Joanna Jędrzejowicz   Andrzej Szepietowski (Eds.)

Mathematical Foundations of Computer Science 2005

30th International Symposium, MFCS 2005
Gdansk, Poland, August 29 – September 2, 2005
Proceedings

Springer

Volume Editors

Joanna Jędrzejowicz
Andrzej Szepietowski
Gdansk University
Institute of Mathematics
Wita Stwosza 57, 80 952 Gdansk, Poland
E-mail: {jj, matszp}@math.univ.gda.pl

Library of Congress Control Number: 2005931473

CR Subject Classification (1998): F.1, F.2, F.3, F.4, G.2, E.1

ISSN   0302-9743
ISBN-10   3-540-28702-7 Springer Berlin Heidelberg New York
ISBN-13   978-3-540-28702-5 Springer Berlin Heidelberg New York

This work is subject to copyright. All rights are reserved, whether the whole or part of the material is concerned, specifically the rights of translation, reprinting, re-use of illustrations, recitation, broadcasting, reproduction on microfilms or in any other way, and storage in data banks. Duplication of this publication or parts thereof is permitted only under the provisions of the German Copyright Law of September 9, 1965, in its current version, and permission for use must always be obtained from Springer. Violations are liable to prosecution under the German Copyright Law.

Springer is a part of Springer Science+Business Media

springeronline.com

© Springer-Verlag Berlin Heidelberg 2005
Printed in Germany

Typesetting: Camera-ready by author, data conversion by Scientific Publishing Services, Chennai, India
Printed on acid-free paper   SPIN: 11549345   06/3142   5 4 3 2 1 0

Lecture Notes in Computer Science 3618

Commenced Publication in 1973
Founding and Former Series Editors:
Gerhard Goos, Juris Hartmanis, and Jan van Leeuwen

Editorial Board

David Hutchison
  Lancaster University, UK
Takeo Kanade
  Carnegie Mellon University, Pittsburgh, PA, USA
Josef Kittler
  University of Surrey, Guildford, UK
Jon M. Kleinberg
  Cornell University, Ithaca, NY, USA
Friedemann Mattern
  ETH Zurich, Switzerland
John C. Mitchell
  Stanford University, CA, USA
Moni Naor
  Weizmann Institute of Science, Rehovot, Israel
Oscar Nierstrasz
  University of Bern, Switzerland
C. Pandu Rangan
  Indian Institute of Technology, Madras, India
Bernhard Steffen
  University of Dortmund, Germany
Madhu Sudan
  Massachusetts Institute of Technology, MA, USA
Demetri Terzopoulos
  New York University, NY, USA
Doug Tygar
  University of California, Berkeley, CA, USA
Moshe Y. Vardi
  Rice University, Houston, TX, USA
Gerhard Weikum
  Max-Planck Institute of Computer Science, Saarbruecken, Germany

Preface

This volume contains the papers presented at the 30th Symposium on Mathematical Foundations of Computer Science (MFCS 2005) held in Gdansk, Poland from August 29th to September 2nd, 2005. Taking place alternately in the Czech Republic, Slovakia and Poland, this year the conference was organized by the Institute of Mathematics of Gdansk University.

From the first meeting in 1972 to this year's 30th event, the MFCS series has provided a basis for theoretical computer scientists to present their latest research results. The scope of the conference, consequently, covers all branches of theoretical computer science ranging from automata, algorithms, data structures, models of computation to complexity theory, also including artificial intelligence, computational biology, computational geometry and cryptography.

The 137 submissions from 22 countries revealed a continued strong interest in the conference as well as the high-quality research results the MFCS series stands for. The Program Committee carefully selected 62 papers for presentation at the conference complemented by 7 invited talks.

The meeting took place at a conference hotel located on Sobieszewo Island, 15 km from Gdansk, offering both a beautiful landscape with sandy beaches and forests and the possibility to explore the old Hanseatic city of Gdansk with its interesting history of over 1000 years.

Last but not least we have the pleasure to thank all the people involved in making the meeting a success: the authors for showing their interest by submitting their research results, the invited speakers for agreeing to attend the conference and for presenting their insights, the members of the Program Committee for thorough discussions during the selection process, and the referees for the effort of reading all the submissions. We would also like to thank the Organizing Committee for the excellent preparation and realization of the symposium, Springer for a smooth cooperation, and of course all the conference participants for creating an inspiring atmosphere.

June 2005

Joanna Jędrzejowicz
Andrzej Szepietowski

Organization

MFCS 2005 was organized by the Institute of Mathematics, University of Gdansk.

Program Committee

Ch. Choffrut	Paris, France
T. Coquand	Goteborg, Sweden
V. Cortier	Nancy, France
V Geffert	Kosice, Slovakia
L A. Hemaspaandra	Rochester, USA
M. Jantzen	Hamburg, Germany
J Jędrzejowicz	Co-chair, Gdansk, Poland
J. Karhumaki	Turku, Finland
M. Karpinski	Bonn, Germany
R. Kralovic	Bratislava, Slovakia
J. Kratochvil	Prague, Czech Republic
M. Kubale	Gdansk, Poland
F. Moller	Swansea, UK
B. Monien	Paderborn, Germany
A.W. Mostowski	Gdansk, Poland
A. Muscholl	Paris, France
S. Radziszowski	Rochester, USA
A. Szepietowski	Co-chair, Gdansk, Poland
W. Thomas	Aachen, Germany
P. Urzyczyn	Warsaw, Poland

Organizing Committee

A. Brekiewicz
T. Dzido
H. Furmanczyk
M. Nadolska
J. Neumann
P. Zyliński

Referees

M. Agnew	G. Andrejkova	A. Beygelzimer
R. Amadio	R. Bartak	F. Blanchard
P. Anderson	J. Berstel	S. Blazy

Organization

Y. Bleischwitz
J. Blömer
L. Boasson
H.J. Böckenhauer
M. Bojańczyk
B. Borchert
T. Borzyszkowski
A.M. Borzyszkowski
O. Bournez
V. Brattka
P. Bürgisser
Z. Butler
P. Campadelli
R. Canosa
D. Catalano
D. Caucal
K. Cechlarova
B. Chlebus
M. Chrobak
R. Clifford
J. Cohen
R. Cole
T. Coquand
V. Cortier
B. Courcelle
F. Cucker
E. Czeizler
R. Dabrow
S. Demri
D. Deng
R. Downey
M. Droste
R. Durikovic
Ch. Durr
J. Dvorakova
R. Elsässer
D. Eppstein
T. Erlebach
G. Exoo
P. Faliszewski
R. Feldmann
J. Fiala
L. Fousse
S. Fratani
R. Freivalds

S. Froeschle
H. Furmańczyk
R.S. Gaborski
M. Gairing
J. Gajdosikova
A. Gambin
C. Gavoille
V. Geffert
K. Giaro
Ch. Glasser
A. Gomolińska
J. Goubault-Larrecq
P. Grabner
E. Graedel
S. Grothklags
F. Hüffner
A. Habel
V. Halava
M.M. Halldorsson
D. Harel
T. Harju
M. Hauptmann
F. Heine
P. Hell
L.A. Hemaspaandra
E. Hemaspaandra
C. de la Higuera
M. Hirvensalo
P. Hlineny
Ch. Homan
J. Honkala
P. Hřyer
L. Istonova
A. Jakoby
A. Janiak
J. Jędrzejowicz
A. Jez
G. Jiraskova
A. Kaminsky
M. Kanazawa
J. Karhumäki
J. Kari
J. Karkkainen
M. Karpinski
M. Kaufmann

G. Kliewer
M. Koehler
P. Kolman
V. Koubek
M.Y. Kovalyov
D. Kowalski
R. Kralovic
E. Kranakis
J. Kratochvil
D. Kratsch
A. Krebs
S. Kremer
M. Kubale
A. Kucera
M. Kudlek
M. Kunc
M. Kutylowski
J.M. Kwon
G. Lakemeyer
W. Lamersdorf
S. Laplante
M. Latteux
M. Lehmann
S. Lietsch
Ch. Löding
S. Lombardy
U. Lorenz
T. Luecking
J. Mairesse
D. Makowiec
M. Malafiejski
A. Malinowski
J. Marciniec
N. Markey
J. Matousek
A. May
B. McKay
J. Meakin
D. Meister
J. Messner
H. Meyerhenke
K. Michael
M. Milkowska
A. Miller
P. Mlynarcik

F. Moller
B. Monien
L. Moss
A.W. Mostowski
A. Muscholl
A. Nadolski
B. Neumann
R. Niedermeier
A. Okhotin
L. Olsen
J. Oravec
P. Orponen
F. Otto
L. Pacholski
P. Pączkowski
I. Paenke
D. Pardubská
A. Pavan
E. Petersen
I Petre
M. Piotrów
K. Piwakowski
T. Plachetka
W. Plandowski
S. Radziszowski
S. Ranise
M. Ren
H. Roeglin
D. Rossin

F. Ruskey
W. Rytter
Z. Sadowski
J. Sakarovitch
P. Sankowski
M. Sauerhoff
S. Schamberger
Ch. Schindelhauer
U.P. Schroeder
Ch. Schwarzweller
L. Segoufin
S. Seibert
O. Serre
J. Sgall
J. Simon
J. Skurczyński
Cz. Smutnicki
J. Sobecki
Ch. Sohler
R. Somla
G. Stachowiak
L. Staiger
H. Stamer
K. Stencel
F. Sur
A. Szabari
B. Szepietowski
A. Szepietowski
T. Tantau

J. Tarhio
G. Tel
L. Tendera
W. Thomas
E. Thomé
S. Tison
J. Toran
E. Torng
S. Travers
R. Tripathi
T. Truderung
M. Turuani
J. Tyszkiewicz
P. Urzyczyn
M. Van Wie
D. Vilenchik
B. Vöcking
K. Volbert
M. Volkov
K. Voß
T. Walen
I. Walukiewicz
G. Weiss
A. Woclaw
A. Wojna
Y. Yao
S. Zemke
M. Zhong
P. Żyliński

Previous MFCS

1972 Jablonna (Poland)
1973 Strbske Pleso (Czechoslovakia)
1974 Jadwisin (Poland)
1975 Marianske Lazne (Czechoslovakia)
1976 Gdansk (Poland)
1977 Tatranska Lomnica (Czechoslovakia)
1978 Zakopane (Poland)
1979 Olomouc (Czechoslovakia)
1980 Rydzyna (Poland)
1981 Strbske Pleso (Czechoslovakia)

1990 Banská Bystrica (Czechoslovakia)
1991 Kazimierz Dolny (Poland)
1992 Praque (Czechoslovakia)
1993 Gdansk (Poland)
1994 Kosice (Slovakia)
1995 Prague (Czech Republic)
1996 Krakow (Poland)
1997 Bratislava (Slovakia)
1998 Brno (Czech Republic)
1999 Szklarska Poreba (Poland)
2000 Bratislava (Slovakia)

1984 Prague (Czechoslovakia)
1986 Bratislava (Czechoslovakia)
1988 Karlovy Vary (Czechoslovakia)
1989 Porabka-Kozubnik (Poland)
2001 Marianske Lazne (Czech Republic)
2002 Warsaw (Poland)
2003 Bratislava (Slovakia)
2004 Prague (Czech Republic)

Table of Contents

Invited Lectures

Page Migration in Dynamic Networks
  Marcin Bienkowski, Friedhelm Meyer auf der Heide 1

Knot Theory, Jones Polynomial and Quantum Computing
  Rūsiņš Freivalds 15

Interactive Algorithms 2005
  Yuri Gurevich 26

Some Computational Issues in Membrane Computing
  Oscar H. Ibarra 39

The Generalization of Dirac's Theorem for Hypergraphs
  Endre Szemerédi, Andrzej Ruciński, Vojtěch Rödl 52

On the Communication Complexity of Co-linearity Problems
  Andrew C. Yao 57

An Invitation to Play
  Wiesław Zielonka 58

Papers

The Complexity of Satisfiability Problems: Refining Schaefer's Theorem
  *Eric Allender, Michael Bauland, Neil Immerman, Henning Schnoor,
  Heribert Vollmer* 71

On the Number of Random Digits Required in MonteCarlo Integration
of Definable Functions
  César L. Alonso, José L. Montaña, Luis M. Pardo 83

Pure Nash Equilibria in Games with a Large Number of Actions
  Carme Àlvarez, Joaquim Gabarró, Maria Serna 95

On the Complexity of Depth-2 Circuits with Threshold Gates
  Kazuyuki Amano, Akira Maruoka 107

Isomorphic Implication
   Michael Bauland, Edith Hemaspaandra 119

Abstract Numeration Systems and Tilings
   Valérie Berthé, Michel Rigo 131

Adversarial Queueing Model for Continuous Network Dynamics
   *María J. Blesa, Daniel Calzada, Antonio Fernández, Luis López,
   Andrés L. Martínez, Agustín Santos, Maria Serna* 144

Coloring Sparse Random k-Colorable Graphs in Polynomial Expected Time
   Julia Böttcher 156

Regular Sets of Higher-Order Pushdown Stacks
   Arnaud Carayol 168

Linearly Bounded Infinite Graphs
   Arnaud Carayol, Antoine Meyer 180

Basic Properties for Sand Automata
   J. Cervelle, E. Formenti, B. Masson 192

A Bridge Between the Asynchronous Message Passing Model and Local Computations in Graphs
   Jérémie Chalopin, Yves Métivier 212

Reconstructing an Ultrametric Galled Phylogenetic Network from a Distance Matrix
   Ho-Leung Chan, Jesper Jansson, Tak-Wah Lam, Siu-Ming Yiu 224

New Resource Augmentation Analysis of the Total Stretch of SRPT and SJF in Multiprocessor Scheduling
   *Wun-Tat Chan, Tak-Wah Lam, Kin-Shing Liu,
   Prudence W.H. Wong* 236

Approximating Polygonal Objects by Deformable Smooth Surfaces
   Ho-lun Cheng, Tony Tan 248

Basis of Solutions for a System of Linear Inequalities in Integers: Computation and Applications
   D. Chubarov, A. Voronkov 260

Asynchronous Deterministic Rendezvous in Graphs
   *Gianluca De Marco, Luisa Gargano, Evangelos Kranakis,
   Danny Krizanc, Andrzej Pelc, Ugo Vaccaro* 271

Zeta-Dimension
  *David Doty, Xiaoyang Gu, Jack H. Lutz, Elvira Mayordomo,
  Philippe Moser* 283

Online Interval Coloring with Packing Constraints
  Leah Epstein, Meital Levy 295

Separating the Notions of Self- and Autoreducibility
  Piotr Faliszewski, Mitsunori Ogihara 308

Fully Asynchronous Behavior of Double-Quiescent Elementary Cellular
Automata
  Nazim Fatès, Michel Morvan, Nicolas Schabanel, Éric Thierry 316

Finding Exact and Maximum Occurrences of Protein Complexes in
Protein-Protein Interaction Graphs
  Guillaume Fertin, Romeo Rizzi, Stéphane Vialette 328

Matrix and Graph Orders Derived from Locally Constrained Graph
Homomorphisms
  Jiří Fiala, Daniël Paulusma, Jan Arne Telle 340

Packing Weighted Rectangles into a Square
  *Aleksei V. Fishkin, Olga Gerber, Klaus Jansen,
  Roberto Solis-Oba* 352

Nondeterministic Graph Searching: From Pathwidth to Treewidth
  Fedor V. Fomin, Pierre Fraigniaud, Nicolas Nisse 364

Goals in the Propositional Horn$^\supset$ Language Are Monotone Boolean
Circuits
  J. Gaintzarain, M. Hermo, M. Navarro 376

Autoreducibility, Mitoticity, and Immunity
  *Christian Glaßer, Mitsunori Ogihara, A. Pavan, Alan L. Selman,
  Liyu Zhang* 387

Canonical Disjoint NP-Pairs of Propositional Proof Systems
  Christian Glaßer, Alan L. Selman, Liyu Zhang 399

Complexity of DNF and Isomorphism of Monotone Formulas
  Judy Goldsmith, Matthias Hagen, Martin Mundhenk 410

The Expressive Power of Two-Variable Least Fixed-Point Logics
  Martin Grohe, Stephan Kreutzer, Nicole Schweikardt 422

Languages Representable by Vertex-Labeled Graphs
  Igor Grunsky, Oleksiy Kurganskyy, Igor Potapov 435

On the Complexity of Mixed Discriminants and Related Problems
  Leonid Gurvits 447

Two Logical Hierarchies of Optimization Problems over the Real Numbers
  Uffe Flarup Hansen, Klaus Meer 459

Algebras as Knowledge Structures
  Bernhard Heinemann 471

Combining Self-reducibility and Partial Information Algorithms
  André Hernich, Arfst Nickelsen 483

Complexity Bounds for Regular Games
  Paul Hunter, Anuj Dawar 495

Basic Mereology with Equivalence Relations
  Ryszard Janicki 507

Online and Dynamic Recognition of Squarefree Strings
  Jesper Jansson, Zeshan Peng 520

Shrinking Restarting Automata
  Tomasz Jurdziński, Friedrich Otto 532

Removing Bidirectionality from Nondeterministic Finite Automata
  Christos Kapoutsis 544

Generating All Minimal Integral Solutions to Monotone ∧, ∨-Systems of Linear, Transversal and Polymatroid Inequalities
  L. Khachiyan, E. Boros, K. Elbassioni, V. Gurvich 556

On the Parameterized Complexity of Exact Satisfiability Problems
  Joachim Kneis, Daniel Mölle, Stefan Richter, Peter Rossmanith 568

Approximating Reversal Distance for Strings with Bounded Number of Duplicates
  Petr Kolman 580

Random Databases and Threshold for Monotone Non-recursive Datalog
  Konstantin Korovin, Andrei Voronkov 591

An Asymptotically Optimal Linear-Time Algorithm for Locally
Consistent Constraint Satisfaction Problems
  Daniel Král', Ondřej Pangrác 603

Greedy Approximation via Duality for Packing, Combinatorial
Auctions and Routing
  Piotr Krysta 615

Tight Approximability Results for the Maximum Solution Equation
Problem over $\mathbf{Z}_p$
  Fredrik Kuivinen 628

The Complexity of Model Checking Higher Order Fixpoint Logic
  Martin Lange, Rafał Somla 640

An Efficient Algorithm for Computing Optimal Discrete Voltage
Schedules
  Minming Li, Frances F. Yao 652

Inverse Monoids: Decidability and Complexity of Algebraic Questions
  Markus Lohrey, Nicole Ondrusch 664

Dimension Is Compression
  María López-Valdés, Elvira Mayordomo 676

Concurrent Automata vs. Asynchronous Systems
  Rémi Morin 686

Completeness and Degeneracy in Information Dynamics of Cellular
Automata
  Hidenosuke Nishio 699

Strict Language Inequalities and Their Decision Problems
  Alexander Okhotin 708

Event Structures for the Collective Tokens Philosophy of Inhibitor Nets
  G. Michele Pinna 720

An Exact 2.9416^n Algorithm for the Three Domatic Number Problem
  Tobias Riege, Jörg Rothe 733

D-Width: A More Natural Measure for Directed Tree Width
  Mohammad Ali Safari 745

On Beta-Shifts Having Arithmetical Languages
  Jakob Grue Simonsen 757

A BDD-Representation for the Logic of Equality and Uninterpreted Functions
  Jaco van de Pol, Olga Tveretina 769

On Small Hard Leaf Languages
  Falk Unger 781

Explicit Inapproximability Bounds for the Shortest Superstring Problem
  Virginia Vassilevska 793

Stratified Boolean Grammars
  Michał Wrona 801

Author Index 813

Page Migration in Dynamic Networks*

Marcin Bienkowski[1] and Friedhelm Meyer auf der Heide[2]

[1] International Graduate School of Dynamic Intelligent Systems,
University of Paderborn, Germany
young@upb.de

[2] Heinz Nixdorf Institute and Computer Science Department,
University of Paderborn, Germany
fmadh@upb.de

1 Introduction

In the last couple of decades, network connected systems have gradually replaced centralized parallel computing machines. To provide smooth operation of network applications, the underlying system has to provide so-called *basic services*. One of the most crucial services is to provide a transparent access to data like variables, databases, memory pages, or files, which are shared by the instances of programs running at nodes of the network.

The traditional approach of storing the shared data in one or a few central repositories does not scale up well with the increase of the network size and is therefore inherently inefficient.

In this paper, we survey data management strategies that try to exploit *topological locality*, i.e., try to migrate the shared data in the network in such a way that a node accessing a data item finds it "nearby" in the network. This problem can be modeled as an online problem; several such models are discussed and will be presented in this survey. We will mainly deal with the classical, most basic of these data management problems, called *Page Migration*.

Our main focus will be on very recent results on page migration in a *dynamic* scenario: Here we assume that the network is no longer static, but it behaves like a mobile ad-hoc network, i.e., the nodes are allowed to move. Thus, we have to deal with two sources of online events, namely the requests from nodes to data items and the movements of the nodes. The new challenges both for modelling and for algorithm design and analysis arising from these *two adversaries* will be the main topic of this paper.

2 Static Networks

In many applications, access patterns to a shared object change frequently. This is common for example in parallel pipelined data processing, where the set of

---

* Partially supported by DFG-Sonderforschungsbereich 376 "Massive Parallelität: Algorithmen Entwurfsmethoden Anwendungen", and by the Future and Emerging Technologies programme of the EU under EU Contract 001907 DELIS "Dynamically Evolving, Large Scale Information Systems".

processors accessing shared variables changes in the runtime. In these cases, any static placement of the object copies is inefficient. Moreover, the knowledge of the future accesses to the objects is in reality either partial or completely non-existing, which renders any solution based on static placement infeasible. Instead, a data management strategy should migrate the copies, to further exploit the locality of accesses. kkeeping overhead small, it is often required that only *one copy* of each object is stored in the system. Additionally, in a typical situation in the parallel environments, shared objects are usually bigger than the part of their data that is being accessed at one time. Usually, processors want to read or change only one single unit of data from the object, or one record from a database. On the other hand, the data of one object should be kept in one place to reduce the maintenance overhead. This leads to a so-called *non-uniform model*, where migrating or copying the whole object is much more expensive than accessing one unit of data from it.

2.1 Page Migration

This traditional paradigm, called *Page Migration* (PM) was introduced by Black and Sleator [16]. It models an underlying network as a connected, undirected graph, where each edge e has an associated cost $c(e)$ of sending one unit of data over the corresponding communication channel. In case of wired networks, this cost might represent the load generated by sending a data through this communication link. The cost of sending one unit of data between two nodes v_a and v_b is defined as the sum of costs of edges on the cheapest path between v_a and v_b. There is one copy of one single object of size D, which is further called a *(memory) page*, stored initially at one fixed node in the network.

A PM problem instance is a sequence of nodes $(\sigma_t)_t$, which want to access (read or write) one unit of data from the page. In one step t, one node σ_t issues a request to the node holding the page and appropriate data is sent back. For such a request, an algorithm for PM is charged a cost of sending one unit of data between σ_t and the node holding the page. Then the algorithm may move the page to an arbitrary node. Such a transaction incurs a cost which is greater by D, the page size factor, than the cost of sending one unit of data between these two nodes.

The goal is to compute a schedule of page movements to minimize the total cost. Furthermore, computing the optimal schedule *offline*, i.e. on the basis of the *whole* input sequence $\mathcal{I} = (\sigma_t)_t$ is an easy task, which can be performed in polynomial time. Thus, the main effort was placed on constructing online algorithms, i.e. ones which have to make decision in time step t solely on the part of the input up to step t.

Competitive Analysis. To evaluate any online strategy, we use competitive analysis [35,17], i.e. we compare the cost of an online solution to the cost of the optimal offline strategy. In the following we assume that an optimal solution is denoted by OPT, and for any algorithm ALG, $C_{\text{ALG}}(\mathcal{I})$ denotes the cost of this algorithm on input sequence $\mathcal{I} = (\sigma_t)_t$.

An online deterministic algorithm ALG is $\mathcal{R}$-competitive, if there exists a constant A, s.t. for any input sequence $\mathcal{I}$ holds

$$C_{\text{ALG}}(\mathcal{I}) \leq \mathcal{R} \cdot C_{\text{OPT}}(\mathcal{I}) + A \ . \tag{1}$$

For a randomized algorithm ALG, we replace its cost in the definition above by its expectation $\mathbf{E}[C_{\text{ALG}}(\mathcal{I})]$. The expected value is taken over all possible random choices made by ALG. Additionally, we have to distinguish between the three adversary types: *oblivious, adaptive-online, adaptive-offline* (see e.g. [10]), depending on their knowledge of the random bits used by ALG.

Results. The PM problem was thoroughly investigated for different types of adversaries. While we shortly state the results below, for a gentle introduction to the algorithms mentioned here, we refer the reader to the survey by Bartal [7].

First randomized solutions presented by Westbrook [36] were a memoryless algorithm which was 3-competitive against an adaptive-online adversary, and a phase-based algorithm whose competitive ratio against an oblivious adversary tends to 2.618 as D goes to infinity. The former result was proven to be tight by Bartal et al. [9,7]. The lower-bound construction was a slight modification of the analogous lower-bound for deterministic algorithms by Black and Sleator [16]. On the other hand, the exact competitive ratio against an oblivious adversary is not a completely settled issue. The currently best known lower-bound, $2 + \frac{1}{2D}$, is due to Chrobak et al. [18]. It is matched only for certain topologies, like trees or uniform networks (see [18] and [22], respectively).

The first deterministic, phase-based, 7-competitive algorithm Move-To-Min was given by Awerbuch et al. [3]. The result was subsequently improved by the Move-To-Local-Min algorithm [8] attaining competitive ratio of 4.086. On the other hand, Chrobak et al. [18] showed a network with a lower bound of approximately 3.148.

2.2 Data Management

In this subsection we give a brief overview of extensions of Page Migration that allow more flexible data management in networks. Let n denote the number of nodes of the network. One of the possible generalizations of PM is allowing more than one copy of an object to exist in the network. This poses new interesting algorithmic questions which have to be resolved by a *data management* scheme.

- How many copies of shared objects should be created?
- Which accesses to shared objects should be handled by which copies?

A basic version of this problem, where only one shared object is present in the system, called *file allocation*, was first examined in the framework of competitive analysis by Bartal et al. [9]. They present a randomized strategy that achieves an optimal competitive ratio of $\mathcal{O}(\log n)$ against an adaptive-online adversary, by a reduction to the online Steiner tree problem. Additionally, they show how to get rid of the central control (which is useful for example for locating the nearest copy of the object) and create $\mathcal{O}(\log^4 n)$-competitive algorithm which works in a distributed fashion. Awerbuch et al. [3] show that the randomization is not crucial by constructing deterministic algorithms (centralized and distributed ones) for file allocation problem attaining asymptotically the same ratios.

For uniform topologies Bartal et al. [9] showed an optimal 3-competitive deterministic algorithm. Lund et al. [22] gave a 3-competitive algorithm for trees based on *work functions* technique.

Memory Constraints. If multiple objects are present in the network and the local memory capacity at nodes is limited, then running file allocation scheme for each single object in the network might encounter some problems. Above all, it is not possible to copy an object into node's memory, if it is already full. Possibly, some other objects' copies have to be dropped, which induces problems if they were the last copies present in the network. This leads to a so called *distributed paging* problem, where file allocation solutions have to be combined with schemes known from *uni-processor paging* (see for example [1,19,24,35]).

For uniform networks, Bartal et al. [9] presented the deterministic $\mathcal{O}(m)$-competitive Distributed-Flush-When-Full algorithm, where m denotes the total number of copies that can be stored within the network. They also proved that this bound is tight by showing $\Omega(m)$ lower bound for competitiveness against an adaptive-online adversary. Awerbuch et al. [4] used randomized uni-processor paging algorithms [1,19,24] to get an up to a constant factor optimal algorithm HEAT & DUMP, which is $\mathcal{O}(\max\{\log(m - f), \log k\})$-competitive against an oblivious adversary. In this context, f is the number of different objects in the network, and k is the maximum number of files that can be stored at any node. If we again restrict the number of copies of object to one, it results in a problem called *page migration with memory constraints*. Albers and Koga [2] presented deterministic and randomized algorithms for this problem, which are much simpler than their distributed paging counterparts, and attain competitive ratios $\mathcal{O}(n)$ and $O(\log n)$, respectively.

For general networks Awerbuch et al. [5] adopted the model suggested primarily for uniprocessor paging [35], which goes beyond pure competitive analysis. In order to compensate the optimal offline algorithm advantage of knowing the future, Sleator and Tarjan [35] proposed limiting the memory capacity that the optimal algorithm has at its disposal. This extension, which is sometimes referred to as *resource augmentation*, allowed authors of [5] to present a deterministic $\mathcal{O}(\mathrm{polylog}\, n)$-competitive algorithm, under the assumption that the online algorithm has $\mathcal{O}(\mathrm{polylog}\, n)$ times more memory than the optimal algorithm.

Optimizing Congestion. In case of wired networks the communication cost between a pair of nodes might be measured in terms of the load generated by sending the data through a communication link. All the algorithms presented above were designed to minimize the total communication load. A more challenging task it to derive fine-grained algorithms, whose objective is to minimize congestion, i.e. the maximum load on each single link.

Maggs et al. [23] developed a distributed data management strategy for tree networks, which was 3-competitive for the uniform model (the size of object equal to 1). The aforementioned 3-competitive algorithm for trees by Lund et al. [22] was proven to be also competitive with respect to congestion minimization, and worked for the non-uniform model. However, as it was based on computing work-functions, it was inherently centralized. Meyer auf der Heide et al. [25] fixed this deficiency, presenting the deterministic 3-competitive distributed strategy for trees.

However, the main result of [23] was *bisimulation technique*. It was shown that for some regular networks like meshes of clustered networks, the original problem instance can be, without enlarging congestion, mapped into a virtual network, a so called *access tree*. As mentioned above, solving the problem on a tree is relatively easy. Finally, the virtual tree was randomly mapped into the original network, so that, with high probability (w.h.p.), the congestion increases at most by a factor of $\mathcal{O}(\log n)$. This yields a randomized algorithm, which is $\mathcal{O}(\log n)$-competitive against an oblivious adversary. Similar results for fat trees and hypercubic networks, as well as $\mathcal{O}(1)$-competitive algorithms for uniform networks, were presented in [26,37] and experimentally evaluated in [21]. Finally, Räcke [27,28] showed that it is possible to construct access trees for any network topology, showing an $\mathcal{O}(\log^3 n)$-competitive algorithm. This was subsequently improved to $\mathcal{O}(\log^2 n \cdot \log \log n)$ by Harrelson et al. [20].

Furthermore, Meyer auf der Heide et al. [26] and later Westermann [37] showed how to extend these strategies to respect the capacity constraints on the local memory modules. Their algorithms also exploit the paradigm of resource augmentation, giving the online algorithm $\mathcal{O}(\log n)$ times more memory than to the offline strategy. The competitive ratios are asymptotically the same as in the case without memory capacity restrictions.

3 Dynamic Networks

Basic services for mobile wireless networks and dynamically changing wired networks are a relatively new area. Some results have been achieved for topology control and routing in dynamic networks (see surveys by Rajaraman [29], and Scheideler [33], as well as the paper of Awerbuch et al. [6]). In comparison, data management solutions in dynamically changing networks are still in their infancy. Till recently, neither theoretical analysis was present in this area, nor a reasonable model of network changes was proposed. In particular, any model similar to described in [6], where we assume adversarial link failures, would give no chance to any data management scheme.

Hence, for theoretical modelling dynamics of networks, we assume that an adversary may modify the costs of point-to-point communication arbitrarily, as long as the pace of these changes is restricted by, say, an additive constant per step. Intuitively, this gives the data management algorithm time to react to the changes. Such a model can be motivated by a reality-close *pedestrian model* by Schindelhauer et al. [34].

The model of slow changes in the communication costs, formally defined in the next section, tries also to capture slow changes in available bandwidth in wired networks, which are inherently induced by other programs running in the network.

In our considerations we do not take into account the dynamics induced by nodes joining and leaving the network. In fact, a model where nodes may become active and inactive was already investigated by Awerbuch et al. [5] in context of file allocation.

3.1 Models and Results

To model the Page Migration problem in dynamic networks we make the following assumptions. The network is modelled as a set of n mobile nodes (processors) labelled $v_1, v_2, \ldots, v_n$. These nodes are placed in a metric space $(\mathcal{X}, d)$, where the distance between any pair of points from $\mathcal{X}$ is given by the metric d.

Time is discrete and slotted into time steps $t = 0, 1, 2, \ldots$. To model dynamics we assume that the position of each node is a function of t, i.e. $p_t(v)$ denotes the position of v in time step t. As a natural consequence, the distance between a pair of nodes may also change with time. The distance between any pair of nodes v_a and v_b in time step t is denoted by

$$d_t(v_a, v_b) := d(p_t(v_a), p_t(v_b)) \ . \tag{2}$$

Note that such a distance can be equal to zero in two different cases. The first one occurs, if v_a and v_b are different nodes occupying the same position in $\mathcal{X}$. The second one is when $a = b$, in which case we are dealing with a single node (and we write $v_a \equiv v_b$).

A tuple describing the positions of all the nodes in time step t is called *configuration* in step t, and is denoted by $\mathcal{C}_t$. A configuration sequence $(\mathcal{C}_t)_{t=0}^T$ contains the configurations in the first $T+1$ time steps, beginning with the *initial configuration* $\mathcal{C}_0$.

The changes in nodes' positions over time are arbitrary, as long as the nodes move with a *bounded speed*, as mentioned in the previous section. Formally, for any node v_i, its positions in two consecutive time steps t and $t+1$ cannot be too far apart, i.e.

$$d(p_t(v_i), p_{t+1}(v_i)) \leq \delta \ , \tag{3}$$

for some fixed constant δ. Furthermore, if $\mathcal{X}$ is a bounded metric space, then let λ denote its *diameter*, i.e. the maximum possible distance between two points from $\mathcal{X}$. For an unbounded space, $\lambda = \infty$.

Any two nodes are able to communicate directly with each other. The cost of sending a unit of data from node v_a to v_b at time step t is defined by a *cost function* $c_t(v_a, v_b)$, defined as

$$c_t(v_a, v_b) = d_t(v_a, v_b) + 1 \ , \tag{4}$$

if v_a and v_b are different nodes. Obviously, the communication within one node is free, i.e. if $v_a \equiv v_b$, then $c_t(v_a, v_b) = 0$. Essentially, the communication cost is proportional to the distance between these two nodes, plus a constant overhead. This overhead represents the startup cost for establishing connection.

Naturally, the changes in the network themselves (described by $(\mathcal{C}_t)_{t=0}^T$ sequence) do not constitute a problem of its own. According to the described model of Page Migration, a copy of memory page of size D is stored at one of the network's nodes, initially at v_1. In each time step t, exactly one node, denoted by σ_t, tries to access one unit of data from the page. Since the model assumes that there is only one copy of the object stored in the system, there is no need of making distinction between between read and write accesses. Further, we refer to them as *accesses* or *requests*. The requests σ_t create the sequence $(\sigma_t)_{t=1}^T$, complementary to the configuration sequence $(\mathcal{C}_t)_{t=0}^T$.[1]

In each step an algorithm for the Page Migration in dynamic networks has to serve the request, and then to decide, whether it wants to migrate the page to some other node. Precisely, for any algorithm ALG the following stages happen in time step $t \geq 1$.

1. The positions of the nodes in the current step are defined by $\mathcal{C}_t$.
2. A node σ_t wants to access one single unit of data from the page. It sends a write or a read request to $P_{\text{ALG}}(t)$, the node holding ALG's page in the current step.
3. ALG serves this request, i.e. it sends a confirmation in case of write, or a requested unit of data in case of read. This transaction incurs a cost $c_t(P_{\text{ALG}}(t), \sigma_t)$.
4. ALG optionally moves the page to another node of its choice. A movement to $P'_{\text{ALG}}(t)$ incurs a cost $D \cdot c_t(P_{\text{ALG}}(t), P'_{\text{ALG}}(t))$.

In fact, the only part which ALG may influence is choosing a new node $P'_{\text{ALG}}(t)$ in the fourth stage. The problem, to which we further refer *Dynamic Page Migration* (DPM) is to construct a schedule of page movements to minimize the total cost of communication for any pair of sequences $(\mathcal{C}_t)_t, (\sigma_t)_t$.

3.2 Competitive Analysis in Different Scenarios

Like in the Page Migration case, the problem of minimizing the total cost incurred is relatively easy, if both $(\mathcal{C}_t)_t$ and $(\sigma_t)_t$ are given in *offline* setting, i.e. if an algorithm may read the whole input beforehand. In fact, an easy algorithm using dynamic programming approach is able to find an optimal schedule of page movements for any instance of the DPM problem consisting of T steps, using $\mathcal{O}(T \cdot n^2)$ operations and $\mathcal{O}(T \cdot n)$ additional space.

However, as mentioned earlier, DPM has to be primarily solved in an *online* scenario, where an algorithm must make its decisions (where to move the page) in time step t solely on the basis of the initial part of the input up to step t, i.e. on the sequence $\mathcal{C}_0, \mathcal{C}_1, \sigma_1, \mathcal{C}_2, \sigma_2, \ldots, \mathcal{C}_t, \sigma_t$. To evaluate any online strategy for the DPM problem, we use competitive analysis. Since the input sequence consists

---
[1] Note that nodes issue requests from the first step. The initial configuration in time step 0 is introduced for simplifying notation only.

of two practically independent streams, one describing the request patterns and one reflecting the changes in network topology, it is reasonable to assume that they are created by two separate adversarial entities, the request adversary and the configuration/network adversary. This separation yields different scenarios depending on ways in which these adversaries interact.

Adversarial (Cooperative) Scenario. The most straightforward modelling, which also creates the most difficult task to solve, arises when both adversaries may cooperate to create the combined input sequence. In fact, this is equivalent to having one adversary capable of constructing the whole input sequence and brings the problem back to the classical formulation of online analysis.

For this scenario, Bienkowski et al. [13] constructed a deterministic strategy, which is $\mathcal{O}(\min\{n \cdot \sqrt{D}, D, \lambda\})$-competitive. Recall that λ denotes the maximum distance that can be achieved between two nodes. Their algorithm is up to a constant factor optimal, due to the matching lower bound for adaptive-online adversaries, given in [15]. Further, they show how to randomize this strategy to get a competitive ratio of $\mathcal{O}(\sqrt{D} \cdot \log n, D, \lambda\})$ against an oblivious adversary. This result is up to a $\mathcal{O}(\sqrt{\log n})$ factor optimal in the common case $D \geq \log^3 n$, due to the lower bound of $\Omega(\min\{\sqrt{D \cdot \log n}, D^{2/3}, \lambda\})$ from [13]. All the presented competitive ratios are strict, which means that the constant A occurring in (1) is equal to zero.

The competitive ratios of the best possible algorithms for DPM problem are large, even against the weakest, oblivious adversaries. It can be inferred that the poor performance of algorithms for this scenario is caused by the fact that the network and request adversaries might combine their efforts in order to destroy our algorithm. If cooperation between them was forbidden, then one might hope for a provably better performance. However, it is semantically not clear what non-cooperativeness means. Therefore, it was proposed in [11,14,15] that the DPM problem could be analyzed in another extreme case, where one of the adversaries is replaced by a stochastic process. This leads to another two scenarios.

Brownian Motion Scenario. In this scenario the mobile nodes perform a random walk on a bounded area of diameter B, and the request adversary dictates which nodes issue requests during runtime. However, the adversary is "oblivious", i.e. it has to create the whole request sequence $(\sigma_t)_t$ in advance, without knowledge of the actual configuration sequence $(\mathcal{C}_t)_t$ induced by a random walk. The definition of competitiveness has to be adapted appropriately to reflect the fact that the input sequence is created both by an adversary and a stochastic process. A deterministic algorithm ALG is $\mathcal{R}$-competitive with probability p, if there exists a constant A, s.t. for all request sequences $(\sigma_t)_t$ holds

$$\Pr\nolimits_{(\mathcal{C}_t)_t} \left[C_{\text{ALG}}((\mathcal{C}_t)_t, (\sigma_t)_t) \leq \mathcal{R} \cdot C_{\text{OPT}}((\mathcal{C}_t)_t, (\sigma_t)_t) + A \right] \geq p , \qquad (5)$$

where the probability is taken over all possible configuration sequences generated by the random movement.

The main result of [14], based on the preliminary result of [15] is an algorithm MAJ, which is $\mathcal{O}(\min\{\sqrt[4]{D}, n\} \cdot \text{polylog}(B, D, n))$-competitive. This result holds for 1-dimensional areas if $B \leq \tilde{\mathcal{O}}(\sqrt{D})$, or for any constant-dimensional areas if $B \geq \tilde{\mathcal{O}}(\sqrt{D})$. The ratio is achieved w.h.p., i.e. the probability p occurring in (5) can be amplified to $1 - D^{-\alpha}$ by setting $A = \alpha \cdot A_0$ for a fixed constant A_0.

Stochastic Requests Scenario. This is the scenario symmetric to the Brownian motion one. It is assumed that requests appear with some given frequencies, i.e. in step t, σ_t is a node chosen randomly according to a fixed probability distribution π. Analogously, a deterministic algorithm ALG is $\mathcal{R}$-competitive with probability p, if there exists a constant A, s.t. for all possible network topology changes (configuration sequences) $(\mathcal{C}_t)_t$ and all possible probability distributions π holds

$$\Pr\nolimits_{(\sigma_t)_t} \left[C_{\text{ALG}}((\mathcal{C}_t)_t, (\sigma_t)_t) \leq \mathcal{R} \cdot C_{\text{OPT}}((\mathcal{C}_t)_t, (\sigma_t)_t) + A \right] \geq p \ , \qquad (6)$$

where the probability is taken over all possible request sequences $(\sigma_t)_t$ generated according to π.

The Move-To-First-Request algorithm presented in [11] achieves strict $\mathcal{O}(1)$-competitive ratio, w.h.p. In this context, high probability means that one can achieve probability $1 - D^{-\alpha}$, if the input sequence is sufficiently long. Moreover, the algorithm can be slightly modified to handle also the following cost function

$$c_t(v_a, v_b) = (d_t(v_a, v_b))^\beta + 1 \ , \qquad (7)$$

for any constant β, still remaining $\mathcal{O}(1)$-competitive. For the case of wireless radio networks, one can choose the parameter β to respect a *propagation exponent* of the medium (see for example [31]). For example by setting $\beta = 2$, the cost definition reflects the energy consumption used the send the message in the ideally free space along a given distance. Thus, this result minimizes, up to a constant factor, the total energy used in the system.

4 Algorithms and Lower Bounds

In this section we give some technically interesting results for the DPM model. First, we present MARK, the main building block of the $\mathcal{O}(\min\{n \cdot \sqrt{D}, D, \lambda\})$ upper bound for competitiveness in the adversarial scenario. Later, we show that this ratio is inherently high by showing a lower bound of $\Omega(\min\{\sqrt{D}, \lambda\})$ (which works even in two-node networks) for a randomized algorithm against an oblivious adversary. Finally, we present a simple majority algorithm, which is $\mathcal{O}(\log n)$-competitive, w.h.p., in a very restricted version of the Brownian motion scenario.

4.1 Algorithm MARK

The $\mathcal{O}(n \cdot \sqrt{D})$-competitive deterministic MARK algorithm [13] for the adversarial scenario of the DPM problem was inspired by the Move-To-Min algorithm [3] for the regular Page Migration problem. Move-To-Min divides the whole input sequence works into chunks of length D. In any chunk, it serves all the requests, and move the page at the end of the chunk to a so called *gravity center*. A gravity center is a node, which would be the best place for a page in this chunk.

MARK works in chunks of length $\sqrt{D}$. This length constitutes a tradeoff – it has to be long enough to amortize the movement of the page against the cost of serving the requests in the chunk, and short enough to make the adversarial network changes negligible. However, it can be shown that any algorithm, which considers *only* gravity centers as candidates for the nodes holding the page, has no chance to be better than $\Omega(D)$-competitive.

On the other hand, keeping the page close to the gravity center is, generally, a desirable thing. Hence, we consider the following marking scheme, which depends only on the input sequence. Chunks are grouped in epochs, each epoch begins with all nodes unmarked. First epoch starts with the beginning of the input. In each epoch we track A_i counters for the part of the epoch seen so far. A_i counter is the cost of an algorithm, which remains at v_i, and does not move. If such a counter exceeds D, then the corresponding node becomes marked. At the end of a chunk, in which all nodes are already marked, the current epoch ends, the scheme unmarks all nodes, and a new epoch begins.

MARK uses this scheme in the following way. It remains in a node till the end of chunk, in which this node gets marked, and then moves to any not yet marked node. Additionally, at the end of the last chunk in epoch, it moves to the gravity center associated with this chunk.

It can be proven, that even considering the adversarial movement of the nodes, if a node remains far away from the gravity center, A_i counter increases rapidly, which leads to marking the node.

Lemma 1 ([13]). *If at the end of a chunk I a node is not marked, then its distance to the gravity center is at most $\mathcal{O}(\sqrt{D})$.*

Thus, if MARK moves, it moves to the neighborhood of the gravity center. Denoting the sequence of chunks between two movements of MARK by *phase*, and using similar kind of amortized analysis (with adequately chosen potential function) as for Move-To-Min algorithm, the following can be shown.

Lemma 2 ([13]). *In each phase the amortized cost of MARK is not greater than the cost of OPT times $\mathcal{O}(\sqrt{D})$, plus an additive term of $\mathcal{O}(D \cdot \sqrt{D})$*

However, we may eradicate this additive term by resorting to the properties of the marking scheme. First, since in each phase at least one new node gets marked, the number of phases in one epoch is at most n. Second, the OPT's cost in one epoch is at least D. It follows from the case analysis: if OPT moves then it is charged at least D, otherwise it remains in one node v_i, and thus its cost is equal to $A_i \geq D$. Hence, the additive terms in one epoch amount to $\mathcal{O}(n \cdot D \cdot \sqrt{D})$, which is at most $\mathcal{O}(n \cdot \sqrt{D})$ times the optimal cost. This concludes the proof of MARK's competitiveness.

Straightforward generalization of MARK, i.e. choosing not any, but a random not yet marked node, reduces the number of phases to $\log n$ and the competitive ratio to $\mathcal{O}(\sqrt{D} \cdot \log n)$. Choosing different chunks' length and a refined randomization presented in [12] yields a competitive ratio of $\mathcal{O}(\sqrt{D \cdot \log n})$ against oblivious adversary.

4.2 A Lower Bound Against an Oblivious Adversary

Let $B_{\text{exp}} = \min\{\sqrt{D}, \lambda\}$. We construct a probability distribution π over inputs of arbitrary length and prove that for any deterministic algorithm DET, which knows this distribution, holds $\mathbf{E}_\pi[C_{\text{DET}}(\mathcal{I})] \geq \Omega(B_{\text{exp}}) \cdot \mathbf{E}_\pi[C_{\text{OPT}}(\mathcal{I})]$. Then, the lower bound of $\Omega(B_{\text{exp}})$ for any randomized algorithm against oblivious adversary follows directly from the Yao min-max principle [38,17].

We divide input into phases, each of length $D + 2 \cdot B_{\text{exp}}$ steps. Each phase consists of *expanding part*, (B_{exp} steps), *main part* (D steps), and *contracting part* (also B_{exp} steps). Each phase begins with v_1 and v_2 occupying the same point in the space. Then within the expanding part, nodes are moved apart, so that in the t-th step of the expanding part the distance between them is $t - 1$. Throughout the whole main part the distance amounts to B_{exp}. Finally, in the contracting part nodes are moved closer to each other, so that at the end of the phase they meet again. Note that the movement of the nodes is fixed deterministically.

In the expanding part all the requests are issued at v_1, and all the requests of the contracting one occur at v_2. Further, in the main part, with probability $1/2$, all the requests are issued at v_1, and, with probability $1/2$, all the requests are issued at v_2.

We concentrate on one single phase P. It is relatively easy to show that OPT pays at most $\mathcal{O}(D)$ in each phase. On the other hand, a deterministic online algorithm DET can base its decisions only on the past requests. In particular, in the last step of the expanding part it has to decide whether to end this step at v_1 or v_2. Independently of DET's choice, with probability $1/2$, all the next D requests in the main part are given at the opposite node. In this case DET has two options. If it moves the page within the main part, then it pays $D \cdot B_{\text{exp}}$. Otherwise, it pays $D \cdot B_{\text{exp}}$ for serving the requests during this part. Hence, the expected cost of DET in one phase is at least $\frac{1}{2} \cdot D \cdot B_{\text{exp}}$.

Thus, $\mathbf{E}_\pi[C_{\text{DET}}(P)] = \Omega(B_{\text{exp}}) \cdot C_{\text{OPT}}(P)$. Since we may construct arbitrarily long input sequences, the lower bound follows by linearity of expected value.

4.3 The Majority Algorithm

We analyze the Brownian motion scenario in a simplified setting, where only two nodes perform a random walk on a discrete ring of size $B = \sqrt{D}$. In each time step the coordinate of a node, with probability $1/3$, increases by 1, decreases by 1, or remains the same.

Algorithm MAJ simply divides the input into phases $P_1, P_2, P_3, \ldots$, each of length $B^2 = D$. At the end of each phase, it moves to the node which issued majority of requests in this phase.

We sketch a proof that MAJ is $\mathcal{O}(\log n)$-competitive, w.h.p. We neglect the cost of MAJ in the first two phases, putting it into additive constant A, occurring in (5). The remaining phases are divided into three disjoint, alternating sets $\mathcal{M}_i = \{P_j : j \equiv i \,(\mathrm{mod}\, 3)\}$. Naturally, there exists a set $\mathcal{M}_\chi$, which incurs at least $1/3$ of the total cost, hence we need to bound $C_{\mathrm{MAJ}}(\mathcal{M}_\chi)$ only.

The crucial part is relating $C_{\mathrm{MAJ}}(P_j)$ to $C_{\mathrm{OPT}}(P_{j-1} \uplus P_j)$, for any phase P_j. We charge MAJ $\mathcal{O}(B)$ for any request issued not at the node holding its page, and $\mathcal{O}(D \cdot B)$ for moving its page.

Lemma 3 ([15,14]). *For each phase P_j there exists a critical subphase $P_j' \subseteq P_{j-1} \uplus P_j$, of length $\Theta(\frac{1}{\log D}) \cdot D$, s.t. P_j' is similar to P_j.*

Similarity means that, under the assumption that within P_j' the distance between nodes is $\Omega(B)$, the cost of any algorithm ALG is $C_{\mathrm{ALG}}(P_j') = \Omega(1/\log D) \cdot C_{\mathrm{MAJ}}(P_j)$. The key observation helping to prove the lemma above is that even if MAJ is at node v_1 within P_j, and all the requests are given at v_2 (and thus $C_{\mathrm{MAJ}}(P_j) = \Omega(B \cdot D)$), then in the previous phase P_{j-1} the majority of requests must have been issued at v_1. Thus, we are able to find a subphase with roughly the same number of requests of v_1 and v_2.

Naturally, in the subphase nodes might by at a distance of $o(B)$. The following lemma assures that this is frequently not the case.

Lemma 4 ([14]). *Let P_{j-3}' and P_j' be two consecutive critical subphases. For any configuration at the end of P_{j-3}', with a constant probability, within P_j' nodes are at the distance $\Omega(B)$.*

The proof utilizes two facts. First, at least B^2 steps separate P_{j-3}' and P_j'. The Markov chain induced by nodes' random walks converges relatively quickly (see [32]), i.e. after B^2 steps the position of nodes are almost uniform. Thus, with a constant probability nodes are at distance $\Omega(B)$ at the beginning of P_j'. Moreover, if their initial distance is $\Omega(B)$, then during $\mathcal{O}(B^2/\log B)$ steps, they approximately maintain this distance.

By Lemma 4 we get $C_{\mathrm{MAJ}}(P_j) \leq \mathcal{O}(\log D) \cdot \mathbf{E}[C_{\mathrm{OPT}}(P_j')]$. Moreover, the expected value of this bound on OPT is taken only on the random walk in phases P_{j-2}, P_{j-1}, and P_j, and thus for different $P_j \in \mathcal{M}_\chi$, $C_{\mathrm{OPT}}(P_j')$ are independent random variables. Hence, we may use Hoeffding inequality [30] to show that $\sum_{P_j \in \mathcal{M}_\chi} C_{\mathrm{OPT}}(P_j')$ is sharply concentrated around its mean value.

References

1. D. Achlioptas, M. Chrobak, and J. Noga. Competitive analysis of randomized paging algorithms. *Theoretical Computer Science*, 234(1–2):203–218, 2000.
2. S. Albers and H. Koga. Page migration with limited local memory capacity. In *Proc. of the 4th Int. Workshop on Algorithms and Data Structures (WADS)*, pages 147–158, 1995.
3. B. Awerbuch, Y. Bartal, and A. Fiat. Competitive distributed file allocation. In *Proc. of the 25th ACM Symp. on Theory of Computing (STOC)*, pages 164–173, 1993.

4. B. Awerbuch, Y. Bartal, and A. Fiat. Heat & dump: Competitive distributed paging. In *Proc. of the 34th IEEE Symp. on Foundations of Computer Science (FOCS)*, pages 22–31, 1993.
5. B. Awerbuch, Y. Bartal, and A. Fiat. Distributed paging for general networks. *Journal of Algorithms*, 28(1):67–104, 1998. Also appeared in *Proc. of the 7th SODA*, pages 574–583, 1996.
6. B. Awerbuch, A. Brinkmann, and C. Scheideler. Anycasting in adversarial systems: routing and admission control. In *Proc. of the 30th Int. Colloq. on Automata, Languages and Programming (ICALP)*, pages 1153–1168, 2003.
7. Y. Bartal. Distributed paging. In *Dagstul Workshop on On-line Algorithms*, pages 97–117, 1996.
8. Y. Bartal, M. Charikar, and P. Indyk. On page migration and other relaxed task systems. *Theoretical Computer Science*, 268(1):43–66, 2001. Also appeared in *Proc. of the 8th SODA*, pages 43–52, 1997.
9. Y. Bartal, A. Fiat, and Y. Rabani. Competitive algorithms for distributed data management. *Journal of Computer and System Sciences*, 51(3):341–358, 1995. Also appeared in *Proc. of the 24nd STOC*, pages 39–50, 1992.
10. S. Ben-David, A. Borodin, R. M. Karp, G. Tardos, and A. Wigderson. On the power of randomization in online algorithms. In *Proc. of the 22nd ACM Symp. on Theory of Computing (STOC)*, pages 379–386, 1990.
11. M. Bienkowski. Dynamic page migration with stochastic requests. In *Proc. of the 17th ACM Symp. on Parallelism in Algorithms and Architectures (SPAA)*, 2005. To appear.
12. M. Bienkowski and J. Byrka. Bucket game with applications to set multicover and dynamic page migration. Unpublished manuscript, 2005.
13. M. Bienkowski, M. Dynia, and M. Korzeniowski. Improved algorithms for dynamic page migration. In *Proc. of the 22nd Symp. on Theoretical Aspects of Computer Science (STACS)*, pages 365–376, 2005.
14. M. Bienkowski and M. Korzeniowski. Dynamic page migration under brownian motion. In *Proc. of the European Conf. in Parallel Processing (Euro-Par)*, 2005. To appear.
15. M. Bienkowski, M. Korzeniowski, and F. Meyer auf der Heide. Fighting against two adversaries: Page migration in dynamic networks. In *Proc. of the 16th ACM Symp. on Parallelism in Algorithms and Architectures (SPAA)*, pages 64–73, 2004.
16. D. L. Black and D. D. Sleator. Competitive algorithms for replication and migration problems. Technical Report CMU-CS-89-201, Department of Computer Science, Carnegie-Mellon University, 1989.
17. A. Borodin and R. El-Yaniv. *Online Computation and Competitive Analysis*. Cambridge University Press, 1998.
18. M. Chrobak, L. L. Larmore, N. Reingold, and J. Westbrook. Page migration algorithms using work functions. In *Proc. of the 4th Int. Symp. on Algorithms and Computation (ISAAC)*, pages 406–415, 1993.
19. A. Fiat, R. M. Karp, M. Luby, L. A. McGeoch, D. D. Sleator, and N. E. Young. Competitive paging algorithms. *Journal of Algorithms*, 12(4):685–699, 1991.
20. C. Harrelson, K. Hildrum, and S. Rao. A polynomial-time tree decomposition to minimize congestion. In *Proc. of the 15th ACM Symp. on Parallelism in Algorithms and Architectures (SPAA)*, pages 34–43, 2003.
21. C. Krick, F. Meyer auf der Heide, H. Räcke, B. Vöcking, and M. Westermann. Data management in networks: Experimental evaluation of a provably good strategy. *Theory of Computing Systems*, 2:217–245, 2002. Also appeared in *Proc. of the 11nd SPAA*, pages 165–174, 1999.

22. C. Lund, N. Reingold, J. Westbrook, and D. C. K. Yan. Competitive on-line algorithms for distributed data management. *SIAM Journal on Computing*, 28(3):1086–1111, 1999. Also appeared as On-Line Distributed Data Management in *Proc. of the 2nd ESA*, pages 202–214, 1994.
23. B. M. Maggs, F. Meyer auf der Heide, B. Vöcking, and M. Westermann. Exploiting locality for data management in systems of limited bandwidth. In *Proc. of the 38th IEEE Symp. on Foundations of Computer Science (FOCS)*, pages 284–293, 1997.
24. L. A. McGeoch and D. D. Sleator. A strongly competitive randomized paging algorithm. *Algorithmica*, 6(6):816–825, 1991.
25. F. Meyer auf der Heide, B. Vöcking, and M. Westermann. Provably good and practical strategies for non-uniform data management in networks. In *Proc. of the 7th European Symp. on Algorithms (ESA)*, pages 89–100, 1999.
26. F. Meyer auf der Heide, B. Vöcking, and M. Westermann. Caching in networks. In *Proc. of the 11th ACM-SIAM Symp. on Discrete Algorithms (SODA)*, pages 430–439, 2000.
27. H. Räcke. Minimizing congestion in general networks. In *Proc. of the 43rd IEEE Symp. on Foundations of Computer Science (FOCS)*, pages 43–52, 2002.
28. H. Räcke. Data management and routing in general networks. PhD thesis, Universität Paderborn, 2003.
29. R. Rajaraman. Topology control and routing in ad hoc networks: a survey. *SIGACT News*, 33(2):60–73, 2002.
30. S. Rajesekaran, P. M. Pardalos, J. H. Reif, and J. Rolim. *Handbook of Randomized Computing*, volume II. Kluwer Academic Publishers, 2001.
31. T. S. Rappaport. *Wireless Communications: Principles and Practices*. Prentice Hall, 1996.
32. J. S. Rosenthal. Convergence rates for Markov chains. *SIAM Review*, 37(3):387–405, 1995.
33. C. Scheideler. Models and techniques for communication in dynamic networks. In *Proc. of the 19th Symp. on Theoretical Aspects of Computer Science (STACS)*, pages 27–49, 2002.
34. C. Schindelhauer, T. Lukovszki, S. Rührup, and K. Volbert. Worst case mobility in ad hoc networks. In *Proc. of the 15th ACM Symp. on Parallelism in Algorithms and Architectures (SPAA)*, pages 230–239, 2003.
35. D. D. Sleator and R. E. Tarjan. Amortized efficiency of list update and paging rules. *Communications of the ACM*, 28(2):202–208, 1985.
36. J. Westbrook. Randomized algorithms for multiprocessor page migration. *DIMACS Series in Discrete Mathematics and Theoretical Computer Science*, 7:135–150, 1992.
37. M. Westermann. *Caching in Networks: Non-Uniform Algorithms and Memory Capacity Constraints*. PhD thesis, Universität Paderborn, 2000.
38. A. C.-C. Yao. Probabilistic computation: towards a uniform measure of complexity. In *Proc. of the 18th IEEE Symp. on Foundations of Computer Science (FOCS)*, pages 222–227, 1977.

Knot Theory, Jones Polynomial and Quantum Computing

Rūsiņš Freivalds

Institute of Mathematics and Computer Science, University of Latvia,
Raiņa bulv. 29, Rīga, Latvia*
Rusins.Freivalds@mii.lu.lv

Abstract. Knot theory emerged in the nineteenth century for needs of physics and chemistry as these needs were understood those days. After that the interest of physicists and chemists was lost for about a century. Nowadays knot theory has made a comeback. Knot theory and other areas of topology are no more considered as abstract areas of classical mathematics remote from anything of practical interest. They have made deep impact on quantum field theory, quantum computation and complexity of computation.

1 Introduction

Scott Aaronson writes in [Aa 05]: "In my (unbiased) opinion, the showdown that quantum computing has forced - between our deepest intuitions about computers on the one hand, and our best-confirmed theory of the physical world on the other - constitutes one of the most exciting scientific dramas of our time.

But why did this drama not occur until so recently? Arguably, the main ideas were already in place by the 1960's or even earlier. First, many scientists see the study of "speculative" models of computation as at best a diversion from more serious work; this might explain why the groundbreaking papers papers by Simon [Si 94] and Bennett et al. [BBBV 97] were initially rejected from the major theory conferences. And second, many physicists see computational complexity as about as relevant to the mysteries of Nature as dentistry or tax law.

Today, however, it seems clear that there is something to gain from resisting these attitudes.

We would do well to ask: *what else* about physics might we have overlooked in thinking about the limits of efficient computation? The goal of this article is to encourage the serious discussion of this question. For concreteness, I will focus on a single sub-question: *can NP-complete "problems be solved in polynomial time using the resources of the physical universe?"*

The rest of my paper is rather far from the topic of S. Aaronson's paper but the problem stays: can NP-complete problems be solved in polynomial time using the resources of the physical universe? We will see that unexpected help

---

* Research supported by Grant No.05.1528 from the Latvian Council of Science and European Commission, contract IST-1999-11234.

to understand this problem comes from one of the most abstract parts of the classical mathematics, namely, from topology.

2 Knots

A knot is just such a knotted loop of string, except that we think of the string as having no thickness, its cross-section being a single point. We do not distinguish between the original closed knotted curve and the deformations of that curve through space that do not allow the curve to pass through itself. All of these deformed curves are considered to be the same knot. We think of the knot as if it were made of easily deformable rubber. The simplest knot of all is just the unnknotted circle, which we call the unknot. The next simplest knot is called a trefoil knot.

Why knots are interesting? Much of the early interest in knot theory came from from chemistry. Lord Kelvin (William Thomson) hypothesized that atoms were merely knots in the fabric of ether [Th 1867, Th 1869]. Different knots would then correspond to different elements. This convinced the Scottish physicist Peter Guthrie Tait that if he could list all of the possible knots, he would be creating a table of the elements. He spent many years tabulating knots.

Unfortunately, Kelvin was wrong. In 1887, the Michelson-Morley experiment demonstrated that there was no such thing as ether. A more adequate model of atomic structure appeared at the end of the nineteenth century and chemists lost interest in knots. But in the meantime the mathematicians developed a mathematical theory of knots.According to Tait [Ta 1898], a knot, being a closed curve in space, could be represented by a planar curve obtained by projecting it perpendicularly on the horizontal plane. This projection could have *crossings*, where the projection of one part of the curve crossed another; the planar representation shows the position in space of two strands that cross each other by interrupting the line that represents the lower strand at the crossing.

Luckily for Tait, he learned that another amateur mathematician, the Reverend Thomas Penyngton Kirkman, had already classified planar curves with minimal crossings [Ki 1883], and all that remained was to eliminate the duplications systematically. For a curve with 10 crossings, for example, there were 2^{10}, or, 1024 possibilities for making a knot. Tait decided to list only *alternating knots,* that is, those in which overpasses and underpasses alternate along the curve. In this way, exactly two alternating knots corresponded to each planar curve. Nonetheless, Tait spent the rest of his life to this task. Nonalternating knots with 10 or fewer crossings were classified by C.N.Little [Li 1900]. A difficult problem arose. How one can find out whether or not two knots are equivalent. Even two projections of the same knot may look very much differently. However, equivalence of two knots can be more complicated. Parts of the knot can be pushed and twisted into many topologically equivalent forms. The existence of innumerable versions of the given knot gives rise to a mathematical problem. Two knots are called (topologically) equivalent if it is possible to deform one smoothly into the other so that all the intermediate stages are loops without self intersections. The key result that makes it possible to begin a (combinatorial) theory of knots is the theorem of Reidemeister [Re 32] that states that two diagrams represent equivalent knots if and only if one diagram can be obtained from the other by a finite sequence of special deformations called the Reidemeister moves. There are three types of Reidemeister moves:

1. Allows us to put in/take out a twist.
2. Allows us to either add two crossings or remove two crossings.
3. Allows us to slide a strand of the knot from one side of a crossing to the other.

3 Knot Invariants

A link is a set of knotted loops all tangled together. Two links are considered to be the same if we can deform the one link to the other link without ever having any one of the loops intersect itself or any of the other loops in the process. If we have two loops knotted with each other, we say that it is a *link of two components*. It is easy to see that no Reidemeister move changes the number of components in a link. Hence the number of components is an invariant of the link. Unfortunately, it is not true that two links are equivalent if and only if their numbers of components are the same. If we wish to distinguish knots by their invariants, we need more invariants.

3.1 Linking Number

The linking number is a way of measuring numerically how linked up two components are. If there are more than two components, add up the link numbers and divide by two. The crossing is called *positive* if you rotate the under-strand clockwise they line up. The crossing is called *negative* if you rotate the understrand counterclockwise they line up. You go along the strand and add +1 if you meet a positive crossing and add -1 if you meet a negative crossing.

The linking number is unaffected by all three Reidemeister moves. Therefore it is an invariant of the oriented link.

3.2 Alexander Polynomial

An early example of a successful knot invariant is the Alexander polynomial, discovered by J. W. Alexander [AL 28]. The Alexander polynomial for the knot called *trefoil* is $-(txt) + t - 1$ and the polynomial for the knot called *the figure-eight* is -(txt)+3t-1.

3.3 Topology Comes in

So the search was on for more sensitive knot invariants that would detect when two knots were different. This led to alternate understandings of the notion of sameness. In particular, it is possible that to a topologist there is no difference between two loops but what is different is the space away from these loops, that is the complement of the knot. Let R^3 be the space in which a knot K sits. Then the space "around" the knot, i.e., everything but the knot itself, is denoted $R^3 - K$ and is called the knot complement of K.

Understanding that the principle object of study is the knot complement places knot theory inside the larger study of 3-manifolds. A 3-manifold is a space which locally (assume you are near sighted) looks like standard xyz-space and knot complements are readily seen as examples of 3-manifolds. It was through the study of 3-manifolds that in the 1970's knot theory began returning to its ancestoral roots in physics. To understand this we have to flashback to the 1860's work of Bernhard Riemann. Riemann was interested in relating geometric structures to the forces in physics. Building on Gauss' work, Riemann investigated three different geometric structures for 3-dimensional spaces—elliptic, Euclidean, and hyperbolic. (Einstein's Theory of Relativity was built on Riemannian geometry.)

Each of these distinct structures can be characterized by the behavior of triangles in planes. In elliptic 3-space, the interior angles of a triangle in a plane have a sum greater than 180 degrees. In Euclidean 3-space, the sum is 180 degrees and in hyperbolic 3-space the sum is less than 180 degrees. In 1977 William Thurston [Thu 77] established sufficient conditions for when a 3-manifold possesses a hyperbolic structure. Surprisingly, except for a well understood subclass of knots, all knot complements possess a complete hyperbolic structure.

Thurston's work on hyperbolic structures firmly re-established knot theory's connections with physics. In the 1980's, through some totally unexpected routes, knot theory made further connections with its ancestral roots. In 1987 Vaughan Jones [Jo 89] discovered a totally different polynomial invariant from that of Alexander using the theory of operator algebras. Within a short period of time, more than five new polynomial invariants generalizing the Jones polynomial were discovered. Moreover, Jones polynomial quickly led to the proofs that established all of Tait's original conjectures on knot projections.

3.4 Kauffman Bracket Polynomial

Kauffman bracket polynomial of the knot (or link) K is denoted as $< K >$. This will be a Laurent polynomial, i.e. it will contain both positive and negative degrees of the variable. Kauffman's first rule says that the bracket of the trivial knot (unknot) is equal to 1. Second, we want a method for obtaining the bracket polynomial of a link in terms of the bracket polynomials of simpler links. We use the following skein relation. Given a crossing in our link projecton, we split it open vertically and horizontally, in order to obtain two new link projections, each of which has one fewer crossing. We make the bracket polynomial of our link projection a linear combination of the bracket polynomials of our two new link projections, where we have not decided on the coefficients, so we just call them A and B. We consider here two equations. The second equation here is just the first equation looked at from a perpendicular view. If you bend your neck so that your head is horizontal and look at the first tangle of the second equation, it will appear the same as the first tangle in the first equation. Finally, we would like a rule for adding in a trivial component to a link (the result of which will always be a split link). So we say

$$< L \cap O > = C < L >$$

Each time we add in an extra trivial component that is not tangled up with the original link, we just multiply the entire polynomial by C. As with A and B, we consider C a variable in the polynomial.

In order to show that a given polynomial is in fact knot/link invariant, it is necessary and sufficient to show that the invariant in question is unchanged under each of the three Reidemeister moves. We need a following notion. *Planar isotopy* is the motion of a diagram in the plane that preserves the graphical structure of the underlying projection. A knot or a link is said to be *ambient isotopic* to another if there is a sequence of Reidemeister moves and planar equivalences between them.

Luckily, the Kauffman bracket is invariant under Reidemeister moves II and III i.e. is an invariant of regular isotopy. Unluckily, the "naive" Kauffman bracket is not an invariant under Reidemeister I. The next task is to find a way to make it work.

Kauffman found a way to add some terms to the bracket, and this solved the problem.

3.5 The Jones Polynomial

The Jones polynomial was discovered by Vaughan F.R. Jones in 1984. Unlike the Alexander polynomial, the Jones polynomial distingushes between a knot and its mirror image. The Jones polynomial is essentially the same as the Kauffman bracket polynomial. In fact, they can be derived from each other through the equation

$$V_L(t) = f[L]t^{-\frac{1}{4}}$$

where $V_L(t)$ is the Jones polynomial and $f[L]$ is the Kauffman polynomial.

The Jones polynomial is derived from an oriented knot diagram through two basic rules:

1. $V_U(t) = 1$
2. $t^{-1}V_{pos}(t) - tV_{neg}(t) = (t^{\frac{1}{2}} - t^{-\frac{1}{2}})V_{zero}(t)$

It was already said above that the Kauffman bracket is invariant under Reidemeister moves II and III but not under Reidemeister move I. To overcome this obstacle Kauffman invented simple but surprising trick that altered his bracket. Let K be a knot or link. Let $|K|$ be the nonoriented diagram obtained from the oriented diagram of K by forgetting its orientation. Let $< . >$ be the Kauffman bracket. Kauffman introduced

$$X(K) = (-a)^{-3w(K)} < |K| >$$

This little additional factor changed everything. Now the modified Kauffman bracket polynomial turns out to be invariant under all the Reidemeister moves. Moreover, this invariant is equivalent to Jones polynomial.

4 New Invariants

Alexander polynomial was the only polynomial invariant of knots for over 50 years. But when Jones discovered his polynomial, many mathematicians introduced their polynomial invariants. The first one in this series was so called HOMFLY polynomial. This is not the name of the inventor. Rather this is an acronym for 6 independent inventors. They are H= Hoste, O=Ocneanu, M=Millet, F=Freyd, L=Lickorish and Y=Yetter. Later several more letters were added to this acronym. The HOMFLY polynomial is a polynomial of two variables.

Vladimir Vassiliev [Va 90] invented a whole class of polynomial invariants. His approach to knots could be called sociological (as Vaughan Jones does). He considers the space of all knots in which knots are only points a nd therefore have lost their intrinsic properties. Moreover, Vassiliev does not go looking for one invariant - he wants to find all of them, to define the entire space of invariants. In the same way that classical sociology makes an abstraction of the personality of the people it studies, focusing only on their position in the social, economic or other stratification. This approch is a trademark of the catastrophe theory but it was used much earlier by David Hilbert, and it is characteristic for category theory.

5 Statistical Mechanics

It is instructive to see how V.Jones discovered his polynomial invariant. The discovery came indirectly by way of a branch of quantum mechanics called Von Neumann algebras. These were developed to handle quantum mechanical observables such as energy, position and momentum. The capacity of the operators representing such quantities to be added or multiplied results in them having the structure of an algebra. Von Neumann algebras can be built out of simpler structures called factors which have the intriguing property that they can have "continuous dimensions" i.e. real numbers such as g or $\frac{1}{27}$. Jones was studying subfactors when he discovered that, rather than having continuous dimensions the only dimensions less than 4 were $4cos2\frac{g}{n}$.

While showing the proof to some friends at Geneva, it was remarked that sections resembled the group of a braid, which is like a knot except that it is a series of threads beginning at the top which are woven over and under before being realigned at the bottom. A braid can be converted into a knot by joining its ends together. But the reverse process is not so easy.

Vaughan Jones ended up having a meeting at Columbia University with knot theorist Joan Birman to see if his work might have some application in knot theory. When the two sat down together, the discovery was almost instantaneous.

Jones proved that Von Neumann algebras are related to knot theory and provide a way to tell very complicated knots apart.

6 Jones Polynomial and Quantum Computing

The Jones polynomial involves a complicated mathematical formula, and although calculating it is easy for simple knots, it is enormously difficult for big knots. Mathematicians have found evidence that the difficulty of computing Jones polynomials rises exponentially with the size of the knot.

Just for the comparison, we consider several invariant polynomials for the same knot (whatever it is).

Alexander polynomial is: $3t^2 - 11t^{-1} + 17 - 11t + 3t^2$.

Convay polynomial of the same knot is: $1 + z^2 + 3z^4$.

HOMFLY-PT polynomial is: $-a^{-2} - 2a^{-2}z^2 + 3 + 4z^2 + 2z^4 - a^2 + a^2z^4 - a^4z^2$.

Kauffman polynomial is: $-2a^{-3}z + 3a^{-3}z^3 + a^{-2} - 6a^{-2}z^2 + 4a^{-2}z^4 + a^{-2}z^6 - 5a^{-1}z + 10a^{-1}z^3 - 6a^{-1}z^5 + 3a^{-1}z^7 + 3 - 9z^2 + 8z^4 - 3z^6 + 2z^8 - 5az + 16az^3 - 17az^5 + 7az^7 + a^2 - 3a^2z^4 - a^2z^6 + 2a^2z^8 - 2a^3z + 7a^3z^3 - 10a^3z^5 + 4a^3z^7 + 3a^4z^2 - 7a^4z^4 + 3a^4z^6 - 2a^5z^3 + a^5z^5$.

However, the Jones polynomial is: $2q^{-15} - 3q^{-14} + q^{-13} + 9q^{-12} - 14q^{-11} - 5q^{-10} + 30q^{-9} - 21q^{-8} - 24q^{-7} + 52q^{-6} - 17q^{-5} - 46q^{-4} + 62q^{-3} - 7q^{-2} - 57q^{-1} + 56 + 4q - 50q^2 + 35q^3 + 11q^4 - 29q^5 + 12q^6 + 8q^7 - 8q^8 + 3q^{10} - q^{-30} + 3q^{-29} - q^{-28} - 5q^{-27} - q^{-26} + 14q^{-25} + 7q^{-24} - 28q^{-23} - 22q^{-22} + 38q^{-21} + 52q^{-20} - 38q^{-19} - 92q^{-18} + 21q^{-17} + 132q^{-16} + 16q^{-15} - 163q^{-14} - 70q^{-13} + 184q^{-12} + 124q^{-11} - 184q^{-10} - 182q^{-9} + 177q^{-8} + 229q^{-7} - 160q^{-6} - 265q^{-5} + 140q^{-4} + 286q^{-3} - 109q^{-2} - 301q^{-1} + 82 + 290q - 38q^2 - 276q^3 + 7q^4 + 229q^5 + 36q^6 - 184q^7 - 53q^8 + 121q^9 + 65q^{10} - 72q^{11} - 51q^{12} + 28q^{13} + 36q^{14} - 5q^{15} - 21q^{16} + q^{17} + 3q^{18} + 4q^{19} - 2q^{20} + 4q^{-50} - 3q^{-49} + q^{-48} + 5q^{-47} - 3q^{-46} + q^{-45} - 17q^{-44} + 5q^{-43} + 31q^{-42} + 3q^{-41} + 5q^{-40} - 81q^{-39} - 33q^{-38} + 80q^{-37} + 78q^{-36} + 105q^{-35} - 170q^{-34} - 204q^{-33} - 11q^{-32} + 149q^{-31} + 424q^{-30} - 46q^{-29} - 381q^{-28} - 378q^{-27} - 76q^{-26} + 765q^{-25} + 413q^{-24} - 219q^{-23} - 780q^{-22} - 682q^{-21} + 775q^{-20} + 932q^{-19} + 323q^{-18} - 898q^{-17} - 1358q^{-16} + 442q^{-15} + 1218q^{-14} + 939q^{-13} - 744q^{-12} - 1820q^{-11} + 26q^{-10} + 1269q^{-9} + 1389q^{-8} - 505q^{-7} - 2034q^{-6} - 327q^{-5} + 1193q^{-4} + 1657q^{-3} - 246q^{-2} - 2039q^{-1} - 635 + 972q + 1746q^2 + 98q^3 - 1771q^4 - 896q^5 + 529q^6 + 1568q^7 + 485q^8 - 1180q^9 - 939q^{10} - 17q^{11} + 1046q^{12} + 664q^{13} - 456q^{14} - 643q^{15} - 333q^{16} + 404q^{17} + 480q^{18} + 4q^{19} - 221q^{20} - 268q^{21} + 26q^{22} + 168q^{23} + 77q^{24} - 6q^{25} - 80q^{26} - 31q^{27} + 16q^{28} + 18q^{29} + 12q^{30} - 5q^{31} - 6q^{32} + q^{345} - q^{-75} + 3q^{-74} - q^{-73} - 5q^{-72} + 3q^{-71} + 3q^{-70} + 2q^{-69} + 5q^{-68} - 7q^{-67} - 26q^{-66} - 7q^{-65} + 28q^{-64} + 42q^{-63} + 39q^{-62} - 17q^{-61} - 102q^{-60} - 127q^{-59} - 17q^{-58} + 149q^{-57} + 250q^{-56} + 183q^{-55} - 105q^{-54} - 419q^{-53} - 474q^{-52} - 114q^{-51} + 460q^{-50} + 842q^{-49} + 622q^{-48} - 218q^{-47} - 1144q^{-46} - 1349q^{-45} - 434q^{-44} + 1098q^{-43} + 2104q^{-42} + 1544q^{-41} - 517q^{-40} - 2612q^{-39} - 2898q^{-38} - $

$691q^{-37} + 2578q^{-36} + 4209q^{-35} + 2408q^{-34} - 1847q^{-33} - 5178q^{-32} - 4376q^{-31} + 507q^{-30} + 5563q^{-29} + 6246q^{-28} + 1318q^{-27} - 5366q^{-26} - 7815q^{-25} - 3267q^{-24} + 4681q^{-23} + 8900q^{-22} + 5146q^{-21} - 3692q^{-20} - 9584q^{-19} - 6731q^{-18} + 2623q^{-17} + 9897q^{-16} + 8002q^{-15} - 1618q^{-14} - 9989q^{-13} - 8957q^{-12} + 737q^{-11} + 9940q^{-10} + 9698q^{-9} + 10q^{-8} - 9816q^{-7} - 10229q^{-6} - 747q^{-5} + 9557q^{-4} + 10715q^{-3} + 1495q^{-2} - 9169q^{-1} - 10991 - 2384q + 8414q^2 + 11191q^3 + 3386q^4 - 7379q^5 - 10965q^6 - 4454q^7 + 5826q^8 + 10402q^9 + 5405q^{10} - 4024q^{11} - 9195q^{12} - 6043q^{13} + 1987q^{14} + 7564q^{15} + 6145q^{16} - 186q^{17} - 5470q^{18} - 5652q^{19} - 1263q^{20} + 3434q^{21} + 4596q^{22} + 1988q^{23} - 1583q^{24} - 3245q^{25} - 2134q^{26} + 325q^{27} + 1932q^{28} + 1728q^{29} + 354q^{30} - 874q^{31} - 1136q^{32} - 553q^{33} + 252q^{34} + 604q^{35} + 404q^{36} + 41q^{37} - 212q^{38} - 248q^{39} - 96q^{40} + 69q^{41} + 79q^{42} + 56q^{43} + 12q^{44} - 30q^{45} - 23q^{46} - q^{47} + 3q^{48} + 2q^{49} + 4q^{50} - 2q^{51}.$

Calculating Jones polynomial for complicated knots is considrered beyond the reach of even the fastest computers. However in the late 1980s physicist Edward Witten [Wi 89] described a physical system that should calculate information about the Jones polynomial. The idea of a physical system calculating something about knots or other loops may sound strange, but in fact examples of such systems exist. In an electrical transformer two loops of wire are coiled around an iron core. The electric current passing through on of the wires generates voltage in the other wire that is proportional to the number of times the second wire twists around the core. Thus, even if you cannot see the wire, you can figure out its number of twists simply measuring the voltage. Witten proposed, in a similar way, that it should be possible to obtain approximate information about the Jones polynomial of a knot by taking appropriate measurements in a more complicated physical system. Topological quantum field theory was created, and this lead to creation of topological quantum computation by Michael H. Freedman, Alexei Kitaev, Michael J.Larsen and Zhenghan Wang [FKLW 01].

Witten gave a heuristic definition of the Jones polynomial in terms of a topological quantum field theory, following the outline of a program proposed by M.Atiyah [At 88]. Specifically, he considered a knot in a 3-manifold and a connection A on some principal G-bundle, with G a simple Lie group. The Chern-Simons functional associates a number $CS(A)$ to A, but it is well-defined only up to an integer, so the quantity $exp(2\pi ikCS(A))$ is well-defined. Also, the holonomy of the connection around the knot is an element of G well-defined up to conjugation, so the trace of it with respect to a given representation is well-defined. Multiplying these two gives a number depending on the knot, the manifold, the representation and the connection. The magic comes when we average over all the connections and all principal bundles. Of course, this makes no sense, since there is no apparent measure on the infinite-dimenional space of connections. But proceeding heuristically, such an average should depend only on the manifold, representation and the isotopy type of the knot. Witten argued using a close correspondence with conformal field theory, that when the manifold is S^3 and the representation is the fundamental one, this invariant had combinatorial properties that forced it to be the analogue of the Jones polynomial for the given group. Needless to say, a long physics tradition of very successful heuristic reasoning along these lines suggested to Witten that this ill-defined average should make sense in this case.

Strangely enough, Freedman had ideas about relations between knot theory and quantum field theory as early as in 1980. No theory of quantum computing existed then. His colleagues physicists considered his ideas too abstract that time. Freedman postponed his reasearch in that direction, got Fields Medal for knot theory, and returned to this topic only after the results by A.Kitaev [FKW 02, FKLW 01]. Topological quantum computer would not be able to provide the exact value of the Jones polynomial, only approximation. However, it would be no less efficient rather than the well-known qubit quantum computer.

More can be said about the possibilities of topological quantum computers. When practically built, they will be able to compute approximate values of Jones polynomials. Hence, computation of Jones polynomial becomes an etalon problem for topological quantum computation. If there is problem for which topological quantum computers can compute in a polynomial time a function not computable in polynomial time by classical computers, then computation of Jones polynomial is also such a problem.

7 Back to NP

Outside of theoretical computer science, parallel computers are sometimes discussed as they were fundamentally more powerful than serial computers. But of course, anything that can be done with 10^{20} processors in time T can also be done with one processor in time $10^{20}T$. When quantum computing came along, it was hoped that we might have a type of parallelism commeasurate with the difficulty of NP-complete problems. For in quantum mechanics, we need a vector of 2^n complex numbers called "amplitudes" just to specify the state of an n-bit computer. Surely, we could exploit this exponentiality inherent in Nature to try out all 2^n possible solutions to an NP-complete problems in parallel.

Unfortunately, we do not know whether this is possible. Let BQP denote the class of problems solvable in polynomial time by a quantum computer. We still do not know whether $NP = BQP$. Moreover, Bennett, Bernstein, Brassard,and Vazirani proved in [BBBV 97] that it is not so for a specific oracle. We have seen advantages in size of quantum finite automata [AF 98] but we do not know what happens in the case of larger memory.

What are the real advantages of quantum computation? This is a challenging problem. Quantum skeptics sometimes argue that we do not really know whether quantum mechanics itself will remain valid in the regime tested by quantum computing. Leonid Levin [Le 03] writes: "The major problem [with quantum computing] is the requirement that basic quantum equations hold to multi-hundredth if not millionth decimal positions where the significant digits of the relevant quantum amplitudes reside. We have never seen a physical law valid to over a dozen decimals."

Scott Aaronson answers to him in [Aa 05]: "The irony is that most of the specific proposals for how quantum mechanics could be wrong suggest a world with more, not less, computational power than BQP. For, as we saw in $[\cdots]$, the linearity of quantum mechanics is what prevents one needle in an exponentially large haystack from shouting above the others."

Who is right, who is wrong? Only the future will show. But today we see that Jones polynomial has something to say in this challenging discussion.

References

[Aa 05]   Scott Aaronson. Guest column: NP-complete problems and physical reality. *ACM SIGACT News,* vol. 36, No. 1, pp. 30–52, 2005.

[Ad 94]   Colin C. Adams. *The Knot Book. An Elementary Introduction to the Mathematical Theory of Knots,* American Mathematical Society, 1994.

[ADO 92]   Y. Akutsu, T. Deguchi, and T. Ohtsuki. Invariants of colored links. *Journal of Knot Theory Ramifications,* vol. 1, No. 2, pp. 161–184, 1992.

[AKN]   Dorit Aharonov, Alexei Kitaev, Noam Nisan. Quantum Circuits with Mixed States.

[AJL 05]   D. Aharonov, V. Jones, Z. Landau. On the quantum algorithm for approximating the Jones polynomial. Unpublished, 2005.

[AL 28]   James Waddell Alexander. Topological invariants of knots and links. *Transactions of American Mathematical Society,* vol. 30, pp. 275–306, 1928.

[AF 98]   Andris Ambainis, Rūsiņš Freivalds. 1-way quantum finite automata: strengths, weaknesses and generalizations. *Proc. IEEE FOCS'98,* pp. 332–341, 1998. Also quant-ph/9802062.

[AKV 02]   Andris Ambainis, Arnolds Ķikusts, Māris Valdats. On the class of languages recognizable by 1-way quantum finite automata.

[At 88]   Michael Francis Atiyah. New invariants of three and four dimensional manifolds. In: *The mathematical heritage of Herman Weyl,* vol. 48 of *Proc. Symp. Pure Math.,* American Mathematical Society, 1988.

[BBBV 97] Charles Bennett, Ethan Bernstein, Gilles Brassard, Umesh Vazirani. Strengths and weaknesses of quantum computing. *SIAM Journal on Computing,* vol. 26, No. 5, pp. 1510–1523, 1997.

[BV 97]   Ethan Bernstein, Umesh Vazirani, Quantum complexity theory. *SIAM Journal on Computing,* 26:1411–1473, 1997.

[BP 99]   Alex Brodsky, Nicholas Pippenger. Characterizations of 1-way quantum finite automata. quant-ph/9903014.

[FKW 02]   Michael H. Freedman, Alexei Kitaev, Zhenghan Wang. Simulation of topological field theories by quantum computers. *Communications in Mathematical Physics,* vol. 227, No. 3, pp. 587–603, 2002.

[FKLW 01] Michael H. Freedman, Alexei Kitaev, Michael J. Larsen, Zhenghan Wang. Topological quantum computation. quant-ph/0101025.

[Jo 89]   Vaughan F.R. Jones. On knot invariants related to some statistical mechanical models. *Pacific Journal of Mathematics,* vol. 137, No. 2, pp. 311–334, 1989.

[Jo 90]   Vaughan F.R. Jones. Knot theory and statistical mechanics. *Scientific American,* vol. 263, No. 5, pp. 98–103, 1990.

[JR 03]   Vaughan F.R. Jones, Sarah A. Reznikoff. Hilbert space representations of the annular Temperley-Lieb algebra. http://math.berkeley.edu/ vfr/hilbertannular.ps

[Ka 87]   Louis H. Kauffman. State models and the Jones polynomial. *Topology*, vol. 26, No. 3, pp. 395–407, 1987.
[Ka 88]   Louis H. Kauffman. New invariants in the theory of knots. *American Mathematical Monthly*, vol. 95, No. 3, pp. 195–242, 1988.
[Ka 94]   Louis H. Kauffman. Knot Automata. *Proc. ISMVL*, pp. 328–333, 1994.
[Ka 03]   Louis H. Kauffman. Review of "Knots" by Alexei Sossinsky, Harvard University Press, 2002, ISBN 0-674-00944-4. http://arxiv.org/abs/math.HO/0312168
[KL 94]   Louis H. Kauffman and S. L. Lins. *Temperley-Lieb recoupling theory and invariant of 3manifolds*, Princeton University Press, Princeton, 1994.
[KS 91]   Louis H. Kauffman and H. Saleur. Free fermions and the Alexander-Conway polynomial. *Comm. Math. Phys.*, vol. 141, No. 2, pp. 293–327, 1991.
[Ki 1883] Thomas Penyngton Kirkman. The enumeration, description and construction of knots with fewer than 10 crossings. *Transactions R.Soc. Edinburgh*, vol. 32, pp. 281–309, 1883.
[Le 03]   Leonid A. Levin. Polynomial time and extravagant machines, in the tale of one-way machines. *Problems of Information Transmission*, vol. 39, No. 1, pp. 92–103, 2003.
[Li 1900] C.N. Little. Non-alternate + - knots. *Transactions R.Soc. Edinburgh*, vol. 39, pp. 771–778, 1900.
[MM 01]   Hitoshi Murakami, Jun Murakami. The colored Jones polynomials and the simplicial volume of a knot. *Acta Mathematica*, vol. 186, pp. 85–104, 2001. Also http://arxiv.org/abs/math/9905075
[Re 32]   Kurt Werner Friedrich Reidemeister. Knotentheorie. *Eregebnisse der Mathematik und ihrer Grenzgebiete (Alte Folge 0, Band 1, Heft 1)*, (Reprint Springer-Verlag, Berlin, 1974.
[Si 94]   D. Simon. On the power of quantum computation. *Proc. IEEE FOCS*, pp. 116–123, 1994.
[So 02]   Alexei Sossinsky. *Knots. Mathematics with a Twist*, Harvard University Press, 2002.
[Ta 1898] Peter Guthrie Tait. On knots I, II, III. in *Scientific papers*, vol. 1, pp. 273–347, London: Cambridge University Press, 1898.
[Th 1867] William Thomson. Hydrodynamics. *Transactions R.Soc. Edinburgh*, vol. 6, pp. 94–105, 1867.
[Th 1869] William Thomson. On vortex motion. *Transactions R.Soc. Edinburgh*, vol. 25, pp. 217–260, 1869.
[Thu 77] William Paul Thurston. *The Geometry and Topology of Three-Manifolds*, Princeton University Lecture Notes, 1977.
[Thu 97] William Paul Thurston. *Three-Dimensional Geometry and Topology*, vol.1, Princeton Lecture Notes, 1997.
[Va 90]   Vladimir A. Vassiliev. Cohomology of Knot Spaces. In *Theory of Singularities and Its Applications* (Ed. V. I. Arnold). Providence, RI: Amer. Math. Soc., pp. 23–69, 1990.
[Wi 89]   Edward Witten. Quantum field theory and the Jones polynomial. *Communications in Mathematical Physics*, vol. 121, No. 3, pp. 351Ű-399, 1989.

Interactive Algorithms 2005

Yuri Gurevich

Microsoft Research, One Microsoft Way,
Redmond, WA 98052, USA

Abstract. A sequential algorithm just follows its instructions and thus cannot make a nondeterministic choice all by itself, but it can be instructed to solicit outside help to make a choice. Similarly, an object-oriented program cannot create a new object all by itself; a create-a-new-object command solicits outside help. These are but two examples of intra-step interaction of an algorithm with its environment. Here we motivate and survey recent work on interactive algorithms within the Behavioral Computation Theory project.

1 Introduction

In 1982, the University of Michigan hired this logician on his promise to become a computer scientist. The logician eagerly wanted to become a computer scientist. But what is computer science? Is it really a science? What is it about?

After thinking a while, we concluded that computer science is largely about algorithms. Operating systems, compilers, programming languages, etc. are all algorithms, in a wide sense of the word. For example, a programming language can be seen as a universal algorithm that applies the given program to the given data. In practice, you may need a compiler and a machine to run the compiled program on, but this is invisible on the abstraction level of the programming language.

A problem arises: What is an algorithm? To us, this is a fundamental problem of computer science, and we have been working on it ever since.

But didn't Turing solve the problem? The answer to this question depends on how you think of algorithms. If all you care is the input-to-output function of the algorithm, then yes, Turing solved the problem. But the behavior of an algorithm may be much richer than its input-to-output function. An algorithm has its natural abstraction level, and the data structures employed by an algorithm are intrinsic to its behavior. The parallelism of a parallel algorithm is an inherent part of its behavior. Similarly, the interactivity of an interactive algorithm is an inherent part of its behavior as well.

Is there a solution à la Turing to the problem what an algorithm is? In other words, is there a state-machine model that captures the notion of algorithm up to behavioral equivalence? Our impression was, and still is, that the answer is yes. In [13], we defined sequential abstract state machines (ASMs) and put forward a sequential ASM thesis: for every sequential algorithm, there is a sequential ASM with the same behavior. In particular, the ASM is supposed to simulate the given algorithm step-for-step. In [14], we defined parallel and distributed abstract state machines and generalized the ASM thesis for parallel and distributed algorithms. Parallel ASMs gave rise to a specification (and high-level programming) language AsmL [2] developed by the group of Foundations of Software Engineering of Microsoft Research.

At this point, the story forks. One branch leads to experimental evidence for the ASM thesis and to applications of ASMs [1,2,12]. Another branch leads to behavioral computation theory. We take the second branch here and restrict attention to *sequential time algorithms* that compute in a sequence of discrete steps.

In §2 we discuss a newer approach to the explication of the notion of algorithm. The new approach is axiomatic, but it also involves a machine characterization of algorithms. This newer approach is used in the rest of the article.

In §3 we sketch our explication of sequential (or small-step) algorithms [15]. We mention also the explication of parallel (or wide-step) algorithms in [3] but briefly. In either case, the algorithms in questions are *isolated step algorithms* that abstain from intra-step interaction with the environment. They can interact with the environment in the inter-step manner, however.

§4 is a quick introduction to the study of intra-step interaction of an algorithm with its environment; much of the section reflects [5]. We motivate the study of intra-step interaction and attempt to demonstrate how ubiquitous intra-step interaction is. Numerous disparate phenomena are best understood as special cases of intra-step interaction. We discuss various forms of intra-step interaction, introduce the query mechanism of [5] and attempt to demonstrate the universality of the query mechanism: the atomic interactions of any mechanism are queries. In the rest of the article, we concentrate on intra-step interaction; by default interaction means intra-step interaction. To simplify the exposition, we consider primarily the small-step (rather than wide-step) algorithms; by default algorithms are small-step algorithms.

§5 is devoted to the explication of *ordinary interactive algorithms* [5,6,7]. Ordinary algorithms never complete a step until all queries from that step have been answered. Furthermore, the only information from the environment that an ordinary algorithm uses during a step is answers to its queries.

§6 is devoted to the explication of general interactive algorithms [8,9,10]. Contrary to ordinary interactive algorithms, a general interactive algorithm can be *impatient* and complete a step without waiting for all queries from that step to have been answered. It also can be *time sensitive*, so that its actions during a step depend not only on the answers to its queries but also on the order in which the answers have arrived. We mention also the explication of general wide-step algorithms [11] but briefly.

§7 is a concluding remark.

Much of this article reflects joint work with Andreas Blass, Benjamin Rossman and Dean Rosenzweig.

2 Explication of Algorithms

The theses mentioned in the introduction equate an informal, intuitive notion with a formal, mathematical notion. You cannot prove such a thesis mathematically but you can argue for it. Both Church and Turing argued for their theses. While their theses are equivalent, their arguments were quite different [4]. The ASM theses, mentioned in the introduction, have the following form.

ASM Thesis Form

1. Describe informally a class **A** of algorithms.
2. Describe the behavioral equivalence of **A** algorithms. Intuitively two algorithms are behaviorally equivalent if they do the same thing in all circumstances. Since **A** is defined informally, the behavioral equivalence may be informal as well.
3. Define a class **M** of abstract state machines.
4. Claim that **M** ⊆ **A** and that every $A \in$ **A** is behaviorally equivalent to some $M \in$ **M**.

The thesis for a class **A** of algorithms explicates algorithms in **A** as abstract state machines in **M**. For example, sequential algorithms are explicated as sequential ASMs. The thesis is open to criticism. One can try to construct an ASM in **M** that falls off **A** or an algorithm in **A** that is not behaviorally equivalent to any ASM in **M**.

Since the ASM thesis for **A** cannot be proven mathematically, experimental confirmation of the thesis is indispensable; this partially explains the interest in applications of ASMs in the ASM community. But one can argue for the thesis, and we looked for the best way to do that. Eventually we arrived at a newer and better explication procedure.

Algorithm Explication Procedure

1. Axiomatize the class **A** of the algorithms of interest. This is the hardest part. You try to find the most convincing axioms (or postulates) possible.
2. Define precisely the notion of behavioral equivalence. If there is already an ASM thesis T for **A**, you may want to use the behavioral equivalence of T or a precise version of the behavioral equivalence of T.
3. Define a class **M** of abstract state machines. If there is already an ASM thesis T for **A**, you may want to use the abstract state machines of T.
4. Prove the following characterization theorem for **A**: **M** ⊆ **A** and every $A \in$ **M** is behaviorally equivalent to some $M \in$ **M**.

The characterization provides a theoretical programming language for **A** and opens a way for more practical languages for **A**. Any instance of the explication procedure is open to criticism of course. In particular, one may criticize the axiomatization and the behavioral equivalence relation.

If an explication procedure for **A** uses (a precise version of) the behavioral equivalence and the machines of the ASM thesis for **A**, then the explication procedure can be viewed as a proof of the thesis given the axiomatization.

A priori it is not obvious at all that a convincing axiomatization is possible. But our experience seems to be encouraging. The explication procedure was used for the first time in [15] where sequential algorithms were axiomatized and the sequential ASM thesis proved; see more about that in the next section. In [3], parallel algorithms were axiomatized and the parallel ASM thesis was proved, except that we slightly modified the notion of parallel ASM. Additional uses of the explication procedure will be addressed in § 4–6.

In both, [15] and [3], two algorithms are behaviorally equivalent if they have the same states, initial states and transition function. It follows that behaviorally equivalent algorithms simulate each other step-for-step. We have been criticized that this behavioral equivalence is too fine, that step-for-step simulation is too much to require, that appropriate bisimulation may be a better behavioral equivalence. We agree that in some applications bisimulation is the right equivalence notion. But notice this: the finer the behavioral equivalence, the stronger the characterization theorem.

3 Isolated-Step Algorithms

As we mentioned above, sequential algorithms were explicated in [15]. Here we recall and motivate parts of that explication needed to make our story self-contained.

Imagine that you have some entity E. What does it mean that E is a sequential algorithm? A part of the answer is easy: every algorithm is a (not necessarily finite-state) automaton.

Postulate 3.1 (Sequential Time). *The entity E determines*

- *a nonempty collection of states,*
- *a nonempty collection of initial states, and*
- *a state-transition function.*

The postulate does not say anything about final states; we refer the interested reader to [15, § 3.3.2] in this connection. This single postulate allows us to define behavioral equivalence of sequential algorithms.

Definition 3.2. Two sequential algorithms are *behaviorally equivalent* if they have the same states, initial states and transition function.

It is harder to see what else can be said about sequential algorithms in full generality. Of course, every algorithm has a program of one kind or another, but we don't know how to turn this into a postulate or postulates. There are so many different programming notations in use already, and it is bewildering to imagine all possible programming notations.

Some logicians, notably Andrey A. Markov [17], insisted that the input to an algorithm should be *constructive*, like a string or matrix, so that you can actually write it down. This excludes abstract finite graphs for example. How would you put an abstract graph on the Turing machine tape? It turned out, however, that the constructive input requirement is too restrictive. Relational databases for example represent abstract structures, in particular graphs, and serve as inputs to important algorithms.

Remark 3.3. You can represent an abstract graph by an adjacency matrix. But this representation is not unique. Note also that it is not known whether there is a polynomial-time algorithm that, given two adjacency matrices, determines whether they represent the same graph.

A characteristic property of sequential algorithms is that they change their state only locally in any one step. Andrey N. Kolmogorov, who looked into this problem, spoke about "steps whose complexity is bounded in advance" [16]. We prefer to speak about bounded work instead; the amount of work done by a sequential algorithm in any one step is bounded, and the bound depends only on the algorithm and not on the state or the input. But we don't know how to measure the complexity of a step or the work done during a step. Fortunately we found a way around this difficulty. To this end, we need two additional postulates.

According to the abstract state postulate, all states of the entity E are structures (that is first-order structures) of a fixed vocabulary. If X is an (initial) state of A and a structure Y is isomorphic to X then Y is an (initial) state of A. The abstract state postulate allows us to introduce an abstract notion of location and to mark locations explored by an algorithm during a given step. The bounded exploration postulate bounds the number of locations explored by an algorithm during any step; the bound depends only on the algorithm and not on the state or the input. See details in [15].

Definition 3.4. A sequential algorithm is any entity that satisfies the sequential-time, abstract-state and bounded-exploration postulates.

A *sequential abstract state machine* is given is by a program, a nonempty isomorphism-closed collection of states and a nonempty isomorphism-closed subcollection of initial states. The program determines the state transition function.

Like a Turing machine program, a sequential ASM program describes only one step of the ASM. It is presumed that this step is executed over and over again. The machine halts when the execution of a step does not change the state of the machine. The simplest sequential ASM programs are assignments:

$f(t_1, \ldots, t_j) := t_0$

Here f is a j-ary *dynamic function* and every t_i is a ground first-order term. To execute such a program, evaluate every t_i at the given state; let the result be a_i. Then set the value of $f(a_1, \ldots, a_j)$ to a_0. Any other sequential ASM program is constructed from assignments by means of two constructs: if-then-else and do-in-parallel. Here is a sequential ASM program for the Euclidean algorithm: given two natural numbers a and b, it computes their greatest common divisor d.

Example 3.5 (Euclidean Algorithm 1).
```
if a = 0 then d := b
else do in-parallel
   a := b mod a
   b := a
```

The do-in-parallel constructs allows us to compose and execute in parallel two or more programs. In the case when every component is an assignment, the parallel composition can be written as a simultaneous assignment. Example 3.5 can be rewritten as

```
if a = 0 then d := b
else a, b := b mod a, a
```

A question arises what happens if the components perform contradictory actions in parallel, for example,

```
do in-parallel
   x := 7
   x := 11
```

The ASM breaks down in such a case. One can argue that there are better solutions for such situations that guarantee that sequential ASMs do not break down. In the case of the program above, for example, one of the two values, 7 or 11, can be chosen in one way or another and assigned to x. Note, however, that some sequential algorithms do break down. That is a part of their behavior. If sequential ASMs do not ever break down, then no sequential ASM can be behaviorally equivalent to a sequential algorithm that does break down.

In the Euclidean algorithm, all dynamic functions are nullary. Here is a version of the algorithm where some of dynamic functions are unary. Initially $mode = s = 0$.

Example 3.6 (Euclidean Algorithm 2).
```
if mode = 0 then a(s), b(s), mode := Input1(s), Input2(s), 1
elseif mode = 1 then
   if a(s) = 0 then d(s), s, mode := b(s), s+1, 0
   else a(s), b(s) := b(s) mod a(s), a(s)
```

Theorem 3.7 (Sequential Characterization Theorem). *Every sequential ASM is a sequential algorithm, and every sequential algorithm is behaviorally equivalent to a sequential ASM.*

We turn our attention to parallel algorithms and quote from [4]: "The term 'parallel algorithm' is used for a number of different notions in the literature. We have in mind sequential-time algorithms that can exhibit unbounded parallelism but only bounded sequentiality within a single step. Bounded sequentiality means that there is an *a priori* bound on the lengths of sequences of events within any one step of the algorithm that must occur in a specified order. To distinguish this notion of parallel algorithms, we call such parallel algorithms *wide-step*. Intuitively the width is the amount of parallelism. The 'step' in 'wide-step' alludes to sequential time." Taking into account the bounded sequentiality of wide-step algorithms, they could be called "wide and shallow step algorithms".

4 Interaction

4.1 Inter-step Interaction

One may have the impression that the algorithms of the previous section do not interact at all with the environment during the computation. This is not necessarily so. They do not interact with the environment during a step; we call such algorithm *isolated step algorithms*. But the environment can intervene between the steps of an algorithm. The environment preserves the vocabulary of the state but otherwise it can change the state in any way. It makes no difference in the proofs of the two characterization theorems whether inter-step interaction with the environment is or is not permitted.

In particular, Euclidean Algorithm 2 could be naturally inter-step interactive; the functions Input1 and Input2 do not have to be given ahead of time. Think of a machine that repeatedly applies the Euclidean algorithm and keeps track of the number s of the current session. At the beginning of session s, the user provides numbers Input1(s) and Input2(s), so that the functions Input1(s) and Input2(s) are *external*. The inter-step interactive character of the algorithm becomes obvious if we make the functions Input1, Input2 nullary.

Example 4.1 (Euclidean Algorithm 3).

```
if mode = 0 then a(s), b(s), mode := Input1, Input2, 1
elseif mode = 1 then
   if a(s) = 0 then d(s), s, mode := b(s), s+1, 0
   else a(s), b(s) := b(s) mod a(s), a(s)
```

4.2 Intra-step Interaction

In applications, however, much of the interaction of an algorithm with its environment is intra-step. Consider for example an assignment

   x := g(f(7))

where $f(7)$ is a remote procedure call whose result is used to form another remote procedure call. It is natural to view the assignment being done within one step. Of course, we can break the assignment into several steps so that interaction is inter-step but this forces us to a lower abstraction level. Another justification of intra-step interaction is related to parallelism.

Example 4.2. This example reflects a real-world AsmL experience. To paint a picture, an AsmL application calls an outside paint applications. A paint agent is created, examines the picture and repeatedly calls the algorithm back: what color for such and such detail? The AsmL application can make two or more such paint calls in parallel. It is natural to view parallel conversations with paint agents happening intra-step.

Proviso 4.3. In the rest of this article, we concentrate on intra-step interaction and ignore inter-step interaction. By default, interaction is intra-step interaction.

4.3 The Ubiquity of Interaction

Intra-step interaction is ubiquitous. Here are some examples.

- Remote procedure calls.
- Doing the following as a part of expression evaluation: getting input, receiving a message, printing output, sending a message, using an oracle.
- Making nondeterministic choices among two or more alternatives.
- Creating new objects in the object-oriented and other paradigms.

The last two items require explanation. First we address nondeterministic choices. Recall that we do not consider distributed algorithms here. A sequential-step algorithm just follows instructions and cannot nondeterministically choose all by itself. But it can solicit help from the environment, and the environment may be able to make a choice for the algorithm. For example, to evaluate an expression

$$\text{any } x \mid x \text{ in } \{0, 1, 2, 3, 4, 5\} \text{ where } x > 1$$

an AsmL program computes the set $\{2, 3, 4, 5\}$ and then uses an outside pseudo-random number generator to choose an element of that set. Of course an implementation of a nondeterministic algorithm may incorporate a choosing mechanism, so that there is no choice on the level of the implementation.

Re new object creation. An object-oriented program does not have the means necessary to create a new object all by itself: to allocate a portion of the memory and format it appropriately. A create-a-new-object command solicits outside help. This phenomenon is not restricted to the object-oriented paradigm. We give a non-object-oriented example. Consider an ASM rule

```
import v
    NewLeaf := v
```

that creates a new leaf say of a tree. The import command is really a query to the environment. In the ASM paradigm, a state comes with an infinite set of so-called reserve elements. The environment chooses such a reserve elements and returns it as a reply to the query.

4.4 Interaction Mechanisms

One popular interaction form is exemplified by the Remote Procedure Call (RPC) mechanism. One can think of a remote procedure call as a query to the environment where the caller waits for a reply to its query in order to complete a step and continue the computation. This interaction form is often called synchronous or blocking. Another popular interaction form is message passing. After sending a message, the sender proceeds with its computation; this interaction form is often called asynchronous or nonblocking. The synchronous/asynchronous and blocking/nonblocking terminologies may create an impression that every atomic intra-step interaction is in one of the two form. This is not the case. There is a spectrum of possible interaction forms. For example, a query may require two replies: first an acknowledgment and then an informative reply. One can think of queries with three, four or arbitrarily many replies.

Nevertheless, according to [5], there a universal form of atomic intra-step interaction: not-necessarily-blocking single-reply queries. In the previous paragraph, we have already represented a remote procedure call as a query. Sending a message can be thought of as a query that gets an immediate automatic reply, an acknowledgment that the query has been issued. Producing an output is similar. In fact, from the point of view of an algorithm issuing queries, there is no principal difference between sending a message and producing an output; in a particular application of course messages and outputs may have distinct formats.

What about two-reply queries mentioned above? It takes two single-reply queries to get two answers. Consider an algorithm A issuing a two-reply query q and think of q as a single-reply query. When the acknowledgment comes back, A goes to a mode where it expects an informative answer to q. This expectation can be seen as implicitly issuing a new query q'. The informative reply ostensibly to q is a usual reply to q'. In a similar way, one can explain receiving a message. It may seem that the incoming message is not provoked by any query. What query is it a reply to? An implicit query. That implicit query manifests itself in A's readiness to accept the incoming message. Here is an analogy. You sleep and then wake up because of the alarm clock buzz. Have you been expecting the buzz? In a way you were, in an implicit sort of way. Imagine that, instead of producing a buzz, the alarm clock quietly produces a sign "Wake up!" This will not have the desired effect, would it?

In general we do not assume that the query issuer has to wait for a reply to a query in order to resume its computation. More about that in §6.

What are potential queries precisely? This question is discussed at length in [5]. It is presumed that potential answers to a query are elements of the state of the algorithm that issued the query, so that an answer makes sense to the algorithm.

5 Ordinary Interactive Small-Step Algorithms

Proviso 5.1. To simplify the exposition, in the rest of the paper we speak primarily about small-step algorithms. By default, algorithms are small-step algorithms.

Informally speaking, an interactive algorithm is *ordinary* if it has the following two properties.

- The algorithm cannot successfully complete a step while there is an unanswered query from that step.
- The only information that the algorithm receives from the environment during a step consists of the replies to the queries issued during the step.

Ordinary interactive algorithms are axiomatized in [5]. Some postulates of [5] refactor those of [15]. One of the new postulates is this:

Postulate 5.2 (Interaction Postulate). *An interactive algorithm determines, for each state X, a causality relation $\vdash_X$ between finite answer functions and potential queries.*

Here an answer function is a function from potential queries to potential replies. An answer function α is *closed* under a causality relation $\vdash_X$ if every query caused by α or by a subfunction of α is already in the domain of α. Minimal answer functions closed under $\vdash_X$ are *contexts* at X.

As before, behaviorally equivalent algorithms do the same thing in all circumstances. To make this precise, we need a couple of additional definitions. Given a causality relation $\vdash_X$ and an answer function α, define an α-*trace* to be a sequence $\langle q_1, \ldots, q_n \rangle$ of potential queries such that each q_i is caused by the restriction α_i of α to $\{q_j : j < k\}$ or by some subfunction of α_i. A potential query q is *reachable* from α under $\vdash_X$ if it occurs in some α-trace. Two causality relations are *equivalent* if, for every answer function α, they make the same potential queries reachable from α.

Definition 5.3. Two ordinary interactive algorithms are *behaviorally equivalent* if

- they have the same states and initial states,
- for every state, they have equivalent causality relations, and
- for every state and context, they both fail or they both succeed and produce the same next state. □

We turn our attention to ordinary abstract state machines. Again, a machine is given by a program, a collection of states and a subcollection of initial states. We need only to describe programs.

The syntax of ordinary ASM programs is nearly the same as that of isolated state algorithms, the algorithms of [15]. The crucial difference is in the semantics of external functions. In the case of isolated step algorithms, an invocation of an external function is treated as a usual state-location lookup; see Euclidean Algorithm 2 or 3 in this connection. In the case of interactive algorithms, an invocation of an external function is a query.

The new interpretation of external functions gives rise to a problem. Suppose that you have two distinct invocations $f(3)$ of an external function $f()$ in your program. Should the replies be necessarily the same? In the case of an isolated-step program, the answer is yes. Indeed, the whole program describes one step of an algorithm, and the state does not change during the step. Two distinct lookups of $f(3)$ will give you the same result. In the case of an interactive program, the replies don't have to be the same. Consider

Example 5.4 (Euclidean Algorithm 4).

```
if mode = 0 then a, b, mode := Input, Input, 1
elseif mode = 1 then
   if a = 0 then d, mode := b, 0
   else a, b := b mod a, a
```

The two invocations of Input are different queries that may have different results. Furthermore, in the object-oriented paradigm, two distinct invocations of the same create-a-new-object command with the same parameters necessarily result in two distinct objects. We use a mechanism of template assignment to solve the problem in question [6,7].

The study of ordinary interactive algorithms in [5,6,7] culminates in

Theorem 5.5 (Ordinary Interactive Characterization Theorem). *Every ordinary interactive ASM is an ordinary interactive algorithm, and every ordinary interactive algorithm is behaviorally equivalent to an ordinary interactive ASM.*

6 General Interactive Algorithms

Call an interactive algorithm *patient* if it cannot finish a step without having the replies to all queries issued during the step. While ordinary interactive algorithms are patient, this does not apply to all interactive algorithms. The algorithm

Example 6.1 (Impatience).

```
do in parallel
    if α or β then x:=1
    if ¬α and ¬β then x:=2
```

issues two Boolean queries α and β. If one of the queries returns "true" while the other query is unanswered, then the other query can be aborted.

Call an interactive algorithm *time insensitive* if the only information that it receives from the environment during a step consists of the replies to the queries issued during the step. Ordinary algorithms are time insensitive. Since our algorithms interact with the environment only by means of queries, it is not immediately obvious what information the algorithm can get from the environment in addition to the replies. For example, time stamps, reflecting the times when the replies were issued, can be considered to be parts of the replies.

The additional information is the order in which the replies come in. Consider for example an automated financial broker with a block of shares to sell and two clients bidding for the block of shares. If the bid of client 1 reaches the broker first, then the broker sells the shares to client 1, even if client 2 happened to issue a bid a tad earlier.

An algorithm can be impatient and time sensitive at the same time. Consider for example a one-step algorithm that issues two queries, q_1 and q_2, and then does the following. If q_i is answered while q_{2-i} is not, then it sets x to i and aborts q_{2-i}. And if the queries are answered at the same time, then it sets x to 0.

The following key observation allowed us to axiomatize general interactive algorithms. Behind any sequential-step algorithm there is a single executor of the algorithm. In particular, it is the executor who gets query replies from the environment, in batches, one after another. It follows that the replies are linearly preordered according to the time or arrival. In [8], we successfully execute the algorithm explication procedure of § 2 in the case of general interactive algorithms.

Theorem 6.2 (Interactive Characterization Theorem). *Every interactive ASM is an interactive algorithm, and every interactive algorithm is behaviorally equivalent to an interactive ASM.*

A variant of this theorem is proved in [9]. The twist is that, instead of interactive algorithms, we speak about their components there.

Patient (but possibly time sensitive) interactive algorithms as well as time insensitive (but possibly impatient) interactive algorithms are characterized in [10].

These variants of the interactive characterization theorem as well as the theorem itself are about small-step algorithms. The interactive characterization theorem is generalized to wide-step algorithms in [11].

7 Finale

The behavioral theory of small-isolated-step algorithms [15] was an after-the-fact explanation of what those algorithms were. Small-isolated-step algorithms had been studied for a long time.

The behavioral theory of wide-isolated-step algorithms was developed in [3]. Wide-isolated-step algorithms had been studied primarily in computational complexity where a number of wide-isolated-step computation models had been known. But the class of wide-isolated-step algorithms of [3] is wider. The theory was used to develop a number of tools [1], most notably the specification language AsmL [2]. Because of the practical considerations of industrial environment, intra-step interaction plays a considerable role in AsmL. That helped us to realize the importance and indeed inevitability of intra-step interaction.

The behavioral theory of intra-step interactive algorithms is developed in [5]–[11]. While intra-step interaction is ubiquitous, it has been studied very little if at all. We hope that the research described above will put intra-step interaction on the map and will give rise to further advances in specification and high-level programming of interactive algorithms.

References

1. ASM Michigan Webpage, http://www.eecs.umich.edu/gasm/, maintained by James K. Huggins.
2. The AsmL webpage, http://research.microsoft.com/foundations/AsmL/.
3. Andreas Blass and Yuri Gurevich, "Abstract state machines capture parallel algorithms," *ACM Trans. on Computational Logic*, 4:4 (2003), 578–651.
4. Andreas Blass and Yuri Gurevich, "Algorithms: A Quest for Absolute Definitions," Bull. Euro. Assoc. for Theor. Computer Science Number 81, October 2003, pages 195–225. Reprinted in *Current Trends in Theoretical Computer Science: The Challenge of the New Century*, Vol. 2, eds. G. Paun et al., World Scientific, 2004, 283–312.
5. Andreas Blass and Yuri Gurevich, "Ordinary Interactive Small-Step Algorithms, I", *ACM Trans. on Computational Logic*, to appear. Microsoft Research Tech. Report MSR-TR-2004-16.
6. Andreas Blass and Yuri Gurevich, "Ordinary Interactive Small-Step Algorithms, II", *ACM Trans. on Computational Logic*, to appear. Microsoft Research Tech. Report MSR-TR-2004-88.
7. Andreas Blass and Yuri Gurevich, "Ordinary Interactive Small-Step Algorithms, III", *ACM Trans. on Computational Logic*, to appear. Microsoft Research Tech. Report MSR-TR-2004-88, mentioned above, covers this article as well; the material was split into two pieces by the journal because of its article-length restriction.
8. Andreas Blass, Yuri Gurevich, Dean Rosenzweig and Benjamin Rossman, "General Interactive Small-Step Algorithms", in preparation.
9. Andreas Blass, Yuri Gurevich, Dean Rosenzweig and Benjamin Rossman, "Composite Interactive Algorithms" (tentative title), in preparation.
10. Andreas Blass, Yuri Gurevich, Dean Rosenzweig and Benjamin Rossman, "Interactive Algorithms: Impatience and Time Sensitivity" (tentative title), in preparation.
11. Andreas Blass, Yuri Gurevich, Dean Rosenzweig and Benjamin Rossman, "Interactive Wide-Step Algorithms", in preparation.
12. Egon Börger and Robert Stärk, "Abstract State Machines: A Method for High-Level System Design and Analysis", Springer-Verlag, 2003.
13. Yuri Gurevich, "Evolving Algebras: An Introductory Tutorial", Bull. Euro. Assoc. for Theor. Computer Science 43, February 1991, 264–284. A slightly revised version is published in *Current Trends in Theoretical Computer Science*, eds. G. Rozenberg and A. Salomaa, World Scientific, 1993, 266-292.
14. Yuri Gurevich, "Evolving Algebra 1993: Lipari Guide," in *Specification and Validation Methods*, ed. E. Börger, Oxford University Press, 1995, 9–36.
15. Yuri Gurevich, "Sequential Abstract State Machines Capture Sequential Algorithms," *ACM Trans. on Computational Logic* 1:1 (2000), 77–111.
16. Andrey N. Kolmogorov, "On the Concept of Algorithm", *Uspekhi Mat. Nauk* 8:4 (1953), 175–176, Russian.
17. Andrey A. Markov, "Theory of Algorithms", Transactions of the Steklov Institute of Mathematics, vol. 42 (1954), Russian. Translated to English by the Israel Program for Scientific Translations, Jerusalem, 1962.

Some Computational Issues in Membrane Computing[*]

Oscar H. Ibarra

Department of Computer Science, University of California,
Santa Barbara, CA 93106, USA
ibarra@cs.ucsb.edu

Abstract. Membrane computing is a branch of molecular computing that aims to develop models and paradigms that are biologically motivated. It identifies an unconventional computing model, namely a P system, from natural phenomena of cell evolutions and chemical reactions. Because of the nature of maximal parallelism inherent in the model, P systems have a great potential for implementing massively concurrent systems in an efficient way that would allow us to solve currently intractable problems (in much the same way as the promise of quantum and DNA computing) once future bio-technology (or silicon-technology) gives way to a practical bio-realization (or chip realization). Here we report on recent results that answer some interesting and fundamental open questions in the field. These concern computational issues such as determinism versus nondeterminism, membrane and alphabet-size hierarchies, and various notions of parallelism.

1 Introduction

There has been a great deal of research activity in the area of membrane computing (a branch of natural computing) initiated by Gheorghe Paun six years ago in his seminal paper [23] (see also [24]). Membrane computing identifies an unconventional computing model, namely a P system, from natural phenomena of cell evolutions and chemical reactions. It abstracts from the way the living cells process chemical compounds in their compartmental structure. Thus, regions defined by a membrane structure contain objects that evolve according to specified rules. The objects can be described by symbols or by strings of symbols, in such a way that multisets of objects are placed in regions of the membrane structure. The membranes themselves are organized as a Venn diagram or a tree structure where one membrane may contain other membranes. By using the rules in a nondeterministic, maximally parallel manner, transitions between the system configurations can be obtained. A sequence of transitions shows how the system is evolving. Various ways of controlling the transfer of objects from a region to another and applying the rules, as well as possibilities to dissolve, divide, or create membranes have been studied. P systems were introduced with the goal to abstract a new computing model from the structure and the functioning of the living cell (as a branch of the general effort of Natural Computing – to explore new models, ideas, paradigms from the way nature computes).

---

[*] This wark was supported in part by NSF Grants CCR-0208595, CCF-0430945, and IIS-0451097. Some of the results reported here were obtained jointly with Zhe Dang, Sara Woodworth, and Hsu-Chun Yen.

Membrane computing has been quite successful: many models have been introduced, most of them Turing complete and/or able to solve computationally intractable problems (NP-complete, PSPACE-complete) in polynomial time, by trading space for time. In fact, the Institute for Scientific Information (ISI) has selected membrane computing as a fast "Emerging Research Front" in Computer Science (*http://esi-topics.com/ erf/october2003.html*. See also the P system website at *http://psystems.disco.unimib.it* for a large collection of papers in the area, and in particular the monograph [25].) Due to the built-in nature of maximal parallelism inherent in the model, P systems have a great potential for implementing massively concurrent systems in an efficient way that would allow us to solve currently intractable problems (in much the same way as the promise of quantum and DNA computing) once future bio-technology (or silicon-technology) gives way to a practical bio-realization (or chip-realization).

In this paper, we report on recent results that answer some interesting and fundamental open questions in the area of membrane computing. These concern computational issues such as determinism versus nondeterminism, membrane and alphabet-size hierarchies, and various notions of parallelism.

2 Determinism Versus Nondeterminism

In the standard semantics of P systems [24,25,27], each evolution step of a system G is a result of applying all the rules in G in a maximally parallel manner. More precisely, starting from the initial configuration, w, the system goes through a sequence of configurations, where each configuration is derived from the directly preceding configuration in one step by the application of a multiset of rules, which are chosen nondeterministically. For example, a catalytic rule $Ca \rightarrow Cv$ in membrane m is applicable if there is a catalyst C and an object (i.e., symbol) a in the preceding configuration in membrane m. The result of applying this rule is the evolution of v from a. The catalyst C remains in membrane m, but each symbol in v has an associated target indicating the membrane where the symbol is to be transported to (of course, if the system has only one membrane, each symbol in v remains in the membrane). If there is another occurrence of C and another occurrence of a, then the same rule or another rule with Ca on the left hand side can be applied. Thus, in general, the number of times a particular rule is applied at anyone step can be unbounded. We require that the application of the rules is maximal: all objects, from all membranes, which *can be* the subject of local evolution rules *have to* evolve simultaneously. Configuration z is reachable (from the starting configuration) if it appears in some execution sequence; z is halting if no rule is applicable on z.

An interesting class of P systems acceptors with symport/antiport rules was studied in [10] – each system is *deterministic* in the sense that the computation path of the system is unique, i.e., at each step of the computation, the maximal multiset of rules that is applicable is unique. It was shown in [10] that any recursively enumerable unary language $L \subseteq o^*$ can be accepted by a deterministic 1-membrane symport/antiport system. Thus, for symport/antiport systems, the deterministic and nondeterministic versions are equivalent. It also follows from the construction in [31] that for communicating P systems, the deterministic and nondeterministic versions are equivalent as both can accept any unary recursively enumerable language. The deterministic-versus-nondeterministic question was left open in [10] for the class of catalytic systems, where the proofs of universality involve a high degree of parallelism [31,8]. In particular, it was an open problem [1] whether there is a class of (universal or nonuniversal) P systems where the nondeterministic version is strictly more powerful than the deterministic version. For a discussion of this open question and its importance, see [1,26].

In this section, we look at three popular models of P systems (catalytic system, symport/antiport system, and communicating P system), and report on recent results that answer some open questions concerning determinism versus nondeterminism.

2.1 Catalytic Systems

First we look at 1-membrane catalytic systems (CSs). A CS has rules of the forms: $Ca \to Cv$ or $a \to v$, where C is a catalyst, a is a noncatalyst symbol, and v is a (possibly null) string of noncatalyst symbols. (Note that we are only interested in the multiplicities of the symbols.) A CS whose rules are only of the form $Ca \to Cv$ is called *purely* CS.

For a catalytic system serving as a *language acceptor*, the system starts with an initial configuration wz, where $z = a_1^{n_1}...a_k^{n_k}$ with $\{a_1, ..., a_k\}$ (the input alphabet) a distinguished subset of noncatalyst symbols, $n_1, ..., n_k$ are nonnegative integers, and w is a fixed string of catalysts and noncatalysts not containing a_i ($1 \le i \le k$). At each step, a maximal multiset of rules are nondeterministically selected and applied in parallel to the current configuration to derive the next configuration (note that the next configuration is not unique, in general). The string z is accepted if the system eventually halts. A CS is *deterministic* if at each step, there is a *unique* maximally parallel multiset of rules applicable.

Before we state the results, we recall the definition of a semilinear set [13]. Let N be the set of nonnegative integers and k be a positive integer. A subset R of N^k is a *linear set* if there exist vectors $v_0, v_1, \ldots, v_t$ in N^k such that

$$R = \{v \mid v = v_0 + m_1 v_1 + \ldots + m_t v_t, \ m_i \in N\}.$$

The vectors v_0 (referred to as the *constant vector*) and $v_1, v_2, \ldots, v_t$ (referred to as the *periods*) are called the *generators* of the linear set R. The set $R \subseteq N^k$ is *semilinear* if it is a finite union of linear sets. The empty set is a trivial (semi)linear set, where the set of generators is empty. Every finite subset of N^k is semilinear – it is a finite union of linear sets whose generators are constant vectors. It is also clear that the semilinear sets are closed under union. It is also known that they are closed under complementation and intersection. A (bounded) language $L \subseteq a_1^*...a_k^*$ is semilinear if its Parikh map, $P(L) = \{(n_1, ..., n_k) \mid a_1^{n_1}...a_k^{n_k} \in L\}$, is a semilinear set.

Unlike nondeterministic 1-membrane catalytic system acceptors (with 2 catalysts) which are universal [8], we were able to show in [19] using a graph-theoretic approach the following:

Theorem 1. *Any language $L \subseteq a_1^*...a_k^*$ accepted by a deterministic catalytic system is effectively semilinear. In fact, L is either empty, or $a_1^{n_1}....a_k^{n_k}$, where $n_i = *$ or 0, $1 \le i \le k$.*

Corollary 1. *Deterministic catalytic systems are not universal.*

The corollary above gives the first example of a P system for which the nondeterministic version is universal, but the deterministic version is not.

For deterministic 1-membrane purely catalytic systems (i.e., the rules of are of the form $Ca \to Cv$), the set of all reachable configurations from a given initial configuration is effectively semilinear. In contrast, the reachability set is no longer semilinear in general if rules of type $a \to v$ are also used.

We also considered in [19] deterministic catalytic systems which allow rules to be prioritized. We investigated three such systems, namely, *totally prioritized, strongly prioritized* and *weakly prioritized* catalytic systems. For totally prioritized systems, rules are divided into different priority groups, and if a rule in a higher priority group is applicable, then no rules from a lower priority group can be used. For both strongly prioritized and weakly prioritized systems, the underlying priority relation is a *strict partial order* (i.e., irreflexive, asymmetric, and transitive). Under the semantics of strong priority, if a rule with higher priority is used, then no rule of a lower priority can be used even if the two rules do not compete for objects. For weakly prioritized systems, a rule is applicable if it cannot be replaced by a higher priority one. For these three prioritized systems, we obtained contrasting results: deterministic strongly and weakly prioritized catalytic systems are universal, whereas totally prioritized systems only accept semilinear sets.

Finally, we note that the results above generalize to multi-membrane catalytic systems where now, in the rules $Ca \to Cv$ or $a \to v$, each symbol in v is associated with a target membrane.

2.2 Symport/Antiport Systems

Another popular model of a P system is called a symport/antiport system, first introduced in [22]. It is a simple system whose rules closely resemble the way membranes transport objects between themselves in a purely communicating manner. Symport/antiport systems (S/A systems) have rules of the form (u, out), (u, in), and $(u, out; v, in)$ where $u, v \in \Sigma^*$. Again, u and v are strings representing multisets. A rule of the form (u, out) in membrane i sends the symbols in u from membrane i out to the membrane directly enclosing i. A rule of the form (u, in) in membrane i transports the symbols in u from the membrane enclosing i into membrane i. Hence this rule can only be used when the symbols in u exist in the outer membrane. A rule of the form $(u, out; v, in)$ simultaneously sends u out of the membrane i while transporting v into membrane i. Hence this rule cannot be applied unless membrane i contains the symbols in u and the membrane surrounding i contains the symbols in v. Formally an S/A system is defined as

$$G = (V, H, \mu, w_1, \ldots, w_{|H|}, E, R_1, \ldots, R_{|H|}, i_o)$$

where V is the set of objects (symbols) the system uses. H is the set of membrane labels. The membrane structure of the system is defined in μ. The initial multiset of objects within membrane i is represented by w_i, and the rules are given in the set R_i.

$E \subseteq V$ is the set of objects that occur abundantly (i.e., each object in E has infinite copies) in the environment; other objects in $V - E$ can occur in the environment, but in bounded number. The designated output membrane is i_o. (When the system is used as a recognizer or acceptor, there is no need to specify i_o.)

A large number of papers have been written concerning symport/antiport systems. In particular, it is known that every (unary) recursively set $L \subseteq o^*$ can be accepted by a deterministic 1-membrane S/A system (hence, such a system is universal) [10].

Small Universal Deterministic S/A Systems

There has been much interest in finding "small" systems (in terms of the number of objects, weights of the rules, etc) that are universal. For a discussion of minimal universal S/A systems, see [28,26]. Here we look at deterministic S/A system acceptors with 1 or 2 membranes and 1, 2, or 3 objects.

Let G be an m-membrane S/A system with alphabet V and $E \subseteq V$ be the set of objects that occur abundantly in the environment. Each object in $V - E$ has bounded number of copies in the environment. If $|E| = k$, then we say the system is an m-membrane k-symbol S/A system.

Let o be a distinguished symbol in E. There is a fixed $w \in (V - E)^*$ such that at the start of the computation, a multiset wo^n for some nonnegative integer n, is placed in the skin membrane. We say that o^n is accepted if the system halts. A deterministic m-membrane k-symbol S/A system is defined as before. We can show the following:

Theorem 2. *Let $L \subseteq o^*$.*

1. *L is accepted by a deterministic 1-membrane 1-symbol S/A system if and only if it is semilinear.*
2. *Let $(m, k) \in \{(1, 3), (3, 1), (2, 2)\}$. Then L is accepted by a deterministic m-membrane k-symbol S/A system if and only if it is recursively enumerable.*
3. *There are recursive sets that cannot be accepted by deterministic 1-membrane 2-symbol S/A systems and deterministic 2-membrane 1-symbol S/A systems.*

One can show that Theorem 2 part 1 holds for the nondeterministic case. Obviously, part 2 holds for the nondeterministic version as well. We believe that part 3 also holds for the nondeterministic case, but we have no proof at this time.

Restricted (Nonuniversal) S/A Systems

In [18], we studied some restricted versions of S/A systems. One model, called *bounded S/A system*, has only one membrane and has rules of the form $(u, out; v, in)$ with the restriction that $|u| \geq |v|$. The environment has an infinite supply of every object in V. An input $z = a_1^{n_1}...a_k^{n_k}$ (each n_i a nonnegative integer) is accepted if the system when started with wz, where w is a fixed string not containing a_i ($1 \leq i \leq k$) eventually halts. We showed the following:

Theorem 3. *Let $L \subseteq a_1^*...a_k^*$. Then the following statements are equivalent:*

1. *L is accepted by a bounded S/A system.*
2. *L is accepted by a $\log n$ space-bounded Turing machine.*
3. *L is accepted by a two-way multihead finite automaton.*

This result holds for both deterministic and nondeterministic versions.

The next result follows from Theorem 3 and the following result in [30]: Deterministic and nondeterministic two-way multihead finite automata over a unary input alphabet are equivalent if and only if deterministic and nondeterministic linear bounded automata (over an arbitrary input alphabet) are equivalent.

Theorem 4. *Deterministic and nondeterministic bounded S/A systems over a unary input alphabet are equivalent if and only if deterministic and nondeterministic linear-bounded automata (over an arbitrary input alphabet) are equivalent. The latter problem is a long-standing open question in complexity theory [29].*

We also considered multi-membrane S/A systems, called special S/A systems, which are restricted in that only rules of the form $(u, out; v, in)$, where $|u| \geq |v|$, can appear in the skin membrane (there are no restrictions on the rules in the other membranes). Thus, the number of objects in the system during the computation cannot increase. *The environment does not contain any symbol initially.* Only symbols exported from the skin membrane to the environment can be retrieved from the environment. (Note that in the bounded S/A system, the environment has an infinite supply of every object in V.)

Theorem 5. *Let $L \subseteq a_1^*...a_k^*$. Then the following statements are equivalent:*

1. *L is accepted by a special S/A system.*
2. *L is accepted by a bounded S/A system.*
3. *L is accepted by a $\log n$ space-bounded Turing machine.*
4. *L is accepted by a two-way multihead finite automaton.*

This result holds for both deterministic and nondeterministic versions.

We also studied in [18] a model of a (one-membrane) bounded S/A system whose alphabet of symbols V contains a distinguished input alphabet Σ. We assume that Σ contains a special symbol $, the (right) end marker. The rules are restricted to be of the forms:

(1) $(u, out; v, in)$
(2) $(u, out; vc, in)$

where u, v are in $(V - \Sigma)^*$ with $|u| \geq |v|$, and c is in Σ. The second type of rule is called a *read-rule*. There is an abundance (i.e., infinite copies) of each symbol from $V - \Sigma$ in the environment. The only symbols from Σ available in the environment are in the input string $z = a_1...a_n$ (where a_i is in $\Sigma - \{$\}$ for $1 \leq i < n$, and $a_n = $), which is provided online externally.

There is a fixed string w in $(V - \Sigma)^*$, which is the initial configuration of the system. Maximal parallelism in the application of the rules is assumed as usual. Hence, in general, the size of the multiset of rules applicable at each step can be unbounded. In particular, the number of instances of read-rules (i.e., rules of the form $(u, out; vc, in)$) applicable in a step can be unbounded. However, if a step calls for reading k input symbols (for some k), these symbols must be consistent with the next k symbols of the input string z that have not yet been processed. Note that rules of first type do not consume any input symbol from z.

The input string $z = a_1...a_n$ (with $a_n = \$$) is accepted if, after reading all the input symbols, the system eventually halts. The language accepted is $\{a_1...a_{n-1} \mid a_1...a_n$ is accepted $\}$ (we do not include the end marker).

We call the system above a *bounded S/A string acceptor*. As described above, the system is nondeterministic. Again, in the deterministic case, the maximally parallel multiset of rules applicable at each step of the computation is unique. In [18] we showed the following:

Theorem 6. *Deterministic bounded S/A string acceptors are strictly weaker than nondeterministic bounded S/A string acceptors. An example of a language accepted by the nondeterministic version that cannot be accepted by the deterministic version is* $L = \{x \# y \mid x, y \in \{0, 1\}^*, x \neq y\}$.

2.3 Communicating P Systems

There is another model that also works on a purely communicating mode, called communicating P system (CPS), first introduced and studied in [31]. It has multiple membranes labeled $1, 2, ...$, where 1 is the skin membrane. The rules are of the form:

1. $a \rightarrow a_x$
2. $ab \rightarrow a_x b_y$
3. $ab \rightarrow a_x b_y c_{come}$

where a, b, c are objects, x, y (which indicate the directions of movements of a and b) can be *here*, *out*, or in_j. The designation *here* means that the object remains in the membrane containing it, *out* means that the object is transported to the membrane directly enclosing the membrane that contains the object (or to the environment if the object is in the skin membrane). The designation in_j means that the object is moved into the membrane, labeled j, that is directly enclosed by the membrane that contains the object. A rule of the form (3) can only appear in the skin membrane. When such a rule is applied, c is imported through the skin membrane from the environment and will become an element in the skin membrane. As usual, in one step, all rules are applied in a maximally parallel manner.

An RCPS [15] is a restricted CPS where the environment does not contain any object initially. The system can expel objects into the environment but only expelled objects can be retrieved from the environment. Hence, at any time during the computation, the objects in the system (including in the environment) are always the same.

Let $a_1, ..., a_k$ be the symbols in the input alphabet $\Sigma \subseteq V$. Assume that an RCPS G has m membranes, with a distinguished *input membrane*. We say that G accepts $z = a_1^{n_1}...a_k^{n_k}$ if G, when started with z in the input membrane initially (with no a_i's in the other membranes), eventually halts. Note that initially (at the start of the computation), each membrane contains a multiset from $(V - \Sigma)^*$. At any time during the computation, the number of each object $b \in (V - \Sigma)$ in the whole system (including the environment) remains the same, although the distribution of the b's among the membranes may change at each step. The language accepted by G is $L(G) = \{z \mid z$ is accepted by $G\}$.

A nondeterministic (deterministic) RCPS is one in which there may be more than one (at most one) maximally parallel multiset of rules that is applicable at each step.

We showed in [16] that RCPSs are equivalent to two-way multihead finite automata (in both the deterministic and nondeterministic cases). Thus, we have:

Theorem 7. *Let $L \subseteq a_1^* ... a_k^*$. Then the following statements are equivalent:*

1. *L is accepted by an RCPS.*
2. *L is accepted by a special S/A system.*
3. *L is accepted by a bounded S/A system.*
4. *L is accepted by a $\log n$ space-bounded Turing machine.*
5. *L is accepted by a two-way multihead finite automaton.*

This result holds for both deterministic and nondeterministic versions.

Corollary 2. *Deterministic and nondeterministic RCPSs over a unary input alphabet are equivalent if and only if deterministic and nondeterministic linear-bounded automata (over an arbitrary input alphabet) are equivalent.*

3 Hierarchies

Various models of P systems have been investigated and have been shown to be universal, i.e., Turing machine complete, even with a very small number of membranes (e.g., 1 or 2 membranes). Not much work has been done on investigating P systems that are nonuniversal. The question of whether there exists a model of P systems where the number of membranes induces an infinite hierarchy in its computational power had been open since the beginning of membrane computing. Clearly, for models that are universal, there cannot be a hierarchy. So the hierarchy question makes sense only for non-universal systems. We resolved this question in the affirmative in [18], where we proved the following result:

Theorem 8. *For every r, there exist an $s > r$ and a unary language L (i.e., subset of o^*) accepted by an s-membrane special S/A system that cannot be accepted by any r-membrane special S/A system. The result holds for both deterministic and nondeterministic versions.*

The proof Theorem 8 reduces the membrane hierarchy to a result in [21] that shows that there is an infinite hierarchy of two-way (non)deterministic multihead finite operating on unary input in terms of the number of heads. We note that the theorem also holds for RCPSs [16].

Similary, the number of symbols in the alphabet V of a bounded S/A system induces an infinite hierarchy [18].

Theorem 9. *For every r, there exist an $s > r$ and a unary language L accepted by a bounded S/A system with an alphabet of s symbols that cannot be accepted by any bounded S/A system with an alphabet of r symbols. This result holds for both deterministic and nondeterministic versions.*

4 The Power of Maximal Parallelism

As already mentioned above, in the standard semantics of P systems, each evolution step of the system is a result of applying the rules in a nondeterministic maximally parallel manner.

Current digital and bio technologies do not permit a direct implementation of a P system (under the parallel semantics). Also, because of the highly parallel and nondeterministic nature of the computation in a P system, simulation and analysis (such as reachability between configurations) are mostly undecidable (i.e., no algorithms exist or combinatorially intractable). Let G be a P system and $R = \{r_1, ..., r_k\}$ be the set of (distinct) rules in all the membranes. Note that r_i uniquely specifies the membrane the rule belongs to. We say that G operates in maximal parallel mode if at each step of the computation, a maximal subset of R is applied, and at most one instance of any rule is used at every step (thus at most k rules are applicable at any step). For example, if r_i is a catalytic rule $Ca \to Cv$ in membrane q and the current configuration has two C's and three a's in membrane q, then only one a can evolve into v. Of course, if there is another rule r_j, $Ca \to Cv'$, in membrane q, then the other a also evolves into v'. In [17], we investigated the computing power of P systems under three semantics of parallelism. For a positive integer $n \leq k$, define:

n-**Max-Parallel:** At each step, nondeterministically select a maximal subset of at most n rules in R to apply (this implies that no larger subset is applicable).

$\leq n$-**Parallel:** At each step, nondeterministically select any subset of at most n rules in R to apply.

n-**Parallel:** At each step, nondeterministically select any subset of exactly n rules in R to apply.

In all three cases, if any rule in the subset selected is not applicable, then the whole subset is not applicable. When $n = 1$, the three semantics reduce to the **Sequential** or **Asynchronous** mode.

Before proceeding further, we need the definition of a vector addition system. An n-dimensional *vector addition system* (VAS) is a pair $G = \langle x, W \rangle$, where $x \in \mathbf{N}^n$ is called the *start point* (or *start vector*) and W is a finite set of vectors in $\mathbf{Z}^n$, where $\mathbf{Z}$ is the set of all integers (positive, negative, zero). The *reachability set* of the VAS $\langle x, W \rangle$ is the set $R(G) = \{z \mid$ for some j, $z = x + v_1 + ... + v_j$, where, for all $1 \leq i \leq j$, each $v_i \in W$ and $x + v_1 + ... + v_i \geq 0\}$. An n-dimensional *vector addition system with states* (VASS) is a VAS $\langle x, W \rangle$ together with a finite set T of transitions of the form $p \to (q, v)$, where p and q are states and v is in W. The meaning is that such a transition can be applied at point y in state p and yields the point $y + v$ in state q, provided that $y + v \geq 0$. The VASS is specified by $G = \langle x, W, T, p_0 \rangle$, where p_0 is the starting state. It is known that n-dimensional VASS can be effectively simulated by $(n+3)$-dimensional

VAS [14]. The *reachability problem* for a VAS (VASS) G is to determine, given a vector y, whether y is in $R(G)$.

In [17], we studied catalytic systems, symport/antiport systems, and communicating P systems with respect to the three modes of parallelism defined above. We showed that for these systems, n-**Max-Parallel** mode is strictly more powerful than any of the following three modes: **Sequential**, $\leq n$-**Parallel**, or n-**Parallel**. For example, it follows from a result in [9] that a maximally parallel communicating P system is universal for $n = 2$. However, under the three limited modes of parallelism, the system is equivalent to a vector addition system (VAS), which are equivalent Petri nets (PN). VAs and PNs are well-known models of concurrent and parallel systems. They are used extensively to analyze properties (e.g. reachability) in concurrent and parallel systems. A fundamental result is that there is a decision procedure for the reachability problem (given two configurations w and w', is w' reachable from w?). Thus, we can decide reachability for the three P systems mentioned operating under the three notions of parallelism. These results show that "maximal parallelism" is key for the model to be universal.

5 Sequential (Asynchronous) P Systems

Note that in a P system operating in sequential mode, at every step, only one nondeterministically chosen rule instance is applied. Sequential P systems (also called asynchronous P systems) have been studied in various places in the literature (see, e.g., [2,3,6,7,11,17]). In a recent paper [20], we showed the following results that complement these earlier results:

1. Any sequential P system with rules of the form $u \rightarrow v$ (where u, v are strings of symbols) with rules for membrane creation and membrane dissolution can be simulated by a vector addition system (VAS), provided the rules are not prioritized and the number of membranes that can be created during the computation is bounded by some fixed integer. Hence the reachability problem (deciding if a configuration is reachable from the start configuration) is decidable. Interestingly, if such cooperative systems are allowed to create an unbounded number of new membranes during the course of the computation, then they become universal.
2. A sequential communicating P system language acceptor (CPA) is equivalent to a partially blind multicounter machine (PBCM) [12]. Several interesting corollaries follow from this equivalence, for example:
   (a) The emptiness problem for CPAs is decidable.
   (b) The class of CPA languages is a proper subclass of the recursive languages.
   (c) The language $\{a^n b^n \mid n \geq 1\}^*$ cannot be accepted by a CPA.
   (d) For every r, there is an $s > r$ and a language that can be accepted by a quasi-realtime CPA with s membranes that cannot be accepted by a quasi-realtime CPA with r membranes. (In a CPA, we do not assume that the CPA imports an input symbol from the environment at every step. Quasi-realtime means that the CPA has to import an input symbol from the environment with delay of no more than k time steps for some nonnegative integer k independent of the computation.)

(e) A quasi-realtime CPA is strictly weaker than a linear time CPA. (Here, linear time means that for some constant c, the CPA accepts an input of length n within cn time.)

(f) The class of quasi-realtime CPA languages is not closed under Kleene + and complementation.

We note that the relationship between PBCMs and sequential symport/antiport P systems (similar to communication P systems) has been studied recently in [11], but only for systems with symbol objects and not as language acceptors. Thus, the results in [11] deal only with tuples of nonnegative integers defined by P systems and counter machines. For example, it was shown in [11] that a set of tuples of nonnegative integers that is definable by a partially blind counter machine can be defined by a sequential symport/antiport system with two membranes. Our results cannot be derived from the results in [11].

3. The results for CPA above generalize to cooperative system acceptors with membrane dissolution and bounded creation rules. Hence, the latter are also equivalent to PBCMs.
4. The reachability problem for sequential catalytic systems with prioritized rules is NP-complete.

6 Some Problems for Future Research

Limited Parallelism in Other P Systems: We believe the results in the previous section concerning sequential P systems can be shown to hold for other more general P systems. We also think that, in fact, similar results hold for $\leq n$-**Parallel** and n-**Parallel** modes of computation.

Characterizations: It would be of interest to study various classes of nonuniversal P systems and characterize their computing power in terms of well-known models of sequential and parallel computation: Investigate language-theoretic properties of families of languages defined by P systems that are not universal (e.g., closure and decidable properties), find P system models that correspond to the Chomsky hierarchy, and in particular, characterize the "parallel" computing power of P systems in terms of well-known models like alternating Turing machines, circuit models, cellular automata, parallel random access machines, develop useful and efficient algorithms for their decision problems.

Reachability Problem in Cell Simulation: Another important research area that has great potential applications in biology is the use of P systems for the modeling and simulation of cells. While previous work on modeling and simulation use continuous mathematics (differential equations), P systems will allow us to use discrete mathematics and algorithms. As a P system models the computation that occurs in a living cell, an important problem is to develop tools for determining reachability between configurations, i.e., how the system evolves over time. Specifically, given a P system and two configurations α and β (a configuration is the number and distribution of the different types of objects in the various membranes in the system), is β reachable from α? Unfortunately, unrestricted P systems are universal (i.e., can simulate a Turing machine), hence all nontrivial decision problems (including reachability) are undecidable. Therefore, it is important to identify special P systems that are decidable for reachability. Some results along this lines have appeared in [4,5].

References

1. C. S. Calude and Gh. Paun. *Computing with Cells and Atoms: After Five Years (new text added to Russian edition of the book with the same title first published by Taylor and Francis Publishers, London, 2001)*. To be published by Pushchino Publishing House, 2004.
2. E. Csuhaj-Varju, O. H. Ibarra, and G. Vaszil. On the computational complexity of P automata. In *Proc. DNA 10* (C. Ferretti, G. Mauri, C. Zandron, eds.), Univ. Milano-Bicocca, 97–106, 2004.
3. Z. Dang and O. H. Ibarra. On P systems operating in sequential and limited parallel modes. In *Pre-Proc. 6th Workshop on Descriptional Complexity of Formal Systems*, 2004.
4. C. Li, Z. Dang, O. H. Ibarra, and H.-C. Yen. Signaling P systems and verification problems. *Proc. ICALP 2005*, to appear.
5. Z. Dang, O. H. Ibarra, C. Li, and G. Xie. On model-checking of P systems. Submitted.
6. R. Freund. Sequential P-systems. Available at *http://psystems.disco.unimib.it*, 2000.
7. R. Freund, Asynchronous P systems, in *Pre-Proc. Fifth Workshop on Membrane Computing*, eds. G. Mauri, Gh. Paun, and C. Zandron (2004).
8. R. Freund, L. Kari, M. Oswald, and P. Sosik. Computationally universal P systems without priorities: two catalysts are sufficient. *Theoretical Computer Science*, 330(2): 251–266, 2005.
9. R. Freund and A. Paun. Membrane systems with symport/antiport rules: universality results. In *Proc. WMC-CdeA2002*, volume 2597 of *Lecture Notes in Computer Science*, pages 270–287. Springer, 2003.
10. R. Freund and Gh. Paun. On deterministic P systems. Available at *http://psystems.disco.unimib.it*, 2003.
11. P. Frisco. About P systems with symport/antiport. *Second Brainstorming Week on Membrane Computing, Sevilla, Spain*, pp.224-236, Feb 2-7, 2004.
12. S. Greibach. Remarks on blind and partially blind one-way multicounter machines. *Theor. Comput. Sci.* 7:311-324, 1978.
13. S. Ginsburg. The Mathematical Theory of Context-Free Languages. New York: McGraw-Hill, 1966.
14. J. Hopcroft and J. J. Pansiot. On the reachability problem for 5-dimensional vector addition systems. *Theor. Comput. Sci.*, 8(2):135–159, 1979.
15. O. H. Ibarra. On Membrane hierarchy in P systems. *Theoretical Computer Science*, 334, 115-129, 2005.
16. O. Ibarra. On determinism versus nondeterminism in P systems. *Theoretical Computer Science*, to appear.
17. O. Ibarra, H. Yen, and Z. Dang. The power of maximal parallelism in P systems. *Proc. Eighth International Conference on Developments in Language Theory (DLT'04) (LNCS 3340)*, pp. 212-224, 2004.
18. O. Ibarra and S. Wood. On bounded symport/antiport systems. *Pre-proceedings of 11th International Meeting on DNA Computing*, 2005, to appear.
19. O. Ibarra and H. Yen. On deterministic catalytic systems. *Pre-proceedings of 10th International Conference on Implementation and Application of Automata*, 2005, to appear.
20. O. Ibarra, S. Woodworth, H.-C. Yen, and Z. Dang. On sequential and 1-deterministic P systems. *Proc. of the Eleventh International Computing and Combinatorics Conference (COCOON)*, 2005, to appear.

21. B. Monien, Two-way multihead automata over a one-letter alphabet, *RAIRO Informatique theorique*, 14(1):67–82, 1980.
22. A. Paun and Gh. Paun. The power of communication: P systems with symport/antiport. *New Generation Computers* 20(3): 295–306, 2002.
23. Gh. Paun. Computing with membranes. *Turku University Computer Science Research Report No. 208*, 1998.
24. Gh. Paun. Computing with membranes. *Journal of Computer and System Sciences*, 61(1):108–143, 2000.
25. Gh. Paun. *Membrane Computing: An Introduction.* Springer-Verlag, 2002.
26. Gh. Paun. Further twenty six open problems in membrane computing. Written for the Third Brainstorming Week on Membrane Computing, Sevilla, Spain, February, 2005. Available at *http://psystems.disco.unimib.it*.
27. Gh. Paun and G. Rozenberg. A guide to membrane computing. *Theoretical Computer Science*, 287(1):73–100, 2002.
28. Gh. Paun, J. Pazos, M. J. Perez-Jimenez, and A. Rodriguez-Paton. Symport/antiport P systems with three objects are universal. *Fundamenta Informaticae*, 64, 1–4, 2004.
29. W. Savitch. Relationships between nondeterministic and deterministic tape complexities. *J. Comput. Syst. Sci.*, 4(2): 177–192, 1970.
30. W. Savitch. A note on multihead automata and context-sensitive languages. *Acta Informatica*, 2:249–252, 1973.
31. P. Sosik. P systems versus register machines: two universality proofs. In *Pre-Proceedings of Workshop on Membrane Computing (WMC-CdeA2002), Curtea de Arges, Romania*, pages 371–382, 2002.

The Generalization of Dirac's Theorem for Hypergraphs

Endre Szemerédi[1], Andrzej Ruciński[2,*], and Vojtěch Rödl[3,**]

[1] Rutgers University, New Brunswick
szemered@cs.rutgers.edu
[2] A. Mickiewicz University, Poznań, Poland
rucinski@amu.edu.pl
[3] Emory University, Atlanta, GA
rodl@mathcs.emory.edu

1 Introduction and Main Result

A substantial amount of research in graph theory continues to concentrate on the existence of hamiltonian cycles and perfect matchings. A classic theorem of Dirac states that a sufficient condition for an n-vertex graph to be hamiltonian, and thus, for n even, to have a perfect matching, is that the minimum degree is at least $n/2$. Moreover, there are obvious counterexamples showing that this is best possible.

The study of hamiltonian cycles in hypergraphs was initiated in [1] where, however, a different definition than the one considered here was introduced. Given an integer $k \geq 2$, a k-uniform hypergraph is a hypergraph (a set system) where every edge (set) is of size k.

By *a cycle* we mean a k-uniform hypergraph whose vertices can be ordered cyclically $v_1, \ldots, v_l$ in such a way that for each $i = 1, \ldots, l$, the set $\{v_i, v_{i+1}, \ldots, v_{i+k-1}\}$ is an edge, where for $h > l$ we set $v_h = v_{h-l}$. A *hamiltonian cycle* in a k-uniform hypergraph H is a spanning cycle in H, that is, a sub-hypergraph of H which is a cycle and contains all vertices of H. A k-uniform hypergraph containing a hamiltonian cycle is called *hamiltonian*.

This notion and its generalizations have a potential to be applicable in many contexts which still need to be explored. An application in the relational database theory can be found in [2]. As observed in [5], the square of a (graph) hamiltonian cycle naturally coincides with a hamiltonian cycle in a hypergraph built on top of the triangles of the graph. More precisely, given a graph G, let $Tr(G)$ be the set of triangles in G. Define a hypergraph $H^{Tr}(G) = (V(G), Tr(G))$. Then there is a one-to-one correspondence between hamiltonian cycles in $H^{Tr}(G)$ and the squares of hamiltonian cycles in G. For results about the existence of squares of hamiltonian cycles see, e.g., [6].

As another potential application consider a seriously ill patient taking 24 different pills on a daily basis, one at a time every hour. Certain combinations

---

* Research supported by KBN grant 2 P03A 015 23. Part of research performed at Emory University, Atlanta.
** Research supported by NSF grant DMS-0300529.

of three pills can be deadly if taken within 2.5 hour. Let D be the set of deadly triplets of pills. Then any safe schedule corresponds to a hamiltonian cycle in the hypergraph which is precisely the complement of D.

A natural extension of Dirac's theorem to k-graphs, $k \geq 2$, has been conjectured in [5], where as a sufficient condition one demands that every $(k-1)$-element set of vertices is contained in at least $\lfloor n/2 \rfloor$ edges. The following construction of a k-uniform hypergraph H_0, also from [5], shows that the above conjecture, if true, is nearly best possible (best possible for $k = 3$).

Let $V = V' \cup \{v\}$, $|V| = n$. Split $V' = X \cup Y$, where, $|X| = \lfloor \frac{n-1}{2} \rfloor$ and $|Y| = \lceil \frac{n-1}{2} \rceil$. The edges of H_0 are all k-element subsets S of V such that $|X \cap S| \neq \lfloor \frac{k}{2} \rfloor$ or $v \in S$. It is shown in [5] that H_0 is not hamiltonian, while every $(k-1)$-element set of vertices belongs to at least $\lfloor \frac{n-k+1}{2} \rfloor$ edges.

In [9] we proved an approximate version of the conjecture from [5] for $k = 3$, and in [11] we give a generalization of that result to k-uniform hypergraphs for arbitrary k.

Theorem 1 ([11]). *Let $k \geq 3$ and $\gamma > 0$. Then, for sufficiently large n, every k-uniform hypergraph on n-vertices such that each $(k-1)$-element set of vertices is contained in at least $(1/2 + \gamma)n$ edges is hamiltonian.*

2 The Idea of Proof

The idea of the proof is as follows. As a preliminary step, we find in H a powerful path A, called *absorbing* which has the property that *every* not too large subset of vertices can be "absorbed" by that path. We also put aside a small subset of vertices R which preserves the degree properties of the entire hypergraph.

On the sub-hypergraph $H' = H - (A \cup R)$ we find a collection of long, disjoint paths which cover almost all vertices of H'. Then, using R we "glue" them and the absorbing path A together to form a long cycle in H. In the final step, the vertices which are not yet on the cycle are absorbed by A to form a hamiltonian cycle in H.

The main tool allowing to cover almost all vertices by disjoint paths is a generalization of the regularity lemma from [12].

Given a k-uniform hypergraph H and k non-empty, disjoint subsets $A_i \subset V(H)$, $i = 1, \ldots, k$, we define $e_H(A_1, \ldots, A_k)$ to be the number of edges in H with one vertex in each A_i, and the *density* of H with respect to $(A_1, \ldots, A_k)$ as

$$d_H(A_1, \ldots, A_k) = \frac{e_H(A_1, \ldots, A_k)}{|A_1| \cdots |A_k|}.$$

A k-uniform hypergraph H is k-*partite* if there is a partition $V(H) = V_1 \cup \cdots \cup V_k$ such that every edge of H intersects each set V_i in precisely one vertex. For a k-uniform, k-partite hypergraph H, we will write d_H for $d_H(V_1, \ldots, V_k)$ and call it the *density of H*.

We say that a k-uniform, k-partite hypergraph H is ϵ-*regular* if for all $A_i \subseteq V_i$ with $|A_i| \geq \epsilon |V_i|$, $i = 1, \ldots, k$, we have

$$|d_H(A_1, \ldots, A_k) - d_H| \leq \epsilon.$$

The following result, called *weak regularity lemma* as opposed to the stronger result in [4], is a straightforward generalization of the graph regularity lemma from [12].

Lemma 1 (Weak regularity lemma for hypergraphs). *For all $k \geq 2$, every $\epsilon > 0$ and every integer t_0 there exist T_0 and n_0 such that the following holds. For every k-uniform hypergraph H on $n > n_0$ vertices there is, for some $t_0 \leq t \leq T_0$, a partition $V(H) = V_1 \cup \cdots \cup V_t$ such that $|V_1| \leq |V_2| \leq \cdots \leq |V_t| \leq |V_1| + 1$ and for all but at most ϵt^k sets of partition classes $\{V_{i_1}, \ldots, V_{i_k}\}$, the induced k-uniform, k-partite sub-hypergraph $H[V_{i_1}, \ldots, V_{i_k}]$ of H is ϵ-regular.*

The above regularity lemma, combined with the fact that every dense ϵ-regular hypergraph contains an almost perfect path-cover, yields an almost perfect path-cover of the entire hypergraph H.

3 Results for Matchings

A perfect matching in a k-uniform hypergraph on n vertices, n divisible by k, is a set of n/k disjoint edges. Clearly, every hamiltonian, k-uniform hypergraph with the number of vertices n divisible by k contains a perfect matching.

Given a k-uniform hypergraph H and a $(k-1)$-tuple of vertices $v_1, \ldots, v_{k-1}$, we denote by $N_H(v_1, \ldots, v_{k-1})$ the set of vertices $v \in V(H)$ such that $\{v_1, \ldots, v_{k-1}, v\} \in H$. Let $\delta_{k-1}(H) = \delta_{k-1}$ be the minimum of $|N_H(v_1, \ldots, v_{k-1})|$ over all $(k-1)$-tuples of vertices in H.

For all integer $k \geq 2$ and n divisible by k, denote by $t_k(n)$ the smallest integer t such that every k-uniform hypergraph on n vertices and with $\delta_{k-1} \geq t$ contains a perfect matching.

For $k = 2$, that is, in the case of graphs, we have $t_2(n) = n/2$. Indeed, the lower bound is delivered by the complete bipartite graph $K_{n/2-1,n/2+1}$, while the upper bound is a trivial corollary of Dirac's condition [3] for the existence of Hamilton cycles.

In [10] we study t_k for $k \geq 3$. As a by-product of our result about hamiltonian cycles in [11] (see Theorem 2 above), it follows that $t_k(n) = n/2 + o(n)$. Kühn and Osthus proved in [7] that

$$\frac{n}{2} - k + 1 \leq t_k(n) \leq \frac{n}{2} + 3k^2 \sqrt{n \log n}.$$

The lower bound follows by a simple construction, which, in fact, for k odd yields $t_k(n) \geq n/2 - k + 2$. For instance, when $k = 3$ and $n/2$ is an odd integer, split the vertex set into sets A and B of size $n/2$ each, and take as edges all triples of vertices which are either disjoint from A or intersect A in precisely two elements.

In [10] we improve the upper bound from [7].

Theorem 2. *For every integer $k \geq 3$ there exists a constant $C > 0$ such that for sufficiently large n,*

$$t_k(n) \leq \frac{n}{2} + C \log n.$$

It is very likely that the true value of $t_k(n)$ is yet closer to $n/2$. Indeed, in [5] it is conjectured that $\delta_{k-1} \geq n/2$ is sufficient for the existence of a Hamilton cycle, and thus, when n is divisible by k, the existence of a perfect matching. Based on this conjecture and on the above mentioned construction from [7], we believe that $t_k(n) = n/2 - O(1)$. In fact, for $k = 3$, we conjecture that $t_3(n) = \lceil n/2 \rceil - 1$.

Our belief that $t_k(n) = n/2 - O(1)$ is supported by some partial results. For example, we are able to show that the threshold function $t_k(n)$ has a stability property, in the sense that hypergraphs that are "away" from the "extreme case" H_0, described in Section 1, contain a perfect matching even when δ_{k-1} is smaller than but not too far from $n/2$.

Interestingly, if we were satisfied with only a partial matching, covering all but a constant number of vertices, then this is guaranteed already with $n/2 + o(n)$ replaced by n/k, that is, when $\delta_{k-1} \geq n/k$.

We have also another related result, about the existence of a fractional perfect matching, which is a simple consequence of Farkas' Lemma (see, e.g.,[8]). A *fractional perfect matching* in a k-uniform hypergraph $H = (V, E)$ is a function $w : E \to [0, 1]$ such that for each $v \in V$ we have

$$\sum_{e \ni v} w(e) = 1.$$

In particular, it follows from our result that if $\delta_{k-1}(H) \geq n/k$ then H has a fractional perfect matching, so, again, the threshold is much lower than that for perfect matchings.

References

1. J. C. Bermond et al., Hypergraphes hamiltoniens, *Prob. Comb. Theorie Graph Orsay* 260 (1976) 39-43.
2. J. Demetrovics, G. O. H. Katona and A. Sali, Design type problems motivated by database theory, *Journal of Statistical Planning and Inference* 72 (1998) 149-164.
3. G. A. Dirac, Some theorems for abstract graphs, *Proc. London Math. Soc.* (3) 2 (1952) 69-81.
4. P. Frankl and V. Rödl, Extremal problems on set systems, *Random Struct. Algorithms* 20, no. 2, (2002) 131-164.
5. Gyula Y. Katona and H. A. Kierstead, Hamiltonian chains in hypergraphs, *J. Graph Theory* 30 (1999) 205-212.
6. J. Komlós, G. N. Sárközy and E. Szemerédi, On the Pósa-Seymour conjecture, *J. Graph Theory* 29 (1998) 167-176.
7. D. Kuhn and D. Osthus, Matchings in hypergraphs of large minimum degree, submited.
8. L. Lovász & M.D. Plummer, *Matching theory*. North-Holland Mathematics Studies 121, Annals of Discrete Mathematics 29, North-Holland Publishing Co., Amsterdam; Akadémiai Kiadó, Budapest, 1986
9. V. Rödl, A. Ruciński and E. Szemerédi, A Dirac-type theorem for 3-uniform hypergraphs, *Combinatorics, Probability and Computing*, to appear.
10. V. Rödl, A. Ruciński and E. Szemerédi, Perfect matchings in uniform hypergraphs with large minimum degree, submitted.

11. V. Rödl, A. Ruciński and E. Szemerédi, An approximative Dirac-type theorem for k-uniform hypergraphs, submitted.
12. E. Szemerédi, Regular partitions of graphs. Problemes combinatoires et theorie des graphes (Colloq. Internat. CNRS, Univ. Orsay, Orsay, 1976), pp. 399–401, Colloq. Internat. CNRS, 260, CNRS, Paris, 1978.

On the Communication Complexity of Co-linearity Problems

Andrew C. Yao*

Tsinghua University, Beijing, China

Abstract. In the k-party simultaneous message model, $k-1$ parties holding respectively $x_1, x_2, \cdots, x_{k-1}$ wish to compute the value of some boolean function $f(x_1, x_2, \ldots, x_{k-1})$, by each sending a stochastically chosen message to a k-th party, the *referee*, who then decides on the value of f with probability at least $2/3$ of being correct. Let $R^{\|}(f)$ be the minimum number of total communication bits needed to compute f by any such algorithm.

The (k, n)-*Co-Linearity Problem* is defined by $CL_{k,n}(x_1, x_2, \ldots, x_{k-1}) = 1$, if and only if $\oplus_{1 \leq i \leq k-1} x_i = 0^n$ (where x_i are n-bit strings). It is well known that, for any fixed $k \geq 3$, $R^{\|}(CL_{k,n}) = O(n^{(k-2)/(k-1)})$, and that the bound is tight for $k=3$. It is an interesting open question whether the bound is tight for $k > 3$. In this talk we present some new results on this question. Specifically, we prove that the above bound is tight in the *linear model*, in which all the transmitted message bits are linear functions of the input bits. We also discuss $CL_{k,n}$'s quantum communication complexity, which also has received considerable attention in recent years.

---

* This research was supported in part by the US National Science Foundation under Grants CCR-0310466 and CCF-0426582.

An Invitation to Play*

Wiesław Zielonka

LIAFA, case 7014, Université Paris 7 and CNRS,
2, Place Jussieu, 75251 Paris Cedex 05, France
zielonka@liafa.jussieu.fr

Abstract. Parity games and their subclasses and variants pop up in various contexts: μ-calculus, tree automata, program verification [3,1,8]. Such games provide only binary information indicating the winning player. However, in classical games theory [12] the emphasis is rather on how much we win or lose. Can we incorporate the information about the profits and losses into parity games?

1 Games

Our games oppose two players, player 1 and player 2. At each moment the game is in some state s and the player controlling s chooses an action available at s which results in issuing an immediate reward r and changing the state to a new one s'. Both the reward and the new state depend deterministically on the executed action, i.e. we can assume without loss of generality that the set of actions A is just a subset of $S \times \Re \times S$, where S is the set of all states and $\Re$ is a set of (immediate) rewards. If $a = (s_1, r, s_2) \in A$ then the state $s_1 = \text{source}(a)$ is the *source* of the action a indicating the state where a is available, $s_2 = \text{target}(a)$ is the *target* state where the game moves upon the execution of a and finally $r = \text{reward}(a) \in \Re$ is the reward associated with a.

The set S of states is partitioned onto two sets, the set S_1 of states controlled by player 1 and the set S_2 of states controlled by player 2. For each state s the set $A(s) = \{a \in A \mid \text{source}(a) = s\}$ is the set of actions *available* at s and we assume that this set is always non-empty for each state s.

The tuple $\mathcal{A} = (S_1, S_2, A)$ satisfying the conditions above is called an *arena* over the set $\Re$ of rewards. Unless otherwise stated, we assume always that an "arena" means in fact a finite arena, i.e. an arena with finite state and action spaces.

A *history* in arena $\mathcal{A}$ is a finite or an infinite sequence $h = a_1 a_2 \ldots$ of actions such that $\forall i, \text{target}(a_i) = \text{source}(a_{i+1})$. The source of the first action a_0 is the source, source(h), of history h. If h is finite then the target of the last action is the target, target(h), of h.

It is convenient to assume that for each state s there is an empty history $\mathbf{1}_s$ with the source and the target s.

---

* This research was supported by European Research Training Network: Games and Automata for Synthesis and Validation and ACI Sécurité Informatique 2003-22 VERSYDIS.

We start the play by putting a token at some initial state s_1 and the players play by moving the token from state to state: at each stage if the token is at a state $s \in S_i$ controlled by player i than player i chooses an action $a \in A(s)$ available at s and moves the token to the state target(a).

Starting from an initial state s_1, the infinite sequence of actions $p = a_1 a_2 \ldots$ executed by the players is called a *play* in the arena $\mathcal{A}$, i.e. *plays* are just infinite histories in $\mathcal{A}$.

Upon the termination of a play p player 1 receives from player 2 a payoff. In this paper we assume that the payoff depends uniquely on the infinite sequence of rewards occurring in the play p.

An infinite sequence $r = r_1 r_2 \ldots$ of elements of $\Re$ is said to be finitely generated if there exists a finite subset X of $\Re$ such that all r_i belong to X.

A *payoff mapping* u over $\Re$ maps finitely generated infinite sequences of rewards $r = r_1 r_2 \ldots$ into $\mathbb{R}$, $u : r \mapsto u(r) \in \mathbb{R}$. Since we are concerned only with plays over finite arenas we do not need to specify what is the payoff for those infinite reward sequences which are not finitely generated.

A *game* over $\Re$ is just a couple $\mathbf{G} = (\mathcal{A}, u)$ consisting of an arena and a payoff mapping. A play $p = a_0 a_1 \ldots$ in the game $\mathbf{G}$ is a play in the underlying arena. Upon completing p player 1 receives from player 2 the amount $u(\text{reward}(p))$, where $\text{reward}(p) := \text{reward}(a_1), \text{reward}(a_2), \ldots$ is the sequence of rewards occurring in p. To avoid clutter we abuse the notation and we write systematically $u(p)$ to denote $u(\text{reward}(p))$ (this can be seen as an extension of payoff mapping to plays).

Parity Games. For parity games the set of rewards $\Re$ is the set $\mathbb{N}$ of non-negative integers and following the tradition we call the elements of $\Re = \mathbb{N}$ priorities rather than rewards.

For any infinite finitely generated sequence of priorities $n = n_1 n_2 \ldots$ let

$$\text{priority}(n) = \limsup_{i \to \infty} n_i \qquad (1)$$

be the maximal priority occurring infinitely often in n. The payoff mapping in the parity games is given by

$$u(n) = \begin{cases} 1 & \text{if priority}(n) \text{ is odd,} \\ 0 & \text{if priority}(n) \text{ is even.} \end{cases} \qquad (2)$$

Two remarks are in order. Usually in parity games we speak about the winning and the losing player, however it is clear that this is equivalent to the binary payoff formulation given above and we prefer payoffs since subsequently we will be interested in profits or losses and not just in the mere information who wins. Secondly, in parity games we usually attach priorities to states not to actions but this has no influence on the game analysis and game theoretists prefer to associate rewards with actions [11].

Discounted and Mean-Payoff Games. Let us compare briefly parity games with other similar games studied in game theory rather than in computer science.

Two types of games are particularly popular, discounted and mean-payoff games [11,4]. In both these games $\Re = \mathbb{R}$, i.e. the rewards are real numbers.

In *mean-payoff* games the payoff for an infinite sequence of real numbers $r = r_1 r_2 \ldots$ is calculated through the formula $\overline{\operatorname{mean}}(r) = \limsup_{n \to \infty} \frac{1}{n} \sum_{i=1}^{n} r_i$. Instead of taking lim sup it is possible to consider the games with the payoff $\underline{\operatorname{mean}}(r) = \liminf_{n \to \infty} \frac{1}{n} \sum_{i=1}^{n} r_i$.

In the case of *discounted games* player 1 receives from 2 the amount $\operatorname{disc}_\lambda(r) = (1-\lambda) \sum_{i=0}^{\infty} \lambda^i r_i$, where $\lambda \in (0,1)$ is a discount factor.

The striking difference between the parity games on the one hand and the mean-payoff or the discounted games on the other hand is that in the later the emphasis is put on the amount of profit/loss while for the parity games the information is just binary, indicating the winner without any attempt to quantify his profit. Obviously for games inspired by economic applications to be able to quantify the profit is essential, after all, the difference between winning or losing 10\$ is hardly noticeable (and both events in themselves are of little interest) while the difference between winning 10\$ and winning 10^6\$ is formidable and of great interest to the player.

Can parity games be adapted to provide a pertinent information about the player's profits/losses instead of just a plain indication who wins? It turns out that in fact several such extensions are possible for parity games and moreover these games preserve the most appealing property of parity games: the existence of optimal memoryless strategies for both players.

1.1 Strategies

A strategy of a player is his plan of action, it tells him which action to take when it is his turn to move. The choice of the action to be executed can depend on the whole sequence of previous moves. Thus a *strategy* for player 1 is a mapping

$$\sigma : \{h \mid h \text{ a finite history with target}(h) \in S_1\} \longrightarrow A \qquad (3)$$

such that if $s = \operatorname{target}(h)$ then $\sigma(h) \in A(s)$.

A strategy σ of player 1 is said to be *positional* or *memoryless* if the chosen action depends only on the last state in the history. It is convenient to view a positional strategy as a mapping

$$\sigma : S_1 \to A \qquad (4)$$

such that $\sigma(s) \in A(s)$, $\forall s \in S_1$.

Strategies and positional strategies for player 2 are defined in the similar way with S_2 replacing S_1.

In the sequel, σ and τ, possibly with subscripts or superscripts, will always denote strategies for players 1 and 2 respectively.

A finite or infinite history $h = a_1 a_2 \ldots$ is said to be *consistent* with a strategy σ of player 1 if for each i such that $\operatorname{target}(a_i) \in S_1$, $a_{i+1} = \sigma(a_0 \ldots a_i)$. Moreover, if $s = \operatorname{source}(a_1) \in S_1$ then we require that $a_1 = \sigma(\mathbf{1}_s)$ (recall that $\mathbf{1}_s$ is a special play of zero length with the source and target s). The consistency with strategies of player 2 is defined similarly.

Given a pair of strategies σ and τ for both players and a state s, there exists in arena $\mathcal{A}$ a unique infinite play p, denoted $p_\mathcal{A}(s, \sigma, \tau)$, consistent with σ and τ and such that $s = \text{source}(p)$.

Strategies $\sigma^\sharp$ and $\tau^\sharp$ of players 1 and 2 are *optimal* in the game $\mathbf{G} = (\mathcal{A}, u)$ if for any state $s \in S$ and any strategies σ and τ

$$u(p_\mathcal{A}(s, \sigma, \tau^\sharp)) \leq u(p_\mathcal{A}(s, \sigma^\sharp, \tau^\sharp)) \leq u(p_\mathcal{A}(s, \sigma^\sharp, \tau)) \ . \tag{5}$$

Thus if both strategies are optimal the players do not have any incentive to change them unilaterally.

Note that for zero sum games that we consider here, where the profit of one player is equal to the loss of his adversary, we have the exchangeability property for optimal strategies: for any other pair of optimal strategies $\tau^\ddagger$, $\sigma^\ddagger$, the couples $(\tau^\ddagger, \sigma^\sharp)$ and $(\tau^\sharp, \sigma^\ddagger)$ are also optimal and $u(p_\mathcal{A}(s, \sigma^\sharp, \tau^\sharp)) = u(p_\mathcal{A}(s, \sigma^\ddagger, \tau^\ddagger))$; this last quantity is called the value of the game $\mathbf{G} = (\mathcal{A}, u)$ at the state s.

The basic problem of game theory is to determine for a given payoff mapping u if for every game $\mathbf{G} = (\mathcal{A}, u)$ both players have optimal strategies.

In computer science we prefer positional strategies since they are particularly easy to implement. For this reason the question that we ask in this paper for every payoff u is whether for each game $\mathbf{G} = (\mathcal{A}, u)$ over a finite arena $\mathcal{A}$ both players have positional optimal strategies.

2 From Parity Games to Games with Profits

2.1 Simple Priority Games

The simplest adjustment of parity games enabling any real-valued payoff consists in associating with each priority a real number by means of a mapping $\alpha : \mathbb{N} \to \mathbb{R}$, we call α a priority valuation. Let $n = n_1 n_2 \ldots$ be any finitely generated infinite sequence of elements of $\mathbb{N}$.

Then the payoff mapping of *simple priority games* is given by

$$u_\alpha(n) = \alpha(\text{priority}(n)), \tag{6}$$

where $\text{priority}(n)$ is defined as in (1). Clearly for different priority valuations α we have different simple priority games, in particular for α that maps even numbers to 0 and odd numbers to 1 we recover the parity game. In fact simple priority games are still very close to parity games. Let $\alpha(\mathbb{N}) = \{x_1 < \ldots < x_k\}$ be all priority values taken in the increasing order[1]. Then to establish if player 1 has a strategy allowing him to win at least x_i in the game with the priority valuation α we solve the game with the binary priority valuation β_i defined by $\beta_i(l) = 1$ if $\alpha(l) \geq x_i$ and $\beta_i(l) = 0$ if $\alpha(l) < x_i$. Games with binary valuations are obviously equivalent to parity games thus both players have optimal positional strategies

---

[1] We can assume without loss of generality that $\alpha(\mathbb{N})$ is finite.

$\sigma_i^\sharp, \tau_i^\sharp$ in the game with the valuation β_i (in fact this is true even for infinite arenas [3,10]). These strategies can used to build optimal positional strategies $\sigma^\sharp, \tau^\sharp$ in the game with the valuation α. For a given state $s \in S$ define the rank of s to be the maximal l such that the strategy β_l of player 1 allows him to win 1 in the binary priority game with the valuation β_l when the initial state is s. Then, for $s \in S_1$, we set $\sigma^\sharp(s) = \sigma_l^\sharp(s)$ while for $s \in S_2$ we set $\tau^\sharp(s) = \tau_{l+1}^\sharp(s)$, where l is the rank of s. Clearly the strategies $\sigma^\sharp$ and $\tau^\sharp$ are positional. Moreover, it is easy to see that in $(\mathcal{A}, u_\alpha)$ for plays starting at a state s with the rank l the strategy $\sigma^\sharp$ assures for player 1 that he will win at least x_l while the strategy $\tau^\sharp$ assures for player 2 that he will pay no more than x_l. This proves the optimality of strategies $\sigma^\sharp$ and $\tau^\sharp$ (also for infinite arenas).

2.2 Mean-Payoff Priority Games

To generalize yet further our games set $\Re = \mathbb{N} \times \mathbb{R}$ as the set of rewards. Each couple $(n, r) \in \Re$ consists now of a non-negative priority n and a real valued reward $r \in \mathbb{R}$. For an infinite finitely generated reward sequence $x = (n_1, r_1), (n_2, r_2), \ldots$ we calculate now the payoff in the following way. Let $n = \text{priority}(n_1 n_2 \ldots)$ be the maximal priority appearing infinitely often in x and let $x(n) = (n_{i_1}, r_{i_1}), (n_{i_2}, r_{i_2}), \ldots$ be the subsequence of x consisting of the elements with priority n, $n = n_{i_1} = n_{i_2} = \cdots$. Then

$$\overline{\text{mean}}(x) = \limsup_{k \to \infty} \frac{r_{i_1} + \cdots + r_{i_k}}{k} \tag{7}$$

defines the payoff for *mean-payoff priority games*. Thus, intuitively, we calculate here mean-payoff of rewards but limited to the subsequence of the maximal priority occurring infinitely often. Note that if there is only one priority then this payoff mapping reduces to the payoff of mean-payoff games (and for this reason we keep the same name). But, on the other hand, if we limit ourselves to reward sequences such that $n_i = n_j$ implies $r_i = r_j$ for all i, j, i.e. to sequences where the reward is constant for each priority, then $\overline{\text{mean}}$ reduces to a simple priority payoff of Sect. 2.1 with an appropriate priority valuation. Thus mean-payoff priority games combine the principal characteristics of mean-payoff and parity games.

Are these games positional?

If the arena is controlled by player 1, i.e. $S = S_1$, then player 1 has an obvious optimal positional strategy that can be found in the following way (we do not pretend that the method given below is the most efficient one). First note that for any play of the form $p = xy^\omega$, where x is a finite history, $y = a_1 a_2 \ldots a_k$ is a simple cycle[2] in the arena $\mathcal{A}$ and $y^\omega = yy\ldots$ is the infinite concatenation of y, we can calculate $\overline{\text{mean}}(p)$ in the following way:
let for $1 \leq i \leq k$, $\text{reward}(a_i) = (n_i, r_i)$, let $l = \max\{n_i \mid 1 \leq i \leq k\}$ be the maximal priority occurring in y, and let $M = \{i \mid 1 \leq i \leq k \text{ and } n_i = l\}$ be the occurrences of l in y, then $\overline{\text{mean}}(p) = \frac{1}{|M|} \sum_{m \in M} r_m$.

---

[2] That means that source$(y) = $ target(y) and source$(a_i) \neq $ source(a_j) for $1 \leq i < j \leq k$.

Let y be a simple cycle such that $\overline{\mathrm{mean}}(y^\omega)$ is maximal. It is easy to see that for any other play p in $\mathcal{A}$, $\overline{\mathrm{mean}}(p) \leq \overline{\mathrm{mean}}(y^\omega)$. Thus to maximize his gain player 1 should arrive at this cycle y, which can be done with a positional strategy, and then he should turn round y forever which is obviously positional. If there are states in $\mathcal{A}$ from which the cycle y of the maximal payoff is not accessible then in the subarena consisting of such states we repeat the procedure described above.

For arenas controlled by player 2 (which means that $S = S_2$) the optimal positional strategy of player 2 can be found in the similar way by finding the simple cycle minimizing the payoff.

The main result of Sect. 3 (Theorem 3) states that the existence of optimal positional strategies for one-player games implies the existence of optimal positional strategies for two-player games and thus we can conclude

Proposition 1. *For all priority mean-payoff games over finite arenas both players have optimal positional strategies.*

This result above was first established by other methods in [7].

2.3 Weighted Reward Games

Yet another extension of parity games can be obtained in the following way. Suppose that for an infinite finitely generated sequence of priorities $n = n_1 n_2 \ldots$, the priorities n_e and n_o are respectively the greatest even and the greatest odd priority occurring infinitely often in n. Then player 1 wins the parity game iff the quantity $n_o - n_e$ is positive. However, intuitively, $n_o - n_e$ gives us a more detailed information of how much the winning player outperforms the losing player in parity games, and we can as well consider the game where $n_o - n_e$ is the payoff obtained by player 1, i.e. the game where player 1 tries now to maximize this value. It is convenient then to replace even priorities by their negatives, i.e. consider finitely generated sequences $m_1 m_2 \ldots$ of integers and then the payoff for player 1 takes the form $\limsup_{i \to \infty} m_i + \liminf_{i \to \infty} m_i$. However, we can then go a step further and take as the set of rewards the set $\mathbb{R}$ of real numbers and for a fixed parameter $\lambda \in [0; 1]$ and a finitely generated sequence $r_1 r_2 \ldots$ of real numbers define a weighted reward payoff mapping

$$u^{\mathrm{wr}}_\lambda(r_1 r_2 \ldots) = (1 - \lambda) \liminf_{i \to \infty} r_i + \lambda \limsup_{i \to \infty} r_i \ . \tag{8}$$

Note that for $\lambda = 1$ the definition above gives just the classical gambling payoff [2,9] while $\lambda = 1/2$ can be used, as explained above, to generalize parity games. In the same way as for mean-payoff priority games one can verify that weighted reward one-player games have optimal positional strategies which implies by Theorem 3 the following result obtained first in [6]:

Proposition 2. *For all weighted reward games over finite arenas both players have optimal positional strategies.*

3 From One-Player Games to Two-Player Games

It turns our that to assure that a payoff mapping u allows optimal positional strategies for all two-player games it suffices to verify whether one-player games with payoff u have optimal positional strategies. In fact a similar result holds also for perfect information stochastic games [5] and the proof below is just an adaptation of the one of [5].

This result is useful in practice since, as we have seen, the verification if a given payoff mapping admits optimal positional strategies can be trivial for one-person games but can require a bit of dexterity for two-person games.

An arena $\mathcal{A} = (S_1, S_2, A)$ is said to be *controlled by player* i if for each state $s \in S_j$ controlled by his adversary j, $j \neq i$ and $i, j \in \{1, 2\}$, there is only one action $a \in A$ with source s. Thus essentially the adversary player j has never any choice, in particular he has only one strategy and this strategy is positional. In this case we can as well put all the states of j under the control of player i and remove player j altogether from our game. A *one-player arena* is just an arena controlled by one of the two players and a *one-player game* is a game on a one-player arena. Note that in one-player games it suffices to exhibit an optimal strategy for the controlling player since the unique strategy of his adversary is trivial.

Theorem 3. *Let u be a payoff mapping over a set $\Re$ of rewards. If for each finite one-player arena $\mathcal{A}$ over $\Re$ the player controlling $\mathcal{A}$ has an optimal positional strategy in the game $\mathbf{G} = (\mathcal{A}, u)$ then for all two-person games over finite arenas with payoff u both players have optimal positional strategies.*

Proof. Suppose that u satisfies the conditions of the theorem. In the sequel whenever we speak about games over arenas $\mathcal{A}$ the payoff u is tacitly assumed.

For any arena $\mathcal{A} = (S_1, S_2, A)$ we call the value $|A| - |S|$ the *rank* of $\mathcal{A}$ ($|X|$ denotes the cardinality of X). Since for each state there is at least one available action the rank is always non-negative. If the rank is 0 then for each state s there is exactly one available action and therefore each player has only one possible strategy and these strategies are positional and optimal.

We shall continue the proof of Theorem 3 by induction over the rank value.

Let $\mathcal{A} = (S_1, S_2, A)$ be an arena with rank $k > 0$ and suppose that both players have optimal positional strategies for all games over the arenas with the ranks smaller than k. We shall construct a pair of optimal strategies $\sigma^\sharp, \tau^\sharp$ for the game over $\mathcal{A}$, the strategy $\sigma^\sharp$ of player 1 will be positional but the strategy $\tau^\sharp$ of player 2 will use some finite memory. In the next step we shall show that also player 2 has an optimal positional strategy.

If one of the players controls $\mathcal{A}$ then both of them have optimal positional strategies and there is nothing to do.

Thus we can assume that there exists a state $x \in S_1$ controlled by player 1 such that the set $A(x) = \{a \in A \mid \text{source}(a) = s\}$ of actions available at x contains more than one element. Let us fix such a state x which we shall call the *pivot*. We fix also a partition of the set $A(x)$ onto two non-empty sets $A_L(x)$ and $A_R(x)$, $A(x) = A_L(x) \cup A_R(x)$, $A_L(x) \cap A_R(x) = \emptyset$.

We define two subarenas $\mathcal{A}_L$ and $\mathcal{A}_R$ of $\mathcal{A}$ which we call respectively the left and the right (sub)arena. In $\mathcal{A}_L$ and $\mathcal{A}_R$ we keep the same states as in $\mathcal{A}$. Also the actions with the source in the states $s \neq x$ are the same in $\mathcal{A}_L$, $\mathcal{A}_R$ and $\mathcal{A}$. The only difference concerns the actions with the source x, in the left arena $\mathcal{A}_L$ we keep only the actions of $A_L(x)$ while in the right arena only the actions of $A_R(x)$ removing all the other actions with source x.

Since the ranks of the arenas $\mathcal{A}_L$ and $\mathcal{A}_R$ are smaller than the rank of $\mathcal{A}$, by the induction hypothesis, there exist optimal positional strategies $\sigma_L^\sharp$ and $\tau_L^\sharp$ in the game $(\mathcal{A}_L, u)$ and optimal positional strategies $\sigma_R^\sharp$ and $\tau_R^\sharp$ in the game $(\mathcal{A}_R, u)$.

We pretend that one of the two strategies $\sigma_L^\sharp$ or $\sigma_R^\sharp$ is also optimal for player 1 in the initial game over $\mathcal{A}$. The situation is more complicated for player 2, usually neither $\tau_L^\sharp$ nor $\tau_R^\sharp$ is optimal for him in the game over $\mathcal{A}$. However, it turns out to be possible to intertwine in some way the strategies $\tau_L^\sharp$ and $\tau_R^\sharp$ to obtain an optimal strategy for player 2 on $\mathcal{A}$.

Using the arena $\mathcal{A}_L$ and the strategy $\tau_L^\sharp$ of player 2 we construct an arena $\mathcal{A}_L[\tau_L^\sharp]$ that has the same states as $\mathcal{A}_L$ but we restrict the actions available to player 2: for each state $s \in S_2$ controlled by 2 we leave in $\mathcal{A}_L[\tau_L^\sharp]$ only one action with the source s, namely the action $\tau_L^\sharp(s)$ provided by the strategy $\tau_L^\sharp$. We do not restrict the moves of player 1, he can take exactly the same actions as in $\mathcal{A}_L$.

In a similar way we construct from the arena $\mathcal{A}_R$ and the optimal strategy $\tau_R^\sharp$ of player 2 in the game on $\mathcal{A}_R$ an arena $\mathcal{A}_R[\tau_R^\sharp]$ by restricting the actions player 2 to those that are provided by the strategy $\tau_R^\sharp$.

Notice that arenas $\mathcal{A}_L[\tau_L^\sharp]$ and $\mathcal{A}_R[\tau_R^\sharp]$ are controlled by player 1.

Next we rename in $\mathcal{A}_L[\tau_L^\sharp]$ and $\mathcal{A}_R[\tau_R^\sharp]$ all the states that are different from the pivot state x.

Let
$$U = S \setminus \{x\} \qquad (9)$$
be the set of states that are different from the pivot x. Let U_L, U_R be two disjoint copies of the set U and let
$$S_L := U_L \cup \{x\} \quad \text{and} \quad S_R := U_R \cup \{x\} \ . \qquad (10)$$

For a state $s \in U$ its left and right copy are denoted respectively s_L and s_R. It is convenient to assume that the pivot x is the only state that is a copy of itself, i.e. $x_L = x = x_R$. By π_L we shall denote the natural bijections
$$\pi_L : S \to S_L \quad \text{and} \quad \pi_R : S \to S_R \qquad (11)$$
$\pi_L : s \mapsto s_L = \pi_L(s)$ and $\pi_R : s \mapsto s_R = \pi_R(s)$, for all $s \in S$. The renaming mappings π_L and π_R are extended in a natural way to actions
$$\pi_L((s,r,t)) = (\pi_L(s), r, \pi_L(t)) \quad \text{and} \quad \pi_R((s,r,t)) = (\pi_R(s), r, \pi_R(t)) \qquad (12)$$
for actions (s, r, t) respectively on the left and the right subarena.

The arenas obtained from $\mathcal{A}_L[\tau_L^\sharp]$ and $\mathcal{A}_R[\tau_R^\sharp]$ by applying the corresponding renaming mappings are denoted $\pi_L(\mathcal{A}_L[\tau_L^\sharp])$ and $\pi_R(\mathcal{A}_R[\tau_R^\sharp])$. Note that $\pi_L(\mathcal{A}_L[\tau_L^\sharp])$ and $\pi_R(\mathcal{A}_R[\tau_R^\sharp])$ have only one common state x and the only common actions are eventually the actions of the form (x, r, x) if such actions with source and target x exist in $\mathcal{A}$. Finally we construct the arena $\mathcal{A}_{LR} = \pi_L(\mathcal{A}_L[\tau_L^\sharp]) \cup \pi_R(\mathcal{A}_R[\tau_R^\sharp])$, where the union means that we take simply the union of state sets and the union of action sets of $\pi_L(\mathcal{A}_L[\tau_L^\sharp])$ and $\pi_R(\mathcal{A}_R[\tau_R^\sharp])$. Let us note that since the only state common to $\pi_L(\mathcal{A}_L[\tau_L^\sharp])$ and $\pi_R(\mathcal{A}_R[\tau_R^\sharp])$ is the pivot x the arena $\mathcal{A}_{LR}$ can be seen informally as the arena obtained from $\mathcal{A}_L[\tau_L^\sharp]$ and $\mathcal{A}_R[\tau_R^\sharp]$ by gluing them together at x.

Obviously, $\mathcal{A}_{LR}$ is a one-player arena controlled by player 1. Intuitively, for each state of $\mathcal{A}_{LR}$ controlled by player 1 he has at his disposition the same actions as in $\mathcal{A}$. On the other hand, player 2 is compelled to use either the strategy $\tau_L^\sharp$ or the strategy $\tau_R^\sharp$ depending on whether the current position is in the left or in the right subarena of $\mathcal{A}_{LR}$. Each time the pivot x is visited player 1 can choose if he prefers to play till the next visit to x against the strategy $\tau_L^\sharp$ or against the strategy $\tau_R^\sharp$ by choosing either a left or a right action at x.

Example 4. Figure 1 illustrates different stages of the construction of $\mathcal{A}_{LR}$. To avoid clutter the rewards associated with actions are omitted. The states controlled by players 1 and 2 are represented respectively by circles and squares. The pivot state x has three outgoing actions and we fix the following left/right partition of $A(x)$: $A_L(x) = \{(x, s^2)\}$, $A_R(x) = \{(x, s^1), (x, s^3)\}$. Suppose now that for the state s^4 the optimal positional strategy $\tau_L^\sharp$ for player 2 in $(\mathcal{A}_L, u)$ chooses the action (s^4, s^1) while in $(\mathcal{A}_R, u)$ the optimal positional strategy $\tau_R^\sharp$ for the same player chooses the action (s^4, s^3) (for the other state s^2 both strategies choose the only available action (s^2, s^4)). The bottom part of Fig. 1 presents the resulting arenas $\mathcal{A}_L[\tau_L^\sharp]$ and $\mathcal{A}_R[\tau_R^\sharp]$. Finally, the upper left part of Fig. 1 shows the arena $\mathcal{A}_{LR}$. Note that in $\mathcal{A}_{LR}$ at the pivot state x player 1 has again three available actions, as in the initial arena $\mathcal{A}$.

Since the game $(\mathcal{A}_{LR}, u)$ is controlled by player 1 he has in this game an optimal positional strategy $\sigma_{LR}^\sharp$. Now let us look which action is chosen by $\sigma_{LR}^\sharp$ at the pivot state, we can have either $\mathrm{target}(\sigma_{LR}^\sharp(x)) \in S_L$ or $\mathrm{target}(\sigma_{LR}^\sharp(x)) \in S_R$. Exchanging if necessary "left" and "right", we can assume without loss of generality that

$$\mathrm{target}(\sigma_{LR}^\sharp(x)) \in S_L \ . \tag{13}$$

Under condition (13) it turns out that the strategy

$$\sigma^\sharp := \sigma_L^\sharp \tag{14}$$

is optimal for player 1 in the game $(\mathcal{A}, u)$

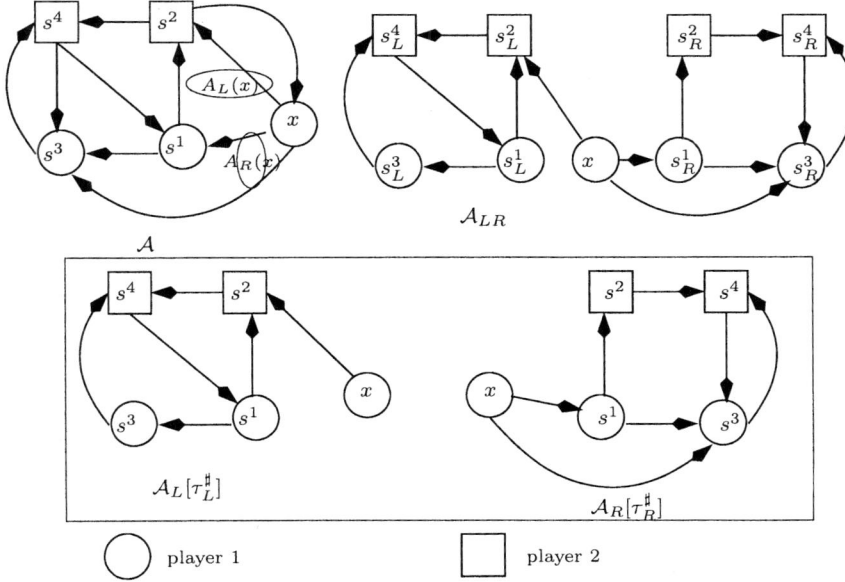

Fig. 1. Construction of $\mathcal{A}_{LR}$

It remains to define an optimal strategy $\tau^\sharp$ for player 2 on $\mathcal{A}$. Let h be a finite history in $\mathcal{A}$ with $\text{target}(h) \in S_2$, then

$$\tau^\sharp(h) = \begin{cases} \tau_L^\sharp(\text{target}(h)) & \text{if either } h \text{ does not contain any action with source } x \\ & \text{or the last such action belongs to } A_L(x), \\ \tau_R^\sharp(\text{target}(h)) & \text{if } h \text{ contains at least one action with source } x \\ & \text{and the last such action belongs to } A_R(x). \end{cases}$$
(15)

Thus player 2 applies in $\tau^\sharp$ either the strategy $\tau_L^\sharp$ or $\tau_R^\sharp$ and which of the two strategies is chosen depends on the action taken by player 1 at the last passage through the pivot state. To implement the strategy $\tau^\sharp$ one needs a memory, albeit a finite memory taking two values L and R is sufficient. The initial memory value is L. Every time the play traverses the pivot state x player 2 observes the action taken by player 1 and updates his memory either to L or to R depending on whether this action belongs to $A_L(x)$ or to $A_R(x)$.

Up to the next visit to x player 2 uses either the strategy $\tau_L^\sharp$ or $\tau_R^\sharp$ depending on the memory value.

We shall show that strategies $\sigma^\sharp$ and $\tau^\sharp$ defined by (13) and (15) are optimal in the game over $\mathcal{A}$.

Let $s \in S$ and consider the play $p_{\mathcal{A}}(s, \sigma^\sharp, \tau^\sharp)$. From (13) it follows that player 1 chooses during this play at each passage through x the same left hand side action from $A_L(x)$, however in this case player 2 plays all the time using the strategy $\tau_L^\sharp$, thus

$$p_{\mathcal{A}}(s, \sigma^\sharp, \tau^\sharp) = p_{\mathcal{A}}(s, \sigma_L^\sharp, \tau_L^\sharp) = p_{\mathcal{A}_L}(s, \sigma_L^\sharp, \tau_L^\sharp) \ . \tag{16}$$

Let τ be any strategy of player 2 on $\mathcal{A}$ and let τ' be the restriction of this strategy to histories in $\mathcal{A}_L$. Obviously, τ' is a valid strategy on $\mathcal{A}_L$ and $p_{\mathcal{A}_L}(s, \sigma_L^\sharp, \tau') = p_\mathcal{A}(s, \sigma_L^\sharp, \tau) = p_\mathcal{A}(s, \sigma^\sharp, \tau)$. On the other hand, $u(p_{\mathcal{A}_L}(s, \sigma_L^\sharp, \tau_L^\sharp)) \leq u(p_{\mathcal{A}_L}(s, \sigma_L^\sharp, \tau'))$ by optimality of $\sigma_L^\sharp, \tau_L^\sharp$ on $\mathcal{A}_L$. These facts and (16) imply

$$u(p_\mathcal{A}(s, \sigma^\sharp, \tau^\sharp)) \leq u(p_\mathcal{A}(s, \sigma^\sharp, \tau)) \ . \tag{17}$$

Now let σ be any strategy of player 1 on $\mathcal{A}$. This strategy can be transformed to a strategy σ_{LR} on $\mathcal{A}_{LR}$ in the following way.

Let

$$\pi : S_L \cup S_R \to S$$

be the mapping from the states of $\mathcal{A}_{LR}$ to the states of $\mathcal{A}$ such that $\pi(s_L) = \pi(s_R) = s$ for all $s \in S$. This mapping can be extended to actions by putting $\pi((y', r, y'')) = (\pi(y'), r, \pi(y''))$ for any states $y', y'' \in S_L \cup S_R$ and next to finite and infinite histories, for a history $h = a_1 a_2 \ldots$ in $\mathcal{A}_{LR}$, $\pi(h) = \pi(a_1) \pi(a_2) \ldots$ is a history in $\mathcal{A}$. Now for any history $h = a_1 \ldots a_n$ in $\mathcal{A}_{LR}$ with the target $y := \text{target}(h)$ controlled by player 1 we define

$$\sigma_{LR}(h) = \begin{cases} \pi_L(\sigma(\pi(h))) & \text{if target}(h) \in S_L = U_L \cup \{x\}, \\ \pi_L(\sigma(\pi(h))) & \text{if target}(h) \in U_R, \end{cases} \tag{18}$$

where π_L and π_R were defined in (11). Thus, intuitively, when playing according to σ_{LR} player 1 takes the projection $\pi(h)$ of the history h onto $\mathcal{A}$, applies the strategy σ which gives him an action $(s, r, t) := \sigma(\pi(h))$ in $\mathcal{A}$. The target state of h, target(h), is either the state $s_L \in U_L \cup \{x\}$ or the state $s_R \in U_R$. In the first case player 1 executes the action $\pi_L((s, r, t)) = (s_L, r, t_L)$, in the second case he executes $\pi_R((s, r, t)) = (s_R, r, t_R)$.

Let $p_{\mathcal{A}_{LR}}(s_L, \sigma_{LR}, \cdot)$ be the play in $\mathcal{A}_{LR}$ starting at a left hand side state $s_L \in S_L$ and consistent with σ_{LR} (we left out here the strategy of player 2 since he has only one strategy on $\mathcal{A}_{LR}$ and therefore it is useless to specify it explicitly). From the construction of $\mathcal{A}_{LR}$ and definitions (18) and (15) of σ_{LR} and $\tau^\sharp$ it follows that for any strategy σ of player 1 on $\mathcal{A}$

$$p_\mathcal{A}(s, \sigma, \tau^\sharp) = \pi(p_{\mathcal{A}_{LR}}(s_L, \sigma_{LR}, \cdot)), \tag{19}$$

Since $\sigma_{LR}^\sharp$ is an optimal positional strategy for player 1 in $(\mathcal{A}_{LR}, u)$ we have

$$u(p_{\mathcal{A}_{LR}}(s_L, \sigma_{LR}, \cdot)) \leq u(p_{\mathcal{A}_{LR}}(s_L, \sigma_{LR}^\sharp, \cdot)). \tag{20}$$

The play $p_{\mathcal{A}_{LR}}(s_L, \sigma_{LR}^\sharp, \cdot)$ starts in the left hand side state s_L and is consistent with the strategy $\sigma_{LR}^\sharp$ that, according to (13), chooses at the pivot state x a left hand side action, therefore this play traverses uniquely the left hand side states S_L. We can define for player 1 a positional strategy $\pi \circ \sigma_{LR}^\sharp \circ \pi_L$ on $\mathcal{A}_L$ that corresponds to the left hand side part of the strategy $\sigma_{LR}^\sharp$: for $s \in S_1$, $\pi \circ \sigma_{LR}^\sharp \circ \pi_L(s) = \pi(\sigma_{LR}^\sharp(s_L))$. That we have defined in this way a valid strategy for player 1 on $\mathcal{A}_L$ is guaranteed by (13). Since player 2 is constrained in $\mathcal{A}_{LR}$ to play on S_L according to the strategy $\tau_L^\sharp$ we can see that applying the strategy $\sigma_{LR}^\sharp$ for a play starting at s_L in $\mathcal{A}_{LR}$ gives, modulo the renaming, the same result as applying the strategies $\pi \circ \sigma_{LR}^\sharp \circ \pi_L$ and $\tau_L^\sharp$ in $\mathcal{A}_L$ for a play staring at s, formally

$$\pi(p_{\mathcal{A}_{LR}}(s_L, \sigma_{LR}^\sharp, \cdot)) = p_{\mathcal{A}_L}(s, \pi \circ \sigma_{LR}^\sharp \circ \pi_L, \tau_L^\sharp). \tag{21}$$

Eq. (19) and (21) imply the equality of corresponding rewards sequences, i.e. also the equality of corresponding payoffs while by optimality $\sigma_L^\sharp$ and $\tau_L^\sharp$ in $\mathcal{A}_L$ we have $u(p_{\mathcal{A}_L}(s, \pi \circ \sigma_{LR}^\sharp \circ \pi_L, \tau_L^\sharp)) \leq u(p_{\mathcal{A}_L}(s, \sigma_L^\sharp, \tau_L^\sharp))$.

Putting together the last inequality and (16), (19), (20), (21) we deduce

$$u(p_\mathcal{A}(s, \sigma, \tau^\sharp)) \leq u(p_\mathcal{A}(s, \sigma^\sharp, \tau^\sharp)).$$

This and (17) imply the optimality of strategies $\sigma^\sharp$ and $\tau^\sharp$ in the game $(\mathcal{A}, u)$.

Our problem is that the optimal strategy of player 2 constructed above is not positional, however, the remedy is simple. Consider the game with the payoff mapping $-u$ and where the roles of players 1 and 2 are permuted, i.e. it is player 1 that pays to player 2 the amount $-u(p)$ after a play p. Thus player 2 wants to maximize the payment while player 2 tries to minimize it. In the new game choose as the pivot a state controlled by player 2 with at least two available actions, then the construction above repeated in this new setting will provide optimal strategies $\sigma^\ddagger$ and $\tau^\ddagger$ for players 1 and 2 in the new game with $\tau^\ddagger$ being positional. However optimal strategies in the new and the old games are the same thus $\tau^\ddagger$ is an optimal positional strategy for player 2 in $(\mathcal{A}, u)$. By exchangeability property for optimal strategies, $\sigma^\sharp$ and $\tau^\ddagger$ constitute a pair of optimal positional strategies in the game $(\mathcal{A}, u)$. □

4 Final Remarks

In Sect. 3 and in Theorem 3 instead of payoff mappings we could use, without any substantial modification, preference relations [12] over infinite reward sequences. Such a relation $\precsim$ is a binary complete transitive relation (where "complete" means that $a \precsim b$ or $b \precsim a$ for all a, b in the domain of $\precsim$). Obviously each payoff mapping u defines a preference relation $\precsim_u$, for infinite finitely generated sequences of rewards r and r', $r \precsim_u r'$ iff $u(r) \leq u(r')$. Although, at least in principle, preference relations can be represented by real valued payoffs, this representation is not always natural and therefore it may be advantageous to reformulate Sect. 3 and trade payoffs for preference relations.

4.1 Nash Equilibria

Suppose that we have a finite set $\{1, \ldots, N\}$ of players and the set of states is partitioned onto N disjoint sets $S = S_1 \cup \ldots \cup S_N$, S_i being the states controlled by player i. Again, if the current state is s then the player controlling s chooses and executes an action available at s. Now each player i has his own payoff mapping u_i that gives for each infinite finitely generated sequence of rewards r the payoff $u_i(r)$ of player i. A strategy profile $\sigma = (\sigma_1, \ldots, \sigma_N)$ is an N-tuple of strategies, where σ_i is a strategy of player i. Fixing a strategy profile σ and an initial state $s \in S$ we have exactly one play $p_\mathcal{A}(s, \sigma)$ starting at s and consistent with all strategies σ. For a strategy profile σ and a strategy σ'_i of player i by (σ^{-i}, σ'_i) we denote the strategy profile obtained from the profile σ by replacing σ_i by σ'_i.

A strategy profile σ is in *Nash equilibrium* if for each i, $1 \leq i \leq N$, and each strategy σ'_i of player i, $u_i(p_\mathcal{A}(s, (\sigma^{-i}, \sigma'_i))) \leq u_i(p_\mathcal{A}(s, \sigma))$. From the result of Sect. 3, using the trigger strategy described in [12], we can deduce that

Proposition 5. *Suppose that for all i, $1 \leq i \leq N$, for one-player games over finite arenas with payoffs u_i and $-u_i$ there exist optimal positional strategies. Then for each N-person game with payoff profile $(u_1, \ldots, u_N)$ over a finite arena there exists a Nash equilibrium profile σ where the strategy σ_i of each player i is a finite memory strategy.*

Acknowledgments. During the preparation of this paper we have profited from numerous conversations with Hugo Gimbert.

References

1. André Arnold and Damian Niwiński. *Rudiments of μ-calculus*, volume 146 of *Studies in Logic and the Foundations of Mathematics*. Elsevier, 2001.
2. Lester E. Dubins and Leaonard J. Savage. *Inequalities for Stochastic Processes (How to gamble if you must)*. Dover Publications, 2nd edition, 1976.
3. E.A. Emerson and C. Jutla. Tree automata, μ-calculus and determinacy. In *FOCS'91*, pages 368–377. IEEE Computer Society Press, 1991.
4. Jerzy Filar and Koos Vrieze. *Competitive Markov Decision Processes*. Springer, 1997.
5. Hugo Gimbert and Wiesław Zielonka. From Markov decision processes to perfect information stochastic games. in preparation.
6. Hugo Gimbert and Wiesław Zielonka. When can you play positionally? In *Mathematical Foundations of Computer Science 2004*, volume 3153 of *LNCS*, pages 686–697. Springer, 2004.
7. Hugo Gimbert and Wiesław Zielonka. Games where you can play optimally without any memory. In *CONCUR 2005*, LNCS. Springer, 2005. to appear.
8. Erich Grädel, Wolfgang Thomas, and Thomas Wilke, editors. *Automata, Logics, and Infinite Games*, volume 2500 of *LNCS*. Springer, 2002.
9. Ashok P. Maitra and William D. Sudderth. *Discrete Gambling and Stochastic Games*. Springer, 1996.
10. A.W. Mostowski. Games with forbidden positions. Technical Report 78, Uniwersytet Gdański, Instytut Matematyki, 1991.
11. Abraham Neyman and Sylvain Sorin, editors. *Stochastic Games and Applications*, volume 570 of *NATO Science Series C, Mathematical and Physical Sciences*. Kluwer Academic Publishers, 2004.
12. Martin J. Osborne and Ariel Rubinstein. *A Course in Game Theory*. The MIT Press, 2002.

The Complexity of Satisfiability Problems: Refining Schaefer's Theorem*

Eric Allender[1], Michael Bauland[2], Neil Immerman[3],
Henning Schnoor[2], and Heribert Vollmer[2]

[1] Department of Computer Science, Rutgers University, Piscataway, NJ 08855
allender@cs.rutgers.edu
[2] Theoretische Informatik, Universität Hannover, Appelstr. 4,
30167 Hannover, Germany
{bauland, schnoor, vollmer}@thi.uni-hannover.de
[3] Department of Computer and Information Science, University of Massachusetts,
Amherst, MA 01003
immerman@cs.umass.edu

Abstract. Schaefer proved in 1978 that the Boolean constraint satisfaction problem for a given constraint language is either in P or is NP-complete, and identified all tractable cases. Schaefer's dichotomy theorem actually shows that there are at most two constraint satisfaction problems, up to polynomial-time isomorphism (and these isomorphism types are distinct if and only if P $\neq$ NP). We show that if one considers AC^0 isomorphisms, then there are exactly six isomorphism types (assuming that the complexity classes NP, P, $\oplus$L, NL, and L are all distinct).

1 Introduction

In 1978, Schaefer classified the Boolean constraint satisfaction problem and showed that, depending on the allowed relations in a propositional formula, the problem is either in P or is NP-complete [Sch78]. This famous "dichotomy theorem" does not consider the fact that different problems in P have quite different complexity, and there is now a well-developed complexity theory to classify different problems in P. Furthermore, in Schaefer's original work (and in the many subsequent simplified presentations of his theorem [CKS01]) it is already apparent that certain classes of constraint satisfaction problems are either trivial (the 0-valid and 1-valid relations) or are solvable in NL (the bijunctive relations) or $\oplus$L (the affine relations), whereas for other problems (the Horn and anti-Horn relations) he provides only a reduction to problems that are complete for P. Is this a complete list of complexity classes that can arise in the study of constraint satisfaction problems? Given the amount of attention that the dichotomy theorem has received, it is surprising that no paper has addressed the question of how to refine Schaefer's classification beyond some steps in this direction in Schaefer's original paper (see [Sch78, Theorem 5.1]).

---

* Supported in part by DFG grant Vo 630/5-1.

Our own interest in this question grew out of the observation that there is at least one other fundamental complexity class that arises naturally in the study of Boolean constraint satisfaction problems that does *not* appear in the list (AC^0, NL, $\oplus L$, P) of feasible cases identified by Schaefer. This is the class SL (symmetric logspace) that has recently been shown by Reingold to coincide with deterministic logspace [Rei05]. (Theorem 5.1 of [Sch78] does already present examples of constraint satisfaction problems that are complete for SL.) Are there other classes that arise in this way? We give a negative answer to this question. If we examine constraint satisfaction problems using AC^0 reducibility $\leq_m^{AC^0}$, then we are able to show that the following list of complexity classes is exhaustive: Every constraint satisfaction problem not solvable in coNLOGTIME is isomorphic to the standard complete set for one of the classes NP, P, $\oplus L$, NL, or L under isomorphisms computable and invertible in AC^0.

Our proofs rely heavily on the connection between complexity of constraint languages and universal algebra (in particular, the theory of *polymorphisms* and *clones*) which has been very useful in analyzing complexity issues of constraints. An introduction to this connection can be found in [Pip97b], and we recall some of the necessary definitions in the next section. One of the contributions of this paper is to point out that, in order to obtain a complete classification of constraint satisfaction problems (up to AC^0 isomorphism) it is necessary to go beyond the partition of constraint satisfaction problems given by their polymorphisms, and examine the constraints themselves in more detail.

2 Preliminaries

An n-ary Boolean relation is a subset of $\{0,1\}^n$. For a set V of variables, a constraint application C is an application of an n-ary Boolean relation R to an n-tuple of variables $(x_1, \ldots, x_n)$ from V. An assignment $I: V \to \{0,1\}$ *satisfies* the constraint application $R(x_1, \ldots, x_n)$ iff $(I(x_1), \ldots, I(x_n)) \in R$. In this paper we use the standard correspondence between Boolean relations and propositional formulas: A formula $\varphi(x_1, \ldots, x_n)$ defines the relation $R_\varphi = \{(\alpha_1, \ldots, \alpha_n) \mid \varphi(\alpha_1, \ldots, \alpha_n) = 1\}$. The meaning should always be clear from the context.

A *constraint language* is a finite set of Boolean relations. The Boolean *Constraint Satisfaction Problem* over a constraint language Γ (CSP(Γ)) is the question if a given set φ of Boolean constraint applications using relations from Γ is simultaneously satisfiable, i.e. if there exists an assignment $I: V \to \{0,1\}$, such that I satisfies every $C \in \varphi$. It is easy to see that the Boolean CSP over some language Γ is the same as satisfiability of conjunctive Γ-formulas. A well-known restriction of the general satisfiability problem is 3SAT, which can be seen as the CSP problem over the language $\Gamma_{3SAT} = \{(x_1 \vee x_2 \vee x_3), (\overline{x_1} \vee x_2 \vee x_3), (\overline{x_1} \vee \overline{x_2} \vee x_3), (\overline{x_1} \vee \overline{x_2} \vee \overline{x_3})\}$.

There is a very useful connection between the complexity of the CSP problem and universal algebra, which requires a few definitions:

A class of Boolean functions is called *closed* or a *clone*, if it is closed under superposition. (As explained in the surveys [BCRV03, BCRV04] being closed under superposition is essentially the same thing as containing all projections (in particular, the identity) and being closed under arbitrary composition.) Since the intersection of clones is again a clone, we can define, for a set B of Boolean functions, $\langle B \rangle$ as the smallest clone containing B.

It is clear that $\langle B \rangle$ is the set of Boolean functions that can be calculated by Boolean circuits using only gates for functions from B [BCRV03, Pip97a].

It is easy to see that the set of clones forms a lattice. For the Boolean case, Emil Post identified all clones and their inclusion structure (Figure 1). A description of the clones and a list of bases for each one can be found in Table 1. The clones are interesting for the study of the complexity of CSPs, because the complexity of CSP(Γ) depends on the closure properties of the relations in Γ, which we will define next.

Definition 2.1. *A k-ary relation R is closed under an n-ary Boolean function f, or f is a polymorphism of R, if for all $x_1, \ldots, x_n \in R$ with $x_i = (x_i[1], x_i[2], \ldots, x_i[k])$, we have*

$$(f(x_1[1], \ldots, x_n[1]), f(x_1[2], \ldots, x_n[2]), \ldots, f(x_1[k], \ldots, x_n[k])) \in R.$$

We denote the set of all polymorphisms of R by $\mathrm{Pol}(R)$, *and for a set Γ of Boolean relations we define* $\mathrm{Pol}(\Gamma) = \{f \mid f \in \mathrm{Pol}(R) \text{ for every } R \in \Gamma\}$. *For a set B of Boolean functions,* $\mathrm{Inv}(B) = \{R \mid B \subseteq \mathrm{Pol}(R)\}$ *is the set of* invariants *of B.*

It is easy to see that every set of the form $\mathrm{Pol}(\Gamma)$ is a clone. The operators Pol and Inv form a "Galois connection" between the lattice of clones and certain sets of Boolean relations, which is very useful for complexity analysis of the CSP problem. The concept of relations closed under certain Boolean functions is interesting, because many properties of Boolean relations can be equivalently formulated using this terminology. For example, a set of relations can be expressed by Horn-formulas if and only if every relation in the set is closed under the binary AND function. Horn is one of the properties that ensures the corresponding satisfiability problem to be tractable. More generally, tractability of formulas over a given set of relations only depends on the set of its polymorphisms. A proof of the following theorem can be found in e.g. [JCG97] and [Dal00]:

Theorem 2.2. *If* $\mathrm{Pol}(\Gamma_2) \subseteq \mathrm{Pol}(\Gamma_1)$, *then every $R \in \Gamma_1$ can be expressed by a formula*

$$R(x_1, \ldots, x_n) \iff \exists y_1, \ldots, y_m R_1(z_{1,1}, \ldots, z_{1,n_1}) \wedge \cdots \wedge R_k(z_{k,1}, \ldots, z_{k,n_k})$$

$$\wedge (x_{i_1} = x_{i_2}) \wedge (x_{i_3} = x_{i_4}) \wedge \cdots \wedge (x_{i_{r-1}} = x_{i_r})$$

for some $R_i \in \Gamma_2$ (where $z_{i,j} \in \{x_1, \ldots, x_n, y_1, \ldots, y_m\}$).

Therefore:

Theorem 2.3. *Let Γ_1 and Γ_2 be sets of Boolean relations such that Γ_1 is finite and* $\mathrm{Pol}(\Gamma_2) \subseteq \mathrm{Pol}(\Gamma_1)$. *Then* $\mathrm{CSP}(\Gamma_1) \leq_m^P \mathrm{CSP}(\Gamma_2)$.

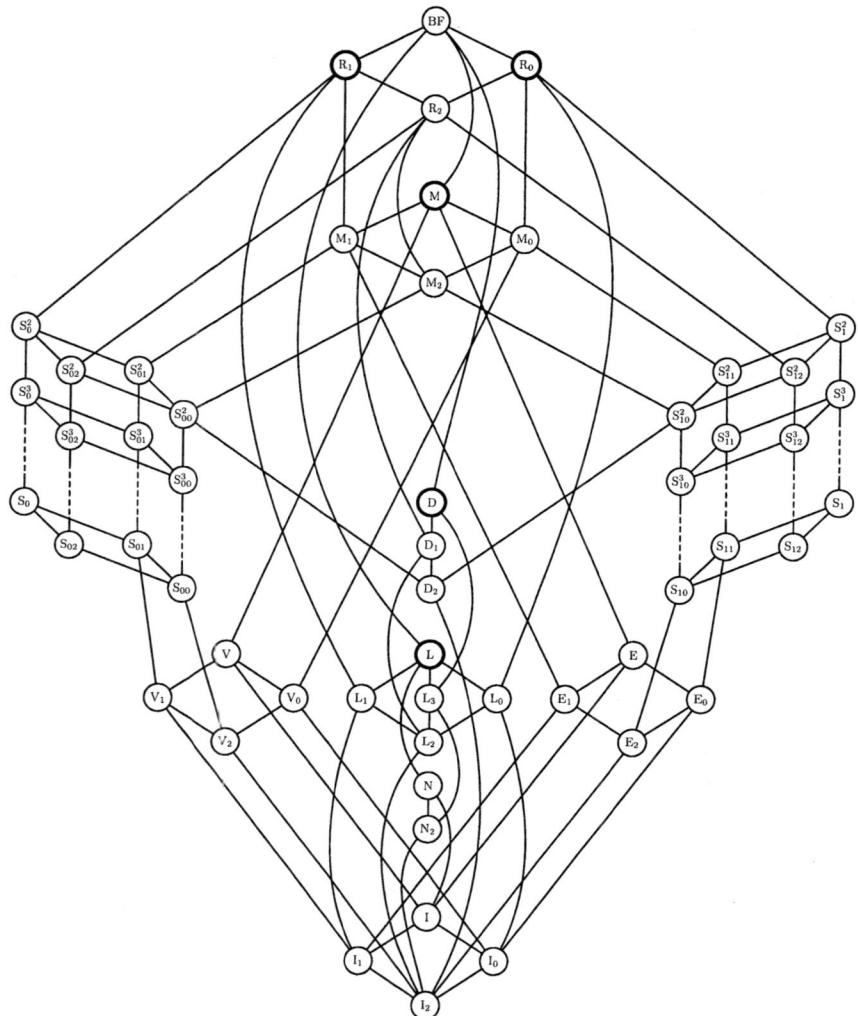

Fig. 1. Graph of all closed classes of Boolean functions

Trivially, the binary equality predicate = is closed under every Boolean function. Thus, = is contained in every set $\text{Inv}(B)$ for a clone B (these sets often are called *co-clones*). On the other hand, every relation is closed under the projection function, $\phi_i^n(x_1, \ldots, x_n) = x_i$. It is clear that when a set of relations is "big", the set of its polymorphisms is "small". So the most general case is a constraint language Γ such that $\text{Pol}(\Gamma)$ only contains the projections, and these cases of the CSP are NP-complete. An example for this is the language $\Gamma_{3\text{SAT}}$ from above: It can be shown that $\text{Pol}(\Gamma_{3\text{SAT}})$ only contains the projections, and therefore 3SAT is NP-complete.

Table 1. List of all closed classes of Boolean functions, and their bases (for definitions of these properties, see e.g. [BCRV03])

Name	Definition	Base
BF	All Boolean functions	$\{\vee, \wedge, \neg\}$
R_0	$\{f \in BF \mid f \text{ is 0-reproducing }\}$	$\{\wedge, \oplus\}$
R_1	$\{f \in BF \mid f \text{ is 1-reproducing }\}$	$\{\vee, \leftrightarrow\}$
R_2	$R_1 \cap R_0$	$\{\vee, x \wedge (y \leftrightarrow z)\}$
M	$\{f \in BF \mid f \text{ is monotonic }\}$	$\{\vee, \wedge, 0, 1\}$
M_1	$M \cap R_1$	$\{\vee, \wedge, 1\}$
M_0	$M \cap R_0$	$\{\vee, \wedge, 0\}$
M_2	$M \cap R_2$	$\{\vee, \wedge\}$
S_0^n	$\{f \in BF \mid f \text{ is 0-separating of degree } n\}$	$\{\rightarrow, \text{dual}(h_n)\}$
S_0	$\{f \in BF \mid f \text{ is 0-separating }\}$	$\{\rightarrow\}$
S_1^n	$\{f \in BF \mid f \text{ is 1-separating of degree } n\}$	$\{x \wedge \overline{y}, h_n\}$
S_1	$\{f \in BF \mid f \text{ is 1-separating }\}$	$\{x \wedge \overline{y}\}$
S_{02}^n	$S_0^n \cap R_2$	$\{x \vee (y \wedge \overline{z}), \text{dual}(h_n)\}$
S_{02}	$S_0 \cap R_2$	$\{x \vee (y \wedge \overline{z})\}$
S_{01}^n	$S_0^n \cap M$	$\{\text{dual}(h_n), 1\}$
S_{01}	$S_0 \cap M$	$\{x \vee (y \wedge z), 1\}$
S_{00}^n	$S_0^n \cap R_2 \cap M$	$\{x \vee (y \wedge z), \text{dual}(h_n)\}$
S_{00}	$S_0 \cap R_2 \cap M$	$\{x \vee (y \wedge z)\}$
S_{12}^n	$S_1^n \cap R_2$	$\{x \wedge (y \vee \overline{z}), h_n\}$
S_{12}	$S_1 \cap R_2$	$\{x \wedge (y \vee \overline{z})\}$
S_{11}^n	$S_1^n \cap M$	$\{h_n, 0\}$
S_{11}	$S_1 \cap M$	$\{x \wedge (y \vee z), 0\}$
S_{10}^n	$S_1^n \cap R_2 \cap M$	$\{x \wedge (y \vee z), h_n\}$
S_{10}	$S_1 \cap R_2 \cap M$	$\{x \wedge (y \vee z)\}$
D	$\{f \mid f \text{ is self-dual}\}$	$\{x\overline{y} \vee x\overline{z} \vee (\overline{y} \wedge \overline{z})\}$
D_1	$D \cap R_2$	$\{xy \vee x\overline{z} \vee y\overline{z}\}$
D_2	$D \cap M$	$\{xy \vee yz \vee xz\}$
L	$\{f \mid f \text{ is linear}\}$	$\{\oplus, 1\}$
L_0	$L \cap R_0$	$\{\oplus\}$
L_1	$L \cap R_1$	$\{\leftrightarrow\}$
L_2	$L \cap R$	$\{x \oplus y \oplus z\}$
L_3	$L \cap D$	$\{x \oplus y \oplus z \oplus 1\}$
V	$\{f \mid f \text{ is constant or a } n - \text{ary OR function}\}$	
V_0	$[\{\vee\}] \cup [\{0\}]$	$\{\vee, 0\}$
V_1	$[\{\vee\}] \cup [\{1\}]$	$\{\vee, 1\}$
V_2	$[\{\vee\}]$	$\{\vee\}$
E	$\{f \mid f \text{ is constant or a } n - \text{ary AND function}\}$	
E_0	$[\{\wedge\}] \cup [\{0\}]$	$\{\wedge, 0\}$
E_1	$[\{\wedge\}] \cup [\{1\}]$	$\{\wedge, 1\}$
E_2	$[\{\wedge\}]$	$\{\wedge\}$
N	$[\{\neg\}] \cup [\{0\}] \cup [\{1\}]$	$\{\neg, 1\}$
N_2	$[\{\neg\}]$	$\{\neg\}$
I	$[\{\text{id}\}] \cup [\{0\}] \cup [\{1\}]$	$\{\text{id}, 0, 1\}$
I_0	$[\{\text{id}\}] \cup [\{0\}]$	$\{\text{id}, 0\}$
I_1	$[\{\text{id}\}] \cup [\{1\}]$	$\{\text{id}, 1\}$
I_2	$[\{\text{id}\}]$	$\{\text{id}\}$

The function h_n is defined as:

$$h_n(x_1, \ldots, x_{n+1}) = \bigvee_{i=1}^{n+1} x_1 \wedge x_2 \wedge \cdots \wedge x_{i-1} \wedge x_{i+1} \wedge \cdots \wedge x_{n+1}$$

As we have seen in the above theorem, the complexity of the CSP problem for a given constraint language is determined by the set of its polymorphisms. At least this is the case when considering gross classifications of complexity (such as whether a problem is in P or is NP-complete). However, when we examine finer complexity classifications, such as determining the circuit complexity of a constraint satisfaction problem, then the set of polymorphisms of a constraint language Γ does *not* completely determine the complexity of $\mathrm{CSP}(\Gamma)$, as can easily be seen in the following important example:

Example 2.4. Let $\Gamma_1 = \{\overline{x}, x\}$, $\Gamma_2 = \Gamma_1 \cup \{=\}$. It is obvious that $\mathrm{Pol}(\Gamma_1) = \mathrm{Pol}(\Gamma_2)$; the set of polymorphisms is the clone R_2. Formulas over Γ_1 only contain clauses of the form x or $\overline{x}$ for some variable x, whereas in Γ_2, we additionally have the binary equality predicate. We will now see that $\mathrm{CSP}(\Gamma_1)$ has very different complexity than $\mathrm{CSP}(\Gamma_2)$.

Satisfiability of a Γ_1-formula φ can be decided in coNLOGTIME. (Such a formula is unsatisfiable if and only if for some variable x, both x and $\overline{x}$ are clauses.)

In contrast, $\mathrm{CSP}(\Gamma_2)$ is complete for L under $\leq_m^{\mathrm{AC}^0}$ reductions: The complement of the graph accessibility problem (GAP) for undirected graphs, which is known to be complete for L [Rei05], can be reduced to $\mathrm{CSP}(\Gamma_2)$. Let $G = (V, E)$ be a finite, undirected graph, and s, t vertices in V. For every edge $(v_1, v_2) \in E$, add a constraint $v_1 = v_2$. Also add $\overline{s}$ and t. It is obvious that there exists a path in G from s to t if and only if the resulting formula is not satisfiable. In fact, it is easy to see that $\mathrm{CSP}(\Gamma_2)$ is not only hard for L, but it also lies within L so it is complete for L under $\leq_m^{\mathrm{AC}^0}$ reductions.

The lesson to learn from this example is that the usual reduction among constraint satisfaction problems arising from the same co-clone is not an $\leq_m^{\mathrm{AC}^0}$ reduction. The following lemma summarizes the main relationships.

Lemma 2.5. *Let Γ_1 and Γ_2 be sets of relations over a finite set, where Γ_1 is finite and $\mathrm{Pol}(\Gamma_2) \subseteq \mathrm{Pol}(\Gamma_1)$. Then $\mathrm{CSP}(\Gamma_1) \leq_m^{\mathrm{AC}^0} \mathrm{CSP}(\Gamma_2 \cup \{=\}) \leq_m^{\log} \mathrm{CSP}(\Gamma_2)$.*

Proof. Since the local replacement from Theorem 2.2 can be computed in AC^0, this establishes the first reducibility relation (note that variables are implicitly existentially quantified and therefore the quantifiers do not need to be written).

For the second reduction, we need to eliminate all of the =-constraints. We do this by identifying variables x_{i_1} and x_{i_2} if there is an =-path from x_{i_1} to x_{i_2} in the formula. By [Rei05], this can be computed in logspace. □

3 Classification

Theorem 3.1. *Let Γ be a finite set of Boolean relations.*

- *If $\mathrm{I}_0 \subseteq \mathrm{Pol}(\Gamma)$ or $\mathrm{I}_1 \subseteq \mathrm{Pol}(\Gamma)$, then every constraint formula over Γ is satisfiable, and therefore $\mathrm{CSP}(\Gamma)$ is trivial.*
- *If $\mathrm{Pol}(\Gamma) \in \{\mathrm{I}_2, \mathrm{N}_2\}$, then $\mathrm{CSP}(\Gamma)$ is $\leq_m^{\mathrm{AC}^0}$-complete for NP.*
- *If $\mathrm{Pol}(\Gamma) \in \{\mathrm{V}_2, \mathrm{E}_2\}$, then $\mathrm{CSP}(\Gamma)$ is $\leq_m^{\mathrm{AC}^0}$-complete for P.*
- *If $\mathrm{Pol}(\Gamma) \in \{\mathrm{L}_2, \mathrm{L}_3\}$, then $\mathrm{CSP}(\Gamma)$ is $\leq_m^{\mathrm{AC}^0}$-complete for $\oplus\mathrm{L}$.*
- *If $\mathrm{S}_{00} \subseteq \mathrm{Pol}(\Gamma) \subseteq \mathrm{S}_{00}^2$ or $\mathrm{S}_{10} \subseteq \mathrm{Pol}(\Gamma) \subseteq \mathrm{S}_{10}^2$ or $\mathrm{Pol}(\Gamma) \in \{\mathrm{D}_2, \mathrm{M}_2\}$, then $\mathrm{CSP}(\Gamma)$ is $\leq_m^{\mathrm{AC}^0}$-complete for NL.*

- If $\text{Pol}(\Gamma) \in \{D_1, D\}$, then $\text{CSP}(\Gamma)$ is $\leq_m^{\text{AC}^0}$-complete for L.
- If $S_{02} \subseteq \text{Pol}(\Gamma) \subseteq R_2$ or $S_{12} \subseteq \text{Pol}(\Gamma) \subseteq R_2$, then either $\text{CSP}(\Gamma)$ is in coNLOGTIME, or $\text{CSP}(\Gamma)$ is complete for L under $\leq_m^{\text{AC}^0}$. There is an algorithm deciding which case occurs.

Theorem 3.1 is a refinement of Theorem 5.1 from [Sch78] and Theorem 6.5 from [CKS01]. It is immediate from a look at Figure 1 that this covers all cases. The proof follows from the lemmas in the following subsections. First, we mention a corollary:

Corollary 3.2. *For any set of relations Γ, $\text{CSP}(\Gamma)$ is AC^0-isomorphic either to $0\Sigma^*$ or to the standard complete set for one of the following complexity classes: $\text{NP}, \text{P}, \oplus\text{L}, \text{NL}, \text{L}$.*

Proof. It is immediate from Theorem 3.1 that if $\text{CSP}(\Gamma)$ is not in AC^0, then it is complete for one of $\text{NP}, \text{P}, \text{NL}, \text{L}$, or $\oplus\text{L}$ under $\leq_m^{\text{AC}^0}$ reductions. By [Agr01] each of these problems is AC^0-isomorphic to the standard complete set for its class. On the other hand, if $\text{CSP}(\Gamma)$ is solvable in AC^0 then it is an easy matter to reduce any problem $A \in \text{AC}^0$ to $\text{CSP}(\Gamma)$ via a length-squaring, invertible AC^0 reduction (by first checking if $x \in A$, and then using standard padding techniques to map x to a long satisfiable instance if $x \in A$, and mapping x to a long syntactically incorrect input if $x \notin A$). AC^0 isomorphism to the standard complete set now follows by [ABI97] (since the standard complete set is complete under invertible, length-squaring reductions). □

3.1 Upper Bounds: Algorithms

First, we state results that are well-known; see e.g. [Sch78, BCRV04]:

Proposition 3.3. *Let Γ be a Boolean constraint language.*

1. *If $\text{Pol}(\Gamma) \in \{I_2, N_2\}$, then $\text{CSP}(\Gamma)$ is NP-complete. Otherwise, $\text{CSP}(\Gamma) \in \text{P}$.*
2. *$L_2 \subseteq \text{Pol}(\Gamma)$ implies $\text{CSP}(\Gamma) \in \oplus\text{L}$.*
3. *$D_2 \subseteq \text{Pol}(\Gamma)$ implies $\text{CSP}(\Gamma) \in \text{NL}$.*
4. *$I_0 \subseteq \text{Pol}(\Gamma)$ or $I_1 \subseteq \text{Pol}(\Gamma)$ implies every instance of $\text{CSP}(\Gamma)$ is satisfiable by the all-0 or the all-1 tuple, and therefore $\text{CSP}(\Gamma)$ is trivial.*

Lemma 3.4. *Let Γ be a constraint language.*

1. *If $S_{02} \subseteq \text{Pol}(\Gamma)$ or $S_{12} \subseteq \text{Pol}(\Gamma)$, then $\text{CSP}(\Gamma) \in \text{L}$.*
2. *If $S_{00} \subseteq \text{Pol}(\Gamma)$ or $S_{10} \subseteq \text{Pol}(\Gamma)$, then $\text{CSP}(\Gamma) \in \text{NL}$.*

Proof. First we consider the cases S_{00} and S_{02}. The following algorithm is based on the proof for Theorem 6.5 in [CKS01]. Observe that there is no finite set Γ such that $\text{Pol}(\Gamma) = S_{00}$ ($\text{Pol}(\Gamma) = S_{02}$, resp.). Therefore, $\text{Pol}(\Gamma) \supseteq S_{00}^k$ ($\text{Pol}(\Gamma) \supseteq S_{02}^k$, resp.) for some $k \geq 2$. Note that $\text{Pol}(\{\text{OR}^k, x, \overline{x}, \rightarrow, =\}) = S_{00}^k$ (OR^k refers to the k-ary OR relation) and $\text{Pol}(\{\text{OR}^k, x, \overline{x}, =\}) = S_{02}^k$ ([BRSV05]), and therefore by Lemma 2.5 we can assume w.l.o.g. $\Gamma = \{\text{OR}^k, x, \overline{x}, \to, =\}$ ($\Gamma = \{\text{OR}^k, x, \overline{x}, =\}$, resp.).

Now the algorithm works as follows: For a given formula φ over the relations mentioned above, consider every positive clause $x_{i_1} \lor \cdots \lor x_{i_k}$. The clause is satisfiable if and only if there is one variable in $\{x_{i_1}, \ldots, x_{i_k}\}$ which can be set to 1 without violating any of the $\overline{x}$ and $x \to y$ clauses (without violating any of the $\overline{x}$, resp.). For a variable $y \in \{x_{i_1}, \ldots, x_{i_k}\}$, this can be checked as follows:

For each clause $\overline{x}$, check if there is an $\to$-=-path (=-path, resp.) from y to x, by which we mean a sequence $y R_1 z_1, z_1 R_2 z_2, \ldots, z_{m-1} R_m x$ for $R_i \in \{\to, =\}$ ($R_i \in \{=\}$, resp.). (This is just an instance of the GAP problem on *directed* graphs (*undirected* graphs, resp.), which is the standard complete problem for NL (L, resp.).) If one of these is the case, then y cannot be set to 1. Otherwise, we can set y to 1, and the clause is satisfiable. If a clause is shown to be unsatisfiable, reject. If no clause is shown to be unsatisfiable in this way, accept.

The S_{10}- and S_{12}-case are analogous; in these cases we have NAND instead of OR. □

Our final upper bound in this section is combined with a hardness result, and thus serves as a bridge to the next two sections.

Lemma 3.5. *Let Γ be a constraint language. If $\text{Pol}(\Gamma) \in \{D_1, D\}$, then $\text{CSP}(\Gamma)$ is $\leq_m^{\text{AC}^0}$-complete for L.*

Proof. Note that $\text{Pol}(\{\oplus\}) = D$ and $\text{Pol}(\{R\}) = D_1$, where $R = x_1 \land (x_2 \oplus x_3)$. Thus by Lemmas 2.5 and 3.7, and Proposition 3.6, we can restrict ourselves to the cases where Γ consists of these relations only. The satisfiability problem for formulas that are conjunctions of clauses of the form x or $y \oplus z$ is complete for L by Problem 4.1 in Section 7 of [AG00], which proves completeness for the case $\text{Pol}(\Gamma) = D_1$ and thus proves membership in L for the case $\text{Pol}(\Gamma) = D$. It suffices to prove hardness in the case $\text{Pol}(\Gamma) = D$.

This can easily be shown by introducing a new variable f and replacing clauses x with $x \oplus f$ (observe that $\oplus$ is closed under negation). □

3.2 Removing the Equality Relation

Lemma 2.5 reveals that polymorphisms completely determine the complexity of a given constraint satisfaction problem only if the equality relation is contained in the corresponding constraint language. In Example 2.4 we saw that this question does lead to different complexity results. We now show that for most constraint languages, we can get equality "for free" and therefore the question of whether we have equality directly or not does not make a difference.

We say a constraint language Γ *can express* the relation $R(x_1, ..., x_n)$ if there is a formula $R_1(z_1^1, \ldots, z_{n_1}^1) \land \cdots \land R_l(z_1^l, \ldots, z_{n_l}^l)$, where $R_i \in \Gamma$ and $z_j^i \in \{y_1, \ldots, y_n, w_1, \ldots, w_r\}$ (the z_j^i's need not be distinct) such that for each assignment of values $(c_1, \ldots, c_n)$ to the variables $y_1, \ldots, y_n$, $R(c_1, ..., c_n)$ evaluates to TRUE if and only if there is an assignment of values to the variables $w_1, \ldots, w_r$ such that all R_i-clauses, with y_i replaced by c_i, evaluate to TRUE.

The following proposition is immediate.

Proposition 3.6. *Let Γ be a constraint language. If Γ can express the equality relation, then $\mathrm{CSP}(\Gamma \cup \{=\}) \leq_m^{\mathrm{AC}^0} \mathrm{CSP}(\Gamma)$.*

Lemma 3.7. *Let Γ be a finite set of Boolean relations where $\mathrm{Pol}(\Gamma) \subseteq \mathrm{M}_2$, $\mathrm{Pol}(\Gamma) \subseteq \mathrm{L}$, or $\mathrm{Pol}(\Gamma) \subseteq \mathrm{D}$. Then Γ can express the equality relation.*

Proof. The relation "$x \to y$" is invariant under M_2. Thus given any such Γ, by Theorem 2.2 we can construct "$x \to y$" with help of new existentially quantified variables that do not appear anywhere else in the formula. Equality clauses between the variables x and y do not appear, since $x = y$ does not hold for every element of the relation (equality involving existentially quantified variables does not appear in the construction given in Theorem 2.2). Hence Γ can express $x = y$ with $x \to y \wedge y \to x$.

For the L-case, apply an analogous argument for the relation R^4_{even}, which consists of all 4-tuples with an even number of 1's. Note that $x = y$ is expressed by $R^4_{\mathrm{even}}(z, z, x, y)$. If $\mathrm{Pol}(\Gamma) \subseteq \mathrm{D}$, then we can express $x \oplus y$, and thus we express equality by $x = y \iff (x \oplus z) \wedge (z \oplus y)$. □

As noted in Example 2.4, for some classes, the question whether equality is contained in the constraint language or not does lead to different complexities, namely complete for L or contained in coNLOGTIME. We now show that there are no intermediate complexity classes arising in these cases. As we saw in the lemmas above, this only concerns constraint languages Γ such that $\mathrm{Pol}(\Gamma) \supseteq \mathrm{S}_{02}^m$ or $\mathrm{Pol}(\Gamma) \supseteq \mathrm{S}_{12}^m$ holds for some $m \geq 2$.

Lemma 3.8. *Let R be a relation such that $\mathrm{Pol}(R) \supseteq \mathrm{S}_{02}$ ($\mathrm{Pol}(R) \supseteq \mathrm{S}_{12}$, resp.). Let $S = \mathrm{OR}^m$ ($S = \mathrm{NAND}^m$, resp.). Then either $\mathrm{CSP}(\{x, \overline{x}, S, R\}) \in \mathrm{coNLOGTIME}$ or R can express equality (in which case $\mathrm{CSP}(\{x, \overline{x}, S, R\})$ is complete for L under AC^0 reductions). There is an algorithm deciding which of the cases occurs.*

Proof. If $\mathrm{Pol}(R) \supseteq \mathrm{S}_{02}$, then as in the proof of Lemma 3.4 we know that $\mathrm{Pol}(R) \supseteq \mathrm{S}_{02}^m$ for some $m \geq 2$. Thus we know from Theorem 2.2 that R can be expressed using equality, literals, and the m-ary OR predicate, since $\mathrm{Pol}(\{x, \overline{x}, \mathrm{OR}^m\}) = \mathrm{S}_{02}^m$ ([BRSV05]). Let φ be a representation of R in this form. We simplify φ as follows (without loss of generality, assume that R is not the empty relation):

1. For any clause $x_1 = x_2$ where x_1 or x_2 appears as a literal, remove this clause and insert the corresponding literals for x_1 and x_2. Repeat until no such clause remains.
2. Remove variables from OR-clauses which appear as negative literals.
3. For an OR-clause containing variables connected with $=$, remove all of them except one.

Note that this does not change the relation represented by the formula. If no $=$-clause remains, then R can be expressed using only OR and literals and therefore leads to a CSP solvable in coNLOGTIME (a CSP-formula using only these relations is unsatisfiable iff there appear two contradictory variables or an OR-clause containing only variables which also appear as a negative literal).

Otherwise, let $x_1 = x_2$ be a remaining clause. We existentially quantify all variables in R except x_1 and x_2, and call the resulting relation R'. We that claim R' is the equality relation. Let $(x_1, x_2) \in R'$. Since $x_1 = x_2$ appears in the defining formula, $x_1 = x_2$ holds. For the other direction, let $x_1 = x_2$. We assign the value 0 to every existentially quantified variable that appears as a negative literal, the same value as x_1 to every variable connected to x_1 via an =-path, and the value 1 to all others. Obviously, all literals are satisfied this way: Remember x_1 and x_2 do not appear as literals due to step 1, and there are no contradictory literals since R is nonempty. All equality clauses are satisfied because none of the variables appearing here also appear as literals. Let $(x_1 \vee \cdots \vee x_j)$ be a clause. None of these variables appear as negative literals due to step 2, and at most one of them can be =-connected to x_1 and x_2 due to step 3. Therefore, the assignment constructed above assigns 1 to at least one of the occurring variables, thus satisfying the formula. Hardness for L now follows with the same construction as in Example 2.4.

It is decidable which of these cases occurs: Since the only way to obtain equality is by existentially quantifying all variables except two, this is a finite number of combinations which can be easily verified by an algorithm. An analogous argument can be applied to the dual case $\text{Pol}(R) \supseteq S_{12}^m$. □

3.3 Lower Bounds: Hardness Results

One technique of proving hardness for constraint satisfaction problems is to reduce certain problems related to Boolean circuits to CSPs. In [Rei01], many decision problems regarding circuits were discussed. In particular, the "Satisfiability Problem for B Circuits" ($\text{SAT}^C(B)$) is very useful for our purposes here. $\text{SAT}^C(B)$ is the problem of determining if a given Boolean circuit with gates from B has an input vector on which it computes output "1".

Lemma 3.9. *Let Γ be a constraint language such that $\text{Pol}(\Gamma) \in \{\text{E}_2, \text{V}_2\}$. Then $\text{CSP}(\Gamma)$ is $\leq_m^{\text{AC}^0}$-hard for P.*

Proof. It is well-known that the satisfiability problem for Horn and anti-Horn formulas is complete for P. To show that this hardness results holds under AC^0 reductions, it is easy to construct a reduction from $\text{SAT}^C(\text{S}_{11})$, by just simulating every gate in the circuit as an anti-Horn clause. The result then follows from [Rei01]. The dual case for Horn-formulas is analogous. □

Lemma 3.10. *Let Γ be a constraint language such that $\text{Pol}(\Gamma) \in \{\text{L}_2, \text{L}_3\}$. Then $\text{CSP}(\Gamma)$ is $\leq_m^{\text{AC}^0}$-hard for $\oplus\text{L}$.*

Proof. Assume that Γ contains =. The proof of the general case then follows from Lemmas 2.5 and 3.7, and Proposition 3.6. For the L_2-case, we show $\text{SAT}^C(\text{L}_0)$ $\leq_m^{\text{AC}^0} \text{CSP}(\Gamma)$ for a constraint language Γ with $\text{Pol}(\Gamma) = \text{L}_2$. The result then follows with [Rei01]. Since we can express x_{out} and $x_1 = x_2 \oplus x_3$ as L_2-invariant relations, we can directly reproduce the given L_0-circuit.

This does not work for L_3, since we cannot express x or $\overline{x}$ in L_3. However, since L_3 is basically L_2 plus negation, we can "extend" a given relation from $\text{Inv}(L_2)$ so that it is invariant under negation, by simply doubling the truth-table. More precisely, given a constraint language Γ such that $\text{Pol}(\Gamma) = L_2$, we show that there is a constraint language Γ' such that $\text{Pol}(\Gamma') = L_3$ and $\text{CSP}(\Gamma) \leq_m^{\text{AC}^0} \text{CSP}(\Gamma')$. For an n-ary relation $R \in \Gamma$, let $\overline{R} = \{(\overline{x_1}, \ldots, \overline{x_n}) \mid (x_1, \ldots, x_n) \in R\}$, and let R' be the $(n+1)$-ary relation $R' = (\{0\} \times R) \cup (\{1\} \times \overline{R})$. It is obvious that R' is closed under N_2 and under L_2, and hence under L_3. Let φ be an instance of $\text{CSP}(\Gamma)$. Let $\Gamma' = \{R' \mid R \in \Gamma\}$. Let $\varphi = \bigwedge_{i=1}^{n} R_n(x_{i_1}, \ldots, x_{i_{n_i}})$. We set $\varphi' = \bigwedge_{i=1}^{n} R'_n(t, x_{i_1}, \ldots, x_{i_{n_i}})$ for a new variable t.

Let $\varphi \in \text{CSP}(\Gamma)$, $I \models \varphi$. Then $I \cup \{t = 0\} \models \varphi'$.

Let $\varphi' \in \text{CSP}(\Gamma)$, $I' \models \varphi'$. Without loss of generality, let $I'(t) = 0$ (otherwise, observe $\overline{I'} \models \varphi'$ holds as well), therefore $I'\{t = 0\} \models \varphi$, and thus $\text{CSP}(\Gamma) \leq_m^{\text{AC}^0} \text{CSP}(\Gamma')$ holds. □

With the same technique as in Example 2.4, we can examine the complexity of CSPs invariant under M_2. The relation $x \to y$ is invariant under M_2, and thus we can model search in directed graphs here.

Lemma 3.11. *Let Γ be a constraint language such that $\text{Pol}(\Gamma) \subseteq M_2$. Then $\text{CSP}(\Gamma)$ is $\leq_m^{\text{AC}^0}$-hard for NL.*

4 Conclusion and Further Research

We have obtained a complete classification for constraint satisfaction problems under AC^0 isomorphisms, and identified six isomorphism types corresponding to the complexity classes $\text{NP}, \text{P}, \text{NL}, \oplus \text{L}, \text{L}$, and AC^0. One can also show that all constraint satisfaction problems in AC^0 are either trivial or are complete for coNLOGTIME (under logtime-uniform projections).

One natural question for further research concerns constraint satisfaction problems over larger domains. In particular, it would be interesting to see if the dichotomy theorem of Bulatov [Bul02] over three-element domains can be refined to obtain a complete classification up to AC^0-isomorphism.

Acknowledgments

The first and third authors thank Denis Thérien for organizing a workshop at Bellairs research institute where Phokion Kolaitis lectured at length about constraint satisfiability problems. We thank Phokion Kolaitis for his lectures and for stimulating discussions. We also thank Nadia Creignou for helpful hints.

References

[ABI97] E. Allender, J. Balcazar, and N. Immerman. A first-order isomorphism theorem. *SIAM Journal on Computing*, 26:557–567, 1997.

[AG00] C. Alvarez and R. Greenlaw. A compendium of problems complete for symmetric logarithmic space. *Computational Complexity*, 9(2):123–145, 2000.

[Agr01] M. Agrawal. The first-order isomorphism theorem. In *Foundations of Software Technology and Theoretical Computer Science: 21st Conference, Bangalore, India, December 13-15, 2001. Proceedings*, Lecture Notes in Computer Science, pages 58–69, Berlin Heidelberg, 2001. Springer Verlag.

[BCRV03] E. Böhler, N. Creignou, S. Reith, and H. Vollmer. Playing with Boolean blocks, part I: Post's lattice with applications to complexity theory. *SIGACT News*, 34(4):38–52, 2003.

[BCRV04] E. Böhler, N. Creignou, S. Reith, and H. Vollmer. Playing with Boolean blocks, part II: Constraint satisfaction problems. *SIGACT News*, 35(1):22–35, 2004.

[BRSV05] E. Böhler, S. Reith, H. Schnoor, and H. Vollmer. Simple bases for Boolean co-clones. *Information Processing Letters*, 2005. to appear.

[Bul02] A. Bulatov. A dichotomy theorem for constraints on a three-element set. In *Proceedings 43rd Symposium on Foundations of Computer Science*, pages 649–658. IEEE Computer Society Press, 2002.

[CKS01] N. Creignou, S. Khanna, and M. Sudan. *Complexity Classifications of Boolean Constraint Satisfaction Problems*. Monographs on Discrete Applied Mathematics. SIAM, 2001.

[Dal00] V. Dalmau. *Computational complexity of problems over generalized formulas*. PhD thesis, Department de Llenguatges i Sistemes Informàtica, Universitat Politécnica de Catalunya, 2000.

[JCG97] P. G. Jeavons, D. A. Cohen, and M. Gyssens. Closure properties of constraints. *Journal of the ACM*, 44(4):527–548, 1997.

[Pip97a] N. Pippenger. Pure versus impure Lisp. *ACM Transactions on Programming Languages and Systems*, 19:223–238, 1997.

[Pip97b] N. Pippenger. *Theories of Computability*. Cambridge University Press, Cambridge, 1997.

[Rei01] S. Reith. *Generalized Satisfiability Problems*. PhD thesis, Fachbereich Mathematik und Informatik, Universität Würzburg, 2001.

[Rei05] Omer Reingold. Undirected st-connectivity in log-space. In *STOC '05: Proceedings of the thirty-seventh annual ACM symposium on Theory of computing*, pages 376–385, New York, NY, USA, 2005. ACM Press.

[Sch78] T. J. Schaefer. The complexity of satisfiability problems. In *Proceedings 10th Symposium on Theory of Computing*, pages 216–226. ACM Press, 1978.

On the Number of Random Digits Required in MonteCarlo Integration of Definable Functions

César L. Alonso[1,*], José L. Montaña[2,**], and Luis M. Pardo[2,***]

[1] Centro de Inteligencia Artificial, Universidad de Oviedo,
Campus de Viesques, 33271 Gijón, Spain
`calonso@aic.uniovi.es`
[2] Departamento de Matemáticas, Estadística y Computación,
Facultad de Ciencias, Universidad de Cantabria, Spain
`{montana, pardo}@matesco.unican.es`

Abstract. Semi-algebraic objects are subsets or functions that can be described by finite boolean combinations of polynomials with real coefficients. In this paper we provide sharp estimates for the the precision and the number of trials needed in the MonteCarlo integration method to achieve a given error with a fixed confidence when approximating the mean value of semi-algebraic functions. Our study extends to the functional case the results of P. Koiran ([7]) for approximating the volume of semi-algebraic sets.

Keywords: MonteCarlo algorithms, discrepancy bounds, learning theory, Chebyshev inequalities, semi-algebraic geometry.

1 Introduction

Numerical methods that are known as MonteCarlo methods can be loosely described as statistical simulation methods, where statistical simulation is defined in quite general terms to be any method that uses sequences of random numbers to perform the simulation. MonteCarlo methods have been used for centuries, but only in the past several decades has the technique gained the status of a full-fledged numerical method capable of addressing the most complex applications in a wide field of areas, including many subfields of physics, like statistical physics, high energy physics, biology or analysis of financial markets.

In a MonteCarlo method many simulations must be performed (multiple "trials" or "histories") and the desired result is taken as an average over the number of observations (which may be a single observation or perhaps millions of them). In many practical applications, one wishes to predict the error in this average result, and hence an estimate of the number of MonteCarlo trials that are needed to achieve a given error is necessary. More precisely, a major goal

---

* Partially supported by the Spanish grant TIC2003-04153.
** Partially supported by the Spanish Programa de Movilidad PR2005-0422.
*** Partially supported by the Spanish grant MTM2004-01176.

whilst studying the theoretical performance of some MonteCarlo method is to estimate as a function of $\epsilon > 0$ and $\delta > 0$ how large the number of trials m needs to be in order to obtain an ϵ- approximation of the approximating object with confidence at least $1 - \delta$.

The random numbers used in computer implementations of basic MonteCarlo integration (described in Sect. 2) are usually referred to as pseudo-random numbers, since they are not truly random. Pseudo-random numbers are produced by the computer deterministically by simple algorithms that generate integer numbers up to a range N. To obtain real numbers in the interval $[0, 1]$ one divides therefore by N. After this reduction they work directly with floating point numbers. In these algorithms (see [20] for a survey exposition) pseudo-random numbers are generated within a given fixed precision depending on characteristics like the programming language used, the computer platform, and other implementation features. On the other hand the theoretical study of the number of MonteCarlo trials needed to achieve a given error bound based on quantitative estimates of the convergence of stochastic processes as Central Limit Theorem or Chebyshev-like estimates, assumes, in general, infinite precision generation of random numbers. This feature has been pointed out in [7] when approximating the volume of semi-algebraic sets and it is mimetic while studying the size of the sample in many learning situations. In the learning context a learning machine is supposed to learn from randomly drawn examples and infinite precision generation of the sample is implicit in the core of the results of learning theory (see [17], [18] and [3]).

Hence, there is a gap between theoretical studies (assuming infinite precision generation of random points) and real-life computer limitations (where only finite precision can be achieved). The aim of this paper is to go inside this gap, providing new mathematical tools to grasp the theoretical analysis of the error and the number of random points needed in MonteCarlo integration and other learning problems taking into account that only finite precision generation of random samples can be assumed.

Our main results (Theorem 2 and Corollary 4) study the convergence of stochastic processes predicting the mean value of a semi-algebraic function. We take as starting point Chebyshev and Hoeffding-like inequalities. We provide sharp estimates not only for the number of examples but also for the minimal precision with which examples must be generated to ensure an approximation of a given error with high confidence. Both lower bounds, on the number of examples and the precision, are given as a function of some suitable parameters: dimension, degree and number of polynomials involved in the description of the semi-algebraic object. The paper is organized as follows. Section 2 describes basic MonteCarlo integration and the statement of the main results. Section 3 and Section 4 develop the main technical tools used in this paper, that is, they provide sharp discrepancy bounds for semi-algebraic subsets of the hypercube $[0, 1]^n$. Section 5 contains the proof of our main results.

As a main conclusion our results (Corollary 4) suggest the idea that even if learning processes involving semi-algebraic objects can require an exponential number of examples in the desired learning precision, these examples can be drawn within a polynomial (linear) precision in the error and confidence precision.

2 Basic MonteCarlo Integration and Statement of Main Results

In this paper we shall be mainly concerned with the MonteCarlo integration method that can be seen as an early instance of learning (see [3]). Suppose that $f : [0,1]^n \to \mathbb{R}$ is a Lebesgue measurable function and that there exists a constant $M \geq 0$ such that $|f(x) - \int_{[0,1]^n} f| \leq M$ almost everywhere. Let $E[f] := \int_{[0,1]^n} f$ be the mean value of f and $\sigma^2(f)$ its variance. A way of approximating the integral $E[f]$ consists in randomly drawing points $x_1, ..., x_m \in [0,1]^n$ and then, computing

$$I_m(f) = \frac{1}{m} \sum_{1 \leq i \leq m} f(x_i) \qquad (1)$$

This method of approximating the value $E[f]$ relies on the following well known fact: $I_m(f) \to E[f]$ in probability; i.e for all $\epsilon > 0$

$$lim_{m \to \infty} prob\{(x_1, ..x_m) \in [0,1]^{n.m} : |I_m(f) - E[f]| \leq \epsilon\} = 1 \qquad (2)$$

Remark 1. Using any quantitative version of the property expressed in Equation 2 above, as for instance Chebyshev inequality

$$prob\{(x_1, ..x_m) \in [0,1]^{n.m} : |I_m(f) - E[f]| < \epsilon\} > 1 - \frac{\sigma^2(f)}{m\epsilon^2} \qquad (3)$$

or exponential Chebyshev-like inequalities like Hoeffding inequality (see [14])

$$prob\{(x_1, ..x_m) \in [0,1]^{n.m} : |I_m(f) - E[f]| < \epsilon\} > (1 - 2e^{-\frac{m\epsilon^2}{2M^2}}), \qquad (4)$$

it is possible to estimate how large m needs to be to ensure that the value $I_m(f)$ defined in Equation 1 is an ϵ-approximation of $E[f]$ with confidence at least $1-\delta$.

Given $\epsilon \in (0,1)$ the precision of ϵ is defined by $pr(\epsilon) := \lceil -\log_{10} \epsilon \rceil$. Our main result combines the following Chebyshev and Hoeffding inequalities with precision.

Theorem 2. *Let $f : [0,1]^n \to \mathbb{R}$ be a measurable semi-algebraic function whose graph can be defined by at most s real polynomials of degree at most d. Assume that $|f(x) - E[f]| \leq M$ a. e.. Let $L := nm$ and $K(n, m, s, d) := mn(4(s+2)d+1)^8$. Let $\mathcal{P}$ be the probability that a random choice of m rational numbers $x_1, \ldots, x_m \in \mathbb{Q}^n \cap [0,1]^n$ given with precision $p > 0$ satisfies*

$$\left|\frac{1}{m}\sum_{i=1}^m f(x_i) - E[f]\right| < \epsilon.$$

Let $\delta \in (0,1)$ be any positive real number. Then

(1) *For L big enough (depending on δ), if $p \in \mathbf{N}$ is a positive integer that satisfies the following inequality:*

$$p \geq log_{10}K(n,m,s,d) - log_{10}\delta + 4log_{10}\epsilon + 2signum(\sigma^2(f)-1)log_{10}(\sigma^2(f)) \quad (5)$$

it holds [C-inequality]

$$\mathcal{P} > 1 - (1+\delta)\frac{\sigma^2(f)}{m\epsilon^2} \quad (6)$$

(2) *For L big enough if $p \in \mathbf{N}$ is a positive integer that satisfies the following inequality:*

$$p \geq log_{10}K(m,n,s,d) - log\delta + \frac{m\epsilon^2}{M^2} \quad (7)$$

it holds [H-inequality]

$$\mathcal{P} > (1 - 2(1+\delta)e^{-\frac{m\epsilon^2}{2M^2}}) \quad (8)$$

Remark 3. The reader should observe that the probability measure in Theorem 2 is defined by the uniform distribution on the finite set of rational numbers $x \in \mathbf{Q}^n \cap [0,1]^n$ given with precision p while the probability in Equations 3 and 4 is just the Lebesgue Borel probability measure.

From Theorem 2 we get the following Corollary.

Corollary 4. *Let $\epsilon, \delta \in (0,1)$ be positive real constants. Let $f : [0,1]^n \to \mathbf{R}$ be a measurable semi-algebraic function whose graph can be defined by at most s real polynomials of degree at most d. Assume that $|f(x) - E[f]| \leq M$ almost everywhere. Let $L := nm$ and $K(m,n,s,d) := mn(4(s+2)d+1)^8$. Let $\mathcal{P}$ be the probability that a random choice of m rational numbers $x_1, \ldots, x_m \in \mathbf{Q}^n \cap [0,1]^n$ given with precision $p > 0$ satisfies*

$$\left| \frac{1}{m} \sum_{i=1}^{m} f(x_i) - E[f] \right| < \epsilon.$$

Then there exit universal constants k_1 and k_2 such that for L big enough it holds

(1) *A sufficient condition for $I_m(f)$ to be an ϵ-approximation of $E[f]$ with confidence at least $1-\delta$ is that*

$$m > \sigma^2(f)10^{k_1(pr(\epsilon)+pr(\delta))}$$

and

$$p > k_2 \left(signum(\sigma^2(f)-1)log_{10}\sigma^2(f) + log_{10}(nsd) + pr(\epsilon) + pr(\delta) \right)$$

(2) *A sufficient condition for $I_m(f)$ to be an ϵ-approximation of $E[f]$ with confidence at least $1-\delta$ is that*

$$m > pr(\delta)10^{k_1 pr(\epsilon)}M^2$$

and

$$p > k_2 \left(pr(\epsilon) + pr(\delta)log_{10}M + log_{10}(nsd) \right)$$

Remark 5. Note that the analysis in Corollary 4 implies that the number of examples needed in a MonteCarlo trial to achieve error less than ϵ and confidence at least $1 - \delta$ can be exponential in the error precision ($pr(\epsilon)$) while the minimal precision p of the examples $x = (x_1, \cdots, x_m) \in [0,1]^n \times \cdots \times [0,1]^n$ is at most linear in $pr(\epsilon)$ and in $pr(\delta)$.

3 Notions from Real Algebraic Geometry

A semi-algebraic set is a subset $\mathcal{W} \subset \mathbb{R}^n$ given by a first order formula in the theory of real closed fields. From Tarski's Principle (c.f. [15]), a subset $\mathcal{W} \subset \mathbf{R}^n$ is semi-algebraic if and only if it is given by a boolean combination of polynomial sign conditions. A semi-algebraic function is a function $f : \mathcal{W} \subset \mathbf{R}^n \to \mathbf{R}$ whose graph $Gr(f) \subset \mathbf{R}^{n+1}$ is a semi-algebraic set. Note that the domain of definition $\mathcal{W}$ of a semi-algebraic function $f : \mathcal{W} \subset \mathbf{R}^n \to \mathbf{R}$ has to be a semi-algebraic set. Next we expose the necessary notions needed to achieve the proof of our main results.

Definition 6. *Let $s, d \in \mathbf{N}$ be two positive integer numbers. Assume that $d \geq 2$ and $s \geq 1$. Let $\mathcal{F} := \{f_1, \ldots, f_s\} \subset \mathbf{R}[x_1, \ldots, x_n]$ be a finite set of s polynomials of degree at most d. A semi-algebraic subset $\mathcal{W} \subset \mathbf{R}^n$ is said to be $\mathcal{F}$-definable if and only if there are positive integers $u, v \in \mathbf{N}$ and sign conditions $\epsilon_{i,j} \in \{>, =, <\}, 1 \leq i \leq u, 1 \leq j \leq v$ such that the following equality holds.*

$$\mathcal{W} := \bigcup_{i=1}^{u} \{x \in \mathbf{R}^n : f_{i,1}(x)\epsilon_{i,1}0, \ldots, f_{i,v}(x)\epsilon_{i,v}0\}, \tag{9}$$

where $f_{i,j} \in \mathcal{F}$ for all $i, j, 1 \leq i \leq u, 1 \leq j \leq v$.

Definition 7. *A semi-algebraic subset $\mathcal{W} \subset \mathbf{R}^n$ is said to be an (s,d)-definable semi-algebraic set if there is a finite set $\mathcal{F} \subset \mathbf{R}[x_1, \ldots, x_n]$ of at most s polynomials of degree at most d such that $\mathcal{W}$ is an $\mathcal{F}$-definable semi-algebraic set.*

Definition 8. *We say that a semi-algebraic subset $\mathcal{W} \subset \mathbf{R}^n$ is the M-projection of some (s,d)-definable semi-algebraic set if there is $M \geq 0$ and an (s,d)-definable semi-algebraic subset $\mathcal{W}' \subset \mathbf{R}^{M+n}$ such that*

$$\mathcal{W} := \{x \in \mathbf{R}^n : \exists y_1 \in \mathbf{R}, \ldots, \exists y_M \in \mathbf{R} \ (x, y_1, \ldots, y_M) \in \mathcal{W}'\}$$

Definition 9. *A semi-algebraic function $f : \mathcal{W} \subset \mathbf{R}^n \to \mathbf{R}$ is called an (s,d)-definable semi-algebraic function if its graph $Gr(f) \subset \mathbf{R}^{n+1}$*

$$Gr(f) := \{(x, y) \in \mathbf{R}^{n+1} : x \in \mathcal{W}, \ y - f(x) = 0\}$$

is an (s,d)-definable semi-algebraic set.

Definition 10. Let $f : [0,1]^n \to \mathbf{R}$ be an (s,d)-definable semi-algebraic function. Let $H \in \mathbf{R}$ be a positive real number. We define the following functions

(1) $f^{(H)} : [0, H]^n \to \mathbf{R}$ the function given by

$$f^{(H)}(x) := f\left(\frac{1}{H} \cdot x\right), \forall x \in [0, H]^n \tag{10}$$

(2) $Im(f) : ([0,1]^n)^m \to \mathbf{R}$ given by

$$Im(f)(x_1, x_2, \ldots, x_m) := \frac{1}{m} \sum_{i=1}^{m} f(x_i) \tag{11}$$

(3) and the function $Im(f^H) : ([0, H]^n)^m \to \mathbf{R}$,

$$Im(f^{(H)})(x_1, x_2, \ldots, x_m) := \frac{1}{m} \sum_{i=1}^{m} f^{(H)}(x_i) \tag{12}$$

(4) Let $a \in \mathbf{R}$ be any real number and let $\epsilon > 0$ be a positive real number. We define $\mathcal{W}(m, f, H, a, \epsilon) \subset ([0, H]^n)^m$ as the set given by the following identity:

$$\mathcal{W}(m, f, H, a, \epsilon) := \left\{(x_1, \ldots, x_m) \in \mathbf{R}^{n \cdot m} : |Im(f^{(H)})(x_1, \ldots, x_m) - a| < \epsilon\right\} \tag{13}$$

We say that $\mathcal{W}(m, f, H, a, \epsilon)$ is a Chebyshev set.

Next Lemma is an easy consequence of the previous definitions.

Lemma 11. If f is an (s,d)-definable semi-algebraic function then

(1) $f^{(H)}$ is an (s,d)-definable semi-algebraic function
(2) $Im(f)$ and $Im(f^H)$ are (sm,d)-definable
(3) $\mathcal{W}(m, f, H, a, \epsilon)$ is the m-projection of an (sm+2, d)-definable semi-algebraic set.

The following statement gives a precise estimate for the number of connected components of a semi-algebraic set. It is a consequence of the estimates in [8], [16], [11],[12], and [19] for the homology of semi-algebraic sets and the fact that projections do not increase the number of connected components. A detailed proof is given in [9] (see also [13]).

Proposition 12. Let $\mathcal{W} \subset \mathbf{R}^n$ be the M-projection of an (s,d)-definable semi-algebraic set. Then,

$$\beta_0(\mathcal{W}) \leq (2sd+1)^2(4sd+1)^{2(n+M+1)} \tag{14}$$

Here $\beta_0(\mathcal{W})$ denotes the number of connected components (0-th Betti number) of $\mathcal{W}$.

4 Lattices, Rational Points of Given Denominator and Discrepancy Bounds

A lattice $L \subset \mathbf{R}^n$ is the free abelian subgroup of the additive group $(\mathbf{R}^n, +)$ generated by a basis of $\mathbf{R}^n$ as real vector space. Among a large class of lattices we discuss here the lattice $\mathbf{Z}^n \left[\frac{1}{H}\right] \subset \mathbf{R}^n$ of all rational points with denominator H. Namely,

$$\mathbf{Z}^n \left[\frac{1}{H}\right] := \left\{ \left(\frac{a_1}{H}, \ldots, \frac{a_n}{H}\right) : a_i \in \mathbf{Z}, \; 1 \leq i \leq n \right\} \tag{15}$$

Remark 13. Assume that $H := 10^p$. In this case, the intersection

$$\mathbf{Z}^n \left[\frac{1}{H}\right] \cap [0,1]^n \tag{16}$$

represents all points $x = (x_1, \ldots, x_n) \in [0,1]^n$ in base 10 where the mantissa is given up to precision p. Namely,

$$\mathbf{Z}^n \left[\frac{1}{H}\right] \cap [0,1]^n = \{(x_1, \ldots, x_n) : x_i = 0.a_1 \ldots a_p, \; a_i \in \{0, \ldots, 9\}\} \tag{17}$$

Note that $\lfloor \log_{10} H \rfloor$ is the number of digits of the mantissa.

Let $\mathcal{W} \subset [0,1]^n$ be a semi-algebraic subset and let $H \in \mathbf{R}$ be a positive real number. We denote by $N(\mathcal{W}, H)$ the following quantity:

$$N(\mathcal{W}, H) := \# \left(\mathcal{W} \cap \mathbf{Z}^n \left[\frac{1}{H}\right] \right) = \sum_{x \in \mathbf{Z}^n \left[\frac{1}{H}\right]} \chi_\mathcal{W}(x), \tag{18}$$

where $\chi_\mathcal{W} : \mathbf{R}^n \to \{0,1\}$ is the characteristic function of $\mathcal{W}$. Namely,

$$\chi_\mathcal{W}(x) := \begin{cases} 1, & \text{if } x \in \mathcal{W} \\ 0, & \text{otherwise} \end{cases}$$

We denote by $\mathcal{W}^{(H)} \subset [0,H]^n$ the following semi-algebraic set:

$$\mathcal{W}^{(H)} := \left\{ (x_1, \ldots, x_n) \in [0,H]^n : \left(\frac{x_1}{H}, \ldots, \frac{x_n}{H}\right) \in \mathcal{W} \right\} \tag{19}$$

Let μ_L on $[0,1]^n$ denote the Lebesgue measure. Next Lemma is an immediate consequence of the definitions.

Lemma 14. *Let $\mathcal{W} \subset [0,1]^n$ be a semi-algebraic set then:*

(1) if $\mathcal{W}$ is an (s,d)-definable semi-algebraic set, also $\mathcal{W}^{(H)}$ is an (s,d)-definable semi-algebraic set.

(2) if $\mathcal{W}$ is the projection of an (s,d)-definable semi-algebraic set, then also $\mathcal{W}^{(H)}$ is the projection of an (s,d)-definable semi-algebraic set.

(3) $\mu_L(\mathcal{W}^{(H)}) = \mu_L(\mathcal{W}) \cdot H^n$
(4) $N(\mathcal{W}, H) = \#(\mathbf{Z}^n \cap \mathcal{W}^{(H)})$

The following technical statement provides sharp estimates for the quantity $|N(\mathcal{W}, H) - \mu_L(\mathcal{W}^{(H)})|$ where $\mathcal{W}$ is the M-projection of some (s,d)-definable semi-algebraic set. This quantity is usually call discrepancy. Results about discrepancies where initiated by Erdos and Turan in [5] and [6]. The proof technique follows the general guidelines in [4] and the refinements obtained in [1] and [2] for discrepancies of semi-algebraic cones in closed balls $B(0,H)$ of projective spaces. The affine case with $M = 1$ has been studied in [7], Theorem 3.

Lemma 15. *Let $\mathcal{W} \subset [0,1]^n$ be the M-projection of some (s,d)-definable semi-algebraic set. Then, for any natural number $H > 0$ the following holds:*

$$|N(\mathcal{W}, H) - \mu_L(\mathcal{W}^{(H)})| \leq T(M, s, d) \cdot n \cdot (H+1)^{n-1}, \tag{20}$$

where $T(M, s, d) \leq (2sd+1)^2 (4sd+1)^{2(M+2)}$.

Next statement estimates the discrepancy of the Chebyshev set associated to a (s,d)-definable function. It represents the main technical contribution of this paper. We briefly sketch its proof.

Proposition 16. *Let $f : [0,1]^n \to \mathbf{R}$ be an (s,d)-definable semi-algebraic function. Let $H > 0$ and $m > 0$ be positive integers and let $\epsilon > 0$ be a positive real number. Let $N(f, m, H, \epsilon)$ be the number of rational points $x = (x_1, \ldots, x_m) \in (\mathbf{Z}^n[\frac{1}{H}])^m$ that satisfy the following in equality*

$$|I_m(f)(x) - E[f]| < \epsilon.$$

Then, the following holds

$$|N(f, m, H, \epsilon) - V \cdot H^{mn}| \leq m \cdot n \left(4(s+2)d+1\right)^8 (H+1)^{m \cdot n - 1} \tag{21}$$

where

$$V = \mu_L(\{x \in [0,1]^{n \cdot m} : |I_m(f)(x) - E[f]| < \epsilon\})$$

is the Lebesgue measure of the Chebyshev set

$$\{x \in [0,1]^{n \cdot m} : |I_m(f)(x) - E[f]| < \epsilon\}$$

Proof. With the same notations as in Section 3, let $a \in \mathbf{R}$ be a real number and let $\mathcal{W}$ be the Chebyshev set given by $\mathcal{W} := \mathcal{W}(m, f, H, a, \epsilon) \subset ([0,H]^n)^m$ (see Definition 10, item 4). Let $W(a) := \mathcal{W}(m, f, 1, a, \epsilon) \subset ([0,1]^n)^m$. Note that $\mathcal{W} = W(a)^{(H)}$. According to Lemma 14, items 3 and 4, we have $|N(f, m, H, \epsilon) - V \cdot H^{mn}| = |N(W(E[f]), H) - H^{nm}\mu_L(W(E[f]))|$. Let $\delta(m, f, H, a, \epsilon)$ be defined as follows:

$$\delta(m, f, H, a, \epsilon) := \frac{1}{(H+1)^{n \cdot m}} |N(W(a), H) - \mu_L(\mathcal{W})| \tag{22}$$

Let $\triangle(m, s, d, H, a, \epsilon)$ be the supremo of all the quantities $\delta(m, f, H, a, \epsilon)$ when f range over the set of all (s,d)-definable semi-algebraic function.

We start with the case $m = 1$. In this case, according to lemma 11, item (3), $\mathcal{W}(m, f, H, a, \epsilon)$ is the 1-projection of an $(s + 2,d)$-definable semi-algebraic set and the result follows from Lemma 15.

We proceed by induction on m. Suppose now $m \geq 2$. We introduce the following two auxiliary quantities:

$$S_1(m, f, H, a, \epsilon) :=$$

$$\frac{1}{(H+1)^{n \cdot m}} |N(W(a), H) - \sum_{x \in \mathbf{Z}^n \cap [0,H]^n} \int_{[0,H]^{n(m-1)}} \chi_{\mathcal{W}_x}(y) dy| \quad (23)$$

where

$$\mathcal{W}_x := \{(x_1, \ldots, x_{m-1}) \in ([0, H]^n)^{m-1} : (x_1, \ldots, x_{m-1}, x) \in \mathcal{W}\} \quad (24)$$

and

$$S_2(m, f, H, a, \epsilon) :=$$

$$\frac{1}{(H+1)^{n \cdot m}} | \sum_{x \in \mathbf{Z}^n \cap [0,H]^n} \int_{[0,H]^{n(m-1)}} \chi_{\mathcal{W}_x}(y) dy - \mu_L(\mathcal{W})| \quad (25)$$

We obviously have

$$\delta(m, f, H, a, \epsilon) \leq S_1(m, f, H, a, \epsilon) + S_2(m, f, H, a, \epsilon).$$

We then estimate both quantities separately. First note that $S_1(m, f, H, a, \epsilon)$ is less than or equal to:

$$\frac{1}{(H+1)^{n \cdot (m-1)}} max_{x \in \mathbf{Z}^n \cap [0,H]^n} | \sum_{y \in (\mathbf{Z} \cap [0,H])^{n(m-1)}} \chi_{\mathcal{W}_x}(y) - \int_{[0,H]^{n(m-1)}} \chi_{\mathcal{W}_x}(y) dy| \quad (26)$$

Let $f_1(x) := \frac{m-1}{m} f(x)$ $\forall x \in [0,1]^n$. We clearly have that f_1 is an (s,d)-definable semi-algebraic function. Define $a_x := a - \frac{1}{m} f^{(H)}(x) \in \mathbf{R}$. Thus, we conclude $\mathcal{W}_x = \mathcal{W}(m-1, f_1, H, a_x, \epsilon)$. Hence there is some $a_1 \in \mathbf{R}$ s.t.

$$S_1(m, f, H, a, \epsilon) \leq \triangle(m-1, s, d, H, a_1, \epsilon) \quad (27)$$

On the other hand, we have that $S_2(m, f, H, a, \epsilon)$ is less than or equal to

$$\frac{1}{(H+1)^{n(m-1)}} \int_{[0,H]^{n(m-1)}} | \sum_{x \in (\mathbf{Z} \cap [0,H])^n} \chi_{\mathcal{W}^y}(x) - \int_{[0,H]^n} \chi_{\mathcal{W}^y}(x) dx| dy, \quad (28)$$

where for every $y \in \mathbf{Z}^{n(m-1)} \cap [0, H]^{n(m-1)}$:

$$\mathcal{W}^y := \{x \in \mathbf{R}^n : (x, y) \in \mathcal{W}(m, f, H, a, \epsilon)\} \quad (29)$$

Defining $f_2(x) := \frac{1}{m} f(x)$ and for each $y := (y_1, \ldots, y_{m-1}) \in ([0, H] \cap \mathbf{Z})^{n(m-1)}$, $a_y := a - \frac{1}{m} \sum_{i=1}^{m-1} f^{(H)}(y_i)$ we conclude that $S_2(m, f, H, a, \epsilon)$ is less than or equal to

$$(\frac{H}{H+1})^{n(m-1)} \max\{\delta(1, f_2, H, a_y, \epsilon) : y \in ([0, H] \cap \mathbf{Z})^{n(m-1)}\} \qquad (30)$$

Since f_2 is an (s,d)-definable semi-algebraic function, we conclude that there exist real numbers $a_1, a_2 \in \mathbf{R}$ such that

$$\triangle(m,, s, d, H, a, \epsilon) \leq \triangle(m-1, s, d, H, a_1, \epsilon) + \triangle(1, s, d, H, a_2, \epsilon) \qquad (31)$$

Since $\triangle(1, s, d, H, a, \epsilon)$ is bounded by a function which is independent of a and ϵ it follows that:

$$\triangle(m, s, d, H, a, \epsilon) \leq \triangle(m-1, s, d, H) + \frac{nT(1, s+2, d)}{(H+1)} \qquad (32)$$

where $T(1, s+2, d) \leq (4(s+2)d+1)^8$ is the quantity introduced in Lemma 15 and the result follows by induction.

Now, combining Proposition 16 with Chebyshev and Hoeffding inequalities (Equations 3 and 4 respectively), one easily gets the following result.

Proposition 17. *Let $f : [0,1]^n \to \mathbf{R}$ be a measurable semi-algebraic function whose graph can be defined by at most s real polynomials of degree at most d. Assume that $|f(x) - E[f]| \leq M$ almost every where. Let $\epsilon > 0$ be a real positive constant. Let $K(m, n, s, d) := mn(4(s+2)d+1)^8$. Let $\mathcal{P}$ be the probability that a random choice of m rational numbers $x_1, \ldots, x_m \in \mathbf{Q}^n \cap [0,1]^n$ given with precision $p > 0$ verifies*

$$\left| \frac{1}{m} \sum_{i=1}^{m} f(x_i) - E[f] \right| < \epsilon$$

For every natural number $p \geq 1$ it holds it holds
[C-inequality]

$$\mathcal{P} > (1 - \frac{\sigma^2(f)}{m\epsilon^2})(1 - \frac{1}{10^p + 1})^{mn} - \frac{K(m,n,s,d)}{10^p + 1} \qquad (33)$$

[H-inequality]

$$\mathcal{P} > (1 - 2e^{-\frac{\epsilon^2}{2M}})(1 - \frac{1}{10^p + 1})^{mn} - \frac{K(m,n,s,d)}{10^p + 1} \qquad (34)$$

5 Proof of Theorem 2

In this section we sketch the proof of Theorem 2. We shall make use of the following technical statement which reflects well known properties of limits.

Lemma 18.
(1) For every positive real number $\theta \in (0,1)$ it holds:

$$\lim_{L \to \infty} \left(-\frac{L}{\ln \theta} \left(1 - \theta^{1/L}\right) \right) = 1 \tag{35}$$

(2) For every real constant $\theta \neq 0$ it holds:

$$\lim_{L \to \infty} \left(-\frac{L}{\theta} \ln\left(1 - \frac{\theta}{L}\right) \right) = 1 \tag{36}$$

Lemma 19. *For every positive real constant $\delta \in (0,1)$ there exists $L_0 \in \mathbf{N}$ such that if $L \geq L_0$ the following property holds: for every natural number $p \in \mathbf{N}$ satisfying*

$$p \geq \log_{10} L + \log_{10} 2 - \log_{10}(-\ln(1-\delta)) \tag{37}$$

it holds

$$\left(\frac{10^p}{1 + 10^p} \right)^L > 1 - \delta \tag{38}$$

Proof. Let $H := 10^p$. Given $\delta \in (0,1)$ condition (38) is equivalent to

$$H + 1 > \frac{1}{1 - (1-\delta)^{1/L}} \tag{39}$$

Due to property (35), for every $t \in (0,1)$ it holds that for L big enough and H satisfying

$$H > \frac{1}{1-t}\left(-\frac{L}{\ln(1-\delta)} \right) \tag{40}$$

condition (38) is satisfied. Finally set $t := \frac{1}{2}$ in inequality (40) and take logarithms.

Let $\delta 1, \delta 2 \in (0,1)$ be positive real numbers. As a direct consequence of Lemma 19 and Proposition 17, with the hypothesis of Theorem 2, the following holds: for L big enough if $p \in \mathbf{N}$ is a positive integer number such that

$$p \geq \max\{\log_{10} L + \log_{10} 2 - \log_{10}(-\ln(1-\delta 1)), \log_{10} K(m,n,s,d) - \log_{10} \delta 2\} \tag{41}$$

it holds
[C-inequality]

$$\mathcal{P} > \left(1 - \frac{\sigma^2(f)}{m\epsilon^2}\right)(1 - \delta 1) - \delta 2 \tag{42}$$

[H-inequality]

$$\mathcal{P} > \left(1 - \frac{2}{e^{\frac{\epsilon^2}{2M^2}}}\right)(1 - \delta 1) - \delta 2 \tag{43}$$

Now we can finish the proof of Theorem 2. We proceed as follows. For the [C-inequality] in Equation 42 take $\delta 2 := \delta 1 \frac{\sigma^2(f)}{m\epsilon^2}$ and $\delta 1 := \delta \frac{\sigma^2(f)}{m\epsilon^2}$. Using Equation 36 to simplify Equation 41 one gets the desired bound on p. For the [H-inequality] in Equation 43 take $\delta 2 := 2\delta 1 e^{-\frac{m\epsilon^2}{2M^2}}$ and $\delta 1 := 2\delta e^{-\frac{m\epsilon^2}{2M^2}}$. Using again Equation 36 to simplify Equation 41 one gets the desired bound on p.

References

1. Castro D., Montaña J. L., Pardo L. M. and San Martin J.: The Distribution of Condition Numbers of Rational Data of Bounded Bit Length, Found. of Comput. Math. Vol. 1, N. 1 1-52. 2002.
2. Castro D., Pardo L.M. and San Martín J.:Systems of rational polynomial equations have polynomial size approximate zeros on the average. J. Complexity 19 (2003), no. 2, 161–209
3. Cucker F., Smale S. :On the Mathematical foundations of learning. Bulletin (New Series) Of the AMS, vol 39, number 1, pp 1–4. 2001.
4. Davenport H.: On a Principle of Lipschitz, J. London Math. Soc. 26. 179-183. 1951.
5. Erdos P., Turan P.: On a Problem in the Theory of Uniform Distribution. I, Indagationes Math. 10 370–378. 57. 1948
6. Erdos P. and Turan P.:, On a Problem in the Theory of Uniform Distribution. II, Indagationes Math. 10 406–413. 1948
7. Koiran P.: Approximating the volume of definable sets. In Proc. 36th IEEE Symposyum on Foundations of Computer Science FOCS95. 134–141. 1995.
8. Milnor J., On the Betti Numbers of Real Varieties, Proc. Amer. Math. Soc. 15 275–280. 1964.
9. Montaña J. L. and Pardo L. M., Lower bounds for Arithmetic Networks, Applicable Algebra in Engineering Communications and Computing 4 (1993) 1–24.
10. Mordell L. J. , On some Arithmetical Results in the Geometry of Numbers, Compositio Mathematica 1 (1934) 248–253.
11. Oleinik O.A.: Estimates of the Betti Numbers of Real Algebraic Hypersurfaces, Mat. Sbornik 70 (1951) 63–640.
12. Oleinik O. A. and Petrovsky I. B.: On the topology of Real Algebraic Surfaces, Izv. Akad. Nauk SSSR (in Trans. of the Amer. Math. Soc.) 1 (1962) 399–417.
13. Pardo L. M.: How Lower and Upper Complexity Bounds Meet in Elimination Theory, in: G. Cohen, M.Giusti and T. Mora, eds., Proc. 11th International Symposium Applied Algebra, Algebraic Algorithms and Error–Correcting Codes, AAECC-11, (Paris, Lect. Notes in Comput. Sci. 948, Springer, 1995) 33–69.
14. Pollard P.: Convergence of Stocastic Processes, Springer-Verlarg. 1984
15. Tarski A.: A Decision Method for Elementary Algebra and Geometry, (RAND Corporation, Santa Monica, Calif., 1948)
16. Thom R.: Sur l'Homologie des Varietes Alg´ebriques R´eelles, in: Differential and Combinatorial Topology (A Symposium in Honor of Marston Morse), (Princeton Univ. Press, 1965) 255–265.
17. Valiant L. G.: A theory of the learneable. Communications of the ACM, vol. 27 pp. 1134–1142. 1984.
18. Vapnik V.: Statistical learning theory, John Willey & Sons. 1998
19. Warren H. E.: Lower Bounds for Approximation by non Linear Manifolds, Trans. A.M.S. 133 (1968) 167–178.
20. Weinzierl S.:Introduction to MonteCarlo methods. Technical Report NIKHEF-00-012 Theory Group. 2000.

Pure Nash Equilibria in Games with a Large Number of Actions[*],[**]

Carme Àlvarez, Joaquim Gabarró, and Maria Serna

ALBCOM Research Group. Universitat Politècnica de Catalunya,
Jordi Girona, 1-3, Barcelona 08034, Spain
{alvarez, gabarro, mjserna}@lsi.upc.edu

Abstract. We study the computational complexity of deciding the existence of a Pure Nash Equilibrium in multi-player strategic games. We address two fundamental questions: how can we represent a game? and how can we represent a game with polynomial pay-off functions? Our results show that the computational complexity of deciding the existence of a pure Nash equilibrium in a strategic game depends on two parameters: the number of players and the size of the sets of strategies. In particular we show that deciding the existence of a Nash equilibrium in a strategic game is NP-complete when the number of players is large and the number of strategies for each player is constant, while the problem is Σ_2^p-complete when the number of players is a constant and the size of the sets of strategies is exponential (with respect to the length of the strategies).

Keywords: Strategic games, Nash equilibria, complexity classes.

1 Introduction

The question that motivates the present work is *which is the complexity of deciding whether a strategic game has a pure Nash equilibrium?* A strategic game is defined by a set of players, each player has a set of possible actions to play and a pay-off function that measures their benefit depending on the actions adopted by each one of the players. A pure Nash equilibrium describes a situation in which each player has selected an action from their set of actions (this is their strategy) and no individual player can derive any benefit from deviating from their strategy. In this context each player chooses to play an action in a deterministic way. Contrasting with this a mixed Nash equilibrium can be defined in a similar way, but now each player chooses a distribution on their set of actions.

The focus of our study is inspired in one of the fundamental open problems in Theoretical Computer Science: *Which is the complexity of finding a Nash*

---

[*] Work partially supported by the EU IST-2001-33116 (Flags) and IST-2004-15964 (AEOLUS) and by Spanish CICYT TIC2002-04498-C05-03 (Tracer).
[**] Due to space restrictions some proofs are omitted, we refer the reader to [1] for further details.

equilibrium?. This question was posed by Papadimitriou [19] and it has initiated a line of research towards understanding the complexity of computing a pure or a mixed Nash equilibrium [17,14,12,4,5,9,10,8,13,6,11].

There have been some results in determining the complexity of Nash equilibria and related questions for particular cases. In the context of strategic games Conitzer et al. in [6] demonstrate the NP-hardness of determining whether a Nash equilibrium with certain natural properties exists. Gottlob et al. in [13] show that determining whether a strategic game has a pure Nash equilibrium is NP-hard in the case that each player of the game has a polynomial-time computable pay-off function. In the case of congestion games Fabrikant et al. in [8] show that a pure Nash equilibrium can be computed in polynomial time in the symmetric network case, while the problem is PLS-hard in general. Fotakis et al. in [9] study several algorithmic problems related to the computation of Nash equilibria for a certain game that models selfish routing over a network consisting of parallel links. However, in any of those references there is a lack of uniformity in the representation of games.

How Can We Represent a Game? To answer this question we have to take into account the main elements that form part of a strategic game. We note that, for any problem on games to be computationally meaningful, the number of players or the number of actions of each player or both should be large, furthermore the set of actions or the payoff functions should be given in some implicit way. For example in [6,13] the set of possible actions of each player is given explicitly by enumerating each of their elements. In [8,10] the set of actions are given implicitly, they are the set of $s-t$ paths of a given network. And with respect to the pay-off functions in general it is assumed that they are polynomial time computable but there is no discussion about how can they be described.

We propose a framework that allow us to represent a strategic game with different levels of succinctness, reflecting the constraints of the natural components. We consider three natural ways of describing a game depending of the succinctness of their representation (see Table 1). We consider explicit descriptions for any of those elements defining a strategic game, by means of listing the set of actions and tabulating the pay-off functions, what we call the *explicit form*. We also consider more succinct representations in which the set of actions and/or the pay-off functions are described in terms of Turing machines. When considering a Turing machine as part of a description, an additional element is needed in it, the computation time allowed to the machine. In this way we obtain succinct descriptions of games that are *non-uniformly* described from Turing machines. We can further describe the actions explicitly, by giving the list of the actions allowed to each player, what we call the *general form*, or succinctly, by giving the length of the actions, what we call the *implicit form*. Observe that the players are not represented in a succinct way, that means in practice "one bit for player". This is a reasonable assumption because any strategy profile $a = (a_1, \ldots, a_n)$ has one strategy for player, therefore we need at least n bits. If we describe the number of players using $\log n$ bits any strategy profile will be exponential in that quantity and that seems an unreasonable additional constraint.

What Means a Game with Polynomial Pay-off Functions? Even though in many papers studying the computational complexity of some specific games, it is assumed that the utilities are computable in polynomial time. This assumption has had different interpretations (see for example [8,13,9,10,11]).

For instance, Gottlob et al. in [13] consider that "each player has a polynomial time computable real valued utility function" however a machine computing such function is not given as part of the description of a game [13]. Fabrikant et al. [8] consider congestion games with a different representation. A congestion game is defined by n players, a set E of resources, and a delay function d mapping $E \times \{1, \ldots, n\}$ to the integers. The action for each player are subsets of E. Setting $f(a_1, \ldots, a_n, e) = |\{i \mid e \in a_i\}|$, the pay-off are $u_i(a_1, \ldots, a_n) = -(\sum_{e \in a_i} d(e, f(a_1, \ldots, a_n, e)))$, and thus can be computed in polynomial time. In this case, they consider a uniform family of games in the sense that the different instances are given by considering different number of players, action sets and delay functions, but in each of them the pay-off functions can be computed by a DTM which works in polynomial time with respect to n and m, being m the maximum length of the actions a_i.

This notion leads us to consider families of games that can be defined uniformly in the sense that there is a DTM that gives the way of computing the utilities when the game is played with different number of players and/or different sets of actions. Hence, each DTM defines a uniform family of games. As in the non-uniform representation of games, in this case we also consider uniform families defined in *general form* or in *implicit forms* depending on the succinctness of the representation of their action sets.

Once we have defined carefully how to represent a strategic game, we study the computational complexity of deciding the existence of a pure Nash equilibrium (the SPN problem) for strategic games. In the case of non-uniform game families we show that the SPN problem for games given in implicit and in general form is hard (Table 1 summarises the results). The SPN problem is Σ_2^p-complete in the case of implicit form description, while it is NP-complete in the the case of general form descriptions. Contrasting with this, when the game is given in explicit form the SPN problem is tractable.

When we consider families of games defined uniformly and implicitly from a polynomial time deterministic Turing machine M, we show that the SPN problem is in Σ_2^p. Furthermore we show that there are Turing machines for which the problem is Σ_2^p-hard. Contrasting with this, when the representation of games is in general form the positive and hardness results are for the NP class instead of Σ_2^p. (Table 2 summarises the results).

Table 1. Degrees of succinctness in *non-uniform* strategic games description and associated complexity results

Non-uniform	Succinctness		Exist PNE?	
representation	actions	utilities	n-players	k-players ($k \geq 2$)
implicit	yes	yes	Σ_2^p-complete	Σ_2^p-complete
general	no	yes	NP-complete	P-complete
explicit	no	no	AC^0	AC^0

Table 2. Degrees of succinctness in *uniform* strategic games described by a polynomial time Turing machine M with associated complexity results

Uniform representation	Succinctness actions	Exist PNE? any M	some M
M-implicit	yes	Σ_2^p	Σ_2^p-complete
M-general	no	NP	NP-complete

Hence we solve the fundamental question of classifying the complexity of the SPN problem. Under the best of our knowledge, all the previous results presented in the literature concerning to this question only solve the problem for restricted cases.

We wish to mention that recently several researchers have independently obtained results related to ours: Daskalakis and Papadimitriou [7] studied the complexity of concisely represented graphical games, and Schoennebeck and Vadhan [20] studied the complexity of circuit games.

Finally, we consider two function problems related to game theory concepts. The first one is that of computing a best response for a player to the actions of the other players. The second one is the computation of a mixed Nash equilibrium. We show that for games given in implicit form both problems can not be solved in polynomial time unless P = NP. Our result hold even in case of two players and even when only one of the two players has a large number of actions. For games in explicit or general form the first problem can be trivially solved in polynomial time while the complexity of the second one remains open. Hence, we show for the first time a fairly general class of games in which the computation of a mixed Nash equilibrium is hard. We also consider two decision version of the above problems. Deciding whether an action is a best response is coNP-hard and deciding whether an action is in the support of some mixed Nash equilibrium is NP-hard.

The paper is organised as follows: Section 2 contains basic definitions. In Section 3 we study the SPN problem for non-uniform families of games. Section 4 contains the results for uniform families of games. Section 5 consider the additional computational problems.

2 Strategic Games

The following definition of a strategic game is borrowed from [16]. A *strategic game* Γ is defined by the following components:

- A set of n players denoted by $N = \{1, \ldots, n\}$.
- A finite set of actions A_i for each player $i \in N$. The elements of $A_1 \times \ldots \times A_n$ are the strategy profiles.
- An *utility* (or *payoff*) function u_i for each player $i \in N$ mapping $A_1 \times \ldots \times A_n$ to the rationals.

Given a strategy profile $a \in A_1 \times \ldots \times A_n$ and given any action $a_i \in A_i$, we denote by (a_{-i}, a_i) the strategy profile obtained by replacing the i-th component of

a by a_i. For any player i, we denote the set of player i's best actions when the profile of the other player's is a_{-i} as $B_i(a_{-i})$. A strategy profile $a^* = (a_1^*, a_2^*, \ldots, a_n^*)$ is a strategic pure Nash (PNE) equilibrium if, for any player i and any $a_i \in A_i$ we have $u_i(a^*) \geq u_i(a_{-i}^*, a_i)$.

We define a strategic game associated to a given property P. Our game has a PNE only in the case that the property is true.

Gadget(P). It has two players whose action sets are the same, $A_1 = A_2 = \{0,1\}$. Given the following functions f_1, f_2 defined on $\{0,1\} \times \{0,1\}$, the pay-off functions are the following.

f_1	0	1
0	1	4
1	2	3

f_2	0	1
0	4	3
1	1	2

	$u_1(a_1, a_2)$	$u_2(a_1, a_2)$
P is true	5	5
P is false	$f_1(a_1, a_2)$	$f_2(a_1, a_2)$

Note that when P is true, the players play a game such that all the strategy profiles are Nash equilibria. When P is false, no strategy profile is a pure Nash equilibrium because $(0,0) <_1 (1,0) <_2 (1,1) <_1 (0,1) <_2 (0,0)$, where $a <_i a'$ denotes $u_i(a) \leq u_i(a')$.

Proposition 1. *Given a property P, the game Gadget(P) has a PNE if and only if the property P is true.*

Determining whether a pure Nash equilibrium exists is a problem that have attracted much research in computer science (see [19]). This problem can be stated as follows:

Strategic Pure Nash (SPN). Given a strategic game Γ, decide whether Γ has a Pure Nash equilibrium.

Now we turn our attention to mixed equilibria. A *mixed strategy* is a probability distribution over the player's actions. We use σ_i to denote mixed strategy for player i such that $\sigma_i(a_i)$ is the probability assigned by player i to a_i. A *mixed strategy profile* $\sigma = (\sigma_1, \ldots \sigma_n)$ is a list of mixed strategies, one for each player. The utility functions are extended as usual taking the expected pay-off. The mixed strategy profile σ^* is a *mixed Nash equilibrium* if, for each player i and every mixed strategy σ_i of player i, we have $U_i(\sigma^*) \geq U_i(\sigma_{-i}^*, \sigma_i)$.

By the theorem of Nash [15] we know that every strategic game in which each player has a finite set of actions has a mixed Nash equilibrium. Therefore the decision version of the problem is trivial and the interesting question is the complexity of computing one.

All through the paper we use standard notation for computational complexity classes. See for example [3,2,18].

3 Non-uniform Families of Games

In the context of computational complexity it is very important to define how an input game Γ is represented. In order to define an instance of the SPN problem we have to make clear how to describe the set of players, and for each player their set of actions and pay-off functions.

All the TMs appearing in the description of games are deterministic. We use the following convention: there is a pre-fixed interpretation of the contents of the output tape of a TM so that, both when the machine stops or when the machine is stopped, it always computes a value. Let us assume that Σ is a pre-fixed alphabet. Hence we can describe the pay-off functions of a game by giving a tuple $\langle M, 1^t \rangle$ where M is a deterministic TM (DTM) and t is a natural number bounding its computation time. The interpretation is that given a strategy profile a and a natural number i, the output of M on input $\langle a, i \rangle$ is the value of the pay-off function of the i-th player on input a.

First, we consider a way of describing the set of actions so that they are not given explicitly and directly, by listing all their actions, but succinctly and implicitly. We are interested in descriptions whose length does not depend dramatically on the number of the actions, but depends on the length of the actions. Such descriptions are exponentially more succinct than the sets they describe. The following definition captures this idea.

Strategic Games in Implicit Form [1]. A game is a tuple $\Gamma = \langle 1^n, 1^m, M, 1^t \rangle$. This game has n players. For each player i, their set of actions is $A_i = \Sigma^m$ and $\langle M, 1^t \rangle$ is the description of the pay-off functions.

The second family of games is defined by considering that the set of actions of each player is given explicitly.

Strategic Games in General Form. A game $\Gamma = \langle 1^n, A_1, \ldots, A_n, M, 1^t \rangle$ has n players, for each player i, their set of actions A_i is given by listing all its elements. The description of their pay-off functions is given by $\langle M, 1^t \rangle$.

Finally, we consider a less succinct description of games. This is the usual description adopted in basic books giving us a complete description in form of a bimatrix or trimatrix (set of bimatrices).

Strategic Games in Explicit Form. A game is a tuple $\Gamma = \langle 1^n, A_1, \ldots, A_m, T \rangle$. It has n players, and for each player i, their set of actions A_i is given explicitly. T is a table with an entry for each strategy profile a and a player i. In this case $u_i(a) = T(a, i)$.

We analyse the complexity of the SPN problem in the different representations of games answering in this way the question posed by Papadimitriou in [19] for the case of strategic games.

First we study the complexity of deciding whether a game in implicit form has a PNE. We show that this problem is really hard since it is complete for the second level of the Polynomial Time Hierarchy. Observe that the proof of Theorem 3.4 of [13] can be rewritten to show that the problem of deciding whether a

---

[1] In the games in implicit form we assume $A_i = \Sigma^m$, this is not a major restriction because we can also consider $A_i \subseteq \Sigma^{\leq m}$ with just small modifications. In this case $\Gamma = \langle 1^n, 1^m, M_1, \ldots M_n, M, 1^t \rangle$ with $M_1, \ldots, M_n, M$ being DTM. The game is played by n players. For each player i, M_i is a succinct description of their set of actions $A_i \subseteq \Sigma^{\leq m}$. We say that $a_i \in A_i$ iff M_i accepts a_i in at most t steps. Given a and i, $u_i(a)$ is the output of $M(a, i)$ after at most t steps.

given strategy (of a game given in implicit form) is a Nash equilibrium is coNP-complete. At first glance this fact seems to imply that the hardness of the SPN problem follows trivially from the coNP-completeness and the additional existential quantification. It is worth noticing that this approach is false in general as it is known that the equivalence problem for circuits is coNP-complete while the isomorphisms for circuits is not Σ_2^p-hard unless the Polynomial Time Hierarchy collapses to the third level [21].

Theorem 1. *The* SPN *problem for strategic games in implicit form is Σ_2^p-complete.*

Proof. Let $\Gamma = \langle 1^n, 1^m, M, 1^t \rangle$ be a strategic game in implicit form, the problem of deciding whether Γ has a PNE can be formalised as follows:

$$\Gamma \in \text{SPN} \Leftrightarrow \exists a_1^* \in A_1 \ldots \exists a_n^* \in A_n \ \forall a_1 \in A_1 \ldots \forall a_n \in A_n$$
$$u_1(a_{-1}^*, a_1) \leq u_1(a_{-1}^*, a_1^*) \wedge \ldots \wedge u_n(a_{-n}^*, a_n) \leq u_n(a_{-n}^*, a_n^*).$$

Hence we can define an Alternating Turing machine that guesses the strategy profile $(a_1^*, \ldots, a_n^*)$ and then using a universal state it can verify that this strategy profile is a Nash equilibrium. Since the length of any action is bounded by m, and for each player i, u_i can be computed in time t, then the computation time of this Alternating Turing machine is bounded by a polynomial with respect to $\max\{n, m, t\}$. Then SPN $\in \Sigma_2^p$.

In order to prove the hardness of the SPN problem let us consider a restricted version of the Quantified Boolean Formula, the Q2SAT problem, which is Σ_2^p-complete.

Q2SAT. Given $\Phi = \exists \alpha_1, \ldots, \alpha_{n_1} \forall \beta_1, \ldots \beta_{n_2} F$ where F is a Boolean formula over the boolean variables $\alpha_1, \ldots, \alpha_{n_1}, \beta_1, \ldots, \beta_{n_2}$, decide whether Φ is valid.

For each Φ we define the following game: $\Gamma(\Phi)$. There are four players:

- Player 1, the *existential player*, assigns truth values to $\alpha_1, \ldots, \alpha_{n_1}$. Their set of actions is $A_1 = \{0, 1\}^{n_1}$ and $a_1 = (\alpha_1, \ldots \alpha_{n_1}) \in A_1$.
- Player 2, the *universal player*, assigns truth values to $\beta_1, \ldots, \beta_{n_2}$ and then their set of actions is $A_2 = \{0, 1\}^{n_2}$ and $a_2 = (\beta_1, \ldots, \beta_{n_2}) \in A_2$.
- Players 3 and 4 avoid entering into a Nash equilibrium when the actions played by players 1 and 2 do not satisfy F. Their set of actions are $A_3 = A_4 = \{0, 1\}$.

Let us denote by $F(a_1, a_2)$ the truth value of F under the assignment given by a_1 and a_2. Now it only remains to define the utility functions in such a way that they guarantee that Φ is valid if and only if $\Gamma(\Phi)$ has a Nash equilibrium. Given a strategy profile $a = (a_1, a_2, a_3, a_4)$ the utilities $u_i(a)$ as defined as follow:

	$u_1(a)$	$u_2(a)$	$u_3(a)$	$u_4(a)$
$F(a_1, a_2) = 1$	1	0	5	5
$F(a_1, a_2) = 0$	0	1	$f_1(a_3, a_4)$	$f_2(a_3, a_4)$

We claim: Φ is valid $\Leftrightarrow \Gamma(\Phi)$ has a PNE. Note that players 3 and 4 play the Gadget game associated to $F(a_1, a_2) = 1$. Furthermore a description of the above game in implicit form can be obtained in polynomial time. □

In the previous proof, for the sake of clarity, we use a game with four players, but a similar game with only two players can also be used.

Corollary 1. *The* SPN *problem for games in implicit form with k players is Σ_2^p-complete, for any $k \geq 2$.*

Contrasting with the previous results, when we allow to describe the set of actions explicitly, although the SPN problem remains hard, it is not as hard as the SPN problem for games where the set of actions are described implicitly.

Theorem 2. *The* SPN *problem for strategic games in general form is* NP*-complete, even in the case that the number of actions of each a player is some constant k, for any $k \geq 2$.*

Proof. Consider $\Gamma = \langle 1^n, A_1, \ldots, A_n, M, 1^t \rangle$, We can conjecture a strategy profile $(a_1^*, \ldots, a_n^*)$ and then check that for any i and any $a_i \in A_i$ $u_i(a^*) \geq u_i(a_{-i}^*, a_i)$. Each computation of M takes time at most t and the overall number of tests to be performed is at most $\sum_{i=1}^{n} |A_i|$. As the sets of actions are given explicitly the Nash equilibrium property can be checked in time polynomial in the input size.

In order to prove the hardness let us reduce the Satisfiability of boolean formulae problem to the SPN problem in general form. Given a formula F in conjunctive normal form on n variables, we consider the following game:

$\Gamma(F)$. We have $n + 2$ players, for each $1 \leq i \leq n + 2$, $A_i = \{0, 1\}$. Therefore the set of strategy profiles coincides with the set of truth assignments with two additional bits. The utilities are defined as follows, where $a = (a_1, \ldots, a_n)$ and $1 \leq j \leq n$:

	$u_j(a, a_{n+1}, a_{n+2})$	$u_{n+1}(a, a_{n+1}, a_{n+2})$	$u_{n+2}(a, a_{n+1}, a_{n+2})$
$F(a) = 1$	5	5	5
$F(a) = 0$	1	$f_1(a_{n+1}, a_{n+2})$	$f_2(a_{n+1}, a_{n+2})$

We can show that F is satisfiable iff $\Gamma(F)$ has a PNE. Here players $n + 1$ and $n + 2$ play the Gadget game associated to $F(a) = 1$. Moreover, $\Gamma(F)$ can be represented in general form by $\langle 1^{n+2}, \{0, 1\} \ldots \{0, 1\}, M^F, 1^{(n+2+|F|)^3} \rangle$ where M^F is a TM that on input (a, a_{n+1}, a_{n+2}, i), evaluates the formula F on input a. Afterwards it implements the utility function of the i-th player. Since we can construct M^F in polynomial time and its computation time is also polynomial, always respect to $|F|$, we have that the representation of $\Gamma(F)$ in general form can be constructed in polynomial time with respect to $|F|$. □

Contrasting with the previous hardness results, in the following two cases the SPN problem becomes tractable.

Theorem 3. *For any $k \geq 2$, the* SPN *problem for strategic games in general form with k-players is P-complete. Furthermore, the* SPN *problem for strategic games in explicit form is in AC^0.*

4 Uniform Families of Strategic Games with Polynomial Time Computable Utilities

In the previous section we have analysed the representations of the strategic games as potential inputs of the SPN problem. Here we are interested in families of strategic games that arise when the utility functions are computable in polynomial time. Thus we are interested in families of games defined uniformly by Turing machines. Following the ideas of Fabrikant et al. in [8], for each DTM M we define uniform families of strategic games in such a way that the pay-off functions of each game in the family are computed by M. Moreover, as in the previous section, we consider further refinements according to the input representation. Let M be a DTM and let us assume that an alphabet Σ is fixed. We define the following uniform families of games associated to M:

M-Implicit Form Family [2]. It is an implicit description of the family of games in which the pay-off functions are computed by the DTM M. Each instance of the family specifies the number of players n and their set of actions in an succinct way. We consider that a description of a set is succinct when the length of the description is at most polynomial with respect to its length. Formally, the M-implicit form family is $\{\langle 1^n, 1^{m_1}, \ldots, 1^{m_n}\rangle \mid n, m_1, \ldots, m_n \in \mathbb{N}\}$. In the game described by $\langle 1^n, 1^m, M_1, \ldots, M_n\rangle$, if a is a strategy profile of such game, and $1 \leq i \leq n$, then the utility of the i-th player on a is defined as $u_i(a) = M(a, i)$.

M-General Form Family. It is a general form description of the family of games in which the pay-off functions are computed by M. Each instance of the family describes a game by giving the number of players n and the set of actions of each player. Here, every set of actions is given by listing all its elements. Formally, the M-general form family is $\{\langle 1^n, A_1, \ldots, A_n\rangle \mid n, m \in \mathbb{N}\}$ where, for all i, A_i is given by listing all its elements. As in the M-implicit form, in the game described by $\langle 1^n, A_1, \ldots, A_n\rangle$, if a is a strategy profile of such game, and $1 \leq i \leq n$, then the utility of the i-th player on a is defined as $u_i(a) = M(a, i)$.

Hence, given a family of games defined from a polynomial time DTM M, we can also pose the question of determining whether a game of this family has a Nash equilibrium.

M-Strategic Pure Nash (M-SPN). Given a strategic game Γ whose pay-off functions are defined by M, decide whether Γ has a Pure Nash equilibrium.

As we have seen in the previous section, depending on whether the games are described in implicit or general form we obtain different hardness results. The following results are obtained by a modification of the corresponding result for non-uniform families. We only have to consider, in the reductions, an additional player whose set of actions is the input formula.

---

[2] In the games in implicit form we assume $A_i = \Sigma^{\leq m_i}$. We can also consider $A_i \subseteq \Sigma^{\leq m_i}$. In this case the machine M has to be able to recognise whether a given action a_i belongs to A_i.

Theorem 4.

(i) For any polynomial time DTM M, *the* M-SPN *problem for games in the M-implicit form family is in* Σ_2^p.

(ii) There exists a polynomial time DTM M *for which the* M-SPN *problem for games in the M-implicit form family is* Σ_2^p-*complete.*

Theorem 5.

(i) For any polynomial time DTM M, *the* M-SPN *problem for games in the M-general form family is in* NP.

(ii) There exists a polynomial time DTM M *for which the* M-SPN *problem for games in the M-general form family is* NP-*complete.*

If we consider the results presented in [13], they propose to study, among many other problems, the complexity of the SPN problem for games in

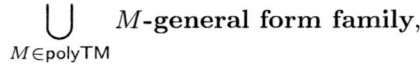
$$\bigcup_{M \in \text{polyTM}} M\text{-general form family,}$$

where polyTM is the class of TM working in polynomial time. They assume that the utility functions of their games are polynomially computable functions and they show that deciding whether a game in general form has a PNE is NP-complete. To prove the membership in NP, they strongly need to make use of the assumption that the utilities are polynomial time computable. However, in their hardness result, they construct polynomial time computable utilities, but the utilities are non-uniform in the sense that for each instance they get a different utility function.

Our contribution is different, for the uniform families our reduction produces a Turing machine for all the game instances. Furthermore, in the previous section, for non-uniform families of games, we give a general way of describing all the games with "computable utilities". In order to prove our complexity results, we do not have to assume that the description of the pay-off functions can be given as polynomial time DTM, we represent any 'computable' pay-off function by giving a DTM and a natural number t (in unary) bounding its computation time.

5 Other Computational Problems Related to Games

In this section we consider two function problems related to game theory concepts. The first one is that of computing a best response for a player given the actions of the other players. The second one is the computation of a mixed Nash equilibrium. For games in explicit or general form the first problem can be trivially solved in polynomial time while the complexity of the second one remains open.

Theorem 6. *Given a game in implicit form, a player i, and a list of the other player's actions a_{-i}, computing a best response of player i to a_{-i} can not be done in polynomial time unless $P = NP$. This is so even in the case of k-players when $k \geq 2$.*

Proof. Given a boolean CNF formula F on n variables we consider the following game:

$\Gamma(F)$. There are two players, player 1 has $A_1 = \{0,1\}^n$ and player 2 has $A_2 = \{0,1\}$. For any $\alpha \in \{0,1\}^n$ and $\beta \in \{0,1\}$, the utility functions are the following

	$u_1(\alpha,\beta)$	$u_2(\alpha,\beta)$
$F(\alpha) = \beta$	1	$F(\alpha)$
$F(\alpha) \neq \beta$	0	$F(\alpha)$

Notice that when F is satisfiable, $B_1(1) = \{\alpha \in \{0,1\}^n \mid F(\alpha) = 1\}$, but when F is not satisfiable $B_1(1) = \{0,1\}^n$ and $F(\alpha) = 0$ for any α. Therefore, given $\alpha \in B_1(1)$ we have $F(\alpha) = 1 \iff F$ is satisfiable.

Using similar arguments as the one presented in Theorem 1, given a formula F, we can construct in polynomial time the implicit representation of the game $\Gamma(F)$. Therefore if we could compute α in $B_1(1)$ in polynomial time, asking for the truth of $F(\alpha)$ we could decide the satisfiability of F in polynomial time. □

Theorem 7. *Given a strategic game Γ in implicit form, a mixed Nash equilibrium for Γ can not be computed in polynomial time unless $P = NP$. This is so even in the case of k-players when $k \geq 2$.*

Proof. Given a boolean CNF formula F on n variables we consider the following two players game $\Gamma(F)$.

$\Gamma(F)$. There are two players whose sets of actions are $A_1 = \{0,1\}^n$ and $A_2 = \{0,1\}$ respectively. For any $\alpha \in \{0,1\}^n$ and $\beta \in \{0,1\}$, the utility functions are the following:

	$u_1(\alpha,\beta)$	$u_2(\alpha,\beta)$
$F(\alpha) = \beta$	$F(\alpha)$	1
$F(\alpha) \neq \beta$	$F(\alpha)$	0

When F is not satisfiable, for any σ_1 the payoff of player 2 is maximised when their action is 0. Therefore, $\sigma = (\sigma_1, \sigma_2)$ is a mixed Nash equilibrium iff (i) σ_1 is any probability distribution on $\{0,1\}^n$ and (ii) $\sigma_2(1) = 0$ and $\sigma_2(0) = 1$.

When F is satisfiable, the payoff of player 1 is maximised when their strategy σ_1 assigns positive probability only to satisfying assignments. In such a case, the best response of player 2 is to play 1. Therefore, $\sigma = (\sigma_1, \sigma_2)$ is a mixed Nash equilibria iff (i) σ_1 is any distribution on the set $\{\alpha \in \{0,1\}^n \mid F(\alpha) = 1\}$ and (ii) $\sigma_2(1) = 1$ and $\sigma_2(0) = 0$.

Therefore, if σ is a mixed Nash equilibrium of Γ the strategy for player 2 is $\sigma_2(1) = 1$ and $\sigma_2(0) = 0$ if and only if F is satisfiable. As before an implicit representation of $\Gamma(F)$ can be obtained in polynomial time, and the result follows. □

We also analyse the complexity of the following problems:

Best Response (BR). Given a strategic game Γ, a player i, and strategy profile $a = (a_{-i}, a_i)$ decide whether $a_i \in B_i(a_{-i})$.

Mixed Nash Support (MNS). Given a strategic game Γ, a player i and $a_i \in A_i$ decide whether there is a mixed Nash equilibrium σ such that $\sigma_i(a_i) > 0$.

Theorem 8. *The* BR *problem is* coNP-*complete and the* MNS *problem is* NP-*hard.*

References

1. C. Àlvarez, J. Gabarró, and M. Serna. Pure Nash equilibria in games with a large number of actions. Technical Report 31, Electronic Colloquium on Computational Complexity, 2005.
2. J. L. Balcazar, J. Díaz, and J. Gabarró. *Structural Complexity II*. Springer-Verlag, 1990.
3. J. L. Balcazar, J. Díaz, and J. Gabarró. *Structural Complexity I*. Springer-Verlag, 2nd. edition, 1995.
4. E. Ben-Porath. The complexity of computing a best response automaton in repeated games with mixed strategies. *Games and Economic Behavior*, 2(1):1–12, 1990.
5. F. Chu and J. Halpern. On the NP-completeness of finding an optimal strategy in games with commons pay-offs. *International Journal of Game Theory*, 2001.
6. V. Conitzer and T. Sandholm. Complexity results about Nash equilibra. In *IJCAI 2003*, pages 765–771, 2003.
7. K. Daskalakis and C. Papadimitriou. The complexity of games on highly regular graphs. Technical report, available at http://www.cs.berkeley.edu/ christos/, 2005.
8. A. Fabrikant, C. Papadimitriou, and K. Talwar. The complexity of pure Nash equilibria. In *STOC 2004*, pages 604–612, 2004.
9. D. Fotakis, S. Kontogiannis, E. Koutsoupias, M. Mavronicolas, and P. Spirakis. The structure and complexity of Nash equilibria for a selfish routing game. In *ICALP 2002*, pages 123–134, 2002.
10. D. Fotakis, S. Kontogiannis, and P. Spirakis. Selfish unsplittable flows. In *ICALP 2004*, pages 593–605, 2004.
11. M. Gairing, T. Lücking, M. Mavronicolas, B. Monien, and M. Rode. Nash equilibria in discrete routing games with convex latency functions. In *ICALP 2004*, pages 645–657, 2004.
12. I. Gilboa and E. Zemel. Nash and correlated equilibria. *Games and Economic Behavior*, 1(1):80–93, 1989.
13. G. Gottlob, G. Greco, and F. Scarcello. Pure Nash equilibria: Hard and easy games. In *Theoretical Aspects of Rationality and Knowledge.*, pages 215–230, 2003.
14. D. Koller and M-Megiddo. The complexity of two-person zero sum games in extensive form. *Games and Economic Behavior*, 4(4):528–552, 1992.
15. J. Nash. Non-cooperative games. *Annals of Mathematics*, pages 286–295., 1951.
16. M.J. Osborne and A. Rubinstein. *A Course in Game Theory*. MIT Press, 1994.
17. C. Papadimitriou. On players with a bounded number of actions. *Games and Economic Behavior*, 4(1):122–131, 1992.
18. C. Papadimitriou. *Computational Complexity*. Addison-Wesley, 1994.
19. C. Papadimitriou. Algorithms, games and the internet. In *STOC 2001*, pages 4–8, 2001.
20. G.R. Schoenebeck and S. Vadham. The complexity of Nash equilibria in concisely represented games. Technical Report 52, Electronic Colloquium on Computational Complexity, 2005.
21. T. Thierauf. *The computational complexity of equivalence and isomorphisms problems*. Springer, 2000.

On the Complexity of Depth-2 Circuits with Threshold Gates

Kazuyuki Amano* and Akira Maruoka

GSIS, Tohoku University, Aoba 6-6-05, Aramaki, Sendai 980-8579, Japan
{ama, maruoka}@ecei.tohoku.ac.jp

Abstract. The paper investigates the complexity of depth-two circuits with threshold gates and consisting of two parts.

First, we develop a method for deriving a lower bound on the size of depth two circuits with a threshold gate at the top and a certain type of gates at the bottom. We apply the method for circuits with symmetric gates at the bottom that compute the "inner product mod 2", and obtain a lower bound of 1.3638^n. Although our lower bound is slightly weaker than the best known lower bound of $\Omega(2^{n/2}/n)$, which was recently proved by Forster et al. [5,6], our method has unique features: A lower bound is obtained by solving a certain linear program, and solving larger linear programs yield higher lower bounds. We also discuss the generalization of the proposed method.

Second, we develop a simplified simulation of a depth-one threshold circuit with unbounded weights by a depth-two threshold circuit with small weights. Precisely, we give an explicit construction of depth-two circuits with small weights consist of $\tilde{O}(n^5)$ gates that compute an arbitrary threshold function. We also give the construction of such circuits with $O(n^3/\log n)$ gates computing the COMPARISON and CARRY functions, and that with $O(n^4/\log n)$ gates computing the ADDITION function. These improve the previously known constructions on its size and simplicity.

1 Introduction

Threshold circuits of depth two are the current borderline for circuit lower bounds. We have strong lower bounds on the size of depth-two circuits consisting of threshold gates with polynomial weights [10], but we have no super-linear lower bounds for depth-two circuits with threshold gates of unbounded weights (see, e.g., [14,20] for surveys). This motivates us to investigate the computational power of various types of depth-two circuits containing threshold gates.

In the first half of the paper (Section 2), we mainly deal with depth-two circuits with a threshold gate of unbounded weights at the top and a certain type of gates at the bottom. Studying this type of circuits is interesting for several reasons. If we place PARITY gates at the bottom, the minimum size of a

---
* Supported in part by Grant-in-Aid for Scientific Research on Priority Areas "New Horizons in Computing" from MEXT of Japan.

circuit that computes a Boolean function f is equal to the minimum number of terms in a polynomial over $GF(2)$ that sign represents f. Such a representation is known as the *polynomial threshold representation*, and has been extensively studied (e.g., [13,15,17,18,21]). If we place threshold gates at the bottom, proving a good lower bound for an explicit function turns out to be quite hard. Hanjal et al. [10] introduced the "discriminator lemma" and proved that the size of a depth-two circuit with unbounded-weight threshold gates at the bottom and a polynomial-weight threshold gate at the top that computes the inner product mod 2 is at least $2^{(1/3-\epsilon)n}$. Recently, Forster et al. [5,6] proved that the size of a depth-two circuit with polynomial-weight threshold gates at the bottom and an unbounded-weight threshold gate at the top that computes the inner product mod 2 is at least $2^{(1/2-o(1))n}$. However, for unrestricted weights, there are no strong lower bounds on the size of a depth-two circuit for an explicit function.

Recently, Basu et al. [3] introduced a new method for deriving lower bounds on the size of depth-two threshold of ANDs circuits. Their method is based on the fact that the minimum size of a threshold of ANDs circuit that computes a function f equals the minimum number of terms in a polynomial over a certain basis that sign represents f. Interestingly, the method is quite simple but yields stronger lower bounds than previously known bounds for several functions. These include a lower bound of 1.5^n for the parity function, and a lower bound of 2^n for the inner product mod 2.

In this paper, inspired by their results, we develop a new method for deriving a lower bound on the size of a certain type of depth-two circuits. We apply our method for depth-two circuits in which the top gate is a threshold gate of unrestricted weights and the bottom level has symmetric gates. Such circuits have been previously considered by, for example, Krause and Pudlák [12], and Forster et al. [6]. We demonstrate that we can obtain a lower bound of 1.3638^n (*and possibly higher*) on the size of such a circuit that computes the inner product mod 2. Although our lower bound is slightly weaker than the best known lower bound of $\Omega(2^{n/2}/n)$, which was recently obtained by using an algebraic method by Forster et al. [5,6], our method has unique features: A lower bound is obtained by solving a certain linear program, and solving larger linear programs yield higher lower bounds. These results together with the discussion of the generalization of the proposed method are described in Section 2.

In the second half of the paper (Section 3), we develop a new and simplified simulation of a depth one threshold circuit with unbounded weights (i.e., a linear threshold function with possibly exponential weights) by a depth two threshold circuit with polynomial weights.

Goldmann, Håstad and Razborov showed in [8] that any linear threshold function can be computed by a depth two threshold circuit of polynomial size and polynomial weights. Goldmann and Karpinski [9] gave an explicit construction of such a circuit. The construction was then simplified by Hofmeister [11]. Unfortunately, the size of the constructed circuit is still quite large. (It seems that their circuit consists of $\tilde{O}(n^4)$ subcircuits each having $O(n^2 p^2)$ gates where p is the $\tilde{O}(n^3)$-th prime number.) In this paper, we further simplify the construction of such a circuit. Precisely, we give an explicit construction of depth two threshold circuit with polynomial weights and $\tilde{O}(n^5)$ gates that computes an arbitrary linear threshold function. Here we use the "$\tilde{O}$" (soft O) notation, which ignores the polylogarithmic factors.

In this paper, we also give explicit constructions of depth two threshold circuits with polynomial weights that compute the "comparison" and "addition" functions. The comparison function is the Boolean function of two n-bit integers X and Y whose output is 1 iff $X > Y$. Note that the comparison function can be computed by a single threshold gate with exponential weights, but not by a gate with polynomial weights. The addition function outputs all the bits of the sum of two n-bit numbers.

Siu and Bruck [22] showed that both functions can be computed by a depth two threshold circuit with polynomial size and polynomial weights. Alon and Bruck [1] presented the constructions of such circuits. In fact, they constructed depth two circuits with a threshold gate at the top, and parity gates at the bottom. The size of their circuit for the comparison is $O(n^4)$, and that for the addition is $O(n^5)$. Since a parity gate can be replaced by $O(n)$ threshold gates with unit weights, their construction yields a depth two threshold circuit of size $O(n^5)$ for the comparison, and that of size $O(n^6)$ for the addition. Subsequently, Bohossian et al. [4] presented a construction of depth two threshold circuit with $\tilde{O}(n^4)$ gates for the comparison.

In this paper, we further improve these constructions on its size and simplicity. The size of our circuit for the comparison is $O(n^3/\log n)$ and that for the addition is $O(n^4/\log n)$.

2 Getting Lower Bounds on Circuit Size by LP

Let $X = (x_1, \ldots, x_n)$ and $Y = (y_1, \ldots, y_n)$ be two binary vectors of length n. The *inner product mod 2* function, denoted by $\mathsf{IP}_n(X, Y)$, is defined to be $\oplus_i x_i y_i$ where $\oplus$ denotes the exclusive-OR operation. In what follows, we consider the size of depth-two circuits with a threshold gate at the top and symmetric gates below that compute IP_n.

Definition 1. *A linear threshold function $f(X)$ is a Boolean function with input $X = (x_1, \ldots, x_n) \in \{0,1\}^n$ such that*

$$f(X) = sgn[F(X)] = \begin{cases} 1, & \text{if } F(X) \geq 0; \\ 0, & \text{otherwise,} \end{cases}$$

where $F(X) = w_0 + \sum_{i=1}^{n} w_i x_i$. The coefficients w_i are called the weights of the threshold function. A gate that computes a threshold function is called a threshold gate. A function $f : \{0,1\}^X \to \mathbb{R}$ is called symmetric if the value of f depends only on the number of inputs that are 1. A gate that computes a symmetric function is called a symmetric gate.

Note that a symmetric gate is usually defined as a binary gate, i.e., it computes a symmetric function $f : \{0,1\}^X \to \{0,1\}$. The reason why we extend the domain from $\{0,1\}$ to reals is to make the set of symmetric functions closed under linear combinations, which we state below as Fact 1.

Fact 1. Any linear combination of two (or more) symmetric functions over the same set of variables is also a symmetric function.

Let TH∘SYM denote a circuit of depth two where threshold gate is the top and symmetric gates are the bottom. For a Boolean function f, the minimum number of gates in a TH∘SYM circuit that computes f is denoted by $s(f)$.

In what follows, we consider a polynomial P of the form

$$P(X,Y) = \sum_{S \subseteq X \cup Y} w_S h_S(X,Y), \tag{1}$$

where w_S is a real number and h_S denotes a symmetric function over the set of variables S. The support of a polynomial P, denoted by supp(P), is defined to be supp$(P) = \{S \subseteq X \cup Y \mid w_S \neq 0\}$. We denote the size of the support of P by $\sharp(P)$, i.e., $\sharp(P) = |\text{supp}(P)|$. We say that a polynomial P *sign represents* a Boolean function f if $P(X) > 0$ whenever $f(X) = 1$ and $P(X) < 0$ whenever $f(X) = 0$. (Without loss of generality, we can assume that $P(X) \neq 0$ for every input X.) Note that $s(f)$ is equal to the minimum size of the support of a polynomial of the form (1) that sign represents f.

2.1 Basic Step

Let f be a (not necessarily Boolean) function on a set of variables X and ρ be a partial assignment of the variables, i.e., ρ is a map from X to the set $\{0, 1, *\}$. For a partial assignment ρ on X, $res(\rho)$ denotes the set of variables that mapped to 0 or 1 by ρ, i.e., $res(\rho) = \{v \in X \mid \rho(v) \neq *\}$. The restriction of f by ρ, denoted by $f|_\rho$, is the function obtained by setting x_i to be $\rho(x_i)$ if $x_i \in res(\rho)$ and leaving x_i as a variable otherwise.

We also define the restriction of a polynomial P of the form (1) by ρ, denoted by $P|_\rho$ as follows: First, replace each h_S in P by $h_S|_\rho$. Note that $h_S|_\rho$ is a symmetric function on $S - res(\rho)$. Then, for every S_1 and S_2 such that $h_{S_1}|_\rho$ and $h_{S_2}|_\rho$ are on the same set of variables S', then replace $w_{S_1} h_{S_1}|_\rho + w_{S_2} h_{S_2}|_\rho$ by an equivalent symmetric function $h'_{S'}$. Fact 1 guarantees that such a replacement is always possible.

Consider a polynomial P that sign represents IP_n and two partial assignments ρ_1 and ρ_2 such that $res(\rho_1) = res(\rho_2) = \{x_1, y_1\}$, $\rho_1(x_1) = 0$, $\rho_1(y_1) = 1$, and $\rho_2(x_1) = \rho_2(y_1) = 1$. It is obvious that $P|_{\rho_1}$ sign represents IP_{n-1}, and $P|_{\rho_2}$ sign represents the *complement* of IP_{n-1}. This implies that the polynomial $P|_{\rho_1} - P|_{\rho_2}$ sign represents IP_{n-1}. The key observation is that if the polynomial $P|_{\rho_1} - P|_{\rho_2}$ has fewer terms than P, then we can obtain a recursive formula on the minimum size of the support of a polynomial for IP_n.

Now we describe the method in detail. Given a polynomial P that sign represents IP_n, we decompose P into P_Ts for each $T \subseteq \{x_1, x_2, y_1, y_2\}$ as

$$P_T(X,Y) = \sum_{\substack{S \in \text{supp}(P) \\ S \cap \{x_1,x_2,y_1,y_2\}=T}} w_S h_S(X,Y).$$

The following fact is easy to prove.

Fact 2. Let ρ_1 and ρ_2 be two restrictions such that $res(\rho_1) = res(\rho_2)$. Then,

(i) $\sharp(P_T|_{\rho_1} - P_T|_{\rho_2}) \leq \sharp(P_T)$ for every T,
(ii) if $\sum_{v \in T \cap res(\rho_1)} \rho_1(v) = \sum_{v \in T \cap res(\rho_2)} \rho_2(v)$, then $P_T|_{\rho_1} - P_T|_{\rho_2} = 0$.

Proof. First, we show the statement (i). Consider a term $w_S h_S$ that appears in P_T. Then a corresponding formula $w_S h_S|_{\rho_1} - w_S h_S|_{\rho_2}$ is appearing in $P_T|_{\rho_1} - P_T|_{\rho_2}$. Since both of $h_S|_{\rho_1}$ and $h_S|_{\rho_2}$ are symmetric functions on the set $S - res(\rho_1)$ of variables, we can simplify this formula to a single term $w_S h'$ with a certain symmetric function h' on $S - res(\rho_1)$ by Fact 1. This completes the proof of (i). The statement (ii) is now obvious since $h_S|_{\rho_1} \equiv h_S|_{\rho_2}$ if $\sum_{v \in T \cap res(\rho_1)} \rho_1(v) = \sum_{v \in T \cap res(\rho_2)} \rho_2(v)$. □

We consider the two types of a pair (ρ_1, ρ_2) of restrictions:

Type 1. Choose $i \in \{1,2\}$ and $v \in \{x_i, y_i\}$. The unchosen variable in $\{x_i, y_i\}$ is denoted by u. Let $res(\rho_1) = res(\rho_2) = \{x_i, y_i\}$, $(\rho_1(v), \rho_1(u)) = (0,1)$ and $(\rho_2(v), \rho_2(u)) = (1,1)$.

Type 2. Choose $v_1 \in \{x_1, y_1\}$ and $v_2 \in \{x_2, y_2\}$. Let u_1 and u_2 be the unchosen variables in $\{x_1, y_1\}$ and in $\{x_2, y_2\}$, respectively. Let $res(\rho_1) = res(\rho_2) = \{x_1, y_1, x_2, y_2\}$, $(\rho_1(v_1), \rho_1(u_1), \rho_1(v_2), \rho_1(u_2)) = (0,1,1,0)$ and $(\rho_2(v_1), \rho_2(u_1), \rho_2(v_2), \rho_2(u_2)) = (1,1,0,0)$.

Note that we can obtain 4 pairs of restrictions of Type 1 and 4 pairs of Type 2. For an arbitrary pair (ρ_1, ρ_2) of restrictions of Type 1, we have $\mathsf{IP}|_{\rho_1} \equiv \mathsf{IP}_{n-1}$ and $\mathsf{IP}|_{\rho_2} \equiv \overline{\mathsf{IP}_{n-1}}$. This implies that $P|_{\rho_1} - P|_{\rho_2}$ sign represents IP_{n-1}. Similarly, for an arbitrary pair (ρ_1, ρ_2) of restrictions of Type 2, we have $\mathsf{IP}|_{\rho_1} \equiv \mathsf{IP}_{n-2}$ and $\mathsf{IP}|_{\rho_2} \equiv \overline{\mathsf{IP}_{n-2}}$. This implies that $P|_{\rho_1} - P|_{\rho_2}$ sign represents IP_{n-2}.

By combining the above argument with Fact 2, we obtain the following.

Fact 3. For every polynomial P that sign represents IP_n, the following are true: For each $v \in \{x_1, y_1, x_2, y_2\}$, $\sum_{T: v \in T} \sharp(P_T) \geq s(\mathsf{IP}_{n-1})$. For each $v_1 \in \{x_1, y_1\}$ and $v_2 \in \{x_2, y_2\}$, $\sum_{T: |\{v_1, v_2\} \cap T| = 1} \sharp(P_T) \geq s(\mathsf{IP}_{n-2})$.

If P is an optimal polynomial for IP_n, then $s(\mathsf{IP}_n) = \sum_T \sharp(P_T)$. It is obvious that $s(\mathsf{IP}_{n-1}) \geq s(\mathsf{IP}_{n-2})$. Putting them altogether, we have:

Fact 4. Let z be the minimum value of the objective function of the following linear program. Then $s(\mathsf{IP}_n) \geq z \cdot s(\mathsf{IP}_{n-2})$.

$$\text{Minimize} \quad \sum_{T \subseteq \{x_1, x_2, y_1, y_2\}} q_T$$

$$\text{subject to} \quad \sum_{T: v \in T} q_T \geq 1 \ (v \in \{x_1, x_2, y_1, y_2\})$$

$$\sum_{T:|\{v_1,v_2\}\cap T|=1} q_T \geq 1 \ (v_1 \in \{x_1,y_1\}, v_2 \in \{x_2,y_2\}),$$
$$q_T \geq 0 \ (T \subseteq \{x_1,x_2,y_1,y_2\}). \tag{2}$$

LP (2) has $2^4 = 16$ variables and $4+4=8$ constraints, and is easy to solve. The minimum value of the objective function is 1.5, and this implies $s(\mathsf{IP}_n) \geq 1.5^{n/2} \sim 1.2247^n$.

2.2 Incremental Step

It seems natural to expect that we can obtain better lower bounds if we consider a larger collection of restrictions. In this subsection, we show that the incremental use of LP methods yields higher lower bounds.

Let $k \geq 3$ be an integer whose value will be chosen later. We now consider a set of restrictions ρ such that $res(\rho) \subseteq \{x_1,y_1,\ldots,x_k,y_k\}$. More specifically, we consider the following two types of pairs of restrictions:

Type 1. Choose $i \in \{1,\ldots,k\}$ and $v \in \{x_i,y_i\}$. The unchosen variable in $\{x_i,y_i\}$ is denoted by u. Let $res(\rho_1) = res(\rho_2) = \{x_i,y_i\}$, $(\rho_1(v),\rho_1(u)) = (0,1)$ and $(\rho_2(v),\rho_2(u)) = (1,1)$.

Type 2. Choose $i,j \in \{1,\ldots,k\}$ such that $i \neq j$. Choose $v_1 \in \{x_i,y_i\}$ and $v_2 \in \{x_j,y_j\}$. Let u_1 and u_2 be the unchosen variables in $\{x_i,y_i\}$ and in $\{x_j,y_j\}$, respectively. Let $res(\rho_1) = res(\rho_2) = \{x_i,y_i,x_j,y_j\}$, $(\rho_1(v_1),\rho_1(u_1),\rho_1(v_2),\rho_1(u_2)) = (0,1,1,0)$ and $(\rho_2(v_1),\rho_2(u_1),\rho_2(v_2),\rho_2(u_2)) = (1,1,0,0)$.

Let P be a polynomial that sign represents IP_n. It is obvious that $P|_{\rho_1} - P|_{\rho_2}$ sign represents IP_{n-i} for a pair (ρ_1,ρ_2) of restrictions of Type $i \in \{1,2\}$. Thus, by arguments analogous to those in the last subsection, we can obtain the following.

Fact 5. Suppose that $k \geq 3$. Let z_{k-1} and z_{k-2} be real numbers such that $s(\mathsf{IP}_n) \geq z_{k-1} \cdot s(\mathsf{IP}_{n-(k-1)})$ and $s(\mathsf{IP}_n) \geq z_{k-2} \cdot s(\mathsf{IP}_{n-(k-2)})$ for every n. Let z_k be the minimum value of the objective function of the following linear program. Then $s(\mathsf{IP}_n) \geq z_k \cdot s(\mathsf{IP}_{n-k})$.

$$\text{Minimize} \sum_{T \subseteq \{x_1,y_1,\ldots,x_k,y_k\}} q_T$$

subject to
$$\sum_{T:v \in T} q_T \geq z_{k-1} \ (v \in \{x_1,y_1,\ldots,x_k,y_k\})$$
$$\sum_{T:|\{v_1,v_2\}\cap T|=1} q_T \geq z_{k-2} \ \begin{pmatrix} i,j \in \{1,\ldots,k\}, i \neq j \\ v_1 \in \{x_i,y_i\}, v_2 \in \{x_j,y_j\} \end{pmatrix}$$
$$q_T \geq 0 \quad (T \subseteq \{x_1,y_1,\ldots,x_k,y_k\}). \tag{3}$$

Note that the constraint matrix of LP (3) is a $(2k+4\binom{k}{2}) \times 2^{2k}$ binary matrix and is almost balanced, i.e., there is about the same number of 1s and 0s. This matrix is easy to generate by a simple computer program. In addition, if the value of k is relatively small, then we can solve this LP by a LP solver program.

Solving LP (3) for $k = 3$ with $z_1 = 1$ and $z_2 = 1.5$ yields $z_3 = 2$. This implies $s(\mathsf{IP}_n) \geq 2^{n/3} \sim 1.2599^n$, which is slightly better than the lower bound obtained by solving LP (2). Solving LP (3) again for $k = 4$ with $z_2 = 1.5$ and $z_3 = 2$ yields $z_4 \sim 2.8333$, which implies better lower bound of $s(\mathsf{IP}_n) \geq 2.8333^{n/4} \sim 1.2974^n$. By repeating this procedure, we can obtain $z_5 \sim 4.0277$, $z_6 \sim 5.7500$, $z_7 \sim 8.2541$ and $z_8 \sim 11.9700$. These imply the lower bounds on $s(\mathsf{IP}_n)$ of 1.3213^n, 1.3384^n, 1.3519^n and 1.3638^n, respectively. We have not succeeded to compute the value of z_k for $k \geq 9$ at the time of writing the paper.

The best possible lower bound obtained by applying Fact 5 may be $s(\mathsf{IP}_n) \geq z_\infty^n$ where $z_\infty = \lim_{k \to \infty} z_k^{1/k}$. So the problem of determining the value of z_∞ seems to be interesting. Note that $s(\mathsf{IP}_n) \leq 2^n$ since $\mathsf{IP}_n(X,Y) = \mathrm{sign}(\sum_{S \subseteq [n]} (-2)^{|S|+1} X_S Y_S)$, where X_S and Y_S denote $\prod_{i \in S} x_i$ and $\prod_{i \in S} y_i$, respectively.

2.3 Discussions

In this subsection, we generalize the arguments in the last subsection in order to analyze the potential of the proposed method.

Let f be a collection of Boolean functions $f = \{f^n : \{0,1\}^n \to \{0,1\}\}_n$. Let $H = \{h : \{0,1\}^n \to \{0,1\}\}$ be a certain set of functions and let $s_H(f^n)$ be the size of a smallest subset $K \subseteq H$ such that f^n can be represented as the sign of a weighted sum of functions in K. Our objective is to lower bound the value of $s_H(f^n)$. Let $P(X) = \sum_{h \in K} w_h h(X)$ be an optimal polynomial that sign represents f^n, i.e., $s_H(f^n) = |K|$. Let k be an arbitrary but fixed integer. Let S be a set consisting of restrictions ρ with $res(\rho) \subseteq \{x_1, \ldots, x_k\}$ and pairs of restrictions (ρ_1, ρ_2) with $res(\rho_1), res(\rho_2) \subseteq \{x_1, \ldots, x_k\}$. For a restriction ρ or a pair of restrictions (ρ_1, ρ_2) of S, define

$$\begin{aligned}
V_\rho &= \{h \in H : h|_\rho \text{ is a constant.}\}, &&\text{if } f|_\rho \equiv f^{n-k} \text{ or } f|_\rho \equiv \overline{f^{n-k}}, \\
V_{(\rho_1,\rho_2)} &= \{h \in H : h|_{\rho_1} \equiv \overline{h|_{\rho_2}}\}, &&\text{if } f|_{\rho_1} \equiv \overline{f|_{\rho_2}} \equiv f^{n-k} \text{ or } \overline{f^{n-k}}, \\
V_{(\rho_1,\rho_2)} &= \{h \in H : h|_{\rho_1} \equiv h|_{\rho_2}\}, &&\text{if } f|_{\rho_1} \equiv f|_{\rho_2} \equiv f^{n-k} \text{ or } \overline{f^{n-k}}, \\
V_\rho, V_{(\rho_1,\rho_2)} &= \emptyset, &&\text{otherwise.}
\end{aligned}$$

Let z_k denotes the optimal value of the objective function of the program:

$$\begin{aligned}
\text{Minimize} \quad & \sum_{h \in H} q_h \\
\text{subject to} \quad & \sum_{h \in H} M_{\beta, h} q_h \geq 1, \ (\beta \in S), \\
& q_h \geq 0, \quad (h \in H),
\end{aligned} \qquad (4)$$

where $M_{\beta,h} = 0$ if $h \in V_\beta$ and $M_{\beta,h} = 1$ if $h \notin V_\beta$.

By the arguments analogous to those in the last subsection, we have $s_H(f^n) \geq z_k \cdot s_H(f^{n-k})$, which implies $s_H(f^n) \geq z_k^{n/k}$. Introducing variables r_β for the constraint corresponding to $\beta \in S$, we get the dual program of (4):

$$\text{Maximize} \sum_\beta r_\beta$$
$$\text{subject to} \sum_\beta M_{\beta,h} r_\beta \leq 1, \ (h \in H), \tag{5}$$
$$r_\beta \geq 0, \quad (\beta \in S).$$

The *LP-duality theorem* says that the optimal values of objective functions of LPs (4) and (5) are identical. We define

$$\alpha = \min_{\mathcal{D}_S} \max_{h \in H} \Pr_{\gamma \sim \mathcal{D}_S}[h \notin V_\gamma],$$

where $\mathcal{D}_S$ denotes a distribution on S. Let $\mathcal{D}_S^\star$ be the distribution that attains the minimum value of α. Then $r_\beta = \alpha^{-1} \mathcal{D}_S^\star(\beta)$, for each β, is a feasible solution of LP (5). This guarantees that the optimal value of the objective function of LP (5) is at least α^{-1}, which implies $s_H(f^n) \geq (\alpha^{-1})^{n/k}$.

Intuitively, the above argument says that if the value of α is shown to be strictly smaller than 1 for a base set H and for reasonable k, then our technique can yield a good lower bound on the size of depth-two circuits where a threshold gate with unbounded weights is the top and gates that can compute a function in H are the bottom. Examples of such base sets are symmetric functions and threshold functions with very small weights. It seems that we need more ideas in order to obtain a good lower bound for depth-two threshold circuits with threshold gates of unbounded weights in this line of work.

3 Simulation of Exponential Weights by Polynomial Weights

In this section, we describe a simplified simulation of a depth one threshold circuit with unbounded weights by a depth two threshold circuit with polynomial weights[1].

3.1 Construction for General Threshold Functions

For two integers $a \leq b$, $[a,b]$ denotes the set of integers $\{a, a+1, \ldots, b\}$. The set $[1,n]$ is simply denoted by $[n]$. Given a linear combination $F(X) = w_0 + \sum_{i \in [n]} w_i x_i$ with $w_i \in \mathbb{Z}$ and $|w_i| \leq 2^{O(n \log n)}$. In the following, we describe the construction of depth two threshold circuit with small weights that computes the sign of $F(X)$. It is well known that the weights of a threshold function can be restricted to integers with absolute values less than $2^{O(n \log n)}$ without changing the set of realizable functions [16] (or see [19, Theorem 3.3.9]).

Let L be the minimum integer such that $|w_i| < 2^L$ for every i. Note that $L = O(n \log n)$. Define $F^{(0)}(X) = F(X)$. For $l \in [L]$, define a linear combinations $F^{(l)}$ and $E^{(l)}$ as follows:

---
[1] In this section, many proofs are omitted due to the space limitation. A technical report [2] describes the results in this section and includes all the proofs.

$$w_i^{(l)} = \begin{cases} \lfloor w_i/2^l \rfloor, & \text{if } w_i \geq 0, \\ \lceil w_i/2^l \rceil, & \text{otherwise,} \end{cases}$$

$$F^{(l)}(X) = w_0^{(l)} + \sum_{i \in [n]} w_i^{(l)} x_i,$$

$$E^{(l)}(X) = F^{(l-1)}(X) - 2F^{(l)}(X).$$

Note that if we represent the weight w_i by a binary sequence $s_i, w_{i,1}, \ldots, w_{i,L}$ such that $w_i = (-1)^{s_i} \sum_{j \in [L]} w_{i,j} 2^{j-1}$, then $E^{(l)}$ can also be represented by

$$E^{(l)}(X) = (-1)^{s_0} w_{0,l} + \sum_{i \in [n]} (-1)^{s_i} w_{i,l} x_i. \tag{6}$$

Let $E_{max} = |\max_{l \in [L]} \max_{X \in \{0,1\}^n} E^{(l)}(X)|$. From Eq. (6), it is obvious that $E_{max} \leq n+1$. For simplicity of presentation, we assume that $|F(X)| \geq E_{max}+1$ for every input X. (The other case can be dealt with easily.) The construction due to Hofmeister [11] is based on the following equality:

$$F(X) \geq 0 \Leftrightarrow \bigvee_{l \in [L]} (F^{(l)}(X) \in [0, E_{max}] \land F^{(l-1)}(X) \notin [-E_{max}, E_{max}]).$$

The following lemma, which is the key to our construction, shows that the sign of $F(X)$ can be computed more efficiently.

Lemma 2. $F(X)$ *is positive iff* $F^{(l)}(X) \in [E_{max}+1, 3E_{max}]$ *for some* $l \in [0, L-1]$. □

The proof of the lemma is omitted and can be found in [2]. The rest of the construction is similar to that of Hofmeister [11]. The following lemma is a slight modification from the lemma used in their construction. This can easily be proved by using the *Chinese Remainder Theorem*.

Lemma 3. *[11, Lemma 2] Let $a \leq b$ be two non-negative integers. Let $b < p_1 < p_2 < \cdots$ be prime numbers and let s be the minimum integer which satisfies $p_1 \cdots p_s \geq 2 \cdot Z_{max} + 1$. Then for every $Z \in \mathbb{Z}$ with $|Z| \leq Z_{max}$, it holds that:*

1. $Z \in [a, b] \Rightarrow Z \bmod p_i \in [a, b]$ *for all* p_i,
2. $Z \notin [a, b] \Rightarrow Z \bmod p_i \in [a, b]$ *for less than* $s \cdot ((b-a)+1)$ *many* p_i. □

Let $p_1 < \ldots < p_r$ be r consecutive prime numbers. The value of r will be chosen later. We choose p_1 such that $3E_{max} < 4n < p_1$ in order to guarantee that no distinct integers in $[E_{max}+1, 3E_{max}]$ can be equivalent modulo p_i for every i. Let s be the smallest integer such that $p_1 \cdots p_s > (n+1)2^L$. Note that $s = O(n)$.

For $l \in [0, L-1]$ and $i \in [r]$, we define a linear combination $F_i^{(l)}$ as follows:

$$F_i^{(l)}(X) = (w_0^{(l)} \bmod p_i) + \sum_{j \in [n]} (w_j^{(l)} \bmod p_i) x_j.$$

Let $\text{TEST}_{l,i}(X)$ be a Boolean function that outputs 1 iff $F_i^{(l)}(X) \bmod p_i \in [E_{max}+1, 3E_{max}]$. By Lemmas 2 and 3, we have

$$F(X) \geq 0 \Rightarrow \sum_{l \in [0, L-1]} \sum_{i \in [r]} \text{TEST}_{l,i}(X) \geq r,$$

$$F(X) < 0 \Rightarrow \sum_{l \in [0, L-1]} \sum_{i \in [r]} \text{TEST}_{l,i}(X) \leq 2E_{max} \cdot L \cdot s.$$

If we choose r such that $r > 2E_{max} \cdot L \cdot s$, e.g., $r = O(E_{max} \cdot n^2 \log n)$ will suffice, then $F(X)$ is positive if and only if the sum of the values of $rL = O(E_{max} \cdot n^3 \log^2 n)$ test functions is at least r. Since $F_i^{(l)}(X) < (n+1)p_i$ for every input X, $\text{TEST}_{l,i}(X)$ can be represented as the sum of $O(n)$ linear threshold functions

$$\sum_{k \in [0,n]} (\text{``}F_i^{(l)}(X) \geq (E_{max}+1) + kp_i\text{''} + \text{``}F_i^{(l)}(X) \leq 3E_{max} + kp_i\text{''} - 1).$$

Here and hereafter, we use the notation of the form "$F(X) \geq a$" that denotes the Boolean function whose value is 1 if $F(X) \geq a$ holds and is 0 otherwise. Putting them all together, we can construct a depth two threshold circuit with at most $O(nrL) = O(E_{max} \cdot n^4 \log^2 n) = \tilde{O}(n^5)$ gates that computes $f(X) = \text{sgn}[F(X)]$. Remark that the total number of wires in the resulting circuit is $\tilde{O}(n^6)$ and the weight of each wire is at most $O(np_r) = O(E_{max} \cdot n^3 \log^2 n) = \tilde{O}(n^4)$. Here we use the prime number theorem, which says that $p_r = O(r \log r)$.

3.2 More Economical Construction for Simple Functions

For $X = (x_n, \ldots, x_1) \in \{0,1\}^n$, we consider X as the integer $\sum_{i \in [n]} 2^{i-1} x_i$. The CARRY function is a Boolean function with two n-bit inputs X and Y that outputs 1 iff $X + Y \geq 2^n$, or equivalently $\sum_{i \in [n]} (x_i + y_i) 2^{i-1} \geq 2^n$. The COMPARISON function is a Boolean function with two n-bit inputs X and Y that outputs 1 iff $X > Y$, or equivalently $\sum_{i \in [n]} (x_i - y_i) 2^{i-1} > 0$. Since $E_{max} = O(1)$ for both functions, the construction described in the previous section yields circuits with $\tilde{O}(n^4)$ gates. In the following, we show that the number of gates in circuits for these functions can be further reduced to $O(n^3 / \log n)$.

First, we describe a construction of a circuit for the CARRY function. Let $n < p_1 < \ldots < p_r$ be r consecutive prime numbers. The value of r will be chosen later. Let s be the smallest integer such that $p_1 \cdots p_s > 2^{n+1}$. Note that $s = O(n/\log n)$.

For $l \in [n]$ and $i \in [r]$, let $m_{l,i}$ be an integer satisfying

$$\sum_{j \in [l,n]} (2^{j-l} \bmod p_i) + 1 - (2^{n+1-l} \bmod p_i) = m_{l,i} p_i.$$

Such an integer always exists since $\sum_{j \in [l,n]} 2^{j-l} + 1 - 2^{n+1-l} = 0$. For $i \in [r]$ and $l \in [n]$, let $\text{CHK}_{l,i}(X, Y)$ be a Boolean function that outputs 1 iff

$$\sum_{j \in [l,n]} (2^{j-l} \bmod p_i)(x_j + y_j) - (2^{n+1-l} \bmod p_i) = m_{l,i} p_i.$$

Then the following are true (see [2] for the proofs).

$$\text{CARRY}(X,Y) = 1 \Rightarrow \sum_{l \in [n]} \sum_{i \in [r]} \text{CHK}_{l,i}(X,Y) \geq r,$$

$$\text{CARRY}(X,Y) = 0 \Rightarrow \sum_{l \in [n]} \sum_{i \in [r]} \text{CHK}_{l,i}(X,Y) \leq sn.$$

If we choose r such that $r > sn$, e.g, some $r = O(n^2/\log n)$ will suffice, then CARRY$(X,Y) = 1$ if and only if the sum of the values of $rn = O(n^3/\log n)$ test functions is at least r. Since a Boolean function of the form "$F(x) = y$" is equal to "$F(x) \geq y$" + "$F(x) \leq y$" $- 1$, we can construct a depth two threshold circuit of size $O(n^3/\log n)$ that computes CARRY. The total number of wires in the resulting circuit is $O(n^4/\log n)$ and the weight of each wire is at most $O(np_r) = O(n^3)$. The construction for the COMPARISON function is almost analogous and is omitted (see [2] for details).

Finally, we sketch the construction of circuit that computes the addition of two n-bit integers based on our circuit for the carry function. For $X = (x_n, \ldots, x_1)$ and $Y = (y_n, \ldots, y_1)$, ADDITION(X,Y) outputs $Z = (z_{n+1}, \ldots, z_1)$ such that $X + Y = Z$, or equivalently $\sum_{i \in [n]}(x_i + y_i)2^{i-1} = \sum_{i \in [n+1]} z_i 2^{i-1}$.

The k-th bit of the output of ADDITION is given by $z_k = x_k \oplus y_k \oplus c_k$ where c_k denotes the output of CARRY$(x_{k-1} \cdots x_1, y_{k-1} \cdots y_1)$. To compute z_k, we slightly modify the definition of our test functions for CARRY. For $t \in [0,2]$, $l \in [k-1]$ and $i \in [r]$, let CHK$_{l,i,t}(X,Y)$ be a Boolean function that outputs 1 iff

$$\sum_{j \in [l,k-1]} (2^{j-l} \bmod p_i)(x_j + y_j) - (2^{k-l} \bmod p_i) + 4kp_i(x_k + y_k) = m_{l,i}p_i + 4kp_i t,$$

where $m_{l,i}$ is an integer satisfying

$$\sum_{j \in [l,k-1]} (2^{j-l} \bmod p_i) + 1 - (2^{k-l} \bmod p_i) = m_{l,i}p_i.$$

Note that if $x_k + y_k \neq t$, then CHK$_{l,i,t}(X,Y) = 0$ for every l and i. It is easy to check that the k-th bit of the output of ADDITION is 1 iff

$$\sum_{t \in [0,2]} \sum_{l \in [k-1]} \sum_{i \in [r]} (-1)^t \text{CHK}_{l,i,t}(X,Y) + (r+sn)\text{“}x_k + y_k = 1\text{”} \geq r.$$

Hence, each bit of the output of ADDITION can be computed by a depth two threshold circuit with polynomial weights and $O(n^3/\log n)$ gates. Thus, the total number of gates in our circuit for ADDITION is $O(n^4/\log n)$.

References

1. N. Alon, J. Bruck, *Explicit Constructions of Depth-2 Majority Circuits for Comparison and Addition*, SIAM J. Disc. Math. **7(1)** (1994) 1–8
2. K. Amano, A. Maruoka, *Better Simulation of Exponential Weights by Polynomial Weights*, Technical Report ECCC TR04-090 (2004)

3. S. Basu, N. Bhatnagar, P. Gopalan and R.J. Lipton, *Polynomials that Sign Represent Parity and Descartes Rule of Signs*, Proc. 19th CCC (2004) 223–235
4. V. Bohossian, M.D. Riedel and J. Bruck, *Trading Weight Size for Circuit Depth: An LT2 Circuit for Comparison*, Tech. Report of PARADISE, ETR028 (1998) (Available at http://www.paradise.caltech.edu/~riedel/research/lt2comp.html)
5. J. Forster, *A Linear Lower Bound on the Unbounded Error Probabilistic Communication Complexity*, J. Comput. Syst. Sci. **65** (2002) 612–625
6. J. Forster, M. Krause, S.V. Lokam, R. Mubarakzjanov, N. Schmitt and H.U. Simon, *Relations Between Communication Complexity, Linear Arrangements, and Computational Complexity*, Proc. 21st FSTTCS, LNCS **2245** (2001) 171–182
7. M. Goldmann, *On the Power of a Threshold Gate at the Top*, Info. Proc. Let., **63** (1997) 287–293
8. M. Goldmann, J. Håstad, A.A. Razborov, *Majority Gates vs. General Weighted Threshold Gates*, Computational Complexity **2** (1992) 277–300
9. M. Goldmann, M. Karpinski, *Simulating Threshold Circuits by Majority Circuits*, SIAM J. Comput. **27(1)** (1998) 230–246
10. A. Hajnal, W. Maass, P. Pudlák, M. Szegedy and G. Turán, *Threshold Circuits of Bounded Depth*, J. Comput. Syst. Sci., **46(2)** (1993) 129–154
11. T. Hofmeister, *A Note on the Simulation of Exponential Threshold Weights*, Proc. 2nd COCOON, LNCS **1090** (1996) 136–141
12. M. Krause and P. Pudlák, *On the Computational Power of Depth-2 Circuits with Threshold and Modulo Gates*, Proc. 25th STOC (1993) 48—57.
13. M. Krause and P. Pudlák, *Computing Boolean Functions by Polynomials and Threshold Circuits*, Comput. Complexity, **7(4)** (1998) 346–370
14. M. Krause and I. Wegener, *Circuit Complexity*, in "Boolean Functions Vol.II", Eds. Y. Crama and P. Hammer (2004)
15. M. Minsky and S. Papert, *Perceptrons (Expanded Ed.)*, MIT Press (1988)
16. S. Muroga, Threshold Logic and its Applications, John Wiley, New York (1971)
17. R. O'Donnell and R. Servedio, *New Degree Bounds for Polynomial Threshold Functions*, Proc. 35th STOC (2003) 325–334.
18. R. O'Donnell and R. Servedio, *Extremal Properties of Polynomial Threshold Functions*, Proc. 18th CCC (2003) 3–12
19. I. Parberry, Circuit Complexity and Neural Networks, The MIT Press, London, England (1994)
20. A.A. Razborov, *On Small Depth Threshold Circuits*, Proc. 3rd SWAT, LNCS **621** (1992) 42–52
21. M. Saks, *Slicing the Hypercube*, Cambridge University Press (1993)
22. K.I. Siu and J. Bruck, *On the Power of Threshold Circuits with Small Weights*, SIAM J. Disc. Math. **4(3)** (1991) 423–435

Isomorphic Implication*

Michael Bauland[1] and Edith Hemaspaandra[2]

[1] Theoretische Informatik, Universität Hannover,
Appelstr. 4, D-30167 Hannover, Germany**
bauland@thi.uni-hannover.de
[2] Department of Computer Science,
Rochester Institute of Technology,
Rochester, NY 14623, USA***
eh@cs.rit.edu

Abstract. We study the isomorphic implication problem for Boolean constraints. We show that this is a natural analog of the subgraph isomorphism problem. We prove that, depending on the set of constraints, this problem is in P, NP-complete, or NP-hard, coNP-hard, and in $P_{||}^{NP}$. We show how to extend the NP-hardness and coNP-hardness to $P_{||}^{NP}$-hardness for some cases, and conjecture that this can be done in all cases.

1 Introduction

One of the most interesting and well-studied problems in complexity theory is the graph isomorphism problem (GI). This is the problem of determining whether two graphs are isomorphic, i.e., whether there exists a renaming of vertices such that the graphs become equal. This is a fascinating problem, since it is the most natural example of a problem that is in NP, not known to be in P, and unlikely to be NP-complete (see [KST93]).

The obvious analog of graph isomorphism for Boolean formulas is the formula isomorphism problem. This is the problem of determining whether two formulas are isomorphic, i.e., whether we can rename the variables such that the formulas become equivalent. This problem has the same behavior as the graph isomorphism problem one level higher in the polynomial hierarchy: The formula isomorphism problem is in Σ_2^p, coNP-hard, and unlikely to be Σ_2^p-complete [AT00].

Note that graph isomorphism can be viewed as a special case of Boolean isomorphism, since graph isomorphism corresponds to Boolean isomorphism for 2-positive-CNF formulas, in the following way: Every graph G (without isolated vertices) corresponds to the (unique) formula $\bigwedge_{\{i,j\} \in E(G)} x_i \vee x_j$. Then two graphs without isolated vertices are isomorphic if and only if their corresponding formulas are isomorphic.

---

* Supported in part by grants NSF-CCR-0311021 and DFG VO 630/5-1.
** Work done in part while visiting the Laboratory for Applied Computing at Rochester Institute of Technology.
*** Work done in part while on sabbatical at the University of Rochester.

One might wonder what happens when we look at other restrictions on the set of formulas. There are general frameworks for looking at all restrictions on Boolean formulas: The most often used is the Boolean constraint framework introduced by Schaefer [Sch78]. Basically (formal definitions can be found in the next section) we look at formulas as CNF formulas (or sets of clauses) where each clause is an application of a constraint (a k-ary Boolean function) to a list of variables. Each finite set of constraints gives rise to a new language, and so there are an infinite number of languages to consider. Schaefer studied the satisfiability problem for all finite sets of constraints. He showed that all of these satisfiability problems are either in P or NP-complete, and he gave a simple criterion to determine which of the cases holds.

The last decade has seen renewed interest in Schaefer's result, and has seen many dichotomy (and dichotomy-like) theorems for problems related to the satisfiability of Boolean constraints. For example, such results were obtained for the maximum satisfiability problem [Cre95], counting satisfying assignments [CH96], the inverse satisfiability problem [KS98], the unique satisfiability problem [Jub99], approximability problems [KSTW01], the minimal satisfying assignment problem [KK01], and the equivalence problem [BHRV02]. For an excellent survey of dichotomy theorems for Boolean constraint satisfaction problems, see [CKS01].

Most of the results listed above were proved using methods similar to the one used by Schaefer [Sch78]. A more recent approach to proving results of this form is with the help of algebraic tools [Jea98, JCG97, BKJ00]. This approach uses the clone (closed classes) structure of Boolean functions called Post's lattice, after Emil Post, who first identified these classes [Pos44]. A good introduction of how this can be used to obtain short proofs can be found in [BCRV04]. However, this approach does not work for isomorphism problems, because it uses existential quantification.

For the case of most interest for this paper, the Boolean isomorphism problem for constraints, Böhler et al. [BHRV02, BHRV04, BHRV03] have shown that this problem is in P, GI-complete, or GI-hard, coNP-hard, and in $P_{||}^{NP}$ (the class of problems solvable in polynomial time with one round of parallel queries to NP). As in Schaefer's theorem, simple properties of the set of constraints determine the complexity.

A problem closely related to the graph isomorphism problem is the subgraph isomorphism problem. This is the problem, given two graphs G and H, to determine whether G contains a subgraph isomorphic to H. In contrast to the graph isomorphism problem, the subgraph isomorphism problem can easily be seen to be NP-complete (it contains, for example, CLIQUE, HAMILTONIAN CYCLE, and HAMILTONIAN PATH).

To further study the relationship between the isomorphism problems for graphs and constraints, we would like to find a relation $\mathcal{R}$ on constraints that is to isomorphism for constraints as the subgraph isomorphism problem is to graph isomorphism.

Such a relation $\mathcal{R}$ should at least have the following properties:

1. A graph G is isomorphic to a graph H if and only if G contains a subgraph isomorphic to H and H contains a subgraph isomorphic to G. We want the

same property in the constraint case, i.e., for S and U sets of constraint applications, S is isomorphic to U if and only if $S\mathcal{R}U$ and $U\mathcal{R}S$.

2. The subgraph isomorphism problem should be a special case of the decision problem induced by $\mathcal{R}$, in the same way as the graph isomorphism problem is a special case of the constraint isomorphism problem. In particular, for G and H graphs, let $S(G)$ and $S(H)$ be their (standard) translations into sets of constraint applications of $\lambda xy.(x \vee y)$, i.e., $S(G) = \{x_i \vee x_j \mid \{i,j\} \in E(G)\}$ and $S(H) = \{x_i \vee x_j \mid \{i,j\} \in E(H)\}$. For G and H graphs without isolated vertices, G is isomorphic to H if and only if $S(G)$ is isomorphic to $S(H)$. We want G to have a subgraph isomorphic to H if and only if $S(G)\mathcal{R}S(H)$.

Borchert et al. [BRS98, p. 692] suggest using the subfunction relations $\gg_v$ and $\gg_{cv}$ as analogs of subgraph isomorphism. These relations are defined as follows. For two formulas ϕ and ψ, $\phi \gg_v \psi$ if and only if there exists a function π from variables to variables such that $\pi(\phi)$ is equivalent to ψ. $\phi \gg_{cv} \psi$ if and only if there exists a function π from variables to variables and constants such that $\pi(\phi)$ is equivalent to ψ [BR93]. Borchert and Ranjan [BR93] show that these relations satisfy our first desirable property, i.e., S is isomorphic to U if and only if $S \gg_v U$ and $U \gg_v S$, and that S is isomorphic to U if and only if $S \gg_{cv} U$ and $U \gg_{cv} S$. They also show that the problem of determining whether $\phi \gg_v \psi$ and the problem of determining whether $\phi \gg_{cv} \psi$, for unrestricted Boolean formulas, are Σ_2^p-complete.

But Borchert et al.'s subfunction relations will not give the second desirable property. Consider, for example, the graphs G and H such that $V(G) = V(H) = \{1,2,3\}$, $E(G) = \{\{1,2\},\{1,3\},\{2,3\}\}$, and $E(H) = \{\{1,2\},\{1,3\}\}$. Clearly, G contains a subgraph isomorphic to H, but $(x_1 \vee x_2) \wedge (x_1 \vee x_3) \wedge (x_2 \vee x_3) \not\gg_{cv} (x_1 \vee x_2) \wedge (x_1 \vee x_3)$.

How could the concept of a subgraph be translated to sets of constraint applications? As a first attempt at translating subgraph isomorphism to constraint isomorphism one might try the following: For sets of constraint applications S and U, does there exist a subset $\widehat{S}$ of S that is isomorphic to U. Certainly, such a definition satisfies the second desired property. But this definition does not satisfy the first desired property, since it is quite possible for sets of constraint applications to be equivalent without being equal.

We show that *isomorphic implication* satisfies both desired properties, and is a natural analog of the subgraph isomorphism problem. For S and U sets of constraint applications over variables X, we say that S isomorphically implies U (notation: $S \Rrightarrow U$) if and only if there exists a permutation π on X such that $\pi(S) \Rightarrow U$. In Section 4, we show that, depending on the set of constraints, the isomorphic implication problem is in P, NP-complete, or NP-hard, coNP-hard, and in $P_{||}^{NP}$. Our belief is that the isomorphic implication problem is $P_{||}^{NP}$-complete for all the cases where it is both NP-hard and coNP-hard. In Section 5, we prove this conjecture for some cases. *Because of space limitations, most of the proofs are omitted; please refer to the full version of this paper [BH04].*

2 Preliminaries

We will mostly use the constraint terminology from [CKS01].

Definition 1.

1. A constraint C *(of arity k) is a Boolean function from $\{0,1\}^k$ to $\{0,1\}$.*
2. *If C is a constraint of arity k, and $z_1, z_2, \ldots, z_k$ are (not necessarily distinct) variables, then $C(z_1, z_2, \ldots, z_k)$ is a* constraint application *of C.*
3. *If C is a constraint of arity k, and for $1 \leq i \leq k$, z_i is a variable or a constant (0 or 1), then $C(z_1, z_2, \ldots, z_k)$ is a* constraint application of C with constants.
4. *If S is a set of constraint applications [with constants] and X is a set of variables that includes all variables that occur in S, we say that S is a* set of constraint applications [with constants] over variables X.

Definition 2. *Let C be a k-ary constraint.*

- *C is* 0-valid *if $C(0, \ldots, 0) = 1$.*
- *C is* 1-valid *if $C(1, \ldots, 1) = 1$.*
- *C is* Horn *(or* weakly negative*) if $C(x_1, \ldots, x_k)$ is equivalent to a CNF formula where each clause has at most one positive literal.*
- *C is* anti-Horn *(or* weakly positive*) if $C(x_1, \ldots, x_k)$ is equivalent to a CNF formula where each clause has at most one negative literal.*
- *C is* bijunctive *if $C(x_1, \ldots, x_k)$ is equivalent to a 2CNF formula.*
- *C is* affine *if $C(x_1, \ldots, x_k)$ is equivalent to an XOR-CNF formula.*
- *C is* 2-affine *(or* affine of width 2*) if $C(x_1, \ldots, x_k)$ is equivalent to an XOR-CNF formula, such that every clause contains at most two literals.*
- *C is* complementive *(or* C-closed*) if for every $s \in \{0,1\}^k$, $C(s) = C(\overline{s})$, where $\overline{s} \in \{0,1\}^k =_{def} (1, \ldots, 1) - s$, i.e., $\overline{s}$ is obtained by flipping every bit of s.*

Let $\mathcal{C}$ be a finite set of constraints. We say $\mathcal{C}$ is 0-valid, 1-valid, Horn, anti-Horn, bijunctive, affine, 2-affine, or complementive, if *every* constraint $C \in \mathcal{C}$ has this respective property. We say that $\mathcal{C}$ is *Schaefer* if $\mathcal{C}$ is Horn, anti-Horn, affine, or bijunctive.

Definition 3 ([BHRV02]). *Let $\mathcal{C}$ be a finite set of constraints.*

1. *ISO($\mathcal{C}$) is the problem, given two sets S and U of constraint applications of $\mathcal{C}$ over variables X, to decide whether S is isomorphic to U (denoted by $S \cong U$), i.e., whether there exists a permutation π of X such that $\pi(S) \equiv U$; Here $\pi(S)$ is the set of constraint applications that results when we simultaneously replace every variable x in S by $\pi(x)$.*
2. *ISO$_c$($\mathcal{C}$) is the problem, given two sets S and U of constraint applications of $\mathcal{C}$ with constants, to decide whether S is isomorphic to U.*

Theorem 4 ([BHRV02, BHRV04, BHRV03]). *Let $\mathcal{C}$ be a finite set of constraints.*

1. If $\mathcal{C}$ is not Schaefer, then ISO($\mathcal{C}$) and ISO$_c$($\mathcal{C}$) are coNP-hard, GI-hard, and in $P_{||}^{NP}$.
2. If $\mathcal{C}$ is Schaefer and not 2-affine, then ISO($\mathcal{C}$) and ISO$_c$($\mathcal{C}$) are polynomial-time many-one equivalent to GI.
3. Otherwise, $\mathcal{C}$ is 2-affine and ISO($\mathcal{C}$) and ISO$_c$($\mathcal{C}$) are in P.

The isomorphic implication problem combines isomorphism with implication in the following way.

Definition 5. *Let $\mathcal{C}$ be a finite set of constraints.*

1. *ISO-IMP($\mathcal{C}$) is the problem, given two sets S and U of constraint applications of $\mathcal{C}$ over variables X, to decide whether S isomorphically implies U (denoted by $S \xRightarrow{\cong} U$), i.e., whether there exists a permutation π of X such that $\pi(S) \Rightarrow U$; Here $\pi(S)$ is the set of constraint applications that results when we simultaneously replace every variable x in S by $\pi(x)$.*
2. *ISO-IMP$_c$($\mathcal{C}$) is the problem, given two sets S and U of constraint applications of $\mathcal{C}$ with constants, deciding whether S isomorphically implies U.*

Definition 6.

1. *The graph isomorphism problem is the problem, given two graphs G and H, to decide whether G and H are isomorphic, i.e., whether there exists a bijection π from $V(G)$ to $V(H)$ such that $\pi(G) = H$. $\pi(G)$ is the graph such that $V(\pi(G)) = \{\pi(v) \mid v \in V(G)\}$ and $E(\pi(G)) = \{\{\pi(v), \pi(w)\} \mid \{v, w\} \in E(G)\}$.*
2. *The subgraph isomorphism problem is the problem, given two graphs G and H, to decide whether G contains a subgraph isomorphic to H, i.e., whether there exists a graph G' such that $V(G') \subseteq V(G)$ and $E(G') \subseteq E(G)$ and G' is isomorphic to H.*

Theorem 7 ([GJ79, Coo71]). *The subgraph isomorphism problem is NP-complete.*

Corollary 8. *The subgraph isomorphism problem for graphs without isolated vertices is NP-complete.*

3 Subgraph Isomorphism and Isomorphic Implication

The following lemma, whose proof can be found in the full version, shows that the isomorphic implication problem is a natural analog of the subgraph isomorphism problem, in the sense explained in the introduction.

Lemma 9.

1. *Let S and U be sets of constraint applications of $\mathcal{C}$ with constants. Then $S \cong U$ if and only if $S \xRightarrow{\cong} U$ and $U \xRightarrow{\cong} S$.*
2. *For graphs G and H without isolated vertices, G contains a subgraph isomorphic to H if and only if $S(G) \xRightarrow{\cong} S(H)$, where S is the "standard" translation from graphs to sets of constraint applications of $\lambda xy.x \vee y$, i.e., for $\widehat{G}$ a graph, $S(\widehat{G}) = \{x_i \vee x_j \mid \{i, j\} \in E(\widehat{G})\}$.*

4 Complexity of the Isomorphic Implication Problem

The following theorem gives a trichotomy-like theorem for the isomorphic implication problem.

Theorem 10. *Let $\mathcal{C}$ be a finite set of constraints.*

1. *If every constraint in $\mathcal{C}$ is equivalent to a constant or a conjunction of literals, then ISO-IMP$(\mathcal{C})$ and ISO-IMP$_c(\mathcal{C})$ are in P.*
2. *Otherwise, if $\mathcal{C}$ is Schaefer, then ISO-IMP$(\mathcal{C})$ and ISO-IMP$_c(\mathcal{C})$ are NP-complete.*
3. *If $\mathcal{C}$ is not Schaefer, then ISO-IMP$(\mathcal{C})$ and ISO-IMP$_c(\mathcal{C})$ are NP-hard, coNP-hard, and in $P_{||}^{NP}$.*

The proofs of the upper bounds and of the coNP lower bound are reasonably straightforward and can be found in the full version. It remains to show the NP lower bounds.

When proving dichotomy or dichotomy-like theorems for Boolean constraints, the proofs of some of the lower bounds are generally most involved. In addition, proving lower bounds for the case without constants is often a lot more involved than the proofs for the case with constants. This is particularly true in the case for isomorphism problems, since here, we cannot introduce auxiliary variables.

The approach taken in [BHRV02, BHRV04, BHRV03], which examine the complexity of the isomorphism problem for Boolean constraints, is to first prove lower bounds for the case with constants, and then to show that all the hardness reductions can be modified to obtain reductions for the cases without constants.

In contrast, in this paper we prove the lower bounds directly for the case without constants. We have chosen this approach since careful analysis of the cases shows that proving the NP lower bounds boils down to proving NP-hardness for ten different cases (far fewer than in the isomorphism paper).

It should be noted that our NP lower bound results do not at all follow from the lower bound results for the isomorphism problem. This is also made clear by comparing Theorems 4 and 10: In some cases, the complexity jumps from P to NP-complete, in other cases we jump from GI-hard to NP-complete.

Lemma 11. *Let C be a k-ary constraint such that $C(x_1, \ldots, x_k)$ is not equivalent to a conjunction of literals. Then there exists a set of constraint applications of C that is equivalent to one of the following ten constraint applications:*

- $t \wedge (x \vee y)$, $\overline{f} \wedge t \wedge (x \vee y)$, $\overline{f} \wedge (\overline{x} \vee \overline{y})$, $\overline{f} \wedge t \wedge (\overline{x} \vee \overline{y})$,
- $x \leftrightarrow y$, $t \wedge (x \leftrightarrow y)$, $\overline{f} \wedge (x \leftrightarrow y)$, $\overline{f} \wedge t \wedge (x \leftrightarrow y)$,
- $x \oplus y$, or $\overline{f} \wedge t \wedge (x \oplus y)$.

The proof of this lemma can be found in the full version.

To prove the NP lower bounds of Theorem 10 it suffices to show that the isomorphic implication problem is NP-hard for each of the ten constraints from Lemma 11.

As shown in Lemma 9(2), the NP-complete subgraph isomorphism problem is closely related to the isomorphic implication problem for sets of constraint applications of $\lambda xy.x \vee y$. We use this observation to prove NP-hardness for the constraints from Lemma 11 that are similar to $\lambda xy.x \vee y$, namely, we reduce the subgraph isomorphism problem to the isomorphic implication problems for $\lambda txy.t \wedge (x \vee y)$, $\lambda ftxy.\overline{f} \wedge t \wedge (x \vee y)$, $\lambda fxy.\overline{f} \wedge (\overline{x} \vee \overline{y})$, and $\lambda ftxy.\overline{f} \wedge t \wedge (\overline{x} \vee \overline{y})$. The actual reductions and the proofs of their correctness can be found in the full version of this paper.

The remaining six constraints from Lemma 11 behave differently. In these cases, the isomorphism problem is in P. Thus, GI does not reduce to these isomorphism problems (unless GI is in P), and there does not seem to be a simple reduction from the subgraph isomorphism problem to the isomorphic implication problem. In these cases, we prove NP-hardness by reduction from a suitable partitioning problem, namely, the unary version of the problem 3-Partition [GJ79, Problem SP15]. Both 3-Partition and the unary version of 3-Partition are NP-complete [GJ79]. The actual reductions and the proofs of their correctness can be found in the full version of this paper. This completes the proof of Theorem 10.

5 Toward a Trichotomy Theorem

The current main theorem (Theorem 10) is not a trichotomy theorem, since for $\mathcal{C}$ not Schaefer, it states that ISO-IMP($\mathcal{C}$) is NP-hard, coNP-hard, and in $P_{||}^{NP}$. The large gap between the lower and upper bounds is not very satisfying. We conjecture that the current lower bounds for ISO-IMP($\mathcal{C}$) for $\mathcal{C}$ not Schaefer can be raised to $P_{||}^{NP}$ lower bounds, which would give the following trichotomy theorem.

Conjecture 12. *Let $\mathcal{C}$ be a finite set of constraints.*

1. *If every constraint in $\mathcal{C}$ is equivalent to a constant or a conjunction of literals, then* ISO-IMP($\mathcal{C}$) *and* ISO-IMP$_c$($\mathcal{C}$) *are in* P.
2. *Otherwise, if $\mathcal{C}$ is Schaefer, then* ISO-IMP($\mathcal{C}$) *and* ISO-IMP$_c$($\mathcal{C}$) *are NP-complete.*
3. *If $\mathcal{C}$ is not Schaefer, then* ISO-IMP($\mathcal{C}$) *and* ISO-IMP$_c$($\mathcal{C}$) *are* $P_{||}^{NP}$-*complete.*

We believe this conjecture for two reasons. First of all, it is quite common for problems that are NP-hard, coNP-hard, and in $P_{||}^{NP}$ to end up being $P_{||}^{NP}$-complete. (For an overview of this phenomenon, see [HHR97].) Secondly, we will prove $P_{||}^{NP}$ lower bounds for some cases in Theorem 16.

To raise NP and coNP lower bounds to $P_{||}^{NP}$ lower bounds, the following theorem by Wagner often plays a crucial role, which it will also do in our case.

Theorem 13 ([Wag87]). *Let L be a language. If there exists a polynomial-time computable function h such that*

$$||\{i \mid \phi_i \in \text{SAT}\}|| \text{ is odd iff } h(\phi_1, \ldots, \phi_{2k}) \in L$$

for all $k \geq 1$ and all Boolean formulas $\phi_1, \ldots, \phi_{2k}$ such that $\phi_i \in$ SAT $\Rightarrow \phi_{i+1} \in$ SAT, then L is $P_{||}^{NP}$-hard.

Wagner's theorem can be used to prove the following lemma, which shows situations in which an NP lower bound and a coNP lower bound can be turned into a $\mathrm{P}_{||}^{\mathrm{NP}}$ lower bound.

Lemma 14. *Let L be a language. If L is NP-hard and coNP-hard, and (L has polynomial-time computable and- and ω-or-functions or L has polynomial-time computable or- and ω-and-functions), then L is $\mathrm{P}_{||}^{\mathrm{NP}}$-hard.*[1]

Agrawal and Thierauf [AT00] proved that the formula isomorphism problem has polynomial-time computable ω-and- and ω-or-functions. Since the formula isomorphism problem is trivially coNP-hard, we obtain the following corollary.

Corollary 15. *If the formula isomorphism problem is NP-hard, then it is $\mathrm{P}_{||}^{\mathrm{NP}}$-hard.*

Unfortunately, Agrawal and Thierauf's ω-or-function does not work for isomorphic implication. Their ω-and-function seems to work for isomorphic implication, but since this function or's two formulas together, it will not work for sets of constraint applications.

To prove the $\mathrm{P}_{||}^{\mathrm{NP}}$ lower bound of the following theorem, we need to come up with a completely new construction.

Theorem 16. *Let $\mathcal{D}$ be a set of constraints that is 0-valid, 1-valid, not complementive, and not Schaefer. Let $\mathcal{C} = \mathcal{D} \cup \{\lambda xy.x \vee y\}$. Then ISO-IMP($\mathcal{C}$) is $\mathrm{P}_{||}^{\mathrm{NP}}$-complete.*

Proof. By Theorem 10, ISO-IMP($\mathcal{C}$) is in $\mathrm{P}_{||}^{\mathrm{NP}}$. Thus it suffices to show that ISO-IMP($\mathcal{C}$) is $\mathrm{P}_{||}^{\mathrm{NP}}$-hard. Let $k \geq 1$ and let $\phi_1, \ldots, \phi_{2k}$ be formulas such that $\phi_i \in \mathrm{SAT} \Rightarrow \phi_{i+1} \in \mathrm{SAT}$. We will construct a polynomial-time computable function h such that

$$||\{i \mid \phi_i \in \mathrm{SAT}\}|| \text{ is odd iff } h(\phi_1, \ldots, \phi_{2k}) \in \mathrm{ISO\text{-}IMP}(\mathcal{C}).$$

By Theorem 13, this proves that ISO-IMP($\mathcal{C}$) is $\mathrm{P}_{||}^{\mathrm{NP}}$-hard.

Note that $||\{i \mid \phi_i \in \mathrm{SAT}\}||$ is odd if and only if there exists an i such that $1 \leq i \leq k$, $\phi_{2i-1} \notin \mathrm{SAT}$, and $\phi_{2i} \in \mathrm{SAT}$. This is a useful way of looking at it, and we will prove that there exists an i such that $1 \leq i \leq k$, $\phi_{2i-1} \notin \mathrm{SAT}$ and $\phi_{2i} \in \mathrm{SAT}$ if and only if $h(\phi_1, \ldots, \phi_{2k}) \in \mathrm{ISO\text{-}IMP}(\mathcal{C})$.

From Theorem 10 we know that ISO-IMP($\mathcal{C}$) is NP-hard and coNP-hard, and thus there exist (polynomial-time many-one) reductions from SAT to ISO-IMP($\mathcal{C}$) and from $\overline{\mathrm{SAT}}$ to ISO-IMP($\mathcal{C}$).

Let f be a polynomial-time computable function such that for all ϕ, $f(\phi)$ is a set of constraint applications of $\mathcal{D}$ and

$$\phi \in \overline{\mathrm{SAT}} \text{ iff } f(\phi) \overset{*}{\Rightarrow} \bigcup_{1 \leq j, \ell \leq n} \{x_j \to x_\ell\}.$$

---

[1] An or-function for a language L is a function f such that for all $x, y \in \Sigma^*$, $f(x, y) \in L$ iff $x \in L$ or $y \in L$. An ω-or-function for a language L is a function f such that for all $x_1, \ldots, x_n \in \Sigma^*$, $f(x_1, \ldots, x_n) \in L$ iff $x_i \in L$ for some i; and-functions are defined similarly [KST93].

Here $x_1, \ldots, x_n$ are exactly all variables in $f(\phi)$. Such a function exists, since $\overline{\text{SAT}}$ is reducible to $\overline{\text{CSP}_{\neq \mathbf{0},\mathbf{1}}(\mathcal{D})}$ ($\text{CSP}_{\neq \mathbf{0},\mathbf{1}}(\mathcal{D})$ is the problem of deciding whether a set of constraint applications of $\mathcal{D}$ has a satisfying assignment other than $\mathbf{0}$ and $\mathbf{1}$), which is reducible to ISO-IMP($\mathcal{D}$) via a reduction that satisfies the properties above. (See the proofs of the coNP lower bound from Theorem 10 and [BHRV02, Claims 19 and 14].)

Let g be a polynomial-time computable function such that for all ϕ, $g(\phi)$ is a set of constraint applications of $\lambda xy.x \vee y$ without duplicates (i.e., if $z \vee z' \in g(\phi)$, then $z \neq z'$) and

$$\phi \in \text{SAT iff } g(\phi) \Rrightarrow \{y_j \vee y_{j+1} \mid 1 \leq j < n\}.$$

Here $y_1, \ldots, y_n$ are exactly all variables occurring in $g(\phi)$. Such a function exists, since SAT is reducible to HAMILTONIAN PATH, which is reducible to ISO-IMP($\{\lambda xy.x \vee y\}$) via a reduction that satisfies the properties above. (Basically, use the standard translation from graphs to sets of constraint applications of $\lambda xy.x \vee y$: For G a connected graph on vertices $\{1, \ldots, n\}$, let $g(G) = \{y_i \vee y_j \mid \{i, j\} \in E(G)\}$.)

Recall that we need to construct a polynomial-time computable function h with the property that there exists an i such that $1 \leq i \leq k$, $\phi_{2i-1} \notin \text{SAT}$, and $\phi_{2i} \in \text{SAT}$ if and only if $h(\phi_1, \ldots, \phi_{2k}) \in \text{ISO-IMP}(\mathcal{C})$.

In order to construct h, we will apply the coNP-hardness reduction f on ϕ_i for odd i, and the NP-hardness reduction g on ϕ_i for even i. It will be important to make sure that all obtained sets of constraint applications are over disjoint sets of variables.

For every i, $1 \leq i \leq k$, we define O_i to be the set of constraint applications $f(\phi_{2i-1})$ with each variable x_j replaced by $x_{i,j}$. Clearly,

$$\phi_{2i-1} \notin \text{SAT iff } O_i \Rrightarrow \bigcup_{1 \leq j, \ell \leq n_i} \{x_{i,j} \to x_{i,\ell}\},$$

where n_i is the n from $f(\phi_{2i-1})$.

For every i, $1 \leq i \leq k$, we define E_i to be the set of constraint applications $g(\phi_{2i})$ with each variable y_j replaced by $y_{i,j}$. Clearly,

$$\phi_{2i} \in \text{SAT iff } E_i \Rrightarrow \{y_{i,j} \vee y_{i,j+1} \mid 1 \leq j < n'_i\},$$

where n'_i is the n from $g(\phi_{2i})$.

Note that the sets that occur to the right of $O_i \Rrightarrow$ are almost isomorphic (apart from the number of variables). The same holds for the sets that occur to the right of $E_i \Rrightarrow$. It is important to make sure that these sets are exactly isomorphic. In order to do so, we simply pad the sets O_i and E_i.

Let $n = \max\{n_i, n'_i + 2 \mid 1 \leq i \leq k\}$. For $1 \leq i \leq k$, let

$$\widehat{O}_i = O_i \cup \{x_{i,1} \to x_{i,j}, x_{i,j} \to x_{i,1} \mid n_i < j \leq n\}.$$

$\widehat{O}_i$ is a set of constraint applications of $\mathcal{D}$, since there exists a constraint application $A(x, y)$ of $\mathcal{D}$ that is equivalent to $x \to y$ (see [BHRV02, Claim 14]). It is immediate that

$$\widehat{O}_i \Rrightarrow \bigcup_{1 \leq j, \ell \leq n} \{x_{i,j} \to x_{i,\ell}\} \text{ iff } O_i \Rrightarrow \bigcup_{1 \leq j, \ell \leq n_i} \{x_{i,j} \to x_{i,\ell}\}.$$

For $1 \leq i \leq k$, let

$$\widehat{E}_i = E_i \cup \{y_{i,j} \vee y_{i,n'_i+1} \mid 1 \leq j \leq n'_i\} \cup \{y_{i,j} \vee y_{i,j+1} \mid n'_i + 1 \leq j < n\}.$$

Then

$$\widehat{E}_i \Rrightarrow \{y_{i,j} \vee y_{i,j+1} \mid 1 \leq j < n\} \text{ iff } E_i \Rrightarrow \{y_{i,j} \vee y_{i,j+1} \mid 1 \leq j < n'_i\}.$$

The right-to-left direction is immediate. The left-to-right to direction can easily be seen if we think about this as graphs. Since $n \geq n'_i + 2$, any Hamiltonian path in $\widehat{E}_i$ contains the subpath $n'_{i+1}, n'_{i+2}, \ldots, n$, where n is an endpoint. This implies that there is a Hamiltonian path in the graph restricted to $\{1, \ldots, n'_i\}$, i.e., in E_i.

So, our current situation is as follows. For all i, $1 \leq i \leq k$, $\widehat{O}_i$ is a set of constraint applications of $\mathcal{D}$ such that

$$\phi_{2i-1} \notin \text{SAT iff } \widehat{O}_i \Rrightarrow \bigcup_{1 \leq j, \ell \leq n} \{x_{i,j} \to x_{i,\ell}\}$$

and $\widehat{E}_i$ is a set of constraint applications of $\lambda xy.x \vee y$ without duplicates such that

$$\phi_{2i} \in \text{SAT iff } \widehat{E}_i \Rrightarrow \{y_{i,j} \vee y_{i,j+1} \mid 1 \leq j < n\}.$$

Our reduction h is defined as follows

$$h(\phi_1, \ldots, \phi_{2k}) = \langle S, U \rangle,$$

where

$$S = \bigcup_{i=1}^{k} \left(\widehat{O}_i \cup \widehat{E}_i \cup \bigcup_{1 \leq j, \ell \leq n} \{x_{i,j} \to y_{i,\ell}\} \right)$$

and

$$U = \bigcup_{1 \leq j, \ell \leq n} \{x_j \to x_\ell\} \cup \bigcup_{j=1}^{n-1} \{y_j \vee y_{j+1}\} \cup \bigcup_{1 \leq j, \ell \leq n} \{x_j \to y_\ell\}.$$

The proof that h is the desired reduction can be found in the full version of this paper. □

We can modify the proof of Theorem 16 to show that ISO-IMP($\mathcal{D} \cup \{C\}$) is $P_{\|}^{NP}$-hard for various other constraints C. Note however that the proof of Theorem 16 crucially uses the fact that $\mathcal{D}$ is 0-valid, 1-valid, and complementive. Thus, new insights and constructions will be needed to obtain $P_{\|}^{NP}$-hardness for all non-Schaefer cases.

6 Open Problems

The most important question left open by this paper is whether Conjecture 12 holds. In addition, the complexity of the isomorphic implication problem for Boolean formulas is still open. This problem is trivially in Σ_2^p, and, by Theorem 16, $P_{||}^{NP}$-hard. Note that an improvement of the upper bound will likely give an improvement of the best-known upper bound (Σ_2^p) for the isomorphism problem for Boolean formulas, since that problem is 2-conjunctive-truth-table reducible to the isomorphic implication problem.

Schaefer's framework is not the only framework to study generalized Boolean problems. It would be interesting to study the complexity of isomorphic implication in other frameworks, for example, for Boolean circuits over a fixed base.

Acknowledgments. The authors thank Henning Schnoor, Heribert Vollmer, and the anonymous referees for helpful comments.

References

[AT00] M. Agrawal and T. Thierauf. The formula isomorphism problem. *SIAM Journal on Computing*, 30(3):990–1009, 2000.

[BCRV04] E. Böhler, N. Creignou, S. Reith, and H. Vollmer. Playing with Boolean blocks, part II: Constraint Satisfaction Problems. *SIGACT News*, 35(1):22–35, 2004.

[BH04] M. Bauland and E. Hemaspaandra. Isomorphic implication. Technical Report cs.CC/0412062, Computing Research Repository, http://www.acm.org/repository/, December 2004. Revised, May 2005.

[BHRV02] E. Böhler, E. Hemaspaandra, S. Reith, and H. Vollmer. Equivalence and isomorphism for Boolean constraint satisfaction. In *Proceedings of the 16th Annual Conference of the EACSL (CSL 2002)*, pages 412–426. Springer-Verlag *Lecture Notes in Computer Science #2471*, September 2002.

[BHRV03] E. Böhler, E. Hemaspaandra, S. Reith, and H. Vollmer. The complexity of Boolean constraint isomorphism. Technical Report cs.CC/0306134, Computing Research Repository, http://www.acm.org/repository/, June 2003. Revised, April 2004.

[BHRV04] E. Böhler, E. Hemaspaandra, S. Reith, and H. Vollmer. The complexity of Boolean constraint isomorphism. In *Proceedings of the 21st Symposium on Theoretical Aspects of Computer Science*, pages 164–175. Springer-Verlag *Lecture Notes in Computer Science #2996*, March 2004.

[BKJ00] A. Bulatov, A. Krokhin, and P. Jeavons. Constraint satisfaction problems and finite algebras. In *Proceedings of the 27th International Colloquium on Automata, Languages and Programming*, pages 272–282. Springer-Verlag, 2000.

[BR93] B. Borchert and D. Ranjan. The ciruit subfunction relations are Σ_2^p-complete. Technical Report MPI-I-93-121, MPI, Saarbrücken, 1993.

[BRS98] B. Borchert, D. Ranjan, and F. Stephan. On the computational complexity of some classical equivalence relations on Boolean functions. *Theory of Computing Systems*, 31(6):679–693, 1998.

[CH96] N. Creignou and M. Hermann. Complexity of generalized satisfiability counting problems. *Information and Computation*, 125:1–12, 1996.

[CKS01] N. Creignou, S. Khanna, and M. Sudan. *Complexity Classifications of Boolean Constraint Satisfaction Problems*. Monographs on Discrete Applied Mathematics. SIAM, 2001.

[Coo71] S. Cook. The complexity of theorem-proving procedures. In *Proceedings of the 3rd ACM Symposium on Theory of Computing*, pages 151–158. ACM Press, 1971.

[Cre95] N. Creignou. A dichotomy theorem for maximum generalized satisfiability problems. *Journal of Computer and System Sciences*, 51:511–522, 1995.

[GJ79] M. Garey and D. Johnson. *Computers and Intractability: A Guide to the Theory of NP-Completeness*. W. H. Freeman and Company, 1979.

[HHR97] E. Hemaspaandra, L. Hemaspaandra, and J. Rothe. Raising NP lower bounds to parallel NP lower bounds. *SIGACT News*, 28(2):2–13, 1997.

[JCG97] P. Jeavons, D. Cohen, and M. Gyssens. Closure properties of constraints. *Journal of the ACM*, 44(4):527–548, 1997.

[Jea98] P. Jeavons. On the algebraic structure of combinatorial problems. *Theoretical Computer Science*, 200(1-2):185–204, 1998.

[Jub99] L. Juban. Dichotomy theorem for generalized unique satisfiability problem. In *Proceedings of the 12th Conference on Fundamentals of Computation Theory*, pages 327–337. Springer-Verlag *Lecture Notes in Computer Science #1684*, 1999.

[KK01] L. Kirousis and P. Kolaitis. The complexity of minimal satisfiability problems. In *Proceedings of the 18th Symposium on Theoretical Aspects of Computer Science*, pages 407–418. Springer-Verlag *Lecture Notes in Computer Science #2010*, 2001.

[KS98] D. Kavvadias and M. Sideri. The inverse satisfiability problem. *SIAM Journal on Computing*, 28(1):152–163, 1998.

[KST93] J. Köbler, U. Schöning, and J. Torán. *The Graph Isomorphism Problem: Its Structural Complexity*. Birkhäuser, 1993.

[KSTW01] S. Khanna, M. Sudan, L. Trevisan, and D. Williamson. The approximability of constraint satisfaction problems. *SIAM Journal on Computing*, 30(6):1863–1920, 2001.

[Pos44] E. Post. Recursively enumerable sets of integers and their decision problems. *Bulletin of the AMS*, 50:284–316, 1944.

[Sch78] T. Schaefer. The complexity of satisfiability problems. In *Proceedings of the 10th ACM Symposium on Theory of Computing*, pages 216–226, 1978.

[Wag87] K. Wagner. More complicated questions about maxima and minima, and some closures of NP. *Theoretical Computer Science*, 51(1–2):53–80, 1987.

Abstract Numeration Systems and Tilings

Valérie Berthé[1] and Michel Rigo[2]

[1] LIRMM, CNRS-UMR 5506, Univ. Montpellier II,
161 rue Ada, 34392 Montpellier Cedex 5, France
berthe@lirmm.fr
[2] Université de Liège, Institut de Mathématiques,
Grande Traverse 12 (B 37), B-4000 Liège, Belgium
M.Rigo@ulg.ac.be

Abstract. An abstract numeration system is a triple $S = (L, \Sigma, <)$ where $(\Sigma, <)$ is a totally ordered alphabet and L a regular language over Σ; the associated numeration is defined as follows: by enumerating the words of the regular language L over Σ with respect to the induced genealogical ordering, one obtains a one-to-one correspondence between $\mathbb{N}$ and L. Furthermore, when the language L is assumed to be exponential, real numbers can also be expanded. The aim of the present paper is to associate with S a self-replicating multiple tiling of ăthe space, under the following assumption: the adjacency matrix of the trimmed minimal automaton recognizing L is primitive with a dominant eigenvalue being a Pisot unit. This construction generalizes the classical constructions performed for Rauzy fractals associated with Pisot substitutions [16], and for central tiles associated with a Pisot beta-numeration [23].

1 Introduction

To any infinite regular language L over a totally ordered alphabet $(\Sigma, <)$, an *abstract numeration system* $S = (L, \Sigma, <)$ is associated in the following way [10]. Enumerating the words of L by increasing genealogical order gives a one-to-one correspondence between $\mathbb{N}$ and L, the non-negative integer n being represented by the $(n+1)$-th word of the ordered language L. Nonnegative integers as well as positive real numbers (under some natural assumptions on L) can thus be expanded in such a numeration system [10,11,12]. In this latter situation, a real number is represented by an infinite word which is the limit of a converging sequence of words in L. These systems generalize in a natural way classical positional systems like the k-ary numeration, the Fibonacci numeration, more generally, the numeration scales built on a sequence of integers satisfying a linear recurrence relation, including the beta-numeration when β is a Parry number [13], as well as the Dumont-Thomas numeration associated with a substitution [6,7]. Many classical properties of such numerations extend in a natural way to abstract numeration systems: see for instance [3,9,11,12,17,18,19].

The aim of this paper is to introduce a self-replicating multiple tiling of the space that can be associated with an abstract numeration system with some pre-scribed algebraic properties: these systems are built upon an exponential regular language such that the adjacency matrix of the trimmed minimal automaton recognizing L is primitive with a dominant eigenvalue being a Pisot unit. We recall that a *Pisot number* is an algebraic integer whose other conjugates have modulus smaller than 1; a Pisot number is a *unit* if its norm is equal to 1, that is, the constant term in its minimal polynomial equals ± 1. The basic tiles are compact sets that are the closure of their interior, that have non-zero measure and a fractal boundary; they are attractors of some graph-directed Iterated Function System. By tiling, we mean here tilings by translation having finitely many tiles up to translation (a tile is assumed to be the closure of its interior); we assume furthermore that each compact set intersects a finite number of tiles. By *multiple tiling*, we mean arrangements of tiles such that almost all points are covered exactly p times for some positive integer p.

The multiple tiling we propose here is directly inspired by the tilings of the space that can be associated with beta-numeration [23] and with substitutions (see e.g., Chap. 7 in [15]). It is conjectured that the corresponding multiple tiling is indeed a tiling in the Pisot case. This conjecture is known as the *Pisot conjecture* and can also be reformulated in spectral terms: the associated dynamical systems have pure discrete spectrum. Notice that the existence of such tilings has applications in Diophantine approximation, or in the study of mathematical quasicrystals, for instance.

Our main motivation for this work is the following. The central tiles in the beta-numeration framework are defined in a natural way [1,2,23]; one can consider the formalism introduced in the substitutive case as a first generalization of the beta-numeration formalism [5]. Indeed the underlying substitutions and automata have a very particular shape in the beta-numeration case. We develop here a further generalization by working directly on the automaton. In particular final acceptance states play a crucial rôle in our study. We thus wish to put to the test the Pisot conjecture in a more general context.

This paper is organized as follows. We first recall in Section 2 a few basic definitions and properties. We focus on the representation of real numbers in abstract number systems in Section 3. We respectively introduce in Section 4 and in Section 5 the central tile and our multiple tiling. We illustrate these notions in Section 6 with two examples. We conclude this paper by mentioning a few natural prospects concerning this work in Section 7.

2 Definitions

An *abstract numeration system* is a triple $S = (L, \Sigma, <)$, where L is an infinite regular language over the totally ordered alphabet $(\Sigma, <)$.

Let $\Sigma = \{s_0 < s_1 < \cdots < s_k\}$ be a finite and totally ordered alphabet. Since Σ is totally ordered, we can order the words of Σ^* using the *genealogical ordering*. Let $u, v \in \Sigma^*$. We say that $u < v$ if $|u| < |v|$ or if $|u| = |v|$ and there exist $p, u', v' \in \Sigma^*$, $s, t \in \Sigma$, $s < t$ such that $u = psu'$ and $v = ptv'$.

The trimmed minimal automaton of L is denoted $\mathcal{M}_L = (Q, q_0, \Sigma, \delta, F)$ where Q is the set of states, q_0 is the initial state, $F \subseteq Q$ is the set of final states and $\delta : Q \times \Sigma \to Q$ is the (partial) transition function. As usual, δ can be extended to $Q \times \Sigma^*$. In this paper $L \subset \Sigma^*$ will always denote an infinite regular language having the property that $\mathcal{M}_L$ is such that

$$\delta(q_0, s_0) = q_0. \tag{1}$$

In other words, $\mathcal{M}_L$ has a loop of label s_0 in the initial state q_0. In particular, this implies that L has the following property: $s_0^* L \subseteq L$.

The entry of index $(p, q) \in Q^2$ of the *adjacency matrix* $\mathbf{M}_L$ of the automaton $\mathcal{M}_L$ is given by the cardinality of the set of letters $s \in \Sigma$ such that $\delta(p, s) = q$. An abstract numeration system is said *primitive* if the matrix $\mathbf{M}_L$ is primitive, that is, there exists a nonnegative integer n such that $\mathbf{M}_L^n$ has only positive entries. According to Perron-Frobenius theorem, the adjacency matrix of a primitive abstract numeration system admits a simple dominating eigenvalue $\beta > 0$.

For any state $q \in Q$, we denote by L_q the regular language accepted by $\mathcal{M}_L$ from state q, by $\mathbf{u}_q(n)$ the number of words of length n in L_q, and by $\mathbf{v}(n)$ the number of words of length at most n in L. In particular, $L = L_{q_0}$ and $\mathbf{u}_q(n) = e_q \mathbf{M}_L^n e_F$ for appropriate row (resp. column) vector e_q (resp. e_F).

Let us introduce several sets of right-sided and left-sided infinite words built upon the abstract numeration system S. We use here the topology induced by the infinite product topology on $\Sigma^{\mathbb{N}}$, $\Sigma^{\mathbb{N}^*}$ and $^{\mathbb{N}}\Sigma$ respectively, where $\mathbb{N}^*$ denotes the set of positive integers, and $^{\mathbb{N}}\Sigma$ the set of left-infinite words over Σ. We use the following notation for elements of $^{\mathbb{N}}\Sigma$: $v = \cdots v_2 v_1 v_0$.

We first define $\mathcal{L}^\omega \subset \Sigma^{\mathbb{N}^*}$ as the set of right-infinite words $w = (w_i)_{i \in \mathbb{N}^*}$ for which there exists a sequence of words $(W_n)_{n \in \mathbb{N}}$ in L converging to w, that is, for all ℓ, there exists N_ℓ such that for all $n \geq N_\ell$, a prefix of length at least ℓ of W_n is a prefix of w. Notice that a main difference with the set $\overrightarrow{L}$ classically encountered in the literature (we refer for instance to [22]) is that if w belongs to $\mathcal{L}^\omega$ then it does not necessarily imply that infinitely many prefixes of w belongs to L (see [11]).

Definition 1. *We define the set $\mathcal{K}^\omega \subset (\Sigma \times Q)^{\mathbb{N}}$ by $(w, r) = (w_0 w_1 \cdots, r_0 r_1 \cdots)$ belongs to $\mathcal{K}^\omega$ if and only if the following conditions hold*

1. *there exists a sequence of words $(W_n)_{n \in \mathbb{N}}$ in $\cup_{q \in Q} L_q$ converging to $w_1 w_2 \cdots$,*
2. *for all $i \geq 0$, $\delta(r_i, w_{i+1}) = r_{i+1}$.*

For a given $q \in Q$, the subset $\mathcal{K}_q^\omega \subset \mathcal{K}^\omega$ is defined as the set of elements $(w, r) \in \mathcal{K}^\omega$ such that $r_0 = q$. One has $\mathcal{K}^\omega = \cup_{q \in Q} \mathcal{K}_q^\omega$.

Definition 2. *We similarly define the set $^\omega \mathcal{K} \subset {^{\mathbb{N}}}(\Sigma \times Q)$. A pair $(v, p) = (\cdots v_2 v_1 v_0, \cdots p_2 p_1 p_0)$ belongs to $^\omega \mathcal{K}$ if and only if the following conditions hold*

1. *there exists a sequence $(V_n)_{n \in \mathbb{N}}$ of words in L converging to v, i.e., $v_0 v_1 v_2 \cdots$ is the limit of the sequence of words $(\widetilde{V_n})_{n \in \mathbb{N}}$, where $\widetilde{W}$ denotes the mirror image of the word W,*
2. *for all $i \geq 0$, $\delta(p_{i+1}, v_i) = p_i$.*

Definition 3. *Finally, $^\omega\mathcal{K}^\omega \subset (\Sigma \times Q)^{\mathbb{Z}}$ is defined as the set of two-sided sequences $((\cdots v_2 v_1 v_0 \cdot w_1 w_2 \cdots), (\cdots p_2 p_1 p_0 \cdot r_1 r_2 \cdots))$ (denoted $((v,w);(p,r))$) that satisfy*

1. $(\cdots v_2 v_1 v_0, \cdots p_2 p_1 p_0)$ *belongs to* $^\omega\mathcal{K}$,
2. $(v_0 w_1 w_2 \cdots, p_0 r_1 r_2 \cdots)$ *belongs to* $\mathcal{K}^\omega$.

These three sets are easily shown to be nonempty, by a classical compactness argument; they have the rôle played by the beta-shift in the beta-numeration case [13, Chap. 7].

3 Expansions of Real Numbers

The abstract numeration system $S = (L, \Sigma, <)$ gives a one-to-one correspondence between $\mathbb{N}$ and L [10]: the representation of the integer n is defined as the $(n+1)$-th word w of L. We conversely define val : $L \to \mathbb{N}$, which maps the $(n+1)$-th word of L onto n.

We want now to expand real numbers. Let us assume that S is a primitive abstract numeration system. Let $\beta > 1$ denote its dominating eigenvalue. Consequently, L is an exponential regular language (i.e., $\mathbf{u}_{q_0}(n) \geq C\beta^n$, for infinitely many n and some $C > 0$) and thanks to [11, Prop. 3], we deduce that the set $\mathcal{L}^\omega$ is uncountable.

We assume moreover that L is a language for which there exist $P \in \mathbb{R}[X]$, and some nonnegative real numbers a_q, $q \in Q$, which are not simultaneously equal to 0, such that for all state $q \in Q$

$$\lim_{n \to \infty} \frac{\mathbf{u}_q(n)}{P(n)\beta^n} = a_q. \qquad (2)$$

The coefficients a_q are defined up to a scaling constant; in fact, the vector $(a_q)_{q \in Q}$ is an eigenvector of $\mathbf{M}_L$ [12]; by Perron-Frobenius theorem, all its entries a_q are positive; we normalize it so that $a_{q_0} = 1 - 1/\beta$, according to [19].

For $q \in Q$ and $s \in \Sigma$, set

$$\alpha_q(s) := \sum_{q' \in Q} a_{q'} \cdot \text{Card}\{t < s \mid \delta(q,t) = q'\} = \sum_{\substack{t < s \\ (q,t) \in \text{dom}(\delta)}} a_{\delta(q,t)}.$$

One has for all $q \in Q$, $0 \leq \alpha_q(s) \leq \beta a_q$, since $(a_q)_{q \in Q}$ is a positive eigenvector of $\mathbf{M}_L$. Notice also that if $s < t$, $s, t \in \Sigma$, then $\alpha_q(s) \leq \alpha_q(t)$.

For any sequence of words $(W_k)_{k \in \mathbb{N}}$ converging to a word $w \in \mathcal{L}^\omega$, let us recall that the limit

$$\lim_{k \to \infty} \frac{\text{val}(W_k)}{\mathbf{v}(|W_k|)}$$

only depends on w, belongs to $[1/\beta, 1]$, and is equal to

$$(1 + \alpha_{q_0}(w_1))\beta^{-1} + \sum_{j=2}^{\infty} \alpha_{\delta(q_0, w_1 \cdots w_{j-1})}(w_j)\beta^{-j},$$

according to [11,19]. Hence it is natural to introduce the map

$$\varphi^\omega : \mathcal{K}^\omega \to [0, \max(a_q)], \ (w, r) = (w_0 w_1 \cdots, r_0 r_1 \cdots) \mapsto \sum_{j=1}^\infty \alpha_{r_{j-1}}(w_j) \beta^{-j}.$$

Conversely, let us expand real numbers by introducing a suitable dynamical system analogous to the β-transformation $T_\beta : \ x \in [0,1] \mapsto \{\beta x\}$, where $\{z\}$ denotes the fractional part of z. The corresponding transformation for abstract dynamical systems has been introduced in [19]. The underlying dynamics depends on each interval $[0, a_q)$, and is defined as follows: we first set for $y \in \mathbb{R}^+$,

$$\lfloor y \rfloor_q = \max\{\alpha_q(s) \mid s \in \Sigma, \ \alpha_q(s) \leq y\};$$

let us recall that $(a_q)_{q \in Q}$ is an eigenvector of $\mathbf{M}_L$ of eigenvalue β, hence

$$\beta a_q = \sum_{r \in Q} a_r \cdot \text{Card}\{s \in \Sigma \mid \delta(q, s) = r\},$$

and one checks that for $y \in [0, a_q)$, then $\beta y - \lfloor \beta y \rfloor_q \in [0, a_{q'})$, with $\lfloor \beta y \rfloor_q = \alpha_q(s)$ and $\delta(q, s) = q'$. Furthermore, s may be not uniquely determined since α_q is nondecreasing. We define

$$T_S : (\cup_{q \in Q}[0, a_q]) \times Q \to (\cup_{q \in Q}[0, a_q]) \times Q,$$
$$(x, q) \mapsto (\beta x - \lfloor \beta x \rfloor_q, q')$$

where q' is determined as follows: let s be the largest letter such that $\alpha_q(s) = \lfloor \beta x \rfloor_q$; then $q' = \delta(q, s)$. To retrieve this information given by the largest letter s, we thus set

$$\rho_S : (\cup_{q \in Q}[0, a_q]) \times Q \to \Sigma,$$
$$(x, q) \mapsto s.$$

We thus can expand any real number $x \in [0, a_{q_0}) = [0, 1 - 1/\beta)$ as follows. Let $(x_i, r_i)_{i \geq 1} := (T_S^i(x, q_0))_{i \geq 1} \in ((\cup_{q \in Q}[0, a_q]) \times Q)^{\mathbb{N}^*}$. Moreover, set $(w_0, r_0) := (s_0, q_0)$ and for every $i \geq 1$, set $w_i := \rho_S(x_{i-1}, r_{i-1})$ where it is assumed that $x_0 := x$. According to [11], one has $x = \sum_{j=1}^\infty \alpha_{r_{j-1}}(w_j) \beta^{-j}$. So we have the following definition.

Definition 4. *Let S be a primitive abstract numeration system satisfying (1) and (2). Every real number $x \in [0, 1 - 1/\beta)$ can be expanded as*

$$x = \sum_{i \geq 1} \alpha_{r_{i-1}}(w_i) \beta^{-i},$$

where (w, r) belongs to $\mathcal{K}^\omega$ and satisfies for every $i \geq 1$, $w_i = \rho_S(x_{i-1}, r_{i-1})$, with $(x_i, r_i)_{i \geq 1} = (T_S^i(x, q_0))_{i \geq 1}$, $(w_0, r_0) = (s_0, q_0)$ and $x_0 = x$. We call $(w, r) = (w_i, r_i)_{i \in \mathbb{N}} \in \mathcal{K}^\omega$ the S-expansion of x and denote it $d_S(x)$.

Similarly, one can expand every real positive number by rescaling. Indeed let $x \geq a_{q_0}$. Let k be the smallest positive integer such that $\beta^{-k}x \in [0, a_{q_0})$ and let us set $(w_i, r_i)_{i \in \mathbb{N}} := d_S(\beta^{-k}x)$. One has $\beta^{-k}x = \sum_{j \geq 1} \alpha_{r_{j-1}}(w_j)\beta^{-j}$. Let $(v, p) \in {}^\omega\mathcal{K}$, with $v = (\cdots s_0 \cdots s_0 v_{k-1} v_{k-2} \cdots v_0)$, $p = (\cdots q_0 \cdots q_0 p_{k-1} \cdots p_0)$, and $v_{k-1} \cdots v_0 = w_1 \cdots w_k$, $p_{k-1} \cdots p_0 = r_1 \cdots r_k$. (Notice that we have explicitly used (1) to define (v, p).) One thus gets

$$x = \alpha_{q_0}(v_{k-1})\beta^{k-1} + \alpha_{p_{k-1}}(v_{k-2})\beta^{k-2} + \cdots + \alpha_{p_1}(v_0)$$
$$+ \alpha_{r_k}(w_{k+1})\tfrac{1}{\beta} + \cdots + \alpha_{r_{k+1}}(w_{k+2})\tfrac{1}{\beta^2} + \cdots$$

with $(v \cdot w_{k+1} w_{k+2} \cdots, p \cdot r_{k+1} r_{k+2} \cdots) \in {}^\omega\mathcal{K}^\omega$.

Definition 5. *Let S be a primitive abstract numeration system and x be a positive real number. If $x \in [0, a_{q_0}) = [0, 1 - 1/\beta)$, then the S-fractional part of x is simply $d_S(x)$. Otherwise, let k be the smallest positive integer such that $\beta^{-k}x \in [0, a_{q_0})$. Using the same notation as above, the S-fractional part of x is $(w_k w_{k+1} w_{k+2} \cdots, r_k r_{k+1} r_{k+2} \cdots) \in \mathcal{K}^\omega$. We denote it $\mathrm{Frac}_S(x)$.*

4 The Central Tile

We have given in Section 3 a geometric representation of the set $\mathcal{K}^\omega$ thanks to the map φ^ω. The aim of the present section is to provide a similar representation for the set ${}^\omega\mathcal{K}$. We follow here the formalism of [1,2,5].

We assume now that S is a primitive abstract numeration system satisfying (1) and (2), whose dominant eigenvalue β is a Pisot unit. Let $\beta^{(2)}$, ..., $\beta^{(r)}$ denote the real conjugates of β, and let $\beta^{(r+1)}$, $\overline{\beta^{(r+1)}}$, ..., $\beta^{(r+s)}$, $\overline{\beta^{(r+s)}}$ be its complex conjugates. If d denotes the degree of β, then $d = r + 2s$. We set $\beta^{(1)} = \beta$. Let $\mathbb{K}^{(k)}$ be equal to $\mathbb{R}$ if $1 \leq k \leq r$, and to $\mathbb{C}$, if $k > r$. We furthermore denote by $\mathbb{K}_\beta$ the *representation space*

$$\mathbb{K}_\beta := \mathbb{R}^{r-1} \times \mathbb{C}^s \simeq \mathbb{R}^{d-1}.$$

Let us note that, according to [19, Lemma 4.1], a_q belongs to $\mathbb{Q}(\beta)$ for all $q \in Q$. Let us consider now the following algebraic embeddings:

– The *canonical embedding* on $\mathbb{Q}(\beta)$ maps a polynomial to all its conjugates

$$\Phi_\beta : \mathbb{Q}(\beta) \to \mathbb{K}_\beta, \ P(\beta) \mapsto (P(\beta^{(2)}), \ldots, P(\beta^{(r)}), P(\beta^{(r+1)}), \ldots, P(\beta^{(r+s)})).$$

– For any $(v, p) \in {}^\omega\mathcal{K}$, the series

$$\lim_{n \to \infty} \Phi_\beta \left(\sum_{i=0}^n \alpha_{p_{i+1}}(v_i)\beta^i \right) = \sum_{i \geq 0} \Phi_\beta(\alpha_{p_{i+1}}(v_i))\Phi_\beta(\beta^i)$$

are convergent in $\mathbb{K}_\beta$. The *representation map* of ${}^\omega\mathcal{K}$ is then defined as

$${}^\omega\varphi : {}^\omega\mathcal{K} \to \mathbb{K}_\beta, \ (v, p) \mapsto \lim_{n \to +\infty} \Phi_\beta \left(\sum_{i=0}^n \alpha_{p_{i+1}}(v_i)\beta^i \right).$$

Definition 6. *Let S be a primitive abstract numeration system satisfying (1) and (2) whose dominant eigenvalue β is a Pisot number. We define the central tile $\mathcal{T}_S$ as*

$$\mathcal{T}_S := {}^\omega\varphi({}^\omega\mathcal{K}).$$

The central tile can be naturally divided into $\operatorname{Card}(Q)$ pieces, called *basic tiles*, as follows:

$$\text{for } q \in Q, \ \mathcal{T}_S(q) := {}^\omega\varphi\Big(\{(v,p) \in {}^\omega\mathcal{K} \mid p_0 = q\}\Big).$$

5 A Self-replicating Multiple Tiling

We introduce the following countable set

$$\mathcal{F}_S := \operatorname{Frac}_S(\mathbb{Z}[\beta]_{>0}) \subset \mathcal{K}^\omega.$$

Let $(w,r) = (w_i, r_i)_{i\in\mathbb{N}} \in \mathcal{F}_S$. By definition of $\mathcal{F}_S$, we can apply Φ_β to $\varphi^\omega(w,r)$ which belongs to $\mathbb{Q}(\beta)$. We define the tile $\mathcal{T}_{(w,r)}$ as

$$\mathcal{T}_{(w,r)} = \Phi_\beta \circ \varphi^\omega(w,r) + {}^\omega\varphi\Big(\{(v,p) \in {}^\omega\mathcal{K} \mid ((v,w);(p,r)) \in {}^\omega\mathcal{K}^\omega\}\Big).$$

One checks that the tiles $\mathcal{T}_{(w,r)}$ are finite unions of translates of the basic tiles $\mathcal{T}_S(q)$ for $q \in Q$ by considering the minimal automaton $\mathcal{M}_L$; furthermore, one proves similarly as in [2] that there are finitely many such tiles.

Definition 7. *The primitive abstract numeration system S for which (1) and (2) hold is said to satisfy the* strong coincidence condition *if for any pair of states $(q, q') \in Q$, there exist a state $q'' \in Q$, a positive integer n and two words $w_1 \cdots w_n$, $w'_1 \cdots w'_n \in \Sigma^n$ such that*

$$\begin{cases} \sum_{1 \leq i \leq n} \alpha_{\delta(q, w_1 \cdots w_{i-1})}(w_i) \beta^{n-i} = \sum_{1 \leq i \leq n} \alpha_{\delta(q', w'_1 \cdots w'_{i-1})}(w'_i) \beta^{n-i} \\ \delta(q, w_1 \cdots w_n) = \delta(q', w'_1 \cdots w'_n) = q''. \end{cases}$$

We have now gathered all the required tools to be able to state and prove the main theorem of the present paper. This theorem and its proof are directly inspired by the corresponding statements in the beta-numeration case [1,2,5], and in the substitutive case [21].

Theorem 1. *Let S be a primitive abstract Pisot numeration system for which (1) and (2) hold and whose dominant eigenvalue β is a Pisot number. The finite (up to translation) set of tiles $\mathcal{T}_{(w,r)}$, for $(w,r) \in \mathcal{F}_S$, covers $\mathbb{K}_\beta$, that is,*

$$\mathbb{K}_\beta = \bigcup_{(w,r)\in\mathcal{F}_S} \mathcal{T}_{(w,r)}. \tag{3}$$

For each (w,r), the tile $\mathcal{T}_{(w,r)}$ has non-empty interior. Hence it has non-zero measure.

We denote by $h_\beta : \mathbb{K}_\beta \to \mathbb{K}_\beta$ the β-multiplication map *that multiplies the coordinate of index i by $\beta^{(i)}$, for $2 \leq i \leq d$. The basic tiles of the central tile $\mathcal{T}_S$ are solutions of the following graph-directed self-affine Iterated Function System:*

$$\forall q \in Q, \ \mathcal{T}_S(q) = \bigcup_{\substack{p \in Q, \ s \in \Sigma, \\ \delta(p,s)=q}} h_\beta(\mathcal{T}_S(p)) + \varPhi_\beta(\alpha_p(s)). \tag{4}$$

If S satisfies the strong coincidence condition, then the basic tiles have disjoint interiors and they are the closure of their interior.

Furthermore, there exists an integer $k \geq 1$ such that the covering (3) is almost everywhere k-to-one.

Proof. We first notice that there exists $C > 0$ such that if

$$\varPhi_\beta \circ \varphi^\omega(\mathrm{Frac}_S(P(\beta))) \neq \varPhi_\beta \circ \varphi^\omega(\mathrm{Frac}_S(P'(\beta))),$$

with $\mathrm{Frac}_S(P(\beta)) \neq \mathrm{Frac}_S(P(\beta))$, then

$$\|\varPhi_\beta \circ \varphi^\omega(\mathrm{Frac}_S((P(\beta)))) - \varPhi_\beta \circ \varphi^\omega(\mathrm{Frac}_S((P'(\beta))))\| > C, \tag{5}$$

where $\|\cdot\|$ denotes a given norm in $\mathbb{K}_\beta$. Indeed $\varPhi_\beta \circ \varphi^\omega(\mathrm{Frac}_S(P-P')(\beta))$ is an algebraic integer: this a direct consequence of the fact that β is a unit and that $a_q \in \mathbb{Q}(\beta)$, for all q. We now conclude by using the fact that for any $C' > 0$, there exist only finitely many algebraic integers x in $\mathbb{Q}(\beta)$ such that $|x| < \max(a_q)$ and $\|\varPhi_\beta(x)\| < C'$.

Let us prove now (3). From β being a Pisot number, we first deduce that $\varPhi_\beta(\mathbb{Z}[\beta]_{\geq 0})$ is dense in $\mathbb{K}_\beta$, according to [1, Prop. 1]. Let $x \in \mathbb{K}_\beta$. There thus exists a sequence $(P_n)_{n \in \mathbb{N}}$ of polynomials in $\mathbb{Z}[X]$ with $P_n(\beta) \geq 0$, for all n, such that $(\varPhi_\beta(P_n(\beta)))_{n \in \mathbb{N}}$ tends towards x. For all n, $\varPhi_\beta(P_n(\beta)) \in \mathcal{T}_{(w,r)^{(n)}}$, with $(w,r)^{(n)} = \mathrm{Frac}_S(P_n(\beta))$. We deduce from (5) that there exist infinitely many n such that $\varPhi_\beta \circ \varphi^\omega((w,r)^{(n)})$ take the same value, say, $\varPhi_\beta \circ \varphi^\omega(w,r)$. Since the tiles are closed, $x \in \mathcal{T}_{(w,r)}$. We now deduce from Baire's theorem that the tiles have non-empty interior.

Let $q \in Q$ be given. Let $(v,p) \in {}^\omega\mathcal{K}$ with $p_0 = q$. One has:

$$\begin{aligned}{}^\omega\varphi(v,p) \ {}^\omega\varphi((v_k,p_k)_{k \geq 1}) + \varPhi_\beta(\alpha_{p_1}(v_0)) \\ = h_\beta \circ {}^\omega\varphi((v_{k-1},p_{k-1})_{k \geq 1}) + \varPhi_\beta(\alpha_{p_1}(v_0)).\end{aligned}$$

One deduces (4) by noticing that $(v_{k-1},p_{k-1})_{k \geq 1}$ belongs to ${}^\omega\mathcal{K}$.

We deduce from the uniqueness of the solution of the IFS [14], that the basic tiles are the closure of their interior, since the interiors of the pieces are similarly shown to satisfy the same IFS equation (4).

We assume that S satisfies the strong coincidence condition. We deduce from this strong coincidence condition that there exist $q'' \in Q$, $w_1 \cdots w_n$, $w'_1 \cdots w'_n \in \Sigma^n$ such that $\mathcal{T}_S(q)$ contains

$$h_\beta^n(\mathcal{T}_S(q)) + \varPhi_\beta\left(\sum_{1 \leq i \leq n} \alpha_{\delta(q,w_1 \cdots w_{i-1})}(w_i) \beta^{n-i}\right)$$

and
$$h_\beta^n(\mathcal{T}_S(q')) + \Phi_\beta\left(\sum_{1\leq i\leq n}\alpha_{\delta(q',w'_1\cdots w'_{i-1})}(w'_i)\beta^{n-i}\right)$$

with
$$\sum_{1\leq i\leq n}\alpha_{\delta(q',w_1\cdots w_{i-1})}(w_i)\beta^{n-i} = \sum_{1\leq i\leq n}\alpha_{\delta(q',w'_1\cdots w'_{i-1})}(w'_i)\beta^{n-i},$$

according to (4), when iterated n times. Indeed $\mathcal{T}_S(q'')$ is equal to the union on the states $p\in Q$ for which there exists a path $a_1\cdots a_n$ of length n in $\mathcal{M}_L$ from p to q'', of
$$h_\beta^n(\mathcal{T}_S(p)) + \Phi_\beta\left(\sum_{1\leq i\leq n}\alpha_{\delta(p,a_1\cdots a_{i-1})}(a_i)\beta^{n-i}\right).$$

We denote by μ the Lebesgue measure of $\mathbb{K}_\beta$: for every Borelian set B of $\mathbb{K}_\beta$, one has $\mu(h_\beta(B)) = \frac{1}{\beta}\mu(B)$, according to [20]: we have used here the fact that β is a Pisot unit. One has for a given $q\in Q$ according to (4)

$$\begin{aligned}\mu(\mathcal{T}_S(q)) &\leq \sum_{p:\delta(p,a)=q}\mu(h_\beta(\mathcal{T}_S(p)))\\ &\leq \frac{1}{\beta}\sum_{p:\delta(p,a)=q}\mu(\mathcal{T}_S(p)).\end{aligned} \quad (6)$$

Let $\mathbf{m} = (\mu(\mathcal{T}_S(q)))_{q\in Q}$ denotes the vector in $\mathbb{R}^d$ of measures in $\mathbb{K}_\beta$ of the basic tiles; we have proved above that $\mathbf{m}$ is a non-zero vector. Since $\mathbf{m}$ has furthermore nonnegative entries, according to Perron-Frobenius theorem the previous inequality implies that $\mathbf{m}$ is an eigenvector of the primitive matrix $\mathbf{M}_L$, and thus of $\mathbf{M}_L^n$. In particular
$$\mu(\mathcal{T}_S(q)) = \sum_{p\in Q}\mathbf{M}_L^n[p,q]\cdot\mu(\mathcal{T}_S(p)),$$

which implies that $\mathcal{T}_S(q)$ and $\mathcal{T}_S(q')$ have disjoint interiors. We thus have proved that the Card(Q) basic tiles are disjoint up to sets of zero measure.

Finally, one deduces from the statement below ([5], Lemma 1) that there exists an integer k such that this covering is almost everywhere k-to-one:

Let $(\Omega_i)_{i\in I}$ be a collection of open sets in $\mathbb{R}^k$ such that $\cup_{i\in I}\overline{\Omega_i} = \mathbb{R}^k$ and for any compact set K, $I_k := \{i\in I;\ \overline{\Omega_i}\cap K\neq\emptyset\}$ is finite. For $x\in\mathbb{R}^k$, let $f(x) := \text{Card}\{i\in I;\ x\in\overline{\Omega_i}\}$. Let $\Omega = \mathbb{R}^k\setminus\cup_{i\in I}\delta(\omega_i)$, where $\delta(\Omega_i)$ denotes the boundary of Ω_i. Then f is locally constant on Ω.

6 Some Examples

Example 1. Let us consider the automaton depicted in Fig. 1. It defines an abstract numeration system over $\Sigma = \{0,1,2\}$. Its adjacency matrix is

$$\mathbf{M}_L = \begin{pmatrix}1 & 1 & 0\\ 1 & 0 & 1\\ 2 & 1 & 0\end{pmatrix}.$$

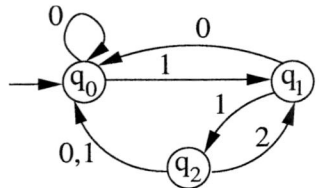

Fig. 1. A trimmed minimal automaton

It is primitive, its characteristic polynomial is $X^3 - X^2 - 2X - 1$. The unique real root of the characteristic polynomial is $\beta \simeq 2,148$ and one of its complex root is $\beta^{(2)} \simeq -0,573 + 0,369\,i$, with $|\beta^{(2)}| < 1$, hence β is a Pisot unit. We have $\mathbb{K}_\beta = \mathbb{C}$. We assume that q_1 and q_2 are final states. One can check that

$$\begin{cases} \alpha_{q_0}(0) = 0, \; \alpha_{q_1}(1) = a_{q_0}, \\ \alpha_{q_1}(0) = 0, \; \alpha_{q_1}(1) = a_{q_0}, \\ \alpha_{q_2}(0) = 0, \; \alpha_{q_2}(1) = a_{q_0}, \; \alpha_{q_2}(2) = 2a_{q_0}. \end{cases}$$

Let us note that on this particular example, the value taken by $\alpha_{q_i}(j)$ does only depend on the letter $j \in \{0, 1, 2\}$. Let us recall that $a_{q_0} = 1 - 1/\beta$, hence $\Phi_\beta(a_{q_0}) = 1 - \frac{1}{\beta^{(2)}}$. We represent the basic tiles (the one associated to q_0, q_1 and q_2 is coloured in red, green and blue respectively) in Fig. 2.

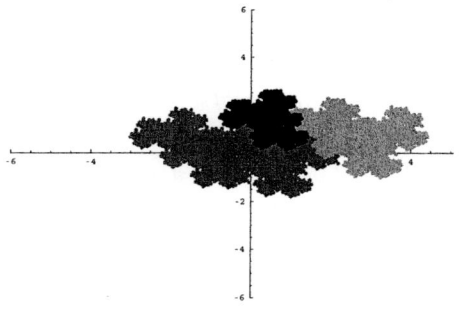

Fig. 2. The basic tiles

Let $\widetilde{\mathcal{M}_L}$ denote the automaton obtained by reversing in $\mathcal{M}_L$ the direction of the arrows. The basic tiles satisfy for $i = 0, 1, 2$:

$$\mathcal{T}_S(q_i) = \left\{ (1 - \tfrac{1}{\beta^{(2)}}) \cdot \left(\sum_{i \geq 0} v_i \cdot (\beta^{(2)})^i \right) \mid (v_0 v_1 \cdots) \text{ being the label} \right.$$

of an infinite path in the automaton $\widetilde{\mathcal{M}_L}$ starting from state $q_i \Big\}$.

The strong coincidence condition is satisfied: for any pair of states (q_i, q_j), the transitions $\delta(q_i, 0) = q_0$, and $\delta(q_j, 0) = q_0$ are suitable. The graph-directed IFS equation satisfies

$$\begin{cases} \mathcal{T}_S(q_0) = (\beta^{(2)} \cdot \mathcal{T}_S(q_0)) \cup (\beta^{(2)} \cdot \mathcal{T}_S(q_1)) \cup (\beta^{(2)} \cdot \mathcal{T}_S(q_2)) \\ \qquad \cup (\beta^{(2)} \cdot \mathcal{T}_S(q_2) + (1 - \frac{1}{\beta^{(2)}})) \\ \mathcal{T}_S(q_1) = (\beta^{(2)} \cdot \mathcal{T}_S(q_0)) \cup (\beta^{(2)} \cdot \mathcal{T}_S(q_2) + (2 - \frac{2}{\beta^{(2)}})) \\ \mathcal{T}_S(q_2) = \beta^{(2)} \cdot \mathcal{T}_S(q_1)) + (1 - \frac{1}{\beta^{(2)}}). \end{cases}$$

The multiple tiling of Theorem 1 is indeed a tiling (up to a set of zero measure). This result can be proved by using the same ideas as in [1,2]: it can be checked in an effective way that every element of $\mathbb{Z}[\beta]$ admits a finite fractional part; this classical property for beta-numeration is called *Finiteness Property* [8], and implies that the covering (3) is a tiling.

Example 2. Let us consider another example given by the automaton depicted in Fig 3. Again we fulfill the Pisot type assumption. Here $\beta \simeq 2,324$ and $\beta^{(2)} \simeq$

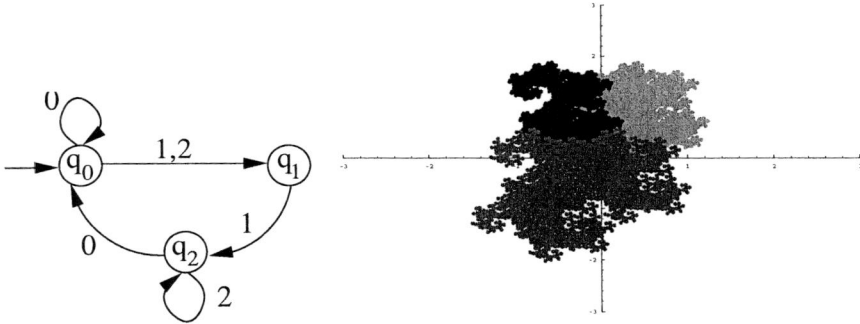

Fig. 3. Another minimal automaton and the corresponding basic tiles

$0,338 + 0,526i$. When considering q_2 as the unique final state, all the α_q's are vanishing except for

$$\alpha_{q_0}(1) = \alpha_{q_2}(2) = 1 - \frac{1}{\beta}, \ \alpha_{q_0}(2) = 1 - \frac{1}{\beta} + \frac{1}{\beta^2}.$$

The corresponding three basic tiles are represented in Fig. 3.

7 Conclusion

All the theory developed in the substitutive and in the beta-numeration case can now be extended in the present framework; mutual insight will by no doubt be brought by handling this more general situation. To mention just but a few prospects, we are planning to study the topological properties of the tiles (connectedness, disklike connectedness), to work out some sufficient tiling conditions in the flavour of the finiteness properties, to define p-adic tiles according to [20], to characterize purely periodic expansions in abstract numeration systems thanks to the central tile as in [4], and to study the geometric representation of the underlying dynamical systems, such as the odometer introduced in [3]: for instance, an exchange of pieces can be performed on the central tile by exchanging the basic tiles; the action of this exchange of pieces can be factorized into a rotation of the torus when the covering (3) is a tiling.

Acknowledgment

We would like to warmly thank Anne Siegel for many useful discussions.

References

1. S. Akiyama, Self affine tiling and Pisot numeration system, *Number theory and its applications (Kyoto, 1997)*, Dev. Math. **2** (1999), 7–17, Kluwer Acad. Publ., Dordrecht.
2. S. Akiyama, On the boundary of self affine tilings generated by Pisot numbers, *J. Math. Soc. Japan* **54** (2002), 2833–308.
3. V. Berthé, M. Rigo, Odometers on regular languages, to appear in *Theory Comput. Syst.*
4. V. Berthé, A. Siegel, Purely periodic expansions in the non-unit case, *preprint* 2004.
5. ăV. Berthé, A. Siegel, Tilings associated with beta-numeration and substitutions, to appear in *Integers: Electronic Journal of Combinatorial Number Theory*.
6. J.-M. Dumont, A. Thomas, Systèmes de numération et fonctions fractales relatifs aux substitutions, *Theoret. Comput. Sci.* **65** (1989), 153–169.
7. J.-M. Dumont, A. Thomas, Digital sum moments and substitutions, *Acta Arith.* **64** (1993), 205–225.
8. C. Frougny and B. Solomyak, Finite beta-expansions, *Ergodic Theory Dynam. Systems* **12**, (1992), 713–723.
9. P. J. Grabner, M. Rigo, Additive functions with respect to numeration systems on regular languages, *Monatsh. Math.* **139** (2003), 205–219.
10. P.B.A. Lecomte, M. Rigo, Numeration systems on a regular language, *Theory Comput. Syst.* **34** (2001), 27–44.
11. P. Lecomte, M. Rigo, On the representation of real numbers using regular languages, *Theory Comput. Syst.* **35** (2002), 13–38.
12. P. Lecomte, M. Rigo, Real numbers having ultimately periodic representations in abstract numeration systems, *Inform. and Comput.* **192** (2004), 57–83.
13. M. Lothaire, *Algebraic Combinatorics on words*, Cambridge University Press, Cambridge, (2002).
14. R.D. Mauldin, S. C. Williams, Hausdorff dimension in graph directed constructions, *Trans. Amer. Math. Soc.* **309** (1988), 811–829.
15. N. Pytheas Fogg, *Substitutions in Dynamics, Arithmetics and Combinatorics*, Lect. Notes in Math. **1794**, Springer-Verlag, Berlin, (2002).
16. G. Rauzy, Nombres algébriques et substitutions, *Bull. Soc. Math. France*, **110** (1982), 147–178.
17. M. Rigo, Numeration systems on a regular language: arithmetic operations, recognizability and formal power series, *Theoret. Comput. Sci.* **269** (2001), 469–498.
18. M. Rigo, Construction of regular languages and recognizability of polynomials, *Discrete Math.* **254** (2002), 485–496.

19. M. Rigo, W. Steiner, Abstract β-expansions and ultimately periodic representations, *J. Théor. Nombres Bordeaux* **17** (2005), 283–299.
20. A. Siegel, Représentation des systèmes dynamiques substitutifs non unimodulaires, *Ergodic Theory Dynam. Systems* **23**, (2003), 1247–1273.
21. V. Sirvent, Y. Wang, Self-affine tiling via substitution dynamical systems and Rauzy fractals, *Pacific J. Math.* **206** (2002), 465–485.
22. W. Thomas, Automata on infinite objects, *Handbook of theoret. comput. sci.*, Vol. B, Elsevier, Amsterdam, (1990), 133–191.
23. W. P. Thurston, Groups, tilings and finite state automata, Lectures notes distributed in conjunction with the Colloquium Series, in *AMS Colloquium lectures*, (1989).

Adversarial Queueing Model for Continuous Network Dynamics[*,**]

María J. Blesa[1], Daniel Calzada[2], Antonio Fernández[3], Luis López[3], Andrés L. Martínez[3], Agustín Santos[3], and Maria Serna[1]

[1] ALBCOM, LSI, Universitat Politècnica de Catalunya, E-08034 Barcelona, Spain
{mjblesa, mjserna}@lsi.upc.edu
[2] ATC, EUI, Universidad Politécnica de Madrid, E-28031 Madrid, Spain
dcalzada@eui.upm.es
[3] LADyR, GSyC, ESCET, Universidad Rey Juan Carlos, E-28933 Madrid, Spain
{anto, llopez, aleonar, asantos}@gsyc.escet.urjc.es

Abstract. In this paper we start the study of generalizing the Adversarial Queueing Theory (AQT) model towards a continuous scenario in which the usually assumed synchronicity of the evolution is not required anymore. We consider a model, named *continuous AQT* (CAQT), in which packets can have arbitrary lengths, and the network links may have different speeds (or bandwidths) and propagation delays. We show that, in such a general model, having bounded queues implies bounded end-to-end packet delays and vice versa. From the network point of view, we show that networks with directed acyclic topologies are universally stable, i.e., stable independently of the protocols and the traffic patterns used in it, and that this even holds for traffic patterns that make links to be fully loaded. Concerning packet scheduling protocols, we show that the well-known LIS, SIS, FTG and NFS protocols remain universally stable in our model. We also show that the CAQT model is strictly stronger than the AQT model by presenting scheduling policies that are unstable under the former while they are universally stable under the latter.

1 Introduction

The Adversarial Queueing Theory (AQT) model [2,3] has been used in the latest years to study the stability and performance of packet-switched networks. The AQT model, (like other adversarial models) allows to analyze the system in a worst-case scenario, since it replaces traditional stochastic arrival assumptions in the traffic pattern by worst-case inputs. In this model, the arrival of packets to the network (i.e., the traffic pattern) is controlled by an adversary that defines,

---

[*] Partially supported by EU IST-2001-33116 (FLAGS), IST-2004-15964 (AEOLUS), COST-295 (DYNAMO), and by Spanish MCyT TIC2002-04498-C05-03 (TRACER), by the Comunidad de Madrid 07T/0022/2003, and by the Universidad Rey Juan Carlos project PPR-2004-42.

[**] We address the reader to the extended technical report version of the paper in [1], for details on the proofs of the theorems.

for each packet, the place and time in which it joins the system and, additionally it might decide the path it has to follow. In order to study non-trivial overloaded situations, the adversary is restricted so that it can not overload any link (in an amortized sense). Under these assumptions, we study the *stability* of network systems $(\mathcal{G}, \mathcal{P}, \mathcal{A})$, which are represented by three elements: the network topology $\mathcal{G}$, the protocol $\mathcal{P}$ used for scheduling the packets at every link, and the adversary $\mathcal{A}$, which defines the traffic pattern. Stability is the property that at any time the maximum number of packets present in the system is bounded by a constant that may depend on system parameters.

The original AQT model assumes a synchronous behavior of the network, that evolves in steps. In each step at most one packet crosses each link. Implicitly, this assumption means that all the packets have the same size and all the links induce the same delay in each packet transmission. There have been generalizations of the AQT model to dynamic networks, like networks with failures [4,5,6,7] and networks with links with different and possibly variable capacities or delays [8,9,10]. These works still assume a synchronous network evolution, to the point that, for instance in [8] all capacities and slow-downs must have an integral value. To the best of our knowledge, the work included in [11] is the only generalization of the AQT model considering packets of arbitrary lengths (up to a maximum) or links of arbitrary (not integral) speeds and propagation delays. In that model the adversary is more powerful than in the AQT model, and a sufficient condition on the adversary injection rate for assuring network stability is presented.

In this paper we propose a generalization of the AQT model allowing arbitrary packet lengths, link speeds (bandwidths), and link propagation delays. The network traffic flow is considered to be continuous in time. Since we do not restrict a synchronous system evolution anymore, we call this model *continuous* AQT (CAQT). Note that all the results for the AQT model which are concerned with instability, also hold for our CAQT model, e.g., the instability of the FIFO protocol at any constant rate [12]. The CAQT model is inspired in the traffic conditions of the session oriented model proposed by Cruz [13], which is widely studied in the communication networks literature. The synchronous assumptions of the AQT model limit the capacity of the adversary as well. In the CAQT model the adversary is more powerful, and any instability result shown in the AQT model can be reproduced in ours.

We show that several results from the AQT model still hold in the CAQT model. First, we show that having bounded queue size implies having bounded packet end-to-end delays and vice versa. Then, we show that networks with a directed acyclic graph (DAG) topology are always stable even if the links are fully loaded. Concerning packet scheduling protocols, we show that the well-known LIS, SIS, FTG and NFS protocols remain universally stable in our model. Finally, we show that some protocols whose policies are based on criteria concerning the length of the packets, the bandwidth of the links or their propagation delay, can configure unstable systems.

2 System Model

Like AQT, the CAQT model represents a network as a finite directed graph $\mathcal{G}$ in which the set of nodes $V(\mathcal{G})$ represent the hosts, and the set of edges $E(\mathcal{G})$ represent the links between those hosts. Each link $e \in E(\mathcal{G})$ in this graph has associated a positive but not infinite transmission speed (a bandwidth), denoted as B_e. The bandwidth of a link establishes how many bits can be transmitted in the link per second. Instead of considering the bandwidth as a synonym for parallel transmission, we relate the bandwidth to the transmission velocity. We consider that only one bit can be put in a link $e \in E(\mathcal{G})$ at each time, and that conceptually the sender puts the associated signal level to the corresponding bit for $1/B_e$ seconds for each bit. This means that a bit can be partially transmitted or partially received at a given time. Let us denote as $B_{\min} = \min_{e \in E(\mathcal{G})} B_e$ and as $B_{\max} = \max_{e \in E(\mathcal{G})} B_e$ the minimum and maximum bandwidth, respectively, of the edges in $\mathcal{G}$.

Each link $e \in E(\mathcal{G})$ has also associated a propagation delay, denoted here as P_e, being $P_e \geq 0$. This delay, measured in seconds, establishes how long it takes for a signal (the start of a bit, for instance) to traverse the link. This parameter has to do with the propagation speed of the changes in the signal that carry the bits along the physical medium used for the transmission. We will denote as $P_{\min} = \min_{e \in E(\mathcal{G})} P_e$ and $P_{\max} = \max_{e \in E(\mathcal{G})} P_e$ the minimum and maximum propagation delay, respectively, of the edges in $\mathcal{G}$.

Like in the AQT model, we assume the existence of an adversary that defines the traffic pattern of the system by choosing when and where to inject packets into the system, and the path to be followed by each of them. We assume that a packet path is edge-simple, in the sense that it does not contain the same edge more than once (it can visit the same vertex several times, though). Again, we restrict the adversary so that it can not trivially overload any link. To do so, we also define two system-wide parameters: the *injection rate* r (with $0 < r \leq 1$), and the *burstiness* b (with $b \geq 1$). For every link $e \in E(\mathcal{G})$, if we denote by $N_e(I)$ the total size (in bits) of the packets injected by the adversary in the interval I whose path contains link e, it must be satisfied that

$$N_e(I) \leq r|I|B_e + b.$$

We call an adversary $\mathcal{A}$ that satisfies this restriction an (r,b)-adversary. The injection rate r is sometimes expressed alternatively as $(1-\varepsilon)$, with $\varepsilon \geq 0$.

Regarding packet injections, we assume that the adversary injects packets instantaneously. From the above restriction, this implies that packets have a maximum size of b bits. In general, we will use L_p to denote the length (in bits) of a packet p, and $L_{\max} = \max_p L_p \leq b$ to denote the maximum packet length. Once a packet p starts being transmitted through a link $e \in E(\mathcal{G})$, it will only take $P_e + L_p/B_e$ units of time more until it crosses it completely.

Let us now look at the packet switching process. We assume that each link has associated an *output queue*, where the packets that have to be sent across the link are held. The still unsent portion of a packet that is being transmitted

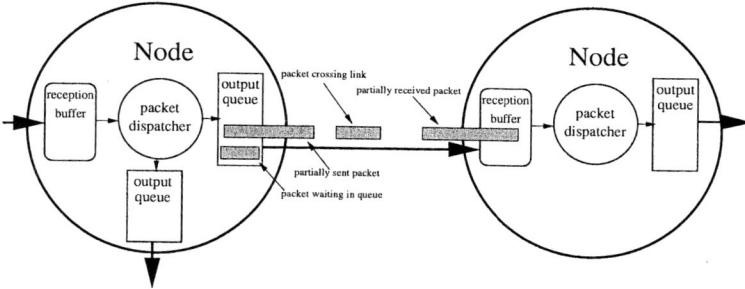

Fig. 1. Elements involved in the nodes and links of the network in the CAQT model

is also held in this queue. In fact, if a bit has only been partially sent, we assume that the still unsent portion of the bit still resides in this queue. A packet can arrive to a node either by direct injection of the adversary or by traversing some incoming link. In the latter case we assume that only full packets are dispatched (moved to an output queue). Hence, we assume that each link has a *reception buffer* in the receiving node where the portion of a partially received packet is held. As soon as the very last bit of a packet is completely received, the packet is dispatched instantaneously (by a packet dispatcher) to the corresponding output queue (or removed, if this is the final node of the packet). Figure 1 shows these network elements.

The definition of *stability* in the CAQT model is analogous to the definitions stated under other adversarial models.

Definition 1. *Let $\mathcal{G}$ be a network with a bandwidth and a propagation delay associated to each link, $\mathcal{P}$ be a scheduling policy, and $\mathcal{A}$ an (r,b)-adversary, with $0 < r \leq 1$ and $b \geq 1$. The system $(\mathcal{G}, \mathcal{P}, \mathcal{A})$ is stable if, at every moment, the total number of packets (or, equivalently, the total number of bits) in the system is bounded by a value C, that can depend on the system parameters.*

We also use common definitions of *universal stability*. We say that a scheduling policy $\mathcal{P}$ is universally stable if the system $(\mathcal{G}, \mathcal{P}, \mathcal{A})$ is stable for each network $\mathcal{G}$ and each (r,b)-adversary $\mathcal{A}$, with $0 < r < 1$ and $b \geq 1$. Similarly, we say that a network $\mathcal{G}$ is universally stable if the system $(\mathcal{G}, \mathcal{P}, \mathcal{A})$ is stable for each *greedy* scheduling policy[1] $\mathcal{P}$ and each (r,b)-adversary $\mathcal{A}$, with $0 < r < 1$ and $b \geq 1$.

Some additional notation is needed to describe the state of the queues and the packets at a specific time step. We will use $Q_t(e)$ to denote the queue size (in bits) of edge $e \in E(\mathcal{G})$ at time t, and define $Q_{\max}(e) = \max_t Q_t(e)$. Similarly,

---

[1] Greedy (or work-conserving) protocols are those forwarding a packet across a link e whenever there is at least one packet waiting to traverse e. Three types of packets may wait to traverse a link in a particular instant of time: the incoming packets arriving from adjacent links, the packets injected directly into the link, and the packets that could not be forwarded in previous steps. At each time step, only one packet from those waiting is forwarded through the link; the rest are kept in a queue.

we will use $R_t(e)$ to denote the number of bits at time t that are crossing link e, or already crossed it but are still in its reception buffer at the target node of e. Then, we define $R_{\max}(e) = \max_t R_t(e)$. Observe that $R_{\max}(e) < P_e B_e + L_{\max}$ and is hence bounded. $A_t(e)$ will denote the number of bits in the system that require to cross e and still have to be transmitted across link e at time t. The bits in $Q_t(e)$ are included in $A_t(e)$, but those in $R_t(e)$ are not.

3 General Results

We point out some general results that apply to every system $(\mathcal{G}, \mathcal{P}, \mathcal{A})$ in the CAQT model, independently of which is the network topology, the protocol used and the traffic pattern.

3.1 Relation Between Maximum Queue Size and Maximum Delay

We show that for injection rate $r < 1$, having bounded queues is equivalent to having bounded end-to-end packet delay. This generalizes a result from the AQT model to the stronger CAQT model.

Theorem 1. *Let $\mathcal{G}$ be a network, $\mathcal{P}$ a protocol, and $\mathcal{A}$ an (r,b)-adversary with $r \leq 1$ and $b \geq 1$. If the maximum end-to-end delay is bounded by D in the system $(\mathcal{G}, \mathcal{P}, \mathcal{A})$, then the maximum queue size of an edge e is bounded by $(D - P_e)B_e$.*

Theorem 2. *Let $\mathcal{G}$ be a network with $m = |E(\mathcal{G})|$ links, $\mathcal{P}$ a greedy protocol, and $\mathcal{A}$ an (r,b)-adversary, with $r = 1 - \varepsilon < 1$ and $b \geq 1$. If the maximum queue size is bounded by Q in the system $(\mathcal{G}, \mathcal{P}, \mathcal{A})$, then the end-to-end delay of a packet p with path $e_1, ..., e_d$ is bounded by*

$$\sum_{i=1}^{d} \frac{mQ + \sum_{e \in E(\mathcal{G})} R_{\max}(e) + b}{\varepsilon B_{e_i}} + P_{e_i}.$$

Then, the following corollary follows from the above two lemmas.

Corollary 1. *Let $\mathcal{G}$ be a network, $\mathcal{P}$ a greedy protocol, and $\mathcal{A}$ an (r,b)-adversary, with $r < 1$ and $b \geq 1$. In the system $(\mathcal{G}, \mathcal{P}, \mathcal{A})$ the maximum end-to-end delay experienced by any packet is bounded if and only if the maximum queue size is bounded.*

3.2 Initial Configurations

The moment in which a system $(\mathcal{G}, \mathcal{P}, \mathcal{A})$ starts its dynamics is usually denoted as t_0, and usually $t_0 = 0$. The system can start either with no packet placed at any element of the network or with some kind of initial configuration. Usually, an initial configuration C_0 consists of a set S of packets located in the output queues of the network links. Trivially, any such initial configuration for a system $(\mathcal{G}, \mathcal{P}, \mathcal{A})$ can be built from an empty initial configuration at time 0 if we allow a large enough burstiness. Thus, any system $(\mathcal{G}, \mathcal{P}, \mathcal{A})$ that starts with a non-empty initial configuration as described can be simulated by another system $(\mathcal{G}, \mathcal{P}, \mathcal{A}')$ that starts with an empty one.

Theorem 3. *Let $A_S = \max_e A_0(e)$ be the maximum number of bits that have to be transmitted across any given edge in the paths of the set S of packets. A system $(\mathcal{G}, \mathcal{P}, \mathcal{A})$, where $\mathcal{G}$ is a network, $\mathcal{P}$ a greedy protocol, and $\mathcal{A}$ an (r, b)-adversary with $r \leq 1$ and $b \geq 1$, that starts with an initial configuration C_0 consisting of a set S of packets in the network output queues can be simulated by a system $(\mathcal{G}, \mathcal{P}, \mathcal{A}')$ starting from an empty configuration, where $\mathcal{A}'$ is an $(r, A_S + b)$-adversary.*

Corollary 2. *A policy or network that is universally stable for systems with empty initial configurations is also universally stable for initial configurations in which there are initially packets in the network output queues.*

4 Stability of Networks

We focus first on the study of stability of networks. We show that networks with a directed acyclic graph topology are universally stable, even when the traffic pattern can fully load the links, i.e., even for the injection rate $r = 1$. Note that this proof is not a direct adaptation of the one in [2] for the corresponding analogous result in the AQT model.

Theorem 4. *Let $\mathcal{G}$ be a directed acyclic graph, $\mathcal{P}$ any greedy protocol, and $\mathcal{A}$ any (r, b)-adversary with $r \leq 1$ and $b \geq 1$. The system $(\mathcal{G}, \mathcal{P}, \mathcal{A})$ is stable.*

Proof: Let us first denote with T_e the node at the tail of link e (i.e., the node that contains the output queue of e), for every edge $e \in E(\mathcal{G})$. Let us also denote with $in(v)$ the set of incoming links to node v, for all $v \in V(\mathcal{G})$. Let us define the function Ψ on the edges of $\mathcal{G}$ as

$$\Psi(e) = Q_0(e) + b + R_{\max}(e) + \sum_{e' \in in(T_e)} \Psi(e').$$

If we call nodes without incoming links *sources*, we will show that $A_t(e) + R_t(e)$ is bounded by $\Psi(e)$, for all e and all $t \geq 0$, by induction on the maximum distance of T_e to a source (i.e., the length of the longest directed path from any source to T_e). Then, stability follows.

The base case of the induction is when T_e is a source. In this case, $A_t(e) = Q_t(e)$ and $\Psi(e) = Q_0(e) + b + R_{\max}(e)$. Let us fix a time t and consider two cases, depending on whether in the interval $[0, t]$ the output queue of e was empty at any time. If it was never empty, then by the restriction on the adversary and the fact that $\mathcal{P}$ is greedy we have that

$$Q_t(e) \leq Q_0(e) + rtB_e + b - tB_e \leq Q_0(e) + b.$$

Otherwise, if time t' was the last time in interval $[0,t]$ that the queue of e was empty (i.e., $Q_{t'}(e) = 0$), by the same facts,

$$Q_t(e) \leq Q_{t'}(e) + r(t-t')B_e + b - (t-t')B_e \leq b.$$

Clearly, in either case,

$$A_t(e) + R_t(e) \leq Q_t(e) + R_{\max}(e) \leq Q_0(e) + b + R_{\max}(e) = \Psi(e).$$

Now, let us assume that the maximum distance of T_e to any source is $k > 0$. Note that for any edge $e' \in in(T_e)$, the maximum distance of $T_{e'}$ to a source is at most $k - 1$. Then, by induction hypothesis, we assume that $(A_t(e') + R_t(e')) \leq \Psi(e')$ for all $t \geq 0$ and all $e' \in in(T_e)$. Note that $A_t(e) \leq Q_t(e) + \sum_{e' \in in(T_e)} (A_t(e') + R_t(e'))$. Again, we fix t and consider separately the case when the output queue of e was never empty in the interval $[0,t]$ and the case when it was. In the first case we have that

$$A_t(e) \leq Q_0(e) + rtB_e + b - tB_e + \sum_{e' \in in(T_e)} (A_0(e') + R_0(e'))$$

$$\leq Q_0(e) + b + \sum_{e' \in in(T_e)} \Psi(e').$$

In the second case, if time t' was the last time in interval $[0,t]$ that the queue of e was empty (i.e., $Q_{t'}(e) = 0$), we have that

$$A_t(e) \leq Q_{t'}(e) + r(t-t')B_e + b - (t-t')B_e + \sum_{e' \in in(T_e)} (A_{t'}(e') + R_{t'}(e'))$$

$$\leq b + \sum_{e' \in in(T_e)} \Psi(e').$$

In either case, we have that

$$A_t(e) + R_t(e) \leq Q_0(e) + b + R_{\max}(e) + \sum_{e' \in in(T_e)} \Psi(e') = \Psi(e). \quad \blacksquare$$

5 Stability of Queueing Policies

Stability can also be studied from the point of view of the protocols. Unstable protocols in the AQT model are also unstable in the CAQT model. In the following, we show that the so-called LIS, SIS, FTG and NFS protocols are universally stable in the CAQT model, as they were in the AQT model [3].

5.1 Universal Stability of LIS

The LIS (*longest-in-system*) protocol gives priority to the packet which was earliest injected in the system. Independently of the network topology and the (r,b)-adversary, any system $(\mathcal{G}, \text{LIS}, \mathcal{A})$ is stable.

Theorem 5. *Let $\mathcal{G}$ be a network, $\mathcal{A}$ an (r,b)-adversary with $r = 1 - \varepsilon < 1$, and d the length of the longest simple directed path in $\mathcal{G}$. Then all packets spend less than $(\frac{b}{B_{\min}} + P_{\max})/(r\varepsilon^d)$ time in the system $(\mathcal{G}, \text{LIS}, \mathcal{A})$.*

Corollary 3. *Let $\mathcal{G}$ be a network, $\mathcal{A}$ an (r,b)-adversary with $r = 1 - \varepsilon < 1$, and d the length of the longest edge-simple directed path in $\mathcal{G}$. Then, the system $(\mathcal{G}, \text{LIS}, \mathcal{A})$ is stable, and there are always less than $(\frac{b}{B_{\min}} + P_{\max})\varepsilon^{-d}B_{\max} + b$ bits trying to cross any edge e.*

5.2 Universal Stability of SIS

The SIS (*shortest-in-system*) protocol gives priority to the packet which was injected the latest in the system. In the case of the SIS protocol, bounding the size of the packets recently injected is related to bounding the time that a packet packet p requires to cross the edge e. The following lemma provides us with such a bound:

Lemma 1. *Let p be a packet that, at time t, is waiting in the queue of edge $e \in E(\mathcal{G})$. At that instant, let $k - 1$ be the total size in bits of the packets in the system that also require e and that may have priority over p (i.e., that were injected later in the system). Then p will start crossing e in at most $(k+b)/(\varepsilon B_e)$ units of time.*

Observe that, once the packet p starts being transmitted through the link e, it will only take $P_e + L_p/B_e$ units of time more until it crosses it completely. Using the bound obtained in Lemma 1 in a recursive way, we can derive more general bounds, thus proving the universal stability of the SIS protocol.

Theorem 6. *Let $\mathcal{G}$ be a network, $\mathcal{A}$ an (r,b)-adversary with $r = 1 - \varepsilon < 1$ and $b \geq 1$, and d the length of the longest edge-simple directed path in $\mathcal{G}$. The system $(\mathcal{G}, \text{SIS}, \mathcal{A})$ is stable and, moreover:*

- *no queue ever contains $k_d + L_{\max}$ bits, and*
- *no packet spends more than $(d(b + \varepsilon L_{\max}) + \sum_{i=1}^{d} k_i)/(\varepsilon B_{\min}) + dP_{\max}$ time in the system.*

where k_i is defined according to the following recurrence:

$$k_i = \begin{cases} b & \text{for } i = 1 \\ k_{i-1} + (1-\varepsilon)\left(\frac{k_{i-1}+b}{\varepsilon B_{\min}} + \frac{L_{\max}}{B_{\min}} + P_{\max}\right)B_{\max} + b & \text{for } 1 < i \leq d \end{cases}$$

5.3 Universal Stability of FTG

The FTG (*farthest-to-go*) protocol gives priority to the packet which still has to traverse the longest path until reaching its destination. We show that FTG is universally stable by using the fact that all the packets have to traverse at least one edge, and that all the packet go at most d edges further.

Theorem 7. *Let $\mathcal{G}$ be a network with $m = |E(\mathcal{G})|$ links, $\mathcal{A}$ an (r,b)-adversary with $r < 1$ and $b \geq 1$, and d the length of the longest edge-simple directed path in $\mathcal{G}$. The system $(\mathcal{G}, \text{FTG}, \mathcal{A})$ is stable and:*

- *there are never more than k_1 bits in the system,*
- *no queue ever contains more than $k_2 + b$ bits, and*
- *no packet spends more than $dP_{\max} + (d(b + \varepsilon L_{\max}) + \sum_{i=2}^{d} k_i)/(\varepsilon B_{\min})$ time in the system.*

where k_i is defined according to the following recurrence:

$$k_i = \begin{cases} 0 & \text{for } i > d \\ mk_{i+1} + mb + \sum_{e \in E(\mathcal{G})} R_{\max}(e) & \text{for } 1 \leq i \leq d \end{cases}$$

5.4 Universal Stability of NFS

The NFS (*nearest-from-source*) protocol gives priority to the packet which is closest to its origin, i.e., which has traversed the less portion of its whole path. We show that NFS is universally stable by using a similar argument as the one used for FTG; however the bounds will be provided now taking the length of the longest path as a reference point.

Theorem 8. *Let $\mathcal{G}$ be a network with $m = |E(\mathcal{G})|$ links, $\mathcal{A}$ an (r,b)-adversary with $r < 1$ and $b \geq 1$, and d the length of the longest edge-simple directed path in $\mathcal{G}$. The system $(\mathcal{G}, \text{NFS}, \mathcal{A})$ is stable and:*

- *there are never more than k_d bits in the system,*
- *no queue ever contains more than $k_{d-1} + b$ bits, and*
- *no packet spends more than $dP_{\max} + (d(b + \varepsilon L_{\max}) + \sum_{i=1}^{d-1} k_i)/(\varepsilon B_{\min})$ time in the system.*

where k_i is defined according to the following recurrence:

$$k_i = \begin{cases} 0 & \text{for } i = 0 \\ mk_{i-1} + mb + \sum_{e \in E(\mathcal{G})} R_{\max}(e) & \text{for } 1 \leq i \leq d \end{cases}$$

6 Instability of Queueing Policies

In this section we introduce some new protocols that base their policies in the main features of the CAQT model, namely, the length of the packets, the edge bandwidths and the edge propagation delays. We show that the CAQT model is strictly stronger than the AQT model by presenting scheduling policies that are unstable under the former while they are universally stable under the latter.

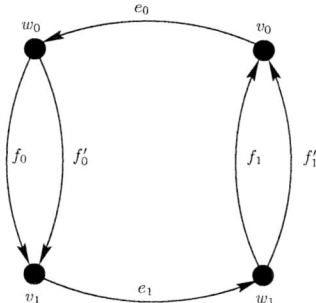

Fig. 2. Baseball network $\mathcal{G}_B$ presented in [3]

6.1 Instability by Difference in Packet Length

Consider the LPL (*longest-packet-length*) protocol which gives priority to the packet with longest length. Let us denote as LPL-LIS the same protocol when ties are broken according to the LIS policy. Note that LPL-LIS is universally stable under the AQT model, since in this model all packets have the same length and hence the policy simply becomes LIS [3]. However, we show here that LPL-LIS is unstable in an extension of AQT with multiple packet lengths just by considering two different packet lengths (1 and 2). For simplicity we will assume that time advances in synchronous steps (as in AQT). Packets of length 2 take 2 steps to cross each link. In the LPL-LIS protocol, these double packets will have priority over the single packets. Note that this model is trivially included in CAQT. To show the instability of the LPL-LIS protocol, we use the baseball network presented in [3] (see Figure 2).

Theorem 9. *Let $\mathcal{G}_B$ be the graph with nodes $V(\mathcal{G}_B) = \{v_0, v_1, w_0, w_1\}$, and edges $E(\mathcal{G}_B) = \{(v_0, w_0), (v_1, w_1), (w_1, v_0), (w_1, v_0), (w_0, v_1), (w_0, v_1)\}$. All the edges in $E(\mathcal{G}_B)$ have bandwidth 1 and null propagation delay. For $r > 1/\sqrt{2}$ there is an (r, b)-adversary $\mathcal{A}$ that makes the system $(\mathcal{G}_B, \text{LPL-LIS}, \mathcal{A})$ to be unstable only with packets of length 1 and 2.*

6.2 Instability by Difference in Bandwidth

Consider the SPL (*slowest-previous-link*) protocol which gives priority to the packet whose last crossed link was the slowest, i.e., had the smallest bandwidth. This policy aims to equilibrate the lost in transmission velocity suffered in previous links. Let us denote as SPL-NFS this protocol, when ties are broken according to the NFS protocol. Observe that the SPL-NFS protocol is equivalent to NFS in the AQT model and thus universally stable [3]. However, we show that in a similar way as shown for the LPL-LIS protocol, the SPL-NFS protocol can be made unstable in the CAQT model.

Theorem 10. *Let $\mathcal{G}_B$ be the graph with nodes $V(\mathcal{G}_B) = \{v_0, v_1, w_0, w_1\}$ and edges $E(\mathcal{G}_B) = \{(v_0, w_0), (v_1, w_1), (w_1, v_0), (w_1, v_0), (w_0, v_1), (w_0, v_1)\}$. Let $\mathcal{G}$ be the graph obtained from $\mathcal{G}_B$ whose set of nodes is $V(\mathcal{G}) = V(\mathcal{G}_B) \cup \{v'_0, v'_1, w'_0, w'_1\}$, and whose set of edges is $E(\mathcal{G}) = E(\mathcal{G}_B) \cup \{(v'_0, v_0), (v'_1, v_1), (w'_0, w_0), (w'_1, w_1)\}$. Those edges inciding to v_0 and v_1 have bandwidth 2, while the rest have bandwidth 1. All the edges have null propagation delays. For $r > 1/\sqrt{2}$ there is an (r, b)-adversary $\mathcal{A}$ that makes the system $(\mathcal{G}, \text{SPL-NFS}, \mathcal{A})$ to be unstable.*

6.3 Instability by Difference in Propagation Delays

Consider the SPP (*smallest-previous-propagation*) protocol which gives priority to the packet whose previously traversed edge had smallest propagation delay, and combine it with NFS to break ties. Let us denote this protocol as SPP-NFS. Observe that the SPP-NFS protocol is equivalent to NFS in the AQT model and thus universally stable [3]. However, we show with the SPP-NFS protocol as example, that just the fact of considering propagation delays can make a policy unstable in CAQT.

Theorem 11. *Let $\mathcal{G}_B$ be the graph with nodes $V(\mathcal{G}_B) = \{v_0, v_1, w_0, w_1\}$ and edges $E(\mathcal{G}_B) = \{(v_0, w_0), (v_1, w_1), (w_1, v_0), (w_1, v_0), (w_0, v_1), (w_0, v_1)\}$. Let $\mathcal{G}$ be the graph obtained from $\mathcal{G}_B$ whose set of nodes is $V(\mathcal{G}) = V(\mathcal{G}_B) \cup \{v'_0, v'_1, w'_0, w'_1\}$, and whose set of edges is $E(\mathcal{G}) = E(\mathcal{G}_B) \cup \{(v'_0, v_0), (v'_1, v_1), (w'_0, w_0), (w'_1, w_1)\}$. Those edges inciding to v_0 and v_1 have propagation delay 1, while the rest have null propagation delay. All the edges have unary bandwidth. For $r > 1/\sqrt{2}$ there is an (r, b)-adversary $\mathcal{A}$ that makes the system $(\mathcal{G}, \text{SPP-NFS}, \mathcal{A})$ to be unstable.*

7 Conclusions and Open Questions

We consider a networking scenario in which packets can have arbitrary lengths, and the network links may have different speeds and propagation delays. Taking into account these features, we have presented a generalization of the well-known Adversarial Queueing Theory (AQT) model which does not assume anymore synchronicity in the evolution of the system, and makes it more appropriate for more realistic continuous scenarios. We called it the CAQT model.

We have shown that, in the CAQT model having bounded queues is equivalent to having bounded packet end-to-end delays. From the network point of view, we show that networks with a directed acyclic topologies are universally stable even when the traffic pattern fully loads the links. From the protocol point of view, we have also shown that the well-known LIS, SIS, FTG and NFS protocols remain universally stable in the CAQT model. New protocols have also been proposed which are universally stable in the AQT model but unstable in the CAQT model.

Many interesting questions remain still open in the CAQT model. More results are needed concerning the stability of networks, starting from simple topologies like the ring, to finally tackle the universal stability of networks. It would be of interest to know the queue sizes to be expected (as is was studied in [3,14] for the AQT model), as well as which conditions guarantee that all the packets are actually delivered to destination (as it was studied in [15] for AQT).

Acknowledgments

The authors would like to thank the unknown referees for their comments and suggestions.

References

1. Blesa, M., Calzada, D., Fernández, A., López, L., Martínez, A., Santos, A., Serna, M.: Adversarial queueing model for continuous network dynamics. Research Report LSI-05-10-R, Dept. Llenguatges i Sistemes Informàtics, UPC (2005)
2. Borodin, A., Kleinberg, J., Raghavan, P., Sudan, M., Williamson, D.: Adversarial queueing theory. Journal of the ACM **48** (2001) 13–38
3. Andrews, M., Awerbuch, B., Fernández, A., Kleinberg, J., Leighton, T., Liu, Z.: Universal stability results for greedy contention-resolution protocols. Journal of the ACM **48** (2001) 39–69
4. Awerbuch, B., Berenbrink, P., Brinkmann, A., Scheideler, C.: Simple routing strategies for adversarial systems. In: 42th. IEEE Symposium on Foundations of Computer Science, IEEE Computer Society Press (2001) 158–167
5. Anshelevich, E., Kempe, D., Kleinberg, J.: Stability of load balancing algorithms in dynamic adversarial systems. In: 34th. Annual ACM Symposium on Theory of Computing, ACM Press (2002) 399–406
6. Àlvarez, C., Blesa, M., Díaz, J., Fernández, A., Serna, M.: Adversarial models for priority-based networks. Networks **45** (2005) 23–35
7. Àlvarez, C., Blesa, M., Serna, M.: The impact of failure management on the stability of communication networks. In: 10th International Conference on Parallel and Distributed Systems, IEEE Computer Society Press (2004) 153–160
8. Borodin, A., Ostrovsky, R., Rabani, Y.: Stability preserving transformations: Packet routing networks with edge capacities and speeds. Journal of Interconnection Networks **5** (2004) 1–12
9. Koukopoulos, D., Mavronicolas, M., Spirakis, P.: Instability of networks with quasi-static link capacities. In: 10th Internaltional Colloquium on Structural Information Complexity. Volume 17 of Proceedings in Informatics., Carleton Scientific (2003) 179–194
10. Koukopoulos, D., Mavronicolas, M., Spirakis, P.: Performance and stability bounds for dynamic networks. In: 7th International Conference on Parallel Architectures, Algorithms and Networks, IEEE Computer Society Press (2004) 239–246
11. Echagüe, J., Cholvi, V., Fernández, A.: Universal stability results for low rate adversaries in packet switched networks. IEEE Communication Letters **7** (2003) 578–580
12. Bhattacharjee, R., Goel, A., Lotker, Z.: Instability of FIFO at arbitrarily low rates in the adversarial queueing model. SIAM Journal on Computing **34** (2004) 318–332
13. Cruz, R.: A calculus for network delay. Part I (network elements in isolation) and II (network analysis). IEEE Transactions on Information Theory **37** (1991) 114–141
14. Weinard, M.: The necessity of timekeeping in adversarial queueing. In: 4th International Workshop on Efficient and Experimental Algorithms. Volume 3503 of Lecture Notes in Computer Science., Springer-Verlag (2005) 440–451
15. Rosén, A., Tsirkin, M.: On delivery times in packet networks under adversarial traffic. In: 16th ACM Symposium on Parallel Algorithms and Architectures, ACM Press (2004) 1–10

Coloring Sparse Random k-Colorable Graphs in Polynomial Expected Time

Julia Böttcher

Humboldt-Universität zu Berlin, Institut für Informatik,
Unter den Linden 6, 10099 Berlin, Germany
boettche@informatik.hu-berlin.de

Abstract. Feige and Kilian [5] showed that finding reasonable approximative solutions to the coloring problem on graphs is hard. This motivates the quest for algorithms that either solve the problem in most but not all cases, but are of polynomial time complexity, or that give a correct solution on all input graphs while guaranteeing a polynomial running time on average only. An algorithm of the first kind was suggested by Alon and Kahale in [1] for the following type of random k-colorable graphs: Construct a graph $\mathcal{G}_{n,p,k}$ on vertex set V of cardinality n by first partitioning V into k equally sized sets and then adding each edge between these sets with probability p independently from each other. Alon and Kahale showed that graphs from $\mathcal{G}_{n,p,k}$ can be k-colored in polynomial time with high probability as long as $p \geq c/n$ for some sufficiently large constant c. In this paper, we construct an algorithm with polynomial expected running time for $k = 3$ on the same type of graphs and for the same range of p. To obtain this result we modify the ideas developed by Alon and Kahale and combine them with techniques from semidefinite programming. The calculations carry over to general k.

1 Introduction

The coloring problem on graphs remains one of the most demanding algorithmic tasks in graph theory. Since it is one of the classical $\mathcal{NP}$-hard problems (see [7]) it is unlikely that efficient coloring algorithms exist. If no exact answer to a problem can be found within a reasonable amount of time, one alternative is to search for approximation algorithms. However, for the coloring problem even this suboptimal approach fails. While the standard greedy heuristic with high probability does not use more than $2\chi(G)$ colors on a random input graph G, guaranteeing a similar performance ratio for all graphs is intractable under reasonable computational assumptions. In fact, Feige and Kilian [5] proved that for all $\epsilon > 0$ it is impossible to approximate the coloring problem within $n^{1-\epsilon}$, provided $\mathcal{ZPP} \neq \mathcal{NP}$, where n is the number of vertices of the input graph. Moreover, Khanna, Linial, and Safra [9] showed that coloring 3-colorable graphs with 4 colors is $\mathcal{NP}$-hard.

Accordingly, different approaches must be pursued. One possibility is to ask for algorithms that work with high probability. While finding algorithms of this type is not too difficult for dense k-colorable graphs [4,11,13], it turns out to be harder for sparse k-colorable graphs. For constructing such sparse graphs set $p = c/n$ for a constant c in the following process: Partition the vertex set V into k sets C_i of equal size and allow only edges between these sets, taking each one independently with probability p. We denote graphs obtained in this way by $\mathcal{G}_{n,p,k}$. The sets C_i are also called the *color classes* of $\mathcal{G}_{n,p,k}$.

In 1997, Alon and Kahale [1] established the following result for $\mathcal{G}_{n,p,k}$.

Theorem 1 (Alon and Kahale [1]). *Let $p > c/n$ for some sufficiently large constant c. Then there is a polynomial time algorithm for k-coloring $\mathcal{G}_{n,p,k}$ with high probability.*

But an algorithm that works with high probability has one drawback: For some inputs it does not provide any solution at all. Alternatively, we could require that the algorithm always gives a correct answer to the problem under study but performs well only on average: An algorithm $\mathcal{A}$ with running time $t_\mathcal{A}(G)$ on input G has *polynomial expected running time* on $\mathcal{G}_{n,p,k}$ if $\sum_G t_\mathcal{A}(G) \cdot \mathbf{P}[\mathcal{G}_{n,p,k} = G]$ remains polynomial. Here, the sum ranges over all graphs on n vertices. Observe that this is a stronger condition than to work correctly with high probability: An algorithm that k-colors $\mathcal{G}_{n,p,k}$ in polynomial expected running time also solves the k-coloring problem with high probability in polynomial time.

In this paper we present an algorithm for coloring sparse 3-colorable graphs with 3 colors in polynomial expected time.

Theorem 2. *If $p > c/n$ for some sufficiently large constant c, then there is an algorithm COLOR that 3-colors $\mathcal{G}_{n,p,3}$ in polynomial expected time.*

This improves on results of Subramanian [12] and Coja-Oghlan [2] and answers a question of Subramanian [12] and Krivelevich [10]. The calculations carry over to general k. The best known previous algorithm is due to Coja-Oghlan [2] and k-colors graphs from $\mathcal{G}_{n,p,k}$ in polynomial expected time if $np \geq c \cdot \max(k \ln n, k^2)$ for a sufficiently large constant c.

The main philosophy of COLOR can be described as follows. On input G, we start by executing a polynomial time algorithm $\mathcal{A}$: In the first step $\mathcal{A}$ determines an initial coloring of G which colors all but a constant fraction of G correctly with high probability. $\mathcal{A}$ then refines this initial coloring by using different combinatorial methods which are modifications of the methods used by Alon and Kahale. With high probability this results in a valid coloring of G. However, since we are interested in returning a valid coloring for all graphs, COLOR also has to take care of exceptional cases. In the case that $\mathcal{A}$ does not produce a valid coloring of G, we therefore proceed by removing a set Y of vertices from G, rerun $\mathcal{A}$ on $G \setminus Y$ and treat Y with brute force coloring methods. In the beginning, Y contains only a single vertex. We repeat this procedure and gradually increase $|Y|$ until G is finally properly colored. We verify that COLOR has polynomial expected running time by showing that $\mathcal{A}$ can handle all but a small number of vertices for most graphs from $\mathcal{G}_{n,p,3}$.

In addition, and in contrast to Alon and Kahale, we apply the concept of semidefinite programming in order to obtain a good initial coloring of $\mathcal{G}_{n,p,3}$ in the first stage of COLOR. To this end we use the semidefinite program $\mathcal{SDP}_3$ introduced by Frieze and Jerrum [6]. The value of $\mathcal{SDP}_3$ on $\mathcal{G}_{n,p}$ has been investigated by Coja-Oghlan, Moore, and Sanwalani [3]. We use their result to show that $\mathcal{SDP}_3$ behaves similarly on $\mathcal{G}_{n,p,3}$. This will then allow us to construct a coloring of $\mathcal{G}_{n,p,3}$ from a solution of $\mathcal{SDP}_3$ that already colors all but a small linear fraction of the input graph correctly. Similar methods have been used by Coja-Oghlan [2].

The remainder of this paper is structured as follows: In Section 2 we investigate the behaviour of $\mathcal{SDP}_3$ on $\mathcal{G}_{n,p,3}$, in Section 3 we give the details of our coloring algorithm, in Section 4 its analysis and in Section 5 some concluding remarks.

2 The Value of $\mathcal{SDP}_3$ on Graphs from $\mathcal{G}_{n,p,3}$

Recall that a *k-cut* of a graph G is a partition of $V(G)$ into k disjoint sets $V_1, \ldots, V_k$, its weight is the total number of edges crossing the cut, and that MAX-k-CUT is the the problem of finding a k-cut of maximum weight.

In the algorithm COLOR we make use of the following SDP relaxation $\mathcal{SDP}_3$ of MAX-3-CUT due to Frieze and Jerrum [6] which provides an upper bound for MAX-3-CUT:

$$\max \sum_{vw \in E(G)} \frac{2}{3} (1 - \langle \mathbf{x}_v \,|\, \mathbf{x}_w \rangle)$$
$$\text{s.t.} \quad \|\mathbf{x}_v\| = 1 \quad \forall v \in V,$$
$$\langle \mathbf{x}_v \,|\, \mathbf{x}_w \rangle \geq -\frac{1}{2} \quad \forall v, w \in V.$$

Here, the maximum runs over all vector assignments $(\mathbf{x}_v)_{v \in V(G)}$ obeying $\mathbf{x}_v \in \mathbb{R}^{|V|}$. Observe that, if G_1 is a subgraph of G_2, then $\mathcal{SDP}_3(G_1) \leq \mathcal{SDP}_3(G_2)$. One way to realize a feasible solution of $\mathcal{SDP}_3$ corresponding to a 3-cut V_1, V_2, V_3 of G is to assign the same vector $\mathbf{s}_i$ to each vertex in C_i in such a way that $\langle \mathbf{s}_i \,|\, \mathbf{s}_j \rangle = -1/2$ for $i \neq j$.

Moreover, there is an obvious connection between maximum 3-cuts and 3-colorings: In the case of a 3-colorable graph G a maximum 3-cut simply contains all edges. Then, we know that each edge contributes exactly 1 to the value of $\mathcal{SDP}_3$. In this case, the special feasible solution to $\mathcal{SDP}_3$ discussed above is optimal. Conversely, if the optimal solution of $\mathcal{SDP}_3(G)$ has this structure, then it is clearly easy to read off a proper 3-coloring of G from this solution. Unfortunately, the position of the vectors $\mathbf{x}_v$ can get "far away" from this ideal picture in general. In this section we show however that with high probability such a scenario does not occur in the case of random 3-colorable graphs from $\mathcal{G}_{n,p,3}$. Although for such graphs the vectors corresponding to vertices of one color class do not necessarily need to be equal, most of them will be comparably close. Similar techniques were used in [2].

In [3] Coja-Oghlan, Moore, and Sanwalani studied the behaviour of $\mathcal{SDP}_3$ on $\mathcal{G}_{n,p}$. They obtained the following result, which will be the key ingredient to our analysis of $\mathcal{SDP}_3$ on graphs from $\mathcal{G}_{n,p,3}$.

Theorem 3 (Coja-Oghlan, Moore & Sanwalani [3]). *If $p \geq c/n$ for sufficiently large c then*

$$\mathcal{SDP}_3(\mathcal{G}_{n,p}) \leq \frac{2}{3}\binom{n}{2}p + \mathcal{O}\left(\sqrt{n^3 p(1-p)}\right) \tag{1}$$

with probability at least $1 - \exp(-3n)$.

Note that $2\binom{n}{2}p/3$ is also the size of a random 3-cut in $\mathcal{G}_{n,p}$. So Theorem 3 estimates the difference of the sizes of a maximum 3-cut and a random 3-cut in $\mathcal{G}_{n,p}$.

Let $G = (V, E) \in \mathcal{G}_{n,p,3}$. We can construct a random graph $G^* \in \mathcal{G}_{n,p}$ from G by inserting additional edges with probability p within each color class. The following lemma investigates the effect of this process on the value of $\mathcal{SDP}_3$.

Lemma 1. *Consider a graph $G^* = (V, E^*)$ from $\mathcal{G}_{n,p}$ with $V = [n]$ and let $G = (V, E)$ be the subgraph of G^* with edges $E = E^* \cap \{vw \mid \lceil 3v/n \rceil \neq \lceil 3w/n \rceil\}$. Then for some constant c' not depending on d*

$$\mathcal{SDP}_3(G^*) - \mathcal{SDP}_3(G) \leq c'\sqrt{n^3 p} \tag{2}$$

with probability at least $1 - \exp(-5n/2)$.

Proof. In order to establish this result we prove that

$$\frac{2}{3}\binom{n}{2}p - \frac{c'}{2}\sqrt{n^3 p} \leq \mathcal{SDP}_3(G) \leq \mathcal{SDP}_3(G^*) \leq \frac{2}{3}\binom{n}{2}p + \frac{c'}{2}\sqrt{n^3 p}$$

holds with the same probability. In fact, the second inequality holds by construction since G is a subgraph of G^* and the third inequality is asserted by Theorem 3 if we choose c' accordingly. Thus it remains to show the first inequality. This is obtained by a straightforward application of the Chernoff bound and the fact that $\mathcal{SDP}_3(G) = |E|$ as mentioned earlier.

Equation (2) asserts that the values of $\mathcal{SDP}_3$ for G and G^* are not likely to differ much, if the additional edges within the color classes C_i of G are chosen at random. It follows that in an optimal solution to $\mathcal{SDP}_3(G)$, most of the vectors corresponding to vertices of C_i for a particular i can not be far apart. This is shown in the next lemma.

If not stated otherwise, we consider $\mathcal{SDP}_3$ on input G from now on. Let $(\mathbf{x}_v)_{v \in V(G)}$ be an optimal solution to $\mathcal{SDP}_3(G)$. Then we call

$$\mathbf{N}^\mu(v) := \{v' \in V \mid \langle \mathbf{x}_v \mid \mathbf{x}_{v'} \rangle > 1 - \mu\}$$

the μ-neighborhood of v.

Lemma 2. *For fixed ϵ with $0 < \epsilon < 1/2$ there is a constant $0 < \mu < 1/2$ such that for any μ' with $\mu \leq \mu' < 1/2$ the following holds with probability greater than $1 - \exp(-7n/3)$: For each $i \in \{1, 2, 3\}$ there is a vertex $v_i \in C_i$ such that the set $\mathbf{N}^\mu(v_i)$ contains at least $(1-\epsilon)n/3$ vertices of the color class C_i and the set $\mathbf{N}^{\mu'}(v_i)$ contains at most $\epsilon n/3$ vertices from other color classes C_j $(j \neq i)$.*

For proving this lemma, we first observe that the edges uv of G^* with u,v in one color class C_i of G and $\langle \mathbf{x}_u \,|\, \mathbf{x}_v \rangle$ small, form a random subgraph of G^*. From Lemma 1 we can then deduce the first statement of Lemma 2. The second statement follows since $\mathbf{N}^{\mu'}(v)$ induces an empty graph in G for $\mu' < 1/2$. We omit the details.

In the following we will call a vertex $v \in C_i$ obeying the properties asserted by Lemma 2 an (ϵ, μ, μ')-*representative* for color class C_i. In the case $\mu' = \mu$ we omit the parameter μ'.

3   The Algorithm

Roughly speaking, there are two basic principles underlying the mechanisms of COLOR. On the one hand a number of steps, also called the *main steps*, aim at constructing a valid 3-coloring of the input graph with sufficiently high probability. These are the *initial step*, the *iterative recoloring step*, the *uncoloring step* and the *extension step*. However, on atypical graphs this approach might fail. For guaranteeing a valid output for each input, COLOR also has to handle this case; possibly by using computationally expensive methods. Accordingly, the purpose of the remaining operations is to fix the mistakes of the main steps on such atypical graphs. This constitutes the second principle, the so-called *recovery procedure* of COLOR.

The details of COLOR are given in Algorithm 1. We now briefly describe the main steps and then turn to the recovery procedure.

The Initial Step: (Steps 1, 5, and 6) This step is concerned with finding an initial coloring Υ_0 of the input graph $G = (V, E)$ such that Υ_0 fails on at most ϵn vertices of G. Here, ϵ is small but constant. To obtain Υ_0 we apply the SDP relaxation $\mathcal{SDP}_3$ of MAX-3-CUT. An optimal solution of this semidefinite program can efficiently be computed within any numerical precision (cf. [8]). This solution gives rise to the coloring Υ_0 by grouping vertices whose corresponding vectors have large scalar product into the same color class. In the algorithm COLOR we use a randomized method for this grouping process.

The Iterative Recoloring Step: (Step 8) This step refines the initial coloring in order to find a valid coloring of a much larger vertex set by repeating the following step at most a logarithmic number of times: Assign to each vertex in G the color that is the least favorite among its neighbors. In Section 4 we will show that this approach is indeed successful, in the sense that with sufficiently high probability at most $\alpha_0 n$ vertices are still colored incorrectly after the iterative recoloring step, where α_0 is of order $\exp(-np)$.

The Uncoloring Step: (Step 9) This step proceeds iteratively as well. In each iteration, the uncoloring step uncolors all vertices that have less than $np/6$ neighbors of some color other than their own. Observe that in a "typical" graph

Algorithm 1: COLOR(G)

Input: a graph $\mathcal{G}_{n,p,3} = G = (V, E)$
Output: a valid coloring of $\mathcal{G}_{n,p,3}$

begin

1. let $(\mathbf{x}_v)_{v \in V(G)}$ be an optimal solution of $\mathcal{SDP}_3(G)$;
2. **for** $0 \le y \le n$ **do**
3.   **foreach** $Y \subset V$ with $|Y| = y$ and **each** valid 3-coloring Υ_Y of Y **do**
4.   **for** $\mathcal{O}(n)$ times **do**
            /** The initial step **/
5.   Randomly choose three vectors $\mathbf{x}_1, \mathbf{x}_2, \mathbf{x}_3 \in \{\mathbf{x}_v | v \in V(G)\}$;
6.   Extend Υ_Y to a coloring Υ_0 of G by setting $\Upsilon_0(v) := i$ for all $v \in G - Y$ where i is such that $\langle \mathbf{x}_v | \mathbf{x}_i \rangle$ is maximal ;
7.   **for** $t = \log n$ **downto** $t = 0$ **do**
               /** The iterative recoloring step **/
8.   **for** $0 \le s < t$ **do**
                  Construct a coloring Υ_{s+1} of G with $\Upsilon_{s+1}(v) := \Upsilon_Y(v)$ for $v \in Y$ and $\Upsilon_{s+1}(v) := i$ for $v \notin Y$ where i minimizes $|\mathbf{N}(v) \cap \Upsilon_s^{-1}(i)|$;
            Set $\Upsilon' := \Upsilon_t$;
            /** The uncoloring step **/
9.   **while** $\exists v \in G - Y$ with $\Upsilon'(v) = i$ and $|\{w | w \in \mathbf{N}(v), \Upsilon'(w) = j\}| < np/6$ for some $j \ne i$ **do** uncolor v in Υ' ;
            /** The extension step **/
10.   **if** each component of uncolored vertices is of size at most $\alpha_0 n$ **then**
                Extend the partial coloring Υ' to a coloring Υ of G by exhaustively trying each coloring of each component in the set of uncolored vertices ;
11.   **if** Υ is a valid coloring of G **then** return $\Upsilon(G)$ and stop;

end

from $\mathcal{G}_{n,p,3}$ a "typical" vertex and its neighbors will not have this property if they are colored correctly.

The Extension Step: (Step 10) Here, an exact coloring method is used to extend the partial coloring obtained to the whole graph $\mathcal{G}_{n,p,3}$. In this process the components induced on the uncolored vertices are treated seperately. On each such component K, the algorithm tries all possible colorings until it finds one that is compatible to the coloring of the rest of G.

The main steps are all we need for 3-coloring $\mathcal{G}_{n,p,3}$ with high probability This will be formally proven in Section 4; the analysis of the recoloring, the uncoloring and the extension step is similar to that of Alon and Kahale [1].

If the main steps do not produce a valid coloring of G the recovery procedure (the loops in Steps 2 to 7) comes into play. The concept is as follows. Assume that for an input graph G the main steps of COLOR produce a correct coloring on the subgraph induced by $V \setminus Y$ for some $Y \subset V$ but "fail" on Y. Then an easy way of "repairing" the coloring obtained is to exhaustively test all valid colorings of Y. Of course, we neither know this set Y nor its size $|Y|$. To deal with these two problems, COLOR proceeds by trying all possible subsets Y of V with $|Y| = y$. Here, we start with $y = 0$ and then gradually increase the value of y until a valid coloring of G is determined. This is performed in Steps 2 and 3 of the recovery procedure. We also call Step 3, where all colorings of all vertex sets of size y are constructed, the *brute force coloring method* or *repair mechanism* for these sets.

Since the size of Y in the recovery procedure (Step 3) is increased until a valid coloring is obtained, the correctness of Algorithm 1 is inherent (ultimately, a proper coloring will be found in the last iteration, when Y contains all vertices of G). An analysis of the expected running time of Algorithm 1 will be presented in the next section.

4 Analysis of the Algorithm

In the following we assume that G is a graph sampled from $\mathcal{G}_{n,p,3}$ where $d := np > c$ for some sufficiently large constant c. Moreover, let **3-COL**$(x) := 3^x$ be the time needed to find all 3-colorings of a graph of order x.

4.1 The Initial Step

Lemma 2 guarantees that, given an optimal solution of $\mathcal{SDP}_3(G)$, we can construct a reasonably good initial coloring Υ_0 via the sets $\mathbf{N}^\mu(v_i)$ by choosing an appropriate representative v_i for each color class C_i and setting $\Upsilon_0(v) := i$ for all $v \in N^\mu(v_i)$. Here, ties are broken arbitrarily and vertices not appearing in any of the sets $N^\mu(v_i)$ get assigned an arbitrary color.

Thus it remains to determine appropriate representatives. The easiest way is to simply try all different triples of vertices from G as representatives. This, however, introduces an extra factor of n^3 in the running time. In order to reduce this factor to a linear one, Algorithm 1 proceeds differently. Let v_1, v_2 and v_3 be (ϵ', μ, μ')-representatives of the color classes C_1, C_2, and C_3, respectively. We will make use of the following lemma.

Lemma 3. *If v_i is a (ϵ', μ, μ')-representative for C_i and $\mu' > 4\mu + \sqrt{2\mu}$, then each vertex $v \in \mathbf{N}^\mu(v_i)$ is an $(\epsilon', 4\mu)$-representative of color class C_i.*

By choosing $\mu' > 4\mu + \sqrt{2\mu}$ (and μ sufficiently small such that $\mu' < 1/2$) we therefore get at least $(1 - \epsilon') \cdot n/3$ representatives per color class. But then the probability of obtaining a set of representatives for G by picking three vertices r_1, r_2, r_3 from V at random is at least $(1 - 3 \cdot \epsilon')/9$. Repeating this process raises the probability of success. More specifically, the probability that in $c'n$ trials none yields a triple of $(\epsilon', 4\mu)$-representatives is smaller than $(8/9 + \epsilon'/3)^{c'n}$. Here, $c' > 0$ is an arbitrary constant. Observe additionally that if r_1, r_2, r_3 form a triple of $(\epsilon', 4\mu)$-representatives for G, then $\langle \mathbf{x}_v | \mathbf{x}_{r_i} \rangle > \langle \mathbf{x}_v | \mathbf{x}_{r_j} \rangle$ for at least $(1 - 2\epsilon')n/3$ vertices $v \in C_i$ if $i \neq j$. Let $2 \cdot \epsilon' = \epsilon$. This guarantees a coloring Υ_0 of G that colors at least ϵn vertices of G correctly by assigning each vertex v the color i such that $\langle \mathbf{x}_v | \mathbf{x}_{r_i} \rangle$ is maximal.

The strategy just described is applied in Step 4 of Algorithm 1. We conclude that this randomized approach gives rise to a valid coloring of an $(1 - \epsilon)$-fraction of the graph with probability at least

$$1 - \exp\left(-\frac{7}{3}n\right) - (8/9 + 2\epsilon/3)^{c'n} \geq 1 - 2\exp\left(-\frac{7}{3}n\right) \geq 1 - 10^{-n} \quad (3)$$

for c' and n sufficiently large and ϵ small enough, e.g. $c' = 30$ and $\epsilon < 1/10$. Here, the probability is taken with respect to the input graphs $\mathcal{G}_{n,p,k}$ and to the random choices of representatives.

Now, consider the coloring Υ_0 constructed in Step 6 of Algorithm 1 and let F_{SDP} be a vertex set of minimal cardinality such that Υ_0 colors at most $\epsilon n/3$ vertices incorrectly in each set $C_i \setminus F_{SDP}$. The following lemma is an immediate implication of Equation (3) and Lemma 2.

Lemma 4. *For all $y > 0$ the following relation holds:* $\mathbf{P}[\,|F_{SDP}| \geq 1\,] \leq 10^{-n}$.

4.2 The Iterative Recoloring Step

After the initial coloring $\Upsilon_0(G)$ is constructed, Step 8 of Algorithm 1 aims at improving this coloring iteratively. We show that this attempt is indeed successful on a large subgraph H of G with high probability.

Let H be the subgraph of G obtained by the following process:

1. Delete all vertices in $\overline{H}^+ := \left\{ v \in V \;\middle|\; v \in C_i, \exists j \neq i : \deg_{C_j}(v) > (1+\delta)\frac{d}{3} \right\}$
2. Delete all vertices in $\overline{H}^- := \left\{ v \in V \;\middle|\; v \in C_i, \exists j \neq i : \deg_{C_j}(v) < (1-\delta)\frac{d}{3} \right\}$
3. Iteratively delete all vertices having more than $\delta d/3$ neighbors that were deleted earlier in C_i for some i, i.e., delete all vertices in $\bigcup_{0 < l} \overline{H}^l$, where $\overline{H}^0 := \overline{H}^+ \cup \overline{H}^-$ and

$$\overline{H}^l : \left\{ v \in V \;\middle|\; \exists i : \mathbf{N}(v) \cap C_i \cap \bigcup_{l' < l} \overline{H}^{l'} > \delta \frac{d}{3} \right\}$$

for $l > 0$.

We also denote $G - H$ by $\overline{H}$. The lemma below shows that H spans a large subgraph of G with high probability.

Lemma 5. *Let $0 < \alpha < \frac{1}{2}$ and $0 < \delta < \frac{1}{2}$ be constant in the definition of H. Then*

$$\mathbf{P}\left[\,|\overline{H}| \geq \alpha n\,\right] \leq \exp(-(\log \alpha + \Omega(d)) \cdot \alpha n) + \exp(\Omega(d \cdot \log \alpha \cdot \alpha n)). \quad (4)$$

This lemma follows from the observation that it is unlikely that many vertices are deleted in the first two steps of the construction of H. But then it is also unlikely that many vertices are deleted in the iterative step.

Now, we can use the structural properties of H to show that the algorithm succeeds on H with high probability. For this, we prove that with high probability the number of vertices in H that are colored incorrectly decreases by more than a factor of 2 in each of the iterations of the recoloring step. If this performance is actually achieved, we call the corresponding iteration *successful*, otherwise we say that it *fails*. Algorithm 1 performs at most $\log n$ of these iterations. Afterwards, either the entire graph H is colored correctly or one of the iterations failed. In the latter case Algorithm 1 runs the iterative recoloring step until just before the iteration, say iteration t, when it fails for the first time. The algorithm then proceeds by exhaustively trying all colorings on all subsets of G of size y and thus fixes the coloring of H in this way. However, since Algorithm 1 can not discover whether a particular iteration of the recoloring step succeeds or fails another iteration is necessary at this point. Algorithm 1 applies the strategy of simply trying to repair each of the iterations of the recoloring step subsequently, starting with the last one and proceeding until it reaches iteration t. This explains the innermost loop of the recovery procedure (Step 7 of Algorithm 1). Once the recovery procedure repaired iteration t, all vertices of H are colored correctly.

Lemma 6. *Consider the first iteration of the recoloring step that fails and let $F_H \subseteq H$ denote the set of vertices of H that were colored incorrectly in H before this iteration. Then*

$$\mathbf{P}[\,|F_H| = \alpha n\,] \leq \exp(\Omega\,(d \cdot \log \alpha \cdot \alpha n))$$

for δ sufficiently small but constant in the definition of H.

This follows from the fact that H has good expansion properties and that vertices in H do not have many neighbors outside of H.

4.3 The Uncoloring Step

Note that, if H is colored correctly before the application of the uncoloring step, no vertex of H gets uncolored by this procedure. Indeed, since each vertex v in H has at least $(1-\delta)d/3$ neighbors in $C_i \cap H$ for each i such that $v \notin C_i$ and all these neighbors are colored with color i, v does not get uncolored as long as $\delta < 1/2$.

The following Lemma shows that vertices $v \notin H$ that were not colored correctly by the iterative recoloring step are likely to get uncolored in the uncoloring step.

Lemma 7. *Let $F_\Upsilon \subset G - H$ be the set of vertices that are colored incorrectly and remain colored after the execution of the uncoloring step. Then*

$$\mathbf{P}[\,|F_\Upsilon| = \alpha n\,] \leq \exp(\Omega\,(d \cdot \log \alpha \cdot \alpha n))$$

for $0 < \alpha < \frac{1}{2}$.

Proof. If a vertex v in C_i is colored incorrectly, say with color j, and remains colored after the uncoloring step, v must have at least $d/6$ neighbors of color i. Since v is not adjacent to any vertex in its own color class C_i, all these neighbors are elements of F_Υ as well. Hence, the lemma follows from an estimation of the probability that there is some set $Y \subset V(G)$ with $|Y| = \alpha n$ and minimum degree at least $d/6$.

4.4 The Extension Step

Knowing that the uncoloring step succeeds in uncoloring all vertices of wrong color with high probability, we are now left with the task of assigning a new color to these uncolored vertices. In Algorithm 1 this is taken care of by the extension step (Step 10). Using similar techniques as those developed by Alon and Kahale in [1], we show that all components induced on the set of uncolored vertices are likely to be rather small. Recall that $\alpha_0 = \exp(-\mathcal{O}(d))$.

Lemma 8. *For $\alpha < \alpha_0$,*

$$\mathbf{P}\big[\text{ there is a component of order } \alpha n \text{ in } \overline{H} \,\big] \leq \left(\frac{d}{\exp(\Omega(d))}\right)^{\alpha n}.$$

From this lemma it follows that with high probability a valid coloring of H can indeed be extended to the whole graph G by Step 10 of Algorithm 1 as long as $\overline{H}$ does not get too large.

Lemma 9. *The extension step (Step 10) of Algorithm 1 has polynomial expected running time.*

Proof. As explained, in Step 10 of Algorithm 1 an exact coloring method is used to extend the partial coloring obtained in earlier steps to the whole graph G. In this process the components induced on the vertices uncolored by the uncoloring step are considered independently. On each such component the algorithm tries all possible colorings until it finds one that is compatible to the coloring of the rest of G. Trivially there are at most n components in the set of uncolored vertices and so it suffices to show that the probability that a component of $G - H$ has αn vertices multiplied by $\mathbf{3\text{-}COL}(\alpha n)$ remains small for all $\alpha < \alpha_0$ since the extension step is only executed for $\alpha < \alpha_0$:

$$\mathbf{P}[\text{there is a component of order } \alpha n \text{ in } G - H \,] \cdot \mathbf{3\text{-}COL}(\alpha n)$$
$$\leq \left(\frac{d}{\exp(\Omega(d))}\right)^{\alpha n} \cdot 3^{\alpha n} \leq \left(\frac{3d}{\exp(\Omega(d))}\right)^{\alpha n} = \mathcal{O}(1).$$

4.5 The Expected Running Time of COLOR

All main steps of Algorithm 1, i.e., the construction of the initial coloring, the recoloring step, and the uncoloring step are executed in polynomial time.

Moreover, Lemma 8 guarantees that the extension step of COLOR has polynomial expected running time. It therefore remains to investigate the recovery procedure consisting of the loops in Steps 2, 3, 4 and 7 of Algorithm 1.

The results derived in the last few subsections estimate the probabilities that one of the main steps of Algorithm 1 fails on a vertex set Y of size y. As explained in Section 3, these vertex sets are taken care of by the recovery procedure. The polynomial expected running time of Algorithm 1 is a consequence of the exponentially small probabilities in the previous lemmas. This is shown in Lemma 10 and it immediately implies Theorem 2.

Lemma 10. *The recovery procedure (i.e., Steps 2, 3, 4 and 7) of Algorithm 1 has polynomial expected running time.*

Proof. Consider the vertex set Y from Algorithm 1 that is colored correctly in Step 3 of the recovery procedure and let $t(y)$ be the time the algorithm needs to execute this step in the case $|Y| = y$. Further, denote by F the set Y used in the iteration when the algorithm finally obtains a valid coloring. The expected running time $\mathbf{E}[t]$ of the repair mechanism can then be written as

$$\mathbf{E}[t] \sum_{y \leq n} \mathbf{P}[\,|F| = y\,] \cdot t(y)$$

$$\leq \mathcal{O}(n) \sum_{y \leq n} \mathbf{P}[\,|F| = y\,] \cdot \binom{n}{y} 3\text{-}\mathbf{COL}(y).$$

Recall the definition of H in subsection 4.2 and that F_H are those vertices in H which are colored incorrectly after the recoloring step. F_{SDP} are those vertices that need to be assigned a different color for obtaining a valid coloring on an $(1-\epsilon)$-fraction of G after the inital phase and F_Υ are those that are colored incorrectly after the uncoloring step. Moreover, $\alpha_0 = \exp(-\mathcal{O}(d))$.

We bound $\mathbf{P}[\,|F| = y\,]$ by rewriting F as sum of F_{SDP}, F_H, F_Υ and possibly $\overline{H}$. For this, observe that $F = F_{SDP} \cup F_H \cup F_\Upsilon$. However, we use this partitioning of F only in the case that $|\overline{H}| \leq \alpha_0 n$. If $|\overline{H}| > \alpha_0 n$ we use $F \subset F_{SDP} \cup F_H \cup \overline{H}$. Since all probabilities involved are monotone decreasing, we get

$$\mathbf{P}[\,|F|=y\,] \leq \sum_{\substack{y_1+y_2+y_3=y \\ y_3 \leq \alpha_0 n}} \mathbf{P}[\,|F_{SDP}|=y_1, |F_H|=y_2, |F_\Upsilon|=y_3\,]$$

$$+ \sum_{\substack{y_1+y_2+y_3=y \\ y_3 > \alpha_0 n}} \mathbf{P}\bigl[\,|F_{SDP}|=y_1, |F_H|=y_2, |\overline{H}|=y_3\,\bigr].$$

One of the vertex sets involved in the conjunctions of this sum certainly contains more than $y/3$ vertices and so

$$\mathbf{P}[\,|F|=y\,]$$

$$\leq n^3 \begin{cases} \max\bigl(\mathbf{P}\bigl[\,|F_{SDP}|=\tfrac{y}{3}\,\bigr], \mathbf{P}\bigl[\,|F_H|=\tfrac{y}{3}\,\bigr], \mathbf{P}\bigl[\,|F_\Upsilon|=\tfrac{y}{3}\,\bigr]\bigr) & \text{if } \tfrac{y}{3} \leq \alpha_0 n, \\ \max\bigl(\mathbf{P}\bigl[\,|F_{SDP}|=\tfrac{y}{3}\,\bigr], \mathbf{P}\bigl[\,|F_H|=\tfrac{y}{3}\,\bigr], \mathbf{P}\bigl[\,|\overline{H}|=\tfrac{y}{3}\,\bigr]\bigr) & \text{otherwise.} \end{cases}$$

The lemma then follows from Lemmas 4, 5, 6, and 7. We omit the details.

5 Concluding Remarks

We proved that random 3-colorable graphs taken from $\mathcal{G}_{n,p,3}$ can be 3-colored in polynomial expected time if $p \geq c/n$, where c is some sufficiently large constant. The same methods can be used for obtaining a similar result for $\mathcal{G}_{n,p,k}$ with values of k other than 3. More precisely, the calculations carry over directly to arbitrary k for $pn \geq c_k$ where c_k is a constant depending on k.

One remaining question is whether it is possible to design an algorithm for coloring $\mathcal{G}_{n,p,k}$ in polynomial expected time for all values of p. In particular, it is not clear how to deal with the case that pn is constant but much smaller than c_k.

Acknowledgements

I am grateful to Amin Coja-Oghlan for many helpful discussions.

References

1. N. Alon and N. Kahale. A spectral technique for coloring random 3-colorable graphs. *SIAM Journal on Computing*, 26(6):1733–1748, 1997.
2. A. Coja-Oghlan. Coloring semirandom graphs optimally. In *Proceedings of the 31st International Colloquium on Automata, Languages and Programming*, pages 383–395, 2004.
3. A. Coja-Oghlan, C. Moore, and V. Sanwalani. MAX k-CUT and approximating the chromatic number of random graphs. In *Proceedings of the 30th International Colloquium on Automata, Languages and Programming*, pages 200–211, 2003.
4. M. E. Dyer and A. M. Frieze. The solution of some random NP-hard problems in polynomial expected time. *Journal of Algorithms*, 10:451–489, 1989.
5. U. Feige and J. Kilian. Zero knowledge and the chromatic number. *Journal of Computer and System Sciences*, 57(2):187–199, 1998.
6. A. M. Frieze and M. Jerrum. Improved approximation algorithms for MAX k-CUT and MAX BISECTION. *Algorithmica*, 18:61–77, 1997.
7. M. R. Garey and D. S. Johnson. *Computers and Intractability*. W.H. Freeman and Company, 1979.
8. M. Grötschel, L. Lovász, and A. Schrijver. *Geometric Algorithms and Combinatorial Optimization*. Springer, Berlin, 1993.
9. S. Khanna, N. Linial, and S. Safra. On the hardness of approximating the chromatic number. *Combinatorica*, 20(3):393–415, 2000.
10. M. Krivelevich. Deciding k-colorability in expected polynomial time. *Information Processing Letters*, 81:1–6, 2002.
11. L. Kučera. Expected behavior of graph colouring algorithms. In *Proceedings of the 1977 International Conference on Fundamentals of Computation Theory*, pages 447–451, 1977.
12. C. R. Subramanian. Algorithms for coloring random k-colorable graphs. *Combinatorics, Probability and Computing*, 9:45–77, 2000.
13. J. S. Turner. Almost all k-colorable graphs are easy to color. *Journal of Algorithms*, 9:253–261, 1988.

Regular Sets of Higher-Order Pushdown Stacks

Arnaud Carayol

IRISA – Campus de Beaulieu – 35042 Rennes Cedex – France
Arnaud.Carayol@irisa.fr

Abstract. It is a well-known result that the set of reachable stack contents in a pushdown automaton is a regular set of words. We consider the more general case of higher-order pushdown automata and investigate, with a particular stress on effectiveness and complexity, the natural notion of regularity for higher-order stacks: a set of level k stacks is regular if it is obtained by a regular sequence of level k operations. We prove that any regular set of level k stacks admits a normalized representation and we use it to show that the regular sets of a given level form an effective Boolean algebra. In fact, this notion of regularity coincides with the notion of monadic second order definability over the canonical structure associated to level k stacks. Finally, we consider the link between regular sets of stacks and families of infinite graphs defined by higher-order pushdown systems.

1 Introduction

Higher-order pushdown automata (hopdas for short) were introduced as a generalization of pushdown automata [Aho69, Gre70, Mas76]. Whereas a pushdown automaton works on a stack of symbols, a pushdown automaton of level 2 (or 2-hopda) works with a stack of level 1 stacks. In addition to the ability to push and to pop a symbol on the top-most level 1 stack, a 2-hopda can copy or remove the entire top-most level 1 stack. The k-hopdas are similarly defined for all level k and have been extensively studied as language recognizers [Dam82, Eng83].

Recently, the infinite structures defined by hopdas have received a lot of attention. First, in [KNU02, Cau02], the families of infinite terms defined by k-hopdas were shown to correspond to the solutions of safe higher-order recursive schemes. This study was later extended to the transition graphs of k-hopdas in [CW03]. Several characterizations of this hierarchy of families of infinite graphs were obtained. In particular, it was shown to coincide with a hierarchy defined using graph transformations by Caucal in [Cau96b] (see [Tho03] for a survey on this hierarchy).

The transition graphs of hopdas are defined in [CW03] by ε-closure of the reachability graphs. The vertices of the reachability graph of an hopda are the configurations reachable by the automaton from the initial configuration, and the edges represent the transition rules of the hopda. A major drawback of this definition is that it does not provide a direct description of the relations defining the edges of the transition graphs.

At level 1, the set of edges of the transition graph of a pushdown automaton can be given by prefix-recognizable relations [Cau96a, Cau03]. This characterization essentially uses the fact that the set of stack contents reachable by a pushdown automaton is regular. At level 2, due to the introduction of the copy operation, the set of words representing the stacks of stacks reachable by a 2-hopda is not a regular set of words. Hence, in order to obtain an internal representation of the transition graphs of k-hopdas, it is necessary to define a notion of regularity for sets of stacks of level k (k-stacks for short).

In this article, we study the notion of regularity for k-stacks induced by k-hopdas: a set of k-stacks is regular if it is obtained by a regular sequence of level k operations applied to the empty k-stack. In Section 3, we study the algebraic and algorithmic properties of this notion. We define a normal form for regular sets of k-stacks and use it to prove that they are closed under complementation. From the algorithmic point of view, the complexity of the normalization algorithm presented in Section 3.3 is a lower bound. We also show that the k-regular sets correspond to the sets definable by monadic second order logic over the canonical infinite structure associated with k-stacks. Finally, in Section 4, we use the notion of k-regularity to define an internal representation of the transition graphs of k-hopdas.

A complete version of this work including proofs can be found in [Car05].

2 Preliminary Definitions

2.1 Regular Parts of a Monoid

A *monoid* is given by a set M together with an associative internal product operation written · that admits a neutral element 1_M. The product operation is extended to subsets of M by $P \cdot Q = \{p \cdot q \mid p \in P \text{ and } q \in Q\}$. For any subset N of M, N^n is defined by $N^0 = \{1_M\}$ and $N^{n+1} = N \cdot N^n$. The iteration of N written N^* is equal to $\cup_{i \in \mathbb{N}} N^i$. Similarly, N^+ is defined as $\cup_{i > 0} N^i$. The set of regular parts of a monoid M noted $\text{Reg}(M)$ is the smallest set containing the finite subsets of M and closed under union, product and iteration.

A common example of monoid is the set of words over a finite alphabet Γ. A finite sequence of symbols (also called letters) in Γ is a word and the set of all words is written Γ^*. The empty word is noted ε.

2.2 Infinite Graphs and Transformations

Infinite Graphs. Given a finite set Σ of edge labels and a countable set V, a Σ-*labeled graph* G is a subset of $V \times \Sigma \times V$. An element (s, a, t) of G is an *edge* of *source* s, *target* t and *label* a, and is written $s \xrightarrow{a}_G t$ or simply $s \xrightarrow{a} t$ if G is understood. The set of all sources and targets of a graph is its *support* V_G. A sequence of edges $s_1 \xrightarrow{a_1} t_1, \ldots, s_k \xrightarrow{a_k} t_k$ with $\forall i \in [2, k]$, $s_i = t_{i-1}$ is a *path* starting from s_1. We write $s_1 \xRightarrow{u} t_k$, where $u = a_1 \ldots a_k$ is the corresponding *path label*.

The *unfolding* $\mathrm{Unf}(G,r)$ of a Σ-labeled graph G from a vertex $r \in V_G$ is the Σ-labeled tree T satisfying for any $a \in \Sigma$ that $\pi \xrightarrow{a} \pi' \in T$ if and only if π and π' are two paths in G starting from r and $\pi' = \pi s \xrightarrow{a} t$.

Inverse Mappings. Let $\bar{\Sigma}$ be a set of symbols disjoint from but in bijection with Σ. For any $x \in \Sigma$, we write $\bar{x}$ the corresponding symbol in $\bar{\Sigma}$. We extend every Σ-labeled graph G to a $(\Sigma \cup \bar{\Sigma})$-labeled graph $\bar{G}$ by adding reverse edges (i.e. $\bar{G} = G \cup \{s \xrightarrow{\bar{x}} t \mid t \xrightarrow{x} s \in G\}$). Let Γ be a set of edge labels, a *rational mapping* is a mapping $h : \Gamma \to \mathrm{Reg}((\Sigma \cup \bar{\Sigma})^*)$ which associates to every symbol from Γ a regular subset of $(\Sigma \cup \bar{\Sigma})^*$. If $h(a)$ is finite for every $a \in \Gamma$, we also speak of a *finite mapping*. We apply a rational mapping h to a Σ-labeled graph G by the inverse to obtain a Γ-labeled graph $h^{-1}(G) = \{s \xrightarrow{a} t \mid s \xRightarrow[\bar{G}]{h(a)} t\}$.

Monadic Second Order Logic. We define the *monadic second-order logic* (MSO for short) over Σ-labeled graphs as usual (see e.g. [EF95]). For any monadic second order formula $\varphi(x_1, \ldots, x_n)$ whose free-variables are first-order variables in $\{x_1, \ldots, x_n\}$ and for any vertices $u_1, \ldots, u_n \in V_G$, we write $G \models \varphi(u_1, \ldots, u_n)$ the fact that the graph G satisfies the formula φ when x_i is interpreted as u_i for all $i \in [1,n]$.

2.3 Higher-Order Stacks and Operations

Stacks. A stack over a finite alphabet Γ is a word over Γ. We write $\mathrm{Stacks}_1(\Gamma) = \Gamma^*$ for the set of all stacks of level 1 and note $[\,]_1$ the empty level 1 stack ε. For all $k > 1$, a level k stack over Γ (or simply a k-stack) is a non-empty sequence of $(k-1)$-stacks over Γ. We write $\mathrm{Stacks}_k(\Gamma) = (\mathrm{Stacks}_{k-1}(\Gamma))^+$ the set of all k-stacks or simply Stacks_k if Γ is understood. The empty stack of level k is the k-stack containing only the empty $(k-1)$-stack and is written $[\,]_k$. The stack $[[AB][ABC][BA]]_2$ designates a 2-stack whose top most 1-stack is $[BA]_1$. The set of all stacks over Γ is written $\mathrm{Stacks}(\Gamma) = \bigcup_{k \in \mathbb{N}} \mathrm{Stacks}_k(\Gamma)$.

Operations. An operation on higher-order stacks is a (partial) function from $\mathrm{Stacks}(\Gamma)$ to $\mathrm{Stacks}(\Gamma)$ which preserves the level of the stack (i.e. the image of a k-stack is a k-stack). The level $|\rho|$ of an operation ρ is the smallest k such that $\mathrm{Dom}(\rho) \cap \mathrm{Stacks}_k \neq \emptyset$. The only operation for which the level is not defined is the empty function $\emptyset$. Note that for any two functions f and g, $f \cdot g$ designates the mapping associating to x the value $g(f(x))$.

The operations, we consider, respect the access mode of higher-order stacks that is to say, in a level $k+1$ stack only the top most level k stack can be accessed. It implies that for any level k operation ρ and for all $k' > k$, we have: $\rho([w_1, \ldots, w_n]_{k'}) = [w_1 \ldots \rho(w_n)]_{k'}$. Hence, it is only necessary to define a level k operation on Stacks_k, its definition for level of stacks greater than k is implicit.

The operations of level 1 for stacks over Γ are the well known push_x and pop_x for all $x \in \Gamma$. The operations added at level $k+1$ are the copy of the top most k-stack written copy_k and the inverse operation which is usually the destruction of the top most k-stack written pop_k. We consider a more symmetric operation $\overline{\text{copy}}_k$ that only destroys the top most k-stack if it is equal to previous one[1]. These operations are formally defined by:

$$\begin{aligned}
\text{push}_x([\,x_1 \ldots x_n\,]_1) &= [\,x_1 \ldots x_n x\,]_1 \\
\text{pop}_x([\,x_1 \ldots x_n x\,]_1) &= [\,x_1 \ldots x_n\,]_1 \\
\text{copy}_k([\,w_1 \ldots w_n\,]_{k+1}) &= [\,w_1 \ldots w_n w_n\,]_{k+1} \\
\text{pop}_k([\,w_1 \ldots w_{n+1}\,]_{k+1}) &= [\,w_1 \ldots w_n\,]_{k+1} \\
\overline{\text{copy}}_k([\,w_1 \ldots w_n w_n\,]_{k+1}) &= [\,w_1 \ldots w_n\,]_{k+1}
\end{aligned}$$

In addition for each level k, we consider an operation written E_k to test whether the top most k-stack is empty (i.e $E_k([\ ]_k) = [\ ]_k$ and is undefined otherwise). This operation is usually avoided by considering a bottom symbol in the definition of the stacks but we wish to remain as general as possible. Moreover, we write id_k the identity seen as a level k operation.

We define $\text{Ops}_1 = \{\text{push}_x, \text{pop}_x \mid x \in \Gamma\} \cup \{E_1\}$. and $\text{Ops}_{k+1} = \text{Ops}_k \cup \{\text{copy}_k, \overline{\text{copy}}_k, E_{k+1}\}$. The set $\text{Ops}_k^* = \{\rho \mid |\rho| = k, \rho = \rho_1 \cdots \rho_n$ for $\rho_1, \ldots, \rho_n \in \text{Ops}_k\} \cup \{\emptyset\}$ is a monoid for composition of functions with neutral element id_k.

Instructions. In order to work in a symbolic manner, we associate to each operation in Ops_k a symbol in an alphabet Γ_k called an *instruction*. Let $\bar{\Gamma}$ be a finite alphabet disjoint from but in bijection with Γ, we write $\bar{x}$ the letter of $\bar{\Gamma}$ corresponding to $x \in \Gamma$. The set of instructions of level k written Γ_k is defined by: $\Gamma_1 = \Gamma \cup \bar{\Gamma} \cup \{\bot_1\}$ and $\Gamma_{k+1} = \Gamma_k \cup \{\bot_{k+1}, k, \bar{k}\}$. We write $\Gamma_k^t = \{\bot_1, \ldots, \bot_k\}$ and $\Gamma_k^o = \Gamma_k - \Gamma_k^t$. For all sequence $w \in \Gamma_k^*$, we designate by $\text{Last}(w)$ (resp. $\text{First}(w)$) the last (resp. first) element of Γ_k^o appearing in w.

We define a morphism[2] of monoid O from Γ_k^* to Ops_k^* associating to any sequence of instruction $w \in \Gamma_k^*$ the corresponding operation $O(w) \in \text{Ops}_k^*$ as follows: $O(\varepsilon) = \text{id}_k, O(x) = \text{push}_x$ and $O(\bar{x}) = \text{pop}_x$ for all $x \in \Gamma$, $O(i) = \text{copy}_i$, $O(\bar{i}) = \overline{\text{copy}}_i$, and $O(\bot_i) = E_i$ for all $i \in [1, k]$. The morphism O is extended to Γ_k^* in the canonical way. For example, the sequence of instructions $m = ab\bar{b}a1\bar{a}$ is evaluated to $O(m) = \text{push}_a\text{push}_b\text{pop}_b\text{push}_a\text{copy}_1\text{pop}_a = \text{push}_a\text{push}_a\text{copy}_1\text{pop}_a$. For any subset R of Γ_k^*, we write $O(R)$ the corresponding set of operations in Ops_k^* and $S(R) = O(R)([\ ]_k)$ the corresponding set of stacks in $\text{Stacks}_k(\Gamma)$.

For each k-stack s, there exists a minimal sequence of instructions $w \in \Gamma_k^*$ such that $S(w) = s$. It is easy to see that if $k = 1$, w belongs to Γ^* and if $k > 1$, w does not contain $\bar{k}$. In fact, a sequence of instructions $w \in (\Gamma_k^o \cup \{k\})^*$ (such that $S(w) \neq \emptyset$) is the minimal sequence of some level $k+1$ stack if and only if it does not contain $x\bar{x}$, $l\bar{l}$ or $\bar{l}l$ for any $x \in \Gamma \cup \bar{\Gamma}$ or any $l < k$. A sequence of instructions that does not contain such sub-sequences will be called *loop-free*. A

---

[1] It is already known from [CW03] that hopdas defined using $\overline{\text{copy}}_k$ recognize the same languages as the ones defined using pop_k (see. Proposition 4.1).

[2] The definition is such that we always obtain a level k operation. So strictly speaking, there should be one evaluation mapping for each level.

k-stack s is a prefix of a k-stack s' (written $s \sqsubseteq s'$) if the minimal sequence of s is a prefix of the minimal sequence of s'.

Higher-Order Pushdown Automata. An higher-order pushdown automaton P over $\mathrm{Stacks}_k(\Gamma)$ (k-hopda for short) with Σ as an input alphabet is a tuple (Q, i, F, δ) where Q is a finite set of states, i is the initial state, F is the set of final states and $\delta \subset Q \times \Sigma \cup \{\varepsilon\} \times \Gamma_k^* \times Q$. The set of configurations of P noted C_P is $Q \times \mathrm{Stacks}_k(\Gamma)$. For each $x \in \Sigma \cup \{\varepsilon\}$, P induces a transition relation $\xrightarrow{x} \subset C_P \times C_P$ defined by $(p, w) \xrightarrow{x} (q, w')$ if (p, x, ρ, q) belongs to δ and $w' = O(\rho)(w)$. For any word $u \in \Sigma^*$, we write $c \xRightarrow{u} c'$ if there exists a sequence $c \xrightarrow{x_1} c_1 \ldots c_{n-1} \xrightarrow{x_n} c'$ and $u = x_1 \ldots x_n$. A word $u \in \Sigma^*$ is accepted by P if $(i, [\,]_k) \xRightarrow{u} (f, w)$ for some $f \in F$.

3 Regular Set of Higher-Order Stacks

We consider the notion of regularity for sets of higher-order stacks that naturally extends what is known at level 1. A set of k-stack is k-*regular* if it is the set of stacks appearing in the reachable final configurations of a k-hopda. In other terms, a k-regular set is obtained by applying a regular set of operations in Ops_k^* to the empty stack of level k. The set of all k-regular subsets of $\mathrm{Stacks}_k(\Gamma)$ is written $\mathrm{Reg}_k(\Gamma) = \mathrm{Reg}(\mathrm{Ops}_k^*(\Gamma))([\,]_k) = S(\mathrm{Reg}(\Gamma_k^*))$.

A normal form for k-regular sets is presented in Section 3.1 and it is proved in Section 3.2 that every k-regular set admits a normalized representation. Complexity related issues are dealt with in Section 3.3. Finally, Section 3.4 establishes that k-regular sets correspond to MSO-definable sets in the canonical infinite structure associated to k-stacks.

3.1 Normal Forms

A regular set of instructions is not *per se* a useful representation of a set of stacks. We therefore define a normal form that gives a *forward* representation of the set of stacks in the sense that the set of instructions produced are loop-free.

At level 1, such a normal form is easily achieved: the set of minimal sequences of a 1-regular set is also regular [Büc64]. Hence, any 1-regular set admits a normalized representation in $\mathrm{Norm}_1 = \mathrm{Reg}(\Gamma^*)$. At level 2, a loop-free set of instruction does not contains $\bar{1}$. As illustrated by the following example, it is not possible to describe all sets in $\mathrm{Reg}(\Gamma_2^*)$ without $\bar{1}$.

Example 3.1. The regular set of instructions $R = \{a, b\}^* 1 \{\bar{a}, \bar{b}\}^* \bar{b} 1 (\bar{a}\bar{a})^+ \bar{b} b a^* \bar{1}$ represents the set of stacks $S(R) = \{[\,[\,wba^{2n}bw'\,][\,wba^{2n}\,]\,]_2 \mid w \in \Gamma^*, w' \in \Gamma^* \text{ and } n \geq 0\}$. It can be proved that R is not equivalent to any set in $\mathrm{Reg}((\Gamma_1 \cup \{1\})^*)$. The problem is that the set $1(\bar{a}\bar{a})^+ \bar{b} b a^* \bar{1}$ correspond to the operation $\mathrm{id}_2|_A$ where $A = \{[\,w_1 \ldots w_n\,]_2 \mid w_n \in \Gamma^* b(aa)^+\}$ which tests, in a non-destructive manner, that the top-most 1-stack belongs to A.

Hence, in order to give a forward presentation of sets in $\mathrm{Reg}(\Gamma_k^*)$, we need to introduce k-regular tests as a new operation. Theorem 3.1 will prove that we do not need additional operations. For any set Q of k-stacks, $\mathrm{id}_k|_Q$ designates the identity function restricted to the set of k'-stack whose top-most k-stack is in Q for all $k' > k$.

Definition 3.1 (Regular tests of level k). *Let T_k be an countable set of symbols with one symbol written T_R for each $R \in \mathrm{Reg}(\Gamma_k)$. We extend the evaluation mapping to $(\Gamma_k \cup T_k)^*$ by defining $O(T_R) = \mathrm{id}_k|_{S(R)}$.*

A subset R of $\mathrm{Reg}((\Gamma_k \cup T_k)^*)$ is loop-free is the set of obtained by removing the tests from R is.

In order to normalize a 2-regular set of stacks, it is necessary to give a normal form for sets of operations in $O(\mathrm{Reg}((\Gamma_k \cup T_k)^*))$. At level 1, a normal form was obtained in [Cau96a, Cau03]. It is proved that any $R \in \mathrm{Reg}((\Gamma_1 \cup T_1)^*)$, there exists a finite union $R' = \cup_{i \in I} U_i \cdot T_{W_i} \cdot V_i$ where $U_i \in \mathrm{Reg}(\overline{\Gamma}^*)$, $V_i, W_i \in \mathrm{Reg}(\Gamma^*)$ for some finite set I with $\mathrm{Last}(U_i) \cap \overline{\mathrm{First}(V_i)} = \emptyset^3$ such that $O(R) = O(R')$. We write Rew_1 the set all $R \in \mathrm{Reg}((\Gamma_1 \cup T_1)^*)$ than can be expressed as such a finite union.

We now define Norm_k and Rew_k for level $k > 1$ as a straightforward extension of what is known at level 1:

- Norm_{k+1} is the set of all finite union of elements in $\mathrm{Norm}_k \cdot \mathrm{Reg}((k\,\mathrm{Rew}_k)^*)$,
- Pop_{k+1} and Push_{k+1} designate respectively the sets $\mathrm{Rew}_k \cdot \mathrm{Reg}((\bar{k}\,\mathrm{Rew}_k)^*)$ and $\mathrm{Rew}_k \cdot \mathrm{Reg}((k\,\mathrm{Rew}_k)^*)$,
- Rew_{k+1} is the set of all finite unions of sets of the form $U \cdot T_W \cdot V$ where $W \in \mathrm{Norm}_{k+1}$, $U \in \mathrm{Pop}_{k+1}$ and $V \in \mathrm{Push}_{k+1}$ with $\mathrm{Last}(U) \cap \overline{\mathrm{First}(V)} = \emptyset$ and $\overline{\mathrm{Last}(W)} \cap (\overline{\mathrm{Last}(U)} \cup \mathrm{First}(V)) = \emptyset$.

An equivalent characterization of the sets in Norm_{k+1} is through finite automata $A = (Q, i, F, \delta)$ labeled by a finite subset $\mathrm{Norm}_k \cup k \cdot \mathrm{Rew}_k$ such the only edges labeled by an element of Norm_k are starting from the initial state i and such that no transition comes back to i. We will call such an automaton a $(k+1)$-automaton. It is obvious that $L(A)$ belongs to Norm_{k+1} and that conversely, all $R \in \mathrm{Norm}_k$ is accepted by a k-automaton. By a slight abuse of language, we will say that a set R of k-stacks is accepted by A if $R = S(L(A))$.

The interest of this notion is that a deterministic version can be defined : a $(k+1)$-automaton labeled by $\{N_1, \ldots, N_n\} \subset \mathrm{Norm}_k$ and $\{R_1, \ldots, R_m\} \subset \mathrm{Rew}_k$ is deterministic if $S(N_i) \cap S(N_j) = \emptyset$ for $i \neq j$, $O(R_i) \cap O(R_j) = \emptyset$ for $i \neq j$ and A is deterministic. In a deterministic k-automaton, if two k-stacks are produced by two different executions (not necessarily successful) then they are different.

Proposition 3.1. *For all level k, any set in $S(\mathrm{Norm}_k)$ can be accepted by a deterministic k-automaton. Moreover, $S(\mathrm{Norm}_k)$ and $O(\mathrm{Rew}_k)$ are effective Boolean algebras[4].*

---

[3] This corresponds to the right-irreducible prefix-recognizable relations in [Cau03].
[4] This result was already obtained in [Cau96a] for $O(\mathrm{Rew}_1)$.

3.2 Normalization

In this part, we prove by induction on the level k that for any set in $R \in \text{Reg}(\Gamma_k^*)$, there exists a set $N \in \text{Norm}_k$ such that $S(R) = S(N)$.

Characterization of Loop Languages. Let $B = (Q, i, F, \delta)$ be an automaton labeled by Γ_{k+1}. For any two states p and $q \in Q$, the automaton B loops on a $(k+1)$-stack w starting in p and ending in q if there exists a sequence $(p, w) \xrightarrow[B]{x_1} (p_1, w_1) \ldots \xrightarrow[B]{x_n} (p_n, w_n)$ such that $q = p_n$, $w_n = w$ and for all $i \in [1, n]$, $w \sqsubseteq w_i$.

It follows from the definition that "looping" behavior of an automaton only depends on the top-most k-stack. Hence, we define the *loop language* $L_{p,q}$ to be the set of k-stacks such that $w \in L_{p,q}$ if and only if for any $(k+1)$-stack w' with top-most k-stack w, B loops on w' starting in p and ending in q. The loop languages allow us to define a loop-free equivalent of $L(A)$.

Proposition 3.2. *For any automaton A labeled by Γ_k, $L(A)$ is equivalent to a loop-free set in $\text{Reg}((\Gamma_k \cup \{T_{L_{p,q}} \mid p, q \in Q\})^*)$.*

The Loop Languages Are Regular. In order to simulate the copy_{k+1} and $\overline{\text{copy}}_{k+1}$ operations on a level k stack, we use *alternation* to simultaneously perform the computation taking place on different copies of the stack.

Definition 3.2. *An alternating automaton A over Γ_k is a tuple (Q, i, Δ) where Q is a finite set of states, i is the initial state and $\Delta \subset Q \times \Gamma_k^t \cup \{\varepsilon\} \times 2^{Q \times \Gamma_k^o}$ is the set of transitions.*

A transition $(p, t, \{(q_1, a_1), \ldots, (q_n, a_n)\}) \in \Delta$ is written $p, t \longrightarrow (q_1, a_1) \wedge \ldots \wedge (q_n, a_n)$. An execution of A is a finite tree T with vertices V_T whose edges are labeled by Γ_k and whose vertices are labeled by $Q \times \text{Stacks}_k(\Gamma)$. We write c the labeling mapping from V_T to $Q \times \text{Stacks}_k(\Gamma)$. An execution satisfies the following conditions:

- $x \xrightarrow{a} y \in T$ implies $c(x) = (q, w)$ and $c(y) = (p, w')$ and $w' = O(a)(w)$.
- for all $x \in V_T$ with children $y_1, \ldots, y_n$ (i.e $x \xrightarrow{a_1} y_1, \ldots, x \xrightarrow{a_n} y_n$), if $c(y_i) = (q_i, w_i)$ then there exists $q, t \longrightarrow (q_1, a_1) \wedge \ldots \wedge (q_n, a_n) \in \Delta$ such that $O(t)(w)$ is defined[5].

An execution T of A is accepting a stack w if the root of T is labeled by (i, w). We write $S(A) \subset \text{Stacks}_k(\Gamma)$ the set of stacks accepted by A.

The following lemma states that the loop languages defined in the previous part are accepted by alternating automata labeled by Γ_k.

Lemma 3.1. *For any automaton B labeled by Γ_{k+1}, there exists an alternating automaton labeled by Γ_k accepting $L_{p,q}$.*

---

[5] The set of final states is implicitly given by transitions of the form $q, t \longrightarrow \emptyset$.

In order to prove that the language accepted by an alternating automaton over Γ_k is k-regular, we define a normal form for alternating automata in which at most one execution can be sent for a given instruction in Γ_k^o and such that no execution contains two vertices labeled by the same stack. More formally, a *normalized* alternating automaton over Γ_k is an alternating automaton with transitions of the form $q, t \longrightarrow (q_1, b_1) \wedge \ldots (q_n, b_n)$ with $b_i \neq b_j$ for $i \neq j$ for which no execution tree T contains $x \xrightarrow[T]{a\bar{a}} y$ for $a \in \Gamma_k^o$. The following proposition establishes that any alternating automaton can be transformed into an equivalent normalized alternating automaton.

Proposition 3.3. *For any alternating automaton A labeled by Γ_k, an equivalent normalized alternating automaton B labeled by Γ_k can be constructed in $\mathcal{O}(2^{p(|B|)})$ for some polynomial p.*

We can now establish that the languages accepted by alternating automata labeled by Γ_k are k-regular languages.

Proposition 3.4. *The sets of k-stacks accepted by alternating automaton labeled by Γ_k are k-regular sets.*

Proof (Sketch). First, we establish that the languages accepted by normalized alternating automaton over Γ_{k+1} are loop-free languages in $\text{Reg}((\Gamma_k^o \cup \{k\} \cup T_k^A)^*)$ where T_k^A designates the tests by languages accepted by alternating automaton over Γ_k. The result follows by induction on the level k combining the above property and Proposition 3.3.

□

Normalization Result. We proceed by induction on the level k of stacks. It follows from the Proposition 3.2, Lemma 3.1 and Proposition 3.4, that any set of instructions in $R \in \text{Reg}(\Gamma_k^*)$ is equivalent to a loop-free subset of $\text{Reg}((\Gamma_k \cup T_k)^*)$.

Proposition 3.5. *For all loop-free set R in $\text{Reg}((\Gamma_k \cup T_k)^*)$ with tests languages in $S(\text{Norm}_k)$, there exists a $R' \in \text{Norm}_k$ such that $S(R) = S(R')$.*

Note that to achieve this normalization we need to determinize the languages appearing in the tests. However, if the languages appearing in the tests are already determinize the transformation is polynomial. The normalization result is obtained by a straightforward induction.

Theorem 3.1 (Normalization). *Every k-regular set can be accepted by a k-automaton. Hence, $\text{Reg}_k = S(\text{Reg}(\Gamma_k^*)) = S(\text{Reg}(\text{Norm}_k))$ is an effective Boolean algebra.*

3.3 Complexity and Lower Bounds

In order to evaluate the complexity of the normalization algorithm, we need a notation for towers of exponentials. We define $2\uparrow^0(n) = n$ and $2\uparrow^{k+1}(n) = 2^{2\uparrow^k(n)}$.

The complexity of the algorithm obtained in the previous section when applied to a k-regular set of stacks is $\mathcal{O}(2\uparrow^{2k+1}(p(n)))$ where p is a polynomial. This complexity can be improved by transforming directly an alternating automaton over Γ_k into a deterministic k-automaton.

Theorem 3.2 (Lower bound).

1. For any alternating automaton A labeled by Γ_k, there exists a deterministic k-automaton accepting $S(A)$ which can be computed in time $\mathcal{O}(2\uparrow^k(p(n)))$ for some polynomial p.
2. For any automaton A labeled by Γ_k, there exists a k-automaton accepting $S(A)$ which can be computed in time $\mathcal{O}((2\uparrow^{k-1}(p(n))))$

It is easy to see that normalization can be used to test the emptiness of a regular set of k-stacks. In fact, for any set $R \in \text{Reg}(\Gamma_k^*)$, the normalized representation of $R \cdot \Gamma_k^* \cdot \bot_k$ contains $[\ ]_k$ if and only if $S(R)$ is not empty. Therefore, the normalization can be used to test the emptiness of the language accepted by a k-hopda (i.e it is equivalent to the emptiness of the set of reachable final configurations).

In [Eng83], the author proves that $\mathcal{O}(2\uparrow^{k-1}(p(n)))$ is a lower bound for the emptiness problem of k-hopda. It follows the complexity of the normalization algorithm obtained in Theorem 3.2 is a lower bound.

3.4 MSO-Definability Over Δ_2^n

In this part, we fix $\Gamma = \{a, b\}$. The canonical infinite structure associated to words in Γ^* is the infinite binary tree Δ_2. A set $X \subset V_G$ (resp. $Y \subset V_G \times V_G$) is MSO-definable in G if there exists a formula $\varphi(x)$ (resp. $\varphi(x,y)$) such that X (resp. Y) is the set of $u \in V_G$ (resp. $(u,v) \in V_G \times V_G$) such that $G \models \varphi(u)$ (resp. $G \models \varphi(u,v)$). It is well known that the MSO-definable sets in Δ_2 are the regular sets of words. Moreover, Blumensath [Blu01] proved that the relations MSO-definable in Δ_2 are the prefix-recognizable relations (i.e $O(\text{Rew}_1)$).

In order to investigate the notion of MSO-definable set of k-stacks, we consider the canonical structure Δ_2^k associated to k-stacks (see Figure 1). The graph Δ_2^k is labeled by $\Sigma_k = \{a, b, 1, \ldots, k\}$, its set of vertices is $\text{Stacks}_k(\Gamma)$ and it edges are defined by: $w \xrightarrow{i} w'$ if $w' = O(i)(w)$ for $i \in \Sigma_k$. For instance, the set of k-stacks whose top-most 1-stack is empty is defined by the formula $\varphi(x) = \neg(\exists y.y \xrightarrow{a} x) \wedge \neg(\exists y.y \xrightarrow{b} x)$.

Proposition 3.6. The set of k-stacks MSO-definable in Δ_2^k are the k-regular sets and the sets of relations MSO-definable in Δ_2^k are the relation in $O(\text{Rew}_k)$.

4 Higher-Order Pushdown Graphs

In this section, we consider infinite graphs associated to hopdas and we show how the notion of k-regularity can be used to study their structure.

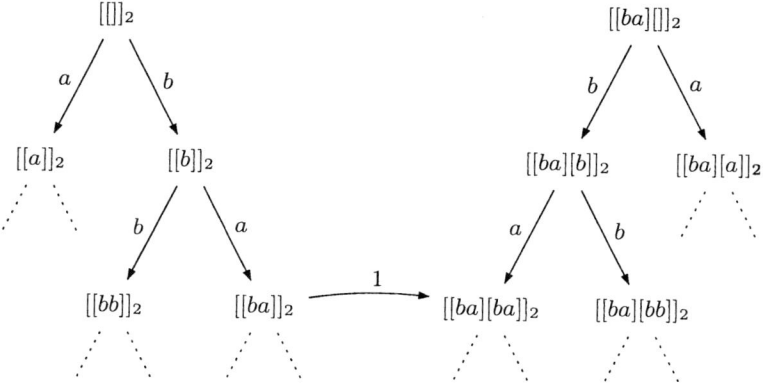

Fig. 1. The canonical structure associated to level 2 stacks

4.1 Reachability Graphs

The most natural way to associate an infinite graph to a k-hopda is to consider its *reachability graph*: the set of vertices is the set of configurations reachable from the initial configuration and the edges are given by the k-hopda. Due to the restriction by reachability, it is possible to encode in the stack a finite information corresponding to rational tests. The idea behind this encoding is taken from [Cau02] and was used in [CW03] to simulate the $\overline{copy}_k$ operation using pop_k.

Proposition 4.1. *The reachability graphs of k-hopdas enriched with k-regular tests and the reachability graphs of the k-hopda defined only with the pop_k operation instead of $\overline{copy}_k$ coincide up to isomorphism.*

From a structural point of view however, this approach is limited as the graphs obtained are necessarily directly connected. For instance, Δ_2^k is not the reachability graph of any hopda.

4.2 Configuration and Transition Graphs

The configuration graph of a k-hopda P is obtained by restricting the transition relation induced by P to a k-regular set of configurations R:

$$\{w \xrightarrow{x} w' \mid x \in \Sigma \cup \{\varepsilon\}, w \xrightarrow[P]{x} w' \text{ and } w, w' \in R\}$$

As the set of k-stacks reachable from the initial configuration is a k-regular set, the reachability graph is a particular case of configuration graph. The following property gives a structural characterization of configurations graphs.

Proposition 4.2. *The configuration graphs of k-hopdas are the graphs obtained by a k-fold iteration of unfolding and finite inverse mapping starting from a finite graph.*

The transition graphs are defined as the ε-closure of the configuration graphs:

$$\{w \xrightarrow{x} w' \mid x \in \Sigma, w \underset{P}{\overset{x}{\Longrightarrow}} w' \text{ and } w, w' \in R\}$$

for some k-regular set of configurations R. The following proposition summarizes various equivalent characterization of the transition graphs of k-hopda.

Proposition 4.3 ([CW03]). *The family of transition graphs of k-hopdas is equal up-to isomorphism to the families of:*

- *graphs whose edges are defined by relations in $O(\text{Rew}_k)$,*
- *graphs obtained by a k-fold iteration of unfolding and inverse rational mapping starting from a finite graph,*
- *graphs MSO-definable in Δ_2^k.*

5 Conclusion

We define a natural notion of regularity for higher-order pushdown stacks shares some of the most important properties of regular sets of words. In fact, we proved that they can be accepted by a deterministic machine and form an effective Boolean algebra. Furthermore, in the same way as regular set of words corresponds to the MSO-definable sets of words over the infinite binary tree, regular sets of k-stacks correspond to the MSO-definable sets over the canonical infinite structure associated to k-stacks. From the algorithmic point of view, we provided a normalization algorithm whose complexity is in fact a lower bound. To demonstrate the usefulness of this notion to work with hopdas, we used it to give a simple characterization of transitions graphs of hopdas similar to prefix-recognizable graphs for level 1.

From the model checking point of view, this notion could be used to extend the work done on pushdown automaton [BEM97] or on sub-families of hopdas [BM04]. From the structural point of view, it would be interesting to obtain an internal characterization of the transition graphs of k-hopdas of bounded degree. At level 1, the prefix recognizable graphs of bounded degree correspond to the configuration graphs of pushdown automata. It would be interesting to know how this property extends to higher-order.

Acknowledgment

The author would like to thank Didier Caucal and Antoine Meyer for their support, Jacques Sakarovitch for his suggestions on this work and the anonymous referees for their detailed and thorough reports.

References

[Aho69] Alfred V. Aho. Nested stack automata. *J. ACM*, 16(3):383–406, 1969.

[BEM97] A. Bouajjani, J. Esparza, and O. Maler. Reachability analysis of pushdown automata: Application to model checking. In *Proc. of CONCUR '97*, volume 1243 of *LNCS*, pages 135–150. Springer Verlag, 1997.

[Blu01] A. Blumensath. Prefix-recognisable graphs and monadic second-order logic. Technical Report AIB-2001-06, RWTH Aachen, 2001.

[BM04] A. Bouajjani and A. Meyer. Symbolic reachability analysis of higher-order context-free processes. In *Proc. of FSTTCS '04*, volume 3328 of *LNCS*, pages 135–147. Springer Verlag, 2004.

[Büc64] R. Büchi. Regular canonical systems. In *Archiv fur Math. Logik und Grundlagenforschung 6*, pages 91–111, 1964.

[Car05] A. Carayol. Regular sets of higher-order pushdown stacks. Extended version. A preliminary version of this article is available at the address http://www.irisa.fr/galion/acarayol/, 2005.

[Cau96a] D. Caucal. On infinite transition graphs having a decidable monadic theory. In *Proc. of ICALP '96*, volume 1099 of *LNCS*, pages 194–205. Springer Verlag, 1996.

[Cau96b] D. Caucal. Sur des graphes infinis réguliers. Habilitation thesis, Université de Rennes 1, 1996.

[Cau02] D. Caucal. On infinite terms having a decidable monadic theory. In *Proc. of MFCS '02*, volume 2420 of *LNCS*, pages 165–176. Springer Verlag, 2002.

[Cau03] D. Caucal. On infinite transition graphs having a decidable monadic theory. *Theor. Comput. Sci.*, 290:79–115, 2003.

[CW03] A. Carayol and S. Wöhrle. The Caucal hierarchy of infinite graphs in terms of logic and higher-order pushdown automata. In *Proc. of FSTTCS '03*, volume 2914 of *LNCS*, pages 112–123. Springer Verlag, 2003.

[Dam82] W. Damm. The OI- and IO-hierarchies. *Theor. Comput. Sci.*, 20(2):95–207, 1982.

[EF95] H.D. Ebbinghaus and J. Flum. *Finite Model Theory*. Springer-Verlag, 1995.

[Eng83] J. Engelfriet. Iterated pushdown automata and complexity classes. In *Proc. of STOC '83*, pages 365–373. ACM Press, 1983.

[Gre70] S. Greibach. Full AFL's and nested iterated substitution. *Inf. Control*, 16(1):7–35, 1970.

[KNU02] T. Knapik, D. Niwiński, and P. Urzyczyn. Higher-order pusdown trees are easy. In *Proc. of FoSSaCS '02*, volume 2303 of *LNCS*, pages 205–222. Springer Verlag, 2002.

[Mas76] A.N. Maslov. Multilevel stack automata. *Problemy Peredachi Informatsii*, 12:55–62, 1976.

[Tho03] W. Thomas. Constructing infinite graphs with a decidable MSO-theory. In *Proc. of MFCS '03*, volume 2747 of *LNCS*, pages 113–124. Springer Verlag, 2003.

Linearly Bounded Infinite Graphs

Arnaud Carayol[1] and Antoine Meyer[1,2]

[1] IRISA, Campus de Beaulieu, 35042 Rennes Cedex, France
{Arnaud.Carayol, Antoine.Meyer}@irisa.fr
[2] LIAFA, Université Denis Diderot, Case 7014, 2 place Jussieu,
F-75251 Paris Cedex 05, France
Antoine.Meyer@liafa.jussieu.fr

Abstract. Linearly bounded Turing machines have been mainly studied as acceptors for context-sensitive languages. We define a natural family of canonical infinite automata representing their observable computational behavior, called linearly bounded graphs. These automata naturally accept the same languages as the linearly bounded machines defining them. We present some of their structural properties as well as alternative characterizations in terms of rewriting systems and context-sensitive transductions. Finally, we compare these graphs to rational graphs, which are another family of automata accepting the context-sensitive languages, and prove that in the bounded-degree case, rational graphs are a strict sub-family of linearly bounded graphs.

1 Introduction

One of the cornerstones of formal language theory is the hierarchy of languages introduced by Chomsky in [Cho59]. It rests on the definition of four increasingly restricted families of grammars, which respectively generate the *recursively enumerable*, *context-sensitive*, *context-free* and *regular* languages. All were extensively studied, and were given several alternative characterizations using different kinds of formalisms (or *acceptors*). For instance, pushdown systems characterize context-free languages, and linearly bounded Turing machines (LBMs) characterize context-sensitive languages. More recently, several authors have related these four families of languages to families of infinite graphs (see for instance [Tho01]). Given a fixed initial vertex and a set of final vertices, one can associate a language to a graph by considering the set of all words labeling a path between the initial vertex and one of the final vertices. In [CK02], a summary of four families of graphs accepting the four families of the Chomsky hierarchy is presented. They are the Turing graphs [Cau03b], rational graphs [Mor00, MS01], prefix-recognizable graphs [Cau96, Cau03a] and finite graphs.

Several approaches exist to define families of infinite graphs, among which we will cite three. The first one is to consider the finite acceptor of a language, and to build a graph representing the structure of its computations: vertices represent configurations, and each edge reflects the observable effect of an input on the configuration. One speaks of the *transition graph* of the acceptor. An interesting consequence is that the language of the graph can be deduced from the language of the acceptor it was built from. A second method proposed in [CK02] is to consider the Cayley-type graphs of some families of word rewriting systems. Each vertex is a normal form for a given rewriting system, and an edge between two vertices represents the addition of a letter and re-normalization by the rewriting system. Finally, a third possibility is to directly define the edge relations in a graph using automata or other formalisms. One speaks of derivation, transduction or computation graphs. In this approach, a path no longer represents a run of an acceptor, but rather a composition of binary relations.

Both prefix-recognizable graphs and Turing graphs have alternative definitions along all three approaches. Prefix-recognizable graphs are defined as the graphs of recognizable prefix relations. In [Sti00], Stirling presented them as the transition graphs of pushdown systems. It was also proved that they coincide with the Cayley-type graphs of prefix rewriting systems. As for Turing graphs, Caucal showed that they can be seen indifferently as the transition and computation graphs of Turing machines [Cau03b]. They are also the Cayley-type graphs of unrestricted rewriting systems. Rational graphs, however, are only defined as transduction graphs (using rational transducers) and as the Cayley-type graphs of left-overlapping rewriting systems, and lack a characterization as transition graphs. In this paper, we are thus interested in defining a suitable notion of transition graphs of linearly bounded Turing machines, and to determine some of their structural properties as well as to compare them with rational graphs.

As in [Cau03b] for Turing machines, we define a labeled version of LBMs, called LLBMs. Their transition rules are labeled either by a symbol from the input alphabet or by a special symbol denoting an unobservable transition. Following an idea from [Sti00], we consider that in every configuration of a LLBM, either internal actions or inputs are allowed, but not both at a time. This way, we can distinguish between internal and external configurations. The transition graph of a LLBM is the graph whose vertices are external configurations, and whose edges represent an input followed by a finite number of silent transitions. This definition is purely structural and associates a unique graph to a given LLBM. For convenience, we call such graphs *linearly bounded graphs*. To our knowledge, the notion of transition graph of a LBM was never considered. A similar work was proposed in [KP99, Pay00], where the family of configuration graphs of LBMs up to weak bisimulation is studied. However, it provides no formal definition associating LBMs to a family of *real-time* graphs (without edges labeled by silent transitions) representing their observable computations.

To further illustrate the suitability of our notion, we provide two alternative definitions of linearly bounded graphs. First, we prove that they are isomorphic to the Cayley-type graphs of length-decreasing rewriting systems. The second alternative definition directly represents the edge relations of a linearly bounded graph as a certain kind of context-sensitive transductions. This allows us to straightforwardly deduce structural properties of linearly bounded graphs, like their closure under synchronized product (which was already known from [KP99]) and under restriction to a context-sensitive set of vertices. To conclude this study, we show that linearly bounded graphs and rational graphs form incomparable families, even in the finite degree case. However, bounded degree rational graphs are a strict sub-family of linearly bounded graphs.

A more complete study of this family of graphs, including the proofs of the results stated in this article can be found in [CM05b].

2 Preliminary Definitions

A labeled, directed and simple *graph* is a set $G \subseteq V \times \Sigma \times V$ with Σ is a finite set of labels and V a countable set of *vertices*. An element (s, a, t) of G is an *edge* of source s, target t and *label* a, and is written $s \xrightarrow{a}_G t$ or simply $s \xrightarrow{a} t$ if G is understood. The set of all sources and targets of a graph is its *support* V_G. A sequence of edges $s_1 \xrightarrow{a_1} t_1, \ldots, s_k \xrightarrow{a_k} t_k$ with $\forall i \in [2, k]$, $s_i = t_{i-1}$ is a *path*. It is written $s_1 \xrightarrow{u} t_k$, where $u = a_1 \ldots a_k$ is the corresponding *path label*. A graph is *deterministic* if it contains no pair of edges with the same source and label. One can relate a graph to a languages by considering its path language, defined as the set of all words labeling a path between two given sets of vertices.

Definition 2.1. *The (path) language of a graph G between two sets of vertices I and F is the set $L(G, I, F) = \{ w \mid s \xrightarrow{w}_G t,\ s \in I,\ t \in F\}$.*

Linearly Bounded Turing Machines. We now recall the definition of context-sensitive languages and linearly bounded Turing machines. A context-sensitive language is a set of words generated by a grammar whose production rules are of the form $\alpha \to \beta$ with $|\beta| \geq |\alpha|$. Such grammars are called *context-sensitive*.

A more operational definition of context-sensitive languages is as the family of languages accepted by *linearly bounded Turing machines* (LBMs). Informally, a LBM is a Turing machine accepting each word w of its language using at most $k.|w|$ tape cells, where k is a fixed constant. Without loss of generality, one usually considers k to be equal to 1. Note that, contrary to unbounded Turing machines, it is sufficient to only consider linearly bounded machines which always terminate, also called *quasi-real time* machines. An interesting open problem raised by Kuroda [Kur64] concerns deterministic context-sensitive languages, which are the languages accepted by deterministic LBMs. It is not known whether they coincide with non-deterministic context-sensitive languages, as is the case for recursively enumerable or rational languages.

Rational Graphs. Consider the product monoid $\Sigma^* \times \Sigma^*$, whose elements are pairs of words (u, v) in Σ^*, and whose composition law is defined by $(u_1, v_1) \cdot (u_2, v_2) = (u_1 u_2, v_1 v_2)$. A finite transducer is an automaton over $\Sigma^* \times \Sigma^*$ with labels in $(\Sigma \cup \{\varepsilon\}) \times (\Sigma \cup \{\varepsilon\})$. Transducers accept the rational subsets of $\Sigma^* \times \Sigma^*$, which are seen as binary relations on words and called rational transductions. We do not distinguish a transducer from the relation it accepts and write $(w, w') \in T$ if the pair (w, w') is accepted by T. Graphs whose vertices are words and whose edge relations is defined by transducers (one per letter in the label alphabet) are called rational graphs.

Definition 2.2 ([Mor00]). *A rational graph labeled by Σ with vertices in Γ^* is given by a tuple of transducers $(T_a)_{a \in \Sigma}$ over Γ. For all $a \in \Sigma$, $(u, a, v) \in G$ if and only if $(u, v) \in T_a$.*

For $w \in \Sigma^+$ and $a \in \Sigma$, we write $T_{wa} = T_w \circ T_a$, and $u \xrightarrow{w} v$ if and only if $(u, v) \in T_w$. In general, there is no bound on the size difference between input and output in a transducer (and hence between the lengths of two adjacent vertices in a rational graph). Interesting subclasses are obtained by enforcing some form of synchronization. The most well-known was defined by Elgot and Mezei [EM65] as follows. A transducer over Σ with initial state q_0 is (left-)synchronized if for every path $q_0 \xrightarrow{x_0/y_0} q_1 \ldots q_{n-1} \xrightarrow{x_n/y_n} q_n$, there exists $k \in [0, n]$ such that for all $i \in [0, k-1]$, x_i and y_i belongs to Σ and either $x_k = \ldots = x_n = \varepsilon$ or $y_k = \ldots = y_n = \varepsilon$. A rational graph defined by synchronized transducers will simply be called a synchronized (rational) graph.

3 Linearly Bounded Graphs

3.1 LBM Transition Graphs

Following [Cau03b], we define the notion of labeled linearly bounded Turing machine (LLBM). As in standard definitions of LBMs, the transition rules can only move the head of the LLBM between the two end markers [and]. In addition, a silent step can decrease the size of the configuration (without removing the markers) and a Σ-transition can increase the size of the configuration by one cell. This ensures that while reading a word of length n, the labeled LBM uses at most n cells.

Definition 3.1. *A labeled linearly bounded Turing machine is a tuple $M = (\Gamma, \Sigma, [,], Q, q_0, F, \delta)$, where Γ is a finite set of tape symbols, $\Sigma \subseteq \Gamma$ is the input alphabet, [and] $\notin \Gamma$ are the left and right end-marker, Q is a finite set (disjoint from Γ) of control states, $q_0 \in Q$ is the unique initial state, $F \subseteq Q$ is a set of final states and δ is a finite set of labeled transition rules of one of the forms:*

$$pA \xrightarrow{\varepsilon} qB\pm \qquad p[\xrightarrow{\varepsilon} q[+ \qquad p] \xrightarrow{\varepsilon} q]-$$
$$pB \xrightarrow{a} qAB \qquad p] \xrightarrow{a} qA] \qquad pA \xrightarrow{\varepsilon} q$$

with $p, q \in Q$, $A, B \in \Gamma$, $\pm \in \{+, -\}$ and $a \in \Sigma$.

The set of configurations C_M of M is the set of words uqv such that $q \in Q$, $v \neq \varepsilon$ and $uv \in [\Gamma^*]$. For all $x \in \Sigma \cup \{\varepsilon\}$, the transition relation $\xrightarrow[M]{x}$ is a subset of $C_M \times C_M$ defined as

$$\xrightarrow[M]{x} = \{\ (upAv, uBqv) \mid pA \xrightarrow{x} qB+ \in \delta\ \}$$
$$\cup\ \{\ (uCpAv, uqCBv) \mid pA \xrightarrow{x} qB- \in \delta\}$$
$$\cup\ \begin{cases} \{\ (upAv, uqv) \mid pA \xrightarrow{x} q \in \delta\} & \text{with } x = \varepsilon \\ \{\ (upAv, uqBAv) \mid pA \xrightarrow{x} qBA \in \delta\} & \text{with } x \in \Sigma. \end{cases}$$

We will simply write $\xrightarrow{x}$ when M is understood. As usual, we define $\xrightarrow{wx}$ as ($\xrightarrow{w} \circ \xrightarrow{x}$) for all $w \in (\Sigma \cup \{\varepsilon\})^*$. The unique initial configuration is $[q_0]$ and a final configuration c_f is a configuration containing a terminal control state. A word w is accepted by M if $[q_0] \xrightarrow{w} c_f$ where c_f is a final configuration. Quite naturally, M is *deterministic* if, from any configuration, either all possible moves are labeled by *distinct* letters of Σ, or there is only one possible move. Formally, it means that for all configurations c, c_1, c_2 with $c_1 \neq c_2$, if $c \xrightarrow{a} c_1$ and $c \xrightarrow{b} c_2$ then $a \neq b$, $a \neq \varepsilon$ and $b \neq \varepsilon$.

Remark 3.1. For convenience, one may consider LBMs whose initial configuration is not of the form $[q_0]$ but is any fixed configuration c_0. This does not add any expressive power, as can be proved by a simple encoding of c_0 into the control state set of the machine.

Let $M = (Q, \Sigma, \Gamma, \delta, q_0, F, [,])$ be a LLBM, we define its configuration graph

$$C_M = \{(c, a, c') \mid c \xrightarrow[M]{a} c' \text{ for } a \in \Sigma \cup \{\varepsilon\}\ \}.$$

The vertices of this graph are all configurations of M, and its edges denote the transitions between them, including ε-transitions. One may wish to only consider the behavior of M from an external point of view, i.e. only looking at the sequence of inputs. This means one has to find a way to conceal ε-transitions without changing the accepted language or destroying the structure. One speaks of the *transition graph* of an acceptor, as opposed to its configuration graph.

In [Sti00], Stirling mentions a normal form for pushdown automata which allows him to consider a structural notion of transition graphs, without relying on the naming of vertices. We first recall this notion of *normalized* systems adapted to labeled LBMs. A labeled LBM is *normalized* if its set of control states can be partitioned in two subsets: one set of *internal* states, noted Q_ε, which can always and only perform ε-rules, and a set of *external* states noted Q_Σ, which can only perform Σ-rules. More formally:

Definition 3.2. *A labeled LBM $M = (Q, \Sigma, \Gamma, \delta, q_0, F, [,])$ is normalized if there are disjoint sets Q_Σ and Q_ε such that $Q = Q_\varepsilon \cup Q_\Sigma$, $F \subseteq Q_\Sigma$, and*

$$pB \xrightarrow{a} qAB \in \delta \implies p \in Q_\Sigma,$$
$$pA \xrightarrow{\varepsilon} qB\pm \in \delta \implies p \in Q_\varepsilon,$$
$$p \in Q_\varepsilon \implies \exists\ pA \xrightarrow{\varepsilon} qB\pm \in \delta.$$

This definition implies in particular that a control states from which there exists no transition must belong to Q_Σ. A configuration is external if its control state is in Q_Σ, and internal otherwise. This makes it possible to *structurally* distinguish between internal vertices, which have one or more outgoing ε-edges, and external ones which only have outgoing Σ-edges or have no outgoing edges. Given any labeled LBM, it is always possible to normalize it without changing the accepted language.

From this point on, unless otherwise stated, we will only consider normalized LLBMs. We can now define our notion of LLBM transition graph as the ε-closure of its configuration graph, followed by a restriction to its set of external configurations (which happens to be a rational set).

Definition 3.3. *Let $M = (\Gamma, \Sigma, [,], Q, q_0, F, \delta)$ be a (normalized) LLBM, and C_Σ be its set of external configurations. The transition graph of M is*

$$G_M = \{(c, a, c') \mid c, c' \in C_\Sigma,\ a \in \Sigma,\ \wedge\ c \xrightarrow[M]{a\varepsilon^*} c'\}.$$

We now define the family of linearly bounded graphs as the closure under isomorphism of transition graphs of labeled LBMs, i.e. as the set of all graphs which can be obtained by renaming the vertices of a LLBM transition graph.

3.2 Alternative Definitions

This section provides two alternative definitions of linearly bounded graphs. In [CK02], it is shown that all previously mentioned families of graphs can be expressed in a uniform way in terms of Cayley-type graphs of certain families of rewriting systems. We show that it is also the case for linearly bounded graphs, which are the Cayley-type graphs of length-decreasing rewriting systems. The second alternative definition we present changes the perspective and directly defines the edges of linearly bounded graphs using incremental context-sensitive transductions. This variety of definitions will allow us to prove in a simpler way some of the properties of linearly bounded graphs.

Cayley-Type Graphs of Decreasing Rewriting Systems. We first give the relevant definitions about rewriting systems and Cayley-type graphs. A *word rewriting system* R over alphabet Γ is a subset of $\Gamma^* \times \Gamma^*$. Each element $(l, r) \in R$ is called a *rewriting rule* and noted $l \to r$. The words l and r are respectively called the left-hand and right-hand side of the rule. The *rewriting relation* of R is the binary relation $\{(ulv, urv) \mid u, v \in \Gamma^*,\ l \to r \in R\}$ which we also denote by R, consisting of all pairs of words (w_1, w_2) such that w_2 can be obtained by replacing (*rewriting*) an instance of a left-hand side l in w_1 with the corresponding right-hand side r. The reflexive and transitive closure R^* of this relation is called the *derivation* of R. Whenever for some words u and v we have uR^*v, we say R rewrites u into v. A word which contains no left-hand side is called a *normal form*. The set of all normal forms of R is written NF(R).

One can associate a unique infinite graph to any rewriting system by considering its *Cayley-type graph* defined as follows:

Definition 3.4. *The Σ-labeled Cayley-type graph of a rewriting system R over Γ, with $\Sigma \subseteq \Gamma$, is the infinite graph*

$$G_R = \{(u, a, v) \mid a \in \Sigma,\ u, v \in \mathrm{NF}(R),\ uaR^*v\}.$$

The family of rewriting systems we consider is the family of *finite length-decreasing word rewriting systems*, i.e. rewriting systems with a finite set of rules of the form $l \to r$ with $|l| \geq |r|$, which can only preserve or decrease the length of the word to which they are applied. The reason for this choice is that the derivation relations of such systems coincide with arbitrary compositions of labeled LBM ε-rules.

Theorem 3.1. *The two families of linearly bounded graphs and of Cayley-type graphs of decreasing rewriting systems are equal up to isomorphism.*

Incremental Context-Sensitive Transduction Graphs. The notion of *computation graph* was first introduced in early versions of [Cau03b] and systematically used in [CK02]. It corresponds to the graphs defined by the *transductions* (i.e. binary relations on words) associated to a family of finite machines. These works prove that for pushdown automata and Turing machines, the classes of transition and computation graphs coincide. We show that it is also the possible to give a definition of linearly bounded graphs as the computation graphs of a certain family of LBMs, or equivalently as the graphs defined by a certain family of context-sensitive transductions.

A relation R is recognized by a LBM M if the language $\{u\#v \mid (u,v) \in R\}$ where $\#$ is a fresh symbol is accepted by M. However, this type of transductions generates more than linearly bounded graphs. Even if we only consider linear relations (i.e relations R such that there exists c and $k \in \mathbb{N}$ such that $(u,v) \in R$ implies $|v| \leq c \cdot |u| + k$), we obtain graphs accepting the languages recognizable in exponential space (EXPSPACE) which strictly contain the context-sensitive languages [Imm88]. We need to consider relations for which the length difference between a word and its image is bounded by a certain constant. Such relations can be associated to LBMs.

Definition 3.5. *A k-incremental context-sensitive transduction T over Γ is defined by a LBM recognizing a language $L = \{u\#v \mid u, v \in \Gamma^*$ and $|v| \leq |u| + k\}$ where $\#$ does not belong to Γ. The relation T is defined as $\{(u,v) \mid u\#v \in L\}$.*

The following proposition states that incremental context-sensitive transductions form a boolean algebra.

Proposition 3.1. *For all k-incremental context-sensitive transductions T and T' over Γ^*, $T \cup T'$, $T \cap T'$ and $\overline{T} = E - T$ (where E is $\{(u,v) \mid 0 \leq |v| \leq |u| + k\}$) are incremental context-sensitive transductions.*

The canonical graph associated to a finite set of transductions is called a *transduction graph*. Relating graphs to a family of binary relations on words was already used to define rational graphs and their sub-families.

Definition 3.6. *The Σ-labeled transduction graph G_T of a finite set of incremental context-sensitive transductions $(T_a)_{a \in \Sigma}$ is*

$$G_T = \{(u, a, v) \mid a \in \Sigma \text{ and } (u, v) \text{ is recognized by } T_a\}.$$

Length-preserving context-sensitive transductions have already been extensively studied in [LST98]. In the rest of this presentation, unless otherwise stated, we will only consider 1-incremental transductions without loss of generality regarding the obtained family of graphs.

Theorem 3.2. *The families of linearly bounded graphs and of incremental context-sensitive transduction graphs are equal up to isomorphism.*

3.3 Structural Properties

Languages. It is quite obvious that the language of the transition graph of a LLBM M between the vertex representing its initial configuration and the set of vertices representing its final configurations is the language of M. In fact, the choice of initial and final vertices has no importance in terms of the family of languages one obtains.

Proposition 3.2. *The languages of linearly bounded graphs between an initial vertex i and a finite set F of final vertices are the context-sensitive languages.*

Remark 3.2. When a linearly bounded graph is explicitly seen as the transition graph of a LLBM, as a Cayley-type graph or as a transduction graph, i.e. when the naming of its vertices is fixed, considering context-sensitive sets of final vertices does not increase the accepted family of languages.

Closure Properties. Linearly bounded graphs enjoy several good properties, which will be especially important when comparing this class to other families of graphs related to LBMs or context-sensitive languages (see Section 4).

Proposition 3.3. *The family of linearly bounded graphs is closed under restriction to reachable vertices from any vertex and under restriction to a context-sensitive set of vertices.*

Since all rational languages are context-sensitive, linearly bounded graphs are also closed under restriction to a rational set of vertices. This shows that it is not necessary to allow arbitrary rational restrictions in the definition of transition graphs of linearly bounded machine, since such a restriction can be directly applied to the set of external configurations of a machine. By a slight adaptation of the proofs used in [KP99], one also gets the result below.

Proposition 3.4 ([KP99]). *Linearly bounded graphs are closed under synchronized product.*

Deterministic Linearly Bounded Graphs. It is straightforward to notice that there exist non-deterministic labeled LBMs whose transition graphs are deterministic, and we do not know whether, for all non-deterministic labeled LBM whose transition graph is deterministic, it is possible to build an equivalent deterministic labeled LBM, possibly having the same transition graph. In fact, we can show that, for any context-sensitive language, it is always possible to build a deterministic linearly bounded graph accepting it.

Proposition 3.5. *For all context-sensitive language L, there exists a deterministic linearly bounded graph G, a vertex i and a rational set of vertices F of G such that $L = L(G, \{i\}, F)$.*

We are of course not able to conclude that the languages of deterministic transition graphs of labeled LBMs are the deterministic context-sensitive languages, because it would imply that deterministic and non-deterministic context-sensitive languages coincide. However, if we only consider quasi real-time linearly bounded machines, which have no infinite run on any given input word, the family of transition graphs we obtain faithfully illustrates the determinism of the languages.

Proposition 3.6. *The languages of deterministic transition graphs of quasi real-time LBMs are the deterministic context-sensitive languages.*

4 Comparison with Rational Graphs

We will now give some remarks about the comparison between linearly bounded graphs and several different sub-families of rational graphs. First note that since linearly bounded graphs have by definition a finite degree, it is more relevant to only consider rational graphs of finite degree. However, even under this structural restriction, rational and linearly-bounded graphs are incomparable, due to the incompatibility in the growth rate of their vertices degrees.

In a rational graph the out-degree at distance n from any vertex can be c^{c^n}, whereas in a linearly bounded graph is at most c^n for some c.

Lemma 4.1. *For any linearly bounded graph L and any vertex x, there exists $c \in \mathbb{N}$ such that the out-degree of L at distance $n > 0$ of x is at most c^n.*

Figure 1 shows a rational graph whose vertices at distance n from the root A have out-degree $2^{2^{n+1}}$. This graph is thus not linearly bounded.

Conversely, in a rational graph of finite degree, the in-degree at distance n from any vertex is at most c^{c^n} for some $c \in \mathbb{N}$, in a linearly bounded graph it can be as large as $f(n)$ for any mapping f from $\mathbb{N}$ to $\mathbb{N}$ recognizable in linear space (i.e. such that the language $\{0^n 1^{f(n)} \mid n \in \mathbb{N}\}$ is context-sensitive).

Lemma 4.2. *For any mapping $f : \mathbb{N} \mapsto \mathbb{N}$ recognizable in linear space, there exists a linearly bounded graph L with a vertex x such that the in-degree at distance $n > 0$ of x is $f(n)$.*

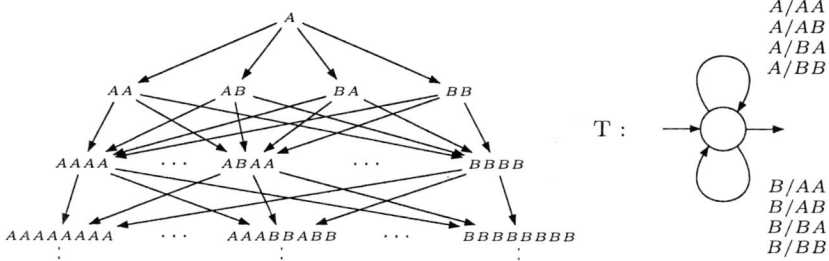

Fig. 1. A finite degree rational graph (together with its transducer) which is not isomorphic to any linearly bounded graph

An instance of such a mapping is $f : n \mapsto 2^{2^{2^n}}$, which is more than the in-degree at distance n of a vertex in any rational graph of finite degree. From these two observations, we get the result below.

Proposition 4.1. *The families of finite degree rational graphs and of linearly bounded graphs are incomparable.*

Since finite-degree rational graphs and linearly bounded graphs are incomparable, we investigate more restricted sub-families of rational graphs. For synchronized graphs of finite out-degree, we have the following result.

Proposition 4.2. *The synchronized graphs of finite degree form a strict sub-family of linearly bounded graphs (up to isomorphism).*

Proof (Sketch). Synchronized transducers of finite image can only map together words whose length difference is at most some constant k. It can thus very easily be seen that synchronized rational relations of finite image are incremental context-sensitive transductions.

For the even more restricted family of bounded-degree rational graphs, we show the following comparison.

Theorem 4.1. *The rational graphs of bounded degree form a strict sub-family of linearly bounded graphs of bounded degree (up to isomorphism).*

Proof (Sketch). The inclusion is based on a uniformization result for rational relations of bounded image due to Weber [Web96], which states that they can be decomposed into a finite union of functional transductions. This allows us to propose a coding of the rational graph's vertices such that the edge relation of the obtained graph is a 1-incremental context-sensitive transduction. The idea of this coding is to identify a vertex either by its name in the rational graph, or by a unique path from another vertex, whichever is shortest. This allows to express the edge relation of the graph as a 1-incremental context-sensitive transduction.

As rational graphs are closed under edge reversal, an equality between the two families would imply that linearly bounded graphs of bounded degree are also closed under edge reversal, which can be proved wrong. □

It may be interesting at this point to recall that all existing proofs that the rational graphs accept the context-sensitive languages break down when the out-degree is bounded. It is thus not at all clear whether rational graphs of bounded degree accept all context-sensitive languages. However, as noted in 3.5, it is still the case for bounded degree linearly bounded graphs, and in particular for deterministic linearly bounded graphs.

5 Conclusion

This paper gives a natural definition of a family of canonical graphs associated to the observable computations of labeled linearly bounded machines. It provides equivalent characterizations of this family as the Cayley-type graphs of length-decreasing term-rewriting systems, and as the graphs defined by a subfamily of context-sensitive transductions which can increase the length of their input by at most a constant number of letters. Although of a sensibly different nature from rational graphs, we showed that all rational graphs of bounded degree are linearly bounded graphs of bounded degree, and that this inclusion is strict. This leads us to consider a more restricted notion of infinite automata, closer to classical finite automata (as was already observed in [CM05a]), and to propose a hierarchy of families of infinite graphs of bounded degree accepting the families of languages of the Chomsky hierarchy (see Fig. 2).

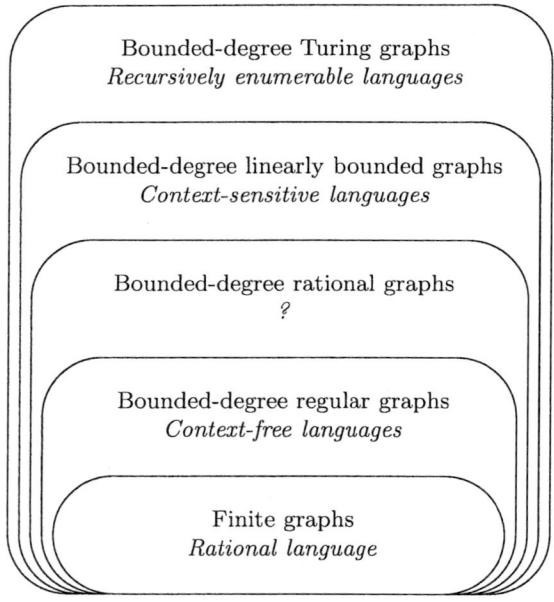

Fig. 2. A Chomsky-like hierarchy of bounded-degree infinite graphs

Acknowledgment

The authors would like to thank Didier Caucal for his support.

References

[Cau96] D. Caucal. On infinite transition graphs having a decidable monadic theory. In *Proc. of ICALP '96*, volume 1099 of *LNCS*, pages 194–205. Springer Verlag, 1996.

[Cau03a] D. Caucal. On infinite transition graphs having a decidable monadic theory. *Theor. Comput. Sci.*, 290:79–115, 2003.

[Cau03b] D. Caucal. On the transition graphs of Turing machines. *Theor. Comput. Sci.*, 296:195–223, 2003.

[Cho59] N. Chomsky. On certain formal properties of grammars. *Inf. Control*, 2:137–167, 1959.

[CK02] D. Caucal and T. Knapik. A Chomsky-like hierarchy of infinite graphs. In *Proc. of MFCS '02*, volume 2420 of *LNCS*, pages 177–187. Springer Verlag, 2002.

[CM05a] A. Carayol and A. Meyer. Context-sensitive languages, rational graphs and determinism. Submitted to publication, preliminary version available at http://www.irisa.fr/galion/acarayol/, 2005.

[CM05b] A. Carayol and A. Meyer. Linearly bounded infinite graphs (extended version). Submitted to publication, preliminary version available at http://www.irisa.fr/galion/acarayol/, 2005.

[EM65] C. Elgot and J. Mezei. On relations defined by finite automata. *IBM J. Res. Develop.*, 9:47–68, 1965.

[Imm88] N. Immerman. Nondeterministic space is closed under complementation. *SIAM J. Comput.*, 17(5):935–938, 1988.

[KP99] T. Knapik and É. Payet. Synchronized product of linear bounded machines. In *Proc. of FCT '99*, volume 1684 of *LNCS*, pages 362–373. Springer Verlag, 1999.

[Kur64] S. Kuroda. Classes of languages and linear-bounded automata. *Inf. Control*, 7(2):207–223, 1964.

[LST98] M. Latteux, D. Simplot, and A. Terlutte. Iterated length-preserving rational transductions. In *Proc. of MFCS '98*, volume 1450 of *LNCS*, pages 286–295. Springer Verlag, 1998.

[Mor00] C. Morvan. On rational graphs. In *Proc. of FoSSaCS '00*, volume 1784 of *LNCS*, pages 252–266. Springer Verlag, 2000.

[MS01] C. Morvan and C. Stirling. Rational graphs trace context-sensitive languages. In *Proc. of MFCS '01*, volume 2136 of *LNCS*, pages 548–559. Springer Verlag, 2001.

[Pay00] É. Payet. *Thue Specifications, Infinite Graphs and Synchronized Product*. PhD thesis, Université de la Réunion, 2000.

[Sti00] C. Stirling. Decidability of bisimulation equivalence for pushdown processes. Technical Report EDI-INF-RR-0005, School of Informatics, University of Edinburgh, 2000.

[Tho01] W. Thomas. A short introduction to infinite automata. In *Proc. of DLT '01*, volume 2295 of *LNCS*, pages 130–144. Springer Verlag, 2001.

[Web96] A. Weber. Decomposing a k-valued transducer into k unambiguous ones. *Inf. Théor. Appl.*, 30(5):379–413, 1996.

Basic Properties for Sand Automata

J. Cervelle[1], E. Formenti[2], and B. Masson[2]

[1] Université de Marne-la-Vallée, Institut Gaspard Monge,
77454 Marne-la-Vallée cedex 2, France
`julien.cervelle@univ-mlv.fr`
[2] Université de Nice-Sophia Antipolis, Laboratoire I3S, Bât. ESSI,
930 Route des Colles, 06903 Sophia Antipolis, France
`{enrico.formenti, benoit.masson}@I3S.unice.fr`

Abstract. We prove several results about the relations between injectivity and surjectivity for sand automata. Moreover, we begin the exploration of the dynamical behavior of sand automata proving that the property of ultimate periodicity is undecidable. We believe that the proof technique used for this last result might turn out to be useful for many other results in the same context.

Keywords: sand automata, reversibility, undecidability, ultimate periodicity.

1 Introduction

Self-organized criticality (SOC) is a notion which tries to explain the peculiar behavior of many natural and physical phenomena. These systems evolve, according to some law, to a "critical state". Any perturbation, no matter how small, of the critical state generates a deep spontaneous re-organization of the system. Thereafter, the system evolves to another critical state and so on.

Examples of SOC systems are: sandpiles, snow avalanches, star clusters in the outer space, earthquakes, forest fires, load balance in operating systems [1,2,3,4,5].

Sandpiles models are a paradigmatic formal model for SOC systems [6,7]. In [8], the authors introduced sand automata as a generalization of sandpiles models and transposed them in the setting of discrete dynamical systems. A key-point of [8] was to introduce a suitable topology and study the dynamical behavior of sand automata $w.r.t.$ this new topology. This resulted in a fundamental representation theorem similar to the well-known Hedlund's theorem for cellular automata [8,9].

This paper continues the study of sand automata starting from basic set properties like injectivity and surjectivity. The decidability of those two last properties is still an open question. In order to simplify the decision problem we study the relations between basic set properties. We prove that many relations between set properties that are true in cellular automata are no more true in the context of sand automata. This allows to conclude that sand automata are a completely new model and not a peculiar "sub-model" of cellular automata as it might seem at a glance.

In particular, we show that injective sand automata are not necessarily reversible but they might have a right inverse automaton which is not a left inverse. This is a completely new situation w.r.t. cellular automata which we think is worthwhile future studies.

Understanding the dynamical behavior of sand automata is in general very difficult. Hence we started from very "simple" behavior: ultimate periodicity. We have proved (Theorem 2) that the problem of establishing if a given automaton is ultimately periodic is undecidable (when considering spatial periodic or finite configurations).

We believe that the proof technique developed for Theorem 2 might be used for proving many other similar results.

The paper is structured as follows. The next section introduces the topology on sandpiles and related known results. Section 3 recalls the definition of sand automata and their representation theorem. Very interesting and useful examples of sand automata are presented in Sect. 4. The main results are in Sect. 5 and 6. In Sect. 7 we draw our conclusions.

Remark that, due to lack of space, some results have no proof. Their proofs can be found in the appendix.

2 The Topology on Sandpiles

A *configuration* represents a set of sand grains, organized in piles and distributed all over a d-dimensional grid. Every point of the grid $\mathbb{Z}^d$ is associated with the *number of grains* i.e. an element of $\widetilde{\mathbb{Z}} = \mathbb{Z} \cup \{-\infty, +\infty\}$. The value $-\infty$ represents a *sink* and $+\infty$ a *source* of sand grains. Hence a configuration is an element of $\widetilde{\mathbb{Z}}^{\mathbb{Z}^d}$. We denote by $c_{i_1,\ldots,i_d}$ or c_i the number of grains in the column of c indexed by the vector $i = (i_1, \ldots, i_d)$. Denote $\mathfrak{C}$ the set of all configurations. Finally, for $i \in \widetilde{\mathbb{Z}}$, $\mathfrak{C}_i$ is the set of configurations whose sand amount at position $(0, \ldots, 0)$ is i. A configuration c is *finite* if $\exists k \in \mathbb{N}$ such that for any vector $i \in \mathbb{Z}^d$, $|i| \geq k$, $c_i = 0$ (we denote by $|\cdot|$ the infinite norm). The set of finite configurations is noted $\mathfrak{F}$. For any finite configuration c, the *size* of c is $|c| = \max_{i,j \in \mathbb{Z}^d} \{|i - j|, c_i \neq 0 \text{ and } c_j \neq 0\}$. A configuration c is (spatially) *periodic* if there is a vector $p \in \mathbb{Z}^d$ such that for any vector $i \in \mathbb{Z}^d$ and any integer $t \in \mathbb{Z}$, $c_i = c_{i+tp}$; $\mathfrak{P}$ denotes the set of (spatially) periodic configurations.

In the remainder of the section, definitions are only given for dimension 1. The generalization to higher dimensions is straightforward.

In [8], the authors equipped $\mathfrak{C}$ with a metric topology defined in two steps. First, one fixes a reference point (for example the column of index 0); then the metric is designed in such a way that two configurations are at small distance if they have "the same" number of grains in a (finite) neighborhood of the reference point. Of course, one should make more precise the meaning of the sentence "have the same grains content". The differences in the number of grains is quantified by a *measuring device* of precision $l \in \mathbb{N}$ and reference height $m \in \mathbb{Z}$

$$\beta_l^m(n) = \begin{cases} +\infty & \text{if } n > m+l \ , \\ -\infty & \text{if } n < m-l \ , \\ n-m & \text{otherwise.} \end{cases}$$

If the difference (in the number grains) between the measured height n and the reference height m is too high (resp. too low), then it is declared to be $+\infty$ (resp. $-\infty$). We assume $\infty - \infty = 0$.

For any configuration $c \in \widetilde{\mathbb{Z}}^{\mathbb{Z}}$, $l \in \mathbb{N}$, $l \neq 0$ and $i \in \mathbb{Z}$, define the following sequence of differences:

$$d_l^i(c) = \begin{cases} (\beta_l^{c_i}(c_{i-l}), \ldots, \beta_l^{c_i}(c_{i-1}), \beta_l^{c_i}(c_{i+1}), \ldots, \beta_l^{c_i}(c_{i+l})) & \text{if } |c_i| \neq \infty \ , \\ (\beta_l^0(c_{i-l}), \ldots, \beta_l^0(c_{i-1}), \beta_l^0(c_{i+1}), \ldots, \beta_l^0(c_{i+l})) & \text{if } c_i = \pm\infty \ . \end{cases}$$

For $l = 0$, define $d_0^i(c)$ as the singleton (c_i). Finally, the distance between two configurations x and y is defined as follows: $d(x,y) = 2^{-l}$, where l is the smallest integer such that $d_l^0(x) \neq d_l^0(y)$.

From now on, $\mathfrak{C}$ is equipped with the metric topology induced by d. The following propositions prove that the structure of the topology on $\mathfrak{C}$ is rich enough to justify the study of dynamical systems on it.

Proposition 1 ([8]). *The space $\mathfrak{C}$ is perfect (i.e. it has no isolated point) and locally compact (i.e. for any point x there is a neighborhood of x whose closure is compact).*

Proposition 2 ([8]). *The space $\mathfrak{C}$ is totally disconnected (i.e. for any points x, y there are two open sets U and V such as $x \in U$, $y \in V$, $U \cap V = \emptyset$ and $U \cup V = \mathfrak{C}$).*

Proposition 3 ([8]). *For any $i \in \widetilde{\mathbb{Z}}$, the set $\mathfrak{E}_i$ is compact.*

The following result completes the characterization of the topological structure of $\mathfrak{C}$.

Proposition 4. *The space $\mathfrak{C}$ is complete.*

3 Sand Automata

A *sand automaton* (SA) is a deterministic automaton acting on configurations. It essentially consists in a local rule which is applied synchronously to each column of the current configuration. The local rule describes how many grains are lost or gained in each column according to the grain content of its neighborhood.

In the sequel, we give the formal definition of sand automaton in dimension 1. Its generalization to higher dimensions is straightforward.

Formally, a sand automaton is a structure $\mathcal{A} \equiv \langle r, \lambda \rangle$ where $\lambda : \widetilde{[\![-r,r]\!]}^{2r} \mapsto [\![-r,r]\!]$ is the local rule and r is the *precision* (sometimes also called the *radius*) of the measuring device. The *global function* $f_{\mathcal{A}} : \mathfrak{C} \mapsto \mathfrak{C}$ of $\mathcal{A}$ is defined as follows

$$\forall c \in \mathfrak{C}\, \forall i \in \mathbb{Z}, \quad f_{\mathcal{A}}(c)_i = \begin{cases} c_i & \text{if } c_i = \pm\infty \ , \\ c_i + \lambda(d_r^i(c)) & \text{otherwise.} \end{cases}$$

In [8], the authors show that sand automata can easily simulate all sandpile models known in literature and even cellular automata. They also obtained the fundamental representation result given in Theorem 1; but let us first introduce a few more useful definitions.

We need two special functions: the *shift map* $\sigma : \mathfrak{C} \mapsto \mathfrak{C}$ defined by $\forall c \in \mathfrak{C}, \forall i \in \mathbb{Z}$, $\sigma(c)_i = c_{i+1}$; and the *raising map* $\rho : \mathfrak{C} \mapsto \mathfrak{C}$ defined by $\forall c \in \mathfrak{C}, \forall i \in \mathbb{Z}$, $\rho(c)_i = c_i + 1$. A function $f : \mathfrak{C} \mapsto \mathfrak{C}$ is *shift-invariant* (resp. *vertical-invariant*) if $f \circ \sigma = \sigma \circ f$ (resp. $f \circ \rho = \rho \circ f$). A function $f : \mathfrak{C} \mapsto \mathfrak{C}$ is *infiniteness conserving* if

$$\forall c \in \mathfrak{C}\, \forall i \in \mathbb{Z}, \quad \begin{cases} f(c)_i = +\infty \Leftrightarrow c_i = +\infty \\ \text{and} \\ f(c)_i = -\infty \Leftrightarrow c_i = -\infty \ . \end{cases}$$

Theorem 1 ([8]). *A function $f : \mathfrak{C} \mapsto \mathfrak{C}$ is the global function of a sand automaton if and only if f is continuous, shift-invariant, vertical-invariant and infiniteness conserving.*

By an abuse of terminology, we will often confuse a sand automaton $\mathcal{A} \equiv \langle r, \lambda \rangle$ with its global function $f_{\mathcal{A}}$. For example, we claim that $\mathcal{A}$ is *surjective* (resp. *injective*) if $f_{\mathcal{A}}$ is surjective (resp. injective). For $\mathfrak{U} \subseteq \mathfrak{C}$, $f_{\mathcal{A}}$ is said to be $\mathfrak{U}$-surjective (resp. injective) if the restriction of f to $\mathfrak{U}$ is surjective (resp. injective).

4 Examples

In this section we introduce a series of worked examples with a twofold purpose: illustrate basic behavior of sand automata and constitute a set of counter-examples for later use. Some examples might seem a bit technical but the underlaying ideas are very useful in the sequel.

Example 1. **The automaton $\mathcal{S}$** .
This automaton is the simulation of SPM (Sand Pile Model) in dimension 1: $\mathcal{S} = \langle 1, \lambda_{\mathcal{S}} \rangle$, where

$$\forall x, y \in \widetilde{[\![-1,1]\!]}, \quad \lambda_{\mathcal{S}}(x,y) = \begin{cases} +1 & \text{if } x = +\infty \text{ and } y \neq -\infty \ , \\ -1 & \text{if } x \neq +\infty \text{ and } y = -\infty \ , \\ 0 & \text{otherwise.} \end{cases}$$

Remark the basic grain movement of $\mathcal{S}$: a grain falls to the column on its right when the height difference is bigger than 2.

Example 2. **The automaton $\mathcal{S}^r$.**
This automaton is defined similarly to $\mathcal{S}$, but grains climb the cliffs instead of falling down. Let $\mathcal{S}^r = \langle 1, \lambda_{\mathcal{S}^r} \rangle$ where

$$\forall x, y \in \widetilde{[\![-1,1]\!]}, \quad \lambda_{\mathcal{S}^r}(x,y) = \begin{cases} -1 & \text{if } x = +\infty \text{ and } y \neq -\infty, \\ +1 & \text{if } x \neq +\infty \text{ and } y = -\infty, \\ 0 & \text{otherwise.} \end{cases}$$

Proposition 5. *The SA $\mathcal{S}$ is $\mathfrak{U}$-surjective for $\mathfrak{U} = \mathfrak{C}, \mathfrak{F}, \mathfrak{P}$. The SA $\mathcal{S}^r$ is $\mathfrak{U}$-injective for $\mathfrak{U} = \mathfrak{C}, \mathfrak{F}, \mathfrak{P}$.*

Proof. It is not difficult to see that $\mathcal{S} \circ \mathcal{S}^r = id$, but $\mathcal{S}^r \circ \mathcal{S} \neq id$ (just use the configuration c defined by $c_i = 2$ if $i = 0$ and $c_i = 0$ otherwise). The first equation implies that $\mathcal{S}$ is surjective and $\mathcal{S}^r$ is injective. Moreover, since the pre-image by $\mathcal{S}$ of a configuration is computed by $\mathcal{S}^r$, another SA, the pre-image of a finite configuration is finite, and periodic if the initial configuration is periodic. Hence we have the first part of the thesis. The second part is a consequence of the injectivity of $\mathcal{S}^r$. □

Proposition 6. *The SA $\mathcal{S}$ is not $\mathfrak{U}$-injective for $\mathfrak{U} = \mathfrak{C}, \mathfrak{F}, \mathfrak{P}$.*

Proposition 7. *The SA $\mathcal{S}^r$ is not $\mathfrak{U}$-surjective for $\mathfrak{U} = \mathfrak{C}, \mathfrak{F}, \mathfrak{P}$.*

Example 3. **The automaton $\mathcal{L}$.**
Consider an automaton $\mathcal{L} = \langle 1, \lambda_{\mathcal{L}} \rangle$ where

$$\forall x, y \in \widetilde{[\![-1,1]\!]}, \quad \lambda_{\mathcal{L}}(x,y) = \begin{cases} -1 & \text{if } x < 0, \\ +1 & \text{if } x > 0, \\ 0 & \text{otherwise.} \end{cases}$$

Remark the basic behavior of $\mathcal{L}$: each column tries to reach the height of its left neighbor.

Proposition 8. *The SA $\mathcal{L}$ is not $\mathfrak{F}$-surjective.*

Proposition 9. *The SA $\mathcal{L}$ is both $\mathfrak{C}$-surjective and $\mathfrak{P}$-surjective.*

Proof. Choose an arbitrary configuration c, we are going to build one of its pre-image c'. There is a unique sequence of strictly increasing indices $(i_n)_{n \in N}$, $N \subset \mathbb{Z}$, such that $\forall i \in [\![i_n, i_{n+1}[\![, \; c_i = c_{i_n}$ and $c_{i_n} \neq c_{i_n - 1}$ (every i_n corresponds to a variation of height in c). The idea is to work on these intervals, amplifying the difference at the border so that an application of the rule corrects it. Formally, for every $n \in N$, suppose that $c_{i_n - 1} < c_{i_n}$ (if it is not the case then the symmetrical operations have to be performed). For every $i_n \leq i < i_{n+1}$, let $c'_i = c_i + 1$ if $i - i_n$ is even, $c'_i = c_i - 1$ if $i - i_n$ is odd. There are two little subtleties if N is not bi-infinite. First if $n_0 = \min N$ exists, then let $c'_i = c_i$ for all $i < n_0$. Second, if $n_1 = \max N$ exists, then the $\pm$ operation has to be performed forever on the right. Note that it is why a finite configuration may not have a finite pre-image.

It is not difficult to see that $f_\mathcal{L}(c') = c$. For every $i \in \mathbb{Z}$, first suppose that there is a $n \in N$ such that $i = i_n$. We have $f_\mathcal{L}(c')_i = c'_i + \lambda_\mathcal{L}(d_1^i(c'))$. Supposing that $c_{i-1} < c_i$ (again, if it is the opposite then the operations are symmetrical), we have $c'_i = c_i + 1 > c_{i-1} + 1$, hence $c'_i > c'_{i-1}$ since $|c_{i-1} - c'_{i-1}| \leq 1$. So $\lambda_\mathcal{L}(d_1^i(c')) = -1$, and $f_\mathcal{L}(c')_i = c_i + 1 - 1 = c_i$. Otherwise if $i \neq i_n$ for all $n \in N$, then by construction we have either:

- $c'_i = c_i + 1$ and $c'_{i-1} = c_{i-1} - 1 = c_i - 1$, because c is constant between the i_n's. Hence $c'_{i-1} = c'_i - 2$, and then $f_\mathcal{L}(c')_i = c_i + 1 - 1 = c_i$;
- or $c'_i = c_i - 1$ and $c'_{i-1} = c_{i-1} + 1$, the same method gives the result.

Therefore $\mathcal{L}$ is surjective. Finally, as the operations we perform on the configuration are deterministic, a periodic configuration would have a periodic pre-image (same transformation of the period everywhere). Hence $\mathcal{L}$ is also surjective over periodic configurations. □

The next example is a bit less intuitive since it uses a special neighborhood: the two nearest left neighbors.

Example 4. **The automaton $\mathcal{X}$.**
Consider the sand automaton $\mathcal{X} = \langle 2, \lambda_\mathcal{X} \rangle$ where

$$\forall x, y, z \in \widetilde{[\![-2, 2]\!]}, \quad \begin{aligned} \lambda_\mathcal{X}(+\infty, x, y, z) &= -1 \ , \\ \lambda_\mathcal{X}(2, x, y, z) &= -1 \ , \\ \lambda_\mathcal{X}(1, -1, x, y) &= -1 \ , \\ \lambda_\mathcal{X}(1, -2, x, y) &= -1 \ , \\ \lambda_\mathcal{X}(1, -\infty, x, y) &= -1 \ , \\ \lambda_\mathcal{X}(0, -2, x, y) &= -1 \ , \\ \lambda_\mathcal{X}(0, -\infty, x, y) &= -1 \ , \end{aligned}$$

and any other value gives 0. The evolutions of $\mathcal{X}$ on arbitrary configurations seem quite hard to describe. Anyway, in the sequel we will need to study its evolutions only on special (simple) configurations.

Proposition 10. *The SA $\mathcal{X}$ is $\mathfrak{F}$-injective but not $\mathfrak{C}$- or $\mathfrak{P}$-injective.*

Example 5. **The automaton $\mathcal{Y}$.**
Consider the following SA $\mathcal{Y} = \langle 2, \lambda_\mathcal{Y} \rangle$, where

$$\forall x, y, z \in \widetilde{[\![-2, 2]\!]}, \quad \begin{aligned} \lambda_\mathcal{Y}(+\infty, x, y, z) &= -1 \ , \\ \lambda_\mathcal{Y}(2, x, y, z) &= -1 \ , \\ \lambda_\mathcal{Y}(1, x, y, z) &= -1 \ , \\ \lambda_\mathcal{Y}(0, x, y, z) &= -1 \ , \\ \lambda_\mathcal{Y}(-1, -\infty, x, y) &= -1 \ , \end{aligned}$$

and everything else returns 0.

Proposition 11. *The SA $\mathcal{Y}$ is $\mathfrak{F}$- and $\mathfrak{P}$-injective, but not injective.*

Proof. Consider the two configurations c and c' defined as follows

$$\forall i \in \mathbb{Z}, \quad \begin{cases} c_{2i} = i \\ c_{2i+1} = i+2 \end{cases}, \quad \begin{cases} c'_{2i} = i \\ c'_{2i+1} = i+3 \end{cases}.$$

It is not difficult to see that $f_{\mathcal{Y}}(c) = f_{\mathcal{Y}}(c') = c$ Hence $\mathcal{Y}$ is not injective. In order to show that $\mathcal{Y}$ is injective over finite and periodic configurations, we need an intermediate result: if c, c' are two distinct configurations such that $f_{\mathcal{Y}}(c) = f_{\mathcal{Y}}(c')$, then there are infinitely many differences, of infinitely many different values. Practically, we show that if $c_i > c'_i$ then $c_{i-2} > c'_{i-2}$ and $c_{i-2} < c_i$.

Assume $c_i > c'_i$ for some i, and let $f(c) = f(c')$. Then, without loss of generality, one can choose $c_i = c'_i + 1$ (the difference cannot be greater than one, because $\lambda_{\mathcal{Y}}$ only returns -1 or 0). Therefore, a rule which returns 0 is applied to c' at position i, which means that $c'_{i-2} \leq c'_i - 1$ (since $\lambda_{\mathcal{Y}}(x, -, -, -)$ returns 0 only if $x \leq -1$). For the same reason, one of the five rules which returns -1 is applied to c at position i, hence $c_{i-2} \geq c_i - 1$. So it holds that

$$c_{i-2} \geq c_i - 1 = c'_i \geq c'_{i-2} + 1 > c'_{i-2}. \tag{1}$$

The first consequence of this inequality is that if there is a difference somewhere, there are infinitely many differences, hence $\mathcal{Y}$ is $\mathfrak{F}$-injective. Indeed two finite configurations cannot have infinitely many differences, so two different finite configurations have a different image.

Moreover, $c_{i-2} = c'_{i-2} + 1$ to ensure $f_{\mathcal{Y}}(c) = f_{\mathcal{Y}}(c')$. So the inequalities (1) above are in reality equalities, in particular $c_{i-2} = c_i - 1$. Therefore it holds $\cdots < c_{i-4} < c_{i-2} < c_i$, which proves that two different periodic configurations also have different images (a periodic configuration contains a finite number of different columns, which is contradicted by the above inequality). As a consequence, $\mathcal{Y}$ is $\mathfrak{P}$-injective. □

5 Basic Set Properties

This section deals with the relations between surjectivity and injectivity, w.r.t. all, finite and periodic configurations, in the same way it was done in [10] for cellular automata. In particular the relation between $\mathfrak{F}$-injectivity and surjectivity was interesting, as for cellular automata it can be used to prove undecidability of surjectivity [11]. Unfortunately, no such relation holds between those two properties in the context of SA (see Propositions 5, 6, 7). In this section we try to analyze these relations deeper hoping this might help for the proof of the decidability result about surjectivity or injectivity.

Proposition 12. $\mathfrak{F}$-*surjectivity implies surjectivity.*

Proof. For any configuration c, let c_n^0 be such that $\forall i \in \mathbb{Z}$, $(c_n^0)_i = c_i$ if $-n \leq i \leq n$ and $(c_n^0)_i = 0$ otherwise. Consider a sand automaton f that is $\mathfrak{F}$-surjective and choose an arbitrary configuration $c \in \mathfrak{C}$. For any $n \in \mathbb{N}$, let $c_n = f^{-1}(c_n^0)$. The pre-images c_n are contained in some set $\mathfrak{E}_i$ for $i \in I$, with

$I \subset [\![c_0 - r, c_0 + r]\!]$ where r is the precision of f. Since $\cup_{i \in I} \mathfrak{E}_i$ is compact and $(c_n)_{n \in \mathbb{N}} \subset \cup_{i \in I} \mathfrak{E}_i$, $(c_n)_{n \in \mathbb{N}}$ contains a converging sub-sequence $(c_{n_i})_{i \in \mathbb{N}}$. Let $c^* = \lim_{i \to \infty} c_{n_i}$. By contradiction, assume that $f(c^*) \neq c$. Then there exists $j \in \mathbb{Z}$ such that $f(c^*)_j \neq c_j$ but $f(c_{n_i})_j = c_j$ for n_i big enough. □

Remark that the result of Proposition 12 is true in any dimension but the converse is false (even in dimension 1), since $\mathcal{L}$ is surjective but not $\mathfrak{F}$-surjective (see Propositions 8 and 9).

Proposition 13. $\mathfrak{P}$-surjectivity implies $\mathfrak{C}$-surjectivity.

Proposition 14. In dimension 1, $\mathfrak{C}$-surjectivity implies $\mathfrak{P}$-surjectivity.

Proof. Let $\mathcal{A}$ be a surjective sand automaton in dimension 1, of radius r, and c^0 a periodic configuration of period $p \in \mathbb{Z}$. Let c be a pre-image of c^0 by $\mathcal{A}$. We build a periodic configuration from c, whose image is c^0. Let $X = \{(c_{k-r}, \ldots, c_{k+r-1}) \mid \exists \alpha \in \mathbb{Z}, k = \alpha p\}$. Since for every $i \in \mathbb{Z}$, $|c_i - c_i^0| \leq r$ (as λ returns an element of $[\![-r, r]\!]$), and because c^0 is p-periodic, there are at most $2r \cdot (2r + 1)$ elements in X. Let $k_1 = \alpha_1 p$ and $k_2 = \alpha_2 p$, $k_1 < k_2$ such that $(c_{k_1-r}, \ldots, c_{k_1+r-1}) = (c_{k_2-r}, \ldots, c_{k_2+r-1})$. Let the $(k_2 - k_1)$-periodic configuration c' where the period is defined by (see Fig. 1 for the construction) $c'_{k_1+i} = c_{k_1+i}$ for all $0 \leq i < k_2 - k_1$. It is easy to see that $f(c') = c^0$, because for every configuration of the period of c', the automaton sees the same neighborhood as for c (due to the construction of c'), so it acts in the same correct way. And as $k_2 - k_1$ is a multiple of p, each period of c' coincides with a period of c^0, so the image of c' is equal to c everywhere: $\mathcal{A}$ is $\mathfrak{P}$-surjective. □

In dimensions greater than 1, the above problem is currently open, we have no direct proof nor counter-example. The problem is due to the fact that in dimension 2 and above, the size of the perimeter of a ball (the $2r$ sequence we

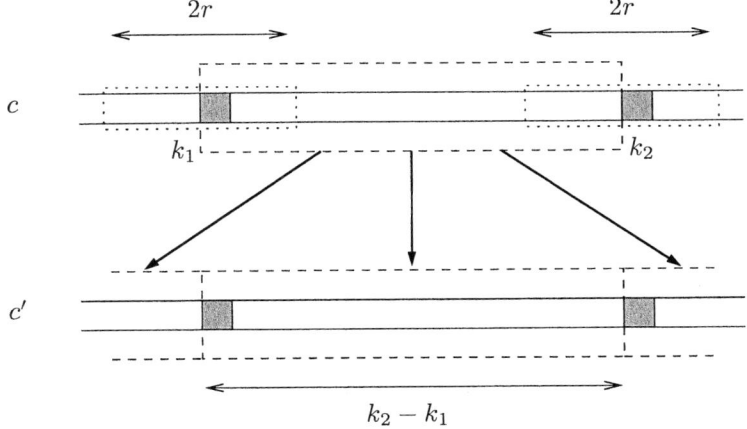

Fig. 1. Construction of c' using c

used in X for the proof in dimension 1) is linked to the size of the ball. Therefore we cannot say that there is a finite number of perimeters, and then stick them together to build the periodic configuration.

Corollary 1. *In dimension 1, $\mathfrak{F}$-surjectivity implies $\mathfrak{P}$-surjectivity.*

The question whether the above corollary is true in dimension 2 and above is still open and its solution appears to be quite difficult.

Note that the opposite implication of Corollary 1 is false in any dimension, thanks to $\mathcal{L}$ which is $\mathfrak{P}$-surjective but not $\mathfrak{F}$-surjective (see Propositions 8 and 9).

Clearly injectivity implies $\mathfrak{F}$-injectivity and $\mathfrak{P}$-injectivity, but the opposite implications are not true. In fact, because of $\mathcal{X}$, $\mathfrak{F}$-injectivity does not imply injectivity (Proposition 10); and $\mathcal{Y}$ shows that $\mathfrak{P}$-injectivity does not mean global injectivity (Proposition 11). The following proposition completes these results.

Proposition 15. $\mathfrak{P}$-*injectivity implies $\mathfrak{F}$-injectivity.*

Proof. This is proved using the contrapositive. Let $\mathcal{A}$ be an automaton not $\mathfrak{F}$-injective. Let x^1, x^2 be the two distinct finite configurations which lead to the same image c. Let $k \in \mathbb{N}$ such that for all $i \in \mathbb{Z}^d$, $|i| > k$, $x_i^1 = x_i^2 = 0$. We are going to build two distinct periodic configurations by surrounding the non-zero part of x^1 and x^2 with a crown of zeros, of thickness r, and repeat this pattern. For $\alpha \in \{1, 2\}$, let y^α be the $(2k + 2r + 1)$-periodic configuration defined by

$$\forall i \in \mathbb{Z}^d, |i| \leq k + r, \quad \begin{cases} y_i^\alpha = x_i^\alpha & \text{if } |i| \leq k \ , \\ y_i^\alpha = 0 & \text{if } k < |i| \leq k + r \ . \end{cases}$$

We have $f(y^1) = f(y^2)$. For every configuration, we can consider the translated configuration whose index is lower in norm than $k + r$ because of the periodicity. This configuration reacts as it did in x^1 and x^2 because its neighborhood is the same : inside the k "circle", it is obvious. If it is inside the crown of 0's, then the only non-zero values it can see are the values located inside the initial pattern. So its behavior is equivalent to the one of the point at the border of the initial finite configuration, and $\mathcal{A}$ is not $\mathfrak{P}$-injective. □

The opposite implication of Proposition 15 is false since $\mathcal{X}$ is $\mathfrak{F}$-injective but not $\mathfrak{P}$-injective (Proposition 10). Figure 2 summarizes the relations between basic set properties.

Fig. 2. Relations between basic set properties. I means injectivity and S surjectivity. $I_\mathfrak{U}$ (resp. $S_\mathfrak{U}$) means injectivity (resp. surjectivity) restricted to $\mathfrak{U}$. The symbol $\stackrel{1}{\Longrightarrow}$ means that the implication it true in dimension 1 and open in higher dimensions.

6 Ultimate Periodicity

Understanding the dynamical behavior of SA seems very difficult. This is confirmed by the main result of this section: ultimate periodicity, one of the simplest dynamical behavior, is undecidable for sand automata.

Recall that given a SA f, a configuration c is ultimately periodic if $\exists p, t \in \mathbb{N}$ such that $\forall i, k \in \mathbb{N}$, $f^{pk+i+t}(c) = f^{i+t}(c)$. A SA f is $\mathfrak{U}$-ultimately periodic if for all $c \in \mathfrak{U}$, c is ultimately periodic for f. Define $\overline{\mathfrak{F}} = \mathfrak{F} \cap \mathbb{Z}^{\mathbb{Z}^d}$ and $\overline{\mathfrak{P}} = \mathfrak{P} \cap \mathbb{Z}^{\mathbb{Z}^d}$, in other words we remove sources and sinks in $\mathfrak{F}$ and $\mathfrak{P}$.

Problem ULT($\mathfrak{U}$)
   INSTANCE: a SA $\mathcal{A} = \langle \lambda, r \rangle$;
   QUESTION: is every configuration in $\mathfrak{U}$ ultimately periodic for $\mathcal{A}$?

Details of the proof of the following result are given in the Appendix.

Theorem 2. *Both problems $ULT(\overline{\mathfrak{P}})$ and $ULT(\overline{\mathfrak{F}})$ are undecidable.*

Proof (Sketch). First of all, remark that it is enough to prove the thesis on $\overline{\mathfrak{F}}$. In fact, from any finite configuration one can obtain a periodic configuration by repeating periodically the non-zero pattern surrounded by a suitable border of zeroes (if necessary). Moreover, we provide the proof for dimension 1 only, since a similar construction can be done for other dimensions. We reduce these problems to the halting problem of a two registers machine with finite control started with both registers at 0.

Each two registers machine $\mathcal{M}$ is associated with a SA $\mathcal{S}_\mathcal{M}$ such that $\mathcal{S}_\mathcal{M}$ is ultimately periodic if and only if $\mathcal{M}$ halts when started with both registers at zero. The idea is that $\mathcal{S}_\mathcal{M}$ uses a certain number of grain stacks for the registers (R) and for the finite control (Q) in order to simulate the iterations of $\mathcal{M}$. For technical reasons we also need a counter (C) which counts the number of iterations of $\mathcal{M}$. Figure 3 illustrates the general "architecture" of $\mathcal{S}_\mathcal{M}$.

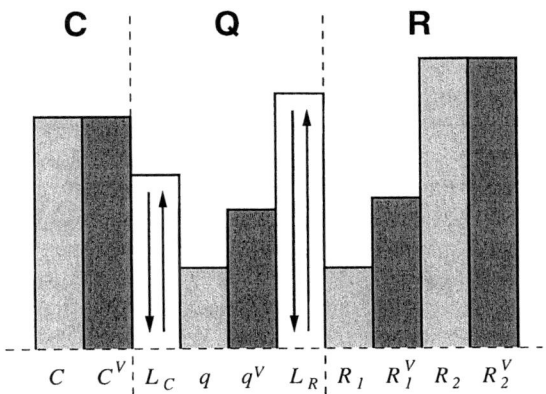

Fig. 3. Simulation of a two registers machine by a SA

Each iteration of $\mathcal{M}$ can be simulated by $\mathcal{S}_\mathcal{M}$ in three main steps:

S. *simulation* of one iteration of $\mathcal{M}$;
V. *verification* from the beginning to the current iteration, in the verification columns (with a ^{V} superscript);
C. *comparison* between the results of the first two steps, to ensure that the simulation is correct.

These three steps are necessary since not all initial configurations of $\mathcal{S}_\mathcal{M}$ represent valid computations of $\mathcal{M}$. For this reason, $\mathcal{S}_\mathcal{M}$ is equipped with a verification part that is able to simulate $\mathcal{M}$ when started with both registers at zero. Then $\mathcal{S}_\mathcal{M}$ compares the current state with the one obtained in the verification part. If they coincide, the counter (C) is increased by one and a new iteration of $\mathcal{M}$ is simulated; otherwise $\mathcal{S}_\mathcal{M}$ evolves to a periodic configuration.

In the sequel, a configuration of $\mathcal{S}_\mathcal{M}$ is *valid* if it represents a computation of $\mathcal{M}$ when started with both registers at 0.

If $\mathcal{M}$ halts when started with both registers at 0, then $\mathcal{S}_\mathcal{M}$ evolves to a periodic point when started with a valid configuration. In fact, when the control of $\mathcal{S}_\mathcal{M}$ reaches a halting state, all the other parts freeze in the current value.

A configuration is *malformed* if it does not respect the "architecture" of $\mathcal{S}_\mathcal{M}$ *i.e.*, for example, the value of the counter is negative *etc*. All these situations are easily checkable by using a suitable (large) radius for $\mathcal{S}_\mathcal{M}$ and very simple rules. If $\mathcal{S}_\mathcal{M}$ detects that the current configuration is malformed then it evolves to a periodic configuration in a finite number of steps.

Finally, if $\mathcal{M}$ does not halt when started with both registers at 0, $\mathcal{S}_\mathcal{M}$ keeps on simulating iterations of $\mathcal{M}$ and at each iteration the counter (C) is increased by one. This fact ensures that the evolution of $\mathcal{S}_\mathcal{M}$ is not periodic.

This last remark concludes the proof. More details are given in the appendix. □

7 Conclusions

In this paper we have seen that the quest for decidability results for basic set properties like injectivity and surjectivity is hardened by the lack of relations between them and their restriction to "easy" computable subsets of configurations (such as $\mathfrak{P}$ or $\mathfrak{F}$). This fact can be considered as a first evidence that the study of dynamical behavior of SA might reveal very difficult.

The second evidence is given by Theorem 2. A very simple dynamical behavior like ultimate periodicity over $\overline{\mathfrak{F}}$ or $\overline{\mathfrak{P}}$ is undecidable. Remark that, in the case of cellular automata, the undecidability of the ultimate periodicity is a powerful tool for proving the undecidability of many other problems in cellular automata theory. We think that this property can play a similar role for sand automata. The authors are currently investigating this subject.

References

1. Bak, P.: How nature works - The science of SOC. Oxford University Press (1997)
2. Bak, P., Tang, C.: Earthquakes as a self-organized critical phenomenon. J. Geophys. Res. 94 (1989) 15635-15637

3. Bak, P., Chen, K., Tang, C.: A forest-fire model and some thoughts on turbulence. Physics letters A 147 (1990) 297-300
4. Bak, P., Tang, C., Wiesenfeld, K.: Self-organized criticality. Physical Review A 38 (1988) 364-374
5. Subramanian, R., Scherson, I.: An analysis of diffusive load-balancing. In: ACM Symposium on Parallel Algorithms and Architecture (SPAA'94), ACM Press (1994) 220-225
6. Goles, E., Kiwi, M.A.: Games on line graphs and sand piles. Theoretical Computer Science 115 (1993) 321-349
7. Goles, E., Morvan, M., Phan, H.D.: Sand piles and order structure of integer partitions. Discrete Applied Mathematics 117 (2002) 51-64
8. Cervelle, J., Formenti, E.: On sand automata. In Alt, H., Habib, M., eds.: STACS 2003: 20th Annual Symposium on Theoretical Aspects of Computer Science. Volume 2607 of Lecture Notes in Computer Science., Springer-Verlag Heidelberg (2003) 642-653
9. Hedlund, G.A.: Endomorphism and automorphism of the shift dynamical system. Mathematical System Theory 3 (1969) 320-375
10. Durand, B.: Global properties of cellular automata. In Goles, E., Martinez, S., eds.: Cellular Automata and Complex Systems. Kluwer (1998)
11. Durand, B.: The surjectivity problem for 2d cellular automata. Journal of Computer and Systems Science 49 (1994) 718-725
12. Goles, E., Morvan, M., Phan, H.D.: The structure of a linear chip firing game and related models. Theoretical Computer Science 270 (2002) 827-841
13. Durand, B., Formenti, E., Varouchas, G.: On undecidability of equicontinuity classification for cellular automata. In Morvan, M., Rémila, É., eds.: Discrete Models for Complex Systems DMCS'03. Volume AB of DMTCS Proceedings., Discrete Mathematics and Theoretical Computer Science (2003) 117-128
14. Eriksson, K.: Reachability is decidable in the numbers game. Theoretical Computer Science 131 (1994) 431-439

Appendix. Proofs of Remaining Results

Proofs of Section 2

Proof (of Proposition 4). Let $(c^n)_{n \in \mathbb{N}}$ be a Cauchy sequence of $\mathfrak{C}^{\mathbb{N}}$. There is a $N \in \mathbb{N}$ such that for all $m, n \geq N$, $d(c^m, c^n) < 1$, in other words for all $n \geq N$, $c_0^n = c_0^N$. Every element of the sequence $(c^{N+n})_{n \in \mathbb{N}}$ is in $\mathfrak{E}_{c_0^N}$, which is compact and hence complete. As this is a Cauchy sequence, it has a limit c in $\mathfrak{E}_{c_0^N} \subset \mathfrak{C}$. c is obviously the limit of the initial sequence (c^n), which gives the result. □

Proofs of Section 4

Proof (of Proposition 6). Consider the following finite configurations c, c' where $c_i = 0$ for $i \in \mathbb{Z}$, $c'_i = 0$ for $i \in \mathbb{Z} \setminus \{0, 1\}$, $c'_0 = 1$, and $c'_1 = -1$. Clearly, $f_S(c) = f_S(c') = c$. Now, consider the periodic configuration c'' with $c''_{2i} = 1$ and $c''_{2i+1} = -1$ for every $i \in \mathbb{Z}$, again $f_S(c) = f_S(c'') = c$. □

Proof (of Proposition 7). Consider the following finite configuration c, where $c_i = 2$ if $i = 0$; $c_i = 0$ otherwise. Assume that c has a pre-image c'. There are only three possibilities for the value of c'_0:

$c'_0 = 3$: then the local rule has to return -1, which implies that $c'_{-1} \geq 5$. But $f_{\mathcal{S}^r}(c')_{-1} = 0$, this value cannot be reached from 5;

$c'_0 = 2$: the column is unchanged, which means that $(c'_{-1} \leq 3$ or $c'_1 \leq 0)$ and $(c'_{-1} \geq 4$ or $c'_1 \geq 1)$. For the same reason as before, c'_{-1} cannot be greater than 4, hence $c'_1 \geq 1$. This means that the local rule applied at position 1 returns -1, in other words that $c'_0 \geq 3$, which contradicts the first hypothesis;

$c'_0 = 1$: $\lambda_{\mathcal{S}^r}$ returns $+1$, so $c_1 \leq -1$. Hence at position 1, $\lambda_{\mathcal{S}^r}$ also returns $+1$. That means, in particular, that $c'_2 \leq -3$, which is impossible if one has to obtain $f_{\mathcal{S}^r}(c')_2 = 0$.

We have found a finite configuration with no pre-image, which means that $\mathcal{S}^r$ is not surjective both on $\mathfrak{C}$ and on $\mathfrak{F}$. To show that $\mathcal{S}^r$ is not $\mathfrak{P}$-surjective, one can consider the configuration c where $c_{4i+1} = 2$ for every $i \in \mathbb{Z}$, and everywhere else $c_k = 0$. The proof is similar to the previous part, since the 4 elements of the period act as if the configuration was finite (radius 1, so they do not "see" farther than one column ahead and one column back). □

Proof (of Proposition 8). Consider the finite configuration c where $c_i = 2$ if $i = 0$ and $c_i = 0$ otherwise. By contradiction assume that c' is the pre-image of c and that $c' \in \mathfrak{F}$. Let i be the greatest integer such that $c'_i \neq 0$. Then since $c'_i \neq 0$ and $c'_{i+1} = 0$, it holds that $f_{\mathcal{L}}(c')_{i+1} = c_{i+1} \neq 0$. This implies that $i = -1$ because c_0 is the only non-zero value in c. But in that case, we have $c'_0 = 0$, and as $\lambda_{\mathcal{L}}$ cannot return more than 1, $c_0 = 2$ cannot be reached. This is a contradiction. □

Proof (of Proposition 10). Consider the two periodic configurations c and c' defined as follows :

$$\forall i \in \mathbb{Z}, \quad \begin{cases} c_{2i} = 0 \\ c_{2i+1} = 1 \end{cases}, \quad \begin{cases} c'_{2i} = 0 \\ c'_{2i+1} = 2 \end{cases}.$$

It can be easily verified that $f_\mathcal{X}(c) = f_\mathcal{X}(c') = c$. Hence $\mathcal{X}$ is not $\mathfrak{P}$-injective and, of course, it is not injective.

Let us prove that $\mathcal{X}$ is $\mathfrak{F}$-injective. Let c and c' be two distinct finite configurations, and suppose that their image by $f_\mathcal{X}$ is identical. As the two configurations are finite, we can define $i \in \mathbb{Z}$ being the least integer such that $c_i \neq c'_i$. As $\lambda_\mathcal{X}$ returns only 0 or -1, we know that $|c_i - c'_i| = 1$, and we can suppose that $c_i = c'_i + 1$. That means that the local rule applied to c at position i is one of the seven rules which return -1:

– if the neighborhood is $(+\infty, -, -, -)$ (to make the notations clearer, $-$ represents any value), then since $c'_i = c_i - 1$ and $c'_{i-2} = c_{i-2}$, the same rule is applied to c', which means that $f_\mathcal{X}(c)_i \neq f_\mathcal{X}(c')_i$ which is a contradiction;
– if the neighborhood is $(2, -, -, -)$, for the same reason the rule for the neighborhood $(+\infty, -, -, -)$ is applied to c', which raises the same contradiction;

- again, if the neighborhood is $(1,-1,-,-), (1,-2,-,-)$ or $(1,-\infty,-,-)$, the rule for the neighborhood $(2,-,-,-)$ is applied to c', making c'_i decrease by 1: same contradiction;
- if the neighborhood is $(0,-2,-,-)$ or $(0,-\infty,-,-)$, because $c'_{i-2} = c_{i-2}$, $c'_{i-1} = c_{i-1}$ and $c'_i = c_i - 1$, one of the rules corresponding to the neighborhoods $(1,-1,-,-), (1,-2,-,-)$ or $(1,-\infty,-,-)$ is applied to c'. There again, we have $f_\mathcal{X}(c)_i \neq f_\mathcal{X}(c')_i$. □

Proof (of Proposition 13). Nearly exactly the same proof as for Proposition 12 can be made. The only change is that it starts with c_n^0 defined as the $(2n+1,\ldots,2n+1)$-periodic configuration with $\forall i \in \mathbb{Z}^d, |i| \leq n, (c_n^0)_i = c_i$. Everything else is unchanged. □

Proofs of Section 5

Proof (of Corollary 1). $\mathfrak{F}$-surjectivity implies surjectivity (Proposition 12), which implies in dimension 1 $\mathfrak{P}$-surjectivity (Proposition 14). □

Proofs of Section 6

The sketch of the proof of Theorem 2 given in Section 6 ommits several technical but fundamental details.

The reduction is made from a two registers machine $\mathcal{M}$ defined by $\mathcal{M} = \langle Q, q_0, q_f, \delta \rangle$, where Q is a finite set of states, $q_0 \in Q$ is the initial state, $q_f \in Q$ the final state. The registers R_1 and R_2 always contain positive integer values. In our case, $\mathcal{M}$ is always started with both registers at 0.

The function $\delta : Q \times \{0,1\} \times \{0,1\} \mapsto Q \times \{1,2\} \times \{-1,0,+1\}$ is the *transition function*. The second and third arguments of δ indicate whether or not the registers are 0 (hence a 1 means that the register contains a value strictly greater than 0). δ returns the new state, the number of the register which is modified (1 or 2), and its modification (decrease by 1, increase by 1, no change). For clarity we denote these transitions by the expression $\delta(q, b_1, b_2) = (q', R_i + j)$.

For example, the rules

$$\begin{cases} \delta(q_0, 0, 0) = (q_1, R_1 + 1) \\ \delta(q_1, 0, -) = (q_f, R_1 + 0) \\ \delta(q_1, 1, -) = (q_2, R_1 - 1) \\ \delta(q_2, -, -) = (q_3, R_2 + 1) \\ \delta(q_3, -, -) = (q_1, R_2 + 1) \end{cases}$$

define a machine which first initializes R_1 to 1, then multiplies it by 2 and puts the result in R_2.

A two registers machine $\mathcal{M}$ (started with both registers at 0) is associated with a SA $\mathcal{S_M}$. Before describing $\mathcal{S_M}$ we need the following "tips and tricks" which will be fundamental in the construction.

The Lifts. The control has to send commands both to the registers (R) and to the counters (C). The point is that the radius of the local rule is finite and the difference of height between the control and the registers or the counters could be much bigger than the radius. Hence, the control cannot deliver commands directly to the registers or to the counters. This problem can be solved by introducing two more columns which we call *lifts* : L_C delivers commands to the counters and L_R delivers commands to the registers (see Fig. 3).

Knowing Themselves. The local rule of $\mathcal{S}_\mathcal{M}$ is formed by several sub-rules. Each sub-rule concerns the evolution of a single column of the simulation zone of $\mathcal{S}_\mathcal{M}$. The point is that each column must know "which it is" in order to apply the right sub-rule. This problem is solved by splitting each column c into two columns (l, r) and the "identity" of the original column is coded by the difference of height between l and r. For example, a difference of 1 says that c is the counter C, 2 stands for C^V and so on. We also use a height difference to code an error symbol E whose meaning will be explained later.

In the sequel, when speaking of height of a column $c = (l, r)$ we will always mean the height of r since l is simply the height of r plus the "identity" number.

Finally, the height of q_r is used as (relative) zero height by all other columns when needed.

Commands, Colors and States. The idea used to code "identity" information in the difference between pairs of successive columns can be used to store additional information which will be useful for the simulation. For example, one can code the following commands for the lifts: C_{+1} which increases the counter C by one; $C^V_{\to 0}$ that resets C^V; $R_{1,-1}$ which decreases R_1 by one; $L_\searrow$ which instruct the lift to go down and so on. Remark that lifts are colored **S**, **V**$_0$, **V** and **C** to indicate the current simulation step as described at page 202; **V**$_0$ is the initialization step of **V**. Clearly, colors can be coded using the height difference as well.

Finally, we need to code the state of the control q (or q^V). Once more, this piece of information can be coded into the height difference.

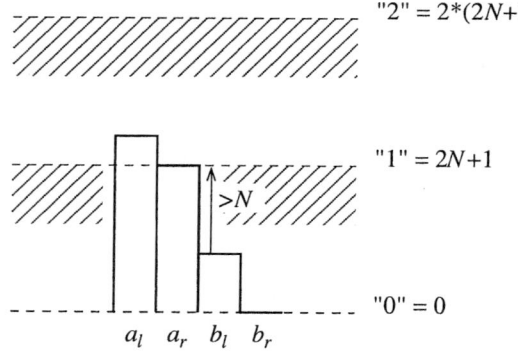

Fig. 4. How to distinguish "the left side from the right side"

Removing Ambiguities. Let N be the biggest difference used to code objects (or actions, see above) needed in the simulation. In the sequel, in order to maintain a strict correspondence between the two registers machine and the simulated model, we prefer to say that "a column $c = (l, r)$ is increased by $t \in \mathbb{N}$" even if in reality in $\mathcal{S}_\mathcal{M}$, r is increased by $(2N+1) \cdot t$ and l is increased by $(2N+1) \cdot t + \alpha$ where $\alpha \in \{-N+1, \ldots, 0, \ldots, N-1\}$ is meant to code the modification of the state of the column or its color. This trick avoids ambiguities in the "identity" of the columns as can be seen in Fig. 4. All rightmost columns with a $_r$ subscript are located at levels $k \cdot (2N+1)$, while the leftmost columns avoid the cross-hatched zone and remains between the line $k \cdot (2N+1)$ and $k \cdot (2N+1) + N$. As a consequence, the difference between any consecutive a_r and b_l exceeds N and cannot be mistaken for a code: a_r is guaranteed to be the right column of a pair, and b_l the left column of another.

Malformed Configurations. In view of the previous considerations, the notion of malformed configuration has to be extended by adding the case in which a column cannot decide if it is a left part or a right part. When the local rule cannot determine if a given column is a right part or a left part then it does add 0 to the current value of the column.

The Beginning. At the beginning of the simulation, C contains the number of simulation steps (w.r.t. $\mathcal{M}$) since the beginning, $q = (q_l, q_r)$ contains the current state of $\mathcal{M}$, the registers R_1 and R_2 contain some value. The lifts L_C and L_R are at 0 (relatively to q_r). Moreover, the lifts are in color **S**.

All other columns contain arbitrary values. They will be reset later on when necessary.

S. Simulation Step. In this step, $\mathcal{S}_\mathcal{M}$ simulates a single iteration of $\mathcal{M}$. For example, assume that R_1 and R_2 contain a strictly positive value and that $\delta(q_1, 1, 1) = (q_2, R_i + j)$ for some $i \in \{1, 2\}$ and $j \in \{-1, 0, +1\}$. Then, $\mathcal{S}_\mathcal{M}$ changes q_1 into q_2 in column q and at the same time fires L_C with the command C_{+1} and L_R with the command $R_{i,j}$. Below we give the local rules of $\mathcal{S}_\mathcal{M}$ for this transition which perform the update of q.

$$\begin{cases} \lambda\Big(\ldots, \mathbf{L_C}, \alpha_{q_1} \Big| \underbrace{-, -}_{q^V}, \mathbf{L_R}, \mathbf{R_1}, \underbrace{-, -}_{R_1^V}, \mathbf{R_2}, \ldots \Big) = 0 & \text{(right)} \\ \lambda\Big(\ldots, \mathbf{L'_C} \Big| -\alpha_{q_1}, \overbrace{-, -}, \mathbf{L'_R}, \mathbf{R'_1}, \overbrace{-, -}, \mathbf{R'_2}, \ldots \Big) = \alpha_{q_2} - \alpha_{q_1} & \text{(left)} \end{cases}$$

with

$$\mathbf{L_C} = \underbrace{\alpha_{L_C}, \mathbf{S}, 0}_{L_C \text{ at } 0, \mathbf{S}\text{-colored}}, \quad \mathbf{L_R} = \underbrace{\alpha_{L_R}, \mathbf{S}, 0}_{L_R \text{ at } 0, \mathbf{S}\text{-colored}},$$

$$\mathbf{L'_C} = \overbrace{\alpha_{L_C}, \mathbf{S} - \alpha_{q_1}, -\alpha_{q_1}}, \quad \mathbf{L'_R} = \overbrace{\alpha_{L_R}, \mathbf{S} - \alpha_{q_1}, -\alpha_{q_1}},$$

and

$$\mathbf{R_1} = \underbrace{> \alpha_{R_1}, > 0}_{R_1 \neq 0}, \qquad \mathbf{R_2} = \underbrace{> \alpha_{R_2}, > 0}_{R_2 \neq 0},$$

$$\mathbf{R'_1} = \overbrace{> (\alpha_{R_1} - \alpha_{q_1}), > -\alpha_{q_1}}, \qquad \mathbf{R'_2} = \overbrace{> (\alpha_{R_2} - \alpha_{q_1}), > -\alpha_{q_1}},$$

where $0 < \alpha_c \leq N$ represents the difference used to code all the characteristics of column c (identity, state, color, *etc.*). In the above formulas, the notation $> x$ means any number greater than x, while $-$ means any number. Moreover, the $|$ symbol is used as a delimiter between the neighborhood on the left and on the right.

Surely, the reader has remarked how involved are the formulas for the local rule of $\mathcal{S}_\mathcal{M}$. For this reason we prefer to describe them by words in the sequel. We stress that translating the descriptions into rules is not difficult.

The next iterations are for the lifts to reach their destination height and deliver the command. As a result, C finally increases by 1 and if necessary one of the registers can also have its value modified. Then, L_C and L_R go down (this can be done by turning into the command $L_\searrow$), changing their color to $\mathbf{V_0}$.

The step $\mathbf{S}$ ends when both lifts have reached the reference height, and are colored in $\mathbf{V_0}$.

$\mathbf{V_0}$. Initialization of the Verification Step. Before starting the verification step one should reset the verification columns (*i.e.* those with the ^{V} superscript in Fig. 3). In $\mathcal{S}_\mathcal{M}$ this is performed by sending $C^V_{\to 0}$ command to L_C and $R^V_{\to 0}$ to L_R. Finally, q^V is set to q_0.

The $R^V_{\to 0}$ command starts a sequence of actions. First, L_R goes up until it is above both registers. Then it goes down forcing the registers to go down with it. The same holds for $C^V_{\to 0}$.

Finally, when the lifts reach the reference height (*i.e.* the height of q_r), they turn into color $\mathbf{V}$ to indicate that the initialization step is complete, and that the verification step can begin.

V. Verification Step. Each time both lifts are on the ground, colored in $\mathbf{V}$, C iterations of $\mathcal{M}$ (started with both registers at 0) are performed in the verification columns. This is done exactly like in step $\mathbf{S}$: the lifts L^V_C and L^V_R deliver commands to the counter C^V and to the registers R^V_i ($i \in \{1, 2\}$), while the current state q^V is modified according to the rules of $\mathcal{M}$.

Moreover, L_C has to detect when $C = C^V$, which corresponds to the end of the verification step. In that case it goes down with color $\mathbf{C}$. When it reaches height 0 (*i.e.* the height of q_r), L_R checks the color of L_C and turns into the same color. At this point, $C = C^V$, q, R_1, R_2 should be equal to q^V, R^V_1, R^V_2 (the next step will determine if this is really the case), and L_C and L_R are at the reference height colored $\mathbf{C}$.

C. Comparison Step. The lift L_C is launched and it goes up until it reaches the highest among $C, C^V, R_1, R^V_1, R_2, R^V_2$. Then, it starts going down, comparing columns two by two when it reaches their height.

If everything is correct *i.e.* L_C reaches 0, then it changes its color into **S**. At this point L_R become **S**-colored also and the comparison step is finished.

If L_C finds that the comparison failed, it changes into the error state E, and does not move anymore: the simulation is blocked forever, since all other columns are waiting for the L_C to go down. Remark that in this last case, $S_\mathcal{M}$ is in an ultimately periodic point.

Concluding the Construction. For all neighborhoods that were not considered above, the local rule of $S_\mathcal{M}$ returns 0. This assumption is essential for several proofs that will follow.

Halting on Errors

When running on a malformed configuration, the sand automaton has to reach a periodic state. To force this, our simulation will freeze after a few iterations when it finds an error.

There are two main categories of errors for a particular column. First, *neighborhood* errors which are not due to the column itself, but to its global situation. For example, a pair of columns which code a register, but containing a negative value. Or any misplaced pair of columns, such as 2 state columns in the same configuration. Another neighborhood error is when two consecutive columns code for the error symbol.

Second, when a pair of consecutive columns do not code for anything, or there is an ambiguity in the coding, the configuration is also invalid. This is called an *identity* error.

Neighborhood Errors. When this type of error occurs one has to prevent any further movement. When a pair of columns finds unexpected values in its neighborhood it changes into the error symbol E. In terms of the local rule, this means that any sub-rule concerning a particular type of column $c = (l, r)$ with an incorrect neighborhood returns 0 for column r, and for l it returns the height difference coding E minus the current identity number.

Identity Errors. For identity errors the local rule returns 0. This concerns both columns whose neighbors do not code for anything (in this case we have that $\lambda(\ldots, x \mid y, \ldots) = 0$, with $x > N$ or $x \leq 0$, and $y < -N$ or $y \geq 0$, see Fig. 5(a)) and columns which cannot decide which column they are paired with (in this case we have $\lambda(\ldots, x \mid y, \ldots) = 0$, with $0 < x \leq N$ and $0 \geq y > -N$, see Fig. 5(b)).

From now on, fix a two registers machine $\mathcal{M}$ and let $S_\mathcal{M}$ be the associated SA given by the above construction. Let f be the global rule of $S_\mathcal{M}$.

Lemma 1. *For any finite configuration c and for any $t \in \mathbb{N}$, $|f^t(c)| \leq |c| + 1$.*

Proof. Let $c \in \overline{\mathfrak{F}}$ and i be its leftmost non-zero value. By construction, we have that $\lambda(-, \ldots, -, 0 \mid -, \ldots, -) = 0$ and hence $\forall j \in \mathbb{Z}\, \forall t \in \mathbb{N},\, j < i \Rightarrow f^t(c)_j = 0$.

Now, let k be the rightmost non-zero value of c. Remark that $f^t(c)_{k+1}$ is always a multiple of $2N + 1$. Since c_{k+1} is either an identity error or the right column of a pair coding for something then, by construction, either it does not

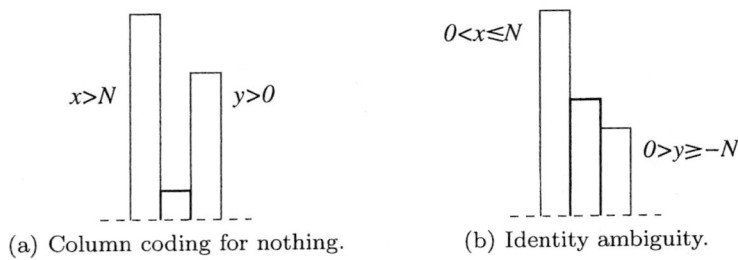

(a) Column coding for nothing.   (b) Identity ambiguity.

Fig. 5. Typical *identity* errors

increase at all (in the case of an identity error) or it changes by multiples of $2N+1$. Again, by construction, for any $j > k+1$, one finds $\forall t \in \mathbb{N}, f^t(c)_j = 0$ since $\lambda(-,\ldots,-,x\,|\,0,-,\ldots,-) = 0$ for $x \neq \alpha_{R_2^V}$, where $\alpha_{R_2^V}$ is the height difference coding for R_2^V. Remark that if $x = \alpha_{R_2^V}$ then the rule corresponding to the register R_2^V has to be applied and may not return 0. Anyway, in the present case, if $j = k+2$ then c_{j-1} is a multiple of $2N+1$ and hence $x \neq \alpha_{R_2^V}$. If $j > k+2$ then $c_{j-1} = 0$, $x = 0 \neq \alpha_{R_2^V}$. □

Lemma 2. *Consider a configuration $c \in \overline{\mathfrak{F}}$. If c is such that the columns (c_i, c_{i+1}) code for an identity I then for all $t \in \mathbb{N}$, the columns $(f^t(c)_i, f^t(c)_{i+1})$ code for the same identity I.*

Proof. Let $c \in \overline{\mathfrak{F}}$ be a configuration containing a symbol I at position $(i, i+1)$. During a valid simulation, this pair evolves according to the local rules which may change its state or color, but preserve its identity I.

The only problem which could occur to change the identity of c_i or c_{i+1} is when a column comes "too close" on the left or on the right of the pair (see Fig. 6).

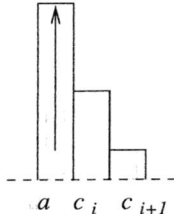

Fig. 6. The identity of (c_i, c_{i+1}) is not modified

When this happens, then c_i becomes an identity error and do not evolve anymore (for instance, in the Fig. 6, this happens when $0 < a - c_i \leq N$). To prevent c_{i+1} from moving and hence maintain the identity I, one should just add the constraint that a local rule returns a non-zero value if and only if both members of the pair do not have ambiguity in their code. This is easy to check, for example $\lambda(\ldots, -\,|\,\alpha_I, x, -, \ldots) = 0$ whenever $0 < x - \alpha_I \leq N$ for the left column. This new constraint does not affect the simulation, as such a situation should not happen in a valid configuration. □

Lemma 3. *Consider a malformed configuration $c \in \overline{\mathfrak{F}}$. If $\mathcal{S}_\mathcal{M}$ does not halt on c, then there is a lift whose color changes infinitely often.*

Proof. Assume $c \in \overline{\mathfrak{F}}$ is malformed, and $\mathcal{S}_\mathcal{M}$ does not halt when started from c. Because of Lemma 1, $\forall t \in \mathbb{N}$, $|f^t(c)|$ is bounded independently from t. So the infinite behavior is due to "vertical" movement in c, i.e. there is a column whose content changes infinitely often. Because of the conservation of the identity shown in Lemma 2, this column is in fact a pair of columns, as its identity cannot be modified. Hence there is a lift in c which evolves infinitely often (otherwise the configuration cannot change, since pairs of columns move only when they have a lift in their neighborhood, at most once every time the lift moves).

Moreover, there are no infinite columns in configurations taken from $\overline{\mathfrak{F}}$, which prevents this lift from keeping increasing or decreasing (lifts never go higher than the maximal value in their neighborhood, nor lower than the minimal one). As a consequence, its color changes infinitely often, otherwise the lift would have either stopped or gone to $\pm\infty$. Indeed, if the color does not change, the lift has no other choice but go towards the same direction after a finite number of steps. □

Proposition 16. *Consider a configuration $c \in \overline{\mathfrak{F}}$. If c contains an error (either identity or neighborhood error) then c is ultimately periodic for $\mathcal{S}_\mathcal{M}$.*

Proof. Let $c \in \overline{\mathfrak{F}}$. By contradiction, assume that c contains an error (no matter if identity or neighborhood error) and is aperiodic.

First of all, Lemma 3 implies that there is a lift in the configuration, whose color changes infinitely often. Hence there are infinitely many simulation steps **S-V-C**, which leads to infinitely many correct comparison steps **C**.

In this step, L_C checks the validity of all columns C, C^V, q, q^V, R_1, R_1^V, R_2, R_2^V. If one of them contains an error, either identity or neighborhood error, the simulation stops. This contradicts the aperiodicity of c. The same holds for L_R. It has to be valid, otherwise the next **S** step cannot be started and the simulation is blocked forever. □

Proof (of Theorem 2). By construction, if c represents a valid computation of $\mathcal{M}$ when $\mathcal{M}$ is started with both registers at 0, then c is ultimately periodic for $\mathcal{S}_\mathcal{M}$ if and only if $\mathcal{M}$ halts (when started with both registers at 0).

If c is malformed then, by Proposition 16, c is ultimately periodic for $\mathcal{S}_\mathcal{M}$. □

A Bridge Between the Asynchronous Message Passing Model and Local Computations in Graphs

(Extended Abstract)

Jérémie Chalopin and Yves Métivier

LaBRI Université Bordeaux 1, ENSEIRB,
351 cours de la Libération, 33405 Talence, France
{chalopin, metivier}@labri.fr

1 Introduction

A distributed system is a collection of processes that can interact. Three major process interaction models in distributed systems have principally been considered: - the message passing model, - the shared memory model, - the local computation model. In each model the processes are represented by vertices of a graph and the interactions are represented by edges. In the message passing model and the shared memory model, processes interact by communication primitives: messages can be sent along edges or atomic read/write operations can be performed on registers associated with edges. In the local computation model interactions are defined by labelled graph rewriting rules; supports of rules are edges or stars. These models (and their sub-models) reflect different system architectures, different levels of synchronization and different levels of abstraction. Understanding the power of various models, the role of structural network properties and the role of the initial knowledge enhances our understanding of basic distributed algorithms. This is done with some typical problems in distributed computing: election, naming, spanning tree construction, termination detection, network topology recognition, consensus, mutual exclusion. Furthermore, solutions to these problems constitute primitive building blocks for many other distributed algorithms. A survey may be found in [FR03], this survey presents some links with several parameters of the models including synchrony, communication media and randomization. An important goal in the study of these models is to understand some relationships between them. This paper is a contribution to this goal; more precisely we establish a bridge between tools and results presented in [YK96] for the message passing model and tools and results presented in [Ang80, BCG+96, Maz97, CM04, CMZ04, Cha05] for the local computation model.

In the message passing model studied by Yamashita and Kameda in [YK96], basic events are: send events, receive events, internal events and transmission events. They have obtained characterizations of graphs permitting a leader election algorithm, a spanning tree construction algorithm and a topology recognition algorithm. For this, they introduced the concept of view. The view from a vertex v of a graph G is an infinite labelled tree rooted in v obtained by considering all labelled walks in G starting from v. The characterizations use also the notion of symmetricity. The symmetricity of a graph depends on the number of vertices that have the same view. The local computation model has been studied intensively since the pioneer work of Angluin [Ang80]. A basic event changes the state attached to one vertex or the states of a group of neighbouring vertices. The new state depends on the state of one neighbour or depends on the states of a group of neighbours (some examples are presented in [RFH72, BV99, BCG+96]). Characterizations of graphs, for the existence of an election algorithm, have been obtained using classical combinatorial material like the notions of fibration and of covering: special morphisms which ensure isomorphism of neighbourhoods of vertices or arcs. Some effective characterizations of computability of relations in anonymous networks using fibrations and views are given in [BV01]. The new state of a vertex must depend on the previous state and on the states of the in-neighbours.

The Election Problem and the Naming Problem. The election problem is one of the paradigms of the theory of distributed computing. It was first posed by LeLann [LeL77]. A distributed algorithm solves the election problem if it always terminates and in the final configuration exactly one process is marked as *elected* and all the other processes are *non-elected*. Moreover, it is supposed that once a process becomes *elected* or *non-elected* then it remains in such a state until the end of the algorithm. Election algorithms constitute a building block of many other distributed algorithms. The naming problem is another important problem in the theory of distributed computing. The aim of a naming algorithm is to arrive at a final configuration where all processes have unique identities. Being able to give dynamically and in a distributed way unique identities to all processes is very important since many distributed algorithms work correctly only under the assumption that all processes can be unambiguously identified. The enumeration problem is a variant of the naming problem. The aim of a distributed enumeration algorithm is to assign to each network vertex a unique integer in such a way that this yields a bijection between the set $V(G)$ of vertices and $\{1, 2, \ldots, |V(G)|\}$.

The Main Results. In Section 3 we introduce a new labelled directed graph which encodes a network in which processes communicate by asynchronous message passing with a symmetric port numbering. The basic events (send, receive, internal, transmission) are encoded by local computations on arcs. From this directed graph, we deduce necessary conditions for the existence of an election (and a naming) algorithm on a network (Proposition 4). The conditions are also sufficient (Theorem 1): we give a naming (and an election) algorithm in Section 5 (Algorithm 1). This algorithm is totally asynchronous (the Yamashita and Kameda algorithm needs a pseudo-synchronization). Furthermore, our algorithm does not need the FIFO property of channels (i.e., it does not require that messages are received in the same order as they have been sent). The size of the buffer does not interfere in the impossibility proof for the election. Moreover, we present a fully polynomial algorithm. Given a graph G with n vertices and m edges, in Yamashita and Kameda algorithm the size of each message can be 2^n

whereas in our algorithm the size is bounded by $O(m \log n)$ and the number of messages is $O(m^2 n)$. Another consequence of this bridge between these models is a direct characterization of graphs having a symmetricity equal to 1 in the sense of Yamashita and Kameda using the notion of covering. The same techniques can be applied to some other problems such as spanning tree computation or the topology recognition problem. We can note also that our algorithm may elect even if the necessary condition is not verified: in this case an interesting problem is the study of the probability of this event.

2 Preliminaries

The notations used here are essentially standard. The definitions and main properties are presented in [BV02]. We consider finite, undirected, connected graphs having possibly self-loops and multiple edges, $G = (V(G), E(G), \text{Ends})$, where $V(G)$ denotes the set of vertices, $E(G)$ denotes the set of edges and Ends is a map assigning to every edge two vertices: its ends. A symmetric digraph (V, A, s, t) is a digraph endowed with a symmetry, that is, an involution $Sym : A \to A$ such that for every $a \in A : s(a) = t(Sym(a))$. Labelled graphs will be designated by bold letters like **G**, **H**, ... If $\mathbf{G} = (G, \lambda)$ is a labelled graph then G denotes the underlying graph and λ denotes the labelling function. The labelling may encode any initial process knowledge. Examples of such knowledge include: (a bound on) the number of processes, (a bound on) the diameter of the graph, the topology, identities or partial identities, distinguished vertices. The notion of fibration and of of covering are fundamental in this work.

Definition 1. *A fibration between the digraphs D and D' is a morphism φ from D to D' such that for each arc a' of $A(D')$ and for each vertex v of $V(D)$ such that $\varphi(v) = v' = t(a')$ there exists a unique arc a in $A(D)$ such that $t(a) = v$ and $\varphi(a) = a'$.*

The arc a is called the lifting of a' at v, D is called the total digraph and D' the base of φ. We shall also say that D is fibred (over D'). In the sequel directed graphs are always strongly connected and total digraphs non empty thus fibrations will be always surjective.

Definition 2. *An opfibration between the digraphs D and D' is a morphism φ from D to D' such that for each arc a' of $A(D')$ and for each vertex v of $V(D)$ such that $\varphi(v) = v' = s(a')$ there exists a unique arc a in $A(D)$ such that $s(a) = v$ and $\varphi(a) = a'$. A covering projection is a fibration that is also an opfibration.*

If a covering projection $\varphi : D \to D'$ exists, D is said to be a covering of D' via φ. Covering projections verify:

Proposition 1. *A covering projection $\varphi : D \to D'$ with a connected base and a nonempty covering is surjective; moreover, all the fibres have the same cardinality. This cardinality is called the number of sheets of the covering.*

A digraph D is covering prime if there is no digraph D' not isomorphic to D such that D is a covering of D' (i.e., D is a covering of D' implies that D is isomorphic to D'). Let D and D' be two digraphs such that D is a surjective covering of D' via φ. If D' has no self-loop then for each arc $a \in A(D) : \varphi(s(a)) \neq \varphi(t(a))$. Finally the following property is a direct consequence of the definitions and it is fundamental in the sequel of this paper :

Proposition 2. *Let D and D' be two digraphs such that D' has no self-loop and D is a surjective covering of D' via φ. If $a_1 \neq a_2$ and $a_1, a_2 \in \varphi^{-1}(a')$ ($a' \in A(D')$) then $Ends(a_1) \cap Ends(a_2) = \emptyset$.*

The notions of fibrations and of coverings extend to labelled digraphs in an obvious way: the morphisms must preserve the labelling. Examples of coverings are given in Figures 1 and 2.

Local Computations on Arcs. In this paper we consider labelled digraphs and we assume that local computations modify only labels of vertices. Digraph relabelling systems on arcs and more generally local computations on arcs satisfy the following constraints, that arise naturally when describing distributed computations with decentralized control: -($C1$) they do not change the underlying digraph but only the labelling of vertices, the final labelling being the result of the computation (*relabelling relations*), -($C2$) they are *local*, that is, each relabelling step changes only the label of the source and the label of the target of an arc, -($C3$) they are *locally generated*, that is, the applicability of a relabelling rule on an arc only depends on the label of the arc, the labels of the source and of the target (locally generated relabelling relation). The relabelling is performed until no more transformation is possible, i.e., until a normal form is obtained. Let $\mathcal{R}$ be a locally generated relabelling relation, $\mathcal{R}^*$ stands for the reflexive-transitive closure of $\mathcal{R}$. The labelled digraph $\mathbf{D}$ is $\mathcal{R}$-*irreducible* (or just irreducible if $\mathcal{R}$ is fixed) if there is no $\mathbf{D_1}$ such that $\mathbf{D}\ \mathcal{R}\ \mathbf{D_1}$.

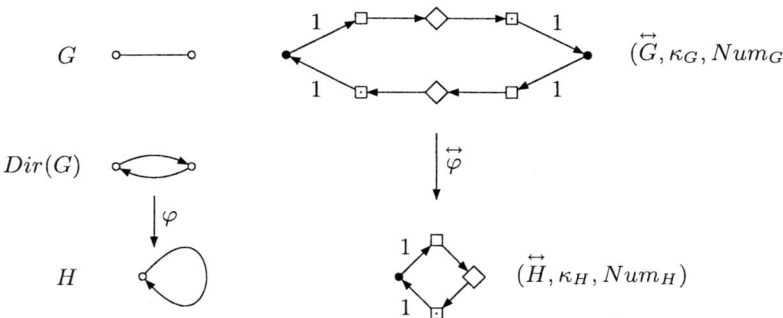

Fig. 1. We adopt the following notation conventions for vertices of $(\overleftrightarrow{G}, \kappa_G, Num_G)$ and $(\overleftrightarrow{H}, \kappa_H, Num_H)$. A black-circle vertex corresponds to the label **process**, a square vertex corresponds to the label **send**, a diamond vertex corresponds to the label **transmission**, and a square-dot vertex corresponds to the label **receive**. The digraph $(\overleftrightarrow{G}, \kappa_G, Num_G)$ is a covering of $(\overleftrightarrow{H}, \kappa_H, Num_H)$ and the port numbering is symmetric. Thus there is no election algorithm for G.

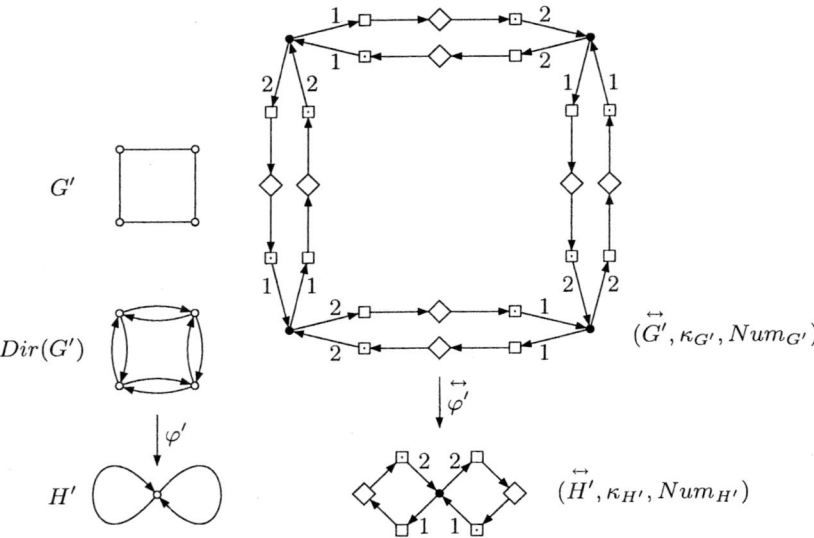

Fig. 2. With the notation conventions of Figure 1, we deduce from the covering relation and the symmetry of the port numbering that there is no election algorithm for the graph G'. With the same argument the same result is obtained for any ring.

3 From Asynchronous Message Passing to Local Computations on Arcs

The Model. Our model follows standard models for distributed systems given in [AW98, Tel00]. The communication model is a point-to-point communication network which is represented as a simple connected undirected graph where vertices represent processes and two vertices are linked by an edge if the corresponding processes have a direct communication link. Processes communicate by message passing, and each process knows from which channel it receives a message or it sends a message. An edge between two vertices v_1 and v_2 represents a channel connecting a port i of v_1 to a port j of v_2. We consider the asynchronous message passing model: processes cannot access a global clock and a message sent from a process to a neighbour arrives within some finite but unpredictable time.

From Undirected Labelled Graphs to Labelled Digraphs. A first approximation of a network, with knowledge about the structure of the underlying graph, is a simple labelled graph $\mathbf{G} = (V(\mathbf{G}), E(\mathbf{G}))$. We associate to this undirected labelled graph a labelled digraph $\overleftrightarrow{\mathbf{G}} = (V(\overleftrightarrow{\mathbf{G}}), A(\overleftrightarrow{\mathbf{G}}))$ defined in the following way. (This construction is illustrated in Figure 1 and in Figure 2).

Let u and v be two vertices of $\mathbf{G}$ such that u and v are neighbours, we associate to the edge $\{u,v\}$ the set $V_{\{u,v\}}$ of 6 vertices denoted $\{outbuf(u,v), t(u,v),$

$inbuf(u,v), outbuf(v,u), t(v,u), inbuf(v,u)\}$, and the set $A_{\{u,v\}}$ of 8 arcs defined by: $\{(u, outbuf(u,v)), (outbuf(u,v), t(u,v)), (t(u,v), inbuf(u,v)), (inbuf(u,v), v), (v, outbuf(v,u)), (outbuf(v,u), t(v,u)), (t(v,u), inbuf(v,u)), (inbuf(v,u), u)\}$.

Finally, $V(\overleftrightarrow{\mathbf{G}}) = V(G) \cup (\bigcup_{\{u,v\}\in E(G)} V_{\{u,v\}})$ and $A(\overleftrightarrow{\mathbf{G}}) = \bigcup_{\{u,v\}\in E(G)} A_{\{u,v\}}$.

The arc $(u, outbuf(u,v))$ is denoted $out(u,v)$, $receiver(out(u,v))$ is the vertex v, and the arc $(inbuf(v,u), u)$ is denoted by $in(v,u)$.

If $\mathbf{G} = (G, \lambda)$ then $\overleftrightarrow{\mathbf{G}} = (\overleftrightarrow{G}, \lambda_{\overleftrightarrow{G}})$ where $\lambda_{\overleftrightarrow{G}}(v) = \lambda(v)$ for each $v \in V(G)$.

In the sequel we consider digraphs obtained by this construction; in general networks are anonymous: vertices have no name. Nevertheless we need to memorize the meaning (semantic) of vertices thus we label vertices of $\overleftrightarrow{\mathbf{G}}$ with a labelling function κ, the set of labels is: {**process, send, receive, transmission**},
- if a vertex x of $V(\overleftrightarrow{\mathbf{G}})$ corresponds to a vertex u of $V(G)$ then $\kappa(x) = $ **process**,
- if a vertex x of $V(\overleftrightarrow{\mathbf{G}})$ corresponds to a vertex of the form $outbuf(u,v)$ then $\kappa(x) = $ **send**, - if a vertex x of $V(\overleftrightarrow{\mathbf{G}})$ corresponds to a vertex of the form $inbuf(u,v)$ then $\kappa(x) = $ **receive**, - if a vertex x of $V(\overleftrightarrow{\mathbf{G}})$ corresponds to a vertex of the form $t(u,v)$ then $\kappa(x) = $ **transmission**. Using a new label *neutral*, κ is extended to $(V(\overleftrightarrow{\mathbf{G}}), A(\overleftrightarrow{\mathbf{G}}))$. We denote by $\mathcal{E}$ the map which associates to a labelled graph $\mathbf{G}$ the labelled digraph $\mathcal{E}(\mathbf{G}) = (\overleftrightarrow{\mathbf{G}}, \kappa)$ described above.

Two adjacent vertices of $\mathcal{E}(\mathbf{G}) = (\overleftrightarrow{\mathbf{G}}, \kappa)$ have different labels thus if the digraph $\mathcal{E}(\mathbf{G}) = (\overleftrightarrow{\mathbf{G}}, \kappa)$ is a covering of a digraph $\mathbf{D}$ then $\mathbf{D}$ has no self-loop.

Remark 1. By the insertion of special vertices in arcs and the labelling of vertices, we define a transformation $\mathcal{E}'$ such that $(\overleftrightarrow{\mathbf{G}}, \kappa)$ can be obtained directly from $Dir(\mathbf{G})$, i.e., $\mathcal{E}'(Dir(\mathbf{G})) = (\overleftrightarrow{\mathbf{G}}, \kappa)$. Furthermore if $Dir(\mathbf{G})$ is a covering of a labelled digraph $\mathbf{D}$ then $(\overleftrightarrow{\mathbf{G}}, \kappa)$ is a covering of $\mathcal{E}'(\mathbf{D})$.

Port Numbering and Symmetric Port Numbering. We can notice, that for a digraph $\mathcal{E}(\mathbf{G}) = (\overleftrightarrow{\mathbf{G}}, \kappa)$, if we consider a vertex x labelled **process** then $deg^+(x) = deg^-(x)$. Each process knows from which channel it receives a message or it sends a message, that is, each process assigns numbers to its ports. Thus we consider a labelling Num of arcs of $\mathcal{E}(\mathbf{G})$ coming into or going out of vertices labelled **process** such that for each vertex x labelled **process** the restriction of Num assigns to each outgoing arc a unique integer of $[1, deg^+(x)]$ and assigns to each arc coming into a unique integer of $[1, deg^-(x)]$, such a labelling is a local enumeration of arcs incident to **process** vertices and it is called a port numbering. In a message passing system the communication is done over communication channels. A channel provides a bidirectional connection between two processes. Finally, the topology is encoded by an undirected graph G where an edge corresponds to a channel. Let v be a vertex of G, the port numbering for the vertex v is defined by an enumeration of edges incident to the vertex v, this enumeration induces an enumeration of the arcs of $(\overleftrightarrow{\mathbf{G}}, \kappa)$. This enumeration is symmetric, i.e., Num verifies for each arc of the form $out(u,v)$:

$Num(out(u,v)) = Num(in(v,u))$; this condition is called the symmetry of the port numbering (or equivalently of Num). Such a port numbering is said symmetric. Again, using the special label *neutral*, Num is considered as a labelling function of $\mathcal{E}(\mathbf{G})$. The graph $(\overleftrightarrow{\mathbf{G}}, \kappa, Num)$ is denoted by $\mathcal{H}(\mathbf{G})$. The hypothesis of the symmetry of the port numbering is done in [YK96] and it corresponds to the complete port awareness model in [BCG+96].

Basic Instructions. As in [YK96] (see also [Tel00] pp. 45-46), we assume that each process, depending on its state, either changes its state, or receives a message via a port or sends a message via a port. Let $Inst$ be this set of instructions. This model is equivalent to the model of local computations on arcs with respect to the initial labelling as it is depicted in the following remark.

Remark 2. Let $\mathbf{G}$ be a labelled graph, let $\mathcal{H}(\mathbf{G}) = (\overleftrightarrow{\mathbf{G}}, \kappa, Num)$ be the labelled digraph obtained from $\mathbf{G}$. The labelled digraph $\mathcal{H}(\mathbf{G})$ enables to encode the following events using local computations on arcs: - an internal event "a process changes its state" can be encoded by a relabelling rule concerning a vertex labelled **process**, - a send event "the process x sends a message via the port i" can be encoded by a relabelling rule concerning an arc of the form (x, y) with $\kappa(x) = $ **process**, $\kappa(y) = $ **send** and $Num((x,y)) = i$, - a receive event "the process y receives a message via the port i" can be encoded by a relabelling rule concerning an arc of the form (x, y) with $\kappa(x) = $ **receive**, $\kappa(y) = $ **process** and $Num((x,y)) = i$, - an event concerning the transmission control can be encoded by a relabelling rule concerning an arc of the form (x, y) or (y, z) with $\kappa(x) = $ **send**, $\kappa(y) = $ **transmission** and $\kappa(z) = $ **receive**.

The Election and the Naming Problems. Consider a network $\mathbf{G}$ with a symmetric port numbering Num. An algorithm $\mathcal{A}$ is an election algorithm for $(\overleftrightarrow{\mathbf{G}}, \kappa, Num)$ if each execution of $\mathcal{A}$ on $\mathbf{G}$ with the port numbering Num successfully elects a process. We are particularly interested in characterizing the networks that admit an election algorithm whatever the symmetric port numbering is. We say that an algorithm $\mathcal{A}$ is an election algorithm for a graph $\mathbf{G}$ if for each symmetric port numbering Num, $\mathcal{A}$ is an election algorithm for $(\overleftrightarrow{\mathbf{G}}, \kappa, Num)$. We will use the same conventions for the naming problem.

4 A Necessary Condition for the Election Problem and the Naming Problem

First, we present a fundamental lemma which connects coverings and locally generated relabelling relations on arcs. It is the natural extension of the Lifting Lemma [Ang80] and it is a direct consequence of Proposition 2.

Lemma 1 (Lifting Lemma). *Let $\mathcal{R}$ be a locally generated relabelling relation on arcs and let $\mathbf{D}_1$ be a covering of the digraph $\mathbf{D}'_1$ via the morphism γ; we assume that $\mathbf{D}'_1$ has no self-loop. If $\mathbf{D}'_1 \mathcal{R}^* \mathbf{D}'_2$ then there exists $\mathbf{D}_2$ such that $\mathbf{D}_1 \mathcal{R}^* \mathbf{D}_2$ and $\mathbf{D}_2$ is a covering of $\mathbf{D}'_2$ via γ.*

As a direct consequence of this lemma and of Proposition 1, if $\mathbf{D}_2$ is a proper covering of $\mathbf{D}'_2$, each label that appears in $\mathbf{D}'_2$ appears at least twice in $\mathbf{D}_2$ and therefore, we have the following result.

Proposition 3. *Let $\mathbf{G}$ be an undirected labelled graph. Let Num be a port numbering of $\mathbf{G}$. If the labelled digraph $(\overleftrightarrow{\mathbf{G}}, \kappa, Num)$ is not covering prime then there is no election algorithm and no naming algorithm for the graph $\mathbf{G}$ with Num as port numbering using $Inst$ as set of basic instructions.*

The election algorithm must work whatever the symmetric port numbering is. Let $(\overleftrightarrow{\mathbf{G}}, \kappa)$ be a covering of $(\overleftrightarrow{\mathbf{G}}', \kappa')$, and let Num be a local enumeration of arcs incident to vertices labelled **process** in the graph $(\overleftrightarrow{\mathbf{G}}', \kappa')$. The labelling Num induces a port numbering of $(\overleftrightarrow{\mathbf{G}}, \kappa)$ which is not necessarily symmetric (see the example in Figure 3).

Before the next propositions we need two definitions [BV02]:

Definition 3. *Let $\mathbf{D}_1$ and $\mathbf{D}_2$ be two symmetric labelled digraphs, let Sym_1 and Sym_2 be symmetric relations of $\mathbf{D}_1$ and $\mathbf{D}_2$, $\mathbf{D}_1$ is a covering of $\mathbf{D}_2$ modulo Sym_1 and Sym_2 if there exists a morphism φ such that $\mathbf{D}_1$ is a covering of $\mathbf{D}_2$ via φ and $\varphi \circ Sym_1 = Sym_2 \circ \varphi$.*

Definition 4. *Let $\mathbf{D}_1$ be a symmetric digraph, $\mathbf{D}_1$ is symmetric covering prime if whenever there exists a symmetric relation Sym_1 of $\mathbf{D}_1$, a symmetric digraph $\mathbf{D}_2$ with a symmetric relation Sym_2 of $\mathbf{D}_2$ such that $\mathbf{D}_1$ is a covering of $\mathbf{D}_2$ modulo Sym_1 and Sym_2 then $\mathbf{D}_1$ is isomorphic to $\mathbf{D}_2$.*

From these definitions, there exists a symmetric port numbering Num of $\mathbf{G}$ such that $(\overleftrightarrow{\mathbf{G}}, \kappa, Num)$ is not covering prime if and only if $Dir(\mathbf{G})$ is not symmetric covering prime. Finally:

Proposition 4. *Let $\mathbf{G}$ be an undirected graph. If the labelled digraph $Dir(\mathbf{G})$ is not symmetric covering prime then there is no election algorithm and no naming algorithm for the graph $\mathbf{G}$ using $Inst$ as set of basic instructions.*

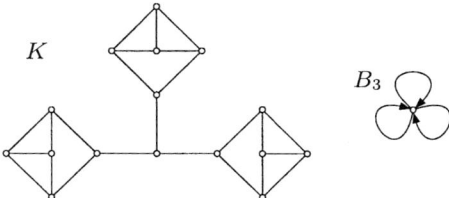

Fig. 3. There exists exactly one digraph B_3 such that $Dir(K)$ is a covering of B_3 : it is the 3-bouquet (the digraph with one node and three self-loops [BV02]). Thus $\mathcal{E}'(Dir(K))$ is a covering of $\mathcal{E}'(B_3)$ (Remark 1). It is easy to verify that no lifting of a local enumeration of arcs of $\mathcal{E}'(B_3)$ gives a symmetric port numbering of $\mathcal{E}'(Dir(K))$. Thus $Dir(K)$ is symmetric covering prime although it is not covering prime (see Definition 4). It follows from Theorem 1 that a naming algorithm exists for the graph K.

As immediate consequences of this result we deduce two classical results: there exists no deterministic election algorithm in an anonymous network of two processes that communicate by asynchronous message passing ([Tel00] p. 316) and more generally there exists no deterministic algorithm for election in an anonymous ring of known size ([Tel00] Theorem 9.5 p. 317) (sketches of the proofs are given in Figure 1 and in Figure 2).

5 A Mazurkiewicz-Like Algorithm

The aim of this section is to prove the main result of this work:

Theorem 1. *Let* **G** *be a graph. There exist an election algorithm and a naming algorithm for* **G** *if and only if* $Dir(\mathbf{G})$ *is symmetric covering prime.*

The necessary part is Proposition 4, the following algorithm proves the other part. In [Maz97] Mazurkiewicz presents a distributed enumeration algorithm for non-ambiguous graphs (see also [GMM04]). The computation model in [Maz97] allows relabelling of all vertices in balls of radius 1. In the following we adapt Mazurkiewicz algorithm to graphs with port numbering and using $Inst$ as set of basic instructions. We shall denote our algorithm $\mathcal{M}$.

Description of $\mathcal{M}$. We first give a general description of the algorithm $\mathcal{M}$ applied to a labelled graph **G** equipped with a port numbering Num. We assume that **G** is connected. Let $\mathbf{G} = (G, \lambda)$ and consider a vertex v_0 of G, and the set $\{v_1, ..., v_d\}$ of neighbours of v_0. During the computation, each vertex v_0 will be labelled by a pair of the form $(\lambda(v_0), c(v_0))$, where $c(v_0)$ is a triple $(n(v_0), N(v_0), M(v_0))$ representing the following information obtained during the computation (formal definitions are given below): $n(v_0) \in \mathbb{N}$ is the *number* of the vertex v_0 computed by the algorithm, $N(v_0) \in \mathcal{N}$ is the *local view* of v_0, this view can be either empty or it is a set of the form: $\{((n(v_i), p_{s,i}, p_{r,i}), \lambda(v_i)) | 1 \leq i \leq d\}$, $M(v_0) \subseteq L \times \mathbb{N} \times \mathcal{N}$ is the *mailbox* of v_0 containing the whole information received by v_0 at previous computation steps. Let $(((n(v_i), p_{s,i}, p_{r,i}), \lambda(v_i)) 1 \leq i \leq d)$ be the local view of v_0. For each i, $(n(v_i), p_{s,i}, p_{r,i})$ encodes a neighbour v_i of v_0, where: $n(v_i)$ is the number of v_i, v_i has sent its number to v_0 via the port $p_{s,i}$, and v_0 has received this message via the port $p_{r,i}$. Each vertex v gets information from its neighbours via messages and then attempts to calculate its own number $n(v)$, which will be an integer between 1 and $|V(G)|$. If a vertex v discovers the existence of another vertex u with the same number, then it compares its own label and its own local view with the label and the local view of u. If the label of u or the local view of u is "stronger", then v chooses another number. Each new number, with its local view, is broadcasted again over the network. At the end of the computation, it is not guaranteed that every vertex has a unique number, unless the graph $(\overleftrightarrow{\mathbf{G}}, \kappa, Num)$ is covering prime. However, all vertices with the same number will have the same label and the same local view.

Algorithm 1: The algorithm $\mathcal{M}$

Var : $n(v_0)$: integer **init** 0 ;
$N(v_0)$: set of local view **init** $\emptyset$;
N : set of local view ;
$M(v_0)$: mailbox **init** $\emptyset$;
M, M_a : mailbox;
$\lambda(v_0), c_a, l$: element of L;
i, x, p, q, n_a : integer;

$\mathbf{I_0}$: $\{n(v_0) = 0$ and no message has arrived at $v_0\}$
begin
$\quad n(v_0) := 1;$
$\quad M(v_0) := \{(\lambda(v_0), 1, \emptyset)\};$
$\quad$ **for** $i := 1$ **to** $deg(v_0)$ **do** send$< (n(v_0), M(v_0)), i >$ via port i ;
end

$\mathbf{R_0}$: $\{$A message $<$ **mes**$= (n_a, M_a)$, $p >$ has arrived at v_0 from port $q\}$
begin
$\quad M := M(v_0);$
$\quad M(v_0) := M(v_0) \cup M_a;$
$\quad$ **if** $((x, p, q) \notin N(v_0)$ *for some* $x)$ **then**
$\quad\quad N(v_0) := N(v_0) \cup \{(n_a, p, q)\};$
$\quad$ **if** $((x, p, q) \in N(v_0)$ *for some* $x < n_a)$ **then**
$\quad\quad N(v_0) := (N(v_0) \setminus \{(x, p, q)\}) \cup \{(n_a, p, q)\};$
$\quad$ **if** $(n(v_0) = 0)$ *or* $(n(v_0) > 0$ *and there exists* $(l, n(v_0), N) \in M(v_0)$
$\quad$ *such that* $(\lambda(v_0) <_L l)$ *or* $((\lambda(v_0) = l)$ *and* $(N(v_0) \prec N))))$ **then**
$\quad\quad n(v_0) := 1 + \max\{n \in \mathbb{N} \mid (l, n, N) \in M(v_0)$ for some $l, N\};$
$\quad\quad M(v_0) := M(v_0) \cup \{(\lambda(v_0), n(v_0), N(v_0))\};$
$\quad$ **if** $(M(v_0) \neq M))$ **then**
$\quad\quad$ **for** *(i := 1 **to** $deg(v_0)$)* **do** send $< (n(v_0), M(v_0)), i >$ via port i;
end

An Order on Local Views. We assume for the rest of this paper that the set of labels L is totally ordered by $<_L$. Consider a vertex v such that the local view $N(v) \in \mathcal{N}$ is the set $\{(n_1, p_{s,1}, p_{r,1}), (n_2, p_{s,2}, p_{r,2}), \ldots, (n_d, p_{s,d}, p_{r,d})\}$. We assume that for each $i < d$, $(n_{i+1}, p_{s,i+1}, p_{r,i+1}) <_{Lex} (n_i, p_{s,i}, p_{r,i})$ where $<_{Lex}$ denotes the usual lexical order. We say that $((n_1, p_{s,1}, p_{r,1}), (n_2, p_{s,2}, p_{r,2}), \ldots, \ldots (n_d, p_{s,d}, p_{r,d}))$ is the ordered representation $N_>(v_0)$ of the local view of v_0. Let $\mathcal{N}_>$ be the set of such ordered tuples. We define a total order $\prec$ on $\mathcal{N}_>$ using the alphabetical order that induces naturally a total order on $\mathcal{N}$. This order can also be defined on $\mathcal{N}$ as follows: $N_1 \prec N_2$ if the maximal element for the lexical order $<_{Lex}$ of the symmetric difference $N_1 \triangle N_2 = N_1 \cup N_2 \setminus N_1 \cap N_2$ belongs to N_2. If $N(u) \prec N(v)$, then we say that the local view $N(v)$ of v is stronger than the one of u.

The Final Labelling. Let $\mathbf{G} = (G, \lambda)$ be a connected labelled graph with the port numbering Num. If v is a vertex of G then the label of v after a run ρ of $\mathcal{M}$

is denoted $(\lambda(v), c_\rho(v))$ with $c_\rho(v) = (n_\rho(v), N_\rho(v), M_\rho(v))$ and (λ, c_ρ) denotes the final labelling. Finally $\mathcal{M}$ verifies:

Proposition 5. *Any run ρ of $\mathcal{M}$ on $\mathbf{G} = (G, \lambda)$, a connected labelled graph with the port numbering Num, terminates and yields a final labelling (λ, c_ρ) verifying the following conditions for all vertices v, v' of G:*

1. *there exists an integer $k \leq V(G)$ such that $\{n_\rho(v) \mid v \in V(G)\} = [1, k]$.*
2. $M_\rho(v) = M_\rho(v')$.
3. $(\lambda(v), n_\rho(v), N_\rho(v)) \in M_\rho(v')$.
4. *Let $(l, n, N) \in M_\rho(v')$. Then $\lambda(v) = l$, $n_\rho(v) = n$ and $N_\rho(v) = N$ for some vertex v if and only if there is no triple $(l', n, N') \in M_\rho(v')$ with $l <_L l'$ or $(l = l'$ and $N \prec N')$.*
5. $n_\rho(v) = n_\rho(v')$ *implies* $(\lambda(v) = \lambda(v')$ and $N(v) = N(v'))$.

Consider a graph $\mathbf{G}$ that is symmetric covering prime. For each port numbering Num, the graph $(\overleftrightarrow{\mathbf{G}}, \kappa, Num)$ is covering prime and then from Proposition 5, at the end of the computation, each vertex $v \in V(G)$ has a unique number $n(v)$. Moreover, once a vertex gets a number $n(v) = |V(G)|$, it knows that all the vertices have a unique identifier, it can take the label *elected* and broadcast the information. Theorem 1 follows from Proposition 5 and the impossibility results of the previous section.

Remark 3. The proof of this proposition uses increasing properties and invariant properties as in [Maz97]. In particular, the number $n(v)$ (resp. the mailbox $M(v)$) can only increase for the order $\leq$ (resp. for $\subseteq$) during the computation. Consequently, if a message $m_1 = (n_1(v), M_1(v), p)$ has been sent before $m_2 = (n_2(v), M_2(v), p)$ by a vertex v to a node w is such that m_2 arrives before m_1, then when the message m_1 is read by w, $M_1(v) \subsetneq M_2(v) \subseteq M(w)$ and $n_1(v) \leq n_2(v)$. Consequently, this message does not modify the state of the vertex w and can be considered as ignored by the vertex w. We can therefore deduce that Algorithm 1 does not require ordering of messages, that is, it does not require that messages are received in the same order that they have been sent.

Remark 4. Note that not all the elements of $M(v)$ are useful during the whole computation. In fact, for all $(n, l_1, N_1), (n, l_2, N_2) \in M(v)$, if $l_1 <_L l_2$ or $l_1 = l_2$ and $N_1 \prec N_2$, we can remove (n, l_1, N_1) from $M(v)$. Consequently, if we remove all such elements of $M(v)$, we get for each number n exactly one element in $M(v)$. If we can encode the labels l of L with $O(\log |V(G)|)$ bits, then the size of the mailbox of v is $O(|E(G)| \log |V(G)|)$ and therefore, the size of the messages is also $O(|E(G)| \log |V(G)|)$. Moreover, we can show that the total number of messages sent during the computation is $O(|E(G)|^2 |V(G)|)$ and therefore the amount of information sent all over the network during the computation is polynomial in the size of the network.

References

[Ang80] D. Angluin. Local and global properties in networks of processors. In *Proceedings of the 12th Symposium on Theory of Computing*, pages 82–93, 1980.

[AW98] H. Attiya and J. Welch. *Distributed computing: fundamentals, simulations, and advanced topics*. McGraw-Hill, 1998.

[BCG+96] P. Boldi, B. Codenotti, P. Gemmell, S. Shammah, J. Simon, and S. Vigna. Symmetry breaking in anonymous networks: Characterizations. In *Proc. 4th Israeli Symposium on Theory of Computing and Systems*, pages 16–26. IEEE Press, 1996.

[BV99] P. Boldi and S. Vigna. Computing anonymously with arbitrary knowledge. In *Proceedings of the 18th ACM Symposium on principles of distributed computing*, pages 181–188. ACM Press, 1999.

[BV01] Paolo Boldi and Sebastiano Vigna. An effective characterization of computability in anonymous networks. In Jennifer L. Welch, editor, *Distributed Computing. 15th International Conference, DISC 2001*, volume 2180 of *Lecture Notes in Computer Science*, pages 33–47. Springer-Verlag, 2001.

[BV02] P. Boldi and S. Vigna. Fibrations of graphs. *Discrete Math.*, 243:21–66, 2002.

[Cha05] J. Chalopin. Election and local computations on closed unlabelled edges (*extended abstract*). In *Proc. of SOFSEM 2005*, number 3381 in LNCS, pages 81–90, 2005.

[CM04] J. Chalopin and Y. Métivier. Election and local computations on edges (*extended abstract*). In *Proc. of Foundations of Software Science and Computation Structures, FOSSACS'04*, number 2987 in LNCS, pages 90–104, 2004.

[CMZ04] J. Chalopin, Y. Métivier, and W. Zielonka. Election, naming and cellular edge local computations (*extended abstract*). In *Proc. of International conference on graph transformation, ICGT'04*, number 3256 in LNCS, pages 242–256, 2004.

[FR03] F. Fich and E. Ruppert. Hundreds of impossibility results for distributed computing. *Distributed computing*, 16:121–163, 2003.

[GMM04] E. Godard, Y. Métivier, and A. Muscholl. Characterization of Classes of Graphs Recognizable by Local Computations. *Theory of Computing Systems*, (37):249–293, 2004.

[LeL77] G. LeLann. Distributed systems: Towards a formal approach. In B. Gilchrist, editor, *Information processing'77*, pages 155–160. North-Holland, 1977.

[Maz97] A. Mazurkiewicz. Distributed enumeration. *Inf. Processing Letters*, 61:233–239, 1997.

[RFH72] P. Rosenstiehl, J.-R. Fiksel, and A. Holliger. Intelligent graphs. In R. Read, editor, *Graph theory and computing*, pages 219–265. Academic Press (New York), 1972.

[Tel00] G. Tel. *Introduction to distributed algorithms*. Cambridge University Press, 2000.

[YK96] M. Yamashita and T. Kameda. Computing on anonymous networks: Part i - characterizing the solvable cases. *IEEE Transactions on parallel and distributed systems*, 7(1):69–89, 1996.

Reconstructing an Ultrametric Galled Phylogenetic Network from a Distance Matrix

Ho-Leung Chan, Jesper Jansson, Tak-Wah Lam, and Siu-Ming Yiu

Department of Computer Science, The University of Hong Kong,
Pokfulam Road, Hong Kong
{hlchan, jjansson, twlam, smyiu}@cs.hku.hk

Abstract. Given a distance matrix M that specifies the pairwise evolutionary distances between n species, the phylogenetic *tree* reconstruction problem asks for an edge-weighted phylogenetic tree that satisfies M, if one exists. We study some extensions of this problem to rooted phylogenetic *networks*. Our main result is an $O(n^2 \log n)$-time algorithm for determining whether there is an ultrametric galled network that satisfies M, and if so, constructing one. In fact, if such an ultrametric galled network exists, our algorithm is guaranteed to construct one containing the minimum possible number of nodes with more than one parent (*hybrid* nodes). We also prove that finding a largest possible submatrix M' of M such that there exists an ultrametric galled network that satisfies M' is NP-hard. Furthermore, we show that given an incomplete distance matrix (i.e., where some matrix entries are missing), it is also NP-hard to determine whether there exists an ultrametric galled network which satisfies it.

1 Introduction

A phylogenetic network is a generalization of a phylogenetic tree which can be used to describe the evolutionary history of a set of species that is non-treelike, for example, due to recombination events such as hybrid speciation or horizontal gene transfer [8, 14, 15, 17] or to represent several conflicting phylogenetic trees at once in order to identify parts where the trees disagree [2, 10].

To develop efficient methods for inferring phylogenetic networks is an important topic in computational biology. In particular, one promising category of methods which includes methods such as Neighbor-Net [2] and several others (see [15] for a survey) is known as *distance-based*. Here, the input consists of a (symmetric and non-negative) distance matrix which specifies the pairwise evolutionary distances between the species. To infer a phylogenetic *tree* from such a matrix is a well-studied problem [3, 5, 6, 16, 18], the basic objective being to construct an edge-weighted phylogenetic tree such that for any two species, the length of the path between them in the tree equals the corresponding entry

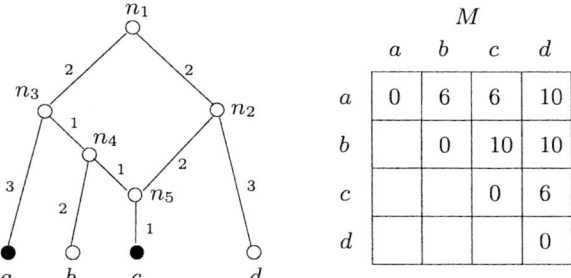

Fig. 1. The (galled and ultrametric) phylogenetic network on the left satisfies the distance matrix M on the right. There are two evolutionary paths (a, n_3, n_4, n_5, c) and $(a, n_3, n_1, n_2, n_5, c)$ with lengths 6 and 10, respectively, connecting a and c. The entry $M(a, c)$ corresponds to the first path. Note that there does not exist any phylogenetic tree that satisfies M.

in the matrix. Note that in a phylogenetic tree, the path between two specified leaves is always unique. On the other hand, due to recombination events, for any two species in a phylogenetic network, there can be more than one path connecting them with different path lengths. The entry in the input matrix may correspond to one of these paths only. Hence, in some cases, there may exist a phylogenetic network that satisfies the given distance matrix (see the definition below) while no such phylogenetic tree exists. See Figure 1 for an example. In this paper, we consider some natural extensions of the distance-based variant of the phylogenetic tree reconstruction problem to phylogenetic networks and present a new algorithm.

Problem Definitions: A *rooted phylogenetic network* for a set S of species is a rooted, connected, directed acyclic graph such that: (1) exactly one node (the *root*) has indegree 0 and all other nodes have indegree 1 or 2; (2) any node with indegree 2 (called a *hybrid node*) has outdegree 1 and all other nodes have outdegree 0 or 2; and (3) each node with outdegree 0 (a *leaf*) is labeled with a distinct species from S. A rooted phylogenetic network is called a *galled phylogenetic network*, or *galled network* for short[1], if all cycles in the underlying undirected graph (i.e., where edge orientations are ignored) are node-disjoint. For example, the phylogenetic network in Figure 1 and the network N_1 in Figure 2 are galled networks. From here on, we only consider phylogenetic networks that are *edge-weighted*, i.e., where each edge has a positive length. In analogy with the standard usage of the term "ultrametric" for phylogenetic trees, we say that a galled network is *ultrametric* if every directed path from the root to a leaf has the same length.

---

[1] Galled networks are also known in the literature as *topologies with independent recombination events* [17], *galled-trees* [8], *gt-networks* [14], and *level-1 phylogenetic networks* [13].

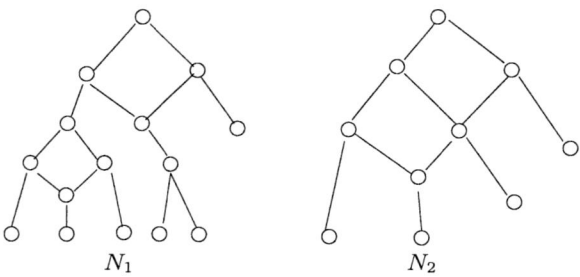

Fig. 2. N_1 is a galled network, while N_2 is not (The leaf labels are omitted for clarity.)

For any rooted phylogenetic network N, an *evolutionary path* between two leaves a and b is a simple path which goes up (i.e., moving in a child-to-parent direction) from a to a common ancestor u of a and b, and then down (i.e., moving in a parent-to-child direction) from u to b. Observe that even if N is galled and ultrametric, there can be more than one evolutionary path between a and b, and moreover, these paths may have different lengths (again, see Figure 1). However, in an ultrametric galled network, there can exist at most two different evolutionary path lengths between each pair of leaves.

A *distance matrix* for a set S of n species is a symmetric, non-negative $(n \times n)$-matrix M such that $M(a,a) = 0$ for every $a \in S$. Intuitively, for each $a, b \in S$, $M(a,b)$ contains the measured evolutionary distance between a and b. A rooted phylogenetic network N for S *satisfies* M if, for every $a, b \in S$, it holds that N contains an evolutionary path between a and b of length equal to $M(a,b)$. In this case, we also say that M is *satisfied by* N. We are now ready to define the problem which is the main focus of this paper.

Problem Statement: Given a distance matrix M for a set S of n species, return an ultrametric galled network for S satisfying M, if one exists; otherwise, return *fail*.

Motivation: The rationale behind the way we define the problem is as follows. There are a number of methods to estimate the evolutionary distance between two species. One common approach is to align the DNA sequences for some related genes from the species. The alignment score usually provides a reasonable estimation on the evolutionary distance between the species. However, if recombination events had occurred, there may exist more than one common ancestor (at different evolutionary distances) for a pair of species. Thus, depending on which common ancestor the selected genes were inherited from, the measured evolutionary distance may reflect only one of the possible evolutionary paths. Therefore, for any two species in the phylogenetic network, we only require one of their evolutionary paths to satisfy the matrix entry.

If there are no restrictions on the topological structure of the constructed phylogenetic network, it may not make sense from a biological point of view.

We therefore concentrate on galled networks, a very useful class of rooted phylogenetic networks which despite their simple structure are powerful enough to describe evolutionary history when the frequency of recombination events is moderate or when most of the recombination events have occurred recently [8]. See [8] for a discussion on the importance of galled networks. Also, the biological meaning of the ultrametric assumption is that the species have evolved according to a constant rate; see, e.g., [3, 5, 6, 18] and the references therein for justification of this assumption.

Finally, there may be more than one ultrametric galled network that satisfies an input matrix. From the biological point of view, it is more reasonable if we could find the simplest explanation that is consistent with the observed distances. So, although recombination events (corresponding to hybrid nodes) may occur, a more reasonable network is the one with the minimum number of hybrid nodes.

Our Contributions: Our main result in this paper is an exact $O(n^2 \log n)$-time algorithm to construct an ultrametric galled network (if one exists) that satisfies a given distance matrix M. When a solution exists, our algorithm always outputs one having as few hybrid nodes as possible. On the other hand, we prove that finding a largest possible submatrix M' of M such that there exists an ultrametric galled network that satisfies M' is an NP-hard problem. We also show that given an incomplete distance matrix (i.e., where some matrix entries are missing), it is NP-hard to determine whether there exists an ultrametric galled network which satisfies it.

Related Works: In the context of reconstructing a phylogenetic network from distance data, the most related work is the *Neighbor-Net* method, developed by Bryant and Moulton [2], which outputs a planar, unrooted phylogenetic network from a given distance matrix. Neighbor-Net is based on the well-known Neighbor-Joining method for trees [16]. Earlier proposed distance-based methods for reconstructing phylogenetic networks include [4] and others described in [15]. However, all of these approaches are heuristics-based and there is no guarantee that the output is a phylogenetic network that satisfies the given matrix exactly, even when a galled network exists. Also, Neighbor-Net runs in $O(n^3)$ time, which is slower than the method we present here.

Some other models of computation for reconstructing phylogenetic networks (i.e., assuming other types of input) are reviewed in [15]. Recently, in addition to distance-based methods, researchers have also studied *character-based* [7, 8, 17] and *supertree-based* [9, 10, 11, 12, 14] methods for inferring phylogenetic networks.

To reconstruct a phylogenetic *tree* with n species consistent with a given distance matrix (if one exists) can easily be done in $O(n^2)$ time (see [5, 6]). Note however, that when an exact solution does not exist, obtaining a tree that is as "close" as possible to the matrix has been shown to be NP-hard on several closeness metrics [3, 5, 18].

2 Preliminaries

Let N be a galled network. In the rest of this paper, we will use the following terminology. A node h in N is a *hybrid node* if the indegree of h is equal to 2. Let s be an ancestor of h such that there are two edge-disjoint paths from s to h. Then s is called *the split node of h*. In a galled network, each split node is a split node of exactly one hybrid node, and each hybrid node has exactly one split node (see Lemma 1 in [13]). The two paths from s to h are *the merge paths of h*, and they form a *galled loop* rooted at s. The galled loop rooted at s is *skew* if one of its two merge paths consists of a single edge from s to h; otherwise, it is *non-skew*. Nodes other than h and s on the merge paths of h are called *side nodes*, and a node is called a *tree node* if it is not on any galled loop. For any node u in N, the *subnetwork* rooted at u is the minimal subgraph of N including all nodes and directed edges reachable from u, and is denoted by N_u. Finally, N_u is a *side network* if the parent of u belongs to a merge path P in N but u itself is not on P.

In a galled network, the smallest possible galled loop is skew and consists of exactly three nodes (a split node, a hybrid node, and a side node). A simple induction can show that a galled network with n leaves contains at most $3n - 3$ internal nodes. This property is useful to our algorithm.

For any internal node u of an ultrametric galled network N, every directed path from u to a leaf under u has the same length. We call this length the *height* of u and denote it by $height(u)$. For any leaf a, $height(a) = 0$. Note that the length of any edge (a, b) can be calculated from $height(a)$ and $height(b)$. Thus, to find a network for M, we only need to determine the heights of all internal nodes and the parent-child relations between nodes.

3 Framework of the Algorithm

Given an $n \times n$ distance matrix M for a set S of n species, we first analyze some properties for the ultrametric galled network satisfying M. For simplicity, we say a *network* to refer to an ultrametric galled network. M is *satisfiable* if there exists a network satisfying it. For any $S' \subseteq S$, if a network N for S' satisfies the submatrix of M induced by the species in S', we say that N satisfies S'.

Consider any two species a and b in S. To satisfy M, the network contains an evolutionary path between a and b with length equal to $M(a, b)$. We notice that this path starts from a, goes up to a common ancestor of height $M(a, b)/2$, and then goes down to b. Let D_S be the maximum distance between two species in S as specified by M. If M is satisfiable, then there is a network satisfying M whose root has height $D_S/2$.

Also, we have the following observation about the internal nodes of N.

Observation 1. Assume that M can be satisfied by a network N. For any node u that is a tree node or a split node, let N_u be the subnetwork rooted at u, and let S_u be the set of species in N_u.

- For any two species $a, b \in S_u$, $M(a, b) = 2 \times height(v)$ for some internal node v in N_u, and hence $M(a, b) \leq 2 \times height(u)$.
- For any species $a \in S_u$ and $c \in S - S_u$, $M(a, c) > 2 \times height(u)$.

Observation 1 motivates us to consider the following definition.

Definition 1. For any set of species $S' \subseteq S$, S' is called a *cluster* if there exists a value x such that for any two species $a, b \in S'$, $M(a, b) \leq x$ and for any species $a \in S'$ and $c \in S - S'$, $M(a, c) > x$.

S itself is the biggest cluster. Note that clusters are nested, i.e., two clusters are always either disjoint or one is a subset of the other. Observation 1 states that every tree node and split node in N corresponds to a cluster. In fact, the reverse is also true.

Lemma 1. Assume that M can be satisfied by some network. Then there exists one such network N such that, for every cluster $S' \subseteq S$, N has a tree node or a split node u such that all species in S' are in the subnetwork N_u, and no species in $S - S'$ are in N_u.

To prove Lemma 1, we let N be any network satisfying M. If N does not satisfy Lemma 1, we can modify it to obtain a network satisfying Lemma 1. Details will be given in the full paper.

We call a network satisfying Lemma 1 a *well-structured* network, which has a very nice property as follows. Consider any $S' \subseteq S$ that is a cluster. Let $S_1, S_2, \ldots, S_t$ be all the maximal clusters which are proper subsets of S'. We call $S_1, S_2, \ldots, S_t$ the *side clusters* of S'. Note that $S' = S_1 \cup S_2 \cup \ldots \cup S_t$.

Lemma 2. Let S' be a cluster with side clusters $S_1, \ldots, S_t$. Let N be any well-structured network satisfying S' (w.r.t. the submatrix of M induced by S'). N consists of a root node u, with the networks satisfying $S_1, \ldots, S_t$ attached to u, or attached to a galled loop rooted at u.

Proof. As N is well-structured, for each side cluster S_i, there is a tree node or a split node v whose subnetwork contains exactly all species in S_i. We notice that on the path from v to the root u, there is no tree node or split node other than u or v (otherwise, let v' be that intermediate node; the species under the subnetwork rooted at v' form a cluster S'' and $S_i \subset S'' \subset S'$, meaning that S_i is not a side cluster of S'). Thus, v is directly attached to u or a galled loop rooted at u. It means that N is formed by attaching the networks for $S_1, \ldots, S_t$ to u, or to a galled loop rooted at u. □

The Algorithm

Lemma 2 states that we can construct a network for a cluster by connecting the networks for its side clusters. Thus, our algorithm takes a bottom-up approach, which continuously identifies subsets of S that are clusters, starting from smaller ones to bigger ones. It maintains an invariant that as soon as a cluster S' is found, a subnetwork satisfying S' is constructed. For the base case, a set containing only a single species is a cluster, and the corresponding network is a single leaf for this species. Since S is the biggest cluster, the algorithm will eventually find a network satisfying S.

To ease the finding of clusters, our algorithm constructs a graph G as follows. Initially, G has n isolated nodes, each representing a species in S. Edges which represent the distance among the species are added in rounds, where two nodes u, v will be connected by an edge of length $M(u, v)$. In the i-th round, all edges with the i-th shortest length are added. Suppose that after the i-th round, a connected component of G becomes a clique. Then the species inside this connected component form a cluster for which a network is built immediately. The algorithm is shown below. Details of Step 2c will be given in the next section.

Algorithm 1. GalledNet

Step 1. Sort the entries in M and let $m_1 < m_2 < ... < m_r$ be the distinct positive values in M. If $r > 3n - 3$, return failure.

Step 2. Build the networks while constructing a graph G. Initially, G contains n isolated nodes representing the n species. For $i = 1, 2, \ldots, r$,

  a. Add all edges of length m_i to G.
  b. Identify all connected components that become a new clique.
  c. For each new clique, let $S' \subseteq S$ be the corresponding cluster. Run the procedure ConnectingSideClusters (shown in the next section) which constructs a network satisfying S', if S' is satisfiable. This is done by creating a new root u, and attaching the networks for the side clusters of S' to u, or to a galled loop rooted at u.

Note that any galled network for n species can contain at most $3n-3$ internal nodes, and the length of any evolutionary path is $2 \times height(u)$ for some internal node u. Thus, if there are more than $3n - 3$ distinct positive values in M, no network can satisfy M.

We analyse the running time of GalledNet as follows. Step 1 takes $O(n^2 \log n)$ time. Step 2a takes $O(n^2)$ time over the whole algorithm. With some straightforward bookkeeping (which takes $O(1)$ time for each edge added), Step 2b can be done in $O(n)$ time in each iteration and $O(n^2)$ time in total. We will show in the next section that Step 2c, which calls ConnectingSideClusters, takes totally $O(n^2)$ time. Thus, the whole algorithm takes $O(n^2 \log n)$ time.

Theorem 1. *Algorithm GalledNet runs in $O(n^2 \log n)$ time.*

4 Attaching Side Clusters to a Galled Loop

This section explains how Step 2c of GalledNet is performed. Let S be a satisfiable cluster with side clusters $S_1, S_2, \cdots, S_t$. Suppose that we have constructed side networks for these side clusters. Below we overload S_i to also denote the corresponding side network. To build a network for S, we need to determine how these side clusters (more precisely, their side networks) are attached to a new root or to a galled loop, and compute the height of the new root and nodes on the loop.

We skip the simple case of $t = 2$ and we consider only the general case that $t \geq 3$, i.e., S has three or more side clusters. We need to build a galled loop to accommodate the corresponding side networks S_i's. We focus on the network N that satisfies S and we show that the structure of N can be determined from the relations between the side clusters. Recall that N has a galled loop at the top. Let S_h be the side cluster attached to the hybrid node. Let LEFT(S_h) be the group of side clusters attached to the side nodes on the left merge path. Define RIGHT(S_h) similarly. The following lemma tells how to identify the side clusters in the two groups. For simplicity, we say a species a is in LEFT(S_h) (resp. RIGHT(S_h)) if a belongs to some side cluster in LEFT(S_h) (resp. RIGHT(S_h)).

Lemma 3 (Partitioning the side clusters to the two merge paths). Let D_S be the maximum distance between two species in S. **(i)** For any two species a, b in LEFT(S_h), $M(a,b) < D_S$; similarly, for any two species a, b in RIGHT(S_h), $M(a,b) < D_S$; and **(ii)** for any species a in LEFT(S_h) and c in RIGHT(S_h), $M(a,c) = D_S$.

Assume that LEFT(S_h) contains ℓ side clusters and their side networks are attached to side nodes $v_1, v_2, \cdots, v_\ell$ on the left merge path of N, where v_i is the i-th node next the hybrid node. Let r be the root. Denote the side cluster (as well as the side network) attached to v_i as $S(v_i)$. That is, LEFT(S_h) = $\{S(v_i) \mid 1 \leq i \leq \ell\}$.

The following lemmas provide some structural characteristics of each side network $S(v_i)$, which allow us to identify each of them easily. For each side cluster S' of S, let $inter\_dist(S')$ denote the minimum distance $M(x,y)$ between a species x in S' and a species y in $S - S'$.

Lemma 4 (Identifying the order of side clusters). **(i)** $inter\_dist(S(v_1)) \leq inter\_dist(S(v_2))$, and $inter\_dist(S(v_2)) < inter\_dist(S(v_3)) < \cdots < inter\_dist(S(v_\ell))$; **(ii)** $height(v_i) = inter\_dist(S(v_i))/2$ for $i = 2, \cdots, \ell$.

Lemma 4(i) allows us to identify which side cluster in LEFT(S_h) is attached to each v_i, except when $inter\_dist(S(v_1)) = inter\_dist(S(v_2))$. In this case, we exploit the relationship with S_h to distinguish the side clusters attached to v_1 and v_2. Note that a species x in $S(v_2)$ and a species y in S_h are connected by two evolutionary paths, with the root r and v_2 as the highest node, respectively. Since N satisfies S, the distance of x and y (i.e., $M(x,y)$) must equal the length of either path, i.e., $2 \times height(r)$ or $2 \times height(v_2)$. The latter value is strictly less than $2 \times height(r) = D_S$.

Lemma 5 (Resolving ambiguity). (i) If $inter\_dist(S(v_1)) = inter\_dist(S(v_2))$, then $S(v_2)$, but not $S(v_1)$, contains a species x whose distance to some species y in S_h (i.e., $M(x,y)$) is less than D_S, and $height(v_1)$ can be any value in the range $(height(S_h), height(v_2))$. (ii) Otherwise, $height(v_1) = inter\_dist(S(v_1))/2$.

The above lemmas explain how the side clusters are attached to the merge paths, once the side cluster under the hybrid node is known. The following lemma shows that we can in fact find the side cluster attached to the hybrid node easily.

Lemma 6 (Finding S_h). (i) $inter\_dist(S_h) \leq inter\_dist(S_i)$ for any side cluster S_i of S; and (ii) there can be at most five side clusters S_i of S such that $inter\_dist(S_i) = inter\_dist(S_h)$.

Based on the above lemmas, we can construct a galled loop to connect the side clusters for S, as follows. By Lemma 6, there are at most five candidates for the side cluster attached to the hybrid node. We try to build the network using each of the candidate according to Lemma 3, 4 and 5. We verify each network constructed and return the one that satisfies S. Details of the algorithm are shown in Algorithm 2. It builds a network for S if and only if S is satisfiable.

Algorithm 2. ConnectingSideClusters $(S, S_1, S_2, \ldots, S_t)$, $t \geq 3$.

Find $inter\_dist(S_i)$ for each side cluster S_i and sort the side clusters according to the $inter\_dist$ value. If there are more than five side clusters having the minimum $inter\_dist$ value, return failure. Otherwise, for each side cluster S_h with the minimum $inter\_dist$ value, try to build a network which attaches S_h to the hybrid node, as follows.

a. Divide the remaining side clusters into two groups LEFT(S_h) and RIGHT(S_h) that satisfy Lemma 3.
b. Sort the side clusters in LEFT(S_h) according to the $inter\_dist$ value and attach the side clusters to the left merge path according to Lemma 4 and 5. Repeat it for the side clusters in RIGHT(S_h).
c. Let h_l and h_r be the height of the lowest side node on the left and right merge path, respectively. Set $height(h)$ to any value in $(height(S_h), \min\{h_l, h_r\})$.
d. Verify that for any two species $a, b \in S$, there is an evolutionary path between a and b with length equal to $M(a,b)$. Return the network if it is true.

Runtime of ConnectingSideClusters. It is straightforward to implement the procedure ConnectingSideClusters in $O(t_S \log t_S + \#_S)$ time, where t_S is the number of side clusters in S and $\#_S$ is the number of species pair (x, y) where x and y are species belonging to two different side clusters of S.

Over the whole execution of GalledNet, $\sum t_S \leq 2n - 1$ and $\sum \#_S \leq \frac{n(n-1)}{2}$. Thus, the total runtime for ConnectingSideClusters is $\sum O(t_S \log t_S + \#_S) = O(n \log n + \frac{n(n-1)}{2}) = O(n^2)$.

The Minimality of Number of Hybrid Nodes. Given a satisfiable matrix M, the network produced by GalledNet has the minimum number of hybrid nodes among all networks satisfying M. Proofs will be given in the full paper.

5 NP-Hardness Results

In the following, we say that a distance matrix M *admits* an ultrametric galled network if there exists such a network which satisfies M. We first prove that finding a maximum submatrix M' of a given distance matrix M such that M' admits an ultrametric galled network is an NP-hard problem. Our proof consists of a reduction from the NP-hard independent set problem.

The Independent Set Problem

Instance: An undirected graph $G = (V, E)$ and a positive integer $I \leq |V|$.
Question: Is there a subset V' of V with $|V'| = I$ such that V' is an independent set, i.e., such that no two vertices in V' are joined by an edge in E?

The Maximum Submatrix Admitting an Ultrametric Galled Network Problem, Decision Problem Version (MSGN-d)

Instance: A set S, a distance matrix M for S, and a positive integer $K \leq |S|$.
Question: Is there a subset S' of S with $|S'| = K$ such that M restricted to S' admits an ultrametric galled network?

The following shows the reduction of the independent set problem to MSGN-d. Let (G, I) be any given instance of the independent set problem. For convenience, write $n = |V|$ and $V = \{v_1, v_2, \ldots, v_n\}$. Construct an instance (S, M, K) of MSGN-d as follows. Let $S = V \cup P \cup Q$, where $P = \{p_1, p_2, \ldots, p_n\}$ and $Q = \{q_1, q_2, \ldots, q_n\}$ are two disjoint sets of elements not in V, and set $K = I + 2n$. Next, let M be a distance matrix for S satisfying, for every $i, j \in \{1, 2, \ldots, n\}$: $M(p_i, p_j) = \max\{i, j\}$; $M(q_i, q_j) = \max\{i, j\}$; $M(p_i, q_j) = n + 1$; $M(v_i, v_j) = \max\{i, j\}$ if the edge $\{i, j\}$ does not belong to E and $M(v_i, v_j) = n + 1$ if the edge $\{i, j\}$ belongs to E; $M(v_i, p_j) = n + 1$; and $M(v_i, q_j) = n + 1$.

Lemma 7. *M has a submatrix of size $K \times K$ which admits an ultrametric galled network if and only if G has an independent set of size I.*

Theorem 2. *MSGN is NP-hard.*

Next, we prove that it is NP-hard to determine whether a given incomplete distance matrix admits an ultrametric galled network. The proof consists of a reduction from the NP-hard 3-coloring problem.

The 3-Coloring Problem

Instance: An connected undirected graph $G = (V, E)$.

Question: Can G be 3-colored, i.e., can V be partitioned into three disjoint subsets in such a way that E contains no edge between two vertices in the same subset?

The Incomplete Distance Matrix Admitting an Ultrametric Galled Network Problem (IDGN)

Instance: A set S and an incomplete distance matrix M (i.e., where some entries are missing) for S.

Question: Is there an ultrametric galled network which satisfies all of the nonempty entries in M?

Let G be any given instance of 3-coloring with at least two vertices. Construct an instance (S, M) of IDGN by setting $S = V$ and defining the ($|S| \times |S|$)-matrix M as follows: for every $i \in V$, let $M(i,i) = 0$; and for every edge $\{i,j\} \in E$, let $M(i,j) = M(j,i) = 1$. For every pair of vertices i, j in V such that $\{i,j\} \notin E$, leave the matrix entries $M(i,j)$ and $M(j,i)$ empty.

Lemma 8. *G is 3-colorable if and only if there exists an ultrametric galled network which satisfies all of the nonempty entries in M.*

Theorem 3. *IDGN is NP-hard.*

Acknowledgement. We thank Wing-Kin Sung for introducing this problem to us and for his useful comments.

References

[1] D. Bryant. *Building Trees, Hunting for Trees, and Comparing Trees: Theory and Methods in Phylogenetic Analysis*. PhD thesis, University of Canterbury, Christchurch, New Zealand, 1997.

[2] D. Bryant and V. Moulton. Neighbor-Net: An agglomerative method for the construction of phylogenetic networks. *Molecular Biology and Evolution*, 21(2):255–265, 2004.

[3] H.-F. Chen and M.-S. Chang. An efficient exact algorithm for the minimum ultrametric tree problem. In *Proceedings of the 15th International Symposium on Algorithms and Computation* (ISAAC 2004), pages 282–293, 2004.

[4] E. Diday and P. Bertrand. An extension of hierarchical clustering: the pyramidal representation. *Pattern Recognition in Practice II*, pages 411–424, 1986.
[5] M. Farach, S. Kannan, and T. Warnow. A robust model for finding optimal evolutionary trees. *Algorithmica*, 13(1/2):155–179, 1995.
[6] D. Gusfield. *Algorithms on Strings, Trees, and Sequences.* Cambridge University Press, New York, 1997.
[7] D. Gusfield and V. Bansal. A fundamental decomposition theory for phylogenetic networks and incompatible characters. In *Proceedings of the 9th Annual International Conference on Research in Computational Molecular Biology* (RECOMB 2005), pages 217–232, 2005.
[8] D. Gusfield, S. Eddhu, and C. Langley. Optimal, efficient reconstruction of phylogenetic networks with constrained recombination. *Journal of Bioinformatics and Computational Biology*, 2(1):173–213, 2004.
[9] Y.-J. He, T. N. D. Huynh, J. Jansson, and W.-K. Sung. Inferring phylogenetic relationships avoiding forbidden rooted triplets. In *Proceedings of the 3rd Asia-Pacific Bioinformatics Conference* (APBC 2005), pages 339–348, 2005.
[10] D. H. Huson, T. Dezulian, T. Klöpper, and M. Steel. Phylogenetic super-networks from partial trees. In *Proceedings of the 4th Workshop on Algorithms in Bioinformatics* (WABI 2004), pages 388–399, 2004.
[11] T. N. D. Huynh, J. Jansson, N. B. Nguyen, and W.-K. Sung. Constructing a smallest refining galled phylogenetic network. In *Proceedings of the 9th Annual International Conference on Research in Computational Molecular Biology* (RECOMB 2005), pages 265–280, 2005.
[12] J. Jansson, N. B. Nguyen, and W.-K. Sung. Algorithms for combining rooted triplets into a galled phylogenetic network. In *Proc. of the 16th Annual ACM-SIAM Symposium on Discrete Algorithms* (SODA 2005), pages 349–358, 2005.
[13] J. Jansson and W.-K. Sung. The maximum agreement of two nested phylogenetic networks. In *Proceedings of the 15th International Symposium on Algorithms and Computation* (ISAAC 2004), pages 581–593, 2004.
[14] L. Nakhleh, T. Warnow, and C. R. Linder. Reconstructing reticulate evolution in species – theory and practice. In *Proceedings of the 8th Annual International Conference on Research in Computational Molecular Biology* (RECOMB 2004), pages 337–346, 2004.
[15] D. Posada and K. A. Crandall. Intraspecific gene genealogies: trees grafting into networks. *TRENDS in Ecology & Evolution*, 16(1):37–45, 2001.
[16] N. Saitou and M. Nei. The neighbor-joining method: A new method for reconstructing phylogenetic trees. *Molecular Biology and Evolution*, 4(4):406–425, 1987.
[17] L. Wang, K. Zhang, and L. Zhang. Perfect phylogenetic networks with recombination. *Journal of Computational Biology*, 8(1):69–78, 2001.
[18] B. Y. Wu, K.-M. Chao, and C. Y. Tang. Approximation and exact algorithms for constructing minimum ultrametric trees from distance matrices. *Journal of Combinatorial Optimization*, 3(2–3):199–211, 1999.

New Resource Augmentation Analysis of the Total Stretch of SRPT and SJF in Multiprocessor Scheduling

Wun-Tat Chan[1], Tak-Wah Lam[1],
Kin-Shing Liu[1], and Prudence W.H. Wong[2]

[1] Department of Computer Science, University of Hong Kong
{wtchan, twlam, ksliu}@cs.hku.hk
[2] Department of Computer Science, University of Liverpool
pwong@csc.liv.ac.uk

Abstract. This paper studies online job scheduling on multiprocessors and, in particular, investigates the algorithms SRPT and SJF for minimizing total stretch, where the stretch of a job is its flow time (response time) divided by its processing time. SRPT is perhaps the most well-studied algorithm for minimizing total flow time or stretch. This paper gives the first resource augmentation analysis of the total stretch of SRPT, showing that it is indeed $O(1)$-speed 1-competitive. This paper also gives a simple lower bound result that SRPT is not s-speed 1-competitive for any $s < 1.5$.

This paper also makes contribution to the analysis of SJF. Extending the work of [4], we are able to show that SJF is $O(1)$-speed 1-competitive for minimizing total stretch. More interestingly, we find that the competitiveness of SJF can be reduced arbitrarily by increasing the processor speed (precisely, SJF is $O(s)$-speed $(1/s)$-competitive for any $s \geq 1$). We conjecture that SRPT also admits a similar result.

1 Introduction

We study the problem of online job scheduling for minimizing total stretch. There is a pool of $m \geq 1$ processors. Jobs arrive at arbitrary times, and their processing times are known when they arrive. Jobs are sequential in nature and can be scheduled on at most one processor at a time. Preemption is allowed. The flow time (or response time) of a job is the amount of time the job spent before it is completed, and the stretch of the job is the ratio of its flow time to its required processing time (see the survey by Pruhs et al. [21]). We are interested in scheduling algorithms that minimize the total stretch (or equivalently the average stretch) of the jobs. Roughly speaking, if the average stretch is λ, a job on average takes λ times the required processing time to complete, i.e., it appears to be processed by a $\frac{1}{\lambda}$-speed processor[1]. Stretch is a useful indicator of system performance, and has received a lot of attention in recent years (see, e.g.,

---
[1] A speed-s processor, where $s > 0$, can process s units of work in one unit of time.

[3, 5–7, 11, 12, 19]). Competitive analysis is often used to measure the performance of an online algorithm with respect to total stretch (or any other objective function). An online scheduling algorithm A is said to be c-competitive for any number $c > 0$ if for any input job sequence, the total stretch of the jobs as defined by A is at most c times that of the optimal offline algorithm.

SRPT (Shortest Remaining Processing Time First) is a popular online algorithm when the concern is the total flow time or total stretch. With respect to total flow time, SRPT is 1-competitive for a single processor [2]. But for multiprocessors ($m \geq 2$), Leonardi and Raz [17] have shown that SRPT is $\Theta(\min(\log P, \log n/m))$-competitive, where n is the number of jobs and P is the ratio of the maximum possible processing time to the minimum possible processing time. To obtain better performance guarantee for multiprocessor scheduling, Phillips et al. [20] applied resource augmentation analysis (which was pioneered by Kalyanasundaram and Pruhs [16]) to SRPT, showing that SRPT is 1-competitive when using processors that are two times faster, or in short, 2-speed 1-competitive. This result means that a modest increase in the processor speed of the online scheduler can compensate its lack of future information. Recently, McCulloguh and Torng [18] further showed that SRPT is s-speed ($\frac{1}{s}$)-competitive for any $s \geq 2 - \frac{1}{m}$. In a wider context, resource augmentation analysis has been found useful in a number of difficult scheduling problems (see, e.g., [8–10, 13–16, 20]).

Muthukrishan et al. [19] were the first to study total stretch. They showed that SRPT is 2-competitive on a single processor and 14-competitive on multiprocessors, and no online algorithm can be 1-competitive. Chekuri et al. [12] proposed a different algorithm (called SG) that is 9.81-competitive on multiprocessors. Existing resource augmentation results are actually based on algorithms like SJF (Shortest Job First), which assigns fixed priorities to jobs independently of the schedule. On a single processor, the work of Phillips (on weighted flow time) [20] implies that an algorithm called Preemptively-Schedule-by-Halves as well as SJF are 2-speed 1-competitive for minimizing total stretch. For multiprocessors, there are two algorithms known to be $O(1)$-speed $O(1)$-competitive (namely, SJF is $(2+2\epsilon)$-speed $(1+\frac{1}{\epsilon})$-competitive [4], and IMD [1] is $(1+\epsilon)$-speed $O(1+\frac{1}{\epsilon})$-competitive [11]). Though SRPT is believed to perform well, it is generally agreed that SRPT is more difficult for resource augmentation analysis (see, e.g., [21]). In this paper we show that SRPT is indeed 2-speed 1-competitive for minimizing total stretch on a single processor. A more elaborate analysis further reveals that SRPT is 5-speed 1-competitive on multiprocessors. This is the first result on exploiting extra speed to achieve 1-competitiveness. Table 1 gives a summary of the performance of SRPT and SJF. We also derive a simple lower bound that for any $s < 1.5$, SRPT is not s-speed 1-competitive.

Technically speaking, our analysis of SRPT is based on an observation that the optimal offline algorithm, at any time, has no more finished jobs than SRPT does, and more interestingly, each finished job of the optimal offline algorithm can be mapped to a unique finished job of SRPT with same or smaller processing time.

Table 1. SRPT and SJF using faster processors can be 1-competitive (or even better) for minimizing total stretch. Results given in this paper are marked with asterisks.

	Single processor	Multiprocessors
SRPT	2-competitive [19] 2-speed 1-competitive *	14-competitive [19] 5-speed 1-competitive *
SJF	$(1+\epsilon)$-speed $(1+\frac{1}{\epsilon})$-competitive [4] 2-speed 1-competitive [20,4]	$(2+2\epsilon)$-speed $(1+\frac{1}{\epsilon})$-competitive [4] $(24s)$-speed $(\frac{1}{s})$-competitive, for $s \geq 1$ *

This paper also makes contribution to the analysis of SJF. It has been known that based on the result on weighted flow time, SJF is 2-speed 1-competitive for minimizing total stretch on a single processor, and $(2 + 2\epsilon)$-speed $(1 + \frac{1}{\epsilon})$-competitive on multiprocessors [4, 20]. We improve the analysis of SJF on multiprocessors to show that SJF is indeed 24-speed 1-competitive and, in general, $(24s)$-speed $(\frac{1}{s})$-competitive for any $s \geq 1$, for minimizing total stretch. We conjecture that SRPT also admits a similar result.

Before moving on to the analysis of SRPT and SJF, we give a definition of these two algorithms. Suppose there are $m \geq 1$ processors. At any time, if there are at most m unfinished jobs, SRPT and SJF both schedule each job to a distinct processor; otherwise, SJF gives priority to the m jobs with the shortest processing times, and SRPT schedules the m jobs with the shortest remaining processing times. A tie is simply broken by job ID.

Formally speaking, we say that SRPT (or SJF) is s-speed c-competitive if for any job sequence, SRPT (or SJF) using m s-speed processors incurs a total stretch at most c times of that of the optimal offline algorithm using m unit-speed processors.

Organization of the Paper: Section 2 gives three useful properties of an SRPT schedule regardless of processor speed. Section 3 presents a resource augmentation analysis of SRPT on multiprocessors, revealing that SRPT is 5-speed 1-competitive for minimizing total stretch. Section 4 shows a lower bound of SRPT. Section 5 analyzes the performance of SJF. Section 6 discusses some future work. The proof of the result that SRPT is 2-speed 1-competitive on a single processor will be given in the full paper.

2 Preliminaries

In this section we give some basic definitions and three useful properties of an SRPT schedule regardless of processor speed. Let I be an input sequence of jobs to be scheduled on $m \geq 2$ processors. For any job $J \in I$, let $p(J)$ and $r(J)$ denote the processing time and release time of J, respectively. Let $x > 0$ be any number. A job J is said to be x-large if $p(J) > x$, and x-small if $p(J) \leq x$. For any set K of jobs, $p(K)$ is defined to be $\sum_{J \in K} p(J)$. Consider a schedule S for I on m processors. We assume that jobs can be preempted and later resumed at the point of preemption. The following notations are concerned with a particular time t in the schedule S.

- Let $w_t^S(J)$ and $rw_t^S(J)$ denote the processed work and the remaining work, respectively, of a job J in S at time t. Note that $w_t^S(J) + rw_t^S(J) = p(J)$. A job J is said to be *partially processed* if $0 < rw_t^S(J) < p(J)$.
- Let Q_t^S denote the set of jobs released at or before time t and unfinished at time t, and let $Q_t^S(x) \subseteq Q_t^S$ denote the set of x-small jobs in Q_t^S.
- Let $\mathrm{Shrink}_t^S(x)$ denote the set of x-large jobs J in Q_t^S such that $rw_t^S(J) \leq x$ (note that any job J in $\mathrm{Shrink}_t^S(x)$ is partially processed because $p(J) > x$).
- Let F_t^S denote the set of jobs finished at or before time t, and let $F_t^S(x) \subseteq F_t^S$ denote the set of x-small jobs in F_t^S.

When the context is clear, we will omit the superscript S in the above notations, which become $w_t(J)$, $rw_t(J)$, Q_t, $Q_t(x)$, F_t, and $F_t(x)$. Using the above definitions, the stretch of a schedule S can be expressed as $\int_0^\infty \sum_{J \in Q_t} \frac{1}{p(J)} dt$.

To ease our discussion, we use c-speed SRPT, for any $c \geq 1$, to denote an online scheduler running SRPT on m c-speed processors, and we let OPT denote an optimal schedule for I on m unit-speed processors. To compare the stretch of c-speed SRPT and OPT, we focus on analyzing the corresponding Q_t and F_t. Hereafter, we use the notations Q_t^*, $Q_t^*(x)$, F_t^* and $F_t^*(x)$ to denote the above concepts for OPT.

Before we move on to the analysis of 5-speed SRPT, we show in the rest of this section three useful properties of an SRPT schedule regardless of processor speed. Precisely, let S denote the schedule defined by c-speed SRPT for any $c \geq 1$. The properties are concerned with three categories of jobs defined at any time in S as follows.

- There are at most m unfinished x-large jobs with remaining work at most x (Lemma 1).
- While there is an x-small job J waiting (i.e., not being processed by an processor), jobs that can be scheduled only include x-small jobs or jobs in $\mathrm{Shrink}_{r(J)}(x)$ (Lemma 2).
- The accumulated work on all unfinished x-small jobs is less than mx (Lemma 3).

Lemma 1. *At any time $t \geq 0$ and for any $x > 0$, $|\mathrm{Shrink}_t(x)| \leq m$.*

Proof. We prove the lemma by contradiction. Suppose that $\mathrm{Shrink}_t(x) = \{J_1, J_2, \cdots, J_{m'}\}$, for some $m' > m$. By definition, $rw_t(J_i) \leq x < p(J_i)$. Let y be a number such that $x < y < \min_{J \in \mathrm{Shrink}_t(x)} \{p(J)\}$. For each job $J_i \in \mathrm{Shrink}_t(x)$, let $t_i < t$ be the latest time such that $rw_{t_i}(J_i) = y$. Note that J_i must be processed by some processor at time t_i and its remaining work is strictly less than y immediately after t_i. Without loss of generality, we assume that $t_1 \leq \cdots \leq t_{m'} < t$.

Suppose that $t_{m'-k} < t_{m'-k+1} = \cdots = t_{m'}$ for some integer $k \in [1, m]$. At $t_{m'}$, the jobs $J_{m'-k+1}, \cdots, J_{m'}$ are each being processed by a processor. And, for $1 \leq i \leq m' - k$, $rw_{t_{m'}}(J_i) < y = rw_{t_{m'}}(J_{m'})$, which implies that all these J_i are also being processed at time $t_{m'}$ (because SRPT always processes jobs with smallest remaining work first). This leads to a contradiction that $m' > m$ jobs are being processed at the same time. □

Lemma 2. *Let J be any x-small job. Whenever J is waiting, S can only schedule other x-small jobs or jobs in $\text{Shrink}_{r(J)}(x)$.*

Proof. Any x-large job $J' \notin \text{Shrink}_{r(J)}(x)$ has $rw_{r(J)}(J') > x \geq p(J)$. Starting from $r(J)$, whenever J' is being processed, J is also being processed. Therefore, the remaining work of J is always less than that of J'. Thus, whenever J is waiting, J' also needs to wait as J' has more remaining work. □

Lemma 3. *At any time $t \geq 0$ and for any $x > 0$, $w_t(Q_t(x)) < mx$, where $w_t(Q_t(x)) = \sum_{J \in Q_t(x)} w_t(J)$.*

Intuitively, at any time t, there may be many unfinished x-small jobs, but the above lemma states that their total processed work up to time t is less than mx. To prove the lemma, we let L be the set of jobs in $Q_t(x)$ that are partially processed. Notice that $w_t(Q_t(x)) = w_t(L)$. Assume that $L = \{J_1, J_2, \cdots, J_{|L|}\}$ where $p(J_1) \geq p(J_2) \geq \cdots \geq p(J_{|L|})$.

Below we show that jobs in L can be partitioned into m disjoint sets $Y_1, Y_2, \cdots, Y_m$ such that for $1 \leq k \leq m$, $w_t(Y_k) < x$. Then $w_t(Q_t(x)) = \sum_{k=1}^{m} w_t(Y_k) < mx$ and the upper bound follows. We construct the partition by adding the jobs in L one by one into the m sets. Denote by $last(Y_k)$ the last job added to Y_k.

- Initially, set $Y_1 = \{J_1\}, Y_2 = \{J_2\}, \cdots, Y_m = \{J_m\}$.
- For $i = m + 1$ to $|L|$, add J_i to the set Y_k with the largest $rw_t(last(Y_k))$ value.

The following lemma gives a property on $rw_t(last(Y_k))$.

Lemma 4. *Whenever a job J_i is added to a set Y_k, $p(J_i) \leq rw_t(last(Y_k))$.*

Proof. Suppose on the contrary that $rw_t(last(Y_k)) < p(J_i)$. Notice that for all $1 \leq z \leq m$, we have (1) $rw_t(last(Y_z)) \leq rw_t(last(Y_k))$ because J_i is added to Y_k; and (2) $p(J_i) \leq p(last(Y_z))$. Therefore, all the $m+1$ jobs including J_i and $last(Y_1), \cdots, last(Y_m)$ have processing time at least $p(J_i)$ but remaining work at time t less than $p(J_i)$. So letting $x = \max\{rw_t(J_i), \max_{1 \leq h \leq m} rw_t(last(Y_h))\}$, we have $x < p(J_i)$ and $|\text{Shrink}_t(x)| \geq m + 1 > m$, which is a contradiction to Lemma 1. □

Consider any Y_k. Suppose $J'_1 (= J_k), J'_2, \cdots, J'_h$, for some $h \geq 1$, are the jobs added to Y_k (in that order). Then by Lemma 4, $w_t(Y_k) = \sum_{1 \leq i \leq h} w_t(J'_i) = \sum_{1 \leq i < h}(p(J'_i) - rw_t(J'_i)) + w_t(J'_h) \leq \sum_{1 \leq i < h}(p(J'_i) - p(J'_{i+1})) + w_t(J'_h) = p(J'_1) - p(J'_h) + w_t(J'_h) < p(J'_1) \leq x$. The second last inequality holds because J'_h is not finished at time t. Therefore, we have $w_t(Q_t(x)) = \sum_{1 \leq k \leq m} w_t(Y_k) < mx$ and Lemma 3 follows.

3 Resource Augmentation Analysis of SRPT

In this section we show that SRPT is 5-speed 1-competitive for minimizing total stretch on multiprocessors. We analyze the schedule of 5-speed SRPT, denoted by S_5 below, against OPT on a given input sequence, and in particular, we show in Lemma 6 that at any time t, S_5 outperforms OPT on finished jobs; precisely, for any $x > 0$, $p(F_t(x)) \geq p(F_t^*(x))$. Then we show in Lemma 7 (Section 3.2) that there is a one-to-one mapping from F_t^* to F_t such that each job $J^* \in F_t^*$ can be mapped to a unique job $J \in F_t$ with $p(J^*) \geq p(J)$. In other words, at any time t, we have $\sum_{J \in F_t} 1/p(J) \geq \sum_{J^* \in F_t^*} 1/p(J^*)$, implying that $\sum_{J \in Q_t} 1/p(J) \leq \sum_{J^* \in Q_t^*} 1/p(J^*)$. It is then easy to see that the total stretch of S_5 is no more than that of OPT (Theorem 1).

3.1 Outperforming the Optimal Schedule on Finished Jobs

In this section we show that 5-speed SRPT outperforms OPT on finished jobs. Consider the schedule S_5 defined by 5-speed SRPT. For any $x > 0$, a time interval is said to be a $\lambda(x)$-*interval* if at any time within the interval, there is an x-small job waiting.

Lemma 5. *Let J be a job with $p(J) = x$. Suppose that S_5 does not complete J at time $t \geq r(J) + x$.*

- *Then during $[r(J), t]$, S_5 schedules at least $3m(t - r(J))$ units of work on x-small jobs; and $p(F_t(x)) \geq p(F_{r(J)}(x)) + 2m(t - r(J))$.*
- *Furthermore, if $r(J)$ is inside a $\lambda(x)$-interval starting from $t' \leq r(J)$, then during $[t', t]$, the work scheduled by S_5 on x-small jobs is at least $3m(t - t')$; and $p(F_t(x)) \geq p(F_{t'}(x)) + 2m(t - t')$.*

Proof. J is not finished in S_5 at time $t \geq r(J) + x$. During $[r(J), t]$, J incurs a waiting time longer than $t - r(J) - \frac{x}{5} \geq \frac{4}{5}(t - r(J))$, and S_5 must process at least $5m \cdot \frac{4}{5}(t - r(J)) = 4m(t - r(J))$ units of work. By Lemma 2, while J is waiting, S_5 can only process other x-small jobs or jobs in $\text{Shrink}_{r(J)}(x)$. By Lemma 1, $|\text{Shrink}_{r(J)}(x)| \leq m$. Each job in $\text{Shrink}_{r(J)}(x)$, by definition, has remaining work at most x at time $r(J)$. Thus, during $[r(J), t]$, the work scheduled by S_5 on x-small jobs is at least $4m(t - r(J)) - mx \geq 3m(t - r(J))$.

Since, by Lemma 3, $w_t(Q_t(x)) < mx$, during $[r(J), t]$, the work scheduled by S_5 on x-small jobs that are completed by time t is at least $3m(t - r(J)) - mx \geq 2m(t - r(J))$. Consider jobs in $F_t(x)$ but not in $F_{r(J)}(x)$. They are all x-small jobs scheduled by S_5 to completion during $[r(J), t]$, and their total processing time is at least the work scheduled by S_5 on them during $[r(J), t]$, i.e., at least $2m(t - r(J))$. Thus, $p(F_t(x)) \geq p(F_{r(J)}(x)) + 2m(t - r(J))$.

Furthermore, if $r(J)$ is inside a $\lambda(x)$-interval starting from $t' \leq r(J)$, we have $\text{Shrink}_{r(J)}(x) \subseteq \text{Shrink}_{t'}(x)$. During $[t', r(J)]$ and the waiting time of J, S_5 can only process x-small jobs or jobs in $\text{Shrink}_{t'}(x)$. Using the same argument above, we can conclude that during $[t', t]$, the work scheduled by S_5 on x-small jobs is at least $3m(t - t')$, and $p(F_t(x)) \geq p(F_{t'}(x)) + 2m(t - t')$. □

Lemma 6. *At any time $t \geq 0$ and for any $x > 0$, $p(F_t(x)) \geq p(F_t^*(x))$.*

Proof. Let $\Psi_u(x)$ denote the set of x-small jobs released before time u. We will use a property of $F_t^*(x)$ that for any time $u < t$, $p(F_t^*(x)) \leq p(\Psi_u(x)) + m(t-u)$.

We prove the lemma by contradiction. Suppose that $t \geq 0$ is the earliest time such that at time t, there is a smallest $x > 0$ such that $p(F_t(x)) < p(F_t^*(x))$. Then there must be an x-small job J with $p(J) = x$ such that at time t, J finishes in OPT but J is unfinished in S_5. Note that $t - r(J) \geq x$.

We consider two cases. First, if $r(J)$ is not within a $\lambda(x)$-interval in S_5, then at time $r(J)$ in S_5, no x-small jobs are waiting, and there are at most m unfinished x-small jobs. Thus, $p(\Psi_{r(J)}(x)) \leq p(F_{r(J)}(x)) + mx$, and $p(F_t^*(x)) \leq p(F_{r(J)}(x)) + mx + m(t - r(J)) \leq p(F_{r(J)}(x)) + 2m(t - r(J))$. By Lemma 5, $p(F_t(x)) \geq p(F_{r(J)}(x)) + 2m(t - r(J))$, and thus $p(F_t(x)) \geq p(F_t^*(x))$. A contradiction occurs.

Second, if $r(J)$ is within a $\lambda(x)$-interval starting from time $t' \leq r(J)$ in S_5, then we can upper bound $p(\Psi_{t'}(x))$ by $p(F_{t'}(x)) + mx$. Then $p(F_t^*(x)) \leq p(F_{t'}(x)) + 2m(t - t')$. Using Lemma 5, we can again derive the contradiction that $p(F_t(x)) \geq p(F_t^*(x))$. □

3.2 5-Speed SRPT is 1-Competitive

Based on Lemma 6, we can prove that 5-speed SRPT is 1-competitive. First, we show that at any time t, there is a one-to-one mapping between F_t^* and F_t.

Lemma 7. *Consider any time $t \geq 0$. Assume that $p(F_t(x)) \geq p(F_t^*(x))$ for any $x > 0$. Then there is a one-to-one mapping from F_t^* to F_t such that each job $J^* \in F_t^*$ is mapped to a unique job $J \in F_t$ with $p(J^*) \geq p(J)$.*

Proof. Suppose that the processing times of the jobs in F_t^* have d distinct values, denoted by $x_1 < x_2 < \cdots < x_d$. We construct a mapping from F_t^* to F_t incrementally, each time we consider jobs in F_t^* with the same processing time.

Consider all jobs in F_t^* that have the smallest processing time (i.e., equal to x_1). Given that $p(F_t(x_1)) \geq p(F_t^*(x_1))$, $F_t(x_1)$ must contain at least as many jobs as $F_t^*(x_1)$. Thus, each job in $F_t^*(x_1)$ can be mapped to a unique job in $F_t(x_1)$ with processing time at most x_1.

Assume that for some $k \geq 1$, we have constructed a mapping from $F_t^*(x_k)$ to F_t as required by Lemma 7. Next, we consider jobs in F_t^* with processing time x_{k+1}. Let $Y \subset F_t$ be the set of jobs in F_t to which jobs in $F_t^*(x_k)$ are mapped. As each job in $F_t^*(x_k)$ is mapped to a job with the same or shorter processing time, we have $p(F_t^*(x_k)) \geq p(Y)$. The number of jobs in F_t^* with processing time x_{k+1} is exactly $(p(F_t^*(x_{k+1})) - p(F_t^*(x_k)))/x_{k+1}$. The number of unmapped jobs in $F_t(x_{k+1})$ is at least $(p(F_t(x_{k+1})) - p(Y))/x_{k+1}$, which is at least $(p(F_t^*(x_{k+1})) - p(F_t^*(x_k)))/x_{k+1}$. Thus, each job in F_t^* with processing time x_{k+1} can be mapped to a unique job in F_t with the same or shorter processing time. □

We are now ready to show our main theorem.

Theorem 1. *SRPT is 5-speed 1-competitive for minimizing total stretch.*

Proof. For any time $t \geq 0$, let Φ_t denote the set of jobs released at or before time t. (Note that Φ_t equals the union of the set of jobs released at time t and Ψ_t, the set of jobs released before time t.) The set of unfinished jobs in S_5 is $Q_t = \Phi_t - F_t$; and the set of unfinished jobs in OPT is $Q_t^* = \Phi_t - F_t^*$.

$$\text{Total stretch of 5-speed SRPT} = \int \sum_{J \in Q_t} \frac{1}{p(J)} dt = \int \sum_{J \in \Phi_t} \frac{1}{p(J)} dt - \int \sum_{J \in F_t} \frac{1}{p(J)} dt$$

By Lemma 7, each job J^* in F_t^* is mapped to a unique job in F_t with processing time at most $p(J^*)$, so we have

$$\int \sum_{J \in F_t} \frac{1}{p(J)} dt \geq \int \sum_{J^* \in F_t^*} \frac{1}{p(J^*)} dt$$

Thus,

$$\text{total stretch of 5-speed SRPT} \leq \int \sum_{J \in \Phi_t} \frac{1}{p(J)} dt - \int \sum_{J^* \in F_t^*} \frac{1}{p(J^*)} dt$$
$$= \int \sum_{J^* \in Q_t^*} \frac{1}{p(J^*)} dt,$$

which is equal to the stretch of OPT. Hence, 5-speed SRPT is 1-competitive. □

3.3 Remark

The analysis given in Sections 3.1 and 3.2 can be easily generalized to show that for any $m' \geq 1$, SRPT using m' processors that are $(m/m' + 4)$-speed gives a total stretch no more than an optimal schedule with m unit-speed processors. In particular, we can generalize Lemma 5 to show that if SRPT (using m' $(m/m' + 4)$-speed processors) does not complete a job J with $p(J) = x$ at time $t \geq r(J) + x$, then $p(F_t(x)) \geq p(F_{r(J)}(x)) + (m + m')(t - r(J))$. And all the other lemmas remain true.

4 Speed Requirement for SRPT to Be 1-Competitive

In this section we give a lower bound on the speed requirement for SRPT to be 1-competitive.

Theorem 2. *For minimizing total stretch, SRPT is not c-speed 1-competitive for any $c < 1.5$.*

Proof. Let $c = 1.5 - \epsilon$ where $0 < \epsilon < 1.5$. We construct a sequence of jobs such that the total stretch of c-speed SRPT schedule is larger than that of OPT. At time 0, m jobs $J_1, J_2, \cdots, J_m$ of equal processing time are released. (The processing time p will be fixed shortly.) Each job is processed in a distinct processor at time 0. Just after c-speed SRPT has started the last unit of work, i.e., at time $(p-1)/c+\delta$, for some small $0 < \delta < 1/c$, m more jobs $J'_1, J'_2, \cdots, J'_m$, all of which have processing time equal to 1, are released. In this case, c-speed SRPT continues processing J_i, and starts processing J'_i only after finishing all J_i.

We analyze the total stretch of c-speed SRPT and OPT. For c-speed SRPT, the stretch of J_i is $1/c$. Since J'_i is processed after J_i finishes, J'_i finishes at $p/c + 1/c$ and thus with stretch $2/c - \delta$. Therefore, the total stretch of c-speed SRPT is $m(3/c - \delta)$. On the other hand, a unit-speed schedule can start processing J'_i immediately after the job is released and then resume processing J_i; the stretch of J'_i and J_i is 1 and $(p+1)/p$, respectively. Therefore, the total stretch of OPT is at most $m(2 + 1/p)$. We can fix the value of $p > 1/(4\epsilon/(3-2\epsilon) - \delta)$, then we have $3/c - \delta = 3/(3/2 - \epsilon) - \delta > 2 + 1/p$. (Notice that $0 < \epsilon < 3/2$ and if we choose $\delta < 4\epsilon/(3-2\epsilon)$, then we can ensure p to be positive.) The total stretch of c-speed SRPT is greater than that of OPT and thus the theorem follows. □

5 Resource Augmentation Analysis of SJF

In this section we analyze the performance of SJF for scheduling $m \geq 2$ processors. We show that the total stretch of the schedule of $(24c)$-speed SJF is at most $1/c$ times the total stretch of an optimal schedule using unit-speed processors. Recall that SJF gives higher priority to jobs with shorter processing times, with tie broken by job ID.

Our analysis makes use of the result by Becchetti et al. [4] that HDF (Highest Density First) is $(2 + 2\epsilon)$-speed $(1 + 1/\epsilon)$-competitive for minimizing weighted flow time. If we define the weight of a job J to be $1/p(J)$, then the stretch of J is equal to the weighted flow time of J. Furthermore, HDF is equivalent to SJF (since the density of a job is defined to be its weight divided by its processing time). Thus, the work of Becchetti et al. [4] implies that SJF is $(2 + 2\epsilon)$-speed $(1 + 1/\epsilon)$-competitive for minimizing total stretch.

The framework of our analysis of SJF is as follows. Let $\tau = (2 + 2\epsilon)$, and let $c \geq 1$ be any number. We compare the schedules of $(c\tau)$-speed SJF and τ-speed SJF. We show that the flow time of each job in the former schedule is at most $3/c$ times of the flow time in the latter schedule. Combining with the result of Becchetti et al., we conclude that SJF is $c(2 + 2\epsilon)$-speed $(3/c)(1+1/\epsilon)$-competitive, or equivalently, $(24c)$-speed $((2+2\epsilon)(1+1/\epsilon)/(8c))$-competitive. Putting $\epsilon = 1$ (so as to minimize $(2+2\epsilon)(1+1/\epsilon)$), we obtain the result that SJF is $(24c)$-speed $(1/c)$-competitive.

Lemma 8. *Consider any real numbers $z \geq z' \geq 1$. Given an input job sequence, denote the schedules of z-speed SJF and z'-speed SJF as S and S', respectively. At any time $t \geq 0$ and for any job J, we have $rw_t^S(J) \leq rw_t^{S'}(J)$.*

Proof. We prove the lemma by contradiction. Let t be the earliest time such that there is a job J with $rw_t^S(J) > rw_t^{S'}(J)$. If there are more than one such J, then we pick the one with the highest priority. Since $z \geq z'$, we can assume that, at time t, J is processed by some processor in S' but not by any processor in S. By the definition of SJF, as J is processed in S' at time t, there are at most

$m-1$ unfinished jobs with priority higher than J. On the other hand, since J is not processed in S at time t, there are at least m unfinished jobs with priority higher than J, and one of these m jobs must be finished in S'. It contradicts the assumption that t is the earliest time and J is the job with highest priority that $rw_t^S(J) > rw_t^{S'}(J)$. □

Corollary 1. *Assume that $z \geq z' \geq 1$. For any job J, the flow time of J in the schedule of z-speed SJF is at most that of J in the schedule of z'-speed SJF.*

Lemma 9. *Consider a schedule S of (any speed) SJF. At the time when a job J finishes, the total remaining work of the unfinished jobs arrived before J finishes and with priority higher than J is at most $(m-1)p(J)$.*

Proof. At the time when J finishes, there are at most $m-1$ unfinished jobs arrived before $r(J)$ and with priority higher than J. Otherwise, J will be preempted and cannot finish at the time. Since a job with priority higher than J has processing time at most $p(J)$, the total remaining work of those jobs is at most $(m-1)p(J)$. □

Let S_τ denote the schedule of τ-speed SJF. We denote the flow time of a job J in S_τ as $flow_\tau(J)$, which can be divided into two parts, $wait_\tau(J)$ and $busy_\tau(J)$, corresponding to the amount of time J is waiting for a processor and J is being processed by a processor, respectively. Similarly, we use the notations $S_{c\tau}$, $flow_{c\tau}(J)$, $wait_{c\tau}(J)$ and $busy_{c\tau}(J)$ for the schedule of $(c\tau)$-speed SJF.

Consider any job J. Our goal is to show that $flow_{c\tau}(J) \leq \frac{3}{c} flow_\tau(J)$. This is done by proving $busy_{c\tau}(J) \leq \frac{1}{c} flow_\tau(J)$ and $wait_{c\tau}(J) \leq \frac{2}{c} flow_\tau(J)$. The former is straightforward because $busy_{c\tau}(J) = p(J)/(c\tau) = busy_\tau(J)/c \leq flow_\tau(J)/c$.

The rest of this section is devoted to showing that the work scheduled by $S_{c\tau}$ while J is waiting, denoted W below, is upper bounded by $2m\tau flow_\tau(J)$. Then it follows that $wait_{c\tau}(J) \leq W/(mc\tau) \leq \frac{2}{c} flow_\tau(J)$. Let G be the set of jobs that have ever been scheduled by $S_{c\tau}$ while J is waiting. Note that jobs in G must arrive before $r(J) + flow_{c\tau}(J)$, and they all have priority higher than J. We partition G into two subsets G_1 and G_2 such that G_1 contains jobs arriving before $r(J)$ and G_2 the rest. The work scheduled by $S_{c\tau}$ while J is waiting, i.e., W, is at most $\sum_{J' \in G_1} rw_{r(J)}^{S_{c\tau}}(J') + p(G_2)$. To relate W with the flow time of J in S_τ, we consider two sets of jobs H_1 and H_2 in the schedule S_τ.

- H_1 contains jobs with priority higher than J that arrive before $r(J)$ and are unfinished at $r(J)$ in S_τ.
- H_2 contains jobs J' with priority higher than J such that $r(J) \leq r(J') < r(J) + flow_\tau(J)$.

It is not difficult to see that $G_1 \subseteq H_1$ and $G_2 \subseteq H_2$ (see Lemma 10), and hence W can be bounded by the remaining work of H_1 in S_τ at $r(J)$ plus the processing time of H_2.

Lemma 10. $G_1 \subseteq H_1$ *and* $G_2 \subseteq H_2$. *Furthermore*, $rw_{r(J)}^{S_{c\tau}}(G_1) \leq rw_{r(J)}^{S_\tau}(H_1)$, *where* $rw_t^S(K) = \sum_{J' \in K} rw_t^S(J')$ *for any schedule S, any time t, and any set K of jobs.*

Proof. Consider any job J' in G_1. By definition, J' is unfinished in $S_{c\tau}$ at $r(J)$ and has priority higher than J. By Corollary 1, J' is also unfinished in S_τ at $r(J)$. Therefore, we have $J' \in H_1$. Furthermore, by Lemma 8, $rw^{S_{c\tau}}_{r(J)}(J') \leq rw^{S_\tau}_{r(J)}(J')$. Together with $G_1 \subseteq H_1$, we have $rw^{S_{c\tau}}_{r(J)}(G_1) \leq rw^{S_\tau}_{r(J)}(H_1)$.

Consider any job J'' in G_2. J'' arrives at or after $r(J)$. Furthermore, J'' must arrive before J finishes in $S_{c\tau}$. That is, $r(J) \leq r(J'') < r(J) + flow_{c\tau}(J)$. By Corollary 1, $flow_{c\tau}(J) \leq flow_\tau(J)$, and hence $r(J'') < r(J) + flow_\tau(J)$. Therefore, $G_2 \subseteq H_2$. □

Corollary 2. $W \leq rw^{S_\tau}_{r(J)}(H_1) + p(H_2)$.

Lemma 11 further shows that the upper bound of W is $2m\tau\, flow_\tau(J)$.

Lemma 11. $rw^{S_\tau}_{r(J)}(H_1) + p(H_2) \leq 2m\tau\, flow_\tau(J)$.

Proof. Let us consider how S_τ schedules the work in H_1 and H_2 starting from time $r(J)$. First, we note that the total amount of such work is exactly $rw^{S_\tau}_{r(J)}(H_1) + p(H_2)$. During $[r(J), r(J) + flow_\tau(J)]$, the work scheduled by S_τ is at most $m\tau\, flow_\tau(J)$. At time $r(J) + flow_\tau(J)$, S_τ may not complete all work in H_1 and H_2; yet, by Lemma 12, at the time when J finishes, all unfinished jobs arriving before J finishes and with priority higher than J have a total remaining work at most $(m-1)p(J)$, and thus, from $r(J) + flow_\tau(J)$ onwards, S_τ can schedule at most $(m-1)p(J)$ units of work on H_1 and H_2. In conclusion, $rw^{S_\tau}_{r(J')}(H_1) + p(H_2) \leq m\tau\, flow_\tau(J) + (m-1)p(J) \leq 2m\tau\, flow_\tau$ because $p(J) \leq \tau\, flow_\tau(J)$. □

The waiting time of the job J in $S_{c\tau}$ (i.e., $wait_{c\tau}(J)$) is at most $W/mc\tau$, which, by Corollary 2 and Lemma 11, is at most $\frac{2}{c} flow_\tau(J)$.

Corollary 3. $flow_{c\tau}(J) \leq \frac{3}{c} flow_\tau(J)$.

Proof. By definition, $flow_{c\tau}(J) = wait_{c\tau}(J) + busy_{c\tau}(J)$. The corollary follows from the facts that $wait_{c\tau}(J) \leq \frac{2}{c} flow_\tau(J)$ and $busy_{c\tau}(J) \leq \frac{1}{c} flow_\tau(J)$. □

The following theorem follows from Corollary 3 and that SJF is $(2+2\epsilon)$-speed $(1 + 1/\epsilon)$-competitive.

Theorem 3. *SJF is $(24c)$-speed $(1/c)$-competitive, for any $c \geq 1$.*

6 Future Work

In this paper we have studied online job scheduling on multiprocessors and showed that, with respect to total stretch, SRPT is 5-speed 1-competitive and SJF is $(24c)$-speed $(1/c)$-competitive, for any $c \geq 1$. We conjecture that SRPT also admits a similar result as SJF, i.e., SRPT is also $O(c)$-speed $(1/c)$-competitive for any $c \geq 1$. It is also interesting to analyze the performance of SRPT and SJF when the online algorithm is given extra processors instead of extra speed, and to determine whether 1-competitiveness can be achieved.

Another open question is to derive a c-speed 1-competitive online algorithm for minimizing weighted flow time on multiprocessors. Note that both SRPT and SJF require job migration. Another direction is to consider non-migratory algorithms, i.e., once a job is assigned to a processor, it cannot be migrated to other processors, though it may be preempted.

References

1. N. Avrahami and Y. Azar. Minimizing total flow time and total completion time with immediate dispatching. In *SPAA*, pages 11–18, 2003.
2. K. R. Baker. *Introduction to Sequencing and Scheduling*. Wiley, New York,, 1974.
3. N. Bansal and K. Pruhs. Server scheduling in the L_p norm: a rising tide lifts all boat. In *STOC*, pages 242–250, 2003.
4. L. Becchetti, S. Leonardi, A. Marchetti-Spaccamela, and K. Pruhs. Online weighted flow time and deadline scheduling. In *RANDOM-APPROX*, pages 36–47, 2001.
5. L. Becchetti, S. Leonardi, and S. Muthukrishnan. Scheduling to minimize average stretch without migration. In *SODA*, pages 548–557, 2000.
6. M. A. Bender, S. Chakrabarti, and S. Muthukrishnan. Flow and stretch metrics for scheduling continuous job streams. In *SODA*, pages 270–279, 1998.
7. M. A. Bender, S. Muthukrishnan, and R. Rajaraman. Improved algorithms for stretch scheduling. In *SODA*, pages 762–771, 2002.
8. M. Brehob, E. Torng, and P. Uthaisombut. Applying extra-resource analysis to load balancing. *J. Scheduling*, 3(5):273–288, 2000.
9. H. L. Chan, T. W. Lam, and K. K. To. Non-migratory online deadline scheduling on multiprocessors. In *SODA*, pages 970–979, 2004.
10. W. T. Chan, T. W. Lam, H. F. Ting, and P. W. H. Wong. A unified analysis of hot video schedulers. In *STOC*, pages 179–188, 2002.
11. C. Chekuri, A. Goel, S. Khanna, and A. Kumar. Multi-processor scheduling to minimize flow time with ϵ resource augmentation. In *STOC*, pages 363–372, 2004.
12. C. Chekuri, S. Khanna, and A. Zhu. Algorithms for minimizing weighted flow time. In *STOC*, pages 84–93, 2001.
13. M. Chrobak, L. Epstein, J. Noga, J. Sgall, R. van Stee, T. Tichý, and N. Vakhania. Preemptive scheduling in overloaded systems. In *ICALP*, pages 800–811, 2002.
14. J. Edmonds. Scheduling in the dark. In *STOC*, pages 179–188, 1999.
15. B. Kalyanasundaram and K. Pruhs. Maximizing job completions online. In *ESA*, pages 235–246, 1998.
16. B. Kalyanasundaram and K. Pruhs. Speed is as powerful as clairvoyance. *J. ACM*, 47(4):617–643, 2000.
17. S. Leonardi and D. Raz. Approximating total flow time on parallel machines. In *STOC*, pages 110–119, 1997.
18. J. McCullough and E. Torng. SRPT optimally utilizes faster machines to minimize flow time. In *SODA*, pages 350–358, 2004.
19. S. Muthukrishnan, R. Rajaraman, A. Shaheen, and J. Gehrke. Online scheduling to minimize average stretch. In *FOCS*, pages 433–442, 1999.
20. C. A. Phillips, C. Stein, E. Torng, and J. Wein. Optimal time-critical scheduling via resource augmentation. In *STOC*, pages 140–149, 1997.
21. K. Pruhs, J. Sgall, and E. Torng. Online scheduling. In J. Leung, editor, *Handbook of Scheduling: Algorithms, Models and Performance Analysis*, pages 15-1–15-41. CRC Press, 2004.

Approximating Polygonal Objects by Deformable Smooth Surfaces

Ho-lun Cheng and Tony Tan

School of Computing, National University of Singapore
{hcheng, tantony}@comp.nus.edu.sg

Abstract. We propose a method to approximate a polygonal object by a deformable smooth surface, namely the t-skin defined by Edelsbrunner for all $0 < t < 1$. We guarantee that they are homeomorphic and their Hausdorff distance is at most $\epsilon > 0$. Such construction makes it possible for fully automatic, smooth and robust deformation between two polygonal objects with different topologies. En route to our results, we also give an approximation of a polygonal object with a union of balls.

1 Introduction

Geometric deformation is a heavily studied topic in disciplines such as computer animation and physical simulation. One of the main challenges is to perform deformation between objects with different topologies, while at the same time maintaining a good quality mesh approximation of the deforming surface.

Edelsbrunner defines a new paradigm for the surface representation to solve these problems, namely the *skin surface* [5] which is a smooth surface based on a finite set of balls. It provides a robust way of deforming one shape to another without any constraints on features such as topologies [2]. Moreover, the skin surfaces possess nice properties such as curvature continuity which provides quality mesh approximation of the surface [3].

However, most of the skin surface applications are still mainly on molecular modeling. The surface is not widely used in other fields because general geometric objects cannot be represented by the skin surfaces easily. This leaves a big gap between the nicely defined surfaces and its potential applications. We are trying to fill this gap in this paper.

1.1 Motivation and Related Works

One of the main goals of the work by Amenta et. al in [1] is to convert a polygonal object into a skin surface. We can view our work here as achieving this goal and the purpose of doing so is to perform deformation between polygonal objects. As noted earlier in some previous works [2,5], deformation can be performed robustly and efficiently if the object is represented by the skin surface.

Moreover, our work here can also be viewed as a step toward converting an arbitrary smooth object into a provably accurate skin surface. In this regard, previous work has been done by Kruithof and Vegter [8]. For input their method requires a so-called r-admissible set of balls B which approximate the object well. Then, it expands all the weights of the balls by a carefully computed constant t, before taking the $\frac{1}{t}$-skin of the expanded balls to approximate the smooth object.

However, we observe that there are at least two difficulties likely to occur in such approach. First, such an r-admissible balls are not trivial to obtain. Furthermore, when the computed factor t is closed to 1, the skin surface is almost the same as the union of balls, thus, does not give much improvement from the union of balls. On the other hand, the approach discussed here allows the freedom to choose any constant $0 < t < 1$ for defining the skin surface.

On top of the skin approximation, we also give an approximation of a polygonal object with a union of balls. Such approximation has potential applications in computer graphics such as collision detection and deformation [7,9,10]. Ranjan and Fournier [9] proposed using a union of balls for object interpolation. Sharf and Shamir [10] also proposed using the same representation for shape matching. Those algorithms require a union of balls which accurately approximate the object as an input and to provide such a good set of balls at the beginning is still not trivial.

A comparison with our Previous Work. In [4], we proposed a method to construct a set of weighted points whose alpha shape is the same as the input simplicial complex in $\mathbb{R}^d$, which we call the *subdividing alpha complex*. Given such alpha complex it is quite straightforward to obtain a set of balls which can be used to approximate the object. However, to construct the subdividing alpha complex, we need to make the assumption that the constrained triangulation of the input is given too.

In this paper the input is a piecewise linear complex which constitutes the boundary of the object. To avoid assuming we are given the constrained triangulation, we make use of the notion of *local gap size*(lgs) in the construction of the subdividing alpha complex.

1.2 Approach and Outline

The first step is to construct a set of balls whose alpha shape is the same as the boundary of the polygonal object, namely, the subdividing alpha complex. The radii of the balls constructed are at most ϵ, for a given real number $\epsilon > 0$.

In the second step we fill the interior of the object with balls according to the Voronoi complex of the balls constructed in the first step, namely, the balls that make up the subdividing alpha complex. Specifically, we consider all the Voronoi vertices which are inside the object. Each Voronoi vertex determines an orthogonal ball. We use the set of all such orthogonal balls to approximate the object. It is shown that that the union of such balls is homeomorphic to the object and furthermore, the Hausdorff distance between them is at most ϵ.

To obtain the skin approximation, we invert the weights of the balls that make up the subdividing alpha complex of the boundary. Those inverted balls, together with the balls in the interior of the object(computed in the second step), generate a skin surface which is homeomorphic to the object. It is also shown that the Hausdorff distance between them is at most ϵ.

Outline. This paper is organized as follows. In the next section we introduce some basic terminologies on piecewise linear complex(PLC) and alpha complex. In Section 3 we describe our method in constructing the subdividing alpha complex of a given PLC and the approximation of a polygonal object with a union of balls. Then we briefly review the definition of the skin surface in Section 4. The object approximation by the skin surface is described in Section 5. Finally, we end with some discussions in Section 6.

2 Notations and Basic Definitions

In this section we introduce a few basic definitions that we use throughout this paper: polygonal objects, piecewise linear complexes and alpha complexes.

Polygonal Objects. A polygonal object $\mathcal{O} \subseteq \mathbb{R}^3$ is a compact 3-manifold whose boundary is a piecewise linear 2-manifold. Our algorithm takes as an input a *piecewise linear complex*(PLC) which constitutes the boundary of $\mathcal{O}$.

Piecewise Linear Complexes. In $\mathbb{R}^3$, a piecewise linear complex is a set $\mathcal{P}$ of vertices, line segments and polygons with the following conditions:
 $i)$ all elements on the boundary of an element in $\mathcal{P}$ also belong to $\mathcal{P}$, and,
 $ii)$ if two elements intersect, the intersection is a lower dimensional element in $\mathcal{P}$.
The underlying space of $\mathcal{P}$ is denoted by $|\mathcal{P}| = \bigcup_{\sigma \in \mathcal{P}} \sigma$.

The *local gap size* is a function $lgs : |\mathcal{P}| \mapsto \mathbb{R}$ where $lgs(x)$ is the radius of the smallest ball centered on x that intersects an element of $\mathcal{P}$ that does not contain x. We remark that lgs is continuous on the interior of every element $\sigma \in \mathcal{P}$.

Alpha Complexes. We describe a *weighted point* $b \in \mathbb{R}^3 \times \mathbb{R}$ by its *location* $z_b \in \mathbb{R}^3$ and its *weight* $w_b \in \mathbb{R}$, written also as $b = (z_b, w_b)$. A weighted point b can also be viewed as a *ball* with center z_b and radius $\sqrt{w_b}$, that is, the *set of points* $\{p \in \mathbb{R}^3 \mid \|p - z_b\|^2 \leq w_b\}$. If w_b is negative then b is an imaginary ball, which is, an empty set. In this paper, we will use the terms *ball* and *weighted point* interchangeably.

The *weighted distance* of a point $p \in \mathbb{R}^3$ to a ball b is defined as
$$\pi_b(p) = \|p - z_b\|^2 - w_b.$$
Two balls b_1 and b_2 are *orthogonal* to each other if $\|z_{b_1} - z_{b_2}\|^2 = w_{b_1} + w_{b_2}$.

Given a finite set of balls B, each ball $b \in B$ defines a *Voronoi cell* ν_b which consists of the points in $\mathbb{R}^3$ with weighted distance to b less than or equal to any other ball in B. For $X \subseteq B$, the *Voronoi cell* of X is
$$\nu_X = \bigcap_{b \in X} \nu_b.$$
If ν_X consists of only one point then it is called a *Voronoi vertex*.

Let $\nu_X = \{p\}$ be a Voronoi vertex. We can associate ν_X with the ball b' where $z_{b'} = p$ and $w_{b'} = \pi_b(p)$ for some $b \in X$. Note that b' is orthogonal to every ball $b \in X$. For this reason, we call b' the *associated orthogonal ball* of the Voronoi vertex ν_X.

The collection of all Voronoi cells is called the *Voronoi complex* of B,

$$V_B = \{\nu_X \mid X \subseteq B \text{ and } \nu_X \neq \emptyset\}.$$

In this paper, we make an important but standard assumption regarding V_B:

General Position Assumption. *Let $B \subseteq \mathbb{R}^3 \times \mathbb{R}$ be a finite number of set of balls and let $X \subseteq B$. Suppose $\nu_X \neq \emptyset$ with respect to the Voronoi complex V_B. Then $1 \leq \mathrm{card}(X) \leq 4$ and the dimension of ν_X is $4 - \mathrm{card}(X)$.*

Such assumption can be achieved by small perturbation on either one of the weights or positions of the balls in X. (See, for example, [6])

For a set of balls X, we abuse the notation z_X to denote the set of the ball centers of X. The *Delaunay complex* of B is the collection of simplices,

$$D_B = \{\mathrm{conv}(z_X) \mid \nu_X \in V_B\}.$$

Note that by the *general position assumption*, the number of tetrahedra in D_B is the same as the number of Voronoi vertices in V_B.

The *alpha complex* of B is a subset of the Delaunay complex D_B which is defined as follow,

$$\mathcal{K}_B = \{\mathrm{conv}(z_X) \mid (\bigcup X) \cap \nu_X \neq \emptyset\}.$$

The *alpha shape* of B is the underlying space of $\mathcal{K}_B$, namely, $|\mathcal{K}_B|$. Note that if $\mathrm{conv}(z_X) \in \mathcal{K}_B$ then $\bigcap X \neq \emptyset$. Conversely, if $\bigcap X = \emptyset$ then $\mathrm{conv}(z_X) \notin \mathcal{K}_B$.

3 Subdividing Alpha Complex

Given a PLC $\mathcal{P}$ and a set of balls B, we say $\mathcal{K}_B$ subdivides $\mathcal{P}$ if $|\mathcal{K}_B| = |\mathcal{P}|$. In this section, we show how to construct B such that $\mathcal{K}_B$ subdivides $\mathcal{P}$. For this we need the following Lemma 1 which is a straightforward generalization of Theorem 1 in [4]. The proof is very similar, thus, we omit it.

Lemma 1. *Let $\mathcal{P}$ be a PLC. If B is a set of balls that satisfies the following two conditions:*

C1. *For $X \subseteq B$, if $\bigcap X \neq \emptyset$ then $\mathrm{conv}(z_X) \subseteq \sigma$ for some $\sigma \in \mathcal{P}$, and,*
C2. *For each $\sigma \in \mathcal{P}$, define $B(\sigma) = \{b \in B \mid b \cap \sigma \neq \emptyset\}$.*
Then we have: $z_{B(\sigma)} \subseteq \sigma \subseteq \bigcup B(\sigma)$,

then $\mathcal{K}_B$ subdivides $\mathcal{P}$.

We call $\mathcal{K}_B$ a *subdividing alpha complex*, or in short SAC, of $\mathcal{P}$. Furthermore, if all the weights in B are less than a real value ϵ, then $\mathcal{K}_B$ is called an ϵ-SAC of $\mathcal{P}$.

The aim is to construct a set of balls B that satisfies Conditions C1 and C2 in Lemma 1 and at the same time all the weights of the balls are bounded above by an input real number $\epsilon > 0$. In the first step we fix a real number $0 < \gamma < 0.5$. Then we construct the set of balls $B(\sigma)$ for each $\sigma \in \mathcal{P}$, starting with those of dimension 0, then dimension 1 and ending with those of dimension 2. Algorithm 1 outlines the sequence of computational steps.

Algorithm 1. Construction of a set of balls B such that $\mathcal{K}_B$ subdivides $\mathcal{P}$

1: Fix a real number $0 < \gamma < 0.5$
2: **for** $i = 0, 1, 2$ **do**
3:    Construct $B(\sigma)$ for all $\sigma \in \mathcal{P}$ of dimension i.
4: **end for**
5: Output $B = \bigcup_{\sigma \in \mathcal{P}} B(\sigma)$.

The construction of $B(\sigma)$ where $\dim(\sigma) = 0$ is trivial. For each vertex v in $\mathcal{P}$, we add a ball with center v and radius $r = \min(\gamma \cdot lgs(v), \sqrt{\epsilon})$. So, $B(v) = \{(v, r^2)\}$. For completeness, we present it as Algorithm 2.

Algorithm 2. To construction $B(\sigma)$ for all $\sigma \in \mathcal{P}$ with dimension 0

1: **for** each vertex $\sigma \in \mathcal{P}$ **do**
2:    $r := \min(\gamma \cdot lgs(v), \sqrt{\epsilon})$
3:    $B(\sigma) := \{(v, r^2)\}$
4: **end for**

To describe the construction of $B(\sigma)$ with σ is of dimension 1 or 2, we need the notations of *restricted Voronoi complex*. The restricted Voronoi complex of a set of balls X on $\sigma \in \mathcal{P}$, denoted by $V_X(\sigma)$, is the complex which consists of $\nu_X \cap \sigma$, for all $\nu_X \in V_X$. A Voronoi vertex u in $V_X(\sigma)$ is called a *positive* vertex if $\pi_b(u) > 0$, for all $b \in X$. Note that such a vertex is outside every ball in X. To determine whether a vertex is positive, it suffices to compute $\pi_{b'}(u)$ where u is the Voronoi vertex in the Voronoi cell of b'.

We construct $B(\sigma)$ where $\dim(\sigma) = 1$ according to Algorithm 3. The basic idea is to add a ball to a positive vertex in an edge until the edge is covered by the balls. To avoid unwanted elements other than the edge itself, we set the radius of every ball to be less than both $\sqrt{\epsilon}$ and γ times the lgs of the ball center. The construction of $B(\sigma)$ where σ is of dimension 2 is similar. For completeness, we present it as Algorithm 4 here.

We claim that our algorithms terminate and the output $B = \bigcup_{\sigma \in \mathcal{P}} B(\sigma)$ satisfies both Conditions C1 and C2. It should be clear that all weights in B are at most ϵ. Since every ball with center p has radius less than $0.5 \times lgs(p)$, it is obvious that Condition C1 is satisfied. Condition C2 follows from Proposition 1 below. Theorem 2 establishes the termination of our algorithm.

Proposition 1. *Let X be a set of balls. Suppose $z_X \subseteq \sigma$. Then $\sigma \subseteq \bigcup X$ if and only if there is no positive vertex in $V_X(\sigma)$.*

Algorithm 3. To construct $B(\sigma)$ for all $\sigma \in \mathcal{P}$ with dimension 1

1: **for all** the edge $\sigma \in \mathcal{P}$ **do**
2:     Let v_1, v_2 be the two vertices of σ.
3:     $X := B(v_1) \cup B(v_2)$
4:     **while** there exists a positive vertex u in $V_X(\sigma)$ **do**
5:         $r := \min(\gamma \cdot lgs(u), \sqrt{\epsilon})$
6:         $X := X \cup \{(u, r^2)\}$
7:     **end while**
8:     $B(\sigma) := X$
9: **end for**

Algorithm 4. To construct $B(\sigma)$ for all $\sigma \in \mathcal{P}$ with dimension 2

1: **for all** each polygon $\sigma \in \mathcal{P}$ **do**
2:     Let $\tau_1, \ldots, \tau_m$ be the edges of σ.
3:     $X := B(\tau_1) \cup \cdots \cup B(\tau_m)$
4:     **while** there exists a positive vertex u in $V_X(\sigma)$ **do**
5:         $r := \min(\gamma \cdot lgs(u), \sqrt{\epsilon})$
6:         $X := X \cup \{(u, r^2)\}$
7:     **end while**
8:     $B(\sigma) := X$
9: **end for**

Proof. The "only if" part is immediate. We will show the "if" part. Suppose there is no positive Voronoi vertex in $V_X(\sigma)$. We claim that $\nu_b(\sigma) \subseteq b$ for all $b \in X$. This claim follows from the fact that $\nu_b(\sigma)$ is the convex hull of its Voronoi vertices and bounded. Thus, by our assumption that all the Voronoi vertices are not positive, it is immediate that $\nu_b(\sigma) \subseteq b$ for any $b \in X$. Since σ is partitioned into $\nu_b(\sigma)$ for all $b \in X$, it follows that $\sigma \subseteq \bigcup X$.

To establish the termination of the algorithm, we need the following fact.

Proposition 2. *Let $\rho \in \mathcal{P}$. Suppose $\Gamma \subset \sigma$ is a closed region such that it does not intersect the boundary of σ. Then there exists a constant $c > 0$ such that for every point $p \in \Gamma$, $\mathrm{lgs}(p) > c$.*

Proof. We observe that lgs is a continuous function on Γ. Moreover, Γ is compact. Thus, there exists $p_0 \in \Gamma$ such that $lgs(p_0) = \min_{p \in \Gamma} lgs(p)$. The value $lgs(p_0) \neq 0$ since p_0 is in the interior of σ. Thus, we can choose $c = \frac{1}{2} lgs(p_0)$ to establish our proposition.

Lemma 2. *Both algorithms 3 and 4 terminate.*

Proof. We prove that Algorithm 3 terminates. It suffices to show that the `while`-loop does not iterate infinitely many times. The proof is by contradiction and it follows from the fact that each element ρ in $\mathcal{P}$ is compact.

Assume to the contrary that for some edge $\sigma = (v_1, v_2) \in \mathcal{P}$ the `while`-loop iterates infinitely many times. That is, it inserts infinitely many balls to $B(\sigma)$

whose centers are in the region $\sigma - (b_1 \cup b_2)$ where $b_i \in B(v_i)$ for $i = 1, 2$. The region $\sigma - (b_1 \cup b_2)$ is a closed region which does not intersect with the boundary of σ. By Proposition 2, there exists a constant $c > 0$ such that all the radii of the balls are greater than c.

Moreover, $\sigma - (b_1 \cup b_2)$ is compact, so if $B(\sigma)$ contains infinitely many balls, then there are two balls b and b' whose centers are at the distance less than c. Without loss of generality, we assume that b was inserted before b'. This is impossible, because at the time b' was inserted, its center would be a negative vertex. Therefore, the while-loop iterates only finitely many times. The proof of the termination of Algorithm 4 is similar.

3.1 Approximating Polygonal Object with a Union of Balls

Let $\mathcal{O}$ be a polygonal object and $\mathcal{P}$ be its boundary, given in the form of piecewise linear complex. Our method to approximate the object $\mathcal{O}$ with a union of balls can be summarized as follows.

1. Construct a set of balls B such that $\mathcal{K}_B$ is an ϵ^2-SAC of $\mathcal{P}$.
2. Compute the Voronoi complex of B.
3. Denote by T, the set of all the Voronoi vertices which are located inside the object $\mathcal{O}$.
4. Let $B^\perp$ be the set of all orthogonal balls associated with all the Voronoi vertices in T.
5. Output $B^\perp$.

Remark 1. We remark that every ball in $B^\perp$ has positive weight, thus, is a real ball. The reasoning is as follows. Because $|\mathcal{K}_B| = |\mathcal{P}|$, there is no tetrahedron in $\mathcal{K}_B$. This means each Voronoi vertex is not inside any ball $b \in B$, thus, has positive weighted distance to the each ball in B. Therefore, the associated orthogonal ball of each Voronoi vertex has positive weight.

We claim that $\bigcup B^\perp$ can be used to approximate the object $\mathcal{O}$ well.

Theorem 1. *The union of balls $\bigcup B^\perp$ is contained inside the object $\mathcal{O}$ and homeomorphic to $\mathcal{O}$. Moreover, the Hausdorff distance between them is at most ϵ.*

Proof. (Sketch) Consider the Delaunay complex D_B. Let Δ be the set of all Delaunay tetrahedra which are located inside the object $\mathcal{O}$. The object $\mathcal{O}$ is decomposable into the tetrahedra of Δ. By the general position assumption, $\text{card}(\Delta) = \text{card}(B^\perp)$.

Furthermore, $\bigcup B^\perp$ is decomposable into $\text{conv}(z_x) \cap b$ where b is the associated orthogonal ball of ν_x for all $\text{conv}(z_x) \in \Delta$. We can establish a homeomorphism between $b \cap \text{conv}(z_x)$ and $\text{conv}(x)$ for each $\text{conv}(z_x) \in \Delta$. By combining all such homeomorphisms, we obtain a homeomorphism between $\bigcup B^\perp$ and $\mathcal{O}$.

The Hausdorff nearness part can be established via the fact that each $b \in B^\perp$ is contained inside the object $\mathcal{O}$. Furthermore, the ball b is orthogonal to some balls $b' \in B$ and all the weights of the balls in B are less than or equal to ϵ.

4 Skin Surface

In this section we briefly review both the algebra of balls and the definition of the skin surface which is based on the algebra of balls [5].

Algebra of Balls. The algebra of balls is based on a bijection $\phi : \mathbb{R}^3 \times \mathbb{R} \mapsto \mathbb{R}^4$ defined as
$$\phi(b) = (z_b, \|z_b\|^2 - w_b).$$
The space $\mathbb{R}^4$ together with the usual componentwise addition and scalar multiplication forms a vector space. The addition and scalar multiplication operations are defined on $\mathbb{R}^3 \times \mathbb{R}$ in such a way that ϕ is a vector space isomorphism, that is,
$$\phi(b_1 + b_2) = \phi(b_1) + \phi(b_2),$$
$$\phi(\gamma \cdot b) = \gamma \cdot \phi(b),$$
where $b_1, b_2, b \in \mathbb{R}^3 \times \mathbb{R}$ and $\gamma \in \mathbb{R}$. One can easily verify that
$$b_1 + b_2 = (z_{b_1} + z_{b_2}, w_{b_1} + w_{b_2} + 2\langle z_{b_1}, z_{b_2}\rangle), \tag{1}$$
$$\gamma b = (\gamma z_b, \gamma w_b + (\gamma^2 - \gamma)\|z_b\|^2). \tag{2}$$

By the two operations above, the convex hull of a set of balls $B = \{b_1, \ldots, b_n\}$ is the set of balls $\mathrm{conv}(B) = \{\sum_i \gamma_i b_i \mid \sum_i \gamma_i = 1 \text{ and } \gamma_i \geq 0 \text{ for all } i = 1, \ldots, n\}$. It is straightforward to verify that if a ball b is orthogonal to every ball $b_i \in \{b_1, \ldots, b_n\}$, then b is orthogonal to every ball $b' \in \mathrm{conv}(b_1, \ldots, b_n)$.

Skin Surfaces. Let b be a weighted point and $t \in \mathbb{R}$, we define $b^t = (z_b, tw_b)$. For a set of balls B, B^t is defined as $B^t = \{b^t \mid b \in B\}$.

For $0 \leq t \leq 1$, the skin body of a set of balls B is defined as
$$\mathrm{body}^t(B) = \bigcup \mathrm{conv}(B)^t,$$
that is, the set of points obtained by shrinking all balls in the convex combination of B. The skin surface is the boundary of the skin body of B, denoted by $\mathrm{skin}^t(B)$. Note that $\bigcup B = \mathrm{body}^1(B)$. We cite here an important relation between a union of balls $\bigcup B$ and the skin body that it generates.

Theorem 2. [5] *The union of balls $\bigcup B$ is homeomorphic to* $\mathrm{body}^t(B)$, *for* $0 < t < 1$.

5 Approximating a Polygonal Object with the Skin Surface

To approximate a polygonal object with a union of balls, we start by constructing a set of balls B such that $\mathcal{K}_B$ is an ϵ^2-SAC of the boundary of the object. Then we use the associated orthogonal balls $B^\perp$ to approximate the object $\mathcal{O}$.

In this section we will show that the set of balls $B^\perp \cup B^{-1}$ will generate a skin body that approximates the object well too, as stated in Theorem 3 below.

Theorem 3. *For all $0 \leq t \leq 1$, the skin body $\mathrm{body}^t(B^\perp \cup B^{-1})$ is homeomorphic to the object $\mathcal{O}$. Moreover, the Hausdorff distance between them is at most ϵ.*

Proof. All balls in B^{-1} have negative weights, thus, $\bigcup(B^\perp \cup B^{-1}) = \bigcup B^\perp$. By Theorem 1, $\bigcup B^\perp \subseteq \mathcal{O}$, thus, it follows that $\mathrm{skin}^t(B^\perp \cup B^{-1}) \subseteq \bigcup(B^\perp \cup B^{-1}) = \bigcup B^\perp \subseteq \mathcal{O}$.

The homeomorphism follows from Theorem 2 that $\mathrm{skin}^t(B^\perp \cup B^{-1})$ is homeomorphic to $\bigcup(B^\perp \cup B^{-1}) = \bigcup B^\perp$ which is homeomorphic to $\mathcal{O}$(Theorem 1). The proof for the Hausdorff nearness part is presented in the next subsection.

5.1 Proof of the Hausdorff Nearness in Theorem 3

Note that for every point p in the object $\mathcal{O}$, there is a weighted point $b \in \mathrm{conv}(B^\perp \cup B^{-1})$ such that $z_b = p$. In other words, $\mathcal{O} \subseteq \mathcal{Z}$ where $\mathcal{Z} = \{z_b \mid b \in \mathrm{conv}(B^\perp \cup B^{-1})\}$. In view of this, it suffices to prove the following lemma.

Lemma 3. *For every ball $b \in \mathrm{conv}(B^\perp \cup B^{-1})$ where $z_b \in \mathcal{O}$, if $w_b < 0$ then there exists a ball $b' \in \mathrm{conv}(B^\perp \cup B^{-1})$ such that $w_{b'} > 0$ and $\|z_b - z_{b'}\| \leq \epsilon$.*

We note that the object $\mathcal{O}$ can be partitioned into tetrahedra of Delaunay complex $D_{B^\perp \cup B}$[1]. We made a few simple observations concerning the tetrahedron of $D_{B^\perp \cup B}$ which is contained inside $\mathcal{O}$.

Fact 1. *Let $X = \{b_1, \ldots, b_4\}$ such that $\mathrm{conv}(z_X)$ is a tetrahedron in $D_{B^\perp \cup B}$ and is contained inside $\mathcal{O}$. Then,*

1. *At least one of the balls in X is a ball in $B^\perp$.*
2. *If $b_i \in X \cap B^\perp$ and $b_j \in X \cap B$ then b_i and b_j are orthogonal to each other.*
3. *The simplex $\mathrm{conv}(z_{B \cap X})$ is a simplex in $\mathcal{K}_B$, i.e. $\mathrm{conv}(z_{B \cap X}) \subseteq |\mathcal{P}|$.*

Statements 1 and 2 are pretty straightforward. The intuition of Statement 3 is as follows. Let $X' = X \cap B$. It is clear when $\mathrm{card}(X') = 1$. For $\mathrm{card}(X') = 2$ or 3, assume to the contrary that $\mathrm{conv}(z_{X'}) \notin \mathcal{K}_B$. Since $|\mathcal{K}_B| = |\mathcal{P}|$, the simplex $\mathrm{conv}(z_{X'})$ is in the interior of $\mathcal{O}$. Then, there exist at least $5 - \mathrm{card}(X')$ balls of $B^\perp$ which are orthogonal to every ball in X'[2]. These balls of $B^\perp$ make $\nu_X = \emptyset$, thus, yields a contradiction that $\mathrm{conv}(z_X)$ is a Delaunay tetrahedron. Therefore, $\mathrm{conv}(z_{X'}) \in \mathcal{K}_B$, where $X' = X \cap B$.

In view of Statement 3 in Fact 1, we categorize the tetrahedra of $D_{B^\perp \cup B}$ within $\mathcal{O}$ into four types according to $\mathrm{card}(X \cap B)$. We illustrate it in Figure 1.

1. Tetrahedron type I is a tetrahedron where $\mathrm{card}(X \cap B) = 1$.
   In Figure 1, $b_1 \in B$ and $b_2, b_3, b_4 \in B^\perp$.

---

[1] Note that $D_{B^\perp \cup B}$ may not be the same as $D_{B^\perp \cup B^{-1}}$. The object $\mathcal{O}$ may not be partitioned into tetrahedra of $D_{B^\perp \cup B^{-1}}$.

[2] That is, if $\mathrm{card}(X') = 2$, then $\dim(\mathrm{conv}(z_{X'})) = 1$. So, $\mathrm{conv}(z_{X'})$ is incident to at least three tetrahedra in D_B and each tetrahedron corresponds to one ball in $B^\perp$. Similarly, if $\mathrm{card}(X') = 3$, then $\mathrm{conv}(z_{X'})$ is incident to two tetrahedra in D_B and each tetrahedron correspond to one ball in $B^\perp$.

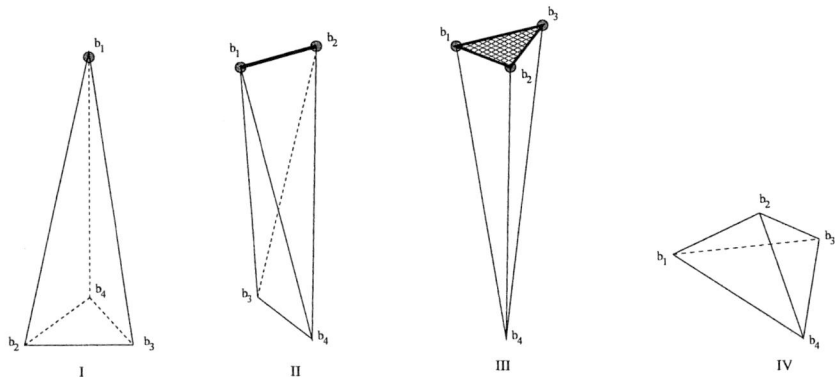

Fig. 1. The bold point in type I, the bold edge in type II and the the shaded triangle in the type III indicate that they are in $\mathcal{K}_B$, thus in the boundary of the object. None of the vertices in the type IV tetrahedron belongs to B.

2. Tetrahedron type II is a tetrahedron where $\mathrm{card}(X \cap B) = 2$. In Figure 1, $b_1, b_2 \in B$ and $b_3, b_4 \in B^\perp$.
3. Tetrahedron type III is a tetrahedron where $\mathrm{card}(X \cap B) = 3$. In Figure 1, $b_1, b_2, b_3 \in B$ and $b_4 \in B^\perp$.
4. Tetrahedron type IV is a tetrahedron where $\mathrm{card}(X \cap B) = 0$. In Figure 1, all $b_1, b_2, b_3, b_4 \in B^\perp$.

In view of this, to prove Lemma 3 it is sufficient to prove the following.

Claim. Let $\mathrm{conv}(z_X) \in D_{B^\perp \cup B}$ and located inside $\mathcal{O}$. For every ball $b \in \mathrm{conv}(X)$, if $w_b < 0$ then there exists a ball $b' \in \mathrm{conv}(X)$ such that $w_{b'} > 0$ and $\|z_b - z_{b'}\| \leq \epsilon$.

We divide the proof of the claim according to $\mathrm{card}(X \cap B)$, that is, the type of the tetrahedron that contains z_b. If $\mathrm{card}(X \cap B) = 4$ then all balls $b \in \mathrm{conv}(X)$ have weights $w_b > 0$.

The following Lemma 4 states that all points in $\mathrm{conv}(b_1^{-1}, b_2, b_3, b_4)$ (i.e. in tetrahedron type I) with negative weights are located within the ϵ-neighborhood of $z_{b_1^{-1}}$. This immediately implies the validity of claim for tetrahedron type I.

Lemma 4. *Let $(p, w) \in \mathrm{conv}(b_1^{-1}, b_2, b_3, b_4)$. If $w \leq 0$ then $\|p - z_{b_1^{-1}}\| \leq \epsilon$.*

Proof. Let
$$(p, w) = \gamma_1 b_1^{-1} + \gamma_2 b_2 + \gamma_3 b_3 + \gamma_4 b_4$$
$$= \gamma_1 b_1^{-1} + (1 - \gamma_1) b',$$

where $b' = \frac{1}{1-\gamma_1} \sum_{i=2}^4 \gamma_i b_i$ and $\sum \gamma_i = 1$ and $\gamma_i \geq 0$, for $i = 1, \ldots, 4$. Since b_2, b_3, b_4 are all orthogonal to b_1, then b' is also orthogonal to b_1, i.e. $w_{b'} + w_{b_1} = \|z_{b_1} - z_{b'}\|^2$. We apply the formula of combination of weighted points:

$$w = (1 - \gamma_1) w_{b'} + \gamma_1 w_{b_1^{-1}} + (\gamma_1^2 - \gamma_1) \|z_{b'} - z_{b_1^{-1}}\|^2.$$

Since $w \leq 0$, we arrange the terms into

$$(\gamma_1^2 - \gamma_1)\|z_{b'} - z_{b_1^{-1}}\|^2 - \gamma_1(w_{b'} + w_{b_1}) + w_{b'} \leq 0 \tag{3}$$

$$\gamma_1^2 \|z_{b'} - z_{b_1^{-1}}\|^2 - 2\gamma_1 \|z_{b'} - z_{b_1^{-1}}\|^2 \leq -w_{b'} \tag{4}$$

$$\gamma_1^2 \|z_{b'} - z_{b_1^{-1}}\|^2 - 2\gamma_1 \|z_{b'} - z_{b_1^{-1}}\|^2 + \|z_{b'} - z_{b_1^{-1}}\|^2 \leq \|z_{b'} - z_{b_1^{-1}}\|^2 - w_{b'} \tag{5}$$

$$(\gamma_1 - 1)^2 \|z_{b'} - z_{b_1^{-1}}\|^2 \leq w_{b_1} \tag{6}$$

$$(1 - \gamma_1)^2 \|z_{b'} - z_{b_1^{-1}}\|^2 \leq \epsilon^2 \tag{7}$$

$$\|p - z_{b_1^{-1}}\| \leq \epsilon \tag{8}$$

From Inequality 3 to Inequality 4 and Inequality 5 to Inequality 6, we apply $w_{b'} + w_{b_1} = \|z_{b_1} - z_{b'}\|^2$. From Inequality 7 to Inequality 8, we apply $\|p - z_{b_1^{-1}}\| = (1 - \gamma_1)\|z_{b'} - z_{b_1^{-1}}\|$.

The validity of the claim for tetrahedra types II and III is presented as Lemmas 5 and 6 below. Lemma 5 states that all points in $\mathrm{conv}(b_1^{-1}, b_2^{-1}, b_3, b_4)$ (i.e. in tetrahedron type II) with negative weights are located within the ϵ-neighborhood of $\mathrm{conv}(z_{b_1^{-1}, b_2^{-1}})$. Similarly, Lemma 6 states that all points in $\mathrm{conv}(b_1^{-1}, b_2^{-1}, b_3^{-1}, b_4)$ (i.e. in tetrahedron type III) with negative weights are located within the ϵ-neighborhood of $\mathrm{conv}(z_{b_1^{-1}, b_2^{-1}, b_3^{-1}})$. Both proofs are just a slight twist of the proof of Lemma 4 and we omit them.

Lemma 5. *Let $(p, w) = \mathrm{conv}(b_1^{-1}, b_2^{-1}, b_3, b_4)$. If $w \leq 0$ then there exists $b' \in \mathrm{conv}(b_1^{-1}, b_2^{-1})$ such that $\|p - z_{b'}\| \leq \epsilon$.*

Lemma 6. *Let $(p, w) = \mathrm{conv}(b_1^{-1}, b_2^{-1}, b_3^{-1}, b_4)$. If $w \leq 0$ then there exists $b' \in \mathrm{conv}(b_1^{-1}, b_2^{-1}, b_3^{-1})$ such that $\|p - z_{b'}\| \leq \epsilon$.*

6 Discussion

One future direction is to implement the same idea in approximating smooth objects with skin surfaces. Amenta et.al [1] showed that given a sufficiently dense sample points on a smooth surface, the set of polar balls obtained can be used to approximate the object well. There is an analogy between such approach with our method here. We can view the ϵ-SAC constructed as the sample points and $B^{\perp}$ as the polar balls.

By appropriately assigning certain weights to the sample points and taking the polar balls, we hope to be able to approximate the smooth object by a skin surface. At this point, the usefulness of this idea is still under investigation.

References

1. N. Amenta and R. K. Kolluri. Accurate and Efficient Unions of Balls. *Proc. 16th Sympos. Computational Geometry*, pages 119-128, ACM-SIAM, 2000.
2. H.-l. Cheng, H. Edelsbrunner and P. Fu. Shape Space from Deformation. *Comput. Geometry: Theory and Applications*, **19**:191-204, 2001.

3. H.-l. Cheng and X.-w. Shi. Guaranteed Quality Triangulation of Molecular Skin Surfaces. *Proc. IEEE Visualization*, pages 481-488, 2004.
4. H.-l. Cheng and T. Tan. Subdividing Alpha Complex. *Proc. 24th Conf. on Foundations of Software Technology and Theoretical Computer Science*, pages 179-190, LNCS 3328 - Springer Verlag, 2004.
5. H. Edelsbrunner. Deformable Smooth Surface Design. *Discrete and Computational Geometry*, **21**:87-115, 1999.
6. H. Edelsbrunner and E. P. Mucke. Simulation of Simplicity: a Technique to Cope with Degenerate Cases in Geometric Algorithms. *ACM Trans. Graphics*, **9**: 66-104, 1990.
7. P. M. Hubbard. Approximating Polyhedra with Spheres for Time-critical Collision Detection. *ACM Transactions on Graphics*, **15**: 3, 179-210, 1996.
8. N. Kruithof and G. Vegter. Approximation by Skin Surfaces. *Proc. 8th Sympos. Solid Modeling and Applications*, pages 86-95, ACM-SIAM, 2003.
9. V. Ranjan and A. Fournier. Matching and Interpolation of Shapes Using Unions of Circles. *Computer Graphics Forum*, **15**(3):129-142, 1996.
10. A. Sharf and A. Shamir. Feature-sensitive 3D Shape Matching. *Proc. Computer Graphics International*, pages 596-599, 2004.

Basis of Solutions for a System of Linear Inequalities in Integers: Computation and Applications

D. Chubarov* and A. Voronkov

The University of Manchester

Abstract. We define a basis of solutions of a system of linear inequalities and present a general algorithm for finding such a basis. Our algorithm relies on an algorithm for finding a Hilbert basis for the set of nonnegative solutions of a system of linear inequalities and can be used in conjunction with any such algorithm.

1 Introduction

It is a classical combinatorial fact that the set of integral solutions of a system of linear inequalities can be represented as the set of nonnegative linear combinations of a finite number of vectors. For instance, the set of nonnegative integral solutions of a system of linear inequalities can be represented by the Hilbert basis as defined in [9].

In this paper we consider bases for the set of *all* integral solutions of a system of linear inequalities.

Definition 1 (Basis of Solutions). A set of vectors H *generates* the set of vectors S if S is the set of all linear combinations of vectors from H with nonnegative integer coefficients. Let S be the set of integral solutions of a system of linear inequalities $A\mathbf{x} \geq \mathbf{0}$. A set of vectors $H \subset S$ is a *basis of solutions* of $A\mathbf{x} \geq \mathbf{0}$ if H generates S and S is not generated by any proper subset of H.

When the system has only non-negative solutions (for example, when it contains the subsystem $\mathbf{x} \geq \mathbf{0}$), the basis of solutions is unique and usually called the Hilbert basis.

A number of algorithms for finding the Hilbert basis have been considered in the literature. We are interested in the problem of finding bases of solutions over integers. This problem has a number of applications, for example in infinite-state model checking or constraint satisfaction over integers.

In this paper we define a general algorithm for finding a basis of solutions. Our algorithm relies on an algorithm for finding the Hilbert basis and can be used in conjunction with any such algorithm. It can also be used in conjunction with an algorithm for finding the basis for systems with a matrix of full column rank [12] or in conjunction with an incremental basis finding algorithm [25].

2 Related Work

Hilbert bases were introduced in computer science in the context of automated reasoning by Stickel [29], and by Giles and Pulleyblank [9] in the context of combinatorial

---

* Supported by an ORSAS grant.

optimization. Giles and Pulleyblank were also the first to introduce the name, which stems from Hilbert's result on the existence of a finite generating set for the set of nonnegative integral solutions of a system of linear equations [16]. Since their introduction the Hilbert bases gave rise to several theoretical questions, such as the question of finding an integral analogue of Carathéodory theorem initially posed by Sebő [4,28] and the question of upper bounds on the elements of a Hilbert basis [13,18,23].

Two major classical results regarding Hilbert bases are the existence of a finite Hilbert basis for the set of solutions of any system of linear inequalities [10] and the uniqueness of the Hilbert basis for any system of linear inequalities in nonnegative integers [30]. The latter is generalized to systems of linear inequalities with a matrix of full column rank.

Recently Hilbert bases found several new applications in computer science. Hilbert bases serve as a representation for the complete solution of the unification problem for terms with associative and commutative function symbols (AC-unification) [29]. Hilbert bases are used in integer programming as test sets [1,27]; in model checkers for infinite state systems to speed up computations [25]; in CLP framework a solver based on Hilbert bases can serve as a truly incremental solver for linear inequality constraints with unbounded integer variables [2]. The main advantage of a solver based on Hilbert bases in CLP setting and in model checking is the way it can be used to specialize entailment tests for systems of non-homogeneous linear inequalities.

Most of the existing algorithms for computing Hilbert bases are defined for computing the Hilbert basis of the set of nonnegative integral solutions of a system of linear equations or inequalities. There are several algorithms available for this case starting from the algorithm for a single linear equation proposed by Huet [17], there is an algorithm for systems of linear equations by Domenjoud [5] and algorithms for finding Hilbert bases for systems of linear inequalities directly without introducing slack variables by Ajili and Contejean [2], Pasechnik [22], Tomas and Filgueiras [7,8]. The algorithm by Hemmecke [12] finds the basis of solutions of a system with a matrix of full column rank. A recent survey of algorithms for Hilbert basis computation can be found in [3].

One reason why nonnegative integral solutions are important comes from applications to AC-unification where integral vectors represent monomials. On the contrary, the systems of linear inequalities that model checkers need to solve do not always have this restriction.

In this paper we present an algorithm for computing a Hilbert basis of the set of integral solutions of a system of linear inequalities with a matrix with nonzero column defect. We show that the bases of solutions can be used to solve the entailment problem for systems of non-homogeneous linear inequalities in integers and therefore are suitable for applications in CLP framework and in model checking.

The algorithm presented in the paper relies significantly on the use of the Hermite normal form theorem. By using it, we avoid introducing new variables and show how existing algorithms for the nonnegative case can be generalised to finding the Hilbert basis of a system of inequalities with a matrix of full column rank.

3 Preliminaries

Unless otherwise stated we use notation and terminology from Schrijver [26].

If **a** is an integral vector and b is a number, then $\mathbf{ax} \geq b$ is called a *linear inequality* and $\mathbf{ax} = b$ is called a *linear equation*. If A is an $m \times n$ matrix and **b** is a vector of dimension m, then $A\mathbf{x} \geq \mathbf{b}$ is called a *system of linear inequalities*. A linear inequality $\mathbf{ax} \geq b$ is called *homogeneous* if $b = 0$ and *non-homogeneous* otherwise.

We only consider *integral matrices* here, that is matrices such that all their entries are integer numbers. If a matrix A has m rows and n columns, we say that A is an $m \times n$ matrix. The *rank* of a matrix A is the number of linearly independent columns in A. If r is the rank of an $m \times n$ matrix A then the number $d = n - r$ is called the *column defect* of A. A matrix such that its column defect is zero is also called a matrix of *full column rank*.

It is sometimes convenient to look at a system of linear inequalities as a conjunction of individual inequalities. Given a system of linear inequalities $A\mathbf{x} \geq \mathbf{b}$ and a new inequality $\mathbf{ax} \geq b$ on the same variables, we write $A\mathbf{x} \geq \mathbf{b} \wedge \mathbf{ax} \geq b$ for the system of linear inequalities with a matrix obtained by adding the row **a** to A and the right hand side obtained by inserting a new component b into the vector **b** at corresponding position.

If S is a set of vectors of the same dimension and A is a matrix of corresponding dimensions then by AS we denote the set $\{A\mathbf{s} \mid \mathbf{s} \in S\}$.

By E_r we denote the $r \times r$ identity matrix and by $O_{m,n}$ we denote the $m \times n$ matrix of zeros.

Non-decomposable Solutions

Let H be a set of vectors. In the rest of this paper we denote the set that the set H generates by $\mathbb{Z}_+(H)$.

Definition 2 (directed set). A set B of integral vectors is a *directed set* if no nontrivial linear combination with nonnegative rational coefficients of vectors from B is equal to **0**.

Let S be a set of integral vectors. A vector $\mathbf{v} \in S$ is called *non-decomposable in* S if **v** cannot be represented as a sum of two nonzero vectors from S. In particular, if S is the set of integral solutions of a system of linear inequalities $A\mathbf{x} \geq \mathbf{0}$, and **v** is non-decomposable in S then **v** is called *non-decomposable solution* of the system of linear inequalities.

If the set of solutions of a system of linear inequalities is not a directed set then it does not contain any non-decomposable solutions. Note that if the system of linear inequalities has a matrix with a nonzero column defect then all solutions are decomposable.

On the other hand, in the case of a system of linear inequalities with a full column rank we have the following result due to van der Corput.

Lemma 1 (van der Corput's lemma). *Let $A\mathbf{x} \geq \mathbf{0}$ be a system of linear inequalities with a matrix A of full column rank, then it has a unique basis of solutions which consists of all non-decomposable solutions.*

Lattices

Our solution to the problem of finding Hilbert bases is based on some classical results from the theory of integral lattices.

A set of integral vectors Λ is called a *lattice* of dimension d if there exist linearly independent integral vectors $\mathbf{a}_1, ..., \mathbf{a}_d$ in $\mathbb{Z}^n$ such that

$$\Lambda = \left\{ \sum_{i=1}^{d} \lambda_i \mathbf{a}_i \,\bigg|\, \lambda_i \in \mathbb{Z} \right\}.$$

The set $\{\mathbf{a}_1, ..., \mathbf{a}_d\}$ is called a *lattice basis* of the lattice lambda Λ.

Let $A\mathbf{x} = \mathbf{0}$ be a system of linear equations with a matrix A of column defect d. It is a classical result of 19th century number theory that the set of all integral solutions of the system of linear equations forms a lattice of dimension d (for modern presentation see [26]).

Lemma 2. *A basis of solutions of a system of linear equations can be found in polynomial time.*

Proof. A lattice basis of a lattice given by a system of linear equations can be found in polynomial time ([26], Chapter 4).

Let $A\mathbf{x} = \mathbf{0}$ be a system of equations with a matrix A of column defect d and let Λ denote the corresponding lattice.

Let $\{\mathbf{a}_1, ..., \mathbf{a}_d\}$ be a lattice basis of L. Every vector in L can be represented as a linear combination of vectors $\mathbf{a}_1, ..., \mathbf{a}_d$ with integral coefficients, therefore it can also be represented as a nonnegative linear combination of the vectors $\mathbf{a}_1, ..., \mathbf{a}_d, -\mathbf{a}_1, ..., -\mathbf{a}_d$. Since the vectors in the lattice basis are linearly independent, these $2d$ vectors form a basis of solution of the system of linear equations.

Observe that there also exists a basis of solutions that consists of only $d+1$ vectors. One can consider, for instance, the set $\{\mathbf{a}_1, ..., \mathbf{a}_d, -\mathbf{a}_1, ..., -\mathbf{a}_d\}$. □

Hermite Normal Form

The algorithmic aspect of the proposed solution to the problem of finding Hilbert bases, as will be seen later, is based on the existence of a polynomial algorithm for transforming a matrix into a lower triangular form by elementary column operations. Existence of such an algorithm is the essence of the following Hermite normal form theorem.

Schrijver ([26], Section 4.1) gives a definition of Hermite normal form only for matrices of full row rank. In this paper we will need a definition of Hermite normal form for matrices with arbitrary row defect. We assume the following definition of Hermite normal form as given by Micciancio and Warinschi [20]. If a matrix is of full row rank, then this definition coincides with the definition of Schrijver.

Definition 3 (Lower triangular matrix, Hermite normal form). An $m \times n$ matrix T of rank r is called a *lower triangular matrix* if $T_{pq} = 0$ for every entry T_{pq} in row p and column q, such that $p < q$.

A lower triangular matrix T is in *Hermite normal form* if all entries are nonnegative and there exists a sequence of indices $1 \leq j_1 < ... < j_r \leq m$ such that the following conditions are satisfied:

a) A row with the index j_i has a unique maximal entry located in ith column.

b) If $p < j_i$ and $q \geq i$ then the entry in row p and column q is equal to zero.

An *elementary column operation* on a matrix is either adding an integral multiple of one column to another column or multiplying a column by -1.

Any sequence of elementary column operations on an $m \times n$ matrix is equivalent to right multiplication by an integral square matrix in a specific form (see Schrijver [26], Theorem 4.3). Such integral square matrices are called *unimodular*.

Here we state the Hermite normal form theorem in a generalised form corresponding to the above definition of Hermite normal form.

Theorem 1 (Hermite normal form theorem). *For any integral matrix A there exists a unimodular matrix U and a unique matrix T in Hermite normal form such that $AU = T$. The matrix T and the matrix U can be found in polynomial time.*

Complexity

The complexity of the problem of computing a basis of solutions can be assessed with at least two complexity measures. One is the time complexity of the algorithm, which outputs the basis in some order, another is the cardinality of the resulting basis.

For Hilbert bases these complexity measures were studied before. The cardinality of the Hilbert basis can be doubly exponential in the number of variables of the system [19]. Hermann, Juban and Kolaitis [15] show that the problem of counting the cardinality of the Hilbert basis is in the counting class #NP.

The problem of testing if a given vector belong to the Hilbert basis is coNP-complete [15] and the problem of recognizing the Hilbert basis is coNP-complete even if the coefficients of the system are given in unary [6].

The above complexity results represent the worst case. There are nontrivial classes of systems of linear inequalities with lower complexity of the problems related to finding the Hilbert basis. One example is the class of systems of linear inequalities with totally unimodular matrices[26].

4 Basis of Solutions. The Case of Nonzero Column Defect

Let A be an $m \times n$ integral matrix and $A\mathbf{x} \geq \mathbf{0}$ be a system of linear inequalities. Let T be the Hermite normal form of A and U be a unimodular matrix such that $T = AU$. From the definition of Hermite normal form it follows that if A is of nonzero column defect d then the last d columns of T are vectors of zeros.

Let $r = n - d$ be the rank of A. Let A^* be the matrix formed by the first r columns of T, Q be the $n \times r$ matrix formed by the first r columns of U and X be the $n \times d$ matrix formed by the last d columns of U.

This can be presented in a more compact form by the following equations:

$$T = AU \quad U = [Q\ X] \quad A^* = AQ \quad AX = O_{m,d} \quad T = [A^*\ O_{m,d}].$$

The matrix A^* is of full column rank, therefore by van der Corput's lemma the basis of solutions of $A^*\mathbf{x} \geq \mathbf{0}$ is the set of all non-decomposable solutions. Let H^* denote this basis of solutions.

This theorem splits the problem of finding a basis of solutions of an arbitrary system of linear inequalities into a problem of finding a basis of solutions H_L of a system of linear equations and a problem of finding a basis of solutions H^* of a system of linear inequalities with a matrix of full column rank.

Theorem 2. *Let Q and H^* be as defined above and let H_L be a basis of solutions of the system of linear equations $A\mathbf{x} = \mathbf{0}$. The set $H = H_L \cup QH^*$ is a basis of solutions for the system of linear inequalities $A\mathbf{x} \geq \mathbf{0}$.*

The proof reduces to routine checking of the two conditions in the definition of the basis of solutions.

Observe, that the columns of the matrix X form a lattice basis of $A\mathbf{x} = \mathbf{0}$, therefore a basis of solutions H_L for this system of linear equations can be found immediately by Lemma 2.

Essentially, this theorem can be considered as an algorithm for computing a basis of solutions which relies on an algorithm for finding a basis of solutions in a special case of a system of linear inequalities with a matrix of full column rank.

A number of algorithms is available in the literature for computing the Hilbert basis of the set of nonnegative solutions of a system of linear inequalities. In the next section we consider the problem of finding a basis of solutions of a system of linear inequalities with a matrix of full column rank.

5 Basis of Solutions. The Case of a Full Column Rank

A Hilbert basis of the set of nonnegative solutions of a system of a linear inequalities $A\mathbf{x} \geq \mathbf{0}$ is the basis of solutions of a system of linear inequalities

$$A\mathbf{x} \geq \mathbf{0}, \quad \mathbf{x} \geq \mathbf{0}.$$

This system of linear inequalities is a system with a matrix of full column rank.

With the following theorem any algorithm for finding the Hilbert basis of the set of nonnegative solutions of a system of linear inequalities can be used to find the basis of solutions of a system of linear inequalities of full column rank.

Theorem 3. *Let A be a matrix of full column rank, then there exists an unimodular matrix B such that a vector $\mathbf{y}$ is a non-decomposable solution of $AB\mathbf{y} \geq \mathbf{0}, \quad \mathbf{y} \geq \mathbf{0}$ if and only if $B\mathbf{y}$ is a non-decomposable solution of $A\mathbf{x} \geq \mathbf{0}$.*

Proof. To prove the theorem we find a matrix B such that the columns of B generate a superset of the set of solutions of $A\mathbf{x} \geq \mathbf{0}$. If B is such a matrix, then every solution to $A\mathbf{x} \geq \mathbf{0}$ is a linear combination of columns of B with nonnegative integral coefficients and therefore can be represented as $\mathbf{x} = B\mathbf{y}$ for some integral vector $\mathbf{y} \geq \mathbf{0}$. The one-to-one correspondence between the sets of solutions follows.

Without loss of generality we may assume that A is initially in Hermite normal form, otherwise we can compute the normal form T and the unimodular matrix U, such that $T = AU$, and let $A = T$ and $B = UB$. We may further assume that the elements in each row of A are co-prime. For simplicity we also assume that A is of full row rank.

Then the first two inequalities are of the form $x_1 \geq 0 \quad ax_1 + bx_2 \geq 0$, where $0 \leq a < b$.

Consider a system of linear inequalities $A'\mathbf{x} \geq \mathbf{0}$, where all inequalities are the same, but the second inequality is replaced by $ax_1 + x_2 \geq 0$. Since $b \geq 1$, it follows that if $\mathbf{x}$ is a solution of the original system, then $bax_1 + bx_2 \geq ax_1 + bx_2 \geq 0$. Therefore the set of solutions of $A'\mathbf{x} \geq \mathbf{0}$ is a superset of solutions of the original system. Let U_1 be a unimodular matrix such that $A'U_1$ is in the Hermite normal form. Note that U_1 corresponds to adding the second column to the first column with coefficient $-a$.

We can repeat this process of widening the set of solutions of the original system as follows. Let A_k be an $m \times n$ matrix in Hermite normal form such that its first k rows are of the form $[E_k \ O_{k,(n-k)}]$ and let $a_1 x_1 + ... + a_k x_k + b x_{k+1} \geq 0$ be the inequality defined by the row $k+1$. Let A'_k be the matrix of a system of linear inequalities where all inequalities are the same, but the inequality defined by row $k+1$ is replaced by $a_1 x_1 + ... + a_k x_k + x_{k+1} \geq 0$. As in the case of two inequalities, the special form of the first k rows guarantees that the set of solutions of this system is a superset of the set of solutions of $A_k \mathbf{x} \geq \mathbf{0}$. Let U_k be a unimodular matrix such that $A'_k U_k$ is in Hermite normal form and let $A_{k+1} = A'_k U_k$. Note that U_k corresponds to adding column $k+1$ to each of the first k columns with coefficient $-a_k$.

Starting with $A_1 = A$, this process terminates with the matrix $A_n = E$ and a sequence of unimodular matrices $U_{n-1}, ..., U_1$.

Observe that if V is a unimodular matrix, A and B are such that $AV = B$ and the vectors $\mathbf{g}_1, ..., \mathbf{g}_n$ generate the set of solutions of $B\mathbf{x} \geq \mathbf{0}$ then the vectors $V\mathbf{g}_1, ..., V\mathbf{g}_n$ generate the set of solutions of $A\mathbf{x} \geq \mathbf{0}$. Using this observation we can see that the non-negative unit vectors $\mathbf{e}_1, ..., \mathbf{e}_n$ generate the set of solutions of $A_n \mathbf{x} \geq \mathbf{0}$ and therefore $U_{n-1}\mathbf{e}_1, ..., U_{n-1}\mathbf{e}_n$ generate the set of solutions of a system of linear inequalities with the matrix A'_{n-1}. By construction, these vectors generate a superset of the set of solutions of $A_{n-1}x \geq \mathbf{0}$.

Let $B = U_1 \cdot ... \cdot U_{n-1}$. It follows that the columns of B generate a superset of the set of solutions of $A\mathbf{x} \geq \mathbf{0}$. □

Note that Theorem 3 provides a polynomial algorithm for finding the matrix B. If A is a matrix of rank r then to find the matrix B it is necessary to multiply r unimodular matrices of a special form. Each unimodular matrix has ones on its diagonal and with the exception of the diagonal contains nonzero entries only in one row.

With the help of theorem 3 any algorithm for computing the Hilbert basis of a system of linear inequalities can be used to compute the basis of solutions of a system of linear inequalities with a matrix of full column rank.

It must be noted that this is not the only available method of computing the basis of solutions. First, there are algorithms capable of computing the set of non-decomposable solutions of a system of linear inequalities with a matrix of full column rank by definition. One example of such algorithms is the algorithm by Hemmecke [12]. For such algorithms Theorem 3 is not necessary.

Among other algorithms we distinguish algorithms that we call *incremental*. The input of an incremental algorithm for finding a Hilbert basis is a directed set of vectors B and a system of linear inequalities $A\mathbf{x} \geq \mathbf{0}$. The algorithm finds all vectors non-decomposable in the set $\{\mathbf{v} \mid \mathbf{v} \in \mathbb{Z}_+(B) \text{ and } A\mathbf{v} \geq \mathbf{0}\}$. To find the Hilbert basis of the set of nonnegative solutions of a system of linear inequalities an incremental algorithm takes as input the set of nonnegative unit vectors and a system of linear inequalities. Incremental algorithms are described in [2] and in [25].

In order to apply an incremental algorithm to find a basis of solutions of a system of linear inequalities with a matrix of full column rank one can find the matrix B as in Theorem 3 and give its columns and the system of linear inequalities as input to the algorithm.

The remaining algorithms are *black box* algorithms for computing the Hilbert basis for the set of solutions of a system of linear inequalities of the form

$$A\mathbf{x} \geq \mathbf{0}, \quad \mathbf{x} \geq \mathbf{0}.$$

With such algorithms Theorem 3 is applied directly.

6 Non-homogeneous Inequalities

Basis of solutions is defined for systems of homogeneous linear inequalities.

For system of non-homogeneous linear inequalities a basis of solutions as defined above does not exist, since the sum of two nonzero solutions to a homogeneous inequality may not itself be a solution. In the case of non-homogeneous inequalities we can define a basis of the set of solutions that consists of two parts.

Definition 4 (Basis of solutions for a system of nonhomogeneous inequalities). A pair of sets $\langle H_0, H_1 \rangle$ is a basis of solutions for a system of non-homogeneous linear inequalities $A\mathbf{x} \geq \mathbf{b}$ if the following two conditions hold.

a) An integral vector $\mathbf{v}$ satisfies $A\mathbf{v} \geq \mathbf{b}$ if and only if $\mathbf{v}$ can be represented as

$$\mathbf{v} = \mathbf{v}_0 + \mathbf{v}_1,$$

where $\mathbf{v}_0$ is a nonnegative integral linear combination of vectors from H_0 and $\mathbf{v}_1 \in H_1$.

b) If a vector $\mathbf{v}$ in H_1 is such that $A\mathbf{v} \geq \mathbf{0}$, then $\mathbf{v} = \mathbf{0}$.

The algorithm for computing a basis of solutions for a system of homogeneous linear inequalities can be applied to find a basis of a system of non-homogeneous linear inequalities.

Given a system of non-homogeneous linear inequalities

$$A\mathbf{x} \geq \mathbf{b} \tag{1}$$

with an $m \times n$ matrix A, consider a system of homogeneous linear inequalities

$$A'\mathbf{y} \geq \mathbf{0}, \tag{2}$$

with the matrix

$$A' = \begin{bmatrix} 1 & 0 \ldots 0 \\ -\mathbf{b} & A \end{bmatrix} \tag{3}$$

Let H be a basis of solutions for the system of linear inequalities in (3). Let $\pi : \mathbb{Z}^{n+1} \to \mathbb{Z}^n$ denote the function that takes the last n components of its argument and let $\rho : \mathbb{Z}^{n+1} \to \mathbb{Z}$ denote the function that takes the first component of its argument.

Define

$$H_0 = \{\pi(\mathbf{h}) \mid \mathbf{h} \in H \text{ and } \rho(\mathbf{h}) = 0\}$$

and

$$H_1 = \{\pi(\mathbf{h}) \mid \mathbf{h} \in H \text{ and } \rho(\mathbf{h}) = 1\}.$$

The pair of sets $\langle H_0, H_1 \rangle$ is a basis of the system of non-homogeneous inequalities as the following lemma shows.

Theorem 4. *A vector $\mathbf{v}$ is a solution to (2) if and only if it can be represented as $\mathbf{v} = \mathbf{v}_0 + \mathbf{v}_1$, where $\mathbf{v}_0$ is a linear combination of vectors from H_0 with nonnegative integral coefficients and $\mathbf{v}_1 \in H_1$.*

The proof is standard.

7 Entailment Problem

To conclude the paper we would like to present an example of the use of bases of solutions in applications. In this section we consider the use of bases of solutions to obtain a specialized algorithm for checking entailment for systems of linear inequalities.

The entailment problem for systems of linear inequalities is to check if the set of integral solutions of one system is contained in the set of solutions of another system. We say that the system of linear inequalities $A_1 \mathbf{x} \geq \mathbf{b}_1$ *entails* the system of linear inequalities $A_2 \mathbf{x} \geq \mathbf{b}_2$ if every integral solution of the second system is a solution to the first system. It easily follows from the result of Papadimitriou [21] on the complexity of the integer programming problem that the entailment problem is co-NP complete.

Let the sets $\langle H_0, H_1 \rangle$ be a basis for the system of linear inequalities $A_2 \mathbf{x} \geq \mathbf{b}_2$. In [24] the following criterion for the entailment problem is stated.

Corollary 1. *The system $A_1 \mathbf{x} \geq \mathbf{b}_1$ entails $A_2 \mathbf{x} \geq \mathbf{b}_2$ if*
  a) $A_1 \mathbf{h} \geq \mathbf{b}_1$ *for every $\mathbf{h}$ in H_1*
  b) $A_1 \mathbf{h} \geq \mathbf{0}$ *for every $\mathbf{h}$ in H_0*

It follows that with the knowledge of a basis, the entailment problem becomes easy.

In a situation where many instances of the entailment problem have to be solved where the second system remains the same, computing a basis once and using this criterion may be more efficient then solving each instance separately. When the number of instances grows, the effect of economy of scale comes into play.

Essentially, specialization of entailment tests based on the use of bases of solutions allows one to perform a linear time (in the size of the basis) computation instead of solving instances of a co-NP complete problem. Rybina and Voronkov [24] use this method in the nonnegative case and present empirical evidence confirming that the method performs much better then other ways of solving instances of entailment problem during the run of a model checker. A related technique is used in model checkers for transition systems with variables that range over the reals [11] and is implemented in state of the art tools like HyTech [14]. Using the results presented in this paper this approach can be applied to systems of linear inequalities that have matrices with nonzero column defect.

Acknowledgement

We would like to thank Tatiana Rybina for comments on a preliminary version of this paper and an anonymous referee for several helpful suggestions.

References

1. Karen Aardal, Robert Weismantel, and Laurence A. Wolsey. Non-standard approaches to integer programming. *Discrete Applied Mathematics*, 123(1-3):5–74, 2002.
2. Farid Ajili and Evelyn Contejean. Avoiding slack variables in the solving of linear Diophantine equations and inequations. *Theoret. Comput. Sci.*, 173:183–208, 1997.
3. Alexander Bockmayr and Volker Weispfenning. Solving numerical constraints. In A. Robinson and A. Voronkov, editors, *Hanbook of Automated Reasoning*, volume I, chapter 12, pages 751–842. Elsevier Science, 2001.
4. Winfried Bruns, Joseph Gubeladze, Martin Henk, Alexander Martin, and Robert Weismantel. A counterexample to an integer analogue of Carathéodory's theorem. *J. Reine Angew. Math.*, 510:179–185, 1999.
5. Eric Domenjoud. Solving systems of linear Diophantine equations: An algebraic approach. In Andrzej Tarlecki, editor, *MFCS*, volume 520 of *Lecture Notes in Computer Science*, pages 141–150. Springer, 1991.
6. Arnaud Durand, Miki Hermann, and Laurent Juban. On the complexity of recognizing the hilbert basis of a linear diophantine system. In Miroslaw Kutylowski, Leszek Pacholski, and Tomasz Wierzbicki, editors, *MFCS*, volume 1672 of *Lecture Notes in Computer Science*, pages 92–102. Springer, 1999.
7. Miguel Filgueiras and Ana Paula Tomás. A fast method for finding the basis of nonnegative solutions to a linear Diophantine equation. *J. Symbolic Comput.*, 19:507–526, 1995.
8. Miguel Filgueiras and Ana Paula Tomás. Solving linear Diophantine equations using the geometric structure of the solution space. In Hubert Comon, editor, *RTA*, volume 1232 of *LNCS*, pages 269–283. Springer, 1997.
9. R. Giles and W.R.Pulleyblank. Total dual integrality and integer polyhedra. *Linear Algebra and Appl.*, 25:191–196, 1979.
10. P. Gordan. Über die auflösung linearer gleichungen mit reellen coefficienten. *Math. Ann.*, 6:23–28, 1873.
11. Nicolas Halbwachs. Delay analysis in synchronous programs. In Costas Courcoubetis, editor, *CAV'93, proceedings*, volume 697 of *LNCS*, pages 333–346. Springer, 1993.
12. Raymond Hemmecke. On the computation of Hilbert bases and extreme rays of cones. E-print arXiv:math.CO/0203105, March 2002.
13. Martin Henk and Robert Weismantel. On minimal solutions of linear diophantine equations. *Contrib. Algebra and Geometry*, 41(1):49–55, 2000.
14. T. Henzinger, Pei-Hsin Ho, and H. Wong-Toi. HyTech: A model checker for hybrid systems. In *CAV'97, proceedings*, volume 1254 of *LNCS*, pages 460–463. Springer, 1997.
15. Miki Hermann, Laurent Juban, and Phokion G. Kolaitis. On the complexity of counting the hilbert basis of a linear diophnatine system. In Harald Ganzinger, David A. McAllester, and Andrei Voronkov, editors, *LPAR*, volume 1705 of *Lecture Notes in Computer Science*, pages 13–32. Springer, 1999.
16. D. Hilbert. Über die endlichkeit des invariantensystems für binäre grundformen. *Math. Ann.*, 33:223–226, 1888. In German.
17. Gérard P. Huet. An algorithm to generate the basis of solutions to homogeneous linear Diophantine equations. *Inform. Process. Lett.*, 7:144–147, 1978.

18. Jean Lambert. Une borne pour générateurs des solutions entière positives d'une équation diophantienne linéaire. (A bound for the minimal positive integer solutions of a linear diophantine equation). (in French). *C. R. Acad. Sci. Paris Sér. I Math.*, 305:39–40, 1987.
19. Ernst W. Mayr. Some complexity results for polynomial ideals. *J. Complexity*, 13(3):303–325, 1997.
20. Daniele Micciancio and Bogdan Warinschi. A linear space algorithm for computing the Hermite normal form. In *ISAAC'01, proceedings*, pages 231–236. ACM, 2001.
21. Christos H. Papadimitriou. On the complexity of integer programming. *J. ACM*, 28(4):765–768, 1981.
22. Dmitrii V. Pasechnik. On computing Hilbert bases via the Elliot–MacMahon algorithm. *Theoret. Comput. Sci.*, 263:37–46, 2001.
23. Loic Pottier. Minimal solutions of linear diophantine systems: Bounds and algorithms. In Ronald V. Book, editor, *RTA*, volume 488 of *LNCS*, pages 162–173. Springer, 1991.
24. Tatiana Rybina and Andrei Voronkov. Using canonical representations of solutions to speed up infinite-state model checking. In Ed Brinksma and Kim Guldstrand Larsen, editors, *CAV'02, proceedings*, volume 2404 of *LNCS*, pages 386–400. Springer, 2002.
25. Tatiana Rybina and Andrei Voronkov. Fast infinite-state model checking in integer-based systems. In Matthias Baaz and Johann A. Makowsky, editors, *CSL'03, proceedings*, volume 2803 of *LNCS*, pages 546–573. Springer, 2003.
26. Alexander Schrijver. *Theory of Linear and Integer Programming*. John Wiley, 1998.
27. A. Schulz and R. Weismantel. An oracle-polynomial time augmentation algorithm for integer programming. In *ACM–SIAM SODA'99, proceedings*, pages 967–968, 1999.
28. András Sebő. Hilbert bases, Caratheodory's theorem and combinatorial optimization. In Ravi Kannan and William R. Pulleyblank, editors, *IPCO'90, proceedings*, pages 431–455. University of Waterloo Press, 1990.
29. Mark E. Stickel. A unification algorithm for associative-commutative functions. *J. ACM*, 28:423–434, 1981.
30. J.G. van der Corput. Über systeme von linear-homogenen Gleichungen und Ungleichungen. (in German). *Proc. Roy. Acad. Amsterdam*, 34:368–371, 1931.

Asynchronous Deterministic Rendezvous in Graphs

Gianluca De Marco[1], Luisa Gargano[1], Evangelos Kranakis[2],
Danny Krizanc[3], Andrzej Pelc[4], and Ugo Vaccaro[1]

[1] Dipartimento di Informatica e Applicazioni, Università di Salerno,
84081 Baronissi (SA), Italy
{demarco, lg, uv}@dia.unisa.it
[2] School of Computer Science, Carleton University, Ottawa,
Ontario, K1S 5B6, Canada*
kranakis@scs.carleton.ca
[3] Computer Science Group, Mathematics Department,
Wesleyan University, Middletown, CT 06459 USA
dkrizanc@wesleyan.edu
[4] Département d'informatique, Université du Québec en Outaouais,
Gatineau, Québec J8X 3X7, Canada**
pelc@uqo.ca

Abstract. Two mobile agents (robots) having distinct labels and located in nodes of an unknown anonymous connected graph, have to meet. We consider the asynchronous version of this well-studied rendezvous problem and we seek fast deterministic algorithms for it. Since in the asynchronous setting meeting at a node, which is normally required in rendezvous, is in general impossible, we relax the demand by allowing meeting of the agents inside an edge as well. The measure of performance of a rendezvous algorithm is its *cost*: for a given initial location of agents in a graph, this is the number of edge traversals of both agents until rendezvous is achieved. If agents are initially situated at a distance D in an infinite line, we show a rendezvous algorithm with cost $O(D|L_{min}|^2)$ when D is known and $O((D + |L_{max}|)^3)$ if D is unknown, where $|L_{min}|$ and $|L_{max}|$ are the lengths of the shorter and longer label of the agents, respectively. These results still hold for the case of the ring of unknown size but then we also give an optimal algorithm of cost $O(n|L_{min}|)$, if the size n of the ring is known, and of cost $O(n|L_{max}|)$, if it is unknown. For arbitrary graphs, we show that rendezvous is feasible if an upper bound on the size of the graph is known and we give an optimal algorithm of cost $O(D|L_{min}|)$ if the topology of the graph and the initial positions are known to agents.

1 Introduction

The Problem. Two mobile agents (robots) initially located in nodes of a network, modeled as an undirected connected graph, have to meet. This task is

---

* Research supported in part by NSERC and MITACS.
** Research supported in part by NSERC grant OGP 0008136 and by the Research Chair in Distributed Computing of the Université du Québec en Outaouais.

known in the literature as the rendezvous problem in graphs. It was mostly studied in the synchronous setting and meeting was required at a node. In this paper we study the asynchronous version of the rendezvous problem. In this setting meeting at a node may be impossible even in the two-node graph, as the adversary can desynchronize the agents and make them visit nodes at different times. Thus we have to relax the requirement and allow agents to meet either in a node or inside an edge. Such a definition of meeting is natural, e.g., when agents are robots traveling in a labyrinth. We seek efficient deterministic algorithms to solve this asynchronous rendezvous problem.

If nodes of the graph are labeled then agents can decide to meet at a predetermined node and the rendezvous problem reduces to graph exploration. However, in many applications, when rendezvous is needed in a network of unknown topology, such unique labeling of nodes may be unavailable, or agents may be unable to perceive such labels. Hence it is important to design rendezvous algorithms for agents operating in *anonymous* graphs, i.e., graphs without unique labeling of nodes. Clearly, the agents have to be able to *locally* distinguish ports at a node: otherwise, an agent may even be unable to visit all neighbors of a node of degree 3 (after visiting the second neighbor, the agent cannot distinguish the port leading to the first visited neighbor from that leading to the unvisited one). Consequently, agents initially located at two nodes of degree 3, might never be able to meet. Hence we make a natural assumption that all ports at a node are locally labeled $1,...,d$, where d is the degree of the node. No coherence between those local labelings is assumed. When an agent leaves a node, it is aware of the port number by which it leaves and when it enters a node, it is aware of the entry port number and of the degree of the node. Unless otherwise stated, we do not assume any knowledge of the topology of the graph or of its size. Likewise, agents are unaware of the distance separating them.

The Model. *The network.* The network is modeled as an undirected connected graph. Since we allow meetings inside an edge, we have to avoid unwanted crossings. Thus, for simplicity, we consider an embedding of the underlying graph in the three-dimensional Euclidean space, with nodes of the graph being points of the space and edges being pairwise disjoint line segments joining them. For any graph such an embedding exists. Mobile agents are modeled as points moving inside this embedding.

Adversarial Decisions, Definition of Rendezvous and Its Cost. An agent, currently located at a node, does not know the other endpoints of yet unexplored incident edges. If the agent decides to traverse such a new edge, the choice of the actual edge belongs to the adversary, as we are interested in the worst-case performance. This choice given to the adversary captures the fact that the topology and the orientation of the network are unknown to agents. Clearly, sometimes an agent may decide to traverse an already known edge, e.g., when it traverses an edge, goes back and then traverses it again. An algorithm for agent with label L depends on L and causes the agent to make the following decision at any node of the graph: either take a specific already explored incident edge, or take a yet unexplored incident edge (in which case the choice of the edge is made by the adversary). There is another important choice given to the adversary, this one capturing the asynchronous characteristics of the process. When the agent, situated at a node v at time t_0 has to traverse an edge modeled as a segment $[v,w]$, the adversary performs the following choice. It selects a time point $t_1 > t_0$ and any continuous function $f : [t_0, t_1] \longrightarrow [v, w]$, with $f(t_0) = v$ and $f(t_1) = w$. This function models the actual movement of the agent inside the line segment $[v, w]$ in the time period $[t_0, t_1]$. Hence this movement can be at arbitrary speed, the agent may go back and forth, as long as it does not leave the segment and the movement is continuous. We say that at time $t \in [t_0, t_1]$ the agent is in point $f(t) \in [v, w]$. Moreover, the adversary chooses the starting time of the agent. Hence an agent's trajectory is represented by the concatenation of the functions chosen by the adversary for consecutive edges that the agent traverses. (Recall that the choice of the edge incident to a current node also belongs to the adversary, whenever the edge is yet unexplored.) For a given algorithm, given starting points of agents and a given sequence of adversarial decisions in an embedding of a graph G, a rendezvous occurs if both agents are at the same point at the same time. We say that rendezvous is *feasible* in a given graph, if there exists an algorithm for agents such that for any embedding of the graph, any starting points and any sequences of adversarial decisions, the rendezvous does occur. The *cost* of rendezvous is defined as the worst-case number of edge traversals by both agents (the last partial traversal counted as a complete one for both agents), where the worst case is taken over all decisions of the adversary.

Labels and Local Knowledge. If agents are identical, i.e., they do not have distinct identifiers, and execute the same algorithm, then deterministic rendezvous is impossible even in the oriented ring: the adversary will make the agents move in the same direction at the same speed, thus they will never meet. Hence we assume that agents have distinct identifiers, called labels, which are two different nonempty binary strings, and that every agent knows its own label. If *both* agents knew *both* labels, the problem could be again reduced to that of graph exploration: the agent with smaller label does not move, and the other agent searches the graph until it finds it. (This strategy is sometimes called "wait for mommy".) However, the assumption that agents know each other may often be unrealistic: agents may be created in different parts of the network in a distributed fashion, oblivious of each other. Hence we assume that each agent knows its own label but does not know the label of the other. The only initial input of a (deterministic) rendezvous algorithm executed by an agent is the agent's label. During the execution of the algorithm, an agent learns the local port number by which it enters a node and the degree of the node.

Notation. L_{min} denotes the shorter label and L_{max} the longer one, with ties broken arbitrarily. If L is a label, $|L|$ denotes its length. n denotes the number of nodes in the graph, and D the distance between initial positions of the agents.

1.1 Our Results

We first look at the case of rendezvous in an infinite line. Besides its intrinsic interest, this case is important, as the results for the infinite line carry over to the case of the ring of unknown size and the cost depends on the initial distance between agents rather than on the size of the ring. For agents initially situated at a distance D in an infinite line, we show a rendezvous algorithm with cost $O(D|L_{min}|^2)$ when D is known and $O((D+|L_{max}|)^3)$ if D is unknown, where $|L_{min}|$ and $|L_{max}|$ are the lengths of the shorter and longer label of the agents, respectively. These results still hold for the case of the ring (even of unknown size) but then we also give an algorithm of cost $O(n|L_{min}|)$ (and this is optimal), if the size n of the ring is known, and of cost $O(n|L_{max}|)$, if it is unknown. In both these algorithms the knowledge of the initial distance D between agents is not assumed, and for D of the order of n, their complexity is better than that of infinite line algorithms. On the other hand, for small D and small labels of agents, the opposite is true. For arbitrary graphs, we show that rendezvous is feasible if an upper bound on the size of the graph is known, and we give an optimal algorithm of cost $O(D|L_{min}|)$ if the topology of the graph and the initial positions are known to agents.

It should be noted that asynchronous rendezvous techniques significantly differ from the synchronous case. An important ingredient in synchronous rendezvous algorithms in graphs is the insertion of idle periods for each of the agents (depending on its label) during which the other agent walks in the graph and has the chance of meeting the standing agent. This is, of course, impossible in the asynchronous setting, as the adversary controls waiting time of the agents. Instead, the algorithm should be designed to force the agents to move on the same path in opposite directions sufficiently far to guarantee meeting. This is complicated by the fact that agents do not have any sense of direction and do not know each other's labels which serve as the algorithm's parameters.

1.2 Related Work

The rendezvous problem has been introduced in [24]. The vast literature on rendezvous (see the book [4] for a complete discussion and more references) can be divided into two classes: papers considering the geometric scenario (rendezvous in the line, see, e.g., [11,12,19], or in the plane, see, e.g., [9,10]), and those discussing rendezvous in graphs, e.g., [2,5]. Most of the papers, e.g., [2,3,7,11,20] consider the probabilistic scenario: inputs and/or rendezvous strategies are random. In [20] randomized rendezvous strategies are applied to study self-stabilized token management schemes. Randomized rendezvous strategies use random walks in graphs, which have been widely studied and applied also, e.g., in graph traversing [1], on-line algorithms [14] and estimating volumes of convex bodies [17]. A natural extension of the rendezvous problem is that of gathering [18,20,23,25], when more than 2 agents have to meet in one location.

Deterministic rendezvous with anonymous agents working in unlabeled graphs but equipped with tokens used to mark nodes was considered e.g., in [22]. In [26] the authors considered rendezvous of many agents with unique labels. In [16,21] deterministic rendezvous in graphs with labeled agents was considered. However, in all the above papers, the synchronous setting was assumed. While asynchronous rendezvous under geometric scenarios has been studied, e.g., in

[18], the present paper is, to the best of our knowledge, the first to consider deterministic asynchronous rendezvous in graphs.

Due to space limitations, proofs are deferred to the journal version.

2 Rendezvous in the Infinite Line

We first observe that a simple modification of the (synchronous) rendezvous algorithm proposed in [16] can be used to perform asynchronous rendezvous in an n-node tree in time $O(n)$. Trees have a convenient feature from the point of view of rendezvous. Every tree has either a *central node*, defined as the unique node minimizing the distance from the farthest leaf, or a *central edge*, defined as the edge joining the only two such nodes. This suggests the following natural rendezvous algorithm: explore the tree, find the central node or the central edge, and try to meet there. Each agent can explore the tree by DFS, keeping a stack for used port numbers. At the end of the exploration, the agent has a map of the tree, can identify the central node or the central edge, and can find its way either to the central node or to one endpoint of the central edge, in the latter case knowing which port corresponds to the central edge. In the first case, rendezvous is accomplished after the other agent gets to the central node. In the second case, both agents traverse the central edge once and they have to meet by the time the later agent performs this traversal.

However, the above method uses the possibility of exploring the entire tree in order to construct a map of it. This is impossible, e.g., in the case of the infinite line. In this section we consider the case when the two agents are initially situated in an infinite line at distance D. The case of the infinite line is also important because the obtained results carry over to a ring of arbitrary unknown size and do not depend on this size.

2.1 Distance D Known to Agents

In this section, we present a rendezvous algorithm with cost $O(D|L_{min}|^2)$. The algorithm assumes that the agents know D and it is formulated for an agent with label L. Each agent has an initial local orientation left-right and these orientations of both agents may be the same or different.

Algorithm Rendezvous-in-Infinite-Line(D) The algorithm consists of two parts: Label Transformation and Label Execution.

Label Transformation. The Label Transformation part takes the label L of an agent and produces the label L^* consisting of a string of $|L|$ zeros, followed by a 1 and then followed by the string L.

Label Execution. For a given agent, we define the execution of the i-th bit of L^* as performing $2iD$ steps left, $(4i+1)D$ steps right and $(2i+1)D$ steps left (resp. $2iD$ steps right, $(4i+1)D$ steps left and $(2i+1)D$ steps right) if $L^*(i) = 0$ (resp. if $L^*(i) = 1$), according to the agent's initial local orientation. For an agent with label L, the Label Execution part consists of consecutive executions of all bits of L^* from left to right. The algorithm stops when rendezvous is achieved.

The following fact is straightforward.

Fact 2.1.

1. *The label execution of* L^* *requires* $\sum_{i=1}^{|L^*|}(8i+2)D = \sum_{i=1}^{2|L|+1}(8i+2)D = O(|L|^2 D)$ *steps;*
2. *For any labels* L_1, L_2, *if* $L_1 \neq L_2$ *then none of the* L_1^*, L_2^* *is a prefix of the other;*

The execution of the p-th bit of L^* can be divided into three segments: the *first segment* consists of the first $2pD$ steps, the *second segment* is formed by the next $(4p+1)D$ steps and finally the *last segment* consists of the last $(2p+1)D$ steps. Let $s_H(p)$ be the time when agent H finishes executing the p-th bit of its transformed label.

Correctness and Analysis. Fix a global left-right orientation of the line. Below we use the terms "left" and "right" according to this fixed orientation. We define the direction of an agent H on bit position p as left (resp. right) if the execution of the first segment of p for H consists of $2pD$ consecutive steps to the left (resp. right) from its initial position. It is clear that depending on their local orientations and on the values of the p-th bit of their transformed labels, the agents may have either the same or different directions on a given bit position. When the directions of the agents differ on a given bit position, two situations are possible during the execution of the *first* segment of p, depending on the agents' initial positions: either in the first step the agents move approaching each other (in this case, we say that the directions are *convergent*), or they move receding farther from each other (in this case, we say that the directions are *divergent*).

Lemma 1. *If the agents have different directions on a given bit position p, they meet by time* $\min\{s_A(p), s_B(p)\}$.

Proof. Two cases may happen: either (a) the directions are convergent or (b) they are divergent.

Case (a). Let X be the agent that first completes the execution of the first segment of the p-th bit of its transformed label. At this time, if Y is already executing the p-th bit, then they have to meet because they are at distance D and move approaching each other for more than D steps. If Y is still executing some previous bit $q \leq p-1$, then it can be at a distance of at most $(2(p-1)+1)D = (2p-1)D$ steps from its initial position (in some direction), while the first segment of the pth bit executed by X carries it at distance $2pD$, starting towards the initial position of Y. Again, recalling that their initial positions are at distance D, they have to meet.

Case (b). Let X be the agent that first completes the execution of the second segment of the p-th bit of its transformed label. During the second segment, agent X moves $(2p+1)D$ steps from its initial position, towards Y's initial position. As in the previous case, if Y is still executing some bit $q \leq p-1$ when X completes the last segment, then Y can be at most $(2(p-1)+1)D = (2p-1)D$ steps (in any direction) from its initial position: the agents have to meet.

Assume now that at the time τ when X completes the second segment of the p-th bit, Y is already executing the p-th bit. In this case, we can observe that during the execution of the second segment of their p-th bit, the two agents move approaching each other. It follows that at time τ, agent Y can be as far from its initial position as permitted by the execution of the first segment of p, i.e., at most $2pD$ steps in some direction. Recalling that X and Y's initial positions are D steps apart one from the other, and that X moves $(2p+1)D$ steps from its initial position towards Y's initial position, agents have to meet by time $\min\{s_X(p), s_Y(p)\}$.

Theorem 2 (Correctness). *Let q be the length of the shortest tranformed label. Agents must meet by time $\min\{s_A(q), s_B(q)\}$.*

Proof. In view of Lemma 1, it is enough to show that there exists a bit position $1 \leq p \leq q$ on which the agents have different directions. In fact, if the orientations of the agents differ, then $p = 1$; otherwise, in view of part 2 of Fact 2.1, there is a bit position $d \leq q$ where the transformed labels differ. In this case, $p = d$.

Theorem 3 (Analysis). *Assume the agents are initially at distance D and D is known to both of them. Then the cost of Algorithm Rendezvous-in-Infinite-Line(D) is $O(D|L_{min}|^2)$.*

Proof. Each agent runs the algorithm proceeding from left to right on the transformed label, executing bit by bit. By Theorem 2 we have that the agents must meet by time $\min\{s_A(d), s_B(d)\}$, where d is the first bit position where their transformed labels differ. Since $d = O(|L_{min}|)$, by part 1. of Fact 2.1, the agents must meet after $O(D|L_{min}|^2)$ steps.

2.2 Distance D Unknown to Agents

In this section, we present a rendezvous algorithm with cost $O((D + L_{max})^3)$. The algorithm does not assume the knowledge of D. It is formulated for an agent with label L. Each agent has an initial local orientation left-right.

Algorithm Rendezvous-in-Infinite-Line The algorithm consists of two parts: Label Transformation and Label Execution.

Label Transformation. The Label Transformation part takes the label L of an agent and produces the label L^* in the following way.

Step 1. Produce label L' as follows: insert a new bit before every bit of L, alternating 0 and 1 (e.g., if L is the string 110, we obtain the new string 101100).

Step 2. Produce label L'' by adding the pattern 11110 at the end of L'.

Step 3. Finally, label L^*, called the *transformed label* of the agent, is obtained as an infinite concatenation of copies of L''.

Label Execution. For a given agent, we define the execution of the i-th bit of L^* as performing i^2 steps left, $2i^2 + i$ steps right and $i^2 + i$ steps left (resp. i^2 steps right, $2i^2 + i$ steps left and $i^2 + i$ steps right), if $L^*(i) = 0$ (resp. if $L^*(i) = 1$) according to the agent's initial local orientation. For an agent with label L, the Label Execution part consists of consecutive executions of all bits of L^*, from left to right, until rendezvous is achieved.

As in the previous subsection, the execution of the p-th bit of L^* can be divided into three segments: the first p^2 steps form the *first segment* of p, the next $2p^2 + p$ form the *second segment* of p and the last $p^2 + p$ steps of p form the *last segment*.

Denote by $L^*(a, b)$ the substring of L^* contained between bit positions a and b, extremities included. The following fact is straightforward.

Fact 2.2.

1. L^* does not contain the substring 0000;
2. The only place in L^* where the substring 11110 occurs is at the end of a copy of L';
3. For any bit position b of L^*, the execution of $L^*(1, b)$ requires $\sum_{i=1}^{b}(4i^2 + 2i) = O(b^3)$ steps;
4. For any labels L_1, L_2, if $L_1 \neq L_2$ then $L'_1 \neq L'_2$;

Correctness and Analysis. Directions of agents on a given bit position are defined analogously as in Section 2.1. Also the notions of convergent and divergent directions on a given bit position are similar.

Lemma 4. *Let X and Y be the two agents. For any $p > 1$, there exists a $q > p$ such that $q - p = O(|L_{max}|)$ and the direction of X on $L_X^*(q)$ is different from that of Y on $L_Y^*(q)$.*

As before, the integer $s_H(p)$ denotes the time when agent H finishes executing the p-th bit of its transformed label.

Lemma 5. *If the agents have different directions on a given bit position $p \geq D$, they meet by time $\min\{s_A(p), s_B(p)\}$.*

Theorem 6 (Correctness and analysis). *Assume the agents are initially at distance D. Algorithm Rendezvous-in-Infinite-Line achieves the rendezvous in $O((D + |L_{max}|)^3)$ steps.*

Proof. Each agent runs the algorithm proceeding from left to right on the transformed label, executing bit by bit. By Lemma 5 the agents must meet by time $\min\{s_A(d), s_B(d)\}$, where $d \geq D$ is the first bit position on which they have different directions. By Lemma 4, there exists a bit position $d \geq D$ on which the agents have different directions. This proves the correctness of the algorithm.

Moreover, Lemma 4 also guarantees that $d = D + O(|L_{max}|) = O(D + |L_{max}|)$. Hence, from part 3 of Fact 2.2, the cost of the algorithm is $\sum_{i=1}^{d}(4i^2 + 2i) = O((D + |L_{max}|)^3)$.

3 Optimal Rendezvous in the Ring

Our results for the infinite line carry over to the case when agents are situated in a ring of unknown size n. However, for the ring there is no danger of "infinite escape" by going in divergent directions: in this case the agents can still meet "on the other side" of the ring. Consequently, for the ring we get rendezvous algorithms whose cost depends on n (which is also an upper bound on the initial distance between agents) and on the size of the labels. The knowledge of D is not assumed and for D of the order of n the bound on the cost of rendezvous is better than that previously established.

We present a rendezvous algorithm with cost of optimal order of magnitude $O(n|L_{min}|)$, working on an arbitrary unoriented ring of known size and an algorithm with cost $O(n|L_{max}|)$, when the size of the ring is unknown. Since the ring is unoriented, each of the agents has a local right/left orientation and these orientations for the two agents may differ.

Algorithm Rendezvous-in-Ring The algorithm consists of two parts: Label Transformation and Label Execution. The Label Transformation part takes the label L of an agent and produces the label L^* in the following way. First produce label L' consisting of a string of $|L|$ zeros, followed by a 1 and then followed by the string L. The label L^*, called the *transformed label* of the agent, is obtained by replacing in L' each 0 by 01 and each 1 by 10.

The Label Execution part is divided into phases numbered 1,2,... For a given agent, we define the execution of bit 0 (resp. 1) in phase a as performing 3^a steps left (resp. right), according to the agent's local orientation. For an agent with label L, phase a consists of consecutive executions of all bits of L^* from left to right. Since the agents do not know the size of the ring, the number of phases is unbounded. The algorithm stops when rendezvous is achieved.

The following fact is straightforward.

Fact 3.1.

1. $|L^*| = O(|L|)$;
2. For any labels L_1, L_2, if $L_1 \neq L_2$ then none of the L_1^*, L_2^* is a prefix of the other;
3. There are no more than two consecutive equal bits in L^*;

Correctness and Analysis. In order to show the correctness of Algorithm Rendezvous-in-Ring, we need to prove that for any size of the ring, any initial position of the agents and any behavior of the adversary, the agents will eventually meet. Let n be the size of the ring and $x = \lceil \log_3 n \rceil + 1$. Consider agents A and B. Let L_A and L_B be the labels of agents A and B, respectively. Let P be the longest common prefix of L_A^* and L_B^*. For the b-th bit of L_A^*, we denote by $L_A^*(b)$ the value of this bit. Similarly for $L_B^*(b)$. If the b-th bit of L_A^* is still in P then we use notation $P(b)$ to denote $L_A^*(b) = L_B^*(b)$. For an agent H and bit position b, let $t'_H(b)$ be the time when agent H starts executing the b-th bit of its transformed label in phase x. Analogously, let $t''_H(b)$ be the time when agent H finishes executing the b-th bit of its transformed label in phase x.

The following fact is straightforward.

Fact 3.2. *If during a time interval $[t_1, t_2]$ one of the agents does at least $c + n$ consecutive steps in one direction and the other agent does at most c steps in this direction, for any $c \geq 0$, then they must meet by time t_2.*

We first state two lemmas and then we prove the correctness and the time complexity of Algorithm Rendezvous-in-Ring.

Lemma 7. *Let X be the agent that first starts the execution of phase x and let Y be the other agent (i.e. $t'_X(1) \leq t'_Y(1)$).*

1. *If the agents don't meet by time $\min\{t''_X(1), t''_Y(1)\}$, then $t'_Y(1) < t''_X(1)$.*
2. *If the orientations of the agents differ, they meet by time $\min\{t''_X(1), t''_Y(1)\}$.*

The next lemma shows that if agents have the same orientation then they are "almost" synchronized in phase x on the common prefix P.

Lemma 8. *Suppose that both agents have the same orientation. Let P be the longest common prefix of L^*_A and L^*_B. Consider the b-th bit of P. Let X_b be the agent that first starts executing this bit in phase x and let Y_b be the other agent (i.e., $t'_{X_b}(b) \leq t'_{Y_b}(b)$). Then $t'_{Y_b}(b) \leq t''_{X_b}(b)$, unless the agents meet by time $t''_{X_b}(b+1)$.*

Theorem 9 (Correctness). *Let n be the size of the ring. Let d be the first position where the transformed labels of the agents differ. Agents must meet by time $\min\{t''_A(d), t''_B(d)\}$.*

Theorem 10 (Analysis). *The Algorithm Rendezvous-in-Ring has $O(n|L_{max}|)$ cost, for the n-node ring. Moreover, if the size n of the ring is known to the agents, then Algorithm Rendezvous-in-Ring can be modified to have time complexity $O(n|L_{min}|)$, which is optimal.*

Proof. It follows from Theorem 9 that by the meeting time none of the agents completed phase $x = \lceil \log_3 n \rceil + 1$. Hence the cost of the algorithm is $O(n(|L_A| + |L_B|)) = O(n|L_{max}|)$.

Suppose that both agents know n. Then they can start executing Algorithm Rendezvous-in-Ring at the beginning of phase x. By the proof of Theorem 9, agents must meet by time $\min\{t''_A(d), t''_B(d)\}$, where d is the first position where their transformed labels differ. Clearly, $d \leq |L_{min}|$. Hence both agents execute at most $|L_{min}|$ bits before meeting, which implies that the cost is $O(n|L_{min}|)$.[1] The optimality of this cost follows from a result in [16], applied to the case when the distance between agents is $\Theta(n)$.

---

[1] The reader may wonder why the bound of $O(n|L_{min}|)$ does not hold also in the case of n unknown. This is due to the fact that the work done by the agents through phases 1 to $x - 1$, where meeting is not yet assured, is already $\Theta(n(|L_A| + |L_B|))$.

4 Rendezvous in Arbitrary Graphs

We start with the following observation.

Proposition 11. *If a map of the graph with labeled ports and indicated initial positions of agents is available to both of them then deterministic asynchronous rendezvous can be done at cost $O(D|L_{min}|)$, which is optimal.*

Proof. Each agent computes the distance D between them and finds the lexicographically smallest path of length D from its own position to the position of the other agent (paths are viewed as sequences of port numbers). Thus both agents identify the same cycle of length $2D$ on which their both initial positions are situated. (Notice that the cycle need not be simple, some edges may be repeated, it may even degenerate to one path considered in both directions.) Then agents apply the modified version of Algorithm Rendezvous-in-Ring to this cycle. The size of the cycle is known, so rendezvous can be achieved at cost $O(D|L_{min}|)$.

We conclude this section with a feasibility result in the case when an upper bound on the size of the graph is known to agents.

Theorem 12. *Suppose that a bound M on the number of nodes in the graph is known to both agents. Then deterministic asynchronous rendezvous is feasible.*

5 Conclusion

The results that we presented for the asynchronous deterministic rendezvous in graphs contribute to understanding the feasibility and complexity of this problem which seems far more complex than its synchronous counterpart. In fact, it remains open if asynchronous deterministic rendezvous is at all feasible in arbitrary graphs of unknown size. Our solution heavily uses the knowledge of the upper bound on the size.

As far as complexity is concerned, the optimal cost of rendezvous remains open even in an infinite line. Our algorithm for known D has time complexity $O(D|L_{min}|^2)$, while the lower bound, following from [16], is $\Omega(D|L_{min}|)$. In the case of the n-node ring, we established upper bounds $O(n|L_{min}|)$ and $O(n|L_{max}|)$, for the known and unknown size, respectively. Is rendezvous with cost $O(n|L_{min}|)$ possible for the ring of unknown size? Finally, our algorithm showing feasibility of rendezvous for arbitrary graphs with known upper bound M on the size is not efficient. Is there a rendezvous algorithm polynomial in the bound M and in the lengths of the agents' labels?

References

1. R. Aleliunas, R.M. Karp, R.J. Lipton, L. Lovász, and C. Rackoff, Random walks, universal traversal sequences, and the complexity of maze problems, Proc. FOCS'1979, 218-223.
2. S. Alpern, The rendezvous search problem, SIAM J. on Control and Optimization 33 (1995), 673-683.

3. S. Alpern, Rendezvous search on labelled networks, Naval Reaserch Logistics 49 (2002), 256-274.
4. S. Alpern and S. Gal, The theory of search games and rendezvous. Int. Series in Operations research and Management Science, Kluwer Academic Publisher, 2002.
5. J. Alpern, V. Baston, and S. Essegaier, Rendezvous search on a graph, Journal of Applied Probability 36 (1999), 223-231.
6. S. Alpern and S. Gal, Rendezvous search on the line with distinguishable players, SIAM J. on Control and Optimization 33 (1995), 1270-1276.
7. E. Anderson and R. Weber, The rendezvous problem on discrete locations, Journal of Applied Probability 28 (1990), 839-851.
8. E. Anderson and S. Essegaier, Rendezvous search on the line with indistinguishable players, SIAM J. on Control and Optimization 33 (1995), 1637-1642.
9. E. Anderson and S. Fekete, Asymmetric rendezvous on the plane, Proc. 14th Annual ACM Symp. on Computational Geometry, 1998.
10. E. Anderson and S. Fekete, Two-dimensional rendezvous search, Operations Research 49 (2001), 107-118.
11. V. Baston and S. Gal, Rendezvous on the line when the players' initial distance is given by an unknown probability distribution, SIAM J. on Control and Optimization 36 (1998), 1880-1889.
12. V. Baston and S. Gal, Rendezvous search when marks are left at the starting points, Naval Res. Log. 48 (2001), 722-731.
13. S.A. Cook and P. McKenzie, Problems complete for deterministic logarithmic space, Journal of Algorithms 8 (5) (1987), 385-394.
14. D. Coppersmith,, P. Doyle, P. Raghavan, and M. Snir, Random walks on weighted graphs, and applications to on-line algorithms, Proc. STOC'1990, 369-378.
15. D. Coppersmith, P. Tetali, and P. Winkler, Collisions among random walks on a graph, SIAM J. on Discrete Math. 6 (1993), 363-374.
16. A. Dessmark, P. Fraigniaud, and A. Pelc, Deterministic rendezvous in graphs, Proc. 11th European Symposium on Algorithms (ESA'2003), LNCS 2832, 184-195.
17. M. Dyer, A. Frieze, and R. Kannan, A random polynomial time algorithm for estimating volumes of convex bodies, Proc. 21st Annual ACM Symposium on Theory of Computing (STOC'1989), 375-381.
18. P. Flocchini, G. Prencipe, N. Santoro, P. Widmayer, Gathering of asynchronous oblivious robots with limited visibility, Proc. STACS'2001, LNCS 2010, 247-258.
19. S. Gal, Rendezvous search on the line, Operations Research 47 (1999), 974-976.
20. A. Israeli and M. Jalfon, Token management schemes and random walks yield self stabilizing mutual exclusion, Proc. PODC'1990, 119-131.
21. D. Kowalski, A. Pelc, Polynomial deterministic rendezvous in arbitrary graphs, Proc. 15th Annual Symposium on Algorithms and Computation (ISAAC'2004), December 2004, Hong Kong.
22. E. Kranakis, D. Krizanc, N. Santoro and C. Sawchuk, Mobile agent rendezvous in a ring, Proc. 23rd International Conference on Distributed Computing Systems (ICDCS'2003), 592-599.
23. W. Lim and S. Alpern, Minimax rendezvous on the line, SIAM J. on Control and Optimization 34 (1996), 1650-1665.
24. T. Schelling, The strategy of conflict, Oxford University Press, Oxford, 1960.
25. L. Thomas, Finding your kids when they are lost, Journal on Operational Res. Soc. 43 (1992), 637-639.
26. X. Yu and M. Yung, Agent rendezvous: a dynamic symmetry-breaking problem, Proc. International Colloquium on Automata, Languages, and Programming (ICALP'1996), LNCS 1099, 610-621.

Zeta-Dimension
(Preliminary Version)

David Doty[1], Xiaoyang Gu[1,*], Jack H. Lutz[1,**],
Elvira Mayordomo[2,***], and Philippe Moser[1,†]

[1] Department of Computer Science, Iowa State University, Ames, IA 50011, USA
{ddoty, xiaoyang, lutz, moser}@cs.iastate.edu
[2] Departamento de Informática e Ingeniería de Sistemas, María de Luna 1,
Universidad de Zaragoza, 50018 Zaragoza, Spain
elvira@unizar.es

Abstract. The *zeta-dimension* of a set A of positive integers is

$$\mathrm{Dim}_\zeta(A) = \inf\{s \mid \zeta_A(s) < \infty\},$$

where

$$\zeta_A(s) = \sum_{n \in A} n^{-s}.$$

Zeta-dimension serves as a fractal dimension on $\mathbb{Z}^+$ that extends naturally and usefully to discrete lattices such as $\mathbb{Z}^d$, where d is a positive integer.

This paper reviews the origins of zeta-dimension (which date to the eighteenth and nineteenth centuries) and develops its basic theory, with particular attention to its relationship with algorithmic information theory. New results presented include a gale characterization of zeta-dimension and a theorem on the zeta-dimensions of pointwise sums and products of sets of positive integers.

1 Introduction

Natural and engineered complex systems often produce structures with fractal properties. These structures may be explicitly observable (e.g., shapes of neurons or patterns created by cellular automata), or they may be implicit in the behaviors of the systems (e.g., strange attractors of dynamical systems, Brownian trajectories in financial data, or Boolean circuit complexity classes). In either

---

* This research was supported in part by National Science Foundation Grant 0344187.
** This research was supported in part by National Science Foundation Grant 0344187.
*** This research was supported in part by Spanish Government MEC project TIC 2002-04019-C03-03.
† This research was supported in part by Swiss National Science Foundation Grant PBGE2–104820.

case, the choice of appropriate mathematical models is crucial to understanding the systems.

Many, perhaps most, fractal structures are best modeled by classical fractal geometry [15], which provides top-down specifications of many useful fractals in Euclidean spaces and other manifolds that support continuous mathematical methods and attendant methods of numerical approximation. Classical fractal geometry also includes powerful quantitative tools, the most notable of which are the *fractal dimensions* (especially Hausdorff dimension [19,15], packing dimension [39,38,15], and box dimension [15]). Theoretical computer scientists have recently developed *effective* fractal dimensions [28,26,27,10,4] that work in complexity classes and other countable settings, but these, too, are best regarded as continuous, albeit effective, mathematical methods.

Some fractal structures are inherently discrete and best modeled that way. To some extent this is already true for structures created by cellular automata. For the nascent theory of nanostructure self-assembly [1,33], the case is even more compelling. This theory models the *bottom-up* self-assembly of molecular structures. The tile assembly models that achieve this cannot be regarded as discrete approximations of continuous phenomena (as cellular automata often are), because their bottom-level units (tiles) correspond directly to discrete objects (molecules). Fractal structures assembled by such a model are best analyzed using discrete tools.

This paper concerns a discrete fractal dimension, called *zeta-dimension*, that works in discrete lattices such as $\mathbb{Z}^d$, where d is a positive integer. Curiously, although our work is motivated by twenty-first century concerns in theoretical computer science, zeta-dimension has its mathematical origins in eighteenth and nineteenth century number theory. Specifically, zeta-dimension is defined in terms of a generalization of Euler's 1737 *zeta-function* [14] $\zeta(s) = \sum_{n=1}^{\infty} n^{-s}$, defined for nonnegative real s (and extended in 1859 to complex s by Riemann [32], after whom the zeta-function is now named). Moreover, this generalization can be formulated in terms of Dirichlet series [12], which were developed in 1837, and one of the most important properties of zeta-dimension (in modern terms, the entropy characterization) was proven in these terms by Cahen [8] in 1894.

Our objectives here are twofold. First, we present zeta-dimension and its basic theory, citing its origins in scattered references, but stating things in a unified framework emphasizing zeta-dimension's role as a discrete fractal dimension in theoretical computer science. Second, we present several results on zeta-dimension and its interactions with classical fractal geometry and algorithmic information theory.

Our presentation is organized as follows. In section 2, we give an intuitive development of zeta-dimension in the positive integers. In section 3, we extend this development in a natural way to the integer lattices $\mathbb{Z}^d$, for $d \geq 1$. In addition to reviewing known properties of zeta-dimension, we prove discrete analogs of two theorems of classical fractal geometry, namely, the dimension inequalities for Cartesian products and the total disconnectivity of sets of dimension less than 1.

In section 4, we discuss relationships between zeta-dimension and classical fractal dimensions. Many discrete fractals in $\mathbb{Z}^d$ have been observed to "look like" corresponding fractals in $\mathbb{R}^d$. The most famous such correspondence is the obvious resemblance between Pascal's triangle modulo 2 and the Sierpinski triangle [36]. We define a version of discrete self similar fractal and its continuous counterpart and use result from [6] to show that the zeta-dimension of the discrete fractal is always the Hausdorff dimension of its continuous version. We will further discuss issues along these lines [6,7,37] in the full version of this paper. We also prove a result relating zeta-dimension in $\mathbb{Z}^+$ to Hausdorff dimension in the Baire space.

Section 5 concerns the relationships between zeta-dimension and algorithmic information theory. We review the Kolmogorov-Staiger characterization [43,35] of the zeta-dimensions of computably enumerable and co-computably enumerable sets in terms of the Kolmogorov complexities (algorithmic information contents) of their elements. We prove a theorem on the zeta-dimensions of sets of positive integers that are defined in terms of the digits, or strings of digits, that can occur in the base-k expansions of their elements. Most significantly, we prove that zeta-dimension, like classical and effective fractal dimensions, can be characterized in terms of gales. Finally, we prove a theorem on the zeta-dimensions of pointwise sums and products of sets of positive integers that may have bearing on the question of which sets of natural numbers are definable by McKenzie-Wagner circuits [29].

Note: Researchers have considered other fractal dimensions in $\mathbb{Z}^d$ that are not equivalent to zeta-dimension, but nevertheless of interest [5,6,16,21]. These will be discussed further in the full version of this paper.

Throughout this paper, $\log t = \log_2 t$, and $\ln t = \log_e t$.

2 Zeta-Dimension in $\mathbb{Z}^+$

A set of positive integers is generally considered to be "small" if the sum of the reciprocals of its elements is finite [2,18]. Easily verified examples of such small sets include the set of nonnegative integer powers of 2 and the set of perfect squares. On the other hand, the divergence of the harmonic series means that the set $\mathbb{Z}^+$ of all positive integers is not small, and a celebrated theorem of Euler [14] says that the set of all prime numbers is not small either.

If a set is small in the above qualitative (yes/no) sense, we are still entitled to ask, "Exactly how small is the set?" This section concerns a natural, quantitative answer to this question. For each set $A \subseteq \mathbb{Z}^+$ and each nonnegative real number s, let

$$\zeta_A(s) = \sum_{n \in A} n^{-s}. \tag{2.1}$$

Note that $\zeta_{\mathbb{Z}^+}$ is precisely ζ, the Riemann zeta-function [32] (actually, Euler's original version [14] of the zeta-function, since we only consider $\zeta_A(s)$ for real s). The *zeta-dimension* of a set $A \subseteq \mathbb{Z}^+$ is then defined to be

$$\mathrm{Dim}_\zeta(A) = \inf\{s | \zeta_A(s) < \infty\}. \tag{2.2}$$

Since $\zeta_{\mathbb{Z}^+}(s) < \infty$ for all $s > 1$, we have $0 \le \text{Dim}_\zeta(A) \le 1$ for every set $A \subseteq \mathbb{Z}^+$. By the results cited in the preceding paragraph, the set of all positive integers and the set of all prime numbers each have zeta-dimension 1. Every finite set has zeta-dimension 0, because $\zeta_A(0)$ is the cardinality of A. It is easy to see that the set of nonnegative integer powers of 2 also has zeta-dimension 0. For a deeper example, Wirsing's $n^{O(\frac{1}{\ln \ln n})}$ upper bound on the number of perfect numbers not exceeding n [42] implies that the set of perfect numbers also has zeta-dimension 0.

The zeta-dimension of a set of positive integers can also lie strictly between 0 and 1. For example, if A is the set of all perfect squares, then $\zeta_A(s) = \zeta(2s)$, so $\text{Dim}_\zeta(A) = \frac{1}{2}$. Similarly, the set of all perfect cubes has zeta-dimension $\frac{1}{3}$, etc. In fact, this argument can easily be extended to show that, for every real number $\alpha \in [0,1]$, there exist sets $A \subseteq \mathbb{Z}^+$ such that $\text{Dim}_\zeta(A) = \alpha$.

Intuitively, we regard zeta-dimension as a fractal dimension, analogous to Hausdorff dimension [19,15] or (more aptly, as we shall see) upper box dimension dimension [39,38,15], on the space $\mathbb{Z}^+$ of positive integers. This intuition is supported by the fact that zeta-dimension has the following easily verified functional properties of a fractal dimension.

1. Monotonicity: $A \subseteq B$ implies $\text{Dim}_\zeta(A) \le \text{Dim}_\zeta(B)$.
2. Stability: $\text{Dim}_\zeta(A \cup B) = \max\{\text{Dim}_\zeta(A), \text{Dim}_\zeta(B)\}$.
3. Translation invariance: For each $k \in \mathbb{Z}^+$, $\text{Dim}_\zeta(k+A) = \text{Dim}_\zeta(A)$, where $k + A = \{k+n | n \in A\}$.
4. Expansion invariance: For each $k \in \mathbb{Z}^+$, $\text{Dim}_\zeta(kA) = \text{Dim}_\zeta(A)$, where $kA = \{kn | n \in A\}$.

Equation (2.1) can be written as a Dirichlet series

$$\zeta_A(s) = \sum_{n=1}^{\infty} f(n) n^{-s} \qquad (2.3)$$

in which f is the characteristic function of A. In the terminology of analytic number theory, (2.2) then says that the zeta-dimension of A is the *abscissa of convergence* of the series (2.3) [23,18,2,3]. In this sense, zeta-dimension was introduced in 1837 by Dirichlet [12]. The following useful characterization of zeta-dimension was proven in this more general setting in 1894.

Theorem 2.1 (entropy characterization of zeta-dimension – Cahen [8]; see also [22,23,18,2,3]). *For all $A \subseteq \mathbb{Z}^+$,*

$$\text{Dim}_\zeta(A) = \limsup_{n \to \infty} \log |A \cap \{1,\ldots,n\}| / \log n. \qquad (2.4)$$

Example 2.2. *The set C', consisting of all positive integers whose ternary expansions do not contain a 1, can be regarded as a discrete analog of the Cantor middle thirds set C, which consists of all real numbers in $[0,1]$ who ternary expansions do not contain a 1. Theorem 2.1 implies immediately that C' has zeta-dimension $\frac{\log 2}{\log 3} \approx 0.6309$, which is exactly the classical fractal (Hausdorff, packing or box) dimension of C. We will see in section 4 that this is not a coincidence, but rather a special case of a general phenomenon.*

By Theorem 2.1 and routine calculus, we have

$$\mathrm{Dim}_\zeta(A) = \limsup_{n \to \infty} \log |A \cap \{1, \ldots, 2^n\}|/n \qquad (2.5)$$

and

$$\mathrm{Dim}_\zeta(A) = \limsup_{n \to \infty} \log |A \cap \{2^n, \ldots, 2^{n+1} - 1\}|/n \qquad (2.6)$$

for all $A \subseteq \mathbb{Z}^+$. The right-hand side of (2.6) has been called the (*channel*) *capacity* of A, the (*topological*) *entropy* (*rate*) of A, and the *upper* (*fractal/mass*) *dimension* of A [34,24,17,13,9,11,20,35,7,30,31,5,6]. In particular, Staiger [35] (see also [20]) rediscovered (2.6) as a characterization of the entropy of A.

The following section shows how to extend zeta-dimension to the integer lattices $\mathbb{Z}^d$, for $d \geq 1$.

3 Zeta-Dimension in $\mathbb{Z}^d$

For each $\vec{n} = (n_1, \ldots, n_d) \in \mathbb{Z}^d$, where d is a positive integer, let $\|\vec{n}\|$ be the Euclidean distance from the origin to $\vec{n}$, i.e.,

$$\|\vec{n}\| = \sqrt{n_1^2 + \cdots + n_d^2}. \qquad (3.1)$$

For each $A \subseteq \mathbb{Z}^d$, define the *A-zeta-function* $\zeta_A : [0, \infty) \to [0, \infty]$ by

$$\zeta_A(s) = \sum_{\vec{0} \neq \vec{n} \in A} \|\vec{n}\|^{-s} \qquad (3.2)$$

for all $s \in [0, \infty)$, and define the *zeta-dimension* of A to be

$$\mathrm{Dim}_\zeta(A) = \inf\{s \mid \zeta_A(s) < \infty\}. \qquad (3.3)$$

Note that, if $d = 1$ and $A \subseteq \mathbb{Z}^+$, then definitions (3.2) and (3.3) agree with definitions (2.1) and (2.2), respectively. The zeta-dimension that we have defined in $\mathbb{Z}^d$ is thus an extension of the one that was defined in $\mathbb{Z}^+$.

Observation 3.1. *For all $d \in \mathbb{Z}^+$ and $A \subseteq \mathbb{Z}^d$, $0 \leq \mathrm{Dim}_\zeta(A) \leq d$.*

We next note that zeta-dimension has key properties of a fractal dimension in $\mathbb{Z}^d$. We state the invariance property a bit more generally than in section 2.

Definition. A function $f : \mathbb{Z}^d \to \mathbb{Z}^d$ is *bi-Lipschitz* if there exists $\alpha, \beta \in (0, \infty)$ such that, for all $\vec{m}, \vec{n} \in \mathbb{Z}^d$, $\alpha \|\vec{m} - \vec{n}\| \leq \|f(\vec{m}) - f(\vec{n})\| \leq \beta \|\vec{m} - \vec{n}\|$.

Observation 3.2 (fractal properties of zeta-dimension). *Let $A, B \subseteq \mathbb{Z}^d$.*

1. *Monotonicity:* $A \subseteq B$ *implies* $\mathrm{Dim}_\zeta(A) \leq \mathrm{Dim}_\zeta(B)$.
2. *Stability:* $\mathrm{Dim}_\zeta(A \cup B) = \max\{\mathrm{Dim}_\zeta(A), \mathrm{Dim}_\zeta(B)\}$.
3. *Lipschitz invariance: If* $f : \mathbb{Z}^d \to \mathbb{Z}^d$ *is bi-Lipschitz, then* $\mathrm{Dim}_\zeta(f(A)) = \mathrm{Dim}_\zeta(A)$.

For $A \subseteq \mathbb{Z}^d$ and $I \subseteq [0, \infty)$, let $A_I = \{\vec{n} \in A \mid \|\vec{n}\| \in I\}$. Then the Dirichlet series

$$\zeta_A^D(s) = \sum_{n=1}^{\infty} |A_{[n,n+1)}| n^{-s} = \sum_{\vec{0} \neq \vec{n} \in A} \lfloor \|\vec{n}\| \rfloor^{-s}, \tag{3.4}$$

converges exactly when $\zeta_A(s)$ converges, so equation (3.3) says that $\text{Dim}_\zeta(A)$ is the abscissa of convergence of this series. Cahen's 1894 characterization of this abscissa thus gives us the following extension of Theorem 2.1.

Theorem 3.3 (entropy characterization of zeta-dimension in $\mathbb{Z}^d$ – Cahen [8]). *For all $A \subseteq \mathbb{Z}^d$,*

$$\text{Dim}_\zeta(A) = \limsup_{n \to \infty} \log |A_{[1,n]}| / \log n. \tag{3.5}$$

As in $\mathbb{Z}^+$, it follows immediately by routine calculus that

$$\text{Dim}_\zeta(A) = \limsup_{n \to \infty} \log |A_{[1,2^n]}| / n \tag{3.6}$$

and

$$\text{Dim}_\zeta(A) = \limsup_{n \to \infty} \log |A_{[2^n, 2^{n+1})}| / n \tag{3.7}$$

for all $A \subseteq \mathbb{Z}^d$. Willson [40] has used (a quantity formally identical to) the right-hand side of (3.6) as a measure of the *growth-rate dimension* of a cellular automaton.

We next note that "subspaces" of $\mathbb{Z}^d$ have the "correct" zeta-dimensions.

Theorem 3.4. *If $\vec{m}_1, \ldots, \vec{m}_k \in \mathbb{Z}^d$ are linearly independent (as vectors in $\mathbb{R}^d$) and $S = \{a_1 \vec{m}_1 + \cdots + a_k \vec{m}_k \mid a_1, \ldots, a_k \in \mathbb{Z}\}$, then $\text{Dim}_\zeta(S) = k$.*

By translation invariance, it follows that "hyperplanes" in $\mathbb{Z}^d$ also have the "correct" zeta-dimensions. inconvenient for calculations. When desirable, the L^1 norm, $\|\vec{n}\|_1 = |n_1| + \cdots + |n_d|$, can be used in its place. That is, if we define the L^1 A-zeta-function $\zeta_A^{L^1}$ by $\zeta_A^{L^1}(s) = \sum_{\vec{0} \neq \vec{n} \in A} \|\vec{n}\|_1^{-s}$, then $2^{-s} \zeta_A(s) \leq \zeta_A^{L^1}(s) \leq \zeta_A(s)$ holds for all $s \in [0, \infty)$, so $\text{Dim}_\zeta(A) = \inf\{s \mid \zeta_A^{L^1}(s) < \infty\}$. The entropy characterizations (3.5), (3.6), and (3.7) also hold with each set A_I replaced by the set $A_I^{L^1} = \{\vec{n} \in A \mid \|\vec{n}\|_1 \in I\}$. Note that other norms can be used to define zeta dimension too.

Example 3.5 (Pascal's triangle modulo 2). Let $A = \{(m, n) \in \mathbb{N}^2 \mid \binom{m+n}{m} \equiv 1 \mod 2\}$. Then it is easy to see that $|A_{[1,2^n]}^{L^1}| = 3^n$ for all $n \in \mathbb{N}$, whence the L^1 version of (3.6) tells us that $\text{Dim}_\zeta(A) = \log 3 \approx 1.5850$. This is exactly the fractal (Hausdorff, packing or box) dimension of the Sierpinski triangle that A so famously resembles [36]. This connection will be further illuminated in section 4.

In order to examine the zeta-dimensions of Cartesian products, we define the *lower zeta-dimension* of a set $A \subseteq \mathbb{Z}^+$ to be

$$\dim_\zeta(A) = \liminf_{n \to \infty} \log |A_{[1,n]}| / \log n. \tag{3.8}$$

By Theorem 3.3, $\dim_\zeta(A)$ is a sort of dual of $\text{Dim}_\zeta(A)$. By routine calculus, we also have
$$\dim_\zeta(A) = \liminf_{n \to \infty} \log|A_{[1,2^n]}|/n, \qquad (3.9)$$
i.e., the dual of equation (3.6) holds. Note, however, that the dual of equation (3.7) does *not* hold in general.

The following theorem is exactly analogous to a classical theorem on the Hausdorff and packing dimensions of Cartesian products [15].

Theorem 3.6. *For all $A \subseteq \mathbb{Z}^{d_1}$ and $B \subseteq \mathbb{Z}^{d_2}$, $\dim_\zeta(A) + \dim_\zeta(B) \leq \dim_\zeta(A \times B) \leq \dim_\zeta(A) + \text{Dim}_\zeta(B) \leq \text{Dim}_\zeta(A \times B) \leq \text{Dim}_\zeta(A) + \text{Dim}_\zeta(B)$.*

Although connectivity properties play an important role in classical fractal geometry, their role in discrete settings like $\mathbb{Z}^d$ will perforce be more limited. Nevertheless, we have the following. Given $d, r \in \mathbb{Z}^+$, and points $\vec{m}, \vec{n} \in \mathbb{Z}^d$, an *r-path* from $\vec{m}$ to $\vec{n}$ is a sequence $\pi = (\vec{p_0}, \ldots, \vec{p_l})$ of points $\vec{p_i} \in \mathbb{Z}^d$ such that $\vec{p_0} = \vec{m}$, $\vec{p_l} = \vec{n}$, and $\|\vec{p_i} - \vec{p_{i+1}}\| \leq r$ for all $0 \leq i < l$. Call a set $A \subseteq \mathbb{Z}^d$ *boundedly connected* if there exists $r \in \mathbb{Z}^+$ such that, for all $\vec{m}, \vec{n} \in A$, there is an r-path $\pi = (\vec{p_0}, \ldots, \vec{p_l})$ from $\vec{m}$ to $\vec{n}$ in which $\vec{p_i} \in A$ for all $0 \leq i \leq l$.

A result of classical fractal geometry says that any set of dimension less than 1 is totally disconnected. The following theorem is an analog of this for zeta-dimension.

Theorem 3.7. *Let $d \in \mathbb{Z}^+$ and $A \subseteq \mathbb{Z}^d$. If $\text{Dim}_\zeta(A) < 1$, then no infinite subset of A is boundedly connected.*

The next section examines the relationships between zeta-dimension and classical fractal dimensions in greater detail.

4 Zeta-Dimension and Classical Fractal Dimension

The following result shows that the agreement between zeta-dimension and Hausdorff dimension noticed in Examples 2.2 and 3.5 are instances of a more general phenomenon: Given any discrete fractal with enough self similarity, its zeta-dimension is equal to the Hausdorff dimension of its classical version. In earlier investigations along these lines, discrete self similar fractals were defined using additive cellular automata [40,41], reverse iterative function system [5,6,37], etc. Here we give a slightly different definition of self similarity.

Definition. Let $c, d \in \mathbb{N}$, $F \subset \mathbb{N}^d$. F is a c-discrete self similar fractal, if there exists a function $S : \{1, 2, \cdots, c\}^d \to \{\text{no}, R_0, R_1, R_2, R_3\}$ such that $S(1, \cdots, 1) = R_0$, and for every integer k and every $(i_1, \cdots, i_d) \in \{1, 2, \cdots, c\}^d$,

$$F \cap C^k_{i_1, i_2, \cdots, i_d} \begin{cases} R_j(C^k_{1,\cdots,1}) & \text{if } S(i_1, \cdots, i_d) = R_j, \\ \varnothing & \text{if } S(i_1, \cdots, i_d) = \text{no} \end{cases}$$

where R_j ($j = 0, \cdots, 3$) is a rotation of angle $j\pi/2$, and $C^k_{i_1, i_2, \cdots, i_d}$ is a d-dimensional cube of side c is defined by $[(i_1 - 1)c^k + 1, i_1 c^k] \times \cdots \times [(i_d - 1)c^k + 1, i_d c^k]$.

Given any c-discrete self similar fractal $F \subset \mathbb{N}^d$, we construct its continuous analogue $\mathbb{F} \subset [0,1]^d$ recursively, via the following contraction $T : x \mapsto \frac{1}{c}x$. $\mathbb{F}_0 = [0,1]$ and $\mathbb{F}_k = T^{(k)}(F \cap [1, c^k]^d)$, where $T^{(k)} = T \circ \cdots \circ T$, denotes k iterations of T. The fractal $\mathbb{F} = \lim_{k \to \infty} \mathbb{F}_k$ obtained by this construction is a self-similar continuous fractal with contraction ratio $1/c$. The following result shows that the zeta-dimension of the discrete fractal is equal to the Hausdorff dimension of its continuous counterpart. See Barlow and Taylor [6] for their more general result that implies this theorem.

Theorem 4.1. *If $c, d, F, \mathbb{F}$ are as above, then* $\text{Dim}_\zeta(F) = \dim_H(\mathbb{F})$.

The following result gives a relationship between zeta-dimension and dimension in the Baire space. We consider the Baire space $\mathbb{N}^\infty$ representing total functions from $\mathbb{N}$ to $\mathbb{N}$ in the obvious way. Given $w \in \mathbb{N}^*$, let $C_w = \{z \in \mathbb{N}^\infty | w \sqsubset z\}$. We define real : $\mathbb{N}^\infty \to [0,1]$ by $\text{real}(z) = \cfrac{1}{(z_0 + 1) + \cfrac{1}{(z_1+1) + \cdots}}$. The cylinder generated by w is the interval $\Delta(w) = \{x \in [0,1] | x = \text{real}(z), w \sqsubset z\}$.

A subprobability supermeasure on $\mathbb{N}^\infty$ is a function $p : \mathbb{N}^* \to [0,1]$ such that $p(\lambda) \leq 1$ and for each $w \in \mathbb{N}^*$, $p(w) \geq \sum_n p(wn)$.

For each subprobability supermeasure p we can define a Hausdorff dimension and a packing dimension on $\mathbb{N}^\infty$, $\dim_p$ and Dim_p, using the metric ρ defined as $\rho(z, z') = p(w)$ for $w \in \mathbb{N}^*$ the longest common prefix of $z, z' \in \mathbb{N}^\infty$.

Gauss measure is defined on each $E \subseteq \mathbb{R}$ as $\gamma(E) = 1/\ln 2 \int_E (1+t)^{-1} dt$. We will abuse notation and use $\gamma(w) = \gamma(\text{real}(C_w))$ for each $w \in \mathbb{N}^*$. Notice that $\gamma(\lambda) = 1$ and therefore γ is a probability measure on $\mathbb{N}^\infty$.

Define $F_A = \{f : \mathbb{N} \to \mathbb{N} | f(\mathbb{N}) \subseteq A \text{ and } \lim_{n \to \infty} f(n) = \infty\}$, for each $A \subseteq \mathbb{Z}^+$. The following result relates zeta-dimension to Gauss-dimension.

Theorem 4.2. $\text{Dim}_\zeta(A) = 2 \cdot \dim_\gamma(F_A) = 2 \cdot \text{Dim}_\gamma(F_A)$.

5 Zeta-Dimension and Algorithmic Information

The entropy characterization of zeta-dimension (Theorem 3.3) already indicates a strong connection between zeta-dimension and information theory. Here we explore further such connections. The first concerns the zeta-dimensions of sets of positive integers that are defined in terms of the digits, or strings of digits, that can appear in the base-k expansions of their elements. We write $\text{rep}_k(n)$ for the base-k expansion $(k \geq 2)$ of a positive integer n. Conversely, given a nonempty string $w \in \{0, 1, \cdots, k-1\}^*$ that does not begin with 0, we write $\text{num}_k(w)$ for the positive integer whose base-k expansion is w.

A *prefix set* over an alphabet Σ is a set $B \subseteq \Sigma^*$ such that no element of B is a proper prefix of another element of B. An *instantaneous code* is a nonempty prefix set that does not contain the empty string.

Theorem 5.1. *Let $\Sigma = \{0, 1, \cdots, k-1\}$, where $k \geq 2$. Assume that $\varnothing \neq \Delta \subseteq \Sigma - \{0\}$ and that $B \subseteq \Sigma^*$ is a finite instantaneous code, and let $A = \{n \in$*

$\mathbb{Z}^+|\mathrm{rep}_k(n) \in \Delta B^*\}$. Then $\mathrm{Dim}_\zeta(A) = s^*$, where s^* is the unique solution of the equation $\sum_{w \in B} k^{-s^*|w|} = 1$.

Corollary 5.2. Let $\Sigma = \{0, 1, \cdots, k-1\}$, where $k \geq 2$. If $\Gamma \subseteq \Sigma$ and $\Gamma \not\subseteq \{0\}$ and $A = \{n \in \mathbb{Z}^+ | \mathrm{rep}_k(n) \in \Gamma^*\}$, then $\mathrm{Dim}_\zeta(A) = \frac{\ln |\Gamma|}{\ln k}$.

Example 5.3. Corollary 5.2 gives a quantitative articulation of the "paradox of the missing digit"[18]. If A is the set of positive integers in whose decimal expansions some particular digit, such as 7, is missing, then a naive intuition might suggest that A contains "most" integers, but A has long been known to be small in the sense that the sum of the reciprocals of its elements is finite (i.e., $\zeta_A(1) < \infty$). In fact, Corollary 5.2 says that $\mathrm{Dim}_\zeta(A) = \frac{\ln 9}{\ln 10} \approx 0.9542$, a quantity somewhat smaller than, say, the zeta-dimension of the set of prime numbers.

The main connection between zeta-dimension and *algorithmic* information theory is a theorem of Staiger [35] relating entropy to Kolmogorov complexity. To state Staiger's theorem in our present framework, we define the *Kolmogorov complexity* $\mathrm{K}(\vec{n})$ of a point $\vec{n} \in \mathbb{Z}^d$ to be the length of a shortest program $\pi \in \{0,1\}^*$ such that, when a fixed universal self-delimiting Turing machine U is run with (π, d) as its input, U outputs $\vec{n}$ (actually, some straightforward encoding of $\vec{n}$ as a binary string) and halts after finitely many computation steps. Detailed discussions of Kolmogorov complexity's definition, fundamental properties, history, significance, and applications appear in the definitive textbook by Li and Vitanyi [25]. As we have already noted, $\mathrm{K}(\vec{n})$ is a measure of the *algorithmic information content* of $\vec{n}$.

For $\vec{0} \neq \vec{n} \in \mathbb{Z}^d$, we write $l(\|\vec{n}\|)$ for the length of the standard binary expansion (no leading zeroes) of the positive integer $\lfloor \|\vec{n}\| \rfloor$.

If $f : \mathbb{Z}^d \to [0, \infty)$ and $A \subseteq \mathbb{Z}^d$, then the *limit superior* of f on A is $\limsup_{\vec{n} \in A} f(\vec{n}) = \lim_{k \to \infty} \sup f(A_{[k, \infty]})$. Note that this is 0 if A is finite.

Theorem 5.4 (Kolmogorov [43], Staiger [35]). For every $A \subseteq \mathbb{Z}^d$, $\mathrm{Dim}_\zeta(A) \leq \limsup_{\vec{n} \in A} \frac{\mathrm{K}(\vec{n})}{l(\|\vec{n}\|)}$, with equality if A or its complement is computably enumerable.

In the case where $d = 1$ and $A \subseteq \mathbb{Z}^+$, Theorem 5.4 says that, if A is Σ_1^0 or Π_1^0, then $\mathrm{Dim}_\zeta(A) = \limsup_{n \in A} \frac{\mathrm{K}(n)}{l(n)}$, where $l(n)$ is the length of the binary representation of A. Kolmogorov [43] proved this for Σ_1^0 sets, and Staiger [35] proved it for Π_1^0 sets. The extension to $A \subseteq \mathbb{Z}^d$ for arbitrary $d \in \mathbb{Z}^+$ is routine.

As Staiger has noted, Theorem 5.4 cannot be extended to Δ_2^0 sets, because an oracle for the halting problem can easily be used to decide a set $B \subseteq \mathbb{Z}^+$ such that, for each $k \in \mathbb{Z}^+$, $B_{[2^k, 2^{k+1}]}$ contains exactly one integer n, and this n also satisfies $\mathrm{K}(n) \geq k$. Such a set B is a Δ_2^0 set satisfying $\mathrm{Dim}_\zeta(B) = 0 < 1 = \limsup_{n \in B} \frac{\mathrm{K}(n)}{l(n)}$.

Classical Hausdorff and packing dimensions were recently characterized in terms of gales, which are betting strategies with a parameter s that quantifies how favorable the payoffs are [26,4]. These characterizations have played a central role in many recent studies of effective fractal dimensions in algorithmic information theory and computational complexity theory [28]. We show here that zeta-dimension also admits such a characterization.

Briefly, given $s \in [0, \infty)$, an *s-gale* is a function $d : \{0,1\}^* \to [0, \infty)$ satisfying $d(w) = 2^{-s}[d(w0) + d(w1)]$ for all $w \in \{0,1\}^*$. For purposes of this paper, an s-gale d *succeeds* on a positive integer n if $d(w) \geq 1$, where w is the standard binary representation of n.

Theorem 5.5 (gale characterization of zeta-dimension). *For all $A \subseteq \mathbb{Z}^+$, $\mathrm{Dim}_\zeta(A) = \inf\{s \mid \text{there is an } s\text{-gale } d \text{ that succeeds on every element of } A\}$.*

Our last result is a theorem on the zeta-dimensions of pointwise sums and products of sets of positive integers. For $A, B \subseteq \mathbb{Z}^+$, we use the notations $A + B = \{a+b \mid a \in A \text{ and } b \in B\}$, $A * B = \{ab \mid a \in A \text{ and } b \in B\}$. The first equality in the following theorem is due to Staiger [35].

Theorem 5.6. *If $A, B \subseteq \mathbb{Z}^+$ are nonempty, then $\mathrm{Dim}_\zeta(A*B) = \max\{\mathrm{Dim}_\zeta(A), \mathrm{Dim}_\zeta(B)\} \leq \mathrm{Dim}_\zeta(A+B) \leq \mathrm{Dim}_\zeta(A) + \mathrm{Dim}_\zeta(B)$, and the inequalities are tight in the strong sense that, for all $\alpha, \beta, \gamma \in [0,1]$ with $\max\{\alpha, \beta\} \leq \gamma \leq \alpha + \beta$, there exist $A, B \subseteq \mathbb{Z}^+$ with $\mathrm{Dim}_\zeta(A) = \alpha$, $\mathrm{Dim}_\zeta(B) = \beta$, and $\mathrm{Dim}_\zeta(A+B) = \gamma$.*

We close with a question concerning circuit definability of sets of natural numbers, a notion introduced recently by McKenzie and Wagner [29]. Briefly, a McKenzie-Wagner *circuit* is a combinational circuit (finite directed acyclic graph) in which the inputs are singleton sets of natural numbers, and each gate is of one of five types. Gates of type $\cup$, $\cap$, $+$, and $*$ have indegree 2 and compute set union, set intersection, pointwise sum, and pointwise product, respectively. Gates of type $^-$ have indegree 1 and compute set complement. Each such circuit *defines* the set of natural numbers computed at its designated output gate in the obvious way. The fact that 0 is a natural number is crucial in this model. Interesting sets that are known to be definable in this model include the set of primes, the set of powers of a given prime, and the set of counterexamples to Goldbach's conjecture. Is there a zero-one law, according to which every set definable by a McKenzie-Wagner circuit has zeta-dimension 0 or 1? Such a law would explain the fact that the set of perfect squares is not known to be definable by such circuits. Theorem 5.6 suggests that a zero-one law, if true, will not be proven by a trivial induction on circuits.

Acknowledgment. The third author thanks Tom Apostol and Giora Slutzki for useful discussions. We thank anonymous referees for their useful comments.

References

1. L. Adleman. Toward a mathematical theory of self-assembly. Technical report, USC, January 2000.
2. T. M. Apostol. *Introduction to Analytic Number Theory*. Undergraduate Texts in Mathematics. Springer-Verlag, 1976.

3. T. M. Apostol. *Modular Functions and Dirichlet Series in Number Theory*, volume 41 of *Graduate Texts in Mathematics*. 1976.
4. K. B. Athreya, J. M. Hitchcock, J. H. Lutz, and E. Mayordomo. Effective strong dimension, algorithmic information, and computational complexity. *SIAM Journal on Computing*. To appear. Preliminary version appeared in *Proceedings of the 21st International Symposium on Theoretical Aspects of Computer Science*, pages 632-643, 2004.
5. M. T. Barlow and S. J. Taylor. Fractional dimension of sets in discrete spaces. *Journal of Physics A*, 22:2621–2626, 1989.
6. M. T. Barlow and S. J. Taylor. Defining fractal subsets of $\mathbb{Z}^d$. *Proc. London Math. Soc.*, 64:125–152, 1992.
7. T. Bedford and A. Fisher. Analogues of the lebesgue density theorem for fractal sets of reals and integers. *Proc. London Math. Soc.*, 64:95–124, 1992.
8. E. Cahen. Sur la fonction $\zeta(s)$ de Riemann et sur des fonctions analogues. *Annales de l'École Normale supérieure*, 1894. (3) **11**, S. 85.
9. W. M. Conner. The dimension of a formal language. *Information and Control*, 29:1–10, 1975.
10. J. J. Dai, J. I. Lathrop, J. H. Lutz, and E. Mayordomo. Finite-state dimension. *Theoretical Computer Science*, 310:1–33, 2004.
11. A. deLuca. On the entropy of a formal language. *Lecture Notes in Computer Science*, (33):103–109, 1975. Automata Theory and Formal Languages (H. Brakhage, Ed.), Proc. 2nd GI Conference.
12. L. Dirichlet. Über den satz: das jede unbegrenzte arithmetische Progression, deren erstes Glied und Differenz keinen gemeinschaftlichen Factor sind, unendlichen viele Primzahlen enthalt. *Mathematische Abhandlungen*, 1837. Bd. 1, (1889) 313-342.
13. S. Eilenberg. *Automata, Languages, and Machines*, volume A. Academic Press, 1974.
14. L. Euler. Variae observationes circa series infinitas. *Commentarii Academiae Scientiarum Imperialis Petropolitanae*, 9:160–188, 1737.
15. K. Falconer. *Fractal Geometry: Mathematical Foundations and Applications*. Wiley, second edition, 2003.
16. H. Furstenberg. Intersections of Cantor sets and transversality of semigroups. *Problems in analysis. Symposium Salomon Bochner*, 1969.
17. G. Hansel, D. Perrin, and I. Simon. Compression and entropy. In *Proceedings of the 9th Annual Symposium on Theoretical Aspects of Computer Science*, pages 515–528, 1992.
18. G. Hardy and E. Wright. *An Introduction to the Theory of Numbers*. Clarendon Press, 5th edition, 1979.
19. F. Hausdorff. Dimension und äusseres Mass. *Mathematische Annalen*, 79:157–179, 1919.
20. J. M. Hitchcock. *Effective fractal dimension: foundations and applications*. PhD thesis, Iowa State University, 2003.
21. I. Hueter and Y. Peres. Self-affine carpets on the square lattice. *Comb. Probab., Computing.*, 6:197–204, 1997.
22. K. Knopp. Über die Abszisse der Grenzgeraden einer Dirichletschen Reihe. *Sitzungsberichte der Berliner Mathematischen Gesellschaft*, 1910.
23. K. Knopp. *Theory and Application of Infinite Series*. Dover Publications, New York, 1990. First published in German in 1921 and in English in 1928.
24. W. Kuich. On the entropy of context-free languages. *Information and Control*, 16(2):173–200, 1970.

25. M. Li and P. M. B. Vitányi. *An Introduction to Kolmogorov Complexity and its Applications.* Springer-Verlag, Berlin, 1997. Second Edition.
26. J. H. Lutz. Dimension in complexity classes. *SIAM Journal on Computing*, 32:1236–1259, 2003.
27. J. H. Lutz. The dimensions of individual strings and sequences. *Information and Computation*, 187:49–79, 2003.
28. J. H. Lutz. Effective fractal dimensions. *Mathematical Logic Quarterly*, 51:62–72, 2005.
29. P. McKenzie and K. Wagner. The complexity of membership problems for circuits over sets of natural numbers. *Proceedings of the Twentieth Annual Symposium on Theoretical Aspects of Computer Science*, pages 571–582, 2003.
30. L. Olsen. Distribution of digits in integers: fractal dimension and zeta functions. *Acta Arith.*, 105(3):253–277, 2002.
31. L. Olsen. Distribution of digits in integers: Besicovitch-Eggleston subsets of N. *Journal of London Mathematical Society*, 67(3):561–579, 2003.
32. B. Riemann. Über die Anzahl der Primzahlen unter einer gegebener Grösse. *Monatsber. Akad. Berlin*, pages 671–680, 1859.
33. P. W. K. Rothemund and E. Winfree. The program-size complexity of self-assembled squares (extended abstract). In *STOC*, pages 459–468, 2000.
34. C. E. Shannon. A mathematical theory of communication. *Bell System Technical Journal*, 27:379–423, 623–656, 1948.
35. L. Staiger. Kolmogorov complexity and Hausdorff dimension. *Information and Computation*, 103:159–94, 1993.
36. I. Stewart. Four encounters with Sierpinski's gasket. *The Mathematical Intelligencer*, 17(1):52–64, 1995.
37. R. S. Strichartz. Fractals in the large. *Canadian Journal of Mathematics*, 50(3):638–657, 1998.
38. D. Sullivan. Entropy, Hausdorff measures old and new, and limit sets of geometrically finite Kleinian groups. *Acta Mathematica*, 153:259–277, 1984.
39. C. Tricot. Two definitions of fractional dimension. *Mathematical Proceedings of the Cambridge Philosophical Society*, 91:57–74, 1982.
40. S. J. Willson. Growth rates and fractional dimensions in cellular automata. *Physica D*, 10:69–74, 1984.
41. S. J. Willson. The equality of fractional dimensions for certain cellular automata. *Physica D*, 24:179–189, 1987.
42. E. Wirsing. Bemerkung zu der arbeit über vollkommene zahlen. *Mathematische Annalen*, 137:316–318, 1959.
43. A. K. Zvonkin and L. A. Levin. The complexity of finite objects and the development of the concepts of information and randomness by means of the theory of algorithms. *Russian Mathematical Surveys*, 25:83–124, 1970.

Online Interval Coloring
with Packing Constraints

Leah Epstein[1,*] and Meital Levy[2]

[1] Department of Mathematics, University of Haifa, 31905 Haifa, Israel
   lea@math.haifa.ac.il
[2] School of Computer Science, Tel-Aviv University, Israel
   levymeit@post.tau.ac.il

Abstract. We study online interval coloring problems with bandwidth. We are interested in some variants motivated by bin packing problems. Specifically we consider open-end coloring, cardinality constrained coloring, coloring with vector constraints and finally a combination of both the cardinality and the vector constraints. We construct competitive algorithms for each of the variants. Additionally, we present a lower bound of 24/7 for interval coloring with bandwidth, which holds for all the above models, and improves the current lower bound for the standard interval coloring with bandwidth.

1 Introduction

We study variants of the online interval coloring problem with bandwidth. In these coloring problems, the intervals are presented one by one and the online algorithm must assign each interval a color before the next interval arrives. In the classical problem the intervals have no bandwidth and two intersecting intervals can not be colored by the same color. We are interested in the case where every interval has an associated bandwidth in (0,1]. This problem (standard coloring of intervals with bandwidth) was introduced by Adamy and Erlebach [1]. A set of intervals can be assigned the same color c, if for any point p on the real line, the sum of the bandwidths of intervals colored c and containing p, does not exceed 1. We refer to a coloring satisfying the above condition as a *proper coloring*.

Online coloring of intervals with bandwidth is a simultaneous generalization of two major problems. The first one Online bin packing, the study of which dates back to the works of Johnson and Ullman in the early 1970's [11,20], see [6] for a survey. If all the presented intervals intersect, colors correspond to bins. The second problem is the classical Online coloring of interval graphs, introduced by Kierstead and Trotter [14].

As mentioned in [1], the problem of coloring intervals with bandwidth arises in many applications. Most of these applications come from the field of networks. Consider a network with a line topology that consists of links, where each link has channels of constant capacity. This can be either an all-optical

---

* Research supported by Israel Science Foundation (grant no. 250/01).

WDM (wavelength-division multiplexing) network or an optical network supporting SDM (space-division multiplexing). A connection request is from one network node a to another node b has a bandwidth associated with it. The set of requests assigned to a channel must not exceed the capacity of the channel on any of the links on the path $[a, b]$. Another network related application is one where requests have constant duration c, and we have to serve all requests as fast as possible. With respect to our online coloring intervals with bandwidth problem, the colors correspond to time slots, and the total number of colors corresponds to the schedule length. The last example comes from scheduling, a requested job has a duration and resource requirement during its execution. Jobs (intervals) arrive online and must be assigned to a machine (color) immediately. All the machines have the same capabilities and the objective is to minimize the number of machines used.

The unweighted (classical) problem is equivalent to coloring an interval graph, where each interval corresponds to a node and an edge between two nodes appears if the corresponding intervals intersect. Interval graphs are perfect, therefore the chromatic number of the graph is the maximum clique size [10], which represents a point where the most intervals intersect. It can be elaborated for the bandwidth case, if we refer to the maximum clique size as the maximum weighted clique. Each node has the weight of the related interval, *i.e.*, its bandwidth, and the clique size is the sum of the weights of the corresponding intervals of the clique. We study online coloring problems in terms of the *asymptotic competitive ratio*. Thus we compare an online algorithm to an optimal offline algorithm OPT that knows the complete sequence of intervals in advance.

Let $B(\sigma)$ (or B, if the sequence σ is clear from the context), be the cost of algorithm B on the request sequence σ. An algorithm A is $\mathcal{R}$-competitive (with respect to the absolute competitive ratio) if for every sequence σ, $A(\sigma) \leq \mathcal{R} \cdot OPT(\sigma)$. The absolute competitive ratio of an algorithm is the infimum value of $\mathcal{R}$ such that the algorithm is $\mathcal{R}$-competitive.

The asymptotic competitive ratio for an online algorithm A is defined to be $\mathcal{R}_A^\infty = \limsup_{n \to \infty} \sup_\sigma \left\{ \frac{A(\sigma)}{OPT(\sigma)} \middle| OPT(\sigma) = n \right\}$. All results given in this paper apply to both the absolute and the asymptotic competitive ratios.

Coloring interval graphs has been intensively studied, Kierstead and Trotter [14] gave an upper and lower bounds of 3 on the competitive ratio. Much research has been done analyzing the performance of the simple First Fit algorithm for the unweighted problem. Upper bounds on the competitive ratio of 40, 25.72 and 10 were given in [12,13,19] respectively. Chrobak and Slusarek [5] showed a lower bound close to 4.5 on the competitive ratio of First Fit.

Coloring intervals with bandwidth was first posed in 2003 in [1] by Adamy and Erlebach, they presented an online algorithm with a competitive ratio of 195. Narayanaswamy [18] presented a new algorithm with a competitive ratio of 10. In [7] we studied several extensions of this problem including coloring unit length intervals.

Motivated by the well known bin packing problem, we investigate four variants studied in the past with respect to bin packing. Namely, Open-end bin packing, Vector packing, Cardinality constrained packing and Vector packing with cardinality constraints. Open-end online bin packing (also called the Ordered open-end problem) was introduced by Young and Leung [21]. Online vector packing was studied by Garey et al. as a scheduling problem with resource constraints [9], this problem was studied also in [15,8,3]. Cardinality constrained bin packing was first studied by Krause, Shen and Schwetman [16,17]. It was also studied in [2]. The vector packing problem with cardinality constraints was mentioned in [4]. In that paper it is treated as a special case of the vector packing problem.

We make adjustments to these variants to suit the interval coloring with bandwidth problem in the following way.

Open-End Interval Coloring: Given a point p and color c, we remove the restriction that all intervals intersecting point p colored with c should have total bandwidth of at most 1. Instead, we require that if the last interval which received color c and intersects p is removed, then the total bandwidth of all such other intervals, is strictly less than 1. A possible application of this model is the situation where the decision on the color of a new interval does not depend on the exact value of its bandwidth, but on the current load of each color. This is consistent with our algorithms which use a partition into classes of bandwidth rather than using the exact bandwidth to classify a new interval and to assign it a color.

Interval Coloring with Vector Constraints: Instead of one dimensional bandwidths, the intervals are associated with d-dimensional vectors. This is a generalization of the standard interval coloring with bandwidth problem. Here each interval has d distinct weights and each color has d corresponding unit capacities. An interval can receive color c if the assignment is valid according to all d components. This variant models a multiple number of available resources that each request needs and all requests must share, rather than a single resource as in the standard problem.

Cardinality Constrained Interval Coloring: The cardinality constrained coloring, or the k-bounded interval coloring with bandwidth problem, additionally imposes the constraint that at each point p, at most k intersecting intervals are allowed to use one color. This variant models applications where only a limited number of requests can be satisfied simultaneously, a restriction that occurs in addition to the bandwidth constraints.

Cardinality and Vector Constrained Interval Coloring: This is a combination of the two previous variants. Each interval is associated with a d-dimensional vector of d distinct bandwidths and each color has d corresponding capacities.

Additionally at most k intersecting intervals are allowed in one color at each point.

Our Results: We present competitive online algorithms for each of the variants. For the open-end coloring model we present an algorithm with competitive ratio of at most 12. For the cardinality constrained variant we suggest an algorithm with competitive ratio of $\min\{10+2\cdot\frac{k}{k-1}, k+3\}$, for odd k and $\min\{12, k+3\}$, for even k.

A $10d$ competitive algorithm is presented for the vector constrained model and a competitive ratio of $\min\{10d+2, 3k\}$, for even k and $\min\{10d+2\frac{k}{k-1}, 3k\}$, for odd k for the combined model of both vector and cardinality constraints. The description of these results is omitted from this version and can be found in the full version of the paper.

We also present a lower bound of $\frac{24}{7} \approx 3.428571$, an improvement of the previously known lower bound of 3.26 for standard interval coloring with bandwidth presented in [7]. The latter lower bound does not apply for cardinality constrained coloring (a simplification of that lower bound can be applied to large values of k) and to the open-end model. However the lower bound of Kierstead and Trotter [14] can be used in both models. By using intervals of bandwidth 1, as done in the construction of [14], two intersecting intervals can not receive the same color in both models. Therefore the best lower bound known for these models is 3. Our lower bound can be easily modified by a simple change of parameter to all the variants considered in this paper.

2 Preliminaries

A weighted interval graph G of a set of intervals S, is a graph where each node corresponds to an interval. The weight of the node is the bandwidth of the interval in S related to it. If two intervals intersect, there is an edge between their related nodes in G. Recall that we denote the optimal coloring of the offline algorithm by OPT.

Let $\omega(G)$ or $\omega(S)$ denote the size of the maximum cardinality clique in G (ω for short), i.e., ignoring the weights. Let $\omega^*(G)$ or $\omega^*(S)$ (ω^* for short) denote the largest weighted clique in G. A weighted clique is the sum of the weights of the vertices in a clique. Note that for the interval coloring problem with bandwidth we have $OPT \geq \lceil \omega^* \rceil$

Below we give the generalized presentation of the algorithm of Kierstead and Trotter [14] presented in [7]. For convenience we include the full presentation and list three relevant lemmas from [7]. We refer the reader to [7] for the proofs of these lemmas.

Let $\sigma = v_1, \ldots, v_n$ be the list of vertices of G, in the order of arrival. Algorithm $KT_{l,b}$ is defined for inputs σ such that, $b(v_i) \in (0,b]$. The algorithm partitions the intervals (i.e. the vertices of G) into sets A_m (for integer values of m, such that $m \geq 1$). We use C_m to denote the set of colors dedicated to A_m. Every set A_m is colored using First Fit, independently of other sets. Therefore the colors have the property $C_x \cap C_y = \emptyset$ for $x \neq y$.

Algorithm 1. $KT_{l,b}$

On a new interval v_i:
1: For every integer $m \geq 1$, let $V_m(v_i)$ and $E_m(v_i)$ be the following subsets of $V(G)$ and $E(G)$ respectively.
$V_m(v_i) = \{v_j \in V(G) : j < i, m(v_j) \leq m\}$;
$E_m(v_i) = \{(u,v) \in E(G) : u, v \in V_m(v_i)\}$;
$G_m(v_i) \cup \{v_i\} = G(V_m(v_i) \cup v_i, E_m(v_i) \cup \{(u, v_i) \in E(G) : u \in V_m(v_i)\})$
$\omega_i^*(H) = $ The maximum weighted clique in graph H that contains the interval v_i
2: Let $G_m(v_i) = G(V_m(v_i), E_m(v_i))$
3: $m(v_i) = $ the smallest m such that $\omega_i^*(G_m(v_i) \cup \{v_i\}) \leq m \cdot l$.
4: $A_{m(v_i)} \Leftarrow A_{m(v_i)} \cup \{v_i\}$
5: Color v_i considering only the intervals of $A_{m(v_i)}$ using First Fit on colors of $C_{m(v_i)}$.

Lemma 1. *For every m, $\omega^*(A_m) \leq 2(b+l)$.*

Lemma 2. *If all intervals have the same bandwidth, b, and l is divisible by b, for every m, $\omega^*(A_m) \leq 2l$.*

Lemma 3.

(i) The largest value of m ever used in $KT_{l,b}$ is $\lceil \frac{\omega^*}{l} \rceil$

(ii) The coloring of $KT_{l,b}$ is at most $\lceil \frac{\omega^*}{l} \rceil (\max_m FF(A_m))$, where $FF(A_m)$ denotes the coloring of the First Fit algorithm on the set A_m of intervals that were presented online.

Note that $KT_{1,1}$ without bandwidth is equivalent to the original algorithm of Kierstead and Trotter [14]. In their algorithm every layer can be colored by First Fit with at most 3 colors. The number of layers equals to the size of the maximum cardinality clique. Therefore the coloring is at most $3OPT$.

3 Upper Bounds

In this section we present algorithms for different models. We denote the optimal offline algorithm for a specific variant A, by OPT_A, i.e., for the open-end model we denote the optimal offline algorithm that follows the restrictions of the model by $OPT_{Open-End}$.

Open End Coloring

In the Open End version, colors can consist of intersecting intervals with a total bandwidth of more than 1. However, for any given point, the removal of the last interval colored with a specific color must bring the color's level back to strictly below 1 at that point.

Theorem 1. *There exists an online algorithm with competitive ratio 12 for the open-end interval coloring.*

Proof. **Algorithm:** Perform an online partition of the intervals into three disjoint subsequences S_1, S_2, and S_3 according to the bandwidth of the intervals. The subsequences are defined as follows. For every interval I, $I \in S_1$ if $b(I) \leq \frac{1}{4}$, $I \in S_2$ if $\frac{1}{4} < b(I) < 1$ and $I \in S_3$ if $b(I) = 1$. Each subsequence is colored by a different set of colors. The colors to be assigned are split into three disjoint classes C_1, C_2, and C_3. Each class is designated to intervals of one subsequence, i.e., C_1 for S_1, C_2 for S_2 and C_3 for S_3.

The classes of colors are built dynamically, when a new color is required, the first unused color is assigned. When a color is assigned to one of the three classes it can no longer be assigned to any of the other classes.
Run in parallel (i.e., independently) the following three sub algorithms:

SubAlgorithm A_{S_1}: Use $KT_{\frac{1}{4},\frac{1}{4}}$ on the intervals of S_1 ignoring the open-end option.

SubAlgorithm A_{S_2}: Use a variant of $KT_{1,1}$ without bandwidth on S_2. In lines 1-4 of the algorithm $KT_{1,1}$, treat all intervals as if they have bandwidth of exactly 1. The change is made in line 5 of the algorithm. Instead of using at most 3 colors for each A_m, use only one color.

SubAlgorithm A_{S_3}: Use the Algorithm of Kierstead and Trotter, i.e., $KT_{1,1}$ without bandwidth.

Analysis of the Competitive Ratio:
We show the following properties. (i) A_{S_1} uses at most $5 \cdot OPT_{Open-End}(S_1)$ colors; (ii) A_{S_2} uses at most $4 \cdot OPT_{Open-End}(S_2)$ colors; (iii) A_{S_3} uses at most $3 \cdot OPT_{Open-End}(S_3)$ colors.

SubAlgorithm A_{S_1}: According to Lemma 3 part (i), the coloring of $KT_{\frac{1}{4},\frac{1}{4}}$ is $\lceil \frac{\omega^*}{\frac{1}{4}} \rceil \max_m FF(A_m)$, where $FF(A_m)$ denotes the coloring of the First Fit algorithm on the set A_m of intervals that were presented online. By Lemma 1 For every m, $\omega^*(A_m) \leq 1$. Therefore $\max_m FF(A_m) = 1$ and we get a coloring of at most $\lceil 4\omega^* \rceil$. In the open-end version, the total bandwidth for each color may exceed 1. Since all the intervals in S_1 have a maximum bandwidth of $\frac{1}{4}$, OPT can use each color for a total bandwidth of at most $\frac{5}{4}$. Therefore it needs at least $\frac{4}{5}\omega^*$ colors. Hence, we get, $A_{S_1}(S_1) \leq \lceil 4\omega^* \rceil = \lceil 5 \cdot \frac{4}{5}\omega^* \rceil \leq 5 OPT_{Open-End}(S_1)$.

SubAlgorithm A_{S_2}: Note that in this variant of $KT_{1,1}$ without bandwidth, all the sets of intervals A_m should contain the same intervals as if we had used the regular $KT_{1,1}$ without bandwidth. The only difference is the coloring of the intervals within these sets.

First we claim that this variant results in a proper coloring. By Lemma 2, the cardinality clique is at most 2 in each A_m. In the Open End variant two intersecting intervals of bandwidth strictly less than 1 can be colored by the same color. Since every interval $I \in S_2$ satisfies $\frac{1}{4} < b(I) < 1$ every A_m can be colored by a single color and the claim is proved.

Next we show that the coloring of $A_{S_2}(S_2)$ is at most $4 \cdot OPT_{Open-End}(S_2)$. Algorithm A_{S_2} uses at most ω colors, where ω is the cardinality clique and not the weighted clique. Since for every $I \in S_2$, $b(I) > \frac{1}{4}$, $OPT_{Open-End}(S_2)$ can use at most four intersecting intervals in a single color. Therefore $OPT_{Open-End}(S_2)$ uses at least $\frac{\omega}{4}$ colors. Thus we get that A_{S_2} uses at most $4 \cdot OPT_{Open-End}(S_2)$.

SubAlgorithm A_{S_3}: According to the analysis of the algorithm of Kierstead and Trotter, the coloring of $A_{S_3}(S_3) \leq 3OPT(S_3)$. Since for every $I \in S_3$, $b(I) = 1$, every two intersecting intervals can not receive the same color. Thus $OPT(S_3) = OPT_{Open-End}(S_3)$.

By combining the competitive ratios of the subalgorithms of A_{S_1}, A_{S_2} and A_{S_3} we get a total of 12 competitive ratio for the complete algorithm.

Coloring with Cardinality Constraints

In the cardinality constrained, or the k-bounded interval coloring with bandwidth problem there is an additional restriction. In this variant at most k intersecting interval are allowed in one color at each point.

Theorem 2. *There exists an online algorithm for cardinality constrained interval coloring with a competitive ratio $\min\{10 + 2 \cdot \frac{k}{k-1}, k+3\}$ for odd k and $\min\{12, k+3\}$ for even k, where k is the cardinality constraint.*

Proof. Algorithm: If $\min\{10 + 2 \cdot \frac{k}{k-1}, k+3\} = 10 + 2 \cdot \frac{k}{k-1}$ for odd k or $\min\{12, k+3\} = 12$ for even k, use the algorithm described in case 1. Otherwise use case 2.

Case 1: Perform an online partition of the intervals into two disjoint subsequences S_1 and S_2 according to the bandwidth of the intervals. The subsequences are defined as follows:
For every interval I, $I \in S_1$ if $b(I) \leq \frac{1}{k}$ and $I \in S_2$ if $\frac{1}{k} < b(I) < 1$.

SubAlgorithm A_{S_1}: For **even** k, take the bandwidth of every interval in S_1 to be exactly $\frac{1}{k}$ and use algorithm $KT_{\frac{1}{2}, \frac{1}{k}}$. For **odd** k, take the bandwidth of every interval in S_1 to be exactly $\frac{1}{k-1}$ and use algorithm $KT_{\frac{1}{2}, \frac{1}{k-1}}$

SubAlgorithm A_{S_2}: Use the algorithm presented in [18] on S_2. We provide the details of this algorithm in the full version of the paper.

Case 2: Perform an online partition of the intervals into two disjoint subsequences R_1 and R_2 according to the bandwidth of the intervals. The subsequences are defined as follows:
For every interval I, $I \in R_1$ if $b(I) \leq \frac{1}{2}$ and $I \in R_2$ if $\frac{1}{2} < b(I) < 1$.

SubAlgorithm A_{R_1}: Use a variant of $KT_{1,1}$ without bandwidth on R_1. In lines 1-4 of the algorithm $KT_{1,1}$, treat all intervals as if they have bandwidth of exactly 1. The change is made in line 5 of the algorithm. Instead of using at most 3 colors for each A_m, use only one color.

SubAlgorithm A_{R_2}: Use $KT_{1,1}$ without bandwidth, treating every interval as if its bandwidth is exactly 1.

Analysis of the Competitive Ratio:
Case 1: SubAlgorithm A_{S_1}:
Since each color can be used for a total of k intersecting intervals, we can treat all interval in S_1 as if they have bandwidth of exactly $\frac{1}{k}$. The value ω^* is computed using this assumption.

Even k. The coloring of $KT_{\frac{1}{2},\frac{1}{k}}$ is $\lceil \frac{\omega^*}{\frac{1}{2}} \rceil \max_m FF(A_m)$, according to Lemma 3 part (ii). Since now all intervals have the same bandwidth $\frac{1}{k}$, and since $\frac{1}{2}$ is divisible by $\frac{1}{k}$, by Lemma 2, we get that for every m, $\omega^*(A_m) \leq 1$. Therefore $\max_m FF(A_m) = 1$ and we get a coloring of at most $\lceil 2\omega^* \rceil \leq 2\lceil \omega^* \rceil$.

Odd k. Similarly to the previous case, we get a competitive ratio of 2 if OPT can use only $k-1$ intersecting intervals for every color. However $OPT_{K-Bounded}$ can use k intersecting intervals for each color. Therefore the $OPT_{K-Bounded}$ uses at least $\frac{k-1}{k}\omega^*$ colors. Hence we get a competitive ratio of $2 \cdot \frac{k}{k-1}$.

SubAlgorithm A_{S_2}: The algorithm presented in [18] has a competitive ratio of 10 on intervals in (0,1]. Since for every $I \in S_2$, $b(I)$ in (0,1], the competitive ratio for this part is also at most 10.

Combining the competitive ratio of this case we get $10 + \frac{2k}{k-1}$ for odd k and 12 for even k.

Case 2: SubAlgorithm A_{R_1}:
Note that in this variant of $KT_{1,1}$ without bandwidth, all the sets of intervals A_m contain the same intervals as if we had used the regular $KT_{1,1}$ without bandwidth. The only difference is the coloring of the intervals within these sets.

First we claim that this variant results in a proper coloring. By Lemma 2, the cardinality clique is at most 2 in each A_m. Since every interval $I \in R_1$ satisfies $b(I) \leq \frac{1}{2}$ every A_m can be colored by a single color and the claim is proved.

We next show that the coloring of $A_{R_1}(R_1)$ is at most $k \cdot OPT_{K-Bounded}(R_1)$. Algorithm A_{R_1} uses at most ω colors, where ω is the cardinality clique of the set R_1 and not the weighted clique. In the cardinality constraint variant $OPT_{K-Bounded}$ can only color k intersecting intervals with the same color. Therefore uses at least $\frac{\omega}{k}$ colors. Thus we get that $A_{R_1}(R_1)$ uses at most $k \cdot OPT_{K-Bounded}(R_1)$ colors.

SubAlgorithm A_{R_2}: The coloring of $A_{R_2}(R_2) \leq 3OPT(R_2)$. Since for every $I \in R_2$, $b(I) > \frac{1}{2}$, every two intersecting intervals can not receive the same color. Thus $OPT(R_2) = OPT_{K-Bounded}(R_2)$.

Combining the competitive ratio of this case we get $k + 3$.

To complete the analysis, since k is known in advance, the algorithm uses the best case for a specified k, thus getting the minimum competitive ratio out of the two cases.

4 Lower Bound

In this section we prove the following theorem, using a single type of construction for all models, that can be achieved by choosing appropriate parameters. The theorem holds for all models studied in this paper. The variant with vector constraints is a generalization of standard coloring with bandwidth, where the vector has dimension 1. Additional dimensions can be added trivially by adding zero components. Similarly, the variant with both vector and cardinality constraints is a generalization of cardinality constrained coloring.

Theorem 3. *Any deterministic online algorithm for interval coloring with bandwidth in the standard model, open-end model, and cardinality constrained model, has competitive ratio of at least $\frac{24}{7} \approx 3.428571$.*

The general structure of the input sequence is as follows. In the first part of the construction, all intervals have bandwidth α and in the second (optional) part all intervals have bandwidth $\beta > \alpha$. The values of α and β are picked depending on the exact problem. The choice is such that it is possible to give the same color to two intersecting intervals of bandwidth α, or even to one interval of bandwidth α and one of bandwidth β, which are intersecting. However, we need to make sure that it is impossible to give the same color to two intersecting intervals of bandwidth β, or to any three intersecting intervals of bandwidth at least α.

Such choices can be e.g. $\alpha = 0.4$ and $\beta = 0.6$ for the standard problem, or to the cardinality constrained problem (for any $k \geq 2$). For the open-end problem, we can take $\alpha = 0.6$ and $\beta = 1$.

Given an integer value s, the first part of the sequence is built so that the largest clique size (ignoring the bandwidth) is $2s$. Therefore it is possible to color the input using s colors, as follows. First, we greedily distribute $2s$ colors so that no two intersecting intervals receive the same color. Then we can partition colors into pairs, and unite every two colors into one. This can be done since at every point there will be at most two intersecting intervals of bandwidth α.

The second part of the sequence is built in a way that the largest clique size (again, ignoring the bandwidth) of intervals introduced in this part is $2s$. The complete sequence can be colored using $2s$ colors, similarly to the explanation above, by coloring each part of the sequence using $2s$ colors. The same palette of $2s$ colors can be used for both parts.

Consider a subset of the input, usually this is a subset of input intervals contained in some mega-interval. A color which was used for at least one interval in the subset is called a "used color". Next, we define the notion of "full colors" and "partial colors" in a the coloring of this subset as follows. If there exists a point p and a color c such that two distinct intervals X and Y such that $p \in X$ and $p \in Y$ received both the color c, or a single interval from the second part, i.e., with bandwidth β then c is a full color. Otherwise, if this does not hold, but c is a used color, then it is a partial color.

The construction of the two parts of the sequence are adaptations of the lower bound in [14]. The first part of the sequence uses intervals of bandwidth α and therefore two intersecting intervals may receive the same color. This is a main difference with the proof in [14], since we need to deal with such a situation, whereas in [14] all intervals have bandwidth 1. Another difference, that we already used in the construction in [7] is the assumption that some information on the optimal cost (which is either s or $2s$ in our case) is known in advance.

The construction of each part works in phases, after a phase we shrink some parts of the line into single points. Given a point p, that is a result of shrinking an interval $[a, b]$. Every interval presented in the past which is contained in $[a, b]$ is also shrunk into p and therefore such a point inherits a list of used (partial and full) colors that some interval received. A partial color can be used again exactly once in some interval containing p. A full color cannot be assigned to any interval that contains the point p. This is done for simplification. In practice it means that for a given point p that is the result of shrinking, every future interval either contains this point or not, i.e., it either contains all intervals that were shrunk into this point, or has no overlap with any of them.

We would like to show that either the number of colors used in the first part is at least $\frac{24s-2}{7}$, or the number of colors used after the second step is at least $\frac{48s-4}{7}$. This would imply the lower bound. Therefore, the sequence construction can clearly stop once $7s$ colors have been used. Therefore we may assume that we are initially given a palette of $7s$ colors, $1, \ldots, 7s$, that can all be used by the algorithm. The ith color ever used is called color number i. As soon as color $7s$ is used, the proof is complete. This is just one stopping condition, we may stop the sequence earlier as well.

The first part of the sequence has intervals of bandwidth α and starts with introducing $S(0)$ non-intersecting intervals, this is phase 1. A bound on the value $S(0)$ is fixed later.

Since the algorithm is using at most $7s$ colors, this means that there exists a set of $\frac{S(0)}{7s}$ intervals that share the exact same color c. We shrink all intervals into single points. Later phases result in additional points. Since there are no intersecting intervals, color c is partial in all points colored with it.

We now define phase i. The phases are constructed in a way that in the beginning of phase i there is a set of at least $S(i-1)$ points that contain a two given subsets of the $7s$ colors. The first subset is of $P(i-1)$ partial colors and the second is of $F(i-1)$ full colors. These points are called points of interest. Note that after phase 1 we have $P(1) = 1$ and $F(1) = 0$.

There exist some other points containing other subsets of full and partial colors. All these points are called void points. At this time, we partition the points of interest into consecutive sets of four. At most three points of interest that do not participate in this become void points.

We next define additional intervals, increasing the size of the largest cardinality clique (with respect to the number of intervals, i.e., ignoring bandwidth) by exactly one. Given a set of four points listed from left to right a_1, a_2, a_3, a_4, let b be the leftmost void point on the right hand side of a_1, between a_1 and a_2. If no such point exists, then let $b = \frac{a_1+a_2}{2}$, i.e., the point which is halfway between a_1 and a_2. Similarly, let d be the rightmost void point between a_3 and

a_4, and if no such point exists then $d = \frac{a_3+a_4}{2}$. Let f be a point between a_2 and a_3 that is not a void point. We introduce the intervals $I_1 = [a_1, \frac{a_1+b}{2}]$ and $I_2 = [\frac{d+a_4}{2}, a_4]$.

If they both receive the same color (used or unused at points a_1 and a_4), we introduce the intervals $I_3 = [\frac{a_1+b}{2}, f]$ and $I_4 = [f, \frac{d+a_4}{2}]$. The interval I_3 intersects with a_2, and with I_1. The second interval I_4 intersects I_3, a_3 and I_2. If at most two distinct colors were used, then there exists a point in the range $[a_1, a_4]$ where two intersecting intervals received the same color, and therefore there is at least one new full color in this interval. If a color that is partial in the point a_1, a_2, a_3, a_4 was used, then this color becomes full in $[a_1, a_4]$. If three unused colors were used, then these colors become additional partial colors in $[a_1, a_4]$.

If I_1, I_2 receive distinct colors (used or unused), we introduce the interval $I_5 = [\frac{a_1+b}{2}, \frac{d+a_4}{2}]$. Interval I_5 intersects with I_1, I_2, a_2, a_3. If it gets the same color as I_1 or I_2 this color becomes full in $[a_1, a_4]$. If a color that is partial in the point a_1, a_2, a_3, a_4 was used, then this color becomes full in $[a_1, a_4]$. If three unused colors were used, then these colors become additional partial colors in $[a_1, a_4]$.

We shrink every such interval $[a_1, a_4]$ into a single point. Each of the new shrunk points received either three new partial colors, or one full (not necessarily new) color.

Note that we do not use more than $7s$ colors, and each new shrunk point receives a number of full and partial colors, which is at most three colors in total. Four intervals are introduced only if the first two received the same color. If the point has no new full colors, then it has exactly three new partial colors. Otherwise, it has at least one new full color, and possibly one or two new partial or full colors. Before the phase, all points of intervals had the exact same subsets of partial and full colors. This gives seven options for the type of new colors (or colors which changed status from partial to full). Let "f" denote full and "p" denote partial, then the options are $(p, p, p), (f, p, p), (f, f, p), (f, f, f), (f, p), (f, f), (f)$. There are less than $(7s)^3$ options for each type, and thus in total, there are less then $7 \cdot (7s)^3$ choices for the updated subsets given the previous ones. We can choose at least $S(i) = \frac{S(i-1)}{4 \cdot 7 \cdot (7s)^3}$ points having the same sets of full and partial colors. The points containing these exact sets of colors become the points of interest of the next phase, and the others become void points of the next phase. Points that are void points of previous phases and are not contained in shrunk intervals remain void points. Note that the only points where the new intervals intersect are points with no previous intervals, and therefore the clique size increases by exactly 1.

After the first $2s$ phases, the sequence may continue with the second part. If $P(2s) + F(2s) \geq \frac{24s-2}{7}$ the sequence stops since the lower bound is obtained. Otherwise, the second part goes on for $2s$ phases, however the intervals have bandwidth β, therefore no new partial colors are introduced, and every phase results in three new full colors. To verify this, it can be checked that in both construction cases, the new intervals must receive three distinct colors, that are either unused or partial. The number of full colors after all phases is at least $F(2s) + 6s$. Let A be the number of phases among the first $2s$ which increased the number of partial colors by 3. Therefore $P(2s) + F(2s) \geq 3A + 1$. In all other phases except the first one and up to phase $2s$, the number of full colors increased. Therefore $F(2s) \geq 2s - 1 - A$. We get $2s - 1 - F(2s) \leq A < \frac{24s-9}{21}$ and therefore $F(2s) > \frac{6s-4}{7}$ and $F(2s) + 6s > \frac{48s-4}{7}$, which proves the lower bound in this case.

Note that in each phase, the number of intervals which can be used for the next phase decreases by a factor of at most $28 \cdot (7s)^3$. To complete the construction, we need $S(4s) \geq 1$. If the initial amount of intervals introduced is $S(0) = (28 \cdot (7s)^3)^{4s}$, this holds and we are done.

References

1. U. Adamy and T. Erlebach. Online coloring of intervals with bandwidth. In *Proc. of te First International Workshop on Approximation and Online Algorithms (WAOA2003)*, pages 1–12, 2003.
2. L. Babel, B. Chen, H. Kellerer, and V. Kotov. Algorithms for on-line bin-packing problems with cardinality constraints. *Discrete Applied Mathematics*, 143(1-3):238–251, 2004.
3. D. Blitz, A. van Vliet, and G. J. Woeginger. Lower bounds on the asymptotic worst-case ratio of online bin packing algorithms. Unpublished manuscript, 1996.
4. A. Caprara, H. Kellerer, and U. Pferschy. Approximation schemes for ordered vector packing problems. *Naval Research Logistics*, 92:58–69, 2003.
5. M. Chrobak and M. Ślusarek. On some packing problems relating to dynamical storage allocation. *RAIRO Journal on Information Theory and Applications*, 22:487–499, 1988.
6. J. Csirik and G. J. Woeginger. On-line packing and covering problems. In *A. Fiat and G. J. Woeginger, editors,* Online Algorithms: The State of the Art, pages 147–177, 1998.
7. L. Epstein and M. Levy. Online interval coloring and variants. In *Proceedings of The 32nd International Colloquium on Automata, Languages and Programming (ICALP'05)*, 2005. to appear.
8. G. Galambos, H. Kellerer, and G. J. Woeginger. A lower bound for online vector packing algorithms. *Acta Cybernetica*, 10:23–34, 1994.
9. M. R. Garey, R. L. Graham, D. S. Johnson, and A. C. C. Yao. Resource constrained scheduling as generalized bin packing. *Journal of Combinatorial Theory (Series A)*, 21:257–298, 1976.
10. T. R. Jensen and B. Toft. *Graph coloring problems*. Wiley, 1995.
11. D. S. Johnson. *Near-optimal bin packing algorithms*. PhD thesis, MIT, Cambridge, MA, 1973.
12. H. A. Kierstead. The linearity of first-fit coloring of interval graphs. *SIAM Journal on Discrete Mathematics*, 1(4):526–530, 1988.
13. H. A. Kierstead and J. Qin. Coloring interval graphs with First-Fit. *SIAM Journal on Discrete Mathematics*, 8:47–57, 1995.
14. H. A. Kierstead and W. T. Trotter. An extremal problem in recursive combinatorics. *Congressus Numerantium*, 33:143–153, 1981.
15. L. T. Kou and G. Markowsky. Multidimensional bin packing algorithms. *IBM Journal on Research and Development*, 21:443–448, 1977.

16. K. L. Krause, V. Y. Shen, and H. D. Schwetman. Analysis of several task-scheduling algorithms for a model of multiprogramming computer systems. *Journal of the ACM*, 22(4):522–550, 1975.
17. K. L. Krause, V. Y. Shen, and H. D. Schwetman. Errata: "Analysis of several task-scheduling algorithms for a model of multiprogramming computer systems". *Journal of the ACM*, 24(3):527–527, 1977.
18. N. S. Narayanaswamy. Dynamic storage allocation and online colouring interval graphs. In *Proc of the 10th Annual International Conference on Computing and Combinatorics (COCOON2004)*, pages 329–338, 2004.
19. S. Pemmaraju, R. Raman, and K. Varadarajan. Buffer minimization using max-coloring. In *Proc. of the Fifteenth Annual ACM-SIAM Symposium on Discrete Algorithms (SODA 2004)*, pages 562–571, 2004.
20. J. D. Ullman. The performance of a memory allocation algorithm. Technical Report 100, Princeton University, Princeton, NJ, 1971.
21. J. Yang and J. Y.-T. Leung. The ordered open-end bin-packing problem. *Oper. Res.*, 51(5):759–770, 2003.

Separating the Notions of Self- and Autoreducibility

Piotr Faliszewski and Mitsunori Ogihara

Department of Computer Science,
University of Rochester,
Rochester, USA
{pfali, ogihara}@cs.rochester.edu

Abstract. Recently Glaßer et al. have shown that for many classes C including PSPACE and NP it holds that all of its nontrivial many-one complete languages are autoreducible. This immediately raises the question of whether all many-one complete languages are Turing self-reducible for such classes C.

This paper considers a simpler version of this question—whether all PSPACE-complete (NP-complete) languages are length-decreasing self-reducible. We show that if all PSPACE-complete languages are length-decreasing self-reducible then PSPACE = P and that if all NP-complete languages are length-decreasing self-reducible then NP = P.

The same type of result holds for many other natural complexity classes. In particular, we show that (1) not all NL-complete sets are logspace length-decreasing self-reducible, (2) unconditionally not all PSPACE-complete languages are logpsace length-decreasing self-reducible, and (3) unconditionally not all EXP-complete languages are polynomial-time length-decreasing self-reducible.

1 Introduction

Self-reducibility [1,2] and autoreducibility [3,4] are among the most frequently used central concepts in complexity theory. Intuitively, these notions refer to the situations in which the membership question about a word in a language can be answered by asking the membership question in the same language about other strings. While autoreducibility essentially permits querying about any word other than the input (within a certain resource constraint), self-reducibility permits querying only those words preceding the input with respect to some partial order (the exact definition of the partial order changes the characteristic of the self-reducibility). It is well-known that the NP-complete problem SAT possesses such a property: given a non-trivial formula φ as input, one can decide whether φ is satisfiable by asking whether at least one of the two formulas constructed by setting the value of the first variable of φ to 0 and to 1 is satisfiable. This is called the disjunctive-self-reducibility of SAT. A rich theory of self-reducibility and autoreducibility has been established by studying the structure of the reductions, that is, how "easier" the queries should be and how powerful the underlying computation is. The theory encompasses such concepts as coherence [5], logspace-self-reducibility [6], random-self-reducibility [7], and word-decreasing self-reducibility [6].

Like SAT, many concrete complete sets are known to be self-reducible or autoreducible. In many cases, their self-reductions or autoreductions are *length-decreasing*, in the sense that the query strings are shorter than the input. It is easy to see that the disjunctive-self-reduction of SAT presented in the above is indeed length-decreasing. The standard complete language QBF for PSPACE has a similar self-reduction in which the membership is queried about the formulas constructed by fixing the first variable to 0 and to 1. This self-reduction is length-decreasing too.

One might then ask whether every complete problem is indeed length-decreasing self-reducible for NP and for PSPACE. This paper shows that that is unlikely to be the case. We show that for a wide variety of classes C, including PSPACE and NP, it holds that if all complete sets for C are length-decreasing self-reducible then $C \subseteq \text{P}$.

The above result about NP can be contrasted with the results about autoreducibility of NP-complete languages. Formally, a language A is autoreducible if it is accepted by a polynomial-time oracle Turing machine M with A as the oracle and for no input x M queries x to the oracle. Beigel and Feigenbaum [8] showed that all languages complete for NP with respect to polynomial-time Turing reductions are (Turing) autoreducible. More recently, Glaßer et al. [9] show that all NP-complete sets and all PSPACE-complete sets are many-one autoreducible. So, from our new result it follows that under the assumption that P $\neq$ PSPACE there are languages that are autoreducible but not length-decreasing self-reducible.

Buhrman and Torenvliet [10] gave evidence that general self-reducibility (as defined by Meyer and Paterson in [1]) differs from autoreducibility on NP. Our result does not follow from their result, since the length-decreasing self-reducibility is a special case of the Meyer–Paterson self-reducibility. Indeed, our results are stronger than the result of Buhrman and Torenvliet.

Our results can be obtained by using a single, simple technique. For each class we are concerned with, we construct a complete language with the property that the language is length-decreasing self-reducible if and only if the language is *easy*. Here by "easy" we mean that the language is polynomial-time decidable in the case where the self-reductions are polynomial-time and that the language is logarithmic-space decidable in the case where the self-reductions are logarithmic-space. With simple modifications, our technique applies to logspace length-decreasing self-reducibility, obtaining collapses to L.

Note that in complexity theory there is no agreement on the definition of self-reducibility. In this paper, following Balcázar [6], we use length-decreasing Turing self-reducibility. However, a more general notion of Meyer and Paterson [1] has been widely used. Köbler and Watanabe [11] use an even more generic notion of self-reducibility, which can be applied beyond PSPACE (all length-decreasing self-reducible sets, and all sets self-reducible with respect to the Meyer–Paterson self-reducibility, are in PSPACE). One may consider our result as an argument against the use of length-decreasing self-reducibility, since we show that sets that are unlikely to be length-decreasing self-reducibility can be easily constructed.

Finally, we note here that the work presented in this paper was motivated by Problem 5.15 in the book by Bovet and Crescenzi [12]. The problem asks to prove that SAT is length-decreasing self-reducible and then asks whether this fact is enough for us to conclude that *all* NP-complete sets are length-decreasing self-reducible. This paper gives a solution to this problem.

The paper is organized as follows. In Section 2 we give basic definitions that will be useful throughout the paper. Section 3 presents our main results along with some simple, but interesting, corollaries.

2 Preliminaries

Throughout this paper we use standard definitions of complexity theory, as in [13,12]. Class PSPACE is the set of all languages that can be decided using polynomial amount of space. L and NL are sets of languages than can be decided using logarithmic amount of space by a deterministic and a nondeterministic Turing Machine respectively. Without the loss of generality, we assume that all languages we consider are over the alphabet $\Sigma = \{0,1\}$. All polynomials we consider here have nonnegative integer coefficients so for all $n \geq 0$ their value is nonnegative.

We define autoreducible sets as follows:

Definition 1. *A language A is* autoreducible *if there exists a polynomial-time oracle Turing machine M such that*

- $L(M^A) = A$ *and*
- *for all inputs x, M does not query x to its oracle.*

It is easy to see that SAT is autoreducible by the reduction provided in the introduction: On input formula φ, if φ is trivial, that is, it contains no variable, accept or reject according to whether φ is true or false; otherwise, construct two formulas, φ_0 and φ_1, by setting the value of the first variable of φ to 0 and 1, respectively, and then accept if and only if either φ_0 or φ_1 belongs to SAT. In this reduction the queries are disjunctive and the query strings are shorter than the input, so it is actually a disjunctive length-decreasing self-reduction. Formally, we define length-decreasing self-reductions as follows:

Definition 2. *We say that a language A is* length-decreasing self-reducible *if there exists a polynomial-time autoreduction M such that for all x, M on input x does not query a string whose length is greater than or equal to $|x|$.*

The above two definitions are with respect to polynomial-time Turing machines. The logarithmic-space versions of those reducibility notions are defined simply by requiring that the machine M runs in logarithmic space. The logarithmic bound does not apply to the oracle tape, the queries can be of up to polynomial size.

Let f be a function from **N** to itself. We say that f is a *padding-length function* if for all $n \geq 0$ $f(n) > n$. We say that f is *polynomially bounded* if there exists a polynomial p such that for all $n \geq 0$ it holds that $f(n) \leq p(n)$. We say that f is *logspace computable* if the mapping $1^n \mapsto 1^{f(n)}$ is computable in logarithmic space.

Definition 3. *We say that a class C is closed under logspace padding if for every nontrivial (neither $\{0,1\}^*$ nor emptyset) $A \in C$ and for every polynomially bounded, logspace computable padding-length function f it holds that the language A' defined by:*

$$A' = \{x10^m \mid x \in A \land 1 + |x| + m = f(|x|)\}$$

belongs to C.

It is easy to see that classes L, NL, NP and PSPACE are all closed under logspace padding.

Definition 4. *We say that a complexity class C is* normal *if it has complete sets (polynomial-time many-one complete sets for classes known to contain P and logspace many-one complete sets for the others) and is closed under logspace padding.*

It is easy to see that all classes that have complete sets and are closed under many-one reductions are normal.

Lemma 1. *Let C be a normal class. Let f be an arbitrary polynomially bounded, logspace computable padding-length function. Let A be an arbitrary C-complete set. Then $A' = \{x10^m \mid x \in A \land |x| + 1 + m = f(|x|)\}$ is C-complete.*

Proof. Let C, f, and A be as in the hypothesis of the lemma. Since C is normal, A' belongs to C. To show that A' is hard for C, define $g(x) = x10^{|f(x)|-|x|-1}$. This function g is logspace computable: our machine simulates the machine for computing f on input x and replaces its first $|x|+1$ bits by $x1$. It is easy to see that g many-one reduces A to A'. So, A' is hard for C. This proves the lemma. □

Note that if A in the lemma above is logspace C-complete then so is its padded version A'.

3 Main Results

In this section we prove our two main results regarding length-decreasing self-reducibility and logspace length-decreasing self-reducibility.

Theorem 1. *Let C be a normal class. If all C-complete languages under polynomial-time many-one reductions are length-decreasing self-reducible then $C \subseteq$ P.*

Proof. Let A be an arbitrary C-complete language under polynomial-time many-one reductions. Let $k \geq 2$ be an integer. Let f be the function from $\mathbf{N}$ to itself defined for all $n \geq 0$ by: $f(n) =$ the smallest integer 2^{k^i} such that i is an integer and $2^{k^i} > n$. Define B as follows:

$$B = \{x10^m \mid x \in A \land |x| + 1 + m = f(|x|)\}.$$

Note that B is simply a padded version of A, in which the length of each member of A is inflated to an integer of the form 2^{k^i}. The function f is clearly a padding-length function. Also, f is polynomially bounded because for all $n \geq 0$ it holds that $f(n) \leq 2 + n^k$. Furthermore, f is logspace computable: On input x, $f(x)$ can be computed by successively calculating in binary $2, 2^k, (2^k)^k, ((2^k)^k)^k, \cdots$ until the value exceeds $|x|$. Thus, by Lemma 1, B is C-complete. Then by our assumption B is length-decreasing self-reducible.

Since B is length-decreasing self-reducible, there is a deterministic Turing machine M such that:

- M runs in polynomial time;
- $L(M^B) = B$;
- On input x, M queries its oracle only about strings shorter than x.

We now describe a polynomial-time algorithm for B that does not need the oracle. On input x our algorithm behaves as follows: If $|x|$ is not of the form 2^{k^i}, then by definition x is clearly a non-member of B, so we immediately terminate the simulation (or return from the simulation if we are dealing with a recursive call) by asserting that $x \notin B$. Otherwise, we simulate M on input x replacing each oracle query by a recursive call to M. Nontrivial simulations of M take place only when the length of the input is of the form 2^{k^i} for some integer i. Since M is length-decreasing self-reducible, we can assume that on inputs of appropriate length all oracle queries regard strings of length at most $|x|^{\frac{1}{k}}$. We also modify M to use a look-up table to decide strings of length at most 2.

Now we are ready to prove that our algorithm runs in polynomial time. Let p be a polynomial such that for all x, the running time of M on input x is $p(|x|)$ assuming that the cost of each oracle query is 1. Note that all polynomials we are concerned with have nonnegative coefficients, so p is strictly increasing and for all nonnegative n it holds that $p(n) > 1$. What is the time complexity of our algorithm? In the recursion tree, there are at most $\lceil \log \log n \rceil$ levels because strings of length at most $n^{\frac{1}{k^{\lceil \log \log n \rceil}}}$ are of length at most 2, and we have a look-up table to decide whether we accept them or not. For each level $i \geq 0$ of the recursion tree, let P_i be the number of computational steps other than processing of recursive calls required to simulate all the level-i simulations. It holds that:

$$P_0 \leq p(n)$$
$$P_1 \leq p(n) \cdot p(n^{\frac{1}{k}})$$
$$P_2 \leq p(n) \cdot p(n^{\frac{1}{k}}) \cdot p(n^{\frac{1}{k^2}})$$

and so on. This is because at the 0-th level we just have one call to M with input of length n. At this level M can make at most $p(n)$ oracle queries, each of length at most $n^{\frac{1}{k}}$, thus at the first level we need to execute at most $p(n) \cdot p(n^{\frac{1}{k}})$ basic steps. The same analysis applies to P_2, P_3, and so on, up to $P_{\lceil \log \log n \rceil}$.

Let us now estimate the value of P_j for some arbitrary j. We can assume without the loss of generality that $p(n) \leq n^d$ for some nonnegative integer d. It holds that:

$$P_j \leq \prod_{i=0}^{j} p(n^{\frac{1}{k^i}}) \leq \prod_{i=0}^{j} n^{\frac{d}{k^i}} = n^{\sum_{i=0}^{j} \frac{d}{k^i}} \leq n^{\sum_{i=0}^{\infty} \frac{d}{k^i}} = n^{\frac{dk}{k-1}}$$

Note that the final result does not depend on j so the time complexity of the whole algorithm is bounded by:

$$\lceil \log \log n \rceil \cdot n^{\frac{dk}{k-1}} \in O(n^h),$$

where h is some nonnegative integer such that $h \geq \frac{dk}{k-1} + 1$. Thus, we have shown that B is decidable in polynomial time. Since B is complete for C, it holds that all languages in C are decidable in polynomial time. This proves the theorem. □

For all C's other than some pathological cases ($C = \{\emptyset\}$ or $C = \{\Sigma^*\}$) we have, by the above theorem, that if C is closed under many-one reductions and all C-complete sets are length-decreasing self-reducible (and C has at least one complete set) then $C =$ P. Thus, we have the following corollaries:

Corollary 1. *If all NP-complete languages are length-decreasing self-reducible, then* P = NP.

Corollary 2. *If all PSPACE-complete languages are length-decreasing self-reducible then* P = PSPACE.

A similar theorem holds for length-decreasing self-reducibility in logarithmic space.

Theorem 2. *Let C be a normal complexity class. If C has complete languages with respect to logspace many-one reduction and all its complete languages are logspace length-decreasing self reducible then $C \subseteq$ L.*

Proof. The proof is essentially the same as the proof of Theorem 1. The only difference is that now we need to make sure that we do not use more than a logarithmic amount of space for our recursive calls.

Let A, k, f, B, and M be as defined in the proof of Theorem 1 with the exception that A is logspace many-one complete for C, and that M is an oracle Turing Machine witnessing that A is logspace length-decreasing self-reducible. We need to implement the following space-preserving strategy. We cannot generate the input for the recursive calls on the tape because that would use more than a logarithmic amount of space. Instead, after each recursive call that completed, we store the contents of the tape and the position of the head (replacing the previously stored one) and run the machine—without writing out the next query string—until finally a subsequent query is to be asked (recursive call is to be performed). Then, in that recursive call we only pass the stored tape so that the recursive call can recreate any of the bits of its input string on demand (using an at most $\log n$ bit counter).

By definition of logspace length-decreasing self-reducibility, excluding the recursive calls, our machine uses at most logarithmic amount of space. Thus, if the input string has length n then for the recursive calls at the first level we only need at most $c \log n$ bits, where c is some constant. Consequently, the total amount of space that we need to handle each branch of recursion is:

$$\sum_{i=0}^{\lceil \log \log n \rceil} c \log \left(n^{\frac{1}{k^i}} \right) \leq \sum_{i=0}^{\infty} c \log \left(n^{\frac{1}{k^i}} \right)$$
$$= c \log n \sum_{i=1}^{\infty} \frac{1}{k^i}$$
$$= c \left(\frac{k}{k-1} \right) \log n.$$

Since we handle recursion branches one at a time, this means that our algorithm requires at most logarithmic space. As B is C-complete (with respect to logspace many-one reductions) it holds that all languages in C can be decided in logarithmic space. □

The above theorem yields the following corollary.

Corollary 3. *If all* NL *sets complete with respect to logspace many-one reductions are logspace length-decreasing self-reducible then* L = NL.

By the Time and Space Hierarchy Theorems, it holds that

- L $\neq$ PSPACE and
- P $\neq$ EXP.

Now by Theorems 1 and 2, it holds that there is a PSPACE-complete language that is not logspace length-decreasing self-reducible, and that there exists an EXP-complete language that is not length-decreasing self-reducible.

We conclude the paper by noting that it follows from Theorem 1 that if P $\neq$ PSPACE then there are PSPACE-complete sets that are not length-decreasing self-reducible. However, it was proved by by Beigel and Feigenbaum [8] (see also the work of Glaßer et al. [9]) that all PSPACE-complete languages are autoreducible. Thus, if P $\neq$ PSPACE then the notions of length-decreasing self-reducibility and autoreducibility are different.

Acknowledgments

We wish to thank Daniel Pierre Bovet and Pierluigi Crescenzi for inspiring this work. We also thank Lane Hemaspaandra, Staszek Radziszowski and Alan Selman for useful discussions. This work is supported in part by NSF Grants EIA-0080124, EIA-0205061, and CCF-0426761.

References

1. Meyer, A., Paterson, M.: With what frequency are apparently intractable problems difficult? Technical Report MIT/LCS/TM-126, Laboratory for Computer Science, MIT, Cambridge, MA (1979)
2. Schnorr, C.: Optimal algorithms for self-reducible problems. In: Proceedings of the 3rd International Colloquium on Automata, Languages, and Programming. (1976) 322–337
3. Ladner, R.: Mitotic recursively enumerable sets. Journal of Symbolic Logic **38** (1973) 199–211
4. Ambos-Spies, K.: P-mitotic sets. In: Logic and Machines. (1983) 1–23
5. Yao, A.: Coherent functions and program checkers. In: Proceedings of the 22nd ACM Symposium on Theory of Computing, ACM Press (1990) 84–94
6. Balcázar, J.: Self-reducibility. Journal of Computer and System Sciences **41** (1990) 367–388
7. Abadi, M., Feigenbaum, J., Kilian, J.: On hiding information from an oracle. Journal of Computer and System Sciences **39** (1989) 21–50
8. Beigel, R., Feigenbaum, J.: On being incoherent without being very hard. Computational Complexity **2** (1992) 1–17
9. Glaßer, C., Ogihara, M., Pavan, A., Selman, A., Zhang, L.: Autoreducibility, mitoticity, and immunity. Technical Report TR05-011, Electronic Colloquium on Computational Complexity, http://www.eccc.uni-trier.de/eccc/ (2005)
10. Buhrman, H., Torenvliet, L.: P-selective self-reducible sets: A new characterization of P. Journal of Computer and System Sciences **53** (1996) 210–217
11. Köbler, J., Watanabe, O.: New collapse consequences of NP having small circuits. SIAM Journal on Computing **28** (1998) 311–324
12. Bovet, D., Crescenzi, P.: Introduction to the Theory of Complexity. Prentice Hall (1993)
13. Papadimitriou, C.: Computational Complexity. Addison-Wesley (1994)

Fully Asynchronous Behavior of Double-Quiescent Elementary Cellular Automata

Nazim Fatès[1], Michel Morvan[1,2], Nicolas Schabanel[1], and Éric Thierry[1]

[1] ENS Lyon - LIP (UMR CNRS - ENS Lyon - UCB Lyon - INRIA 5668),
46 allée d'Italie, 69 364 Lyon Cedex 07, France
[2] Institut universitaire de France,
École des hautes études en sciences sociales and Santa Fe Insitute
{Nazim.Fates, Michel.Morvan, Nicolas.Schabanel, Eric.Thierry}@ens-lyon.fr

Abstract. In this paper we propose a probabilistic analysis of the fully asynchronous behavior (i.e., two cells are never simultaneously updated, as in a continuous time process) of elementary finite cellular automata (i.e., $\{0,1\}$ states, radius 1 and unidimensional) for which both states are quiescent (i.e., $(0,0,0) \mapsto 0$ and $(1,1,1) \mapsto 1$). It has been experimentally shown in previous works that introducing asynchronism in the global function of a cellular automaton may perturb its behavior, but as far as we know, only few theoretical work exist on the subject. The cellular automata we consider live on a ring of size n and asynchronism is introduced as follows: at each time step one cell is selected uniformly at random and the transition rule is applied to this cell while the others remain unchanged. Among the sixty-four cellular automata belonging to the class we consider, we show that fifty-five other converge almost surely to a random fixed point while nine of them diverge on all non-trivial configurations. We show that the convergence time of these fifty-five automata can only take the following values: either 0, $\Theta(n \ln n)$, $\Theta(n^2)$, $\Theta(n^3)$, or $\Theta(n2^n)$. Furthermore, the global behavior of each of these cellular automata can be guessed by simply reading its code.

1 Introduction

The aim of this article is to analyze theoretically the asynchronous behavior of unbounded finite cellular automata. During the last two decades, several empirical studies [3,12,9,1,13,4] have shown that certain cellular automata behavior change drastically under asynchronous behavior. In particular, [1,5] observe that finite size Game of Life space-time diagrams under synchronous and asynchronous updating differ qualitatively. For instance, fixed size Game of Life exhibits convergence to cycles of arbitrary length under synchronous updating, while appears to converge towards a random fixed point under asynchronous dynamics [1].

Cellular automata are widely used to model systems involving a huge number of interacting elements such as agents in economy, particles in physics, proteins in biology, etc. In most of these applications, in particular in many real system models, agents are not synchronous. Interestingly enough, in spite of this lack of synchronism, real living systems are very resilient over time. One might then expect the cellular automata used to model these systems to be robust to asynchronism and other kind of failure as well (such as misreading the state of the neighbors). Surprisingly enough, it turns out that the resilience to asynchronism widely varies from one automata to another (e.g., [1,4]). In particular, the aspect of asynchronous space-time diagrams of cellular automata may differ radically from their synchronous ones.

As far as we know, the question of the importance of perfect synchrony on the behavior of a cellular automaton is not yet understood theoretically. To our knowledge, only Gács shows in [8] undecidability results on the invariance with respect to the update history. Studies have also been led in the more general context of probabilistic cellular automata regarding the question of the existence of stationary distribution on infinite configurations (see [10] for a state of the art).

In this paper, we quantify the convergence time and describe the space-time diagrams for a class of cellular automata under fully asynchronous updating,

Table 1. Behavior of DQECA under fully asynchronous dynamics. WECT stands for worst expected convergence time. See Section 2 for explanations.

Behavior	ECA (#)	Rule	01	10	010	101	WECT
Identity	204 (1)	∅	·	·	·	·	0
Coupon collector	200 (2)	E	·	·	+	·	$\Theta(n \ln n)$
	232 (1)	DE	·	·	+	+	
Monotonic	206 (4)	B	←	·	·	·	$\Theta(n^2)$
	222 (2)	BC	←	→	·	·	
	234 (4)	BDE	←	·	+	+	
	250 (2)	BCDE	←	→	+	+	
	202 (4)	BE	←	·	+	·	
	192 (4)	EF	→	·	+	·	
	218 (2)	BCE	←	→	+	·	
	128 (2)	EFG	→	←	+	·	
Biased Random Walk	242 (4)	BCDEF	↭	→	+	+	
	130 (4)	BEFG	↭	←	+	·	
Random Walk	226 (2)	BDEF	↭	·	+	+	$\Theta(n^3)$
	170 (2)	BDEG	←	←	+	+	
	178 (1)	BCDEFG	↭	↭	+	+	
	194 (4)	BEF	↭	·	+	·	
	138 (4)	BEG	←	←	+	·	
	146 (2)	BCEFG	↭	↭	+	·	
Biased Random Walk	210 (4)	BCEF	↭	→	+	·	$\Theta(n 2^n)$
Divergent	198 (2)	BF	↭	·	·	·	Divergent
	142 (2)	BG	←	←	·	·	
	214 (4)	BCF	↭	→	·	·	
	150 (1)	BCFG	↭	↭	·	·	

where two cells are not updated simultaneously. This asynchronous regime, also known as step-driven asynchronous dynamics [13], arises for instance in continuous time updating processes. We focus on double-quiescent elementary automata. We show that among these sixty-four automata, nine diverge on all non-trivial configurations (see Theorem 13), and the fifty-five other converge almost surely to a random fixed point (see Theorem 1). Furthermore, the convergence time

Fig. 1. Examples of space-time diagrams under fully asynchronous and synchronous dynamics for each type of convergence, with $n = 50$. For each automaton, the larger left and the smaller right diagrams are respectively examples of asynchronous and synchronous dynamics. White and black pixels respectively stand for states 0 and 1. The k-th line from bottom is the configuration at time $t = 50\,k$ for the asynchronous dynamics, and at time $t = k$ for the synchronous one. Note that automata (a) and (c) are respectively the classic Majority and Shift rules. Each automata is described by two codes: a number, which is the classic Wolfram's number, and a sequence of letters, which will be introduced later in the paper.

of these fifty-five automata on (spatially) periodic configurations, can only take the following values: either 0, $\Theta(n \ln n)$, $\Theta(n^2)$, $\Theta(n^3)$, or $\Theta(n2^n)$, where n is the size of the configurations. One of the most striking results is that the fully asynchronous global behavior of double quiescent elementary automata is obtained simply by reading the code of their local transition rules (see Tab. 1), which is known to be a difficult problem in general. Moreover, the asynchronous behavior of all automata is in a certain sense *characterized* by this convergence time: all automata within the same convergence time present the same kind of space-time diagrams (see Tab. 1 and Fig. 1). Remark that the asynchronous behavior of some very simple automata like the shift (Wolfram rule code 170) actually simulates intricate stochastic processes that are currently under investigation in mathematics and physics, such as annihilating random walks, studied for instance in [11]. Our results rely on coupling the automata with a proper random process.

Definitions and our main result are given in Section 2. In section 3, we present basic but useful properties of the automata we consider. Section 4 is a technical section that develops probabilistic tools used to analyze the automata. Section 5 finally analyzes in details the asynchronous behavior of each automaton.

2 Definitions, Notations and Main Results

In this paper, we consider two-state cellular automata on finite size configurations.

Definition 1. *An* Elementary Cellular Automata (ECA) *is given by its* transition function $\delta : \{0, 1\}^3 \to \{0, 1\}$. *We denote by* $Q = \{0, 1\}$ *the set of* states. *A state q is* quiescent *if $\delta(q, q, q) = q$. An ECA is* double-quiescent (DQECA) *if both states 0 and 1 are quiescent.*

We denote by $U = \mathbb{Z}/n\mathbb{Z}$ the set of cells. *A finite configuration with periodic boundary conditions $x \in Q^U$ is a word indexed by U with letters in Q. For a given pattern $w \in Q^U$, we denote by $|x|_w = \#\{i \in U : x_{i+1} \ldots x_{i+|w|} = w\}$ the number of occurrences of w in configuration x.*

We consider two kinds of dynamics for ECAs: the synchronous dynamics and the fully asynchronous dynamics. The synchronous dynamics is the classic dynamics of cellular automata, where the transition function is applied at each (discrete) time step on each cell simultaneously.

Definition 2 (Synchronous Dynamics). *The* synchronous dynamics $S_\delta : Q^U \to Q^U$ *of an ECA δ, associates to each configuration x the configuration y, such that for all i in U, $y_i = \delta(x_{i-1}, x_i, x_{i+1})$.*

The asynchronous regime studied here can be seen as the most extreme asynchronous regime as two cells are never updated simultaneously.

Definition 3 (Fully Asynchronous Dynamics). *The* fully asynchronous dynamics AS_δ *of an ECA δ associates to each configuration x a random configuration y, such that $y_j = x_j$ for $j \neq i$, and $y_i = \delta(x_{i-1}, x_i, x_{i+1})$, where i is* uniformly chosen at random in U. AS_δ could equivalently be seen as a function with two arguments, the configuration x and the random index $i \in U$. For a given ECA δ, we denote by x^t the random variable for the configuration obtained by t applications of the asynchronous dynamics function AS_δ on configuration x, i.e., $x^t = (AS_\delta)^t(x)$.

Definition 4 (Fixed point). *We say that a configuration x is a fixed point for δ under fully asynchronous dynamics if $AS_\delta(x) = x$ whatever the choice of i (the cell to be updated) is. $\mathfrak{F}_\delta$ denotes the set of fixed points for δ.*

The set of fixed points of the asynchronous dynamics is clearly identical to $\{x : S_\delta(x) = x\}$ the set of fixed points of the synchronous dynamics. Note that every DQECA admits two *trivial fixed points*, 0^n and 1^n.

Definition 5 (Worst Expected Convergence Time). *Given an ECA δ and a configuration x, we denote by $T_\delta(x)$ the random variable for the time to reach a fixed point from configuration x under fully asynchronous dynamics, i.e., $T_\delta(x) = \min\{t : x^t \in \mathfrak{F}_\delta\}$. The worst expected convergence time T_δ of ECA δ is :*

$$T_\delta = \max_{x \in Q^U} \mathbb{E}[T_\delta(x)].$$

We can now state our main theorem.

Theorem 1 (Main result). *Under fully asynchronous dynamics, among the sixty-four DQECAs,*

- *fifty-five converge almost surely to a random fixed point on any initial configuration, and the worst expected convergence times of these fifty-five convergent DQECAs are 0, $\Theta(n \ln n)$, $\Theta(n^2)$, $\Theta(n^3)$, and $\Theta(n2^n)$;*
- *the nine others diverge almost surely on any initial configuration that is neither 0^n, nor 1^n nor, when n is even, $(01)^{n/2}$.*

Furthermore, the behaviors of the different DQECAs are similar within each class, and are obtained by simply reading its code as illustrated in Tab. 1.

Figure 1 gives examples of the asynchronous space-time diagrams of a representative of each class (but Identity). It is interesting to notice that except for the first diagram (Fig. 1(a)), the asynchronous space-time diagrams (the larger ones) considerably differ from the corresponding synchronous ones (the smaller ones).

3 Basic properties of DQECAs

The transition function δ of an ECA is given by the set of its eight transitions $\delta(000), \delta(001), \ldots, \delta(111)$, traditionally written $\frac{000}{\delta(000)}, \ldots, \frac{111}{\delta(111)}$. The following code describes each ECA by its differences to the Identity automaton. We use this notation rather than the classic Wolfram's one [14] since it is not immediate to infer the local behavior of the cellular automaton just by looking at its Wolfram code. In order to allow comparison with other work we still indicate the classic Wolfram number in Tab. 1.

Notation 1. We say that a transition is *active* if it changes the state of the cell where it is applied. Each ECA is fully determined by its active transitions. We label each active transition by a letter as follow:

A	B	C	D	E	F	G	H
000	001	100	101	010	011	110	111
1	1	1	1	0	0	0	0

We label each ECA by the set of its active transitions.

Note that with these notations, the DQECAs are exactly the ECAs having a label containing neither A nor H. By 0/1 and horizontal symmetries of configurations, we shall w.l.o.g. only consider the 24 DQECAs listed in Tab. 1 among the 64 DQECAs. For each of these 24 DQECAs, the number of the equivalent automata under symmetries is written within parentheses after their classic ECA code in the table.

From now on, we only consider the fully asynchronous dynamics (with uniform choice); this will be implicit in all the following propositions. Our results rely on the study of the evolution of the "regions" in the space-time diagram (i.e., of the intervals of consecutive 0s or 1s in configuration x^t). The key observation is that for DQECAs, under fully asynchronous dynamics, the number of regions is non-increasing since no new region can be created; furthermore, only regions of length one can disappear (see Fig. 1). We denote by $Z(x) = |x|_{01}$ ($= |x|_{10}$) the *number of alternations* from 0 to 1 in configuration x, which will be our counter for the number of regions.

Fact 2. *For any DQECA, $Z(x^t)$ is a non-increasing function of time. Furthermore, $Z(x^{t+1}) < Z(x^t)$ if and only if x^{t+1} is obtained from x^t by applying a transition D or E at time t, and then $Z(x^{t+1}) = Z(x^t) - 1$.*

On the one hand, transitions D and E are thus responsible for decreasing the number of regions in the space-time diagram: D "erases" the 1-regions and E the 0-regions. On the other hand, transitions B and F act on patterns 01. Intuitively, transition B moves a pattern 01 to the left, and transition F moves it to the right. In particular, patterns 01 perform a kind of random walk for DQECA with both transitions B and F. Similarly, transitions C and G act on patterns 10. Transition C moves a pattern 10 to the right, and transition G moves it to the left. The arrows in Tab. 1 represent the different behavior of the patterns: ← or →, for left or right moves of the patterns 01 or 10; ⤳, for random walks of these patterns.

The following lemma characterizes the fixed points of a given DQECA according to its code.

Fact 3. *If a DQECA δ admits a non-trivial fixed point x, then:*
- *if δ contains transition B or C, then all 0s in x are isolated;*
- *if δ contains transition F or G, then all 1s in x are isolated;*
- *if δ contains transition D, then none of the 0s in x is isolated;*
- *if δ contains transition E, then none of the 1s in x is isolated.*

The next section is devoted to analyzing particular random walk-like processes that will be used as tools to obtain our bounds on the convergence time.

4 Probabilistic Toolbox

Notation 2. For a given random sequence $(X_t)_{t \in \mathbb{N}}$, we denote by $(\Delta X_t)_{t>0}$ the random sequence $\Delta X_t = X_t - X_{t-1}$.

Quadratic DQECA Toolbox. Consider $\epsilon > 0$, a non-negative integer m, and $(X_t)_{t \in \mathbb{N}}$ a sequence of random variables with values in $\{0, \ldots, m\}$ given with a suitable filtration $(\mathcal{F}_t)_{t \in \mathbb{N}}$. In probability theory, $\mathcal{F}_t$ represents intuitively the σ-algebra (the "set") of the events that happened up to time t and is the formal tool to condition relatively to the past (see [7, Chap. 7]). In the sequel, $\mathcal{F}_t$ will either be the values of the previous random variables $X_0, \ldots, X_t$, or in some cases, the set of past configurations $x^0, \ldots, x^t$. The following lemma bounds the convergence time of a random variable that decreases by a constant on expectation.

Lemma 4. *Assume that if $X_t > 0$, then $\mathbb{E}[\Delta X_{t+1} | \mathcal{F}_t] \leqslant -\epsilon$. Let $T = \min\{t : X_t \leqslant 0\}$ denote the random variable for the first time t where $X_t \leqslant 0$. Then, if $X_0 = x_0$,*

$$\mathbb{E}[T] \leqslant \frac{m + x_0}{\epsilon}.$$

Cubic DQECA Toolbox. Let $\epsilon > 0$ and $(X_t)_{t \in \mathbb{N}}$ a sequence of random variables with values in $\{0, \ldots, m\}$, given with a suitable filtration $(\mathcal{F}_t)_{t \in \mathbb{N}}$.

Definition 6. *The following two types of process will be extensively used in the next section:*

- *We say that $(X_t)_{t \in \mathbb{N}}$ is of type I if for all t:*
  - $\mathbb{E}[X_{t+1} | \mathcal{F}_t] = X_t$ *(i.e., (X_t) is a martingale), and*
  - *if $0 < X_t < m$, then $\Pr\{\Delta X_{t+1} \geqslant 1\} = \Pr\{\Delta X_{t+1} \leqslant -1\} \geqslant \epsilon$.*

- *We say that $(X_t)_{t \in \mathbb{N}}$ is of type II if for all t:*
  - *if $X_t < m$, then $\mathbb{E}[X_{t+1}] = X_t$ (i.e., (X_t) behaves as a martingale when $X_t < m$), and*
  - *if $0 < X_t < m$, then $\Pr\{\Delta X_{t+1} \geqslant 1\} = \Pr\{\Delta X_{t+1} \leqslant -1\} \geqslant \epsilon$, and*
  - *if $X_t = m$, then $\Pr\{X_{t+1} \leqslant m - 1\} \geqslant \epsilon$ (i.e., X_t "bounces on m").*

Note that when (X_t) is of type I, if for some t, $X_t \in \{0, m\}$, then $X_{t'} = X_t$ for all $t' \geqslant t$, because (X_t) is a martingale bounded between 0 and m. Thus, $\{0, m\}$ are the (only) fixed points of any type I sequence. When (X_t) is of type II, if for some t, $X_t = 0$, then $X_{t'} = X_t$ for all $t' \geqslant t$, because (X_t) is a martingale lower bounded by 0. Thus, 0 is the (only) fixed point of any type II sequence.

Definition 7. *The convergence time of a type I sequence (X_t) is defined as the random variable $T = \min\{t : X_t \in \{0, m\}\}$. The convergence time of a type II sequence (X_t) is similarly defined as the random variable $T = \min\{t : X_t = 0\}$.*

The following lemmas bound the convergence time of these two types of random processes.

Lemma 5. *For sequence* (X_t), *if* $X_0 = x_0$, *the expectation of* T *satisfies:*

$$\mathbb{E}[T] \leq \frac{x_0(m - x_0)}{2\epsilon} \quad \text{if } (X_t) \text{ is of type I,}$$

$$\mathbb{E}[T] \leq \frac{x_0(2m + 1 - x_0)}{2\epsilon} \quad \text{if } (X_t) \text{ is of type II.}$$

5 Convergence

In this section, we evaluate the worst expected convergence time for each of the twenty-four representative automata in Tab. 1. Our results rely on studying the evolution of quantities computed on the random configurations (x^t), whose convergence implies the convergence of the automaton. The upper bounds on the convergence time of these quantities are obtained by coupling them with one of the integer random processes analyzed in the previous section. The lower bounds are obtained by analyzing the exact expected convergence time for a particular initial configuration (most of the time, a configuration with a single 0-region and a single 1-region). This involves building suitable variants measuring progress towards fixed points. One of the main difficulties is to handle correctly the mergings of the regions, i.e., the applications of transitions D and E.

We introduce the following convenient functions that simplify the evaluation of the quantities that are used to bound the convergence time. These function will spare us tedious parsings of the patterns in the configurations. For a given configuration x, we denote by $a(x), \ldots, h(x)$ the number of cells where transitions A, ..., H are applicable, i.e.:

$$a(x) = |x|_{000}, \ b(x) = |x|_{001}, \ c(x) = |x|_{100}, \ d(x) = |x|_{101},$$
$$e(x) = |x|_{010}, \ f(x) = |x|_{011}, \ g(x) = |x|_{110}, \ h(x) = |x|_{111}.$$

For instance, consider rule BCG. For convenience, we denote by $p = 1/n$ the probability that a given cell is updated under fully asynchronous dynamics. Applying the transitions A, ..., D increases the number of 1s by one and applying E, ..., H decreases it by one. The expected variation of the number of 1s for configuration x in one step is then immediately $p \cdot (b(x) + c(x) - g(x))$. When the context is clear, the argument x will be omitted. Clearly, parsing properly configuration x gives the following useful relationships.

Fact 6. *For all configurations* $x \in Q^U$, *the following equalities hold:*

$$|x|_{01} = b + d = e + f = c + d = e + g = |x|_{10},$$
$$|x|_{001} = b = c = |x|_{100},$$
$$|x|_{011} = f = g = |x|_{110}.$$

Let us now analyze the worst expected convergence time for DQECAs. Due to space constraints, most of the proofs are omitted and can be found in [6].

5.1 "Coupon collector" DQECAs

The behavior of the DQECAs in this class (see Fig. 1(a)) is similar to the classic Coupon Collector random process (e.g., [7]).

Theorem 7. *Under fully asynchronous dynamics, DQECAs E and DE converge a.s. to a fixed point on any initial configuration. Their worst expected convergence time is $\Theta(n \ln n)$. The fixed points for E and DE respectively are the configurations without isolated 1 and the configurations without isolated 0 and 1.*

Proof. These rules simply erase either isolated 0s, isolated 1s or both. They never create any of them (by Fact 2), and reach a fixed point as soon as no more 0 or 1 are isolated (by Fact 3). These processes are then similar to a coupon collector process that has to collect all the isolated 0s or 1s, by drawing at each time step a random location uniformly in $\{1, \ldots, n\}$ (see e.g., [7]). If the number of remaining isolated 0s and 1s is i, the probability to draw one of them is i/n, and then, one of them is drawn on expectation after n/i steps. The expected convergence time is then bounded by $n(1 + \frac{1}{2} + \cdots + \frac{1}{n}) = O(n \ln n)$.

Finally, configuration $(010)^{\lfloor n/3 \rfloor} 0^{n \bmod 3}$, which is a proper coupon collector process, provides a lower bound of $\Omega(n \ln n)$ for both rules.

5.2 Quadratic DQECAs

Figure 1(b) illustrates the typical space-time diagram in this class. All the results of this section are obtained by finding a proper *variant* whose convergence implies the convergence of the DQECA, and which decreases by more than a given constant on expectation.

Lemma 8. *Given an initial configuration x, for each DQECA B, BC, BDE, BCDE, BCDEG, BE, EF, BCE, EFG, BCEFG, and BEFG, there exists a sequence (X_t) of random variables with values in $\{0, \ldots, n\}$ (the variant), such that:*
(a) if $X_t = 0$, then x^t is a fixed point.
(b) for all t such that x^t is not a fixed point, $\mathbb{E}[\Delta X_{t+1} | X_t] \leqslant -p$.

Proof. **Rules B and BC.** Set $X_t = |x^t|_0$ the number of 0s in x^t. (a) is clear since $X_t = 0$ implies that $x^t = 1^n$. We obtain (b) by noticing that each application of transitions B or C decreases X_t by one, and that for any non fixed-point configuration, an active transition is performed with probability greater or equal to p. Similarly, $X_t = |x^t|_1$ is suitable for rules EF and EFG.

Remaining Rules. We need to take into account the presence of isolated 0s and 1s. We set $X_t = |x^t|_0 + Z(x^t)$ for rules BDE, BCDE, BE, BCE, and BCDEG; and $X_t = |x^t|_1 + Z(x^t)$ for rule BEFG. Consider automaton BEFG. Clearly, $X_t \in \{0, \ldots, n\}$, and we have (a) $X_t = 0$ implies that $x^t = 0^n$. For this rule,

$$\mathbb{E}[\Delta X_{t+1} | x^t] = p \cdot (b - e - f - g)(x^t) - p \cdot e(x^t),$$

since only transition E acts on $Z(x^t)$. By Fact 6, one can rewrite

$$\mathbb{E}[\Delta X_{t+1} | x^t] = -p \cdot (d + e + g)(x^t).$$

Second, if x is not a fixed point, then $(b + e + f + g)(x) > 0$. But by Fact 6, if $d + e = 0$, then $b = f = g$. Thus, $b + e + f + g > 0$ implies $d + e + g > 0$. We conclude that if x^t is not a fixed point, we have (b). The proof is similar for all the remaining automata. We can now state the theorem.

Theorem 9. *Under fully asynchronous dynamics, DQECAs B, BC, BDE, BCDE, BCDEG, BE, EF, BCE, EFG, BCEFG, and BEFG converge almost surely to a fixed point on any initial configuration. Their worst expected convergence time is $\Theta(n^2)$. Only the DQECAs B, BC, BE, and BCE have non-trivial fixed points, which are the configurations where all the 0s are isolated.*

Proof. The property on the fixed points is a direct application of Fact 3. Consider now one of the rules. Let X_t be the variant given by Lemma 8. X_t does not exactly verify the hypotheses of Lemma 4: X_t needs to be extended beyond a fixed point if it is reached before $X_t = 0$. We consider the random sequence X'_t defined as follow: $X'_t = X_t$ if x^t is not a fixed point, and $X'_t = 0$ otherwise. Thus, $X'_t = 0$ if and only if x^t is a fixed point, and we can now apply Lemma 4 with $m = n$ and $\epsilon = p$ and we obtain $\mathbb{E}[T] \leqslant X_0/p = O(n^2)$.

The lower bound $\Omega(n^2)$ on the convergence time is simply given by considering the following initial configuration $x = 0^{\lceil n/2 \rceil} 1^{\lfloor n/2 \rfloor}$. Note that $X_t = |x^t|_1$ works for all the rules on initial configuration x and its exact expected convergence time is straightforward to compute by first step analysis (see [2]).

Observe that we can divide this class into two subcategories: the automata that are monotonic, for which the variant is a non-increasing function of time, and the non-monotonic, for which the variant follows a biased random walk (see Tab. 1). Interestingly enough, this distinction is observed on the space-time diagrams.

5.3 Cubic DQECAs

Figure 1(c) and 1(d) illustrate the typical behaviors in this class: one can observe that the dynamics of the regions in the space-time diagram are similar to unbiased random walks. Furthermore, one can observe that the process of the frontiers between regions is similar to annihilating random walks (e.g.,[11]): each frontier follow a random walk and two frontiers vanish when they meet.

All the results of this section are obtained by coupling the process with a suitable unbiased bounded random walk, such that the DQECA is guaranteed to reach a fixed point before the walk reaches a (or one distinguished) boundary.

The upperbounds in Theorem 11 are straightforward applications of the following lemma 10 in combination with the probabilistic lemma 5. The lower bounds are again obtained by analyzing the expected convergence time on the initial configuration $x = 0^{\lceil n/2 \rceil} 1^{\lfloor n/2 \rfloor}$ with variant $X_t = |x^t|_1$.

Lemma 10. *Given an initial configuration x,*

- *for each DQECA BDEF, BDEG, and BCDEFG, there exists an integer $m \leqslant 2n$ and a random integer sequence (X_t) of type I (see section 4) with values in*

$\{0, \ldots, m\}$, such that: for all t, if $X_t = 0$ or $X_t = m$, then x^t is a fixed point.
- for each DQECA BEF, BEG, and BCEFG, there exists an integer $m \leqslant 2n$ and a random integer sequence (X_t) of type II (see section 4) with values in $\{0, \ldots, m\}$, such that for all t, if $X_t = 0$, then x^t is a fixed point.

Theorem 11. *Under fully asynchronous dynamics, DQECAs BDEF, BDEG, BCDEFG, BEF, BEG, and BCEFG converge almost surely to a fixed point on any initial configuration. Their worst expected convergence time is $\Theta(n^3)$. All of them admit only 0^n and 1^n as fixed point.*

For DQECAs BDEF, BDEG, and BCDEFG, the fixed points 0^n and 1^n can be reached from any configuration (respectively distinct from 1^n and 0^n). For DQECAs BEF, BEG, and BCEFG, any configuration distinct from 1^n converges almost surely to 0^n.

5.4 Exponential DQECA

Figure 1(e) illustrates the typical behavior of this class. The illustrated process will eventually converge to 0^n. The trajectory of the 0-regions is similar to a coalescing random walk : the 0-regions follow a kind of coalescing random walk and merge when they meet, until only one 0-region remains. The size of the remaining 0-region then follows a random walk, biased towards 1, that will eventually converge to n after an exponential time (note that a 0-region cannot disappear for rule BCEF). This result is obtained by coupling the process with a process applying the same rule on a suitable single 0-region configuration. The following lemma analyzes the latter process first, from which we deduce the theorem. Note that the expected convergence time is independent of the initial (non-fixed point) configuration, up to a multiplicative constant.

Theorem 12. *The fixed points of DQECA BCEF are 0^n and 1^n. From any non-fixed point initial configuration, DQECA BCEF converges almost surely to 0^n and its expected convergence time is exactly $\Theta(n2^n)$.*

5.5 Diverging DQECAs

Figure 1(f) illustrates the typical behavior of a divergent DQECA: the number of regions is conserved, and all reachable configurations from a given initial configuration are accessed an infinite number of times almost surely. The proof of the following result relies essentially on applying Fact 3.

Theorem 13. *Under fully asynchronous dynamics, the DQECAs BF, BG, BCF, and BCFG diverge almost surely on any initial configuration that is not one of the three following fixed points 0^n, 1^n and, if n is even, $(01)^{n/2}$. Furthermore, given an initial configuration, all reachable configurations are accessed an infinite number of times almost surely.*

References

1. H. Bersini and V. Detours. Asynchrony induces stability in cellular automata based models. In Brooks, Maes, and Pattie, editors, *Proceedings of the 4th International Workshop on the Synthesis and Simulation of Living Systems ArtificialLifeIV*, pages 382–387. MIT Press, July 1994.
2. P. Brémaud. *Markov chains, Gibbs fileds, Monte Carlo simulation, and queues.* Springer, 1999.
3. R.L. Buvel and T.E. Ingerson. Structure in asynchronous cellular automata. *Physica D*, 1:59–68, 1984.
4. N. Fatès and M. Morvan. An experimental study of robustness to asynchronism for elementary cellular automata. Submitted, arxiv:nlin.CG/0402016, 2004.
5. N. Fatès and M. Morvan. Perturbing the topology of the game of life increases its robustness to asynchrony. In *LNCS Proc. of 6th Int. Conf. on Cellular Automata for Research and Industry (ACRI 2004)*, volume 3305, pages 111–120, Oct. 2004.
6. N. Fatès, M. Morvan, N. Schabanel, and E. Thierry. Fully asynchronous behavior of double-quiescent elementary cellular automata. Research report LIP RR2005-04, ENS Lyon, 2005.
7. G. Grimmet and D. Stirzaker. *Probability and Random Process.* Oxford University Press, 3rd edition, 2001.
8. P. Gács. Deterministic computations whose history is independent of the order of asynchronous updating. http://arXiv.org/abs/cs/0101026, 2003.
9. B. A. Huberman and N. Glance. Evolutionary games and computer simulations. *Proceedings of the National Academy of Sciences, USA*, 90:7716–7718, Aug. 1993.
10. P.-Y. Louis. *Automates Cellulaires Probabilistes : mesures stationnaires, mesures de Gibbs associées et ergodicité.* PhD thesis, Université de Lille I, Sep. 2002.
11. M. Mattera. Annihilating random walks and perfect matchings of planar graphs. *Discrete Mathematics and Theoretical Computer Science*, AC:173–180, 2003.
12. M. A. Nowak and R. M. May. Evolutionary games and spatial chaos. *Nature (London)*, 359:826–829, 1992.
13. B. Schönfisch and A. de Roos. Synchronous and asynchronous updating in cellular automata. *BioSystems*, 51:123–143, 1999.
14. S. Wolfram. Universality and complexity in cellular automata. *Physica D*, 10:1–35, 1984.

Finding Exact and Maximum Occurrences of Protein Complexes in Protein-Protein Interaction Graphs

Guillaume Fertin[1], Romeo Rizzi[2], and Stéphane Vialette[3]

[1] Laboratoire d'Informatique de Nantes-Atlantique (LINA), FRE CNRS 2729,
Université de Nantes, 2 rue de la Houssinière, 44322 Nantes Cedex 3 - France
`fertin@lina.univ-nantes.fr`
[2] Università degli Studi di Trento,
Facoltà di Scienze - Dipartimento di Informatica e Telecomunicazioni,
Via Sommarive, 14 I-38050 Povo - Trento (TN) - Italy
`romeo.rizzi@unitn.it`
[3] Laboratoire de Recherche en Informatique (LRI), UMR CNRS 8623,
Faculté des Sciences d'Orsay, Université Paris-Sud, 91405 Orsay - France
`vialette@lri.fr`

Abstract. In the context of comparative analysis of protein-protein interaction graphs, we use a graph-based formalism to detect the preservation of a given protein complex G in the protein-protein interaction graph H of another species with respect to (w.r.t.) orthologous proteins. Two problems are considered: the Exact-(μ_G, μ_H)-Matching problem and the Max-(μ_G, μ_H)-Matching problem, where μ_G (resp. μ_H) denotes in both problems the maximum number of orthologous proteins in H (resp. G) of a protein in G (resp. H). Following [FLV04], the Exact-(μ_G, μ_H)-Matching problem asks for an injective homomorphism of G to H w.r.t. orthologous proteins. The optimization version is called the Max-(μ_G, μ_H)-Matching problem and is concerned with finding an injective mapping of a graph G to a graph H w.r.t. orthologous proteins that matches as many edges of G as possible. For both problems, the emphasis here is clearly on bounded degree graphs and extremal small values of parameters μ_G and μ_H.

1 Introduction

High-throughput analysis makes possible the study of protein-protein interactions at a genome-wise scale [Gav02, HG02, Uet00], and comparative analysis tries to determine the extent to which protein networks are conserved among species. Indeed, mounting evidence suggests that proteins that function together in a pathway or a structural complex are likely to evolve in a correlated fashion, and, during evolution, all such functionally linked proteins tend to be either preserved or eliminated in a new species [PMT+99].

Protein interactions identified on a genome-wide scale are commonly visualized as protein interaction graphs, where proteins are vertices and interactions are edges [TSU04]. Experimentally derived interaction networks can be extremely complex, so that it is a challenging problem to extract biological functions or pathways from them (even if some global features have been found). However, biological systems are hierarchically organized into functional modules. Several methods have been proposed for identifying functional modules in protein-protein interaction graphs. As observed in [PLEO04], cluster analysis is an obvious choice of methodology for the extraction of functional modules from protein interaction networks. Comparative analysis of protein-protein interaction graphs aims at finding complexes that are common to different species. Kelley et al. [KSK+03] developed the program PathBlast, which aligns two protein-protein interaction graphs combining topology and sequence similarity. Sharan et al. [SIK+04] studied the conservation of complexes[1] that are conserved in *Saccharomyces cerevisae* and *Helicobacter pylori*, and found 11 significantly conserved complexes (several of these complexes match very well with prior experimental knowledge on complexes in yeast only). They actually recasted the problem of searching for conserved complexes as a problem of searching for heavy subgraphs in an edge- and node-weighted graph, whose vertices are orthologous protein pairs. A promising computational framework for alignment and comparison of more than one protein network together with a three-way alignment of the protein-protein interaction networks of *Caenorhabditis elegans*, *Drosophila melanogaster* and *Saccharomyces cerevisae* is presented in [SSK+05].

Following the line of research presented in [FLV04], we consider here the problem of finding an occurrence of a given complex in the protein-protein interaction graph of another species. Notice that we do not make any assumption about the topology of the complex, such as clique-like structure. In [FLV04], this is formulated as the problem of searching for a list injective homomorphism, *i.e.*, an injective homomorphism with respect to orthologous links, of the complex (viewed as a graph) to a protein-protein interaction graph. Roughly speaking, the rationale of this is as follows. First, graph homomorphism only preserves adjacency, and hence can deal with interaction datasets that are missing many true protein interactions. Second, injectivity is required in order to establish a bijective relationship between proteins in the complex and proteins in the occurrence. Finally, graph homomorphism with respect to orthologous links can be easily recasted as list homomorphism: a list of putative orthologs is associated to each protein (vertex) of the complex, and each such protein can only be mapped by the homomorphism to a protein occurring in its list. In the context of comparative analysis of protein-protein interaction graphs, we need to impose *drastic restrictions* on the size of the lists. We will make the following important assumption (referred hereafter as the parameters μ_G and μ_H): no protein has an unbounded number of orthologs in the other species, *i.e.*, each list has a constant size (upper bounded by parameter μ_G) and each protein has a constant number of occurrences among the lists (upper bounded by parameter μ_H). The present paper is devoted to analyzing the complexity of this problem (the Exact-(μ_G, μ_H)-Matching problem) together with its natural optimization

---

[1] They focused on dense, clique-like interaction patterns.

version (the Max-(μ_G, μ_H)-Matching problem) in case of bounded degree graphs and extremal small values of parameters μ_G and μ_H.

The paper is organized as follows: Section 2 introduces formally the two problems. We prove in Section 3 new tight complexity results for the Exact-(μ_G, μ_H)-Matching problem for bounded degree graphs. In Section 4, it is shown that the Exact-(μ_G, μ_H)-Matching for bounded degree graphs is **APX**-hard. That result is complemented in Section 5 by showing that the Exact-(μ_G, μ_H)-Matching problem for bounded degree graphs is in **APX**. Finally, we prove in Section 6 that the Exact-(μ_G, μ_H)-Matching problem for bounded degree graphs parameterized by the number of matched edges is fixed-parameter tractable. Due to space constraints, several details and proofs are not presented in this paper.

2 Preliminaries

Let G be a graph. We write $\mathbf{V}(G)$ for the set of vertices and $\mathbf{E}(G)$ for the set of edges edges, and abbreviate $\#\mathbf{V}(G)$ to $\mathbf{n}(G)$ and $\#\mathbf{E}(G)$ to $\mathbf{m}(G)$. The maximum degree $\Delta(G)$ of a graph G is the largest degree over all vertices. Let G and H be two graphs. For any injective mapping $\theta : \mathbf{V}(G) \to \mathbf{V}(H)$, let us denote by Match$(G, H, \theta)$ the edges of G that are matched by θ, i.e., Match$(G, H, \theta) = \{\{u, v\} \in \mathbf{E}(G) : \{\theta(u), \theta(v)\} \in \mathbf{E}(H)\}$. An *homomorphism* of G to H is a mapping $\theta : \mathbf{V}(G) \to \mathbf{V}(H)$ such that $\{u, v\} \in \mathbf{E}(G)$ implies $\{\theta(u), \theta(v)\} \in \mathbf{E}(H)$. Clearly, an injective mapping θ is a homomorphism of G to H if $\#$Match$(G, H, \theta) = \mathbf{m}(G)$. Given lists $\mathcal{L}(u) \subseteq \mathbf{V}(H)$, $u \in \mathbf{V}(G)$, a *list homomorphism* of G to H with respect to the lists $\mathcal{L}(u)$, $u \in \mathbf{V}(G)$, is a homomorphism θ with the additional constraint that $\theta(u) \subseteq \mathcal{L}(u)$ for all $u \in \mathbf{V}(G)$. Mappings of G to H with respect to the lists $\mathcal{L}(u)$, $u \in \mathbf{V}(G)$, are defined in a similar way. For simplicity of notation, given lists $\mathcal{L}(u) \subseteq \mathbf{V}(H)$, $u \in \mathbf{V}(G)$, we abbreviate $\{u : v \in \mathcal{L}(u)\}$ to $\mathcal{L}^{-1}(v)$, $v \in \mathbf{V}(H)$. Let G and H be two graphs. Lists $\mathcal{L}(u) \subseteq \mathbf{V}(H)$, $u \in \mathbf{V}(G)$, are called (μ_G, μ_H)-*bounded* if the two following conditions hold true: (1) $\max\{\#\mathcal{L}(u) : u \in \mathbf{V}(G)\} \leq \mu_G$ and (2) $\max\{\#\mathcal{L}^{-1}(v) : v \in \mathbf{V}(H)\} \leq \mu_H$.

We consider here the problem of finding an occurrence of a given complex in the protein-protein interaction graph of another species. Finding an occurrence with respect to orthologous links can easily be reformulated as a list injective homomorphism problem: a list of putative orthologs is associated to each protein (vertex) of the complex, and each such protein can only be mapped by the homomorphism to a protein occurring in its list. The problem, called the Exact-(μ_G, μ_H)-Matching problem, is defined formally as follows.

Exact-(μ_G, μ_H)-Matching
Input : Two graphs G and H, and (μ_G, μ_H)-bounded lists $\mathcal{L}(u) \subseteq \mathbf{V}(H)$, $u \in \mathbf{V}(G)$.
Question : Is there an injective list homomorphism of G to H w.r.t. lists $\mathcal{L}(u)$, $u \in \mathbf{V}(G)$?

In the context of comparative analysis of protein-protein interaction graphs, we need to impose strong restrictions on the size of the lists we consider. We thus assume, throughout the paper, that both μ_G and μ_H are constant, i.e., $\mu_G = \mathcal{O}(1)$ and $\mu_H = \mathcal{O}(1)$.

It is proved in [FLV04] that the Exact-$(2, \mu_H)$-Matching problem is linear-time solvable for any constant $\mu_H \geq 1$, and that the Exact-(3,1)-Matching problem is **NP**-complete even if both G and H are bipartite graphs or split graphs. A first contribution in this paper is to complete the determination of the precise border between tractable and intractable cases for the Exact-(μ_G, μ_H)-Matching problem. Moreover, we begin here the analysis of optimization versions of the problem. Indeed, requiring an injective homomorphism, i.e., an injective mapping that preserves *all* edges of G, might result in an over-constrained problem, though it may exist good approximate solutions, i.e., solutions that match many edges of G. This suggests the following maximization problem for practical applications.

Max-(μ_G, μ_H)-Matching
Input : Two graphs G and H, and (μ_G, μ_H)-bounded lists $\mathcal{L}(u) \subseteq \mathbf{V}(H)$, $u \in \mathbf{V}(G)$.
Solution : An injective mapping $\theta : \mathbf{V}(G) \mapsto \mathbf{V}(H)$ w.r.t. lists $\mathcal{L}(u)$, $u \in \mathbf{V}(G)$.
Measure : #Match(G, H, θ), i.e., #$\{\{u,v\} \in \mathbf{E}(G)\} : \{\theta(u), \theta(v)\} \in \mathbf{E}(H)\}$.

Of particular importance is the fact that θ is no longer required to be a homomorphism in the Max-(μ_G, μ_H)-Matching problem. Furthermore, the present paper mainly focuses on a particular case of the optimization problem, i.e., the Max-$(\mu_G, 1)$-Matching problem.

Let $(G, H, \mathcal{L})$ be an instance of the Max-(μ_G, μ_H)-Matching. An edge $\{u, v\} \in \mathbf{E}(G)$ is called a *bad edge* if there does not exist distinct $u' \in \mathcal{L}(u)$ and $v' \in \mathcal{L}(v)$ such that $\{u', v'\} \in \mathbf{E}(H)$. Clearly, if we remove from G its bad edges, this does not affect the optimal solutions for the Max-(μ_G, μ_H)-Matching problem, since bad edges can never be matched. Notice that we can tell bad edges apart in $\mathcal{O}(\mu_G{}^2 \mathbf{m}(G)) = \mathcal{O}(\mathbf{m}(G))$ time, provided μ_G is a constant. Furthermore, by resorting on classical bipartite matching techniques, we can check in $\mathcal{O}(\mathbf{n}(H) + \mathbf{m}(G) \sqrt{\mathbf{n}(G)})$ time whether there exists at least an injective mapping of G to H w.r.t. lists $\mathcal{L}(u)$, $u \in \mathbf{V}(G)$. Moreover, before solving the problem, we can surely remove from H all those nodes u' with #$\mathcal{L}^{-1}(u') = 0$. Therefore, throughout the paper, we will consider only trim instances as defined in the following.

Definition 1 (Trim instance). *An instance $(G, H, \mathcal{L})$ of the Max-(μ_G, μ_H)-Matching problem is a trim instance provided that (i) there exists an injective mapping of G to H w.r.t. lists $\mathcal{L}(u)$, $u \in \mathbf{V}(G)$, (ii) #$\mathcal{L}^{-1}(u') > 0$ for all $u' \in \mathbf{V}(H)$ and (iii) G does not contain any bad edges.*

3 Exact Matching

This section is devoted to completing the determination of the precise border between tractable and intractable cases for the Exact-(μ_G, μ_H)-Matching problem [FLV04]. We begin by giving an easy algorithm for the Exact-$(\mu_G, 1)$-Matching problem in case $\Delta(G) \leq 2$.

Proposition 1. *The* Exact-$(\mu_G, 1)$-Matching *problem for* $\Delta(G) \leq 2$ *is solvable in* $\mathcal{O}(\mathbf{n}(G))$ *time for any constant* μ_G.

One may argue that the above proposition is too constrained to be of interest. Unfortunately, despite the simplicity of Proposition 1, the result is quite tight - taking into consideration both $\Delta(G)$ and $\Delta(H)$ - as shown in the two following propositions (recall also that the Exact-$(2, \mu_H)$-Matching problem is polynomial-time solvable for any constant μ_H [FLV04]).

Proposition 2. *The* Exact-$(3, 2)$-Matching *problem is* **NP**-*complete even if both G and H are bipartite graphs with* $\Delta(G) = 1$ *and* $\Delta(H) = 2$.

Proof. The reduction is from the 3-Sat problem. We assume the additional restriction that each variable appears in at most 3 of the clauses, counting together both positive and negative occurrences. It is known that the 3-Sat problem is **NP**-complete even when restricted as above [GJ79]. Notice furthermore that we can always assume that each negated literal and each positive literal occurs at most twice, since otherwise there would be a variable without positive or without negative occurrences, and hence a self-reduction would apply. Assume given an input ϕ to the 3-Sat problem. Let $X = \{x_1, \ldots, x_n\}$ denote the set of variables and $C = \{c_1, \ldots, c_m\}$ denote the set of clauses. We now describe how to construct the corresponding instance of the Exact-$(3, 2)$-Matching problem.

To ϕ we associate a bipartite graph, denoted G_ϕ - which in fact is a matching - as follows. For each variable $x_i \in X$, we introduce two vertices $x_i^G[1]$ and $x_i^G[2]$, and one edge $\{x_i^G[1], x_i^G[2]\}$. For each clause $c_j \in C$, we introduce two vertices $c_j^G[1]$ and $c_j^G[2]$, and one edge $\{c_j^G[1], c_j^G[2]\}$. To ϕ we also associate a second bipartite graph, denoted H_ϕ, as follows. For each variable $x_i \in X$, we introduce four vertices $x_i^H[T, 1]$, $x_i^H[T, 2]$, $x_i^H[F, 1]$ and $x_i^H[F, 2]$, and the two edges $\{x_i^H[T, 1], x_i^H[T, 2]\}$ and $\{x_i^H[F, 1], x_i^H[F, 2]\}$. For each clause $c_j \in C$, we introduce three vertices $c_j^H[1]$, $c_j^H[2]$ and $c_j^H[3]$, and also three edges defined as follows. For $\ell \in \{1, 2, 3\}$, let $\hat{x}_i$ be the ℓ-th literal of the clause c_j. Assume $\hat{x}_i$ is the p-th positive (or, resp., negative) occurrence of variable x_i, where $p \in \{1, 2\}$. Then we introduce the edge $\{c_j^H[\ell], x_i^H[T, p]\}$ (or, resp., $\{c_j^H[\ell], x_i^H[F, p]\}$). Notice that for each $j \in \{1, 2, \ldots, m\}$ and $\ell \in \{1, 2, 3\}$, vertex $c_j^H[\ell]$ has a unique neighbor in H_ϕ. For ease of exposition, we denote by $N(c_j^H[\ell])$ this unique neighbor. We now turn to describing the associated lists. To each $x_i^G[p] \in \mathbf{V}(G)$, $1 \leq p \leq 2$, we associate the list $\mathcal{L}(x_i^G[p]) = \{x_i^H[T, p], x_i^H[F, p]\}$. To each $c_j^G[2] \in \mathbf{V}(G)$, we associate the list $\mathcal{L}(c_j^G[2]) = \{c_j^H[\ell] : 1 \leq \ell \leq 3\}$. Finally, to each $c_j^G[1] \in \mathbf{V}(G)$, we associate the list $\mathcal{L}(c_j^G[1]) = \{N(c_j^H[\ell]) : 1 \leq \ell \leq 3\}$.

Clearly, $\mu_G = 3$, $\mu_H = 2$, $\Delta(G_\phi) = 1$, *i.e.*, G_ϕ is a matching, and $\Delta(H_\phi) = 2$ (H_ϕ is indeed made of paths of length at most 3). We claim that there exists a satisfying truth assignment for ϕ if and only if there exists an injective list homomorphism of G_ϕ to H_ϕ w.r.t. lists $\mathcal{L}(u)$, $u \in \mathbf{V}(G_\phi)$.

Let $f : X \mapsto \{\text{true}, \text{false}\}$ be a truth assignment for ϕ that satisfies all clauses. If $f(x_i) = \text{true}$, then define $\theta(x_i^G[1]) = x_i^H[F, 1]$ and $\theta(x_i^G[2]) = x_i^H[F, 2]$, else define $\theta(x_i^G[1]) = x_i^H[T, 1]$ and $\theta(x_i^G[2]) = x_i^H[T, 2]$. For every clause c_j, take an

$\ell \in \{1, 2, 3\}$ such that the ℓ-th literal of c_j evaluates to true under f, and define $\theta(c_j^G[2]) = c_j^H[\ell]$ and $\theta(c_j^G[1]) = N(c_j^H[\ell])$. It can be easily verified that θ is an injective homomorphism of G_ϕ to H_ϕ w.r.t. lists $\mathcal{L}(u)$, $u \in \mathbf{V}(G_\phi)$.

Conversely, suppose that there is an injective list homomorphism θ of G_ϕ to H_ϕ w.r.t. lists $\mathcal{L}(u)$, $u \in \mathbf{V}(G_\phi)$. We first observe that, by construction, we must have $\theta(x_i^G[1]) = x_i^H[T, 1]$ and $\theta(x_i^G[2]) = x_i^H[T, 2]$, or $\theta(x_i^G[1]) = x_i^H[F, 1]$ and $\theta(x_i^G[2]) = x_i^H[F, 2]$, for all $1 \leq i \leq n$, since $\{x_i^G[1], x_i^G[2]\} \in \mathbf{E}(G_\phi)$. Define a truth assignment $f : X \mapsto \{\text{true}, \text{false}\}$ as follows: If $\theta(x_i^G[1]) = x_i^H[F, 1]$ then $f(x_i) = \text{true}$, else define $f(x_i) = \text{false}$, for all $1 \leq i \leq n$. We claim that f is a satisfying truth assignment for ϕ. Indeed, for any clause c_j, let $\ell \in \{1, 2, 3\}$ be such that $c_j^H[\ell] = \theta(c_j^G[1])$. Clearly, the ℓ-th literal of ϕ evaluates to true under the truth assignment f. □

Proposition 3. *The* Exact-$(3, 1)$-Matching *problem is* **NP**-*complete even when* $\Delta(G) = 3$ *and* $\Delta(H) = 4$.

The remainder of this section is devoted to the Exact-$(\mu_G, 1)$-Matching problem. For each trim instance $(G, H, \mathcal{L})$ of the Exact-$(\mu_G, 1)$-Matching problem, define the *correspondence number* $C(G, H, \mathcal{L})$ of $(G, H, \mathcal{L})$ by

$$C(G, H, \mathcal{L}) = \min_{\{u,v\} \in \mathbf{E}(G)} \frac{\#\{\{u', v'\} : u' \in \mathcal{L}(u) \wedge v' \in \mathcal{L}(v) \wedge \{u', v'\} \in \mathbf{E}(H)\}}{\#\mathcal{L}(u)\, \#\mathcal{L}(v)}$$

Clearly, $0 \leq C(G, H, \mathcal{L}) \leq 1$. Furthermore, if $C(G, H, \mathcal{L}) = 1$, then there exists an injective homomorphism θ of G to H w.r.t. lists $\mathcal{L}(u)$, $u \in \mathbf{V}(G)$; any injective mapping of G to H w.r.t. lists $\mathcal{L}(u)$, $u \in \mathbf{V}(G)$, is indeed a solution. We now turn to proving a better lower bound for the correspondence number (the proof is by the Lovasz local lemma [AS92]).

Proposition 4. *Let* $(G, H, \mathcal{L})$ *be a trim instance of the* Exact-$(\mu_G, 1)$-Matching *problem. If*

$$C(G, H, \mathcal{L}) > \frac{2\Delta(G) - 1 - e^{-1}}{2\Delta(G) - 1}$$

then there exists an injective homomorphism θ *of* G *to* H *w.r.t. lists* $\mathcal{L}(u)$, $u \in \mathbf{V}(G)$.

4 Hardness of the Max-(μ_G, μ_H)-Matching Problem

The present and following sections are concerned with the optimization version of the problem. First, it follows from Proposition 2 that the Max-$(3, 2)$-Matching problem is **NP**-complete even if both G and H are bipartite graphs with $\Delta(G) = 1$ and $\Delta(H) = 3$. Moreover, by Proposition 3, we know that the Max-$(3, 1)$-Matching problem is **NP**-complete even when $\Delta(G) = 3$ and $\Delta(H) = 4$. In this section, we strengthen these results by showing that the Max-$(2, 1)$-Matching problem for bounded degree graphs G and H is **APX**-complete (membership to **APX** is in fact deferred to the next section). This has to be compared with the

Exact-$(2, \mu_H)$-Matching problem, which is linear-time solvable for any constant μ_H [FLV04].

We propose a reduction from the Max-2-Sat-3 problem. The input to the Max-2-Sat-3 problem is a boolean formula ϕ in conjunctive normal form in which each clause contains at most 2 literals and each variable appears in at most 3 of the clauses, counting together both positive and negative occurrences. The optimization problem calls for a truth assignment that satisfies as many clauses as possible. It is known that the Max-2-Sat-3 problem is **APX**-hard [BK99, ACG+99]. Notice furthermore that we can always assume that each negated literal and each positive literal occurs at most twice, since otherwise there would be a variable without positive or without negative occurrences, hence a self-reduction would apply.

Assume given an input ϕ to the Max-2-Sat-3 problem. Let $X = \{x_1, x_2, \ldots, x_n\}$ denote the set of variables and $C = \{c_1, c_2, \ldots, c_m\}$ the set of clauses. We now detail the construction of the corresponding instance of the Max-$(2, 1)$-Matching problem. To ϕ, we associate a first bipartite graph G_ϕ defined as follows. The set of vertices is $\mathbf{V}(G_\phi) = V_X^G \cup V_C^G$ where $V_X^G = \{x_i^G : 1 \leq i \leq n\}$ and $V_C^G = \{c_j^G : 1 \leq j \leq m\}$, and $\{x_i^G, c_j^G\}$ is an edge in $\mathbf{E}(G_\phi)$ if and only if the clause c_j contains a literal on x_i. To ϕ, we also associate a second bipartite graph H_ϕ defined as follows. The set of vertices is $\mathbf{V}(H_\phi) = V_X^H[T] \cup V_X^H[F] \cup V_C^H[1] \cup V_C^H[2]$ where $V_X^H[T] = \{x_i^H[T] : 1 \leq i \leq n\}$, $V_X^H[F] = \{x_i^H[F] : 1 \leq i \leq n\}$, $V_C^H[1] = \{c_j^H[1] : 1 \leq j \leq m\}$ and $V_C^H[2] = \{c_j^H[2] : 1 \leq j \leq m\}$. Now, $\{x_i^H[T], c_j^H[\ell]\}$ is an edge in $\mathbf{E}(H_\phi)$ if and only if the $(3-\ell)$-th literal of c_j is a literal of x_i or the ℓ-th literal of c_j is the positive literal x_i ($\ell \in \{1, 2\}$). Similarly, $\{x_i^H[F], c_j^H[\ell]\}$ is an edge in $\mathbf{E}(H_\phi)$ if and only if the $(3-\ell)$-th literal of c_j is a literal of x_i or the ℓ-th literal of c_j is the negative literal $\overline{x_i}$. We now turn to defining the associated lists. To each $x_i^G \in V_X^G$ we associate the list $\mathcal{L}(x_i^G) = \{x_i^H[T], x_i^H[F]\}$. To each $c_j^G \in V_C^G$ we associate the list $\mathcal{L}(c_j^G) = \{c_j^H[1], c_j^H[2]\}$. Clearly, $\mu_G = 2$, $\mu_H = 1$, $\Delta(G) = 3$, $\Delta(H) = 5$ (since each variable has both positive and negative occurrences), and both G_ϕ and H_ϕ are bipartite graphs. We now turn to proving correctness of the approximation-preserving reduction.

Lemma 1. *Every truth assignment for ϕ that satisfies k clauses can be transformed, in polynomial-time, into an injective mapping $\theta : \mathbf{V}(G_\phi) \mapsto \mathbf{V}(H_\phi)$ w.r.t. lists $\mathcal{L}(u)$, $u \in \mathbf{V}(G_\phi)$, such that #Match$(G_\phi, H_\phi, \theta) = m + k$.*

It follows from Lemma 1 that, for any injective mapping $\theta : \mathbf{V}(G_\phi) \mapsto \mathbf{V}(H_\phi)$ w.r.t. lists $\mathcal{L}(u)$, $u \in \mathbf{V}(G_\phi)$, we have #Match$(G_\phi, H_\phi, \theta) \geq m$.

Lemma 2. *Given an injective mapping $\theta : \mathbf{V}(G_\phi) \mapsto \mathbf{V}(H_\phi)$ w.r.t. lists $\mathcal{L}(u)$, $u \in \mathbf{V}(G_\phi)$, such that #Match$(G_\phi, H_\phi, \theta) = m + k$, we can construct, in polynomial-time, a truth assignment for ϕ that satisfies k clauses.*

Combining Lemma 1 and Lemma 2, we obtain the following proposition.

Proposition 5. *The Max-$(2, 1)$-Matching problem is **APX**-hard even if both G and H are bipartite graphs with $\Delta(G) \leq 3$ and $\Delta(H) \leq 5$.*

By slightly complicating the proof, we can strengthen the above proposition.

Proposition 6. *The Max-$(2,1)$-Matching problem is **APX**-hard even if both G and H are bipartite graphs with $\Delta(G) \leq 3$ and $\Delta(H) \leq 3$.*

5 Approximating the Max-$(\mu_G, 1)$-Matching Problem

We proved in the preceding section that the Max-$(2,1)$-Matching problem is **APX**-hard even if both G and H are bipartite graphs with $\Delta(G) \leq 3$ and $\Delta(H) \leq 3$. We show in this section that the Max-$(\mu_G, 1)$-Matching problem for bounded degree graphs G belongs to **APX** for any constant μ_G, thereby proving that the Max-$(2,1)$-Matching problem is **APX**-complete. In addition, we give a fast randomized algorithm for the Max-$(\mu_G, 1)$-Matching problem that achieves a ratio $\mu_G{}^2$ for any constant μ_G.

Recall first that a *matching* in a graph G is a subset of pairwise vertex disjoint edges of G. The *matching number* $\nu(G)$ of G is the size of a largest matching of G. A *linear forest* is a forest, *i.e.*, an acyclic simple graph, in which every connected component is a path. The *linear arboricity* $\mathsf{la}(G)$ of a graph G is the minimum number of linear forests in G, whose union is the set of all edges of G.

Conjecture 1 (The linear arboricity conjecture [AEH81]). The linear arboricity of every d-regular graph is $\lceil(d+1)/2\rceil$.

This conjecture was shown to be asymptotically correct as $d \to \infty$ [Alo88]. Although the linear arboricity conjecture received a considerable amount of attention, the best general result concerning it is that $\mathsf{la}(G) \leq \lceil 3\Delta(G)/5 \rceil$ for even $\Delta(G)$ and that $\mathsf{la}(G) \leq \lceil(3\Delta(G)+2)/5\rceil$ for odd $\Delta(G)$ [AS92].

Lemma 3. *Let G be a graph. Then, $\nu(G) \geq \mathbf{m}(G)(2\,\mathsf{la}(G))^{-1}$.*

Lemma 4. *For any trim instance, the Max-$(\mu_G, 1)$-Matching problem is approximable within ratio $2\,\mathsf{la}(G)$ in $\mathcal{O}(\mathbf{n}(G) + \mathbf{m}(G)\sqrt{\mathbf{n}(G)})$ time for any constant $\mu_G \geq 1$.*

Proof. Let $(G, H, \mathcal{L})$ be a trim instance of the Max-$(\mu_G, 1)$-Matching problem. Now, let $\mathcal{M} \subseteq \mathbf{E}(G)$ be any maximum matching in G. Consider the mapping $\theta : \mathbf{V}(G) \mapsto \mathbf{V}(H)$ defined as follows. For each edge $\{u, v\} \in \mathcal{M}$, let $u' \in \mathcal{L}(u)$ and $v' \in \mathcal{L}(v)$ be two vertices of H such that $\{u', v'\} \in \mathbf{E}(H)$ (such vertices exist since the instance is supposed to be trim). We then set $\theta(u) = u'$ and $\theta(v) = v'$. For any vertex $u \in \mathbf{V}(G)$ which is not incident to any edge in $\mathcal{M}$ (in case $\mathcal{M}$ is not a perfect matching), we set $\theta(u) = v$, where v is any vertex in $\mathcal{L}(u)$. Clearly, θ is well-defined and is injective since $\mu_H = 1$.

So, if we let θ be our solution mapping, it is a simple matter to check that $\#\mathsf{Match}(G, H, \theta) \geq \#\mathcal{M}$, and hence

$$\frac{\mathsf{opt}(G, H, \mathcal{L})}{\#\mathsf{Match}(G, H, \theta)} \leq \frac{\mathsf{opt}(G, H, \mathcal{L})}{\#\mathcal{M}}$$

Combining this with $\mathsf{opt}(G,H,\mathcal{L}) \leq \mathsf{m}(G)$ and $\#\mathcal{M} = \nu(G) \geq \mathsf{m}(G)(2\,\mathsf{la}(G))^{-1}$ (Lemma 3), we obtain

$$\frac{\mathsf{opt}(G,H,\mathcal{L})}{\#\mathsf{Match}(G,H,\theta)} \leq \mathsf{m}(G)\,\frac{2\,\mathsf{la}(G)}{\mathsf{m}(G)} = 2\,\mathsf{la}(G)$$

and the approximation ratio is proved. We now turn to proving the time complexity. For simplicity, let us assume that $(G, H, \mathcal{L})$ is a trim instance. Finding a maximum matching in G is an $\mathcal{O}(\mathsf{m}(G)\sqrt{\mathsf{n}(G)})$ time procedure [MV80]. Since constructing θ is an $\mathcal{O}(\mu_G{}^2\,\nu(G) + \mathsf{n}(G) - 2\nu(G)) = \mathcal{O}(\mathsf{m}(G) + \mathsf{n}(G))$ time procedure, the algorithm, as a whole, runs in $\mathcal{O}(\mathsf{n}(G) + \mathsf{m}(G)\sqrt{\mathsf{n}(G)})$ time. □

Proposition 7. *The* Max-$(\mu_G, 1)$-Matching *problem is approximable within ratio* $2\lceil 3\Delta(G)/5\rceil$ *for even* $\Delta(G)$ *and ratio* $2\lceil(3\Delta(G)+2)/5\rceil$ *for odd* $\Delta(G)$, *for any* $\Delta(H)$ *and any constant* μ_G.

Corollary 1. *The* Max-$(2,1)$-Matching *problem is* **APX**-*complete even if both G and H are bipartite graphs with* $\Delta(G) \leq 3$ *and* $\Delta(H) \leq 3$.

Corollary 2. *If the linear arboricity conjecture is true, then the* Max-$(\mu_G, 1)$-Matching *problem is approximable within ratio* $\Delta(G) + 1$ *if* $\Delta(G)$ *is odd, and* $\Delta(G) + 2$ *if* $\Delta(G)$ *is even, for any* $\Delta(H)$ *and any constant* μ_G.

We now turn to giving a fast randomized algorithm for the Max-$(\mu_G, 1)$-Matching problem. The proof makes use of the *probabilistic method* [AS92], a powerful tool for demonstrating the existence of combinatorial objects.

Lemma 5. *Let* $(G, H, \mathcal{L})$ *be a trim instance of the* Max-$(\mu_G, 1)$-Matching *problem. For any* μ_G, *there exists an injective mapping* $\theta : \mathbf{V}(G) \mapsto \mathbf{V}(H)$ *w.r.t. lists* $\mathcal{L}(u)$, $u \in \mathbf{V}(G)$, *such that* $\#\mathsf{Match}(G, H, \theta) \geq \mu_G{}^{-2}\,\mathsf{m}(G)$.

Proof. The proof is by the probabilistic method [AS92]. For each $u \in \mathbf{V}(G)$ with $\mathcal{L}(u) = \{v_1, v_2, \ldots, v_q\}$, $q \leq \mu_G$, suppose that $\theta(u)$ is set to $v_1, v_2, \ldots,$ or v_q independently and equiprobably. Since $\mu_H = 1$, it follows that θ is an injective mapping from $\mathbf{V}(G)$ to $\mathbf{V}(H)$ w.r.t. lists $\mathcal{L}(u)$, $u \in \mathbf{V}(G)$. For each $\{u,v\} \in \mathbf{E}(G)$, let $\mathcal{E}(\{u,v\}) = 1$ if $\{\theta(u), \theta(v)\} \in \mathbf{E}(H)$, and 0 otherwise. For any edge $\{u,v\} \in \mathbf{V}(G)$, the probability that $\{u,v\}$ is matched by the injective mapping θ is at least $\mu_G{}^{-2}$ (since $(G, H, \mathcal{L})$ is a trim instance), implying $\mathbf{Exp}[\mathcal{E}(\{u,v\})] \geq \mu_G{}^{-2}$. The expected number of edge matches by this random injective mapping θ is $\sum_{\{u,v\}\in\mathbf{E}(G)} \mathbf{Exp}[\mathcal{E}(\{u,v\})] \geq \mu_G{}^{-2}\mathsf{m}(G)$. Thus, there exists at least one injective mapping $\theta : \mathbf{V}(G) \mapsto \mathbf{V}(H)$ w.r.t. lists $\mathcal{L}(u)$, $u \in \mathbf{V}(G)$, such that $\sum_{\{u,v\}\in\mathbf{E}(G)} \mathcal{E}(\{u,v\}) \geq \mu_G{}^{-2}\mathsf{m}(G)$, and hence $\#\mathsf{Match}(G, H, \theta) \geq \mu_G{}^{-2}\,\mathsf{m}(G)$. □

Corollary 3. *There is linear-time randomized algorithm that achieves a performance ratio* $\mu_G{}^2$ *for the* Max-$(\mu_G, 1)$-Matching *problem restricted to trim instances with unbounded degree graphs G and H.*

6 Fixed-Parameter Tractability

Parameterized complexity [DF99] is an approach to complexity theory which offers a means of analyzing algorithms in terms of their tractability. For many hard problems, the seemingly unavoidable combinatorial explosion can be restricted to a *small part* of the input, the *parameter*, so that the problems can be solved in polynomial-time when the parameter is fixed. The parameterized problems that have algorithms of $f(k)\, n^{\mathcal{O}(1)}$ time complexity are called *fixed-parameter tractable*, where k is the parameter, f can be an arbitrary function depending only on k, and n denotes the overall input size. In the last decade, parameterized complexity has proved to be extremely useful in computational molecular biology, see for example [BDF+95, GGN02, AGGN02].

We follow here this trend by showing in this section that the Max-$(\mu_G, 1)$-Matching problem for bounded degree graph G is fixed-parameter tractable parameterized by the number of matched edges, *i.e.* #Match(G, H, θ). For this, we adopt here a two-step procedure: we first define a new graph representation of the problem, and next use that graph to derive fixed-parameter tractability. At the heart of the algorithm is the *incompatibility graph* of any instance $(G, H, \mathcal{L})$ which is later shown to be a compact representation of the problem.

Definition 2 (Incompatibility graph). *Let $(G, H, \mathcal{L})$ be a trim instance of the* Max-$(\mu_G, 1)$-Matching *problem and $<$ be an arbitrary total order on $\mathbf{V}(G)$. The incompatibility graph of $(G, H, \mathcal{L})$, denoted briefly by $I[G, H, \mathcal{L}]$, is defined by $\mathbf{V}(I[G, H, \mathcal{L}]) = \{(u, v, u', v') : u < v \wedge \{u, v\} \in \mathbf{E}(G) \wedge \{u', v'\} \in \mathbf{E}(H) \wedge u' \in \mathcal{L}(u) \wedge v' \in \mathcal{L}(v)\}$ and $\mathbf{E}(I[G, H, \mathcal{L}]) = \bigcup_{1 \leq i \leq 5} E_i$ where $E_1 = \{\{(u, v, u', v'), (x, y, x', y')\} : u = x \wedge v = y \wedge (u' \neq x' \vee v' \neq y')\}$, $E_2 = \{\{(u, v, u', v'), (x, y, x', y')\} : u = x \wedge v \neq y \wedge u' \neq x'\}$, $E_3 = \{\{(u, v, u', v'), (x, y, x', y')\} : u \neq x \wedge v = y \wedge v' \neq y'\}$, $E_4 = \{\{(u, v, u', v'), (x, y, x', y')\} : u = y \wedge u' \neq y'\}$ and $E_5 = \{\{(u, v, u', v'), (x, y, x', y')\} : v = x \wedge v' \neq x'\}$*

Observe that in E_4 (resp. E_5), $u = y$ (resp. $v = x$) implies $v \neq x$ (resp. $u \neq y$) since $x < y = u < v$ (resp. $u < v = x < y$) by definition of $\mathbf{V}(I[G, H, \mathcal{L}])$. Most of the interest in $I[G, H, \mathcal{L}]$ stems from the following lemma.

Lemma 6. *Let $(G, H, \mathcal{L})$ be a trim instance of the* Max-$(\mu_G, 1)$-Matching *problem. There exists an injective mapping $\theta : \mathbf{V}(G) \to \mathbf{V}(H)$ w.r.t. lists $\mathcal{L}(u)$, $u \in \mathbf{V}(G)$, such that #Match$(G, H, \theta) \geq k$ if and only if there exists an independent set of size at least k in the incompatibility graph $I[G, H, \mathcal{L}]$.*

Thus, finding an injective mapping θ of G to H w.r.t. $\mathcal{L}(u)$, $u \in \mathbf{V}(G)$, that maximizes the number of matched edges (*i.e.*, #Match(G, H, θ)) reduces to finding a maximum independent set in $I[G, H, \mathcal{L}]$. This equivalence gains in interest if we realize that, for any constant μ_G, if G is a bounded degree graph, then so is the incompatibility graph $I[G, H, \mathcal{L}]$.

Lemma 7. *Let $(G, H, \mathcal{L})$ be an instance of the* Max-$(\mu_G, 1)$-Matching *problem. Then, $I[G, H, \mathcal{L}]$ has maximum degree at most $(\mu_G - 1)(2\mu_G \Delta(G) - \mu_G + 1)$.*

It follows from the above lemma that $\Delta(I[G, H, \mathcal{L}]) = \mathcal{O}(\Delta(G))$ when $\mu_G = \mathcal{O}(1)$, and hence if G is a bounded degree graph, then so is $I[G, H, \mathcal{L}]$. Having disposed of these preliminaries steps, we now turn to proving fixed-parameter tractability of the Max-$(\mu_G, 1)$-Matching problem.

Proposition 8. *The* Max-$(\mu_G, 1)$-Matching *problem is solvable in* $\mathcal{O}(\mathbf{m}(G)(D+1)^k)$ *time, where k is the number of matched edges, i.e.,* #Match(G, H, θ), *and* $D = \Delta(I[G, H, \mathcal{L}]) = (\mu_G - 1)(2\mu_G \Delta(G) - \mu_G + 1) = \mathcal{O}(\Delta(G))$, *and hence is fixed-parameter tractable for parameter k, provided that G is a bounded degree graph and μ_G is a constant.*

7 Conclusion

In the context of comparative analysis of protein-protein interaction graphs, we considered the problem of finding an occurrence of a given complex in the protein-protein interaction graph of another species. We proved the Exact-$(3, 2)$-Matching problem and the Max-$(2, 1)$-Matching problem for bounded degree bipartite graphs to be **NP**-complete and **APX**-complete, respectively. The latter problem was shown to be fixed-parameter tractable parameterized by the number of matched edges.

We mention here some possible directions for future works. First, an interesting line of research is to further investigate the approximation of the Max-(μ_G, μ_H)-Matching problem for bounded degree graphs G and H. For example, is the Max-$(2, 2)$-Matching problem for bounded degree graphs G and H in **APX** ? Parameterized complexity of the Max-(μ_G, μ_H)-Matching problem is almost unexplored in the case $\mu_H > 1$. In particular, is the Max-(μ_G, μ_H)-Matching problem for bounded degree graphs G and H fixed-parameter tractable for any constant μ_G and μ_H ?

References

[ACG+99] G. Ausiello, P. Crescenzi, G. Gambosi, V. Kann, A. Marchetti-Spaccamela, and M. Protasi, *Complexity and Approximation: Combinatorial optimization problems and their approximability properties*, Springer-Verlag, 1999.

[AEH81] J. Akiyama, G. Exoo, and F. Harary, *Covering and packing in graphs IV: Linear arboricity*, Networks **11** (1981), 69–72.

[AGGN02] J. Alber, J. Gramm, J. Guo, and R. Niedermeier, *Towards optimally solving the longest common subsequence problem for sequences with nested arc annotations in linear time*, Proc. of the 13th Annual Symposium on Combinatorial Pattern Matching (CPM 2002), LNCS, vol. 2373, Springer-Verlag, 2002, pp. 99–114.

[Alo88] N. Alon, *The linear arboricity of graphs*, Israel Journal of Mathematics **62** (1988), no. 3, 311–325.

[AS92] N. Alon and J.H. Spencer, *The probabilistic method*, Wiley, 1992.

[BDF+95] H.L. Bodlaender, R.G. Downey, M.R. Fellows, M.T. Hallett, and H.T. Wareham, *Parameterized complexity analysis in computational biology*, Computer Applications in the Biosciences **11** (1995), 49–57.

[BK99] P. Berman and M. Karpinski, *On some tighter inapproximability results*, Proc. of the 26th International Colloquium on Automata, Languages and Programming (ICALP) (P. van Emde Boas J. Wiedermann and M. Nielsen, eds.), LNCS, vol. 1644, Springer, 1999, pp. 200–209.

[DF99] R. Downey and M. Fellows, *Parameterized complexity*, Springer-Verlag, 1999.

[FLV04] I. Fagnot, G. Lelandais, and S. Vialette, *Bounded list injective homomorphism for comparative analysis of protein-protein interaction graphs*, Proc. of the 1st Algorithms and Computational Methods for Biochemical and Evolutionary Networks (CompBioNets), KCL publications, 2004, pp. 45–70.

[Gav02] A.C. Gavin, M. Boshe et al, *Functional organization of the yeast proteome by systematic analysis of protein complexes*, Nature **414** (2002), no. 6868, 141–147.

[GGN02] J. Gramm, J. Guo, and R. Niedermeier, *Pattern matching for arc-annotated sequences*, Proc. of the the 22nd Conference on Foundations of Software Technology and Theoretical Computer Science (FSTTCS), LNCS, vol. 2556, 2002, pp. 182–193.

[GJ79] M.R. Garey and D.S. Johnson, *Computers and intractability: a guide to the theory of NP-completeness*, W.H. Freeman, San Franciso, 1979.

[HG02] Y. Ho and A. Gruhler et al, *Systematic identification of protein complexes in Saccharomyces cerevisae by mass spectrometry*, Nature **415** (2002), no. 6868, 180–183.

[KSK+03] B.P. Kelley, R. Sharan, R.M. Karp, T. Sittler, D. E. Root, B.R. Stockwell, and T. Ideker, *Conserved pathways within bacteria and yeast as revealed by global protein network alignment*, PNAS **100** (2003), no. 20, 11394–11399.

[MV80] S. Micali and V.V. Vazirani, *An $O(\sqrt{|V|}|E|)$ algorithm for finding maximum matching in general graphs*, Proc. of the 21st Annual Symposium on Foundation of Computer Science (FOCS), IEEE, 1980, pp. 17–27.

[PLEO04] J.B. Pereira-Leal, A.J. Enright, and C.A. Ouzounis, *Detection of functional modules from protein interaction networks*, Proteins **54** (2004), no. 1, 49–57.

[PMT+99] M. Pellegrini, E.M. Marcotte, M.J. Thompson, D. Eisenberg, and T.O. Yeates, *Assigning protein functions by comparative genome analysis: protein phylogenetic profiles*, PNAS **96** (1999), no. 8, 4285–4288.

[SIK+04] R. Sharan, T. Ideker, B. Kelley, R. Shamir, and R.M. Karp, *Identification of protein complexes by comparative analysis of yeast and bacterial protein interaction data*, Proc. of the 8th annual international conference on Computational molecular biology (RECOMB 2004), ACM Press, 2004, pp. 282–289.

[SSK+05] R. Sharan, S. Suthram, R.M. Kelley, T. Kuhn, S. McCuin, P. Uetz T. Sittler, R. Karp, and T. Ideker, *Conserved patterns of protein interaction in multiple species*, PNAS **102** (2005), no. 6, 1974–1979.

[TSU04] B. Titz, M. Schlesner, and P. Uetz, *What do we learn from high-throughput protein interaction data?*, Expert Review of Anticancer Therapy **1** (2004), no. 1, 111–121.

[Uet00] P. Uetz, L.Giot et al, *A comprehensive analysis of protein-protein interactions in Saccharomyces cerevisae*, Nature **403** (2000), no. 6770, 623–627.

Matrix and Graph Orders Derived from Locally Constrained Graph Homomorphisms

Jiří Fiala[1], Daniël Paulusma[2], and Jan Arne Telle[3]

[1] Charles University, Faculty of Mathematics and Physics,
DIMATIA and Institute for Theoretical Computer Science (ITI)*,
Malostranské nám. 2/25, 118 00, Prague, Czech Republic
fiala@kam.mff.cuni.cz
[2] Department of Computer Science, University of Durham,
Science Laboratories, South Road, Durham DH1 3EY, England
daniel.paulusma@durham.ac.uk
[3] Department of Informatics, University of Bergen, N-5020 Bergen, Norway
telle@ii.uib.no

Abstract. We consider three types of locally constrained graph homomorphisms: bijective, injective and surjective. We show that the three orders imposed on graphs by existence of these three types of homomorphisms are partial orders. We extend the well-known connection between degree refinement matrices of graphs and locally bijective graph homomorphisms to locally injective and locally surjective homomorphisms by showing that the orders imposed on degree refinement matrices by our locally constrained graph homomorphisms are also partial orders. We provide several equivalent characterizations of degree (refinement) matrices, e.g. in terms of the dimension of the cycle space of a graph related to the matrix. As a consequence we can efficiently check whether a given matrix M is a degree matrix of some graph and also compute the size of a smallest graph for which it is a degree matrix in polynomial time.

1 Introduction

By graph homomorphisms we mean edge-preserving mappings, i.e. vertex mappings where images of two adjacent vertices are also adjacent in the target graph. Relating pairs of graphs by the existence of a graph homomorphism defines a quasi-order on the class of all graphs, which can be further factorized into a partial order. For a comprehensive survey of these structures see the recent monograph [15].

In this paper we study similar structural properties derived from *locally constrained* graph homomorphisms [9], where for any vertex u the mapping f induces a function from the neighborhood of u to the neighborhood of $f(u)$ which is required to be either *bijective* [5,17], *injective* [8,9], or *surjective* [18,12]. See [18] for a more general model of locally constrained conditions.

---

* Supported by the Ministry of Education of the Czech Republic as project 1M0021620808.

Locally bijective homomorphisms (also known as local isomorphisms or full covers) have important applications, for example in distributed computing [5], in recognizing graphs by networks of processors [1,2], or in constructing highly transitive regular graphs [4]. Locally injective homomorphisms (local epimorphisms or partial covers) are used in distance constrained labelings of graphs [10] and as indicators of the existence of homomorphisms of derivate graphs (line graphs) [20]. Locally surjective homomorphisms (role assignments) are of interest in social network theory where individuals of the same social role relate to other individuals in the same way [7]. Just as in a graph isomorphism, a locally bijective homomorphism maintains vertex degrees and degrees of neighbors and degrees of neighbors of neighbors and so on. The existence of such a mapping from G to H therefore implies equality of the so-called degree refinement matrices of G and H. Since these are easy to compute, they provide both an important necessary condition and a heuristic for the graph isomorphism problem (cf. [19]).

Our Results

Degree refinement matrices belong to the class of degree matrices corresponding to degree partitions of the vertex set of a graph. Degree partitions of graphs are also known under the name of equitable partitions [14,3]. In Sect. 3 we present four equivalent characterizations of degree matrices, e.g. by conditions on the dimension of the cycle space of some matrix-related graph. Given the rather long history and fame of the graph isomorphism problem it is surprising that no characterization of degree (refinement) matrices had been shown previously. As a consequence we can efficiently check whether a given matrix M is a degree matrix or not. We also prove that the size of a smallest graph corresponding to some degree matrix can be computed in polynomial time. In Sect. 4 we prove that the problem whether a given (degree) matrix M is a degree refinement matrix can be solved in polynomial time. In Sect. 5 we introduce three orderings, in which a graph H is smaller than a graph G if a homomorphism from G to H exists, locally constrained to be respectively bijective, injective or surjective. We prove that these are partial orderings and in Sect. 6 we show that these partial orders can be further extended to degree matrices of graphs. These results generalize the use of degree refinement matrices to locally injective and locally surjective homomorphisms. We emphasize that such a relationship was not originally expected, since such degree conditions are not obvious for the non-bijective local constraints.

2 Preliminaries

If not stated otherwise graphs considered in this paper are finite and *simple*, i.e. without loops and multiple edges. For graph terminology not defined below we refer to [6].

For a function $f : V_G \to V_H$ and a set $S \subseteq V_G$ we use the shorthand notation $f(S)$ to denote the image set of S under f, i.e., $f(S) = \{f(u) \mid u \in S\}$. For any $x \in V_H$, the set $f^{-1}(x)$ is equal to $\{u \in V_G \mid f(u) = x\}$.

For a vertex $u \in V_G$ we denote its *neighborhood* by $N_G(u) = \{v \mid (u,v) \in E_G\}$. A *k-regular* graph is a graph, where all vertices have k neighbors (i.e. are of *degree* k). A (k,l)-*regular bipartite* graph is a bipartite graph where vertices of one class of the bi-partition are of degree k and all others are of degree l.

A *graph homomorphism* from $G = (V_G, E_G)$ to $H = (V_H, E_H)$ is a vertex mapping $f : V_G \to V_H$ satisfying the property that for any edge (u,v) in E_G, we have $(f(u), f(v))$ in E_H as well, i.e., $f(N_G(u)) \subseteq N_H(f(u))$ for all $u \in V_G$. Two graphs G and G' are called *isomorphic*, denoted by $G \simeq G'$, if there exists a one-to-one mapping $f : V_G \to V_{G'}$, where both f and f^{-1} are homomorphisms.

Definition 1. *For graphs G and H we denote:*

- $G \xrightarrow{B} H$ *if there exists a so-called* locally bijective homomorphism *$f : V_G \to V_H$ satisfying:*

$$\text{for all } u \in V_G : f(N_G(u)) = N_H(f(u)) \text{ and } |f(N_G(u))| = |N_G(u)|.$$

- $G \xrightarrow{I} H$ *if there exists a so-called* locally injective homomorphism *$f : V_G \to V_H$ satisfying:*

$$\text{for all } u \in V_G : |f(N_G(u))| = |N_G(u)|.$$

- $G \xrightarrow{S} H$ *if there exists a so-called* locally surjective homomorphism *$f : V_G \to V_H$ satisfying:*

$$\text{for all } u \in V_G : f(N_G(u)) = N_H(f(u)).$$

Note that a locally bijective homomorphism is both locally injective and surjective. Hence, any result valid for locally injective or for locally surjective homomorphisms is also valid for locally bijective homomorphisms. We provide an alternative definition of these three kinds of mappings via subgraphs induced by preimages of edges. As far as we know this quite natural definition has not previously appeared in the literature.

Observation 1. *Let $f : G \to H$ be a graph homomorphism. For every edge (u,v) of H, the subgraph of G induced by $f^{-1}(u) \cup f^{-1}(v)$ is a*

- perfect matching *if and only if f is locally bijective,*
- matching and possibly isolated vertices *if and only if f is locally injective,*
- bipartite graph without isolated vertices *if and only if f is locally surjective.*

Note that for locally bijective homomorphisms the preimage classes all have the same size and for locally surjective homomorphisms all the preimage classes have size at least one. This yields the following observation:

Observation 2. *If $G \xrightarrow{S} H$, for H connected and finite, then either $|V_G| > |V_H|$ or else $G \simeq H$.*

For a connected graph G the *universal cover* is defined in [1] as the only (possibly infinite) tree T_G that allows a locally bijective homomorphism $T_G \xrightarrow{B} G$. The vertices of T_G can be represented as walks in G starting in a fixed vertex u that do not traverse the same edge in two consecutive steps. Edges in T_G connect those walks that differ in the presence of the last edge. The mapping $f_0 : T_G \xrightarrow{B} G$ sending a walk in V_{T_G} to its last vertex is a locally bijective homomorphism.

Proposition 1 ([11]). *Let G and H be two connected graphs. From any function $f : G \xrightarrow{I} H$ a locally injective homomorphism $f' : T_G \to T_H$ can be derived. From any function $g : G \xrightarrow{S} H$ a locally surjective homomorphism $g' : T_G \to T_H$ can be derived.*

In the sequel we consider all isomorphism classes of connected simple graphs. We assume that each of these classes is represented by one of its elements, and these representatives form the set $\mathcal{C}$, called the *set of connected graphs*.

3 Degree Matrices

Any locally bijective graph homomorphism, with graph isomorphism as a special case, preserves not only vertex degrees but also degrees of neighbors and degrees of neighbors of these neighbors and so on. To capture this property the following notions have been defined (cf. [14,19]).

Definition 2. *A degree partition of a graph G is a partition of the vertex set V_G into blocks $\mathcal{B} = \{B_1, \ldots, B_k\}$ such that whenever two vertices u and v belong to the same block B_i, then for any $j \in \{1, \ldots, k\}$ we have $|N_G(u) \cap B_j| = |N_G(v) \cap B_j| = m_{i,j}$. The $k \times k$ matrix M such that $(M)_{i,j} = m_{i,j}$ is a degree matrix.*

Observe that a graph G can allow several degree matrices, with an adjacency matrix itself being the largest one. *Degree refinement matrices*, which will be considered in the next section, are on the other extreme.

Observation 3. *The vertex set V_G of any graph G that has a $k \times k$ matrix M as one of its degree matrices can be partitioned into $B_1 \cup \ldots \cup B_k$ such that $m_{i,j}|B_i| = m_{j,i}|B_j|$ holds for all $1 \le i < j \le k$.*

This immediately implies that for any degree matrix M of size k,

$$m_{i,j} > 0 \text{ if and only if } m_{j,i} > 0 \text{ for all } 1 \le i < j \le k.$$

We call integer matrices that have the above property *well-defined*. It is easy to see that there exist well-defined matrices that are not degree matrices of a finite graph. This makes the following decision problem interesting.

DEGREE MATRIX DETERMINATION (DMD)
Instance: A square matrix M.
Question: Is M a degree matrix of a finite graph G?

To determine the complexity of DMD we introduce the following definitions. A directed graph $D = (V_D, E_D)$ is called *symmetric* if there exists an arc $(j, i) \in E_D$ whenever there exists an arc $(i, j) \in E_D$. Let $w : E_D \to \mathbb{N}$ be a positive weight function defined on the arc set of D. We call such a graph with positive arc weights a *symmetric directed product graph (sdp-graph)*. We say that a cycle $v_0, v_1, \ldots, v_c, v_0$ in an sdp-graph D has the *cycle product identity* if

$$1 = \prod_{i=0}^{c} \frac{w(v_i, v_{i+1})}{w(v_{i+1}, v_i)},$$

where the subscript of v_{i+1} is computed modulo $c+1$. In other words, a cycle has the cycle product identity if the product of arc weights going clockwise around the cycle is the same as the product counter-clockwise. We say that *the sdp-graph D has the cycle product identity* if every cycle of D has the cycle product identity. Using induction on the cycle length immediately yields:

Observation 4. *An sdp-graph D has the cycle product identity if and only if every induced cycle of D has the cycle product identity.*

For a square matrix M we define the weighted directed graph F_M as follows. Its vertex set V_{F_M} consists of vertices $\{1, \ldots, k\}$. There is an arc from i to $j \neq i$ with weight $m_{i,j}$ if and only if $m_{i,j} \geq 1$. Note that F_M is an sdp-graph if and only if M is well-defined.

Let F'_M be the underlying undirected graph of F_M, i.e., $V_{F'_M} = V_{F_M} = \{1, \ldots, k\}$ and (i, j) is an undirected edge of F'_M, whenever both (i, j) and (j, i) are directed arcs of F_M. We define the *weighted incidence matrix IM* to be the $|E_{F'_M}| \times k$ matrix whose rows are indexed by edges $e = (i, j) \in E_{F'_M}$, $i < j$ and its only non-zero entries in the e-th row are $(IM)_{e,i} = m_{i,j}$ and $(IM)_{e,j} = -m_{j,i}$.

The kernel and rank of a matrix M are denoted by $\ker(M)$ and $\mathrm{rank}(M)$ respectively. The transpose of a matrix M is denoted by M^T. We represent each $e \in E_G$ by a unit vector in the vector space $\mathbb{R}^{|E_G|}$, called the *edge space* $\mathcal{E}_G$ of a graph G. The *cycle space* $\mathcal{S}_G$ of G is the linear subspace of $\mathcal{E}_G$ generated by all cycles in G. We denote the dimension of a linear subspace $\mathcal{D}$ by $\dim(\mathcal{D})$. For every edge e not in a spanning tree T of G there is a unique cycle C_e in the graph $T + e$. Since there are $|E_G| - |V_G| + 1$ of these edges, it is clear that $\dim(\mathcal{S}_G) = |E_G| - |V_G| + 1$.

We now present our characterization of degree matrices.

Theorem 1. *The following statements are equivalent:*

(i) M is a degree matrix of a graph $G \in \mathcal{C}$.
(ii) F_M is a connected sdp-graph satisfying the cycle product identity.
(iii) M is well-defined and $\dim(\ker(IM)) = 1$.
(iv) M is well-defined and $\dim(\ker(IM^T)) = \dim(\mathcal{S}_{F'_M})$.

Proof. $(i) \Rightarrow (ii)$ Since M is a degree matrix, M is well-defined. Hence, F_M is an sdp-graph. Obviously, F_M is connected. Let $C = i_0, \ldots, i_c, i_0$ be a cycle in F_M, where vertex v_i corresponds to block B_i. Use Observation 3 for pairs $(i_0, i_1), \ldots, (i_c, i_0)$ to show that C satisfies the cycle product identity.

$(ii) \Rightarrow (iii)$ Since F_M is an sdp-graph, M is well-defined. Consider a path P_{1i} in F_M from the vertex 1 to any vertex i corresponding to the i-th row of M. We apply Observation 3 for consecutive pairs on P_{1i}. Combining these equalities yields a rational $b_i > 0$ such that $|B_i| = b_i |B_1|$ for the blocks B_i and B_1 of any possible graph G with degree matrix M. Because F_M satisfies the cycle product identity, taking another path P'_{1i} between vertices 1 and i would lead to exactly the same equality $|B_i| = b_i |B_1|$. Define $b_1 = 1$. Then any solution of $\ker(IM)$ is a multiple of the vector $\mathbf{b} = (b_1, \ldots, b_k)$.

$(iii) \Rightarrow (i)$ We first determine the block sizes of a candidate graph G. We do this with respect to the following two facts. (1) For $p \geq 1$ there exists a p-regular graph on n vertices if and only if $n \geq p+1$ and np is even. (2) There exists a (p,q)-regular bipartite graph with the degree-p side having m vertices and the degree-q side having n vertices if and only if $m \geq q, n \geq p$ and $mp = nq$. We now choose an integer solution $\mathbf{s}$ of $\ker(IM)$ such that

- $s_i \geq m_{i,i} + 1$ for all i.
- $s_i m_{i,i}$ is even for all i. $\hspace{3cm} (*)$
- $s_i \geq m_{j,i}$ for all i and all $j \neq i$.

Then the following graph G has M as one of its degree matrices. Its vertex set V_G can be partitioned into blocks $B_1 \cup \cdots \cup B_k$ with $|B_i| = s_i$ for all $1 \leq i \leq k$. Its edge set E_G can be chosen such that:

- The subgraph induced by B_i is $m_{i,i}$-regular for $1 \leq i \leq k$.
- The induced bipartite subgraph between vertices of blocks B_i and B_j is $(m_{i,j}, m_{j,i})$-regular for all $1 \leq i < j \leq k$.

$(iii) \Leftrightarrow (iv)$ Note that $\dim(\ker(IM)) = 1$ if and only if $\text{rank}(IM^T) = \text{rank}(IM) = k - 1$ if and only if $\dim(\ker(IM^T)) = |E_{F'_M}| - \text{rank}(IM^T) = |E_{F'_M}| - k + 1 = \dim(\mathcal{S}_{F'_M})$. $\square$

Corollary 1. *The DMD problem can be solved in polynomial time.*

Proof. First we check whether the matrix M is well-defined. If it is, we construct the graph F_M. Let $M_1, \ldots, M_p$ be the submatrices of M corresponding to the components of F_M. For each M_i we compute $\ker(IM_i)$ and use Theorem 1. $\square$

In this paper we only consider matrices that are the degree matrix of some finite connected graph. If we allow infinite graphs, then we only have to check whether a matrix M is finite and has connected F_M. This is since for any such matrix M we can construct its *universal cover* T_M by taking as root of the (possibly infinite) tree T_M a vertex corresponding to row 1, thus of row-type 1, and inductively adding a new level of vertices while maintaining the property that each vertex of row-type i has exactly $m_{i,j}$ neighbors of row-type j.

Theorem 1 and Corollary 1 immediately imply that for examining whether an sdp-graph has the cycle product identity we do not have to check all (induced) cycles explicitly.

Corollary 2. *The problem whether a symmetric directed graph with positive edge weights has the cycle product identity can be solved in polynomial time.*

Corollary 3. *For any degree matrix M the block sizes of a smallest graph G that has M as one of its degree matrices can be computed in polynomial time.*

Proof. Let $m = \max\{m_{i,j} \mid 1 \leq i, j \leq k\}$. Let $\langle m \rangle$ be the number of bits required to encode m. Then the input size of a $k \times k$ matrix M can be defined as $k^2 \langle m \rangle$.

If we compute coefficients b_i as in the proof of Theorem 1, then we find that both nominator and denominator of each b_i have size at most $k \langle m \rangle$. Let α be the product of all denominators of elements b_i. Let b' be a solution of $\ker(IM)$ with entries $b'_i = \alpha b_i$ for all $1 \leq i \leq k$. We divide each b'_i by the greatest common divisor of $b'_1, \ldots, b'_k$. This way we have obtained the smallest integer solution $\mathbf{b}^*$ of $\ker(IM)$ in polynomial time. Now we choose the integer γ such that $\gamma \geq \max_{1 \leq i, j \leq k} \{\frac{m_{i,i}+1}{b^*_i}, \frac{m_{j,i}}{b^*_i}\}$, where γ is required to be even if for some i the product $b^*_i m_{i,i}$ is odd. Then $\mathbf{b} = \gamma \mathbf{b}^*$ satisfies all three conditions $(*)$, i.e., it yields the block sizes of a smallest graph G in the same way as in the proof of Theorem 1. (The size of G itself might be exponential in $\langle \mathbf{b} \rangle$.) □

4 Degree Refinement Matrices

For many pairs of graphs (G, H) we can easily determine that a locally bijective homomorphism from G to H does not exist.

Definition 3. *The degree refinement matrix $\mathrm{drm}(G)$ of G is the degree matrix corresponding to the canonical (as explained below) coarsest degree partition of G, i.e., with the fewest number of blocks.*

If $\mathrm{drm}(G) \neq \mathrm{drm}(H)$ then no locally bijective homomorphism exists between G and H, and this condition can be checked by computing both minimum degree partitions by procedure MDP CONSTRUCTION that runs in $\mathcal{O}(n^3)$ time (cf. [1]).

MDP CONSTRUCTION
Input: A graph G.
Output: The minimal degree partition $\mathcal{B}$.

0. Set $\mathcal{B}^0 = \{B^0_1\} = \{V_G\}$, $t = 1$.
1. For each vertex u compute the degree vector
$$\overrightarrow{d(u)} := \Big(|N(u) \cap B^t_1|, |N(u) \cap B^t_2|, \ldots\Big).$$
2. Set $t := t+1$ and define the new partition $\mathcal{B}^t$ of V_G such that
   - $u, v \in B^t_i$ if and only if $\overrightarrow{d(u)} = \overrightarrow{d(v)}$,
   - $u \in B^t_i$, $v \in B^t_{i'}$ with $i < i'$ if and only if
     * either $u \in B^{t-1}_j$, $v \in B^{t-1}_{j'}$ with $j < j'$,
     * or $u, v \in B^{t-1}_j$ and $\overrightarrow{d(u)} >_{\mathrm{Lex}} \overrightarrow{d(v)}$,

     where $>_{\mathrm{Lex}}$ is the lexicographic order on integer sequences.
3. If $\mathcal{B}^t = \mathcal{B}^{t-1}$ then set $\mathcal{B} = \mathcal{B}^t$ and stop, otherwise continue by step 1.

We modify this procedure into the efficient algorithm DRM CONSTRUCTION. Given a degree matrix M it computes a matrix M' such that $M' = \mathrm{drm}(G)$ for any graph G with degree matrix M. Moreover, given a graph G it computes the degree refinement matrix of G when we take an adjacency matrix of G as its input. Note that in steps **2** and **3** the canonical order of the blocks is defined.

DRM CONSTRUCTION

Input: A degree partition matrix M.
Output: A degree refinement matrix M' that encodes all graphs with degree matrix M.

0. Set $\mathcal{R}^0 = \{R_1^0\} = \{1,\ldots,k\}$, $t = 1$.
1. For each row $r = 1,\ldots,k$ compute the row-degree vector
$$\overrightarrow{d(r)} := \left(\sum_{i \in R_1^t} m_{r,i}, \sum_{i \in R_2^t} m_{r,i}, \ldots \right).$$
2. Set $t := t+1$ and define the new partition $\mathcal{R}^t$ of $\{1,\ldots,k\}$ such that
  - $r, s \in B_i^t$ if and only if $\overrightarrow{d(r)} = \overrightarrow{d(s)}$,
  - $r \in B_i^t$, $s \in B_{i'}^t$ with $i < i'$ if and only if
    * either $r \in B_j^{t-1}$, $s \in B_{j'}^{t-1}$ with $j < j'$,
    * or $r, s \in B_j^{t-1}$, and $\overrightarrow{d(r)} >_{\mathrm{Lex}} \overrightarrow{d(s)}$.
3. If $\mathcal{R}^t = \mathcal{R}^{t-1}$ then set $M' = \begin{pmatrix} \overrightarrow{d(r)} : r \in R_1^t \\ \overrightarrow{d(r)} : r \in R_2^t \\ \vdots \end{pmatrix}$ and stop,

otherwise continue by step 1.

By applying the above algorithm and Corollary 1 we immediately obtain the following.

Theorem 2. *Checking if a given matrix M is a degree refinement matrix can be done in polynomial time.*

5 Partial Orders on Graphs

It is well-known that graph homomorphisms define a quasiorder on the class of all graphs, which can be factorized into a partial order. For an overview of these results see the recent monograph [16]. We show that a similar interesting structure exists on the class of connected graphs $\mathcal{C}$ for locally constrained homomorphisms. For this purpose we will view $\xrightarrow{B}$, $\xrightarrow{I}$ and $\xrightarrow{S}$ as binary relations on $\mathcal{C}$, denoted by $(\mathcal{C}, \xrightarrow{*})$ if necessary, where $*$ will indicate the appropriate local constraint. We show that $(\mathcal{C}, \xrightarrow{*})$ is a partial order for any of the three local constraints $* = B, I, S$.

Observe first that for any $G \in \mathcal{C}$ the identity mapping $i : V_G \to V_G$ clarifies that all three relations $\xrightarrow{*}$ are *reflexive*.

The composition of two graph homomorphisms of the same kind of local constraint (B, I, S) is again a graph homomorphism of the same kind. Hence each $\xrightarrow{*}$ is also *transitive*.

For antisymmetry, suppose for $G, H \in \mathcal{C}$ that $f : G \xrightarrow{*} H$, $g : H \xrightarrow{*} G$, where f, g are of the same local constraint. For $* = B, S$ we can invoke Observation 2 to conclude that $G \simeq H$. For $* = I$ we use the following result.

Theorem 3 ([11]). *Let G be a (possibly infinite) graph and let H be a graph in $\mathcal{C}$. If G allows both a locally injective and a locally surjective homomorphism to H, then both these homomorphisms are locally bijective.*

For $* = I$ we have $g \circ f : G \xrightarrow{I} G$ and $G \xrightarrow{S} G$ by the identity mapping. By Theorem 3 the mapping $g \circ f$ is locally bijective. Since G is connected, $(g \circ f)(V_G) = V_G$ implying that f is (globally) injective. By the same kind of arguments we deduce that g is injective. This means that f is surjective, and hence f is a graph isomorphism from G to H. Hence, all three relations are *antisymmetric*. We would like to mention that the antisymmetry of $\xrightarrow{I}$ also follows from an iterative argument of [20].

Combining the results above with Theorem 3 yields the following.

Theorem 4. *All three relations $(\mathcal{C}, \xrightarrow{B}), (\mathcal{C}, \xrightarrow{I})$ and $(\mathcal{C}, \xrightarrow{S})$ are partial orders with $(\mathcal{C}, \xrightarrow{B}) = (\mathcal{C}, \xrightarrow{I}) \cap (\mathcal{C}, \xrightarrow{S})$.*

6 Partial Orders on Degree Refinement Matrices

We again recall the fact that a locally bijective homomorphism from a graph G to a graph H may exist only if G and H have the same degree refinement matrix.

Theorem 5 ([19]). *Two graphs $G, H \in \mathcal{C}$ have a common degree refinement matrix if and only if their universal covers are isomorphic as well as if and only if there exists a graph $F \in \mathcal{C}$ allowing locally bijective homomorphisms to both G and H.*

In view of this theorem we can also define the universal cover T_M associated with a degree refinement matrix M as the universal cover $T_G = T_M$ of any graph G with $\mathrm{drm}(G) = M$. This implies that the symmetric and transitive closure of the partial order $(\mathcal{C}, \xrightarrow{B})$ is an equivalence relation whose classes can be naturally represented by degree refinement matrices. It is natural to ask if the other two kinds of locally constrained homomorphisms are also conditioned by the existence of a well-defined relation on the degree refinement matrices. Here, we prove that such a relation exists and moreover, that it is a partial order.

Definition 4. *We denote the set of all degree refinement matrices of graphs in $\mathcal{C}$ by $\mathcal{M}$. We define three relations $\xrightarrow{B}, \xrightarrow{I},$ and $\xrightarrow{S}$ respectively, on $\mathcal{M}$ as follows. For two matrices $M, N \in \mathcal{M}$ we have $M \xrightarrow{*} N$ if there exist graphs $G \in \mathcal{C}$ with $\mathrm{drm}(G) = M$ and $H \in \mathcal{C}$ with $\mathrm{drm}(H) = N$ such that $G \xrightarrow{*} H$ holds for the appropriate local constraint.*

As stated above $(\mathcal{M}, \xrightarrow{B})$ is a trivial order where no two distinct elements are comparable. For the other two relations, the *reflexivity* of the relation follows directly from the existence of the identity mapping on any underlying graph. Antisymmetry and transitivity require more effort.

For proving *antisymmetry* we involve the notion of universal cover. Assume that $M \xrightarrow{I} N$ and $N \xrightarrow{I} M$. By Proposition 1, there exist locally injective homomorphisms $f' : T_M \to T_N$ and $g' : T_N \to T_M$. Recall from Sect. 2 that there exist a locally bijective homomorphism $f_0 : T_M \to G_1$. As in the previous section we now invoke Theorem 3 to conclude that $f_0 \circ g' \circ f' : T_M \xrightarrow{I} G_1$ is locally bijective. This implies that both f' and g' are locally bijective, and consequently the universal covers T_M and T_N are isomorphic. Hence $M = N$ due to Theorem 5. The antisymmetry of $\xrightarrow{S}$ can be proven according to exactly the same arguments.

The transitivity property of $\xrightarrow{I}$ follows directly from the next lemma.

Lemma 1. *Let $G_1, G_2, H_1, H_2 \in \mathcal{C}$ be such that $G_1 \xrightarrow{I} H_1$ and $G_2 \xrightarrow{I} H_2$, where H_1 and G_2 share the same degree refinement matrix. Then there exists a graph $F \in \mathcal{C}$ such that $F \xrightarrow{I} H_2$ and $F \xrightarrow{B} G_1$.*

Proof. Using Theorem 5 we first construct a finite graph F' such that $F' \xrightarrow{B} H_1$ and $F' \xrightarrow{B} G_2$. The projection $\pi_2 : F' \xrightarrow{B} G_2$ composed with the locally injective homomorphism $g : G_2 \xrightarrow{I} H_2$ gives that $F' \xrightarrow{I} H_2$. See Fig. 1.

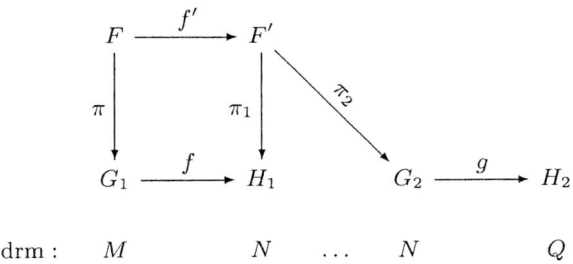

Fig. 1. Commutative diagram for transitivity of $\xrightarrow{I}$ where horizontal mappings are injective and others are bijective

As $F' \xrightarrow{B} H_1$ via projection π_1, by Observation 1 the preimage $\pi_1^{-1}(x)$ has the same size for all vertices $x \in V_{H_1}$, say k. We assume that all vertices of F' that map onto a vertex x are labeled $\{x_1, x_2, \ldots, x_k\}$.

The vertex set of the desired graph F is the Cartesian product $V_{G_1} \times \{1, \ldots, k\}$. For simplicity we abbreviate (u, i) as u_i. Define the edges of F as follows:

$$(u_i, v_j) \in E_F \Leftrightarrow (u, v) \in E_{G_1} \text{ and } (f(u)_i, f(v)_j) \in E_{F'}.$$

We define two mappings $f' : u_i \to f(u)_i$ and $\pi : u_i \to u$. According to Observation 1, f' is a locally injective homomorphism from F to F' and π is a locally bijective homomorphism from F to G_1. The mapping $g \circ \pi_2 \circ f'$ is a locally injective homomorphism $F \xrightarrow{I} H_2$. □

The same assertion can be proven for the order $\xrightarrow{S}$ with exactly the same arguments, the only difference is that the preimage in F of any edge $(x_i, y_j) \in E_{F'}$ is a spanning bipartite graph.

Theorem 6. *For any constraint $* = B, I, S$ the relation $(\mathcal{M}, \xrightarrow{*})$ is a partial order. It arises as a factor of the order $(\mathcal{C}, \xrightarrow{*})$, when we unify the graphs that have the same degree refinement matrices.*

Any locally injective homomorphism $G \xrightarrow{I} H$ can be extended to a locally bijective homomorphism $G' \xrightarrow{B} H$, where $G \subseteq G'$ [17]. This yields an alternative definition of the order $(\mathcal{M}, \xrightarrow{I})$: For matrices M, N holds $M \xrightarrow{I} N$ if and only if there exists graphs G and H with degree refinement matrices M and N, respectively, such that G is a subgraph of H. This straightforwardly implies the first claim of the observation below. The second claim (and the first claim as well) follows by Proposition 1 and a simple inductive argument on the two trees T_M and T_N.

Observation 5. *For any degree refinement matrices $M, N \in \mathcal{M}$ it holds that if $M \xrightarrow{I} N$ then $T_M \subseteq T_N$, and if $M \xrightarrow{S} N$ then $T_N \subseteq T_M$.*

The reverse is not true: for $\xrightarrow{S}$ take $M = \mathrm{drm}(P_4)$ and $N = \mathrm{drm}(P_3)$. The counterexample for $\xrightarrow{I}$ requires a bit more effort (see [13]).

Theorem 3 can now be translated to matrices. If $M \xrightarrow{I} N$ and $M \xrightarrow{S} N$, then $M \xrightarrow{B} N$, i.e., $M = N$.

Corollary 4. $(\mathcal{M}, \xrightarrow{B}) = (\mathcal{M}, \xrightarrow{I}) \cap (\mathcal{M}, \xrightarrow{S}) = (\mathcal{M}, \{(M, M) : M \in \mathcal{M}\})$.

Proof. Suppose $G_1 \xrightarrow{I} H_1$ and $G_2 \xrightarrow{S} H_2$ hold with $\mathrm{drm}(G_i) = M$ and $\mathrm{drm}(H_i) = N$ ($i = 1, 2$). By Observation 5, we have that $T_M \subseteq T_N$ and $T_N \subseteq T_M$. We represent these inclusions by locally injective homomorphisms $f' : T_M \to T_N$ and $g' : T_N \to T_M$. Then we may conclude $M = N$ by the same arguments as in the proof of antisymmetry of $\xrightarrow{I}$. □

7 Conclusion

We have proved that graph homomorphisms with local constraints between finite graphs impose interesting orders on the class of degree matrices. We have also shown that such matrices can be easily detected and, moreover, a canonical representative of a class of equivalent matrices can be computed by an efficient algorithm. The generalization of these concepts beyond the class of degree matrices of finite graphs and their applications in theoretical computer science is subject of further study.

References

1. ANGLUIN, D. Local and global properties in networks of processors. In *Proceedings of the 12th ACM Symposium on Theory of Computing* (1980), 82–93.
2. ANGLUIN, D., AND GARDINER, A. Finite common coverings of pairs of regular graphs. *Journal of Combinatorial Theory B 30* (1981), 184–187.

3. BASTERT, O. Computing equitable partitions of graphs. *Communications in Mathematical and in Computer Chemistry* No. 40, (1999), 265–272.
4. BIGGS, N. Constructing 5-arc transitive cubic graphs. *Journal of London Mathematical Society II. 26* (1982), 193–200.
5. BODLAENDER, H. L. The classification of coverings of processor networks. *Journal of Parallel Distributed Computing 6* (1989), 166–182.
6. BONDY, J.A., AND MURTY. U.S.R. *Graph Theory with Applications,* Macmillan, London and Elsevier, New York (1976).
7. EVERETT, M. G., AND BORGATTI, S. Role coloring a graph. *Mathematical Social Sciences 21,* 2 (1991), 183–188.
8. FIALA, J., AND KRATOCHVÍL, J. Complexity of partial covers of graphs. In *Algorithms and Computation, 12th ISAAC '01, Christchurch, New Zealand* (2001), no. 2223 in Lecture Notes in Computer Science, Springer Verlag, pp. 537–549.
9. FIALA, J., AND KRATOCHVÍL, J. Partial covers of graphs. *Discussiones Mathematicae Graph Theory 22* (2002), 89–99.
10. FIALA, J., KRATOCHVÍL, J., AND KLOKS, T. Fixed-parameter complexity of λ-labelings. *Discrete Applied Mathematics 113,* 1 (2001), 59–72.
11. FIALA J., AND MAXOVÁ, J. Cantor-Bernstein type theorem for locally constrained graph homomorphisms. preprint, http://kam.mff.cuni.cz/~fiala/papers/cantor.ps, 2003.
12. FIALA, J., AND PAULUSMA, D. A complete complexity classification of the role assignment problem. To appear in *Theoretical Computer Science.*
13. FIALA, J., PAULUSMA, D., AND TELLE, J. A. Algorithms for comparability of matrices in partial orders imposed by graph homomorphisms. Accepted to WG'05, *31st Workshop on Graph-Theoretic Concepts in Computer Science,* Metz, June 2005, to appear in Lecture Notes in Computer Science, Springer Verlag.
14. GODSIL, C. *Algebraic Combinatorics* Chapman and Hall, 1993.
15. HELL, P., AND NESETRIL, J. On the complexity of H-colouring. *Journal of Combinatorial Theory, Series B, 48* (1990), 92–110.
16. HELL, P., AND NEŠETŘIL, J. *Graphs and Homomorphisms.* Oxford University Press, 2004.
17. KRATOCHVÍL, J., PROSKUROWSKI, A., AND TELLE, J. A. Covering regular graphs. *Journal of Combinatorial Theory B 71,* 1 (1997), 1–16.
18. KRISTIANSEN, P., AND TELLE, J. A. Generalized H-coloring of graphs. In *Algorithms and Computation, 11th ISAAC '01, Taipei, Taiwan* (2000), no. 1969 in Lecture Notes in Computer Science, Springer Verlag, pp. 456–466.
19. LEIGHTON, F. T. Finite common coverings of graphs. *Journal of Combinatorial Theory B 33* (1982), 231–238.
20. NEŠETŘIL, J. Homomorphisms of derivative graphs. *Discrete Mathematics 1,* 3 (1971), 257–268.

Packing Weighted Rectangles into a Square*

Aleksei V. Fishkin[1], Olga Gerber[2],**,
Klaus Jansen[2], and Roberto Solis-Oba[3],***

[1] Max-Planck-Institut für Informatik, Stuhlsatzenhausweg 85,
66123 Saarbrücken, Germany
`avf@mpi-sb.mpg.de`
[2] University of Kiel, Olshausenstr. 40, 24118 Kiel, Germany
`{oge, kj}@informatik.uni-kiel.de`
[3] Department of Computer Science, The University of Western Ontario,
London, Ontario, N6A 5B7, Canada
`solis@csd.uwo.ca`

Abstract. We consider the problem of packing a set of weighted rectangles into a unit size square frame $[0,1] \times [0,1]$ so as to maximize the total weight of the packed rectangles. We present polynomial time approximation schemes (PTASs) that, for any $\varepsilon > 0$, find $(1-\varepsilon)$-approximate solutions for two special cases of the problem. In the first case we pack a set of squares whose weights are equal to their areas. In the second case we pack a set of weighted rectangles into an augmented square frame $[0, 1+3\varepsilon] \times [0, 1+3\varepsilon]$.

1 Introduction

Two-dimensional packing problems have attracted much attention in the literature since the 80s. A series of approximation results have been obtained for strip packing [10,11,13], 2-dimensional bin packing [3,4,5,12], and rectangle packing [1,2,7,9]. These problems play an important role in a variety of applications in Computer Science and Operations Research, e.g. cutting stock, VLSI design, image processing, and multiprocessor scheduling, just to name a few.

In this paper we address the problem of packing a set of weighted rectangles into a unit size square frame. That is, given a set of weighted rectangles we wish to pack a subset of them into a unit size square frame $[0,1] \times [0,1]$ so that the total weight of the packed rectangles is maximized. In contrast to the above mentioned strip and 2-dimensional bin packing problems, there are only a few known approximation results for our problem. For a long time the only known result was an asymptotic $(4/3)$-approximation algorithm for packing unit-weight squares into a rectangle [2]. Only very recently [9], several approximation

---

* Supported by EU-Project CRESCCO "Critical Resource Sharing for Cooperation in Complex Systems", IST-2001-33135.
** Supported by DFG-Graduiertenkolleg 357.
*** Partially supported by the Natural Sciences and Engineering Research Council of Canada grant R3050A01.

algorithms have been presented for the general problem. The best algorithm in [9] finds packings with weight $(1/2-\varepsilon)$ times the optimum, for any $\varepsilon > 0$. In [7], we considered the problem of packing a set of weighted squares into a unit size square frame, and presented an algorithm which outputs a packing of the squares within an augmented square region $[0, 1+\varepsilon] \times [0, 1+\varepsilon]$ whose weight is at least $(1-\varepsilon)$ times the maximum weight that can be achieved by packing squares into the original unit size square frame $[0,1] \times [0,1]$. We call this the *square packing problem with augmentation*. In [8] the problem of packing weighted rectangles into a rectangular frame of width 1 and height at least ε^{-4} was studied and a $(1-\varepsilon)$-approximation algorithm for the problem was presented.

Here we present algorithms for two special cases of the problem of packing weighted rectangles. First, we consider the case of packing a set of squares when their weights and areas coincide. Then, we consider the problem of packing a set of weighted rectangles with augmentation, i.e., we are allowed to increase the size of the enclosing frame to $[0, 1+O(\varepsilon)] \times [0, 1+O(\varepsilon)]$, for $\varepsilon > 0$.

The problem of packing a set of rectangles into the minimum number of unit size square bins was studied by Correa and Kenyon [6]. They give an algorithm that packs the rectangles into the minimum number of square bins, assuming that the size of each bin can be slightly augmented to $[0, 1+\varepsilon] \times [0, 1+\varepsilon]$. Note that this algorithm cannot be used to pack weighted rectangles into an augmented square frame since the algorithm in [6] does not consider weighted rectangles, and in our problem not all rectangles need to be packed. It is not easy to find a set of rectangles of nearly optimal weight and which can be packed into the augmented square frame, but we show that this can be done in polynomial time. Once this set of rectangles has been found, we can use a slight modification of the algorithm in [6] to find a packing for them.

The problem of finding near-optimal, $(1-\varepsilon)$-approximate solutions for the general problem of packing a set of weighted rectangles into a square frame without augmentation remains a challenging open problem. However, we make some progress towards solving it.

1.1 Our Results

Covering the Maximum Area by Squares. In this problem we wish to pack a set of squares whose weights and areas are the same, i.e. we are interested in covering the maximum area with a subset of squares. Formally, we are given a set Q of n squares S_i ($i = 1, \ldots, n$) with side lengths $s_i \in (0, 1]$. For a given subset $Q' \subseteq Q$, a *packing* of Q' into a unit size square frame $[0,1] \times [0,1]$ is a positioning of the squares in Q' within the frame such that their interiors are disjoint. The goal is to find a subset $Q' \subseteq Q$, and a packing of Q' within $[0,1] \times [0,1]$ of maximum area, $\sum_{S_i \in Q'} (s_i)^2$. Our first main result can be stated as follows.

Theorem 1. *For any set Q of n squares and any accuracy $\varepsilon > 0$, there exists an algorithm A_ε which finds a subset of Q and its packing within the unit square frame $[0,1] \times [0,1]$, with area*

$$A_\varepsilon(Q) \geq (1-\varepsilon)\text{OPT},$$

where OPT *is the maximum area that can be covered by packing any subset of* Q. *The running time of* A_ε *is polynomial in n for fixed* ε.

This result can be extended to the case of packing d-dimensional cubes into a unit d-dimensional square cube, for $d \geq 2$.

Packing Weighted Rectangles. In this problem the goal is to pack a set of weighted rectangles into a unit size square frame. Formally, we are given a set R of n rectangles R_i ($i = 1, \ldots, n$) with widths $a_i \in (0, 1]$, heights $b_i \in (0, 1]$, and weights $w_i \geq 0$. For a given subset $R' \subseteq R$, a *packing* of R' into a unit size square frame $[0, 1] \times [0, 1]$ is a positioning of the rectangles in R' within the frame such that their interiors are disjoint. The goal is to find a subset $R' \subseteq R$, and a packing of R' within the frame of maximum weight, $\sum_{R_i \in R'} w_i$. Our second main result can be stated as follows.

Theorem 2. *For any set* R *of* n *rectangles and any accuracy* $\varepsilon > 0$, *there is an algorithm* W_ε *which finds a subset of* R *and its packing within the augmented unit square frame,* $[0, 1 + 3\varepsilon] \times [0, 1 + 3\varepsilon]$, *with weight*

$$W_\varepsilon(R) \geq (1-\varepsilon)\text{OPT},$$

where OPT *is the maximum weight that can be obtained by packing any subset of* R *into a unit size square frame* $[0, 1] \times [0, 1]$. *The running time of* W_ε *is polynomial in n for fixed* ε.

By scaling, this algorithm can be used for packing a set of rectangles into a rectangular frame. The techniques that we use for designing our algorithms have been used before for solving other problems. The contribution of this work is to show how to combine these techniques with a few new ideas to obtain our results. Some of our lemmas (or close variations of them) have been already proven in the literature.

In the following sections we give our proofs for Theorems 1 and 2. Section 2 describes the algorithm for packing squares and Section 3 describes our algorithm for packing rectangles.

2 Packing Squares

Let Q be a set of n squares S_i ($i = 1, \ldots, n$) with side lengths $s_i \in (0, 1]$. The goal is to find a subset $Q' \subseteq Q$, and a packing of Q' within $[0, 1] \times [0, 1]$, of maximum area, $\sum_{S_i \in Q'} (s_i)^2$.

If all squares S_i in Q are small (their side lengths s_i are at most ε, for some small $\varepsilon > 0$), then we can apply the Next-Fit-Increasing-Height (NFIH) heuristic to pack the squares of Q within a unit square frame $[0, 1] \times [0, 1]$ (see Section 2.1); the total area covered by this solution is at least $\min\{area(Q), 1 - 4\varepsilon - 2\varepsilon^2\}$ for any $\varepsilon > 0$. That is, we either pack all squares or obtain a packing which covers at least a fraction $(1 - 6\varepsilon)$ of the total area of the frame.

For the case of squares of arbitrary sizes, we partition Q into two sets formed by small and large squares, respectively. If we define these set properly, then any feasible packing of the squares in $[0,1] \times [0,1]$ will only contain $O(1)$ large squares. So, in $O(1)$ time we can enumerate all possible *tight packings* for the large squares, where a tight packing does not allow a large square to move to the left or down. For each tight packing of the large squares, we then try to fill up all empty gaps with small squares: we take the small squares one by one in non-decreasing order of size s_i, and use the NFIH heuristic. Among all packings found we select one with the maximum area. The main problem is to define the sets of large and small squares so that the area covered is nearly optimal.

For a subset of squares $Q' \subseteq Q$, we use $area(Q')$ to denote its area, $\sum_{S_i \in Q'} s_i^2$. In addition, we use Q^{opt} to denote an optimal subset of Q that can be packed in the unit square $[0,1] \times [0,1]$. So, $area(Q^{opt}) = \text{OPT}$ and $area(Q^{opt}) \leq 1$. For the rest of the paper, we assume w.l.o.g. that the value of $1/\varepsilon$ is integral.

2.1 The NFIH Heuristic

We consider the following simplified version of the square packing problem: given a positive value β, a set S of squares S_i with side lengths $s_i \leq \varepsilon^\beta$, and a rectangular frame $[0,a] \times [0,b]$ ($a,b \in [0,1]$), pack a subset of S into the frame such that the area covered by the squares is maximized.

First, we sort the squares $S_i \in S$ non-decreasingly by size. Then, we place the squares within $[0,a] \times [0,b]$ by using NFIH; this packs the squares into a sequence of sublevels. The first sublevel is the bottom of the frame. Each subsequent sublevel is defined by a horizontal line drawn at the top of the largest square placed on the previous sublevel. The squares are packed one by one in a left-justified manner, until the next square cannot fit within the current sublevel. At that moment, the current sublevel is closed and a new one is started. The packing procedure runs as above until there are no more squares in S or the next square in the sequence would cross the top b of the frame. For an illustration see Fig. 1.

The following result is a slightly tighter bound on the performance of NFIH than the one that can be derived from [6].

Lemma 1. *Let S be any set of squares S_i with sizes $s_i \leq \varepsilon^\beta$, and let $[0,a] \times [0,b]$ ($a,b \in [0,1]$) be a rectangular frame. The NFIH heuristic, which selects squares S_i in non-decreasing size, outputs a packing of a subset of S whose area is at least $\min\{area(S), ab - 2\varepsilon^\beta(a+b) - 2\varepsilon^{2\beta}\}$.*

Proof. Let q be the number of sublevels and let h_i be the height of the first square on the ith sublevel. Let H be the height of the packing. If no square in S is left unpacked, then the area covered is $area(S)$. Hence, assume that some squares in S are left unpacked. Since all side lengths $s_i \leq \varepsilon^\beta$, then $b - H \leq \varepsilon^\beta$. Furthermore, on each sublevel i, $i = 1, \ldots, q-1$, the area covered by the squares is at least $(a - \varepsilon^\beta)h_i$. Thus, since $h_q \leq \varepsilon^\beta$, the total area covered is at least $(H - h_q)(a - \varepsilon^\beta) \geq (b - 2\varepsilon^\beta)(a - \varepsilon^\beta) \geq a \cdot b - 2\varepsilon^\beta(a+b) - 2\varepsilon^{2\beta}$. □

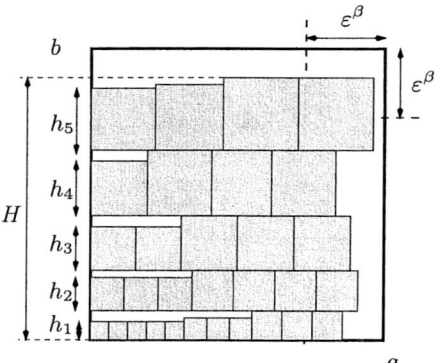

Fig. 1. NFIH for small squares

Corollary 1. *If all squares S_i in Q have sizes s_i at most ε, then the NFIH heuristic packs a subset of Q within $[0,1] \times [0,1]$ of total area at least $(1 - 6\varepsilon)\mathrm{OPT}(Q)$. The running time of the algorithm is $O(n \log n)$.*

2.2 Partitioning the Squares

We define the group $Q^{(0)}$ of squares $S_i \in Q$ with side lengths s_i in $(\varepsilon^4, 1]$, and for $j \in \mathbb{Z}_+$ we define the group $Q^{(j)}$ of squares with side lengths in $(\varepsilon^{2^{j+1}+3}, \varepsilon^{2^j}]$. Then,

$$\cup_{j=0}^{\infty} Q^{(j)} = Q \quad \text{and} \quad Q^{(\ell)} \cap Q^{(j)} = \emptyset, \text{ for} |\ell - j| > 1.$$

Lemma 2. *There is a group $Q^{(k)}$ with $0 \leq k \leq 2/\varepsilon^2 - 1$ such that its contribution to the optimum is*

$$\mathrm{area}(Q^{opt} \cap Q^{(k)}) \leq \varepsilon^2 \mathrm{OPT},$$

where Q^{opt} is an optimal subset of squares.

Proof. Each square belongs to at most two consecutive groups. Therefore,

$$\cup_{k=0}^{2/\varepsilon^2-1} \mathrm{area}(Q^{opt} \cap Q^{(k)}) \leq 2 OPT,$$

and so, there must be a group $Q^{(k)}$ as indicated in the lemma. □

Let $Q^{(k)}$ be a group such that $\mathrm{area}(Q^{opt} \cap Q^{(k)}) \leq \varepsilon^2 OPT$. We drop the squares $Q^{(k)}$ from consideration. Then, an optimal packing for $Q \setminus Q^{(k)}$ must cover area at least $(1 - \varepsilon^2)\mathrm{OPT}$, i.e. this makes a loss of at most a factor of ε^2 in the optimum.

Next, we partition the squares in $Q \setminus Q^{(k)}$ into two groups: $L = \{S_i \mid s_i > \varepsilon^{2^k}\}$ and $S = \{S_i \mid s_i \leq \varepsilon^{2^{k+1}+3}\}$. The squares in L and S are called *large* and *small*, respectively.

2.3 Outline of the Algorithm

Here we give a high level description of the algorithm. The individual steps of the algorithm are analyzed in the next section.

Algorithm A_ε:
INPUT: Set Q of squares, accuracy $\varepsilon > 0$.
OUTPUT: A packing of a subset of Q within $[0,1] \times [0,1]$.

1. For each $k \in \{0, 1 \ldots, 2/\varepsilon^2 - 1\}$ form the group $Q^{(k)}$ as described above.
   (a) Let $\alpha = 2^k$ and $\beta = 2^{k+1} + 3$.
   (b) Partition $Q \setminus Q^{(k)}$ into L and S, the sets of large and small squares with sides larger than ε^α and at most ε^β, respectively.
   (c) Compute the set $FEASIBLE$, containing all subsets of L with at most $1/\varepsilon^{2\alpha}$ large squares.
   (d) For every set in $FEASIBLE$, find all possible *tight packings* of its large squares. For each tight packing use the modified NFIH to pack the small squares in the empty gaps left by the large squares until no further small squares can be packed.
2. Among all packings produced, output one with the maximum area covered.

2.4 Analysis of Algorithm A_ε

Large Squares. The set $FEASIBLE$ which contains all subsets of at most $1/\varepsilon^{2\alpha}$ large squares has polynomial size, $O(n^{\varepsilon^{-2\alpha}})$. Observe also that the optimal set $L \cap Q^{opt}$ of large squares belongs to $FEASIBLE$.

Lemma 3. *For any set $L' \in FEASIBLE$ of large squares, we can find in $O(1)$ time all possible tight packings of its large squares.*

Small Squares. We sort the small squares non-decreasingly by size. Assume that we have a tight packing of some set $L' \in FEASIBLE$. We define a *sliced structure* for this packing as follows. We draw a vertical line at each position where a large square starts or ends. The space between any two consecutive vertical lines is called a *slice*. Looking into each slice we can see that the horizontal boundaries of the large squares cut some slices out. We work with the empty rectangular gaps inside the slices.

We place the small squares from S in the gaps by using the NFIH heuristic: we consider slices one by one, filling the gaps in a bottom-up manner using small squares. To fill a gap, we take the small squares $S_i \in S$ in order of non-decreasing size, and apply the NFIH heuristic. We can prove the following result.

Lemma 4. *For any feasible set $L' \in FEASIBLE$ which has a tight packing within the frame $[0,1] \times [0,1]$, the modified NFIH heuristic adds small squares to the packing in such a way that the area covered is at least $\min\{area(L') + area(S), 1 - \varepsilon^2\}$, for any $0 < \varepsilon \leq 1/16$.*

Proof. Recall that $\alpha = 2^k$, $\beta = 2^{k+1}+3$, and $|L'| \le 1/\varepsilon^{2\alpha}$. The number of slices in a packing of L' is at most $2|L|$. The widths of all slices add up to 1. The heights of all empty gaps in each slice add up to at most 1.

Assume that some small squares are left unpacked. Let q be the number of gaps, and let $x_1 * y_1, x_2 * y_2, \ldots, x_q * y_q$ be their areas. Then, $q \le (2|L|)^2$,

$$\sum_{j=1}^{q} x_j * y_j = 1 - area(L'), \quad \sum_{j=1}^{q} y_j \le 2|L| \text{ and } \sum_{j=1}^{q} x_j \le 2|L|.$$

To see that $\sum_{j=1}^{q} y_j \le 2|L|$, note that all rectangular gaps are inside the slices, so the sum of the lengths of their vertical boundaries is at most $2|L|$, the total length of all the slices. The last inequality follows from a symmetry argument, i.e., if we draw horizontal slices instead of vertical ones, we obtain a similar figure but with respect to the widths x_j.

Remember that each small square in S has side length at most ε^β. Thus, using Lemma 1, we can bound the area covered by the small squares as follows

$$\begin{aligned} AREA &= \sum_{j=1}^{q}(x_j * y_j - 2\varepsilon^\beta(x_j + y_j) - 2\varepsilon^{2\beta}) \\ &\ge (1 - area(L')) - 2\varepsilon^\beta(4|L|) - 2\varepsilon^{2\beta}q \\ &\ge (1 - area(L')) - 2\varepsilon^\beta(4/\varepsilon^{2\alpha}) - 2\varepsilon^{2\beta}(4/\varepsilon^{4\alpha}) \\ &\ge (1 - area(L')) - 8\varepsilon^{\beta-2\alpha} - 8\varepsilon^{2\beta-4\alpha} \\ &\ge (1 - area(L')) - 8\varepsilon^3 - 8\varepsilon^6 \ge (1 - area(L')) - \varepsilon^2 \end{aligned}$$

since $8\varepsilon^3 - 8\varepsilon^6 \le \varepsilon^2$ for $\varepsilon \in (0, 1/16]$.

2.5 Proof of Theorem 1

Algorithm A_ε considers all values $k \in \{0, 1\ldots, 2/\varepsilon^2 - 1\}$ and groups $Q^{(k)}$. By Lemma 2 at least for one of these groups $Q^{(k)}$,

$$area(Q^{opt} \setminus Q^{(k)}) \ge (1 - \varepsilon^2) \text{OPT}.$$

Consider one such group $Q^{(k)}$ and let $\alpha = 2^k$ and $\beta = 2^{k+1}+3$. Partition $Q \setminus Q^{(k)}$ into the sets of large and small squares, L and S, where the side length of each large square is larger than ε^α and the size of each small square is at most ε^β.

We know that $Q^{opt} \cap L$ belongs to the set $FEASIBLE$, which consists of all sets with at most $1/\varepsilon^{2\alpha}$ large squares. Since Q^{opt} can be packed within the frame $[0, 1] \times [0, 1]$, there exists a tight packing for $Q^{opt} \cap L$ as well. For each such a tight packing, the NFIH heuristic adds small squares to the packing such that the total area covered by the squares is at least

$$\min\{area(Q^{opt} \cap L) + area(S), 1 - \varepsilon^2\}.$$

Since OPT ≤ 1, $1 - \varepsilon^2 \geq (1 - \varepsilon^2)$OPT. On the other hand, since $area(Q^{(k)}) \leq \varepsilon^2$OPT, then

$$area(Q^{opt} \cap L) + area(S) \geq area(Q^{opt} \setminus Q^{(k)}) \geq (1 - \varepsilon^2)\text{OPT}.$$

We also know that the set $FEASIBLE$ and all possible tight packings of large squares can be found in $O(n^{O(1)})$ time. The NFIH heuristic runs in time polynomial in the number of squares, n. Hence, the overall running time of the algorithm is polynomial in n for fixed ε.

2.6 Packing d-Dimensional Cubes

Our algorithm can be easily extended to the problem of packing d-dimensional cubes into a unit d-dimensional cubic frame so as to maximize the total volume of the cubes packed. As in the 2-dimensional case, we partition the set of cubes into two sets L and S containing large and small cubes, respectively. Since only a constant number of large cubes can be packed into the frame, we can enumerate all feasible subsets of L that can be packed in the frame in polynomial time. We can prove the following generalization of Lemma 1 (see also [6]).

Lemma 5. *Let S be any set of d-dimensional cubes S_i with sizes $s_i \leq \varepsilon^\beta$, and let $[0, a_1] \times [0, a_2] \times \cdots \times [0, a_d]$ ($a_i \in [0, 1]$) be a parallelepiped. The generalization of the NFIH heuristic to d dimensions outputs a packing of a subset of S whose volume is at least $\min\{volume(S), (a_1 - \varepsilon^\beta)(a_2 - 2\varepsilon^\beta) \cdots (a_d - 2\varepsilon^\beta)\}$.*

This lemma shows that the generalization of NFIH to d dimensions can be used to pack the small cubes in the empty space left by a tight packing of the large cubes so that the total empty space left is only an ε fraction of the total volume of the frame.

3 Packing Weighted Rectangles with Augmentation

Let R be a set of n rectangles, R_i ($i = 1, \ldots, n$) with widths $a_i \in (0, 1]$, heights $b_i \in (0, 1]$, and weights $w_i \geq 0$. The goal is to find a subset $R' \subseteq R$, and a packing of R' within $[0, 1] \times [0, 1]$ of maximum weight, $\sum_{R_i \in R'} w_i$.

We partition the rectangles R into four sets: L, H, V, and S. The rectangles in L have large widths and heights, so only $O(1)$ of them can be packed in the unit square frame. The rectangles in H (V) have large width (height). We round the sizes of these rectangles in order to reduce the number of distinct widths and heights. Then, we use enumeration and a fractional strip-packing algorithm to select the best subsets of H and V to include in our solution. The rectangles in S have very small width and height, so as soon as we have selected near-optimal subsets of rectangles from $L \cup H \cup V$ we add rectangles from S to the set of rectangles to be packed in a greedy way. Once we have selected the set of rectangles to be packed into the frame, we use a slight modification of the algorithm of Correa and Kenyon [6] to pack them.

For a subset of rectangles $R' \subseteq R$, we use $weight(R')$ to denote its weight, $\sum_{R_i \in R'} w_i$. We use R^{opt} to denote an optimal subset of R that can be packed into the unit square frame $[0,1] \times [0,1]$. So,

$$weight(R^{opt}) = \text{OPT} \text{ and } area(R^{opt}) \leq 1.$$

3.1 Partitioning the Rectangles

We define the group $G^{(0)}$ of rectangles $R_i \in R$ with either widths $a_i \in (\varepsilon^3, 1]$ or heights $b_i \in (\varepsilon^3, 1]$. For $j \in Z_+$ we define the group $G^{(j)}$ of rectangles R_i with either widths $a_i \in [\varepsilon^{3(j+1)}, \varepsilon^{3j})$ or heights $b_i \in [\varepsilon^{3(j+1)}, \varepsilon^{3j})$. Each rectangle belongs to at most 2 groups.

Lemma 6. *There is a group $G^{(k)}$ with $0 \leq k \leq 2/\varepsilon^2 - 1$ such that*

$$weight(G^{(k)} \cap R^{opt}) \leq \varepsilon^2 \cdot \text{OPT},$$

where R^{opt} is the subset of rectangles selected by an optimum solution.

We drop the rectangles in group $G^{(k)}$, as described in Lemma 6, from consideration. Then, an optimal packing for $R^{opt} \setminus G^{(k)}$ must have weight at least $(1 - \varepsilon^2)\text{OPT}$, i.e. this causes a loss of at most a factor of ε^2 in the optimum. We partition R into four groups according to their side lengths, as follows:

$$L = \{R_i \mid a_i \geq \varepsilon^{3k} \text{ and } b_i \geq \varepsilon^{3k}\}$$
$$S = \{R_i \mid a_i < \varepsilon^{3k+3} \text{ and } b_i < \varepsilon^{3k+3}\}$$
$$H = \{R_i \mid a_i \geq \varepsilon^{3k} \text{ and } b_i < \varepsilon^{3k+3}\}$$
$$V = \{R_i \mid a_i < \varepsilon^{3k+3} \text{ and } b_i \geq \varepsilon^{3k}\}$$

Lemma 7. *The subset $R^{opt} \setminus G^{(k)}$ of rectangles can be packed within the frame $[0, 1+2\varepsilon] \times [0, 1+2\varepsilon]$ in such a way that*

- *each rectangle $R_i \in H \cup L$ is positioned so that its upper left corner is at an x-coordinate that is a multiple of ε^{3k+1},*
- *each rectangle $R_i \in V \cup L$ is positioned so that its upper left corner is at a y-coordinate that is a multiple of ε^{3k+1},*

Furthermore, any width $a_i \geq \varepsilon^{3k}$ or height $b_i \geq \varepsilon^{3k}$ can be rounded up to the nearest multiple of ε^{3k+1} without affecting the feasibility of the packing.

Due to space limitations we do not include the proof of the lemma.

3.2 Outline of the Algorithm

Algorithm W_ε:
INPUT: Set R of rectangles, accuracy $\varepsilon > 0$.
OUTPUT: A packing of a subset of R within $[0, 1+3\varepsilon] \times [0, 1+3\varepsilon]$.

1. For each $k \in \{0, 1 \ldots, 2/\varepsilon^2 - 1\}$, form the group $G^{(k)}$ as described above.
2. Let $\alpha = 1/\varepsilon^{3k+1}$.
   (a) Partition $R \setminus G^{(k)}$ into sets L, S, H, and V as described above.
   (b) Round the sizes of the rectangles $L \cup H \cup V$ as indicated in Lemma 7.
   (c) Compute the set FL containing all subsets of L with at most $1/(\varepsilon^{3k})^2$ rectangles.
   (d) Compute the set FH containing all *feasible* subsets of H with *profiles* $(h_1, h_2, \ldots, h_\alpha)$ where each entry $h_q \le 1$ ($q = 1, \ldots, \alpha$) is a multiple of ε^{3k+2}. (See next section for details.)
   (e) Compute the set FV containing all *feasible* subsets of V with *profiles* $(v_1, v_2, \ldots, v_\alpha)$ where each entry $v_q \le 1$ ($q = 1, \ldots, \alpha$) is a multiple of ε^{3k+2}. (See next section for details.)
3. For each set $L' \in FL$, $H' \in FH$, and $V' \in FV$ do:
   (a) Try all possible packings for L' in the frame $[0, 1 + 2\varepsilon] \times [0, 1 + 2\varepsilon]$, positioning the rectangles as indicated in Lemma 7.
   (b) For each packing of L' try to pack the rectangles in H' and V' by solving a fractional strip-packing problem as described in the next section.
   (c) Find a subset $S' \subseteq S$ which is *feasible* for L', H' and V' (see next section).
   (d) Increase the size of the frame to $[1 + 3\varepsilon] \times [1 + 3\varepsilon]$ and use the Next Fit Increasing Height algorithm to pack the rectangles S' with the empty gaps left by $L' \cup H' \cup V'$.
4. Among all packings produced, output one having maximum weight.

3.3 Analysis of the Algorithm W_ε

Computing FL. Recall that for each rectangle $R_i \in L \cap R^{opt}$, both sides, $a_i, b_i \in (\varepsilon^{3k}, 1]$. Since $area(L \cap R^{opt})$ is at most 1, there cannot be more than $(1/\varepsilon^{3k})^2$ rectangles in $L \cap R^{opt}$.

Lemma 8. *In $O(n^{\varepsilon^{-6k}})$ time we can find the set FL consisting of all subsets of L with at most ε^{-6k} squares. The optimal subset, $L \cap R^{opt}$, belongs to FL.*

Computing FH. Recall that for each rectangle $R_i \in H$, its width, $a_i \in (\varepsilon^{3k}, 1]$ was rounded to a multiple of ε^{3k+1}. Hence, there are at most $\alpha = 1/\varepsilon^{3k+1}$ distinct widths, $\bar{a}_1, \bar{a}_2, \ldots, \bar{a}_\alpha$, in H. We use $H(\bar{a}_q)$ to denote the subset of H consisting of all rectangles with width $\bar{a}_q$. Let $H' \subseteq H$. The *profile* of H' is an α-tuple $(h_1, h_2, \ldots, h_\alpha)$ such that each entry $h_q \in (0, 1]$ ($q = 1, \ldots, \alpha$) is the total height of the rectangles in $H' \cap H(\bar{a}_q)$.

Consider the profile $(h_1^*, h_2^*, \ldots, h_\alpha^*)$ of $H \cap R^{opt}$. Note that if each value h_i^* is rounded up to the nearest multiple of ε/α, this might increase the height of the frame where the rectangles are packed by at most $\alpha(\varepsilon/\alpha) = \varepsilon$. The advantage of doing this, is that the number of possible values for each entry of the profile of $H \cap R^{opt}$ is only constant, i.e. α/ε. Therefore, the total number of profiles is also constant, $\alpha^{\alpha/\varepsilon}$.

By trying all possible profiles with entries that are multiples of (ε/α) we ensure to find one that is identical to the rounded profile for $H \cap R^{opt}$. However, the profile itself does not yield the set of rectangles in $H \cap R^{opt}$. Fortunately, we do not need to find this set, since any set of rectangles with the same rounded profile as $H \cap R^{opt}$ can be packed in a frame of height $1 + \varepsilon$ by solving a fractional strip-packing problem: the strips are the empty rectangular gaps left by the rectangles in $L \cap R^{opt}$, and inside these strips we try to pack rectangles of width a_i and height h_i^*. Rounding the solution for the fractional strip-packing problem to get an integral solution, increases the height of the packing by at most $(\varepsilon/\alpha)\alpha = \varepsilon$. (For a more detailed explanation, the reader is referred to [6].)

Thus, we just need to find a set of rectangles from H with nearly-maximum weight and with the same rounded profile as $H \cap R^{opt}$.

We say that a subset of $H' \subseteq H$ is *feasible* if

- each entry $h_q \in (0,1]$ ($q = 1, \ldots, \alpha$) in the profile of H' is a multiple of $\varepsilon/\alpha = \varepsilon^{3k+2}$, and
- each subset $H' \cap H(\bar{a}_q)$ ($q = 1, \ldots, \alpha$) is a $(1 - \varepsilon)$-approximate solution of an instance of the knapsack problem where h_q is the knapsack capacity and each rectangle $R_i \in H(\bar{a}_q)$ is an item of size b_i and profit w_i.

Lemma 9. *In $O(n \log n)$ time we can find the set FH consisting of all feasible subsets of H.*

Computing FV. We use similar ideas as above to define *profiles* and to find the set FV consisting of all *feasible* subsets of V.

Selecting the Small Rectangles. Assume that we are given feasible subsets $L' \in FL$, $H' \in FH$, $V' \in FV$ such that $area(L' \cup H' \cup V')$ is at most $(1+2\varepsilon)^2$. A subset $S' \subseteq S$ is feasible for the selection L', H', V', if S' is a $(1-\varepsilon)$-approximate solution for the instance of the knapsack problem where $(1 + 2\varepsilon)^2 - area(L' \cup H' \cup V')$ is the knapsack's capacity, and each rectangle $R_i \in S$ is an item of size $a_i b_i$ and profit w_i.

Proposition 1. *Given sets $L' \subseteq FL, H' \subseteq FH$, and $V' \subseteq FV$, a feasible subset S' of S can be found in $O(n \log n)$ time.*

3.4 Proof of Theorem 2

Lemma 10. *There exist a selection of feasible subsets $L' \in FL, H' \in FH$, $V' \in FV$, and $S' \subseteq S$, such that*

- *$weight(L' \cup H' \cup V' \cup S')$ is at least $(1 - \varepsilon)\mathrm{OPT}$,*
- *algorithm W_ε outputs a packing of $(L' \cup H' \cup V' \cup S')$ within an augmented unit size square frame $[0, 1 + 3\varepsilon] \times [0, 1 + 3\varepsilon]$.*

Due to space limitations we do not include the proof of this lemma. Algorithm W_ε considers all values $k \in \{0, 1 \ldots, 2/\varepsilon^2 - 1\}$. For at least one of these values it must find a group $G^{(k)}$ such that

$$weight(R^{opt} \setminus G^{(k)}) \geq (1 - \varepsilon^2)\mathrm{OPT}.$$

For this group, the rest of the rectangles $R \setminus G^{(k)}$ is partitioned into sets L, S, H, and V.

By Lemma 10 there exist a selection of feasible subsets $L' \in FL, H' \in FH$, $V' \in FV$, and $S' \subseteq S$, such that

$$weight(L' \cup H' \cup V' \cup S') \geq (1 - \varepsilon)\text{OPT},$$

and such that algorithm W_ε outputs a packing of $(L' \cup H' \cup V' \cup S')$ within an augmented unit size square frame $[0, 1 + 3\varepsilon] \times [0, 1 + 3\varepsilon]$. Since algorithm W_ε tries all feasible sets in FL, FH, and FV, and all packing for them, W_ε must find the required solution.

All feasible subsets can be found in $O(n \log n)$ time. The algorithm in [6] for fractional strip-packing also runs in time polynomial in n. Furthermore, there is only a constant number of possible packings for any set of large rectangles from FL. Hence, the running time of algorithm W_ε is polynomial in n for fixed ε.

References

1. B.S. Baker, D.J. Brown, and H.P. Katseff. A 5/4 algorithm for two dimensional packing. *Journal of Algorithms*, 2:348–368, 1981.
2. B.S. Baker, A.R. Calderbank, E.G. Coffman, and J.C. Lagarias. Approximation algorithms for maximizing the number of squares packed into a rectangle. *SIAM Journal on Algebraic and Discrete Methods*, 4:383–397, 1983.
3. N. Bansal and M. Sviridenko. New approximability and inapproximability results for 2-dimensional bin packing. In *15th Annual ACM-SIAM Symposium on Discrete Algorithms (SODA)*, pages 189–196, 2004.
4. A. Caprara. Packing 2-dimensional bins in harmony. In *43rd Annual Symposium on Foundations of Computer Science(FOCS)*, pages 490–499, 2002.
5. F.R.K. Chung, M.R. Garey, and D.S. Johnson. On packing two-dimentional bins. *SIAM Journal on Algebraic and Discrete Methods*, 3:66–76, 1982.
6. J.R. Correa and C. Kenyon. Approximation schemes for multidimensional packing. In *Proceedings 15th Annual ACM-SIAM Symposium on Discrete Algorithms (SODA)*, pages 179–188, 2004.
7. A. V. Fishkin, O. Gerber, K. Jansen, and R. Solis-Oba. On packing squares with resource augmentation: maximizing the profit. In *Computing: The Australasian Theory Symposium (CATS)*, 2005.
8. A.V. Fishkin, O. Gerber, and K. Jansen. On weighted rectangle packing with large resources. In *3rd IFIP International Conference on Theoretical Computer Science*, pages 237–250, 2004.
9. K. Jansen and G. Zhang. On rectangle packing: maximizing benefits. In *15th Annual ACM-SIAM Symposium on Discrete Algorithms (SODA)*, pages 197–206, 2004.
10. C. Kenyon and E. Rémila. Approximate strip-packing. In *37th Annual Symposium on Foundations of Computer Science (FOCS)*, pages 31–36, 1996.
11. I. Schiermeyer. Reverse fit : a 2-optimal algorithm for packing rectangles. In *2nd European Symposium on Algorithms (ESA)*, pages 290–299, 1994.
12. S. Seiden and R. van Stee. New bounds for multi-dimentional packing. *Algorithmica*, 36(3):261–293, 2003.
13. A. Steinberg. A strip-packing algorithm with absolute performance bound 2. *SIAM Journal on Computing*, 26(2):401–409, 1997.

Nondeterministic Graph Searching: From Pathwidth to Treewidth

Fedor V. Fomin[1,*], Pierre Fraigniaud[2,**], and Nicolas Nisse[2,***]

[1] Department of Informatics, University of Bergen,
PO Box 7800, 5020 Bergen, Norway
fomin@ii.uib.no
[2] CNRS, Lab. de Recherche en Informatique, Université Paris-Sud,
91405 Orsay, France
{pierre, nisse}@lri.fr

Abstract. We introduce nondeterministic graph searching with a controlled amount of nondeterminism and show how this new tool can be used in algorithm design and combinatorial analysis applying to both pathwidth and treewidth. We prove equivalence between this game-theoretic approach and graph decompositions called q-*branched* tree decompositions, which can be interpreted as a parameterized version of tree decompositions. Path decomposition and (standard) tree decomposition are two extreme cases of q-branched tree decompositions. The equivalence between nondeterministic graph searching and q-branched tree decomposition enables us to design an exact (exponential time) algorithm computing q-branched treewidth for all $q \geq 0$, which is thus valid for both treewidth and pathwidth. This algorithm performs as fast as the best known exact algorithm for pathwidth. Conversely, this equivalence also enables us to design a lower bound on the amount of nondeterminism required to search a graph with the minimum number of searchers.

Keywords: treewidth, pathwidth, graph searching.

1 Introduction

Treewidth and pathwidth are among the most key parameters in graph algorithms, also playing important roles in structural graph theory. Both parameter serve as the important tools in Robertson and Seymour's Graph Minors project [17]. Many intractable problems can be solved in polynomial time when the input is restricted to graphs of bounded treewidth. (See Bodlaender's survey [4] for a

---

* Additional support by Norges forskningsråd projects 162731/V00 and 160778/V30.
** Additional supports from the INRIA Project "Grand Large", and from the Project PAIRAPAIR of the ACI "Masse de Données".
*** Additional supports from the Project FRAGILE of the ACI "Sécurité & Informatique".

comprehensive overview.) Treewidth also plays a crucial role in Downey & Fellows parameterized complexity theory (Chapter 6 in [9]). Moreover, treewidth is the basic ingredient for many applications in artificial intelligence, databases and logical-circuit design. To mention just a few of these applications: Exact inference in Bayesian networks, reasoning with structured constraint-satisfaction problems, propositional satisfiability and first-order logic. See [1] for further references.

In this paper we introduce the new notion of q-branched treewidth which can be interpreted as a parameterized version of treewidth. Loosely speaking, a rooted tree decomposition is q-branched if every path from the root of the tree to a leaf contains at most q branching nodes (nodes with at least two children). This notion is a natural generalization of path and tree decompositions: For $q = \infty$, q-branched treewidth is equivalent to the treewidth, and, for $q = 0$, q-branched treewidth is equivalent to the pathwidth of a graph.

Both parameters, pathwidth and treewidth, have nice game-theoretic interpretations. (See a survey of Bienstock [3].) Pathwidth can be described as a search game where searchers, looking for a fugitive, are successively placed to and removed from vertices of the graph. (Kirousis and Papadimitriou [13] called this version of searching by node searching.) The purpose of searching is to capture the fugitive that is invisible and moves arbitrarily fast along paths in the graph. The fugitive is not allowed to run through the vertices currently occupied by searchers. So the fugitive is caught when a searcher is placed on the vertex it occupies, and it has no possibility to leave the vertex because all the neighbors are occupied (guarded) by searchers. The goal of search games is to find a search strategy that guarantees the fugitive's capture. The pathwidth of a graph G is equal to the minimum number of searchers needed for a successful search strategy on G, minus one.

Treewidth also can be described as a search game, where a team of searchers are trying to catch a *visible* fugitive. It was shown by Seymour and Thomas [18] that the minimum number of searchers required to catch the fugitive on a graph G in this game is equal to the treewidth of G plus one. (An alternative game-theoretic interpretation of treewidth in terms of searching was given by Dendris et al. [6] who restrict the ability of the (called *inert*) fugitive to move only when a searcher is placed at the vertex where the fugitive currently stands.)

Game theoretic interpretation of width parameters is interesting not only in its own. Very often it provides a deeper insight to the problem yielding new structural and algorithmic results. Good examples are proofs of min-max theorems on treewidth by Seymour and Thomas [18], the polynomial time algorithm computing branch-width of a planar graph in [19], the linear time algorithm on trees for computing cutwidth in [15], as well as the computation of the topological bandwidth in [14], and the vertex separation number in [7]. It is therefore natural to ask if there is a game theoretic interpretation of the q-branched treewidth.

Our Results. To answer the question above, we introduce a new game model providing a unique approach to both search models of Kirousis-Papadimitriou, and Seymour-Thomas. In our search game the searchers can query an oracle which possesses information about the position of the fugitive. However the number of times the searchers can query the oracle is limited. This situation can be interpreted as using powerful but expensive intelligence service with limited resources. More formally, q-limited graph searching is a graph searching game in which the search program is allowed to perform nondeterministic search steps. In the same spirit as in the field of complexity theory addressing limited nondeterminism (cf., e.g., [11] for a survey), the number of nondeterministic steps of the search program is however limited. The parameter q limits the program to at most q nondeterministic steps.

We first show a formal equivalence between q-limited graph searching and q-branched treewidth. Precisely, we prove that a graph G has a q-branched treewidth $\leq k$ if and only if it can be searched with at most $k+1$ searchers by a search strategy using at most q nondeterministic steps. Moreover, we establish a one-to-one correspondence between the q-branched tree decompositions of G of width $\leq k$ and the q-limited search strategies using $\leq k+1$ searchers.

Then we use q-limited graph searching to design an exact (exponential-time) algorithm computing the q-branched treewidth of a graph. The interest in exact and fast exponential-time algorithms solving hard problems dates back to the sixties and seventies [12,20]. The last decade has led to much research in fast exponential-time algorithms. We refer to Woeginger's survey [21] for an overview. However despite of the importance of treewidth and pathwidth, and despite the fact that much progress on exponential-time solutions to other graph problems have been made, the only worst-case bound known so far for finding pathwidth is $2^n \cdot n^{O(1)}$. This can be obtained by adopting classical TSP dynamic programming approach [12]. For treewidth, the fastest known exponential algorithm is an $O(1.96^n)$ algorithm due to Fomin et al. [10]. In this paper we design an algorithm computing q-branched treewidth of a graph on n vertices in time $2^n \cdot n^{O(1)}$ for any $q \geq 0$.

Finally, the equivalence between q-limited graph searching and q-branched tree decomposition enables us to design a lower bound on the amount of nondeterminism required to search a graph with the minimum number of searchers. Precisely, we prove that, for any graph G of treewidth $\mathbf{tw}(G) = k$, the smallest $q \geq 0$ such that G can be searched by $k+1$ searchers using a q-limited search program is at least $\log_2(\mathbf{pw}(G)/\mathbf{tw}(G))$ where $\mathbf{pw}(G)$ is the pathwidth of G.

2 Formal Definitions

In this section, we formally define the two notions of q-branched treewidth and q-limited graph searching. Later on, these two notions will be shown to be equivalent.

Branched Treewidth. A *tree decomposition* of graph G is a pair $(T, \mathcal{X})$ where T is a tree of node set I, and $\mathcal{X} = \{X_i, i \in I\}$ is a collection of subsets of $V(G)$ satisfying the following three conditions:

1. $V(G) = \cup_{i \in I} X_i$;
2. For any edge e of G, there is a set $X_i \in \mathcal{X}$ containing both end-points of e;
3. For any triple i_1, i_2, i_3 of nodes of T, if i_2 is on the path from i_1 to i_3 in T, then $X_{i_1} \cap X_{i_3} \subseteq X_{i_2}$.

The *width* of a tree decomposition is defined as width$(T, \mathcal{X}) = \max_{i \in I} |X_i| - 1$. A *rooted* tree decomposition of a graph G is a tree decomposition $(T, \mathcal{X})$ of G where T is rooted at some node $r \in I$. It is denoted by $(T, \mathcal{X}, r)$. A *branching node* of a rooted tree decomposition is a node with at least 2 children.

For any $q \geq 0$, a *q-branched tree decomposition* of a graph G is a rooted tree decomposition $(T, \mathcal{X}, r)$ of G such that every path in T from the root r to a leaf contains at most q branching nodes.

Thus a path decomposition rooted at one of its extremities is a 0-branched tree decomposition, and a (standard) tree decomposition is a ∞-branched tree decomposition.

For any graph G, the *q-branched treewidth* $\mathbf{tw}_q(G)$ of G is the minimum width of over all q-branched tree-decomposition of G.

Therefore, $\mathbf{pw}(G) = \mathbf{tw}_0(G)$ and $\mathbf{tw}(G) = \mathbf{tw}_\infty(G)$. Figure 1 displays the "spectral width" of a graph G, i.e., the graph of the function $f_G : \mathbf{N} \to \mathbf{N}$ such that $f_G(q) = \mathbf{tw}_q(G)$. In this figure, $\tau(G) = \min\{q \geq 0 \mid \mathbf{tw}_q(G) = \mathbf{tw}(G)\}$ and $\pi(G) = \max\{q \leq \tau(G) \mid \mathbf{tw}_q(G) = \mathbf{pw}(G)\}$.

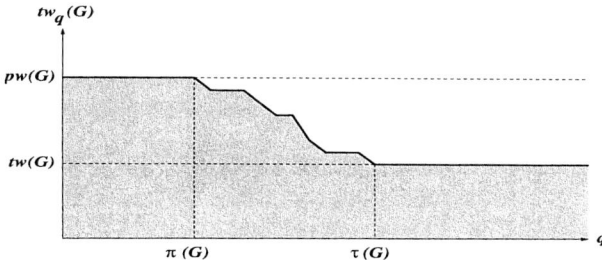

Fig. 1. Spectral width of graph G

2.1 Graph Searching

Search games are games between a *fugitive* and *searchers* in a graph. The fugitive and the searchers occupy vertices of the graphs. The goal of the fugitive is to escape from the searchers. It is caught when a searcher is placed on the vertex currently occupied by the fugitive. The fugitive permanently knows where the searchers are, and moves arbitrarily fast, but it cannot meet a searcher without being caught. The searchers do not know the position of the fugitive.

More formally, a *search program* is a (deterministic) program that takes as input a graph G and an integer $k \geq 1$, and returns an ordered sequence of *search steps*. This sequence is called the *search strategy* for G. Each step is an operation that consists in either "placing a searcher at $v \in V(G)$" or "removing a searcher from $v \in V(G)$". After a searcher s has been placed at v, and before s is removed from v, vertex v is said to be *occupied* by searcher s. When a vertex has been occupied by a searcher, it becomes *clear*. Vertices that have not been cleared yet are called *contaminated*. The search program is correct if it satisfies the following constraints:

1. no more than k searcher are simultaneously occupying vertices of G;
2. a step "place a searcher at v" occurs at most once, for every vertex v;
3. when a searcher is removed from a vertex v, for any path P between v and contaminated vertices, there is a searcher occupying a vertex of P.

A *fugitive program* in a graph G is a deterministic automaton F whose states are all possible triples (S, X, v) where $S \subset V(G)$, $X \subseteq S$, and $v \in V(G) \setminus S$. If the automaton is in state (S, X, v), then the fugitive is currently occupying vertex v, the searchers are occupying vertices in X, and S is the set of clear vertices. Given a state (S, X, v) of the automaton, the transition function of the fugitive program returns a new state (S, X, v') where v and v' are constrained to be in the same connected component of $G \setminus S$. Then the fugitive moves in G from vertex v to vertex v'. The initial state of the fugitive program is the state $(\emptyset, \emptyset, v_0)$ for some $v_0 \in V(G)$.

A *search game* is then a game between the fugitive program and the search program. A *configuration* of the game is a tripple (S, X, v) where S is the set of clear vertices, X is the set of vertices occupied by searchers, and v is the position of the fugitive. From constraint 3 of the search program, we always have $\delta(S) \subseteq X$, where $\delta(S)$ denotes the set of vertices in S that have a neighbor in $G \setminus S$. Initially, the fugitive is placed in v_0, where $(\emptyset, \emptyset, v_0)$ is the initial state of the fugitive program. I.e., the initial configuration of the game is $(\emptyset, \emptyset, v_0)$. Then the search program and the fugitive program play alternatively. Each step of the search program transforms the current configuration (S, X, v) of the game into a configuration $(S \cup \{u\}, X \cup \{u\}, v)$ (in case of a step "place a searcher at u"), or into a configuration $(S, X \setminus \{u\}, v)$ (in case of a step "remove a searcher from u").

The search program *wins* the game if the game reaches a configuration in which $v \in X$. Otherwise the fugitive wins. If the search program wins, then the fugitive is said to be caught. Note that the fugitive wins if the search program cannot carry on without violating one of its three constraints.

The search program that places a different searcher on every vertex of the graph wins against any fugitive. It however requires n searchers in n-node graphs.

Definition 1. *The* search number *of G, denoted by* $s(G)$, *is the minimum number of searchers required by a search program to win against any fugitive in G.*

2.2 Nondeterministic Graph Searching

A *nondeterministic* search program is a search program that can do nondeterministic steps. Each nondeterministic step consists in a *query operation*. Given the set $S \subset V(G)$ of clear vertices, a query returns a connected component C of $G \setminus S$, and all vertices in $G \setminus C$ are cleared. The choice of C is nondeterministic. Alternatively, it can be viewed as given by an oracle answering on a query by letting the searchers know in which component is the fugitive. A nondeterministic search program is thus a nondeterministic program that takes as input a graph G and an integer $k \geq 1$, and returns an ordered sequence of search steps, each of them being one of the following three operations:

- place a searcher at $v \in V(G)$;
- remove a searcher from $v \in V(G)$;
- query the oracle.

Of course, the program must satisfy the same three constraints as any (deterministic) search program.

A nondeterministic search program *wins* the game against a fugitive F if there exists an execution of the program which results in clearing the node currently occupied by the fugitive.

We are interested in the tradeoff between the number of searchers used by a search program and the number of query steps performed by the program. For any $q \geq 0$, a *q-limited* nondeterministic search program is a nondeterministic search program that performs at most q query steps. Therefore, a q-limited nondeterministic search program wins the game against a fugitive F if it can catch the fugitive by querying at most q times an oracle that returns the connected component where is currently hidden the fugitive.

Definition 2. *The q-limited nondeterministic search number $s_q(G)$ of a graph G, or simply the q-limited search number of G, is the minimum number of searchers required by a q-limited nondeterministic search program to win against any fugitive in G.*

Therefore, the 0-limited nondeterministic search number of a graph is its search number, i.e., $s_0(G) = s(G)$. We will prove in the next section that the ∞-limited nondeterministic search number $s_\infty(G)$ of a graph G is equal to its visible-search number.

3 Branched Treewidth vs. Limited Graph Searching

In this section, we show that the q-branched treewidth and the q-limited search number are actually the same, up to 1. This equality will be later shown to be useful for the design of algorithms and for the computation of combinatorial bounds.

Theorem 1. *For any $q \geq 0$, for any graph G, $\mathbf{tw}_q(G) = \mathbf{s}_q(G) - 1$.*

Proof. Let $(T, \mathcal{X}, r)$ be a q-branched decomposition of width k. For a node i of T let $d(i)$ be the set of descendants of i in T. We define the search program of $k+1$ searchers querying the oracle at most q times as follows. Initially the searchers are placed on the vertices of X_r. Suppose that, at some step of searching, for some node i of T the searchers are on vertices X_i and the set of contaminated vertices is $\cup_{j \in d(i)} X_j \setminus X_i$. Note that if i is a leaf, G is cleared. Let i be a non-leaf node of T. Depending on the number of children of i we choose different strategy for the searchers.

Case A. i has only one child l. We remove first the searchers from $X_i \setminus X_l$ and then place searchers to X_l. Since the cardinality of X_i and X_l is at most

$k+1$ we use at most $k+1$ searchers. By properties of tree decompositions, for every contaminated vertex $v \in \cup_{j \in d(i)} X_j \setminus X_i$ and every cleared vertex u, every (u,v)-path contains a vertex from $X_i \cap X_l$. Thus after removing the searchers from $X_i \setminus X_l$ no recontamination occurs and we arrive at the situation when the searchers are at X_l and the set of contaminated vertices is $\cup_{j \in d(l)} X_j \setminus X_l$.

Case B. i has more than one child. In this case the searchers query the oracle. Let C be the connected component of $G[\cup_{j \in d(i)} X_j \setminus X_i]$ returned by the oracle. Then there is a unique child l of i such that $C \cap X_l \neq \emptyset$. We remove the searchers from $X_i \setminus X_l$ and then place searchers to X_l. Again, in this case we arrive at the situation when the searchers are at X_l and the set of contaminated vertices is $\cup_{j \in d(l)} X_j \setminus X_l$.

Eventually, the searchers reach the situation when they are placed on the vertices X_i where i is leaf of T and thus the whole graph is cleared. The number of searchers used is at most $\max_{j \in V(T)} |X_j| \leq k+1$. Since for every leaf i, the path from r to i contains at most q branches, the case B occurs at most q times, thus the searchers query the oracle at most q times. Hence $\mathbf{s}_q(G) \leq \mathbf{tw}_q(G) + 1$.

We prove $\mathbf{tw}_q(G) \leq \mathbf{s}_q(G) - 1$ by proving a slightly stronger claim.

Claim. Suppose that there is a search program of $k+1$ searchers on G with at most q queries and such that, initially, searchers are placed on vertices $X \subseteq V(G)$. Then there is a q-branched tree decomposition $(T, \mathcal{X}, r)$ with $X_r = X$ and of width $\leq k$.

To prove the claim we proceed by induction on q. For $q = 0$, the required path decomposition $P = (X_0, X_1, \ldots, X_m)$ is constructed by taking $X_0 = X$, and, for $i \geq 1$, X_i to be the vertex set occupied by searchers after the ith step. To check that P is the path decomposition we observe that every vertex should be at some step occupied by a searcher and thus is contained in some node of P. Every pair of adjacent vertices $\{u, v\}$ is contained in some node of P because otherwise fugitive can avoid capture by choosing u or v at every step of searching. The third property of tree decompositions follows from the constraints 2 and 3 of graph searching.

Let $q \geq 1$ and suppose that for all $q' < q$, the claim is correct. Consider a winning search program of $k+1$ searchers with at most q queries to the oracle. Suppose that the first time the searchers query the oracle occurs at step $t \geq 0$. Let X be the set of vertices occupied by the searchers and S be the cleared vertices at this step. Let $G_1, G_2, \ldots, G_p$ be the subgraphs of G obtained from the connected components of $G \setminus S$ by adding X. Each of these subgraphs is searchable by $k+1$ searchers with at most $q-1$ queries with the search starting from X. By induction assumption, for each $1 \leq i \leq p$, there is a rooted tree decomposition $(T^{(i)}, \mathcal{Y}^{(i)}, r_i)$ of G_i with at most $q-1$ branches and with the root r_i of $T^{(i)}$ satisfies $Y_{r_i} = X$.

We construct a tree decomposition $(T, \mathcal{Y}, r)$ of G as follows. Let $X_1, \ldots, X_t$ be the vertices occupied by the searchers at the first t steps of searching. In particular $X_t = X$. We construct the path decomposition $(X_1, \ldots, X_t)$ rooted at X_1. Then we add the tree decompositions $(T^{(i)}, \mathcal{Y}^{(i)}, r_i)$, $1 \leq i \leq p$, and identify every r_i to the node t of the path decomposition. The resulting tree decomposition is a q-branched tree decomposition of width $\leq k$. □

4 Exact Exponential Algorithm

For any $q \geq 0$, the decision problem that takes as input a graph G and an integer $k \geq 1$, and returns whether or not $\mathbf{tw}_q(G) \leq k$, is NP-complete. Indeed, it is known [2] that the problem of deciding whether $\mathbf{tw}(G) \leq k$ is NP-complete, even when restricted to the co-bipartite graphs, i.e., the complements of bipartite graphs. Since, for any co-bipartite graph G, $\mathbf{tw}(G) = \mathbf{pw}(G)$, the NP-completeness of deciding $\mathbf{tw}_q(G) \leq k$ follows from the fact that $\mathbf{tw}(G) \leq \mathbf{tw}_q(G) \leq \mathbf{pw}(G)$ for any $q \geq 0$. It is known [5] that $\mathbf{tw}$ can be approximated up to multiplicative factor $O(\log \mathbf{tw})$, in polynomial time. (This bound has been recently improved to $O(\sqrt{\log \mathbf{tw}})$, cf. [8].) However, for pathwidth, no approximation algorithm is known (except by combining the ones for treewidth with the fact that $\mathbf{pw}(G) \leq O(\log n) \cdot \mathbf{tw}(G)$ for any n-node graph G). On the other hand, as mentioned in the introduction, several exact (exponential) algorithms have been designed for treewidth, and for pathwidth as well. In this section, we show that one can design an exact algorithm that applies to q-branched treewidth, for all $q \geq 0$. This algorithm uses the correspondence between q-branched treewidth and q-limited search number.

Theorem 2. *There exists an algorithm that, for any n-node graph G, computes $\mathbf{tw}_q(G)$ and an optimal q-branched tree decomposition of G, in time $O(2^n n \log n)$.*

Proof. Based on Theorem 1, we design an algorithm that computes $\mathbf{s}_q(G)$, and an optimal q-limited search strategy for G. This startegy can be then transformed into a q-branched tree decomposition using the arguments in the proof of Theorem 1. Let G be a graph, and fix $k \geq 1$. We define the configuration digraph H as follows.

$$V(H) = \{S \subseteq V(G) \text{ s.t. } |\delta(S)| \leq k\}.$$

A set S of clear vertices for which $|\delta(S)| > k$ are unreachable by a search program using k searchers, and thus it is not included in $V(H)$. The nodes in H are called H-configurations, to avoid confusion with the configurations of the search game. The edge-set of H has two types of directed edges: *place* edges, and *query* edges. A place edge, or simply p-edge, is an edge (S, S') where $|\delta(S)| < k$ and $S' = S \cup \{v\}$, $v \notin S$. Clearly, a p-edge corresponds to the placement of a searcher at node v. A query edge, or simply q-edge, is an edge (S, S') where $S' = G \setminus C$ and C is a connected component of $G \setminus S$. It is assumed that there is a q-edge $(S, G \setminus C)$ only if $G \setminus S$ has at least two connected components (i.e., there is no self-loop in H). Thus, a q-edge $(S, G \setminus C)$ corresponds to a query to the oracle that returns C. The objective of our algorithm is to find a path in the configuration digraph H from $S = \emptyset$ to $S = V(G)$ that can be put in correspondence with a search strategy.

For the purpose of finding such a path, we label every node of $V(H)$ by a nonnegative integer. The labeling starts from the H-configuration $V(G)$ and goes backwards. The H-configuration $V(G)$ is labeled 0. All the other H-configurations are labeled ∞. All the H-configurations without any outgoing edge are declared *finished*. (In particular $V(G)$ is finished.) All the other H-configurations are declared *pending*. We proceed as long as there is at least one pending H-configuration S satisfying one of the two following conditions:

Case 1. S has an outgoing p-edge e connecting to an H-configuration S' that is finished. (Informally, this case occurs if the labeling has not yet considered the game configuration $(S, \delta(S), v)$, $v \notin S$, from which the next search step is: place a searcher at $S' \setminus S$.)

Case 2. S has all its outgoing q-edges $e_1, \ldots, e_d$ connecting to H-configurations $S'_1, \ldots, S'_d$ that are finished. (Informally, this case occurs if the labeling has not yet considered the game configuration $(S, \delta(S), v)$, $v \notin S$, from which the next search step is: query the oracle.)

In case 1, we update the label of S by:

$$\text{label}(S) = \min\{\text{label}(S), \text{label}(S')\}$$

and the edge e is removed from H. In case 2, we update the label of S by:

$$\text{label}(S) = \min\{\text{label}(S), 1 + \max_{1 \leq i \leq d} \text{label}(S'_i)\}$$

and all the edges $e_1, \ldots, e_d$ are removed from H. In both cases, if the pending H-configuration S has no more outgoing edges because of the edge(s) removal, then S is declared finished.

Claim. The labeling process terminates.

To prove the claim, notice that H is a directed acyclic graph because every edge goes from an H-configuration S to another H-configuration S' with $|S'| > |S|$ (recall that we did not allowed self-loops in H). Removing edges from H preserves this property. Therefore every node becomes eventually finished and thus the labeling process terminates.

Claim. The H-configuration $\emptyset$ is labeled $q < \infty$ if and only if q is the smallest number of queries required to clear G using $\leq k$ searchers. The H-configuration $\emptyset$ is labeled ∞ if one cannot clear G using $\leq k$ searchers, independently of the number of queries to the oracle.

We prove that claim by proving a slightly more general result: for any H-configuration $S \neq V(G)$, S is finished and labeled $\text{label}(S) \leq q \neq \infty$ if and only if one can clear G starting from S with $\leq k$ searchers and performing $\leq q$ queries. By starting from S, it is meant that we assume an initial configuration of the search game in which S is clear, $|\delta(S)|$ searchers are placed at the vertices of $\delta(S)$, and the fugitive is at some vertex of $G \setminus S$. We proceed by induction on $q = \text{label}(S)$.

If $q = 0$, then there is a path P in H from S to $V(G)$ using only p-edges. Let $(S', S'') \in P$, with $S'' = S' \cup \{v\}$. The portion of the search strategy corresponding to that edge consists in removing one by one all searchers occupying vertices $\notin \delta(S')$, and placing a searcher at v. Hence one can catch the fugitive without performing queries by starting from S and following the edges of P until one reaches the configuration $V(G)$. Conversely, if one can clear G starting from S with $\leq k$ searchers and performing no queries, then there is a path in H from S to $V(G)$ composed on only p-edges. These edges are defined by placement steps in the search strategy.

Assume now that the result holds for q, and consider S such that label$(S) = q+1$. We define a *good* edge as a p-edge (S', S'') such that label$(S) = $ label$(S') = $ label(S''). From S, start traveling in H by using good edges only, until one reaches a configuration S^* with

$$\text{label}(S) = \text{label}(S^*) = 1 + \max_{i=1,\ldots,d} \text{label}(S_i^*)$$

where the edges (S^*, S_i^*), $i = 1, \ldots, d$, are all the q-edges out-going from S^*. This configuration S^* exists because (1) a good edge (S', S'') satisfies $|S''| > |S'|$, and (2) label$(S') = $ label$(S'') = $ label$(S) < \infty$. Therefore, if a configuration S^* as specified above would not be met, then, by (1) an H-configuration with outdegree 0 would eventually be reached, and by (2) this H-configuration could only be $V(G)$ since otherwise its label would be ∞. Since label$(S) = q + 1 > 0$, by induction this would contradict the fact that there is no path from S to $V(G)$ composed of p-edges only. So S^* is well defined.

For all $i = 1, \ldots, d$, label$(S_i^*) \leq q$. Therefore, by induction hypothesis, one can clear G starting from any S_i^* using $\leq k$ searchers, and quering $\leq q$ times the oracle. The search strategy from S starts by performing place and remove steps according to the path in H from S to S^*. The search then queries the oracle at S^*, and gets into one of the configurations S_i^*. The rest of the search follows from the induction hypothesis.

Conversely, assume that one can clear G starting from S with $\leq k$ searchers and performing $q + 1$ queries. Consider a corresponding search strategy in G, and assume that the first query to the oracle occurs at step t. The $t - 1$ first steps can be put in correspondence with a path P in H starting at S, and that contains good edges only. Let (S^*, X^*, v^*) be the configuration of the game after step $t - 1$. P connects S with the H-configuration S^*. The query at step t corresponds to the outgoing q-edges (S^*, S_i^*), $i = 1, \ldots, d$, of S^*. From each of the S_i^*s, the search proceeds with at most q queries. Hence, by induction, label$(S_i^*) \leq q$. Therefore, label$(S^*) \leq q+1$. Since P is a path of good edges, we get label$(S) = $ label$(S^*) \leq q + 1$.

For each k, the running time of the labeling procedure is linear in the number of edges of H, which is $O(2^n n)$. Thus by binary search we can find the search number and $\mathbf{tw}_q(G)$ in $O(2^n n \log n)$. □

5 Bounding the Nondeterminism

In this section, we compute a lower bound on the number of nondeterministic steps a search program must perform in a graph G in order to clear the graph with the minimum possible number of searchers, i.e., $\mathbf{tw}(G)+1$ searchers.

Theorem 3. *For any $q \geq 1$, for any graph G, $\mathbf{tw}_{q-1}(G) \leq 2\,\mathbf{tw}_q(G)$.*

Due to the space restrictions the proof of this theorem is omitted.

Note that the bound of Theorem 3 is tight, as witnessed by the graphs consisting in a complete binary tree (all non-leaf vertices, including the root, are of degree 3) of depth q, in which every vertex u is replaced by a complete graph K_u of k vertices, and every edge $\{u,v\}$ is replaced by a perfect matching between the two complete graphs K_u and K_v. Theorem 3 has important corollaries:

Corollary 1. *For any graph G, the smallest $q \geq 0$ such that $\mathbf{tw}_q(G) = \mathbf{tw}(G)$ satisfies $\tau(G) \geq \log_2(\mathbf{pw}(G)/\mathbf{tw}(G))$.*

Remark. There exist n-node graphs G such that $\mathbf{tw}(G) = \mathbf{pw}(G)+1$ and the greatest $q \geq 0$ such that $\mathbf{tw}_q(G) = \mathbf{pw}(G)$ satisfies $\pi(G) \geq \Omega(n)$.

Rephrasing Corollary 1, we get:

Corollary 2. *For any graph G, the smallest number of nondeterministic steps of a nondeterministic search program that clears G with $\mathbf{tw}(G)+1$ searchers is at least $\log_2(\mathbf{pw}(G)/\mathbf{tw}(G))$.*

6 Conclusion

In this paper, we introduced a nondeterministic graph searching game, with a controlled amount of nondeterminism. The objective of this concept was to unify pathwidth and treewidth, at least as far as the design of algorithms, and the computation of combinatorial bounds in concerned. We believe that this is a promising field of investigations, as illustrated by the design of an exact algorithm for q-branched treewidth, valid for any $q \geq 0$. Still, a lot of work has to be done before being able to design common tools for both pathwidth and treewidth.

In particular, it would be particularly interesting to design a polynomial-time algorithm computing the q-limited search number (or equivalently the q-branched treewidth) of trees. As far as algorithm design is concerned, it would also be quite interesting to design an $O(c^n)$-time exact algorithm for the q-branched treewidth of arbitrary graphs, with $c < 2$. Such an algorithm is known [10] in the case of treewidth (i.e., $q = \infty$), but not for pathwidth (i.e., $q = 0$). Last but not least, it is known that, for node-search (i.e., 0-limited graph searching) and visible-search (i.e., ∞-limited graph searching), removing constraint 3 of the search program does not enable to decrease the number of searchers. It would be important to know whether this is true for any $q \geq 0$, i.e., whether or not "recontamination helps" for q-limited graph searching, for any $q \geq 0$.

References

1. E. AMIR, *Efficient approximation for triangulation of minimum treewidth*, in Uncertainty in Artificial Intelligence: Proceedings of the Seventeenth Conference (UAI-2001), San Francisco, CA, 2001, Morgan Kaufmann Publishers, pp. 7–15.
2. S. ARNBORG, D. G. CORNEIL, AND A. PROSKUROWSKI, *Complexity of finding embeddings in a k-tree*, SIAM J. Algebraic Discrete Methods, 8 (1987), pp. 277–284.
3. D. BIENSTOCK, *Graph searching, path-width, tree-width and related problems (a survey)*, DIMACS Ser. in Discrete Mathematics and Theoretical Computer Science, 5 (1991), pp. 33–49.
4. H. L. BODLAENDER, *A partial k-arboretum of graphs with bounded treewidth*, Theoret. Comput. Sci., 209 (1998), pp. 1–45.
5. V. BOUCHITTÉ, D. KRATSCH, H. MÜLLER, AND I. TODINCA. *On treewidth approximations*. Discrete Applied Mathematics 136(2-3), pages 183-196, 2004.
6. N. D. DENDRIS, L. M. KIROUSIS, AND D. M. THILIKOS, *Fugitive-search games on graphs and related parameters*, Theor. Comp. Sc., 172 (1997), pp. 233–254.
7. J. A. ELLIS, I. H. SUDBOROUGH, AND J. TURNER, *The vertex separation and search number of a graph*, Information and Computation, 113 (1994), pp. 50–79.
8. U. FEIGE, M. HAJIAGHAYI, J. LEE . *Improved approximation algorithms for minimum-weight vertex separators*. In 37th ACM Symposium on Theory of Computing (STOC 2005).
9. R. G. DOWNEY AND M. R. FELLOWS, *Parameterized Complexity*, Springer-Verlag, New York, 1999.
10. F. V. FOMIN, D. KRATSCH, AND I. TODINCA, *Exact algorithms for treewidth and minimum fill-in*, Proceedings of the 31st International Colloquium on Automata, Languages and Programming (ICALP 2004), LNCS vol. 3142, Springer, Berlin, 2004, pp. 568–580.
11. J. GOLDSMITH, M. LEVY, AND M. MUNHENK, *Limited Nondeterminism*. SIGACT News, Introduction to Complexity Theory Column 13, June 1996.
12. M. HELD AND R. KARP, *A dynamic programming approach to sequencing problems*, J. Soc. Indust. Appl. Math., 10 (1962), pp. 196–210.
13. L. M. KIROUSIS AND C. H. PAPADIMITRIOU, *Searching and pebbling*, Theor. Comp. Sc., 47 (1986), pp. 205–218.
14. F. S. MAKEDON, C. H. PAPADIMITRIOU, AND I. H. SUDBOROUGH, *Topological bandwidth*, SIAM J. Alg. Disc. Meth., 6 (1985), pp. 418–444.
15. F. S. MAKEDON AND I. H. SUDBOROUGH, *On minimizing width in linear layouts*, Disc. Appl. Math., 23 (1989), pp. 243–265.
16. N. ROBERTSON AND P. D. SEYMOUR, *Graph minors. II. Algorithmic aspects of tree-width*, J. Algorithms, 7 (1986), pp. 309–322.
17. N. ROBERTSON AND P. D. SEYMOUR, *Graph minors. X. Obstructions to tree-decomposition*, J. Combin. Theory Ser. B, 52 (1991), pp. 153–190.
18. P. SEYMOUR AND R. THOMAS, *Graph searching and a min-max theorem for tree-width*, J. Combin. Theory Ser. B, 58 (1993), pp. 22–33.
19. P. SEYMOUR AND R. THOMAS, *Call routing and the ratcatcher*, Combinatorica, 14 (1994), pp. 217–241.
20. R.E. TARJAN AND A.E. TROJANOWSKI, *Finding a maximum independent set*, SIAM J. Computing, 6 (1977), pp. 537–546.
21. G. J. WOEGINGER, *Exact algorithms for NP-hard problems: a survey*, in Combinatorial Optimization: "Eureka, you shrink", LNCS vol. 2570, Springer, Berlin, 2003, pp. 185–207.

Goals in the Propositional Horn$^\supset$ Language Are Monotone Boolean Circuits*

J. Gaintzarain, M. Hermo, and M. Navarro

Dpto de L.S.I., Facultad de Informática,
P.O. Box 649, 20080-San Sebastián, Spain

Abstract. $Horn^\supset$ is a logic programming language which extends usual $Horn$ clauses by adding intuitionistic implication in goals and clause bodies. This extension can be seen as a form of structuring programs in logic programming. Restricted to the propositional setting of this language, we prove that any goal in $Horn^\supset$ can be translated into a monotone Boolean circuit which is linear in the size of the goal.

1 Introduction

In logic programming, some approaches for extending $Horn$ clauses consider to incorporate into the language a new implication symbol, $\supset$, with the aim of structuring logic programs in some blocks with local clauses [2,5,7,8,12,13,14,15,16]. These extensions can also be seen as a sort of inner modularity in logic programming (see [4] for a survey on modularity).

We consider a particular extension named $Horn^\supset$. This programming language has been formally studied in [8,7,2,10,16]. In [2] a natural extension of classical first order logic $\mathcal{FO}$ with the intuitionistic implication ($\supset$), named $\mathcal{FO}^\supset$, was presented as the underlying logic of the programming language $Horn^\supset$.

Model semantics of $\mathcal{FO}^\supset$ is based on Kripke structures consisting of a non-empty partially ordered set of worlds, each world associated to an interpretation. However, to deal with $Horn^\supset$ programs, Kripke structures can be restricted to those with (1) Herbrand interpretations associated to their worlds, (2) a unique minimal world and (3) closure with respect to superset. Moreover, each interpretation I univocally determines a Kripke structure (formed with all the supersets of I) and, conversely, each Kripke structure satisfying conditions (1), (2) and (3) is univocally determined by (the interpretation associated to) its minimal world.

Other "good properties" that verify $Horn$ clauses (as a programming language) with respect to its underlying logic $\mathcal{FO}$ are also conserved by $Horn^\supset$ clauses with respect to $\mathcal{FO}^\supset$: each program has a canonical model, the operational semantics is an effective subcalculus of a complete calculus for $\mathcal{FO}^\supset$ and the goals satisfied in the canonical model are the goals that can be derived from the program in such calculus. The formalization about what are "good properties

---

* This work has been partially supported by the projects TIN2004-07925-C03-03 and UPV 00141.226-E-15965/2004.

of a programming language" is borrowed from [11] and a complete calculus for $\mathcal{FO}^\supset$ is introduced in [10].

More related to implementation issues, a way to proceed is to translate the extended logic programs into the language of some well-known logic ([15,7,3,16]). In particular, [16] introduces a translation preserving the operational semantics from $Horn^\supset$ programs into $Horn$ programs. To be efficient, this translation needs to obtain the $Horn$ program in an extended signature with new predicate symbols. In fact, if we want to preserve the model semantics any translation from a $Horn^\supset$ program into an equivalent $Horn$ program obtains, in general, a $Horn$ program with an exponential number of clauses [6]. Our aim is to study possible correct and efficient translations from $Horn^\supset$ programs into some representation type that, preserving the model semantics, allows a suitable implementation.

It is well-known that boolean circuits are data structures for representing boolean functions. In general, the description of a boolean formula should be rather short and efficient; support the evaluation and manipulation of the function; make particular properties of the function visible; suggest ideas for a technical realization. The boolean circuit is a representation type which satisfies all the properties above, but mainly the first one: the fact that the outdegree of the gates in the boolean circuits can be greater than 1, often allows very compact representation.

The study made in this paper shows, in the propositional setting of $Horn^\supset$ language, that clauses and goals can be represented efficiently by boolean circuits. The paper is organized as follows: In Section 2 the programming language $Horn^\supset$ (restricted to the propositional case) is introduced. In Section 3 some preliminary notions about boolean circuits are given. The core of the paper is Section 4 where we introduce a translation from $Horn^\supset$ goals to monotone circuits and we prove that this transformation is correct and efficient. We conclude, in Section 5, by summarizing our results and by showing further work to do.

2 The Programming Language $Horn^\supset$

In this section we introduce the programming language $Horn^\supset$ by showing its syntax and its model semantics. Although the language is in general a first order language (see [8,2]), in this paper we shall restrict our presentation only to this language in the propositional setting.

2.1 The Syntax

The syntax is an extension of the (propositional) $Horn$ clause language by adding the intuitionistic implication $\supset$ in goals and clause bodies. Let Σ be a fixed set of propositional variables (or signature). The clauses, named D, and the goals, named G, are recursively defined as follows (where v stands for any propositional variable in Σ):

$$G ::= v \mid G_1 \wedge G_2 \mid D \supset G \qquad D ::= v \mid G \to v \mid D_1 \wedge D_2$$

A *Horn*$^\supset$ program is a finite set (or conjunction) of clauses. The main difference with respect to *Horn* clauses is the use of a "local" clause set D in goals of the kind $D \supset G$ (and therefore also in clause bodies).

Example 1. The following set with three clauses is a *Horn*$^\supset$ program over signature $\Sigma = \{a, b, c, d\}$

$$\{((b \to c) \supset c) \to a, \ b, \ ((a \land (b \to c)) \supset (((b \to c) \land (a \to d)) \supset a)) \to d\}$$

The second clause is simply b. The first and the third program clauses are of the form $G \to v$. In the first clause, the goal G is $(b \to c) \supset c$. That is, it contains a local set with one clause. In the third clause, the goal G is of the form $D_1 \supset (D_2 \supset G_3)$, where $D_1 = a \land (b \to c)$ and $D_2 = (b \to c) \land (a \to d)$ are both local sets with two clauses, and $G_3 = a$.

2.2 The Model Semantics

Definition 1. *Given a signature Σ, the model semantics for the propositional Horn$^\supset$ language is given by the set of all Σ-interpretations* $\underline{\text{Mod}}(\Sigma) = \{I | I \subseteq \Sigma\}$.

In the underlying logic of propositional *Horn*$^\supset$ language, well-formed formulas are built from propositional variables in Σ, using constants (*True* and *False*), classical connectives ($\neg, \land, \lor,$ and $\to$) and the intuitionistic implication ($\supset$). The satisfaction relation (or forcing relation) $\Vdash_\Sigma$ (or simply $\Vdash$ if there is no confusion about the signature) between an interpretation I and a formula φ in the underlying logic is given below. Clauses and goals are particular formulas in this logic.

Definition 2. *Let $I \in \underline{\text{Mod}}(\Sigma)$ and φ a formula. We say that*

(a) *I is a model of φ (or φ is forced in I) if $I \Vdash \varphi$*
(b) *The binary forcing relation $\Vdash$ is inductively defined as follows:*
    $I \not\Vdash False$
    $I \Vdash v$ iff $v \in I$ for $v \in \Sigma$
    $I \Vdash \neg \varphi$ iff $I \not\Vdash \varphi$
    $I \Vdash \varphi \land \psi$ iff $I \Vdash \varphi$ and $I \Vdash \psi$
    $I \Vdash \varphi \lor \psi$ iff $I \Vdash \varphi$ or $I \Vdash \psi$
    $I \Vdash \varphi \to \psi$ iff if $I \Vdash \varphi$ then $I \Vdash \psi$
    $I \Vdash \varphi \supset \psi$ iff for all $J \subseteq \Sigma$ such that $I \subseteq J$: if $J \Vdash \varphi$ then $J \Vdash \psi$

Note that the satisfaction of a formula $\varphi \supset \psi$ depends on the satisfaction of ψ in all the interpretations J containing I that satisfy φ. If the formula does not contain the connective $\supset$, then the usual satisfaction relation in classical logic, denoted $\models$, coincides with $\Vdash$.

Example 2. Let φ be the formula (in this case a goal) $((a \land c) \to b) \supset (c \land b)$. $I \Vdash \varphi$ for $I = \{a, b, c\}$, $I = \{a, c\}$ and $I = \{b, c\}$. $I \not\Vdash \varphi$ for $I = \{a, b\}$, $I = \{a\}$, $I = \{b\}$, $I = \{c\}$ and $I = \emptyset$. Note, for instance, that $\{a, b\} \Vdash (a \land c) \to b$ and $\{a, b\} \not\Vdash (c \land b)$.

2.3 Persistency and Equivalence of Formulas

$\underline{\mathsf{Mod}}(\Sigma)$ is partially ordered by the inclusion relation. The forcing relation does not behave monotonically with respect to this relation.
For instance, $a \to b$ is forced in the interpretation $I = \emptyset$ but it is not forced in $J = \{a\}$. We say that a formula is *persistent* whenever the forcing relation behaves monotonically for it.

Definition 3. *A formula φ is* persistent *when for each interpretation I, if $I \Vdash \varphi$ then $J \Vdash \varphi$ for any interpretation J such that $I \subseteq J$.*

Proposition 1. *[2] Any $v \in \Sigma$ is persistent. Any formula $\varphi \supset \psi$ is persistent. If φ and ψ are persistent then $\varphi \vee \psi$ and $\varphi \wedge \psi$ are persistent.*

Proof. For variables and formulas of the form $\varphi \supset \psi$ the property is a trivial consequence of the forcing relation (Definition 2). The other two cases are easily proved, by induction, using the forcing relation definition for $\wedge$ and $\vee$. ∎

From this proposition we obtain the two following results. The second one is a consequence of the former and can be proved by induction on definition of D.

Corollary 1. *Any goal G is a persistent formula.*

Corollary 2. *For any clause D and interpretations I_1, I_2, if $I_1 \Vdash D$ and $I_2 \Vdash D$ then $I_1 \cap I_2 \Vdash D$.*

Definition 4. *Two formulas φ and ψ are* (semantically) equivalent *if both have the same meaning in each I in $\underline{\mathsf{Mod}}(\Sigma)$. In other words, if both are forced in the same interpretations.*

The properties given below will be useful in next sections.

Proposition 2. *The formula $True \supset G$ is equivalent to the formula G.*

Proof. $I \Vdash True \supset G \Leftrightarrow$ for all $J \supseteq I$, $J \Vdash G \Leftrightarrow I \Vdash G$. The last step uses the persistency of G. ∎

Proposition 3. *The formula $((G_1 \to v) \wedge D) \supset G_2$ is equivalent to the formula $((D \supset G_1) \to v) \supset (D \supset G_2)$.*

Proof. ($\Rightarrow$) If $I \nVdash ((D \supset G_1) \to v) \supset (D \supset G_2)$ then there exists J such that $J \supseteq I$, $J \Vdash (D \supset G_1) \to v$ and $J \nVdash D \supset G_2$. Moreover, there exists J_1 such that $J_1 \supseteq J$, $J_1 \Vdash D$ and $J_1 \nVdash G_2$. We distinguish two cases:
- If $v \in J$ also $v \in J_1$. Then $J_1 \Vdash (G_1 \to v) \wedge D$ and $J_1 \nVdash G_2$. Therefore $I \nVdash ((G_1 \to v) \wedge D) \supset G_2$
- If $v \notin J$ then $J \nVdash D \supset G_1$. That is, there exists J_2 such that $J_2 \supseteq J$, $J_2 \Vdash D$ and $J_2 \nVdash G_1$. By using Corollaries 1 and 2, it is easy to prove that the interpretation $J_3 = J_1 \cap J_2$ verifies: $J_3 \Vdash D$, $J_3 \nVdash G_1$ and $J_3 \nVdash G_2$. Then $J_3 \Vdash (G_1 \to v) \wedge D$, $J_3 \nVdash G_2$ and $J_3 \supseteq I$. Therefore $I \nVdash ((G_1 \to v) \wedge D) \supset G_2$.

($\Leftarrow$) If $I \not\Vdash ((G_1 \to v) \wedge D) \supset G_2$ then there exists J such that $J \supseteq I$, $J \Vdash (G_1 \to v)$, $J \Vdash D$ and $J \not\Vdash G_2$. Again two cases:
- If $v \in J$ then trivially $J \Vdash (D \supset G_1) \to v$ and $J \not\Vdash D \supset G_2$. Therefore $I \not\Vdash ((D \supset G_1) \to v) \supset (D \supset G_2)$.
- If $v \notin J$ then $J \not\Vdash G_1$. Since $J \Vdash D$ then $J \not\Vdash D \supset G_1$ and from here trivially $J \Vdash (D \supset G_1) \to v$. Also we have $J \not\Vdash D \supset G_2$. Therefore $I \not\Vdash ((D \supset G_1) \to v) \supset (D \supset G_2)$. ∎

3 Boolean Circuits

In this section we revise from [17] the notion of Boolean circuit.

3.1 The Syntax

A Boolean circuit over signature Σ is a graph $C = (V, E)$, where the nodes $V = \{1, 2, \ldots, n\}$ are called the *gates* of C. Graph C has a rather special structure. First, there are no cycles in the graph, so we can assume that all edges are of the form (i, j) where $i < j$. All nodes in the graph have indegree equal to $0, 1$ or 2. Also, each gate $i \in V$ has a sort $s(i)$ associated with it, where $s(i) \in \{1, 0, \wedge, \vee, \neg\} \cup \Sigma$.

If $s(i) \in \{1, 0\} \cup \Sigma$, then the indegree of i is 0, that is, i must have no incoming edges. Gates with no incoming edges are called the *inputs* of C. If $s(i) = \neg$ then i has indegree one. If $s(i) \in \{\wedge, \vee\}$, then the indegree of i must be two. Finally, node n (the largest numbered gate in the circuit, which necessarily has no outgoing edges) is called the *output gate* of the circuit.

Circuits without gates of the sort $\neg$ are called *monotone Boolean circuits*.

3.2 The Semantics

Given a signature Σ, each $I \subseteq \Sigma$ can be seen as a Σ-interpretation where, for every $v \in \Sigma$, $I(v) = True$ if and only if $v \in I$.

The semantics of a circuit $C = (V, E)$ specifies a truth value $I(C)$ for each interpretation $I \subseteq \Sigma$. The *truth value of gate* $i \in V$, $I(i)$, is defined by induction as follows: If $s(i) = 1$ then $I(i) = True$ and similarly if $s(i) = 0$ then $I(i) = False$. If $s(i) \in \Sigma$ then $I(i) = I(s(i))$. If $s(i) = \neg$ then there is a unique gate $j < i$ such that $(j, i) \in E$. By induction we know $I(j)$, and then $I(i) = True$ if and only if $I(j) = False$. If $s(i) = \vee$ then there are two edges (j, i) and (j', i) entering i. $I(i)$ is then $True$ if and only if at least one of $I(j), I(j')$ is $True$. If $s(i) = \wedge$, then $I(i) = True$ if and only if both $I(j), I(j')$ are $True$, where (j, i) and (j', i) are the incoming edges. Finally, the *value* of the circuit $I(C)$ is $I(n)$, where n is the output gate.

Given a Boolean circuit C (over Σ), a Σ-interpretation I is a Σ-model of C, denoted $I \models_\Sigma C$ or $I \models C$ for short, if the value $I(C)$ is $True$.

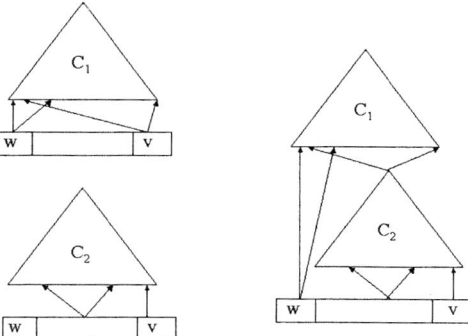

Fig. 1.

3.3 Notation and Properties

Given two circuits $C_1 = (V_1, E_1)$ and $C_2 = (V_2, E_2)$ over the same signature Σ and given $v \in \Sigma$, the new circuit $C_1|_v^{C_2}$ is obtained by changing C_2 for v in C_1. That is, $C_1|_v^{C_2}$ is a pair (V, E) which is the result of combining C_1 and C_2 as follows: V is an adequate enumeration for the union of V_1 and V_2. The edges of the new circuit are the union of E_1 and E_2, according to such enumeration, except those outgoing edges from v in E_1 that now come out from the output gate of C_2. Figure 1 shows $C_1|_v^{C_2}$ from two given circuits C_1 and C_2.

Note that many circuits can compute the same Boolean function, but we are interested in those circuits that have minimum size. Therefore we can assume that the input gates only appear once in the Boolean circuits.

We use some properties on the circuits that are described in the next lemmas. From now on, we consider Boolean circuits over signature Σ.

Lemma 1. *Given two monotone Boolean circuits C_1 and C_2, $I \subseteq \Sigma$, and $v \in \Sigma$. The following holds:*

a. *(Monotonicity)* $I \models C_1 \implies$ *for every* $J \supseteq I$, *it holds* $J \models C_1$
b. $I \models C_1 \implies I \models C_1|_v^{C_2 \vee v}$

Lemma 2. *Given a Boolean circuit C, $I \subseteq \Sigma$, and $v \in \Sigma$. The following holds:*

a. $I \cup \{v\} \models C \iff I \models C|_v^1$
b. $I - \{v\} \models C \iff I \models C|_v^0$

4 A Translation from Goals to Monotone Circuits

We present here how to transform a goal into a monotone Boolean circuit by means of the following function f. Its definition is given by induction on definition

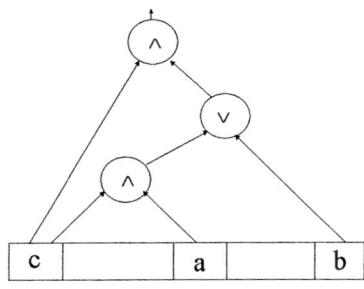

Fig. 2.

of G (on the three cases v, $G_1 \wedge G_2$, and $D \supset G$), but splitting as well the third case $D \supset G$ depending on D.

$$f(G) = \begin{cases} v & \text{if } G = v \quad (1) \\ f(G_1) \wedge f(G_2) & \text{if } G = G_1 \wedge G_2 \quad (2) \\ f(G_2)|_v^1 & \text{if } G = v \supset G_2 \quad (3) \\ f(G_2)|_v^{f(G_1) \vee v} & \text{if } G = (G_1 \to v) \supset G_2 \quad (4) \\ f(D \supset G_2)|_v^1 & \text{if } G = (v \wedge D) \supset G_2 \quad (5) \\ f(D \supset G_2)|_v^{f(D \supset G_1) \vee v} & \text{if } G = ((G_1 \to v) \wedge D) \supset G_2 \quad (6) \end{cases}$$

Figure 2 shows the transformation of the goal $((a \wedge c) \to b) \supset (c \wedge b)$ by f.

This transformation is correct, in the sense that both formulas are equivalent, and efficient, since it obtains a circuit whose size is linear with respect to the goal. In the next sections we prove, respectively, the correctness and the efficiency of the transformation.

4.1 The Correctness of the Transformation

Theorem 1. *Let G be a goal. G is equivalent to $f(G)$, that is, for all $I \subseteq \Sigma$:*

$$I \Vdash G \Longleftrightarrow I \models f(G)$$

Proof. By structural induction on G. Case (1) is trivial and so is case (2) by using induction on G_1 and G_2. Also note that (3) and (5) are respectively particular cases of (4) and (6) because v is equivalent to $True \to v$. Let us see cases (4) and (6).

case (4) $G = (G_1 \to v) \supset G_2$
  ($\Rightarrow$) Let $I \Vdash (G_1 \to v) \supset G_2$.
  – If $I \Vdash G_2$ then, by induction hypothesis on G_2, $I \models f(G_2)$ and then, by Lemma 1 b, $I \models f(G_2)|_v^{f(G_1) \vee v}$.

- If $I \not\Vdash G_2$ then $I \Vdash G_1$, $I \not\Vdash v$ and $I \cup \{v\} \Vdash G_2$. By induction hypothesis on G_1 and G_2: $I \models f(G_1)$ and $I \cup \{v\} \models f(G_2)$. Now by Lemma 2 a, $I \models f(G_2)|_v^1$ and then also $I \models f(G_2)|_v^{f(G_1) \vee v}$ since $I \models f(G_1) \vee v$.

($\Leftarrow$) Let $I \not\Vdash (G_1 \to v) \supset G_2$. There exists J such that $J \supseteq I$, $J \Vdash G_1 \to v$ and $J \not\Vdash G_2$. By induction hypothesis on G_2: $J \not\models f(G_2)$.
  - If $v \in J$ then, by Lemma 2 a, $J \not\models f(G_2)|_v^1$. Then $J \not\models f(G_2)|_v^{f(G_1) \vee v}$ since $J \models f(G_1) \vee v$. And, by monotonicity (Lemma 1 a), $I \not\models f(G_2)|_v^{f(G_1) \vee v}$.
  - If $v \notin J$ then $J \not\Vdash G_1$. By induction hypothesis on G_1, $J \not\models f(G_1)$ and then $J \not\models f(G_1) \vee v$. On other hand, by Lemma 2 b, $J \not\models f(G_2)|_v^0$. Then $J \not\models f(G_2)|_v^{f(G_1) \vee v}$ and as before, by monotonicity, $I \not\models f(G_2)|_v^{f(G_1) \vee v}$.

case (6) $G = ((G_1 \to v) \wedge D) \supset G_2$

By Proposition 3, G is equivalent to the formula

$$G' = ((D \supset G_1) \to v) \supset (D \supset G_2)$$

which is a formula $(G'_1 \to v) \supset G'_2$, in case (4), for $G'_1 = D \supset G_1$ and $G'_2 = D \supset G_2$. Then for all $I \subseteq \Sigma$: $I \Vdash G' \Leftrightarrow I \models f(G')$. But, by definition of f, $f(G') = f((G'_1 \to v) \supset G'_2) = f(G'_2)|_v^{f(G'_1) \vee v} = f(D \supset G_2)|_v^{f(D \supset G_1) \vee v} = f(G)$. Then for all $I \subseteq \Sigma$: $I \Vdash G \Leftrightarrow I \models f(G)$. ∎

4.2 The Complexity of the Transformation

Now we show that the size of any monotone Boolean circuit $f(G)$ with respect to the size of its original goal G is linear. The size of a Boolean circuit is defined as the number of its gates. Respectively, the size of a goal is the number of its connectives ($\wedge, \to, \supset$) and variables.

Theorem 2. *Let G be a goal. The size of $f(G)$ is linear in the size of G.*

Proof. The proof is by induction on the construction of $f(G)$. The cases $(1), (2), (3), (4)$, and (5) are trivial. The case (4) can be seen in figure 3 which shows the transformation of $f(G_2)$ when v is changed by $f(G_1) \vee v$.

We study the transformation in the case (6) and give a sketch of the proof. We have the easiest nontrivial situation depending on D when the goal to transform is the following:

$$G = ((G_{11} \to v_1) \wedge \underbrace{(G_{12} \to v_2)}_{D}) \supset G_2$$

Applying Proposition 3,

$$\underbrace{((G_{11} \to v_1) \wedge (G_{12} \to v_2)) \supset G_2}_{G} \equiv \underbrace{([(G_{12} \to v_2) \supset G_{11}] \to v_1)}_{G'} \supset \underbrace{[(G_{12} \to v_2) \supset G_2]}_{G''}$$

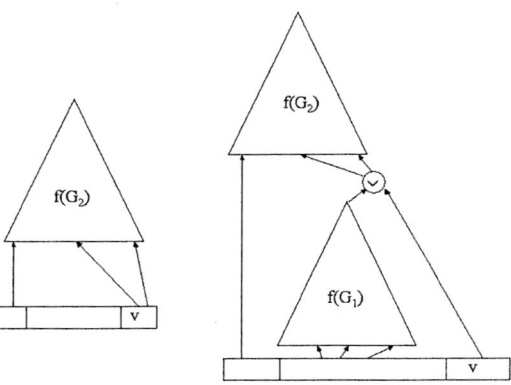

Fig. 3.

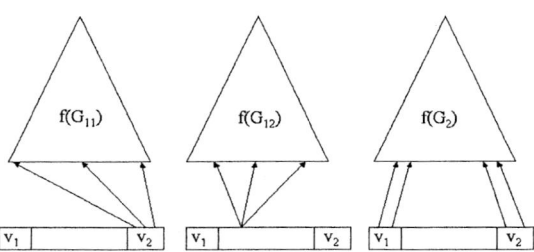

Fig. 4.

and using the case (4) of f,

$$f(G) = f(G'')|_{v_1}^{f(G')\vee v_1}$$
$$= f((G_{12} \to v_2) \supset G_2)|_{v_1}^{f(G')\vee v_1}$$
$$= f(G_2)|_{v_2}^{f(G_{12})\vee v_2}|_{v_1}^{f(G')\vee v_1}$$
$$= f(G_2)|_{v_2}^{f(G_{12})\vee v_2}|_{v_1}^{f((G_{12}\to v_2)\supset G_{11})\vee v_1}$$
$$= f(G_2)|_{v_2}^{f(G_{12})\vee v_2}|_{v_1}^{f(G_{11})|_{v_2}^{f(G_{12})\vee v_2}\vee v_1}$$

Let us see this circuit graphically. Figure 4 represents three Boolean circuits $f(G_{11})$, $f(G_{12})$, and $f(G_2)$.

From these three Boolean circuits, the corresponding $f(G)$ is shown in figure 5. Since the substitution $|_{v_2}^{f(G_{12})\vee v_2}$ is shared by $f(G_2)$ and by $f(G_{11})$, the size of the circuit $f(G)$ is linear with respect to the size of G.

This reasoning can be extended to any D in the goal $((G_1 \to v) \wedge D) \supset G_2$. The idea is that D always induces a substitution σ_D such that $f(G) = f(G_2)\sigma_D|_v^{f(G_1)\sigma_D \vee v}$ and σ_D is shared by $f(G_2)$ and by $f(G_1)$. ∎

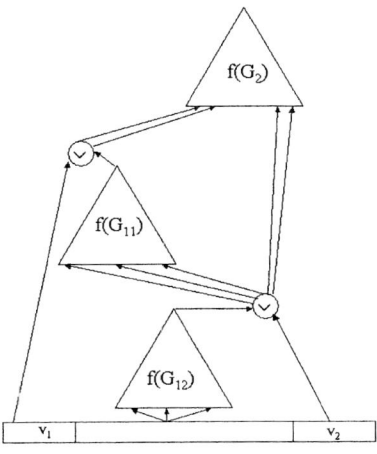

Fig. 5.

5 Conclusions and Open Problems

We have presented an efficient transformation from goals in the propositional $Horn^{\supset}$ language into monotone Boolean circuits. Since the representation of Boolean formulas by Boolean circuits is well established, our transformation allows to work with $Horn^{\supset}$ clauses in an easy and compact way.

On the other hand, in [6] is proved that learning the class of propositional $Horn^{\supset}$ clauses (denoted by $DHorn^{\supset}$) is at least as difficult as learning Boolean formulas, which is a hard problem in the usual learning models (see [1,9]). In the same paper is left open whether both problems of learning are equivalent.

A corollary of our transformation is the fact that learning Boolean circuits is at least as difficult as learning $DHorn^{\supset}$. Therefore, the problem of learning $DHorn^{\supset}$ yields between learning Boolean formulas and Boolean circuits.

A way to prove the equivalence of the three learning problems consists of showing that goals can be transformed into monotone Boolean formulas of polynomial size. In fact, this is a stronger result, because it would mean that the open question of whether Boolean circuits have equivalent polynomial size Boolean formulas would be positively solved for the subclass of formulas $DHorn^{\supset}$.

References

1. Angluin, D. and Kharitonov, M. *When won't Membership Queries Help?*. Journal of Computer and System Sciences, 50(2): 336–355, April 1995.
2. Arruabarrena, R., Lucio P. and Navarro, M. *A Strong Logic Programming View for Static Embedded Implications.* In: Proc. of FOSSACS'99, Springer-Verlag Lect. Notes in Comput. Sciences 1578: 56–72 (1999).

3. Baldoni, M., Giordano, L., and Martelli, A. *Translating a Modal Language with Embedded Implication into Horn Clause Logic.* In: Proc. of 5ž Int. Workshop of Extensions of Logic Programming, ELP'96, Springer-Verlag Lect. Notes in Comput. Sciences 1050 (1996).
4. Bugliesi, M., Lamma, E. and Mello, P. *Modularity in Logic Programming.* Journal of Logic Programming, (19-20): 443–502, (1994).
5. Gabbay, D. M. *N-Prolog: An Extension of Prolog with Hypothetical Implications. II. Logical Foundations and Negation as Failure.* Journal of Logic Programming 2(4): 251–283 (1985).
6. Gaintzarain, J., Hermo, M., and Navarro M. *On Learning Conjunctions of Horn$^\supset$ Clauses.* CiE 2005, New Computational Paradigms. Amsterdam, June 8-12, (2005).
7. Giordano, L., and Martelli, A. *Structuring Logic Programs: A Modal Approach.* Journal of Logic Programming 21: 59–94 (1994).
8. Giordano, L., Martelli, A., and Rossi, G. *Extending Horn Clause Logic with Implication Goals.* Theoretical Computer Science 95: 43–74, (1992).
9. Kearns, M. and Valiant, L. *Criptographic Limitations on Learning Boolean Formulae and Finite Automata.* Journal of the ACM 41(1): 67–95, (1994).
10. Lucio, P. *Structured Sequent Calculi for Combining Intuitionistic and Classical First-Order Logic.* In: Proc. of FroCoSS'2000,Springer-Verlag Lect. Notes in Artificial Intelligence 1794: 88–104 (2000).
11. Meseguer, J. *Multiparadigm Logic Programming.* In: Proc. of ALP'92, Springer-Verlag Lect. Notes in Comput. Sciences 632: 158–200, (1992).
12. Miller, D. *A Logical Analysis of Modules in Logic Programming.* In: Journal of Logic Programming 6: 79–108, (1989).
13. Miller, D., Nadathur, G., Pfenning, F. and Scedrov, A. *Uniform Proofs as a Foundation for Logic Programming.* Annals of Pure and App. Logic 51: 125–157, (1991).
14. Monteiro, L., Porto, A. *Contextual Logic Programming.* In: Proc. 6th International Conf. on Logic Programming 284–299, (1989).
15. Moscowitz, Y., and Shapiro, E. *Lexical Logic Programs.* In: Proc. 8th International Conf. on Logic Programming 349–363, (1991).
16. Navarro, M. *From Modular Horn Programs to Flat Ones: a Formal Proof for the Propositional Case.* In: Proc. of ISIICT 2004 (Int. Symp. on Innovation in Information and Communication Technology), Amman, Jordan. April 2004.
Technical Report UPV-EHU/ LSI/ TR 01-2004.
17. Papadimitriou, Christos H. *Computational Complexity.* Addison-Wesley Publishing Company, Inc. (1995).

Autoreducibility, Mitoticity, and Immunity*

Christian Glaßer[1], Mitsunori Ogihara[2,**], A. Pavan[3,***],
Alan L. Selman[4,†], and Liyu Zhang[4]

[1] Universität Würzburg
glasser@informatik.uni-wuerzburg.de
[2] University of Rochester
ogihara@cs.rochester.edu
[3] Iowa State University
pavan@cs.iastate.edu
[4] University at Buffalo
{selman, lzhang7}@cse.buffalo.edu

Abstract. We show the following results regarding complete sets.
- NP-complete sets and PSPACE-complete sets are many-one autoreducible.
- Complete sets of any level of PH, MODPH, or the Boolean hierarchy over NP are many-one autoreducible.
- EXP-complete sets are many-one mitotic.
- NEXP-complete sets are weakly many-one mitotic.
- PSPACE-complete sets are weakly Turing-mitotic.
- If one-way permutations and quick pseudo-random generators exist, then NP-complete languages are m-mitotic.
- If there is a tally language in NP ∩ coNP − P, then, for every $\epsilon > 0$, NP-complete sets are not $2^{n(1+\epsilon)}$-immune.

These results solve several of the open questions raised by Buhrman and Torenvliet in their 1994 survey paper on the structure of complete sets.

1 Introduction

We solve several open questions identified by Buhrman and Torenvliet in their 1994 survey paper on the structure of complete sets [12]. It is important to study the computational structure of complete sets, because they, by reductions of all the sets in the class to the complete sets, represent all of the structure that a class might have. For this reason, complete sets might have simpler computational structure than some other sets in the class. Here we focus attention primarily on autoreducibility, mitoticity, and immunity.

---

* A full version of this paper is available as ECCC Technical Report TR05-011.
** Research supported in part by NSF grants EIA-0080124, EIA-0205061, and NIH grant P30-AG18254.
*** Research supported in part by NSF grants CCR-0344817 and CCF-0430807.
† Research supported in part by NSF grant CCR-0307077.

Trakhtenbrot [26] introduced the notion of autoreducibility in a recursion theoretic setting. A set A is *autoreducible* if there is an oracle Turing machine M such that $A = L(M^A)$ and M on input x never queries x. Ladner [21] showed that there exist Turing-complete recursively enumerable sets that are not autoreducible. Ambos-Spies [2] introduced the polynomial-time variant of autoreducibility, where we require the oracle Turing machine to run in polynomial time. Yao [30] introduced the notion of coherence, which coincides with probabilistic polynomial-time autoreducibility. In this paper, we assume that all reductions are polynomial-time computable. In particular, we write "autoreducible" to mean "polynomial-time autoreducible."

The question of whether complete sets for various classes are autoreducible has been studied extensively [30,7,10], and is currently an area of active research [15]. Beigel and Feigenbaum [7] showed that Turing complete sets for the classes that form the polynomial-time hierarchy, Σ_i^P, Π_i^P, and Δ_i^P, are Turing autoreducible. Thus, all Turing complete sets for NP are Turing autoreducible. Buhrman et al. [10] showed that Turing complete sets for EXP and Δ_i^{EXP} are autoreducible, whereas there exists a Turing complete set for EESPACE that is not Turing auto-reducible. They showed that answering questions about autoreducibility of intermediate classes results in interesting separation results. Regarding NP, Buhrman et al. [10] showed that all truth-table complete sets for NP are probabilistic truth-table autoreducible. Thus, all NP-complete sets are probabilistic truth-table autoreducible.

Buhrman and Torenvliet [12] asked whether all NP-complete sets are many-one autoreducible and whether all PSPACE-complete sets are many-one autoreducible. We resolve these questions positively: all NP-complete sets and PSPACE-complete sets are (unconditionally) many-one autoreducible. We generalize these two results to show that for each class in MODPH [18] (the hierarchy constructed by applying to P a constant number of operators chosen from $\{\exists \cdot, \forall \cdot\} \cup \{\text{MOD}_k\cdot, \text{coMOD}_k\cdot \mid k \geq 2\}$), all of its nontrivial m-complete sets are m-autoreducible. We obtain as a corollary that for each class C chosen from $\{\text{NP}, \text{PSPACE}\} \cup \text{MODPH}$ it holds that no m-complete sets for C are $2n$-generic.

Autoreducible sets can be thought of as sets having some redundant information. For example, if A is m-autoreducible by the reduction f, then x and $f(x)$ both contain the same information concerning whether x belongs to A. How much redundancy is there in complete sets? Informally, an infinite set is *mitotic* if it can be partitioned into two equivalent parts. Thus, both parts contain the same information as the original set. Ladner [21] introduced and studied mitoticity of recursively enumerable sets. Ambos-Spies [2] formulated two notions in the polynomial time setting, mitoticity and weak mitoticity. Also, he showed that every mitotic set is autoreducible. Here we settle some questions about mitoticity that Buhrman and Torenvliet raised in their 1994 survey paper. First, Buhrman, Hoene, Torenvliet [11] proved that all EXP-complete sets are weakly many-one mitotic, and showed a partial result concerning mitoticity of NEXP-complete sets. Here we strengthen these two results. We prove that all EXP-complete sets are many-one mitotic. This result was also obtained independently by Kurtz [15].

We also prove that all NEXP complete sets are weakly many-one mitotic. In addition, we show that PSPACE-complete sets are weakly Turing mitotic. Also, we show that if one-way permutations and quick pseudo-random generators exist, then NP-complete sets are many-one mitotic.

In Section 5, we study the question of whether NP-complete sets have easy subsets. Berman [9] showed that EXP-complete sets are not P-immune, and Tran [27] showed that NEXP-complete sets are not P-immune. However, we do not have such unconditional results for NP-complete sets. Glaßer et al. [17] showed that if one-way permutations exist, then NP-complete sets are not 2^{n^ϵ}-immune. Here we provide another partial result in this direction. In Section 5 we show that if there exists a tally language in NP ∩ coNP − P, then every NP-complete set includes an infinite subset that is recognizable in time $2^{n(1+\epsilon)}$.

We conclude this paper with results on a few additional properties of complete sets. In Section 6 we show, under a reasonable hypothesis about the complexity class UP, that every NP-complete set is exponentially-honest complete. Exponentially-honest reductions were defined and studied by Ganesan and Homer [16]. Section 7 studies robustness of complete sets. We prove that if a set A is Turing complete for NP and S is a log-dense set, then $A - S$ remains Turing complete for NP. This result is easier to prove than Buhrman and Torenvliet's [14] result about EXP.

2 Preliminaries

We use standard notation and assume familiarity with standard reductions. Words are considered in lexicographic order. All used reductions are polynomial-time computable.

Definition 1 ([2]). *A set A is* polynomially T-autoreducible *(T-autoreducible, for short) if there exists a polynomial-time-bounded oracle Turing machine M such that $A = L(M^A)$ and for all x, M on input x never queries x. A set A is* polynomially m-autoreducible *(m-autoreducible, for short) if $A \leq_m^p A$ via a reduction function f such that for all x, $f(x) \neq x$.*

Definition 2 ([2]). *A recursive set A is* polynomial-time m-mitotic (T-mitotic) *(m-mitotic and T-mitotic, for short) if there exists a set $B \in$ P such that $A \equiv_{m(T)}^p A \cap B \equiv_{m(T)}^p A \cap \overline{B}$. A recursive set A is* polynomial-time weakly m-mitotic (T-mitotic) *(weakly m-mitotic and weakly T-mitotic, for short) if there exist disjoint sets A_0 and A_1 such that $A_0 \cup A_1 = A$, and $A \equiv_{m(T)}^p A_0 \equiv_{m(T)}^p A_1$.*

In general, for a reducibility type r, r-autoreducible sets are sets that are autoreducible with respect to $\leq_r^p$-reductions. The same convention is used for mitotic sets.

A language is DTIME$(T(n))$-*complex* if L does not belong to DTIME$(T(n))$ almost everywhere; that is, every Turing machine M that accepts L runs in time greater than $T(|x|)$, for all but finitely many words x. A language L is *immune* to a complexity class $\mathcal{C}$, or $\mathcal{C}$-*immune*, if L is infinite and no infinite subset of

L belongs to $\mathcal{C}$. A language L is *bi-immune* to a complexity class $\mathcal{C}$, or $\mathcal{C}$-*bi-immune*, if both L and $\overline{L}$ are $\mathcal{C}$-*immune*. Balcázar and Schöning [6] proved that for every time-constructible function T, L is DTIME($T(n)$)-complex if and only if L is bi-immune to DTIME($T(n)$).

3 Autoreducibility

Since EXP-complete sets are complete with respect to length-increasing reductions [9], they are m-autoreducible. Ganesan and Homer [16] showed that NEXP-complete sets are complete under 1-1 reductions. This implies that all NEXP-complete sets are also m-autoreducible. To see this, consider a 1-1 reduction from $0L \cup 1L$ to L, where L is the given NEXP-complete set. These techniques cannot be applied to NP-complete sets, as we do not know any unconditional result on the degree structure of NP-complete sets. Some partial results are known for NP-complete sets. Beigel and Feigenbaum [7] showed that Turing complete sets for NP are T-autoreducible. Buhrman et al. [10] showed that all truth-table complete sets for NP are probabilistic tt-autoreducible. It has not been known whether NP-complete sets are m-autoreducible. Buhrman and Torenvliet raised this question in their survey papers [12,13]. Below, we resolve this question.

Note that neither singletons nor their complements can be m-autoreducible. Therefore, in connection with m-autoreducibility, a set L is called *nontrivial* if $|L| > 1$ and $|\overline{L}| > 1$.

The following theorem is our first main result, which shows that every nontrivial NP-complete set is m-autoreducible. The proof used the left-set technique of Ogihara and Watanabe [22].

Theorem 1. *All nontrivial NP-complete sets are m-autoreducible.*[1]

Proof. Let L be NP-complete and let M be a nondeterministic machine that accepts L. For a suitable polynomial p we can assume that on input x, all computation paths of M have length $p(|x|)$. Since L is nontrivial, there exist different words $y_1, y_2 \in L$ and $\overline{y}_1, \overline{y}_2 \in \overline{L}$. By way of the left set technique [22], let

$$\text{Left}(L) \stackrel{df}{=} \{\langle x, u \rangle \mid |u| = p(|x|) \text{ and } \exists v, |v| = |u|, \text{ such that}$$
$$u \leq v \text{ and } M(x) \text{ accepts along path } v\}.$$

Left(L) $\in$ NP and L is NP-complete. Let $f \in$ PF reduce Left(L) to L. The algorithm below defines function g which is an m-autoreduction for L. Let x be an input. Define $n \stackrel{df}{=} |x|$ and $m \stackrel{df}{=} p(|x|)$.

```
1     if f(⟨x, 0ᵐ⟩) ≠ x then output f(⟨x, 0ᵐ⟩)
2     if f(⟨x, 1ᵐ⟩) = x then
3         if M(x) accepts along 1ᵐ then
4             output a string from {y₁, y₂} − {x}
5         else
```

---
[1] Note that all NP-complete sets are nontrivial, unless P = NP.

```
6            output a string from {ȳ₁, ȳ₂} − {x}
7         endif
8      endif
9      // here f(⟨x, 0ᵐ⟩) = x ≠ f(⟨x, 1ᵐ⟩)
10     determine z of length m such that f(⟨x, z⟩) = x ≠ f(⟨x, z + 1⟩)
11     if M(x) accepts along z then output a string from {y₁, y₂} − {x}
12     else output f(⟨x, z + 1⟩)
```

Note that step 10 can be achieved by simple binary search: Start with $z_1 := 0^m$ and $z_2 := 1^m$. Let z' be the middle element between z_1 and z_2. If $f(z') = x$ then $z_1 := z'$ else $z_2 := z'$. Again, choose the middle element between z_1 and z_2, and so on. This shows $g \in$ PF. Clearly, $g(x) \neq x$, so it remains to show $L \leq^p_m L$ via g.

If the algorithm stops in step 1, then

$$x \in L \Leftrightarrow \langle x, 0^m \rangle \in \text{Left}(L) \Leftrightarrow g(x) = f(\langle x, 0^m \rangle) \in L.$$

If the algorithm stops in step 4 or step 6, then $f(\langle x, 0^m \rangle) = f(\langle x, 1^m \rangle)$. Hence

$$x \in L \Leftrightarrow \langle x, 1^m \rangle \in \text{Left}(L) \Leftrightarrow M(x) \text{ accepts along } 1^m \Leftrightarrow g(x) \in L.$$

Assume we reach step 9. There it holds that $f(\langle x, 0^m \rangle) = x \neq f(\langle x, 1^m \rangle)$. If the algorithm stops in step 11, then $x \in L$ and $g(x) \in L$. Assume we stop in step 12. So $M(x)$ does not accept along z.

$$x \in L \Leftrightarrow f(\langle x, 0^m \rangle) = f(\langle x, z \rangle) \in L \Leftrightarrow g(x) = f(\langle x, z + 1 \rangle) \in L.$$

□

Using the same technique we obtain the following results:

Corollary 1. *For every $k \geq 1$, all nontrivial $\leq^p_{k\text{-}dtt}$-complete sets for NP are $\leq^p_{k\text{-}dtt}$-autoreducible.*

Corollary 2. *All nontrivial $\leq^p_{dtt}$-complete sets for NP are $\leq^p_{dtt}$-autoreducible.*

Corollary 3 below is the analog of Corollaries 1 and 2 for $\leq^p_{1\text{-}tt}$-reductions. It is not known whether the analog holds for $\leq^p_{2\text{-}tt}$-reductions, even for $\leq^p_{2\text{-}ctt}$-reductions.

Corollary 3. *All nontrivial $\leq^p_{1\text{-}tt}$-complete sets for NP are $\leq^p_{1\text{-}tt}$-autoreducible.*

By extending the left-set-based proof of Theorem 1, we can show that the same result holds for PSPACE.

Theorem 2. *All nontrivial m-complete sets for PSPACE are m-autoreducible.*

We generalize the above two results using the concept of polynomial-time bit-reductions [18]. We show that every class that is polynomial-time bit-reducible to a regular language has the property that all of its nontrivial m-complete sets are m-autoreducible. As a corollary to this, we show that for every class in the MODPH hierarchy [18], all of its nontrivial m-complete sets are m-autoreducible.

Definition 3. *[18] A language A is polynomial-time bit-reducible to a language B if there exists a pair of polynomial-time computable functions, (f,g), $f : \Sigma^* \times \mathbf{N}^+ \to \Sigma$, $g : \Sigma^* \to \mathbf{N}$, such that for all x, $x \in A \Leftrightarrow f(x,1) \cdots f(x,g(x)) \in B$.*

Theorem 3. *Let $B \notin \{\emptyset, \Sigma^*\}$ be a regular language recognized by a finite automaton $M = (Q, \{0,1\}, \delta, q_0, q_f)$. Let $\mathcal{C}$ be the polynomial-time bit-reduction closure of B. Then each nontrivial m-complete set for $\mathcal{C}$ is m-autoreducible.*[2]

Definition 4. *[25,29] Let $\mathcal{C}$ be a language class. Define $\exists \cdot \mathcal{C}$ (and analogously $\forall \cdot \mathcal{C}$) to be the set of all languages L for which there exist a polynomial p and an $A \in \mathcal{C}$ such that for all x,*

$$x \in L \Leftrightarrow (\exists y, |y| = p(|x|))[\langle x, y \rangle \in A].$$

Definition 5. *[8,24] Let $k \geq 2$ be an integer. Define $\mathrm{MOD}_k \cdot \mathcal{C}$ to be the set of all languages L for which there exist a polynomial p and a language $A \in \mathcal{C}$ such that for all x,*

$$x \in L \Leftrightarrow \|\{y \mid |y| = p(|x|) \wedge \langle x, y \rangle \in A\}\| \not\equiv 0 \pmod{k}.$$

Definition 6. *[18] MODPH is the hierarchy consisting of the following classes:*

- P *belongs to* MODPH.
- *If $\mathcal{C}$ is a class belonging to* MODPH *then $\exists \cdot \mathcal{C}$ and $\forall \cdot \mathcal{C}$ belong to* MODPH.
- *For each integer $k \geq 2$, if $\mathcal{C}$ is a class belonging to* MODPH *then $\mathrm{MOD}_k \cdot \mathcal{C}$ and $\mathrm{coMOD}_k \cdot \mathcal{C}$ belong to* MODPH.

Proposition 1. *[18] Each class in* MODPH *is the polynomial-time bit-reduction closure of some regular language.*

Combining Theorem 3 with the above proposition, we obtain the following result.

Theorem 4. *For every class in* MODPH *it holds that all of its nontrivial m-complete sets are m-autoreducible.*

Corollary 4. *Every nontrivial set that is m-complete for one of the following classes is m-autoreducible.*

- *the levels Σ_k^P, Π_k^P, and Δ_k^P of the polynomial-time hierarchy*
- *1NP* [3]
- *the levels of the Boolean hierarchy over* NP

---

[2] In terms of leaf languages, this theorem reads as follows: If B is a nontrivial regular language, then m-complete sets for $\mathrm{Leaf}_b^p(B)$ are m-autoreducible. (The latter denotes the balanced leaf-language class defined by B [18,28].)

[3] The set of languages accepted by nondeterministic polynomial-time-bounded Turing machines that accept an input if and only if there exists exactly one accepting path.

Now we obtain corollaries about the genericity of complete sets. The notion of resource bounded genericity was defined by Ambos-Spies, Fleischhack, and Huwig [3]. We use the following equivalent definition [5,23].

Definition 7. *Let t be a polynomial. A set L is $t(n)$-generic if for every oracle Turing machine M, if $M^{L|x}(x) = L(x)$ for all x, then the running time of M is at least $t(2^{|x|})$ for all but finitely many x. Recall that $L|x = \{y \in L \mid y < x\}$.*

We obtain the following corollary regarding genericity of NP-complete sets. Earlier, it was known that there exists a $k > 0$, such that NP-complete sets are not $O(n^k)$-generic. This follows from the work on small span theorems [20,4]. Below we improve this.

Corollary 5. *NP-complete sets are not $2n$-generic.*

Corollary 6. *Let $\mathcal{C}$ be either PSPACE or a class belonging to MODPH. Then no m-complete sets for $\mathcal{C}$ are $2n$-generic.*

3.1 Relativization

We show relativized separation results in this section.

Theorem 5. *For any $k \geq 2$, there is an oracle A such that relative to A there is a set B that is $\leq_{k\text{-}dtt}^p$ complete for NP but not $\leq_{(k-1)\text{-}T}^p$ autoreducible.*

This result suggests that it may not be possible to improve Corollary 1. The theorem immediately yields the following result, which is of interest on its own.

Corollary 7. *There is an oracle A such that relative to A there exists a $\leq_{2\text{-}dtt}^p$-complete set for NP that is not m-mitotic.*

Corollary 7 gives a relativized, partial negative answer to an open question of Buhrman and Torenvliet [12] as to whether all T-complete sets for NP are (weakly) $m(T)$-mitotic. It remains open whether there is an oracle relative to which not all NP-complete sets are m-mitotic. Another related question raised by Buhrman and Torenvliet is whether all tt-complete sets for NP are tt-autoreducible. The following theorem gives a partial answer to this question in a relativized world.

Theorem 6. *For each $k \geq 2$, there is an oracle A relative to which a set exists that is $\leq_{dtt}^p$-complete for NP but not $\leq_{btt}^p$-autoreducible.*

4 Mitoticity

Buhrman, Hoene, and Torenvliet [11] showed that all EXP-complete sets are weakly m-mitotic. We improve this to show that all EXP-complete sets are m-mitotic. We note here that Kurtz independently obtained the same result, which is reported in a recent survey by Buhrman and Torenvliet [15].

Theorem 7. *All* EXP-*complete sets are m-mitotic.*

The proof of the above theorem, which appers in the full paper version, actually shows that for any class $\mathcal{C}$, if L is complete with length-increasing reductions, then L is m-mitotic. Agrawal [1] shows that if one-way permutations and quick pseudo-random generators exist, then every NP-complete language is complete with respect to length increasing reductions. Thus, we have:

Corollary 8. *If one-way permutations and quick pseudo-random generators exist, then* NP-*complete languages are m-mitotic.*

If L is NP-complete with respect to honest many-one reductions, then L is complete with respect to length increasing reductions. Thus, we have:

Corollary 9. *If L is* NP-*complete with respect to honest reductions, then L is m-mitotic.*

Buhrman, Hoene, and Torenvliet [11] showed that every NEXP-complete set can be partitioned into infinitely many sets such that each one of them is NEXP-complete. They asked whether this can be improved to show that every NEXP-complete set is weakly m-mitotic. Theorem 9 below resolves this question affirmatively. The proof uses the following result by Ganesan and Homer [16], which shows that all NE-complete sets are complete with respect to 1-1 reductions. It is easy to see that their proof works for NEXP-complete sets too.

Theorem 8 ([16]). *All $\leq_m^p$-complete sets for* NEXP *are complete under 1-1 reductions.*

Now we are ready to prove the affirmative resolution.

Theorem 9. *All $\leq_m^p$-complete sets for* NEXP *are weakly m-mitotic.*

Proof. Let K be the standard NEXP-complete set. Let L be any given NEXP-complete set. We show that L is weakly m-mitotic.

$$K' \stackrel{df}{=} 0K \cup 1K$$

is NEXP-complete. K' reduces to L via some $f \in \text{PF}$. L reduces to K (really K, not K') via some $g \in \text{PF}$. Choose k such that f and g can be computed in time $O(n^k)$. By Theorem 8, we can assume f and g to be 1-1.

$$L_0 \stackrel{df}{=} \{y \mid \exists x, |x| \leq |y|^k, f(0x) = y\}$$
$$L_1 \stackrel{df}{=} \{y \mid \exists x, |x| \leq |y|^k, f(1x) = y\}$$

Observe that $L_0, L_1 \in \text{EXP}$. Define the following function:

$$f_0(x) \stackrel{df}{=} \begin{cases} f(0x) & : \text{if } |f(0x)|^k \geq |x|, \\ f_0(g(f(0x))) & : \text{otherwise.} \end{cases}$$

Note that if $|f(0x)|^k < |x|$ then $|g(f(0x))| \leq |f(0x)|^k < |x|$. So, for some y, $|y| < |x|$, $f_0(x) = f_0(y)$. So, the recursion terminates. Thus, $f_0 \in$ PF. Note that for every x there exists some y such that $f_0(x) = f(0y)$ and $|f(0y)|^k \geq |y|$. This implies that for all x, $f_0(x) \in L_0$.

Similarly define f_1 as

$$f_1(x) \stackrel{df}{=} \begin{cases} f(1x) & : \text{if } |f(1x)|^k \geq |x|, \\ f_1(g(f(1x))) & : \text{otherwise.} \end{cases}$$

By following an argument similar to the one above we can show that $f_1 \in$ PF and that $f_1(\Sigma^*) \subseteq L_1$.

We first show $K \leq_m^p L$ via reduction f_0. This is done by induction on the number of recursion steps r in the definition of $f_0(x)$. If $r = 0$, then $f_0(x) = f(0x)$ and therefore,

$$x \in K \Leftrightarrow 0x \in 0K \Leftrightarrow 0x \in K' \Leftrightarrow f_0(x) = f(0x) \in L.$$

If $r \geq 1$ then $f_0(x) = f_0(g(f(0x)))$. Let $y = g(f(0x))$. By our induction hypothesis, $y \in K \Leftrightarrow f_0(y) \in L$. So we obtain

$$x \in K \Leftrightarrow 0x \in 0K \Leftrightarrow 0x \in K' \Leftrightarrow f(0x) \in L$$
$$\Leftrightarrow y = g(f(0x)) \in K \Leftrightarrow f_0(y) = f_0(g(f(0x))) = f_0(x) \in L.$$

Thus, $K \leq_m^p L$ via f_0. Analogously we show $K \leq_m^p L$ via f_1.

Since $K \leq_m^p L$ via f_0 and $f_0(\Sigma^*) \subseteq L_0$ we have $K \leq_m^p (L \cap L_0)$ via f_0. Since f is 1-1, L_0 and L_1 are disjoint. Since $f_1(\Sigma^*) \subseteq L_1$, we have $f_1(\Sigma^*) \subseteq \overline{L_0}$. Combining this with $K \leq_m^p L$ via f_1, we have $K \leq_m^p (L \cap \overline{L_0})$ via f_1. Therefore $(L \cap L_0)$ and $(L \cap \overline{L_0})$ are each NEXP-hard. Both sets are NEXP-complete since $L_0 \in$ EXP and $L \in$ NEXP. □

Note that the above proof actually shows that each NEXP-complete set can be split into m-equivalent NEXP-complete m-mitotic parts with a set in EXP.

The following theorem settles another question raised by Buhrman and Torenvliet [12].

Theorem 10. *Every $\leq_m^p$-complete set for* PSPACE *is weakly T-mitotic.*

5 Immunity

In Glaßer et al. [17], the authors proved immunity results for NP-complete sets under the assumption that certain average-case hardness conditions are true. For example, they show that if one-way permutations exist, then NP-complete sets are not 2^{n^ϵ}-immune. Here we obtain a non-immunity result for NP-complete sets under the assumption that the following worst-case hardness hypothesis holds.

Hypothesis T: There is an NP machine N that accepts 0^* and no P-machine can compute its accepting computations. This means that for every polynomial-time machine M there exist infinitely many n such that $M(0^n)$ is not an accepting computation of $N(0^n)$.

Though the hypothesis looks verbose, we note that it is implied by a simply stated and believable hypothesis.

Observation 1. *If there is a tally language in* NP$\cap$coNP$-$P, *then Hypothesis T is true.*

Theorem 11. *If Hypothesis T holds, then, for every $\epsilon > 0$, NP-complete languages are not $2^{n(1+\epsilon)}$-immune.*

Corollary 10. *If there is a tally language in* NP $\cap$ coNP $-$ P, *then, for every $\epsilon > 0$, NP-complete languages are not $2^{n(1+\epsilon)}$-immune.*

Next we consider the possibility of obtaining an unconditional result regarding non-immunity of NP-complete languages. We show that for certain type of NP-complete languages we get an unconditional result.

Definition 8. *A language L does not have superpolynomial gaps, if there exists $k > 0$ such that for all but finitely many n, there exists a string x in L such that $n \leq |x| \leq n^k$.*

We show that NP-complete languages that have no superpolynomial gaps are not immune.

Theorem 12. *If L is an NP-complete language that has no superpolynomial gaps, then for every $\epsilon > 0$, L is not $2^{n(1+\epsilon)}$-immune.*

The above theorem prompts the following question: Are there NP-complete languages with superpolynomial gaps? We have the following result. Given two complexity classes $\mathcal{A}$ and $\mathcal{B}$, we say $\mathcal{A} \subseteq$ io-$\mathcal{B}$, if for every language $A \in \mathcal{A}$, there exists a language $B \in \mathcal{B}$ such that for infinitely many n, $A^n = B^n$.

Theorem 13. *If NP has a complete language with superpolynomial gaps, then for every $\epsilon > 0$, NP $\subseteq$ io-DTIME(2^{n^ϵ}).*

Combining Theorems 12 and 13 we have the following corollary.

Corollary 11. *If for some $\delta > 0$, NP $\not\subseteq$ io-DTIME(2^{n^δ}), then for every $\epsilon > 0$, no NP-complete language is $2^{n(1+\epsilon)}$-immune.*

6 Exponentially Honest Reductions

In the previous section we showed that if NP $\cap$ coNP $-$ P has a tally language, then there exist reductions from 0^* to NP-complete sets that are infinitely often exponentially honest. In this section, we consider a stronger hypothesis and show that if this hypothesis holds, then every NP-complete set is complete with respect to exponentially honest reductions. We consider the following hypothesis.

UP-machine Hypothesis: For some $\epsilon > 0$, there is a UP-machine M that accepts 0^* such that no 2^{n^ϵ}-time-bounded machine can compute infinitely many accepting computations of M.

We first make the following observation.

Observation 2. *If the UP-machine hypothesis holds, then, there exists a UP-machine M that accepts 0^* such that no 2^{2n}-time-bounded machine can compute infinitely many accepting computations of M.*

Theorem 14. *Assume the UP-machine hypothesis holds. Let L be any NP-complete language. For every $S \in$ NP and every $k > 0$, there is a many-one reduction f from S to L such that for every x, $|f(x)| > k \log |x|$.*

7 Robustness

Recently Buhrman and Torenvliet [14] proved that Turing-complete sets for EXP are robust against log-dense sets in P. Using similar ideas, we can prove the same for NP. The proof is easier though, due to the fact that search reduces to decision for all Turing-complete sets for NP [19, Corollary 7.3, p.152]. A set S is *log-dense* if there is a constant $c > 0$ such that for all n, $\|S^{\leq n}\| \leq c \log n$.

Theorem 15. *If a set A is Turing-complete for NP and S is a log-dense set in P, then $A - S$ is Turing-complete for NP.*

References

1. M. Agrawal. Pseudo-random generators and structure of complete degrees. In *17th Annual IEEE Conference on Computational Complexity*, pages 139–145, 2002.
2. K. Ambos-Spies. P-mitotic sets. In E. Börger, G. Hasenjäger, and D. Roding, editors, *Logic and Machines, Lecture Notes in Computer Science 177*, pages 1–23. Springer-Verlag, 1984.
3. K. Ambos-Spies, H. Fleischhack, and H. Huwig. Diagonalizations over polynomial time computable sets. *Theoretical Computer Science*, 51:177–204, 1987.
4. K. Ambos-Spies, H. Neis, and A. Terwijn. Genericity and measure for exponential time. *Theoretical Computer Science*, 168(1):3–19, 1996.
5. J. Balcazar and E. Mayordomo. A note on genericty and bi-immunity. In *Proceedings of the Tenth Annual IEEE Conference on Computational Complexity*, pages 193–196, 1995.
6. J. Balcázar and U. Schöning. Bi-immune sets for complexity classes. *Mathematical Systems Theory*, 18(1):1–18, 1985.
7. R. Beigel and J. Feigenbaum. On being incoherent without being very hard. *Computational Complexity*, 2:1–17, 1992.
8. R. Beigel and J. Gill. Counting classes: Thresholds, parity, mods, and fewness. *Theoretical Computer Science*, 103:3–23, 1992.

9. L. Berman. *Polynomial Reducibilities and Complete Sets*. PhD thesis, Cornell University, Ithaca, NY, 1977.
10. H. Buhrman, L. Fortnow, D. van Melkebeek, and L. Torenvliet. Using autoreducibility to separate complexity classes. *SIAM Journal on Computing*, 29(5):1497–1520, 2000.
11. H. Buhrman, A. Hoene, and L. Torenvliet. Splittings, robustness, and structure of complete sets. *SIAM Journal on Computing*, 27:637–653, 1998.
12. H. Buhrman and L. Torenvliet. On the structure of complete sets. In *Proceedings 9th Structure in Complexity Theory*, pages 118–133, 1994.
13. H. Buhrman and L. Torenvliet. Complete sets and structure in subrecursive classes. In *Proceedings of Logic Colloquium '96*, pages 45–78. Springer-Verlag, 1998.
14. H. Buhrman and L. Torenvliet. Separating complexity classes using structural properties. In *Proceedings of the 19th IEEE Conference on Computational Complexity*, pages 130–138, 2004.
15. H. Buhrman and L. Torenvliet. A Post's program for complexity theory. *Bulleting of the EATCS*, 85:41–51, 2005.
16. K. Ganesan and S. Homer. Complete problems and strong polynomial reducibilities. *SIAM Journal on Computing*, 21:733–742, 1992.
17. C. Glaßer, A. Pavan, A. Selman, and S. Sengupta. Properties of NP-complete sets. In *Proceedings of the 19th Annual IEEE Conference on Computational Complexity*, pages 184–197, 2004.
18. U. Hertrampf, C. Lautemann, T. Schwentick, H. Vollmer, and K. W. Wagner. On the power of polynomial time bit-reductions. In *Proceedings 8th Structure in Complexity Theory*, pages 200–207, 1993.
19. S. Homer and A. Selman. *Computability and Complexity Theory*. Texts in Computer Science. Springer, New York, 2001.
20. D. W. Juedes and J. H. Lutz. The complexity and distribution of hard problems. *SIAM Joutnal on Computing*, 24:279–295, 1995.
21. R. Ladner. Mitotic recursively enumerable sets. *Journal of Symbolic Logic*, 38(2):199–211, 1973.
22. M. Ogiwara and O. Watanabe. On polynomial-time bounded truth-table reducibility of NP sets to sparse sets. *SIAM Journal of Computing*, 20(3):471–483, 1991.
23. A. Pavan and A. Selman. Separation of NP-completeness notions. *SIAM Journal on Computing*, 31(3):906–918, 2002.
24. U. Schöning. Probabilistic complexity classes and lowness. *Journal of Computer and System Sciences*, 39:84–100, 1989.
25. L. Stockmeyer. The polynomial-time hierarchy. *Theoretical Computer Science*, 3:1–22, 1977.
26. B. Trahtenbrot. On autoreducibility. *Dokl. Akad. Nauk SSSR*, 192, 1970. Translation in Soviet Math. Dokl. 11: 814– 817, 1970.
27. N. Tran. On P-immunity of nondeterministic complete sets. In *Proceedings of the 10th Annual Conference on Structure in Complexity Theory*, pages 262–263. IEEE Computer Society Press, 1995.
28. K. W. Wagner. Leaf language classes. In *Proceedings International Conference on Machines, Computations, and Universality*, volume 3354 of *Lecture Notes in Computer Science*. Springer Verlag, 2004.
29. C. Wrathall. Complete sets and the polynomial-time hierarchy. *Theoretical Computer Science*, 3:23–33, 1977.
30. A. Yao. Coherent functions and program checkers. In *Proceedings of the 22nd Annual Symposium on Theory of Computing*, pages 89–94, 1990.

Canonical Disjoint NP-Pairs of Propositional Proof Systems*

Christian Glaßer[1], Alan L. Selman[2,**], and Liyu Zhang[2]

[1] Lehrstuhl für Informatik IV, Universität Würzburg, Am Hubland,
97074 Würzburg, Germany
glasser@informatik.uni-wuerzburg.de
[2] Department of Computer Science and Engineering,
University at Buffalo, Buffalo, NY 14260
{selman, lzhang7}@cse.buffalo.edu

Abstract. We prove that every disjoint NP-pair is polynomial-time, many-one equivalent to the canonical disjoint NP-pair of some propositional proof system. Therefore, the degree structure of the class of disjoint NP-pairs and of all canonical pairs is identical. Secondly, we show that this degree structure is not superficial: Assuming there exist P-inseparable disjoint pairs, there exist intermediate disjoint NP-pairs. That is, if (A, B) is a P-separable disjoint NP-pair and (C, D) is a P-inseparable disjoint NP-pair, then there exist P-inseparable, incomparable NP-pairs (E, F) and (G, H) whose degrees lie strictly between (A, B) and (C, D). Furthermore, between any two disjoint NP-pairs that are comparable and inequivalent, such a diamond exists.

1 Introduction

One reason it is important to study the class DisjNP of all disjoint NP-pairs is its relationship to the theory of proof systems for propositional calculus. Specifically, Razborov [Raz94] defined the canonical disjoint NP-pair, (SAT^*, REF_f), for every propositional proof system f, and he showed that if there exists an optimal propositional proof system f, then its canonical pair is a complete pair for DisjNP. (We will explain this notation later.) In the same paper he asked for evidence of existence of a propositional proof system whose canonical disjoint NP-pair is not separable by a set belonging to the complexity class P, and, relatedly, he asked whether it is possible to reduce to canonical pairs (SAT^*, REF_f), another disjoint NP-pair that we believe to be hard (i.e., not separable by a set in P). We answer these questions in the strongest possible way. We prove that every disjoint NP-pair is polynomial-time, many-one equivalent to the canonical disjoint NP-pair of some propositional proof system. It follows immediately that every disjoint NP-pair we believe to be P-inseparable (cannot be separated

---

* The full paper is available at the Electronic Colloquium on Computational Complexity (ECCC), TR04-106.
** Research partially supported by NSF grant CCR-0307077.

by a set in P) is many-one equivalent to some pair (SAT^*, REF_f) that is also P-inseparable.

The interest in knowing whether a canonical pair of a propositional proof system f is P-separable or P-inseparable arises from the following considerations. First, if the canonical pair is P-inseparable, then f is not trivial (i.e., there exist infinitely many tautologies that have subexponential proofs in f). In other words, for infinitely many inputs, the proof system f beats the brute force algorithm in proving tautologies. Second, as shown by Razborov [Raz94], if the canonical pair is P-separable, then (under the assumption $P \neq NP$) it follows that f is not polynomial bounded.

This paper does not address the question of whether P-inseparable disjoint NP-pairs exist, but we mention that there is evidence for their existence, for example, if $P \neq UP$ or if $P \neq NP \cap coNP$. On the other hand, the hypothesis that $P \neq NP$ does not seem to be sufficient to obtain P-inseparable disjoint NP-pairs. Homer and Selman [HS92] constructed an oracle relative to which $P \neq NP$ and all disjoint NP-pairs are P-separable.

It is easy to see that if proof system f simulates proof system g, then the pair (SAT^*, REF_g) is many-one reducible to the pair (SAT^*, REF_f). A proof system is *optimal* if it simulates every other propositional proof system. Although it is an open question whether optimal proof systems exist, as we stated above, Razborov showed that if there exists an optimal propositional proof system f, then its canonical pair is a complete pair for DisjNP. We obtain this result of Razborov as a corollary of our result above.

Glaßer et al. [GSSZ04] constructed an oracle relative to which the converse of Razborov's result does not hold; i.e., relative to this oracle, using our current result, there is a propositional proof system f whose canonical pair is complete, but f is not optimal. Hence, there is a propositional proof system g such that the canonical pair of g many-one reduces to the canonical pair of f, but f does not simulate g. Our theorem presents a tight connection between disjoint NP-pairs and propositional proof systems. Nevertheless, relative to this oracle, the relationship is not as tight as we might hope for.

In light of our result above, by examining the degree structure of the class DisjNP, we can understand the degree structure of canonical pairs (SAT^*, REF_f). Thus, we should try to understand the degree structure of DisjNP. We prove that between any two comparable and inequivalent disjoint NP-pairs (A, B) and (C, D) there exist P-inseparable, incomparable NP-pairs (E, F) and (G, H) whose degrees lie strictly between (A, B) and (C, D). Our result is an analogue of Ladner's result for NP [Lad75], and our proof is based on Schöning's formulation [Sch82]. Thus, assuming that P-inseparable disjoint NP-pairs exist, the class DisjNP has a rich, dense, degree structure—and each of these degrees contains a canonical pair.

2 Preliminaries

A disjoint NP-pair is a pair (A, B) of nonempty sets A and B such that $A, B \in$ NP and $A \cap B = \emptyset$. Let DisjNP denote the class of all disjoint NP-pairs.

Given a disjoint NP-pair (A,B), a *separator* is a set S such that $A \subseteq S$ and $B \subseteq \overline{S}$; we say that S separates (A,B). Let $Sep(A,B)$ denote the class of all separators of (A,B). For disjoint NP-pairs (A,B), the fundamental question is whether $Sep(A,B)$ contains a set belonging to P. In that case the pair is P-*separable*; otherwise, the pair is P-*inseparable*. The following proposition summarizes known results about P-separability.

Proposition 1 ([GS88]).

1. $P \neq NP \cap coNP$ *implies* DisjNP *contains* P-*inseparable pairs*.
2. $P \neq UP$ *implies* DisjNP *contains* P-*inseparable pairs*.
3. *If* DisjNP *contains* P-*inseparable pairs, then* DisjNP *contains* P-*inseparable pairs of* NP-*complete sets*.

While it is probably the case that DisjNP contains P-inseparable pairs, there is an oracle relative to which $P \neq NP$ and P-inseparable pairs do not exist [HS92]. So $P \neq NP$ probably is not a sufficiently strong hypothesis to show existence of P-inseparable pairs in DisjNP.

We review the natural notions of reducibilities between disjoint pairs. The original notions are nonuniform [GS88]. Here we state only the known equivalent uniform versions [GS88, GSSZ04].

Definition 2. *Let (A,B) and (C,D) be disjoint pairs.*

1. (A,B) *is* many-one reducible in polynomial-time *to* (C,D), $(A,B)\leq_m^{pp}(C,D)$, *if there exists a polynomial-time computable function f such that $f(A) \subseteq C$ and $f(B) \subseteq D$.*
2. (A,B) *is* Turing reducible in polynomial-time *to* (C,D), $(A,B)\leq_T^{pp}(C,D)$, *if there exists a polynomial-time oracle Turing machine M such that for every separator S of (C,D), $L(M,S)$ is a separator of (A,B).*

If $(A,B)\leq_m^{pp}(C,D)$ and $(C,D)\leq_m^{pp}(A,B)$, then we write $(A,B)\equiv_m^{pp}(C,D)$; if $(A,B)\leq_T^{pp}(C,D)$ and $(C,D)\leq_T^{pp}(A,B)$, then we write $(A,B)\equiv_T^{pp}(C,D)$. Since we are interested only in comparing disjoint NP-pairs, it is convenient for us to define the Turing-degree of a pair $(A,B) \in$ DisjNP as follows:

$$\mathbf{d}(A,B) = \{(C,D) \in \text{DisjNP} \mid (A,B)\equiv_T^{pp}(C,D)\}.$$

Let TAUT denote the set of tautologies. Cook and Reckhow [CR79] defined a *propositional proof system* (proof system for short) to be a function $f : \Sigma^* \to$ TAUT such that f is onto and $f \in$ PF, where PF denotes the class of polynomial-time computable functions. The canonical pair of f [Raz94, Pud01] is the disjoint NP-pair (SAT^*, REF_f) where

$$SAT^* = \{(x,0^n) \mid x \in SAT\} \quad \text{and}$$
$$REF_f = \{(x,0^n) \mid \neg x \in TAUT \text{ and } \exists y[|y| \leq n \text{ and } f(y) = \neg x]\}.$$

Let f and f' be two propositional proof systems. We say that f *simulates* f' if there is a polynomial p and a function $h : \Sigma^* \to \Sigma^*$ such that for every $w \in \Sigma^*$, $f(h(w)) = f'(w)$ and $|h(w)| \leq p(|w|)$. A proof system is *optimal* if it simulates every other proof system.

3 Canonical Pairs of Proof Systems

Now we state the main result of this paper. We show that for every disjoint NP-pair (A, B) there exists a proof system f such that $(\text{SAT}^*, \text{REF}_f) \equiv_m^{pp} (A, B)$. This shows that disjoint NP-pairs and canonical pairs of proof systems have identical degree structures.

Theorem 3. *For every disjoint NP-pair (A, B) there exists a proof system f such that $(\text{SAT}^*, \text{REF}_f) \equiv_m^{pp} (A, B)$.*

Proof. Let $\langle \cdot, \cdot \rangle$ be a polynomial-time computable, polynomial-time invertible pairing function such that $|\langle v, w \rangle| = 2|vw|$. Choose g that is polynomial-time computable and polynomial-time invertible such that $A \leq_m^p \text{SAT}$ via g. Let M be an NP-machine that accepts B in time p. Define the following function f.

$$f(z) \stackrel{df}{=} \begin{cases} \neg g(x) & : \text{ if } z = \langle x, w \rangle,\ |w| = p(|x|),\ M(x) \text{ accepts along path } w \\ x & : \text{ if } z = \langle x, w \rangle,\ |w| \neq p(|x|),\ |z| \geq 2^{|x|},\ x \in \text{TAUT} \\ \text{true} & : \text{ otherwise} \end{cases}$$

The function is polynomial-time computable, since in the second case, $|z|$ is large enough so that $x \in \text{TAUT}$ can be decided in deterministic time $O(|z|^2)$. In the first case of f's definition, $x \in B$ and so $g(x) \notin \text{SAT}$. It follows that $f: \Sigma^* \to \text{TAUT}$. The mapping is onto, since for every tautology y,

$$f(\langle y, 0^{2^{|y|}} \rangle) = y.$$

Therefore, f is a propositional proof system.

Claim 4. $(\text{SAT}^*, \text{REF}_f) \leq_m^{pp} (A, B)$.

Choose elements $a \in A$ and $b \in B$. The reduction function h is as follows.

```
1  input (y,0ⁿ)
2  if n ≥ 2^|y| then
3      if y ∈ SAT then output a else output b
4  endif
5  if g⁻¹(y) exists then output g⁻¹(y)
6  output a
```

The exhaustive search in line 3 is possible in quadratic time in n. So $h \in \text{PF}$.

Assume $(y, 0^n) \in \text{SAT}^*$. If we reach line 3, then we output $a \in A$. Otherwise we reach line 5. If $g^{-1}(y)$ exists, then it belongs to A. Therefore, in either case (output in line 5 or in line 6) we output an element from A.

Assume $(y, 0^n) \in \text{REF}_f$ (in particular $\neg y \in \text{TAUT}$). So there exists z such that $|z| \leq n$ and $f(z) = \neg y$. If we reach line 3, then we output b. Otherwise we reach line 5 and so it holds that $|z| \leq n < 2^{|y|}$ and $\neg y$ syntactically differs from the expression true. Therefore, $f(z) = \neg y$ must be due to line 1 in the definition of f. It follows that $g^{-1}(y)$ exists. So we output $g^{-1}(y)$ which belongs to B (again by line 1 of f's definition). This shows Claim 4.

Claim 5. $(A,B) \leq_m^{pp} (\text{SAT}^*, \text{REF}_f)$.

The reduction function is $h'(x) \stackrel{df}{=} \langle g(x), 0^{2(|x|+p(|x|))}\rangle$. If $x \in A$, then $g(x) \in$ SAT and therefore, $h'(x) \in \text{SAT}^*$. Otherwise, let $x \in B$. Let w be an accepting path of $M(x)$ and define $z \stackrel{df}{=} \langle x, w \rangle$. So $|w| = p(|x|)$ and $|z| = 2(|x|+p(|x|))$. By line 1 in f's definition, $f(z) = \neg g(x)$. Therefore, $h'(x) \in \text{REF}_f$. This proves Claim 5 and finishes the proof of Theorem 3. □

Corollary 6. *Disjoint NP-pairs and canonical pairs for proof systems have identical degree structures.*

The following easy to prove proposition also states a strong connection between proof systems and disjoint NP-pairs:

Proposition 7. *Let f and g be propositional proof systems. If g simulates f, then*

$$(\text{SAT}^*, \text{REF}_f) \leq_m^{pp} (\text{SAT}^*, \text{REF}_g).$$

Proof. By assumption there exists a total function $h: \Sigma^* \to \Sigma^*$ and a polynomial p such that for all x, $g(h(x)) = f(x)$ and $|h(x)| \leq p(|x|)$. We claim that $(\text{SAT}^*, \text{REF}_f) \leq_m^{pp} (\text{SAT}^*, \text{REF}_g)$ via reduction r where $r(x, 0^n) \stackrel{df}{=} (x, 0^{p(n)})$. Clearly, if $(x, 0^n) \in \text{SAT}^*$, then $(x, 0^{p(n)}) \in \text{SAT}^*$ as well. Let $(x, 0^n) \in \text{REF}_f$, i.e., $\neg x$ is a tautology and there exists y such that $|y| \leq n$ and $f(y) = \neg x$. So for $y' \stackrel{df}{=} h(x)$ it holds that $|y'| \leq p(n)$ and $g(y') = \neg x$ which shows $(x, 0^{p(n)}) \in \text{REF}_g$. □

The following result of Razborov [Raz94] is an immediate consequence of Theorem 3 and Proposition 7.

Corollary 8 (Razborov). *If there exists an optimal propositional proof system f, then $(\text{SAT}^*, \text{REF}_f)$ is a complete NP-pair.*

We remind the reader that it is known neither whether there exists an optimal propositional proof system nor whether there exist complete NP-pairs. Now it is appropriate to repeat a comment we stated in the introduction. Glaßer et al. [GSSZ04] constructed an oracle relative to which the converse of Corollary 8 does not hold; i.e., relative to this oracle, by Theorem 3, there is a propositional proof system f whose canonical pair is complete, but f is not optimal. Hence, there is a propositional proof system g such that the canonical pair of g many-one reduces to the canonical pair of f, but f does not simulate g. The results of this section present tight connections between disjoint NP-pairs and propositional proof systems. Nevertheless, relative to this oracle, the relationship is not as tight as one might hope for.

4 Degree Structure of Disjoint NP-Pairs

Let $\{M_i\}_i$ be a standard effective enumeration of Turing machines. We require the following definition and theorems:

Definition 9. *We define a class $\mathcal{C}$ of nonempty disjoint NP-pairs to be effectively presentable if there exists a total computable function $f : \mathbb{N} \to \mathbb{N} \times \mathbb{N}$ such that*

1. *for all $(i,j) \in \mathrm{range}(f)$, M_i and M_j halt on all inputs, and*
2. $\mathcal{C} = \{(L(M_i), L(M_j)) \mid (i,j) \in \mathrm{range}(f)\}$.

Note that in item 1 we do not demand halting in polynomial-time.

Theorem 10. *For all $(A,B), (C,D) \in \mathrm{DisjNP}$, the following classes are effectively presentable.*

$$\mathcal{C}_1 \stackrel{df}{=} \{(X,Y) \in \mathrm{DisjNP} \mid (C,D) \leq_T^{pp} (X \oplus A, Y \oplus B)\}$$
$$\mathcal{C}_2 \stackrel{df}{=} \{(X,Y) \in \mathrm{DisjNP} \mid (X,Y) \leq_T^{pp} (A,B)\}$$

A disjoint pair (A', B') is called a *finite variation* of the pair (A, B) if $\|(A \triangle A') \cup (B \triangle B')\|$ is finite. A class $\mathcal{C}$ of disjoint pairs is *closed under finite variations* if for all disjoint pairs (A, B) and (A', B') it holds that if $(A, B) \in \mathcal{C}$, A' and B' are nonempty, and (A', B') is a finite variation of (A, B), then $(A', B') \in \mathcal{C}$.

For any function, define $f^n(x)$ to be the n-fold iteration of f on x ($f^0(x) = x$, $f^1(x) = f(x)$, and $f^{n+1}(x) = f(f^n(x))$). For any function f defined on the set of natural numbers, define

$$G[f] = \{x \in \Sigma^* \mid \text{there exists an even } n \text{ such that } f^n(0) \leq |x| < f^{n+1}(0)\}.$$

The following theorem is a version of Schöning's method [Sch82] for uniform diagonalization, applied to disjoint NP-pairs.

Theorem 11. *Let A, B, C, and D be infinite decidable sets such that (A, B) and (C, D) are disjoint pairs. Let $\mathcal{C}_1$ and $\mathcal{C}_2$ be classes of disjoint pairs with the following properties:*

- $(A, B) \notin \mathcal{C}_1$ *and* $(C, D) \notin \mathcal{C}_2$;
- $\mathcal{C}_1$ *and* $\mathcal{C}_2$ *are effectively presentable; and*
- $\mathcal{C}_1$ *and* $\mathcal{C}_2$ *are closed under finite variations.*

Then there exists a set $T \in \mathrm{P}$ such that the disjoint pair (E, F), where $E = (T \cap A) \cup (\overline{T} \cap C)$ and $F = (T \cap B) \cup (\overline{T} \cap D)$, has the following properties:

- $T \cap A$, $\overline{T} \cap A$, $T \cap B$, $\overline{T} \cap B$, $T \cap C$, $\overline{T} \cap C$, $T \cap D$, $\overline{T} \cap D$ *are infinite,*
- $(E, F) \notin \mathcal{C}_1 \cup \mathcal{C}_2$, *and*
- *if (A, B) is P-separable, then $(E, F) \leq_m^{pp} (C, D)$.*

Proof. Since $\mathcal{C}_1$ and $\mathcal{C}_2$ are effectively presentable, there exist total computable functions f_1 and f_2 such that

- for all $(i,j) \in \mathrm{range}(f_1) \cup \mathrm{range}(f_2)$, M_i and M_j halt on all inputs,
- $\mathcal{C}_1 = \{(L(M_i), L(M_j)) \mid (i,j) \in \mathrm{range}(f_1)\}$, and
- $\mathcal{C}_2 = \{(L(M_i), L(M_j)) \mid (i,j) \in \mathrm{range}(f_2)\}$.

Canonical Disjoint NP-Pairs of Propositional Proof Systems   405

Define the following functions:

$$g_1(n) = \max\left\{\left|\min\{z \mid (i,j) = f_1(k), |z| \geq n, \text{ and } z \in L(M_i) \triangle A \cup L(M_j) \triangle B\}\right| \mid k \leq n\right\}$$

$$g_2(n) = \max\left\{\left|\min\{z \mid (i,j) = f_2(k), |z| \geq n, \text{ and } z \in L(M_i) \triangle C \cup L(M_j) \triangle D\}\right| \mid k \leq n\right\}$$

$$g_3(n) = \min\{m \mid m \geq n \text{ and } \exists u,v,w,x \in \Sigma^{\geq n} \cap \Sigma^{\leq m} \text{ such that } u \in A,\ v \in B,\ w \in C,\text{ and } x \in D\}$$

We prove that g_1, g_2, and g_3 are total computable functions. Since $(A,B) \notin \mathcal{C}_1$, for all $(i,j) \in \text{range}(f_1)$, $(A,B) \neq (L(M_i), L(M_j))$. As $\mathcal{C}_1$ is closed under finite variations, $L(M_i) \triangle A \cup L(M_j) \triangle B$ is an infinite set. Thus, for all k, and for all $n \geq k$, there is a string z such that $|z| \geq n$ and $z \in L(M_i) \triangle A \cup L(M_j) \triangle B$, where $(i,j) = f_1(k)$. Observe that the relation defined by "$z \in L(M_i) \triangle A \cup L(M_j) \triangle B$ and $(i,j) = f_1(k)$" is decidable, because both A and B are decidable, both M_i and M_j halt on all inputs and f_1 is total computable. So the minimal such z is computable and hence g_1 is computable. Similar arguments show that g_2 and g_3 are total and computable (for g_3 we need A, B, C, and D to be infinite).

Since $\max(g_1, g_2, g_3) + 1$ is a total computable function, there exists a fast function[1] g such that for all n, $g(n) > \max(g_1(n), g_2(n), g_3(n))$ (We refer to Proposition 7.3 of the text by Homer and Selman [HS01].) Also, $G[g] \in \text{P}$ (Lemma 7.1, [HS01]). Now take $T = G[g]$. We prove that the pair (E,F), where $E = (T \cap A) \cup (\overline{T} \cap C)$ and $F = (T \cap B) \cup (\overline{T} \cap D)$, has the desired properties.

Suppose $T \cap A$ is finite. Choose an even integer n such that all words in $T \cap A$ are of length less than $g^n(0)$. Substituting $g^n(0)$ for n in the definition of g_3 implies that there exists a word $u \in A$ such that $g^n(0) \leq |u| \leq g_3(g^n(0)) < g^{n+1}(0)$. So $u \in A \cap T$ which contradicts the choice of n. Hence $T \cap A$ must be infinite. Similar arguments show that $\overline{T} \cap A$, $T \cap B$, $\overline{T} \cap B$, $T \cap C$, $\overline{T} \cap C$, $T \cap D$, $\overline{T} \cap D$ are infinite.

We turn to the second consequence. The definition of g_1 implies the following:

$$k \leq n \Rightarrow \exists z[n \leq |z| \leq g_1(n) \text{ and } z \in L(M_i) \triangle A \cup L(M_j) \triangle B, \qquad (1)$$
$$\text{where } f_1(k) = (i,j)]$$

Suppose $(E,F) \in \mathcal{C}_1$. Then, there exists k such that $(E,F) = (L(M_i), L(M_j))$, where $f_1(k) = (i,j)$. Select n to be an even positive integer such that $g^n(0) \geq k$. Substituting $g^n(0)$ for n in Equation (1), there is a string z such that $g^n(0) \leq |z| \leq g_1(g^n(0)) < g^{n+1}(0)$ and $z \in L(M_i) \triangle A \cup L(M_j) \triangle B$. Thus, $z \in T$ and $z \in L(M_i) \triangle A \cup L(M_j) \triangle B$, which implies, using the definition of (E,F), that $z \in L(M_i) \triangle E \cup L(M_j) \triangle F$. This is a contradiction. We conclude that $(E,F) \notin \mathcal{C}_1$. A similar argument shows that $(E,F) \notin \mathcal{C}_2$.

---

[1] A function $g : \mathbb{N} \to \mathbb{N}$ is called *fast* if (i) for all $n \in \mathbb{N}$, $f(n) > n$, and (ii) there is a Turing machine M that computes f in unary notation such that M writes a symbol on its output tape every move of its computation.

Now we show that the third consequence holds. Suppose (A, B) is P-separable. Let S be a separator of (A, B) that belongs to P. Let c and d be fixed words that belong to C and D, respectively. Consider the following function h:

$$h(x) = \begin{cases} x \text{ if } x \in \overline{T}, \\ c \text{ if } x \in T \text{ and } x \in S, \\ d \text{ if } x \in T \text{ and } x \notin S. \end{cases}$$

We claim that $(E, F) \leq_m^{pp} (C, D)$ via h. First it is clear that h is polynomial time computable since both T and S belong to P. Now suppose $x \in E$. If $x \in \overline{T}$, then $x \in C$. Hence, $h(x) = x \in C$. Otherwise $x \in A \subseteq S$. Hence, $h(x) = c \in C$. So in either case we have $h(x) \in C$. Therefore, $h(E) \subseteq C$. Similarly we can show that $h(F) \subseteq D$. □

Now we apply Theorem 11 to obtain the following result about the degree structure of disjoint NP-pairs. Observe that the premise of the following theorem is true as long as there exist P-inseparable disjoint NP-pairs. For under this hypothesis, we can take (A, B) to be P-separable and (C, D) to be P-inseparable. We obtain the full generality of the theorem, in which we do not assume that (A, B) is P-separable, by using a technique of Regan [Reg83, Reg88].

Theorem 12. *Suppose there exist disjoint* NP*-pairs* (A, B) *and* (C, D) *such that* A, B, C, *and* D *are infinite,* $(A,B) \leq_T^{pp} (C,D)$, *and* $(C,D) \not\leq_T^{pp} (A,B)$. *Then there exist incomparable, strictly intermediate disjoint* NP*-pairs* (E, F) *and* (G, H) *between* (A, B) *and* (C, D) *such that* E, F, G, *and* H *are infinite. Precisely, the following properties hold:*

- $(A,B) \leq_m^{pp} (E,F) \leq_T^{pp} (C,D)$ *and* $(C,D) \not\leq_T^{pp} (E,F) \not\leq_T^{pp} (A,B)$;
- $(A,B) \leq_m^{pp} (G,H) \leq_T^{pp} (C,D)$ *and* $(C,D) \not\leq_T^{pp} (G,H) \not\leq_T^{pp} (A,B)$;
- $(E,F) \not\leq_T^{pp} (G,H)$ *and* $(G,H) \not\leq_T^{pp} (E,F)$.

Proof. Define

$$\mathcal{C}_1 = \{(X, Y) \in \text{DisjNP} \mid (C,D) \leq_T^{pp} (X \oplus A, Y \oplus B)\} \quad \text{and}$$
$$\mathcal{C}_2 = \{(X, Y) \in \text{DisjNP} \mid (X,Y) \leq_T^{pp} (A, B)\}.$$

Clearly, $(A,B) \notin \mathcal{C}_1$ and $(C,D) \notin \mathcal{C}_2$. By Theorem 10, both $\mathcal{C}_1$ and $\mathcal{C}_2$ are effectively presentable. Also, it is obvious that $\mathcal{C}_1$ and $\mathcal{C}_2$ are closed under finite variations. Thus by Theorem 11, there exists a set $T \in P$ such that $(E', F') \notin \mathcal{C}_1 \cup \mathcal{C}_2$, where $E' = (T \cap A) \cup (\overline{T} \cap C)$ and $F' = (T \cap B) \cup (\overline{T} \cap D)$ are infinite sets. Clearly, $(E', F') \in \text{DisjNP}$, since both (A, B) and (C, D) belong to DisjNP and $T \in P$. Define $E = E' \oplus A$ and $F = F' \oplus B$. It is straightforward to see that (E, F) also belongs to DisjNP and $(A,B) \leq_m^{pp} (E,F)$. By the definition of $\mathcal{C}_1$ and $\mathcal{C}_2$, $(C,D) \not\leq_T^{pp} (E,F)$ and $(E,F) \not\leq_T^{pp} (A,B)$. In addition, we have the following claim:

Claim 13. $(E, F) \leq_T^{pp} (C, D)$.

Proof. Let S be a separator of (C, D). Since $(A, B) \leq_T^{pp} (C, D)$, there is a separator S_1 of (A, B) such that $S_1 \leq_T^p S$. Then $S_2 = (S_1 \cap T) \cup (S \cap \overline{T})$ is a separator of (E', F') and $S_2 \leq_T^p S_1 \oplus S \leq_T^p S$. Thus $S_3 = S_2 \oplus S_1$ is a separator of (E, F) and $S_3 \leq_T^p S$. □

The following summarizes the properties we proved so far:

- $(A, B) \leq_m^{pp} (E, F) \leq_T^{pp} (C, D)$;
- $(C, D) \not\leq_T^{pp} (E, F) \not\leq_T^{pp} (A, B)$.

Now we define the pair (G, H). It follows from the proof of Theorem 11 that if we take $T' = \overline{T} = \overline{G[g]}$, then all the consequences of the theorem are satisfied as well. So we define

$$G' = (T' \cap A) \cup (\overline{T'} \cap C) = (\overline{T} \cap A) \cup (T \cap C)$$

and

$$H' = (T' \cap B) \cup (\overline{T'} \cap D) = (\overline{T} \cap B) \cup (T \cap D).$$

Then we have $(G', H') \notin \mathcal{C}_1 \cup \mathcal{C}_2$. Similarly we define $G = G' \oplus A$ and $H = H' \oplus B$. By the same arguments as above, the following properties hold for (G, H):

- $(A, B) \leq_m^{pp} (G, H) \leq_T^{pp} (C, D)$;
- $(C, D) \not\leq_T^{pp} (G, H) \not\leq_T^{pp} (A, B)$.

It remains to show $(E, F) \not\leq_T^{pp} (G, H)$ and $(G, H) \not\leq_T^{pp} (E, F)$. We show only that $(E, F) \not\leq_T^{pp} (G, H)$ since the proof of the latter is identical. The proof follows from the following two claims:

Claim 14. $(C, D) \leq_m^{pp} (E \oplus G, F \oplus H)$.

Proof. We define the reduction f as follows: On input x, if $x \in T$, then $f(x) = 10x$, and, if $x \notin T$, then $f(x) = 00x$. We need to prove that $f(C) \subseteq E \oplus G$ and $f(D) \subseteq F \oplus H$. Suppose that $x \in C$. Consider the case that $x \in T$. By definition of G', $x \in G'$. So $0x \in G$. Hence, $f(x) = 10x \in E \oplus G$. In the case that $x \notin T$, we have $x \in E'$. So $0x \in E$. Hence, $f(x) = 00x \in E \oplus G$. Thus, $f(C) \subseteq E \oplus G$. The proof that $f(D) \subseteq F \oplus H$ is similar. □

Claim 15. If $(E, F) \leq_T^{pp} (G, H)$ then $(E \oplus G, F \oplus H) \leq_T^{pp} (G, H)$

Proof. Let S be a separator of (G, H). By the hypothesis, there is a separator S' of (E, F) such that $S' \leq_T^p S$. Then $S' \oplus S$ is a separator of $(E \oplus G, F \oplus H)$ and $S' \oplus S \leq_T^p S$. □

Now we see that if $(E, F) \leq_T^{pp} (G, H)$, then $(C, D) \leq_m^{pp} (E \oplus G, F \oplus H) \leq_T^{pp} (G, H)$, which is a contradiction. □

Corollary 16. *Suppose there exists a P-inseparable disjoint NP-pair (C, D). Let (A, B) be a P-separable disjoint NP-pair such that A and B are infinite. Then there exist incomparable, P-inseparable, strictly intermediate disjoint NP-pairs (E, F) and (G, H) between (A, B) and (C, D) that satisfy all of the consequences of Theorem 12, and in addition, satisfy the following conditions:*

- $(A, B) \leq_m^{pp} (E, F) \leq_m^{pp} (C, D)$, and
- $(A, B) \leq_m^{pp} (G, H) \leq_m^{pp} (C, D)$.

The proof follows readily.

From Corollary 16, Theorem 3, and Proposition 7 it follows that if there exist P-inseparable disjoint NP-pairs, then there exist propositional proof systems f and g so that f does not simulate g and g does not simulate f. However, Messner [Mes00, Mes02] unconditionally proved the existence of propositional proof systems f and g such that f does not simulate g and g does not simulate f. Messner further shows that the simulation order of propositional proof systems is dense. However, as the following argument shows, these results do not replace our study of the degree structure of disjoint NP-pairs. There exist infinite, strictly increasing chains of propositional proof systems (using simulation as the order relation $\leq$) such that all canonical pairs of these proofs systems belong to the same many-one degree of disjoint NP-pairs.

First, observe that for every non-optimal propositional proof system f there is a proof system g such that g simulates f, but f does not simulate g (i.e., $f < g$). (For example, for some h that is not simulated by f, let $g(x) = f(x/2)$ if x is even and $g(x) = h((x-1)/2)$ otherwise.) Glaßer et al. [GSSZ04] constructed an oracle O_2 relative to which many-one complete disjoint NP-pairs exist, but optimal propositional proof systems do not exist. So relative to this oracle, there is a proof system f_0 whose canonical pair is complete, but optimal proof systems do not exist. By our observation, there exists an infinite, strictly increasing chain of proof systems $f_0 < f_1 < \cdots$. However, by Proposition 7, the canonical pair of each f_i is many-one complete.

Acknowledgements. The authors thank Jochen Messner and Kenneth W. Regan for informing them of the methods in their papers [mes00, mes02, reg83, reg88].

References

[CR79]   S. Cook and R. Reckhow. The relative efficiency of propositional proof systems. *Journal of Symbolic Logic*, 44:36–50, 1979.

[GS88]   J. Grollmann and A. Selman. Complexity measures for public-key cryptosystems. *SIAM Journal on Computing*, 17(2):309–335, 1988.

[GSSZ04]   C. Glaßer, A. Selman, S. Sengupta, and L. Zhang. Disjoint NP-pairs. *SIAM Journal on Computing*, 33(6):1369–1416, 2004.

[HS92]   S. Homer and A. Selman. Oracles for structural properties: The isomorphism problem and public-key cryptography. *Journal of Computer and System Sciences*, 44(2):287–301, 1992.

[HS01]   S. Homer and A. Selman. *Computability and Complexity Theory.* Texts in Computer Science. Springer, New York, 2001.
[Lad75] R. Ladner. On the structure of polynomial-time reducibility. *Journal of the ACM*, 22:155–171, 1975.
[Mes00] J. Messner. *On the Simulation Order of Proof Systems.* PhD thesis, Universität Ulm, 2000.
[Mes02] J. Messner. On the structure of the simulation order of proof systems. In *Proceedings 27rd Mathematical Foundations of Computer Science*, Lecture Notes in Computer Science 1450, pages 581–592. Springer-Verlag, 2002.
[Pud01] P. Pudlák. On reducibility and symmetry of disjoint NP-pairs. In *Proceedings 26th International Symposium on Mathematical Foundations of Computer Science*, volume 2136 of *Lecture Notes in Computer Science*, pages 621–632. Springer-Verlag, Berlin, 2001.
[Raz94] A. Razborov. On provably disjoint NP-pairs. Technical Report TR94-006, Electronic Colloquium on Computational Complexity, 1994.
[Reg83] K. Regan. On diagonalization methods and the structure of language classes. In *Proceedings Foundations of Computation Theory*, volume 158 of *Lecture Notes in Computer Science*, pages 368–380. Springer Verlag, 1983.
[Reg88] K. Regan. The topology of provability in complexity theory. *Journal of Computer and System Sciences*, 36:384–432, 1988.
[Sch82] U. Schöning. A uniform approach to obtain diagonal sets in complexity classes. *Theoretical Computer Science*, 18:95–103, 1982.

Complexity of DNF
and Isomorphism of Monotone Formulas

Judy Goldsmith[1], Matthias Hagen[2,*], and Martin Mundhenk[2]

[1] University of Kentucky, Dept. of Computer Science, Lexington, KY 40506–0046
goldsmit@cs.uky.edu
[2] Friedrich-Schiller-Universität Jena, Institut für Informatik, D–07737 Jena
{hagen, mundhenk}@cs.uni-jena.de

Abstract. We investigate the complexity of finding prime implicants and minimal equivalent DNFs for Boolean formulas, and of testing equivalence and isomorphism of monotone formulas. For DNF related problems, the complexity of the monotone case strongly differs from the arbitrary case. We show that it is DP-complete to check whether a monomial is a prime implicant for an arbitrary formula, but checking prime implicants for monotone formulas is in L. We show PP-completeness of checking whether the minimum size of a DNF for a monotone formula is at most k. For k in unary, we show the complexity of the problem to drop to coNP. In [Uma01] a similar problem for arbitrary formulas was shown to be Σ_2^p-complete. We show that calculating the minimal DNF for a monotone formula is possible in output-polynomial time if and only if P = NP. Finally, we disprove a conjecture from [Rei03] by showing that checking whether two formulas are isomorphic has the same complexity for arbitrary formulas as for monotone formulas.

1 Introduction

Monotone formulas are Boolean formulas that contain only conjunction and disjunction as connectives, but no negation. To solve the satisfiability problem for monotone formulas is trivial. Every satisfiable monotone formula is satisfied by the assignment that sets all variables to true. Hence, the computational complexity of the satisfiability problem for monotone formulas is very much simpler than the NP-complete satisfiability problem for arbitrary formulas. On the other hand, counting the number of satisfying assignments has the same complexity for monotone and for arbitrary formulas [Val79]. Hence, it is interesting to compare the complexity of problems for arbitrary and for monotone formulas.

In the first part of this paper, we investigate the complexity of calculating smallest equivalent Disjunctive Normal Forms (DNF). The smallest equivalent DNF for a formula consists of prime implicants of the formula. For arbitrary formulas, it is hard to find the smallest choice of prime implicants. It is still open whether a smallest equivalent DNF can be calculated in polynomial

---

* The author is supported by a Landesgraduiertenstipendium Thüringen.

space. For monotone formulas, the smallest DNF consists of all prime implicants. We consider problems of checking and finding prime implicants (Section 3). We show that checking whether a monomial is a prime implicant for a formula is DP-complete for arbitrary formulas, whereas it is in L for monotone formulas. DP [PY84] contains both NP and coNP and is contained in Σ_2^p in the Polynomial Time Hierarchy. The question, whether a prime implicant of a certain size exists for a given formula, was shown to be Σ_2^p-complete in [Uma01]. We show that the same question is only NP-complete for monotone formulas. The complexity of calculating the size of a smallest DNF depends on the representation of the problem. Umans [Uma01] showed that given a formula φ in DNF and an integer k, it is Σ_2^p-complete to decide whether φ has a DNF of size at most k. Notice that the length of the input DNF is greater than the size of the DNF that is searched for (except trivial cases). This seems necessary to allow the problem to be decided within a non-deterministic polynomial time bound, because the smallest DNF of a (monotone) formula may have size exponential in the length of the formula. The exact complexity of this problem for arbitrary formulas is open. It is Σ_2^p-hard (which follows from [Uma01]) and in EXPTIME. For monotone formulas, we exactly characterize the complexity of this problem by showing it to be PP-complete. If one encodes the upper bound of the DNF length in unary instead — i.e. given formula φ and string 1^k, decide whether φ has a DNF of size $\leq k$ — we prove the problem to be Σ_2^p-complete for arbitrary formulas, and coNP-complete for monotone formulas.

In Section 4 we consider the hardness of calculating the smallest DNF for a monotone formula. It is clear that the smallest DNF is not polynomial time computable. Therefore, we consider the notion of output-polynomial time. A function is in output-polynomial time if it can be computed in time polynomial in the length of the input plus the length of the function value [Pap97]. We show that the DNF for a monotone formula is output-polynomial time computable if and only if P = NP. Even calculating the size of a minimal DNF is shown to be PP-complete. In Section 5 we consider equivalence and isomorphism problems. The problem of deciding whether monotone formulas φ and ψ are equivalent is known to be coNP-complete [Rei03]. For arbitrary formulas the same completeness holds. If φ is in Conjunctive Normal Form (CNF) and ψ is in DNF, the equivalence problem remains coNP-complete for arbitrary formulas, but for monotone formulas an upper bound between P and coNP-complete was settled in [FK96]. In the case that one of the input formulas consists only of terms of bounded length, we give an L-algorithm improving results from [EG95, BEGK00]. Finally, we refute a conjecture from [Rei03], by showing that checking whether two given formulas are isomorphic has exactly the same complexity for arbitrary as for monotone formulas.

2 Definitions

We consider Boolean formulas with connectives ∧ (conjunction), ∨ (disjunction), and ¬ (negation). We assume that the negations appear directly in front of variables. (Other connectives are used as abbreviations, whereas we use the $\leftrightarrow$ only once because of the doubling of the formula length.) Actually this is no limitation because every formula may be transformed in polynomial time to fulfill these assumptions. A *monotone formula* is a Boolean formula without negations. A *term* is a conjunction or a disjunction of *literals*, i.e. of variables and negated variables; a conjunction is called *monomial*, and a disjunction is called *clause*. The empty clause is unsatisfiable, and the empty monomial, denoted λ, is valid. A *monotone term* is a term without negations. Terms are also considered as sets of literals. Term T_1 *covers* term T_2 if $T_1 \subseteq T_2$.

An *assignment* $\mathcal{A}$ for a formula φ is a mapping of the variables of φ to the truth values true and false. An assignment $\mathcal{A}$ is said to *satisfy* formula φ if φ evaluates to true under $\mathcal{A}$. For monotone formulas we regard $\mathcal{A}$ also as a set $\mathcal{A}_m$ where variable x is in $\mathcal{A}_m$ if and only if x gets value true under $\mathcal{A}$. Notice that in this way every monotone monomial can also be interpreted as an assignment.

An *implicant* of a formula φ is a monomial C such that $C \to \varphi$ is valid. A monomial C is a *prime implicant* of φ iff (1) C is an implicant of φ and (2) for every proper subset $S \subset C$ holds that S is not an implicant of φ. Notice that in order to check condition (2) it suffices to check for $C = \{\ell_1, \ell_2, \ldots, \ell_k\}$ whether for each $\ell_i \in C$ it holds that $C - \{\ell_i\}$ is not an implicant of φ. Every proper subset S of C is a subset of $C - \{\ell_i\}$ for some i. For $S \subseteq C - \{\ell_i\}$ holds that $(C - \{\ell_i\}) \to S$ is valid. If S is an implicant of φ, then $S \to \varphi$ is valid. Both valid formulas together yield that $(C - \{\ell_i\}) \to \varphi$ is valid too, inducing that $C - \{\ell_i\}$ is an implicant of φ. Hence, if no $C - \{\ell_i\}$ is an implicant of φ, then no proper subset of C is an implicant of φ.

A formula is in *conjunctive normal form (CNF)* if it is a conjunction of clauses. Similarly a formula is in *disjunctive normal form (DNF)* if it is a disjunction of monomials. It is said to be in k-CNF (k-DNF), if all clauses (monomials) consist of at most k literals. A monotone formula φ in normal form is *irredundant* if and only if no term of φ covers another term of φ. For a monotone formula, the disjunction of all its prime implicants yields an equivalent monotone DNF. On the other hand, every prime implicant must appear in every equivalent DNF for a monotone formula. Hence, the smallest DNF for a monotone formula is unique and equals the disjunction of all its prime implicants. This is not the case for non-monotone formulas, where the smallest DNF is a subset of the set of all prime implicants. It is NP-hard to select the right prime implicants [Mas79]. See also [Czo99] for an overview on the complexity of calculating DNFs.

We use complexity classes L (logarithmic space), P, NP, coNP, DP (difference polynomial time, which appears to be the class for "exact cost" optimization), Σ_2^p (NP with NP oracle), PP (probabilistic polynomial time), and PSPACE. The inclusion structure is $\mathsf{L} \subseteq \mathsf{P} \subseteq \begin{smallmatrix}\mathsf{NP}\\\mathsf{coNP}\end{smallmatrix} \subseteq \mathsf{DP} \subseteq \begin{smallmatrix}\Sigma_2^p\\\mathsf{PP}\end{smallmatrix} \subseteq \mathsf{PSPACE}$. All considered classes except L are closed downwards under $\leq_m^p$-reduction, and PP is closed under complement. Closely related to PP is the function class #P. See [Pap94] for definitions of these classes. As natural complete problems for NP, coNP and PP we consider SAT (is the Boolean formula φ satisfiable?), UNSAT (is φ unsatisfiable?), and MAJSAT (do at least half of the assignments satisfy φ?). We assume that the input formulas for these problems contain only $\wedge$, $\vee$, and $\neg$ (placed right in front of variables) as connectives.

3 Size of Disjunctive Normal Forms

In this section we concentrate on computing the size of minimal DNFs. Therefore we first analyze the complexity of finding prime implicants.

A valid formula has the empty monomial λ as its only prime implicant. An unsatisfiable formula has no prime implicant at all. In general, a formula φ has a prime implicant if and only if φ is satisfiable. Therefore, the question of whether a formula has a prime implicant is NP-complete, and it is in L for monotone formulas.

Next we consider the problem of deciding whether a monomial is a prime implicant of a formula.

> ISPRIMI : *instance:* Boolean formula φ and monomial C
> *question:* is C a prime implicant of φ ?

The complexity of ISPRIMI is intermediate between NP $\cup$ coNP and Σ_2^p.

Theorem 3.1. ISPRIMI *is* DP-*complete.*

Proof. The standard DP-complete problem is SAT-UNSAT $= \{(\varphi,\psi) \mid \varphi \in$ SAT, $\psi \in$ UNSAT$\}$. We show that SAT-UNSAT $\leq_m^p$ ISPRIMI. The reduction function is the mapping $(\varphi,\psi) \mapsto (\neg\varphi \vee (\neg\psi \wedge z), z)$, where z is a new variable that neither appears in φ nor in ψ. It is clear that this mapping is polynomial time computable.

If $(\varphi,\psi) \in$ SAT-UNSAT, then $\neg\psi$ is valid, and therefore $\neg\psi \wedge z$ has z as prime implicant. Hence z is an implicant of $\neg\varphi \vee (\neg\psi \wedge z)$. Because $\varphi \in$ SAT, its negation $\neg\varphi$ is not valid. Therefore, the empty monomial λ is not an implicant of $\neg\varphi$. Hence, z is a prime implicant of $\neg\varphi \vee (\neg\psi \wedge z)$. Next we consider the case that $(\varphi,\psi) \notin$ SAT-UNSAT. If $\varphi \notin$ SAT, then $\neg\varphi \vee (\neg\psi \wedge z)$ is valid and λ is the only prime implicant of this formula. If $\varphi \in$ SAT and $\psi \notin$ UNSAT, then $\neg\psi$ is not valid and therefore z is not an implicant of $\neg\psi \wedge z$. Because $\neg\varphi$ is not valid, it follows that z is neither an implicant of $\neg\varphi \vee (\neg\psi \wedge z)$.

This proves the DP-hardness of ISPRIMI. We now show that ISPRIMI is in DP, by proving ISPRIMI $\leq_m^p$ SAT-UNSAT. Let (φ,C) be an instance for ISPRIMI, where $C = \{\ell_1,\ell_2,\ldots,\ell_k\}$ is a monomial. Let $C(i) = C - \{\ell_i\}$. First, a necessary condition for C being a prime implicant for φ is that no proper subset of C is an implicant for φ (C is called *prime* in this case). This condition is equivalent to every $C(i)$ being not an implicant for φ (as argued in Section 2). I.e., for every i, $C(i) \to \varphi$ is not valid, and equivalently $\neg(C(i) \to \varphi)$ is in SAT. Summarized, if C is a prime implicant for φ, then $p(\varphi, C) = \bigwedge_{i=1}^k \neg(C(i) \to \varphi)$ is in SAT. Second, the other necessary condition for C being a prime implicant for φ is that C is an implicant for φ. I.e. $C \to \varphi$ is valid, and equivalently $n(\varphi, C) = \neg(C \to \varphi)$ is in UNSAT.

C is a prime implicant for φ if and only if C is prime and C is an implicant for φ. This is equivalent to $p(\varphi, C) \in$ SAT and $n(\varphi, C) \in$ UNSAT. Eventually, this yields that $(\varphi, C) \in$ IsPRIMI if and only if $(p(\varphi, C), n(\varphi, C)) \in$ SAT-UNSAT. Since p and n are both polynomial time computable, the function f with $f(\varphi, C) = ((p(\varphi, C), n(\varphi, C))$ is a polynomial time reduction function for IsPRIMI $\leq_m^p$ SAT-UNSAT. Since DP is closed downwards under polynomial time many-one reduction, IsPRIMI $\in$ DP follows. □

For monotone formulas, the same problem is much easier. A monomial is an implicant of a monotone formula, if and only if the assignment that corresponds to the monotone monomial satisfies the formula. It can be checked in logarithmic space whether an assignment satisfies a monotone formula.

IsPRIMI$_{\text{mon}}$: *instance:* monotone formula φ and monotone monomial C
  question: is C a prime implicant of φ ?

Theorem 3.2. IsPRIMI$_{\text{mon}}$ *is in* L.

The problem of checking whether a formula φ has a prime implicant of size at most k was shown to be Σ_2^p-complete [Uma01]. We show, that the same problem for monotone formulas is NP-complete only.

PRIMISIZE$_{\text{mon}}$: *instance:* monotone Boolean formula φ and integer k
  question: does φ have a prime implicant consisting of at most k variables?

In the following, we will define reductions that transform formulas into monotone formulas, such that satisfying assignments of the basic formula are similar to prime implicants of the monotone formula.

Definition 3.3. *Let φ be a Boolean formula with connectives $\wedge$, $\vee$ and $\neg$, and variables $x_1, \ldots, x_n$. Remember that all negation signs appear directly in front of variables. Then $r(\varphi)$ denotes the formula obtained by replacing all appearances of $\neg x_i$ in φ by the new variable y_i (for $i = 1, 2, \ldots, n$). Let $c(\varphi)$ denote the conjunction $\bigwedge_{i=1}^{n}(x_i \vee y_i)$ and $d(\varphi)$ denote the disjunction $\bigvee_{i=1}^{n}(x_i \wedge y_i)$. The formulas φ^c and φ^{cd} are defined as $\varphi^c = r(\varphi) \wedge c(\varphi)$ and $\varphi^{cd} = \varphi^c \vee d(\varphi) = (r(\varphi) \wedge c(\varphi)) \vee d(\varphi)$.*

Since φ^c and φ^{cd} do not contain any negation signs, they are monotone formulas. Let $\mathcal{A}$ be an assignment for φ. Then $\mathcal{A}'_m$ denotes the assignment $\mathcal{A}'_m = \{x_i \mid \mathcal{A} \text{ maps } x_i \text{ to true}\} \cup \{y_i \mid \mathcal{A} \text{ maps } x_i \text{ to false}\}$. Such an assignment, which contains exactly one from x_i and y_i, is called *conform*. Notice that there is a one-to-one relation between assignments to φ and conform assignments to φ^c and φ^{cd}.

Theorem 3.4. PRIMISIZE$_{\text{mon}}$ *is* NP-*complete*.

Proofsketch: PRIMISIZE$_{\text{mon}}$ is easily seen to be in NP. NP-hardness follows from a reduction SAT $\leq_m^p$ PRIMISIZE$_{\text{mon}}$. The reduction function maps every SAT instance φ with variables $x_1, \ldots, x_n$ to (φ^c, n). This reduction is polynomial time computable. □

Since minimal DNFs of a formula are disjunctions of prime implicants, a natural question arises. How hard is it to calculate the size of a minimal DNF? The respective problem

MINDNFSIZE$_{\text{dnf}}$: *instance:* Boolean formula φ in DNF and integer k
question: does φ have a DNF with at most k occurences of variables?

for arbitrary DNF formulas was shown to be Σ_2^p-complete in [Uma01].

The monotone version — input is a monotone DNF-formula φ — is in L since counting the length of the irredundant part of φ suffices and testing irredundancy can be managed in logarithmic space.

If the input is an arbitrary formula, the problem is Σ_2^p-hard (which follows from the latter result from [Uma01]). It is clear that the problem is in EXPTIME, but it is even not known whether the problem is in PSPACE. We show PP-completeness when the input is monotone.

MINDNFSIZE$_{\text{mon}}$: *instance:* monotone Boolean formula φ and integer k
question: does φ have a DNF with at most k occurences of variables?

Theorem 3.5. MINDNFSIZE$_{\text{mon}}$ *is* PP-*complete.*

Proof. A set A is in PP, if there exists a polynomial time bounded non-deterministic machine M that on input x has at least as many accepting as rejecting computation paths iff $x \in A$. The machine M is allowed to have accepting, rejecting, and non-deciding computation paths. Our polynomial time machine M that decides MINDNFSIZE$_{\text{mon}}$ roughly works as follows. Consider input (φ, k). Let l be the maximum length of a monomial with variables from φ. Then M guesses a sequence w of $l + 1$ bits. If the first bit of w equals 0, then it accepts, if the remaining bits encode an integer $< k$ — otherwise it halts non-deciding. In this way, k accepting computation paths are produced. If $w = 1v$ has first bit 1, then M checks in polynomial time (Theorem 3.2) whether v encodes a prime implicant (with variables in increasing order) for φ. If not, then it halts undecided. If yes, then this computation path splits in that many rejecting paths as the monomial v has variables. The smallest DNF of a monotone formula consists of all prime implicants of the formula. Hence, M on input (φ, k) has at least as many accepting as rejecting computation paths if and only if φ has a DNF with at most k occurences of variables. This shows that MINDNFSIZE$_{\text{mon}}$ is in PP.

To show PP-hardness, we give a reduction MAJSAT $\leq_m^p$ MINDNFSIZE$_{\text{mon}}$. Consider an instance φ of MAJSAT where φ has n variables, and let φ^{cd} be as in Definition 3.3. Observe that every prime implicant of φ^{cd} either is conform and hence consists of n variables, or it is not conform and consists of two variables x_i, y_i. If $\varphi \in $ MAJSAT, then there are at least 2^{n-1} satisfying assignments to φ. Every satisfying assignment of φ induces a conform prime implicant of φ^{cd}. Every $i \in \{1, 2, \ldots, n\}$ induces the non-conform prime implicant $x_i \wedge y_i$. Hence, there are at least 2^{n-1} conform and n non-conform prime implicants of φ^{cd}. Because the minimum DNF consists exactly of all prime implicants, it follows that the minimum DNF of φ^{cd} has size at least $n \cdot 2^{n-1} + 2 \cdot n$. If $\varphi \notin \text{MajSat}$, then the minimum DNF of φ^{cd} has size at most $n \cdot (2^{n-1} - 1) + 2 \cdot n$. The function that maps φ to $(\varphi^{cd}, n \cdot (2^{n-1} - 1) + 2 \cdot n)$ is polynomial time computable, and by the above observations it reduces $\overline{\text{MajSat}}$ to $\text{MinDnfSize}_{\text{mon}}$. Since PP is closed under complement, the PP-hardness of $\text{MinDnfSize}_{\text{mon}}$ follows. □

Accordingly, we can show that the function that on input a monotone formula φ outputs the size of the smallest DNF of φ is #P-complete. In [Val79] it is shown that computing the number of prime implicants of a monotone formula is #P-complete. Our result extends the latter since it additionally takes the size of the prime implicants into account.

Using a similar approach, one can show that counting satisfying assignments for monotone formulas and counting prime implicants for monotone formulas both are PP-complete. Notice that in [Val79] it is shown that given a monotone formula in 2CNF (all clauses consist of at most two variables) the function that calculates the number of satisfying assignments is #P-complete. From this result, it only follows that the problem to decide whether a monotone formula in 2CNF with n variables has at least 2^{n-1} satisfying assignments is PP-complete under polynomial time *Turing* reductions. Our approach yields PP-completeness under the stronger polynomial time many-one reduction.

One of the main reasons that an analogue to Theorem 3.5 for arbitrary formulas is unknown is the fact that polynomial time does not allow on input φ, k to guess a candidate for a DNF of length k. Therefore, we consider a variant of MinDnfSize where k is given in unary.

$\text{MinDnfSize}'$: *instance:* Boolean formula φ and string 1^k
  question: does φ have a DNF with at most k occurences of variables?

Theorem 3.6. $\text{MinDnfSize}'$ *is* Σ_2^p-*complete.*

Proof. $\text{MinDnfSize}_{\text{dnf}}$ reduces to $\text{MinDnfSize}'$ by the following function f. Let $|\varphi|$ denote the number of occurences of variables in φ. If $k \geq |\varphi|$, then $(\varphi, k) \in \text{MinDnfSize}_{\text{dnf}}$ and $f(\varphi, k)$ is some fixed element in $\text{MinDnfSize}'$. If $k < |\varphi|$, then $f(\varphi, k) = (\varphi, 1^k)$. Clearly, f is polynomial time computable and reduces the problem $\text{MinDnfSize}_{\text{dnf}}$ to $\text{MinDnfSize}'$. $\text{MinDnfSize}' \in \Sigma_2^p$ can be shown using the standard guess-and-check approach. □

If we restrict the input to be monotone the complexity is lower.

$\text{MinDnfSize}'_{\text{mon}}$: *instance:* monotone Boolean formula φ and string 1^k
  question: does φ have a DNF with at most k occurences of variables?

Theorem 3.7. $\text{MinDnfSize}'_{\text{mon}}$ *is* coNP-*complete.*

Proof. $\text{MinDnfSize}'_{\text{mon}}$ is coNP-hard: A formula φ is unsatisfiable if and only if φ^{cd} has $(x_i \wedge y_i)$ as its only prime implicants (where $i = 1, 2, \ldots, n$ for $x_1, \ldots, x_n$ are the variables of φ). Hence, φ is unsatisfiable if and only if $(\varphi^{cd}, 1^{2n}) \in \text{MinDnfSize}'_{\text{mon}}$. This shows that $\text{MinDnfSize}'_{\text{mon}}$ is coNP-hard.

MINDNFSIZE$'_{\text{mon}}$ ∈ coNP: Consider the problem $A = \{(\varphi, 1^k)|$ the monotone formula φ has a minimal DNF of size $> k\}$. Note that A is the complement of MINDNFSIZE$'_{\text{mon}}$. A is in NP since one has to guess a disjunction D of monomials of size greater than k and less than $k + |\varphi|$ and check that all are different prime implicants for φ. If so, then the minimal DNF for φ has at least the size of D. Both the guess and the check are polynomial time computable. Hence, MINDNFSIZE$'_{\text{mon}}$ ∈ coNP. □

4 Computing DNFs

A DNF of a formula is a disjunction of (prime) implicants. For monotone formulas, the minimal DNF is unique and it is the disjunction of *all* prime implicants. In order to investigate the complexity of the search for all prime implicants, we use the following problem MOREPRIMI$_{\text{mon}}$. It has instances (φ, S), where φ is a formula and S is a set of monomials. A pair (φ, S) belongs to MOREPRIMI$_{\text{mon}}$ if S is a proper subset of a minimal DNF of φ. I.e., every monomial in S is a prime implicant for φ, but there is at least one more prime implicant for φ that must be added to S in order to make S a DNF for φ.

MOREPRIMI$_{\text{mon}}$:
> *instance:* monotone Boolean formula φ and set S of monomials
> *question:* is S a set of prime implicants of φ and $\varphi \not\equiv S$?

Theorem 4.1. *MOREPRIMI$_{\text{mon}}$ is* NP*-complete.*

There are monotone formulas whose minimal DNF have size exponential in the size of the formula. Therefore it is clear that the DNF cannot be computed in time polynomial in the length of the *input*. For such problems one would like to have algorithms that run in time polynomial in the length of the input plus the length of the output.

Definition 4.2. *[Pap97] A function f can be computed in* output-polynomial time, *if there is an algorithm A that for all x on input x outputs $f(x)$ and there is a polynomial q such that for all x, A on input x has running time $q(|x|+|f(x)|)$.*

An algorithm that cycles through all monomials and outputs those that are prime implicants of the monotone input formula, eventually outputs the minimal DNF of its input. For the special case of formulas that have long DNFs, this algorithm can be seen to have running time polynomial in the length of the output. For formulas with short DNFs, the running time of this straightforward algorithm is exponential in the length of the output. Anyway, we show that we cannot expect to find an algorithm that behaves significantly better than this straightforward approach.

Theorem 4.3. *The function that on input a monotone formula φ outputs the smallest DNF for φ is in output-polynomial time if and only if* P = NP.

Proof. Assume that A is an output-polynomial time algorithm for the considered problem, and let q be the polynomial bounding the run time of A. We show how to solve MOREPRIMI$_{\text{mon}}$ in polynomial time. For an instance (φ, S) of MOREPRIMI$_{\text{mon}}$, first check whether S is a set of prime implicants for φ, and reject if this is not the case. Then start A on input φ for $q(|\varphi| + |S|)$ steps. If A does not halt after $q(|\varphi|+|S|)$ steps, then S does not contain all prime implicants of φ, and our algorithm accepts. If A halts after $q(|\varphi| + |S|)$ steps, then accept if and only if S is a proper subset of the output of A. It is clear that this algorithm decides MOREPRIMI$_{\text{mon}}$. Its run time is bounded by the polynomial q, plus some polynomial overhead. Since MOREPRIMI$_{\text{mon}}$ is NP-complete (Theorem 4.1), it solves an NP-complete problem in polynomial time, and therefore P = NP.

For the other proof direction, assume that P = NP. The set $V = \{(w, S, \varphi) \mid w$ is a prefix of a prime implicant C for φ and $C \notin S\ \}$ is in NP. Our algorithm that computes a minimal DNF of a monotone input formula φ starts with S being the empty set, and uses V as an oracle to make S the set of all prime implicants of φ — and hence the minimal DNF of φ — using a prefix search technique. Intuitively spoken, every query to V yields one bit for the output. From P = NP it then follows that the algorithm runs in output-polynomial time. □

Notice that a similar result is not known for arbitrary formulas.

As a final remark we return to the complexity of MOREPRIMI$_{\text{mon}}$ (Theorem 4.1). We have seen that the complexities of ISPRIMI$_{\text{mon}}$ (Theorem 3.2) and MOREPRIMI$_{\text{mon}}$ differ. This is not the case for the corresponding non-monotone problems ISPRIMI (Theorem 3.1) and MOREPRIMI.

Theorem 4.4. MOREPRIMI *is* DP-*complete*.

5 Equivalence and Isomorphism of Monotone Formulas

Deciding equivalence for arbitrary Boolean formulas is coNP-complete. The same holds for monotone formulas [Rei03]. However, if the monotone input formulas are given in k-CNF and DNF it is known that the problem is in P [EG95], even in RNC [BEGK00]. We improve these results by showing that logarithmic space suffices.

MONE$_{\text{const}}$: *instance:* irredundant, monotone Boolean formulas φ in
k-CNF for a constant k and ψ in DNF
question: are φ and ψ equivalent?

Theorem 5.1. MONE$_{\text{const}} \in$ L.

Two Boolean formulas φ and ψ are *isomorphic* if and only if there exists a permutation — a bijective renaming — π of the variables such that φ and $\pi(\psi)$ are equivalent. Two Boolean formulas are *congruent* if they are isomorphic after negating some of the variables. For example $x_1 \wedge x_2$ and $\neg x_3 \wedge x_4$ are congruent. Such a negation of some variables with the bijective renaming of the variables is called *n-permutation*. A witness for the congruence of the above example is the n-permutation π that exchanges $\neg x_3$ and x_1 as well as x_4 and x_2.

We want to compare the problem of testing isomorphism for monotone Boolean formulas to the case of abitrary Boolean formulas. This provides a negative answer to a conjecture from [Rei03].

BOOLISO$_{mon}$: *instance:* monotone Boolean formulas φ and ψ
question: are φ and ψ isomorphic?

BOOLISO: *instance:* Boolean formulas φ and ψ
question: are φ and ψ isomorphic?

BOOLCON: *instance:* Boolean formulas φ and ψ
question: are φ and ψ congruent?

Note that BOOLCON is polynomially equivalent to BOOLISO [BRS98].

Theorem 5.2. BOOLISO$_{mon}$ $\equiv_m^p$ BOOLISO.

Proof. To show BOOLISO$_{mon}$ $\leq_m^p$ BOOLISO we can choose the identy function as reduction function. We now show BOOLISO $\leq_m^p$ BOOLISO$_{mon}$. In [BRS98] it was shown that BOOLISO $\leq_m^p$ BOOLCON. Therefore, it suffices to show that BOOLCON $\leq_m^p$ BOOLISO$_{mon}$. The reduction function maps the instance (φ, ψ) of BOOLCON to the pair $(\varphi^{cd}, \psi^{cd})$ (cf. Definition 3.3). We have to show $(\varphi, \psi) \in$ BOOLCON $\Leftrightarrow (\varphi^{cd}, \psi^{cd}) \in$ BOOLISO$_{mon}$.

$(\varphi, \psi) \in$ BOOLCON $\Rightarrow (\varphi^{cd}, \psi^{cd}) \in$ BOOLISO$_{mon}$: Let $(\varphi, \psi) \in$ BOOLCON by an n-permutation π. Hence, φ and $\pi(\psi)$ are equivalent. We derive a permutation $\tilde{\pi}$ for $(\varphi^{cd}, \psi^{cd})$ from the n-permutation π in an elementary way. If π exchanges x_i with x_j, then $\tilde{\pi}$ exchanges x_i with x_j as well as y_i with y_j. And if π exchanges x_i with $\neg x_j$, then $\tilde{\pi}$ exchanges x_i with y_j as well as y_i with x_j. Note that $\tilde{\pi}$ does not make any remarkable changes on the $c(\psi)$- and $d(\psi)$-part of ψ^{cd} other than rearranging the terms in $c(\psi)$ and $d(\psi)$. We have to prove that φ^{cd} and $\tilde{\pi}(\psi^{cd})$ are equivalent and proceed by case differentiation of all possible monotone assignments.

$\exists i[x_i, y_i \in \mathcal{A}_m]$: Such assignments satisfy φ^{cd} and $\tilde{\pi}(\psi^{cd})$ by satisfying the conjunction $(x_i \wedge y_i)$.

$(\neg \exists i[x_i, y_i \in \mathcal{A}_m]) \wedge (\exists j[x_j, y_j \notin \mathcal{A}_m])$: None of the conjunctions of $d(\varphi)$ and $d(\psi)$ are satisfied by $\mathcal{A}_m$. Furthermore the disjunction $(x_j \vee y_j)$ in $c(\varphi)$ and $c(\psi)$ is not satisfied by $\mathcal{A}_m$ and consequently φ^{cd} and $\tilde{\pi}(\psi^{cd})$ are not satisfied.

It remains to verify the conform assignments: These are assignments that contain only one of the variables x_i and y_i for every $i \leq n$. They do not satisfy $d(\varphi)$ and $d(\psi)$ but do satisfy $c(\varphi)$ and $c(\psi)$. It remains to check $r(\varphi)$ and $\tilde{\pi}(r(\psi))$. From the facts that φ and $\pi(\psi)$ are equivalent and a conform assignment for φ^{cd} and $\tilde{\pi}(\psi^{cd})$ just simulates an assignment for φ and $\pi(\psi)$ it follows that the truth tables of φ^{cd} and $\tilde{\pi}(\psi^{cd})$ are identical in this case. Thus the truth tables of φ^{cd} and $\tilde{\pi}(\psi^{cd})$ are identical with respect to all possible assignments and therefore φ^{cd} and ψ^{cd} are isomorphic.

$(\varphi, \psi) \in$ BOOLCON $\Leftarrow (\varphi^{cd}, \psi^{cd}) \in$ BOOLISO$_{mon}$: A permutation $\tilde{\pi}$ for $(\varphi^{cd}, \psi^{cd}) \in$ BOOLISO$_{mon}$ is called *proper* if and only if (1) whenever x_i and

x_j are exchanged, then so are y_i and y_j, and (2) whenever x_i and y_j are exchanged, then so are y_i and x_j.

Claim. For all $(\varphi^{cd}, \psi^{cd}) \in \text{BOOLISO}_{\text{mon}}$ with more than two x-variables there is a proper permutation $\tilde{\pi}_p$ that ensures the equivalence of φ^{cd} and $\tilde{\pi}_p(\psi^{cd})$.

Proof. Suppose that the proposition of the claim does not hold. Then there exists a pair of formulas $(\varphi^{cd}_{im}, \psi^{cd}_{im}) \in \text{BOOLISO}_{\text{mon}}$ with more than two x-variables for which no proper permutation exists. As a consequence φ^{cd}_{im} and $\tilde{\pi}_{im}(\psi^{cd}_{im})$ are equivalent for some improper permutation $\tilde{\pi}_{im}$. We distinguish between the two cases of $\tilde{\pi}_{im}$ being improper.

$\exists i[\tilde{\pi}_{im}$ exchanges x_i with x_j but not y_i with $y_j]$: Hence, $\tilde{\pi}_{im}$ exchanges y_i with $b \in \{x_k : k \leq n, k \neq j\} \cup \{y_k : k \leq n, k \neq j\}$. We examine the assignment $\mathcal{A}_m = \{x_j, b\}$. The conjunction $(x_j \wedge b)$ in $\tilde{\pi}_{im}(d(\psi))$ is satisfied by $\mathcal{A}_m$ and so is $\tilde{\pi}_{im}(\psi^{cd}_{im})$. But $\mathcal{A}_m$ does not satisfy φ^{cd}_{im}. Note that the conjunction $(x_j \wedge b)$ is not present in $d(\varphi)$ and therefore $\mathcal{A}_m$ cannot satisfy $d(\varphi)$. Furthermore not all of the disjunctions of $c(\varphi)$ contain x_j or b because there are more than two x-variables in φ^{cd}_{im} and ψ^{cd}_{im}. Thus the two formulas φ^{cd}_{im} and $\tilde{\pi}_{im}(\psi^{cd}_{im})$ cannot be equivalent. This is a contradiction to our assumption.

$\exists i[\tilde{\pi}_{im}$ exchanges x_i with y_j but not y_i with $x_j]$: An analogous argumentation as above shows that the formulas φ^{cd}_{im} and $\tilde{\pi}_{im}(\psi^{cd}_{im})$ cannot be equivalent. This is a contradiction to our assumption. Hence, the claim follows.

As a consequence, there is a proper permutation $\tilde{\pi}_p$ for every $(\varphi^{cd}, \psi^{cd}) \in \text{BOOLISO}_{\text{mon}}$. A proper permutation only works on the $r(\psi)$-part of the ψ^{cd}-formula and only rearranges the terms in $c(\psi)$ and $d(\psi)$. Given a proper permutation $\tilde{\pi}_p$ we can easily derive an n-permutation π for (φ, ψ). If $\tilde{\pi}_p$ exchanges x_i with x_j as well as y_i with y_j, then π exchanges x_i with x_j. And if $\tilde{\pi}_p$ exchanges x_i with y_j as well as y_i with x_j, then π exchanges x_i with $\neg x_j$. Since the y-variables are placeholders for the negative literals, we see that π ensures $(\varphi, \psi) \in \text{BOOLCON}$. This concludes the proof of $\text{BOOLCON} \leq^p_m \text{BOOLISO}_{\text{mon}}$.

Thus we have established $\text{BOOLISO} \equiv^p_m \text{BOOLISO}_{\text{mon}}$. □

In [AT00] it is shown that BOOLISO is not complete for Σ^p_2 unless the Polynomial Time Hierarchy collapses. As a consequence of Theorem 5.2, this holds for $\text{BOOLISO}_{\text{mon}}$ as well.

6 Concluding Remarks

We compared the complexity of problems related to the construction of Disjunctive Normal Forms for non-monotone and monotone formulas. We proved that finding an algorithm that computes a minimal DNF for a monotone formula in output-polynomial time is the same as solving P = NP. A similar result for arbitrary formulas is still open. Anyway, we assume that at least P = PSPACE is the consequence. Although we proved that calculating the size of a minimal DNF for a monotone formula is PP-complete (resp. #P-complete), even a PSPACE upper bound for the non-monotone case is open.

Some problems for formulas are easier to decide in the monotone case than for arbitrary formulas. Among them are finding prime implicants (NP- vs. Σ_2^p-complete) and calculating the size of a smallest equivalent DNF (PP-complete vs. unknown). On the other hand, there are problems whose complexity stays the same for monotone formulas. We could show this polynomial time equivalence for isomorphism testing and counting satisfying assignments.

Deciding equivalence for monotone formulas is coNP-complete [Rei03] like it is for Boolean formulas. Nevertheless we were able to prove a log-space upper bound for the special case MONE$_{\text{const}}$ of equivalence testing. The complexity of the general problem MONE without a constant bound for the clause size (which is equivalent to MOREPRIMI$_{\text{mon}}$ for instances (φ, S) with φ in CNF) remains open.

References

[AT00] M. Agrawal and T. Thierauf. The formula isomorphism problem. *SIAM Journal on Computing*, 30(3):990–1009, 2000.

[BEGK00] E. Boros, K. Elbassioni, V. Gurvich, and L. Khachiyan. An efficient incremental algorithm for generating all maximal independent sets in hypergraphs of bounded dimension. *Parallel Processing Letters*, 10(4):253–266, 2000.

[BRS98] B. Borchert, D. Ranjan, and F. Stephan. On the computational complexity of some classical equivalence relations on Boolean functions. *Theory of Computing Systems*, 31(6):679–693, 1998.

[Czo99] S.L.A. Czort. The complexity of minimizing Disjunctive Normal Form formulas. Technical Report IR-130, Dept. of Computer Science, University of Aarhus, January 1999.

[EG95] T. Eiter and G. Gottlob. Identifying the minimal transversals of a hypergraph and related problems. *SIAM Journal on Computing*, 24(6):1278–1304, 1995.

[FK96] M.L. Fredman and L. Khachiyan. On the complexity of dualization of monotone Disjunctive Normal Forms. *Journal of Algorithms*, 21:618–628, 1996.

[Mas79] W.J. Masek. Some NP-complete set covering problems. Unpublished manuscript, 1979.

[Pap94] C.H. Papadimitriou. *Computational Complexity*. Addison-Wesley, 1994.

[Pap97] C.H. Papadimitriou. NP-completeness: a retrospective. In *Proceedings of 24th ICALP*, volume 1256 of *Lecture Notes in Computer Science*, pages 2–6, 1997.

[PY84] C.H. Papadimitriou and M. Yannakakis. The complexity of facets (and some facets of complexity). *Journal of Computer and System Sciences*, 28:224–259, 1984.

[Rei03] S. Reith. On the complexity of some equivalence problems for propositional calculi. In *Proceedings of MFCS 2003*, volume 2747 of *Lecture Notes in Computer Science*, pages 632–641, 2003.

[Uma01] C. Umans. The minimum equivalent DNF problem and shortest implicants. *Journal of Computer and System Sciences*, 63(4):597–611, 2001.

[Val79] L.G. Valiant. The complexity of enumeration and reliability problems. *SIAM Journal on Computing*, 8(3):410–421, 1979.

The Expressive Power of Two-Variable Least Fixed-Point Logics

Martin Grohe, Stephan Kreutzer, and Nicole Schweikardt

Institut für Informatik, Humboldt-Universität, Berlin
{grohe, kreutzer, schweika}@informatik.hu-berlin.de

Abstract. The present paper gives a classification of the expressive power of two-variable least fixed-point logics. The main results are:

1. The two-variable fragment of monadic least fixed-point logic with parameters is as expressive as full monadic least fixed-point logic (on binary structures).
2. The two-variable fragment of *monadic* least fixed-point logic without parameters is as expressive as the two-variable fragment of *binary* least fixed-point logic without parameters.
3. The two-variable fragment of binary least fixed-point logic with parameters is strictly more expressive than the two-variable fragment of monadic least fixed-point logic with parameters (even on finite strings).

1 Introduction

In the fields of mathematical logic and finite model theory it has always been an important issue to compare the expressive power of different logics. Among the logics that received particular attention in theoretical computer science, extensions of first-order logic by mechanisms that allow to define relations *by induction* play a prominent role. Formalising such inductive definitions in a logical language usually involves some kind of fixed-point construction. In particular, *least fixed-point logic*, LFP, is the extension of first-order logic by least fixed-point operators, whereas M-LFP is the fragment of LFP where fixed-point operators are *monadic*, i.e., have arity at most 1.

From a well-known theorem due to Immerman and Vardi [13,19] it is known that on ordered finite structures the logic LFP precisely characterises the complexity class PTIME. Apart from describing complexity classes, logics – and in particular fixed-point logics – are used, e.g., as languages for hardware and process specification and verification, and as query languages for expressing queries against databases. As observed, e.g., in [2,18], the size of intermediate results that occur while evaluating a query (i.e., a logical formula) over a database (i.e., a finite structure) crucially depends on the number of first-order variables that occur in the formula. If the number of such variables is bounded by a constant k, the size of intermediate results remains polynomial in the size of the input structure. Therefore, the combined complexity of the *model checking problem* apparently is much smaller when considering the *k-variable fragment* of a logic instead of the entire logic.

The research community's considerable interest in bounded variable logics (cf., e.g., [3,12,18,4,6,14,16,10,9,8]) can partly be explained by the comparably low combined complexity of the model checking problem for these logics. Further motivations for studying, in particular, *two*-variable logics are the decidability of FO² and the fact that *modal logics* can be embedded into two-variable logics. For example, plain modal logic ML can be embedded into the two-variable fragment of first-order logic, FO², the modal iteration calculus MIC [5] can be embedded into the two-variable fragment of monadic inflationary fixed-point logic, and the modal μ-calculus L_μ can be embedded into the two-variable fragment of monadic least fixed-point logic, M-LFP², which, in turn, can be embedded into two-variable infinitary logic $L^2_{\infty\omega}$. In particular the logics FO² and M-LFP² have received a lot of attention in the past (cf. [9,8,16]) in an attempt to explain the nice model-theoretical and computational properties of modal logics such as ML and L_μ. An overview of what is known about bounded variable logics and, in particular, *two*-variable logics, can be found in [8,16].

The present paper's aim is to study the expressive power of two-variable fragments of least fixed-point logic LFP. As observed in [6,9], defining these fragments requires some care, because allowing or forbidding the use of parameters (i.e., free first-order variables) in least fixed-point operators crucially changes the expressive power of the logic under consideration. In the literature, LFP^k usually refers to the parameter-free fragment of LFP where k first-order variables and second-order variables of arity at most k are available, cf. [8].

The logics we consider are

- M-LFP²$_{param}$ and LFP²$_{param}$, the two-variable fragments of monadic and binary least fixed-point logic, respectively, where the use of parameters is allowed in fixed-point operators, and
- M-LFP² and LFP², the two-variable fragments of monadic and binary least fixed-point logic, respectively, where least fixed-point operators are *not* allowed to have parameters.

We only consider fixed-point operators of arity at most two, since fixed-point definitions of higher arity already syntactically involve more than just two first-order variables. The presence of only two first-order variables furthermore renders it reasonable to restrict attention to *binary structures*, i.e., structures over a signature that consists of constant symbols and of relation symbols of arity at most two.

The logics M-LFP, M-LFP² and M-LFP²$_{param}$, in particular, have explicitly been considered before [6,8,9,17,11]. E.g., M-LFP² coincides with the logic called FP² in [8,9] and $\widehat{FP}^2$ in [6], whereas M-LFP²$_{param}$ coincides with the logic called FP² in [6].

It is known that on finite structures (or, more generally, on classes of structures of bounded cardinality), LFP² can be embedded into infinitary logic $L^2_{\infty\omega}$ [9,16]. Furthermore, it has been observed in various places (cf., e.g., [16,7]) that M-LFP²$_{param}$ can express the transitive closure of a binary relation and therefore cannot be embedded into $L^2_{\infty\omega}$. Consequently, one obtains that already on the class of finite graphs, M-LFP^2 $\lneq$ M-LFP$^2_{param}$ and M-LFP$^2_{param}$ $\not\leq$ LFP^2. (Here we write $L \lneq L'$ to denote that a logic L is strictly less expressive than a logic L', and we write $L \not\leq L'$ to denote that there are problems that can be expressed in L, but not in L'.)

From [6] it is known that the combined complexity of the model checking problem for M-LFP$^2_{param}$ and LFP$^2_{param}$ is PSPACE-complete, whereas the combined complexity of the model checking problem for LFP^2 is closely related to that of the modal μ-calculus and therefore belongs to NP $\cap$ Co-NP and is PTIME-hard.

When restricting attention to the class of finite strings, one obtains an entirely different picture. Due to Büchi's theorem (cf., e.g., [7]), monadic second-order logic MSO can describe exactly the *regular* string languages which, in turn, can already be described by the modal μ-calculus L_μ. Consequently, on finite strings the logics MSO, M-LFP, M-LFP$^2_{param}$, and L_μ all have the same expressive power. Furthermore, it is known (cf., [9,7]) that the two-variable fragment of monadic *inflationary* fixed-point logic, M-IFP^2, is capable of describing non-regular string-languages, and therefore M-LFP^2 $\lneq$ M-IFP^2.

The present paper's contribution is to complete the picture of the expressive power of the two-variable fragments of least fixed-point logics. Our main results are

1. M-LFP$^2_{param}$ = M-LFP$_{param}$ on binary structures. I.e. the two-variable fragment of monadic least fixed-point logic with parameters is as expressive as full monadic least fixed-point logic with parameters. Here, of course, the restriction to binary structures is crucial, as M-LFP contains full first-order logic.
2. LFP^2 = M-LFP^2, i.e., parameter-free two-variable *binary* least fixed-point logic has the same expressive power as parameter-free two-variable *monadic* least fixed-point logic.
3. M-LFP$^2_{param}$ $\lneq$ LFP$^2_{param}$, and the inclusion is strict already on the class of finite strings. We prove this result by showing that the non-regular string-language $\{a^n b^n \mid n \in \mathbb{N}\}$ is expressible in LFP$^2_{param}$.

Altogether this leads to the following inclusion structure of the expressive power of the two-variable fragments of least fixed-point logic:

The paper is organised as follows: After fixing some basic notation in section 2, we formally introduce the two-variable fragments of least fixed-point logic in section 3. The equivalence of M-LFP$^2_{param}$ and M-LFP$_{param}$ is proved in section 4.

M-LFP^2 = LFP^2 $\lneq$ M-LFP$^2_{param}$ = M-LFP$_{param}$ $\lneq$ LFP$^2_{param}$

on the class of binary structures,

and

M-LFP^2 = LFP^2 = M-LFP$^2_{param}$ = M-LFP$_{param}$ $\lneq$ LFP$^2_{param}$

on the class of finite strings.

Fig. 1. Expressive power of the two-variable fragments of least fixed-point logic

Afterwards, in section 5 we show that LFP² is equivalent to M-LFP² and that LFP²$_{param}$ can express non-regular string-languages and therefore is strictly more expressive than M-LFP²$_{param}$. Detailed proofs of the results presented here can be found in the full version of the paper.

2 Preliminaries

As usual, we write Ord for the class of ordinals and ω for the set of finite ordinals (i.e., non-negative integers). By Pow(S) we denote the power set of a set S. A *signature* is a finite set of relation and constant symbols. We call a signature τ *binary* if the arity of every relation symbol occurring in τ is at most two. Thus, structures of a binary signature are essentially coloured graphs.

In this paper we deal primarily with two-variable logics – logics that only allow for two distinct first-order variables. As with only two variables we cannot take advantage of relations of higher arity, we will only consider binary signatures throughout this paper. In most cases, this restriction has no impact on the validity of our statements. In the few places where it does, we will state this explicitly.

We use German letters $\mathfrak{A}, \mathfrak{B}, \ldots$ to denote structures and the corresponding Roman letters $A, B, \ldots$ to denote their universes.

We assume that the reader is familiar with *first-order logic* (FO). We use FO(τ) to denote the class of all first-order formulae of signature τ. Besides first-order variables we also allow free second-order variables in the formulae (but no second-order quantification). We write $\varphi(R_1, \ldots, R_k, x_1, \ldots, x_n)$ to indicate that the free first-order variables of the formula φ are $x_1, \ldots, x_n$ and the free relation variables are $R_1, \ldots, R_k$. We use $\bar{x}$ and $\bar{R}$ as abbreviations for sequences $x_1, \ldots, x_n$ and $R_1, \ldots, R_k$ of variables. Finally, we write $\varphi(R_1, \ldots, R_k, \bar{x}_1, \ldots, \bar{x}_k, \bar{z})$ to indicate that the free first-order variables of φ are the variables in the tuples $\bar{x}_i$ and $\bar{z}$ and that the arity of a tuple $\bar{x}_i$ is the same as the arity of the relation variable R_i.

3 Finite Variable Fragments of Least Fixed-Point Logic

In this section we give a brief introduction to least fixed-point logic and its two-variable fragments. For a detailed exposition see [7].

Least and greatest fixed points of monotone operators. Let τ be a signature and let $\varphi(R, \bar{x})$ be a formula of signature τ which is *positive* in the k-ary relation variable R, i.e. every atom of the form $R\bar{t}$ in φ occurs within the scope of an *even* number of negation symbols. φ defines for every τ-structure $\mathfrak{A}$ a monotone operator[1] $F_{\mathfrak{A},\varphi} : \text{Pow}(A^k) \to \text{Pow}(A^k)$ via $F_{\mathfrak{A},\varphi}(P) := \{\bar{a} \in A^k : (\mathfrak{A}, P) \models \varphi[\bar{a}]\}$, for every $P \subseteq A^k$. In cases where $\mathfrak{A}$ is understood from the context, we simply write F_φ for $F_{\mathfrak{A},\varphi}$.

A set P is called a *fixed point* (a *pre fixed point*) of φ in $\mathfrak{A}$ if, and only if, $F_{\mathfrak{A},\varphi}(P) = P$ ($F_{\mathfrak{A},\varphi}(P) \subseteq P$). P is called the *least fixed point* of φ if P is a fixed point of φ and $P \subseteq Q$ for every fixed point Q of F_φ. We write $\mathbf{lfp}(F_{\mathfrak{A},\varphi})$ for the least fixed point of $F_{\mathfrak{A},\varphi}$. Further, as the intersection of all pre fixed points is itself a fixed point, we get the following characterisation of least fixed points:

$$\mathbf{lfp}(F_{\mathfrak{A},\varphi}) \;=\; \bigcap \{Q : F_{\mathfrak{A},\varphi}(Q) = Q\} \;=\; \bigcap \{Q : F_{\mathfrak{A},\varphi}(Q) \subseteq Q\} \tag{1}$$

There is also the corresponding notion of a *greatest fixed point* of φ which is the fixed point that contains all other fixed points. Least and greatest fixed points are dual to each other, in the sense that for every monotone operator $F : \text{Pow}(M) \to \text{Pow}(M)$ we have $\mathbf{lfp}(F) = \overline{\mathbf{gfp}(\overline{F})}$, where $\overline{F}$ is defined as $\overline{F}(U) := (F(U^c))^c$ (where U^c denotes the complement of U).

Least (and also greatest) fixed points of monotone operators can also be built up inductively. For this we define for all ordinals α sets $R^\alpha_{\mathfrak{A},\varphi} \subseteq A^k$ inductively as $R^0_{\mathfrak{A},\varphi} := \emptyset$, $R^{\alpha+1}_{\mathfrak{A},\varphi} := F_{\mathfrak{A},\varphi}(R^\alpha_{\mathfrak{A},\varphi})$, and $R^\lambda_{\mathfrak{A},\varphi} := \bigcup_{\gamma < \lambda} R^\gamma_{\mathfrak{A},\varphi}$ for infinite limit ordinals λ. In cases where $\mathfrak{A}$ and φ are understood, we simply write R^α. Since $F_{\mathfrak{A},\varphi}$ is monotone we have $R^\alpha \subseteq R^{\alpha+1}$ for all α. Hence the sequence $(R^\alpha)_{\alpha \in \text{Ord}}$ eventually reaches a fixed point, i.e. there is an ordinal α such that $R^\alpha = R^{\alpha+1} = R^\gamma$ for all $\gamma > \alpha$. We refer to this fixed point as R^∞. It is easily seen that if the structure $\mathfrak{A}$ is finite then α is finite too. A theorem due to Knaster and Tarski establishes the equivalence $R^\infty_{\mathfrak{A},\varphi} = \mathbf{lfp}(F_{\mathfrak{A},\varphi})$ for all structures $\mathfrak{A}$ and formulae φ positive in the variable R. Thus, the sequence $(R^\alpha_{\mathfrak{A},\varphi})_{\alpha \in \text{Ord}}$ approximates the least fixed point of $F_{\mathfrak{A},\varphi}$ from below. The sets $R^\alpha_{\mathfrak{A},\varphi}$ are called the *stages* of the least fixed-point induction on φ in $\mathfrak{A}$. A similar induction can be used to define greatest fixed points, where we start with $R^0 := A^k$ and take intersections instead of unions to define the higher stages.

Least fixed-point logic. The logic $\text{LFP}(\tau)$ is the extension of $\text{FO}(\tau)$ by least fixed-point operators. Precisely: $\text{LFP}(\tau)$ contains $\text{FO}(\tau)$ and is closed under Boolean connectives and first-order quantification; and if $\varphi(R, \bar{x}, \bar{z}, \bar{Q})$ is an $\text{LFP}(\tau)$-formula which is positive in the k-ary relation variable R then for every k-tuple $\bar{t}$ of terms $[\mathbf{lfp}_{R,\bar{x}}\,\varphi](\bar{t})$ is an $\text{LFP}(\tau)$-formula such that for every $(\tau \dot\cup \{\bar{z}, \bar{Q}\})$-structure $\mathfrak{A}$ and every tuple $\bar{a} \in A^k$ we have $\mathfrak{A} \models [\mathbf{lfp}_{R,\bar{x}}\,\varphi](\bar{a})$ if, and only if, $\bar{a} \in \mathbf{lfp}(F_{\mathfrak{A},\varphi})$. Similarly, we allow formulae $[\mathbf{gfp}_{R,\bar{x}}\,\varphi](\bar{t})$ defining the greatest fixed point of $F_{\mathfrak{A},\varphi}$. The variables in $\bar{z}$ that are not contained in $\bar{x}$ are called the *parameters* of the fixed-point induction. They will play an important role in later sections.

Due to the above mentioned duality of least and greatest fixed points, **gfp**-operators can easily be replaced by **lfp**-operators with additional negation symbols. On the other hand, the use of **lfp**- *and* **gfp**-operators allows to transform every formula into a formula in *negation normal form*, i.e., a formula where negation symbols only occur directly in front of *atomic* sub-formulae.

We continue with the definition of some important fragments of least fixed-point logic. The *monadic least fixed-point logic* (M-LFP) is defined as the fragment of LFP where all fixed-point variables are unary, i.e. of arity ≤ 1. Analogously

---

[1] An operator $F : \text{Pow}(M) \to \text{Pow}(M)$ is *monotone* iff $F(A) \subseteq F(B)$ for all $A \subseteq B \subseteq M$.

we define *binary least fixed-point logic* (Bin-LFP) as the fragment of LFP where all fixed-point variables are of arity ≤ 2.

We are primarily interested in fragments of M-LFP and Bin-LFP where the number of available first-order variables is restricted to two. Recall from above that the variables $\bar{z}$ occurring free in $\varphi(R, \bar{x}, \bar{z})$ other than $\bar{x}$ are called parameters of the fixed-point formula $[\mathbf{lfp}_{R,\bar{x}}\ \varphi](\bar{t})$. It is well know in finite model theory that parameters can be eliminated by increasing the arity of the fixed-point variables, i.e. for every LFP-formula $[\mathbf{lfp}_{R,\bar{x}}\ \varphi(R,\bar{x},\bar{z})](\bar{t})$ there is an equivalent LFP-formula $[\mathbf{lfp}_{R',\bar{x}'}\ \varphi'(R',\bar{x}')](\bar{t}')$ which is parameter-free. However, this translation does not only require fixed-point variables of higher arity, it also requires the introduction of fresh first-order variables. Thus the standard translation of formulae with parameters into formulae without parameters does not apply to the two-variable fragments defined above. And indeed, as we will see later on, in this restricted setting, parameters increase the expressive power of the logics. We therefore introduce a separate notation for logics with and without parameters.

The logics M-LFP² and LFP² are defined as the parameter-free fragment of M-LFP and Bin-LFP resp. where only two distinct first-order variables may be used in the formulae. Analogously, the logics M-LFP²$_{param}$ and LFP²$_{param}$ are defined as the fragment of M-LFP and Bin-LFP resp. where only two distinct first-order variables may be used in the formulae but where the fixed-point operators may have parameters.

Simultaneous fixed-point inductions. Simultaneous inductions can simplify the formalisation of properties significantly, but as we will see below, they do not add to the expressive power of the logics.

Definition 1 (Simultaneous least fixed-point logic). Let $R_1, \ldots, R_k$ be relation symbols of arity $r_1, \ldots, r_k$, respectively. Simultaneous formulae are formulae of the form $\psi(\bar{x}) := [\mathbf{lfp}\ R_i : S](\bar{x})$, where

$$S := \begin{cases} R_1\bar{x}_1 \leftarrow \varphi_1(R_1, \ldots, R_k, \bar{x}_1) \\ \quad\vdots \\ R_k\bar{x}_k \leftarrow \varphi_k(R_1, \ldots, R_k, \bar{x}_k) \end{cases}$$

is a system of LFP-formulae φ_i which are positive in all variables $R_1, \ldots, R_k$. On any structure $\mathfrak{A}$, the system S induces an operator

$$F_S : \text{Pow}(A^{r_1}) \times \cdots \times \text{Pow}(A^{r_k}) \to \text{Pow}(A^{r_1}) \times \cdots \times \text{Pow}(A^{r_k})$$

defined as $F_S(P_1, \ldots, P_k) = (F_{\varphi_1}(\bar{P}), \ldots, F_{\varphi_k}(\bar{P}))$, where F_{φ_i} is the operator induced by φ_i in S defined as

$$F_{\varphi_i} : \text{Pow}(A^{r_1}) \times \cdots \times \text{Pow}(A^{r_k}) \longrightarrow \text{Pow}(A^{r_i})$$
$$(R_1, \ldots, R_k) \longmapsto \{\bar{a} : (\mathfrak{A}, R_1, \ldots, R_k) \models \varphi_i[\bar{a}]\}.$$

The stages S^α of an induction on such a system S of formulae are k-tuples of sets $(R_1^\alpha, \ldots, R_k^\alpha)$ defined as $R_i^0 := \varnothing$, $R_i^{\alpha+1} := F_{\varphi_i}(R_1^\alpha, \ldots, R_k^\alpha)$, and $R_i^\lambda := \bigcup_{\xi < \lambda} R_i^\xi$ for infinite limit ordinals λ.

For every structure $\mathfrak{A}$ and any tuple $\bar{a}$ from A, $\mathfrak{A} \models \psi[\bar{a}]$ if, and only if, $\bar{a} \in R_i^\infty$, where R_i^∞ denotes the i-th component of the simultaneous least fixed point of S.

Let S-LFP denote the class of LFP-formulae with simultaneous inductions.

We show next that allowing simultaneous fixed points does not increase the expressive power of LFP, i.e. S-LFP = LFP.

Theorem 2. *For any parameter-free system S of formulae in LFP, positive in their free fixed-point variables, $\varphi := [\mathbf{lfp}\ R_i : S](\bar{t})$ is equivalent to a formula φ^* in LFP (without simultaneous inductions). Further, φ and φ^* use the same set of first and second-order variables. In particular, the arity of the involved fixed-point operators does not increase.*

The theorem follows immediately from the following lemma – sometimes called the *Bekič-principle* – which is part of the folklore of the community. (See e.g. [1, Lemma 1.4.2], [15, Lemma 10.9].)

Lemma 3. *Let*

$$S := \begin{cases} R_1 \bar{x}_1 & \leftarrow \varphi_1(R_1, \ldots, R_k, \bar{x}_1) \\ \vdots & \\ R_{k-1} \bar{x}_{k-1} & \leftarrow \varphi_{k-1}(R_1, \ldots, R_k, \bar{x}_{k-1}) \\ R_k \bar{x}_k & \leftarrow \varphi_k(R_1, \ldots, R_k, \bar{x}_k) \end{cases}$$

be a system of formulae in LFP such that $[\mathbf{lfp}\ R_1 : S](\bar{x}_1)$ is parameter-free. Then $[\mathbf{lfp}\ R_1 : S]$ is equivalent to the parameter-free formula $[\mathbf{lfp}\ R_1 : T]$, where

$$T := \begin{cases} R_1 \bar{x}_1 & \leftarrow \varphi_1'(R_1, \ldots, R_{k-1}, \bar{x}_1) \\ \vdots & \\ R_{k-1} \bar{x}_{k-1} & \leftarrow \varphi_{k-1}'(R_1, \ldots, R_{k-1}, \bar{x}_k). \end{cases}$$

Here $\varphi_i' := \varphi_i(R_1, \ldots, R_{k-1}, R_k \bar{u}/[\mathbf{lfp}_{R_k, \bar{x}_k}\ \varphi_k](\bar{u}), \bar{x}_1)$ is obtained from φ_i by replacing every occurrence of an atom $R_k \bar{u}$ by the formula $[\mathbf{lfp}_{R_k, \bar{x}_k}\ \varphi_k](\bar{u})$.
Note that this lemma cannot be applied in cases where parameters are allowed.

4 Monadic Two-Variable Fixed-Point Logic

As already mentioned in section 1, it is known that M-LFP^2 is strictly less expressive than M-LFP$^2_{param}$ on the class of finite graphs. In fact, M-LFP^2 can be embedded into two-variable infinitary logic $L^2_{\infty\omega}$, whereas M-LFP$^2_{param}$ can not. Due to this, it has been claimed by several authors that allowing parameters in a two-variable fixed-point logic does not yield a logic that behaves as a "proper two-variable logic" (cf., [9,8,6]) . The next theorem gives additional backup to this claim by showing that – subject to the obvious restriction to binary structures – the two-variable fragment of monadic least fixed-point logic with parameters is as expressive as full monadic least fixed-point logic.

Theorem 4. M-LFP$^2_{param}$ = M-LFP$_{param}$ *on binary structures. That is, for every binary signature τ the following is true: For every* M-LFP(τ)-*sentence φ there is an* M-LFP$^2_{param}(\tau)$-*sentence φ' that is equivalent to φ on all τ-structures.*

The proof is based on the simple idea of replacing every first-order quantification by a new monadic second-order variable and a fixed point construction.

5 Binary Two-Variable Fixed-Point Logic

In this section we concentrate on the expressive power of two-variable *binary* least fixed-point logic with and without parameters, respectively. First we show that parameter-free two-variable binary least fixed-point logic is no more expressive than parameter-free two-variable *monadic* least fixed-point logic. Afterwards, we prove that (already on the class of finite strings) two-variable binary least fixed-point logic *with* parameters is strictly more expressive than two-variable *monadic* least fixed-point logic with parameters.

Parameter-free two-variable binary least fixed-point logic.

Theorem 5. LFP^2 = M-LFP^2 *on binary structures. That is, for every binary signature τ the following is true: For every* LFP$^2(\tau)$-*formula φ there is an* M-LFP$^2(\tau)$-*formula φ' that is equivalent to φ on all τ-structures.*

Proof. By definition, every M-LFP^2-formula is also a valid formula in LFP^2. Hence, M-LFP$^2 \leq$ LFP^2. Towards the converse, we show by induction on the number n of binary fixed-point operators that every formula in LFP^2 is equivalent to a formula in M-LFP^2. For $n = 0$ this is trivial. Let λ' be a formula with $n > 0$ binary fixed-point operators and let $\lambda(x, y)$ be a sub-formula of λ' of the form $\lambda(x, y) := [\mathbf{lfp}_{R,x,y}\,\varphi](t_1, t_2)$ such that φ is in M-LFP^2. φ can be decomposed into a positive Boolean combination of the following formulae:

- A quantifier-free formula $\theta(x, y)$ with free variables x and y. Here, by "quantifier-free" we mean absence of fixed-point operators too.
- Formulae $\psi_1(x), \ldots, \psi_s(x)$ where x and only x occurs as a free variable.
- Formulae $\chi_1(y), \ldots, \chi_r(y)$ where y and only y occurs as a free variable.
- A formula ϑ without any free variables

The crucial observation is that as φ is in M-LFP^2 and we do not allow parameters to the fixed-point operators, the only sub-formulae with two free variables are atoms or negated atoms.

The formula φ is a positive Boolean combination of the sub-formulae described above and all sub-formulae are positive in the fixed-point variable R. Hence, the system

$$S := \begin{cases} Rxy \leftarrow \hat\varphi(R, x, y, \bar{X}, \bar{Y}, T) & \\ X_i x \leftarrow \psi_i(x) & \text{for all } i \in \{1, \ldots, s\} \\ Y_i y \leftarrow \chi_i(y) & \text{for all } i \in \{1, \ldots, r\} \\ T \leftarrow \vartheta & \end{cases}$$

is positive in all fixed-point variables. Here $\hat{\varphi}$ is obtained from φ by replacing the sub-formulae $\psi_i(x)$ by $X_i x$, the $\chi_i(y)$ by $Y_i y$ and ϑ by T. Note that T is a nullary second-order variable, i.e. it can only take the values $\emptyset$ or $\{()\}$. A simple induction on the stages of the fixed-point induction establishes the next lemma.

Lemma 6. $\lambda(x, y) \equiv [\mathbf{lfp}\ R : S](x, y)$. □

Now we can treat $[\mathbf{lfp}_{R,x,y}\ \hat{\varphi}](x, y)$ as a fixed-point formula over the extended signature $\tau \cup \{\bar{X}, \bar{Y}, T\}$ and consider the formulae $\hat{\varphi}^\alpha$ of the unravelling of $\hat{\varphi}$ defined as $\hat{\varphi}^0(x) := \neg x = x$ and $\hat{\varphi}^{n+1}(x) := \hat{\varphi}(Rt_1t_2/\hat{\varphi}^n(t_1, t_2))$. Here, $\hat{\varphi}(Rt_1t_2/\hat{\varphi}^n(t_1, t_2))$ is the formula obtained from $\hat{\varphi}$ by replacing every occurrence of an atom Rt_1t_2 by the result of substituting in $\hat{\varphi}^n$ x by t_1 and y by t_2. As $\hat{\varphi}$ is quantifier-free the formulae $\hat{\varphi}^n$ are quantifier-free as well. Further, there are (up to equivalence) only finitely many quantifier-free formulae for a fixed (and finite) relational signature. Thus, there is an $n < \omega$ which only depends on the signature and not on a particular interpretation of the relation variables X_i, Y_i, and T such that $\hat{\varphi}^n \equiv \hat{\varphi}^{n+1}$. (Precisely, there are $n, q < \omega$ such that $\hat{\varphi}^n \equiv \hat{\varphi}^{n+q}$. But as $\hat{\varphi}$ is monotone in R, this implies $\hat{\varphi}^n \equiv \hat{\varphi}^{n+1}$.) Consequently, for $\hat{\hat{\varphi}} := \hat{\varphi}^n$,

$$[\mathbf{lfp}_{R,x,y}\ \hat{\varphi}](x, y) \equiv \hat{\hat{\varphi}} \qquad (*)$$

on all structures over the signature $\tau \cup \{X_i, Y_i, T\}$. Note that in $\hat{\hat{\varphi}}$ the variable R does no longer occur.

The next step is to (a) eliminate in S the rule $Rxy \leftarrow \hat{\varphi}$ by applying the construction of Lemma 3 and (b) to replace every occurrence of $[\mathbf{lfp}_{R,x,y}\ \hat{\varphi}](t_1, t_2)$ by $\hat{\hat{\varphi}}(t_1, t_2)$. This construction yields the system

$$S' := \begin{cases} X_i x \leftarrow \psi_i(Rt_1t_2/\hat{\hat{\varphi}}(t_1, t_2)) \text{ for all } i \in \{1, \ldots, s\} \\ Y_i x \leftarrow \chi_i(Rt_1t_2/\hat{\hat{\varphi}}(t_1, t_2)) \text{ for all } i \in \{1, \ldots, r\} \\ T \leftarrow \vartheta(Rt_1t_2/\hat{\hat{\varphi}}(t_1, t_2)) \end{cases}$$

where $\psi_i(Rt_1t_2/\hat{\hat{\varphi}}(t_1, t_2))$ denotes the formula obtained from ψ_i by replacing every occurrence of an atom Rt_1t_2 by the formula $\hat{\hat{\varphi}}(t_1, t_2)$ – the result of substituting in $\hat{\hat{\varphi}}$ t_1 for x and t_2 for y. By Lemma 3 and the equivalence $(*)$, the systems S and S' are equivalent in the sense that for all $i \in \{1, \ldots, s\}$ we have

$$[\mathbf{lfp}\ X_i : S](x) \equiv [\mathbf{lfp}\ X_i : S'](x) \qquad (**)$$

and likewise for T and all Y_i. Let $(R^\infty, X_i^\infty, Y_i^\infty, T^\infty)$ be the simultaneous least fixed point of F_S. Further, $(**)$ implies that $(X_i^\infty, Y_i^\infty, T^\infty)$ is also the simultaneous least fixed point of $F_{S'}$. By definition, $R^\infty = \{(a, b) : (\mathfrak{A}, R^\infty, X_i^\infty, Y_i^\infty, T^\infty) \models \hat{\varphi}(a, b)\}$. We claim that

$$R^\infty = \{(a, b) : (\mathfrak{A}, X_i^\infty, Y_i^\infty, T^\infty) \models [\mathbf{lfp}_{R,x,y}\ \hat{\varphi}](a, b)\}. \qquad (***)$$

We let $R'^\infty := \{(a, b) : (\mathfrak{A}, X_i^\infty, Y_i^\infty, T^\infty) \models [\mathbf{lfp}_{R,x,y}\ \hat{\varphi}](a, b)\}$. Clearly, $R'^\infty \subseteq R^\infty$, as R^∞ is a fixed point of $\hat{\varphi}$ (with the given interpretation $X_i^\infty, Y_i^\infty, T^\infty$ of the other free variables) and R'^∞ is its least fixed point. Conversely, the sequence $(R'^\infty, X_i^\infty, Y_i^\infty, T^\infty)$ is a fixed point of F_S and thus $R^\infty \subseteq R'^\infty$.

Now we can put the various parts together to obtain the following chain of equalities:

$$\begin{aligned} R^\infty &= \{(a,b) : (\mathfrak{A}, X_i^\infty, Y_i^\infty, T^\infty) \models [\mathbf{lfp}_{R,x,y}\, \hat\varphi](a,b)\} & (\text{ by } (***)\text{ }) \\ &= \{(a,b) : (\mathfrak{A}, X_i^\infty, Y_i^\infty, T^\infty) \models \hat\varphi(a,b)\} & (\text{ by } (*)\text{ }) \\ &= \{(a,b) : \mathfrak{A} \models \varphi^*(a,b)\}, \end{aligned}$$

where φ^* is the formula derived from $\hat\varphi$ by replacing every occurrence of an atom $X_i t$ by $[\mathbf{lfp}\, X_i : S'](t)$ and likewise for the relations Y_i and T. By construction, the formula φ^* only contains monadic fixed-point operators and is equivalent to the formula $\lambda(x,y)$ from the beginning of the proof. Thus, we can replace the occurrence of λ in λ' by φ^*. The resulting formula has fewer binary fixed-point operators as λ' and, by induction hypothesis, is therefore equivalent to a formula without any binary fixed-point operators. This concludes the proof of the theorem. □

Remark. The theorem naturally extends to the k-variable fragment of LFP, that is, every parameter-free formula of LFP with at most k distinct first-order variables and k-ary fixed-point operators is equivalent to a parameter-free k-variable formula of LFP with fixed-point relations of arity at most $k-1$.

Two-variable binary least fixed-point logic with parameters. We show next that LFP^2_{param} is strictly more expressive than M-LFP^2_{param}. We prove this by showing that the non-regular string-language $\{a^n b^n \mid n \in \mathbb{N}\}$ can be defined by an LFP^2_{param}-sentence. In order to give a detailed proof, we need some additional notation: A non-empty string w over an alphabet Σ is represented by a structure $\mathfrak{W}$ over the binary signature $\tau_\Sigma := \{min, max, succ, <\} \cup \{Q_\sigma \mid \sigma \in \Sigma\}$ in the usual way: If $w = w_1 \cdots w_n$ with $w_i \in \Sigma$, then $\mathfrak{W}$ is the τ_Σ-structure with universe $W = \{1, \ldots, n\}$, $min^\mathfrak{W} = 1$, $max^\mathfrak{W} = n$, $succ^\mathfrak{W} = \{(i, i+1) \mid i < n\}$, $<^\mathfrak{W}$ is the natural linear order on $\{1, \ldots, n\}$, and $Q_\sigma^\mathfrak{W} = \{i \leq n \mid w_i = \sigma\}$. We say that a string-language $L \subseteq \Sigma^*$ *is expressible* in a logic $\mathcal{L}$, if there is an $\mathcal{L}(\tau_\Sigma)$-sentence φ_L such that for all non-empty strings $w \in \Sigma^*$ we have $w \in L \iff \mathfrak{W} \models \varphi_L$.

Lemma 7. *(i)* $\{a^n b^n \mid n \in \mathbb{N}\}$ *is expressible in* LFP^2_{param}.
(ii) $\{a^n b^n c^n \mid n \in \mathbb{N}\}$ *is expressible in* LFP^2_{param}. *In particular,* LFP^2_{param} *is capable of defining a non-context-free string-language.*

Proof. We use x and y to denote the two first-order variables available in the logic LFP^2_{param}.

A binary relation E over $\{1, \ldots, n\}$ is called a *pairing* iff the following is true for all $(i,j), (i',j') \in E$: (1) $i < j$, (2) $i = i' \iff j = j'$, and (3) $i < i' \iff j' < j$.

Let $\varphi_y(x, y, R)$ be the following M-LFP^2_{param}-formula

$$\varphi_y(x, y, R) := \big[\mathbf{lfp}_{X,x}\, succ(y,x) \vee \exists y\, (Xy \wedge R(x,y)) \vee \\ \exists y\, (Xy \wedge succ(y,x) \wedge \exists x\, (Xx \wedge R(y,x)))\,\big](x).$$

Claim 1. *For every $n \in \mathbb{N}$, every pairing $E \subseteq \{1,..,n\}^2$, and all $i,j \in \{1,..,n\}$ the following is true: $\langle \{1,..,n\}, succ \rangle \models \varphi_y(i,j,E)$ if, and only if, $i \in \{j{+}1,\ i',\ i'{+}1 \mid (i',j{+}1) \in E\}$. (An illustration is given in Figure 2.)*

Due to space limitation, we have to omit the proof of this and the following two claims.

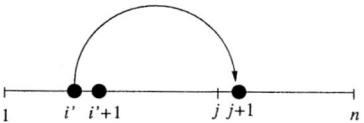 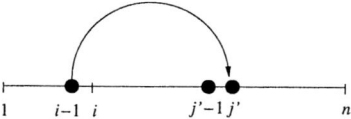

Fig. 2. The unary least fixed-point defined by the formula φ_y in case that y is interpreted by some j for which there exists an i' such that $(i',j{+}1) \in E$. The nodes that belong to this fixed-point are marked by black circles.

Fig. 3. The unary least fixed-point defined by the formula φ_x in case that x is interpreted by some i for which there exists a j' such that $(i{-}1,j') \in E$. The nodes that belong to this fixed-point are marked by black circles.

Analogously to the formula $\varphi_y(x,y,R)$ we define an M-LFP$^2_{param}$-formula $\varphi_x(x,y,R)$ as follows:

$$\varphi_x(x,y,R) := \big[\, \mathrm{lfp}_{Y,y}\ succ(y,x) \vee \exists x\,(Yx \wedge R(x,y)) \vee \\ \exists x\,(Yx \wedge succ(y,x) \wedge \exists y\,(Yy \wedge R(y,x)))\,\big](y).$$

Claim 2. *For every $n \in \mathbb{N}$, every pairing $E \subseteq \{1,..,n\}^2$, and all $i,j \in \{1,..,n\}$ the following is true: $\langle \{1,..,n\}, succ \rangle \models \varphi_x(i,j,E)$ if, and only if, $j \in \{i{-}1,\ j',\ j'{-}1 \mid (i{-}1,j') \in E\}$. (An illustration is given in Figure 3.)*

Finally, we define the LFP$^2_{param}$-formulae $\chi(x,y,R) := x<y\ \wedge\ \big((x = min\ \wedge\ y = max) \vee (\varphi_x(x,y,R) \wedge \varphi_y(x,y,R))\big)$ and $\psi(x,y) := \big[\,\mathrm{lfp}_{R,xy}\,\chi(x,y,R)\,\big](x,y)$.

Claim 3. *For every $n \in \mathbb{N}$ and all $i,j \in \{1,..,n\}$, the following is true: $\langle \{1,..,n\}, min, max, succ, < \rangle \models \psi(i,j) \iff i<j$ and $j = n-i+1$.*

We are now ready to present the LFP$^2_{param}$-sentence $\varphi_{a^n b^n}$ that defines the string-language $\{a^n b^n \mid n \in \mathbb{N}\}$ via

$$\varphi_{a^n b^n} := \exists x \exists y\ \psi(x,y) \wedge succ(x,y) \wedge Q_a(x) \wedge Q_b(y)\ \wedge \\ \forall y\,(y<x \to Q_a(y)) \wedge \forall x\,(y<x \to Q_b(x)).$$

Using Claim 3 it is straightforward to see that $\varphi_{a^n b^n}$ indeed defines the language $\{a^n b^n \mid n \in \mathbb{N}\}$. Thus, the proof of part *(i)* of Lemma 7 is complete.

The proof of part *(ii)* of Lemma 7 uses a similar construction. Now, however, the formula $\varphi_{a^n b^n c^n}$ defining the string-language $\{a^n b^n c^n \mid n \in \mathbb{N}\}$ is given via $\varphi_{a^n b^n c^n} := \varphi_{a^* b^* c^*} \wedge \varphi_{a^n b^n c^*} \wedge \varphi_{a^* b^n c^n}$ where

- $\varphi_{a^* b^* c^*}$ is an FO^2-sentence expressing that the underlying string belongs to the language defined by the regular expression $a^* b^* c^*$.

- $\varphi_{a^n b^n c^*}$ is a LFP^2_{param}-sentence which, for an underlying string of the form $a^* b^* c^*$ expresses that the number of as is equal to the number of bs. This sentence can be obtained in a similar way as the sentence $\varphi_{a^n b^n}$ from the proof of part *(i)* of Lemma 7.
- $\varphi_{a^* b^n c^n}$ is a LFP^2_{param}-sentence which, for an underlying string of the form $a^* b^* c^*$ expresses that the number of bs is equal to the number of cs. Again, this sentence can be obtained in a similar way as the sentence $\varphi_{a^n b^n}$ from the proof of part *(i)* of Lemma 7. □

Using Lemma 7, one easily obtains

Theorem 8. M-$\text{LFP}^2_{param} \lneq \text{LFP}^2_{param}$ *on finite strings. That is, already on the class of finite strings, the two-variable fragment of binary least fixed-point logic where parameters are allowed is strictly more expressive than the two-variable fragment of monadic least fixed-point logic where parameters are allowed.*

Proof. It is well-known that the string-language $\{a^n b^n \mid n \in \mathbb{N}\}$ is not regular, i.e., due to Büchi's theorem, not expressible in monadic second-order logic MSO. As M-$\text{LFP}^2_{param} \leq$ MSO, we therefore obtain that $\{a^n b^n \mid n \in \mathbb{N}\}$ is not expressible in M-LFP^2_{param}. From Lemma 7 we obtain that $\{a^n b^n \mid n \in \mathbb{N}\}$ is expressible in LFP^2_{param}. □

References

1. A. Arnold and D. Niwiński. *Rudiments of μ-calculus*. North Holland, 2001.
2. S. S. Cosmadakis. The complexity of evaluating relational queries. *Information and Control*, 58:101–112, 1983.
3. A. Dawar. *Feasible computation through model theory*. PhD thesis, Univ. of Pennsylvania, 1993.
4. A. Dawar, S. Lindell, and S. Weinstein. Infinitary logic and inductive definability over finite structures. *Information and Computation*, 119:160 – 175, 1995.
5. A. Dawar, E. Grädel, and S. Kreutzer. Inflationary fixed points in modal logic. *ACM Transactions on Computational Logic*, 5:282–315, 2004.
6. S. Dziembowski. Bounded-variable fixpoint queries are pspace-complete. In *Proc. of CSL'96*, volume 1258 of *Lecture Notes in Computer Science*, pages 89–105. Springer, 1997.
7. H.-D. Ebbinghaus and J. Flum. *Finite model theory*. Springer, New York, second edition, 1999.
8. E. Grädel and M. Otto. On Logics with Two Variables. *Theoretical Computer Science*, 224:73–113, 1999.
9. E. Grädel, M. Otto, and E. Rosen. Undecidability Results for Two-Variable Logics. *Archive for Mathematical Logic*, 38:313–354, 1999. Journal version of STACS'97 paper.
10. M. Grohe. Finite variable logics in descriptive complexity theory. *Bulletin of Symbolic Logic*, 4:345–398, 1998.
11. M. Grohe and N. Schweikardt. Comparing the succinctness of monadic query languages over finite trees. *RAIRO - Theoretical Informatics and Applications (ITA)*, 38:343–373, 2004. Journal version of CSL'03 paper.

12. I. Hodkinson. Finite variable logics. *Bull. Europ. Assoc. Theor. Comp. Sci.*, 51:111–140, 1993.
13. N. Immerman. Relational queries computable in polynomial time. *Information and Control*, 68:86–104, 1986.
14. Ph. Kolaitis and M. Vardi. On the expressive power of variable-confined logics. In *Proc. of LICS'96*, pages 348–359, 1996.
15. L. Libkin. *Elements of Finite Model Theory*. Springer, 2004.
16. M. Otto. *Bounded variable logics and counting – A study in finite models*, volume 9 of *Lecture Notes in Logic*. Springer-Verlag, 1997. IX+183 pages.
17. N. Schweikardt. On the expressive power of monadic least fixed point logic. In *Proc. of ICALP'04*, Lecture Notes in Computer Science, pages 1123–1135. Springer, 2004.
18. M. Y. Vardi. On the complexity of bounded-variable queries. In *PODS'95: 14th ACM Symposium on Principles of Database Systems*, pages 266–276, 1995.
19. M. Y. Vardi. The complexity of relational query languages. In *STOC'82: 14th Annual ACM Symposium on the Theory of Computing*, pages 137–146, 1982.

Languages Representable by Vertex-Labeled Graphs

Igor Grunsky[1], Oleksiy Kurganskyy[1], and Igor Potapov[2]

[1] Institute of Applied Mathematics and Mechanics,
Ukrainian National Academy of Sciences,
74 R. Luxemburg St, Donetsk, Ukraine
grunsky@iamm.ac.donetsk.ua, kurgansk@gmx.de
[2] Department of Computer Science,
University of Liverpool, Chadwick Building,
Peach St, Liverpool L69 7ZF, UK
igor@csc.liv.ac.uk

Abstract. In this paper we study the properties of undirected vertex-labeled graphs and the limitations on the languages that they represent. As a main result of this paper we define the necessary and sufficient conditions for the languages to be representable by a class of undirected vertex-labeled graphs and its subclasses. We assume that all obtained results and techniques are transferable to the case of undirected edge-labeled graphs and might give us similar results. The simplicity of necessary conditions emphasizes the naturalness of the result. The proof of their sufficiency is quite non-trivial and it is based on a new notion of quasi-equivalence, that is significantly different from Myhill-Nerode equivalence and might not be reduced to it.

1 Introduction

A formal language is a set of finite-length words that can be specified in a great variety of ways and be associated with different combinatorial objects [2,4,6,7]. For example we can consider a set of paths in a labeled graph that define a language in an alphabet of graph labels. Actually, different classes of labeled graphs can place different limitations on the languages that they represent. It is a well known fact that the necessary and sufficient condition for the language to be representable by finite edge-labeled graphs, is its regularity. The vertex-labeled graphs place another structural limitation:

Theorem 1. *[6]. A language L can be represented by a vertex-labeled graph iff the following two base properties hold:*

1) all words in L start from the same symbol and
2) language L is closed under all nonempty prefixes.

It is also known from [6] that a language L can be represented by *finite* vertex-labeled graph iff L satisfies to the *base properties* and it is regular.

In this paper we study the properties of undirected vertex-labeled graphs with one initial vertex and all final vertices and the limitations on the languages that they represent. The language that we consider is just a set of labels for all paths (in a graph) starting from the initial vertex. As a main result of this paper we define the necessary and sufficient conditions for the languages to be representable by a class of undirected vertex-labeled graphs and its subclasses. In principal the presented results and techniques are transferable to the case of undirected edge-labeled graphs as well and it is most likely that they can give us similar results.

The simplicity of necessary conditions emphasizes the naturalness of the result. The proof of their sufficiency is quite non-trivial and it is based on a new notion of *quasi-equivalence* of vertices, that is significantly different from Myhill-Nerode equivalence and might not be reduced to it. An example of a graph with three quasi-equivalent vertices (labeled by "b") is shown on the Figure 1.

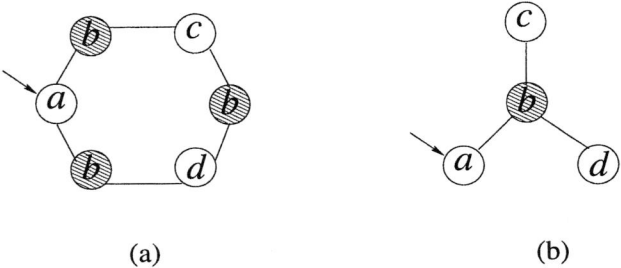

Fig. 1. An example of undirected graph with the initial vertex "a" (before and after the gluing of three quasi-equivalent vertices)

The only common property of equivalent and quasi-equivalent vertices in an undirected vertex-labeled graph is that the gluing of equivalent as well as quasi-equivalent vertices does not change the language (representable by this graph).

This paper is organised as follows. Next section contains preliminaries. In the third section we give a technical proofs of intermediate results about graph modifications, that we use in Section 4 and 5. Next we introduce a new notion of quasiequivalent vertices and prove some of their properties, which are the core elements of proofs from Section 5. The Section 5 contains the main results of this paper. We present here the characterization theorems for classes of languages representable by finite undirected graphs, infinite undirected graphs and finite undirected deterministic graphs. The paper ends with some conclusions.

2 Preliminaries

Let X be a finite alphabet and X^* be a free monoid on the set X. We denote an empty word by e and $X^* - \{e\}$ by X^+. The length of a word $w = x_1 x_2 \ldots x_k$, where $x_i \in X$, $1 \leq i \leq k$, is denoted by $|w|$ and w^{rev} is the reverse word

$x_k x_{k-1} \ldots x_1$. We also introduce a partial binary associative operation of *splicing* on a set X^+ denoted by $\circ$.

Definition 1. *The splicing of two words is defined as follows:*

$$wx \circ yu \begin{cases} wxu, & x = y \\ undefined, & x \neq y \end{cases}$$

where $x, y \in X$ and $w, u \in X^*$.

Now we define other notations in terms of splicing. Let w be a word that can be represented as $u_1 \circ u_2 \circ u_3$, where $u_1, u_2, u_3 \in X^+$. We call u_1 as a prefix of the word w and u_3 as a suffix of w. A language $L \subseteq X^+$ is closed under all non-empty prefixes if $u \in L$ follows from $u \circ w \in L$, where $u, w \in X^+$.

Definition 2. *Let $L, L' \subseteq X^+$. The basic language operations of splicing and prefix contraction are defined as follows:*

- *a splicing of two languages: $L \circ L' = \{w \circ u | w \in L, u \in L'\}$*
- *a prefix contraction of a word w from a language L: $w^{-1} \circ L = \{u | w \circ u \in L\}$*
- *a prefix contraction of a language L' from a language L: $(L')^{-1} \circ L = \bigcap_{w \in L'} w^{-1} \circ L$*

Definition 3. *Let $G = (S, E, X, \mu)$ be a simple (i.e without multiple edges), directed vertex-labeled graph, where S is a (possibly infinite, but counted) set of vertices, E is a set of directed edges, X is a finite set of labels, $\mu : S \to X$ is a mapping from set of vertices to the set of labels and $\mu(s)$ is a label of a vertex $s \in S$.*

The vertices s and t are *adjacent* if they are connected by an edge in G that we denote by st or (s,t). A graph G is *undirected* if for any edge $st \in E$ there is the edge $ts \in E$. A pair of edges st and ts is an undirected edge that we denote by $\{s,t\}$. A graph G is called initial if in the set of vertices there is the initial vertex s_0. In order to indicate that a vertex s_0 is the initial vertex in a graph G we denote it by $G(s_0)$. From now on we only consider initial graphs and assume that intersection of set of vertices for different graphs is empty.

A finite sequence of vertices $p = s_1 s_2 \ldots s_k$ such that $s_i s_{i+1} \in E, 1 \leq i < k$ is a path of length $k-1$ in a graph G. Let us say that the path p begins in the vertex s_1 and ends in s_k. Then the distance between two vertices in a graph is the length of the shortest path between them.

The label $\mu(p)$ of a path $p = s_1 s_2 \ldots s_k$ in a graph is a word

$$\mu(s_1)\mu(s_2)\ldots\mu(s_k),$$

that is concatenation of vertex labels on p.

Definition 4. *Let P and Q are two subsets of the set of vertices S, i.e. $P, Q \subseteq S$. By $P^{-1}Q$ we denote the set of labels for each paths in a graph G that begins in a vertex from the set P and ends in a vertex from the set Q.*

Now we introduce several special cases for the notation $P^{-1}Q$. Let us consider the case where $P = \{s\}$ and $Q = \{t\}$ then we have that $P^{-1}Q = s^{-1}t$. If s is the initial vertex in a graph G then we denote $s^{-1}t$ by **t**. Thus, the notation **t** in an initial graph corresponds to the set of labels of all paths from the initial vertex to the vertex t. The notation $\mathbf{s^{-1}}$ stands for the set $P^{-1}Q$, where $P = \{s\}$ and $Q = S$. The notation $\mathbf{s^{-1}}$ denotes a set of labels of all paths from a vertex s that we call a language generated by the vertex $s \in S$ in a graph G.

Definition 5. *By $L(G)$ we denote a language generated from the initial vertex s_0 in an initial graph G, i.e. $L(G) = \mathbf{s_0}^{-1}$.*

By sw, where $s \in S$, $w \in X^+$, we denote the set of all reachable vertices from a vertex s by any path with a label w, i.e. $sw = \{t \in S | \exists p = s_1 s_2 \ldots s_k \in S^* : s_1 = s, s_k = t, \mu(p) = w\}$.

Definition 6. *Two vertices $s, t \in S$ are equivalent, if $\mathbf{s}^{-1} = \mathbf{t}^{-1}$. This equivalence relation is denoted as ϵ.*

Definition 7. *A graph G is reduced if it does not contain any equivalent vertices.*

Definition 8. *A graph G is deterministic if any vertex $s \in S$ does not have any two adjacent vertices with the same label.*

Let us given some equivalence relation $\rho \subseteq S \times S$ (reflexive, transitive and symmetric relation), such that two vertices s_1 and s_2 have the same labels if they are in the same equivalence class generated by ρ. By $\rho(s)$, where $s \in S$, we denote a set $\{t | (s,t) \in \rho\}$.

Definition 9. *The graph $G/\rho = (G', E', X', \mu')$ is a factor graph of a graph G if $G' = \{\rho(s) | s \in S\}$, $\rho(s)\rho(s') \in E'$ for all edges $ss' \in E$, $\mu'(\rho(s)) = \mu(s)$, $\rho(s_0)$ and s_0 are initial vertices in G/ρ and G, respectively.*

A graph G/ρ from G can be constructed by gluing all vertices in equivalence class $\rho(s)$ for each $s \in S$. In particular we say that G' is constructed from G by gluing two vertices s and t if G' and G/ρ are isomorphic and the equivalence relation ρ partitions the set of vertices of a graph G into one-element equivalence classes excepting a single two-element equivalence class $\rho(s) = \{s, t\}$.

Now we define two kind of simple graphs that we call *Line* and *Yarn*. Let $w = x_1 x_2 \ldots x_k$ is a word of a length k. By $Line(w)$ we denote an undirected graph with k vertices $s_1, \ldots, s_k$, where s_1 is the initial vertex, $\{\{s_i, s_{i+1}\} | 1 \le i < k\}$ is a set of edges and each vertex s_i has a label x_i, $1 \le i \le k$. An example for a graph $Line(w)$ where $w = abcbb$ is shown on the Figure 2.

Definition 10. *By $Yarn(L)$ for a language L, in which all words start from the same symbol, we denote an undirected graph that is constructed from graphs $\{Line(w) | w \in L\}$ by their direct sum and by gluing the initial vertices of all graphs $Line(w)$ into one initial vertex of $Yarn(L)$.*

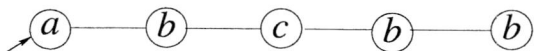

Fig. 2. An example for a $Line(w)$ where $w = abcbb$

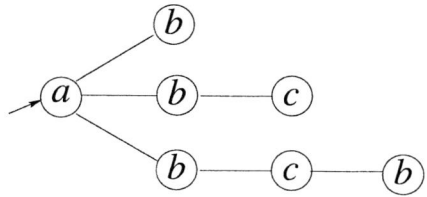

Fig. 3. An example for a $Yarn(L)$, where $L = \{ab, abc, abcb\}$

An example for $Yarn(L)$, where $L = \{ab, abc, abcb\}$, is shown on the Figure 3.

A class of regular languages we denote by *Reg*. We say that a language $L \subseteq X^+$ is representable by a graph (that is denoted by $L \in Graph$) if there exist an initial graph G, such that $L(G) = L$.

We also use the following notations for languages generated by different classes of graphs: *FGraph* is a class of languages generated by finite graphs, *DGraph* is a class of languages generated by deterministic graphs, *UGraph* is a class of languages generated by undirected graphs, *FUGraph* is a class of languages generated by finite undirected graphs, *FUDGraph* is a class of languages generated by finite undirected deterministic graphs. For example, $L \in FUGraph$ means that there is a finite undirected graph G such that $L(G) = L$.

3 Intermediate Results

In this section we show a number of intermediate results that we use for characterization of classes *UGraph*, *FUGraph* and *FUDGraph*. Note, that the languages and graph labels are defined in the same finite alphabet X.

Lemma 1. *Let G be a directed vertex-labeled graph and s, t are two vertices in G with labels x and y, respectively and the following two properties:*

- *an edge ts does not belong to the graph G and*
- $(yx)^{-1} \circ \mathbf{t}^{-1} \supseteq \mathbf{s}^{-1}$.

If G' is a graph constructed from G by adding an edge ts then the language generated by G' does not differ from a language generated by G, i.e. $L(G) = L(G')$.

Proof. Let us add one edge ts to a directed vertex-labeled graph G, that is obviously can only extend language $L(G)$. Thus, it is enough to show that for any path p which goes via an edge ts from the initial vertex we can find a path

p' with the same label that also starts from the initial vertex without passing an edge ts, i.e. $\mu(p) = \mu(p')$.

Let us consider a path $p = s_1 s_2 \ldots s_k$ from the initial vertex such that passes an edge ts in a graph G'. Let $s_1 s_2, s_2 s_3, \ldots, s_{k-1} s_k$ be a sequence of edges that corresponds to the path p and contains an edge ts. Let i be a maximum number, such that $ts = s_i s_{i+1}$. According to the initial condition $(yx)^{-1} \circ \mathbf{t}^{-1} \supseteq \mathbf{s}^{-1}$ there is a path $s_i s'_{i+1} s'_{i+2} \ldots s'_k$ with a label $x_i x_{i+1} x_{i+2} \ldots x_k$, where $x_i x_{i+1} = yx$, but which does not pass an edge ts. Thus there is a path $r = s_1 \ldots s_i s'_{i+1} s'_{i+2} \ldots s'_k$ in G' with a label $x_1 x_2 \ldots x_k$ such that a number of ts-edges in r is less than in p by one. From it follows that for any path p in G' we can construct a path p' in G', that does not contain any ts edges, by repeating the above reduction a finite number of times. In such case we have that for any path p in G there is a path p' in G such that $\mu(p) = \mu(p')$ and therefore $L(G) = L(G')$.

Lemma 2. *For any (finite or infinite) graph G the identity $L(G) = L(G/\epsilon)$ holds.*

Proof. Let G be an initial vertex-labeled graph and let s and t be an equivalent vertices in G. Let us assume that for some vertex s' there is an edge ss' and there is no edge ts' in a graph G. In this case we meet initial condition of Lemma 1, so the language represented by a graph G would not be changed by extending G with an edge ts'. Moreover, a language generated by any vertex in the extended graph would not be changed as well.

Let us extend the graph G in the following way: for any equivalent vertices s and t (i.e. $(s,t) \in \epsilon$) and some vertex s' we add an edges ts' iff an edge ss' belongs to the graph.

From it follows that the sequence of vertices $\epsilon(s_1), \epsilon(s_2), \ldots, \epsilon(s_k)$ is path starting from the initial vertex in a factor graph G/ϵ iff there exist a sequence of vertices $r_1, r_2, \ldots, r_k$, where $r_i \in \epsilon(s_i)$, $1 \leq i \leq k$, that is a path in a graph G and r_1 is its initial vertex. Above facts show that for any graph G the equality $L(G) = L(G/\epsilon)$ holds.

4 Quasiequivalent Vertices

Let us define a quasi-equivalence relation that is significantly different from ϵ.

Definition 11. *Given a graph G that generates a language L. Two vertices s and t from a graph G are quasiequivalent if $(\mathbf{s})^{-1} \circ L = (\mathbf{t})^{-1} \circ L$.*

Let us denote the quasiequivalence relation of vertices by α.

Definition 12. *A graph G is irreducible if it does not contain any quasiequivalent vertices.*

It is clear that equivalence and quasiequivalence relations coincide in deterministic graphs. Lemma 4 shows us another property of quasiequivalence relation that $L(G) = L(G/\alpha)$.

Lemma 3. *Given a graph G and a language L with the following base properties*
- *all words in L start from the same symbol and*
- *language L is closed under all nonempty prefixes.*

The language $L(G) \subseteq L$ iff we have that $\mathbf{s}^{-1} \subseteq (\mathbf{s})^{-1} \circ L$ for any vertex s in a graph G.

Proof. Let us remind, that in expression "$(\mathbf{s})^{-1}$" from "$\mathbf{s}^{-1} \subseteq (\mathbf{s})^{-1} \circ L$" by $\mathbf{s}$ we denote the set of labels of all paths that start from the initial vertex of a graph G and end in the vertex s, which are words from X^+. The above lemma follows from the fact that $\mathbf{s}^{-1} \subseteq (\mathbf{s})^{-1} \circ L$ is equivalent to $\mathbf{s} \circ \mathbf{s}^{-1} \subseteq L$.

Lemma 4. *For any (finite or infinite) graph G the identity $L(G) = L(G/\alpha)$ holds.*

Proof. Let G be an initial vertex-labeled graph. Note that the operation of gluing any two vertices in a vertex-labeled graph can only extend the language that it represents. Let us assume that:

- $L = L(G)$,
- s and t are quasiequivalent vertices and
- $W = (\mathbf{s})^{-1} \circ L = (\mathbf{t})^{-1} \circ L$.

From Lemma 3 follows that $\mathbf{s}^{-1} \subseteq W$ and $\mathbf{t}^{-1} \subseteq W$.

Let us construct an extension of graph G by adding new vertices and edges in such a way that $\mathbf{s}^{-1} = \mathbf{t}^{-1} = W$ in an extended graph holds. In particular we can do it as follows. Let us add by direct sum a graph $Line(w)$ to a graph G for the vertex s and each word $w \in W$ and glue a vertex s with the initial vertex of $Line(w)$. Next we can repeat the same construction with the vertex t. According to the Lemma 3 the language represented by a graph G after its extension is not changed.

Since a pair of vertices s and t are now equivalent in the extended graph G we can glue it without changing the language represented by this graph. Now we can delete all introduced edges and nodes to get a graph that can be constructed by just gluing quasiequivalent vertices s and t.

Lemma 5. *If G be infinite graph and $L(G)$ be a language such that $L(G) \in FGraph$ then a graph G/α is finite.*

Proof. Since the power of a set $\{(W)^{-1} \circ L(G) | W \subseteq L(G)\}$ is bounded by a number 2^N, where $N = |\{w^{-1} \circ L(G) | w \in L(G)\}|$ we have a finite number of classes with quasiequivalent vertices.

Note, that the proposition in Lemma 5 does not hold for the relation ϵ.

The example of an infinite directed graph G with quasiequivalent vertices is shown on the Figure 4. It is clear that $L(G) = a(aa)^*b$. The vertices with the same pattern in the Figure 4 are quasiequivalent. This graph has no equivalent vertices except those that labeled with b. The graph G/ϵ on the Figure 5, is infinite and the graph G/α on the Figure 6 is finite. However it follows from Lemma 2 and Lemma 5 that $L(G) = L(G/\epsilon) = L(G/\alpha)$.

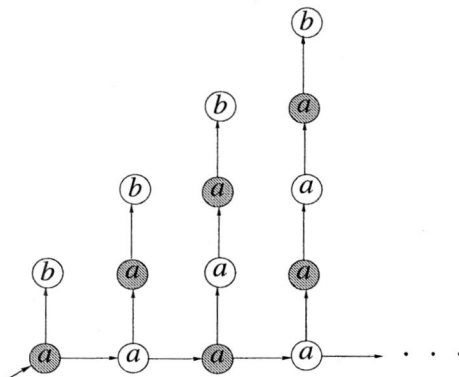

Fig. 4. An example of quasiequivalent vertices. The vertices with the same pattern are quasiequivalent, but not equivalent.

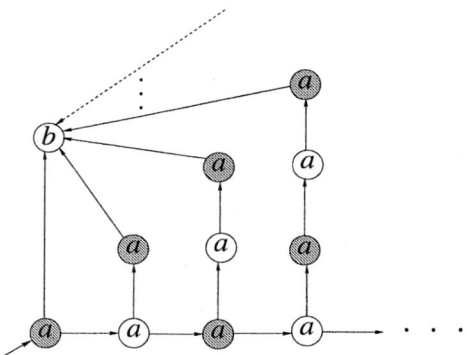

Fig. 5. The graph G/ϵ, where G is shown on Figure 4

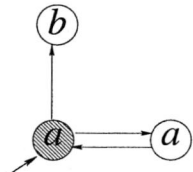

Fig. 6. The graph G/α, where G is shown on Figure 4

5 Characterization of Graph Representable Languages

In this section we present our main results about characterization of languages that can be represented by undirected graphs.

Definition 13. *Symmetric closure of a language $L \in Graph$ is a minimal language $[L] \in UGraph$ such that $L \subseteq [L]$ (note that $L \in UGraph$ iff $L = [L]$).*

Theorem 2. $L(Yarn(L)) = [L]$.

Proof. By Definition 10, $Yarn(L)$ is an undirected graph that represents a language that include all words from a language L. From it follows that $[L] \subseteq L(Yarn(L))$. Let us prove the inverse proposition $L(Yarn(L)) \subseteq [L]$.

Let G be a graph such that $L(G) = [L]$ and $p = s_1 s_2 \ldots s_k$ is a path from G, that starts in the initial vertex and $\mu(p) = w$. Let a graph $Line(w)$ has a set of vertices $\{t_1, t_2, \ldots, t_k\}$, and set of edges $\{t_i t_{i+1}\}$, $1 \leq i < k$ and t_1 is the initial vertex. Now we define a mapping ϕ such that $\phi(t_i) = s_i$, $1 \leq i \leq k$, which is a homomorphism of graph $Line(L)$ to G. From the above mapping and the definition of $Yarn(L)$ follow that there is a homomorphism from $Yarn(L)$ to G. Thus, we have that $L(Yarn(L)) \subseteq [L]$.

Now we have a straightforward but useful corollary.

Corollary 1. $L \in UGraph$ iff $L = L(Yarn(L))$.

Theorem 3. *A language L can be represented by undirected graph, i.e. $L \in UGraph$, iff it satisfies to the following properties:*

1) all words of the language L are starting from the same symbol
2) the language L is closed under prefixes
3) if $w, u \in L$ then $w \circ w^{rev} \circ u \in L$
4) if $w \circ u \circ v \in L$ then $w \circ u \circ u^{rev} \circ u \circ v \in L$.

Proof. The straightforward check of the above four properties give us the necessity of conditions. Let us prove that these four properties are also sufficient. Namely we prove that $L = L(Yarn(L))$ follows from the properties 1-4.

In the first part of the proof we show that if $w = x_1 x_2 \ldots x_k \in L$ then $L(Line(w)) \subseteq L$. A graph $Line(w)$ has k vertices (be definition) that we denote by s_i, $1 \leq i \leq k$, with initial vertex s_1 and a set of edges $\{\{s_i, s_i + 1\} | 1 \leq i < k\}$, where s_i has a label x_i. We will also say that i is the coordinate of a vertex s_i.

Let p be a path from a vertex s_1 in a graph $Line(w)$ which differs from $s_1 s_2 \ldots s_k$. Let us prove that the label of the path p belongs to the language L, i.e. $\mu(p) \in L$. From property 2 of this theorem follows that if extension of any path belongs to a language L the original path also belongs to L, so without loss of generality we can assume that p ends in a vertex s_k. Let us represent p in the form

$$r_1 \circ l_1 \circ r_2 \circ l_2 \circ \ldots \circ r_{n-1} \circ l_{n-1} \circ r_n,$$

where l_i is a path in $Line(w)$ with decreasing coordinates (movement to the left) and r_i is a path in $Line(w)$ with increasing coordinates (movement to the right). Now we can show that there exist a path p' in $Line(w)$, where p' starts and ends at the same vertex as p and $|p'| < |p|$, such that from $\mu(p') \in L$ follows $\mu(p) \in L$. Note that we can get p' from p by deleting some of its subpaths. If we continue the process of deletion we show that from $\mu(s_1 s_2 \ldots s_k) \in L$ follows $\mu(p) \in L$.

Let a path r_i, $1 \leq i \leq n$ starts and ends in vertices with coordinates u_i and o_i, respectively. Actually there are only two cases: either $u_1 < u_2 < \ldots < u_n$ or $u_{i+1} \geq u_{i+2}$ for some i, $0 < i < n-1$.

Case 1. Let us assume that $u_1 < u_2 < \ldots < u_n$ for a set of subpaths $r_1, \ldots, r_n$ in path p. It is obvious that the property $o_1 > o_2 > \ldots > o_n$ does not hold for p since r_n ends with a maximal coordinate k, i.e. $o_n = k$. Thus, there is i, $1 \leq i < n$, such that $o_i \leq o_{i+1}$. From it follows that a path r_i can be represented as $r_i = r \circ l_i^{rev}$ and r_{i+1} as $r_{i+1} = l_i^{rev} \circ r'$. Thus we have that

$$p = r_1 \circ l_1 \circ \ldots \circ r \circ l_i^{rev} \circ l_i \circ l_i^{rev} \circ r' \circ \ldots \circ r_n,$$

and therefore

$$p' = r_1 \circ l_1 \circ \ldots \circ r \circ l_i^{rev} \circ r' \circ \ldots \circ r_n.$$

is a path in $Line(w)$. It is easy to check that $\mu(p) \in L$ if $\mu(p') \in L$, since we can construct $\mu(p)$ from $\mu(p')$ by applying the property 4 of the theorem.

Case 2. Let us assume that $u_{i+1} \geq u_{i+2}$ for some i, $0 \leq i < n-1$. There are also two subcases: $o_i \leq o_{i+1}$ and $o_i > o_{i+1}$. In the proof of the first subcase we mimic the case 1 argumentation. So we consider here only the second subcase.

If $o_i > o_{i+1}$ then a subpath l_i can be represented as $l_i = l \circ r_{i+1}^{rev}$ and a subpath l_{i+1} can be represented as $l_{i+1} = r_{i+1}^{rev} \circ l'$. From it follows that

$$p = r_1 \circ \ldots \circ l \circ r_{i+1}^{rev} \circ r_{i+1} \circ r_{i+1}^{rev} \circ l' \circ \ldots \circ r_n$$

and therefore there is a path p' in a graph $Line(w)$ such that

$$p' = r_1 \circ \ldots \circ l \circ r_{i+1}^{rev} \circ l' \circ \ldots \circ r_n.$$

So we have that $\mu(p) \in L$ if $\mu(p') \in L$, since we can construct $\mu(p)$ from $\mu(p')$ by applying the property 4 of the theorem. Now we can apply the same method a number of times to transfer path p into a path $s_1 s_2 \ldots s_k$, that has a label from a language L. Thus $\mu(p) \in L$ and $L(Line(w)) \subseteq L$.

It is clear that the deletion of all incoming edges for the initial vertex of a graph $Yarn(L)$ give us a graph that generate L. In the second part of the proof we show that the property 3 allows us to restore the deleted edges without any changes in a language generated by that graph. It implies the fact that $L = L(Yarn(L))$.

Let us consider a graph G that generates a language L and let a vertex t be an adjacent to the initial vertex s. It follows from the property 3 of the Theorem that any vertex t satisfies to Lemma 1. It means that any edge, outgoing from the initial vertex in a graph G can be transformed into undirected edges without any changes of the language that G generates. Now we can come to the conclusion that $L = L(Yarn(L))$, since we can restore all edges (incoming to the initial vertex) that we deleted from $Yarn(L)$.

The Theorem 3 shows the characterization of languages generated by undirected vertex-labeled graphs (i.e $UGraph$). Now we consider the class $FUGraph$, which is a subclass of $UGraph$.

Theorem 4. $FUGraph = FGraph \cap UGraph$

Proof. Let L be a language, such that $FGraph \cap UGraph$. From Lemma 5 follows that any infinite undirected graph has a finite number of quasiequivalent classes. By gluing all quasiequivalent vertices in each class we have a finite graph that generates L. Since the graph obtained from an undirected graph by gluing quasiequivalent vertices is undirected, we have that $FUGraph = FGraph \cap UGraph$.

Since $FUGraph = FGraph \cap UGraph$ is equivalent to $FUGraph = Reg \cap UGraph$ we have the following corollary:

Corollary 2. *The langauge L is representable by a finite undirected graph, i.e $L \in FUGraph$ iff L satisfies to the properties of Theorem 3 and $L \in Reg$.*

Since Lemma 5 gives us an upper bound on a number vertices in any irreducible graph that generate a language $L \in FUGraph$, the following corollary holds:

Corollary 3. *The membership problem for $FUGraph$, i.e. checking whether a language L belongs to the class $FUGraph$, is decidable.*

Proof. From Lemma 5 follows that the number of quasieqivalent vertices is bounded by a constant k. Thus we can run over the whole set of undirected detrministic and nondetrministic graphs with a number of verices less then k and check if the language L can be represented by some of these graphs, which is just the comparison of regular languages. If we can find at least one such graph then the language L is in $FUGraph$ and $L \notin FUGraph$ otherwise.

As a last result of this section we present a characterization of languages generated by finite undirected deterministic graphs.

Theorem 5. *Given a reduced deterministic graph G which generates a language L. The language $L \in FUDGraph$ iff G is undirected.*

Proof. If a reduced deterministic graph G, which generates L, is undirected then $L \in FUDGraph$ by definition. Let us prove the necessary part of this proposition.

Let G be a finite undirected deterministic graph, which generates a language L. Since $L \in FUDGraph$ we can state that such graph exists. Since equivalence and quasiequivalent relations coincide in deterministic graphs we have that G/α is a deterministic graph, which generates L. On the other hand G/α is undirected graph, which required to be proved.

Theorem 6. *The langauge L is representable by a finite undirected deterministic graph, i.e $L \in FUDGraph$ iff L satisfies to the following properties:*

1) $L \in Reg$;
2) all words in L starting from the same symbol;

3) language L is closed under all nonempty prefixes;
4) if $wxy \in L$, where $x,y \in X$ and $w \in X^+$ then $(wxyx)^{-1} \circ L = (wx)^{-1} \circ L$.

Proof. The necessity of the above properties is obvious, so we only prove that they are sufficient. Let G be a irreducible graph and $L = L(G)$. Then we have that the set $\{\mathbf{s}^{-1}|s \in G\}$ is equal to the set $\{w^{-1} \circ L | w \in L\}$. From property 4 of this Theorem follows that if from a vertex s to a vertex t in G there exist an edge st then there exist an edge from t to s. Thus, the graph G is undirected, q.e.d.

6 Conclusion

In this work we gave a characterization of languages that can be represented by different types of undirected graphs. Note that all results are ultimate in terms that both necessary and sufficient propositions were proved. The topic is also related to exploration of graphs by software entity (i.e. finite automata, mobile robots, software agents etc.) and map validation problems that have received a widespread attention in recent years [1,5,7,8].

References

1. Dudek, G., Jenkin, M., Milios, E., and Wilkes, D. Map Validation and Robot Self-Location in a Graph-Like World. Robotics and Autonomous Systems, Vol. 22(2), 159-178 (1997)
2. Samuel Eilenberg. Automata, Languages, and Machines. Academic Press, Inc. Orlando, FL, USA (1974)
3. I.S. Grunsky, Oleksiy Kurganskyy. Properties of languages that generated by undirected vertex-labeled graps. In proceeding of 8-th International Seminar on Discrete mathematics and its applications, Moscow State University, 267-269 (2004) [in Russian].
4. Matthias Jantzen, Alexy Kurganskyy: Refining the hierarchy of blind multicounter languages and twist-closed trios. Inf. Comput. 185(2): 159-181 (2003)
5. Tod S. Levitt, Daryl T. Lawton. Qualitative Navigation for Mobile Robots. Artif. Intell. 44(3): 305-360 (1990)
6. Yu. V. Kapitonova and A.A. Letichevsky. Mathematical theory of computer systems design, Nauka, 286 pages (1988) [in Russian]
7. V. B. Kudryavtsev, Sh. M. Ushchumlich and G. Kilibarda. On behaviour of automata in labyrinths. Discrete Mathematics and Applications Volume 3, No. 1, 1-28 (1993)
8. S.V. Sapunov. An equivalence of labeled graphs. Tr. Inst. Prikl. Mat. Mekh. 7, 162-167 (2002) [in Russian]

On the Complexity of Mixed Discriminants and Related Problems

Leonid Gurvits

Los Alamos National Laboratory, Los Alamos, NM 87545
gurvits@lanl.gov

Abstract. We prove that it is #P-hard to compute the mixed discriminant of rank 2 positive semidefinite matrices. We present poly-time algorithms to approximate the "beast". We also prove NP-hardness of two problems related to mixed discriminants of rank 2 positive semidefinite matrices. One of them, the so called Full Rank Avoidance problem, had been conjectured to be NP-Complete in [23] and in [25]. We also present a deterministic poly-time algorithm computing the mixed discriminant $D(A_1, .., A_N)$ provided that the linear (matrix) subspace generated by $\{A_1, .., A_N\}$ is small and discuss randomized algorithms approximating mixed discriminants within absolute error.

1 Introduction

1.1 Mixed Discriminant, Permanent and Mixed Volume

Permanent:
Let $A = (a_{ij})$ be an $n \times n$ matrix. The number

$$per(A) = \sum_{\sigma \in S_n} \prod_{i=1}^{n} a_{i\sigma(i)},$$

where S_n is the symmetric group on n elements, is called the permanent of A. For a $0, 1$ matrix A, $per(A)$ counts the number of perfect matchings in G, the bipartite graph represented by A.

It is #P-hard to compute the permanent of a nonnegative (even $0, 1$) matrix [22], and so it is unlikely to be efficiently computable exactly for all matrices. The realistic goal, then, is to try and *approximate* the permanent efficiently as well as possible, for large classes of matrices. Efficient poly-time probabilistic algorithms that approximate the permanent extremely tightly (($1+\epsilon$)-factor) were developed first for several classes of graphs (e.g. [16]). Eventually, a fully-polynomial probabilistic algorithm that approximate the permanent up to ($1+\epsilon$)-factor was developed in [17].

The decade between [16] and [17] produced many remarkable results. First, the technique from [16] (Markov Chain Monte Carlo Method together with conductance-based upper bounds on the second eigenvalue) was used to approximate volume of convex sets. This line of research naturally led to mixed volumes and mixed discriminants (to be defined below) [9]. Both mixed volumes and mixed discriminants can be viewed as "noncommutative" versions of permanents and have fundamental importance as in convex geometry as well in combinatorics.

In order to partially solve one open problem from [9], A. Barvinok ([3]) developed a probabilistic polynomial algorithm to approximate mixed discriminants of positive-definite n by n matrices up to a factor c^n, with constant being improved in [5].

Let us recall some basic definitions. Let $A_1...A_n$ be $n \times n$ real symmetric matrices. It is well known (and easily seen) that the value of $\det(x_1 A_1 + \cdots x_n A_n)$ is a homogeneous polynomial of degree n in variables $x_1...x_n$. The number

$$D(A_1...A_n) = \frac{\partial^n}{\partial x_1...\partial x_n} \det(x_1 A_1 + \cdots x_n A_n) \qquad (1)$$

is called the *mixed discriminant* of $A_1...A_n$.

The mixed discriminant is known [2] to be monotone, that is, $0 \preceq A_i \preceq B_i$, for $i = 1...n$ implies $D(A_1...A_n) \leq D(B_1...B_n)$. [1] In particular, if the matrices $A_1...A_n$ are positive semidefinite, the mixed discriminant $D(A_1...A_n)$ is nonnegative.

From now on, we will be interested in the special case where the matrices $A_1...A_n$ are positive semidefinite. The number

$$V(K_1...K_n) = \frac{\partial^n}{\partial x_1...\partial x_n} Vol(x_1 K_1 + \cdots x_n K_n)$$

is called the *mixed volume* of convex sets $K_1...K_n$. Mixed discriminants and mixed volumes generalize permanents: If the matrices $A_1...A_n$ are diagonal, namely $A_j = diag(b_{1j}...b_{nj})$, for $j = 1...n$, let $B = (b_{ij})$. Then $per(B) = D(A_1...A_n)$. If entries of matrix B are nonnegative then $per(B)$ can be realized as a mixed volume of the corresponding "boxes". It follows that the computation of a mixed discriminant of n positive semidefinite matrices is #P-hard, since it is at least as hard as computing the permanent of a nonnegative matrix. On the other hand, the first efficient probabilistic algorithm that provides a $2^{O(n)}$-factor approximation for permanents of *arbitrary positive matrices* was obtained by Barvinok in [3] via mixed discriminants of positive definite matrices. We recall that the idea (observation) to look at permanents as mixed discriminants was the only new step in Egorychev's proof of van der Waerden conjecture [11]. It is clear in view of Barvinok's inequality ([3]), which connects mixed volumes of ellipsoids with mixed discriminants of their matrices, that if one can approximate mixed discriminants within a multiplicative constant which depends only on dimension, then it is possible to approximate mixed volumes of ellipsoids (and thus so called well presented convex bodies) within a multiplicative constant depending only on dimension. The question whether exists such deterministic approximating algorithm was posed in [9]. Barvinok results provided such probabilistic algorithm.

---

[1] Here and henceforth the sign $\preceq$ denotes the partial ordering induced by the cone of positive semidefinite matrices, namely $A \preceq B$ iff $B - A$ is positive semidefinite.

Finally, in [18] a deterministic polynomial algorithm that computes the mixed discriminant of an n-tuple of positive semidefinite matrices to within a multiplicative factor of e^n has been developed. Surprisingly, this is the same bound which was achieved in [21] for permanents of arbitrary nonnegative matrices.

Anyway, it might seem that computing mixed discriminants is almost the same as computing permanents. We will try to indicate in this paper that it is not so. We don't say though that it is impossible to get for mixed discriminants the same result as for permanents in [17].

Recall that if $A_i = \sum_j x_{i,j} x_{i,j}^* (1 \leq i,j \leq n)$ then

$$D(A_1...A_n) = \sum_{j_1,...,j_n} Det(x_{1,j_1} x_{i,j_1}^* + ... + x_{n,j_n} x_{i,j_n}^*). \tag{2}$$

There exists a poly-time algorithm to decide whether exist $j_1, ..., j_n$ such that

$$D_{j_1,...,j_n} =: Det(x_{1,j_1} x_{1,j_1}^* + ... + x_{n,j_n} x_{i,j_n}^*) > 0. \tag{3}$$

This algorithm is just a particular case of a standard two matroids intersection problem [14]. It is analogous to checking whether $per(B)$ of nonnegative matrix B is nonzero.

But for the permanents the following problems are easy.

P1. To compute maximum over all permutations of $B(1, \pi(i))...B(n, \pi(n))$.
P2. To compute minimum over all permutations of $B(1, \pi(i))...B(n, \pi(n))$.
P3. To compute $per(B)$ provided each row of B has at most two nonzero entries.
P4. To compute $per(B)$ provided $rank(B)$ is "small".

Natural generalizations of these problems to mixed discriminants are:

D1. To compute maximum over all choices of $j_1, ..., j_n$ of $D_{j_1,...,j_n}$.
D2. To compute minimum over all choices of $j_1, ..., j_n$ of $D_{j_1,...,j_n}$, or to check whether there exist $j_1, ..., j_n$ such that $D_{j_1,...,j_n} = 0$.
D3. To compute $D(A_1...A_n)$ provided that $Rank(A_i) \leq 2$, i.e. $j_i \leq 2 (1 \leq i \leq n)$.
D4. To compute $D(A_1...A_n)$ provided that the linear (matrix) space spanned by $\{A_1, ..., A_n\}$ has "small" dimension.

We will show in the following sections that problems D1 and D2 are NP-HARD and that D3 is #P-HARD even for rank 2 matrices ; and that problem D4 is "easy" for an arbitrary matrices. Notice that if D1 were easy it would produce a deterministic poly-time algorithm to approximate mixed discriminants within n^n multiplivative bound and this is sufficient to answer positively the question from [9]. Before our proof in [15] of Bapat's conjecture [2] the (unpublished) bound was about n^{n^2}.

2 Basic Formulas for Mixed Discriminants of Rank 2 Matrices

Let $A_1...A_n$ be $n \times n$ positive semidefinite matrices of rank 2. I.e., $A_i = x_{i,0} x_{i,0}^* + x_{i,1} x_{i,1}^*, 1 \leq i \leq n$.

As we recalled above, it is "easy" to check whether $D(A_1...A_n) > 0$. Thus let us assume wlog that the vectors $\{x_{1,0}, ..., x_{n,0}\}$ form a basis (are linearly independent).

Then $x_{i,0} = C(e_i)$ for corresponding nonsingular matrix C and a standard canonical basis $\{e_1, ..., e_n\}$. Therefore

$$D(A_1...A_n) = Det(C)^2 D(e_1 e_1^* + y_1 y_1^*, ..., e_n e_n^* + y_n y_n^*), y_i = C^{-1}(x_{i,1}). \quad (4)$$

Definition 1. *Let A be $n \times n$ matrix. For a subset $S \subset \{1,..,n\}$ we define the principal submatrix $A_S = \{A_S(i,j) : i, j \in S\}$.*

$$\phi(A, k, m) = \sum_{|S|=k} Det(A_S)^m, \psi(A, k, m) = \sum_{|S|=k} |Det(A_S)|^m. \quad (5)$$

Correspondingly assuming that $\phi(A, 0, m) = \psi(A, 0, m) \equiv 1$ we define

$$\phi(A, m) = \sum_{0 \leq k \leq n} \phi(A, k, m), \psi(A, m) = \sum_{0 \leq k \leq n} \psi(A, k, m). \quad (6)$$

We also define $A_{S,T} = \{A_{S,T}(i,j) : i \in S, j \in T\}$. ∎

We recall that if the characteristic polynomial $Det(A - xI) = \sum_{0 \leq k \leq n} (-1)^k a(k) x^{n-k}$ then $a(k) = \phi(A, k, 1)$. Also, $\phi(A, 1) = Det(I + A)$.

Proposition 1. *Let A be a $n \times n$ matrix with ith column equal to y_i. Then*

$$D(e_1 e_1^* + y_1 y_1^*, ..., e_n e_n^* + y_n y_n^*) = \phi(A, 2). \quad (7)$$

The mixed volume $V(P_1, ..., P_n) = \psi(A, 1)$, where P_i is a 2-dimensional parallelogram with axis e_i and y_i.

The proof is based on the following observation : associate with an n-dimensional boolean vector $b = (b_1, .., b_n)$ (a subset S) n vectors $(z_1, .., z_n)$, where $z_i = e_i$ if $b_i = 0$ and $z_i = y_i$ if $b_i = 1$. Then $Det([z_1|..|z_n]) = Det(A_S)$.

A particular instance of problem D2 for rank two matrices is to compute $\min |Det(A_S)|$ over all subsets S. In [25] what we call "problem D2" for rank k matrices is called **Full Rank Avoidance Problem** for n blocks consisting of k n-dimensional vectors.

Consider $A = I - xe^T$, where x is $n \times 1$ matrix and e is $1 \times n$ matrix of all ones. It follows that

$$Det(A_S) = 1 - \sum_{i \in S} x(i). \quad (8)$$

Thus to check whether $min|Det(A_S)|$ over all subsets S is equal to zero is equivalent to the **KNAPSACK PROBLEM**. We just proved the NP-hardness of D2 even for rank two matrices.

In other words, we proved that given n pairs $\{(X_{i,0}, X_{i,1}) : 1 \leq i \leq n\}$ of integer n-dimensional vectors it is NP-HARD to decide if there exist $b(1), ..., b(n)$ such that the vectors $X_{1,b(1)}, X_{2,b(2)}, ..., X_{n,b(n)}$ are linearly dependent : **Full Rank Avoidance Problem** for n blocks consisting of 2 n-dimensional vectors is NP-HARD over the field Q of rational numbers. (It had been proved in [27] that the same problem is "easy" for the field $GF(2)$.)

To prove NP-hardness of D1 is also not difficult. We use a reduction from the exact covering by 3-sets. Consider the following symmetric block matrix:

$$D = \begin{pmatrix} 0 & X^* \\ X & 0 \end{pmatrix},$$

where X is some rectangular matrix. It is easy to see that $max|Det(D_S)|$ over all subsets S is equal to $Max(1, M(X))$, where $M(X) = max_{|S_1|=|S_2|\geq 1} |Det(X_{S_1,S_2})|^2$.

Consider a family $F = \{F_1, ..., F_k\}$ of 3-subsets of $\{1, 2, ..., n\} : F_i = \{1 \leq j_{i,1} < j_{i,2} < j_{i,3} \leq n\}$. For a 3-dimensional vector $z = (z_1, z_2, z_3)$ and a subset F_i define a n-dimensional vector

$$(f_{z,i}(1), ..., f_{z,i}(n)) : f_{z,i}(j_{i,l}) = z_l (1 \leq l \leq 3) \text{ and zero otherwise }.$$

Notice that if $F_i \cap F_j = \emptyset$ then for any two 3-dimensional vectors u and v vectors $f_{u,i}$ and $f_{v,j}$ are orthogonal. We need three orthogonal 3-dimensional vectors $z^{(1)}, z^{(2)}, z^{(3)}$ satisfying the following property :

$f_{z^{(r)},i}$ is orthogonal to $f_{z^{(d)},j} (1 \leq r, d \leq 3; i \neq j)$ iff $F_i \cap F_j = \emptyset$.

The following vectors satisfy the condition :

$$z^{(1)} = (1, 1, 1), z^{(2)} = (1, 2, 3), z^{(3)} = (5, -4, -1).$$

Put $u^{(r)} = a_r z^{(r)}$ in such a way that $\|u^{(r)}\| = 2 (1 \leq r \leq 3)$. Consider now the folowing $n \times 3k$ matrix $X = [X_1, X_2, X_3]$, where X_r is $n \times k$ matrix $(1 \leq r \leq 3)$ and the ith column of X_r is equal to $f_{u^{(r)},i}(1 \leq i \leq k; 1 \leq r \leq 3)$.

Notice that all columns of matrix X have norm equal to 2. Using Hadamard's determinantal inequality we get that

$$max_S|Det(D_S)| = 2^n$$

if and only if there exists an exact covering of $\{1, 2, ..., n\}$ by some $\frac{n}{3}$ subsets from F. As this problem is known to be NP-Complete, we have proven that the problem D1 is NP-HARD.

Lemma 1.

$$\phi(A, 2) = E(det(D + A))^2 = E(det(I + DA))^2 = E(det(I + AD))^2. \quad (9)$$

Here D is $Diag(\xi_1,...,\xi_n)$, ξ_i are independent symmetric random variables with first moment equal to zero, and second moment equal to 1.

Proof: We adopt to this rank two situation the formula (a straightforward generalization of Godzil-Gutman formula) from [3] :

$$\phi(A,2) = D(e_1 e_1^* + y_1 y_1^*, ..., e_n e_n^* + y_n y_n^*) =$$
$$E(Det([\mu_i e_i + \xi_i y_i|...|\mu_n e_n + \xi_n y_n]))^2 \tag{10}$$

Here $(\mu_1,...,\mu_n)$ is a random vector independent of $(\xi_1,...,\xi_n)$ and uniformly distrubuted on $\{1,-1\} \times ... \times \{1,-1\}$ (a Rademacher sequence) ; $A = [y_1|...|y_n]$.
It follows from symmetricity and independence that

$$E(Det([\mu_i e_i + \xi_i y_i|...|\mu_n e_n + \xi_n y_n]))^2 =$$

$$E(Det([e_i + \mu_i \xi_i y_i|...|e_n + \mu_n \xi_n y_n]))^2 =$$

$$E(Det([e_i + \xi_i y_i|...|e_n + \xi_n y_n]))^2.$$

This proves that $\phi(A,2) = E(det(I+AD))^2$. The other identities are proved in the same way. ∎

We will use below only the identity $\phi(A,2) = E(det(D+A))^2$ for Rademacher sequences. Our first proof of this identity was based on the "boolean" Fourier transform (Walsh-Hadamard transform) . Indeed, define the following function on the boolean cube :

$$f(S) = Det(A_S).$$

Then its Walsh-Hadamard transform $F(\omega) = \sqrt{2}^{-n} Det(I + Diag(\omega)A)$, where ω is $(+1,-1)$ vector.

It remains to use the unitarity of the Walsh-Hadamard transform. As we explained above, $\phi(A,k,1)$ and $\phi(A,1)$ can be computed in polynomial time and they depend only on the characteristic polynomial of the matrix A. If all $Det(A_S)$ have the same sign for fixed cardinality of S then also as $\psi(A,k,1)$ as well $\psi(A,1)$ can be computed in poly-time. But if all $Det(A_S)$ have the same sign for fixed cardinality of S then

$$|\phi(A,k,1)|^2 \frac{n!}{k!(n-k)!} \leq \phi(A,k,2) \leq |\phi(A,k,1)|^2. \tag{11}$$

We conclude that if all $Det(A_S)$ have the same sign for fixed cardinality of S then there is a strongly polynomial deterministic algorithm to approximate $\phi(A,2)$ within a multiplicative bound $\frac{2^n}{\sqrt{n}}$.

We will prove in the next section that even in this case $\phi(A,2)$ is #P-hard to compute.

This is a bit of an unusual situation: in all papers on the subject mixed volumes are approximated via corresponding mixed discriminants. We just described above a case where it might be better to proceed in the opposite order.

3 Proof of #P-Hardness

Let B be a $n \times n$ nonsingular matrix and α a real scalar. We define the following block $n+1 \times n+1$ matrix $G(B, \alpha)$, with each block being a $n \times n$ matrix:

$$G(B, \alpha)(k, l) = 0 \text{ if } 1 \leq k, l \leq n;$$

$$G(B, \alpha)(n+1, l) = B^{-1} \text{ if } 1 \leq l \leq n \text{ and } G(B, \alpha)(n+1, n+1) = 0;$$

$$G(B, \alpha)(k, n+1) = \alpha B e_k e_k^*.$$

$$G(B, \alpha) = \begin{pmatrix} 0 & 0 & \ldots & 0 & \alpha B e_1 e_1^* \\ 0 & 0 & \ldots & 0 & \alpha B e_2 e_2^* \\ \ldots & \ldots & \ldots & \ldots & \ldots \\ 0 & 0 & \ldots & 0 & \alpha B e_n e_n^* \\ B^{-1} & B^{-1} & \ldots & B^{-1} & 0 \end{pmatrix}$$

Definition 2. *For two permutations π_1 and π_2 we define $I(\pi_1, \pi_2) =: |\{i : \pi_1(i) = \pi_2(i)\}|$. For a $n \times n$ matrix B and integer $r = 0, 1, ..., n-2, n$ we define*

$$\gamma(B, r) =: \sum_{I(\pi_1, \pi_2) = r} (-1)^{sign(\pi_1)} (-1)^{sign(\pi_2)}$$
$$\prod_{i=1}^{n} B(i, \pi_1(i)) \prod_{i=1}^{n} B(i, \pi_2(i)).$$

∎

Notice that $\gamma(B, n) = Per(\{B(i,j)^2\})$. The main technical result of this section is the following theorem.

Theorem 1. *Let B be an $n \times n$ matrix. Then the following identity holds*

$$\phi(G(B, \alpha), 2) = \sum_{r=0,1,\ldots,n-2,n} (1 + \alpha^2)^r \gamma(B, r) Det(B)^{-2}. \tag{12}$$

Proof: We use the first identity in formula (9) with Rademacher diagonal matrix D, i.e. $D = Diag(\mu(i,j); \gamma(k) : 1 \leq i, j, k \leq n)$. Using the specially designed structure of $G(B, \alpha)$, we get that

$$\phi(G(B, \alpha), 2) = E(Det(D + G(B, \alpha))^2) =$$
$$= E(Det(Diag(\gamma) - B^{-1}C)^2)$$

where $(n \times n)$ matrix $C = \{\mu(i,j)B(i,j) : 1 \leq i,j \leq n\}$, $\{\mu(i,j)\}$ is a Rademacher $(n \times n)$ matrix, $\{\gamma(j)\}$ is a Rademacher (n)-dimensional vector; and γ and μ are independent.

As $E(Det(Diag(\gamma) - B^{-1}C)^2) = Det(B)^{-2} E(Det(BDiag(\gamma) - C)^2)$ and all random variables involved are symmetric, it follows that

$$\phi(G(B, \alpha), 2) = Det(B)^{-2} E(Det(\{B(i,j)(\gamma(j) + \alpha\mu(i,j))\})^2).$$

But

$$E(Det(\{B(i,j)(\gamma(j) + \alpha\mu(i,j))\}))^2) =$$
$$E_\gamma(E_\mu(Det(\{B(i,j)(\gamma(j) + \alpha\mu(i,j))\})^2)).$$

Again using symmetricity we get that $E_\mu(Det(\{B(i,j)(\gamma(j) + \alpha\mu(i,j))\}))^2) = E(Det(\{B(i,j)(1 + \alpha\mu(i,j))\})^2)$ for any (non-random) $(+1, -1)$ vector γ. Thus

$$\phi(G(B,\alpha), 2) = Det(B)^{-2} E_\mu(Det(\{B(i,j)(1 + \alpha\mu(i,j))\})^2).$$

After that we get the final identity using the additivity of the first moment. ∎

Remark 1. Notice that we used above the adapted "Godzil-Gutman" formula (9) not for a computation (as it usually used) but rather as a convenient tool to prove the formula (12). ∎

Now everything is ready to prove that $\phi(A, 2)$ (and thus mixed discriminants of rank 2 matrices) are #P-hard to compute.

Let $A = (B(i,j)^2)$ be a $n \times n$ matrix with nonnegative entries. It is "easy" to check whether $Per(A)$ is positive and to find out a "positive permutation". We can assume therefore WLOG that $Per(A) > 0$ and $A(i,i) \equiv 1$.

We need nonsingularity of B. Though the matrix B can be singular but $B + \lambda I$ is nonsingular for all sufficiently large λ. As $Per(A + \lambda I)$ is a monic polynomial of degree n, we can do everything for nonsingular case and "return" to possibly singular situation via a standard (and easy) interpolation.

So, suppose that B is nonsingular. Also, assume that we have an oracle which can compute $\phi(G(B,\alpha), 2)$. We run this oracle n times to obtain $\beta_i = \phi(G(B,\alpha_i), 2), 1 \leq i \leq n$ for distinct positive (rational,integer) $\alpha_1 < ... < \alpha_n$. As a result we get the following nonsingular system of linear equations:

$$\sum_{r=0,1,\ldots,n-2,n} (1 + \alpha_i^2)^r \gamma(B,r) Det(B)^{-2} = \beta_i (1 \leq i \leq n). \quad (13)$$

We solve it, say in $O(n^3)$, operations. But $\gamma(B,n) = Per(\{B(i,j)^2\}) = Per(A)$. This proves our main theorem.

Theorem 2. *Computing $\phi(A, 2)$ and thus mixed discriminants of rank 2 matrices is #P-hard .*

Corollary 1. *Computing $\phi(A, 2)$ is #P-hard even when $Det(A_S) > 0$ for all subsets S.*

Proof: We will use the same interpolational argument as above. It is clear that $\phi(A + \lambda I, 2)$ is a monic polynomial of degree $2n$. For large λ we have that $Det((A + \lambda I)_S)) > 0$ for all subsets S. Taking $n^2 - 1$ distinct large λ will allow to interpolate in poly time. Thus at least one of them is hard. ∎

4 Polynomial Time Algorithm Computing Hyperdeterminants of Small *Prank*

Suppose the linear (matrix) space spanned by $\{A_1, ..., A_N\}$ has dimension $m =: Rank(A_1, ..., A_N)$; A_i is $N \times N$ complex matrix, $1 \leq i \leq N$.

Let us consider 4-dimensional tensor $\rho(i_1, i_2, i_3, i_4), 1 \leq i_1, i_2, i_3, i_4 \leq N$. Define its 4-dimensional Pascal's determinant as

$$QP(\rho) = \frac{1}{N!} \sum_{\tau_1, \tau_2, \tau_3, \tau_4 \in S_N} (-1)^{sign(\tau_1 \tau_2 \tau_3 \tau_4)} \prod_{i=1}^{N} \rho(\tau_1(i), \tau_2(i), \tau_3(i), \tau_4(i)). \tag{14}$$

(See more on those hyperdeterminants in [26] and in numerous refs. from [26].) It is clear that permuting indices of ρ does not change $QP(\rho)$. Partitioning 4 indices of ρ into into 2 groups of two, we get 6 $N^2 \times N^2$ matrices. Define $Prank(\rho)$ as the minimal rank of those $N^2 \times N^2$ matrices.

(The Schmidt Rank, used in Quantum information literature [20], is equal to the maximal rank of those $N^2 \times N^2$ matrices.)

Let us associate with the tuple $\{A_1, ..., A_N\}$ the following 4-dimensional tensor:

$$\rho_A(i_1, i_2, i_3, i_4) = I(i_3, i_4) A_{i_3}(i_1, i_2),$$

where $\{I(i,j) : 1 \leq i,j \leq N\}$ is the identity matrix. It is easy to see that $QP(\rho_A) = D(A_1...A_N)$ and $Prank(\rho_A) \leq m = Rank(A_1, ..., A_N)$.

Theorem 3. *There exists a deterministic algorithm computing $QP(\rho)$ in $O(N^{2m})$ multiplications and additions.*

Proof: (Sketch). Suppose that without loss of generality that (standard) rank of $N^2 \times N^2$ matrix $\rho(i_1, i_2; i_3, i_4) = Prank(\rho) = m$ (here the corresponding partition is $(1,2) \cup (3,4)$). Thus there exist two m-tuples of $N \times N$ matrices $(A_1, ..., A_m)$ and $(B_1, ..., B_m)$ such that

$$\rho(i_1, i_2, i_3, i_4) = \sum_{1 \leq l \leq m} A_l(i_1, i_2) B_l(i_3, i_4) \tag{15}$$

Consider the following two determinantal polynomials :

$$P_A(x_1, ..., x_m) = Det(\sum_{1 \leq l \leq m} x_l A_l),$$

$$P_B(x_1, ..., x_m) = Det(\sum_{1 \leq l \leq m} x_l B_l) \tag{16}$$

Then

$$P_A(x_1, ..., x_m) = \sum_{r_1+...+r_m=N} a_{r_1,...,r_m} x_1^{r_1}...x_m^{r_m},$$

where $a_{r_1,..,r_m} = \frac{D(\mathbf{A}_r)}{r_1!r_2!...r_k!}$ and the N-tuple $\mathbf{A}_r$ consists of r_l copies of $A_l, 1 \leq l \leq m$.

It follows from [7], [4] that

$$QP(\rho) = \sum_{r_1+...+r_m=N} D(\mathbf{A}_r)D(\mathbf{B}_r)(r_1!r_2!...r_m!)^{-1} \quad (17)$$

Now each mixed discriminant $D(\mathbf{A}_r), D(\mathbf{B}_r)$ can be computed as a sum of $O(N^m)$ determinants [3] ; finally, we can compute the right-hand side of (17) in $O(N^{2m})$ multiplications and additions. ∎

5 Randomized Algorithms Computing Mixed Discriminants Within Absolute Error

Let us consider a norm $||(x_1, x_2, .., x_N)||$ in C^N. Assume that this norm is permutation invariant and $||(|x_1|, |x_2|, .., |x_N|)|| = ||(x_1, x_2, .., x_N)||$ for all vectors in C^N. Call such a norm "good". It is well known that such "good" norms induce matrix norms in $M(N)$, i.e. define for complex $N \times N$ matrix A its norm as $||A|| = ||(\sigma_1, \sigma_2, ..., \sigma_N)||$, where $\sigma_i, 1 \leq i \leq N$ are singular values of A, i.e. positive square roots of eigenvalues of AA^*. Associate with an N-tuple $(A_1, ..., A_N)$ of complex $N \times N$ matrices the following linear operator

$$T_A : C^N \to M(N),$$
$$T_A(x_1, x_2, .., x_N) =: \sum_{1 \leq i \leq N} x_i A_i .$$

We define $||T_A|| = \max_{||x||=1} ||T_A(x)||$. The following simple proposition is easily proved using the arithmetic/geometric means inequality.

Proposition 2. *Consider a norm $||(x_1, x_2, .., x_N)||$ in C^N. Assume that it is invariant respect to some transitive subgroup of permutations ,$||(1, 1, ..., 1)|| = 1$,and $||(|x_1|, |x_2|, .., |x_N|)|| = ||(x_1, x_2, .., x_N)||$ for all vectors in C^N. Then $|x_1 x_2 ... x_N| \leq ||(x_1, x_2, .., x_N)||^N$.*

Theorem 4. *If the norm $||.||$ in C^N is "good" then the mixed discriminant $|D(A_1, ..., A_N)| \leq ||T_A||^N$. Moreover there exists computable in poly-time unbiased estimator $F(b_1, .., b_N)$, where $(b_1, .., b_N)$ is uniformly distributed on $\{-1, 1\}^N$, such that*

$$E_b(F(b_1, .., b_N)) = D(A_1, ..., A_N)$$

and $|F(b_1, .., b_N)| \leq ||T_A||^N$.

Proof: One such estimator is $F(b_1, .., b_N) = Det(\sum_{1 \leq i \leq N} b_i A_i) b_1 b_2 ... b_N$. ∎

Corollary 2. *If $||T_A|| \leq 1 + O(\frac{\log(N)}{N})$ then there is a randomized polynomial in N and ϵ^{-1} algorithm computing $D(A_1, ..., A_N) + \delta$ with $|\delta| \leq \epsilon$.*

6 Conclusion and Acknowledgments

We hope that this work will stimulate new approaches to compute/approximate mixed discriminants of rank 2 matrices. As we noticed above, the quantity $\psi(A, 1) = \sum_S |Det(A_S)|$ is equal to the mixed volume $V(P_1, ..., P_n)$, where P_i is a 2-dimensional parallelogram with axis e_i and y_i. We indicated that in certain cases this mixed volume can be computed in polynomial time, say when $A \succeq 0$ or $-A \succeq 0$. We conjecture that, in general, this problem is also #P-hard.

Recall that a tuple $(A_1, ..., A_n)$ of $n \times n$ matrices called doubly stochastic if

$$A_i \succeq 0, tr(A_i) = 1 (1 \leq i \leq n); A_1 + ... + A_n = I.$$

It was proved in [15] that mixed disciminant $D(A_1...A_n) \geq \frac{n!}{n^n}$ for doubly stochastic tuples and the corresponding minimum $\frac{n!}{n^n}$ is uniquely attained at the tuple $(\frac{1}{n}I, ..., \frac{1}{n}I)$.

This result and many other things led to a deterministic polynomial algorithm in [18] to approximate mixed discriminants of positive semidefinite matrices within multiplicative factor $\frac{n^n}{n!}$.

Conjecture 1. Suppose that $(A_1, ..., A_n)$ is doubly stochastic tuple and $Rank(A_i) = 2, 1 \leq i \leq n$. Then $D(A_1...A_n) \geq 2^{(-n+1)}$. ∎

This conjecture is very easy to prove when matrices $(A_1, ..., A_n)$ commute, i.e. for permanents of doubly stochastic matrices with only two nonzero entries in each column. The conjecture would imply, using scaling algorithm from [18], that the mixed discriminant of positive semidefinite $n \times n$ rank two matrices can be approximated in deterministic polynomial time within multiplicative factor $2^{(n-1)}$, which is almost the same as via using inequality (11) for special cases.

Theorem 4 was motivated by a quantum algorithm due to Vwani Roychowdhury and Farrokh Vatan which computes in poly-time $|per(A)|^2$ for complex matrices which are contractions respect to l_2 norm. Theorem 4 allows to do it classically.

References

1. A. Aleksandrov, On the theory of mixed volumes of convex bodies, IV, Mixed discriminants and mixed volumes (in Russian), *Mat. Sb. (N.S.)* 3 (1938), 227-251.
2. R. B. Bapat, Mixed discriminants of positive semidefinite matrices, *Linear Algebra and its Applications* 126, 107-124, 1989.
3. A. I. Barvinok, Computing Mixed Discriminants, Mixed Volumes, and Permanents, *Discrete & Computational Geometry*, 18 (1997), 205-237.
4. L.Gurvits, Classical complexity and Quantum Entanglement, *Jour. of Comp. and Sys. Sciences (JCSS)* 69 (2004) 448-484 .
5. A. I. Barvinok, Polynomial time algorithms to approximate permanents and mixed discriminants within a simply exponential factor, *Random Structures & Algorithms*, 14 (1999), 29-61.
6. A. I. Barvinok, Two algorithmic results for the Traveling Salesman Problem, *Math. Oper. Res.* 21 (1996), 65-84 (2001 version from researchindex.com).

7. L.Gurvits, Classical deterministic complexity of Edmonds' problem and Quantum entanglement, Proc. of 35 annual ACM symposium on theory of computing, San Diego, 2003.
8. L. Gurvits and A. Samorodnitsky, A deterministic algorithm for approximating mixed discriminant and mixed volume, and a combinatorial corollary, Discrete and Computational Geometry 27 : 531 -550, 2002.
9. M. Dyer, P. Gritzmann and A. Hufnagel, On the complexity of computing mixed volumes, *SIAM J. Comput.*, 27(2), 356-400, 1998.
10. L .Gurvits, Classical deterministic complexity of Edmonds' problem and Quantum entanglement, to appear in Proc. of 35 annual ACM symposium on theory of computing (STOC2003), San Diego, 2003.
11. G.P. Egorychev, The solution of van der Waerden's problem for permanents, *Advances in Math.*, 42, 299-305, 1981.
12. D. I. Falikman, Proof of the van der Waerden's conjecture on the permanent of a doubly stochastic matrix, *Mat. Zametki* 29, 6: 931-938, 957, 1981, (in Russian).
13. C. D. Godsil, **Algebraic Combinatorics**, Chapman and Hall, 1993.
14. M. Grötschel, L. Lovasz and A. Schrijver, **Geometric Algorithms and Combinatorial Optimization**, Springer-Verlag, Berlin, 1988.
15. L. Gurvits, Van der Waerden Conjecture for Mixed Discriminants, Advances in Mathematics , 2005 . (Available at the journal web page .)
16. M. Jerrum and A. Sinclair, Approximating the permanent, *SIAM J. Comput.*, 18, 1149-1178, 1989.
17. M. Jerrum, A. Sinclair and E. Vigoda, A polynomial time approximation algorithm for the permanent of a matrix with non-negative entries, *ECCC* , Report No. 79, 2000.
18. L.Gurvits and A.Samorodnitsky, A deterministic polynomial-time algorithm for approximating mixed discriminant and mixed volume, *Proc. 32 ACM Symp. on Theory of Computing*, ACM, 2000.
19. N. Karmarkar, R. Karp, R. Lipton, L. Lovasz and M. Luby, A Monte-Carlo algorithm for estimating the permanent, *SIAM J. Comput.*, 22(2), 284-293, 1993.
20. G. Vidal , Efficient classical simulation of slightly entangled quantum computations , 2003 , available at http://arxiv.org/abs/quant-ph/0301063
21. N. Linial, A. Samorodnitsky and A. Wigderson, A deterministic strongly polynomial algorithm for matrix scaling and approximate permanents, *Proc. 30 ACM Symp. on Theory of Computing*, ACM, New York, 1998.
22. L. G. Valiant, The complexity of computing the permanent, *Theoretical Computer Science*, 8(2), 189-201, 1979.
23. K. Mulmuley and M. Sohoni, Geometric complexity theory 1 : an approach to the P vs. NP and related problems, SIAM J. Comput. , vol. 31, No. 2, pp. 496 - 526.
24. L. Valiant, Quantum computers that can be simulated classically in polynomial time, *Proc. 33 ACM Symp. on Theory of Computing*, ACM, 2001.
25. K. Regan, Understanding the Mulmeley-Sohoni approach to P vs. NP, unpublished manuscript, 2002.
26. J.-G. Luque, J.-Y. Thibon, Hankel hyperdeterminants and Selberg integrals, J.Phys.A: Math. Gen. 36 (2003) 5267-5292.
27. L. Gurvits , On the complexity of mixed discriminants and related problems , Los Alamos Unclassified Technical Report ,2005.

Two Logical Hierarchies of Optimization Problems over the Real Numbers

Uffe Flarup Hansen and Klaus Meer*

Department of Mathematics and Computer Science,
Syddansk Universitet, Campusvej 55, 5230 Odense M, Denmark
FAX: 0045 6593 2691
{flarup, meer}@imada.sdu.dk

Abstract. We introduce and study certain classes of optimization problems over the real numbers. The classes are defined by logical means, relying on metafinite model theory for so called $\mathbb{R}$-structures (see [9], [8]). More precisely, based on a real analogue of Fagin's theorem [9] we deal with two classes MAX-$NP_\mathbb{R}$ and MIN-$NP_\mathbb{R}$ of maximization and minimization problems, respectively, and figure out their intrinsic logical structure. It is proven that MAX-$NP_\mathbb{R}$ decomposes into four natural subclasses, whereas MIN-$NP_\mathbb{R}$ decomposes into two. This gives a real number analogue of a result by Kolaitis and Thakur [10] in the Turing model. Our proofs mainly use techniques from [13]. Finally, approximation issues are briefly discussed.

1 Introduction

Many important problems in mathematics and computer science appear as optimization problems. In the framework of complexity theory a huge number of such problems is studied in relation with the class NPO of combinatorial optimization problems with an exponential search space.

There are at least two directions with respect to studying problems in NPO. The first deals with the study of approximability properties of NP-hard problems in NPO, leading to the consideration of important subclasses of NPO such as $APX, PTAS, FPTAS$, see [1]. The other direction is descriptive complexity theory [11]. Here, optimization problems are studied from a logical definability standpoint. Concerning combinatorial optimization this line was started in a paper by Papadimitriou and Yannakakis [15]. They defined a subclass MAX-NP of NPO by logical means and studied (among other things) this class with respect to approximation algorithms.

A main aspect of all these problems dealt with in Turing (descriptive) complexity theory is that the space of feasible solutions for a given problem instance has at most exponential cardinality and thus is finite. However, many important

---
* Partially supported by the IST Programme of the European Community, under the PASCAL Network of Excellence, IST-2002-506778 and by the Danish Natural Science Research Council SNF. This publication only reflects the authors' views.

optimization problems involving real numbers have an uncountable search space. A typical example is given by determining from a given set of real multivariate polynomials the maximal number of polynomials that share a common real zero. This problem, for example, plays a crucial role in fundamental algorithms for semi-algebraic problems like quantifier elimination, see [16,2].

It is natural to ask whether a similar logical framework to the one mentioned above can be developed for such real number problems. A real number model of computation together with a complexity theory was developed by Blum, Shub and Smale, see [3]. A descriptive complexity theory for the Blum-Shub-Smale (shortly: BSS) model was studied in [9] and further in [6]. It is based on so-called meta-finite model theory introduced in [8]. Most of the fundamental real number complexity classes have been expressed logically using that approach. On the other side, a real number analogue of the class NPO has not been studied thoroughly so far (see some related remarks in [14]).

In this paper we start such an investigation from a logical point of view. We define a class $NPO_\mathbb{R}$ of certain real optimization problems. Then we focus on the definition and analysis of real number analogues of the hierarchies given in [10]. The techniques and problems we use to separate the different classes in these hierarchies are closely related to those used for the study of real number counting problems in [13]. Let us mention that the complexity of such counting problems in the BSS model recently was analyzed in a series of papers by Bürgisser and Cucker, see [4].

Our paper is organized as follows. Section 2 gives a short introduction of the BSS model and meta-finite model theory for $\mathbb{R}$-structures. A real number version of NPO is defined. Using descriptive complexity theory over $\mathbb{R}$ we then introduce the central logical problem classes of this paper: MAX-$\Sigma_{i,\mathbb{R}}$, MAX-$\Pi_{i,\mathbb{R}}$ for maximization problems and MIN-$\Sigma_{i,\mathbb{R}}$, MIN-$\Pi_{i,\mathbb{R}}$ for minimization problems ($i = 0, 1, 2$). Some basic properties of these classes are listed. Sections 3 and 4 build the main part of the paper. We consider several natural real number optimization problems, express them in a logical manner and use them to separate the classes of our hierarchies. In Section 3 this is done for maximization problems, in Section 4 for minimization. Finally, we discuss briefly approximation issues for some of the problem classes.

2 Basic Definitions and Results

In this section we introduce the problem classes we are interested in together with some of their basic properties.

2.1 The BSS Model; $\mathbb{R}$-Structures and Their Logics

We assume the reader to be familiar with the BSS model of computations over the real numbers and descriptive complexity for that model.

In the BSS model over $\mathbb{R}$ real numbers are considered as entities. The basic arithmetic operations $+, -, *, :$ can be performed at unit costs, and there is a test-operation "is $x \geq 0$?" reflecting the underlying ordering of the reals. Decision problems now are subsets $L \subseteq \mathbb{R}^\infty := \bigcup_{n=1}^{\infty} \mathbb{R}^n$. The (algebraic) size of a point $x \in \mathbb{R}^n$ is n. Having fixed these notions it is easy to define real analogues $P_\mathbb{R}$ and $NP_\mathbb{R}$ of the classes P and NP as well as the notion of $NP_\mathbb{R}$-completeness. For more details on the BSS model we refer to [3].

Descriptive complexity for the BSS model over $\mathbb{R}$ was introduced in [9]. It gives a logical presentation of decision problems via so called $\mathbb{R}$-structures and characterizes real number complexity classes in terms of logics for these structures.

For the characterization of our class MAX-$NP_\mathbb{R}$ introduced below we need the following extension of Fagin's theorem to $\mathbb{R}$-structures.

Theorem 1 ([9]). *Let (F, F^+) be a decision problem of $\mathbb{R}$-structures. Then $(F, F^+) \in NP_\mathbb{R}$ iff there is an $\exists SO_\mathbb{R}$-formula ψ such that $F^+ = \{\mathfrak{D} \in F | \mathfrak{D} \models \psi\}$.*

2.2 The Classes $NPO_{\mathbb{R},max}$ and $NPO_{\mathbb{R},min}$

The problems we deal with in this paper are certain optimization problems over $\mathbb{R}$-structures. As a starting point we consider the well known definition of combinatorial optimization problems in class NPO. We extend this definition in order to introduce a similar class in the BSS model. The main additional aspect of this new class $NPO_\mathbb{R}$ is a potentially uncountable set of feasible solutions among which the optimum is searched for. This makes some changes in the definitions necessary, one of which is the lacking requirement of explicitly computing a feasible solution. Moreover, we consider functions with integer values in order to avoid (numerical) approximation issues. Though interesting in its own, the latter is expected to result in completely different issues, see [5].

Let us start with introducing a real analogue of the class NPO of combinatorial optimization problems. Note that generalizations of NPO to arithmetical structures (i.e. metafinite structures with the natural numbers as infinite part) were studied in [12].

The definition below is tailored for the logical framework we study thereafter. It can easily be re-translated to a purely complexity theoretic definition.

Definition 1. *a) A maximization problem $\mathcal{P} := (\mathcal{I}, \{Sol(\mathfrak{D})\}_{\mathfrak{D} \in \mathcal{I}}, m)$ in class $NPO_{\mathbb{R},max}$ is a problem over $\mathbb{R}$-structures that consists of three parts:*

i) A set $\mathcal{I}$ of $\mathbb{R}$-structures (over a fixed vocabulary) as instances for the problem;
ii) for every instance $\mathfrak{D} \in \mathcal{I}$ a set $Sol(\mathfrak{D})$ of feasible solutions. Elements in $Sol(\mathfrak{D})$ are tuples $\mathfrak{S} = (S_1, \ldots, S_r)$ of functions from the universe A of $\mathfrak{D}$ to the reals. Each S_i has a fixed arity $d_i \in \mathbb{N}$, so $S_i : A^{d_i} \mapsto \mathbb{R}$;
iii) a measure function $m : \{(\mathfrak{D}, \mathfrak{S}) | \mathfrak{D} \in \mathcal{I}, \mathfrak{S} \in Sol(\mathfrak{D})\} \to \mathbb{N}$. The value $m(\mathfrak{D}, \mathfrak{S})$ is called the value of the feasible solution $\mathfrak{S}$;
iv) for any input $\mathfrak{D} \in \mathcal{I}$ and for any arbitrary $\mathfrak{S} \in \mathbb{R}^\infty$ of size at most $p(|\mathfrak{D}|)$ (for a fixed polynomial p) it is decidable in polynomial time in the BSS model whether $\mathfrak{S} \in Sol(\mathfrak{D})$;

v) the function m is computable in polynomial time with respect to $|\mathfrak{D}|$ in the BSS model.

We are looking for the maximal solution value $\max\limits_{\mathfrak{S} \in Sol(\mathfrak{D})} m(\mathfrak{D}, \mathfrak{S})$. Finally, we require that the maximum always exists.

b) The class $NPO_{\mathbb{R},min}$ of minimization problems is defined similarly.

Remark 1. a) An important $NP_{\mathbb{R}}$-hard problem is the minimization of a polynomial p, over all possible variable assignments $x \in \mathbb{R}$. Notice that given the remarks before the previous definition this problem is not in class $NPO_{\mathbb{R},min}$ as the value of m is not integer.

b) The logical framework we use to define $NPO_{\mathbb{R}}$ guarantees that a feasible solution has a polynomial (algebraic) size. Note that the set of feasible solutions can be uncountable. In many examples of such problems it might be impossible to even compute a feasible solution; therefore, we do not require it.

c) Due to the structure of classes we consider in this paper it is natural to require the measure function m to yield values in $\mathbb{N}$. This is done because below we define classes via the number of satisfying assignments in some A^u for certain formulas. For a general theory of a class $NPO_{\mathbb{R}}$ of optimization problems over $\mathbb{R}$ not inspired by a logical framework it certainly makes sense to remove this condition. We refer to [14] for some further ideas on a more general definition.

d) For sake of simplicity in many of our proofs we consider $NPO_{\mathbb{R}}$ problems where $\mathcal{S}$ consists of a single function, only. For all proofs it can easily be seen that this is no serious restriction. An extension to the general case can always be done (almost) word by word and just increases the notational complexity.

Definition 2. *An optimization problem $\mathcal{P}$ in classes $NPO_{\mathbb{R},max}$ or $NPO_{\mathbb{R},min}$ is polynomially bounded if for all $\mathfrak{D}$ the value $m(\mathfrak{D}, \mathfrak{S})$ is bounded by a polynomial $p(|\mathfrak{D}|)$ for all feasible solutions $\mathfrak{S}$.*

2.3 The Classes MAX-$NP_{\mathbb{R}}$ and MIN-$NP_{\mathbb{R}}$

We next define a class MAX-$NP_{\mathbb{R}}$ as a subclass of problems in $NPO_{\mathbb{R},max}$ that is defined by certain logical conditions. These classes are extensions of the corresponding discrete ones introduced in [15].

Definition 3. *a) Let $\mathcal{P}$ be an optimization problem in $NPO_{\mathbb{R},max}$ whose instances are given as $\mathbb{R}$-structures over a fixed vocabulary V. Then $\mathcal{P}$ belongs to MAX-$NP_{\mathbb{R}}$ iff for every instance $\mathfrak{D}$ of $\mathcal{P}$ having universe A the maximum can be expressed as*

$$\max_{\mathfrak{S} \in Sol(\mathfrak{D})} |\{x \in A^u \mid \mathfrak{D} \models \phi(x, \mathfrak{S})\}|.$$

Here, $\mathfrak{S}$ is a finite sequence of functions each of some fixed arity from the universe to $\mathbb{R}$ and ϕ is a first-order formula over $V \cup \{\mathfrak{S}\}$; $u \in \mathbb{N}$ is some fixed natural number.

b) We obtain the subclasses MAX-$\Sigma_{0,\mathbb{R}}$, MAX-$\Sigma_{1,\mathbb{R}}$, MAX-$\Pi_{1,\mathbb{R}}$, MAX-$\Sigma_{2,\mathbb{R}}$ and MAX-$\Pi_{2,\mathbb{R}}$, respectively, by restricting ϕ above to be of the corresponding format. For example, MAX-$\Sigma_{1,\mathbb{R}}$ is the class of maximization problems whose maximum is expressible via

$$\max_{\mathfrak{S}\in Sol(\mathfrak{D})} |\{x \in A^u \mid \mathfrak{D} \models \exists y \in A^s \text{ such that } \psi(x,y,\mathfrak{S})\}|,$$

where ψ is first-order quantifier free.

c) In the same way the MIN classes are defined.

Example 1. Let us consider a typical optimization problem in our framework and express it logically. The input are natural numbers n, m together with m polynomials $p_1, \ldots, p_m \in \mathbb{R}[x_1, \ldots, x_n]$. Each p_i has degree 2 and depends on precisely three variables among $\{x_1, \ldots, x_n\}$. The task is to compute the maximal number of p_i's that have a common zero in $\mathbb{R}^n$. This task is $NP_\mathbb{R}$-hard. We give a logical description that places the problem (or better: its used logical version) in MAX-$\Sigma_{0,\mathbb{R}}$.

As vocabulary we choose a nullary function $\mathbf{0}$, a unary relation Pol and a function $C : A^3 \mapsto \mathbb{R}$ (where A is the universe). For an $\mathbb{R}$-structure representing a polynomial system as above the interpretations of these symbols are as follows: The universe A splits into two parts. The first part $A_1 = \{0, \ldots, n\}$ represents the variables $x_1, \ldots, x_n$ plus an additional one x_0 used for homogenization; the second part $A_2 := \{n+1, \ldots, n+m\}$ represents the indices for the polynomials p_i (i.e. $n+i$ stands for p_i). If $\ell \in A$ satisfies $Pol(\ell)$, then ℓ is the index of a polynomial $p_{\ell-n}$. The nullary function $\mathbf{0}$ represents $0 \in A$. Finally, $C : A_1^2 \times A_2 \subset A^3 \mapsto \mathbb{R}$ stands for the coefficients of the p_i; for $(i,j) \in A_1^2, \ell \in A_2$ the coefficient of $x_i \cdot x_j$ in $p_{\ell-n}$ is $C(i,j,\ell)$. Here, we use $x_0 = 1$ as homogenization variable. Note that C is supposed to be symmetric in the first two components. For other arguments not mentioned above we define C to be 0.

The maximal number of polynomials having a common zero can be described as follows: A root is coded via a function $X : A_1 \mapsto \mathbb{R}$. Then we look for

$$\max_{X:A_1 \mapsto \mathbb{R}} |\{(i,j,k,\ell) | Pol(\ell) \wedge X(0) = 1 \wedge \phi_1(i,j,k,\ell) \wedge \phi_2(i,j,k,\ell)\}|.$$

Here, $\phi_1(i,j,k,\ell)$ is a formula guaranteeing that x_i, x_j, x_k are the three variables $p_{\ell-n}$ depends on, i.e.

$$\phi_1(i,j,k,\ell) \equiv \{C(i,0,\ell) \neq 0 \vee C(i,i,\ell) \neq 0 \vee C(i,j,\ell) \neq 0 \vee C(i,k,\ell) \neq 0\}$$
$$\wedge \{C(j,0,\ell) \neq 0 \vee C(j,j,\ell) \neq 0 \vee C(j,i,\ell) \neq 0 \vee C(j,k,\ell) \neq 0\}$$
$$\wedge \{C(k,0,\ell) \neq 0 \vee C(k,k,\ell) \neq 0 \vee C(k,i,\ell) \neq 0 \vee C(k,j,\ell) \neq 0\}$$

Similarly, $\phi_2(i,j,k,\ell)$ expresses that the evaluation of $p_{\ell-n}$ in the point represented by X will give result 0. Note that the knowledge coded in ϕ_1 can be used to design ϕ_2 as quantifier free formula:

$$\phi_2(i,j,k,\ell) \equiv C(0,0,\ell) + C(i,0,\ell) \cdot X(i) + C(j,0,\ell) \cdot X(j)$$
$$+ C(k,0,\ell) \cdot X(k) + C(i,i,\ell) \cdot X(i)^2 + \ldots = 0$$

Altogether, we see that the problem lies in MAX-$\Sigma_{0,\mathbb{R}}$.

Next, we list some basic properties of problems in MAX-$NP_\mathbb{R}$. They mirror the corresponding properties over finite structures [10] and can basically be shown by similar techniques using additional results for $\mathbb{R}$-structures in [9].

First, it is clear that the objective function m of a problem $\mathcal{P}$ in MAX-$NP_\mathbb{R}$ or in MIN-$NP_\mathbb{R}$ is polynomially bounded. Vice versa, the following proposition shows that the converse holds as well. It can be proved along the similar line of the corresponding results for the Turing model in [10] using the extension of Fagin's theorem to the BSS model in [9].

Proposition 1. *a) If $\mathcal{P}$ is a polynomially bounded problem in $NPO_{\mathbb{R},max}$, then $\mathcal{P} \in$ MAX-$NP_\mathbb{R}$.*
b) MAX-$NP_\mathbb{R}$ = MAX-$\Pi_{2,\mathbb{R}}$ and MAX-$\Sigma_{2,\mathbb{R}}$ = MAX-$\Pi_{1,\mathbb{R}}$.
c) If $\mathcal{P}$ is a polynomially bounded problem in $NPO_{\mathbb{R},min}$, then $\mathcal{P} \in$ MIN-$NP_\mathbb{R}$.
d) MIN-$NP_\mathbb{R}$ = MIN-$\Sigma_{2,\mathbb{R}}$, MIN-$\Pi_{1,\mathbb{R}}$ = MIN-$\Sigma_{2,\mathbb{R}}$, MIN-$\Pi_{0,\mathbb{R}}$ = MIN-$\Sigma_{1,\mathbb{R}}$.

3 The 4-Level Maximization Hierarchy

We turn to the main results of this paper. First we prove that MAX-$NP_\mathbb{R}$ can be decomposed into a hierarchy of four distinct levels. More precisely:

Theorem 2. *MAX-$\Sigma_{0,\mathbb{R}} \subsetneq$ MAX-$\Sigma_{1,\mathbb{R}} \subsetneq$ MAX-$\Pi_{1,\mathbb{R}} \subsetneq$ MAX-$\Pi_{2,\mathbb{R}}$ = MAX-$NP_\mathbb{R}$*

The theorem is a real number version of Theorem 2 in [10]. However, the problems we use (as well as our proofs) to establish it are different since they have to involve meta-finite structures. Consider the following real number problems.

Definition 4. *a) For fixed $d \in \mathbb{N}$ the MAX-$HNS_\mathbb{R}(d)$ problem (maximal Hilbert Nullstellensatz) is given as:*
INPUT: $n, m \in \mathbb{N}$, polynomials $p_1, \ldots, p_m$ of degree $\leq d$ in variables $x_1, \ldots, x_n$.
QUESTION: *What is the maximal number of polynomials $p_i, 1 \leq i \leq m$ that have a common zero $x \in \mathbb{R}^n$?*
b) The sign-changes problem is given as:
INPUT: $n \in \mathbb{N}$ together with a sequence of n ordered reals $(x_1, \ldots, x_n)$.
QUESTION: *Find the number of components i with $x_i \neq 0$ for which there exists a j such that $x_j \neq 1$ and $x_i \cdot x_j < 0$.*

The above problems in our framework first become interesting after we formalize them as problems for meta-finite structures. There are several ways to do so depending on which information we include in the structure. This will in particular have impact on the question to which MAX$_\mathbb{R}$-classes the problems belong. Moreover, it will be crucial for our separation results.

3.1 The Non-ordered Version of the MAX-$HNS_\mathbb{R}(d)$ Problem

Let us start with MAX-$HNS_\mathbb{R}(d), d \in \mathbb{N}$. In the first formalization we take, similar as in Example 1, an $\mathbb{R}$-structure with universe $A := \{0, \ldots, n\} \cup \{n +$

$1, \ldots, n+m\}$. The vocabulary includes one unary relation $Pol \subseteq A$ indicating whether an $x \in A$ is a polynomial or a variable. It as well includes a function $C : A^{d+1} \mapsto \mathbb{R}$ that is interpreted as representing the coefficients of the monomials in the corresponding polynomials. Thus, we consider a polynomial system given as $\mathbb{R}$-structure $\mathfrak{D} = (A, Pol, C)$, where $A = \{0, \ldots, n+m\}, Pol(i) \Leftrightarrow i \in \{n+1, \ldots, n+m\}$ and

$$C(i_1, \ldots, i_d, k) = \begin{cases} 0 \text{ if } \neg Pol(k) \\ \text{coefficient of } x_{i_1} \cdot x_{i_2} \cdot \ldots \cdot x_{i_d} \text{ in equation } k \text{ if } Pol(k). \end{cases}$$

In order to also represent monomials of degree strictly less than d we again guarantee in all our formulas $x_0 := 1$.

Most important, we do not include a linear ordering on A in the vocabulary. We therefore denote this formalization of the Hilbert-Nullstellensatz problem by NORD-MAX-$HNS_\mathbb{R}(d)$. To see that NORD-MAX-$HNS_\mathbb{R}(d)$ belongs to MAX-$NP_\mathbb{R}$, and thus by Proposition 1 to MAX-$\Pi_{2,\mathbb{R}}$, is almost straightforward.

Theorem 3. *NORD-MAX-$HNS_\mathbb{R}(4) \notin$ MAX-$\Pi_{1,\mathbb{R}}$.*

Proof. Suppose the claim were false. Then for a corresponding input structure $\mathfrak{D} = (A, Pol, C)$ of the problem NORD-MAX-$HNS_\mathbb{R}(4)$ the maximum can be expressed as

$$\max_{S:A^t \mapsto \mathbb{R}} |\{x \in A^u \mid \mathfrak{D} \models \forall y \in A^s \; \phi(x, y, C, Pol, S)\}|, \qquad (1)$$

where ϕ is first-order quantifier free, $s, t, u \in \mathbb{N}$ fixed. For an even n consider the following polynomials $f_i, 1 \leq i \leq \frac{n}{2}$ in variables $x_1, \ldots, x_n$ and of degree 4 which will be important for constructing an appropriate input-structure:

$$f_1(x_1, x_2) = (x_1 \cdot x_2 - 1)^2 + x_1^2 \; , \; f_2(x_3, x_4) = (x_3 \cdot x_4 - 1)^2 + x_3^2$$

$$\vdots$$

$$f_i(x_{2i-1}, x_{2i}) = (x_{2i-1} \cdot x_{2i} - 1)^2 + x_{2i-1}^2$$

$$\vdots$$

$$f_{\frac{n}{2}}(x_{n-1}, x_n) = (x_{n-1} \cdot x_n - 1)^2 + x_{n-1}^2$$

Define a new polynomial $p(x_1, \ldots, x_n) := \sum_{i=1}^{\frac{n}{2}} f_i(x_{2i-1}, x_{2i}) - \epsilon$ of degree four, where ϵ is an arbitrary, fixed real number in $(0,1)$. It is the (single) polynomial equation $p = 0$ that we now consider as input for NORD-MAX-$HNS_\mathbb{R}(4)$. As input $\mathbb{R}$-structure it is represented as $\mathfrak{D} = (A, Pol, C)$ with $A := \{0, 1, \ldots, n\} \cup \{n+1\}, Pol(i) \Leftrightarrow i = n+1$ and $C(i,j,k,\ell, n+1)$ gives the coefficient of $x_i \cdot x_j \cdot x_k \cdot x_\ell$ in p, where $x_0 := 1$ once again is used to represent monomials of degree less than four. For example, the constant part $\frac{n}{2} - \epsilon$ is given as $C(0,0,0,0,n+1)$.

We claim that p has a real zero for any choice $\epsilon > 0$. In order to see this note that each of the polynomials f_i satisfies $f_i(x_{2i-1}, x_{2i}) > 0$. Non-negativity is obvious by definition whereas strict positivity follows from the fact that $x_{2i-1} := 0$

results in the function value $= 1$. Moreover, $\inf_{x_{2i-1}, x_{2i}} f_i(x_{2i-1}, x_{2i}) = 0$ by choosing $x_{2i-1} := x_{2i}^{-1}$ for $x_{2i} > 0$ and now considering the limit $\lim_{x_{2i} \to \infty} f_i(\frac{1}{x_{2i}}, x_{2i})$.

For any $\epsilon > 0$ if we choose the x_{2i}'s large enough such that $f_i(\frac{1}{x_{2i}}, x_{2i}) < \frac{2\epsilon}{n}$ we get a negative function value for p. Since p clearly has positive values continuity implies the claim.

Thus, the above formula (1) has to give the result 1. Let $X^* : A^t \mapsto \mathbb{R}, x^* \in A^u$ be an assignment such that $\mathfrak{D} \models \forall y \in A^s \phi(x^*, y, X^*, Pol, C)$. We construct a substructure $\mathfrak{D}'$ of $\mathfrak{D}$ that still gives a result of at least 1 for (1) but codes a polynomial without real zeros.

Let i_0 be such that the particular $x^* \in A^u$ chosen above does not depend on x_{2i_0}. For n large enough (f.e. $n > 2u + 1$) such an i_0 exists since u is fixed (independent of n). Define a new input structure $\mathfrak{D}'$ by deleting $2i_0$ from the universe (and identifying the other elements of A with those in the new universe A' correspondingly). Furthermore, Pol' and C' are defined as for $\mathfrak{D}$ on the remaining arguments. The related polynomial p' then is given by

$$p(x_1, \ldots \underbrace{|}_{\neq x_{2i_0}} \ldots, x_n) = \sum_{\substack{i=1, \\ i \neq 2i_0}}^{\frac{n}{2}} f_i - \epsilon + (-1)^2 + x_{2i_0 - 1}^2 .$$

The universal formula $\forall y \in (A')^s \ \phi(x^*, y, X'^*, Pol', C')$ still is satisfied by the substructure $\mathfrak{D}'$, therefore

$$\max_{X' : (A')^t \mapsto \mathbb{R}} |\{x \in (A')^u | \forall y \in (A')^s \ \phi(x, y, X', Pol', C')\}| \geq 1 .$$

But $\quad p(x_1, \ldots \underbrace{|}_{\neq x_{2i_0}} \ldots, x_n) \underbrace{\geq}_{f_i \geq 0} x_{2i_0 - 1}^2 + 1 - \epsilon \geq 1 - \epsilon > 0.$

Thus, p has no real zero but formula ϕ counts at least one. Contradiction. $\square$

3.2 The Ordered Version of the MAX-$HNS_\mathbb{R}(d)$ Problem

In order to separate MAX-$\Sigma_{1,\mathbb{R}}$ from MAX-$\Pi_{1,\mathbb{R}}$ we represent instances from MAX-$HNS_\mathbb{R}(d)$ in a different manner as $\mathbb{R}$-structures. This will push the problem into class MAX-$\Pi_{1,\mathbb{R}}$. For a system $p_1, \ldots, p_m$ over $x_1, \ldots, x_n$ the universe again is $A_1 \cup A_2$, where $A_1 = \{0, \ldots, n\}$ and $A_2 = \{n+1, \ldots, n+m\}$. A unary relation Pol again identifies the polynomials: $Pol(i) \Leftrightarrow i \in A_2$. As before, $C : A^{d+1} \mapsto \mathbb{R}$ denotes the coefficients of the p_i's. In addition, the vocabulary will contain a linear ordering $\rho : A_1 \mapsto \{0, \ldots, n\} \subset \mathbb{R}$ as well as nullary relations $\mathbf{0}$ and $\mathbf{n}$ giving the first and the last element in A_1 (w.r.t. ρ). Note that $\mathbf{0}$ and $\mathbf{n}$ as well as extensions $\rho^4, \mathbf{0}^4, \mathbf{n}^4$ of $\rho, \mathbf{0}, \mathbf{n}$ to A_1^4 can be defined by a universally quantified first-order formula, see [6].

The presence of this linear ordering is the reason why the proof of Theorem 3 cannot be applied in this setting: If we remove an element from the universe, the ordering will be invalid.

Representing the MAX-$HNS_\mathbb{R}(d)$ problem that way we obtain its ordered version which we denote by ORD-MAX-$HNS_\mathbb{R}(d)$.

Theorem 4. ORD-MAX-$HNS_\mathbb{R}(4) \in$ MAX-$\Pi_{1,\mathbb{R}} \setminus$ MAX-$\Sigma_{1,\mathbb{R}}$.

Proof. Concerning membership in MAX-$\Pi_{1,\mathbb{R}}$ let $\rho^4, \mathbf{0}^4, \mathbf{n}^4$ denote the above mentioned extensions of the ranking ρ and $\mathbf{0}, \mathbf{n}$ to A^4 (see [6] for expressibility in $\Pi_{1,\mathbb{R}}$). For an input $\mathbb{R}$-structure $\mathfrak{D}(A, Pol, C, \rho, \mathbf{0}, \mathbf{n})$ of ORD-MAX-$HNS_\mathbb{R}(4)$ the maximal number of polynomials having a common real zero is given as

$$\max_{\substack{X: A_1 \mapsto \mathbb{R}, \\ Y: A_1^4 \times A_2 \mapsto \mathbb{R}}} |\{i \in A_2 \mid \mathfrak{D} \models Pol(i) \wedge \phi_1 \wedge \phi_2 \wedge \phi_3\}|, \text{ where}$$

$\phi_1 \equiv Y(\mathbf{0}^4, i) = C(\mathbf{0}^4, i)$, $\phi_2 \equiv Y(\mathbf{n}^4, i) = 0$ and
$\phi_3 \equiv \forall u, v \in A^4 \ \{\rho^4(u) = \rho^4(v) + 1$
$\implies Y(u, i) = Y(v, i) + C(u, i) \cdot X(u_1) \cdot X(u_2) \cdot X(u_3) \cdot X(u_4)\}$

Here, X is interpreted as a zero giving the maximum, $Y(\bullet, i)$ describes the intermediate sum when evaluating polynomial i in X by cycling through all monomials given by $u \in A_1^4$ (expressed by ϕ_3). Finally, ϕ_1 and ϕ_2 guarantee to start and finish the evaluation process with the correct values.

In order to establish non-expressibility within class MAX-$\Sigma_{1,\mathbb{R}}$ assume to the opposite that the maximum is computed by

$$\max_{S: A^t \mapsto \mathbb{R}} |\{x \in A^u \mid \mathfrak{D} \models \exists y \in A^s \phi(x, y, S)\}|, \quad (2)$$

where ϕ is first-order quantifier free. Consider once more the polynomial system (consisting of a single polynomial) used to prove Theorem 3. This time we represent the system by an ordered $\mathbb{R}$-structure $\mathfrak{D} = (A, Pol, C, \rho, \mathbf{0}, \mathbf{n})$. Let $x^* \in A^u, y^* \in A^s, S^* : A^t \mapsto \mathbb{R}$ satisfy $\mathfrak{D} \models \phi(x^*, y^*, S^*)$ according to our assumption that formula (2) works correctly and the earlier proven fact that polynomial p has a real zero. ϕ is quantifier-free, so $\phi(x^*, y^*, S^*)$ contains at most r elements from A, where r is a constant independent from $\mathfrak{D}$. Choose the size n of universe A such that n is even and $r < \frac{n}{2}$. Then there is a variable among $\{x_2, x_4, \ldots, x_n\}$ which does not occur in $\phi(x^*, y^*, S^*)$. Without loss of generality let x_n be that variable. Now define a new structure $\mathfrak{D}'$ representing a polynomial p' which is generated as before by the polynomials $f_i, 1 \leq i \leq \frac{n}{2}$. The only difference between p and p' is the polynomial $f_{\frac{n}{2}}$ which now has the form $f_{\frac{n}{2}}(x_{n-1}, x_n) := (x_{n-1} \cdot x_n - 1)^2 + x_{n-1}^2 + x_n^2$. Therefore, contrary to p the polynomial p' contains the monomial x_n^2 with coefficient 1, whereas in p the coefficient was 0. The other $f_i's$ remain unchanged.

Since x_n was not occuring in $\phi(x^*, y^*, S^*)$ the new structure $\mathfrak{D}'$ as well satisfies $\mathfrak{D}' \models \phi(x^*, y^*, S^*)$. This implies $\max_{S: A^t \mapsto \mathbb{R}} |\{x \in A^u \mid \mathfrak{D}' \models \exists y \in A^s \ \phi(x, y, S)\}| \geq 1$. But

$$p' = \sum_{i=1}^{\frac{n}{2}-1} f_i(x_{2i-1}, x_{2i}) + (x_{n-1} \cdot x_n - 1)^2 + x_{n-1}^2 + x_n^2 - \epsilon$$
$$> x_{n-1}^2 \cdot x_n^2 - 2x_{n-1} \cdot x_n + 1 + x_{n-1}^2 + x_n^2 - \epsilon \geq (x_{n-1} - x_n)^2 + 1 - \epsilon > 0.$$

That is p' has no real zero and (2) does not give the correct result. □

3.3 Separation Between MAX-$\Sigma_{0,\mathbb{R}}$ and MAX-$\Sigma_{1,\mathbb{R}}$

The separation will be established using the sign-changes problem introduced in part b) of Definition 4. We represent its instances as $\mathbb{R}$-structures $\mathfrak{D} = (A, C)$, where $A = \{1, \ldots, n\}, C : A \mapsto \mathbb{R}$. The number we are looking for is given as

$$|\{i \in A \mid \exists \ell \ C(\ell) \neq 1 \wedge C(i) \cdot C(\ell) < 0\}|. \tag{3}$$

We express this problem a bit artificially as a problem in MAX-$\Sigma_{1,\mathbb{R}}$ by noting that the value given in (3) equals

$$\max_{S: A^t \mapsto \mathbb{R}} |\{i \in A \mid \exists \ell \ C(\ell) \neq 1 \wedge C(i) \cdot C(\ell) < 0\}| \tag{4}$$

(since the $\Sigma_{1,\mathbb{R}}$-formula does not at all depend on S). However, the particular form of the problem is useful for separating the two lowest classes of our hierarchy.

Theorem 5. *MAX-$\Sigma_{0,\mathbb{R}} \subsetneq$ MAX-$\Sigma_{1,\mathbb{R}}$.*

4 The 2-Level Minimization Hierarchy

In this section we show that the hierarchy of polynomially bounded minimization problems consists of the two classes remaining from Proposition 1.

Definition 5. *We define the problem MIN QPS VALUES as follows:*
INPUT: *$n, m \in \mathbb{N}$ and polynomials $p_1, \ldots, p_m$ each of degree 2, and each depends on precisely 3 of the variables $x_1, \ldots, x_n$.*
QUESTION: *Minimizing over $x \in \mathbb{R}^n$ what is the minimal number of different function values we can obtain when we evaluate the polynomials $p_i, 1 \leq i \leq m$ in x?*
We present polynomial systems by $\mathbb{R}$-structures as done in Example 1.

The problem is $NP_\mathbb{R}$-hard: Given a system from Example 1 together with the additional variables x, y and z and additional polynomials $x^2 + y^2 + z^2$ and $-(x^2 + y^2 + z^2)$, the new system yields result 1 for MIN QPS VALUES iff the original system has a zero. The problem trivially is polynomially bounded and thus belongs to class MIN-$\Pi_{1,\mathbb{R}}$ using Proposition 2. The interested reader might try to design a corresponding formula directly without using that proposition. The following is a real number version of Theorem 4 in [10].

Theorem 6. *MIN QPS VALUES $\notin$ MIN-$\Pi_{0,\mathbb{R}}$*

Proof. Let $\mathcal{H}_1$ be an instance with optimal value $opt(\mathcal{H}_1) = k$, where $k \geq 2$. We denote the corresponding interpretations of relation and function symbols by $\mathbf{0}_1, Pol_1, C_1$, compare Example 1. Let $\mathcal{H}_2$ with $\mathbf{0}_2, Pol_2, C_2$ be an isomorphic copy of $\mathcal{H}_1$. Clearly $opt(\mathcal{H}_2) = k$ as well. We define a new structure $\mathcal{H}$ representing a polynomial system consisting of the original set of polynomials and its copy (in new variables). An $\mathbb{R}$-structure $\mathcal{H}$ is obtained as follows. The universe A of

$\mathcal{H}$ is the disjoint union of the universes of $\mathcal{H}_1$ and $\mathcal{H}_2$ except that we identify the elements 0_1 and 0_2 in the new structure (we use the same homogenization variable). The interpretation of the relation Pol as well as of the function C in $\mathcal{H}$ is by using the obvious extension of the corresponding interpretations in $\mathcal{H}_1, \mathcal{H}_2$. At those points where C_1, C_2 are undefined we define C to have the value 0. Clearly $opt(\mathcal{H}) = k$ since the two sets of polynomials in the union have no variables in common and each set has optimal value k.

Assume there is a quantifier free formula over $\mathbb{R}$-structures, ψ, which gives the optimal value of MIN QPS VALUES by minimizing over a real valued function S. For the structure $\mathcal{H}$ let X denote an assignment to this function that realizes the minimal value k : $k = |\{w \in A^t|(\mathcal{H}, X) \models \psi(w, X)\}|$.

Let X_1, X_2 be the restriction of X to the structures $\mathcal{H}_1, \mathcal{H}_2$, respectively.

For both structures the minimal value is k and we obtain

$$k \leq |\{w \in A^t_{\mathcal{H}_1}|(\mathcal{H}_1, X_1) \models \psi(w, X_1)\}|, \ k \leq |\{w \in A^t_{\mathcal{H}_2}|(\mathcal{H}_2, X_2) \models \psi(w, X_2)\}| \ .$$

Since $A_{\mathcal{H}_1}$ and $A_{\mathcal{H}_2}$ contain only the element 0 in common, at most one element w (namely the one with all components 0) can occur as satisfying assignment in both above formulas. All the others are different when considered as elements in the disjoint union A^t.

Now ψ is quantifier free; therefore, all w satisfying the formula in one of the two structures $\mathcal{H}_1, \mathcal{H}_2$ yield (with respect to X) a satisfying assignment in $\mathcal{H}$ as well, and at most a single w can occur twice. Because $k \geq 2$ it follows $k < 2k - 1 \leq |\{w|(\mathcal{H}, X) \models \psi(w, X)\}|$. We arrive at a contradiction. □

5 Conclusions

In this paper we introduced and studied from a logical point of view classes of optimization problems over the reals. Using tools from descriptive complexity theory over the reals two logical hierarchies of such problems were obtained. One for maximization problems consisting of four distinct levels and one for minimization problems containing two levels. Our results provide a real number analogue of corresponding results for the Turing model given in [10].

A most interesting future way to continue in our opinion is to study the relation between the logical description of real number maximization problems and approximation issues. For the Turing setting this line of research was started in [15] by showing that all problems in the discrete version of MAX-$\Sigma_{1,\mathbb{R}}$ have approximation algorithms in class APX. However, in the real number model no serious investigation of approximation classes has been performed so far. Some initial ideas can be found in [14] in relation with probabilistically checkable proofs, but a concise theory is waiting to be developed here. At a first sight, the techniques used in [15] to combine descriptive complexity with approximation seems far from being transferable to the real number world. The same seems to be the case for the techniques used in [12]. The reason might be that on the real number side fixing certain values for variables in a formula results in much stronger backtracking problems than on finite structures. We consider it to be an interesting future research area to develop approximation concepts in the BSS model as well as its relation to descriptive complexity theory for real number maximization problems. For minimization problems we can show a result which, however, is a negative one (compare with the discrete analogue in [10]).

Theorem 7. *There exist problems in MIN-$\Pi_{0,\mathbb{R}}$ that cannot be approximated in polynomial time by any constant factor unless $P_\mathbb{R} = NP_\mathbb{R}$.*

References

1. Ausiello, G., Crescenzi, P., Gambosi, G., Kann, V., Marchetti-Spaccamela, A., Protasi, M.: Complexity and Approximation: Combinatorial Optimization Problems and Their Approximability Properties. Springer (1999).
2. S. Basu, R. Pollack, M.-F. Roy: Algorithms in Real Algebraic Geometry. Springer (2003).
3. L. Blum, F. Cucker, M. Shub, S. Smale: Complexity and Real Computation. Springer, 1998.
4. P. Bürgisser, F. Cucker: Counting Complexity Classes for Numeric Computations II: Algebraic and Semialgebraic Sets, Proc. 36th Symposium on Theory of Computing STOC, 475–485, 2004.
5. T. Chadzelek, G. Hotz: Analytic machines. Theoretical Computer Science 219, 151–167 (1999).
6. F. Cucker, K. Meer: Logics which capture complexity classes over the reals. Journal of Symbolic Logic, Vol. 64, Nr. 1, 363–390 (1999).
7. H.D. Ebbinghaus, J. Flum: Finite Model Theory. Springer-Verlag (1995).
8. E. Grädel, Y. Gurevich: Metafinite Model Theory. In: Logic and computational complexity (D. Leivant, ed.), Springer, 313–366 (1996).
9. E. Grädel, K. Meer: Descriptive complexity theory over the real numbers. Lectures in Applied Mathematics, 32:381–403 (1996).
10. P.G. Kolaitis, M.N. Thakur: Logical definability of NP optimization problems. Information and Computation 115(2), 321–353 (1994).
11. N. Immerman: Descriptive Complexity. Springer (1999).
12. A. Malmström: Optimization problems with approximation schemes. Annual Conference for Computer Science Logic, CSL'96, Springer LNCS 1258, 316–333 (1997).
13. K. Meer: Counting problems over the reals. Theoretical Computer Science 242, 41–58 (2000).
14. K. Meer: On some relations between approximation problems and PCPs over the real numbers. To appear in: Proc. Computability in Europe 2005: New Computational Paradigms, Springer LNCS, 2005.
15. C.H. Papadimitriou, M. Yannakakis, M.: Optimization, approximation and complexity classes. Journal of Computer and System Sciences 43, pp. 425–440 (1991).
16. J. Renegar: On the computational Complexity and Geometry of the first-order Theory of the Reals , I - III. J. of Symbolic Computation, 13, pp. 255–352 (1992).

Algebras as Knowledge Structures

Bernhard Heinemann

Fachbereich Informatik, FernUniversität in Hagen,
58084 Hagen, Germany,
Phone: + 49 2331 987 2714, Fax: + 49 2331 987 319
Bernhard.Heinemann@fernuni-hagen.de

Abstract. We start investigating set algebras from a knowledge theoretical point of view. To this end, we suit hybrid logic to the context of knowledge. The common modal approach is extended in this way, which gives us the necessary expressive power. The main issues of the paper are a completeness and a decidability result for the arising logic of knowledge on algebras.

Keywords: reasoning about knowledge, modal and hybrid logic, topological reasoning, knowledge and algebras, decidability

1 Introduction

In this paper, we look upon set spaces $(S, \mathcal{O})$, where S is a non-empty set and $\mathcal{O} \subseteq \mathcal{P}(S)$ a subset of the powerset of S, as *knowledge structures*. This view is in accordance with the common description of knowledge in terms of modal logic; cf [1], or [2], Ch. 5. In fact, S is to represent the set of all *states of the world* an agent considers possible, whereas $\mathcal{O}$ is intended to model the *knowledge states* of the agent, i.e., the sets of states indistinguishable to the agent by its own knowledge at a time.

The question comes up naturally now, in which way *time* is really visible from such knowledge structures. Presumably, a common answer will be: not at all; if one wants to speak about the change of knowledge in the course of time, then the latter notion has to be added explicitly. This was done for the usual logic of knowledge by means of certain mappings from the domain of time into the set of all states, so-called *runs;* cf [1]. But, actually, time *is* present in knowledge structures, although only implicitly. To see this, note that gaining knowledge means that fewer states are indistinguishable to the agent, i.e., the knowledge state of the agent has been *shrunk*. In other words, the set inclusion relation on $\mathcal{O}$ can serve the modelling of time in knowledge structures, at least as far as knowledge acquisition is concerned.

This is the starting point of the paper [3] (and of the one at hand as well). In that paper, Moss and Parikh designed a bi-modal system, of which one operator corresponds with knowledge and the other with computational *effort;* the latter models some knowledge acquisition procedure thus. Interestingly enough, they called their system *topologic,* as it makes perspicuous the *spatial* content of the concept of knowledge. In fact, knowledge comes along with certain elementary properties of points in space since knowledge states can be viewed as *neighbourhoods,* measuring qualitatively the amount of closeness to complete knowledge. That is why gaining knowledge means *approximating points* (viz states of complete knowledge). Thus, notions from topology enter the realm of knowledge in a natural way.

Several basic classes of spaces have been investigated from this topological view of knowledge, eg, ordinary topological ones, cf [4], as well as 'treelike' and 'directed' ones, cf [5] and [6], respectively. However, more powerful tools are needed to capture other spatial structures arising in computer science or artificial intelligence here and there. In the paper [7], a preliminary formal system was developed, supporting corresponding topological reasoning to some extent. Subsequently, this system is considerably modified and extended in order to make it applicable to *algebras*.

An algebra is a set space $(S, \mathcal{O})$ satisfying the following properties: 1. $S \in \mathcal{O}$, and 2. $\mathcal{O}$ is closed under the formation of complements and unions; see [8], Theorem 1.4. (We take this characterization as a definition because it is most appropriate for our purposes.) The ubiquitous Boolean algebras provide the best-known examples of such structures in computer science. In addition, σ–algebras, which are fundamental to measure theory, have been investigated with regard to effectivity quite recently; see [9], in particular, Definition 2.2. We take this computational interest in algebras as a justification for studying these also from a knowledge theoretical point of view. (Another motivating example, which may convince the reader even more, follows in the next section.)

To this end, we have to enrich the language of knowledge (as we mentioned already above). We do this in a way that preserves the usual modal approach for the most part, viz by means of *hybrid logic.*

Hybrid logic partly makes up for the 'internal, local perspective' of modal logic on relations between states (cf [10], Preface), by allowing among other things *naming* and *jumping to named objects;* see [11] for a dedicated introduction to the subject and an overview of the state of the art. We suit the hybrid methodology to the specific appearance of the semantics of *topologic,* where both points *and* sets are taken into account. In accordance with this, a new hybrid operator for algebras comes into play, leading us to a system which we obtain some nice meta-theorems for.

We give now an overview of the content of this paper. In the next section, we define the just indicated hybrid language for algebras precisely. Afterwards, in Section 3, we introduce the accompanying logic. The main result of this part of the paper is a soundness and completeness theorem for a corresponding logical system. The most important issue of the whole paper is the *decidability* of the hybrid logic of knowledge on algebras. This result and an outline of its proof are contained in Section 4. In the concluding Section 5 we summarize and assess the outcome of the paper.

Finally, it should be remarked that hybrid logics for knowledge or topology (and, in particular, algebras) have hardly been considered up to now. At least a standard hybridization of the classical topological semantics of modal logic (going back to [12] and having been revitalized recently; cf [13]), was briefly studied in [14] with regard to expressiveness.

2 Defining the Language

In this section, we define first a new hybrid language for set algebras. Then we quote a couple of valid formula schemata. Finally, we present a 'generic' example.

We extend the language of *topologic* by two sets of *nominals* on the one hand, and two further modalities on the other hand. The denotation of every nominal is either a unique state or a distinguished knowledge state. The additional connectives are the *global modality*, cf [10], Sec. 7.1, and a new one modelling *complementation*.

Let PROP = $\{p, q, \ldots\}$, $N_{stat} = \{a, b, \ldots\}$ and $N_{sets} = \{X, Y, \ldots\}$ be three mutually disjoint denumerable sets of symbols called *proposition letters*, *names of states*, and *names of sets*, respectively. We define the set WFF of all *well-formed formulas* by the rule

$$\alpha ::= p \mid a \mid X \mid \neg \alpha \mid \alpha \wedge \beta \mid K\alpha \mid \Box\alpha \mid \mathsf{A}\alpha \mid \mathsf{D}\alpha.$$

Here K represents *knowledge* and $\Box$ *effort,* as it is usual for *topologic*. A is the global modality, and D is called the *complementation operator*. The global modality will really reveal the power of names. The missing boolean connectives $\top, \bot, \vee, \to, \leftrightarrow$ are treated as abbreviations, as needed. The duals of K, $\Box$, A and D are denoted L, $\Diamond$, E and C, respectively.

We now turn to semantics. To begin with, we define the domains where formulas will be interpreted in. We let $\mathcal{P}(S)$ designate the powerset of a given set S.

Definition 1 (Set algebras with names).

1. Let S be a non-empty set and $\mathcal{O} \subseteq \mathcal{P}(S)$ a set of subsets of S such that $\mathcal{A} := (S, \mathcal{O})$ is an algebra (as defined in Sec. 1). The set of all neighbourhood situations of $\mathcal{A}$ is defined by $\mathcal{N}_\mathcal{A} := \{(s, U) \mid s \in U \text{ and } U \in \mathcal{O}\}$.[1]
2. Let $\mathcal{A}$ be an algebra as above. An $\mathcal{A}$-valuation is a mapping

$$V : \text{PROP} \cup N_{stat} \cup N_{sets} \longrightarrow \mathcal{P}(S)$$

   such that
   - $V(p) \subseteq \mathcal{P}(S)$ for all $p \in \text{PROP}$,
   - $V(a)$ is a singleton subset of S for all $a \in N_{stat}$, and
   - $V(X) \in \mathcal{O}$ for all $X \in N_{sets}$.
3. A set algebra with names (or, in short, an SAN) is a triple $\mathcal{M} := (S, \mathcal{O}, V)$, where $\mathcal{A} := (S, \mathcal{O})$ is an algebra and V an $\mathcal{A}$-valuation; $\mathcal{M}$ is then called based on $\mathcal{A}$.

---
[1] Neighbourhood situations are always written without brackets below.

Note that the denotation of a set name may be empty. – In Proposition 1 below, we state some formulas which are characteristically valid in every SAN.

As it is common for *topologic,* cf [3], the relation of satisfaction is now defined between neighbourhood situations and formulas.

Definition 2 (Satisfaction and validity). *Let* $\mathcal{M} = (S, \mathcal{O}, V)$ *be an SAN and* s, U *a neighbourhood situation of* $\mathcal{A} = (S, \mathcal{O})$. *Then*

$$
\begin{aligned}
s, U &\models_{\mathcal{M}} p &&:\iff s \in V(p) \\
s, U &\models_{\mathcal{M}} a &&:\iff s \in V(a) \\
s, U &\models_{\mathcal{M}} X &&:\iff V(X) = U \\
s, U &\models_{\mathcal{M}} \neg\alpha &&:\iff s, U \not\models_{\mathcal{M}} \alpha \\
s, U &\models_{\mathcal{M}} \alpha \wedge \beta &&:\iff s, U \models_{\mathcal{M}} \alpha \text{ and } s, U \models_{\mathcal{M}} \beta \\
s, U &\models_{\mathcal{M}} K\alpha &&:\iff t, U \models_{\mathcal{M}} \alpha \text{ for all } t \in U \\
s, U &\models_{\mathcal{M}} \Box\alpha &&:\iff \forall U' \in \mathcal{O} : (s \in U' \subseteq U \Rightarrow s, U' \models_{\mathcal{M}} \alpha) \\
s, U &\models_{\mathcal{M}} \mathsf{A}\alpha &&:\iff t, U' \models_{\mathcal{M}} \alpha \text{ for all } t, U' \in \mathcal{N}_{\mathcal{A}} \\
s, U &\models_{\mathcal{M}} \mathsf{D}\alpha &&:\iff \forall t \in S \setminus U : t, S \setminus U \models_{\mathcal{M}} \alpha,
\end{aligned}
$$

where $p \in \mathrm{PROP}$, $a \in \mathrm{N}_{stat}$, $X \in \mathrm{N}_{sets}$ *and* $\alpha, \beta \in \mathrm{WFF}$. *In case* $s, U \models_{\mathcal{M}} \alpha$ *is true we say that* α *holds in* $\mathcal{M}$ *at the neighbourhood situation* s, U. *– A formula* α *is called* valid *in* $\mathcal{M}$ *(*'$\mathcal{M} \models \alpha$'*) iff it holds in* $\mathcal{M}$ *at every neighbourhood situation of* $\mathcal{A}$.

The following comments may light up this definition in some respects.

Remark 1 (Peculiarities of the just defined language).

1. The meaning of both proposition letters and names of states is independent of neighbourhoods by definition, thus 'stable' with respect to $\Box$. This fact is reflected in two special axioms below (Axioms 6 and 11).
2. The just defined language is fairly expressive. In fact, the formulas of the form $a \wedge X$, where $a \in \mathrm{N}_{stat}$ and $X \in \mathrm{N}_{sets}$, can be taken as names for elements of $\mathcal{N}_{\mathcal{A}}$. And having the global modality on hand, the hybrid *satisfaction operator* @... associated with such a name (cf [10], Sec. 7.3) reads then $\mathsf{E}(a \wedge X \wedge \ldots)$. (Note that E is the dual of A.) I.e., pairs of the form (a, X) act like 'proper' nominals in set algebras with names. It follows from this, in particular, that several important frame properties can be captured by the new language; cf [15].
3. The last clause of Definition 2 explains how the modality D operates: D forces a jump outside the actual neighbourhood U for evaluating α with respect to the complement of U. Note that this cannot be captured with the aid of A on its own.

We express now the characteristic properties of algebras by suitable formula schemata of our hybrid language.

Proposition 1 (Algebra formulas). *Let $\mathcal{M}$ be any SAN based on some algebra $\mathcal{A} = (S, \mathcal{O})$. Then we have that*

$$\mathcal{M} \models X \wedge \mathsf{E}Y \to \mathsf{E}(\Diamond X \wedge L\Diamond Y),$$
$$\mathcal{M} \models \Diamond X \wedge L\Diamond Y \to \Diamond(\Diamond X \wedge L\Diamond Y \wedge K\Diamond(X \vee Y)), \text{ and}$$
$$\mathcal{M} \models \Diamond\mathsf{C}(a \wedge X) \wedge \mathsf{C}(a \wedge Y) \to @_{a \wedge X}\Diamond Y,$$

for all $a \in \mathrm{N}_{stat}$ and $X, Y \in \mathrm{N}_{sets}$.

The abbreviation $@_{a \wedge X}$ is taken as in Remark 1.2. – In Proposition 1, the first formula schema holds due to the property '$S \in \mathcal{O}$', the second corresponds to the closure under unions, and the third is valid because $\mathcal{O}$ is closed under complementation. The proof of this proposition is straightforward and, therefore, omitted.

In the following, we give an example arising quite naturally from potentially infinite computations.

Example 1 (A subalgebra of the Cantor space). Let $\mathcal{C}$ be the set of all infinite 0–1–sequences. A basis $\mathcal{B}$ for the distinguished topology on $\mathcal{C}$ is determined by the set of all finite initial segments of elements of $\mathcal{C}$. Let

$$\mathcal{O} := \left\{ \bigcup \mathcal{A} \mid \mathcal{A} \subseteq \mathcal{B} \text{ finite} \right\}.$$

Then, $\mathcal{A} := (\mathcal{C}, \mathcal{O})$ is an algebra, actually. Note that $(\mathcal{C}, \mathcal{B})$ can be depicted as the full infinite binary tree such that every $U \in \mathcal{B}$ is associated with the node by which it is determined. With the aid of suitable formulas of our language we can now specify certain properties of programs computing binary streams. For example, if a procedure P computes some real number ρ (i.e., the output of P encodes a fast-converging Cauchy sequence having limit ρ; cf [16]) and ρ is different from, eg, π, then *one will know this eventually* (and forever afterwards). Thus the formula $\Diamond K \mathsf{C} \pi$ is valid in a suitable SAN based on $\mathcal{A}$.

Concluding this section, we comment on the relevance of names and accompanying hybrid operators to the context of knowledge. We confine ourselves to names of sets here since the general usefulness of names of states has been sufficiently demonstrated elsewhere; cf, above all, [11]. As mentioned in the introduction already, the elements of $\mathcal{O}$ can be viewed as knowledge states of an agent, for any correspondingly given set space $(S, \mathcal{O})$. Thus the new language supplies one with *names of knowledge states*. And the hybrid operator D allows to switch over to a complementary knowledge state. This additional means of expression is, therefore, quite in accord with the common 'external' view of knowledge in multi-agent systems, where knowledge is 'ascribed' to the agents; cf [1], Ch. 4.

3 A Hybrid Logic for Algebras

Our starting point to this section is a system of axioms for set spaces. Following, some comments on these formulas are given. Then we add suitable schemata for names and the complementation operator D, respectively. The resulting axiomatization leads us to a logical system which turns out to be sound and complete with respect to the class of all set algebras with names. We point to some crucial steps of the completeness proof in the second part of this section.

The usual axioms for arbitrary set spaces read as follows:

1. All instances of tautologies.
2. $K(\alpha \to \beta) \to (K\alpha \to K\beta)$
3. $K\alpha \to \alpha$
4. $K\alpha \to KK\alpha$
5. $L\alpha \to KL\alpha$
6. $(p \to \Box p) \wedge (\Diamond p \to p)$
7. $\Box (\alpha \to \beta) \to (\Box \alpha \to \Box \beta)$
8. $\Box \alpha \to \alpha$
9. $\Box \alpha \to \Box \Box \alpha$
10. $K\Box \alpha \to \Box K \alpha$,

where $p \in$ PROP and $\alpha, \beta \in$ WFF. In this way, it is expressed that

- the accessibility relation $\xrightarrow{K}$ belonging to the knowledge operator is an equivalence,
- the accessibility relation $\xrightarrow{\Box}$ belonging to the effort operator is reflexive and transitive,
- proposition letters are stable with respect to $\xrightarrow{\Box}$ (see Remark 1.1 above), and
- knowledge and effort commute as described by Axiom 10.

The latter schema, which is characteristic of every logic of knowledge and effort, is usually called the *Cross Axiom;* cf [3]. – The next group of axioms concerns names:

11. $(a \to \Box a) \wedge (\Diamond a \to a)$
12. $a \wedge \alpha \to K(a \to \alpha)$
13. $X \to KX$
14. $\mathsf{A}\,(X \wedge L\alpha \to L\beta) \vee \mathsf{A}\,(X \wedge L\beta \to L\alpha)$
15. $K(\Diamond Y \to \Diamond X) \wedge L\Diamond Y \to \Box\,(X \to L\Diamond Y)$
16. $K\Diamond X \to X$,

where $a \in \mathrm{N}_{stat}$, $X, Y \in \mathrm{N}_{sets}$ and $\alpha, \beta \in$ WFF. The meaning of the formulas of this group is not easy to understand at first glance. Actually, these provide for both the right behaviour of names and the necessary properties of the relation $\xrightarrow{K} \circ \xrightarrow{\Box}$ on the canonical model so that really a structure of set space can be ensured there; cf [17]. (Axiom 14 had to be suited to the present case.)

Apart from the schemata 14 and 20 (see below), the global modality is clearly involved in further axioms as well. These say that $\xrightarrow{A}$ is an equivalence relation, too, which includes each of the relations belonging to the other modalities; cf [10], p 417. We do not explicitly list the corresponding formulas here.

The complementation operator is axiomatized by the schemata of the third group:

17. $\mathsf{D}(\alpha \to \beta) \to (\mathsf{D}\alpha \to \mathsf{D}\beta)$ 18. $La \vee Ca$ 19. $\neg\,(La \wedge Ca)$,

where $\alpha, \beta \in$ WFF and $a \in \mathrm{N}_{stat}$. Note that Axiom 18 expresses that $\xrightarrow{K} \cup \xrightarrow{D}$ is exhaustive in a sense, and Axiom 19 that both relations mutually exclude each other.

The final group of axioms goes together with the structure of algebra. In fact, this block consists of the schemata from Proposition 1.

20. $X \wedge \mathsf{E} Y \to \mathsf{E}(\Diamond X \wedge L \Diamond Y)$
21. $\Diamond X \wedge L \Diamond Y \to \Diamond(\Diamond X \wedge L \Diamond Y \wedge K \Diamond (X \vee Y))$
22. $\Diamond \mathsf{C}(a \wedge X) \wedge \mathsf{C}(a \wedge Y) \to @_{a \wedge X} \Diamond Y$,

where $X, Y \in \mathsf{N}_{sets}$ and $a \in \mathsf{N}_{stat}$. Note that the interplay between D and $\square$ is settled by Axiom 22.

A logical system called KA (designating **k**nowledge on **a**lgebras), is now obtained from this list by adding appropriate proof rules. Apart from the standard rules from modal logic (*modus ponens* and *necessitation*) we have also some unorthodox ones which are typical of hybrid logic.

Definition 3 (The logic). *Let* KA *be the smallest set of formulas containing all of the above axiom schemata and closed under application of the following rules:*

$$(\text{MODUS PONENS}) \quad \frac{\alpha \to \beta, \alpha}{\beta} \qquad (\Delta\text{-NECESSITATION}) \quad \frac{\alpha}{\Delta \alpha}$$

$$(\text{NAME}_{stat}) \quad \frac{b \to \beta}{\beta} \qquad (\text{NAME}_{sets}) \quad \frac{Y \to \beta}{\beta}$$

$$(\mathsf{E}_\nabla\text{-ENRICHMENT}) \quad \frac{\mathsf{E}(a \wedge X \wedge \nabla(b \wedge Y \wedge \alpha)) \to \beta}{\mathsf{E}(a \wedge X \wedge \nabla \alpha) \to \beta},$$

where $\alpha, \beta \in \text{WFF}$, $a, b \in \mathsf{N}_{stat}$, $X, Y \in \mathsf{N}_{sets}$, Δ *is contained in* $\{K, \square, \mathsf{A}, \mathsf{D}\}$, $\nabla \in \{L, \Diamond, \mathsf{E}, \mathsf{C}\}$, *and* b, Y *are* new *each time (i.e., do not occur in any other syntactic building block of the respective rule).*

The effect of the NAME and ENRICHMENT rules, respectively, will soon become apparent. – The following result is the first of the main issues of this paper.

Theorem 1 (Completeness). *A formula* $\alpha \in \text{WFF}$ *is valid in all SANs, iff it is* KA*-derivable.*

We cannot present a detailed proof of this theorem here, due to limited space. And we want to put the major emphasis on decidability besides; see Sec. 4. But we fix the starting point and give the broad outlines so that the reader can imagine how the machinery developed so far is working.

The soundness part of Theorem 1 is easy to prove. Towards completeness, the properties of the *canonical model* $\mathcal{M}_{\mathsf{KA}}$ of the system KA have to be exploited. Starting from a given non-derivable formula γ which we want to find a model for, the part of $\mathcal{M}_{\mathsf{KA}}$ accessible from some fixed maximal consistent set determined by γ has to be *named* first. And second, all demands for the existence of points realizing ∇–formulas have to be fulfilled. To this end, we call a maximal consistent set s of formulas

- *named* iff s contains some $a \in \mathrm{N}_{stat}$ and some $X \in \mathrm{N}_{sets}$, and
- *enriched* iff, for every $\nabla \in \{L, \Diamond, \mathsf{E}, \mathsf{C}\}$, we have that

$$\mathsf{E}\,(a \wedge X \wedge \nabla \alpha) \in s \Rightarrow \exists\, b \in \mathrm{N}_{stat}, Y \in \mathrm{N}_{sets} : \mathsf{E}\,(a \wedge X \wedge \nabla(b \wedge Y \wedge \alpha)) \in s,$$

where $a \in \mathrm{N}_{stat}$, $X \in \mathrm{N}_{sets}$, and $\alpha \in \mathrm{WFF}$.

Let N'_{stat} and N'_{sets} be two denumerable sets of *new* symbols, and WFF' the set of formulas extended accordingly. Then, we obtain the following *Modified Lindenbaum Lemma*.

Lemma 1. *Every maximal consistent set $s \subseteq \mathrm{WFF}$ can be extended to a named and enriched maximal consistent set $s' \subseteq \mathrm{WFF}'$.*

As it is clear from the definition of a named and enriched set, the hybrid rules for naming and enrichment are used decisively in the proof of the Modified Lindenbaum Lemma.

Furthermore, the following *Existence Lemma* can be proved for the structure $\mathcal{M}'$ of which the domain D consists of all named points that are reachable (via the canonical accessibility relations) from s' and the relations are the induced ones.

Lemma 2. *Let $\nabla \in \{L, \Diamond, \mathsf{E}, \mathsf{C}\}$. Assume that $s \in D$ contains the formula $\nabla \alpha$. Then there exists some $t \in D$ which is accessible from s with respect to $\xrightarrow{\Delta}$, where Δ is the dual of ∇, and contains α.*

Now, for every $s \in D$ we let $[s] := \{t \in D \mid s \xrightarrow{K} t\}$ be the $\xrightarrow{K}$-equivalence class of s (see Axioms 3 – 5 above), and we consider the set $\mathcal{Q} := \{[s] \mid s \in D\}$ of all such classes. Particularly the axioms of the second group enable us to define a set space structure on a suitable space of partial functions $f : \mathcal{Q} \longrightarrow D$. The axioms of the last two groups make sure that this set space is an algebra, even an SAN, actually. (Clearly, all the other axioms are needed as well.) All in all, it can be guaranteed that the modal operators behave correctly as set space modalities (according to Definition 2) on the model we have just pointed to. Thus we obtain in fact completeness of the system KA, via an appropriate *Truth Lemma*.

4 Decidability

In this section, we show that KA is a decidable set of formulas. This is done by proving the *finite model property (fmp)* for KA. Unfortunately, this property does *not* hold with respect to the intended class of structures; cf [3], Sec. 1.3. But the *fmp* will be established for a suitable subclass of the class of all Kripke models instead. To this end, we define first certain auxiliary structures occurring at an intermediate step of our proof and realizing already a large part of the above list of axioms.

Definition 4 (KA–models). *A quintupel* $\mathfrak{M} := \left(W, \xrightarrow{K}, \xrightarrow{\Box}, \xrightarrow{A}, V \right)$ *is called a* KA–*model, iff the following conditions are satisfied:*

1. W is a non-empty set,
2. the relation $\xrightarrow{K} \subseteq W \times W$ (belonging to the knowledge operator K) is an equivalence,
3. the relation $\xrightarrow{\Box} \subseteq W \times W$ (belonging to the effort operator $\Box$) is reflexive and transitive,
4. for all $u, v, w \in W$ such that $u \xrightarrow{\Box} v \xrightarrow{K} w$ there exists $t \in W$ such that $u \xrightarrow{K} t \xrightarrow{\Box} w$,
5. the relation $\xrightarrow{A} \subseteq W \times W$ (belonging to the global modality A) is universal,
6. there is some $u_0 \in W$ such that $W = \{v \mid (u_0, v) \in (\xrightarrow{K} \circ \xrightarrow{\Box})^*\}$ (i.e., in particular, W is generated by u_0 with respect to $\xrightarrow{K} \circ \xrightarrow{\Box}$),
7. $V : \mathrm{PROP} \cup \mathrm{N}_{stat} \cup \mathrm{N}_{sets} \longrightarrow \mathcal{P}(W)$ is a mapping satisfying
    (a) for all $c \in \mathrm{PROP} \cup \mathrm{N}_{stat}$ and $u, v \in W$: if $u \xrightarrow{\Box} v$, then $u \in V(c) \iff v \in V(c)$,
    (b) for all $a \in \mathrm{N}_{stat}$ and $u \in W$ there is at most one $v \in W$ such that $u \xrightarrow{K} v$ and $v \in V(a)$,
    (c) for all $X \in \mathrm{N}_{sets}$, the set $V(X)$ equals either $\emptyset$ or a unique $\xrightarrow{K}$ - equivalence class.

We give a couple of comments on this definition. First, note that the fourth item above corresponds to the aforementioned Cross Axiom and is, therefore, called the *Cross Property*. This property can be taken as a certain diagram property of the relations $\xrightarrow{K}$ and $\xrightarrow{\Box}$, actually. Second, item 7 suits the notion of $\mathcal{A}$-valuation from Definition 1 to KA–models. And finally, the reader might wonder why Definition 4 does not refer to the relation $\xrightarrow{D}$; the reason for this is that the crucial properties of $\xrightarrow{D}$ are not passed to filtrations; see below. Thus this relation requires an extra treatment.

KA–models are closely related to *SAN*s. In fact, by taking neighbourhood situations as points and defining the accessibility relations and the valuation in a way suggesting itself (in accordance with the proceeding for *topologic*, cf [3], Section 2.3), every SAN induces a KA–model.

Unfortunately, not every KA–model validates necessarily all of the above axioms. But, for a start, we get at least the following partial result.

Proposition 2. *Apart from, at most, Axioms 15 – 19 and 21 – 22, each of the above schemata is valid in every* KA*–model.*

The proof of Proposition 2 is rather straightforward. – We introduce now an ad hoc notation by calling a KA–model $\mathfrak{M}$ *faithful*, iff all the axioms except those containing D (i.e., 17 – 19 and 22) are valid in $\mathfrak{M}$. We want to show next that the D–free fragment of KA satisfies the *Strong Finite Model Property* (cf [10], Def. 6.6) with respect to the class of all faithful KA–models.

To establish this finite model property we use the method of filtration, followed by an appropriate model surgery. So, let $\alpha \in \mathrm{WFF}$ be a consistent formula for which we want to find a model of size at most $f(|\alpha|)$, where f is some computable function and $|\alpha|$ denotes the length of α. Let us assume temporarily that α is D–free. In the following, we have to consider the set sf(α) of all subformulas of α. We construct a suitable filter set Σ from sf(α) as follows. We first let

$$\Sigma_0 := \mathrm{sf}(\alpha) \cup \{\Box\neg X \mid X \in \mathrm{N}_{sets} \text{ occurs in } \alpha\},$$

and secondly $\Sigma^\neg := \{\neg\beta \mid \beta \in \Sigma_0\}$. Then we take the set Σ' of all finite conjunctions of pairwise distinct elements of $\Sigma_0 \cup \Sigma^\neg$. Afterwards, we close Σ' under single applications of the operator L and take, finally, the set of all subformulas of the formulas contained in the resulting set. Let Σ be the union of all these intermediate sets of formulas. Then Σ is subformula closed, and $2^{c\cdot|\alpha|}$ is an upper bound of the cardinality of Σ (for some constant c).

Let $\mathcal{C}$ be the submodel of the canonical model of KA generated by a maximal consistent set realizing α. Moreover, let a Kripke model

$$\mathfrak{M} = \left(W, \xrightarrow{K}, \xrightarrow{\Box}, \xrightarrow{A}, V\right)$$

be obtained from $\mathcal{C}$ as follows:

- W is the filtration of the carrier set C of $\mathcal{C}$ with respect to Σ,
- $\xrightarrow{K}, \xrightarrow{\Box}$ and $\xrightarrow{A}$ are the *smallest* filtrations (cf [10], 2.40) of the accessibility relations of $\mathcal{C}$ belonging to the respective modalities, and
- V is induced by the canonical valuation.

Then we have the following lemma.

Lemma 3. *The just defined structure $\mathfrak{M}$ is a KA–model.*

And we get even more, by exploiting the structure of the filter set Σ and the fixings for the filtrations of the relations to a greater extent.

Proposition 3. *The model $\mathfrak{M}$ can be turned into a faithful KA–model $\mathfrak{M}'$ which is semantically equivalent to $\mathfrak{M}$ with respect to α.*

Proof. The valuation V has to be altered suitably, and Axioms 15, 16 and 21 have to be verified accordingly. A detailed reasoning for Axioms 15 and 16 can be found in [7], proof of Proposition 3. As to Axiom 21, we can argue in a similar manner. Note that $K\Diamond(X \vee Y)$ is really an element of the filter set Σ, if both X and Y occur in α (otherwise there is nothing to prove).

Now we drop our intermediate assumption that the complementation operator D does not occur in α. This will result in a change of the filter set. But since the 'building plan' of this set will be retained, Proposition 3 remains valid.

The following treatment of D reminds one of the handling of the modal *difference operator* in [10], proof of Theorem 7.8.[2] That is, D will be replaced temporarily and reinserted after carrying out the filtration.

---

[2] The complementation operator can, in fact, be viewed as a generalized difference operator: it enables one to jump to points *outside the equivalence class of the actual one* for evaluating a given formula there (instead of jumping to a point *different from the actual one* for the same purpose).

Let Σ be defined as above. We have to take a closer look at the filtration relation induced by Σ on the domain C of $\mathcal{C}$. Let $\sim_\Sigma$ denote this relation, and $\bar{s}$ the equivalence class with respect to $\sim_\Sigma$ of a point $s \in C$. We choose an injective mapping ι from the (finite) set of all such classes into the set of all proposition letters not occurring in Σ. Furthermore, a new proposition letter p_β is assigned to every formula $\mathsf{D}\beta \in \Sigma_0$ in such a way that $p_\beta \neq p_\gamma$ whenever $\beta \neq \gamma$. For every $\delta \in \Sigma_0$, let δ' be the result of replacing every subformula $\mathsf{D}\beta$ of δ with p_β. Then we define

$$\widetilde{\Sigma}_0 := \{\delta' \mid \delta \in \Sigma_0\} \cup \{p_{\iota(\bar{s})}, K p_{\iota(\bar{s})} \mid s \in C\} \cup \{p_\beta \mid \mathsf{D}\beta \in \Sigma_0\},$$

and we let $\widetilde{\Sigma}$ be built from $\widetilde{\Sigma}_0$ as Σ was built from Σ_0; see above. Obviously, $\widetilde{\Sigma}$ is D–free.

$\widetilde{\Sigma}$ will be used as a new filter set in a moment. However, the model $\mathcal{C}$ has to be modified beforehand to the effect that the valuation makes $p_{\iota(\bar{s})}$ true at exactly one point of $\bar{s}$, and p_β at exactly the points where $\mathsf{D}\beta$ is true in model $\mathcal{C}$. Let this variant of $\mathcal{C}$ be designated $\mathcal{C}'$.

Now, $\mathcal{C}'$ is filtrated through $\widetilde{\Sigma}$. Let $\sim_{\widetilde{\Sigma}}$ denote the filtration relation induced by $\widetilde{\Sigma}$ on C. Then every equivalence class with respect to $\sim_\Sigma$ is divided into exactly two equivalence classes with respect to $\sim_{\widetilde{\Sigma}}$ (if the former class consists of more than one point). Moreover, one can prove that these two smaller classes are not $\xrightarrow{K}$–connected. This gives us quite a good behaviour of $\xrightarrow{\mathsf{D}}$ when interpreted appropriately in the filtrated model, denoted $\widetilde{\mathfrak{M}}$. In particular, Axioms 17 – 19 are forced to be valid there.

Furthermore, it can be shown by a suitable induction that $\widetilde{\mathfrak{M}}$ is semantically equivalent to $\mathcal{C}$ with respect to Σ. This is almost that what we would like to have. Only Axiom 22 has still to be established. To this end, all the formulas contained in the set

$$\{\Box\mathsf{D}\neg(a \wedge X) \mid a \in \mathrm{N}_{stat} \text{ and } X \in \mathrm{N}_{sets} \text{ occur in } \alpha\}$$

have to be added to the source set Σ_0. Then one can in fact continue as described above.

All in all, we conclude that α is satisfiable in a finite model of the axioms, of which the size is in $O\left(2^{2^{c \cdot |\alpha|}}\right)$ (where $c \in \mathbb{N}$ is some constant). This yields the main result of this section.

Theorem 2 (Decidability). *The hybrid logic* KA *for knowledge on algebras is decidable.*

We do not know whether or not it is possible to improve the arising upper bound for the complexity of the logic KA.

5 Summary

In the present paper, we developed the fundamental matters of a two-sorted hybrid logic, KA, for set algebras with names, which can be viewed as appropriate knowledge structures in various contexts. A new operator expressing complementation was introduced for this purpose, in particular. We obtained a Completeness as well as a Decidability Theorem for KA. These results may be taken as an indication that our approach to knowledge on algebras is a promising one, which must, however, still prove to be suitable for practical use.

References

1. Fagin, R., Halpern, J.Y., Moses, Y., Vardi, M.Y.: Reasoning about Knowledge. MIT Press, Cambridge, MA (1995)
2. Huth, M., Ryan, M.: Logic in Computer Science: Modelling and Reasoning about Systems. Cambridge University Press, Cambridge (2000)
3. Dabrowski, A., Moss, L.S., Parikh, R.: Topological reasoning and the logic of knowledge. Annals of Pure and Applied Logic **78** (1996) 73–110
4. Georgatos, K.: Knowledge theoretic properties of topological spaces. In Masuch, M., Pólos, L., eds.: Knowledge Representation and Uncertainty, Logic at Work. Volume 808 of Lecture Notes in Artificial Intelligence., Springer (1994) 147–159
5. Georgatos, K.: Knowledge on treelike spaces. Studia Logica **59** (1997) 271–301
6. Weiss, M.A., Parikh, R.: Completeness of certain bimodal logics for subset spaces. Studia Logica **71** (2002) 1–30
7. Heinemann, B.: A hybrid logic of knowledge supporting topological reasoning. In Rattray, C., Maharaj, S., Shankland, C., eds.: Algebraic Methodology and Software Technology, AMAST 2004. Volume 3116 of Lecture Notes in Computer Science., Berlin, Springer (2004) 181–195
8. Bauer, H.: Measure and Integration Theory. Volume 26 of de Gruyter Studies in Mathematics. de Gruyter, New York (2001)
9. Wu, Y., Weihrauch, K.: A computable version of the Daniell-Stone Theorem on integration and linear functionals. In Brattka, V., Staiger, L., Weihrauch, K., eds.: CCA 2004. Number 320 in Informatik Berichte, Hagen, Germany (2004) 195–207
10. Blackburn, P., de Rijke, M., Venema, Y.: Modal Logic. Volume 53 of Cambridge Tracts in Theoretical Computer Science. Cambridge University Press, Cambridge (2001)
11. Blackburn, P.: Representation, reasoning, and relational structures: a hybrid logic manifesto. Logic Journal of the IGPL **8** (2000) 339–365
12. McKinsey, J.C.C.: A solution to the decision problem for the Lewis systems S2 and S4, with an application to topology. Journal of Symbolic Logic **6** (1941) 117–141
13. Aiello, M., van Benthem, J., Bezhanishvili, G.: Reasoning about space: The modal way. Journal of Logic and Computation **13** (2003) 889–920
14. Gabelaia, D.: Modal definability in topology. Master's thesis, ILLC, Universiteit van Amsterdam (2001)
15. Heinemann, B.: Axiomatizing modal theories of subset spaces (an example of the power of hybrid logic). In: HyLo@LICS, Proceedings, Copenhagen, Denmark (2002) 69–83
16. Weihrauch, K.: Computable Analysis. Springer, Berlin (2000)
17. Heinemann, B.: Extended canonicity of certain topological properties of set spaces. In Vardi, M., Voronkov, A., eds.: Logic for Programming, Artificial Intelligence, and Reasoning. Volume 2850 of Lecture Notes in Artificial Intelligence., Berlin, Springer (2003) 135–149

Combining Self-reducibility and Partial Information Algorithms

André Hernich[1] and Arfst Nickelsen[2]

[1] Humboldt-Universität zu Berlin, Germany
hernich@informatik.hu-berlin.de
[2] Technische Universität Berlin, Germany
nicke@cs.tu-berlin.de

Abstract. A partial information algorithm for a language A computes, for some fixed m, for input words $x_1, \ldots, x_m$ a set of bitstrings containing $\chi_A(x_1, \ldots, x_m)$. E.g., p-selective, approximable, and easily countable languages are defined by the existence of polynomial-time partial information algorithms of specific type. Self-reducible languages, for different types of self-reductions, form subclasses of PSPACE.

For a self-reducible language A, the existence of a partial information algorithm sometimes helps to place A into some subclass of PSPACE. The most prominent known result in this respect is: P-selective languages which are self-reducible are in P [9].

Closely related is the fact that the existence of a partial information algorithm for A simplifies the type of reductions or self-reductions to A. The most prominent known result in this respect is: Turing reductions to easily countable languages simplify to truth-table reductions [8].

We prove new results of this type. We show:

1. Self-reducible languages which are easily 2-countable are in P. This partially confirms a conjecture of [8].
2. Self-reducible languages which are $(2m-1, m)$-verbose are truth-table self-reducible. This generalizes the result of [9] for p-selective languages, which are $(m+1, m)$-verbose.
3. Self-reducible languages, where the language and its complement are strongly 2-membership comparable, are in P. This generalizes the corresponding result for p-selective languages of [9].
4. Disjunctively truth-table self-reducible languages which are 2-membership comparable are in UP.

Topic: Structural complexity.

1 Introduction

A *partial information algorithm* for a language A computes on input of m words $x_1, \ldots, x_m$ some information on membership of these words in A. It excludes some of the 2^m bitstrings a priori possible for $\chi_A(x_1, \ldots, x_m)$, where χ_A is the characteristic function for A, by computing a set $D \subset \{0,1\}^m$ with $\chi_A(x_1, \ldots, x_m) \in D$. We call such sets D m-pools. Sets $\mathcal{D}$ of m-pools that may occur as outputs of a specific partial information algorithm are called m-families. In the following we only treat polynomially time bounded partial information algorithms. This line of research started in the seventies with the introduction of p-selective languages (due to [28], see [16]). Many other types of partial information have been studied since then, most prominently cheatability (due to [5]), and membership comparability (due to [2,12], also known as approximability or non-superterseness, see [8,27]); as well as verboseness, strong membership comparability, frequency computations, easily countable languages, multi-selectivity and sortability (for detailed definitions see e.g. [2,3,7,8,15,17,18]). A general theory for polynomial-time partial information classes was developed in [22,23]. For a recent survey on partial information see [26].

From the start the interplay of partial information with reducibility and self-reducibility was investigated. Languages positively Turing reducible to a p-selective language are in fact many-one reducible to that language [10], and therefore are p-selective, as well. But for non-positive reductions the reduction closures of p-selective sets form a strict hierarchy [30,14]. For recent results on reductions and polynomial-time partial information classes see [25,6,4].

A language A is self-reducible if membership for a word x in A can be determined by computing membership for smaller words in A. This property is quite typical for many computational problems. E.g., natural NP-complete problems like SAT are self-reducible. The question whether SAT has partial information algorithms has been studied extensively in the literature, see e.g., [1,8,27,29].

Self-reducible languages are in PSPACE. In some cases, if a self-reducible language additionally has a certain type of polynomial-time partial information algorithm, one can show a better complexity bound than PSPACE. Self-reducible p-selective languages are in P [9]. Similar results hold for p-cheatable languages [13] and frequency computations [8]. We look for new and more general results of that type. A general solution for all types of partial information seems out of reach: An unconditional negative answer would imply P $\neq$ PSPACE; on the other hand there are relativized worlds where the answer is negative for some types of partial information [8].

For two types of partial information we show that in combination with self-reducibility one gets membership in P. Moreover, we give an interesting example where the combination of partial information with a restricted type of self-reduction places languages not into P, but into the class UP.

In particular, we prove the following:
- Easily 2-countable self-reducible languages are in P. In [8] it was conjectured that this even holds for easily m-countable languages for every $m \geq 2$. We think that similar ideas that we use to prove the conjecture for $m = 2$ may turn out helpful to attack the general conjecture.
- If a self-reducible language and its complement are strongly 2-membership comparable, then the language is in P. This is a strict improvement on the result of [9] for self-reducible p-selective languages.
- If a language is 2-membership comparable and disjunctively truth-table self-reducible, then it is in UP. This is especially interesting since there is a relativized world where this cannot be improved to membership in P [8].

In proofs that place self-reducible languages with partial information algorithms into P one often starts as follows: One shows that for languages in the partial information class a certain reducibility simplifies to a more restricted type of reducibility. This is e.g. the case in our proof for easily 2-countable languages. In [9] the essential proof step is to simplify self-reductions to truth-table self-reductions for p-selective languages. We address the question to which types of partial information this simplification result can be extended. We show:

- Self-reducible languages where the m-fold characteristic function is $(2m-1)$-enumerable are truth-table self-reducible.

This generalizes the result for p-selective languages of [9]. Note that for p-selective languages the m-fold characteristic function is $(m+1)$-enumerable. We will apply this result in our proof for languages where the language and its complement are strongly 2-membership comparable.

2 Basics on Partial Information and Self-reducibility

Languages, Bitstrings, Complexity Classes. Languages are subsets of $\Sigma^* = \{0,1\}^*$. The *characteristic function* $\chi_A \colon \Sigma^* \to \{0,1\}$ for a language A is defined by $\chi_A(x) = 1 \iff x \in A$. We extend χ_A to tuples by $\chi_A(x_1, \ldots, x_m) := \chi_A(x_1) \cdots \chi_A(x_m)$. $\overline{A}$ denotes the complement of A. $\#_1(b)$ denotes the number of 1's in a bitstring b, $b[i]$ is the i-th bit of b, and $b[i_1, \ldots, i_k] := b[i_1] \cdots b[i_k]$. For background on Turing machines and complexity classes see e.g. [11]. FP denotes the class of polynomial-time computable functions. A nondeterministic Turing machine is called unambiguous if for every input x there is at most one accepting computation. A language A is in UP if there is an unambiguous polynomial time bounded nondeterministic Turing machine that accepts A.

Reducibility, Self-reducibility. We define several types of reductions and self-reductions:

1. A language A is polynomial-time Turing reducible to a language B if there is a deterministic polynomial-time oracle Turing machine M with $A = L(M, B)$.
2. If the machine is non-adaptive, we say that A is polynomial-time truth-table (or tt, for short) reducible to B. Equivalently, tt-reducibility can be defined in terms of generator and evaluator: A is polynomial-time tt-reducible to B if there exists a generator $g \in \mathrm{FP}$ and an evaluator $\alpha \in \mathrm{FP}$ such that on input x, $g(x)$ is a tuple $\langle q_1, \ldots, q_s \rangle$ and $\chi_A(x) = \alpha(x, \chi_A(q_1, \ldots, q_s))$. If we define $\alpha_x(b_1, \ldots, b_s) := \alpha(x, b_1, \ldots, b_s)$, α_x is the boolean function that evaluates the oracle answers on input x.
3. For $k \in \mathbb{N}$, A is k-*tt-reducible* if on every input x there are at most k words in $g(x)$. We say that A is *dtt-reducible* if for every x the boolean function α_x is a disjunction. A is *nor-nand-reducible* to B if for every x the boolean function α_x is either a nor- or a nand-function. (Note that one can decide which of the two cases for α_x holds without knowing the oracle answers.)

4. A language A is self-reducible if A is polynomial-time reducible to A by some machine M such that all words queried by M on input x are shorter than x. Corresponding to each restricted type of reducibility we also have a restricted type of self-reducibility.

We use here self-reducibility in the narrower sense where queries have to be smaller than the input word *with respect to the word length*. All our results also hold for the more liberal definition of self-reducibility in [21]. The following facts are due to Ko [19].

Fact 1.

1. *Every self-reducible language is in* PSPACE.
2. *Every dtt-self-reducible language is in* NP.
3. *Every 1-tt-self-reducible language is in* P.

Partial information classes. We now introduce the concept of partial information classes. We state facts from [24] (see also [26]).

Definition 1 (Pool, Family, Partial Information Class). *Let $m \geq 1$.*

1. *A subset $D \subseteq \{0,1\}^m$ is called an m-pool.*
2. *A set $\mathcal{D} = \{D_1, \ldots, D_r\}$ of m-pools is called an m-family if*
   (a) $\mathcal{D}$ covers $\{0,1\}^m$, that is $\bigcup_{i=1}^r D_i = \{0,1\}^m$, and
   (b) $\mathcal{D}$ is closed under subsets, that is $D_1 \in \mathcal{D}$ and $D_2 \subseteq D_1$ implies $D_2 \in \mathcal{D}$.
3. *For an m-family $\mathcal{D}$, a language A is in $\mathrm{P}[\mathcal{D}]$ if and only if there is an $f \in \mathrm{FP}$ such that $f(x_1, \ldots, x_m) \in \mathcal{D}$ and $\chi_A(x_1, \ldots, x_m) \in f(x_1, \ldots, x_m)$ for all words $x_1, \ldots, x_m$.*

In general, different m-families may yield the same partial information class. However, to produce all partial information classes we need only consider families in so called *normal form*. Such families are closed under permuting positions in bitstrings of pools in the family, replacing bits of some bitstring position by constant 0 (or 1), and copying bits from one bitstring position to another. For every m-family there is a unique m-family in normal form that produces the same partial information class. Given m-pools $D_1, \ldots, D_s$, $\langle D_1, \ldots, D_s \rangle$ denotes the minimal m-family in normal form that contains these pools. Inclusion of partial information classes corresponds to inclusion of families in normal form.

We next give names and informal descriptions to several pools which we will use to describe types of partial information. We define several families as well, mostly by listing a set of pools generating them.

Definition 2 (Some Pools).

1. equ_2 := $\{00, 11\}$ *The words are equivalent wrt. membership in A.*
2. xor_2 := $\{01, 10\}$ *Exactly one of the words is in A.*
3. sel_2 := $\{00, 01, 11\}$ *If the first word is in A, then also the second.*
4. sel_m := $\{0^i 1^{m-i} \mid i = 0, \ldots, m\}$ *If x_i is in A, then also x_{i+1}.*
5. bottom_m := $\{b \mid |b| = m, \#_1(b) \leq 1\}$ *At most one word is in A.*
6. top_m := $\{b \mid |b| = m, \#_1(b) \geq m - 1\}$ *At least $(m-1)$ words are in A.*

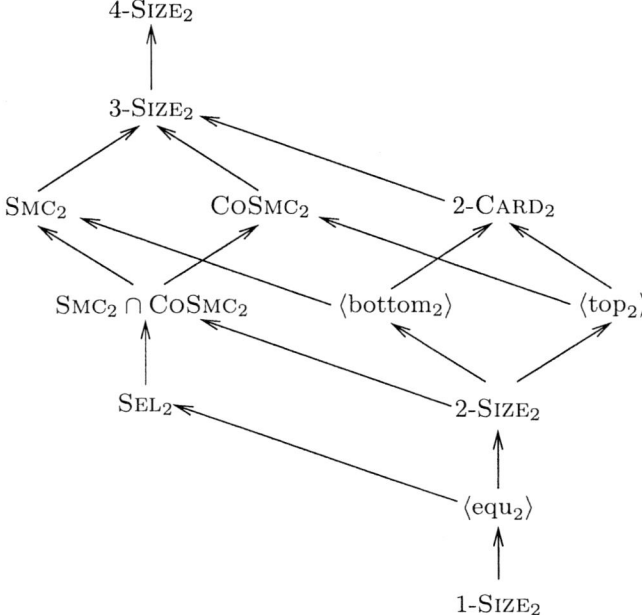

Fig. 1. Inclusion structure of all 2-families in normal form. Arrows stand for strict inclusion.

Definition 3 (Some Families).

1. $k\text{-}\mathrm{SIZE}_m := \{D \subseteq \{0,1\}^m \mid |D| \leq k\}$
2. $\mathrm{SEL}_2 := \langle \mathrm{sel}_2 \rangle$
3. $m\text{-}\mathrm{CARD}_m := \{D \subseteq \{0,1\}^m \mid \exists i \in \{0,\ldots,m\} \, \forall b \in D : \#_1(b) \neq i\}$
4. $\mathrm{SMC}_m := \langle \{0,1\}^m \setminus \{1^m\}, \{0,1\}^m \setminus \{01^{m-1}\} \rangle$
5. $\mathrm{COSMC}_m := \langle \{0,1\}^m \setminus \{0^m\}, \{0,1\}^m \setminus \{10^{m-1}\} \rangle$

Languages in $P[m\text{-}\mathrm{SIZE}_m]$ are called m-cheatable, as defined in [8]. Languages in $P[\mathrm{SEL}_2]$ are called p-selective, as defined in [28]. Languages in $P[m\text{-}\mathrm{CARD}_m]$ are called easily m-countable, as defined in [18]. Languages in $P[(2^m - 1)\text{-}\mathrm{SIZE}_m]$ are called m-approximable or m-membership comparable, as defined in [5]. Languages in $P[k\text{-}\mathrm{SIZE}_m]$ are called (k,m)-verbose. Languages in $P[\mathrm{SMC}_m]$ are called strongly m-membership comparable, as defined in [20].

We close this section with Figure 1 that shows the inclusion structure of all 2-families in normal form, and hence the inclusion structure of all partial information classes produced by 2-families.

3 Self-reducible Easily 2-Countable Languages Are in P

Beigel, Kummer, and Stephan ask in [8] whether every self-reducible easily m-countable language is in P. In this section, we show that this holds at least for the case of easily 2-countable languages:

Theorem 1. *Every self-reducible language in* $P[2\text{-}\textsc{Card}_2]$ *is in* P.

For the proof of Theorem 1 we use a result from [8] on simplification of reductions:

Fact 2. *Every language A reducible to an easily m-countable language B is tt-reducible to B.*

Because in the proof of Fact 2 the truth-table queries are a subset of the queries in the Turing reduction tree, that result also tells us that self-reducible languages in $P[2\text{-}\textsc{Card}_2]$ are tt-self-reducible. Furthermore, we apply the following two lemmas. The first one breaks tt-self-reducibility down to nor-nand-self-reducibility, whereas the second one shows membership in P.

Lemma 1. *Every tt-self-reducible language in* $P[2\text{-}\textsc{Card}_2]$ *is nor-nand-self-reducible.*

Proof. Let A be tt-self-reducible via M, and let $A \in P[2\text{-}\textsc{Card}_2]$ via $f \in FP$. Without loss of generality, $f(x,y) \in \{\text{bottom}_2, \text{equ}_2, \text{top}_2\}$ for all words x, y. The following algorithm decides A in a nor-nand-self-reducing fashion.

On input x, let Q be the set of queries of $M(x)$. If there is a word $q \in Q$ with $f(x,q) = \text{equ}_2$, then $\chi_A(x) = \chi_A(q)$. Replace Q with the set of queries of $M(q)$. Iterate the above process until there is no word $q \in Q$ with $f(x,q) = \text{equ}_2$, or until M accepts or rejects (in which case we accept or reject x).

Assume we end up with a set Q with $f(x,q) \in \{\text{bottom}_2, \text{top}_2\}$ for all $q \in Q$. Then the following procedure computes sets IN, OUT and X such that IN $\cup$ OUT $\cup X = Q$, IN $\subseteq A$, OUT $\subseteq \overline{A}$ and either for all $q \in X$, $f(x,q) = \text{bottom}_2$ or for all $q \in X$, $f(x,q) = \text{top}_2$. At the beginning, set IN = OUT = $\emptyset$ and $X = Q$. While there are two words $q_1, q_2 \in X$ such that $f(x,q_1) = \text{bottom}_2$ and $f(x,q_2) = \text{top}_2$, compute $D = f(q_1,q_2)$. If $D = \text{bottom}_2$, we have $q_1 \notin A$ (otherwise $x \notin A$ and $q_2 \notin A$ which contradicts $f(x,q_2) = \text{top}_2$) and we move q_1 from X into OUT. If $D = \text{top}_2$, we have $q_2 \in A$ (otherwise $x \in A$ and $q_1 \in A$ which contradicts $f(x,q_1) = \text{bottom}_2$) and we move q_2 from X into IN. If $D = \text{equ}_2$, then $x \in A$ if and only if $q_1 \notin A$. In this case, the algorithm stops and $\chi_A(x)$ is determined by querying q_1.

Now we have to deal with the two cases

- $f(x,q) = \text{bottom}_2$ for all $q \in X$, or
- $f(x,q) = \text{top}_2$ for all $q \in X$.

We only treat Case 1. Case 2 is analogous, but with the nand-function instead of the nor-function. Suppose $f(x,q) = \text{bottom}_2$ for all $q \in X$. If at least one query from X is in A, then $x \notin A$. Compute whether M accepts x in case all queries in X are not in A. If *no*, then M rejects x for all oracle answers. We can reject x without using oracle queries at all. If *yes*, then x is in A iff $\neg \bigvee_{q \in X} q \in A$. This means we can query the words in X and evaluate the answers with the nor-function. All computations can be done in polynomial time. □

For membership in P, it suffices to show, by Fact 1, that every nor-nand-self-reducible language in P[2-CARD$_2$] is 1-tt-self-reducible:

Lemma 2. *Every nor-nand-self-reducible language in* P[2-CARD$_2$] *is 1-tt-self-reducible.*

Proof. Let A be nor-nand-self-reducible via generator $g \in$ FP and evaluator $\alpha \in$ FP. Let $A \in$ P[2-CARD$_2$] via $f \in$ FP. Without loss of generality, $f(x,y) \in$ {bottom$_2$, equ$_2$, top$_2$} for all words x, y. The following algorithm decides $x \in A$ in polynomial time with at most one query to A which is shorter than x.

On input x, let Q be the set of queries of $g(x)$, and let α_x be the boolean evaluation function for x. We consider the case that α_x is a nor-function, i.e. $x \in A$ if and only if for all $q \in Q$: $q \notin A$; the case that α_x is a nand-function is analogous. For each $q \in Q$, do the following: compute $g(q) = \langle q_1, \ldots, q_k \rangle$ and determine, whether α_q is a nand- or a nor-function. Consider the following two cases:

Case 1. α_q is a nand-function, i.e. $q \in A$ if and only if $\exists i : q_i \notin A$. Then, for each $i \in \{1, \ldots, k\}$ we can decide either $x \in A$ or $q_i \in A$ (clearly, if we know $\chi_A(q_i)$ for all i, we know $\chi_A(q)$, too). For all $i = 1, \ldots, k$, compute $D = f(x, q_i)$ and consider the following cases. If $D =$ bottom$_2$, then $x \notin A$, since otherwise $q_i \notin A \Rightarrow q \in A \Rightarrow x \notin A$ (because α_x is a nor- and α_q is a nand-function) which contradicts $x \in A$. If $D =$ equ$_2$, then $x \in A \Leftrightarrow q_i \in A$ and we query q_i in order to compute $\chi_A(x)$. If $D =$ top$_2$, then $q_i \in A$, since otherwise $q \in A \Rightarrow x \notin A \Rightarrow q_i \in A$ which contradicts $q_i \notin A$.

Case 2. α_q is a nor-function, i.e. $q \in A$ if and only if for all i: $q_i \notin A$. For all $i = 1, \ldots, k$, compute $D = f(x, q_i)$ and consider the following cases until we can compute $\chi_A(x)$ or $\chi_A(q)$. If $D =$ equ$_2$, then $x \in A \Leftrightarrow q_i \in A$ and we query q_i in order to compute $\chi_A(x)$. If $D =$ top$_2$, then $q \notin A$ since otherwise $x \notin A \Rightarrow q_i \in A \Rightarrow q \notin A$ which contradicts $q \in A$. In the worst case we have $D =$ bottom$_2$ for all i. But then $x \notin A$, since otherwise $q_i \notin A$ for all $i \in \{1, \ldots, k\}$ (because $f(x, q_i) =$ bottom$_2$ for all such i) and therefore $q \in A$ which implies $x \notin A$; a contradiction to $x \in A$.

We have seen that in all cases either we can decide membership in A for all words $q \in Q$, and hence for x, or we find a query q_i such that $x \in A \Leftrightarrow q_i \in A$. This means that A is 1-tt-self-reducible. Moreover, it is easy to see that the whole procedure runs in time polynomial in $|x|$. □

We thus have shown that every self-reducible easily 2-countable language is contained in P.

4 Simplifying Turing to Truth-Table Self-reductions for $(2m-1, m)$-Verbose Languages

In order to show that self-reducible languages in a certain partial information class are contained in P, it is convenient to show first that they are tt-self-reducible. We have already used such a result from [8] in the case of easily countable languages, see Fact 2. We now show a similar result for $P[(2m-1)\text{-SIZE}_m]$ for all $m \in \mathbb{N}$. This is a generalization of Theorem 4 in [9] on p-selective languages.

Theorem 2. *For all $m \in \mathbb{N}$, if $A \in P[(2m-1)\text{-SIZE}_m]$ is self-reducible, then A is tt-self-reducible.*

Proof. Let A be self-reducible via M and $A \in P[(2m-1)\text{-SIZE}_m]$ via $f \in FP$. On input $\langle x_1, \ldots, x_m \rangle$, f computes a set $D \subseteq \{0,1\}^m$ with $|D| \leq 2m-1$ and $\chi_A(x_1, \ldots, x_m) \in D$.

The query tree $T_M(x)$ reflects the computations of M on input x for all possible answers to oracle queries. Each inner node of $T_M(x)$ is labeled with a query q and has two children, one for the oracle answer 'yes' and one for 'no'. Leaf nodes are labeled with '$\chi_A(x) = 1$' if the path of queries and answers from the root to that leaf node describes an accepting computation of M on input x, otherwise they are labeled '$\chi_A(x) = 0$'. We process $T_M(x)$ with a breadth-first extend-and-prune algorithm. Pruning means removing a subtree and replacing it by a leaf node labeled '$\chi_A(x) = 1$' or '$\chi_A(x) = 0$'. Extending means taking in children of nodes labeled with queries. We thus always keep a (pruned) subtree of $T_M(x)$ satisfying the following invariants:

1. On every path in T there are at most $m-2$ nodes having two inner nodes as children.
2. Leaf nodes are labeled '$\chi_A(x) = 1$' if and only if M accepts x for all oracles consistent with the query-answer path leading to that node.

Initially, let T be the full query tree up to depth $m-1$. Suppose T has been extended such that there is a path on which $m-1$ nodes, labeled with $q_1, \ldots, q_{m-1}$, have two inner nodes as children. Let $a = a_1 \ldots a_{m-2}$ be the bitstring encoding the answers to queries along the path: $a_i = 1$ if and only if the answer to query q_i is 'yes'. Compute $D := f(x, q_1, \ldots, q_{m-1})$. Split D into two pools by defining $D_c := \{b \in D \mid b[1] = c\}$ for $c \in \{0, 1\}$. We have $|D_0| + |D_1| = |D| \leq 2m-1$. Hence, for at least one c we have $|D_c| \leq m-1$. Fix such a value for c. Now there must be an $i \in \{1, \ldots, m-1\}$ such that $ca[1, \ldots, i-1]\,0$ is not a prefix of a string in D_c or $ca[1, \ldots, i-1]\,1$ is not a prefix of a string in D_c. Say, $ca[1, \ldots, i-1]\,0$ is not a prefix of a string in D_c. This means that if $\chi_A(x) = c$, then 'no' is not the answer to query q_i. This means: If 'no' is the answer to query q_i, then $\chi_A(x) = 1 - c$. Prune the tree at the 'no'-edge leaving the node labeled q_i. Replace the subtree at this edge by a leaf node labeled '$\chi_A(x) = 1 - c$'.

When $T_m(x)$ has been processed up to some depth, only a constant number of nodes of this depth with two inner nodes as children remain in T. Therefore the whole processing can be completed in polynomial time and in the end only polynomially many queries are left in T. These are the queries that now are given in parallel to the oracle. If the path in T determined by the oracle answers leads to a node labeled '$\chi_A(x) = 1$', then accept x, else reject. We conclude that A is tt-self-reducible. □

5 On Strongly 2-Membership Comparable Languages

In this section, we consider 2-families that strictly include SEL_2, in particular the families $\text{SMC}_2 \cap \text{CoSMC}_2$ and 3-SIZE_2. It turns out that self-reducible languages in the partial information class produced by the first family are in P. It is unlikely that this result can be extended to larger 2-families. However, we can at least show that dtt-self-reducible languages in $\text{P}[3\text{-SIZE}_2]$ are in UP.

5.1 Self-reducibility and P

By Theorem 2, every self-reducible language in $\text{P}[\text{SMC}_2 \cap \text{CoSMC}_2]$ is tt-self-reducible. The proof that every tt-self-reducible language in $\text{P}[\text{SMC}_2 \cap \text{CoSMC}_2]$ is 1-tt-self-reducible, and hence in P, is an easy generalization of Theorem 3 in [9], together with the following fact (Lemma 3.24 from [24]):

Fact 3. *For every language* $A \in \text{P}[\text{SMC}_2 \cap \text{CoSMC}_2]$ *we can compute on input* $\langle x_1, \ldots, x_m \rangle$ *in polynomial time a partition of* $X = \{x_1, \ldots, x_m\}$ *into disjoint sets of one of the following two types:*

1. $X = \text{IN} \cup \text{OUT} \cup S_1 \cup \ldots \cup S_r$ *with* $r \leq m$ *such that:*
   - $\text{IN} \subseteq A$ *and* $\text{OUT} \subseteq \overline{A}$,
   - *For* $1 \leq i \leq r$, $S_i \neq \emptyset$ *and* $S_i \subseteq A$ *or* $S_i \subseteq \overline{A}$.
   - *For* $1 \leq i < j \leq r$, $x \in S_i$ *and* $y \in S_j$: $\chi_A(x) \leq \chi_A(y)$.
2. $X = \text{IN} \cup \text{OUT} \cup X_1 \cup X_2$ *such that:*
   - $\text{IN} \subseteq A$ *and* $\text{OUT} \subseteq \overline{A}$,
   - *For* $i \in \{1, 2\}$: $X_i \neq \emptyset$ *and* $X_i \subseteq A$ *or* $X_i \subseteq \overline{A}$.
   - *For* $x \in X_1$ *and* $y \in X_2$: $\chi_A(x) \neq \chi_A(y)$.

Given a tt-self-reducible language in $\text{P}[\text{SMC}_2 \cap \text{CoSMC}_2]$, we can compute a partition of the words generated in the tt-self-reduction according to Fact 3. The case that the partition is of the first type is handled in the proof of Theorem 3 in [9]. If it is of the second type and the input word x is contained in X_i, then membership of x can be determined by querying some word in X_j, $j \neq i$. We have thus established the following theorem.

Theorem 3. *Every self-reducible language in* $\text{P}[\text{SMC}_2 \cap \text{CoSMC}_2]$ *is in* P. □

5.2 Disjunctive Truth-Table Self-reducibility and UP

We do not know whether every self-reducible language in $\text{P}[\text{SMC}_2]$ is in P. Indeed, it seems that results in this respect are hard to obtain: In the proof of Theorem 7.1 in [8], an oracle is constructed such that relative to it there exists a dtt-self-reducible language in $\text{P}[\text{SMC}_2]$ that is not in P.

To get some results for families above $\text{SMC}_2 \cap \text{CoSMC}_2$ we have to be satisfied to show inclusion in a class larger than P. The following result is of that kind:

Theorem 4. *Every dtt-self-reducible language in* P[3-SIZE$_2$] *is in* UP.

Proof. Let A be dtt-self-reducible via M, and let $A \in$ P[3-SIZE$_2$] via $f \in$ FP. We construct a polynomial-time unambiguous nondeterministic Turing machine N which accepts A.

The dtt-self-reducing tree of M on input x is created by iteratively computing for each query q the queries that M would ask on input q. This tree has polynomial height, but in general contains exponentially many nodes. On an input x we walk a path in the dtt self-reducing tree of M on input x. We use f to decide whether the actual query is in A (the path then ends at this point), or to choose the next node to visit. We accept x if and only if the last node on the path is in A. Note that $x \in A$ if and only if there exists such a path. Assume we already visited $x_1, \ldots, x_l$. If $M(x_l)$ queries $q_1, \ldots, q_s$, we compute $f(q_i, q_j)$ for all $1 \leq i, j \leq s$, $i \neq j$. If for such a pair $f(q_i, q_j) = $ sel$_2$, we can remove q_i from the list of queries. Let $q_{i_1}, \ldots, q_{i_r}$ be the queries remaining after these removals. If $r = 1$, we set $x_{l+1} := q_{i_1}$. If $r > 1$, and $f(q_{i_j}, q_{i_k}) = $ top$_2$ for some pair, then at least one query is in A, and hence x_l is in A. It remains the case that $f(q_{i_j}, q_{i_k}) = $ bottom$_2$ for all $1 \leq j, k \leq r$, $j \neq k$. This means: At most one of these queries is in A. We choose a j nondeterministically and set $x_{l+1} := q_{i_j}$. Clearly, N has on input x at most one accepting computation. □

6 Conclusion

We have improved on previous results on combining partial information and self-reducibility. The improvement from SEL$_2$ to SMC$_2 \cap$ CoSMC$_2$, Theorem 3, may be not so surprising since these two families share many properties and are even linked by a reducibility (which needs access to an NP-oracle, see [4]). But the result on easily 2-countable languages, Theorem 1, is, in our view, an important step forward. It partially solves the conjecture of Beigel, Kummer, and Stephan [8]. Maybe, in the future similar proof ideas may help to answer their question "Are all easily countable self-reducible languages in P?" positively.

An important tool in proofs, but also interesting in their own right, are theorems on the simplification of reductions or self-reductions like Theorem 2. Can it be extended? Our proof is by an iterative query tree pruning procedure which leaves only constantly many branches in the tree. We think that our proof exploits this type of local pruning technique optimally. On the other hand, the simplification result from [8] for easily countable languages is also proved by a pruning technique involving trees of bounded rank. In their case the number of branches is not bounded by a constant. But their technique does not seem to apply to $(2m - 1, m)$-verbose languages. Is there a proof technique powerful enough to subsume both results?

The last result on membership in UP, Theorem 4, suggests that one should not only look for conditions on the type of partial information and type of self-reducibility that yield membership in P. It would be nice if someone came up with characterizations of several classes between P and PSPACE by such combinations. Even for UP such a characterization still has to be found.

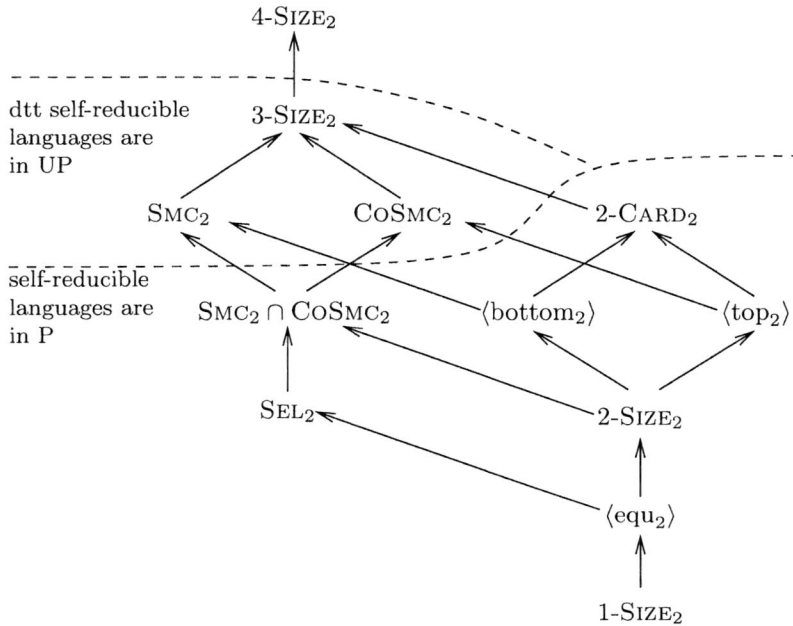

Fig. 2. This figure shows all 2-families in normal form as in Figure 1. Dashed lines separate families $\mathcal{D}$ for which (dtt-)self-reducible languages in P[$\mathcal{D}$] are in P (UP) from those ones where this is not known.

References

1. M. Agrawal and V. Arvind. Polynomial time truth-table reductions to P-selective sets. In *Proc. 9th Structure in Complexity Theory*, 1994.
2. A. Amir, R. Beigel, and W. Gasarch. Some connections between bounded query classes and non-uniform complexity. In *Proc. 5th Struct. in Complexity Th.*, 1990.
3. A. Amir and W. Gasarch. Polynomial terse sets. *Inf. and Computation*, 77, 1988.
4. S. Bab and A. Nickelsen. One query reducibilities between partial information classes. In *Proc. MFCS*, volume 31543 of *LNCS*, pages 404–415. Springer, 2004.
5. R. Beigel. *Query-Limited Reducibilities*. PhD thesis, Stanford University, 1987.
6. R. Beigel, L. Fortnow, and A. Pavan. Membership comparable and p-selective sets. Technical Report 2002-006N, NEC Research Institute, 2002.
7. R. Beigel, M. Kummer, and F. Stephan. Quantifying the amount of verboseness. In *Proc. Logical Found. of Comput. Sci.*, volume 620 of *LNCS*. Springer, 1992.
8. R. Beigel, M. Kummer, and F. Stephan. Approximable sets. *Inf. and Computation*, 120(2), 1995.
9. Buhrman, van Helden, and Torenvliet. P-selective self-reducible sets: A new characterization of P. In *Conference on Computational Complexity*, volume 8, 1993.
10. H. Buhrman, L. Torenvliet, and P. van Emde Boas. Twenty questions to a P-selector. *Information Processing Letters*, 4(4):201–204, 1993.
11. D.-Z. Du and K.I.Ko. *Theory of Computational Complexity*. Wiley & Sons, 2000.

12. W. Gasarch. Bounded queries in recursion theory: A survey. In *Proc. 6th Structure in Complexity Theory*, 1991.
13. J. Goldsmith, D. Joseph, and P. Young. Self-reducible, P-selective, near-testable, and P-cheatable sets: The effect of internal structure on the complexity of a set. In *2nd Structure in Complexity Theory*. IEEE, 1987.
14. L. Hemaspaandra, A. Hoene, and M. Ogihara. Reducibility classes of p-selective sets. *Theoretical Computer Science*, 155:447–457, 1996.
15. L. Hemaspaandra, Z. Jiang, J. Rothe, and O. Watanabe. Polynomial-time multi-selectivity. *J. of Universal Comput. Sci.*, 3(3), 1997.
16. L. Hemaspaandra and T. Torenvliet. *Theory of semi-feasible Alg.* Springer, 2002.
17. M. Hinrichs and G. Wechsung. Time bounded frequency computations. In *Proc. 12th Conf. on Computational Complexity*, 1997.
18. A. Hoene and A. Nickelsen. Counting, selecting, and sorting by query-bounded machines. In *Proc. STACS 93*, volume 665 of *LNCS*. Springer, 1993.
19. K. Ko. On self-reducibility and weak p-selectivity. *Journal of Computer and System Sciences*, 26:209–221, 1983.
20. J. Köbler. On the structure of low sets. In *Proc. 10th Structure in Complexity Theory*, pages 246–261. IEEE Computer Society Press, 1995.
21. A. R. Meyer and M. S. Paterson. With what frequency are apparently intractabel problems difficult? Technical Report MIT/LCS/TM-126, MIT, Cambridge, 1979.
22. A. Nickelsen. On polynomially $\mathcal{D}$-verbose sets. In *Proc. STACS 97*, volume 1200 of *LNCS*, pages 307–318. Springer, 1997.
23. A. Nickelsen. Partial information and special case algorithms. In *Proc. MFCS 01*, pages 573–584. Springer LNCS 2136, 2001.
24. A. Nickelsen. *Polynomial Time Partial Information Classes*. W&T Verlag, 2001. Dissertation, TU Berlin, 1999, also available at www.tal.cs.tu-berlin.de/nickelsen/.
25. A. Nickelsen and T. Tantau. Closure of polynomial time partial information classes under polynomial time reductions. In *Proc. FCT 01*, volume 2138 of *LNCS*, pages 299–310. Springer, 2001.
26. A. Nickelsen and T. Tantau. Partial information classes. Complexity Theory Column,*SIGACT News*, 34, 2003.
27. M. Ogihara. Polynomial-time membership comparable sets. *SIAM Journal on Computing*, 24(5):1068–1081, 1995.
28. A. Selman. P-selective sets, tally languages and the behaviour of polynomial time reducibilities on NP. *Math. Systems Theory*, 13:55–65, 1979.
29. D. Sivakumar. On membership comparable sets. *Journal of Computer and System Sciences*, 59(2):270–280, 1999.
30. Seinosuke Toda. On polynomial-time truth-table reducibility of intractable sets to p-selective sets. *Mathematical Systems Theory*, 24:69–82, 1991.

Complexity Bounds for Regular Games

(*Extended Abstract*)

Paul Hunter and Anuj Dawar

University of Cambridge Computer Laboratory, Cambridge CB3 0FD, UK
{paul.hunter, anuj.dawar}@cl.cam.ac.uk

Abstract. We consider the complexity of infinite games played on finite graphs. We establish a framework in which the expressiveness and succinctness of different types of winning conditions can be compared. We show that the problem of deciding the winner in Muller games is PSPACE-complete. This is then used to establish PSPACE-completeness for Emerson-Lei games and for games described by Zielonka DAGs. Adaptations of the proof show PSPACE-completeness for the emptiness problem for Muller automata as well as the model-checking problem for such automata on regular trees. We also show co-NP-completeness for two classes of union-closed games: games specified by a basis and superset Muller games.

1 Introduction

Recent years have seen an increasing use of two-player infinite games as a means of modelling reactive and concurrent systems. Games have emerged as essential tools for the analysis, synthesis and verification of such systems with a close connection to logic and to automata on infinite objects. The general framework consists of games played on finite or infinite graphs (whose vertices represent a state space) with players moving a token along the edges of the graph. The (possibly infinite) sequence of vertices that is visited constitutes a *play* of the game with the winner of a play being defined by some predetermined condition.

When we are concerned with algorithmic issues surrounding such games, we need to restrict ourselves to games that can be described in a finite fashion. This does not mean that the graph on which the game is played is necessarily finite as it is possible to finitely describe an infinite graph. Nor does having a finite game graph by itself guarantee that the game can be finitely described. Even with two nodes in a graph, the number of distinct plays can be uncountable and there are more possible winning conditions than one could possibly describe. In this paper, we are concerned with *regular* games played on finite graphs. These are games in which the graph is finite and the winner of a play is determined by the set of vertices of the graph that are visited infinitely often in the play (see Section 2 for formal definitions). This category of games is wide enough to include most kinds of game winning conditions that are considered in the literature, including Muller, Streett, Rabin, Büchi and parity games.

Specifically, we are concerned with the problem of deciding, given a game and a starting position, which player has a strategy for winning the game. It is well-known that regular games are determined, i.e. one of the players has a winning strategy and the problem of determining which player has such a strategy is decidable [1]. We are interested in the computational complexity of deciding the winner. Since the complexity is measured as a function of the length of the description, this in turn depends on how exactly the game is described. In general, a regular game is defined by a graph (V, E), where $E \subseteq V \times V$, and a winning condition $\mathcal{F} \subseteq \mathcal{P}(V)$ consisting of a set of subsets of V. One could specify $\mathcal{F}$ by listing all its elements explicitly (we call this an *explicit* presentation) but one could also adopt a formalism which allows one to specify $\mathcal{F}$ more succinctly. In the latter case, there are two possibilities. Either the formalism is general enough that any winning condition $\mathcal{F} \subseteq \mathcal{P}(V)$ can be expressed in it or there is only a restricted class of winning conditions that can be expressed. Muller games are an example of the first case while Rabin, Streett, Büchi and parity games are all examples of the second case. Since the number of possible winning conditions $\mathcal{F}$ is $2^{2^{|V|}}$, if the formalism is general enough to describe any regular winning condition then the description of the game must, in general, be exponential in the size of the game graph. However, some presentations may still be more succinct than the explicit presentation. On the other hand, if the formalism is restricted in its expressive power, it may be possible that the length of a description of the game is always bounded by a polynomial in the size of the graph. We investigate these two dimensions of variation in the description of games – the expressive power of the formalism on the one hand and its succinctness on the other – in the results we establish.

As an example, consider a min-parity winning condition. Here, the winning condition is specified by a priority function $\Omega : V \to \{0, \dots, d\}$. This is treated as a specification of the set $\mathcal{F}$ consisting of those sets $I \subseteq V$ such that the smallest number in $\Omega(I)$ is even. It is clear that the description of Ω is bounded in length by a polynomial (indeed, linear) function of $|V|$. It is also clear that not every set $\mathcal{F} \subseteq \mathcal{P}(V)$ can be described in this way. On the other hand, there are such sets $\mathcal{F}$ for which the description using a priority function is exponentially more succinct than an explicit presentation.

The exact computational complexity of deciding the winner of a parity game is a central open question in the theory of regular games. It is known to be in NP ∩ CO-NP [3] and conjectured by some to be in P. However, lower bounds on the complexity of any class of games are hard to come by. It is known that deciding games specified by the Rabin condition is NP-complete [3] and for the Streett condition the problem is CO-NP-complete. Both of these are condition types that are restricted in that they cannot express all regular games. No lower bounds are previously known for formalisms that are expressive enough to specify all regular games, though algorithms for such games have been studied which establish, for instance, that the games are decidable in PSPACE.

We consider five general-purpose formalisms. Our main result is that the problem of deciding the winner of a Muller game is PSPACE-complete. We then use this to establish PSPACE-completeness for two further general-purpose representations: Emerson-Lei games, where the winning condition is presented as a Boolean formula over the vertices of the graph; and the case where the winning condition is represented as a Zielonka DAG. The latter is a data structure (defined in Section 2) based on the Zielonka trees of [12]. We define a notion of polynomial-time *translatability* between formalisms. A formalism is translatable into another if the representation of a game in the first can be transformed into a representation *of the same game* in the second. This is stronger than polynomial-time reducibility of the corresponding decision problems. We show that Muller games are translatable to Zielonka DAGs which are in turn translatable to Emerson-Lei games, but the reverse translations do not hold. Our hardness result for Muller games is based on the presentation of these games which includes a colouring of the vertices. This allows for more succinct descriptions than the explicit presentation of sets. Indeed, we show that there is a translation in one direction but not the other. The complexity of deciding the winner of the games where the sets are explicitly presented remains an open question. An adaptation of the PSPACE-completeness result shows that two important problems related to Muller *automata* are also PSPACE-complete. These are the emptiness problem and the model-checking problem on regular trees. As an aside, we also show that the PSPACE-completeness result for Muller games holds even when the game arenas are restricted to small tree-width.

We also consider the restriction to games where the winning condition $\mathcal{F}$ is closed under unions. The question of lower-bounds for *union-closed* games was posed by Khoussainov (see [6]). It is known that deciding whether or not Player 0 wins such a game is decidable in CO-NP. The precise formalism used to describe the set $\mathcal{F}$ is not relevant to this upper bound as the non-deterministic algorithm runs in time polynomial in the size of the game graph. We show, for two particular formalisms that the problem of deciding the winner is CO-NP-complete. One such formalism is what we call *Basis* games while the other is the *superset Muller* games defined in [7]. The former is expressive enough to define all union-closed games while the latter is restricted to expressing sets $\mathcal{F}$ that are upward-closed. Both are, as we show, more succinct than an explicit representation of $\mathcal{F}$.

2 Background and Definitions

In this section, we present the basic definitions of games as well as particular winning conditions. Many of the definitions presented here are standard. Where this is the case, we follow terminology and notation from [5].

An *arena* $\mathcal{A}$ is a directed graph on a set of vertices V which is partitioned into two sets V_0 and V_1, i.e. $\mathcal{A} = (V, E)$ where $V = V_0 \cup V_1$, $E \subseteq V \times V$ and V_0 and V_1 are disjoint. For the results we establish in this paper, there is no loss of generality in assuming that the graph is bipartite in the sense that $E \subseteq (V_0 \times V_1) \cup (V_1 \times V_0)$ and that for each $v \in V$, there is a $v' \in V$ such that $(v, v') \in E$. For instance, there is an easy transformation that maps any game to a bipartite game by inserting a new V_0 (resp. V_1) vertex in every edge that connects two V_1 (resp. V_0) vertices. This transformation does not change the existence of winning strategies for either player from any of the original vertices. Thus, wherever it is convenient, we will assume that the arena satisfies the above assumptions.

A *game* G is an arena $\mathcal{A}$ together with a winning set of sequences Win $\subseteq V^\omega$. Informally, we think of the game as played between two players, Player 0 and Player 1, with a token that sits on a vertex v in V. If $v \in V_0$, it is Player 0's turn to move and she[1] moves it to some v' such that there is an edge (v, v') in E and similarly for Player 1 when the token is on a vertex in V_1. The infinite sequence of such moves determines a *play* π which is the sequence in V^ω of vertices visited. We say Player 0 wins the play if $\pi \in$ Win and Player 1 wins otherwise.

A *strategy* (for Player i) is a function f from V^*V_i to V with $f(v_0 v_1 \ldots v_n) \in v_n E$. Given a sequence of vertices visited, ending with a vertex in V_i, a strategy gives the vertex that Player i should then play to. A play is *consistent* with a strategy if every move made by Player i is determined by the strategy, and a strategy is *winning* if every play consistent with it is winning for Player i. If a strategy f has the property that for some fixed m, $f(w) = f(w')$ if w and w' agree on their last m letters, then we say that the strategy requires *finite-memory* (of size $m - 1$). If $m = 1$, we say the strategy is *memoryless*.

A game (V, E, Win) is *regular* if there is a set $\mathcal{F} \subseteq \mathscr{P}(V)$ such that for any $\pi \in V^\omega$, $\pi \in$ Win if, and only if, the set $\{v : v$ occurs infinitely often in $\pi\}$ is in $\mathcal{F}$. In the remainder of the paper, we are concerned with games that are finite (i.e. V is a finite set) and regular. Regular games are known to be determined, that is, for each game and each initial vertex v, either Player 0 or Player 1 has a winning strategy.

We say that a regular game $(V, E, \mathcal{F})$ is *union-closed* if whenever $I, J \in \mathcal{F}$, then $I \cup J \in \mathcal{F}$. A regular game is *upward-closed* if for any $I \in \mathcal{F}$ and $I \subseteq J$, we have $J \in \mathcal{F}$. Clearly any upward-closed game is also union-closed.

The games used in the literature in the study of logics and automata are generally regular games (though not necessarily finite). In these games, the set $\mathcal{F}$ is often not explicitly given but is specified by means of a *condition*. Different types of condition lead to various different types of games. We do not give a formal definition of a *condition type* but we will define specific instances of such types.

The most straightforward presentation of a regular game $(V, E, \mathcal{F})$ is given by listing all elements of $\mathcal{F}$. We call this an *explicit condition*. Games specified by such a condition type are sometimes called Muller games in the literature, but we reserve that term for the more commonly used presentation in terms of colours given next.

A *Muller condition* on an arena (V, E) is given by a set of colours C, a colouring function $\chi : V \to C$ and a set $\mathcal{C} \subseteq \mathscr{P}(C)$. The set $\mathcal{F}$ specified by such a condition is the set $\{I \subseteq V : \chi(I) \in \mathcal{C}\}$.

---

[1] For ease of reference we use the feminine pronoun for Player 0 and the masculine for Player 1.

An *Emerson-Lei condition* [4] on an arena (V, E) is given by a Boolean formula φ with variables from the set V. The set $\mathcal{F}$ specified is the collection of sets $I \subseteq V$ such that the truth assignment that maps each element of I to true and each element of $V \setminus I$ to false satisfies φ.

In [12], Zielonka introduced a representation of a winning set $\mathcal{F} \subseteq \mathscr{P}(V)$ in terms of a labelled tree, where the labels on the nodes are subsets of V. The *Zielonka tree* of the set $\mathcal{F}$, $\mathcal{Z}_{\mathcal{F},V}$, is defined inductively as:

1. If $V \notin \mathcal{F}$ then $\mathcal{Z}_{\mathcal{F},V} = \mathcal{Z}_{\overline{\mathcal{F}},V}$, where $\overline{\mathcal{F}} = \mathscr{P}(V) \setminus \mathcal{F}$.
2. If $V \in \mathcal{F}$ then the root of $\mathcal{Z}_{\mathcal{F},V}$ is labelled with V. Let $M_0, M_1, \ldots, M_{k-1}$ be the maximal sets in $\overline{\mathcal{F}}$, and let $\mathcal{F}|_{M_i} = \mathcal{F} \cap \mathscr{P}(M_i)$. The children of the root are the subtrees $\mathcal{Z}_{\mathcal{F}|_{M_i}, M_i}$, for $0 \leq i \leq k-1$.

A *Zielonka DAG* is constructed as a Zielonka tree except nodes labelled by the same set are identified, making it a directed acyclic graph. A Zielonka tree (DAG) condition is one which uses a Zielonka tree (respectively, DAG) presentation. Nodes of $\mathcal{Z}_{\mathcal{F},V}$ labelled by elements of $\mathcal{F}$ are called *0-level nodes*, and other nodes are *1-level nodes*. In the sequel, we use terms such as *children* and *leaves* when referring to DAGs as well as trees, where the meaning is clear.

From a more practical perspective, when considering applications of these types of games it may be the case that there are vertices whose appearance in any infinite run is irrelevant. This leads to the definition of a *win-set condition*, which is given by $W \subseteq V$ and $\mathcal{W} \subseteq \mathscr{P}(W)$. The sets described by this condition are $\{I \subseteq V : I \cap W \in \mathcal{W}\}$. Win-set games are the type of games considered by McNaughton in [9] where he presents an algorithm to decide the winner of such games.

The five condition types defined above are general purpose in that any regular game can be specified by any one of the condition types. We now look at some less general types of conditions.

A *basis condition* on an arena (V, E) is given by a set $\mathcal{B} \subseteq \mathscr{P}(V)$. This specifies the collection $\mathcal{F}$ of sets $I \subseteq V$ such that there are $B_1, \ldots, B_n \in \mathcal{B}$ with $I \bigcup_{1 \leq i \leq n} B_i$. It is clear that a regular game can be specified using a Basis condition if, and only if, it is union-closed.

A *superset condition* (also called a superset Muller condition in [7]) is given by a set $\mathcal{M} \subseteq \mathscr{P}(V)$ which specifies the set $\mathcal{F} = \{I \subseteq V : J \subseteq I \text{ for some } J \in \mathcal{M}\}$. Only upward-closed games can be specified in this way.

Our main concern is with the complexity of the following decision problem for a fixed condition type: given a game consisting of a finite arena, a condition of the given type and an initial vertex, which of the two players has a winning strategy? We often refer to this as the problem of deciding the winner of a game. This problem has been investigated for condition types other than the ones considered here. For example, in [3] it was shown that deciding games with a winning condition expressed in Rabin form is NP-complete; and the complexity of deciding Parity games is a question that has been the focus of intensive research. However, lower bound proofs for any games are hard to come by.

3 Translations

We begin by considering the five ways we have defined of specifying a winning condition that are *general purpose*, i.e. expressive enough to describe any regular game. These are the explicit presentation, the win-set condition, the Muller condition, the Zielonka DAG and the Emerson-Lei condition. We show that this list is strictly increasing in order of succinctness. That is, any game specified using a condition of one of these types can also be specified using a type later in the list with only a polynomial increase in size. However, for each type, there are specifications of games for which any description of a type earlier in the list is necessarily exponentially longer. We formalise the notion of succinctness through the following definition. Note that this definition is somewhat informal as we have not given a formal definition of a "condition type". It suffices for our present purposes if we take A and B in this definition to range over the types defined in the previous section.

Definition 3.1. *Given two condition types A and B, we say that A is polynomially translatable to B if there is an algorithm, running in polynomial time which, given a game with condition of type A produces a condition of type B which describes the same game.*

As we are only interested in polynomial translations, we simply say A is *translatable* to B to mean that it is polynomially translatable. Clearly, if condition type A is translatable to B then the problem of deciding the winner for games of type A is reducible in polynomial time to the corresponding problem for games of type B.

If condition type A is not translatable to B this may be for one of three reasons. Either A is more expressive than B in that there are sets $\mathcal{F}$ that can be expressed using A but not B; or there are some sets for which the representation of type A is necessarily more succinct; or the translation while not size increasing can not be computed in polynomial time. We are primarily interested in the second situation. Formally, we say that A is *more succinct* than B if B is translatable to A but A is not translatable to B.

It is straightforward to show that win-set conditions are more succinct than explicit presentations. To translate an explicitly presented game $(V, E, \mathcal{F})$ to a win-set condition, simply take $W = V$ and $\mathcal{W} = \mathcal{F}$. To show that win-set conditions are not translatable to explicit presentations, consider a game where $W = \emptyset$ and $\mathcal{W} = \{\emptyset\}$. The set $\mathcal{F}$ described consists of all subsets of V and an explicit presentation must be exponential in length.

The next three theorems show that Emerson-Lei games are more succinct than Zielonka DAG games, which are in turn more succinct than Muller games, which are more succinct than win-set games. In Section 5 we also show that basis and superset games are more succinct than explicit presentations of union-closed and upward-closed games respectively.

Theorem 3.2. *The Muller condition type is more succinct than the win-set condition type.*

Proof. Given a win-set game $(V, E, W, \mathcal{W})$, we construct a Muller condition describing the same set of subsets as $(W, \mathcal{W})$. For the set of colours we use $C = W \cup \{c\}$, where c is distinct from any element of W. The colouring function $\chi : V \to C$ is then defined as:

- $\chi(w) = w$ for $w \in W$,
- $\chi(v) = c$ for $v \notin W$.

The family $\mathcal{C}$ of subsets of C is the set $\{X, X \cup \{c\} : X \in \mathcal{W}\}$. For $I \subseteq V$, if $I \subseteq W$, then $\chi(I) = I$ otherwise $\chi(I) = \{c\} \cup I$. Either way, $I \cap W$ is in $\mathcal{W}$ if and only if $\chi(I) \in \mathcal{C}$.

To show that there is no translation in the other direction, consider a Muller game on (V, E), where half of V, V_r, is coloured red, the other half coloured blue, and the family of sets of colours is $\mathcal{C} = \{\{\text{red}\}\}$. The family $\mathcal{F}$ described by this condition consists of the $2^{|V|/2} - 1$ non-empty subsets of V_r. Now consider trying to describe this family using a win-set condition. In general, if $\mathcal{G}$ is the family of subsets of V described by the win-set condition $(W, \mathcal{W})$, then for any $v \notin W$ and $X \subseteq V$ we have $\{v\} \cup X \in \mathcal{G} \Leftrightarrow X \in \mathcal{G}$. Observe that in our game no vertex has this latter property (if $v \in V_r$, then $\{v\} \in \mathcal{F}$, but $\emptyset \notin \mathcal{F}$; and if $v \notin V_r$ then $\{v\} \cup V_r \notin \mathcal{F}$, but $V_r \in \mathcal{F}$). Thus our win-set must be V, and $\mathcal{W}$ is the explicit listing of the $2^{|V|/2} - 1$ subsets of V_r. Thus $(W, \mathcal{W})$ cannot be produced in polynomial time.

The proofs of the following two theorems are omitted due to lack of space.

Theorem 3.3. *The Zielonka DAG condition type is more succinct than the Muller condition type.*

Theorem 3.4. *The Emerson-Lei condition type is more succinct than the Zielonka DAG condition type.*

4 PSPACE-Completeness

In this section we establish the complexity of deciding the winner for the four main condition types considered in the previous section. McNaughton [9], and later Nerode, Remmel and Yakhnis [10] describe an algorithm for deciding win-set games. An analysis of this algorithm shows it requires space $O(|V|^2)$. Moreover, the algorithm is easily adapted to the case where the winning condition is presented explicitly, or as a Muller condition, a Zielonka DAG or an Emerson-Lei condition without significant increase in the space requirements. Thus, each of these classes of games is decidable in PSPACE. We now show corresponding lower bounds. By the results of the previous section, it suffices to establish the hardness result for the win-set condition type.

Theorem 4.1. *Deciding win-set games is* PSPACE-*complete.*

Proof. (sketch) By the above comments, we only need to show PSPACE-hardness. For this, we reduce the problem of QSAT (satisfiability of a quantified boolean formula [QBF]) to the problem of deciding the winner of a win-set game.

We assume, without loss of generality that we are given a QBF,

$$\Phi = Q_{k-1} x_{k-1} \ldots \forall x_1 \exists x_0 \varphi$$

in which quantifiers are strictly alternating and φ is in disjunctive normal form with 3 literals per clause. We then define a win-set game $\mathcal{G}_\Phi$ as follows:

- $V_0 = \{\varphi\} \cup \{x, \neg x\ :\ \text{for all variables } x\}$
- $V_1 = \{C_0, \ldots, C_{m-1}\}$, the set of clauses in φ.
- E is given by:
  - $(\varphi, C_j) \in E$ for $0 \leq j < m$;
  - If $C_j = (l_0 \wedge l_1 \wedge l_2)$, then $(C_j, l_0), (C_j, l_1), (C_j, l_2) \in E$;
  - $(x_i, x_{i-1}), (x_i, \neg x_{i-1}) \in E$ for $0 < i < k$;
  - $(\neg x_i, x_{i-1}), (\neg x_i, \neg x_{i-1}) \in E$ for $0 < i < k$; and
  - $(x_0, \varphi), (\neg x_0, \varphi) \in E$.
- $W = V_0 \setminus \{\varphi\}$, and $\mathcal{W}$ is

$$\mathcal{W} = \big\{ S_i, S_i \cup \{x_i\}, S_i \cup \{\neg x_i\}\ :\ 0 \leq i < k,\ i \text{ even} \big\}$$

where $S_i = \{x_j, \neg x_j\ :\ 0 \leq j < i\}$.

Note that as this is a win-set game, we are only interested in vertices of W that are visited infinitely often. Observe that the winning condition ensures that Player 0 can win if, and only if, the minimum i such that at most one of x_i and $\neg x_i$ is visited infinitely often is even. The idea behind Player 0's strategy is to perpetually verify φ. The choice of strategies by both players then dictates the choices of the truth values for each of the variables, and the winning condition guarantees a winning strategy for Player 0 if, and only if, Φ is true.

A detailed proof that this construction works is deferred to the full paper.

Corollary 4.2. *Deciding Muller games is* PSPACE-*complete.*

Proof. We have already indicated that the problem is in PSPACE. PSPACE-hardness follows from Theorem 3.2 and Theorem 4.1.

Corollary 4.3. *Deciding Zielonka DAG games is* PSPACE-*complete.*

Proof. From Theorem 3.3 and Theorem 4.1.

Corollary 4.4. *Deciding Emerson-Lei games is* PSPACE-*complete.*

Proof. From Theorem 3.4 and Theorem 4.1.

It can be verified that an explicit presentation of the winning condition constructed in the proof of Theorem 4.1 would be exponentially larger than the presentation using a win-set. Thus, the proof cannot be used to provide a PSPACE-hardness result for the explicitly presented games. The exact complexity of deciding the winner of such games remains open. Indeed, it is conceivable (though it appears unlikely) that the problem is in P.

Infinite tree automata. One of the original motivations for studying Muller and related games was to establish decidability results for problems such as non-emptiness and model checking for infinite tree automata [8]. A reduction to non-emptiness of infinite tree automata is used in some of the most effective algorithms for deciding satisfiability of formulas in logics such as $S2S$, μ-calculus and CTL^* – logics useful for reasoning about non-terminating, branching computation. Furthermore, determining if a structure satisfies a formula in any of these logics reduces to determining if a certain automaton accepts a particular tree.

By adapting the proof of Theorem 4.1 we are able to show that the non-emptiness problem for Muller automata as well as the problem of determining whether a given automaton accepts a given regular tree are both PSPACE-complete. The detailed definitions and proofs are deferred to a full paper.

Theorem 4.5. *The non-emptiness problem for Muller tree automata is* PSPACE-*complete.*

The model checking problem (does a given automaton accept a given tree?) also reduces to deciding which player wins an infinite game. However, depending on how the tree is presented, the resulting arena may be of infinite size. If the tree is regular, a game with finite arena can be constructed, and we can apply Theorem 4.5 to obtain the following corollary.

Corollary 4.6. *Given a regular, infinite, k-ary branching tree t and a Muller automaton $\mathcal{A} = (Q, \Sigma, \delta, q_0, \mathcal{F})$, asking if $\mathcal{A}$ accepts t is* PSPACE-*complete.*

Bounded tree-width arenas. Tree-width is a measure of how closely a graph resembles a tree. It has proved useful in the design of algorithms as many problems that are intractable on general graphs are known to have polynomial time solutions when restricted to graphs of bounded tree-width. In the context of regular games, Obdržálek [11] exhibited a polynomial-time algorithm for deciding the winner in parity games on arenas of bounded tree-width. We show that this is not the case for Muller games (and neither, therefore, for Zielonka DAG games and Emerson-Lei games). The proof of Theorem 4.1 can be modified so that the arenas constructed all have tree-width two provided we allow ourselves to specify the winning condition as a Muller condition rather than a win-set.

Theorem 4.7. *Deciding Muller games on arenas of tree-width 2 is* PSPACE-*complete.*

5 Complexity Bounds for Union-Closed Games

We now turn our attention to games where the winning condition $\mathcal{F}$ is a union-closed set. Among games studied in the literature Streett games and parity games are examples of condition types that can only specify union-closed games. Union-closed games were also studied as a class in [6]. One consideration that makes them an interesting case to study is that they admit memoryless strategies for Player 1 [2]. That is, on a game with a union-closed winning condition, if Player 1 has a winning strategy then he has a strategy which is a function only of the current position. One consequence of this fact is that the problem of deciding whether Player 0 wins such a game is in CO-NP. This is because once a memoryless strategy for Player 1 is fixed, the problem of deciding whether Player 0 wins against that fixed strategy is in P. Indeed, it is a version of the alternating reachability problem. Thus, to decide whether Player 1 has a winning strategy we can nondeterministically guess such a strategy and then verify that Player 0 cannot defeat it. Hence, determining whether Player 1 wins is in NP and therefore deciding whether Player 0 wins is in CO-NP. In this section, we aim to establish a corresponding lower bound for two condition types that can only represent union-closed games, namely the Basis and Superset condition types.

The Basis condition type is a succinct way of describing union-closed types. It is not even known if it is translatable to the Emerson-Lei condition type, the most succinct type considered above. However, the following result shows that the bounds obtained cannot easily be derived from the known completeness results of Streett games.

Theorem 5.1. *The Basis and Streett condition types are incomparable with respect to translatability. That is, neither is translatable to the other.*

Nevertheless, deciding Basis games is still in CO-NP.

Proposition 5.2. *Deciding Basis games is in* CO-NP.

Proof. From the comments above, it suffices to show that if we fix a memoryless strategy for Player 1 then we can decide the resulting single player Basis game in polynomial time.

The algorithm is as follows. Let $\mathcal{B}$ be the basis for the winning condition. Initially let $\mathcal{B}_0 = \mathcal{B}$, and repeat the following:

1. Let $X_i = \bigcup_{B \in \mathcal{B}_i} B$.
2. Partition X_i into strongly connected components (SCCs).
3. Remove any element of $\mathcal{B}_i$ which is not wholly contained in a SCC to obtain $\mathcal{B}_{i+1}$,

until $\mathcal{B}_i = \mathcal{B}_{i-1}$, at which point, let $X = X_i$. This takes at most $O(|\mathcal{B}|(|V|+|E|))$ time using a standard SCC-partitioning algorithm. At this point, every SCC of X is a union of basis elements (all x in X are members of basis elements, and any basis elements not contained in any SCC of X is removed at step 3). Furthermore, any strongly connected set of V which is a union of basis elements is a subset (of an SCC) of X, because the algorithm preserves such sets. Thus, Player 0 can win from any node from which she can reach X (play to X and then visit every node within an SCC of X forever); and Player 0 cannot win if she cannot reach X (there is no union of basis elements for which Player 0 can visit every vertex infinitely often). Thus the set of nodes from which Player 0 wins can be computed in $O(|\mathcal{B}|(|V|+|E|)+|E|)$ time.

It should be clear that the Superset condition type is translatable to the Basis condition type.

We now obtain the lower bounds we seek on Superset games.

Theorem 5.3. *Deciding Superset games is* CO-NP-*complete.*

Proof. Membership of CO-NP follows from the previous two propositions. To show CO-NP-hardness, we use a reduction from validity of DNF formulas.

Given a formula $\varphi(x_0, x_1, \ldots, x_{k-1})$ in DNF, consider the Superset game defined as follows:

- for every variable x_i we include three vertices, $x_i, \neg x_i \in V_0$ and $x'_i \in V_1$;
- for each i we have the edges $(x'_i, x_i), (x'_i, \neg x_i), (x_i, x'_{i+1}), (\neg x_i, x'_{i+1})$, where addition is taken modulo k; and
- the winning condition is specified by the set

$$\mathcal{M} = \big\{ \{l_i \in V_0 \,:\, l_i \text{ is a literal of } C\} \text{ for every clause } C \text{ of } \varphi \big\},$$

Take x_0 to be the initial vertex.

As the Superset condition is closed under union, if Player 1 has a winning strategy he has a memoryless winning strategy. Note that any memoryless strategy for Player 1 effectively chooses a truth value for each variable. The set of vertices visited infinitely often is a superset of an element of $\mathcal{M}$ if, and only if, the truth assignment chosen by Player 1 makes one clause of φ (and hence φ) true. Thus Player 0 wins this game if, and only if, there is no truth assignment which makes φ false.

Corollary 5.4. *Deciding Basis games is* CO-NP-*complete.*

Succinctness Results We finish this section with two succinctness results.

Proposition 5.5. *The Superset condition type is more succinct than an explicit presentation of an upward-closed set.*

Proof. Given an explicitly presented upward-closed game $(V, E, \mathcal{F})$, the set $\mathcal{F}$, viewed as a Superset condition, clearly describes the same set of subsets of V. Conversely, for the Superset game $\big(V, E, \{\{v\} \,:\, v \in V\}\big)$, the set described by the winning condition is of size $2^{|V|} - 1$, and therefore cannot be explicitly presented in polynomial time.

Corollary 5.6. *The Basis condition type is more succinct than an explicit presentation of a union-closed set.*

Proof. The fact that the basis condition type is not translatable to an explicit presentation follows from Proposition 5.5. The other direction is straightforward, the explicit presentation itself suffices as a basis.

We note in conclusion that the exact complexity of deciding union-closed games when they are explicitly presented remains an open problem. It is clearly in CO-NP but the above arguments do not establish lower bounds for it.

6 Conclusion

We have considered the complexity of deciding the winner in a variety of regular games. We establish a framework, through the notion of polynomial translatability, within which the expressive power and the succinctness of types of winning conditions can be considered. We used this, along with an encoding of QBF in win-set conditions to establish PSPACE-completeness for four different condition types that can be used to describe regular games and to establish the PSPACE-completeness of the non-emptiness and model-checking problems for Muller automata. We also showed CO-NP-completeness results for two different condition types describing union-closed games.

References

1. J. Richard Büchi and Lawrence H. Landweber. Solving sequential conditions by finite-state strategies. *Trans. Amer. Math. Soc.*, 138:295–311, 1969.
2. Stefan Dziembowski, Marcin Jurdziński, and Igor Walukiewicz. How much memory is needed to win infinite games? In *Proceedings of the 12th Annual IEEE Symposium on Logic in Computer Science*, pages 99–110, 1997.
3. E. Allen Emerson and Charanjit S. Jutla. The complexity of tree automata and logics of programs (extended abstract). In *Proceedings for the 29th IEEE Symposium on Foundations of Computer Science*, pages 328–337, 1988.
4. E. Allen Emerson and Chin-Laung Lei. Modalities for model checking: Branching time strikes back. In *Proceedings of the 12th Annual ACM Symposium on Principles of Porgramming Languages*, pages 84–96, 1985.
5. Erich Grädel, Wolfgang Thomas, and Thomas Wilke, editors. *Automata Logics, and Infinite Games*, volume 2500 of *Lecture Notes in Computer Science*. Springer, 2002.
6. Hajime Ishihara and Bakhadyr Khoussainov. Complexity of some infinite games played on finite graphs. In *Proceedings of the 28th International Workshop on Graph Theoretical Concepts in Computer Science*, volume 2573 of *Lecture Notes in Computer Science*. Springer, 2002.
7. Salvatore La Torre, Aniello Murano, and Margherita Napoli. Weak Muller acceptance conditions for tree automata. In Agostino Cortesi, editor, *3rd International Workshop on Verification, Model Checking and Abstract Interpretation*, volume 2294 of *Lecture Notes in Computer Science*, pages 240–254. Springer, 2002.
8. Robert McNaughton. Finite-state infinite games. Technical report, Project MAC, Massachusetts Institute of Technology, USA, 1965.
9. Robert McNaughton. Infinite games played on finite graphs. *Annals of Pure and Applied Logic*, 65(2):149–184, 1993.
10. Anil Nerode, Jeffery B. Remmel, and Alexander Yakhnis. McNaughton games and extracting strategies for concurrent programs. *Annals of Pure and Applied Logic*, 78(1-3):203–242, 1996.
11. Jan Obdržálek. Fast mu-calculus model checking when tree-width is bounded. In Warren A. Hunt Jr. and Fabio Somenzi, editors, *Proceedings of 15th International Conference on Computer Aided Verification*, volume 2725 of *Lecture Notes in Computer Science*, pages 80–92. Springer, 2003.
12. Wieslaw Zielonka. Infinite games on finitely coloured graphs with applications to automata on infinite trees. *Theoretical Computer Science*, 200:135–183, 1998.

Basic Mereology with Equivalence Relations[*]

Ryszard Janicki

Department of Computing and Software,
McMaster University, Hamilton, ON, L8S 4K1 Canada
janicki@mcmaster.ca

Abstract. The traditional theory of *"part of"* relations (i.e. *mereology*) is enriched by adding the formal concept of *equivalent* and *exchangeable parts*. Various possible axioms and their roles are discussed. An approach is focused on application to model software structures.

1 Introduction

Mereology is the branch of science of analysing the relation between *the part* and *the whole* based on suitable logical systems. Modern attempts to formalize the concept of "part of" go back to S. Leśniewski (1916-37,[12,19]), and H. Leonard, N. Goodman (1940-50,[3,11]). Leśniewski's systems were invented as an alternative to what is now called "standard set theory" (i.e. based on Zermelo-Fraenkel axioms) [17], and translation of his ideas into the language of standard set theory is not obvious and often problematic [16,17,19], so more practical applications are difficult. Leonard and Goodman's Calculus, formulated within standard set theory, was invented to provide a formal model for a universal concept of parthood [2,3,17]. Both models have been substantially extended, however none of them has been substantially applied outside philosophy, cognitive science and pure logic [2,17,19]. Only in the last decade some serious attempts to apply mereological ideas to industrial engineering [15], knowledge engineering [1], approximate reasoning [13], software engineering [7,10], databases [1], and others [1,15], have been made.

The problem with standard mereology is that it assumes a universal concept of parthood, while for serious specific applications we need to consider many different kinds of part/whole relations [1,13]. We might perhaps even need different kind of mereology for different areas of applications. A mereology more suitable for computer science was proposed in [6]. It was motivated by a formal semantics for tabular expressions[1] [8]. A more operational algebraic version, with "part of" defined by sets of constructors and destructors was proposed in [7]. The mereology of [6,7] has been applied to model tabular expressions [6,7,8] and to define and detect formal discrepancies between two requirement scenarios [10].

---

[*] Partially supported by NSERC of Canada.
[1] Tabular expressions, invented by Dave Parnas [5,9] are *relational* means to represent the complex relations that are used to specify and document software systems. The technique is quite popular in software industry [5].

Standard mereology lacks a formal treatment of equivalent parts [2,17]. In this paper, a Leonard and Goodman style mereology is augumented with formal concepts of *equivalent* and *exchangable* parts, however algebraic structure induced by constructors and destructors (see [7]) will not be discussed.

2 Standard Mereology

In this section we shall present basic concepts of "standard mereology" as described in [2,17]. This is a mereology based on the Calculus of Individuals by Leonard and Goodman [3,11]. From a mathematical point of view it is a part of the theory of partially ordered sets [2,17]. We will not discuss here the mathematics of Leśniewski's systems as they cannot easily be formulated within standard set theory. Conceptually the ideas of Leśniewski and Leonard-Goodman appear to be similar (see [16,17]), but mathematical results about this relationship are hard to find, and a major one [4] has not been widely accepted among Leśniewski's disciples[2] [19].

To make the papers selfsufficient, we start with a survey the principal properties of *partial orders* (see [14] for details).

Let X is a set. A relation $\preceq \subseteq X \times X$ is called a *partial order* iff it is *reflexive* ($x \preceq x$), *anti-symmetric* ($x \preceq y \wedge y \preceq x \Rightarrow x = y$), and *transitive* ($x \preceq y \wedge y \preceq z \Rightarrow x \preceq z$). If $\preceq$ is a partial order then the pair $(X, \preceq)$ is called a *partially ordered set* or *poset*. A relation $\prec$ defined as $x \prec y \iff x \preceq y \wedge x \neq y$ is called a *strict partial order*.

Let $(X, \preceq)$ be a poset and let $A \subseteq X$. An element $a \in X$ is called an *upper bound* (a *lower bound*) of A iff $\forall x \in A.\ x \preceq a$ ($\forall x \in A.\ a \preceq x$). The sets of all *upper bounds* and *lower bounds* of A are denoted by $ub(A)$ and $lb(A)$ respectively.

The element $\top \in X$ satisfying $\forall x \in X.\ x \preceq \top$ is called the *top* of X, and the element $\bot$ satisfying $\forall x \in X.\ \bot \preceq x$ is called the *bottom* of X.

An element $a \in A$ is a *minimal* (*maximal*) element of A iff $\forall x \in A.\ \neg(x \prec a)$ ($\forall x \in A.\ \neg(a \prec x)$). The set of all minimal (maximal) elements of A will be denoted by $min(A)$ ($max(A)$).

An element $a \in X$ is called the *least upper bound* (*supremum*) of A, denoted $sup(A)$, iff $a \in ub(A)$ and $\forall x \in ub(A).\ a \preceq x$, and it is called the *greatest lower bound* (*infimum*) of A, denoted $inf(A)$, iff $a \in lb(A)$ and $\forall x \in lb(A).\ x \preceq a$.

The minimal elements of the set $X \setminus \{\bot\}$ are called *atoms* of the poset $(X, \preceq)$, and *Atoms* denotes the set of all atoms of X.

Now we will start adding mereological axioms. Let $(X, \preceq)$ be a poset (with or without $\bot$). The relation $\preceq$ is now interpreted as *"part of"*, a is a *part of* b if $a \preceq b$, and a is a *proper part of* b if $a \prec b$. The element $\bot$ is interpreted as an *empty part*.

---

[2] The paper [4] is right in the case of elementary mereology of Leśniewski, however Leśniewski's non-elementary mereology is a much richer theory. Therefore, the paper [4] is often considered heresy for some of Leśniewski's students [19].

The relations $\circ$ and $\dagger$ on $X \setminus \{\bot\}$ defined as

$$x \circ y \iff \exists z \in X \setminus \{\bot\}.\ z \preceq x \wedge z \preceq y \qquad \text{[overlap]}$$
$$x \dagger y \iff \neg(x \circ y) \qquad \text{[disjoint]}$$
$$x \diamond y \iff \exists z \in X \setminus \{\bot\}.\ x \preceq z \wedge y \preceq z \qquad \text{[underlap]}$$

are called *overlapping*, *disjointness* and *underlapping* respectively (see [2,3,17]). Two elements x and y *overlap* if they have a common non-empty part, they are *disjoint* if they do not have a common non-empty part, and they *underlap* if they are both parts of another element (see [2,17] for details and more properties).

A partially ordered set $(X, \preceq)$ is called a *Minimal Mereology* [2,17] if the following condition is satisfied:

(WSP) $\qquad x \prec y \Rightarrow (\exists z \in X.\ z \prec y \wedge x \dagger z) \vee x = \bot.$

The axiom WSP, called *Weak Supplementation Principle*, is a part of all known mereologies. Among others, it guarantees that if an element has a proper non-empty part, it has more than one. For example, a totally ordered set is not a minimal mereology. It is widely believed that any reasonable mereology must conform to this axiom [2,17]. From now on, we assume that any partially ordered set that is called "mereology" satisfies the property WSP.

A partially ordered set $(X, \preceq)$ is called an *Extensional Mereology* [2,17] if the following condition is satisfied:

(SSP) $\qquad \neg(y \preceq x) \Rightarrow \exists z.\ (z \preceq y \wedge x \dagger z).$

The axiom SSP is called *strong supplementation principle*. It implies quite regular properties, among others *it guarantees that different objects have diferrent sets of proper parts*. Detailed discussion of SSP and its consequences can be found in [2,17]. We think SSP is too strong restriction for our purposes. In mathematics and computer Science different objects are very often built from the same proper parts. For instance, $A \times B \neq B \times A$ (unless $A = B$), and both $A \times B$ and $B \times A$ are intuitively built from the same parts. The mereological systems for direct products and relations designed to deal with tabular expressions and analysed in [6,7] do not satisfy SSP, either.

A mereology $(X, \preceq)$ is called *Atomistic, Mereology with an Empty Part, Mereology with Universe*, if the axioms ATM, BOT, TOP below are satisfied respectively:

(ATM) $\qquad \forall x \in X \setminus \{\bot\}.\exists y \in \mathcal{A}toms.\ y \preceq x,$
(BOT) $\qquad \bot \in X,$
(TOP) $\qquad \top \in X.$

The axiom BOT simply says that the empty part does exist. Most of mereological theories assume the empty part does not exist. The argument is that the empty part (empty element) is not needed except for completeness properties [2,17]. We believe, that empty part, as empty set or empty string, is a very useful concept that eventually will make our theory simpler. The axiom ATM says that all objects (except the empty part) are built from elementary elements called *atoms*. The top $\top$ is called *Universe* in mereology, and it plays either the role of "universe of concourse" or it simply represents the most complex object [2,17].

Let $\alpha : X \to 2^{Atoms}$ and $\mu : X \to 2^X$ be mappings defined by:

$$\alpha(x) = \{a \mid a \in Atoms \land a \preceq x\} \qquad \text{[atoms]}$$
$$\mu(x) = \{y \mid y \prec x \land y \neq \bot\} \qquad \text{[proper parts]}$$

The set $\alpha(x)$ is interpreted as the set of all atoms from which the element x is built. The set $\mu(x)$ is the set of all proper parts of x, excluding the empty part. For each $A \subseteq X$, we define standardly $\alpha(A) = \bigcup_{x \in A} \alpha(x)$, and $\mu(A) = \bigcup_{x \in A} \mu(x)$. The mapping α can be used to characterize the relations overlapping and disjointness.

The theorem below characterises some relationship between the axioms defined above.

Theorem 1 ([2,17]).

(1) $SSP \Rightarrow [WSP \land (a = b \Leftrightarrow \mu(a) = \mu(b))]$.
(2) $ATM \Rightarrow [SSP \Leftrightarrow (x = y \Leftrightarrow \alpha(x) = \alpha(y))]$.
(3) $ATM \Rightarrow [\forall x, y \in X \setminus \{\bot\}.\ x \circ y \iff \alpha(x) \cap \alpha(y) \neq \emptyset]$.
(4) $ATM \Rightarrow [\forall x, y \in X \setminus \{\bot\}.\ x \dagger y \iff \alpha(x) \cap \alpha(y) = \emptyset]$. □

Posets satisfying ATM and SSP are often called *Hyperextensional Mereologies* [2,3,17]. The name follows from the fact that in such cases the objects are identical if they are built from the identical sets of atoms (Theorem 1.2). Hyperextensional mereologies usually lead to elegant theories, but they appear to be too much restricted for our purposes.

The operations $\oplus$ and $\odot$ defined by

$$z = x \oplus y \iff (\forall w \in X.\ w \circ z \Leftrightarrow w \circ x \lor w \circ y), \qquad \text{[sum]}$$
$$z = x \odot y \iff (\forall w \in X.\ w \preceq z \Leftrightarrow w \preceq x \land w \preceq y) \land z \neq \bot \qquad \text{[product]}$$

are called the *mereological sum* and *mereological product* respectively [2,3,17]. Both concepts can easily be extended from two elements to any set [2,17]. The sum of elements of the set A (if exists) will be denoted by $\bigoplus A$, and the product of elements of the set A (if exists) will be denoted by $\bigodot A$.

The below result shows the similarities between *sum* and *least upper bound*, and between *product* and *gratest lower bound* (compare comments in [17]).

Proposition 1.

(1) If $\bot \notin X$ then $\bigodot A = inf(A)$.
(2) $z = \bigoplus A \iff z = sup(A) \land \mu(z) = \mu(A)$. □

Many mereologies assume that $x \odot y$ implies the existence of $x \oplus y$ [3,17], which results in a very elegant model similar to semi-lattices or, when additional assumptions are made, to quasi boolean algebras [4,17]. However, for our purposes such assumption is too strong, most of the models we are interested in do not have this property (including systems from [6,7]). If different objects are allowed to have identical proper parts, then the sum $x \oplus y$ often does not exist.

The last property we will discuss seems to be implicitly assumed in many mereological theories, however it not usually openly discussed [2,17].

Let $(X, \preceq)$ be any poset. The relation $\widehat{\prec}$ defined as

$$x \widehat{\prec} y \iff x \prec y \wedge \neg(\exists z.\ x \prec z \prec y) \qquad \text{[cover]}$$

is called the *cover relation* for $\preceq$ (c.f. [14], y is here an *upper cover* of x).

A poset $(X, \preceq)$ will be called *cover-closed* if

$$\text{(CCL)} \qquad x \preceq y \iff x(\widehat{\prec})^* y,$$

where $(\widehat{\prec})^*$ is the reflexive and transitive closure of $\widehat{\prec}$, i.e. $(\widehat{\prec})^* = \bigcup_{i=0}^{\infty} (\widehat{\prec})^i$.

It appears most authors assume implicitly their mereologies are cover-closed [2,17]. Clearly, any finite poset satisfies CCL.

3 Mereology with Equivalence and Exchange Relations

Many parts can be seen as equivalent in some circumstances. One can change Michelin tires in a car to Goodyear tires, or Quick Sort to Merge Sort in a program, etc. One can also exchange front tires with rear tires, etc. The outcomes are equivalent but not identical entities. Surprisingly, a formal theory of such equivalences has so far been neglected [2,3,17,19] in the context of mereological theories. To the author's knowledge, this is the first paper that proposes some formal approach. We will consider two kinds of equivalence type relations, *equivalence* denoted by $\equiv$ and *exchange* denoted by $\leftrightarrow$.

To make this paper selfsufficient, we start with formal definition of an *equivalence* relation and recall its basic properties (see [14] for details).

A relation $\equiv \subseteq X \times X$ is called an *equivalence* iff it is *reflexive* ($x \equiv x$), *symmetric* ($x \equiv y \iff y \equiv x$), and *transitive* ($x \equiv y \wedge y \equiv z \Rightarrow x \equiv z$). An *equivalence class* of $\equiv$ containing x, denoted as $[x]_\equiv$, is defined as $[x]_\equiv = \{y \mid x \equiv y\}$. The set of all equivalence classes of $\equiv$ in X is denoted as $X/_\equiv$.

Let $\equiv$ be an equivalence relation on X. We need to define two new concepts: a set $X\wr_\equiv \subseteq 2^X$ and a relation $\iota(\equiv) \subseteq X\wr_\equiv \times X\wr_\equiv$. We define them as follows:

$$A \in X\wr_\equiv \iff (A \in 2^X \wedge (\forall P \in X/_\equiv.\ |A \cap P| \leq 1)) \qquad \text{[cuts]}$$
$$A\ \iota(\equiv)\ B \iff \forall\ P \in X/_\equiv.\ (A \cap P \neq \emptyset \Leftrightarrow B \cap P \neq \emptyset) \qquad [\iota(\equiv)]$$

An element of $X\wr_\equiv$ just has *at most one* element in common with each equivalence class of $\equiv$. The elements of $X\wr_\equiv$ will be called *cuts* of $\equiv$. Two cuts A and B

satisfy $A \iota(\equiv) B$, if they either both have common elements with an equivalence class of $\equiv$, or none of them has.

Corollary 1. *For every equivalence relation $\equiv$ on X, the relation $\iota(\equiv)$ is an equivalence relation on $X\wr_\equiv$.* □

For instance, if $X \stackrel{df}{=} \{a, b, c\}$ and $\equiv \stackrel{df}{=} \{(a, b), (b, a), (a, a), (b, b), (c, c)\}$, then $X\wr_\equiv = \{\emptyset, \{a\}, \{b\}, \{c\}, \{a, c\}, \{b, c\}\}$, and $\iota(\equiv)$ minus identity is: $\{a\} \iota(\equiv) \{b\}$, $\{a, c\} \iota(\equiv) \{b, c\}$.

We also need the following property of partial orders. A partially ordered set $(X, \preceq)$ is *weakly upper bounded* if it satisfies:

(WUB) $\qquad\qquad \forall x \in X. \exists y \in max(X). \ x \preceq y.$

In principle it means the set $max(X)$ is a roof that covers the whole poset.

Definition 1. *A partially ordered set $(X, \preceq)$ will be called a* **Basic Mereology** *if it satisfies BOT, WSP, CCL, ATM and WUB.* □

We start adding the concepts of *equivalent* and *exchangeable parts* (*elements*) to our theory. Let us consider a quadruple $(X, \preceq, \equiv, \leftrightarrow)$, where X is a non-empty set, $\preceq$ is a *partial order* relation on X, $\equiv$ is an *equivalence* relation on X, and $\leftrightarrow$ is an *irreflexive* ($\neg(x \leftrightarrow x)$) and *symmetric* ($x \leftrightarrow y \Leftrightarrow y \leftrightarrow x$) relation on X. We need the following definition.

Let $\widehat{\mu} : X \to 2^X$ be defined as

$$\widehat{\mu}(x) = \{y \mid y \widehat{\prec} x \wedge y \neq \bot\}. \qquad\qquad \text{[maximal parts]}$$

Definition 2. *We say that the relations $\preceq$, $\equiv$, and $\leftrightarrow$ are* **mereologically coherent** *if and only if*

(COH.1) $\qquad\qquad x \equiv y \Rightarrow \neg(x \diamond y),$
(COH.2) $\qquad\qquad x \prec y \wedge y \equiv z \Rightarrow \exists u \in X. \ x \equiv u \wedge u \prec z,$
(COH.3) $\qquad\qquad x \prec y \wedge x \equiv u \Rightarrow \exists z \in X. \ y \equiv z \wedge u \prec z,$
(COH.4) $\quad x \leftrightarrow y \Rightarrow \exists! z_1, z_2 \in X. \ z_1 \equiv z_2 \wedge \{x, y\} \subseteq \widehat{\mu}(z_1) \cap \widehat{\mu}(z_2) \wedge \widehat{\mu}(z_1) \setminus \{x, y\} = \widehat{\mu}(z_2) \setminus \{x, y\}.$ □

The above axioms describe the desired relationship between *parthood*, i.e. $\preceq$, *equivalence of parts*, i.e. $\equiv$, and *exchange of parts*, i.e. $\leftrightarrow$. The relation $\equiv$ models, for instance, a replacement of a Michelin tire by a Goodyear tire, while $\leftrightarrow$ models exchanging front wheels with rear wheels.

The axioms COH.2 and COH.3 say that equivalent objects have equivalent parts and vice versa (drawing appropriate Hasse diagrams helps to understand). The axiom COH.1 simply says that two equivalent elements cannot be both parts of the same object. The axiom COH.4 simply says that the elements x and y are exchangeable, and their exchange results in equivalent objects.

The relation $\leftrightarrow$ may, but does not have to, be transitive. It can be interpreted as a kind of *internal* equivalence (in a common meaning of this word), as $x \leftrightarrow y$

implies x and y are both parts of the same z. The relation $\equiv$ can then be interpreted as *external* equivalence, since $x \equiv y$ implies x, y are not parts of the same object, but they are parts of equivalent objects.

Lemma 1. *If $\preceq, \equiv$ and $\leftrightarrow$ are mereologically coherent, then we have*

1. $\equiv \cap \prec \ = \ \equiv \cap \leftrightarrow \ = \ \leftrightarrow \cap \prec \ = \ \emptyset$
2. $\exists x, y \in X.\ x \equiv y \wedge x \neq y \ \Rightarrow \ \top \notin X$
3. $x \leftrightarrow y \ \Rightarrow \ x \notin max(X) \wedge y \notin max(X)$. $\square$

Point (1) of the above lemma says that the relations $\prec, \equiv$ and $\leftrightarrow$ are mutually disjoint, which seems to be rather obvious property. (2) says that if the equivalence relation is not identity, then the top $\top$ does not exist. If there are x and y that are equivalent and different, there must be at least two equivalent but different maximal elements, one which has a part x and another which has a part y. Point (3) says that maximal elements cannot be exchanged.

In the interpretation we assume, the maximal elements represent equivalent wholes, so we assume additional property:

(EQM) $\qquad\qquad x \in \max X \wedge y \in \max X \ \Rightarrow \ x \equiv y.$

The property EQM cannot be derived from the axioms COH.1-COH.4.
We are now ready to provide the main definition of this Chapter.

Definition 3. *A quadruple $\mathcal{B} = (X, \preceq, \equiv, \leftrightarrow)$ where $\preceq$ is a partial order relation on X, $\equiv$ is an equivalence relation on X, and $\leftrightarrow$ is an irreflexive and symmetric relation on X, is called a* **Mereological Base** *if and only if the following conditions are satisfied*

(1) $(X, \preceq)$ *is a basic mereology,*
(2) $\preceq, \equiv$ *and $\leftrightarrow$ are mereologically coherent,*
(3) $x \in max(X) \wedge y \in max(X) \ \Rightarrow \ x \equiv y.$ $\square$

In other words, $(X, \preceq, \equiv, \leftrightarrow)$ is a mereological base if and only if the axioms BOT, WSP, CCL, ATM, WUB, COH.1-4, and EQM are satisfied.

A natural, next question is "What kind of mereology the equivalence classes of $\equiv$ might constitute?"

Define the following relation $\sqsubseteq$ on $X/_\equiv$:

$$\forall A, B \in X/_\equiv.\ A \sqsubseteq B \iff A = B \vee (\forall x \in A. \exists y \in B.\ x \prec y). \qquad [\sqsubseteq]$$

In general the relation $\sqsubseteq$ may not be a partial order, but if $\preceq$ and $\equiv$ ($\leftrightarrow$ is not involved in the definition of $\sqsubseteq$) are mereologically coherent, it is.

Lemma 2. *If the relations $\preceq$ and $\equiv$ are mereologically coherent, then the tuple $(X/_\equiv, \sqsubseteq)$ is a partially ordered set.* $\square$

For mereologically coherent $\preceq$ and $\equiv$, let $\top_{\equiv}$ denote the top of $(X/_{\equiv}, \sqsubseteq)$ and let $\bot_{\equiv}$ denote the bottom of $(X/_{\equiv}, \sqsubseteq)$. The result below characterises the properties of $(X/_{\equiv}, \sqsubseteq)$ when $(X, \preceq, \equiv, \leftrightarrow)$ is a mereological base.

Proposition 2. *If $(X, \preceq, \equiv, \leftrightarrow)$ is a Mereological Base, then $(X/_{\equiv}, \sqsubseteq)$ is a Basic Mereology with a Universe, and with $\bot_{\equiv} = \{\bot\}$ and $\top_{\equiv} = max(X)$.* □

Definition 4. *A mereological base $\mathcal{B} = (X, \preceq, \equiv, \leftrightarrow)$ is called* **plain** *if the relation $\leftrightarrow$ is empty. In such case we will write $\mathcal{B} = (X, \preceq, \equiv)$.* □

Intuitively, it appears $\mathcal{B} = (X, \preceq, \equiv)$ should be uniquely reconstructed from $(X/_{\equiv}, \sqsubseteq)$. This is true indeed, however not immediately obvious.

Assume for the rest of this chapter that $(X, \preceq, \equiv)$ is a plain mereological base.

A set of atoms $A \subseteq \mathcal{A}toms$ is said to be *cut-proper* if

$$\forall P \in \mathcal{A}toms/_{\equiv} \quad |A \cap P| = 1 \qquad \text{[cut-proper]}$$

Cut-proper sets of atoms are just the maximal (w.r.t. inclusion) elements of the set $\mathcal{A}toms\!\downarrow_{\equiv}$. Let $\mathcal{CPA} \subseteq \mathcal{A}toms\!\downarrow_{\equiv}$ denote *the set of all cut-proper sets of atoms*. We are often interested in objects built from a specific subset of atoms. Below we introduce some concepts to deal with this problem.

For each $A \subseteq X$, let the set $X \!\downarrow_A \subseteq X$ and the relations $\preceq\!\downarrow_A, \equiv\!\downarrow_A$ be defined as follows:

$$X \!\downarrow_A \stackrel{df}{=} \{x \in X \mid \alpha(x) \subseteq \alpha(A)\} \qquad [X\!\downarrow_A]$$
$$x \preceq\!\downarrow_A y \iff x \preceq y \land x \in X\!\downarrow_A \land y \in X\!\downarrow_A \qquad [\preceq\!\downarrow_A]$$
$$x \equiv\!\downarrow_A y \iff x \equiv y \land x \in X\!\downarrow_A \land y \in X\!\downarrow_A \qquad [\equiv\!\downarrow_A]$$

The set $X\!\downarrow_A$ is a subset of X containing all elements built from some atoms from A, the relations $\preceq\!\downarrow_A$ and $\equiv\!\downarrow_A$ are just $\preceq$ and $\equiv$ restricted to $X\!\downarrow_A$. The lemma below states that $\mathcal{CPA}$ has two desired properties.

Lemma 3.

(1) If $A \subseteq \mathcal{CPA}$, then we have $(x \equiv\!\downarrow_A y \iff x = y)$.
(2) For each $x \in X$, there exists $A \in \mathcal{CPA}$ such that $x \in X\!\downarrow_A$. □

Theorem 2.

(1) For each set of atoms $A \in \mathcal{CPA}$, the posets $(X\!\downarrow_A, \preceq\!\downarrow_A)$ and $(X/_{\equiv}, \sqsubseteq)$ are isomorphic.
(2) $x \preceq y \iff [x]_{\equiv} \sqsubseteq [y]_{\equiv} \land (\exists A \in \mathcal{CPA}. \{x,y\} \subseteq X\!\downarrow_A)$. □

Intuitively, the only difference between $\preceq$ and $\sqsubseteq$ is that $\sqsubseteq$ is defined on equivalence classes, while $\preceq$ on ordinary elements, otherwise they both express "the same" relationship. Theorem 2.1 states that this intuition is hold, so the definition of $\sqsubseteq$ is correct. Theorem 2.2 states that $\preceq$ can uniquely be derived from $\sqsubseteq$. The whole theorem justifies the following definitions.

Definition 5.

(1) If $\mathcal{B} = (X, \preceq, \equiv, \leftrightarrow)$ is a mereological base, then every poset isomorphic to $(X/_{\equiv}, \sqsubseteq)$ is called a **representation** of $\mathcal{B}$.
(2) If $\mathcal{B} = (X, \preceq, \equiv)$ is a mereological base, then $\mathcal{B}_{\equiv} = (X/_{\equiv}, \sqsubseteq)$ is called a **quotient mereological base**. □

Among others, the concept of equivalence should allow a replacement of a part by an equivalent one. Before giving a formal definition of such an operation we need the following result.

Lemma 4. Let $x_1, x_2, y_1 \in X$ with $x_1 \equiv x_2$ and $x_1 \prec y_1$.
There is **exactly one** $y_2 \in X$ such that

$$y_1 \equiv y_2 \quad \text{and} \quad x_2 \prec y_2, \quad \text{and} \quad \alpha(y_2) = (\alpha(y_1) \setminus \alpha(x_1)) \cup \alpha(x_2).$$ □

Lemma 4 simply states that the definition below is correct, i.e. *replace* is really a function.

Definition 6. Let $replace : X^3 \to X$ be a partial function such that

(1) $(x_1, x_2, y_1) \in domain(replace) \iff x_1 \equiv x_2 \ \land \ x_1 \prec y_1$
(2) $y_2 = replace(x_1, x_2, y_1) \iff y_1 \equiv y_2 \ \land \ x_2 \prec y_2$
$\qquad \qquad \land \ \alpha(y_2) = (\alpha(y_1) \setminus \alpha(x_1)) \cup \alpha(x_2).$ □

The statement $y_2 = replace(x_1, x_2, y_1)$ reads "y_2 is the outcome of replacing the part x_1 in y_1 by x_2". The first point of the definition states that we can replace only equivalent elements and what is replaced must be a part. The second point describes the result of a replacement. The result is equivalent to a whole in which a part has been replaced. By Lemma 4 the result of replacement is unique.

Definition 7. Let $remove : X^2 \to 2^X$ be a total function defined by

$$remove(x, y) = \begin{cases} \{y\} & \text{if } \neg(x \preceq y) \text{ or } x = \bot \\ max(\{z \mid z \prec y \land z \dagger x\}) & \text{otherwise} \end{cases}$$ □

The above definition describes the result of removing a part. The statement $remove(x, y)$ describes what happens when we remove a part x from y. Nothing happens if x is not a part of y or if x is the empty part, otherwise it is a set of *maximal* remaining parts. The set $remove(x, y)$ might be empty.

The relation $\leftrightarrow$ allows us to define correctly the following *rotate* function.

Definition 8. Let $rotate : X^3 \to X$ be a partial function such that

(1) $(x_1, x_2, y_1) \in domain(rotate) \iff x_1 \leftrightarrow x_2 \ \land \ \{x_1, y_1\} \subseteq \widehat{\mu}(y_1)$
(2) $y_2 = rotate(x_1, x_2, y_1) \iff y_1 \equiv y_2 \ \land \ \{x_1, y_1\} \subseteq \widehat{\mu}(y_1)$
$\qquad \qquad \land \ \widehat{\mu}(y_1) \setminus \{x_1, x_2\} = \widehat{\mu}(y_2) \setminus \{x_1, x_2\}.$ □

The statement $y_2 = rotate(x_1, x_2, y_1)$ reads "y_2 is the outcome of exchanging the part x_1 in y_1 with the part x_2".

Example 1. For every set A, let $\widehat{A} = \{\{a\} \mid a \in A\}$ be the set of all singletons generated by A, i.e. if $A = \{a, b\}$, then $\widehat{A} = \{\{a\}, \{b\}\}$. For every relation R, let R^{eq} denotes the reflexive, symmetric and transitive closure of R, i.e. R^{eq} is the smallest equivalence relation containing R.

Let $D_1 = \{a, b, \mathbf{a}, \mathbf{b}\}$, $D_2 = \{1, 2, \mathbf{1}, \mathbf{2}\}$ be sets and let $\doteq_1, \doteq_2$ be the relations on $D_1 \times D_1$ and $D_2 \times D_2$ given by $\doteq_1 = (\{(a, \mathbf{a}), (b, \mathbf{b})\})^{eq}$ and $\doteq_2 = (\{(1, \mathbf{1}), (2, \mathbf{2})\})^{eq}$. The above relations just say that a is equivalent to $\mathbf{a}$, b is equivalent to $\mathbf{b}$, 1 is equivalent to $\mathbf{1}$, 2 is equivalent to $\mathbf{2}$. Let $\doteq_{1,2}$ be a relation on $(D_1 \times D_2)^2$ defined as: $(x_1, y_1) \doteq_{1,2} (x_2, y_2) \iff x_1 \doteq_1 x_2 \wedge y_1 \doteq_2 y_2$. For instance $(a, 1) \doteq_{1,2} (a, \mathbf{1}) \doteq_{1,2} (\mathbf{a}, 1) \doteq_{1,2} (\mathbf{a}, \mathbf{1})$.

Let $X = D_1 \wr_{\doteq_1} \cup\, D_2 \wr_{\doteq_2} \cup\, (D_1 \times D_2)\wr_{\doteq_{1,2}}$ (see the equation [cuts] for an appropriate definition), $\mathcal{A}toms = \widehat{D_1} \cup \widehat{D_2}$ and $\bot = \emptyset$.

In other words $D_1 \wr_{\doteq_1} = \{\emptyset, \{a\}, \{b\}, \{\mathbf{a}\}, \{\mathbf{b}\}, \{a, b\}, \{\mathbf{a}, \mathbf{b}\}\}$, $D_2 \wr_{\doteq_2} = \{\emptyset, \{1\}, \{2\}, \{\mathbf{1}\}, \{\mathbf{2}\}, \{1, 2\}, \{\mathbf{1}, \mathbf{2}\}\}$, and for instance $\{(a, 1), (\mathbf{a}, 1)\} \notin (D_1 \times D_2)\wr_{\doteq_{1,2}}$ while $\{(a, 1), (\mathbf{a}, 2), (b, \mathbf{1})\} \in (D_1 \times D_2)\wr_{\doteq_{1,2}}$.

Let us fold the three relations $\doteq_1, \doteq_2$ and $\doteq_{1,2}$ into one by defining $\doteq$ in $X \times X$ as the set union of $\doteq_1, \doteq_2$ and $\doteq_{1,2}$, i.e. $\doteq \stackrel{df}{=} \doteq_1 \cup \doteq_2 \cup \doteq_{1,2}$. Since the relations $\doteq_1, \doteq_2$ and $\doteq_{1,2}$ are all disjoint, the relation $\doteq$ is also an equivalence relation.

Define the relations $\preceq$ and $\equiv$ in $X \times X$ as follows:

$$A \preceq B \iff A \subseteq B \vee A \subseteq \pi_i(B),\ i = 1, 2,$$
$$A \equiv B \iff A\iota(\doteq)B \quad \text{(see the equation } [\iota(\equiv)] \text{ in Chapter 2)},$$

where $\pi_i(B)$ is the projection of B on i-th coordinate, i.e. $\pi_1(B) = \{x_1 \mid (x_1, x_2) \in B\}$, $\pi_2(B) = \{x_2 \mid (x_1, x_2) \in B\}$.

One can show by inspection that the quadruple $(X, \preceq, \equiv, \emptyset)$ with the components as defined above is a *plain mereological base*. This is a simple case of a mereological base for direct products. Note that $(X, \preceq)$ does not satisfy SSP, however it satisfies $x = y \iff \mu(x) = \mu(y)$.

The partial order $(X/_{\equiv}, \sqsubseteq)$ is presented (as a Hasse diagram) in Figure 1. The nodes of the graphs correspond to the equivalence classes of $\equiv$. For instance $[\{(a, 1), (a, 2)\}]_\equiv = \{(a, 1), (\mathbf{a}, 1), (a, \mathbf{1}), (\mathbf{a}, \mathbf{1}), (a, 2), (\mathbf{a}, 2), (a, \mathbf{2}), (\mathbf{a}, \mathbf{2})\}$.

This example can be extended to non-empty $\leftrightarrow$ by having X defined as $X \stackrel{df}{=} D_1\wr_{\doteq_1} \cup D_2\wr_{\doteq_2} \cup (D_1 \times D_2)\wr_{\doteq_{1,2}} \cup (D_2 \times D_1)\wr_{\doteq_{1,2}}$, and $\leftrightarrow \stackrel{df}{=} (D_1 \times D_2) \cup (D_2 \times D_1)$, and extend $\equiv$ accordingly. This will make symmetric elements of $(D_1 \times D_2)\wr_{\doteq_{1,2}} \cup (D_2 \times D_1)\wr_{\doteq_{1,2}}$ equivalent, for instance, $\{(a, 2), (b, 1)\} \equiv \{(2, a), (b, 1)\} \equiv \{(2, a), (1, b)\}$, etc. In this case, the property $x = y \iff \mu(x) = \mu(y)$ does *not* hold.

Mereology for direct products but without an equivalence relation was analysed in details, using a different formal model in [6,7]. □

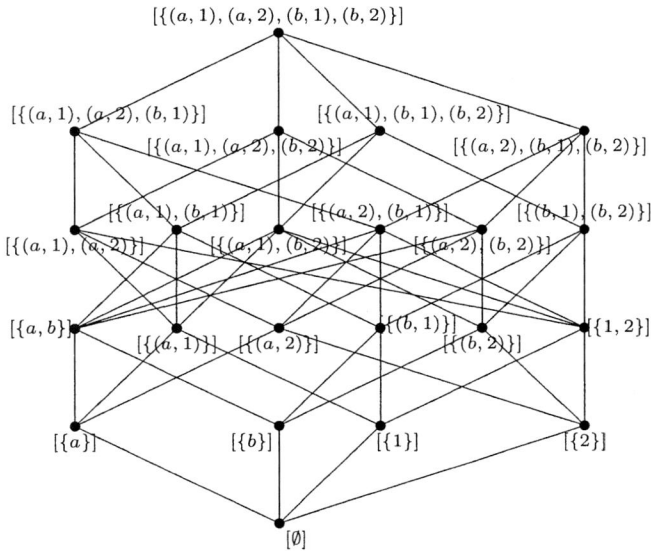

Fig. 1. The relation $\sqsubseteq$ from Example 1. $[A]$ denotes the equivalence class $[A]_\equiv$.

4 Generalisation of Mereological Sum

The mereological sum $\bigoplus A$, if exists, can be interpreted as the most complex object built from the set of parts A only. What if $\bigoplus A$ does not exist? Intuitively, there still might exist the most complex object (or a set of equivalent most complex objects) that can be built from the parts A. We will try to define such an object formally. Attempts to define such a construction, which could be seen as a generalisation of a mereological sum, were made in [6,7], however they all are, on one hand, too complex and on the other hand, too restrictive. In this chapter we will not deal with the relations $\equiv$ and $\leftrightarrow$, and we assume that the poset analysed $(X, \preceq)$ is a representation of a mereological base (Definition 4).

Definition 9. *Let $A, C \subseteq X$. A set C is a **cone** over A if and only if*

$$A \subseteq C \subseteq A \cup ub(A) \quad \wedge \quad sup(C) \in C.$$
□

If $sup(A)$ exists, then $A \cup \{sup(A)\}$ is a cone over A. Consider Example 1 and Figure 1. If $A = \{\{(b,2)\}, \{1,2\}\}$ or $A = \{\{b\}, \{1\}, \{2\}\}$ then $A \cup \{\{(b,1),(b,2)\}\}$, $A \cup \{\{(a,2),(b,1)\}\}$, and $A \cup \{\{(a,1),(b,2)\}\}$ are examples of cones over A in both cases. Note that $sup(A)$ does not exist for either A.

Definition 10. *A set C is a **proper cone** over A if and only if*

$$C \text{ is a cone over } A \text{ and } \mu(C) = \mu(A).$$
□

If $\bigoplus A$ exists, then $A \cup \{\bigoplus A\}$ is a proper cone over A. For the sets A above, $\mu(A) = \{\{b\}, \{1\}, \{2\}\}$, so only $A \cup \{\{(b,1),(b,2)\}\}$ is a proper cone over A (for both A).

Let $PCones(A)$ denote the set of all proper cones over A, and let $\mathbf{PC}_A$ be a partially ordered set defined as $\mathbf{PC}_A = (PCones(A), \subseteq)$. Each cone C (proper or not) can be interpreted as a partially ordered set $(C, \preceq_C)$, where $\preceq_C$ is the partial order $\preceq$ restricted to the subset C of X.

Definition 11. *An element $z \in X$ is called the* **mereological supremum** *of a set $A \subseteq X$, denoted $msup(A)$, if and only if $z = sup_{\preceq_C}(C)$, where $C = sup_\subseteq(min_\subseteq(PCones(A)))$.* □

For both sets A considered above, $msup(A) = \{(b,1),(b,2)\}$. The definition above is a generalisation and a simplification of the similar concepts discussed in [6,7]. The property WSP is necessary, otherwise the above definitions are not sound.

Proposition 3.

(1) $\mathbf{PC}_A$ has a bottom $\perp_{\mathbf{PC}_A}$ if and only if $\bigoplus A$ exists.
(2) If $\perp_{\mathbf{PC}_A}$ exists then $msup(A) = sup_\preceq(A) = sup_\preceq(\perp_{\mathbf{PC}_A})$. □

Definition 12.

(1) A basic mereology $(X, \preceq)$ is **mereologically complete** *if for each $A \in X$, $msup(A)$ does exist.*
(2) A mereological base $(X, \preceq, \equiv, \leftrightarrow)$ is **regular** *if $(X/_\equiv, \sqsubseteq)$ is mereologically complete.* □

If a mereological base is regular, then for each set of parts A a single most complex object can be built from these parts. One can verify by inspection that the mereological system from Example 1 is regular. It can also be proven that the mereological systems discussed in [6,7] are regular.

5 Final Comment

The results of this paper can immediately be applied to mereological models of direct products and relations of [6,7] i.e. to tabular expressions. There are many problems that we have not yet dealt with. The most urgent seems to be that, as pointed out in [15], there are many structurally different kinds of part/whole relations, and without making this distiction clear, we might get unreasonable conclusions (see [1]). Abstraction/refinement relationship between various mereological models has not been analysed either.

References

1. A. Artale, G. Franconi, E. Guarina, L. Pazzi, Part-whole relations in object-centered systems, *Data and Knowledge Engineering*, 20 (1996), 347-383.
2. R. Casati, A. C. Varzi, *Parts and Places*, MIT Press, 1999.
3. N. Goodman, *The Structure of Appearance*, 3rd edition, D. Reidel, 1977.

4. A. Grzegorczyk, The System of Leśniewski in Relation to Contemporary Logical Research, *Studia Logica* 3 (1955), 77-95.
5. D.M. Hoffman, D.M. Weiss (eds.), *Collected Papers by David L. Parnas*, Addison-Wesley, 2001.
6. R. Janicki, Remarks on Mereology of Direct Products and Relations, in J. Desharnais, M. Frappier, W. MacCaull (eds.), *Relational Methods in Computer Science*, Methodos Publ. 2002, pp. 65-84.
7. R. Janicki, On a Mereological System for Relational Software Specifications, Proc. of MFCS'2002, *Lecture Notes in Comp. Science* 2420, Springer 2002, pp. 375-386.
8. R. Janicki, R. Khédri, On a Formal Semantics of Tabular Expressions, *Science of Computer Programming*, 39 (2001) 189-214.
9. R. Janicki, D. L. Parnas, J. Zucker, Tabular Representations in Relational Documents, in C. Brink, W. Kahl, G. Schmidt (eds.), *Relational Methods in Computer Science*, Springer 1997, pp. 184-196.
10. R. Khédri, L. Wang, L. Situ, Requirements Specification Decomposition: A System Testing Driven Approach, *Proc. 7th Int'l Seminar On Relational Methods in Computer Science*, Bad Melente, Germany, 2003,pp. 97-104.
11. H. Leonard, N. Goodman, The calculus of individuals and its uses, *Journal of Symbolic Logic*, 5 (1940), 45-55.
12. S. Leśniewski, Grundzüge eines neuen Systems der Grundlagen der Mathematik, *Fundamenta Matematicae* 24 (1929), 1-81.
13. L. Polkowski, A. Skowron, Rough Mereology; A New Paradigm for Approximate Reasoning, *Journal of Approximate Reasoning*, 15, 4 (1997), 316-333.
14. K. H. Rosen, *Discrete Mathematics and its Applications*, McGraw-Hill 1991.
15. F. A. Salustri, J. C. Lockledge, Towards a formal theory of products including mereology, *Proc. 12th Int'l Conf. on Engn. Design*, Munich, 1999, pp. 1125-30.
16. P. Simons, On Understanding Leśniewski, *Hist. and Phil. of Logic* 3 (1982),165-191.
17. P. Simons, *Parts. A Study in Ontology*, Claredon Press, 1987.
18. C. Szyperski, *Component Software*, Addison-Wesley, 1997.
19. J. T. J. Srzednicki, V. F. Rickey (eds.), *Leśniewski's Systems*, Kluwer, 1984.

Online and Dynamic Recognition of Squarefree Strings

Jesper Jansson and Zeshan Peng

Department of Computer Science,
The University of Hong Kong,
Pokfulam Road, Hong Kong
{jjansson, zspeng}@cs.hku.hk

Abstract. The online squarefree recognition problem is to detect the first occurrence of a square in a string whose characters are provided as input one at a time. We present an efficient algorithm to solve this problem for strings over arbitrarily ordered alphabets. Its running time is $O(n \log n)$, where n is the ending position of the first square, which matches the running times of the fastest known algorithms for the analogous offline problem. We also present a very simple algorithm for a dynamic version of the problem over general alphabets in which we are initially given a squarefree string, followed by a series of updates, and the objective is to determine after each update if the resulting string contains a square and if so, report it and stop.

1 Introduction

A classic problem in computer science is to determine whether a given string T contains a *square*, defined as a substring of T which can be split into two identical parts. Since a square is one of the simplest possible types of patterns in a string, methods for detecting squares efficiently have a wide range of applications in diverse areas such as string algorithms and combinatorics [2, 6, 8, 9, 11, 16, 18], automata and formal language theory [6, 12], data compression [4, 8, 17], coding theory [4], and computational biology [3, 5, 11].

Many people have studied this problem and its variants (see, e.g., [1, 2, 4, 6–9, 11, 12, 15–18] and the numerous references therein). However, previous work has mainly focused on the *offline* version in which the entire string T is available at once. This offline property is not desirable in certain applications. For example, suppose we need to determine whether a string of one million characters contains a square. If we use an offline algorithm, we will have to scan through all the one million characters, which may be very inefficient if a square appears at the very beginning of the string. In some applications, such as online data compression, the offline property is even unacceptable; we need to be able to report a square whenever a new character arrives. The online squarefree recognition problem is also motivated by the local search method for solving the constraints satisfaction problem in [13, 14, 19]; to guarantee that the method will not be trapped in some infinite loop, one can encode the successive states of the search as characters in a growing string and terminate the method if a square is formed at the end of this string [15].

Our main result is an efficient algorithm for the online squarefree recognition problem over an arbitrarily ordered alphabet. This is a reasonable assumption for most applications because when the symbols are encoded as binary numbers in a computer, this will induce a lexicographical ordering among them. Our algorithm is based on the work of Leung, Peng, and Ting [15]. We also introduce and study a dynamic version of the problem.

1.1 Problem Definitions

For any string T, let $|T|$ be the length of T. For every $1 \leq i \leq j \leq |T|$, denote the substring of T starting at position i and ending at position j by $T[i..j]$, and define $T[i] = T[i..i]$. A substring of the form $T[i..(i + 2k - 1)]$ is called a *square* (also known in the literature as a *tandem repeat*) if for every $x \in \{0, 1, \ldots, (k - 1)\}$, it holds that $T[i + x] = T[i + k + x]$. If T does not contain a square, then T is *squarefree*.

We distinguish between the *offline*, *online*, and *dynamic* versions of the squarefree recognition problem. In the offline version, the entire string T is provided as input directly, and the objective is to determine whether or not T contains a square. In the online version, the characters of the string T arrive one at a time in sequential order, and the objective is to determine after receiving each character if the string obtained so far contains a square; if so, report it and stop. Finally, in the dynamic version, a squarefree string T is provided as the initial input and then followed by a series of updates of the form "replace the symbol on position q of T by the symbol x", and the objective is to decide after each update if the resulting T contains a square and if so, report it and stop. In this paper, we also consider a combination of the online and dynamic versions of the problem that also allows updates of the form "append the symbol x to the end of T".

The alphabet of the input string determines how efficiently the various squarefree recognition problems can be solved. Under the least restrictive assumption, the symbols in T cannot be relatively ordered; a comparison between two symbols only tells us if they are equal or not. We call this type of alphabet a *general* alphabet. If the symbols in T admit some arbitrary lexicographical ordering so that any comparison between two symbols yields one of the three outcomes $<$, $=$, and $>$, then the alphabet is called *ordered*.[1] Next, in an *integer* alphabet, all symbols are integers in the range $\{1, 2, \ldots, |T|\}$. Finally, if the size of the alphabet is bounded by a constant, then we say that the alphabet is *constant*.

---

[1] As an example to illustrate the difference between general and ordered alphabets, consider the element uniqueness problem which has a lower bound of $\Omega(n^2)$ for general alphabets but admits an $O(n \log n)$-time solution for ordered alphabets (see [4]).

1.2 Previous Results

For the offline and general alphabet case, Main and Lorentz [17] gave an algorithm that can be used to report all s occurrences of squares in a string T of length n in $O(n \log n + s)$ time, or just the longest square in T in $O(n \log n)$ time. This is optimal because to determine if T is squarefree takes $\Omega(n \log n)$ time for general alphabets [17]. (For the offline and non-general alphabet case, other efficient algorithms for finding squares were presented earlier in [1] and [7].) However, it is still not known if the lower bound $\Omega(n \log n)$ for determining squarefreeness holds for ordered alphabets. For the offline and constant alphabet case, there exist algorithms that determine if T is squarefree in optimal $O(n)$ time [8, 18]. Parallel algorithms for finding squares offline have also been developed (see [4]).

For the online case, the only previously known result is the algorithm by Leung, Peng, and Ting [15] for general alphabets which has a running time of $O(n \log^2 n)$, where n is the ending position in T of the first square. (This is just a factor of $O(\log n)$ worse than the optimal offline algorithm for general alphabets mentioned above.) The algorithm of Leung, Peng, and Ting is outlined in Section 3.1.

1.3 Our Results

We first present an algorithm for the online squarefree recognition problem over arbitrarily ordered alphabets. It reads the successive characters of T until a square has been formed, then reports the occurrence of this square and stops. The running time is $O(n \log n)$, where n is the ending position in T of the square; in other words, if n is the smallest integer such that $T[1..n]$ contains a square, our algorithm correctly determines whether $T[1..h]$ contains a square after reading $T[h]$ for every $h \in \{1, 2, \ldots, n\}$. Note that this matches the running times of the fastest known offline algorithms for determining squarefreeness of strings over ordered alphabets [1, 7, 17].

Next, we give a very simple algorithm for the dynamic version of the squarefree recognition problem. It works for general alphabets and uses $O(n)$ time per update, where n is the length of the input string. The algorithm can easily be extended to also solve the combination of the online and dynamic versions of the problem in which every update either modifies an existing character or adds a new character to the end of T.

The table below summarizes our results.

Alphabet type	Online algorithm	Dynamic algorithm	Online + dynamic
General	$O(n \log^2 n)$ (See [15])	$O(n)$ per update (Theorem 3, Section 4)	$O(n)$ per update (Theorem 4, Section 4)
Ordered	$O(n \log n)$ (Theorem 2, Section 3)	$O(n)$ per update (Theorem 3, Section 4)	$O(n)$ per update (Theorem 4, Section 4)

2 Preliminaries

2.1 Suffix Trees

Let A be a string of length k. A *suffix of A* is a substring of A of the form $A[x..k]$, where $x \in \{1, 2, \ldots, k\}$. A *suffix tree for A* (see, e.g., [10, 11]) is a rooted tree with $O(k)$ nodes which represents each suffix of A as a unique path from the root to a leaf. Every edge in the suffix tree for A encodes a particular substring of A whose starting and ending positions in A are specified by two integers which label that edge. For any two leaves x and y, the unique path from the root to the lowest common ancestor of x and y encodes the longest common prefix of the two suffixes represented by x and y.

3 An Efficient Online Squarefree Recognition Algorithm for Arbitrarily Ordered Alphabets

In this section, we present an algorithm for the online squarefree recognition problem for arbitrarily ordered alphabets. Our algorithm is based on the algorithm of Leung, Peng, and Ting [15] for the general alphabet case, but faster.

3.1 LPT: The Algorithm of Leung, Peng, and Ting

Here, we briefly review the algorithm of Leung, Peng, and Ting [15], henceforth referred to as LPT. LPT reports the first square in the online input string T in $O(n \log^2 n)$ time, where n is the position in T where the square ends.

Algorithm LPT is listed in Fig. 1. It reads the string T one character at a time, starting with $T[1]$. After reading a new position h, LPT immediately checks if $T[1..h]$ contains a square; if so then it reports the square and stops. Otherwise, $T[1..h]$ is squarefree, and the algorithm proceeds to read the character at the next position from T. To efficiently do the checking, LPT makes use of a procedure called $\text{DHangSq}(i,j)$ which solves the following subproblem: for every $h \in \{(j+1), (j+2), \ldots, (2j-i+1)\}$, after $T[h]$ is read, determine if T has a square ending at position h whose first half lies entirely in the interval $T[i..j]$ (such a

---

For $h \in \{1, 2, \ldots\}$, after reading $T[h]$, do the following:
if there is a square in $T[(h-3)..h]$, or any of the running $\text{DHangSq}(i,j)$ detects a square in $T[1..h]$, **then** report it and stop.
$j = h; \ell = 1$;
while $(j \geq 2^\ell)$ **do**
    if $j = q \cdot 2^\ell$ for some integer q **then**
        $i = \max\{1, q \cdot 2^\ell - 4 \cdot 2^\ell + 1\}$;
        start $\text{DHangSq}(i,j)$;
    $\ell = \ell + 1$;

Fig. 1. Algorithm LPT

square is said to be "hanging in $T[i..j]$"). When LPT reaches certain values of h, it starts a new DHangSq process so that at any point of its execution, it will have a number of DHangSq(i,j) processes running (for various values of i and j). Refer to [15] for more details as well as correctness proofs for the algorithm.

For any $1 \leq i \leq j$, the pair (i,j) is called a *level-ℓ pair* if there exists an integer q such that $j = q \cdot 2^\ell$ and $i = \max\{1, q \cdot 2^\ell - 4 \cdot 2^\ell + 1\}$. (Hence, $j - i + 1 \leq 4 \cdot 2^\ell$.) The analysis in [15] of Algorithm LPT can be summarized and expressed as:

Theorem 1. *[15] Suppose n is the smallest integer such that $T[1..n]$ contains a square. For every $h \in \{1, 2, \ldots, n\}$, LPT correctly determines whether $T[1..h]$ contains a square after reading $T[h]$. The total running time of LPT is $\sum_{\ell=1}^{\lceil \log n \rceil} O(\frac{n}{2^\ell}) \cdot t(\ell)$, where $t(\ell)$ is the running time of DHangSq(i,j) for a level-ℓ pair (i,j).*

Leung, Peng, and Ting [15] described how to implement DHangSq(i,j) for general alphabets to run in $O((j-i+1) \cdot \log(j-i+1))$ time, i.e., $t(\ell) = O(2^\ell \cdot \ell)$ above. Using this implementation, it follows from Theorem 1 that the total running time of LPT is $O(n \log^2 n)$.

3.2 Speeding Up DHangSq

Recall that DHangSq(i,j) needs to solve the following problem: for every $h \in \{(j+1), (j+2), \ldots, (2j-i+1)\}$, after $T[h]$ is read, determine if T has a square ending at position h whose first half lies entirely in the interval $T[i..j]$. Section 4 in [15] shows that this problem can in fact be reduced to the following problem (stated slightly differently in [15]) at an additional cost of $O(j-i+1)$ time, where the parameter k in the new problem is equal to $j-i+1$:

The Minimum-Suffix-Centers Checking Problem (MSCC):
Let A be a given string of length k and let L be a given list of pairs of integers of the form $(1, e(1)), (2, e(2)), \ldots, (k, e(k))$, where for each $s \in \{1, 2, \ldots, k\}$ it holds that $1 \leq s \leq e(s) \leq k$. Next, let B be a string of length k which arrives online, one character at a time. Return the smallest possible $h \in \{1, 2, \ldots, k\}$ for which there is a pair $(s, e(s))$ in L such that $A[s..e(s)]$ is equal to $B[1..h]$; if no such h exists then return *fail*.

This means that if we could solve MSCC in $O(k)$ time then we could improve the running time of DHangSq and hence Algorithm LPT; see Theorem 1. More precisely:

Lemma 1. *If we have an $O(k)$-time algorithm for MSCC then $t(\ell) = O(2^\ell)$ in Theorem 1.*

3.3 Solving MSCC for Integer Alphabets

We now give an algorithm for solving MSCC in $O(k)$ time under the additional constraint that A is a string over an integer alphabet $\{1, 2, \ldots, m\}$ with $m \leq k$.

(In the next section, we show how to deal with this extra constraint efficiently for ordered alphabets by using an input alphabet mapping technique.)

The main idea of our algorithm for MSCC for integer alphabets is to store the given A in a suffix tree T_A, and match the successive characters of B along a unique path from the root in T_A until either enough characters match so that $A[s..e(s)]$ equals a prefix of B for some s, or the current character of B fails to match any outgoing edge at the current position in T_A. Our algorithm consists of a preprocessing phase and a matching phase:

Phase I (Preprocessing Phase): Construct a suffix tree T_A for A. For convenience, let s for any $s \in \{1, 2, \ldots, k\}$ also refer to the leaf in T_A that represents the suffix $A[s..k]$. Augment T_A with additional information as follows. For every edge f in T_A, define $v(f)$ as the minimum value of $e(s) - s + 1$ taken over all leaves s belonging to the subtree of T_A below f (note that $v(f) \le k$). Obtain and store $v(f)$ for every edge f in T_A by doing a bottom-up traversal of T_A.

Phase II (Matching Phase): For successive values of $h \in \{1, 2, \ldots, k\}$, check if $B[1..h]$ equals $A[s..e(s)]$ for any $(s, e(s)) \in L$ with the following method. Match the successive characters in B along the unique path in T_A starting at the root by following edges labeled by $B[1], B[2], \ldots$ (to traverse an edge in T_A that represents x characters, we need to match it to x characters from B) until either h reaches the value $v(f)$ for the edge f being traversed (success; return h), or the current character in B does not match any edge at the current position in T_A (failure; return *fail*).

Correctness: In Phase I, the algorithm builds a suffix tree T_A for A. In Phase II, the algorithm starts at the root of T_A and follows a path whose labels match the successive characters of B. Suppose that the algorithm has received $B[1..h]$ for any $h \in \{1, 2, \ldots, k\}$. By the properties of a suffix tree, the set of leaves descending from the current location in T_A encode all prefixes of suffixes (i.e., all substrings) of A having length h that are identical to the string $B[1..h]$ received so far. Now, if there is such a substring $A[s..(s+h-1)]$ that also satisfies $(s, s+h-1) \in L$, then the edge f being traversed will have $v(f) = h$, and since the length of the path from the root is exactly h, the algorithm will succeed and return h.

To see that the algorithm will stop for the smallest possible h, suppose $B[1..h] = A[s..(s+h-1)]$ as well as $B[1..h'] = A[t..(t+h'-1)]$ for some $h < h'$ and $(s, (s+h-1)), (t, (t+h'-1)) \in L$. Then the algorithm must have terminated after $B[1..h]$ has been processed because the corresponding path of length h in T_A from the root will have reached the lowest common ancestor of the two leaves s and t, and the edge f leading to that node satisfies the stopping condition $v(f) \le \min\{h, h'\} = h$.

Running Time: To implement the algorithm above, we use the method of Farach-Colton et al. [10] for constructing suffix trees over integer alphabets to build T_A in $O(k)$ time. Next, the bottom-up traversal to compute $v(f)$ for every edge f in T_A takes $O(k)$ time. Then, in the matching phase, the total time for finding which outgoing edges to follow in T_A from internal nodes is upper-bounded by the number of edges in T_A since each edge is examined at most once; thus, these computations take $O(k)$ time. The rest of the computations in the matching phase take $O(1)$ time per read character and the algorithm reads at most k characters from B. Therefore, the total running time of our algorithm is $O(k)$.

Lemma 2. *MSCC for integer alphabets can be solved in $O(k)$ time.*

3.4 LPT*: An Online Squarefree Recognition Algorithm for Arbitrarily Ordered Alphabets

Our solution for the subproblem MSCC in Section 3.3 requires the alphabet of the input string A to be an integer alphabet $\{1, 2, \ldots, m\}$, where $m \leq |A|$. However, the input T to the online squarefree recognition problem for an arbitrarily ordered alphabet does not necessarily meet this requirement. Therefore, we will modify Algorithm LPT so that before starting DHangSq for any required pair of indices (i, j), it translates $T[i..j]$ into an equivalent string $T''_{i..j}$ over the alphabet $\{1, 2, \ldots, (j - i + 1)\}$. Similarly, when a symbol is read from T, the algorithm will translate that symbol into the corresponding integer alphabet for each currently active DHangSq for checking. For this purpose, the modified LPT will translate the input string T online to a string T' over a growing integer alphabet that is subsequently used to construct all the necessary $T''_{i..j}$-strings. In this section, we demonstrate how these extra steps can be performed without increasing the overall asymptotic running time of LPT. Below, the new version of LPT is referred to as LPT*.

For any positive integer h, denote the set of symbols occurring in $T[1..h]$ by Σ_h. By our assumptions, each Σ_h is arbitrarily ordered; except for this fact, we have no information about the alphabet of T in advance.

Translating T to T': As the characters of T arrive online, LPT* first translates them to obtain a string T' such that for each positive integer h, the alphabet of $T'[1..h]$ is precisely $\{1, 2, \ldots, |\Sigma_h|\}$. To do this, it stores the distinct symbols read from T so far in a balanced binary search tree $\mathcal{B}$ and associates a unique integer with each symbol inserted into $\mathcal{B}$. Since the number of nodes in $\mathcal{B}$ while reading $T[1..h]$ is always less than or equal to h and because Σ_h is ordered, the total time used to translate $T[1..h]$ to $T'[1..h]$ is $O(h \log h)$.

Translating T' to $T''_{i..j}$: Next, whenever LPT* starts DHangSq for some pair of indices (i, j), it also constructs an injective mapping $f_{i..j}$ from the set of symbols occurring in $T'[i..j]$ to the set $\{1, 2, \ldots, (j - i + 1)\}$ and applies $f_{i..j}$ to each position in $T'[i..j]$ to obtain a string $T''_{i..j}$ over $\{1, 2, \ldots, (j - i + 1)\}$. Furthermore, for each such (i, j), until DHangSq(i, j) is terminated, LPT* keeps track of $f_{i..j}$ so that it can translate online the characters in $T'[(j+1)..(2j-i+1)]$ to the same alphabet.

The mapping $f_{i..j}$ is implemented as an array $F_{i..j}$ such that for any $x \in \{1, 2, \ldots, j\}$ occurring as a symbol in $T'[i..j]$, the entry x in $F_{i..j}$ contains the value $f_{i..j}(x)$; the other entries of $F_{i..j}$ are left undefined. For efficiency reasons explained below, LPT* will reuse the array $F_{i..j}$ for a terminated DHangSq, and therefore also associates a "timestamp" of the form (i,j) with each entry of $F_{i..j}$ to directly tell whether an entry is valid or contains old information. Suppose LPT* needs to start a new DHangSq(i,j) for some i immediately after reading a character $T[j]$ and translating it to $T'[j]$. Let c be a counter, initially set to 0, and scan the substring $T'[i..j]$. For each $s \in \{i,(i+1),\ldots,j\}$, first check if entry $T'[s]$ in $F_{i..j}$ already has been set by checking its timestamp: if no then increment c by one, set entry $F_{i..j}(T'[s])$ to c, and update the timestamp of $f_{i..j}(T'[s])$. Clearly, this takes only $O(j-i+1)$ time.

Next, for any DHangSq(i,j) process started by LPT*, say that it is *on level ℓ* if (i,j) is a level-ℓ pair. We make the following crucial observation:

Lemma 3. *At any point during the execution of LPT*, there are at most four active DHangSq processes on each level.*

Proof. Suppose LPT* has just read $T[h]$. Consider any level $\ell \le \log h$. Let a be the largest multiple of 2^ℓ which is less than h, and write $a = q \cdot 2^\ell$, i.e., $q \cdot 2^\ell < h \le (q+1) \cdot 2^\ell$. If $q < 4$ then less than four DHangSq processes on level ℓ have been started and the lemma follows directly. Hence, assume $q \ge 4$. Each DHangSq(i,j) is active while at most $j-i+1 = 4 \cdot 2^\ell$ positions of T are being read. This means that right after $T[h]$ is read, the only active DHangSq(i,j) processes on level ℓ are those that were started for $j \in \{(q-3)\cdot 2^\ell, (q-2)\cdot 2^\ell, (q-1)\cdot 2^\ell, q\cdot 2^\ell\}$. □

By Lemma 3, we only need to keep track of four $F_{i..j}$ arrays for each level reached. This means we can reuse the array $F_{i..j}$ used for storing $f_{i..j}$ after DHangSq(i,j) terminates to store $f_{i'..j'}$ for another DHangSq(i',j') on the same level. By using timestamps, we do not need to reinitialize all the positions of the array. However, note that for any such (i',j'), the array $F_{i..j}$ might not be large enough to store j' entries. To handle this issue, whenever LPT* reaches a position of the input string which equals a power of two, we let it double the size of every existing $F_{i..j}$, (e.g., for each existing $F_{i..j}$, initialize a new array with twice as many entries and copy the contents of the old $F_{i..j}$ into the first half of the new array). Thus, after reading h characters from T, every $F_{i..j}$ contains $O(h)$ entries.

Supposing that LPT* terminates after reading $T[1..n]$ for some positive integer n, the time needed for all these operations is bounded by $\sum_{r=1}^{\lfloor \log n \rfloor} O(r) \cdot 4 \cdot O(2^r) = O(n \log n)$. (LPT* doubles the arrays after reaching position 2^r of T for every integer r, i.e., not more than $\lfloor \log n \rfloor$ times. Every time, there are $O(r)$ levels and at most four active DHangSq on each level, and the doubling of an array uses time proportional to the number of positions read from T so far.)

Total Running Time of LPT*: Suppose n is the smallest integer such that $T[1..n]$ contains a square. The total running time of LPT* is equal to the time needed to do all the string translation operations to integer alphabets plus the running time of LPT using the faster DHangSq for integer alphabets. By the above, the translation operations take a total of $O(n \log n)$ time. By Theorem 1, the running time of LPT is given by $\sum_{\ell=1}^{\lceil \log n \rceil} O(\frac{n}{2^\ell}) \cdot t(\ell)$, and according to Lemmas 1 and 2, we have $t(\ell) = O(2^\ell)$. Adding everything together yields:

Theorem 2. *The online squarefree recognition problem for arbitrarily ordered alphabets can be solved in $O(n \log n)$ time, where n is the ending position of the first square.*

4 An Algorithm for Dynamic Squarefree Recognition over General Alphabets

We now present a simple algorithm for the dynamic squarefree recognition problem over general alphabets. Its input is a squarefree string T of length n, followed by a series of updates of the form $T[q] := $ 'x' (where $1 \leq q \leq n$) which means "replace the symbol on position q of T by the symbol x". After each update, our algorithm uses $O(n)$ time to check if the modified T contains a square, and if so, reports it and stops.

The key observation is that after each update $T[q] := $ 'x', any newly formed square in T must include the position q along with a (possibly empty) substring ending immediately before q and a (possibly empty) substring starting immediately after q, which limits the total number of comparisons we need to make.

For any two positions i, j of T with $1 \leq i < j \leq n$, define $LCSu^{-1}(i, j)$ as the longest common suffix of $T[1..(i-1)]$ and $T[1..(j-1)]$ and $LCPr^{+1}(i, j)$ as the longest common prefix of $T[(i+1)..n]$ and $T[(j+1)..n]$. We have the following.

Lemma 4. *Suppose that T is a squarefree string of length n and we perform an update $T[q] := $ 'x', where $1 \leq q \leq n$. The resulting string T contains a square if and only if there exists a $q' \in \{1, 2, \ldots, n\}$ with $q' \neq q$ such that $T[q] = T[q']$ and $|LCSu^{-1}(q, q')| + |LCPr^{+1}(q, q')| + 1 \geq |q - q'|$.*

Proof. $\Longrightarrow$) Suppose the resulting T contains a square $S = T[p..(p+2k-1)]$. Then we know by the key observation above that $p \leq q \leq p + 2k - 1$. Define the *twin of q* as $q' = q + k$ if $q \leq p + k - 1$ and as $q' = q - k$ if $p + k \leq q$. It is easy to see that $q \neq q'$, $T[q] = T[q']$, and $|LCSu^{-1}(q, q')| + |LCPr^{+1}(q, q')| \geq k - 1 = |q - q'| - 1$.

$\Longleftarrow$) Suppose there exists a $q' \in \{1, 2, \ldots, n\}$ with $q' \neq q$ such that $T[q] = T[q']$ and $|LCSu^{-1}(q, q')| + |LCPr^{+1}(q, q')| + 1 \geq |q - q'|$. Assume without loss of generality that $q < q'$. Define $p = q - |LCSu^{-1}(q, q')|$ and $r = q' - |LCSu^{-1}(q, q')|$. By the definition of $LCSu^{-1}$, we have $T[p..(q-1)] = T[r..(q'-1)]$. Next, we rewrite the inequality as $|LCPr^{+1}(q, q')| \geq -q + r - 1$, which yields $T[(q+1)..(r-1)] = T[(q'+1)..(q'-q+r-1)]$ by the definition of $LCPr^{+1}$. Putting everything together, we have $T[p..(r-1)] = T[r..(q'-q+r-1)]$, i.e., T contains a square. See Fig. 2 for an illustration. The case $q > q'$ is symmetric. □

Now, to determine if T contains a square after performing an update $T[q] := $ 'x', apply Lemma 4. More precisely: for each $q' \in \{1, 2, \ldots, n\}$ with $q' \neq q$,

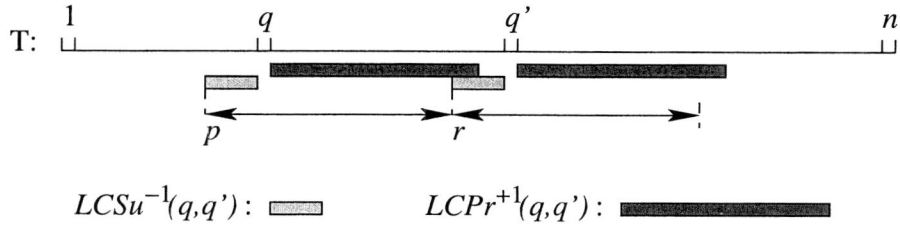

Fig. 2. Illustrating the second part of the proof of Lemma 4

check if the two conditions $T[q] = T[q']$ and $|LCSu^{-1}(q,q')| + |LCPr^{+1}(q,q')| + 1 \geq |q - q'|$ hold. If yes, then T contains a square; report it and stop. If no, then T is still squarefree.

To implement the above, we use an $O(n)$-time method to obtain the values of $|LCSu^{-1}(q,q')|$ and $|LCPr^{+1}(q,q')|$ for all $q' \in \{1, 2, \ldots, n\}$ with $q' \neq q$ as follows. First create a string $S = T[(q+1)..n] \circ T[1..(q-1)]$, where $\circ$ denotes concatenation, of length $n - 1$. Then, for all $j \in \{1, 2, \ldots, (n-1)\}$, compute the length of the longest common prefix of $S[j..(n-1)]$ and $S[1..(n-q)]$ in $O(n)$ total time based on the method on p. 8 in [11] for computing the length of the longest common prefix of $S[j..(n-1)]$ and $S[1..(n-1)]$ for every j. Clearly, this will give us all the values of $|LCPr^{+1}(q,q')|$ for $q' \neq q$. To compute the $|LCSu^{-1}(q,q')|$-values, we repeat the above steps but create $S = T[1..(q-1)]^R \circ T[(q+1)..n]^R$ instead, where A^R means the reverse of string A.

Theorem 3. *The dynamic squarefree recognition problem for general alphabets can be solved in $O(n)$ time per update, where n is the length of the input string.*

We end this section by describing how the above algorithm can be extended to the *online dynamic* squarefree recognition problem that also allows characters to be appended to the current T. Given any update $T[q] :=$ 'x', where $q \in \{1, 2, \ldots, (n+1)\}$, if $1 \leq q \leq n$ then perform the same steps as above. If $q = n+1$ then position $n+1$ must be the endpoint of any possible newly formed square according to the key observation. In this case, calculate $|LCSu^{-1}(n+1,j)|$ for all $j \in \{\lceil \frac{n+1}{2} \rceil, \ldots, n\}$ and note that the resulting T contains a square if and only if $T[n+1] = T[j]$ and $|LCSu^{-1}(n+1,j)| \geq n-j$ for some j as in Lemma 4; use this fact to report any newly formed square. As above, the time needed for one update is $O(n)$, where n is the length of the current T. We obtain:

Theorem 4. *The online dynamic squarefree recognition problem for general alphabets can be solved in $O(n)$ time per update, where n is the current length of the input string.*

5 Concluding Remarks

We have presented an efficient algorithm for the online version of the squarefree recognition problem for arbitrarily ordered alphabets which runs in $O(n \log n)$

time. In comparison, the fastest known offline algorithms for determining if a string of length n over an ordered alphabet is squarefree [1, 7, 17] also run in $O(n \log n)$ time. Moreover, we have provided a simple algorithm for a dynamic version of the problem for general alphabets with $O(n)$ time per update.

Some interesting open questions are:

- Is the running time of our algorithm optimal, i.e., does there exist a lower bound of $\Omega(n \log n)$ for determining squarefreeness of strings over ordered alphabets? Note that the $\Omega(n \log n)$ bound in [17] assumes a *general* alphabet; for ordered alphabets, no lower bound (except for the trivial $\Omega(n)$ bound) has been proved for the offline case.
- Can the online squarefree recognition problem for *constant* alphabets be solved in $O(n)$ time?
- Can the running time of the LPT algorithm [15] be reduced to $O(n \log n)$ for general alphabets?
- How efficiently can the online and dynamic versions of the *cube* (and higher orders of repetitions) detection problem be solved?

References

1. A. Apostolico and F. P. Preparata. Optimal off-line detection of repetitions in a string. *Theoretical Computer Science*, 22(3):297–315, 1983.
2. D. R. Bean, A. Ehrenfeucht, and G. F. McNulty. Avoidable patterns in strings of symbols. *Pacific Journal of Mathematics*, 85(2):261–294, 1979.
3. G. Benson. Tandem repeats finder: A program to analyze DNA sequences. *Nucleic Acids Research*, 27(2):573–580, 1999.
4. D. Breslauer. *Efficient String Algorithmics*. PhD thesis, Columbia University, 1992.
5. A. T. Castelo, W. Martins, and G. R. Gao. TROLL – tandem repeat occurrence locator. *Bioinformatics*, 18(4):634–636, 2002.
6. C. Choffrut and J. Karhumäki. Combinatorics of words. In A. Salomaa and G. Rozenberg, editors, *Handbook of Formal Languages*, volume 1, pages 329–438. Springer-Verlag, 1997.
7. M. Crochemore. An optimal algorithm for computing the repetitions in a word. *Information Processing Letters*, 12(5):244–250, 1981.
8. M. Crochemore. Transducers and repetitions. *Theoretical Computer Science*, 45:63–86, 1986.
9. M. Crochemore and W. Rytter. *Jewels of Stringology*. World Scientific Publishing, 2002.
10. M. Farach-Colton, P. Ferragina, and S. Muthukrishnan. On the sorting-complexity of suffix tree construction. *Journal of the ACM*, 47(6):987–1011, 2000.
11. D. Gusfield. *Algorithms on Strings, Trees, and Sequences*. Cambridge University Press, New York, 1997.
12. J. Karhumäki. Automata on words. In *Proceedings of the 8th International Conference on Implementation and Application of Automata* (CIAA 2003), volume 2759 of *LNCS*, pages 3–10. Springer, 2003.
13. V. Kumar. Algorithms for constraint satisfaction problems: A survey. *AI Magazine*, 12(1):32–44, 1992.

14. J. H. M. Lee, H.-F. Leung, and H. W. Won. Performance of a comprehensive and efficient constraint library based on local search. In *Proceedings of the 11th Australian Joint Conference on Artificial Intelligence*, pages 191–202, 1998.
15. H.-F. Leung, Z. Peng, and H.-F. Ting. An efficient online algorithm for square detection. In *Proceedings of the 10th International Computing and Combinatorics Conference* (COCOON 2004), volume 3106 of *LNCS*, pages 432–439. Springer, 2004.
16. M. G. Main, W. Bucher, and D. Haussler. Applications of an infinite square-free co-CFL. *Theoretical Computer Science*, 49(2–3):113–119, 1987.
17. M. G. Main and R. J. Lorentz. An $O(n \log n)$ algorithm for finding all repetitions in a string. *Journal of Algorithms*, 5(3):422–432, 1984.
18. M. G. Main and R. J. Lorentz. Linear time recognition of squarefree strings. In A. Apostolico and Z. Galil, editors, *Combinatorial Algorithms on Words*, volume F 12 of *NATO ASI Series*, pages 271–278. Springer-Verlag, 1985.
19. J. H. Y. Wong and H.-F. Leung. Solving fuzzy constraint satisfaction problems with fuzzy GENET. In *Proceedings of the 10th IEEE International Conference on Tools with Artificial Intelligence*, pages 184–191, 1998.

Shrinking Restarting Automata*

Tomasz Jurdziński[1] and Friedrich Otto[2]

[1] Institute of Computer Science, University of Wrocław,
51-151 Wrocław, Poland
tju@ii.uni.wroc.pl
[2] Fachbereich Mathematik/Informatik, Universität Kassel,
34109 Kassel, Germany
otto@theory.informatik.uni-kassel.de

Abstract. Restarting automata are a restricted model of computation that is motivated by the so-called *analysis by reduction*. A computation of a restarting automaton consists of a sequence of cycles such that in each cycle the automaton performs exactly one rewrite step, which replaces a small part of the tape content by another, even *shorter* word. Here we consider a natural generalization of this model, called *shrinking restarting automaton*, where we require that there exists a *weight function* such that each rewrite step decreases the weight of the tape content with respect to that function. While it is still unknown whether the two most general types of restarting automata, the RWW-automaton and the RRWW-automaton, differ in their expressive power, we will see that the classes of languages accepted by the shrinking RWW-automaton and the shrinking RRWW-automaton coincide. Further, we will relate shrinking RRWW-automata to finite-change automata, which leads to new insights into the relationships between the classes of languages characterized by (shrinking) restarting automata and some well-known time and space complexity classes.

1 Introduction

The restarting automaton was introduced by Jančar et. al. as a formal tool to model the *analysis by reduction*, which is a technique used in linguistics to analyze sentences of natural languages [8]. This technique consists in a stepwise simplification of a given sentence in such a way that the syntactical correctness or incorrectness of the sentence is not affected. It is applied primarily in languages that have a free word order. Already several programs used in Czech and German (corpus) linguistics are based on the idea of restarting automata [17,19].

A (one-way) restarting automaton, RRWW-automaton for short, is a device M that consists of a finite-state control, a flexible tape containing a word delimited by sentinels, and a read/write window of a fixed size. This window moves from left to right along the tape until the control decides (nondeterministically)

---

* This work was supported by a grant from the Deutsche Forschungsgemeinschaft. It was performed while T. Jurdziński was visiting the University of Kassel.

that the content of the window should be rewritten by some *shorter* string. In fact, the new string may contain auxiliary symbols that do not belong to the input alphabet. After a rewrite, M can continue to move its window to the right until it either halts and accepts, or halts and rejects, or restarts, that is, it places its window over the left end of the tape, and reenters the initial state. Thus, each computation of M can be described through a sequence of cycles. It is easily seen that M can be simulated by a nondeterministic single-tape Turing machine that runs in quadratic time using only linear space, that is, the language $L(M)$ accepted by M belongs to the complexity class CSL ∩ NP.

By requiring that a restarting automaton must always perform a restart step immediately after executing a rewrite operation, we obtain the so-called RWW-automaton. Although the definition of the RWW-automaton is clearly much more restricted than that of the RRWW-automaton, it is a long-standing open problem whether the class of languages $\mathcal{L}$(RWW) accepted by RWW-automata is a proper subclass of the class of languages $\mathcal{L}$(RRWW) accepted by RRWW-automata.

Many well-known classes of formal languages admit characterizations in terms of restricted variants of the restarting automaton. For example, the class of Church-Rosser languages CRL of McNaughton et al. [14] coincides with the class of languages that are accepted by the deterministic variant of the RWW- and the RRWW-automaton [15,16], the class of context-free languages CFL is characterized by the monotone variants of the RWW- and the RRWW-automaton [9], and the class of deterministic context-free languages DCFL is characterized by several different variants of monotone deterministic RWW- and RRWW-automata [9]. In addition, the class of growing context-sensitive languages GCSL considered by Dahlhaus and Warmuth [7] coincides with the class of languages that are accepted by the weakly monotone variant of the RWW- and the RRWW-automaton [10]. Observe that in all these particular cases the considered variant of the RWW-automaton is just as powerful as the corresponding variant of the RRWW-automaton. On the other hand, it is known that for some types of restarting automata without auxiliary symbols, the RWW-variant is strictly less powerful than the RRWW-variant [9]. For a recent survey on restarting automata see [18].

In the present paper we consider a generalization of the restarting automaton, called *shrinking restarting automaton*. A shrinking restarting automaton M is defined just as a restarting automaton with the one exception that it is no longer required that each rewrite step $u \to v$ of M must be length-reducing. Instead there must exist a weight function ω that assigns a positive integer $\omega(a)$ to each letter a of M's tape alphabet Γ such that, for each rewrite step $u \to v$ of M, $\omega(u) > \omega(v)$ holds. Here the function ω is extended to a morphism $\omega : \Gamma^* \to \mathbb{N}$ as usual. Obviously, a shrinking restarting automaton can still be simulated by a nondeterministic single-tape Turing machine in quadratic time and linear space. Observe that similar generalizations have been considered for other types of automata [6], for grammar systems [5], and for string-rewriting systems [2].

The shrinking restarting automaton was introduced in [11], where it was shown that monotone (as well as left-monotone) shrinking restarting automata still characterize the class of context-free languages, and that deterministic shrinking RRWW-automata that are left-monotone are not more expressive than left-monotone deterministic RWW-automata.

Here we study the expressive power of the (nondeterministic) shrinking restarting automaton in general, where we only consider those variants that admit auxiliary symbols in addition to the input alphabet. After restating the basic definitions in Section 2, we establish our first main result in Section 3, which states that the shrinking RWW-automaton is just as expressive as the shrinking RRWW-automaton. As a corollary of our proof, we obtain a reduction by injective morphisms from the language class $\mathcal{L}$(RRWW) to the language class $\mathcal{L}$(RWW). This in itself is a major improvement of the reduction from $\mathcal{L}$(RRWW) to $\mathcal{L}$(RWW) presented in [10]. It clearly indicates that these classes may be very difficult to separate if they are indeed different.

In Section 4 we establish our second main result by giving a characterization of the class of languages accepted by shrinking RWW-automata in terms of the class of finite-change automata as introduced by von Braunmühl and Verbeek [3]. This characterization implies that this class of languages is actually contained in the class of deterministic context-sensitive languages DCSL, thus improving on the best previously known upper bound for $\mathcal{L}$(RRWW), and that it contains the class Q of quasi-realtime languages [1], which coincides with the complexity class NTIME(lin) (the class of languages that are accepted by nondeterministic multi-tape Turing machines in linear time). In the concluding section we summarize our results and state some open problems.

2 Definitions

Throughout the paper ε will denote the empty word, and $\mathbb{N}_+$ will denote the set of all positive integers.

A (one-way) *restarting automaton*, RRWW-automaton for short, is a nondeterministic machine M with a finite-state control Q, a finite tape alphabet Γ containing the input alphabet Σ, a flexible tape, and a read/write window of a fixed size $k \geq 1$. The work space is limited by the left sentinel ¢ and the right sentinel \$, which cannot be removed from the tape. The behaviour of M is described by a transition relation δ that associates to a pair (q, u) consisting of a state q and a possible content u of the read/write window a finite set of possible transition steps. There are four types of transition steps:

1. A *move-right step* (MVR) causes M to shift the read/write window one position to the right and to change the state. However, the read/write window cannot move across the right sentinel \$.
2. A *rewrite step* causes M to replace the content u of the read/write window by a word $v \in \Gamma^*$ satisfying $|v| < |u|$, thereby shortening the tape, and to change the state. Further, the read/write window is placed immediately to the right of the string v.
3. A *restart step* causes M to place its read/write window over the left end of the tape, and to reenter the initial state q_0.
4. An *accept step* causes M to halt and accept.

If $\delta(q,u) = \emptyset$ for some pair (q,u), then M necessarily halts, and we say that M *rejects* in this situation. In addition, it is required that rewrite steps and restart steps occur alternatingly within each computation of M, beginning with a rewrite step.

A *configuration* of M is a string $\alpha q \beta$, where $q \in Q$, and either $\alpha = \varepsilon$ and $\beta \in \{\updownarrow\} \cdot \Gamma^* \cdot \{\$\}$ or $\alpha \in \{\updownarrow\} \cdot \Gamma^*$ and $\beta \in \Gamma^* \cdot \{\$\}$; here q represents the current state, $\alpha\beta$ is the current content of the tape, and it is understood that the window contains the first k symbols of β or all of β when $|\beta| \leq k$. A *restarting configuration* is of the form $q_0 \updownarrow w\$$, where $w \in \Gamma^*$; if $w \in \Sigma^*$, then $q_0 \updownarrow w\$$ is an *initial configuration*. Thus, initial configurations are a particular type of restarting configurations. Each computation of M can be described by a sequence of *cycles*, where a cycle begins with a restarting configuration and ends with the next restarting configuration. The part of the computation after the last restart operation is called the *tail* of the computation.

An input word $w \in \Sigma^*$ is *accepted* by M, if there is a computation which, starting with the initial configuration $q_0 \updownarrow w\$$, finishes by executing an Accept instruction. By $L(M)$ we denote the language consisting of all words accepted by M; we say that M accepts (*recognizes*) the language $L(M)$.

In general, an RRWW-automaton is nondeterministic, that is, for some pairs (q,u), there may be more than one applicable transition step. If that is not the case, then the automaton is deterministic.

Also some restricted classes of restarting automata have been studied. An RWW-*automaton* is an RRWW-automaton that is required to execute a restart step immediately after performing a rewrite step. An RRW-automaton is an RRWW-automaton which does not use any auxiliary symbols, that is, its tape alphabet coincides with its input alphabet. Finally, an RR-automaton is an RRW-automaton whose rewrite instructions can be viewed as deletions, that is, if $(q',v) \in \delta(q,u)$, then v is a scattered subword of u. Obviously, the restrictions on the rewrite operation can be combined with the restriction on the restart operation, which leads to the RW-automaton and the R-automaton.

Finally, we come to the main topic of this paper, the shrinking restarting automaton. A *shrinking restarting automaton* $M = (Q, \Sigma, \Gamma, \updownarrow, \$, q_0, k, \delta)$ is defined in the same way as a 'standard' restarting automaton with one exception. Namely, it is not required that a rewrite operation reduces the length of the tape. Instead, there must exist a *weight function* $\omega : \Gamma \to \mathbb{N}_+$ such that, for each rewrite step $(q',v) \in \delta(q,u)$ of M, $\omega(u) > \omega(v)$ holds. Here ω is extended to a morphism $\omega : \Gamma^* \to \mathbb{N}$ by taking $\omega(\varepsilon) := 0$ and $\omega(wa) := \omega(w) + \omega(a)$ for all $w \in \Gamma^*$ and $a \in \Gamma$. Obviously, the length function $w \mapsto |w|$ is a particular weight function.

To be precise the weight function should also be defined for the delimiters $\updownarrow$ and \$, which are not elements of Γ. However, as the restarting automaton is not allowed to remove $\updownarrow$ or \$ from the tape nor to create new occurrences of these symbols, their weight does not influence the difference $\omega(u) - \omega(v)$ for any rewrite step $(q',v) \in \delta(q,u)$. Therefore, in order to simplify the notation, we do not assign weights to $\updownarrow$ and \$.

If an automaton M is shrinking with respect to a weight function ω, we say that ω is a weight function *compatible* with M. In order to distinguish the 'standard' variant of the restarting automaton from the shrinking restarting automaton, we sometimes denote the former as *length-reducing* restarting automaton.

Notation. For any class A of automata, $\mathcal{L}(A)$ will denote the class of languages that can be accepted by automata from A. The class of shrinking RRWW-automata is denoted by sRRWW, and similarly the class of shrinking RWW-automata is denoted by sRWW.

Let $\omega : \Gamma \to \mathbb{N}_+$ be a weight function, where Γ is a finite alphabet, and let $\Diamond$ and $\triangle$ be two new symbols not contained in Γ. Then $r_\omega : \Gamma^* \to (\Gamma \cup \{\Diamond\})^*$ denotes the morphism that is induced by defining $r_\omega(a) := a\Diamond^{\omega(a)-1}$ for each $a \in \Gamma$. Thus, for each word $u \in \Gamma^*$, $|r_\omega(u)| = \omega(u)$. Further, for $i \in \mathbb{N}_+$, we take r_i to denote the morphism $r_i : (\Gamma \cup \{\Diamond\})^* \to (\Gamma \cup \{\Diamond, \triangle\})^*$ that is defined by $r_i(a) := a\triangle^{i-1}$ for all $a \in \Gamma \cup \{\Diamond\}$. Thus, for each word $u \in (\Gamma \cup \{\Diamond\})^*$, $|r_i(u)| = i \cdot |u|$. Observe that r_ω as well as r_i ($i \geq 1$) are encodings, that is, injective morphisms.

3 sRWW-Automata Versus sRRWW-Automata

In this section we will see that the families of languages accepted by sRWW- and sRRWW-automata coincide. Moreover, we present a reduction by injective morphisms from the language class $\mathcal{L}(\text{RRWW})$ to the language class $\mathcal{L}(\text{RWW})$.

We begin our investigation by establishing a reduction from shrinking to length-reducing restarting automata.

Lemma 1. *If M is an sR(R)WW-automaton, and if ω is a weight function that is compatible with M, then $r_\omega(L(M)) \in \mathcal{L}(\text{R(R)WW})$.*

Proof. Let M be an sR(R)WW-automaton accepting a language $L \subseteq \Sigma^*$, and let $\omega : \Gamma \to \mathbb{N}_+$ be a weight function that is compatible with M. Then an R(R)WW-automaton M' for the language $r_\omega(L)$ is obtained by simply simulating the computation of M on a tape content $¢x\$$ on the corresponding tape content $¢r_\omega(x)\$$. As each rewrite instruction $u \to v$ of M is weight-reducing with respect to ω, the corresponding rewrite instruction $r_\omega(u) \to r_\omega(v)$ of M' is length-reducing. □

Next we come to the technical main result of this section relating sRRWW-automata to sRWW-automata.

Lemma 2. *Let M be an sRRWW-automaton that accepts a language $L \subseteq \Sigma^*$, and let ω be a weight function compatible with M. Then there exists an sRWW-automaton M' such that $L(M') = L(M)$, and M' is compatible with a weight function ω' that satisfies the equality $\omega'(a) = 54 \cdot \omega(a)$ for each input letter $a \in \Sigma$.*

We postpone the proof of Lemma 2 to Section 3.1. The next theorem, which is our first main result, is an immediate consequence of that lemma.

Theorem 1. $\mathcal{L}(\mathsf{sRWW}) = \mathcal{L}(\mathsf{sRRWW})$.

In addition, we obtain the following reduction from the language class $\mathcal{L}(\mathsf{RRWW})$ to the class $\mathcal{L}(\mathsf{RWW})$.

Theorem 2. *For each language* $L \in \mathcal{L}(\mathsf{RRWW})$, $r_{54}(L) \in \mathcal{L}(\mathsf{RWW})$.

Proof. Let M be an RRWW-automaton that accepts a language $L \subseteq \Sigma^*$. Then M can be interpreted as a shrinking RRWW-automaton that is compatible with the weight function that associates the weight 1 to each symbol. Thus, by Lemma 2, there exists an sRWW-automaton M' such that $L(M') = L$, and M' is compatible with a weight function ω' that assigns the weight 54 to each input symbol $a \in \Sigma$. Now Lemma 1 implies that $r_{\omega'}(L) \in \mathcal{L}(\mathsf{RWW})$. As $r_{\omega'}$ maps each symbol $a \in \Sigma$ onto the word $a\Diamond^{53}$, while r_{54} maps a onto the word $a\triangle^{53}$, it is clear that with $r_{\omega'}(L)$ also $r_{54}(L)$ is accepted by some RWW-automaton. □

Theorem 2 is a remarkable improvement over the reduction from $\mathcal{L}(\mathsf{RRWW})$ to $\mathcal{L}(\mathsf{RWW})$ presented in [10] (see Theorem 3 below), which, although being computable in linear-time, even maps regular languages to non-context-free languages, indicating that it is not well-behaved from a language theoretical point of view.

3.1 Proof of Lemma 2

First, we recall the reduction from $\mathcal{L}(\mathsf{RRWW})$ to $\mathcal{L}(\mathsf{RWW})$ from [10] mentioned above.

Theorem 3. *Let L be a language over Σ, let $a, \triangle, c \notin \Sigma$ be three additional symbols, and let the mapping $\varphi : \Sigma^* \to (\Sigma \cup \{a, \triangle, c\})^*$ be defined by*

$$\varphi(x) := a^{3 \cdot |x|} \cdot r_3(x) \cdot c^{3 \cdot |x|}$$

for all $x \in \Sigma^*$. If $L \in \mathcal{L}(\mathsf{RRWW})$, then $\varphi(L) := \{\, \varphi(x) \mid x \in L \,\} \in \mathcal{L}(\mathsf{RWW})$.

Let M be an sRRWW-automaton with input alphabet Σ that accepts a language $L \subseteq \Sigma^*$, and let ω be a weight function that is compatible with M. From Lemma 1 we see that $r_\omega(L) \in \mathcal{L}(\mathsf{RRWW})$, and Theorem 3 implies that $\varphi(r_\omega(L))$ is accepted by some RWW-automaton M_φ. Further, as φ and r_ω are both injective mappings, it follows that $\varphi(r_\omega(x)) \neq \varphi(r_\omega(y))$ for all $x, y \in \Sigma^*$, $x \neq y$. Thus, we have the following equivalence for all $x \in \Sigma^*$:

$$x \in L \text{ if and only if } \varphi(r_\omega(x)) \in \varphi(r_\omega(L)). \tag{1}$$

In order to prove Lemma 2 we will now construct an sRWW-automaton that, given a word $x \in \Sigma^*$ as input, first transforms x into the word $z := \varphi(r_\omega(x))$, and then simulates the computation of the (length-reducing) RWW-automaton M_φ on the input z. Thus, this sRWW-automaton will accept on input x if and only if z belongs to the language $\varphi(r_\omega(L))$, that is by (1), if and only if x belongs to the language L.

Our method of transforming x into $\varphi(r_\omega(x))$ is nondeterministic. Thus, for a given x, there are several different words z that can be produced, only one of them the intended result $\varphi(r_\omega(x))$. Unfortunately we cannot possibly verify the correctness of the word z produced without destroying it, but we can at least guarantee that $z = \varphi(r_\omega(x))$ if z belongs at all to the set $\varphi(r_\omega(L))$. As the RWW-automaton M_φ accepts only inputs from this set, it follows that this property is sufficient for our purposes.

We give a high level description of the algorithm that is realized by the intended sRWW-automaton M'. The details of the implementation can be found in [12].

Let $x \in \Sigma^*$ be the given input. Our algorithm proceeds in three stages, which will be illustrated by an example below:

1. The word x is rewritten deterministically from left to right into $y := r_9(r_\omega(x))$, using a new alphabet A_0 of auxiliary letters. These letters are interpreted as describing two 'tracks,' the first of which now contains the word $y = r_9(r_\omega(x))$, while the second track is empty (we use the symbol $\perp$ to denote 'empty' content).
2. Now the content of the second track is rewritten into $a^p \cdot r_3(r_\omega(x)) \cdot c^p$, where $p := |r_3(r_\omega(x))|$. This is achieved by performing the following steps that all preserve the length of the tape:
   (a) $y' := r_3(x')$ is written as a prefix of the content of the second track for some word x' that is a supersequence of $r_\omega(x)$ (that is, $r_\omega(x)$ is a scattered subsequence of x').
   (b) The second track, containing a word of the form $y' \perp^m$ for some integer m, is rewritten into a word of the form $y'y'' \perp^p$, where $y'' := r_3(x'')$ for a scattered subsequence x'' of x', and $p := m - |y''| > 0$. In order to mark the border between y' and y'', a new subalphabet is used for y''.
   (c) The current content of the second track, $y'y'' \perp^p$, is rewritten deterministically from left to right into $z := a^{|y'|}y''c^p$.

Input	b d
Stage 1	b △△△△△△△△ d △△△△△△△△◇△△△△△△△△ ⊥⊥⊥⊥⊥⊥⊥⊥⊥⊥⊥⊥⊥⊥⊥⊥⊥⊥⊥⊥⊥⊥⊥⊥⊥
Stage 2(a)	b △△△△△△△△ d △△△△△△△△◇△△△△△△△△ b △△ d △△◇△△⊥⊥⊥⊥⊥⊥⊥⊥⊥⊥⊥⊥⊥⊥⊥⊥⊥
Stage 2(b)	b △△△△△△△△ d △△△△△△△△◇△△△△△△△△ b △△ d △△◇△△ b △△ d △△◇△△⊥⊥⊥⊥⊥⊥⊥⊥
Stage 2(c)	b △△△△△△△△ d △△△△△△△△◇△△△△△△△△ a a a a a a a a a b △△ d △△◇△△ c c c c c c c c

Fig. 1. Stages 1 to 2(c) of the computation of M' on input bd.

3. The computation of the RWW-automaton M_φ on input z is simulated on the second track of the tape. The automaton M' accepts if and only if this computation of M_φ is accepting.

Example
Let M be an sRRWW-automaton with input alphabet $\Sigma = \{b, d\}$, and let ω be a weight function compatible with M such that $\omega(b) = 1$ and $\omega(d) = 2$. Given the input word bd, in Stages 1–2(c) M' can execute the transformations displayed in Figure 1.

Here we have omitted some extra information that is stored in symbols, and that is needed to 'coordinate' the computation (see the implementation details in [12]). □

4 Restarting Automata Versus Finite-Change Automata

In [3] von Braunmühl and Verbeek introduced a model of the Turing machine that they called *finite-change automaton*. A finite-change automaton is a nondeterministic single-tape Turing machine A that is parameterized by a constant $k \in \mathbb{N}_+$ and a function $f : \mathbb{N} \to \mathbb{N}$ satisfying $f(n) \geq n$ for all $n \in \mathbb{N}$. Given an input of length n (as the initial inscription of its tape), A must not visit more than $f(n)$ cells, and it must not change the content of any cell more than k times during any accepting computation on the given input. By $k\mathsf{C}(f)$ we denote the class of finite-change automata meeting these restrictions, and by $\mathsf{FC}(f)$ we denote the union

$$\mathsf{FC}(f) := \bigcup_{k>0} k\mathsf{C}(f).$$

For the special case of the identity function $f(n) = n$, we denote the corresponding classes of finite-change automata by $k\mathsf{C}$ and FC, respectively.

In order to enable the finite-change automaton to recognize the left end and the right end of the given input, we assume here that the initial tape content for a given input $x \in \Sigma^*$ is of the form $\mathfrak{c}x\$$, where $\mathfrak{c}$ and $\$$ are special markers not contained in Σ. Hence, the length of the initial tape inscription is $n := |x| + 2$. Another option would be to use particularly marked symbols for the first and the last letter of the input word. That these approaches are equivalent follows from the following technical result of [3].

Lemma 3. *Let $f : \mathbb{N} \to \mathbb{N}$ be a function satisfying $f(n) \geq n$ for each $n \in \mathbb{N}$. Then $\mathcal{L}(\mathsf{FC}(f(n))) = \mathcal{L}(\mathsf{FC}(c \cdot f(n)))$ for each constant $c \geq 1$.*

Now we can state the announced characterization.

Theorem 4. $\mathcal{L}(\mathsf{FC}) = \mathcal{L}(\mathsf{sRRWW})$.

Proof. Let M be an sRRWW-automaton accepting a language $L \subseteq \Sigma^*$. By Lemma 1, there exist an injective morphism r_ω and a (length-reducing) RRWW-automaton M' with input alphabet $\Sigma \cup \{\Diamond\}$ such that, for each word $x \in \Sigma^*$,

$x \in L$ if and only if $r_\omega(x) \in L(M')$. According to [13], Theorems 1 and 2, there exist an injective morphism φ and an RR-automaton M'' such that, for all words $x' \in (\Sigma \cup \{\Diamond\})^*$, $x' \in L(M')$ if and only if $\varphi(x') \in L(M'')$. Thus, we see that, for all words $x \in \Sigma^*$, $x \in L$ if and only if $\varphi'(x) \in L(M'')$, where φ' is the morphism obtained by the composition of r_ω and φ.

Now the finite-change automaton A can proceed as follows:

1. First the image $\varphi'(x)$ of the input word x is written onto the part of the tape immediately to the right of the given input.
2. Then the computation of the RR-automaton M'' on $\varphi'(x)$ is simulated.

Obviously, A accepts the language L with space bound $(c+1) \cdot n$, where $c := \max\{ |\varphi'(a)| \mid a \in \Sigma \}$. During the first part of the computation the content of each cell is changed at most once. During the second part of the computation the content of each cell is again changed at most once by marking those symbols that are deleted by the RR-automaton.

Hence, A is a two-change automaton for the language L that uses only linear space. From Lemma 3 it follows that $L = L(A) \in \mathcal{L}(\mathsf{FC})$. Thus, $\mathcal{L}(\mathsf{sRRWW}) \subseteq \mathcal{L}(\mathsf{FC})$.

To prove the converse inclusion, let $A \in \mathsf{FC}$ be a finite-change automaton which changes each tape cell at most j times, and that meets the following restrictions during accepting computations:

- rewrite operations and head movements of A are performed by different steps;
- the working alphabet Γ of A consists of $j+1$ disjoint subalphabets $\Gamma_0, \ldots, \Gamma_j$, where Γ_0 is the input alphabet, and the i-th change of the content of a cell produces an element of Γ_i for all $i = 1, \ldots, j$.

Now we present a simulation of the finite-change automaton A by an sRRWW-automaton M. If Q is the set of states of A, then we take $\Lambda := \Gamma \cup (\Gamma \times Q \times \{0,1\})$ to be the tape alphabet of M. The automaton M works as follows:

1. If there are no auxiliary symbols on the tape, M simulates that part of a computation of A that starts with an initial configuration and that ends when A changes the content of a tape cell for the first time, or that ends with A accepting or rejecting without changing the content of any tape cell. In the latter case M also accepts or rejects, respectively, in a tail computation. In the former case assume that A executes the instruction $\delta(q,a) = (q',b)$. M determines this tape cell, and then M rewrites the symbol a into the symbol $(b, q', 0)$. There is a slight problem due to the fact that M is a one-way device, while A' can move its head both to the left and to the right. However, in each step M can simply guess the corresponding crossing sequence of A' and verify the correctness of its guess, using the standard method of simulating a two-way finite-state acceptor by a one-way finite-state acceptor.
2. If the tape content is of the form $wa'z$ for some words $w, z \in \Gamma^*$ and a symbol $a' = (a, q, 0) \in \Gamma \times Q \times \{0\}$, then M simulates that part of a computation

of A on waz that starts in state q at the position of the symbol a and that ends when A changes the content of the next tape cell, or that ends with A accepting or rejecting, if no more tape cells are changed. Assume that A executes the instruction $\delta(\hat{q}, b) = (q', c)$. Then M replaces the symbol b by the symbol $b' = (c, q', 1)$. Should the positions of the symbols b and a' coincide, then M replaces the symbol a' by b'.
3. If the tape content is of the form $wa'y$ for a symbol $a' = (a, q, 0) \in \Gamma \times Q \times \{0\}$ and some words w, y satisfying $wy \in \Gamma^* \cdot (\Gamma \times Q \times \{1\}) \cdot \Gamma^*$, then M simply rewrites the symbol a' into the symbol a.
4. If the tape content is of the form $wa'z$ for some words $w, z \in \Gamma^*$ and a symbol $a' = (a, q, 1) \in \Gamma \times Q \times \{1\}$, then M replaces the symbol a' by the symbol $(a, q, 0) \in \Gamma \times Q \times \{0\}$.

Note that the third coordinate of the symbols from $\Gamma \times Q \times \{0, 1\}$ is used to distinguish between the positions of the last and the last but one change operation of A in the current computation. One can easily verify that each accepting computation of M corresponds to an accepting computation of A. Further, for each accepting computation of A, there exists an accepting computation of M. Thus, we conclude that $L(M) = L(A)$ holds. Also it is easily verified that each rewrite operation of M is weight-reducing with respect to an appropriately chosen weight function. Hence, M is indeed an sRRWW-automaton for the language $L(A)$. □

This characterization allows us to establish some new relationships between (shrinking) restarting automata on the one hand and some more classical language classes on the other hand. Let DCSL denote the class of deterministic context-sensitive languages, and let Q denote the class of *quasi-realtime languages*. This is the class of languages that are accepted by nondeterministic multi-tape Turing machines in realtime, that is, Q = NTIME(n). It properly contains the class GCSL of growing context-sensitive languages [4], and it coincides with the complexity class NTIME(lin) [1].

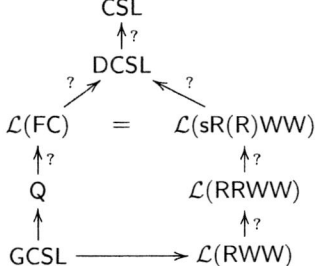

Fig. 2. An unmarked arrow indicates that the inclusion is proper, while a question mark indicates that it is an open problem whether the corresponding inclusion is proper. For those classes that are not connected via directed paths in the diagram it is open whether any inclusions hold.

From Theorems 2 and 3 of [3] the inclusions $\mathsf{Q} \subseteq \mathcal{L}(\mathsf{FC}) \subseteq \mathsf{DCSL}$ follow, which yield our second main result.

Corollary 1. $\mathsf{Q} \subseteq \mathcal{L}(\mathsf{sRRWW}) \subseteq \mathsf{DCSL}$.

Hence, we obtain the relationships depicted in Figure 2. This improves on the previously known results, as it was open whether the language classes $\mathcal{L}(\mathsf{RRWW})$ and DCSL are at all comparable under inclusion.

5 Concluding Remarks

We have investigated the expressive power of the shrinking restarting automaton, which is a rather straightforward generalization of the restarting automaton. For this generalization the model with combined restart and rewrite operations (the shrinking RWW-automaton) is as powerful as the general shrinking RRWW-automaton, and even the additional capabilities of changing the internal state in a restart transition (as opposed to resetting the state to the initial state) and to perform multiple rewrite operations in each cycle do not increase the expressive power of this model [12].

However, it remains open whether the shrinking RRWW-automaton is at all more powerful than the standard RRWW-automaton. Further, although we have obtained a new simplified reduction from the language class $\mathcal{L}(\mathsf{RRWW})$ to the language class $\mathcal{L}(\mathsf{RWW})$, we still have no clue whether or not these two classes coincide. Finally, it is not known whether either of the inclusions $\mathsf{Q} \subseteq \mathcal{L}(\mathsf{sRRWW})$ or $\mathcal{L}(\mathsf{sRRWW}) \subseteq \mathsf{DCSL}$ is proper.

Here we have only studied shrinking restarting automata with auxiliary letters. An obvious direction for future research is the study of the expressive power and the properties of shrinking restarting automata without auxiliary symbols.

Acknowledgements. The authors want to thank F. Mráz and M. Plátek for many fruitful discussions regarding restarting automata and for pointing them to the work of von Braunmühl and Verbeek.

References

1. R.V. Book and S. Greibach, Quasi-realtime languages. *Mathematical Systems Theory*, 4 (1970) 97–111.
2. R.V. Book and F. Otto, *String-Rewriting Systems*. Springer, New York, 1993.
3. B. von Braunmühl and R. Verbeek, Finite-change automata. In K. Weihrauch (ed.), *4th GI Conference, Proc., Lecture Notes in Computer Science* 67, Springer, Berlin, 1979, 91–100.
4. G. Buntrock, *Wachsende kontext-sensitive Sprachen*. Habilitationsschrift, Fakultät für Mathematik und Informatik, Universität Würzburg, 1996.
5. G. Buntrock and K. Loryś, On growing context-sensitive languages. In W. Kuich (ed.), *Automata, Languages and Programming, ICALP'92, Proc., Lecture Notes in Computer Science* 623, Springer, Berlin, 77–88.

6. G. Buntrock and F. Otto, Growing context-sensitive languages and Church-Rosser languages. *Information and Computation*, 141 (1998) 1–36.
7. E. Dahlhaus and M. Warmuth, Membership for growing context-sensitive grammars is polynomial. *Journal of Computer and System Sciences*, 33 (1986) 456–472.
8. P. Jančar, F. Mráz, M. Plátek and J. Vogel, Restarting automata. In H. Reichel (ed.), *FCT'95, Proc.*, Lecture Notes in Computer Science 965, Springer, Berlin, 1995, 283–292.
9. P. Jančar, F. Mráz, M. Plátek and J. Vogel, On monotonic automata with a restart operation. *Journal of Automata, Languages and Combinatorics*, 4 (1999) 287–311.
10. T. Jurdziński, K. Loryś, G. Niemann and F. Otto, Some results on RWW- and RRWW-automata and their relationship to the class of growing context-sensitive languages. *Mathematische Schriften Kassel* 14/01, Universität Kassel, 2001. Also: *Journal of Automata, Languages and Combinatorics*, to appear.
11. T. Jurdziński and F. Otto, On left-monotone restarting automata. *Mathematische Schriften Kassel* 17/03, Universität Kassel, 2003.
12. T. Jurdziński and F. Otto, Shrinking restarting automata. *Mathematische Schriften Kassel* 1/05, Universität Kassel, 2005.
13. T. Jurdziński, F. Otto, F. Mráz and M. Plátek, On the complexity of 2-monotone restarting automata. *Mathematische Schriften Kassel* 4/04, Universität Kassel, 2004. An extended abstract appeared in C.S. Calude, E. Calude and M.J. Dinneen (eds.), *Developments in Language Theory, DLT 2004, Proc.*, Lecture Notes in Computer Science 3340, Springer, Berlin, 2004, 237–248.
14. R. McNaughton, P. Narendran and F. Otto, Church-Rosser Thue systems and formal languages. *Journal of the Association for Computing Machinery*, 35 (1988) 324–344.
15. G. Niemann and F. Otto, Restarting automata, Church-Rosser languages, and representations of r.e. languages. In G. Rozenberg and W. Thomas (eds.), *Developments in Language Theory - Foundations, Applications, and Perspectives, DLT 1999, Proc.*, World Scientific, Singapore, 2000, 103–114.
16. G. Niemann and F. Otto, Further results on restarting automata. In M. Ito and T. Imaoka (eds.), *Words, Languages and Combinatorics III, Proc.*, World Scientific, Singapore, 2003, 352–369.
17. K. Oliva, K. Květoň and R. Ondruška, The computational complexity of rule-based part-of-speech tagging. In V. Matousek and P. Mautner (eds.), *TSD 2003, Proc.*, Lecture Notes in Computer Science 2807, Springer, Berlin, 2003, 82–89.
18. F. Otto, Restarting automata and their relations to the Chomsky hierarchy. In Z. Esik and Z. Fülöp (eds.), *Developments in Language Theory, DLT 2003, Proc.*, Lecture Notes in Computer Science 2710, Springer, Berlin, 2003, 55-74.
19. M. Plátek, M. Lopatková and K. Oliva, Restarting automata: Motivations and applications. In M. Holzer (ed.), *Workshop "Petrinets" und 13. Theorietag "Automaten und Formale Sprachen"*, Institut für Informatik, Technische Universität München, Garching, 2003, 90–96.

Removing Bidirectionality from Nondeterministic Finite Automata

Christos Kapoutsis

Computer Science and Artificial Intelligence Laboratory,
Massachusetts Institute of Technology
cak@mit.edu

Abstract. We prove that every *two-way* nondeterministic finite automaton with n states has an equivalent *one-way* nondeterministic finite automaton with at most $\binom{2n}{n+1}$ states. We also show this bound is exact.

1 Introduction

Converting an arbitrary *one-way nondeterministic finite automaton* (1NFA) to an equivalent *one-way deterministic finite automaton* (1DFA) has long been the archetypal problem of descriptional complexity. Rabin and Scott [1][2] proved that starting with an n-state 1NFA one can always construct an equivalent 1DFA with at most $2^n - 1$ states;[1] later observations [3,4][5,6,7,8] established the tightness of this upper bound, in the strong sense that, for all n, some n-state 1NFA has no equivalent 1DFA with fewer than $2^n - 1$ states. So, we often say that the *trade-off* from 1NFAs to 1DFAs is *exactly* $2^n - 1$. (Fig. 1a.)

The fact that this problem initiated the discussion on issues of descriptional complexity is only one aspect of its significance. A more interesting aspect is that its solution fully uncovered and elegantly described the relationship between the computations of the two types of machines. This is supported not only by the fact that the demonstrated upper and lower bounds match *exactly* (as opposed to merely asymptotically), but also —and more crucially— by the central role that a well-understood set-theoretic object plays in the associated proof: what the theorem really tells us is that every 1NFA N can be simulated by a 1DFA that has one distinct state for each *non-empty subset of states of N* which (as an instantaneous description of N) is both realizable and irreplaceable. From this, the demonstrated trade-off is then only a counting argument away, plus a clever search for 1NFAs that indeed manage to keep all of their instantaneous descriptions realizable and irreplaceable.

In the present study we offer a similar analysis for the conversion of an arbitrary *two-way nondeterministic finite automaton* (2NFA) to a one-way equivalent: we prove that the trade-off from 2NFAs to 1NFAs is exactly $\binom{2n}{n+1}$. As above, we first identify the correct set-theoretic object that 'lives' in the relation between 2NFA and 1NFA computations, and then easily extract the trade-off.

---

[1] In this article, all finite automata are allowed to be *incomplete*: their transition functions may be partial, and thus computations may hang inside the input.

Two-way finite automata were introduced in the late 50's [9,2,10] and shown equivalent to 1DFAs. Originally, their definition included neither endmarkers nor nondeterminism, so they were just *single-pass two-way deterministic finite automata* (ZDFAs). However, they soon grew into full-fledged *two-way deterministic finite automata* (2DFAs) and nondeterministic versions (ZNFAs and 2NFAs), which all remained equivalent to 1DFAs. Since then, the cost of the 2NFA-to-1NFA conversion has been addressed sporadically.

Shepherdson's proof [10] implied every n-state 2NFA can be converted into a 1NFA with at most $n2^{n^2}$ states. A cheaper simluation via crossing sequences [11, Sect. 2.6] has also been known, with only $O(2^{2n \lg n})$ states. However, a straightforward elaboration on [10] shows the cost can be brought down to even $n(n+1)^n$. Which still wastes exponentially many states, as Birget [12] showed $8^n + 2$ are always enough, via an argument based on length-preserving homomorphisms. Here, we establish the still exponentially smaller upper bound of $\binom{2n}{n+1}$.

On the other hand, exponential lower bounds have also been known, even when the starting automaton is deterministic [7,12] and single-pass [6,13]. For example, Damanik [13] gives a language that costs $\leq 4n + 2$ on ZDFAs, but $\geq 2^n$ on 1NFAs. Here, we give a lower bound that matches $\binom{2n}{n+1}$, even when we start from a ZDFA. Hence, *the ability of a* 2NFA *to move its head bidirectionally* strictly inside the input *can alone cause all the hardness a simulating* 1NFA *must subdue.*

The conversions from 1NFAs to 1DFAs and from 2NFAs to 1NFAs are only two of a dozen different conversions among the four basic automata models (1DFAs, 1NFAs, 2DFAs, and 2NFAs) for the regular languages. Each of the 12 arrows in Fig. 1 represents one of these conversions and defines the associated problem of finding the exact trade-off. Of the 12 problems, some are little harder than clever exercises, but others are decades-old open questions on the power of nondeterminism —a surprising range in difficulty. We present a quick review.

The arguments of [2,3,4] and this study establish that $a = 2^n - 1$ and $d = e = \binom{2n}{n+1}$, while an argument of [14] shows the trade-off for every conversion from a weaker to a stronger model (dotted arrows) is exactly $f = n$. From 2DFAs and 2NFAs to 1DFAs, it has been known that the trade-offs are exponential [10,15,4,5,7,16,12], although the exact values remained elusive; refining [12] and following the rational of this study, we can show them to be as in Fig. 1 (the proofs to appear in the full version of this article). This leaves only the questions for the conversions from 1NFAs and 2NFAs to 2DFAs (dashed arrows), which remain wide open: more than 30 years after they were first asked [6], not only are the exact trade-offs unkown, but we cannot even confirm the conjecture that they are exponential (see [17] for a discussion).

Finally, we should note that Fig. 1 shows only four of the numerous automata that solve exactly the regular languages. Bidirectionality, nondeterminism, alternation, randomness, pebbles, and other enhancements, alone or combined, limited or unrestricted, give rise to a long list of variants and to the corresponding descriptional-complexity questions. See [18] for a comprehensive overview.

The next section defines the basic concepts. Section 3 establishes the upper bound, while Sect. 4 proves that it is exact. We conclude in Sect. 5.

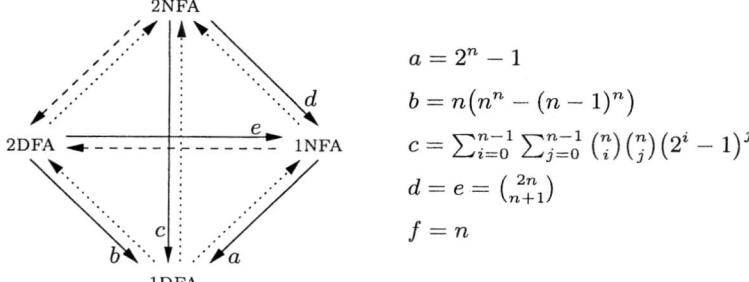

Fig. 1. Trade-off summary ($f = n$ on all dotted arrows; dashed arrows are open)

2 Preliminaries

We write $[n]$ for the set $\{1, 2, \ldots, n\}$. The special objects $\mathtt{l}, \mathtt{r}, \perp$ are used for building the *disjoint union* of two sets and the *augmentation* of a set

$$A \uplus B = (A \times \{\mathtt{l}\}) \cup (B \times \{\mathtt{r}\}) \quad \text{and} \quad A_\perp = A \cup \{\perp\}.$$

When A and B are disjoint, $A \cup B$ is also written as $A + B$. The size of A is denoted by $|A|$, while $(A \to B)$ denotes the set of functions from A to B.

For Σ an alphabet, we use Σ^* for the set of all finite strings over Σ and Σ_e for $\Sigma + \{\vdash, \dashv\}$, where $\vdash$ and $\dashv$ are special endmarking symbols. If w is a string, $|w|$ is its length and w_i is its i-th symbol, for $i = 1, \ldots, |w|$. The 'i-th *boundary* of w' is the boundary between w_i and w_{i+1}, if $i = 1, \ldots, |w| - 1$; or the leftmost (rightmost) boundary of w, if $i = 0$ ($i = |w|$). (Fig. 2a.) We also write w_e for the extension $\vdash w \dashv$ of w and $w_{e,i}$ for $(w_e)_i$. The empty string is denoted by ϵ.

We present 2NFAs and 1NFAs as variations of the more natural model of a 2DFA. The next paragraph defines this model and some basic relevant concepts.

Two-Way Deterministic Finite Automata. We assume the reader is familiar with the intuitive notion of a 2DFA. Formally, this is a triple $M = (s, \delta, f)$, where δ is the *transition function*, partially mapping $Q \times \Sigma_e$ to $Q \times \{\mathtt{l}, \mathtt{r}\}$, for a set Q of *states* and an alphabet Σ, while s, f are the *start* and *final* states.

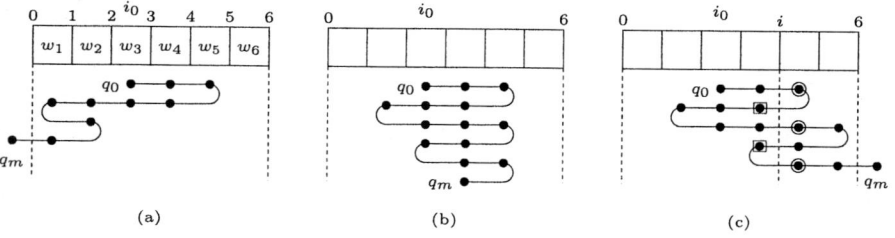

Fig. 2. (a) Cells and boundaries on a 6-long w; a computation that hits left. (b) One that hangs. (c) One that hits right, and its i-th frontier: R_i^c in circles and L_i^c in boxes.

Here, we insist the automaton *accepts* only if it moves past the right endmarker $\dashv$ into f, this being the only case in which an endmarker may be violated.[2]

Computations. Although M is typically started at s and on $\vdash$, many other possibilities exist: for any w, i, q, the *computation of M when started at state q on the i-th symbol of string w* is the unique sequence

$$\text{COMP}_{M,q,i}(w) = \big((q_t, i_t)\big)_{0 \leq t \leq m}$$

where $(q_0, i_0) = (q, i)$, $0 \leq m \leq \infty$, every pair is derived from its predecessor via δ and w, every pair is within w ($1 \leq i_t \leq |w|$) except possibly for the last one, and the last pair is within w iff δ is undefined on the corresponding state and symbol. We say (q_t, i_t) is the *t-th point* and m the *length* of this computation. If $m = \infty$, the computation *loops*. Otherwise, it *hits left* into q_m, if $i_m = 0$; or *hangs*, if $1 \leq i_m \leq |w|$; or *hits right* into q_m, if $i_m = |w| + 1$ (Fig. 2). When $i = 1$ (respectively, $i = |w|$) we get the *left* (*right*) *computation of M from q on w*:[3]

$$\text{LCOMP}_{M,q}(w) ::= \text{COMP}_{M,q,1}(w) \quad \text{and} \quad \text{RCOMP}_{M,q}(w) ::= \text{COMP}_{M,q,|w|}(w).$$

The *computation of M on w* refers to the typical $\text{COMP}_M(w) ::= \text{LCOMP}_{M,s}(w_e)$, so that M accepts $w \in \Sigma^*$ iff $\text{COMP}_M(w)$ hits right (into f). Note that, since M can violate an endmarker only when it moves past $\dashv$ into f, a computation of M on an endmarked string can only loop, or hang, or hit right into f.

Frontiers. Fix a computation $c = ((q_t, i_t))_{0 \leq t \leq m}$ and consider the i-th boundary of the input (Fig. 2c). This is crossed ≥ 0 times. Collect into a set R_i^c (respectively, L_i^c) all states that result from a left-to-right (right-to-left) crossing,

$$R_i^c = \{q_{t+1} \mid 0 \leq t < m \ \& \ i_t = i \ \& \ i_{t+1} = i+1\},$$
$$L_i^c = \{q_{t+1} \mid 0 \leq t < m \ \& \ i_t = i+1 \ \& \ i_{t+1} = i\},$$

also requiring that $R_{i_0-1}^c$ contains q_0.[4] The pair (L_i^c, R_i^c) partially describes the behavior of c over the i-th boundary and we call it the *i-th frontier of c*. Note that the description is indeed partial, as the pair contains no information on the order in which c exhibits the states around the boundary or on number of times each state is exhibited. For a full description one would need instead the *i-th crossing sequence of c* [11]. However, for our purposes, the extra information provided by the complete description is redundant.

Variations. If in the definition of M above more than one next moves are allowed at each step, we say the automaton is *nondeterministic* (a 2NFA). This

---

[2] So, on $\vdash$, δ moves right or hangs. On $\dashv$, it moves left, hangs, or moves right into f.
[3] Note that, when w is the empty string, the left computation of M from q on w is just $\text{LCOMP}_{M,q}(\epsilon) = ((q,1))$ and therefore hits *right* into q, whereas the corresponding right computation $\text{RCOMP}_{M,q}(\epsilon) = ((q,0))$ hits *left* into q.
[4] This reflects the convention that the starting state of any computation is considered to be the result of an 'invisible' left-to-right step.

formally means that δ *totally* maps $Q \times \Sigma_e$ to the *powerset* of $Q \times \{\mathtt{l}, \mathtt{r}\}$ and implies that $C = \text{COMP}_{M,q,i}(w)$ is now a *set* of computations. If then

$$P = \{p \mid \text{some } c \in C \text{ hits right into } p\},$$

we say C *hits right into* P; and w is accepted iff $\text{COMP}_M(w)$ hits right into $\{f\}$.

If the head of M never moves to the left, we say M is *one-way* (a 1NFA).[5] If no computation of M 'continues after arriving at an endmarker', we say M is *single-pass* (a ZNFA; or a ZDFA, if M is deterministic).

3 The Upper Bound

Fix an n-state 2NFA $N = (s, \delta, f)$ over alphabet Σ and state set Q. In this section we build an equivalent $\binom{2n}{n+1}$-state 1NFA N' via an optimal construction.

Frontiers. Assume momentarily that N is deterministic and $c = \text{COMP}_N(w)$ is accepting, for some l-long input w. Consider the i-th frontier (L_i^c, R_i^c) of c, for some $i \neq 0, l+2$. The number of states in R_i^c equals the number of times c left-to-right crosses the i-th boundary: each crossing contributes a state into R_i^c and no two crossings contribute the same state, or else c would be looping. Similarly, $|L_i^c|$ equals the number of times c right-to-left crosses the i-th boundary. Now, since c accepts, it goes from $\vdash$ all the way past $\dashv$, forcing the rightward crossings on every boundary to be exactly 1 more than the leftward crossings. Hence,

$$|L_i^c| + 1 = |R_i^c|,$$

which remains true even on the leftmost boundary ($i = 0$, under our convention from Footn. 4) and also on the rightmost one ($i = l+2$). So, the equality holds over every boundary and motivates the following definition.

Definition 1. *A* frontier *of N is any (L, R) with $L, R \subseteq Q$ and $|L| + 1 = |R|$.*

So, *if the computation of a* 2DFA *on a particular input is accepting, then all frontiers of the computation are frontiers of this* 2DFA.

For our nondeterministic N, though, the argument breaks, as a state repetition under a cell may not imply looping. However, it does imply a cycle. So, let us call a computation *minimal* if it contains no cycles (i.e., if every two of its points are distinct) and repeat the previous argument to establish the following.

Lemma 1. *All frontiers of an* accepting minimal *computation of N on some input are frontiers of N.*

Compatibilities Among Frontiers. Suppose c is an accepting minimal computation of N on an l-long w and let $F_i^c = (L_i^c, R_i^c)$ be its i-th frontier, for each $i = 0, 1, \ldots, l+2$ (Fig. 3). Note that the first and last frontiers are always

$$F_0^c = (\emptyset, \{s\}) \quad \text{and} \quad F_{l+2}^c = (\emptyset, \{f\}),$$

as c starts at s, ends in f, and never right-to-left crosses an outer boundary.

---
[5] Note that our 1NFAs work on *endmarked* inputs, a deviation from typical definitions.

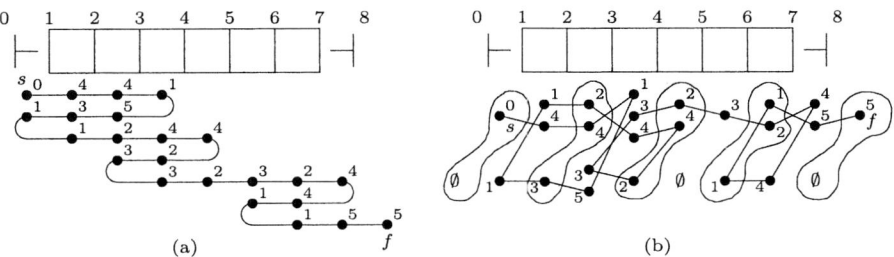

Fig. 3. (a) An accepting minimal $c \in \text{COMP}_N(w)$, for $|w| = 6$, state set $\{0, 1, \ldots, 5\}$, $s = 0$, $f = 5$. (b) The same c arranged in frontiers; the even-indexed ones are circled.

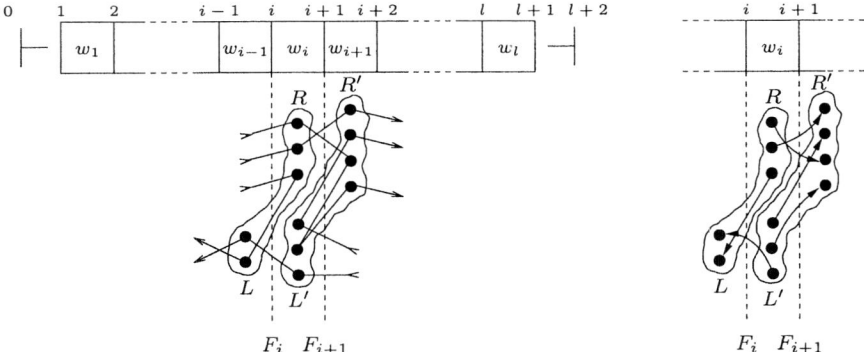

Fig. 4. (a) Two successive frontiers, on the left. (b) The associated ρ, on the right.

Also note that, for $(L, R) = (L_i^c, R_i^c)$ and $(L', R') = (L_{i+1}^c, R_{i+1}^c)$ two successive frontiers (Fig. 4a), it should always be that $R \cap L' = \emptyset$: otherwise, c would be repeating a state under w_i and would not be minimal. Hence, $R + L'$ contains as many states as many (occurences of) states there are in L and R' together:

$$|R + L'| = |R| + |L'| = (|L| + 1) + (|R'| - 1) = |L| + |R'| = |L \uplus R'|.$$

So, bijections can be found from $R + L'$ to $L \uplus R'$. Among them, a very natural one (Fig. 4b): for each $q \in R + L'$ find the unique step in c that produces q under w_i (this is either a rightward crossing of boundary i or a leftward crossing of boundary $i + 1$; the minimality of c guarantees uniqueness); the next step left-to-right crosses boundary $i + 1$ into some state $p \in R'$ or right-to-left crosses boundary i into some $p \in L$; depending on the case, map q to $(p, \mathbf{r})$ or $(p, \mathbf{l})$ respectively. If $\rho : R + L' \to L \uplus R'$ is this mapping, it is easy to verify that it is injective (because c is minimal) and therefore bijective, as promised. In addition, ρ clearly respects the transition function: $\rho(q) \in \delta(q, w_i)$, for all $q \in R + L'$.[6]

Overall, we see that the sequence of frontiers exhibited by an accepting minimal $c \in \text{COMP}_N(w)$ obeys some restrictions. We now formally summarize them.

---

[6] Throughout this argument, w_i really refers to $w_{c,i+1}$. This is w_i only when $i \neq 0, l+1$.

Definition 2. Let (L, R), (L', R') be frontiers of N and $a \in \Sigma_e$. We say that (L, R) is a-compatible to (L', R') iff $R \cap L' = \emptyset$ and some bijection $\rho: R + L' \to L \uplus R'$ respects the transition function on a: for all $q \in R + L'$: $\rho(q) \in \delta(q, a)$.

Definition 3. Suppose $w \in \Sigma^*$ is l-long and $F_0, F_1, \ldots, F_{l+2}$ is a sequence of frontiers of N. We say the sequence fits w iff
 1. $F_0 = (\emptyset, \{s\})$,
 2. for all $i = 0, 1, \ldots, l + 1$: F_i is $w_{e, i+1}$-compatible to F_{i+1},
 3. $F_{l+2} = (\emptyset, \{f\})$.

Lemma 2. If $\text{COMP}_N(w)$ contains an accepting computation, then some sequence of frontiers of N fits w.

Proof. Every accepting computation gives rise to a *minimal* accepting one. □

The Main Observation. The converse of Lemma 2 is also true: *if a sequence of frontiers of N fits w, then $\text{COMP}_N(w)$ contains an accepting computation.*

To prove this, fix an l-long w and assume some sequence of frontiers of N

$$F_0 = (L_0, R_0), \quad F_1 = (L_1, R_1), \quad \ldots, \quad F_{l+2} = (L_{l+2}, R_{l+2})$$

fits w. We show the stronger claim that, for every i, the states of R_i can be produced by $|R_i|$ right-hitting computations on $\vdash w_1 \cdots w_{i-1}$: one starting at s and on $\vdash$, each of the others starting at some $q \in L_i$ and on w_{i-1}.

Claim. For all $i = 0, 1, \ldots, l + 2$, some bijection $\pi_i : (L_i)_\perp \to R_i$ is such that
 1. some $c \in \text{LCOMP}_{N,s}(\vdash w_1 \cdots w_{i-1})$ hits right into $\pi_i(\perp)$, and
 2. for all $q \in L_i$, some $c \in \text{RCOMP}_{N,q}(\vdash w_1 \cdots w_{i-1})$ hits right into $\pi_i(q)$.

(Here, we take w_0 and w_{l+1} to mean the endmarkers $\vdash$ and $\dashv$, respectively.) Note that our main observation follows from this claim for $i = l + 2$.

To prove the claim, we use induction on i. The base case $i = 0$ is trivial. For the inductive step (Fig. 5a), assume $i < l + 2$, let $(L, R) = (L_i, R_i)$, $(L', R') = (L_{i+1}, R_{i+1})$, $a = w_{e,i+1}$, and consider the bijections guaranteed by the inductive hypothesis, $\pi = \pi_i : L_\perp \to R$, and the fact that (L, R) is a-compatible to (L', R'), $\rho : R + L' \to L \uplus R'$. Based on π, ρ and a third function σ, we build a bijection $\pi' = \pi_{i+1} : (L')_\perp \to R'$ that satisfies (1), (2) of the claim. First, we introduce σ.

Definition of σ: pick some $q \in R$ and take a trip around under $\vdash w_1 w_2 \cdots w_{i-1} a$

$$\underset{r_0}{q}, \; \underset{r_1}{\rho(q)}, \; \underset{r_2}{\pi\rho(q)}, \; \underset{r_3}{\rho\pi\rho(q)}, \; \underset{r_4}{\pi\rho\pi\rho(q)}, \; \ldots \tag{1}$$

by alternately following ρ and π, until the first time 'ρ fails to return a state in L'.[7] Let $r_0, r_1, r_2, \ldots$ be the states that we visit. We distinguish two cases.

---
[7] Note that we abuse notation here. Bijection ρ can only return a *pair* of the form $(p, \mathbf{1})$ or $(p, \mathbf{r})$. So, in the description (1) above, $\rho(\cdot)$ really means 'the first component of $\rho(\cdot)$, if the second component is $\mathbf{1}$'. Similarly, 'ρ fails to return a state in L' means 'ρ returns a pair of the form $(p, \mathbf{r})$'. Hopefully, the abuse does not confuse.

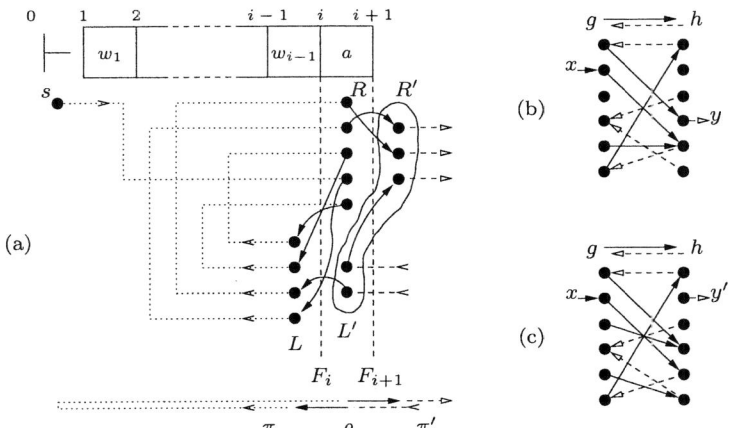

Fig. 5. (a) An example for the inductive step for the main observation. E.g., note that σ maps the 3rd and 5th (from top) states of R to $\bot$, while the 4th state is mapped to the 1st state of R'. (b) A nice input, that has a path. (c) A nice input with no path.

Case 1: ρ does eventually fail to return a state in L. Then the trip is finite: $r_0, r_1, \ldots, r_k$, for an even $k \geq 0$ and an $r_k \in R$ that is ρ-mapped to some $q' \in R'$.

Case 2: ρ always returns a state in L. Then the trip is infinite and (since all even-indexed r_i and all odd-indexed r_i are respectively inside the finite sets R and L) there exist repetitions of states both on the even and on the odd indices. Let k be the first index for which some earlier index $j < k$ of the *same parity* points to the same state: $r_j = r_k$. If k is odd, j is also odd and so $j \geq 1$; then $r_j = r_k \implies \rho^{-1}(r_j) = \rho^{-1}(r_k) \implies r_{j-1} = r_{k-1}$ and $k-1$ also has the property that k is the earliest one to have, a contradiction. So, k must be even, and so must j. In fact, j must be 0 —or we reach a contradiction, as before, with π^{-1} instead of ρ^{-1}. Hence, the first state to be revisited is $r_0 = q$ and the trip consists of infinitely many copies of a list $r_0, r_1, \ldots, r_{k-1}$, for some even $k \geq 2$ and with no two states being both equal and at indices of the same parity.

Overall, either we reach a state $r_k \in R$ that is ρ-mapped to a state $q' \in R'$ or we return to the starting state $q \in R$ having repeated no state in L and no state in R. We define $\sigma : R \to (R')_\bot$ to encode exactly this information: in Case 1, $\sigma(q) = q'$; in Case 2, $\sigma(q) = \bot$. In either case, our trip respects π and ρ, which in turn respect the behavior of N on $\vdash w_1 w_2 \ldots w_{i-1} a$. So, clearly: $\sigma(q) = q'$ implies some $c \in \text{RCOMP}_{N,q}(\vdash w_1 \cdots w_{i-1} a)$ respects π, ρ and *hits right* into q'; $\sigma(q) = \bot$ implies some *looping* $c \in \text{RCOMP}_{N,q}(\vdash w_1 \cdots w_{i-1} a)$ respects π, ρ and visits only states from R when under a. This concludes the definition of σ. □

We can now define π'. We examine three cases about its argument.

(a) some $p \in L'$ that is ρ-mapped to an $r \in R'$. Then we just let $\pi'(p) = r$.
(b) some $p \in L'$ that is ρ-mapped to an $r \in L$. Then we consider $q = \pi(r)$. We know N can start at p under a and eventually reach q under a, so we ask what can happen after that if we keep following ρ and π. We examine $\sigma(q)$.

If $\sigma(q) = \bot$, then we will return to q after a cycle of length ≥ 2, having visited only states of R when under a. But can this happen? If it does, then the next-to-last and last steps in this cycle will follow ρ and π respectively, ending in q. Since ρ, π are bijections, the last two states (before q) in this cycle must respectively be p and r. In particular, p must appear in the cycle under a. But, since the cycle stays within R whenever under a, we must have $p \in R$, and hence R and L' intersect. But then $(L, R), (L', R')$ are not compatible, a contradiction. Hence, we know $\sigma(q) = q' \in R'$. And we can safely set $\pi'(p) = q'$.

(c) the special value $\bot$. The reasoning is similar to the previous case. We consider $q = \pi(\bot)$ and examine $\sigma(q)$. Again, $\sigma(q) = \bot$ is impossible, as it would imply $\bot \in L$. Hence, $\sigma(q) = q'$ for some $q' \in R'$ and we can safely set $\pi'(\bot) = q'$.

This concludes the definition of π'. It is easy to check π' satisfies the conditions of the claim. Hence, the inductive step is complete, as is the overall proof.

The Construction. We now describe the 1NFA N' simulating N. By Lemma 2 and the main observation, N' need simply check if some sequence of frontiers of N fits the input. So, N' just 'guesses' such a sequence. This needs 1 state per frontier, and a standard argument shows N has exactly $\binom{2n}{n+1}$ of them.

4 The Lower Bound

In this section, we exhibit an n-state 2NFA N that has no equivalent 1NFA with fewer than $\binom{2n}{n+1}$ states. In fact, N will even be *deterministic* and *single-pass*.

The Witness. Fix $n \geq 1$ and consider the alphabet $\Gamma = ([n] + ([n] \to [n])) \times \{\mathtt{l}, \mathtt{r}\}$. Of all strings in Γ^*, we will only care about the ones following the pattern

$$(x, \mathtt{l})(g, \mathtt{l})(h, \mathtt{r})(y, \mathtt{r}) \tag{2}$$

where $x, y \in [n]$, g and h are partial functions from $[n]$ to $[n]$, and $h(y)$ is undefined. We call these strings *nice inputs*. Intuitively, given a nice input as (2), we think of the graph of Fig. 5b, where the columns are two copies of $[n]$, the arrows between them are determined by g (left-to-right) and h (right-to-left), and the two special nodes by x (entry point) and y (exit). In this graph, a path from x to y may or may not exist; if it does, we say the nice input 'has a path'.

Letting Π_{yes} (respectively, Π_{no}) be the set of nice inputs that (do not) have a path, we can easily see that the promise problem[8] $\Pi = (\Pi_{\text{yes}}, \Pi_{\text{no}})$ can be solved by a single-pass 2DFA with state set $[n]$. This is our witness, N.

Intuition. Consider an arbitrary frontier $F = (L, R)$ of N and list the elements of $L, R \subseteq [n]$ in increasing order, $L = \{x_1, \ldots, x_m\}$ and $R = \{y_1, \ldots, y_{m+1}\}$, for

---

[8] By a (*promise*) *problem* over Σ we mean a pair $\Pi = (\Pi_{\text{yes}}, \Pi_{\text{no}})$ of disjoint subsets of Σ^*. An automaton *solves* Π iff it accepts *every* $w \in \Pi_{\text{yes}}$ but *no* $w \in \Pi_{\text{no}}$, while arbitrary behavior is allowed on strings outside $\Pi_{\text{yes}} + \Pi_{\text{no}}$.

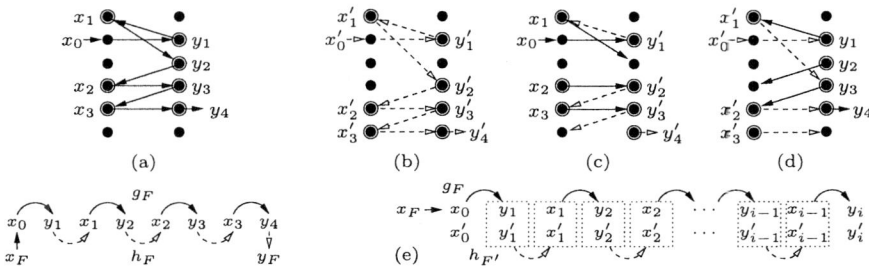

Fig. 6. (a) The input w_F when $n = 6$ and $F = (\{1,4,5\},\{2,3,4,5\})$, and how to derive it from the list $2,2,1,3,4,4,5,5$. (b) A new input $w_{F'}$, for $F' = (\{1,5,6\},\{2,4,5,6\})$. (c,d) Inputs $w_{F,F'}$ and $w_{F',F}$. (e) Proving that at most one of them has a path.

the appropriate $0 \leq m < n$. Since $m < n$, we know $L \neq [n]$ and we can name an element outside L, say $x_0 = \min \overline{L}$. Then the combined list

$$x_0 \, y_1 \, x_1 \, y_2 \, x_2 \, \cdots \, y_m \, x_m \, y_{m+1} \tag{3}$$

gives rise to the nice input $w_F = (x_F, \mathtt{l})(g_F, \mathtt{l})(h_F, \mathtt{r})(y_F, \mathtt{r})$ (see Fig. 6a), where $x_F = x_0$, function g_F maps every x in the list (3) to the following y, function h_F maps every $y \neq y_{m+1}$ to the following x, and $y_F = y_{m+1}$:

$$\begin{aligned} x_F &= \min \overline{L}, & y_F &= \max R, \\ g_F &= \{(x_i, y_{i+1}) \mid 0 \leq i \leq m\}, & h_F &= \{(y_i, x_i) \mid 1 \leq i \leq m\}. \end{aligned} \tag{4}$$

It is easy to verify that, *for any frontier F of N, the computation of N on w_F is accepting and its frontier under the middle boundary is exactly F*. This implies that, if N' is the 1NFA constructed for N as in Sect. 3, then *every* state of N' is used in *some* accepting computation. Which suggests N' is minimal.

The Proof. Every two frontiers F, F' of N give rise to the nice input (Fig. 6c)

$$w_{F,F'} = (x_F, \mathtt{l})(g_F, \mathtt{l})(h_{F'}, \mathtt{r})(y_{F'}, \mathtt{r}),$$

where x_F, g_F, $h_{F'}$, $y_{F'}$ are defined by (4). Crucially, in the $\binom{2n}{n-1} \times \binom{2n}{n+1}$ matrix $W = [w_{F,F'}]_{F,F'}$ containing all such inputs, two distinct strings at cells symmetric with respect to the main diagonal cannot both have a path.

Claim. For F, F' two frontiers of N: $w_{F,F'}, w_{F',F} \in \Pi_{\text{yes}} \iff F = F'$.

Proof. [$\Leftarrow$] Trivial. [$\Rightarrow$] Suppose $F = (L, R)$ and $F' = (L', R')$. We assume that $F \neq F'$ and prove that at least one of $w_{F,F'}$, $w_{F',F}$ lacks a path.

Let $m = |L|$, $m' = |L'|$ and consider the lists defined by F and F', as in (3):

$$x_0 \, y_1 \, x_1 \, y_2 \, x_2 \, \cdots \, y_m \, x_m \, y_{m+1} \quad \text{and} \quad x'_0 \, y'_1 \, x'_1 \, y'_2 \, x'_2 \, \cdots \, y'_{m'} \, x'_{m'} \, y'_{m'+1}.$$

If these were identical *after* their first elements, they would agree in their lengths, their x's (except possibly x_0, x'_0), and their y's, forcing $F = F'$, a contradiction. So, there must be positions of disagreement after 0. Consider the earliest one.

If this position is occupied by y's, say y_i and y'_i, then *either* $y_i < y'_i$ (Case 1) *or* $y_i > y'_i$ (Case 2). If it is occupied by x's, say x_i and x'_i, then *either* $x_i < x'_i$ or x'_i is not present at all[9] (Case 3) *or* $x_i > x'_i$ or x_i is not present at all (Case 4).

We present the argument for Case 1 —the rest are similar. So, suppose the first disagreement is $y_i < y'_i$. Then all previous positions after 0 contain identical elements (Fig. 6e). Also, y_i is not in R': if it were, then it would be in the sublist $y'_1, \ldots, y'_{i-1}$ (since $y_i < y'_i$), and hence in $y_1, \ldots, y_{i-1}$ (the two sublists coincide), a contradiction (since $y_1, \ldots, y_{i-1} < y_i$). So $y_i \notin R'$. Therefore $y_i \neq y_{F'}$ and $h_{F'}(y_i)$ is undefined. But then, searching for a path in $w_{F,F'}$, we travel

$$x_0 \xrightarrow{g_F} (y_1 = y'_1) \xrightarrow{h_{F'}} (x'_1 = x_1) \xrightarrow{g_F} (y_2 = y'_2) \xrightarrow{h_{F'}} \cdots \xrightarrow{h_{F'}} (x'_{i-1} = x_{i-1}) \xrightarrow{g_F} y_i$$

reaching a node which is neither the exit $y_{F'}$ nor the start of an $h_{F'}$-arrow. □

Now suppose a 1NFA A solves Π with fewer than $\binom{2n}{n+1}$ states. For each frontier F of N, we know $w_F = w_{F,F}$ is in Π_{yes}, so A accepts it. Pick an accepting $c_F \in \text{COMP}_A(w_F)$ and let q_F be the state right after the middle boundary is crossed. By the small size of A, we know $q_F = q_{F'}$ for some $F \neq F'$. But then, a standard cut-and-paste argument on $c_F, c_{F'}$ shows A also accepts $w_{F,F'}$ and $w_{F',F}$. Since both are nice inputs, we have $w_{F,F'}, w_{F',F} \in \Pi_{\text{yes}}$, contradicting the last claim.

5 Conclusion

We have shown the exact trade-off in the conversion from 2NFAs to 1NFAs. Our argument complemented that of Birget [12] by carefully removing some redundancies in its constructions. Crucially, the simulation performed by our optimal 1NFA is as 'meaningful' as the simulation given in [2] for the removal of nondeterminism from 1NFAs: each state corresponds to a (realizable and irreplaceable, as an instantaneous description) set-theoretic object that naturally 'lives' in the computations of the simulated 2NFA. Frontiers also allowed a set-theoretic *characterization of* 2NFA *acceptance* (already present in [12], essentially) that complements the set-theoretic *characterization of* 2NFA *rejection* given by Vardi [19]. Finally, by applying the concept of promise problems even to regular languages, we nicely confirmed its reputation for always leading us straight to the combinatorial core of the hardness of a computational task.

We do not know if the large alphabet size of problem Π is neccessary for the exactness of this trade-off. Also, it would be interesting to have the exact trade-offs in the conversions towards and from other types of automata (e.g., alternating, probabilistic) or more powerful machines (e.g., pushdown automata).

Many thanks to J.C. Birget for his help with some references; also, for raising our understanding of the subject to a level from which exact solutions could be seen.

---

[9] This happens if the list for F' stops at y'_i.

References

1. Rabin, M.O., Scott, D.: Remarks on finite automata. In: Proceedings of the Summer Institute of Symbolic Logic, Cornell (1957) 106–112
2. Rabin, M.O., Scott, D.: Finite automata and their decision problems. IBM Journal of Research and Development **3** (1959) 114–125
3. Ott, G.: On multipath automata I. Research report 69, SRRC (1964)
4. Meyer, A.R., Fischer, M.J.: Economy of description by automata, grammars, and formal systems. In: Proceedings of the Symposium on Switching and Automata Theory. (1971) 188–191
5. Moore, F.R.: On the bounds for state-set size in the proofs of equivalence between deterministic, nondeterministic, and two-way finite automata. IEEE Transactions on Computers **20** (1971) 1211–1214
6. Seiferas, J.I.: Manuscript communicated to M. Sipser (October 1973)
7. Sakoda, W.J., Sipser, M.: Nondeterminism and the size of two-way finite automata. In: Proceedings of the Symposium on the Theory of Computing. (1978) 275–286
8. Leung, H.: Separating exponentially ambiguous finite automata from polynomially ambiguous finite automata. SIAM Journal of Computing **27** (1998) 1073–1082
9. Rabin, M.O.: Two-way finite automata. In: Proceedings of the Summer Institute of Symbolic Logic, Cornell (1957) 366–369
10. Shepherdson, J.C.: The reduction of two-way automata to one-way automata. IBM Journal of Research and Development **3** (1959) 198–200
11. Hopcroft, J.E., Ullman, J.D.: Introduction to automata theory, languages, and computation. Addison-Wesley, Reading, MA (1979)
12. Birget, J.C.: State-complexity of finite-state devices, state compressibility and incompressibility. Mathematical Systems Theory **26** (1993) 237–269
13. Damanik, D.: Finite automata with restricted two-way motion. Master's thesis, J. W. Goethe-Universität Frankfurt (1996) (In German.)
14. Birget, J.C.: Two-way automata and length-preserving homomorphisms. Report 109, Department of Computer Science, University of Nebraska (1990)
15. Barnes, B.H.: A two-way automaton with fewer states than any equivalent one-way automaton. IEEE Transactions on Computers **C-20** (1971) 474–475
16. Sipser, M.: Lower bounds on the size of sweeping automata. Journal of Computer and System Sciences **21** (1980) 195–202
17. Kapoutsis, C.: Deterministic moles cannot solve liveness. (In preparation.)
18. Goldstine, J., Kappes, M., Kintala, C.M.R., Leung, H., Malcher, A., Wotschke, D.: Descriptional complexity of machines with limited resources. Journal of Universal Computer Science **8** (2002) 193–234
19. Vardi, M.Y.: A note on the reduction of two-way automata to one-way automata. Information Processing Letters **30** (1989) 261–264

Generating All Minimal Integral Solutions to Monotone ∧, ∨-Systems of Linear, Transversal and Polymatroid Inequalities*

L. Khachiyan[1,**], E. Boros[2], K. Elbassioni[3], and V. Gurvich[2]

[1] Department of Computer Science, Rutgers University, 110 Frelinghuysen Road, Piscataway NJ 08854-8003
[2] RUTCOR, Rutgers University, 640 Bartholomew Road, Piscataway NJ 08854-8003
{boros, gurvich}@rutcor.rutgers.edu
[3] Max-Planck-Institut für Informatik, Saarbrücken, Germany
elbassio@mpi-sb.mpg.de

Abstract. We consider monotone ∨, ∧-formulae ϕ of m atoms, each of which is a monotone inequality of the form $f_i(x) \geq t_i$ over the integers, where for $i = 1, \ldots, m$, $f_i : \mathbb{Z}^n \mapsto \mathbb{R}$ is a given monotone function and t_i is a given threshold. We show that if the ∨-degree of ϕ is bounded by a constant, then for linear, transversal and polymatroid monotone inequalities all minimal integer vectors satisfying ϕ can be generated in incremental quasi-polynomial time. In contrast, the enumeration problem for the disjunction of m inequalities is NP-hard when m is part of the input. We also discuss some applications of the above results in disjunctive programming, data mining, matroid and reliability theory.

1 Introduction

Consider a system of linear inequalities

$$\sum_{j=1}^{n} a_{ij} x_j \geq t_i \text{ for } i = 1, \ldots, m, \qquad (1)$$

where a_{ij} are given non-negative reals, and where we assume that the variables can take only binary values. Due to the non-negativity of the coefficients, if a vector $x \geq 0$ satisfies some of these inequalities and $y \geq x$, then y satisfies the same inequalities (and possibly some others as well), i.e., the system (1) is *monotone*. We say that $x \in \{0,1\}^n$ is a *minimal feasible solution* for a subset $I \subseteq \{1, \ldots, m\}$ of the inequalities (1), if x satisfies all inequalities belonging to I, and any binary vector $y \neq x$ such that $y \leq x$ violates at least one of these

---

* This research was partially supported by the National Science Foundation (Grant IIS-0118635), and by DIMACS, the National Science Foundation's Center for Discrete Mathematics and Theoretical Computer Science.
** Our friend and colleague, Leo Khachiyan passed away with tragic suddenness while we were finalizing this manuscript.

inequalities. Lawler, Lenstra and Rinnooy Kan [23] considered the problems of generating all minimal feasible solutions satisfying

(**P1**) all m inequalities of (1), and
(**P2**) at least one of the m inequalities of (1).

These or equivalent problems arise in a number of areas, including integer programming, scheduling, and polyhedral combinatorics (see e.g., [5,6,23,26,28]). Note that the number of minimal feasible solutions may not be limited by a polynomial of n and m. A generation algorithm is said to be *incrementally polynomial* (or quasi-polynomial[1], or exponential) if for an arbitrary subset $\mathcal{X}$ of minimal feasible solutions it can find an additional minimal feasible solution $x \notin \mathcal{X}$, or recognize that $\mathcal{X}$ contains all such solutions, in time polynomially (or quasi-polynomially, or exponentially) limited in n, m and $|\mathcal{X}|$. Equivalently, an algorithm is incrementally polynomial (quasi-polynomial, or exponential) if for arbitrary integer k, it can generate k minimal feasible solutions, or all of them if k is too large, in time polynomial (quasi-polynomial, exponential) in n, m and k.

When $m = 1$, problems (P1) and (P2) coincide, and incrementally efficient generation of all minimal feasible solutions is possible (see e.g., [23]). For the general case, when $m > 1$, only incrementally exponential algorithms were proposed for (P1) and (P2) in [23], and it was conjectured that no polynomial time algorithm can solve these problems, unless P=NP (i.e., that no algorithm can recognize the completeness of a subset $\mathcal{X}$ of the minimal feasible solutions in polynomial time, unless P=NP, and hence no algorithm can generate all minimal feasible solutions in time limited by a polynomial of n, m, and the number of minimal feasible vectors). These problems were reconsidered recently in [9], and contrary to the conjecture of [23], (P1) was shown to be tractable in incremental quasi-polynomial time (which makes it very unlikely to be NP-hard), while (P2) was shown to be tractable in incremental polynomial time for fixed m, but NP-hard for the general case.

This motivated us to study more complex monotone systems, and generalize the above results in three directions.

First, we generalize the above enumeration problems to *integer* variables. More precisely, we consider the inequalities (1) with the variables running over an arbitrary integer box $\mathcal{C} = \{x \in \mathbb{Z}^n \mid 0 \leq x \leq c\}$, where $c \in \mathbb{Z}_+^n$ is a given integer vector with possibly infinite coordinates.

Second, we consider not only conjunctions (like in (P1)), or disjunctions (like in (P2)), but *arbitrary monotone expressions* of linear inequalities. Specifically, let $Y_i : \mathcal{C} \to \{0, 1\}$ be the characteristic variables corresponding to the inequalities of (1), i.e., $Y_i(x) = 1$ iff $\sum_{j=1}^n a_{ij} x_j \geq t_i$, $i = 1, ..., m$. Then we associate to any given monotone $\vee, \wedge$- formula ϕ in m propositional variables a system Σ_ϕ of inequalities for which a vector $x \in \mathcal{C}$ is feasible iff $\phi(Y_1(x), Y_2(x), ..., Y_m(x)) = 1$. Since Σ_ϕ is a monotone system, the notion of *minimal feasible solutions* of Σ_ϕ is well defined, and we can consider the corresponding enumeration problem:

(**P3**) Generate all minimal feasible solutions of Σ_ϕ.

---
[1] A function $f(x)$ is *quasi-polynomial* if $f(x) = O(2^{polylog(x)})$.

Note that when $\phi = Y_1 \wedge Y_2 \wedge \cdots \wedge Y_m$, then (P3) is the integer variant of (P1), while for $\phi = Y_1 \vee Y_2 \vee \cdots \vee Y_m$, problem (P3) is the integer variant of (P2).

Let us further associate to each monotone $\vee, \wedge$-formula $\phi = \phi(Y_1, \ldots, Y_m)$ a polynomial $P_\phi \in \mathbb{Z}[y_1, \ldots, y_m]$ defined by replacing logical conjunctions by arithmetic additions, and logical disjunctions by arithmetic multiplications. For instance, if $\phi = Y_1 \vee (Y_2 \wedge (Y_3 \vee Y_4))$, then we have $P_\phi(y_1, y_2, y_3, y_4) = y_1(y_2 + (y_3 y_4)) = y_1 y_2 + y_1 y_3 y_4$. We call P_ϕ the *evaluation polynomial of* ϕ.

Theorem 1. *If the degree of the evaluation polynomial P_ϕ is bounded, then we can generate all minimal feasible solutions to system Σ_ϕ in incremental quasi-polynomial time in terms of n and m.*

Let us note that if the degree of F_ϕ is not bounded, then generating minimal feasible solutions for Σ_ϕ is NP-hard already for the Boolean case $\mathcal{C} = \{0,1\}^n$ of (P2), see [9].

Finally, in a third direction, we extend Theorem 1 to *transversal* and *polymatroid* inequalities.

Given a subset $\mathcal{H} \subseteq \mathcal{C}$ and non-negative real weights $w : \mathcal{H} \to \mathbb{R}_+$, we call the function

$$f_{\mathcal{H},w}(x) = \sum \{w(a) \mid a \in \mathcal{H},\ a \not\leq x\}$$

a *(weighted) transversal function* over $\mathcal{C}$. Note that for any given $x \in \mathcal{C}$ we can compute the value $f_{\mathcal{H},w}(x)$ in $O(n|\mathcal{H}|)$ time, and that for the Boolean case $\mathcal{C} = \{0,1\}^n$ we can equivalently define $f_{\mathcal{H},w}(x)$ as the total weight of all hyperedges of the hypergraph $\mathcal{H}$, whose complements intersect the support of x.

An integer-valued monotone function $f : \mathcal{C} \to \mathbb{Z}_+$ is called *polymatroid* if $f(0,\ldots,0) = 0$ and f is submodular, i.e., $f(x \vee y) + f(x \wedge y) \leq f(x) + f(y)$ holds for all vectors $x, y \in \mathcal{C}$, where $x \vee y = (\max\{x_j, y_j\} \mid j = 1,\ldots,n)$ and $x \wedge y = (\min\{x_j, y_j\} \mid j = 1,\ldots,n)$.

Let us note that transversal functions are both monotone and submodular, thus they are also polymatroid, whenever they take only integer values. In what follows we will be dealing with polymatroid functions whose values at any point $x \in \mathcal{C}$ can be evaluated in polynomial time. Some applications of monotone systems defined via transversal and polymatroid functions are discussed in Section 4.

Analogously to the case of linear inequalities, given a system of inequalities

$$f_i(x) \geq t_i, \quad i = 1, \ldots, m, \tag{2}$$

where $f_i(x)$, are monotone functions over $\mathcal{C}$ and $t_i \in \mathbb{R}$ for $i = 1, \ldots, m$, and given a monotone $\vee, \wedge$-formula ϕ in m variables, we can associate to (2) and ϕ a monotone system Σ_ϕ: A vector $x \in \mathcal{C}$ is called *feasible* for Σ_ϕ if $\phi(Y_1(x), \ldots, Y_m(x)) = 1$, where $Y_i(x) = 1$ iff $f_i(x) \geq t_i$, $i = 1, \ldots, m$. Let $\mathcal{F}_\phi$ denote the set of all minimal feasible solutions to system Σ_ϕ.

Theorem 2. *If (2) involves transversal inequalities, i.e., if we have $f_i = f_{\mathcal{H}^i, w^i}$ for $i = 1, \ldots, m$, and if the degree of the evaluation polynomial P_ϕ is bounded, then we can generate $\mathcal{F}_\phi$ in incremental quasi-polynomial time in terms of n, m and $\max_{1 \leq i \leq m} |\mathcal{H}^i|$.*

Note that for the Boolean case $\mathcal{C} = \{0,1\}^n$, linear inequalities are transversal inequalities corresponding to the hypergraph $\mathcal{H}$ consisting of the complements of n singletons $\{1\}, \{2\}, \ldots, \{n\}$. In particular, if the degree of P_ϕ is not bounded, then no efficient generation of minimal feasible solutions for Σ_ϕ is possible, unless P=NP.

Theorem 3. *If (2) involves polymatroid inequalities, and if the degree of the evaluation polynomial P_ϕ is bounded, then we can generate $\mathcal{F}_\phi$ in incremental quasi-polynomial time in terms of n, m and $\max_{1 \le i \le m} t_i$.*

Let us remark that the above theorem provides efficient bounds only if $\max_{1 \le i \le m} t_i$ is bounded by a polynomial or quasi-polynomial expression of n and m. In fact, generating all minimal feasible solutions to a single polymatroid inequality over $\{0,1\}^n$ is already NP-hard, if the right hand side is not bounded [8]. Due to this fact, integrality of polymatroid functions is essential in our analysis.

2 Our Approach and Further Results

Our approach is to utilize a general enumeration method for minimal elements of a monotone system, the so called *joint generation*, proposed first in [7,22], and analyzed at greater detail in [13]. For a subset $\mathcal{X} \subseteq \mathcal{C}$ let us denote by $\mathcal{I}(\mathcal{X})$ the set of all maximal vectors not above any vectors of $\mathcal{X}$, i.e., $\mathcal{I}(\mathcal{X}) = \{\text{maximal } y \in \mathcal{C} \mid \not\exists x \in \mathcal{X} : x \le y\}$. For instance, if $\mathcal{F}_\phi$ denotes the family of all minimal feasible solutions for the system Σ_ϕ, as before, then $\mathcal{I}(\mathcal{F}_\phi)$ is the set of all maximal infeasible solutions.

The method jointly generates $\mathcal{F} \cup \mathcal{I}(\mathcal{F})$ by iteratively extending two partial sets $\mathcal{X} \subseteq \mathcal{F}$ and $\mathcal{Y} \subseteq \mathcal{I}(\mathcal{F})$. This is done by solving the so-called *dualization problem*: either produce a point in $\mathcal{I}(\mathcal{X}) \setminus \mathcal{Y}$ or halt because $\mathcal{Y} = \mathcal{I}(\mathcal{X})$ (see e.g., [19]). In particular, incrementally extending a given subset $\mathcal{X} \subseteq \mathcal{F}$ requires solving at most $|\mathcal{I}(\mathcal{F}) \cap \mathcal{I}(\mathcal{X})|$ dualization problems each of size at most $|\mathcal{X}| + |\mathcal{I}(\mathcal{F}) \cap \mathcal{I}(\mathcal{X})|$ with n variables. Let us denote by $\delta(n, M)$ the complexity of solving a dualization problem in n variables and of size $M = |\mathcal{X}| + |\mathcal{Y}|$. Even though no polynomial upper bound on $\delta(n, M)$ is currently available, it is known that $\delta(n, M) = poly(n) M^{o(\log M)}$, see [9,20].

Note that the joint generation method utilizes the given inequalities in (2) only to check the feasibility of a given vector $x \in \mathcal{C}$, and that any method for generating $\mathcal{F} = \mathcal{F}_\phi$ that uses only such feasibility tests has to perform at least $|\mathcal{F}| + |\mathcal{I}(\mathcal{F})|$ tests (and in fact has to generate both $\mathcal{F}$ and $\mathcal{I}(\mathcal{F})$), see e.g.[21].

The above discussion shows that the quantity $q_\mathcal{F}(k)$ defined by

$$q_\mathcal{F}(k) = \max_{\mathcal{X} \subseteq \mathcal{F},\ |\mathcal{X}|=k} |\mathcal{I}(\mathcal{X}) \cap \mathcal{I}(\mathcal{F})| \qquad (3)$$

is an important parameter intimately related to the generation of the monotone system $\mathcal{F}$. We shall call it the *duality index* of the monotone system $\mathcal{F}$. Using this notion, let us then summarize the above in the following statement.

Proposition 1. *Given a system of monotone inequalities (2), and a monotone Boolean formula ϕ, let Σ_ϕ denote the monotone system associated to these, and let $\mathcal{F}_\phi$ denote the set of all minimal feasible solutions to Σ_ϕ (as above). Then, for an arbitrary subset $\mathcal{X} \subseteq \mathcal{F}_\phi$, we can find $x \in \mathcal{F}_\phi \setminus \mathcal{X}$, or recognize that $\mathcal{X} = \mathcal{F}_\phi$ in time $\ell\delta(n, k+\ell)$ time, where $k = |\mathcal{X}|$ and $\ell = q_{\mathcal{F}_\phi}(|\mathcal{X}|)$.* □

In particular, $\mathcal{F}$ can be generated in incremental quasi-polynomial time whenever the duality index $q_\mathcal{F}(k)$ is bounded by a quasi-polynomial. Hence in order to derive Theorems 1, 2, and 3 from Proposition 1, it now suffices to bound the duality index of $\mathcal{F}_\phi$. In the rest of the paper we focus on the duality index of various monotone systems, and in particular on obtaining the duality index of complex monotone systems from the duality indices of their component systems.

Given a monotone subset $\mathcal{A} \subseteq \mathcal{C}$, i.e., for which $x \in \mathcal{A}$, $y \geq x$ imply $y \in \mathcal{A}$, let us denote by $\min \mathcal{A}$ the set of all minimal vectors of $\mathcal{A}$. Next, if $\mathcal{A}$ is an arbitrary subset of $\mathcal{C}$ then $\mathcal{A}^+ = \{x \in \mathcal{C} \mid x \geq y \text{ for some } y \in \mathcal{A}\}$ denotes the minimum monotone subset (ideal) of $\mathcal{C}$ containing $\mathcal{A}$. Thus, if $\mathcal{F}$ is the set of all minimal feasible solutions of a monotone inequality $f(x) \geq t$ over $\mathcal{C}$, then $\mathcal{F}^+$ is the set of all feasible solutions for the same inequality.

If $\mathcal{A}^+$ and $\mathcal{B}^+$ are monotone subsets of $\mathcal{C}$, then their *conjunction* and *disjunction* are defined as $\mathcal{A}^+ \wedge \mathcal{B}^+ = \mathcal{A}^+ \cap \mathcal{B}^+$ and $\mathcal{A}^+ \vee \mathcal{B}^+ = \mathcal{A}^+ \cap \mathcal{B}^+$. Let us note that the same operations can naturally be extended to their sets of minimal elements, $\mathcal{A} = \min \mathcal{A}^+$ and $\mathcal{B} = \min \mathcal{B}^+$, by defining $\mathcal{A} \wedge \mathcal{B} = \min\{a \vee b \mid a \in \mathcal{A}, b \in \mathcal{B}\}$ and $\mathcal{A} \vee \mathcal{B} = \min \mathcal{A} \cup \mathcal{B}$. In particular, if $\mathcal{A}$ and $\mathcal{B}$ are the minimal feasible solutions of monotone inequalities over $\mathcal{C}$, then $\mathcal{A} \wedge \mathcal{B}$ consists of all minimal vectors satisfying both inequalities, while $\mathcal{A} \vee \mathcal{B}$ consists of all minimal vectors satisfying at least one of the inequalities. Using these definitions, we can thus talk about monotone formulae of arbitrary monotone systems.

Our first result shows that the duality index of conjunctions of monotone systems can effectively be limited in terms of the duality indices of the component systems.

Theorem 4. *Let $\mathcal{F}_i \subseteq \mathcal{C}$ be monotone systems for $i = 1, ..., m$, and let $\phi = Y_1 \wedge Y_2 \wedge \cdots \wedge Y_m$, i.e., $\mathcal{F}_\phi = \mathcal{F}_1 \wedge \mathcal{F}_2 \wedge \cdots \wedge \mathcal{F}_m$. Then, we have*

$$q_{\mathcal{F}_\phi}(k) \leq P_\phi(q_{\mathcal{F}_1}(k), \ldots, q_{\mathcal{F}_m}(k)) = \sum_{i=1}^m q_{\mathcal{F}_i}(k).$$

The above theorem implies that if (2) involves monotone inequalities the duality indices of which are (quasi)-polynomially bounded, then, so is the duality index of the entire system, implying a quasi-polynomially efficient generation of minimal feasible solutions, by Proposition 1.

In fact, an efficient bound on the duality index is known for several types of monotone inequalities. Let us denote temporarily by $\mathcal{A}$ the set of all minimal feasible solutions to a single monotone inequality $f(x) \geq t$. For a monotone linear function $f(x) = \sum_{j=1}^n a_j x_j \geq t$, where $a_j \geq 0$ for $j = 1, ..., n$, we have by [9] that

$$q_\mathcal{A}(k) \leq nk. \tag{4}$$

If $f(x) = f_{\mathcal{H},w}(x) \geq t$ is a weighted transversal function, then it was shown in [14] that
$$q_\mathcal{A}(k) \leq |\mathcal{H}|k, \tag{5}$$
regardless of the weights. Note that both of these bounds are sharp, up to a constant factor. Finally, if f is a polymatroid function, then it was shown in [10] that
$$q_\mathcal{A}(k) \leq max\{n, k^{(\log t)/c(n,k)}\}, \tag{6}$$
where $c(n, \beta)$ is the unique positive root of the equation $2^c(n^{c/\log \beta} - 1) = 1$ (note that the above bound implies that $q_\mathcal{A} \leq (nk)^{\log t}$ and that $c(n,k) \approx \log \log k$ for large k). The exponent of (6) is asymptotically tight [8].

Corollary 1. *If the conjunction of inequalities (2) consists of linear, weighted transversal and/or polymatroid functions (the latter ones with quasi-polynomially limited right-hand sides), then all minimal feasible solutions of (2) can be generated in quasi-polynomial incremental time.* □

Unfortunately, as the following claim shows, no result analogous to Theorem 4 can hold unconditionally for disjunctions of monotone sets.

Theorem 5. *For each $\ell \geq 3$ there exist monotone systems $\mathcal{A}, \mathcal{B} \subseteq \{0,1\}^{4\ell}$, for which $q_\mathcal{A}(k) \leq (2\ell + 1)k$ and $q_\mathcal{B}(k) \leq (2\ell + 1)k$, while $q_{\mathcal{A} \vee \mathcal{B}}(|\mathcal{A} \vee \mathcal{B}|) = 2^{|\mathcal{A} \vee \mathcal{B}|}$.*

However, for the special cases of linear, transversal or polymatroid systems the following result can be regarded as an analogue of Theorem 4.

Theorem 6. *Assume that either all functions $f_i(x)$, $i = 1, ..., m$ in (2) are transversal and/or linear, or all of them are polymatroid, and let $\phi = Y_1 \vee Y_2 \vee \cdots \vee Y_m$. Then*
$$q_{\mathcal{F}_\phi}(k) \leq P_\phi(q_1(k), \ldots, q_m(k)) = \prod_{i=1}^m q_i(k),$$
where P_ϕ is the evaluation polynomial of ϕ, and where $q_i(k)$ are the upper bounds on $q_{\mathcal{F}_i}(k)$ stated in (4), (5), and (6).

We claim next that even though we cannot mix different types of inequalities in Theorem 6, still a result analogous to Theorem 1 can be derived from Theorems 4 and 6.

Theorem 7. *If the system of inequalities (2) consists of either m linear and/or transversal inequalities or m polymatroid inequalities, and ϕ is an arbitrary monotone $\vee, \wedge$-formula in m propositional variables, then the inequality*
$$q_{\mathcal{F}_\phi}(k) \leq P_\phi(q_1(k), q_2(k), ..., q_m(k))$$
holds, where P_ϕ is the evaluation polynomial of ϕ, and where, as in Theorem 6, the functions $q_i(k)$ are the upper bounds on the duality indices of the individual inequalities stated in (4), (5), and (6).

Theorems 1, 2 and 3 readily follow from Theorem 7 in view of Proposition 1 and the bounds of (4), (5) and (6).

In the next section we state the main lemmas from which Theorems 6 and 7 can be derived, and which may be of interest on their own. Proofs omitted here, due to space limitations, can be found in [12]. In the last section we list several applications of monotone systems defined by linear, transversal and polymatroid inequalities.

3 Main Lemmas

Aggregating Polymatroid Inequalities

Let $f_i : \mathcal{C} \to \mathbb{Z}_+$, be a polymatroid function, $t_i \in \mathbb{Z}_+$ be a given positive integer threshold, and denote by $\mathcal{F}_i \subseteq \mathcal{C}$ the set of all minimal feasible solutions to the polymatroid inequality $f_i(X) \geq t_i$, $i = 1, ..., m$. Let us further define

$$(f_1 \wedge \cdots \wedge f_m)(x) = \sum_{i=1}^{m} \min\{f_i(x), t_i\}, \quad \text{and} \tag{7}$$

$$(f_1 \vee \cdots \vee f_m)(x) = \prod_{i=1}^{m} t_i - \prod_{i=1}^{m} (t_i - \min\{f_i(x), t_i\}) \tag{8}$$

for all $x \in \mathcal{C}$.

Lemma 1. *Both functions, $g = f_1 \wedge \cdots \wedge f_m$ and $h = f_1 \vee \cdots \vee f_m$ are polymatroid. Furthermore, the sets $\mathcal{F}_1 \wedge \cdots \wedge \mathcal{F}_m$ and $\mathcal{F}_1 \vee \cdots \vee \mathcal{F}_m$ respectively consist of all minimal feasible solutions of the polymatroid inequalities*

$$g(x) \geq \sum_{i=1}^{m} t_i \quad \text{and} \quad h(x) \geq \prod_{i=1}^{m} t_i.$$

Lemma 1 implies that any monotone $\vee, \wedge$-formula of polymatroid inequalities can be replaced by an equivalent polymatroid inequality:

Corollary 2. *Let ϕ be a monotone $\vee, \wedge$-formula in m variables, $f_i(x)$ be a polymatroid function, $t_i \in \mathbb{Z}_+$ be non-negative integral threshold, and let $\mathcal{F}_i$ denote the set of all minimal feasible solutions of the polymatroid inequality $f_i(x) \geq t_i$, for $i = 1, ..., m$. Then, $\mathcal{F}_\phi = \phi(\mathcal{F}_1, \mathcal{F}_2, ..., \mathcal{F}_m)$ is the set of all minimal feasible solutions of the system Σ_ϕ, and also of the single polymatroid inequality*

$$(\phi(f_1, f_2, ..., f_m))(x) \geq P_\phi(t_1, ..., t_m),$$

where P_ϕ is the evaluation polynomial of ϕ.

Monotone Composition of Transversal Inequalities

Let us associate to a system of nonnegative weights $w(i,z)$, $i = 1,...,n$, $z = 0,...,c_i$ a separable function $f_w : \mathcal{C} \longrightarrow \mathbb{R}_+$ defined by

$$f_w(x) = \sum_{i=1}^{n} \sum_{z=0}^{x_i} w(i,z). \tag{9}$$

Clearly, the mapping f_w is a special, separable transversal function.

Lemma 2. *Let f_w be a nonnegative separable mapping as in (9), and let $t \in \mathbb{R}_+$. Assume further that $\mathcal{X}, \mathcal{Y} \subseteq \mathcal{C}$ are subsets of vectors for which $\mathcal{X} \neq \emptyset$, $(0,...,0) \notin \mathcal{X}$, and which satisfy the following separation constraints:*

(i) $f_w(x) \geq t$ holds for all $x \in \mathcal{X}$;
(ii) $f_w(y) < t$ for all $y \in \mathcal{Y}$.

Then, there exists a subfamily $\mathcal{Y}' \subseteq \mathcal{Y}$ such that $x \not\leq y \vee y'$ for all $y, y' \in \mathcal{Y}'$ and $x \in \mathcal{X}$, and

$$|\mathcal{Y}'| \geq \frac{|\mathcal{Y}|}{(\sum_{x \in \mathcal{X}} \nu(x))} \geq \frac{|\mathcal{Y}|}{n|\mathcal{X}|},$$

where $\nu(x)$ denotes the number of nonzero entries of $x \in \mathcal{C}$.

Corollary 3. *Assume that $f^i_{w^i}$ is a nonnegative separable mapping as in (9), and that $t^i \in \mathbb{R}_+$, for $i = 1,...,m$. Assume further that $\mathcal{X}, \mathcal{Y} \subseteq \mathcal{C}$ are nonempty collections of vectors satisfying the following separation condition:*

(i) For all $x \in \mathcal{X}$ we have $f^i_{w^i}(x) \geq t^i$ for at least one of the indices $i \in \{1,...,m\}$.
(ii) For all $y \in \mathcal{Y}$ we have $f^i_{w^i}(y) < t^i$ for all indices $i \in \{1,...,m\}$.

Then, there exists a subset $\mathcal{Y}' \subseteq \mathcal{Y}$ such that $x \not\leq y \vee y'$ for all $x \in \mathcal{X}$ and $y, y' \in \mathcal{Y}'$, and

$$|\mathcal{Y}'| \geq \frac{|\mathcal{Y}|}{(n|\mathcal{X}|)^m}.$$

Let us remark that the bound in Corollary 3 cannot be improved by more than a factor of $O(m^{2m})$, i.e., it is tight whenever m is constant. Let us add that this bound, and consequently all bounds in our previous theorems (that are driven from this bound) can be improved by a factor of m^m: In our claims the evaluation polynomial P_ϕ associated to a Boolean expression ϕ could be replaced by Q_ϕ obtained from ϕ by replacing conjunctions by arithmetic addition, and disjunctions by the arithmetic $\diamond$ operation defined by $a_1 \diamond a_2 \diamond \cdots \diamond a_r = a_1 a_2 \cdots a_r / r^r$.

4 Applications

Disjunctive Programming. Let $A_1 \in \mathbb{R}_+^{r_1 \times n},...,A_m \in \mathbb{R}_+^{r_m \times n}$ be non-negative real matrices, and $b^1 \in \mathbb{R}_+^{r_1},...,b^m \in \mathbb{R}_+^{r_m}$ be positive real vectors. Consider the following disjunctive normal form (DNF) of linear monotone inequalities [5]:

$$\bigvee_{i=1}^{m}(A_i x \geq b_i), \quad x \in \mathbb{R}_+^n. \tag{10}$$

It follows from Theorem 1 that, if m is bounded by a constant then all minimal integer solutions of (10) can be enumerated in incremental quasi-polynomial time. In contrast, when m is unbounded but $\max\{r_1, \ldots, r_m\} \leq \mathit{const}$, Theorem 1 implies a quasi-polynomial incremental algorithm for enumerating all maximal infeasible vectors for (10).

Data Mining. Given a database $\mathcal{D} = \{Y_1, Y_2, \ldots, Y_d\}$, $Y_i \subseteq [n]$ for $i = 1, \ldots, d$ (i.e., a multi-hypergraph), and an integer threshold t, a subset $X \subseteq [n]$ is said to be t-frequent if $s(X) = |\{i : Y_i \supseteq X\}| \geq t$ and t-infrequent if $s(X) < t$. It is easy to see that the function $f(X) = d - s(X)$ is a transversal function for the hypergraph $\mathcal{H} = \{Y : Y \in \mathcal{D}\}$. Let $\mathcal{D}_1, \ldots, \mathcal{D}_m$ be m binary databases, $t_1, \ldots, t_m$ be real thresholds, and consider the family

$$\mathcal{F} = \min\{X \subseteq [n] \mid \exists i \in [m] : X \text{ is } t_i\text{-infrequent with respect to } \mathcal{D}_i\}.$$

For instance, each database $\mathcal{D}_i$ may represent the set of items purchased in each weekday $i = 1, \ldots, m = 7$, with $\mathcal{F}$ representing the family of minimal collections of items that lie below a specified purchase threshold in at least one of the 7 days of the week. Clearly, $\mathcal{F}$ is the family of minimal true vectors for the disjunction of transversal inequalities, and thus, for constant m, the elements of $\mathcal{F}$ can be enumerated in incremental quasi-polynomial time by Theorem 2. The generation of maximal frequent and minimal infrequent sets arises in the generation of association rules in data mining applications, see e.g. [2,3,21].

Sparse Boxes. Another notion related to data mining applications is that of sparse boxes. Let $\mathcal{S}$ be a set of points in n dimensions, and $t \leq |\mathcal{S}|$ be a given integer. A *maximal t-box* is a closed n-dimensional interval which contains at most t points of $\mathcal{S}$ in its interior, and which is maximal with respect to this property (i.e., cannot be extended in any direction without strictly enclosing more points of $\mathcal{S}$). Typically, the set of points $\mathcal{S}$ represents the set of records in a quantitative database, and maximal sparse boxes correspond to empty or nearly empty regions in the data space [4,16,18,24,25]. It is not difficult to see that the family $\mathcal{F}_{\mathcal{S},t}$ of maximal sparse boxes, with respect to a given set of n-dimensional points $\mathcal{S}$ and a given threshold $t \in \mathbb{Z}_+$, can be represented as the set of minimal feasible vectors of a transversal inequality over a $2n$-dimensional box $\mathcal{C}$ [11]. Given m databases $\mathcal{S}_1, \ldots, \mathcal{S}_m$, and thresholds $t_1, \ldots, t_m$, one may be interested in finding all maximal regions in space which are sparse with respect to at least one of the databases, i.e. finding the disjunction $\mathcal{F} = \bigvee_{i=1}^{m} \mathcal{F}_{\mathcal{S}_i, t_i}$. Theorem 2 implies that the family $\mathcal{F}$ can be generated in quasi-polynomial time if the number of databases m is constant. In contrast, mining all maximal boxes that are sparse for all m databases can be done in incremental quasi-polynomial time regardless of whether m is bounded by a constant or not. Let us add that only exponential algorithms were previously known in the literature for mining sparse boxes.

Matroid Intersections. Given m matroids $M_1, \ldots, M_m$, defined on the common ground set V by m independence oracles, Lawler, Lenstra and Rinnooy Kan [23] considered the problem of enumerating all maximal sets $X \subseteq V$ independent in all the matroids. They gave an exponential-time enumeration algorithm whose running time is $O(|V|^{m+2})$ per each generated maximal independent set. Since a set $X \subseteq V$ is independent in a matroid if and only if $rank^*(V \setminus X) \geq rank^*(V)$, where $rank^*(\cdot)$ is the rank function for the dual matroid, Theorem 3 implies that the above problem can be solved in incremental quasi-polynomial time regardless of m. (Specifically, k maximal sets independent in $M_1, \ldots, M_m$ can be generated in $K^{o(\log K)}$ time and $poly(K)$ independence tests, where $K = \max\{k, |V|, m\}$.) When m is fixed, Theorem 3 also implies an incremental quasi-polynomial-time for generating all maximal sets X independent in at least one of the matroids $M_1, \ldots, M_m$. Our next example deals with graphic matroids.

Reliability. Let R be a finite set of vertices, $R_1, \ldots, R_m \subseteq R$ be m possibly intersecting subsets of R, and $E_1, \ldots, E_n \subseteq \binom{R}{2}$ be a collection of n sets of edges on R. Given a set $X \subseteq [n]$ and $i \in [m]$, define $c_i(X)$ to be the number of connected components of the graph $(R_i, \bigcup_{i \in X}(E_i \cap \binom{R_i}{2}))$. Then for any integral threshold t_i, the inequality $f_i(X) = |R_i| - c_i(X) \geq t_i$ is polymatroid. In particular, if $t_i = |R_i| - 1$ and $c_i([n]) = 1$ then the family $\mathcal{F}_i$, of minimal feasible solutions to this inequality, is the set of all minimal collections of the input sets of edges $E_1, \ldots, E_n$ which interconnect all vertices in R_i. In network reliability applications (see e.g., [1,15,17,27]), the sets of edges $E_1, \ldots, E_n$ correspond to relays, each controlled by a single switch which may work or fail, and the sets of vertices $R_1, \ldots, R_m$ correspond to regions, or sets of nodes in the network, whose connectivity is to be observed. It may be the case that the connectivity of the whole network is measured by the connectivity of these regions, e.g. the network is considered working properly if at least one of the regions R_i is connected, or more generally if a certain monotone Boolean expression ϕ on the connectivity of these regions is satisfied. It follows from Theorem 3 that if the number of prime implicants of ϕ is bounded by a constant, then all minimal collections of relays maintaining the connectivity of the network, as defined by the Boolean expression ϕ, can be enumerated in incremental quasi-polynomial time.

Statistics. Let $(S, 2^S, \mu_1), \ldots, (S, 2^S, \mu_m)$ be m probability spaces defined on some finite sample space S. Given a set $\mathcal{H} \subseteq 2^S$ of events, we are interested in finding all minimal collections $\mathcal{X} \subseteq \mathcal{H}$ of events the probability of the union of which exceeds some threshold t, with respect to at least one of the measures $\mu_1, \ldots, \mu_m$, i.e.

$$\left(\Pr_{\mu_1}[\bigcup_{X \in \mathcal{X}} X] \geq t\right) \vee \cdots \vee \left(\Pr_{\mu_m}[\bigcup_{X \in \mathcal{X}} X] \geq t\right).$$

The above condition is an example of the disjunction of transversal functions, and for constant m, the family of minimal such collections can be enumerated in quasi-polynomial time. This is also true for arbitrary monotone $\vee, \wedge$ -conditions of bounded $\vee$-degree.

References

1. U. Abel and R. Bicker, Determination of All Cutsets Between a Node Pair in an Undirected Graph, *IEEE Transactions on Reliability* 31 (1986), pp. 167–171.
2. R. Agrawal, T. Imielinski and A. Swami, Mining association rules between sets of items in massive databases, in *Proc. the 1993 ACM-SIGMOD Int. Conf. Management of Data*, pp. 207–216.
3. R. Agrawal, H. Mannila, R. Srikant, H. Toivonen and A. I. Verkamo, Fast discovery of association rules, in *Advances in Knowledge Discovery and Data Mining* (U. M. Fayyad, G. Piatetsky-Shapiro, P. Smyth and R. Uthurusamy, eds.), pp. 307–328, AAAI Press, Menlo Park, California, 1996.
4. M. J. Atallah and G. N. Fredrickson, A note on finding a maximum empty rectangle, *Discrete Applied Mathematics* 13 (1986), pp. 87–91.
5. E. Balas, Disjunctive Programming, *Annals of Discrete Mathematics* 5, (1979), pp. 3–51.
6. E. Balas and E. Zemel, All the facets of zero-one programming polytopes with positive coefficients, Management Science Research Report 374, Carnegie Mellon University, Pittsburgh, 1975.
7. J. C. Bioch and T. Ibaraki, Complexity of identification and dualization of positive Boolean functions, *Information and Computation* 123 (1995), pp. 50–63.
8. E. Boros, K. Elbassioni, V. Gurvich, and L. Khachiyan, Generating dual-bounded hypergraphs, *Optimization Methods and Software*, 17 (2002), pp. 749–781.
9. E. Boros, K. Elbassioni, V. Gurvich, L. Khachiyan and K. Makino, Dual-bounded generating problems: All minimal integer solutions for a monotone system of linear inequalities, *SIAM Journal on Computing*, 31 (5) (2002) pp. 1624–1643.
10. E. Boros, K. Elbassioni, V. Gurvich and L. Khachiyan, An inequality for polymatroid functions and its applications, *Discrete Applied Mathematics* 131 (2)(2003), pp. 255–281.
11. E. Boros, K. Elbassioni, V. Gurvich, L. Khachiyan and K. Makino, An Intersection Inequality for Discrete Distributions and Related Generation Problems, in *Automata, Languages and Programming, 30-th International Colloquium, ICALP 2003, Lecture Notes in Computer Science (LNCS)* 2719 (2003), pp. 543–555.
12. E. Boros, K. Elbassioni, V. Gurvich and L. Khachiyan, Generating all minimal integral solutions to AND-OR systems of monotone inequalities: conjunctions are easier than disjunctions, DIMACS Technical Report 2005-12, Rutgers University, New Brunswick, New Jersey, USA, ("http://dimacs.rutgers.edu/TechnicalReports/2005.html").
13. E. Boros, V. Gurvich, L. Khachiyan and K. Makino, Dual-bounded generating problems: partial and multiple transversals of a hypergraph, *SIAM Journal on Computing* 30 (6) (2001), pp. 2036–2050.
14. E. Boros, V. Gurvich, L. Khachiyan and K. Makino, Generating weighted transversals of a hypergraph, Dual-Bounded Generating Problems: Weighted Transversals of a Hypergraph, *Discrete Applied Mathematics* 142(1-3) (2004), pp. 1–15.
15. V.K. Bansal, K.B. Misra and M.P. Jain, Minimal Pathset and Minimal Cutset Using Search Technique, *Microelectr. Reliability* 22 (1982) pp. 1067–1075.
16. B. Chazelle, R. L. (Scot) Drysdale III and D. T. Lee, Computing the largest empty rectangle, *SIAM Journal on Computing*, 15(1) (1986), pp. 550–555.
17. C. J. Colburn, The Combinatorics of Network Reliability, Oxford Univ. Press, New York, 1987.

18. J. Edmonds, J. Gryz, D. Liang and R. J. Miller, Mining for empty rectangles in large data sets, in *Proc. 8th Int. Conf. on Database Theory (ICDT)*, Jan. 2001, *Lecture Notes in Computer Science* 1973, pp. 174–188.
19. T. Eiter and G. Gottlob, Identifying the minimal transversals of a hypergraph and related problems, *SIAM Journal on Computing*, 24 (1995), pp. 1278–1304.
20. M. L. Fredman and L. Khachiyan, On the complexity of dualization of monotone disjunctive normal forms, *Journal of Algorithms*, 21 (1996), pp. 618–628.
21. D. Gunopulos, R. Khardon, H. Mannila, and H. Toivonen, Data mining, hypergraph transversals and machine learning, in *Proc. the 16th ACM-SIGACT-SIGMOD-SIGART Symp. Principles of Database Systems*, (1997) pp. 12-15.
22. V. Gurvich and L. Khachiyan, On generating the irredundant conjunctive and disjunctive normal forms of monotone Boolean functions, *Discrete Applied Mathematics*, 96-97 (1999), pp. 363–373.
23. E. Lawler, J. K. Lenstra and A. H. G. Rinnooy Kan, Generating all maximal independent sets: NP-hardness and polynomial-time algorithms, *SIAM Journal on Computing*, 9 (1980), pp. 558–565.
24. B. Liu, L.-P. Ku and W. Hsu, Discovering interesting holes in data, In *Proc. 15th International Joint Conference on Artificial Intelligence*, pp. 930–935, Nagoya, Japan, 1997.
25. B. Liu, K. Wang, L.-F. Mun and X.-Z. Qi, Using decision tree induction for discovering holes in data, In *Proc. 5th Pacific Rim International Conference on Artificial Intelligence*, pp. 182-193, 1998.
26. M. Pfetsch, The Maximum feasible Subsystem Problem and Vertex-Facet Incidences of Polyhedra, Dissertation, TU Berlin, 2002.
27. J. S. Provan and M. O. Ball, Computing Network Reliability in Time Polynomial in the Number of Cuts, *Operations Research*, 32(1984), pp. 516–526.
28. F. Stork and M. Uetz, On the Generation of Circuits and Minimal Forbidden Sets, *Mathematical Programming, Series A*, 102(1) (2005), pp. 185-203.

On the Parameterized Complexity of Exact Satisfiability Problems

Joachim Kneis, Daniel Mölle, Stefan Richter, and Peter Rossmanith

Dept. of Computer Science, RWTH Aachen University, Germany

Abstract. For many problems, the investigation of their parameterized complexity provides an interesting and useful point of view. The most obvious natural parameterization for the maximum satisfiability problem—the number of satisfiable clauses—makes little sense, because at least half of the clauses can be satisfied in any formula. We look at two optimization variants of the exact satisfiability problem, where a clause is only said to be fulfilled iff exactly one of its literals is set to *true*. Interestingly, these variants behave quite differently. In the case of RESMAXEXACTSAT, where over-satisfied clauses are entirely forbidden, we show fixed parameter tractability. On the other hand, if we choose to ignore over-satisfied clauses, the MAXEXACTSAT problem is obtained. Surprisingly, it is W[1]-complete. Still, restricted variants of the problem turn out to be tractable.

1 Introduction

Satisfiability of boolean formulæ in conjunctive normal form — abbreviated to SAT — is one of the most preeminent problems in computer science to this day. In its most common form, it asks for an assignment to a set of boolean variables $x_1, \ldots, x_n$ that satisfies the formula, which is a conjunction of disjunctive clauses of literals over $x_1, \ldots, x_n$. However, there are many important variations, some of which will be discussed below. All of them are NP-complete just as the original problem [4,7,10,14].

A simple restriction is to bound the length of the clauses by a natural number q, resulting in a problem called q-SAT. It is well-known that 2-SAT is in P whereas 3-SAT is NP-complete.

In the exact satisfiability problem — EXACTSAT — a clause is only counted as satisfied if exactly one of its literals is fulfilled. In this case we say that the clause is *exact-satisfied*. This variety, often combined with restrictions on the length of clauses as above, has been investigated by many authors [1,2,8,10,13,14].

MAXSAT, on the other hand, is the natural generalization of SAT as an optimization problem. That is, given a formula F, the question is *how many* clauses can be satisfied simultaneously. Here, too, restrictions on the length of clauses have been analyzed in addition to the problem itself [4,5,15].

Of course, EXACTSAT can be transformed into an optimization problem as well. There are, however, two natural ways to do this. In the first case, over-satisfied clauses are ignored, and the derived problem is called MAXEXACTSAT.

In the second case, over-satisfied clauses are forbidden, yielding a problem called RESMAXEXACTSAT [9,10]. The first variety turns out to be a generalization of the well-known problem MAXCUT.

In this paper, we analyze both MAXEXACTSAT and RESMAXEXACTSAT in terms of parameterized complexity [3], where the number of clauses that can be satisfied simultaneously gives the most natural parameter. Note that, in opposition to MAXSAT, both problems are non-trivial for small values of this parameter. In MAXSAT, either an arbitrary assignment or its complement already satisfy at least half of the clauses. Hence, MAXSAT belongs to FPT [11].

There are formulæ with arbitrary many clauses, such that no clause can be exact-satisfied. Therefore, MAXEXACTSAT is not trivially solvable for small parameters. Similarly, we can easily construct an instance for RESMAXEXACTSAT such that all assignments are illegal.

The key results achieved in this paper are as follows. MAXEXACTSAT is complete for W[1], but in FPT when restricted to clauses of bounded length or to monotonicity (MONOMAXEXACTSAT). In contrast, RESMAXEXACTSAT is in FPT even on unrestricted clauses.

2 Preliminaries

In parameterized complexity every input instance has an associated natural number, called the *parameter*. Often, the parameter is part of the input, otherwise it is a — usually simple — function of the input. The time complexity of a parameterized problem is measured as a function in both the input length n and the parameter k. The idea behind parameterized complexity theory is that hard problems can be easy in practice if hard instances do not occur. The parameter measures how hard an instance is; the complexity explodes in terms of the parameter only. See the influential monograph by Downey and Fellows [3] for an introduction to parameterized complexity.

Definition 1. $\mathcal{L}$ *is a parameterized language iff* $\mathcal{L} \subseteq \Sigma^* \times \mathbf{N}$ *for some alphabet* Σ. *Let* $(x, k) \in \mathcal{L}$, *then* x *is called instance and* k *is called parameter. The language* $\mathcal{L}$ *is fixed parameter tractable iff there is an algorithm that decides whether* $(x, k) \in \mathcal{L}$ *in* $O(f(k)p(|x|))$ *for some function* $f : \mathbf{N} \to \mathbf{N}$ *and a polynomial* p.

FPT is the complexity class consisting of all fixed parameter tractable languages. To prove the intractability of problems, we need a reduction:

Definition 2. *Let* $\mathcal{L} \subseteq (\Sigma^* \times \mathbf{N}), \mathcal{L}' \subseteq (\Sigma'^* \times \mathbf{N})$, *be parameterized languages. A mapping* $f : \mathcal{L} \to \mathcal{L}'$ *is an fpt-reduction iff*

1. $(x, k) \in \mathcal{L} \Leftrightarrow f(x, k) \in \mathcal{L}'$
2. *there is a computable function* $g : \mathbf{N} \to \mathbf{N}$, *such that* $k' \leq g(k)$ *for every* $(x, k) \in \mathcal{L}$ *with* $f(x, k) = (x', k')$
3. *there is a constant* c *and a computable function* f', *such that* $f(x, k)$ *can be computed in time* $f'(k)|x|^c$

In short, we write $\mathcal{L} \leq^{\text{fpt}} \mathcal{L}'$ if there is an fpt-reduction from $\mathcal{L}$ to $\mathcal{L}'$. Downey and Fellows used this reduction to establish a class hierarchy for parameterized languages. Each of the classes in this hierarchy is defined by a key language which is complete for its class. Results on membership and hardness of other languages are then shown by fpt-reduction. We obtain the classes

$$\text{FPT} \subseteq \text{W}[1] \subseteq \text{W}[2] \subseteq \ldots \subseteq \text{W}[\text{SAT}] \subseteq \text{W}[\text{P}].$$

Note that none of the inclusions is known to be strict. Problems that are hard for any of the classes above FPT are called intractable as they probably admit no FPT algorithm.

Some examples for parameterized languages are the well-known VERTEX-COVER, INDEPENDENTSET and DOMINATINGSET. The first is known to be in FPT, while the last is W[2]-complete. Later, we will use INDEPENDENTSET, which is complete for W[1], to prove our hardness results.

We use the following notation for boolean formulæ: Let V be a set of boolean variables. The negation of a variable x is denoted by $\bar{x}$. *Literals* can be variables or their negations. If l denotes a negated variable $\bar{x}$, then $\bar{l}$ denotes the variable x.

Algorithms for finding the exact solution of MAXSAT are usually designed for the unweighted MAXSAT problem. However, we represent formulæ by multisets to account for positive integer weights.

Definition 3. A CNF formula is a multiset of clauses. A clause is a multiset of literals.

We call a formula *monotone* if each of its variables appears only in positive or only in negative form. Monotone formulæ are uninteresting for traditional MAXSAT, but become challenging if at most one literal per clause may be satisfied.

Definition 4. An assignment *exact-satisfies* a clause C iff it sets exactly one literal in c to *true*. A clause C is *over-satisfied* by an assignment iff two or more of its literals are set to *true*.

This yields the two problems:

Definition 5.

1. The MAXEXACTSAT problem is the question whether there is an assignment that exact-satisfies at least k clauses in a given formula F.
2. The RESMAXEXACTSAT problem asks for an assignment that exact-satisfies at least k clauses under the restriction that no clause is over-satisfied.

Let us analyze the following example:

$$F = \{\{y\}, \{z\}, \{x, y\}, \{x, z\}, \{x, y, z\}\}$$

In the case of MAXSAT, we can satisfy every clause of F. An optimal assignment for MAXEXACTSAT satisfies four clauses ($x = 0, y = 1, z = 1$) while RESMAXEXACTSAT only admits an assignment that satisfies three clauses ($x = 1, y = 0, z = 0$).

Another difference between MAXSAT and these new problems is the way double occurences of variables are handled. The clause $C = \{x, x, y, z\}$ is equivalent to $\{x, y, z\}$ in traditional MAXSAT. In RESMAXEXACTSAT, the clause $C = \{x, x, y, z\}$ translates to $\{y, z\}$ and $x = 0$, because $x = 1$ is an invalid assignment. In MAXEXACTSAT, however, C is not equivalent to any shorter clause. All we can say is that if x is *true*, then C cannot be satisfied.

3 Maximum Exact Satisfiability

The upcoming lemma establishes W[1]-hardness for MAXEXACTSAT. Observe that the proof does not employ variables that occur twice or more in a clause. That is, the result does hold even for the restricted variant where clauses are sets rather than multisets.

Lemma 1. INDEPENDENTSET $\leq^{\text{fpt}}$ MAXEXACTSAT.

Proof. Let $G = (V, E)$ a graph with $V = \{v_1, \ldots, v_n\}$. We construct a formula $F = \{C_1, \ldots, C_n\}$ over variables $x_1, \ldots, x_n, z$ such that C_i is satisfied iff v_i belongs to an independent set. Moreover, we ensure that C_i can only be exact-satisfied by $x_i = 1$ and $x_j = 0$ for all $v_j \in N(v_i)$, where $N(v_i)$ denotes the neighborhood of v_i, excluding v_i itself. For every node v_i, define the clause

$$C_i := \{z, \bar{z}, \bar{x}_i\} \cup \bigcup_{v_j \in N(v_i)} \{x_j\}.$$

Note that C_i is exact-satisfied if and only if $x_i = 1$ and $x_j = 0$ for every j with $v_j \in N(v_i)$. Let $F := \{C_1, \ldots, C_n\}$. We now show that G contains an independent set of size k if and only if it is possible to exact-satisfy k clauses in F:

Let first $I = \{v_{i_1}, \ldots, v_{i_k}\} \subseteq V$ be an independent set in G. Set $x_j = 1$ for every $v_j \in I$ and $x_j = 0$ otherwise. This assignment exact-satisfies C_j if $v_j \in I$ because only x_j is set to *true* in C_j.

Conversely, let $X = \{x_{a_1}, \ldots, x_{a_l}\}$ be a set of variables set to *true*, such that at least k clauses are exact-satisfied. We obviously have $l \geq k$, because $x_i = 1$ holds for every exact-satisfied C_i. Now set $I = \{v_i \mid C_i \text{ is satisfied}\}$. By construction, $x_i = 1$ for every $v_i \in I$. Then I is an independent set, since if $v_i \in N(v_j)$ for some $v_i \in I$, then at least two literals are satisfied in

$$C_j = \{z, \bar{z}, \bar{x}_j, x_i\} \cup \bigcup_{\substack{v_r \in N(v_j) \\ r \neq i}} \{x_r\}. \qquad \square$$

For example, applying the reduction from the proof of Lemma 1 to the graph depicted in Figure 1 constructs the formula

$$\{\{z, \bar{z}, \bar{x}_1, x_2, x_3\}, \{z, \bar{z}, \bar{x}_2, x_1, x_3\}, \{z, \bar{z}, \bar{x}_3, x_1, x_2, x_4\}, \{z, \bar{z}, \bar{x}_4, x_3\}\}.$$

It turns out that MAXEXACTSAT is hard even for monotone formulæ. However, our proof uses clauses that are multisets of literals rather than sets of literals. Later, we will see that this is unavoidable.

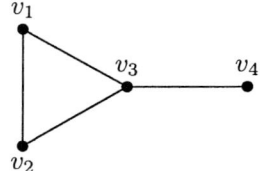

Fig. 1. A graph with an independent set of size two

Lemma 2. INDEPENDENTSET $\leq^{\text{fpt}}$ MONOMAXEXACTSAT

Proof. Let $G = (V, E)$ a graph with $V = \{v_1, \ldots, v_n\}$. We construct a formula F over variables $x_1, \ldots, x_n$ such that C_i is satisfied iff v_i belongs to an independent set. Furthermore, we ensure that C_i can only be exact-satisfied by x_i. For every node v_i, define the clause

$$C_i := \{x_i\} \cup \bigcup_{v_j \in N(v_i)} \{x_j, x_j\}.$$

As in the proof of Lemma 1, C_i is exact-satisfied if and only if $x_i = 1$ and $x_j = 0$ for every $j \neq i$ with $v_j \in N(v_i)$. Let $F := \{C_1, \ldots, C_n\}$. It is easy to see that G has an independent set of size k if and only if k clauses in F can be satisfied simultaneously. □

If, for instance, we apply this reduction to the graph given in Figure 1, we obtain the formula

$$\{\{x_1, x_2, x_2, x_3, x_3\}, \{x_2, x_1, x_1, x_3, x_3\}, \{x_3, x_1, x_1, x_2, x_2, x_4, x_4\}, \{x_4, x_3, x_3\}\}.$$

Let us now further investigate the complexity of MAXEXACTSAT. In order to obtain completeness for W[1], we need to add containment to hardness.

Lemma 3. MAXEXACTSAT $\leq^{\text{fpt}}$ INDEPENDENTSET

Proof. The idea of the proof is to construct a graph G for a formula F, such that the nodes corresponding to literal occurences that satisfy clauses in F form an independent set in G. The graph is built from subgraphs S_i representing clauses, where each literal occurence in a clause C_i is translated to a node in the respective subgraph S_i. Clauses C_i that are oversatisfied for every possible assignment can be detected easily and are ignored in the following reduction. A critical property of the construction is that the parameter does not change at all.

For the moment, assume that the clauses are but sets, rather than multisets. Given the formula $F = \{C_1, \ldots, C_m\}$ with $C_i = \{l_{i,1}, \ldots, l_{i,t_i}\}$, we construct the graph $G = (V, E)$ as follows:

$$V = \{ v_{i,j} \mid 1 \leq i \leq m, 1 \leq j \leq t_i \}$$
$$E = E_1 \cup E_2 \cup E_3 \cup E_4$$

with

$$E_1 := \{\, \{v_{i,j}, v_{a,b}\} \mid i = a\,\} \tag{1}$$
$$E_2 := \{\, \{v_{i,j}, v_{a,b}\} \mid l_{i,j} = \bar{l}_{a,b}\,\} \tag{2}$$
$$E_3 := \{\, \{v_{i,j}, v_{a,b}\} \mid l_{i,j} = l_{a,t}, t \neq b\,\} \tag{3}$$
$$E_4 := \{\, \{v_{i,j}, v_{a,b}\} \mid l_{i,s} = \bar{l}_{a,t}, s \neq j, t \neq b\} \tag{4}$$

The edges from E_1 make every S_i a clique, thus guaranteeing that at most one node in S_i is selected for the independent set, preventing C_i from being over-satisfied. By E_2 it is impossible for l and $\bar{l}$ to satisfy clauses simultaneously. Further consistency between variables occuring together in a clause is ensured by E_3 and E_4.

In order to prove the claim $(F, k) \in$ MAXEXACTSAT $\Leftrightarrow (G = (V, E), k) \in$ INDEPENDENTSET, let first $\lambda = (\lambda_1, \ldots, \lambda_n)$ be an assignment exact-satisfying the clauses $C_1, \ldots, C_k$. Denote by l_{i,t_i} the unique literal satisfying C_i, and let $\mathcal{L}$ be the set of these literals. We want to show that $I = \{\, v_{i,t_i} \mid l_{i,t_i} \in \mathcal{L}\,\}$ is an independent set of size k. Note that I can be larger than $\mathcal{L}$, as $\mathcal{L}$ collects literals, versus nodes depicting literal occurences in I.

Clearly, $|I| = |L| = k$. It remains to show that no two nodes v_{i,t_i} and v_{j,t_j} from I are neighbors in G:

- Since l_{i,t_i} and l_{j,t_j} are from different clauses, we have $\{v_{i,t_i}, v_{j,t_j}\} \notin E_1$.
- Since l_{i,t_i} and l_{j,t_j} satisfy different clauses, we have $l_{i,t_i} \neq l_{j,t_j}$ and thus $\{v_{i,t_i}, v_{j,t_j}\} \notin E_2$.
- Since l_{i,t_i} exact-satisfies C_i and l_{j,t_j} exact-satisfies C_j, l_{i,t_i} cannot appear in C_j and l_{j,t_j} cannot appear in C_i. It follows that $\{v_{i,t_i}, v_{j,t_j}\} \notin E_3$.
- Since l_{i,t_i} exact-satisfies C_i and l_{j,t_j} exact-satisfies C_j, there cannot be a third literal l such that l appears in C_i and $\bar{l}$ in C_j. It follows that $\{v_{i,t_i}, v_{j,t_j}\} \notin E_4$.

Conversely, let $I \subseteq V$ be an independent set in G, $|I| = k$. Let $\mathcal{L} = \{\, l_{i,j} \mid v_{i,j} \in I\,\}$ be the respective literals and $X = \{x_1, \ldots, x_{k'}\}$, $k' \leq k$ the respective variables.

Due to (1), we have $i \neq a$ for any two nodes $v_{i,j}$ and $v_{a,b}$ from I. Hence, for $\mathcal{C} = \{C_i \mid \exists j.\ v_{i,j} \in I\}$ we get $|\mathcal{C}| = k$. By (2), we furthermore have $l_{i,j} \neq \bar{l}_{a,b}$ for any two $l_{i,j}$ and $l_{a,b}$ from $\mathcal{L}$. It is thus possible to satisfy both $l_{i,j}$ and $l_{a,b}$. According to (3), no clause from $\mathcal{C}$ contains more than one literal from $\mathcal{L}$. Therefore, every clause from $\mathcal{C}$ is exact-satisfied by $\mathcal{L}$.

Altogether, the rules (1) through (3) guarantee a sound assignment to the variables from X that satisfies exactly one literal per clause from $\mathcal{C}$. It remains to show that there is an assignment to the rest of the variables that does not interfere with this goal.

For any literal $l \notin \mathcal{L}$ from any clause in $\mathcal{C}$, (4) guarantees that $\bar{l}$ does not appear in $\mathcal{C}$. This can easily be seen by contradiction: assume the opposite. Then, there are clauses C_i and C_a with $l \in C_i$ and $\bar{l} \in C_a$. By construction, C_i is satisfied by l_{i,t_i}, and C_a is satisfied by l_{a,t_a}. In that case, however, we have $\{v_{i,t_i}, v_{a,t_a}\} \in E_4$, implying that I is not an independent set. This, of course,

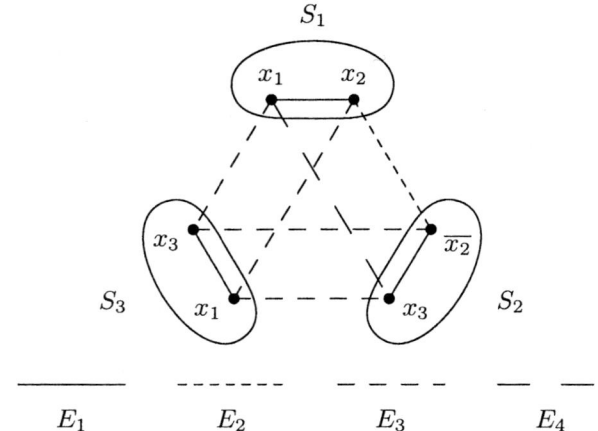

Fig. 2. Construction for $F = \{\{x_1, x_2\}, \{\bar{x_2}, x_3\}, \{x_1, x_3\}\}$

means that we can assign a value to l such that l does not satisfy any of the clauses from $\mathcal{C}$.

When the clauses are allowed to be multisets, caution must be taken. Consider for instance the subgraph S_i generated from a clause $C_i = \{x_1, x_1, x_2\}$. While the first two nodes in S_i may contribute to an independent set, the assignment $x_1 = 1$ leads to over-satisfaction of C_i. Fortunately, there is an easy way out: Following the above reduction, for each i, remove the nodes from S_i that originate from literals that occur multiple times in C_i. □

Figure 2 illustrates the construction on a small example. Combining Lemma 1 and Lemma 3 we obtain:

Theorem 1. MAXEXACTSAT *is complete for* W[1].

By Lemma 2 the problem remains W[1]-hard for monotone formulæ, if a literal can occur more than once in a clause. The complexity for monotone formulæ without repetition of literals will be dealt with in Section 4.

4 Monotone Formulæ

Theorem 2. MONOMAXEXACTSAT $\in$ FPT, *if clauses are sets.*

Proof. Let F be a monotone formula. If F contains a variable x that occurs in k or more clauses, we can simply set x to *true* and all other variables to *false* in order to exact-satisfy at least k clauses. Also, if F consists of no more than $k^2 + k$ clauses, there are at most 2^{k^2+k} variables, and the problem can be solved using exhaustive search.

In the following, we thus assume that no variable is used k times, and that the instance has more than $k^2 + k$ clauses. In this case, the algorithm stated below always finds an assignment exact-satisfying at least k clauses. Hence, any instance with more than $k^2 + k$ clauses is in MONOMAXEXACTSAT.

1. Initially, construct the hypergraph $G(F)$ by introducing a node for each clause and a hyperedge for each variable, where each such hyperedge contains exactly the clauses in which the respective variable occurs.
2. Find a maximal matching M greedily. If the union of the hyperedges in M consists of k or more nodes, assign *true* to the respective variables. This yields a solution exact-satisfying at least k clauses, and we are done.
3. Otherwise, we distinguish two cases. In the first case, $G(F)$ contains a hyperedge of size at least three. Then, we assign *false* to all the variables represented by the edges in M and remove all the clauses contained therein. As a consequence, since M is a matching, every remaining hyperedge shrinks by one. Go to (2).
4. In the second case, $G(F)$ is but a graph. Then, the nodes not covered by M form an independent set I. Step (3) was performed at most $k-3$ times, and each time, we erased at most $k-1$ clauses. Therefore, less than $k^2 - k$ of the clauses have been removed, and more than $2k$ remain. On the other hand, M spans less than k nodes, and thus, we have that $|I| > k$. Because the clauses represented by I are disjoint, we can easily find an assignment that exact-satisfies all of them. □

The above result also affects another problem. While HITTINGSET is W[2]-complete, EXACTHITTINGSET is in FPT when the size of the hitting set is the parameter. If the number of exact-hit sets is chosen instead, we get MAXEXACTHITTINGSET: Given a family $\mathcal{S} = \{S_1, \ldots, S_m\}$ of sets from a universe S, is there an $H \subseteq S$ such that at least k sets from $\mathcal{S}$ share exactly one element with H? Now MAXEXACTHITTINGSET $\in$ FPT, for it is isomorphic to MONOMAXEXACTSAT.

5 Clauses of Bounded Length

There is an FPT algorithm for MAXEXACTSAT if it is restricted to clauses with length two and no negated literals are allowed [10]. In the following we will improve this result to arbitrary formulæ with clause length bounded by a fixed q.

Lemma 4. *Let F be a formula in q-CNF. If some literal x occurs in at least $k \cdot 2^{q-1}$ clauses, an assignment of the variables exists that exact-satisfies at least k clauses.*

Proof. Let X be the set of clauses containing x, and let X^- denote the set of clauses derived from X by removing x from each clause. Observe that k clauses in X can be exact-satisfied by x iff there is an assignment of the variables that leaves k clauses from X^- unsatisfied. By linearity of expectation, the expected number of clauses from X^- exact-satisfied by a random assignment to the variables is

$$E \leq \left(1 - \frac{1}{2^{q-1}}\right) \cdot |X^-|.$$

By the first moment method [12], there is an assignment that satisfies at most E clauses. Thus, there are at least

$$|X^-| - E \geq \left(1 - \left(1 - \frac{1}{2^{q-1}}\right)\right) \cdot |X^-| \geq \frac{1}{2^{q-1}} \cdot k \cdot 2^{q-1} = k$$

clauses not fulfilled by this assignment. □

We can use this lemma to prove that every formula with more than $f(k)$ clauses has an assignment that exact-satisfies at least k clauses.

Lemma 5. *Let F be a formula in q-CNF with $|F| \geq k^2 q 2^q$. Then there is an assignment that exact-satisfies at least k clauses.*

Proof. If a literal occurs in at least $k2^{q-1}$ clauses, k clauses can be exact-satisfied according to Lemma 4.

Otherwise, choose an arbitrary clause $(l_1, \ldots, l_q)$ to exact-satisfy. Remove all clauses containing any of l_1 through l_q. Each literal occurs in no more than $k \cdot 2^{q-1}$ clauses. Thus, each variable is in at most $k \cdot 2^q$ clauses, so we remove no more than $q \cdot k \cdot 2^q$ of the clauses.

Iterating this process k times yields an assignment that exact-satisfies at least k clauses. □

Theorem 3. *There is a linear problem kernel for* MAXEXACT-q-SAT.

Proof. Consider the following algorithm. If $|F| \geq k^2 q 2^q$, return true. Otherwise, exhaustively search for a solution. This takes $O(2^{q^2 k^2 2^q} \cdot |F|)$ time, which is linear in $|F|$. □

6 Restricted Maximum Exact Satisfiability

We will now prove that RESMAXEXACTSAT is fixed parameter tractable. Our approach is based on an idea that was originally used in an algorithm for SET-PACKING [6]. We will see in the end that the only critical case is monotone formulæ. Let us begin with the concept of an *approximator*.

Definition 6. *Let F be a monotone formula. For any variable v_i, let O_i denote the set of clauses from F satisfied by v_i. We say that $(M_1, \ldots, M_l)$ is a $(k_1, \ldots, k_l)$-approximator for F if there is a set of variables $v_1, \ldots, v_l$, such that $O_1, \ldots, O_l$ are pairwise disjoint and for all $1 \leq i \leq l$ we have that $|O_i| = k_i$ and $M_i \subseteq O_i$. If $M_i = O_i$ for all $1 \leq i \leq l$, we call the approximator saturated.*

The following lemma shows how the subroutine from Figure 3 is capable of improving approximators.

Lemma 6. *Let F be a monotone formula. If $(M_1, \ldots, M_l)$ is an unsaturated $(k_1, \ldots, k_l)$-approximator for F, then Improve$(F, M_1, \ldots, M_l, k_1, \ldots, k_l)$ adds elements to $M_1, \ldots, M_l$ such that $(M_1, \ldots, M_l)$ is still a $(k_1, \ldots, k_l)$-approximator for F, in at least one nondeterministic branch.*

Parameters: A monotone formula F, sets of clauses $M_1, \ldots, M_l$,
numbers $k_1, \ldots, k_l$

$P_i := M_i$ for every $i = 1, \ldots, l$;
for $i = 1, \ldots, l$ **do**
  if there is a variable x such that
    - x is in no clause of P_j for any $j \neq i$
    - x is in every clause of M_i
    - x occurs in exactly k_i clauses
  then $P_i := clauses(x)$;
  else guess a clause $C \in P_1 \cup \cdots \cup P_l - (M_1 \cup \cdots \cup M_l)$;
    $M_i := M_i \cup \{C\}$;
    return;
  fi
od;
$M_i := P_i$ for every $i = 1, \ldots, l$;
return;

Fig. 3. Subroutine *Improve*

Proof. There are two ways in which *Improve* adds elements to the M_i's, namely in the lines preceding the two **return** statements.

In the case that the second **return** statement is used, it is easy to see that $(P_1, \ldots, P_l)$ is a saturated $(k_1, \ldots, k_l)$-*approximator* for F.

If the first **return** statement is used in the i-th iteration of the **for**-loop, we get the following situation:

As $(M_1, \ldots, M_l)$ is a $(k_1, \ldots, k_l)$-approximator for F, there exist variables $v_1, \ldots, v_l$ and respective disjoint sets $O_1, \ldots, O_l$ of clauses, such that $M_i \subseteq O_i$. Observe that v_i cannot fulfill the three conditions of the **if**-statement, because the subroutine has entered the **else**-branch. However, v_i obviously fulfills the second and third condition and thus fails the first one. Therefore, there is a clause C that is in both O_i and some P_j, $j \neq i$, but not in $M_i = P_i$, because $P_1, \ldots, P_l$ are pairwise disjoint.

If the guess is correct, this C is added to M_i. Since we still have $M_i \subseteq O_i$, $(M_1, \ldots, M_l)$ remains a $(k_1, \ldots, k_l)$-approximator for F. □

The above subroutine may be profitably used in a nondeterministic algorithm that finds a feasible set $\{v_1, \ldots, v_l\}$ of variables such that each v_i satisfies exactly k_i many clauses, unless no such set exists. As before, no two of these variables are allowed to occur in the same clause.

Lemma 7. *Algorithm A returns "no" in all nondeterministic branches, if no solution exists. Otherwise, it returns correct solutions in at least one branch and "no" in all other branches.*

Input: A monotone formula F, numbers $k_1, \ldots, k_l$
Output: A set $\{v_1, \ldots, v_l\}$ of variables that corresponds
to a saturated $(k_1, \ldots, k_l)$-approximator for F

$M_i := \emptyset$ for every $i = 1, \ldots, l$;
while $|M_1| + \cdots + |M_l| < k_1 + \cdots + k_l$ **do**
  $\text{Improve}(F, M_1, \ldots, M_l, k_1, \ldots, k_l)$
od;
if there are variables $v_1, \ldots, v_l$ such that v_i occurs exactly in the clauses M_i
then return $\{v_1, \ldots, v_l\}$
else return *no* **fi**

Fig. 4. Algorithm A

Proof. The algorithm always checks the correctness of a solution before returning it. Therefore, even on computation paths with wrong guesses no incorrect answer is possible.

By Lemma 6, the fact that $(M_1, \ldots, M_l)$ is a $(k_1, \ldots, k_l)$-approximator for F obviously constitutes an invariant of the **while**-loop, as long as all guesses are correct. □

In order to check whether exactly k clauses can be exact-satisfied, we need to test all possible partitions of the number k using Algorithm A. Unfortunately, there are formulæ that allow for satisfying, say, $k+1$ clauses, but not exactly k. The following observation helps to overcome this difficulty in the final theorem.

Proposition 1. *Let F be a monotone formula in which no variable occurs in more than k clauses. If there is an assignment that exact-satisfies at least k clauses of F, then there is an assignment that exact-satisfies between k and $2k$ clauses.*

Theorem 4. *There is an algorithm that decides* RESMAXEXACTSAT *in time* $O((2k)^{4k} \cdot poly(|F|))$. *Hence,* RESMAXEXACTSAT *is in* FPT.

Proof. If F is not monotone, we can branch on a variable that occurs as a positive and a negative literal. This satisfies at least one clause in every step. The hard case is monotone formulæ, which we assume in what follows.

Try Algorithm A on F and all combinations of $k_1, \ldots, k_l$, such that $k \leq k_1 + \ldots + k_l \leq 2k$, $k \leq l \leq 2k$, and $k_i > 0$ for all $1 \leq i \leq l$. The number of these combinations is bounded by $(2k)^{2k}$. By Proposition 1, this approach suffices to find the solution as long as all guesses are correct.

For up to $2k$ times, Algorithm A guesses a clause from a set of at most $2k$ elements. Resolving the nondeterminism thus involves an additional factor of at most $(2k)^{2k}$ in the running time. □

Acknowledgements. We would like to thank several anonymous referees for helpful comments, especially regarding the final section.

References

1. J. M. Byskov, B. A. Madsen, and B. Skjernaa. New algorithms for exact satisfiability. Technical Report RS-03-30, BRICS, Oct. 2003.
2. V. Dahllöf, P. Johnson, and R. Beigel. Algorithms for four variants of the exact satisfiability problem. *Theoretical Comp. Sci.*, 320(2-3):373–394, 2004.
3. R. G. Downey and M. R. Fellows. *Parameterized Complexity.* Springer-Verlag, 1999.
4. M. Garey, D. Johnson, and L. Stockmeyer. Some simplified np-complete graph problems. *Theoretical Computer Science*, 1:237–267, 1976.
5. J. Gramm, E. A. Hirsch, R. Niedermeier, and P. Rossmanith. New worst-case upper bounds for MAX-2-SAT with application to MAX-CUT. *Discrete Applied Mathematics*, 130(2):139–155, 2003.
6. W. Jia, C. Zhang, and J. Chen. An efficient parameterized algorithm for m-set packing. *Journal of Algorithms*, 50(1):106–117, 2004.
7. R. Karp. Reducibility among combinatorial problems. In R. Miller and J. Thatcher, editors, *Complexity of Computer Communications*, pages 85–103. Plenum Press, 1972.
8. A. S. Kulikov. An upper bound $O(2^{0.16254n})$: A simpler proof. *Zapiski nauchnyh seminarov POMI*, 293:118–128, 2002.
9. B. Madsen. An algorithm for exact satisfiability analysed with the number of clauses as parameter. Technical Report RS-04-18, BRICS, 2004.
10. B. A. Madsen and P. Rossmanith. Maximum exact satisfiability: NP-completeness proofs and exact algorithms. Technical Report RS-04-19, BRICS, Oct. 2004.
11. M. Mahajan and V. Raman. Parameterizing above guaranteed values: MaxSat and MaxCut. *Journal of Algorithms*, 31:335–354, 1999.
12. R. Motwani and P. Raghavan. *Randomized Algorithms.* Cambridge University Press, 1995.
13. S. Porschen, B. Randerath, and E. Speckenmeyer. Exact 3-satisfiability is decidable in time $O(2^{0.16254n})$. In *Proc. 5th Int. Symp. on the Theory and Appl. Satisfiabilty testing (SAT2002)*, pages 231–235, 2002.
14. C. P. Schnorr. Satisfiability is quasilinear complete in NQL. *J. ACM*, 25:136–145, 1978.
15. R. Williams. A new algorithm for optimal constraint satisfaction and its implications. In *Proc. of 31st ICALP*, number 3142 in LNCS, pages 1227–1237. Springer, 2004.

Approximating Reversal Distance for Strings with Bounded Number of Duplicates

Petr Kolman*

Charles University in Prague, Faculty of Mathematics and Physics,
Department of Applied Mathematics
kolman@kam.mff.cuni.cz

Abstract. For a string $A = a_1 \ldots a_n$, a *reversal* $\rho(i,j)$, $1 \le i < j \le n$, transforms the string A into a string $A' = a_1 \ldots a_{i-1} a_j a_{j-1} \ldots a_i a_{j+1} \ldots a_n$, that is, the reversal $\rho(i,j)$ reverses the order of symbols in the substring $a_i \ldots a_j$ of A. In a case of signed strings, where each symbol is given a sign $+$ or $-$, the reversal operation also flips the sign of each symbol in the reversed substring. Given two strings, A and B, signed or unsigned, *sorting by reversals* (SBR) is the problem of finding the minimum number of reversals that transform the string A into the string B.

Traditionally, the problem was studied for permutations, that is, for strings in which every symbol appears exactly once. We consider a generalization of the problem, k-SBR, and allow each symbol to appear at most k times in each string, for some $k \ge 1$. The main result of the paper is a simple $O(k^2)$-approximation algorithm running in time $O(k \cdot n)$. For instances with $3 < k \le O(\sqrt{\log n \log^* n})$, this is the best known approximation algorithm for k-SBR and, moreover, it is faster than the previous best approximation algorithm. In particular, for $k = O(1)$ which is of interest for DNA comparisons, we have a linear time $O(1)$-approximation algorithm.

Keywords: Approximation algorithms, String comparison, Edit distance, Sorting by reversals, Minimum common string partition.

1 Introduction

For a string $A = a_1 \ldots a_n$, a *reversal* $\rho(i,j)$, $1 \le i < j \le n$, transforms the string A into a string $A' = a_1 \ldots a_{i-1} a_j a_{j-1} \ldots a_i a_{j+1} \ldots a_n$, that is, the reversal $\rho(i,j)$ reverses the order of symbols in the substring $a_i \ldots a_j$ of A. In a case of signed strings, where each symbol is given a sign $+$ or $-$, the reversal operation also flips the sign of each symbol in the reversed substring. Given two strings, A and B, signed or unsigned, *sorting by reversals* (SBR) is the problem of finding the minimum number of reversals that transform the string A into the string B; this number, denoted by $\mathsf{SBR}(A,B)$, is called the *reversal distance* of A and B.

---

* Research done while visiting University of California at Riverside. Supported by NSF grants CCR-0208856 and ACI-0085910 and by project 1M0021620808 of MŠMT ČR.

A necessary and sufficient condition for A and B to have a finite reversal distance is that each letter appears the same number of times in A and B (for the signed version, we count together the occurrences of a letter with positive and negative signs). We call such strings *related*.

To give an example, $A = abcabc$ and $B = bcbaac$ are related strings and $\rho(3,6), \rho(1,4)$ is a sequence of reversals that turns A to B, therefore $\mathsf{SBR}(A, B) \leq 2$. Similarly, $\rho(1,4), \rho(4,4)$ turns $A' = +a - c - b - a + b + c$ to $B' = +a + b + c + a + b + c$ and thus, $\mathsf{SBR}(A', B') \leq 2$.

In this paper we study a variant of the problem, denoted by k-SBR, in which each symbol is allowed to appear at most k times in each string. Our particular interest is in the case that $k = O(1)$. The main contribution is a simple $O(k^2)$-approximation algorithm for k-SBR running in time $O(k \cdot n)$.[1] Thus, for $k = O(1)$, we have a linear time $O(1)$-approximation algorithm.

1.1 Terminology

For notational simplicity, we allow a few symbols to have slightly different meanings for signed and unsigned strings. For a string $P = a_1 \ldots a_n$, we denote by $-P$ the result of reversal $\rho(1, n)$ of P (e.g., for $P = +a + b - d$, we have $-P = +d - b - a$). We use two different equivalence relations. Two strings $A = a_1 a_2 \ldots a_n$ and $B = b_1 b_2 \ldots b_n$, signed or unsigned, are *identical*, $A = B$, if $a_i = b_i$ for each $i \in [n]$. In a case of signed strings, by $a_i = b_i$ we mean also equality of the signs. Signed or unsigned strings A and B are *congruent*, $A \cong B$, if $A = B$ or $A = -B$.

The length of a string A is denoted by $|A|$. A *partition* of a string A is a sequence $\mathcal{P} = (P_1, P_2, \ldots, P_m)$ of strings whose concatenation is equal to A, that is, $P_1 P_2 \ldots P_m = A$. The strings P_i are called the *blocks* of $\mathcal{P}$ and their number is the *size* of the partition. Given a partition $\mathcal{P} = (P_1, P_2, \ldots, P_m)$, of a string A, a pair $l, l+1$ is a *break* of the partition $\mathcal{P}$ if $l = \sum_{j=1}^{i} |P_j|$ for some $i \in [m-1]$. Informally, a break of a partition $\mathcal{P}$ of A is a pair of letters that are consecutive in A but are not consecutive in $\mathcal{P}$.

For two strings A and B, we say that S is a *common substring with respect to the relation* $=$ or $\cong$, respectively, if S is a substring of A and there exists a substring R of B such that $S = R$ or $S \cong R$, respectively. When not necessary, we will often avoid specifying the relation and will talk only about a common substring. If S is a common substring of A and B, we use notations S^A and S^B to distinguish between the occurrences of S (or $-S$) in A and B; if S occurs more than once in A then S^A refers to arbitrary but fixed occurrence of S in A and analogous convention applies for the string B. Given two partitions $\mathcal{A} = (A_1, \ldots, A_m)$ and $\mathcal{B} = (B_1, \ldots, B_{m'})$, a common substring of $\mathcal{A}$ and $\mathcal{B}$ is a string S such that S is a common substring of A_i and B_j, for some indices i, j.

---

[1] Tomasz Waleń [15] recently described how to implement the algorithm in time $O(n)$ for $k = O(\log n)$.

1.2 Related Work

String comparison is a fundamental problem in computer science with applications in text processing, data compression or computational biology. The problem of sorting by reversals drew a lot of attention in the last years as a useful tool for DNA comparison [4,12,6,1]. In that application, the letters in the strings represent different genes and the reversal distance measures the similarity of two genomic sequences. A common assumption that a genome contains only one copy of each gene is unwarranted for genomes with multi-gene families such as the human genome [13]. On the other hand, a weaker assumption that a genome contains at most $k = O(1)$ copies of each gene is often warranted (cf. [9]). That is why k-SBR is of interest. In this subsection we will briefly mention the most relevant known results.

Under the assumption that every symbol appears in each input string exactly once, we have the well known problem of permutation sorting by reversals. The problem 1-SBR is solvable in polynomial time for strings with signs [12,1] but is NP-hard [4] and even MAX-SNP hard [3] for strings without signs; the best known approximation ratio for the unsigned 1-SBR is 1.375 by an algorithm of Berman et al. [2]. A recent result of Chen et al. [5] shows that the signed k-SBR is NP-hard even for $k = 2$ (the unsigned k-SBR is obviously NP-hard for all $k \geq 2$). There are $O(1)$-approximation algorithms for signed 2-SBR and 3-SBR [5,7,11]. The best approximation ratio for the general signed SBR is $O(\log n \log^* n)$, using an $O(n \log^* n)$-time algorithm of Cormode and Muthukrishnan [8].

Instead of bounding the number of duplicates, there is another way to restrict the general problem of sorting by reversals with duplicates: bound the size of the alphabet. Unsigned SBR with unary alphabet is trivial; the NP-hardness of unsigned SBR with binary alphabet was proved by Christie and Irving [6].

Closely related is a *minimum common string partition* problem (MCSP). Given a partition $\mathcal{P}$ of a string A and a partition $\mathcal{Q}$ of a string B, we say that the pair $\pi = \langle \mathcal{P}, \mathcal{Q} \rangle$ is a *common partition* of A and B with respect to the relation Rel $\in \{=, \cong\}$, if there exists a permutation σ on $[m]$ such that for each $i \in [m]$, $(P_i, Q_{\sigma(i)}) \in$ Rel. The minimum common string partition problem is to find a common partition of A, B with the minimum size, denoted by MCSP(A, B). The restricted version of MCSP, where each letter occurs at most k times in each input string, is denoted by k-MCSP.

Similarly as for SBR, there is a signed and an unsigned variant of the problem. In *unsigned* MCSP, the input consists of two unsigned strings, and the relation $=$ is used; in *signed* MCSP, the input consists of two signed strings and the relation $\cong$ is used. For unsigned strings, we define yet another variant of the problem, *reversed* MCSP (RMCSP), in which the (unsigned) strings are compared by the relation $\cong$.

The signed MCSP problem was introduced by Chen et al. [5] as a tool for dealing with SBR. They observed that for any two related signed strings A and B, MCSP(A, B) and SBR(A, B) differ only by a constant multiplicative factor: given a partition $(P_1, \ldots, P_m)$ of A, $(Q_1, \ldots, Q_m)$ of B and the permutation σ on $[m]$ such that $P_i \cong Q_{\sigma(i)}$ for each $i \in [m]$, it is possible to move the block

$P_{\sigma^{-1}(1)} \cong Q_1$ to the beginning of A by one reversal and then, if necessary, to reverse it by one more reversal; similarly it is possible to move the block $P_{\sigma^{-1}(2)} \cong Q_2$ to its right position by at most two reversals without affecting the block Q_1 at the beginning of the string, etc. On the other hand, a reversal "breaks" at most two pairs of consecutive letters in the string and thus, from a sequence of m reversals, we derive a common partition with at most $2m$ breaks. Analogous observation applies for related unsigned strings and the problems RMCSP and SBR.

For $k \geq 2$, k-MCSP is NP-hard, and even APX-hard [11]. Due to the close relation between signed SBR and signed MCSP, the known approximation ratios for signed MCSP are within a constant factor of the approximation ratios for signed SBR: $O(1)$ approximation ratios for 2-MCSP and 3-MCSP [7,11], $O(\log n \log^* n)$ approximation ratio for the general MCSP [8].

Chrobak et al. [7] analyzed the behavior of a natural greedy heuristic for MCSP: start with the two strings A and B and iteratively, find the longest common substring of A and B that does not overlap previously marked substrings, and mark this substring. They showed that though GREEDY is a 3-approximation algorithm for 2-MCSP, even for 4-MCSP its approximation ratio is $\Omega(\log n)$. For general MCSP, both signed and unsigned, the approximation ratio is between $\Omega(n^{0.43})$ and $O(n^{0.67})$. It is worth noting that the algorithms described in this paper are simple modifications of GREEDY, yet their approximation ratios for k-MCSP are better, namely $O(k^2)$, in contrast to the $\Omega(\log n)$ of GREEDY for $k \geq 4$.

In the *edit distance* (ED) problem, a set of string operations is given (e.g., DELETE, INSERT or CHANGE a character, SUBSTRING_MOVE or SUBSTRING_REVERSAL) and the task is to find the minimum number of operations needed to convert one string to the other. SBR can be also viewed as an edit distance problem where the only operation is SUBSTRING_REVERSAL and the input strings are related. For any two related strings A and B, $\mathsf{MCSP}(A, B)$ differs by a constant multiplicative factor from the edit distance of A and B with only SUBSTRING_MOVE operations, and the edit distance using only SUBSTRING_MOVE operations differs also by a constant multiplicative factor from the edit distance with operations {INSERT, DELETE a character, SUBSTRING_MOVE} [14].

On the other hand, MCSP can be utilized for approximating the the edit distance even for *unrelated* strings. To give an example, consider ED with operations {INSERT, DELETE a character, SUBSTRING_MOVE}: given strings A and B, let $B - A$ denote the multiset of letters that have more occurrences in B than in A (i.e., if x has x_A occurrences in A and x_B occurrences in B then there are $\max\{0, x_B - x_A\}$ copies of x in $B - A$) and analogously for $A - B$. Then $|A - B| + |B - A|$ is a lower bound on the edit distance $\mathsf{ED}(A, B)$. Let A' denote a concatenation of the string A with all letters from $B - A$ (in any order), and similarly, let B' denote a concatenation of B with all letters from $A - B$; we observe that $\mathsf{ED}(A', B') \leq 2\mathsf{ED}(A, B)$. Exploiting the above mentioned relation between ED and MCSP for related strings we obtain $\mathsf{ED}(A, B) = \Theta(1) \cdot (|A - B| + |B - A| + \mathsf{MCSP}(A', B'))$.

For the edit distance problem with operations {INSERT, DELETE a character, SUBSTRING_MOVE}, Cormode and Muthukrishnan [8] describe an $O(n \log^* n)$-time $O(\log n \log^* n)$-approximation algorithm which yields, by the relations described above, the $O(\log n \log^* n)$-approximation for SBR mentioned earlier in this subsection.

The edit distance problem with a different set of string operations was studied by Ergun et al. [10]. For several edit distance problems that allow SUBSTRING_DELETION, they describe an $O(1)$ approximation algorithm. This is in contrast to the above mentioned known approximations of edit distance *without* SUBSTRING DELETION where the best approximation ratio is of order $\Omega(\log n \log^* n)$.

The rest of the paper is organized as follows. In Section 2.1 we describe how to modify GREEDY to get the $O(k^2)$ approximation for (reversed) k-MCSP and thus, for k-SBR. Section 2.2 explains how to implement the algorithm in time $O(k \cdot n)$.

2 Algorithms

2.1 REFINED GREEDY: $O(k^2)$-Approximation

In the previous section, we briefly described GREEDY algorithm and we recalled that its approximation ratio for k-MCSP and k-SBR, for any $k \geq 4$, is $\Omega(\log n)$. In this section, we show that a simple modification of GREEDY, called REFINED GREEDY, has an $O(k^2)$ approximation ratio for k-MCSP, which implies also an $O(k^2)$ approximation ratio for k-SBR.

A few more terms are needed. A *duo* is a string of length two. To *cut* a duo $a_i a_{i+1}$ of a block $P = a_j \ldots a_k$ of a partition of A, for some $j \leq i < k$, means to replace the block P in the partition by two blocks $P_1 = a_j \ldots a_i$ and $P_2 = a_{i+1} \ldots a_k$. For a substring $S = a_i \ldots a_j$ of $A = a_1 \ldots a_n$, if $i > 1$ we say that $a_{i-1} a_i$ is a *(left) boundary* duo of S, and similarly, if $j < n$ $a_j a_{j+1}$ is a *(right) boundary* duo of S.

For unsigned k-MCSP the algorithm is the following:

Algorithm REFINED GREEDY
Input: two related strings A and B
    $\mathcal{A} \leftarrow (A)$, $\mathcal{B} \leftarrow (B)$
    while there are unmarked blocks in $\mathcal{A}$ and $\mathcal{B}$ **do**
        $S \leftarrow$ longest common substring of $\mathcal{A}, \mathcal{B}$ that does not overlap
            previously marked blocks
        mark S^A in $\mathcal{A}$ and S^B in $\mathcal{B}$
        cut the boundary duos of S^A in $\mathcal{A}$ and the boundary duos of S^B in $\mathcal{B}$
        cut in unmarked blocks of $\mathcal{A}$ and $\mathcal{B}$ *all* occurrences of duos $\delta \in \Phi$,
            where Φ is the set of boundary duos of S^A and S^B
Output: $(\mathcal{A}, \mathcal{B})$

To extend the algorithm for signed k-MCSP and for k-RMCSP, apart from considering common substrings with respect to the other equivalence relation $\cong$, the difference is that in the cutting steps, we cut not only all occurrences of $\delta \in \Phi$ but also all occurrences of $-\delta$.

To give an example, consider an instance of MCSP,

$$A = abcccca f ccccddddhe f ccccebccccgggg \; ,$$
$$B = abccccddddda f cccche f ccccggggebcccc \; .$$

REFINED GREEDY first marks substring $S_1 = ccccdddd$ (we use overline to denote marking in this example) and cuts all unmarked occurrences of duos from $\Phi = \{fc, dh, bc, da\}$. In the second iteration, REFINED GREEDY looks for the longest unmarked substring in partitions $\mathcal{A} = (ab, cccca f, \overline{ccccdddd}, he f, cccceb, ccccgggg)$ and $\mathcal{B} = (ab, \overline{ccccdddd}, a f, cccche f, ccccggggeb, cccc)$, marks substring $S_2 = ccccgggg$ and cuts duos from $\Phi = \{ge\}$. In the third iteration, the algorithm looks for the longest unmarked substring in partitions $\mathcal{A} = (ab, cccca f, \overline{ccccdddd}, he f, cccceb, \overline{ccccgggg})$ and $\mathcal{B} = (ab, \overline{ccccdddd}, a f, cccche f, \overline{ccccgggg}, eb, cccc)$, marks substring $S_3 = cccc$ and cuts duos from $\Phi = \{ca, ch\}$. Eventually, REFINED GREEDY outputs the common partition

$$\mathcal{P} = \langle (ab, cccc, a f, ccccdddd, he f, cccc, eb, ccccgggg),$$
$$(ab, ccccdddd, a f, cccc, he f, ccccgggg, eb, cccc) \rangle \; .$$

The optimal common partition has six blocks:

$$\mathcal{P}_{\text{OPT}} = \langle (abccc c, a f cccc, dddd, he f cccc, ebcccc, gggg),$$
$$(abcccc, dddd, a f cccc, he f cccc, gggg, ebcccc) \rangle \; .$$

Before analyzing REFINED GREEDY, let us briefly look on the behavior of GREEDY on the same instance. The longest common substrings of A and B are $ccccdddd$ and $ccccgggg$, therefore GREEDY starts by matching these substrings in the first two iterations. We observe that there exists a common partition of A and B that uses $ccccdddd$ and $ccccgggg$ as blocks:

$$\mathcal{P}' = \langle (ab, cccc, a f, ccccdddd, he f, cccc, eb, ccccgggg),$$
$$(ab, ccccdddd, a f, cccc, he f, ccccgggg, eb, cccc) \rangle \; .$$

Every common partition induces a matching between the letters (positions) of $\mathcal{A}$ and $\mathcal{B}$. We note that the common partition $\mathcal{P}'$ matches many of the letters of $\mathcal{A}$ and $\mathcal{B}$ in the same way as the optimal partition $\mathcal{P}_{\text{OPT}}$ does. However, after several steps GREEDY will find another common partition:

$$\mathcal{P}_{\text{GR}} = \langle (a, bcccc, a, f, ccccdddd, he, f cccc, e, b, ccccgggg),$$
$$(a, b, ccccdddd, a, f cccc, he, f, ccccgggg, e, bcccc) \rangle \; .$$

Intuitively, the problem of GREEDY is that a wrong decision in *one* iteration can force the use of *several* additional iterations, and in each of them GREEDY may do another wrong decision, and so on. In other words, a deviation from the optimal solution in one iteration encourages deviations in later iterations. In our instance, after the first two iterations, it is still desirable, for example, to match the first b from A with the first b from B, as the common partition $\mathcal{P}'$ does. However, since $bcccc$ is the longest common substring at this point, GREEDY will decide to use the wrong match between the first b from A and the third b from B.

To improve the performance of the algorithm, the idea is to prevent it from propagating "mistakes" from one iteration to later iterations. In our example, the first mistake was to use the substrings $ccccdddd$; a consequence of this was the use of the substrings $bcccc$, another mistake. REFINED GREEDY attempts to suppress this problem by cutting a few additional duos that are related to the current longest common substring, in each iteration. These breaks will constrain later iterations and will confine the propagation of mistakes.

Theorem 1. REFINED GREEDY *is a $2k^2$-approximation algorithm for unsigned and signed k-MCSP and $2(2k-1)^2$-approximation for k-RMCSP.*

Proof. The output of the algorithm is clearly a common partition. We only have to prove the bound on its quality. For simplicity of the presentation, we prove the claim in detail for the unsigned k-MCSP and then we briefly outline the necessary modifications for signed k-MCSP and for k-RMCSP.

For technical reasons, it will be convenient to extend the notions of a partition and a common partition from strings to sequences of strings. A *partition of the sequence* of strings $\mathcal{A} = (A_1, \ldots, A_l)$ is a sequence of strings $A_{1,1}, \ldots, A_{1,k_1}, A_{2,1}, \ldots, A_{2,k_2}, \ldots, A_{l,1}, \ldots, A_{l,k_l}$, such that $A_i = A_{i,1} \ldots A_{i,k_i}$ for $i \in [l]$. For two sequences of strings, the common partition is defined analogously as for pairs of strings.

Observation 2. *Let $(\mathcal{Q}, \mathcal{R})$ be a common partition of sequences of strings $\mathcal{A}$ and $\mathcal{B}$, and let δ be any duo that appears in $\mathcal{Q}$ and $\mathcal{R}$. Let $\mathcal{Q}'$ denote the partition of $\mathcal{A}$ that is obtained from $\mathcal{Q}$ by cutting all occurrences of the duo δ, and let $\mathcal{R}'$ denote the partition of $\mathcal{B}$ that is obtained from $\mathcal{R}$ by cutting all occurrences of the duo δ. Then, $(\mathcal{Q}', \mathcal{R}')$ is a common partition of $\mathcal{A}$ and $\mathcal{B}$.*

Proof. Since $\mathcal{Q}$ is a permutation of $\mathcal{R}$, every block P from $\mathcal{Q}$ that contains δ appears also in $\mathcal{R}$, and vice versa. Thus, if we cut all occurrences of δ in $\mathcal{Q}$ and $\mathcal{R}$, the resulting new partitions $\mathcal{Q}'$ and $\mathcal{R}'$ will be again permutations of each other. □

Let $\pi = (\mathcal{P}, \mathcal{Q})$ be a minimum common partition of A and B, m be its size and let Δ be the set of all boundary duos of blocks in $\mathcal{P}$ and in $\mathcal{Q}$. We are going to iteratively construct common partitions π_i of A and B that will help us to estimate the size of the common partition found by REFINED GREEDY. We define π_1 as the common partition derived from π by cutting *all* occurrences of all duos in Δ (the fact that π_1 is a partition follows from Observation 2). For k-MCSP instances, the number of blocks increases at most k times. The breaks in π_1 are called *initial* breaks. Let S_i denote the substring that REFINED GREEDY used in iteration i and let Φ_i be the set of boundary duos of S_i^A and S_i^B. For iteration $i \geq 1$ of REFINED GREEDY, we define π_{i+1} as the common partition derived from π_i by cutting all occurrences of all duos in Φ_i.

We are going to compare the blocks used by REFINED GREEDY with the blocks in π_i. For ease of reference, we denote the sets $\mathcal{A}$ and $\mathcal{B}$ at the beginning of iteration i by $\mathcal{A}_i$ and $\mathcal{B}_i$, and by s_i the first position of S_i^A in A, by t_i the last position of S_i^A in A, by s_i' the first position of S_i^B in B, and by t_i' the last position of S_i^B in B.

Observation 3. *For every iteration i and for every $0 \leq l < |S_i|$: the pair $s_i + l, s_i + l + 1$ is an initial break of A if and only if the pair $s_i' + l, s_i' + l + 1$ is an initial break of B.*

Proof. The observations follow from the definition of π_1: if one occurrence of a duo is cut in π_1, then all occurrences of this duo are cut. □

Given a break $l, l+1$ of a partition of A, and a substring $S = a_i \ldots a_j$ of A, we say that the substring S *goes over the break* $l, l+1$ if $i \leq l < j$. Observation 3 can be informally stated like this: If the block S_i^A goes over one or more initial breaks, then the block S_i^B goes over the same number of initial breaks, and, moreover, the relative positions of the initial breaks in S_i^A and S_i^B are the same.

Let $\mathcal{A}_i' \subseteq \mathcal{A}_i$ and $\mathcal{B}_i' \subseteq \mathcal{B}_i$ denote the subsets of unmarked strings of $\mathcal{A}_i$ and $\mathcal{B}_i$, resp., at the beginning of phase i, and let π_i' denote the restriction of π_i to $\mathcal{A}_i'$ and $\mathcal{B}_i'$. Observation 3 implies the following important claim.

Observation 4. *For every i, π_i' is a common partition of $\mathcal{A}_i'$ and $\mathcal{B}_i'$.*

Proof. The proof is by induction. For $i = 1$, nothing is marked, $\mathcal{A}_1' = \{A\}$, $\mathcal{B}_1' = \{B\}$, $\pi_1' = \pi_1$ and the claim is obvious. For $i > 1$, Observations 3 and 2 imply that the blocks from π_i corresponding to the newly marked block S_{i-1}^A are the same as the blocks from π_i corresponding to the newly marked block S_{i-1}^B. Observing that outside S_{i-1}^A and S_{i-1}^B, cuts of the same duos (i.e., duos from Φ_{i-1}) are used to obtain π_i' from π_{i-1}' and $(\mathcal{A}_i', \mathcal{B}_i')$ from $(\mathcal{A}_{i-1}', \mathcal{B}_{i-1}')$, the proof is completed. □

Lemma 1. *For every i,*

- *the block $S_i = a_{s_i} \ldots a_{t_i}$ is an entire block in $\mathcal{A}_i'$ and $\mathcal{B}_i'$, or*
- *S_i goes over an initial break or*
- *$s_i - 1, s_i$ is an initial break or $s_i = 1$, and at the same time $t_i, t_i + 1$ is an initial break or $t_i = n$.*

Proof. The lemma follows from Observation 4 and from the greedy nature of REFINED GREEDY: for every common substring S of $\mathcal{A}_i'$ and $\mathcal{B}_i'$ not satisfying any of the conditions in the lemma, there exists another common longer substring S' of $\mathcal{A}_i'$ and $\mathcal{B}_i'$ such that S is a proper substring of S'. □

We are ready to finish the proof of Theorem 1. In every iteration, the number of duos in $\mathcal{A}$ that REFINED GREEDY cuts, is at most $2k$. If REFINED GREEDY chooses for S an entire block of $\mathcal{A}'_i$, then there are no new cuts introduced in this iteration. If REFINED GREEDY chooses for S a string that is not an entire block of $\mathcal{A}'_i$, then, by Lemma 1, S either goes over an initial break or (roughly) S starts and ends at an initial break. In the former case, we charge all cuts done by REFINED GREEDY in this iteration to this initial break; in the later case, we charge half of the new cuts to each of these two new breaks (in the special case that $s_i = 1$ or $t_i = n$, we charge half of the new cuts to the string A itself). In this way each cut done by REFINED GREEDY (except for the $2k$ cuts charged to the string A itself) is charged to one initial break, and the total number of breaks charged to one initial break is not more than $2 \cdot k$. Since there are at most $k \cdot (m-1)$ initial breaks, there are at most $2 \cdot k^2 \cdot (m-1) + 2k$ breaks in the final partition found by REFINED GREEDY. The total number of blocks used by REFINED GREEDY is at most $2 \cdot k^2 \cdot (m-1) + 2k + 1 \leq 2 \cdot k^2 \cdot m$.

For signed k-MCSP and k-RMCSP we only need to adjust the proof to reflect the thing that now a substring S from A can be matched with a substring R from B even if $S \neq R$ but $S = -R$. Thus, in Observation 2 we cut not only all occurrences of duo δ but also all occurrences of duo $-\delta$. To get the common partition π_1 from π, for each $\delta \in \Delta$ we cut all occurrences of δ as well as all occurrences of $-\delta$; for signed k-MCSP the number of breaks in π_1 increases again at most k times, for k-RMCSP it increases at most $2k-1$ times. In Observation 3, we distinguish whether $S_i^A = S_i^B$ or $S_i^A = -S_i^B$. In the later case, we count the relative positions of the initial breaks in S_i^B backwards (i.e., the claim is: $s_i + l, s_i + l + 1$ is an initial break of A if and only if the pair $t'_i - l - 1, t'_i - l$ is an initial break of B); the former case is as before. For signed k-MCSP, the number of duos cut in $\mathcal{A}$ in one iteration is at most $2k$, for k-RMCSP it is at most $2(2k-1)$. □

Considering the relation between signed MCSP and signed SBR, and between RMCSP and unsigned SBR, we get the following theorem.

Theorem 5. *There exists a polynomial time $4k^2$-approximation algorithm for signed k-SBR, and $8(2k-1)^2$-approximation algorithm for unsigned k-SBR.*

Concerning the running time of REFINED GREEDY, we just note that a naive straightforward implementation of the algorithm requires quadratic time in the worst case (e.g., consider $A = abcde \ldots xyz$ and $B = zyxw \ldots cba$).

2.2 EDUCATED GREEDY: $O(k^2)$-Approximation in Time $O(k \cdot n)$

In the previous analysis we never used the fact that S was the *longest* common substring; we only used that it was not possible to extend S^A and still have a matching substring in B (proof of Observation 1). Based on this observation, here we present more efficient implementation of the algorithm. As in the case of REFINED GREEDY, we describe EDUCATED GREEDY in detail for unsigned k-MCSP; the necessary modifications for signed k-MCSP and k-RMCSP are the same as before.

Algorithm EDUCATED GREEDY
Input: two related strings $A = a_1 \ldots a_n$ and $B = b_1 \ldots b_n$
$\quad \mathcal{A} \leftarrow (A), \mathcal{B} \leftarrow (B)$
$\quad i = 1$
$\quad$ **while** $i \leq n$ **do**
$\quad\quad\quad S \leftarrow$ longest common substring of $\mathcal{A}, \mathcal{B}$ that starts in $\mathcal{A}$ on position i
$\quad\quad\quad\quad\quad$ and does not overlap previously marked blocks
$\quad\quad\quad$ mark S^A in $\mathcal{A}$ and S^B in $\mathcal{B}$
$\quad\quad\quad$ cut the boundary duos of S^A in $\mathcal{A}$ and the boundary duos of S^B in $\mathcal{B}$
$\quad\quad\quad$ cut in $\mathcal{A}$ and $\mathcal{B}$ all unmarked occurrences of duos $\delta \in \Phi$, where Φ is
$\quad\quad\quad\quad\quad$ the set of boundary duos of S^A and S^B
$\quad\quad\quad i \leftarrow i + |S|$
Output: $(\mathcal{A}, \mathcal{B})$

Theorem 6. *There exist an $O(k^2)$-approximation algorithms for unsigned and signed k-MCSP, k-RMCSP and k-SBR running in time $O(k \cdot n)$.*

Proof. The proof of Lemma 1 is the only place in the proof of Theorem 1 that refers to the choice of the common substring S used by REFINED GREEDY. However, as mentioned above, the proof only needs the fact that S cannot be extended on either side. Thus, Lemma 1 holds also for the choices of EDUCATED GREEDY and the $O(k^2)$ approximation ratio follows by the same reasoning as for REFINED GREEDY.

Concerning the running time, EDUCATED GREEDY goes once through A from left to right, and in every iteration, there are at most k possibilities (resp., $2k$ for k-RMCSP) where to look for the common substring S_j. EDUCATED GREEDY spends at most $k \cdot |S_j|$ (resp., $2k \cdot |S_j|$) steps in iteration j and advances by $|S_j|$ positions to the right in A. Thus, the common partition is computed in time $O(k \cdot n)$ and the proof is completed. □

3 Conclusion

We presented simple, $O(k^2)$-approximation algorithms for k-MCSP and k-SBR, running in time $O(k \cdot n)$. For instances with $3 < k \leq O(\sqrt{\log n \log^* n})$, this is the best approximation ratio and, moreover, EDUCATED GREEDY is faster than the previous best approximation algorithm.

We conclude with a few challenging open problems. The approximation ratio of REFINED GREEDY is between $\Omega(k)$ and $O(k^2)$; what is the exact value of the ratio? A related questions are whether there is a (simple) $O(k)$-approximation algorithm for k-SBR and what is the best possible approximation ratio for the general SBR? Is it possible to get below the $O(\log n \log^* n)$ upper bound? Is it NP-hard to approximate better than within $\Omega(\log n)$?

Acknowledgment

We would like to thank Jiří Sgall for his suggestion to implement REFINED GREEDY REFINED GREEDY more efficiently.

References

1. A. Bergeron, J. Mixtacki, and J. Stoye. Reversal distance without hurdles and fortresses. In *Proceedings of 15th Annual Combinatorial Pattern Matching Symposium (CPM)*, volume 3109 of *LNCS*, pages 388–399, 2004.
2. P. Berman, S. Hannenhalli, and M. Karpinski. 1.375-approximation algorithm for sorting by reversals. In *Proceedings of the 10th Annual European Symposium on Algorithms (ESA)*, volume 2461 of *LNCS*, pages 200–210, 2002.
3. P. Berman and M. Karpinski. On some tighter inapproximability results. In *Proceedings of the 26th International Colloquium on Automata, Languages and Programming (ICALP)*, volume 1644 of *LNCS*, pages 200–209, 1999.
4. A. Caprara. Sorting by reversals is difficult. In *Proceedings of the First International Conference on Computational Molecular Biology*, pages 75–83, 1997.
5. X. Chen, J. Zheng, Z. Fu, P. Nan, Y. Zhong, S. Lonardi, and T. Jiang. Computing the assignment of orthologous genes via genome rearrangement. In *Proceedings of 3rd Asia-Pacific Bioinformatics Conference*, pages 363–378, 2005.
6. D. A. Christie and R. W. Irving. Sorting strings by reversals and by transpositions. *SIAM Journal on Discrete Mathematics*, 14(2):193–206, 2001.
7. M. Chrobak, P. Kolman, and J. Sgall. The greedy algorithm for the minimum common string partition problem. In *Proceedings of the 7th International Workshop on Approximation Algorithms for Combinatorial Optimization Problems (APPROX)*, volume 3122 of *LNCS*, pages 84–95, 2004.
8. G. Cormode and S. Muthukrishnan. The string edit distance matching problem with moves. In *Proceedings of the 13th Annual ACM-SIAM Symposium On Discrete Mathematics (SODA)*, pages 667–676, 2002.
9. N. El-Mabrouk. Reconstructing an ancestral genome using minimum segments duplications and reversals. *Journal of Computer and System Sciences*, 65(3):442–464, 2002.
10. F. Ergun, S. Muthukrishnan, and S. C. Sahinalp. Comparing sequences with segment rearrangements. In *Proceedings of the 23rd Annual Conference on Foundations of Software Technology and Theoretical Computer Science (FSTTCS)*, volume 2914 of *LNCS*, pages 183–194, 2003.
11. A. Goldstein, P. Kolman, and J. Zheng. Minimum Common String Partition Problem: Hardness and Approximations. In *Proceedings of the 15th International Symposium on Algorithms and Computation (ISAAC)*, LNCS, pages 473–484, 2004.
12. S. Hannenhalli and P. A. Pevzner. Transforming cabbage into turnip: polynomial algorithm for sorting signed permutations by reversals. *Journal of the ACM*, 46(1):1–27, Jan. 1999.
13. D. Sankoff and N. El-Mabrouk. Genome rearrangement. In T. Jiang, Y. Xu, and M. Q. Zhang, editors, *Current Topics in Computational Molecular Biology*. The MIT Press, 2002.
14. D. Shapira and J. A. Storer. Edit distance with move operations. In *Proceedings of the 13th Symposium on Combinatorial Pattern Matching (CPM)*, volume 2373 of *LNCS*, pages 85–98, 2002.
15. T. Walen. Personal communication, 2005.

Random Databases and Threshold for Monotone Non-recursive Datalog

Konstantin Korovin and Andrei Voronkov*

School of Computer Science, The University of Manchester
{korovin, voronkov}@cs.man.ac.uk

Abstract. In this paper we define a model of randomly generated databases and show how one can compute the threshold functions for queries expressible in monotone non-recursive datalog$^{\neq}$. We also show that monotone non-recursive datalog$^{\neq}$ cannot express any property with a sharp threshold. Finally, we show that non-recursive datalog$^{\neq}$ has a $0 - 1$ law for a large class of probability functions, defined in the paper.

1 Introduction

In this paper we consider random databases, in which relations are generated based on some standard probability distributions. Namely, we assume that every tuple belongs to a relation with an equal probability p, called the *tuple probability*. Similar probabilistic models have been intensively studied in the theory of random graphs (see e.g. [2,11]), but also in the database theory (see e.g. [15,10,4]). We are interested in investigating the behavior of queries expressible in database query languages on the random databases. There are several interesting questions that can be asked about such queries, for example, calculating the probability of a query to be true as a function of p and the domain size. One of the most important characteristics of probabilistic behavior of monotone properties of structures is their *threshold functions*. The threshold functions can be used for characterizing asymptotic probabilistic behavior of queries, that is, probabilistic behavior when the database is growing. Intuitively, if a database grows faster than the threshold function, then the probability of the query to be true converges to 1; likewise, if a database grows slower than the threshold function, then the probability of the query to be true converges to 0. Although it is known that every monotone property has a threshold function, the problem of determining threshold functions for particular properties of random structures is an active research area in combinatorics. In this paper we show how one can compute the threshold functions for queries expressible in monotone non-recursive datalog$^{\neq}$. We also introduce a notion of *density* of queries expressible in this language and show the exact relationship between densities and threshold functions of the queries.

Another natural question is the power of databases query languages to express some phenomena related to randomly generated structures. For example, it is known that

---

* The authors are partially supported by grants from EPSRC and the Faculty of Science and Technology.

some monotone properties actually have a *sharp threshold* (see, e. g.,[8]). Finding interesting properties with a sharp threshold or identifying whether some natural properties have a sharp threshold are questions intensively studied in combinatorics, and in particular in the theory of random graphs. We are interested in studying the expressive power of query languages on random databases with respect to the threshold behavior. In particular, we show that monotone non-recursive datalog$^{\neq}$ cannot express any property with a sharp threshold. It is not hard to give an example of a first-order property having a sharp threshold, which shows that relatively simple extensions of monotone non-recursive datalog$^{\neq}$ can express such properties. However, the exact relation between query languages and expressibility of properties with a sharp threshold remains an open question.

The main contributions of this paper are the following.

1. We define a probabilistic model of randomly generated databases.
2. We show how one can compute the threshold functions for queries expressible in monotone non-recursive datalog$^{\neq}$.
3. We introduce the notion of density of queries expressible in monotone non-recursive datalog$^{\neq}$, and show the exact relationship between densities and threshold functions of the queries.
4. We show that monotone non-recursive datalog$^{\neq}$ has a $0 - 1$ law w.r.t. every probability function $p(n)$ satisfying the following condition: for every rational $q > 0$ either $p \gg n^{-q}$ or $p \ll n^{-q}$.
5. We show that monotone non-recursive datalog$^{\neq}$ cannot express any property with a sharp threshold.

For the future research, it would be interesting to consider different probabilistic models for some big existing and evolving databases such as WEB, and study behavior of properties expressible by database query languages in such models (see e.g. [13]).

2 Preliminaries

In this paper we study asymptotic properties of finite relational structures with constants. In particular, we are interested in properties expressible by existential formulas in the language with equality in which negation can only be applied to equalities. In this section we introduce definitions of structures and the query language datalog$^{\neq}$.

Generally speaking, we are dealing with randomly generated finite structures. For simplicity, we assume that the structure contains only one relation symbol. Even in the case of a single relation symbol the proofs are quite involved. We only consider boolean queries, that is, queries with a yes-no answer. In order to deal with some standard queries as boolean queries, we introduce constants in the language. For example, for an input binary relation R, reachability can be formulated as follows: given two elements (x, y), is there a path from x to y in which arcs are pairs belonging to R? Normally, such a query would be formulated using a binary output relation T that is the transitive closure of R. However, since we are restricted to boolean queries, we cannot use a binary output relation. Instead, we can introduce two constants c_1 and c_2 and use $T(c_1, c_2)$ to represent reachability of c_2 from c_1.

Structures. We consider signatures with one relation symbol R and constants. Consider a finite signature $\Sigma = \{R, c_1, \ldots, c_l\}$, where R is a relation symbol of some arity $r > 0$ and each c_i is a constant. Denote by $\mathbb{N}$ the set of natural numbers and by $[n]$ the finite set $\{1, \ldots, n\}$. For every $n \geq l$ denote by $\mathbb{M}_n$ the set of all structures of the signature Σ with the domain $[n]$ in which the constants $c_1, \ldots c_l$ are interpreted as the numbers $1, \ldots, l$ respectively. For $n < l$ we let $\mathbb{M}_n = \emptyset$. Without loss of generality we can restrict ourselves to structures in $\cup_{n>0} \mathbb{M}_n$. For a structure M we denote the interpretation of R by R_M.

For simplicity of the presentation we require all constants to be interpreted by distinct elements of the domain. However, our main results remain valid if the interpretations of constants can coincide.

Properties of Structures. A *property of structures* is a parametrized family $\{A_n\}_{n=0}^{\infty}$ of structures such that for each $n \in \mathbb{N}$ the set A_n consists of structures in $\mathbb{M}_n$. A property $\{A_n\}_{n=0}^{\infty}$ is called *monotone* if for each $n \in \mathbb{N}$ and $M, M' \in \mathbb{M}_n$ such that $R_M \subseteq R_{M'}$ and $M \in A_n$ we have $M' \in A_n$. In other words, if $M \in A_n$, then by adding a tuple to R_M we obtain a structure in A_n.

For example, if R is binary and we have no constants then each structure is a directed graph; it is easy to see that the following properties are monotone: to contain a given graph, to have no isolated vertices, to be a non-planar graph, and connectivity.

Language. Evidently, every formula φ of the signature Σ defines a property: we let $M \in A_n$ if $M \in \mathbb{M}_n$ and $M \models \varphi$. We consider the subset of all formulas of Σ built from formulas $R(\bar{s})$, $s = t$ and $s \neq t$ using only $\vee$, $\wedge$, and $\exists$. Denote this set of formulas by $\Sigma_1^{\neq}$. For example, if R is ternary and $c_1 \in \Sigma$ then $\exists x, y(R(x, c_1, y) \wedge x \neq y \wedge R(x, x, y))$ is a formula in $\Sigma_1^{\neq}$. It is easy to see that all properties definable by formulas in $\Sigma_1^{\neq}$ are monotone. It is also not hard to argue that $\Sigma_1^{\neq}$ has the same expressive power as non-recursive datalog$^{\neq}$.

Theorem 1. *Every query defined by a formula in $\Sigma_1^{\neq}$ is definable by a non-recursive datalog$^{\neq}$-program and vice versa.*

The proof is standard.

Monotone properties of undirected random graphs have been intensively studied (see e.g. [2,11]). One of the most important characteristics of a monotone property of a random graph is to have the *threshold function*. To define this notion for structures, we first introduce a probabilistic model of structures, which is similar to the binomial model of the random graphs.

Random Structures. Let us introduce the probability space $(\mathbb{M}_n, \mathcal{F}, p, \mu)$ on structures, where $\mathcal{F}$ is the set of all subsets of $\mathbb{M}_n$, $0 \leq p \leq 1$ and the probability function μ is defined as follows:

$$\mu(M) = p^m (1-p)^{n^r - m},$$

where $M \in \mathbb{M}_n$, m is the number of tuples satisfying the relation R in M (remember that r is the arity of R). This can be viewed as a result of n^r independent coin flippings with the probability p of success: every tuple is included in R_M with the probability p. We will refer to p as the *tuple probability*. We denote by $M_{n,p}$ the random structure corresponding to this probability distribution, that is a random element of this probability space. When we consider asymptotic behavior of properties of structures we assume that p is a function of n. Let us note that the presence of constants in the language introduces certain peculiarities for the probabilistic analysis. For example, it is well-known that the first-order logic on random graphs (without constants) has a $0-1$ law for the constant distribution, e.g. $p(n) = 1/2$, (see [7,9]). However, when we consider the first-order logic with constants then the logic fails to have a $0-1$ law (see Section 6 for more on $0-1$ laws for datalog$^{\neq}$).

Notation. Asymptotics. For functions f, g, we write $f(n) \asymp g(n)$ if $f(n) = O(g(n))$ and $g(n) = O(f(n))$, also we write $f(n) = \Theta(g(n))$ if $f(n) \asymp g(n)$, and write $f(n) \ll g(n)$ if $f(n) = o(g(n))$.

Probability. The expected value of a random variable X is denoted by $\mathbb{E}(X)$. The indicator variable of a property A will be denoted as I_A.

Thresholds. Let $\{A_n\}_{n=0}^{\infty}$ be a monotone property. A function $p'(n)$ is called the *threshold function*, or simply *threshold*, for this property if

$$\lim_{n \to \infty} \mu(M_{n,p} \in A_n) = \begin{cases} 0 \text{ if } p \ll p'; \\ 1 \text{ if } p \gg p'. \end{cases}$$

Bollobás and Thomason [3] prove that every monotone property has a threshold. *One of our main results is finding threshold functions for all properties of structures expressible by formulas in $\Sigma_1^{\neq}$. Namely, we show that for every formula $\varphi \in \Sigma_1^{\neq}$ the threshold function is either constant or of the form n^{-q_φ}, where q_φ is a non-negative rational number, and give an algorithm for computing q_φ from a given formula φ.*

The problem of determining threshold functions for properties of random structures is an active research area in combinatorics. In particular, the problem of finding thresholds for graph containment has a long history. Starting from the paper [6] where this problem was solved for a special case of balanced undirected graphs, and culminating 21 years later in [1] where this problem is solved for arbitrary undirected graphs. Let us note that the corresponding *structure containment problem*, defined in Section 3, is expressible in $\Sigma_1^{\neq}$.

Sharp Thresholds. Some monotone properties of structures can possess a *sharp threshold*. We call the threshold p' for a property $\{A_n\}_{n=0}^{\infty}$ sharp if for every $\epsilon > 0$

$$\lim_{n \to \infty} \mu(M_{n,p} \in A_n) = \begin{cases} 0 \text{ if } p \leq (1-\epsilon)p'; \\ 1 \text{ if } p \geq (1+\epsilon)p'. \end{cases}$$

If the threshold for a monotone property is not sharp, then we say that the property has a *coarse threshold*. We will show that *for every property definable by a formula in $\Sigma_1^{\neq}$ the threshold is coarse*. We will use the following reformulation of the notion of sharp threshold (see [11]). Consider a property $\{A_n\}_{n=0}^{\infty}$. For each ϵ such that $0 < \epsilon < 1$ define $p_\epsilon(n)$ be the tuple probability such that $\mu(M_{n,p_\epsilon} \in A_n) = \epsilon$. The property $\{A_n\}_{n=0}^{\infty}$ has a sharp threshold if and only if $\lim_{n \to \infty} p_\epsilon(n)/p_{1/2}(n) = 1$ for every ϵ such that $0 < \epsilon < 1$.

Outline. The rest of this paper is structured as follows. In Section 3 we introduce and study the structure containment and the weak structure containment problems. We prove that every sentence in $\Sigma_1^{\neq}$ is equivalent to a weak structure containment problem. In Section 4 we show how one can calculate the threshold for any structure containment problem and show that the threshold is coarse. In Section 5 we prove similar results for the weak structure containment problems. Finally, in Section 6 we present the main results of this paper.

3 Structure Containment

In this section we introduce the structure containment and the weak structure containment properties. We prove that every formula in $\Sigma_1^{\neq}$ is equivalent to a weak structure containment property.

Structure Containment. Consider an arbitrary but fixed structure $\mathcal{M}$. We say that a structure M *contains* $\mathcal{M}$, denoted by $M \sqsupseteq \mathcal{M}$, if there is an injective homomorphism from $\mathcal{M}$ into M. Note that structure containment is *language-dependent*, since every homomorphism maps an integer m into itself if $c_m \in \Sigma$. We define the *structure containment property*, (also referred to as $\mathcal{M}$-*containment property* when we want to emphasise the structure $\mathcal{M}$), to be the set of all structures that contain $\mathcal{M}$. Evidently, for every structure $\mathcal{M}$ the structure containment property is monotone.

Weak Structure Containment. Let S be a finite family of structures. We say that a structure M *weakly contains* S, denoted by $M \sqsupseteq_w S$, if it contains at least one structure from S. We define the *weak structure containment property*, (also referred to as *weak S-containment property* when we want to emphasise the family S), to be the set of all structures that weakly contain S. Evidently, for every finite family S of structures the weak structure containment property is monotone.

Theorem 2. *Given a sentence A of $\Sigma_1^{\neq}$, one can effectively find a finite family of structures S such that for every structure M we have $M \models A$ if and only if $M \sqsupseteq_w S$.*

The proof can be found in the full version of this paper [12].

4 Threshold for the Structure Containment

In this section we show how to compute the threshold function for any structure containment property and show that every such property has a coarse threshold.

Let us consider $\mathcal{M}$-containment property for an arbitrary but fixed structure $\mathcal{M}$. For a structure $M \in \mathbb{M}_n$, we write $M \sqsupset \mathcal{M}$ if $M \sqsupseteq \mathcal{M}$ and $\mathcal{M}$ is not isomorphic to M. Since we are studying the asymptotic behavior of the properties we can always assume that n is greater than the number of elements of $\mathcal{M}$ and therefore for every structure M from $\mathbb{M}_n$ we have $M \sqsupseteq \mathcal{M}$ if and only if $M \sqsupset \mathcal{M}$. We denote by $\sqsubseteq$ and $\sqsubset$ the relations inverse to $\sqsupseteq$ and $\sqsupset$, respectively.

Notation. For a structure M denote by d_M the number of elements of the domain of M and by r_M the number of tuples in R_M. We say that a tuple of elements of M is a *constant tuple* if all its elements are constants (from Σ), and a *non-constant tuple* otherwise. Let $\bar{d}_M$ denote the number of non-constant elements of M, $\bar{r}_M$ denote the number of non-constant tuples in R_M, and $\hat{r}_M$ denote the number of constant tuples in R_M.

First we show exponential bounds for the probability of the random structure $M_{n,p}$ to contain $\mathcal{M}$. In proofs we will use some techniques developed in the theory of random graphs (see [11]).

Exponential Bounds. Let us note that if $R_\mathcal{M}$ is the empty relation then all structures with a sufficiently large domain contain $\mathcal{M}$, so the containment problem is trivial. Furthermore if all tuples in $R_\mathcal{M}$ are constant (i.e. $\hat{r}_\mathcal{M} = r_\mathcal{M}$) then $\mu(M_{n,p} \sqsupseteq \mathcal{M}) = p^{r_\mathcal{M}}$. In this case, trivially, the threshold function for the containment property is just a constant function and the threshold is coarse. We consider the case when $R_\mathcal{M}$ has at least one constant tuple later in Corollary 3.

Now assume that $R_\mathcal{M}$ is non-empty and contains no constant tuples. Define

$$\Phi_\mathcal{M} = \Phi_\mathcal{M}(n,p) = \min_{Q \sqsubseteq \mathcal{M}, r_Q > 0} n^{\bar{d}_Q} p^{r_Q}. \quad (1)$$

Then the following theorem holds.

Theorem 3. *Let $\mathcal{M}$ be a structure in the signature Σ such that $R_\mathcal{M}$ is non-empty and without constant tuples. Then, for every sufficiently large n and for every sequence $p = p(n) < 1$ the following holds:*

$$1 - \exp\{-\Theta(\Phi_\mathcal{M})\} \leq \mu(M_{n,p} \sqsupseteq \mathcal{M}) \leq 1 - \exp\left\{-\frac{\Theta(\Phi_\mathcal{M})}{1-p}\right\}. \quad (2)$$

Proof. Let us rewrite (2) in a more convenient for us form:

$$\exp\left\{-\frac{\Theta(\Phi_\mathcal{M})}{1-p}\right\} \leq \mu(M_{n,p} \not\sqsupseteq \mathcal{M}) \leq \exp\{-\Theta(\Phi_\mathcal{M})\}. \quad (3)$$

We will use the following inequality (4) (see e.g. [11] where this inequality is proved in a more general setting). We say that a structure M *extends* a structure M' if they have the same domain and $R_{M'} \subseteq R_M$. Consider a family S of structures on the domain $[n]$ and such that for each structure $M \in S$ the relation R_M is non-empty. For each structure $M \in S$, let I_M be the indicator variable denoting that the random structure $M_{n,p}$ extends M. Let $X_S = \sum_{M \in S} I_M$, so X_S denotes the number of structures from S which are extended by $M_{n,p}$. Then the following holds:

$$\exp\left\{-\frac{\mathbb{E}(X_S)}{1-p}\right\} \leq \mu(X_S = 0) \leq \exp\left\{-\frac{(\mathbb{E}(X_S))^2}{\sum_{M',M'' \in S \wedge R_{M'} \cap R_{M''} \neq \emptyset} \mathbb{E}(I_{M'} I_{M''})}\right\}. \quad (4)$$

Now let us consider a structure M such that $d_M \leq n$. We call an *n-copy* of M any structure M' with the domain $[n]$ which contains M and has a minimal $R_{M'}$, i.e., after removing any tuple from $R_{M'}$, M' will not contain M. When n is clear from the context we say a copy of M instead of an n-copy. It is clear that a structure with the domain $[n]$ contains M if and only if it contains an n-copy of M. Let X_M be the random variable denoting the number of different n-copies of M in the random structure $M_{n,p}$. Let us calculate $\mathbb{E}(X_M)$. There are exactly $f(n, M) = \binom{n-l}{\bar{d}_M} \bar{d}_M! / aut(M) = \Theta(n^{\bar{d}_M})$ different n-copies of M, where $aut(M)$ is the number of automorphisms of M, (remember that l is the number of constants in the signature). Using the linearity of the expectation we have

$$\mathbb{E}(X_M) = f(n, M) p^{r_M} = \Theta(n^{\bar{d}_M} p^{r_M}). \tag{5}$$

From this it follows that

$$\Phi_{\mathcal{M}} \asymp \min_{Q \sqsubseteq \mathcal{M}, r_Q > 0} \mathbb{E}(X_Q). \tag{6}$$

We will prove the left-hand side of (3) using the left-hand side of (4). To do so we take a structure $H \sqsubseteq \mathcal{M}$ such that $\mathbb{E}(X_H) = \min_{Q \sqsubseteq \mathcal{M}, r_Q > 0} \mathbb{E}(X_Q)$ and consider the family S_H of all n-copies of H. Then, it is easy to see that $X_{S_H} = X_H$. From (4) using (6) we have

$$\exp\left\{-\frac{\Theta(\Phi_{\mathcal{M}})}{1-p}\right\} \leq \exp\left\{-\frac{\mathbb{E}(X_H)}{1-p}\right\} \leq \mu(X_H = 0) = \mu(M_{n,p} \not\sqsupseteq H). \tag{7}$$

It is obvious that if a structure does not contain a copy of H then it does not contain a copy of $\mathcal{M}$ and hence $\mu(M_{n,p} \not\sqsupseteq H) \leq \mu(M_{n,p} \not\sqsupseteq \mathcal{M})$. Therefore the left-hand side of (3) follows from (7).

Let us prove the right-hand side of (3). To this end, we use the right-hand side of (4) for the family $S_{\mathcal{M}}$ of all n-copies of $\mathcal{M}$. Again we have $X_{S_{\mathcal{M}}} = X_{\mathcal{M}}$. Now we estimate the sum in the denominator of the exponent in (4). Let M', M'' be structures with the domain $[n]$, then we can define a new structure $M' \cap M''$ to be a structure with the domain $[n]$ and the relation $R_{M' \cap M''} = R_{M'} \cap R_{M''}$. For each structure $Q \sqsubseteq \mathcal{M}$ there are $\Theta(n^{\bar{d}_Q} n^{2(\bar{d}_{\mathcal{M}} - \bar{d}_Q)}) = \Theta(n^{2\bar{d}_{\mathcal{M}} - \bar{d}_Q})$ pairs (M', M'') such that $M', M'' \in S_{\mathcal{M}}$ and $M' \cap M''$ is isomorphic to an n-copy of Q. So using (6) we have

$$\sum_{M', M'' \in S_{\mathcal{M}} \wedge R_{M'} \cap R_{M''} \neq \emptyset} \mathbb{E}(I_{M'} I_{M''}) \asymp \sum_{Q \sqsubseteq \mathcal{M}, r_Q > 0} n^{2\bar{d}_{\mathcal{M}} - \bar{d}_Q} p^{2r_{\mathcal{M}} - r_Q}$$

$$\asymp \max_{Q \sqsubseteq \mathcal{M}, r_Q > 0} \frac{(\mathbb{E}(X_{\mathcal{M}}))^2}{\mathbb{E}(X_Q)} \asymp \frac{(\mathbb{E}(X_{\mathcal{M}}))^2}{\Phi_{\mathcal{M}}}. \tag{8}$$

Direct substitution of (8) into (4) gives us the right-hand side of (3).

Let $\mathcal{M}$ be a structure such that $R_{\mathcal{M}}$ is non-empty and without constant tuples. Then, we define the *density* of the structure $\mathcal{M}$ to be

$$m(\mathcal{M}) = \max_{Q \sqsubseteq \mathcal{M}, r_Q > 0} r_Q / \bar{d}_Q.$$

Corollary 1. *Let $\mathcal{M}$ be a structure such that $R_{\mathcal{M}}$ is non-empty and without constant tuples. Then the threshold function for the $\mathcal{M}$-containment property is $n^{-1/m(\mathcal{M})}$, i.e., the following holds:*

$$\lim_{n \to \infty} \mu(M_{n,p} \sqsupseteq \mathcal{M}) = \begin{cases} 0, & \text{if } p \ll n^{-1/m(\mathcal{M})}, \\ 1, & \text{if } p \gg n^{-1/m(\mathcal{M})}. \end{cases}$$

Proof. To prove this corollary it is sufficient to prove that (i) if $pn^{1/m(\mathcal{M})} \to \infty$ then $\Phi_\mathcal{M} \to \infty$ and (ii) if $pn^{1/m(\mathcal{M})} \to 0$ then $\Phi_\mathcal{M} \to 0$. Assume that $pn^{1/m(\mathcal{M})} \to \infty$. Then for every structure $Q \sqsubseteq \mathcal{M}$ with $r_Q > 0$ we have $1/m(\mathcal{M}) \leq \bar{d}_Q/r_Q$ and hence

$$n^{\bar{d}_Q} p^{r_Q} = (pn^{\bar{d}_Q/r_Q})^{r_Q} \to \infty.$$

Therefore $\Phi_\mathcal{M} = \min_{Q \sqsubseteq \mathcal{M}, r_Q > 0} n^{\bar{d}_Q} p^{r_Q} \to \infty$, which proves (i).

Now let H be a structure contained in $\mathcal{M}$ such that $r_H/\bar{d}_H = m(\mathcal{M})$. To prove (ii) notice that

$$\Phi_\mathcal{M} = \min_{Q \sqsubseteq \mathcal{M}, r_Q > 0} n^{\bar{d}_Q} p^{r_Q} \leq n^{\bar{d}_H} p^{r_H} = (pn^{1/m(\mathcal{M})})^{r_H} \to 0.$$

Corollary 2. *Let $\mathcal{M}$ be a structure such that $R_\mathcal{M}$ is non-empty and without constant tuples. Then the threshold for the $\mathcal{M}$-containment property is coarse.*

Proof. If $\mathcal{M}$-containment would have a sharp threshold then $\lim_{n \to \infty} p_\epsilon(n)/p_{1/2}(n) = 1$ for every ϵ such that $0 < \epsilon < 1$. Let us show that this is not the case. Take an arbitrary ϵ such that $0 < \epsilon < 1$. Let $p_\epsilon(n)$ be the tuple probability such that $\mu(M_{n,p_\epsilon} \sqsupset \mathcal{M}) = \epsilon$. Corollary 1 implies that $p_\epsilon(n) \to 0$ and therefore $p_\epsilon(n)$ is bounded away form 1. Using this and (2) we obtain that for some constants $A > 0$ and $B > 0$,

$$1 - \exp\{-A\Phi_\mathcal{M}\} \leq \mu(M_{n,p_\epsilon} \sqsupset \mathcal{M}) = \epsilon \leq 1 - \exp\{-B\Phi_\mathcal{M}\} \qquad (9)$$

for all sufficiently large n.

Now consider an ϵ such that

$$0 < \epsilon < 1 - 2^{-2^{-r_\mathcal{M}} A/B}. \qquad (10)$$

It is straightforward to check that $0 < \epsilon < 1$. Let us show that for this ϵ we have $p_\epsilon(n)/p_{1/2}(n) \not\to 1$. To avoid technicalities with $\Phi_\mathcal{M}(p_\epsilon, n) = \min_{Q \sqsubseteq \mathcal{M}, r_Q > 0} n^{\bar{d}_Q} p_\epsilon^{r_Q}$ we consider a structure $W \sqsubseteq \mathcal{M}$ on which the minimum is reached infinitely often. So let $\{n_i | i \in \mathbb{N}\}$ be an infinite subset of $\mathbb{N}$ such that $\Phi_\mathcal{M}(p_\epsilon, n_i) = n_i^{\bar{d}_W} p_\epsilon^{r_W}$. It is clear that for our goal it is enough to prove that $p_\epsilon(n_i)/p_{1/2}(n_i)$ is bounded away from 1. In order to prove it we show lower and upper bounds for $p_\epsilon(n_i)$.

From the right-hand side of (9) we have that

$$\epsilon = \mu(M_{n_i,p_\epsilon} \sqsupset \mathcal{M}) \leq 1 - \exp\{-B(\Phi_\mathcal{M})\}.$$

Straightforward calculations yield

$$\ln(1-\epsilon)^{-1/B} \leq n_i^{\bar{d}_W} p_\epsilon^{r_W}.$$

Finally, the lower bound is

$$\frac{\left(\ln(1-\epsilon)^{-1/B}\right)^{1/r_W}}{n_i^{\bar{d}_W/r_W}} \leq p_\epsilon(n_i). \qquad (11)$$

Obtaining an upper bound for $p_\epsilon(n_i)$ is similar.

The left-hand side of (9) yields

$$1 - \exp\{-A\Phi_{\mathcal{M}}\} \leq \mu(M_{n_i,p_\epsilon} \sqsupset \mathcal{M}) = \epsilon.$$

After straightforward calculations we obtain

$$n_i^{\bar{d}w} p_\epsilon^{rw} \leq \ln(1-\epsilon)^{-1/A},$$

so we have an upper bound

$$p_\epsilon(n_i) \leq \frac{\left(\ln(1-\epsilon)^{-1/A}\right)^{1/rw}}{n_i^{\bar{d}w/rw}}. \tag{12}$$

Now from (11),(12) and (10) we have

$$p_\epsilon(n_i)/p_{1/2}(n_i) \leq \left(\frac{\ln(1-\epsilon)^{-1/A}}{\ln(1/2)^{-1/B}}\right)^{1/rw} = \left(\frac{B\ln(\frac{1}{1-\epsilon})}{A\ln(2)}\right)^{1/rw} \leq$$

$$\left(\frac{B\ln\left(\frac{1}{1-\left(1-2^{-2^{-r_\mathcal{M} A/B}}\right)}\right)}{A\ln(2)}\right)^{1/rw} = (1/2)^{r_\mathcal{M}/rw} \leq 1/2,$$

therefore $p_\epsilon(n_i)/p_{1/2}(n_i)$ is bounded away from 1.

Corollary 3. *Let $\mathcal{M}$ be a structure such that $R_\mathcal{M}$ is non-empty and contains at least one constant tuple. Then the threshold function for the $\mathcal{M}$-containment property is constant and the threshold is coarse.*

Proof. Let $p_\epsilon(n)$ be the tuple probability such that $\mu(M_{n,p_\epsilon} \sqsupset \mathcal{M}) = \epsilon$. Let us show that for every $0 < \epsilon < 1$ we have $\lim_{n \to \infty} p_\epsilon(n) = \epsilon^{1/\hat{r}_\mathcal{M}}$. From this, the corollary easily follows.

Let $\hat{\mathcal{M}}$ be a structure with the same domain as $\mathcal{M}$ and the relation $R_{\hat{\mathcal{M}}}$ consisting of all constant tuples of $R_\mathcal{M}$; likewise let $\bar{\mathcal{M}}$ be a structure with the same domain as $\mathcal{M}$ and the relation $R_{\bar{\mathcal{M}}}$ consisting of all non-constant tuples of $R_\mathcal{M}$. Then we have $R_{\hat{\mathcal{M}}} \cap R_{\bar{\mathcal{M}}} = \emptyset$ and $R_{\hat{\mathcal{M}}} \cup R_{\bar{\mathcal{M}}} = R_\mathcal{M}$. It is easy to see that $M_{n,p}$ contains $\mathcal{M}$ if and only if $M_{n,p}$ contains both $\hat{\mathcal{M}}$ and $\bar{\mathcal{M}}$. Also, since $R_{\hat{\mathcal{M}}}$ and $R_{\bar{\mathcal{M}}}$ are disjoint, the containment of $\hat{\mathcal{M}}$ is independent from the containment of $\bar{\mathcal{M}}$ and therefore

$$\mu(M_{n,p} \sqsupset \mathcal{M}) = \mu(M_{n,p} \sqsupset \hat{\mathcal{M}})\mu(M_{n,p} \sqsupset \bar{\mathcal{M}}).$$

Since all tuples in $R_{\hat{\mathcal{M}}}$ constant, we have $\mu(M_{n,p} \sqsupset \hat{\mathcal{M}}) = p^{\hat{r}_\mathcal{M}}$, then the formula above gives

$$\mu(M_{n,p} \sqsupset \mathcal{M}) = p^{\hat{r}_\mathcal{M}} \mu(M_{n,p} \sqsupset \bar{\mathcal{M}}). \tag{13}$$

There are two possible cases.

1. *The relation $R_{\bar{\mathcal{M}}}$ contains no tuples (that is, all tuples in $R_\mathcal{M}$ are constant).* In this case $p_\epsilon(n) = \epsilon^{1/\hat{r}_\mathcal{M}}$, so the threshold function is constant and the threshold is coarse.
2. *$R_{\bar{\mathcal{M}}}$ contains at least one tuple.* In this case we can apply Theorem 3 in the following way. First, from (13) we have

$$\epsilon = \mu(M_{n,p_\epsilon} \sqsupset \mathcal{M}) = p_\epsilon^{\hat{r}_\mathcal{M}} \mu(M_{n,p_\epsilon} \sqsupset \bar{\mathcal{M}}) \le p_\epsilon^{\hat{r}_\mathcal{M}}.$$

Therefore
$$\epsilon^{1/\hat{r}_\mathcal{M}} \le p_\epsilon(n) \tag{14}$$

for all sufficiently large n. From the left-hand side of (2) and (13) we have

$$p_\epsilon^{\hat{r}_\mathcal{M}} (1 - \exp\{-\Theta(\Phi_{\bar{\mathcal{M}}})\}) \le p_\epsilon^{\hat{r}_\mathcal{M}} \mu(M_{n,p_\epsilon} \sqsupset \bar{\mathcal{M}}) = \mu(M_{n,p_\epsilon} \sqsupset \mathcal{M}) = \epsilon. \tag{15}$$

Since $p_\epsilon(n)$ is bounded from below we have $\lim_{n \to \infty} (1 - \exp\{-\Theta(\Phi_{\bar{\mathcal{M}}})\}) = 1$. Define $g(n) = (1 - \exp\{-\Theta(\Phi_{\bar{\mathcal{M}}})\})^{-1/\hat{r}_\mathcal{M}}$. It is clear that $\lim_{n \to \infty} g(n) = 1$. From (14) and (15) we have

$$\epsilon^{1/\hat{r}_\mathcal{M}} \le p_\epsilon(n) \le \epsilon^{1/\hat{r}_\mathcal{M}} g(n).$$

Therefore $\lim_{n \to \infty} p_\epsilon(n) = \epsilon^{1/\hat{r}_\mathcal{M}}$. So the threshold function for the $\mathcal{M}$-containment property is constant. The threshold for the $\mathcal{M}$-containment is trivially coarse since for $\epsilon \ne 1/2$ we have $\lim_{n \to \infty} p_\epsilon(n)/p_{1/2} = (2\epsilon)^{1/\hat{r}_\mathcal{M}} \ne 1$.

5 Threshold for the Weak Containment

In this section we study the weak containment property for an arbitrary but fixed finite family of structures S, show how to calculate the threshold function for this property, and show that it has a coarse threshold.

It is clear that if there is a structure $M \in S$ such that R_M is empty then the weak S-containment property is trivial since all structures with a sufficiently large domain weakly contain S. Therefore, we assume that for each structure $M \in S$ the relation R_M is non-empty. Let $\bar{S}$ denote the set of all structures M from S such that R_M contains no constant tuples.

Let $\bar{S} \ne \emptyset$. Define *density of S* to be
$$m(S) = \min_{\mathcal{M} \in \bar{S}} m(\mathcal{M}).$$

Then the following holds.

Corollary 4. *Let S be a finite set of structures such that for each $M \in S$ the relation R_M is non-empty. Suppose that $\bar{S} \ne \emptyset$. Then the threshold function for the weak S-containment property is $n^{-1/m(S)}$ and the threshold is coarse.*

Corollary 5. *Let S be a finite set of structures such that for each $M \in S$ the relation R_M is non-empty. Suppose that $\bar{S} = \emptyset$. Then the threshold function for the weak S-containment property is constant and the threshold is coarse.*

The proofs of Corollaries 4 and 5 can be found in the full version of this paper [12].

6 Main Results

We can now put together all the presented results as follows. First we introduce the key notion of *density of a query*. Let φ be a monotone non-recursive datalog$^{\neq}$ query. Then, by Theorem 1 it is equivalent to a sentence in the language $\Sigma_1^{\neq}$. By Theorem 2 such a query is also equivalent to the weak S_φ-structure containment problem for a finite family of structures S_φ. Moreover such a family can be found effectively from φ. If every structure in S_φ contains a constant tuple ($\bar{S}_\varphi = \emptyset$) then the threshold function for the weak S-containment property is constant and the threshold is coarse. Otherwise we define *density* of φ, denoted $m(\varphi)$, to be the density of S_φ, (see Section 5 for the definition of the density for families of structures). From the above it follows that the density of a query can be calculated effectively. Now we are ready to formulate our main theorem.

Theorem 4. *Given a monotone non-recursive datalog$^{\neq}$-query φ one can effectively find the threshold function for the property defined by this query. This threshold function is either constant or has the form $n^{-1/m(\varphi)}$, where $m(\varphi)$ is the density of φ. The density of a query is always a positive rational constant which can be effectively calculated from φ. For every monotone non-recursive datalog$^{\neq}$-query the threshold is coarse.*

Now we give a simple application of Theorem 4 to $0-1$ laws for non-recursive monotone datalog$^{\neq}$. Let us fix a function $0 < p(n) < 1$. We say that non-recursive monotone datalog$^{\neq}$ has a $0-1$ law w.r.t. $p(n)$ if for every boolean query φ expressible in it, $\lim_{n\to\infty} \mu(M_{n,p} \models \varphi)$ equals either to 0 or 1.

Theorem 5. *Monotone non-recursive datalog$^{\neq}$ has a $0-1$ law w.r.t. every probability function $p(n)$ satisfying the following condition: for every rational $q > 0$ either $p \gg n^{-q}$ or $p \ll n^{-q}$ holds.*

Proof. Indeed, from Theorem 4 it follows that for every non-recursive monotone datalog$^{\neq}$ query φ, the threshold function is either constant or has the form n^{-q}, for a rational $q > 0$. Therefore, from the definition of the threshold function follows that for any such query φ, and $p(n)$ as in the statement of the theorem, $\lim_{n\to\infty} \mu(M_{n,p} \models \varphi)$ is either 0 or 1.

For example the theorem holds for $p(n) = n^{-\alpha}$ where $\alpha > 0$ is irrational and also for functions like $ln(n)n^{-t}$ where $t > 0$.

Let us note that $0-1$ laws w.r.t. irrational powers of $1/n$ are proved for the full first-order logic on random graphs in [16] see also [17]. For general accounts on $0-1$ laws for various logics see [5,14], for some applications of $0-1$ laws to the database theory see, e.g., [15].

Acknowledgements. We are grateful to Evgeny Dantsin for introducing us to the fascinating area of randomness and to Leonid Libkin and Moshe Vardi for providing useful references.

References

1. B. Bollobás. Random graphs. In *Combinatorics (Swansea, 1981)*, volume 52 of *London Math. Soc. Lecture Note Ser.*, pages 80–102. Cambridge Univ. Press, 1981.
2. B. Bollobás. *Random graphs*. Academic Press Inc., London, 1985.
3. B. Bollobás and A. Thomason. Threshold functions. *Combinatorica*, 7:35–38, 1987.
4. M. de Rougemont. The reliability of queries. In *Proceedings of the Fourteenth ACM SIGACT-SIGMOD-SIGART Symposium on Principles of Database Systems, May 22-25, 1995, San Jose, California*, pages 286–291. ACM Press, 1995.
5. H.-D. Ebbinghaus and J. Flum. *Finite Model Theory*. Perspectives in Mathematical Logic. Springer, 1999.
6. P. Erdős and A. Rényi. On the evolution of random graphs. *Magyar Tud. Akad. Mat. Kutató Int. Közl.*, 5:17–61, 1960.
7. R. Fagin. Probabilities on finite models. *Journal of Symbolic Logic*, 41:50–58, 1976.
8. E. Friedgut. Sharp thresholds of graph properties, and the k-sat problem. *J. Amer. Math. Soc.*, 12(4):1017–1054, 1999.
9. Y. Glebskii, M. Kogan, M. Liogonkii, and V. Talanov. Range and degree of realizability of formulas in the restricted predicate calculus. *Kibernetika*, 5:17–27, 1969. (in Russian); English translation in Cybernetics 5, 142–154, 1969.
10. E. Grädel, Y. Gurevich, and C. Hirsch. The complexity of query reliability. In *Proceedings of the Seventeenth ACM SIGACT-SIGMOD-SIGART Symposium on Principles of Database Systems, June 1-3, 1998, Seattle, Washington*, pages 227–234. ACM Press, 1998.
11. S. Janson, T. Łuczak, and A. Ruciński. *Random graphs*. John Wiley & Sons, Inc., 2000.
12. K. Korovin and A. Voronkov. Random databases and threshold for monotone non-recursive datalog. Preprint, School of Computer Science, The University of Manchester, 2005.
13. R. Kumar, P. Raghavan, S. Rajagopalan, D. Sivakumar, A. Tomkins, and E. Upfal. The web as a graph. In *Proceedings of the Nineteenth ACM SIGMOD-SIGACT-SIGART Symposium on Principles of Database Systems, May 15-17, 2000, Dallas, Texas, USA*, pages 1–10. ACM, 2000.
14. L. Libkin. *Elements of Finite Model Theory*. Texts in Theoretical Computer Science. Springer, 2004.
15. S. Lifschitz and V. Vianu. A probabilistic view of Datalog parallelization. *Theoretical Computer Science*, 190(2):211–239, 1998.
16. S. Shelah and J. Spencer. Zero one laws for sparse random graphs. *Journal of the AMS*, 1(1):97–115, 1988.
17. J. Spencer. *The Strange Logic of Random Graphs*, volume 22 of *Algorithms and Combinatorics*. Springer, 2001.

An Asymptotically Optimal Linear-Time Algorithm for Locally Consistent Constraint Satisfaction Problems

Daniel Král'[1,2] and Ondřej Pangrác[2]

[1] Institute for Mathematics, Technical University Berlin,[*]
Strasse des 17. Juni 136, D-10623 Berlin, Germany
kral@math.tu-berlin.de

[2] Department of Applied Mathematics,[**]
Faculty of Mathematics and Physics, Charles University,
Malostranské náměstí 25, 118 00 Prague, Czech Republic
{kral, pangrac}@kam.mff.cuni.cz

Abstract. An instance of a constraint satisfaction problem is l-consistent if any l constraints of it can be simultaneously satisfied. For a set Π of constraint types, $\rho_l(\Pi)$ denotes the largest ratio of constraints which can be satisfied in any l-consistent instance composed by constraints from the set Π. We study the asymptotic behavior of $\rho_l(\Pi)$ for sets Π consisting of Boolean predicates. The value $\rho_\infty(\Pi) := \lim_{l \to \infty} \rho_l(\Pi)$ is determined for all such sets Π. Moreover, we design a robust deterministic algorithm (for a fixed set Π of predicates) running in time linear in the size of the input and $1/\varepsilon$ which finds either an inconsistent set of constraints (of size bounded by the function of ε) or a truth assignment which satisfies the fraction of at least $\rho_\infty(\Pi) - \varepsilon$ of the given constraints. Most of our results hold for both the unweighted and weighted versions of the problem.

1 Introduction

Constraint satisfaction problems form an important computational model for problems arising in practice. This is witnessed by an enormous interest in the computational complexity of various variants of constraint satisfaction problems [3,5,6,17]. However, some real instances do not require all the constraints to be satisfied but it is enough to satisfy a large fraction of them. In order to maximize this fraction, the input can be usually pruned at the beginning by removing small sets of contradictory constraints in such a way that the input instance is "locally" consistent. Formally, an instance of the constraint satisfaction problem is *l-consistent* if any l constraints can be simultaneously satisfied.

---

[*] The author was a postdoctoral fellow at TU Berlin within the framework of the European training network COMBSTRU from October 2004 till July 2005.
[**] This research was partly supported by Institute for Theoretical Computer Science (ITI). The Institute is funded by Ministry of Education of Czech Republic as projects LN00A056 and 1M0021620808.

In a weighted version of the problem the constraints are assigned weights and the goal is to maximize the total weight of satisfied constraints. In this paper, we design a robust linear-time asymptotically optimal algorithm for l-consistent constraint satisfaction problems with constraints being Boolean predicates.

If Π is a set of predicate types, then $\rho_l(\Pi)$ is the fraction of the constraints which can be satisfied in each l-consistent instance with constraints from Π. Similarly, $\rho_l^w(\Pi)$ denotes this maximum for the weighted version. Let further $\rho_\infty(\Pi) = \lim_{l \to \infty} \rho_l(\Pi)$ and $\rho_\infty^w(\Pi) = \lim_{l \to \infty} \rho_l^w(\Pi)$. We express $\rho_\infty^w(\Pi)$ for all finite sets of predicates Π and $\rho_\infty(\Pi)$ for all such sets of predicates Π of arities at least two as the minimum of a certain functional Ψ on a convex hull of a finite set $\pi(\Pi)$ of polynomials derived from Π (Corollary 2). Formal definitions of the functional Ψ and the set $\pi(\Pi)$ are provided in Section 2. Some of our results also hold for infinite sets Π.

The main algorithmic result is designing, for any fixed set Π of predicates, a deterministic algorithm which given $\varepsilon > 0$ and a sufficiently locally consistent instance of the constraint satisfaction problem with total weight w_0 finds a truth assignment which satisfies the constraints whose weight is at least $(\rho_\infty^w(\Pi) - \varepsilon) w_0$. The running time of the algorithm is linear in the number of constraints and $1/\varepsilon$. The algorithm is robust in the sense that if it fails to find the desired truth assignment, then it outputs an inconsistent set of constraints whose size is bounded by a function of ε. However, it can find a good truth assignment even if the input instance is not sufficiently locally consistent (in particular, it does not determine the local consistency). Finally, the presented algorithm is asymptotically optimal in the sense that the ratio of the weights of satisfied constraints can be made arbitrarily close to $\rho_\infty^w(\Pi)$ by choosing a sufficiently small ε.

1.1 Previous Results and Their Relation to Our Results

Constraint satisfaction problems with Boolean predicates can be traced to the late 1970's. Schaefer [14] provided a dichotomy result on the complexity of the decision problem. But even if the decision problem is efficiently solvable, the problem to maximize the number of satisfied predicates (if all of them cannot be satisfied) can still be hard, e.g., Håstad [7] showed that there is no $(2 - \varepsilon)$-approximation algorithm for a single-predicate set Π containing $P(x_1, x_2, x_3) = (x_1 + x_2 + x_3) \bmod 2$ unless $P = NP$. Since $\rho_\infty(\Pi) = 1/2$ in this case [4], the ratio of our algorithm is the best possible.

Locally consistent constraint satisfaction problems for constraints which are Boolean predicates were first studied by Trevisan [15] who proved that if Π is the set of all the predicates of arity k, then $\rho_\infty^w(\Pi) = \rho_\infty(\Pi) = 2^{1-k}$. Dvořák et al. [4] showed that if Π is a set containing a single 1-extendable (see Section 2 for the definition) predicate P of arity k, then $\rho_l^w(\Pi) = \rho_l(\Pi) = \sigma(P)/2^k$ for all $l \geq 1$ where $\sigma(P)$ is the number of possible combinations of arguments which satisfy P. In particular, $\rho_\infty^w(\Pi) = \rho_\infty(\Pi) = \sigma(P)/2^k$. In [4], all the values $\rho_l^w(\Pi)$ has also been determined for sets Π consisting of a single Boolean predicate with arity at most three, e.g., it was shown in [4] that $\rho_\infty^w(\Pi^3) = 3/4$ where Π^k is comprised of a single (non-1-extendable) predicate $P^k(x_1,\ldots,x_k) = x_1 \wedge (x_2 \vee \cdots \vee x_k)$. Our results imply that $\rho_\infty^w(\Pi) = 3/4$ for $\Pi = \{P^k\}$ for all $3 \leq k \leq 6$. However, surprisingly, $\rho_\infty^w(\Pi^k) > 3/4$ for all $k \geq 7$ as shown in Example 3.

The most studied variant of the problem are locally consistent CNF formulas. The corresponding set Π_{SAT} of the predicates is the set of all disjunctions. Similarly, $\Pi_{2-\text{SAT}}$ denotes the set $\{(x_1),(x_1 \vee x_2)\}$ of the predicates corresponding to 2-SAT formulas. Locally consistent CNF formulas can be found, e.g., in a recent monograph by Jukna [9]. The exact values of $\rho_l^w(\Pi_{\text{SAT}})$ and $\rho_l^w(\Pi_{2-\text{SAT}})$ are known only for small values of l: clearly, $\rho_1^w(\Pi_{\text{SAT}}) = \rho_1^w(\Pi_{2-\text{SAT}}) = 1/2$. Lieberherr and Specker [11,12] showed that $\rho_2^w(\Pi_{\text{SAT}}) = \rho_2^w(\Pi_{2-\text{SAT}}) = \frac{\sqrt{5}-1}{2} \approx 0.6180$ and $\rho_3^w(\Pi_{\text{SAT}}) = \rho_3^w(\Pi_{2-\text{SAT}}) = 2/3$. Both the proofs were simplified by Yannakakis [18] using a probabilistic argument. The case of 4-locally consistent CNF formulas surprisingly differs from the previous ones: first, $\rho_4^w(\Pi_{\text{SAT}}) \approx 0.6992$ but $\rho_4^w(\Pi_{2-\text{SAT}}) > 0.6992$. Second, the values $\rho_l^w(\Pi_{\text{SAT}})$ for $l = 1,2,3$ coincide with the values defined for a "fractional" version of the problem (which are known for all $l \geq 1$ [10] and are equal to Usiskin's numbers [16]) but the value $\rho_4^w(\Pi_{\text{SAT}})$ differs from the corresponding value 0.6920.

The asymptotic behavior of $\rho_l^w(\Pi_{\text{SAT}})$ was addressed by Huang and Lieberherr [8] who proved that $\rho_\infty^w(\Pi_{\text{SAT}}) \leq 3/4$. The limit was settled by Trevisan [15] by showing $\rho_\infty^w(\Pi_{\text{SAT}}) = \rho_\infty^w(\Pi_{2-\text{SAT}}) = 3/4$. The last equality can be easily derived from our general expression for $\rho_\infty^w(\Pi)$ as shown in Examples 1 and 2.

2 Notation

In the paper, we only deal with constraints which are Boolean predicates and so we prefer to call them *predicates* to emphasize their kind. For a fixed set Π of (types of) predicates, we consider sets Σ of predicates with types from Π. The arguments of the predicates can be both positive and negative literals, but a single variable cannot be contained in two arguments of the same predicate. The latter does not decrease generality: if a single variable is allowed to be contained in several distinct arguments of a single predicate, enhance the set Π by predicates obtained from the predicates of Π by identifying their arguments.

The goal is to find a truth assignment satisfying the largest fraction $\rho(\Sigma)$ of the predicates of Σ. Hence, $\rho_l(\Pi) = \inf \rho(\Sigma)$ where the infimum is taken over all l-consistent sets Σ of predicates of types from Π. In the weighted version, $\rho(\Sigma)$ is the ratio between the weights of the predicates which can be simultanously satisfied and the total weight of the predicates and $\rho_l^w(\Pi) = \inf \rho(\Sigma)$ where the infimum is taken over all l-consistent weighted sets Σ. In the unweighted case, Σ is a set, not a multiset (otherwise, ρ_∞ and ρ_∞^w would coincide).

A predicate P is *1-extendable* if it has the following property: after fixing one of its arguments, the remaining ones can be chosen so that the predicate is satisfied. In particular, the 0-ary predicate which is constantly true is 1-extendable. A *restriction* of a predicate P is a predicate P' obtained from P by fixing some of its arguments, e.g., $P'(x_1,x_2) = (x_1 \wedge x_2)$ is a restriction of $P(x_1,x_2,x_3) = (x_1 \wedge x_2 \wedge x_3) \vee (\neg x_3)$ obtained by setting $x_3 = $ true. A restriction of a k-ary predicate can be described by a vector $\tau \in \{0, 1, \star\}^k$ where 0 and 1 denote arguments fixed to false or true, and $\star$ denotes unfixed arguments. Let $\pi_{P,\tau}(p) : \langle 0, 1 \rangle \to \langle 0, 1 \rangle$ be equal to the probability that the predicate $P(x_1, \ldots, x_k)$ is satisfied if x_i is set to true randomly and independently with the probability $1 - p$, p and $1/2$, if τ_i is 0, 1 and $\star$, respectively. Note that $\pi_{P,\tau}(p)$ is a polynomial in p of degree at most k. For a set Π of predicates, $\pi(\Pi)$ is the set of all $\pi_{P,\tau}$ where $P \in \Pi$ and the restriction of P described by τ is 1-extendable.

Example 1. Let Π be the set consisting of two predicates $P_1(x_1) = (x_1)$ and $P_2(x_1, x_2) = (x_1 \vee x_2)$. There is a single restriction of the predicate P_1 which is 1-extendable and this corresponds to the vector 1. There are five restrictions of the predicate P_2 which are 1-extendable, those corresponding to 11, 10, 1$\star$, $\star$1 and $\star\star$. Hence, the set $\pi(\Pi)$ consists of the following four functions:

$$\pi_{P_1,1}(p) = p \qquad \pi_{P_2,11}(p) = 2p - p^2 \qquad \pi_{P_2,10}(p) = 1 - p + p^2$$

$$\pi_{P_2,1\star}(p) = \pi_{P_2,\star 1}(p) = (p+1)/2 \qquad \pi_{P_2,\star\star}(p) = 3/4.$$

Let Ψ be the functional which assigns a continuous function $f(x) : \langle 0, 1 \rangle \to \langle 0, 1 \rangle$ its maximum for $x \in \langle 0, 1 \rangle$. If F is a family of functions $f : \langle 0, 1 \rangle \to \langle 0, 1 \rangle$, then $\Psi(F)$ is defined to be the infimum $\Psi(f)$ where f ranges over all convex combinations of the functions of F. The infimum is attained if the set F is a finite set of polynomials (which is the case of $\pi(\Pi)$ for any set Π). As mentioned in Section 1, one of our results is that the limit $\rho_\infty(\Pi) = \lim_{l \to \infty} \rho_l(\Pi)$ is equal to $\Psi(\pi(\Pi))$ for any set Π of predicates with arities at least two and $\rho_\infty^w(\Pi)$ is equal to $\Psi(\pi(\Pi))$ for any set Π of predicates (see Corollary 2 and Examples 2–3).

3 The Algorithm and the Lower Bound

The proof of the next auxiliary technical lemma is omitted due to space limitations:

Lemma 1. *Let Π be a set of predicates of arity at most K and let $f(p)$ be any convex combination of functions contained in $\pi(\Pi)$. The derivative of the function $f(p)$ for $p \in \langle 0, 1 \rangle$ takes values from the interval $\langle -K, +K \rangle$.*

We now establish the main result of this section:

Theorem 1. *Fix a set Π of predicates whose arity does not exceed K. There exists an algorithm which given $\varepsilon > 0$ and a set of weighted predicates Σ of total weight w_0 either finds a truth assignment satisfying predicates of Σ of weight at least $(\Psi(\pi(\Pi)) - \varepsilon) w_0$ or finds a set of at most $2K^{\lceil 2K/\varepsilon \rceil - 1}$ inconsistent predicates. The algorithm runs in time linear in $|\Sigma|$ and $1/\varepsilon$.*

Proof. The algorithm consists of three steps:

1. Labeling variables according to the depth of "forcing" their values by the input predicates (or finding at most $2K^{\lceil 2K/\varepsilon \rceil - 1}$ inconsistent predicates).

2. Finding a probability distribution on truth assignments such that the expected weight of the satisfied predicates is at least $(\Psi(\pi(\Pi)) - \varepsilon)w_0$.
3. Construction of a truth assignment which satisfies predicates whose weight is at least $(\Psi(\pi(\Pi)) - \varepsilon)w_0$.

The third step is an easy application of a standard linear-time derandomization technique proposed by Yannakakis [18] formulas nowadays known as the method of conditional expectations (the reader is referred to [1,2,13] for additional details). We focus on the first two steps in the rest of the proof.

In the first step, we construct a sequence of $1 + \lceil 2K/\varepsilon \rceil$ partial truth assignments $\mu_0, \ldots, \mu_{\lceil 2K/\varepsilon \rceil}$ and subsets $\Sigma_1, \ldots, \Sigma_{\lceil 2K/\varepsilon \rceil}$ of Σ. The partial truth assignment μ_0 is the empty one, i.e., no variable is fixed by μ_0. Let i be an integer between 1 and $\lceil 2K/\varepsilon \rceil$ and assume that $\mu_0, \ldots, \mu_{i-1}$ have been constructed. Let Σ_i be the set of the predicates whose restrictions with respect to μ_{i-1} are not 1-extendable. If there is a predicate whose restriction is constantly false, we stop. Otherwise, the assignment μ_{i-1} is extended to the partial truth assignment μ_i by setting the values of the forced variables. The value of a variable x is *forced* if there exists a predicate that can be satisfied only if either x is false or x is true. If the value of a single variable is forced to be both true and false, we also stop.

Let us say few comments on the implementation of this step of the algorithm. Each variable x is labeled by the smallest i such that μ_i fixes x_i. The variables whose values are forced by previously fixed variables are stored in a FIFO queue. When a variable is dequeued, the algorithm checks whether there are some new variables forced after fixing the value of the dequeued variable. If so, the newly forced variables are added to the queue. In addition, in order to quickly find inconsistent sets of clauses, we store for each variable which of the predicates forced its value. Such predicate is also included to the set Σ_i. Note that the labels of the variables correspond to "depths" of derivations forcing their values and each predicate is included to at most K sets $\Sigma_1, \ldots, \Sigma_{\lceil 2K/\varepsilon \rceil}$.

If we stop in the first step because we find an unsatisfied predicate or a variable forced to two different values, we easily find an inconsistent set of at most $2(K^{\lceil 2K/\varepsilon \rceil - 1} + 1)$ predicates: if an unsatisfied predicate is found, let A be the set consisting of this predicate, all the (at most K) predicates forcing the values of the variables contained in its arguments, all the (at most $K(K-1)$) predicates forcing the values of the variables contained in the "first-level" predicates, etc. Since there are at most $\lceil 2K/\varepsilon \rceil$ levels, the size of A does not exceed:

$$1 + K + K(K-1) + \cdots + K(K-1)^{\lceil 2K/\varepsilon \rceil - 2} \leq K^{\lceil 2K/\varepsilon \rceil - 1} + 1.$$

If we stop because there is a variable which is forced to two different values, we include to the set A the two predicates which force it to have opposite values, all the (at most $2(K-1)$) predicates forcing the values of the variables contained in their arguments, etc. In this case, the size of A is bounded by:

$$2 + 2(K-1) + 2(K-1)^2 + \cdots + 2(K-1)^{\lceil 2K/\varepsilon \rceil - 2} \leq 2K^{\lceil 2K/\varepsilon \rceil - 1}.$$

In either of the cases, the number of the predicates contained in A is at most $2K^{\lceil 2K/\varepsilon \rceil - 1}$ and the set A can be constructed in time linear in $|A|K \leq |\Sigma|K$.

If for each variable x, a list of predicates that contain x is formed at the beginning of the computation (which can be simultaneously done for all the variables in linear time), the entire first step of the algorithm can be performed in time $O(|\Sigma|K) = O(|\Sigma|)$ including the construction of an inconsistent set.

We now focus on the second step. Since each predicate of Σ is contained in at most K sets $\Sigma_1, \ldots, \Sigma_{\lceil 2K/\varepsilon \rceil}$, the total weight of all the predicates contained in the sets $\Sigma_1, \ldots, \Sigma_{\lceil 2K/\varepsilon \rceil}$ (counting multiplicities) is at most Kw_0. By an averaging argument, there exists $1 \leq i \leq \lceil 2K/\varepsilon \rceil$ for which the weight of the predicates of Σ_i is at most $\varepsilon w_0/2$. Let w_0' be the total weight of the predicates contained in $\Sigma \setminus \Sigma_i$. Note that $w_0' \geq (1 - \varepsilon/2)w_0$.

Let $f(p)$ be the expected weight of the satisfied predicates of $\Sigma \setminus \Sigma_i$ divided by w_0' where each variable fixed by μ_{i-1} gets the value assigned by μ_{i-1} with the probability p and the remaining variables are set to be true with the probability $1/2$ (all the choices are mutually independent). The coefficients of the polynomial $f(p)$ (of degree at most K) can be computed in time linear in $|\Sigma|$. Since the restriction of each predicate of $\Sigma \setminus \Sigma_i$ with respect to μ_{i-1} is 1-extendable, the function $f(p)$ is a convex combination of functions from $\pi(\Pi)$. In particular, the absolute value of its derivative does not exceed K by Lemma 1.

Evaluate $f(p)$ for the following values of p: $0, \frac{\varepsilon}{K}, \frac{2\varepsilon}{K}, \ldots, \lfloor \frac{K}{\varepsilon} \rfloor \frac{\varepsilon}{K}, 1$. The largest of the computed values differs from the maximum of $f(p)$ for $p \in \langle 0, 1 \rangle$ by at most $\varepsilon/2$ because the absolute value of the derivative of f is at most K. Since for each of the $\lfloor K/\varepsilon \rfloor + 2$ values of p, the function $f(p)$ can be evaluated in time $O(K)$, the algorithm needs time linear in $O(1/\varepsilon)$ to determine p_0.

Consider the probability distribution on the values of the variables for $p = p_0$. The expected weight of the satisfied clauses is clearly at least $f(p_0)w_0'$:

$$f(p_0)w_0' \geq \left(\max_{p \in \langle 0,1 \rangle} f(p) - \varepsilon/2\right)(1 - \varepsilon/2)w_0 \geq$$

$$(\Psi(\pi(\Pi)) - \varepsilon/2)(1 - \varepsilon/2)w_0 \geq (\Psi(\pi(\Pi)) - \varepsilon)w_0.$$

This finishes the second step of the algorithm. Note that the algorithm provides no estimate of $\Psi(\pi(\Pi))$.

An immediate corollary of Theorem 1 is the following:

Corollary 1. *Let Π be a set of Boolean predicates. For each $\varepsilon > 0$, there exists an integer $l \geq 1$ such that*

$$\rho_l(\Pi) \geq \rho_l^w(\Pi) \geq \Psi(\pi(\Pi)) - \varepsilon.$$

4 The Upper Bound

We first introduce some notation. For a set Σ of predicates and a partial truth assignemnt μ, let Σ' be the set of restrictions of the predicates of Σ with respect to μ. Let $G(\Sigma')$ be the multigraph whose vertices are predicates of Σ' and the number of edges between two predicates P_1 and P_2 of Σ' is equal to the number of variables which appear in arguments of both P_1 and P_2 (regardless whether they appear as positive or negative literals). The predicates with all arguments fixed by μ are isolated vertices. A *semicycle* of length l of Σ with respect to μ is a set Γ of l predicates such that the the predicates form a cycle of length l in $G(\Sigma')$ and each edge of the cycle correspond to a different variable. The next lemma relates non-existence of short semicycles and the local consistency of Σ:

Lemma 2. *Let Σ be a set of predicates, μ a partial truth assignment, Σ' the restrictions of the predicates of Σ with respect to μ and $l \geq 2$ an integer. If each predicate of Σ' is 1-extendable and Σ has no semicycle of length at most l with respect to μ, then Σ is l-consistent.*

Proof. We prove by induction on i that any i predicates of Σ' can be simultaneously satisfied. This implies the statement of the lemma because a truth assignment for Σ' can be viewed as an extension of μ.

The claim trivially holds for $i = 1$. Assume that $i > 1$ and let $P_1, \ldots, P_i$ be i predicates of Σ. Since Σ' contains no semicycle of length at most l, there is a predicate that shares at most a single variable with the remaining ones. We assume without loss of generality that P_i is such a predicate. Let $y_1, \ldots, y_n$ be the variables contained in the first $i - 1$ predicates which are not set by μ. By the induction, there is a truth assignment for $y_1, \ldots, y_n$ which satisfies all the predicates $P_1, \ldots, P_{i-1}$. Since P_i has at most one variable in common with the predicates $P_1, \ldots, P_{i-1}$, the truth assignment for $y_1, \ldots, y_n$ can be extended to a truth assignment which satisfies all the predicates $P_1, \ldots, P_i$ (note that the restriction P_i with respect to μ is 1-extendable)..

In the proof of Theorem 2, Markov's inequality and Chernoff's inequality are used to bound the probability of large deviations from the expected value:

Proposition 1. *Let X be a non-negative random variable with the expected value E. The following holds for every $\alpha \geq 1$:*

$$\mathrm{Prob}(X \geq \alpha) \leq \frac{E}{\alpha}.$$

Proposition 2. *Let X be a random variable equal to the sum of N zero-one independent random variables such that each of them is equal to 1 with the probability p. Then, the following holds for every $0 < \delta \leq 1$:*

$$\mathrm{Prob}(X \geq (1+\delta)pN) \leq e^{-\frac{\delta^2 pN}{3}} \quad \text{and} \quad \mathrm{Prob}(X \leq (1-\delta)pN) \leq e^{-\frac{\delta^2 pN}{2}}.$$

We are now ready to prove our upper bounds on $\rho_\infty^w(\Pi)$ and $\rho_\infty(\Pi)$:

Theorem 2. *Let Π be a set of Boolean predicates. For any integer $l \geq 1$ and any real $\varepsilon > 0$, there exists an l-consistent set Σ_0 of weighted predicates whose types are from the set Π such that:*

$$\rho^w(\Sigma_0) \leq \Psi(\pi(\Pi)) + \varepsilon.$$

Moreover, if the arity of each predicate Π is at least two, then there exists such a set Σ_0 of unweighted predicates.

Proof. We can assume that $\varepsilon < 1$ is the inverse of a power of two. Let $f_1, \ldots, f_K$ be all the different functions contained in $\pi(\Pi)$ and let $\sum_{i=1}^{K} \alpha_i f_i$ be their convex combination with $\Psi(\sum_{i=1}^{K} \alpha_i f_i) = \Psi(\pi(\Pi))$. Let further P^i be a predicate of Π whose restriction with respect to a vector τ^i is 1-extendable and $\pi_{P^i, \tau^i} = f_i$. Since f_i are distinct, there are no two indices $i \neq i'$ such that $P^i = P^{i'}$ and $\tau^i = \tau^{i'}$. Finally, let K_0 be the maximum arity of a predicate of Π.

We consider a random set Σ of predicates whose arguments contain variables $x_1, \ldots, x_n$ and $y_1, \ldots, y_n$ where n is a sufficiently large power of two which will be fixed later in the proof. Fix $i \in \{1, \ldots, K\}$ and let k be the arity of P^i and k' the number of stars contained in τ^i. At this point, we abandon the condition that each variable can appear in at most one of the arguments of a predicate. Later, we prune Σ to obey this constraint. If $k > 1$, each of the $n^k 2^{k'}$ predicates P^i whose j-th argument, $1 \leq j \leq k$, is a positive literal containing one of the variables $x_1, \ldots, x_n$ if $\tau_j^i = 1$, a negative literal containing one of the variables $x_1, \ldots, x_n$ if $\tau_j^i = 0$ and a positive or negative literal containing one of the variables $y_1, \ldots, y_n$ if $\tau_j^i = \star$, is included to Σ randomly and independently with the probability $\alpha_i 2^{-k'} n^{-(k-1)+1/2l}$. The weights of the predicates are set to one.

If $k = 1$, each predicate P^i whose only argument is a positive literal containing one of the variables $x_1, \ldots, x_n$ if $\tau_1^i = 1$, a negative literal containing one of the variables $x_1, \ldots, x_n$ if $\tau_1^i = 0$ and a positive or negative literal containing one of the variables $y_1, \ldots, y_n$ if $\tau_1^i = \star$, is included to Σ with the weight $\alpha_i 2^{-k'} n^{1/2l}$. Note that if the arity of each predicate of Π is at least two, the obtained system Σ consists of unweighted predicates.

Let Σ^i be the predicates of Σ corresponding to P^i and τ^i. We prove the following three statements (under the assumption that n is sufficiently large):

1. The total weight of the predicates of Σ^i is at least $\alpha_i(1 - \frac{\varepsilon}{8})n^{1+1/2l}$ with the probability greater than $1 - 1/4K$.
2. With the probability greater than $1 - 1/4K$, each truth assignment which assigns true to exactly n' of the variables $x_1, \ldots, x_n$ satisfies the predicates of Σ^i whose total weight is at most $\alpha_i(f_i(n'/n) + \frac{\varepsilon}{4})n^{1+1/2l}$.
3. The total weight of the predicates whose arguments do not contain different variables is at most $\alpha_i \frac{\varepsilon}{8} n^{1+1/2l}$ with the probability greater than $1 - 1/4K$.

If the arity k of P^i is one or $\alpha_i = 0$, then all the three statements hold with the probability one. In the rest, we consider the case that the arity of P^i is at least two, i.e., $k \geq 2$, and $\alpha_i > 0$.

The probability that the total weight of the predicates of Σ^i is smaller than $\alpha_i(1 - \frac{\varepsilon}{8})n^{1+1/2l}$ is bounded by Proposition 2 from above by the following:

$$e^{-\frac{(\varepsilon/8)^2 (\alpha_i 2^{-k'} n^{-(k-1)+1/2l})(n^k 2^{k'})}{2}} = e^{-\frac{\varepsilon^2 \alpha_i n^{1+1/2l}}{128}}$$

Since ε, α_i, l and K do not depend on n, the probability that the total weight of the predicates of Σ^i exceeds $\alpha_i(1 - \frac{\varepsilon}{8})n^{1+1/2l}$ is smaller than $1/4K$ if n is sufficiently large.

Let μ be any of the 2^{2n} truth assignments for $x_1,\ldots,x_n$ and $y_1,\ldots,y_n$; let n' be the number of variables $x_1,\ldots,x_n$ which are set to be true by μ. A predicate is *good* if it is satisfied by μ. There are exactly $f_i(n'/n)n^k 2^{k'}$ good predicates (i is still fixed). If $f_i(n'/n) \leq \frac{\varepsilon}{8}$, then mark additional predicates corresponding to P^i and τ^i as good so that the total number of good predicates is $\frac{\varepsilon}{8}n^k 2^{k'}$ (note that since ε is the inverse of a power of two, this expression is an integer if n is a sufficiently large).. Hence, the expected number of good predicates included to Σ^i is exactly $\max\{f_i(n'/n), \varepsilon/8\}n^k 2^{k'} \cdot \alpha_i n^{-(k-1)+1/2l} 2^{-k'}$. Using the fact that $f_i(n'/n) \leq 1$ and Proposition 2, we infer the following:

$$\text{Prob}(\mu \text{ satisfies more than } \alpha_i(f_i(n'/n) + \frac{\varepsilon}{4})n^{1+1/2l} \text{ predicates of } \Sigma^i) \leq$$

$$\text{Prob}(\Sigma^i \text{ contains more than } \alpha_i(f_i(n'/n) + \frac{\varepsilon}{4})n^{1+1/2l} \text{ good predicates}) \leq$$

$$\text{Prob}(\Sigma^i \text{contains} > (1+\varepsilon/8)\alpha_i \max\{f_i(n'/n), \varepsilon/8\}n^{1+1/2l} \text{good predicates}) \leq$$

$$e^{-\frac{\varepsilon^2 \alpha_i \max\{f_i(n'/n),\varepsilon/8\} n^{1+1/2l}}{192}} \leq e^{-\frac{\varepsilon^3 \alpha_i n^{1+1/2l}}{1536}}$$

Since there are 2^{2n} possible truth assignments μ, the probability that there exists one which satisfies more than $\alpha_i(f_i(n'/n) + \frac{\varepsilon}{4})n^{1+1/2l}$ clauses of Σ^i is at most $2^{2n} \cdot e^{-\frac{\varepsilon^3 \alpha_i n^{1+1/2l}}{1536}}$. Since ε, α_i and K are fixed, this probability is smaller than $1/4K$ if n is sufficiently large.

It remains to establish our third claim on Σ^i. At most $\binom{k}{2}n^{k-1}2^{k'}$ out of all the $n^k 2^{k'}$ predicates which can be included to Σ^i contain one variable in several arguments. Therefore, the expected number of such predicates contained in Σ^i is at most $\binom{k}{2}n^{k-1}2^{k'}\alpha_i 2^{-k'}n^{-(k-1)+1/2l} = \alpha_i\binom{k}{2}n^{1/2l}$. By Markov's inequality (Proposition 1), the probability that the number of such predicates in Σ^i exceeds $\alpha_i\frac{\varepsilon}{8}n^{1+1/2l}$ is at most the following fraction:

$$\frac{\alpha_i\binom{k}{2}n^{1/2l}}{\alpha_i\frac{\varepsilon}{8}n^{1+1/2l}} = \binom{k}{2}\frac{8}{\varepsilon n}.$$

Since ε, k and K are independent of n, the probability of this event is smaller than $1/4K$ if n is sufficiently large.

We conclude that with the probability greater than $1/4$ the following three statements hold for the set Σ and a sufficiently large n (recall that $\sum_{i=1}^{K}\alpha_i = 1$):

1. The total weight of the predicates of Σ is at least $(1 - \frac{\varepsilon}{8})n^{1+1/2l}$.
2. Any truth assignment which assigns true to exactly n' of the variables $x_1,\ldots,x_n$ satisfies the predicates of Σ whose total weight does not exceed $(\sum_{i=1}^{K}\alpha_i f_i(n'/n) + \frac{\varepsilon}{4})n^{1+1/2l}$.
3. The total weight of the predicates whose arguments do not contain different variables is at most $\frac{\varepsilon}{8}n^{1+1/2l}$.

We now estimate the number of semicycles of length at most l in Σ with respect to the partial truth assignment μ_0 which sets all the variables $x_1,\ldots,x_n$

to be true. Note that all the restrictions of the predicates contained in Σ with respect to μ_0 are 1-extendable. Consider a semicycle corresponding to predicates $P'_1, \ldots, P'_{l'}$, $2 \leq l' \leq l$, described by $\tau'_1, \ldots, \tau'_{l'}$. Let k_i be the arity of P'_i and k'_i the number of stars in τ'_i. The number of all semicycles corresponding to the restrictions of $P'_1, \ldots, P'_{l'}$ determined by $\tau'_1, \ldots, \tau'_{l'}$ is at most $\prod_{i=1}^{l'} n^{k_i - k'_i} n^{k'_i - 1} 2^{k'_i} k'_{i-1}$ (the indices are taken modulo l', i.e., $k'_0 = k'_{l'}$). The probability of including any such particular sequence to Σ is $\prod_{i=1}^{l'} \alpha'_i n^{-(k_i-1)+1/2l} 2^{-k'_i}$ where α'_i is the coefficient α_i corresponding to P'_i and τ'_i. Therefore, the expected number of semicycles in Σ corresponding to the restrictions of $P'_1, \ldots, P'_{l'}$ determined by $\tau'_1, \ldots, \tau'_{l'}$ is at most $\prod_{i=1}^{l'} k'_i n^{1/2l} \leq K_0^{l'} n^{1/2}$ (recall that $0 \leq \alpha'_i \leq 1$ for all $1 \leq i \leq l'$ and K_0 is the maximum arity).

Since there are at most $K^{l'}$ ways how to choose $P'_1, \ldots, P'_{l'}$ and $3^{K_0 l'}$ possible choices of the vectors $\tau'_1, \ldots, \tau'_{l'}$, the expected number of semicycles of Σ of length l' does not exceed $(KK_0 3^{K_0})^{l'} n^{1/2}$. By Proposition 1, the probability that Σ contains more than $\frac{\varepsilon}{8l} n^{1+1/2l}$ semicycles of length at most l is at most:

$$\frac{l(KK_0 3^{K_0})^l n^{1/2}}{\frac{\varepsilon}{8l} n^{1+1/2l}} \leq \frac{8l^2 (KK_0 3^{K_0})^l}{\varepsilon n^{1/2}}$$

Since the numbers l, K, K_0 and ε do not depend on n, this probability is smaller than $1/4$ if n is sufficiently large. Therefore with positive probability, the set Σ has the properties 1–3 stated above and the number of its semicycles of length at most l with respect to μ_0 is at most $\frac{\varepsilon}{8l} n^{1+1/2l}$. In particular, there exists a set Σ' with these properties. Fix such a set Σ' for the rest of the proof.

Remove from Σ' the predicates contained in semicycles of length at most l with respect to μ_0 and the predicates which contain the same variable in several arguments. Let Σ_0 be the resulting set. Note that there are at most at most $l \cdot \frac{\varepsilon}{8l} n^{1+1/2l} = \frac{\varepsilon}{8} n^{1+1/2l}$ predicates contained in semicycles of length at most l. Since each of the predicates of Σ' which is contained in a semicycle must contain one of the variables $y_1, \ldots, y_n$, its arity is at least two. Consequently, its weight is equal to one. Hence, the total weight of the predicates removed from Σ' is at most $\frac{\varepsilon}{8} n^{1+1/2l} + \frac{\varepsilon}{8} n^{1+1/2l} = \frac{\varepsilon}{4} n^{1+1/2l}$ and the total weight of the predicates of Σ_0 is at least $(1 - \frac{3\varepsilon}{8}) n^{1+1/2l}$. Clearly, the total weight of the predicates of Σ_0 which can be simultaneously satisfied by a truth assignment is at most the total weight of such predicates of Σ'. We conclude that the following holds for each truth assignment which sets n' ($0 \leq n' \leq n$) of the variables $x_1, \ldots, x_n$ to true:

$$\rho^w(\Sigma_0) \leq \frac{(\sum_{i=1}^{K} \alpha_i f_i(n'/n) + \frac{\varepsilon}{4}) n^{1+1/2l}}{(1 - \frac{3\varepsilon}{8}) n^{1+1/2l}} \leq \frac{\Psi(\pi(\Pi)) + \frac{\varepsilon}{4}}{1 - \frac{3\varepsilon}{8}} \leq$$

$$\Psi(\pi(\Pi)) \frac{1 + \frac{\varepsilon}{4}}{1 - \frac{3\varepsilon}{8}} \leq \Psi(\pi(\Pi))(1 + \varepsilon) \leq \Psi(\pi(\Pi)) + \varepsilon$$

Since Σ_0 contains no semicycles of length at most l with respect to μ_0 and all the restrictions of the predicates of Σ_0 with respect to μ_0 are 1-extendable, the set Σ_0 is l-consistent by Lemma 2. Consequently, $\rho_l^w(\Pi) \leq \Psi(\pi(\Pi)) + \varepsilon$.

Moreover, if the arity of each predicate of Π is at least two, the weights of all the predicates of Σ are one and $\rho_l(\Pi) \leq \Psi(\pi(\Pi)) + \varepsilon$.

We immediately infer from Corollary 1 and Theorem 2 the following:

Corollary 2. *Let Π be a finite set of predicates. The following holds:*

$$\rho_\infty^w(\Pi) = \Psi(\pi(\Pi)).$$

Moreover, if the arity of each predicate of Π is at least two, then it holds:

$$\rho_\infty(\Pi) = \Psi(\pi(\Pi)).$$

As an application, we compute the values $\rho_\infty^w(\Pi)$ for several sets Π:

Example 2. Let Π be the set of predicates from Example 1. Since $\pi_{P_2,\star\star}(p)$ equals to $3/4$ for all $0 \leq p \leq 1$, we infer $\Psi(\pi(\Pi)) \leq \Psi(\pi_{P_2,\star\star}) = 3/4$. On the other hand, the value of each of $\pi_{P_1,1}$, $\pi_{P_2,11}$, $\pi_{P_2,10}$, $\pi_{P_2,1\star}$ and $\pi_{P_2,\star\star}$ for $p = 3/4$ is at least $3/4$. Thus, the value of any convex combination of them for $p = 3/4$ is also at least $3/4$ and $\Psi(\pi(\Pi)) \geq 3/4$. Hence, $\rho_\infty^w(\Pi) = 3/4$.

Example 3. Let Π^k be the set containing a single predicate $P^k(x_1, \ldots, x_k) = x_1 \wedge (x_2 \vee \cdots \vee x_k)$ for an integer $k \geq 7$. Consider the vector $\tau = 10 \cdots 0 \star \star$. Clearly, the restriction of P^k determined by τ is 1-extendable. It is easy to show that the maximum of the function $\pi_{P^k,\tau}$ is attained for $p_0 = \sqrt[k-3]{\frac{4}{k-2}}$ and it is strictly larger than $3/4$. Moreover, the value $\pi_{P^k,\tau}(p_0)$ is smaller or equal to the value $\pi_{P^k,\tau'}(p_0)$ for any τ' corresponding to a 1-extendable restriction of P. We infer that $\rho_\infty^w(\Pi^k) = \Psi(\pi(\Pi^k)) \geq \Psi(\pi_{P^k,\tau}) > 3/4$.

5 Conclusion

We settled almost completely the case of finite sets Π of predicates: it only remains open to determine $\rho_\infty(\Pi)$ for sets Π containing a predicate of arity one. The case of infinite sets Π seems to be also interesting, but rather from the theoretical point of view than the algorithmic one: in most cases, it might be difficult to describe the input if the set Π is not "nice". For an infinite set Π, one can also define the set $\pi(\Pi)$ and then $\Psi(\pi(\Pi))$ as the infimum of Ψ taken over all convex combinations of finite number of functions from $\pi(\Pi)$. It is not hard to verify that the proof of Theorem 2 translates to this setting. In particular, $\rho_\infty^w(\Pi) \geq \Psi(\pi(\Pi))$ for every infinite set Π. However, the proof of Theorem 1 cannot be adopted since the arity of the predicates of Π is not bounded. We suspect that the equality $\rho_\infty^w(\Pi) = \Psi(\pi(\Pi))$ does not hold for all (infinite) sets Π..

Acknowledgement

The authors would like to thank Gerhard Woeginger for attracting their attention to locally consistent formulas and for pointing out several useful references, Dimitrios M. Thilikos for suggesting the version of the problem considered in the paper and Zdeněk Dvořák for fruitful discussions on the subject.

References

1. N. Alon, J. Spencer: The probabilistic method. 2nd edition, John Wiley, New York (2000).
2. B. Chazelle: The discrepancy method: Randomness and complexity. Cambridge University Press, 2000.
3. S. Cook, D. Mitchell: Finding Hard Instances of the Satisfiability Problem: A Survey. In: Satisfiability Problem: Theory and Applications. DIMACS Series in DMTCS Vol. 35 AMS (1997).
4. Z. Dvořák, D. Král', O. Pangrác: Locally consistent constraint satisfaction problems. In: Proc. of 31st International Colloquium on Automata, Languages and Programming (ICALP), LNCS Vol. 3142, Springer-Verlag Berlin (2004) 469-480.
5. D. Eppstein: Improved Algorithms for 3-coloring, 3-edge-coloring and Constraint Satisfaction. In: Proc. of the 12th ACM-SIAM Symposium on Discrete Algorithms. SIAM (2001) 329–337.
6. T. Feder, R. Motwani: Worst-case Time Bounds for Coloring and Satisfiability Problems. J. Algorithms 45(2) (2002) 192-201.
7. J. Håstad: Some optimal inapproximability results. Journal of ACM 48(4) (2001) 798–859 (2001). A preliminary version appeared in: Proc.. 28th ACM Symposium on Theory of Computing (STOC). ACM Press (1997) 1–10.
8. M. A. Huang, K. Lieberherr: Implications of Forbidden Structures for Extremal Algorithmic Problems. Theoretical Computer Science 40 (1985) 195–210.
9. S. Jukna: Extremal Combinatorics with Applications in Computer Science. Springer, Heidelberg (2001).
10. D. Král': Locally Satisfiable Formulas. In: Proc. of the 15th Annual ACM-SIAM Symposium on Discrete Algorithms (SODA). SIAM (2004) 323-332.
11. K. Lieberherr, E. Specker: Complexity of Partial Satisfaction. J. of the ACM, 28(2) (1981) 411–422.
12. K. Lieberherr, E. Specker: Complexity of Partial Satisfaction II. Technical Report 293, Dept. of EECS, Princeton University (1982).
13. R. Motwani, P. Raghavan: Randomized Algorithms. Cambridge University Press, 1995.
14. T. J. Schaefer: The complexity of satisfiability problems. In: Proc. of the 10th Annual ACM Symposium on Theory of Computing (STOC). ACM Press (1978) 216–226.
15. L. Trevisan: On Local versus Global Satisfiability. SIAM J. Disc. Math. (to appear). A preliminary version available as ECCC report TR97-12.
16. Z. Usiskin: Max-min Probabilities in the Voting Paradox. Ann. Math. Stat. 35 (1963) 857–862.
17. G. J. Woeginger: Exact Algorithms for NP-hard Problems: A Survey. In: Proc. 5th Int. Worksh. Combinatorial Optimization - Eureka, You Shrink. LNCS Vol. 2570. Springer-Verlag Berlin (2003) 185-207.
18. M. Yannakakis: On the Approximation of Maximum Satisfiability. J. Algorithms 17 (1994) 475–502. A preliminary version appeared in: Proc. of the 3rd Annual ACM-SIAM Symposium on Discrete Algorithms (SODA). SIAM (1992) 1–9.

Greedy Approximation via Duality for Packing, Combinatorial Auctions and Routing

Piotr Krysta*

Department of Computer Science, Dortmund University,
Baroper Str. 301, 44221 Dortmund, Germany
piotr.krysta@cs.uni-dortmund.de

Abstract. We study simple greedy approximation algorithms for general class of integer packing problems. We provide a novel analysis based on the duality theory of linear programming. This enables to significantly improve on the approximation ratios of these greedy methods, and gives a unified analysis of greedy for many packing problems. We show matching lower bounds on the ratios of such greedy methods. Applications to some specific problems, including mechanism design for combinatorial auctions, are also shown.

1 Introduction

Combinatorial auctions (CAs) is the canonical problem motivated by applications in electronic commerce and game theoretical treatment of the Internet. A seminal paper of Lehmann et al. [24] identified a class of greedy approximation algorithms for the set packing problem as having certain monotonicity properties. These properties proved crucial in obtaining approximate non-VCG mechanisms for truthful CAs. This is one of our main motivations to study greedy algorithms.

Greedy algorithms for combinatorial optimization problems are very simple and efficient. However, their performance analysis can be difficult, and there are no general, unified tools known. We study simple greedy approximation algorithms for general integer packing problems, and provide a technique for analyzing their performance via the duality theory of linear programming. This significantly improves upon known approximation ratios of greedy methods for a class of integer packing problems.

We are not aware about any existing work of analyzing greedy approximation algorithms for integer packing problems via duality (an exception is [6] – see end of Sec. 1.1). The situation is completely different for the integer covering problems, where starting from the seminal work of Lovász [25] and Chvátal [8], such analyzes were performed and generalized [9,29]. One of our purposes is to initiate filling this gap. In fact our technique is fairly general in that it can even be extended to some routing problems.

A class of packing integer programs [31], (PIP), reads: $\max\{cx : Ax \leq b, x \in \{0,1\}^n\}$, where A is an $m \times n$ matrix with non-negative entries, $b \in \mathsf{R}^m_{\geq 1}$, $c \in \mathsf{R}^n_{\geq 0}$. Our results can be extended to allow $x \in \{0,1,\ldots,u\}^n$ for some $u \in \mathsf{N}$. When all

---

* The author is supported by DFG grant Kr 2332/1-1 within "Aktionsplan Informatik" of the Emmy Noether program.

entries in A are $0/1$, (PIP) is called $(0,1)$-PIP, and generalizes many weighted problems, e.g., maximum weighted independent sets, max-cliques, hypergraph b-matching, k-dimensional matching, (multi-) set (multi-) packing, edge-disjoint paths, etc. If A has non-negative entries, we can also model multicommodity unsplittable flow (UFP), and multi-dimensional knapsack problems.

We reformulate (PIP) as a generalized set packing. Let U be a set of m elements, and $\mathcal{S} \subseteq 2^U$ a family of n subsets of U. Each set $S \in \mathcal{S}$ has cost c_S, and can in fact be a multi-set: let $q(e, S) \in \mathsf{N}_{\geq 0}$ be the number of copies of e in S. Each element $e \in U$ has an upper bound $b_e \in \mathsf{N}$ on the number of times it can appear in the packing. A feasible packing is a subfamily of $\mathcal{S}$, where the total number of copies of each element $e \in U$ in all sets of the subfamily is at most b_e. The problem is to find a feasible packing with maximum total cost. These assumptions can be relaxed to $q(e, S) \in \mathsf{R}_{\geq 0}$ and $b_e \in \mathsf{R}_{\geq 1}$. In terms of (PIP), if $A = (a_{ij})$ and $S \in \mathcal{S}$, then $a_{eS} = q(e, S)$. Thus, (PIP) is now:

$$\max \quad \sum_{S \in \mathcal{S}} c_S x_S \tag{1}$$
$$\text{s.t.} \sum_{S: S \in \mathcal{S}, e \in S} q(e, S) \cdot x_S \leq b_e \;\; \forall e \in U \tag{2}$$
$$x_S \in \{0, 1\} \quad \forall S \in \mathcal{S}. \tag{3}$$

Let $b_{min} = \min\{b_e : e \in U\}$, and $d = \max\{|S| : S \in \mathcal{S}\}$, i.e., $d = $ max. number of non-zero entries in any column of A. Let $\phi = \max\{b_e/b_f : \exists S \in \mathcal{S} \text{ s.t. } e, f \in S\}$.

1.1 Our Contributions in General

We present a novel analysis of greedy algorithms for general PIPs by using the duality theory of linear programming (LP). We employ dual-fitting, and many new ingredients. Two fractional, possibly infeasible, dual solutions are defined. One during the execution of the algorithm, and the second one after it stops. By the dual LP, the solutions must have high values on any set in the problem instance. To achieve this we treat one of these solutions as a *back-up*, and prove that if the first solution is not high enough, the second one is. We combine these two solutions by taking their convex combination, and prove that a suitable scaling gives a feasible dual solution. By weak duality, this combined dual solution is an upper bound on the value of an optimal primal integral solution, thus implying the approximation ratio. An interesting aspect of this analysis is that we do not lose any constant factor when the two dual solutions are combined. Thus, we emphasize here that our constant in front of the approximation ratio is precisely 1. Our analysis results in provably best possible approximation ratios for a large class of integer packing problems. We also show that it gives best possible approximation ratios in the natural class of *oblivious* greedy algorithms (cf. Sec. 2).

Our analysis gives significant improvements on the approximation ratios known for simple (monotone, cf. Sec. 3) greedy methods for general PIPs, and improves the approximation ratios for many specific packing problems by constant factors. It is quite flexible and general. The largest improvements are obtained for general PIPs, and for b-matching in hypergraphs. We also slightly improve (by constant factors) the approximation ratios for truthful CAs and routing UFP problems. An additional advantage of our analysis is that it also implies bounds on the integrality gaps of the LP relaxations.

An LP duality-based (dual-fitting) analysis was previously known for the weighted set cover problem due to Chvátal [8], and for generalized set cover by Rajagopalan & Vazirani [29]. We are not aware of such analyzes for PIPs. (Except a recent primal-dual algorithm and analysis by Briest, Krysta and Vöcking [6], but this analysis does not apply to the simple greedy algorithms that we study.)

1.2 Previous Work and Our Improvements

To compare the previous and our results, we discuss them for (0,1)-PIP and assume $b_{min} \geq 1$. (In fact our bounds are more general – see further sections for details.)

The best approximation ratios, $O\left(\min\left\{m^{1/(b_{min}+1)}, d^{1/b_{min}}\right\}\right)$ [28,31,30], for (0,1)-PIP were obtained by solving an LP relaxation and performing randomized rounding of the fractional solution. They are better than the ratios of combinatorial methods, but the main drawback is the need of solving LPs. This is inefficient in many cases, and also does not guarantee monotonicity properties (cf. Sec. 3). (A monotone randomized rounding algorithm is known [2], but it applies to restricted bidders and supplies of goods, and is truthful only in a probabilistic sense – see Sec. 3 for the definitions.)

Our Main Results. Lehmann et al. [24] have analyzed the following simple greedy algorithm for (0,1)-PIP assuming $b_e = 1$ for all $e \in U$.

```
P := ∅; let S = {S₁, S₂, ..., Sₙ} s.t.  cs₁/√|S₁| ≥ cs₂/√|S₂| ≥ ... ≥ csₙ/√|Sₙ|
for S' = S₁, S₂, ..., Sₙ do if P∪{S'} fulfills (2) & (3) then P := P ∪ {S'}
output packing P
```

(A more general version of this algorithm, called Greedy-2, is given in (14).) Lehmann et al. proved that it gives a $\sqrt{m}$-approximation for (0,1)-PIP with all $b_e = 1$. Gonen & Lehmann [10] have shown that Greedy-2 is a $\sqrt{\sum_{e \in U} b_e}$-approximation algorithm for (0,1)-PIP. We use our duality-based analysis to improve this ratio significantly, showing that Greedy-2 is a $(\sqrt{\sum_{e \in U} b_e/b_{min}} + 1)$-approximation algorithm for (0,1)-PIP. This, for instance, implies a ratio of $(\sqrt{m} + 1)$ for (0,1)-PIP when all $b_e = b$ for some value b, which is not necessarily one. We also show a corresponding lower bound, by proving that this ratio is essentially best possible for this problem in the class of oblivious greedy algorithms – which basically captures all natural greedy algorithms for (0,1)-PIP.

We give another greedy algorithm for (0,1)-PIP, called Greedy-1, and show it is a $(\sqrt{\phi m} + 1)$-approximation using our duality-based technique. A third presented algorithm, Greedy-3, is a $(d+1)$-approximation for (0,1)-PIP. Obtaining a ratio better than $\sqrt{m}$ (even if $\phi = 1$) is not possible, unless NP = ZPP [17], and it is NP-hard to obtain a ratio of $O(d/\log d)$ [18]. Thus, our analysis implies essentially best possible ratios.

It is possible to modify a combinatorial greedy algorithm for the unsplittable flow problem, presented by Kolman and Scheideler [23], to obtain an $O(\sqrt{m})$-approximation algorithm for (0,1)-PIP. However, our greedy algorithms and analysis have the following advantages over that modified algorithm: our algorithm is monotone, which is needed for mechanism design (cf. Sec. 3); our duality-based analysis implies bounds on the integrality gaps of PIPs; and, finally, we do not lose constant factors in the ratio.

The integrality gaps of LP relaxations for (0,1)-PIPs were proved before by Aharoni et al. [1] (for the unweighted (0,1)-PIP), Raghavan [27], and Srinivasan [31]. Our duality-based analysis improves these bounds by constant factors.

Further Known Results Versus Ours. Some combinatorial algorithms are known for (0,1)-PIP. E.g., an algorithm of Bartal, Gonen and Nisan [4], following ideas from [3]. If $b_e = b, \forall e \in U$, the algorithm of [4] achieves the best ratio of $O(b \cdot (m)^{1/(b-1)})$ for (0,1)-PIP. A very recent result is a primal-dual $O(m^{1/(b+1)})$-approximation algorithm for (0,1)-PIP, by Briest, Krysta and Vöcking [6]. Our contribution here is the greedy $(\sqrt{m} + 1)$-approximate algorithm, which is faster and much simpler than the other two combinatorial algorithms in [4,6], and tightens the big-O constants.

An additional motivation for simple greedy method here comes from the branch-and-bound heuristics for CAs. Gonen and Lehmann [10] proved that algorithm Greedy-2 (see (14)) gives the best method of ordering the bids in the branch and bound heuristics for CAs. Their experiments [11] support this claim in practice. Our improved ratio for Greedy-2 might be a step towards theoretical explanation of this good performance.

Combinatorial approximation algorithms are known for special problems modeled by PIPs. Halldórsson et al. [13] gave a greedy $b\sqrt{m}$-approximation for unweighted b-matching in hypergraphs (all $c_S = 1$ and all $b_e = b$). For the same weighted problem [10] gives a greedy $\sqrt{bm}$-approximation. Thus, we improve the ratio of a greedy approximation for the problem to $\sqrt{m} + 1$. A simple greedy $\sqrt{m}$-approximation for unweighted set packing ($b = 1$) [13], and a $2\sqrt{m}$-approximation for weighted set packing [15] are known. Our ratio $\sqrt{m} + 1$ applies here as well. For weighted set packing ($c_S \geq 0$; all $b_e = 1$), Hochbaum [19] gave a greedy d-approximation, and Berman & Krysta [5] show a local search $\frac{2}{3}d$-approximation. Our ratio $d + 1$ applies to a more general problem. See survey [14] for other results on related approximations.

Further Consequences of Our Analysis. Our results can be applied to obtain truthful mechanisms for combinatorial auctions with slightly improved approximation factors. This is discussed in detail in Section 3.

A related problem is the routing unsplittable flow problem (UFP). We can cast UFP by slightly generalizing PIP, but in fact UFP is less general than PIPs. There are greedy and other combinatorial approximation algorithms for UFP and related problems, e.g., [20,12,7,23,22,3]. Their ratios, usually, look like $\sqrt{m}$, sometimes with additional logarithmic factors or factors depending on capacities and demands, with m denoting the number of edges in the graph. Our duality-based analysis can be extended to these problems. See Section 4 for the definitions and details on our improvements.

2 Oblivious Greedy Algorithms

We study a class of algorithms for (PIP) that we call *oblivious greedy algorithms*. This class was studied in context of truthful CAs in [24]. Besides their simplicity, truthful CAs are the main motivation to study these algorithms. The crucial feature of this class is that of *monotonicity*, which implies a truthful mechanism for CAs (cf. Section 3).

An *oblivious greedy algorithm* for (PIP) is a polynomial time algorithm $\mathcal{A}$ with a rank function, say $\rho : \mathcal{S} \times \mathsf{R}_{\geq 0} \longrightarrow \mathsf{R}_{\geq 0}$, assigning a real number $\rho(S, c_S)$ to each pair (S, c_S), $S \in \mathcal{S}$. We assume, after [24], that having any $S \in \mathcal{S}$ fixed, $\rho(S, x)$ is strictly increasing as a function of x. Having $\rho(\cdot, \cdot)$, $\mathcal{A}$ sorts all the sets in $\mathcal{S}$ w.r.t. non-increasing numbers $\rho(S, c_S)$. Then $\mathcal{A}$ scans the sets in this order once, and picks them to the solution one by one, maintaining feasibility.

The monotonicity of $\rho(S, \cdot)$ is a natural assumption for CAs: Let a buyer offer some amount of money for a product, so that the seller wants to sell the product for that money. If, now, the buyer offers even more money, the seller, obviously, is also willing to sell the same product. We will write $\rho(S)$ for short, instead of $\rho(S, c_S)$.

2.1 The Upper Bound

We give a greedy algorithm, Greedy-1, for (PIP), having the best possible approximation ratio for (0,1)-PIP when $b = \min_e b_e$ is small. Greedy-1 will also be shown to have the best possible ratio in the class of all oblivious greedy algorithms for (0,1)-PIP.

Generic Greedy Algorithm and Dual-Fitting Analysis. We can assume w.l.o.g., that given $S \in \mathcal{S}$, we have $q(e, S) \leq b_e$ for each $e \in S$. Let, for any $S \in \mathcal{S}$, a rank value $\rho(S) \in \mathbb{R}_{\geq 0}$ be given. Let $\mathcal{P} \subseteq \mathcal{S}$ be a given packing. We say that element $e \in U$ is *saturated* or *sat* w.r.t. $\mathcal{P}$ if there is a set $S' \in \mathcal{S} \setminus \mathcal{P}$ with $e \in S'$, such that $q(e, S') + \sum_{S: S \in \mathcal{P}, e \in S} q(e, S) > b_e$. Our generic greedy algorithm, Greedy, is then:

```
01. P := ∅
02. let S = {S₁, S₂, ..., Sₙ} s.t. ρ(S₁) ≥ ρ(S₂) ≥ ··· ≥ ρ(Sₙ)
03. for S' = S₁, S₂, ..., Sₙ do
04.     if P ∪ {S'} fulfills (2) and (3) then
05.                 P := P ∪ {S'}
06. output packing P
```

We now present our analysis via dual fitting. Our original primal problem is given by the integer linear program (1)–(3). The LP relaxation of this integer program is:

$$\max \quad \sum_{S \in \mathcal{S}} c_S x_S \qquad (4)$$
$$\text{s.t.} \sum_{S: S \in \mathcal{S}, e \in S} q(e, S) x_S \leq b_e \quad \forall e \in U \qquad (5)$$
$$0 \leq x_S \leq 1 \qquad \forall S \in \mathcal{S}. \qquad (6)$$

Its corresponding dual linear program can be written as:

$$\min \quad \sum_{e \in U} b_e y_e + \sum_{S \in \mathcal{S}} z_S \qquad (7)$$
$$\text{s.t.} \; z_S + \sum_{e \in S} q(e, S) y_e \geq c_S \quad \forall S \in \mathcal{S} \qquad (8)$$
$$z_S, y_e \geq 0 \qquad \forall S \in \mathcal{S}, e \in U. \qquad (9)$$

Dual variable z_S corresponds to $x_S \leq 1$. We present below a performance proof of a greedy with a specific rank function ρ_1, which will be used as *generic proof*.

$$\text{Given a set } S \in \mathcal{S}, \text{ let:} \quad \rho_1(S) = \frac{c_S}{\sqrt{\sum_{e \in S} \frac{q(e,S)}{b_e}}}.$$

We call algorithm Greedy with the rank function $\rho = \rho_1$, Greedy-1.

Theorem 1. *Algorithm* Greedy-*1 has an approximation ratio of $\frac{q_{max}}{q_{min}}\sqrt{\phi m} + 1$ for the generalized set packing problem, and for* (PIP), *assuming* $\phi = \max\{b_e/b_f : e, f \in S, S \in \mathcal{S}\}$, *and for each* $S \in \mathcal{S}, e \in S$, *we have* $q_{min} \leq q(e, S) \leq q_{max}$ *or* $q(e, S) = 0$.

Due to the generality of (PIP), there is no better approximation ratio than $\sqrt{m}$, unless NP = ZPP [17]. This hardness holds even when $\frac{q_{max}}{q_{min}} = 1$, and $\phi = 1$.

Proof. (Theorem 1) This is a sketch of the **generic proof**. Suppose Greedy-1 terminated and output solution $\mathcal{P}$. Let $SAT_\mathcal{P} = \{e \in U : e$ is sat w.r.t. $\mathcal{P}\}$. Notice, for each set $S \in \mathcal{S} \setminus \mathcal{P}$, there is an $e \in SAT_\mathcal{P}$, called a *witness*: That is, when S was considered in line 04, and $\mathcal{P}$ was the current (partial) solution, $e \in S$ was an element such that $q(e, S) + \sum_{S':S'\in\mathcal{P},e\in S'} q(e, S') > b_e$. For each set $S \in \mathcal{S} \setminus \mathcal{P}$ we keep in $SAT_\mathcal{P}$ one (arbitrary) witness for S, and discard the remaining elements from $SAT_\mathcal{P}$.

Defining Two Dual Solutions. We define two fractional dual solutions, y^1 and y^2. y^1 is defined after Greedy-1 has terminated. Let us define the following:

$$\sigma = \sum_{S\in\mathcal{P}} \frac{c_S \cdot q_S}{\sqrt{\sum_{e'\in\mathcal{P}(S)} \frac{q(e',S)\cdot \max_{e''\in\mathcal{P}(S)}\{b_{e''}\}}{b_{e'}}}}, \quad \text{where}$$

$\mathcal{P}(S) = S \cap SAT_\mathcal{P}$ if $S \cap SAT_\mathcal{P} \neq \emptyset$ and $\mathcal{P}(S) = S$ if $S \cap SAT_\mathcal{P} = \emptyset$; and $q_S = \sqrt{|\mathcal{P}(S)|}$. For each $e \in U$, define y^1 as: $y^1_e = \frac{\sigma}{b_e \cdot m}$. Solution y^2 is defined during the execution of Greedy-1. We need to know $\mathcal{P}$ to define y^2, which is needed only for analysis. In line 01 of Greedy-1 we initialize: $y^2_e := 0, z^2_S := 0$, for all $e \in U$, and $S \in \mathcal{S}$. The following is added in line 05 of Greedy-1:

$$y^2_e := y^2_e + \Delta^{S'}_e, \quad \text{for all } e \in \mathcal{P}(S'), \quad \text{where}$$

$$\Delta^{S'}_e = \frac{c_{S'}}{b_e \cdot q_{S'} \cdot \sqrt{\sum_{e'\in\mathcal{P}(S')} \frac{q(e',S')\max_{e''\in\mathcal{P}(S')}\{b_{e''}\}}{b_{e'}}}}, \quad \text{for } e \in \mathcal{P}(S').$$

Note, that for $e \in S' \setminus SAT_\mathcal{P}$ the value of y^2_e is not updated and remains zero. We also add the following in line 05 of Greedy-1: $z^2_{S'} := z^2_{S'} + c_{S'}$. Obviously, if for an $e \in U$, values of y^1, y^2 or z^1, z^2 have not been defined, they are zero.

Dual Lower Bound on the Solution. We now argue that both dual solutions (appropriately scaled) provide lower bounds on the cost of the output solution:

$$\sum_{e\in U} b_e \cdot y^i_e \leq \frac{1}{\sqrt{q_{min}}} \sum_{S\in\mathcal{P}} c_S, \quad \text{for} \quad i = 1, 2. \tag{10}$$

Final Dual Solution. We will now show that there exists a dual solution y such that the scaled solution $(\frac{q_{max}}{\sqrt{q_{min}}}\sqrt{\phi m} \cdot y, z)$ is feasible for the dual LP, i.e., constraints (8) are fulfilled for all sets $S \in \mathcal{S}$. Thus, we have to show that, for *each* set $S \in \mathcal{S}$,

$$z_S + \frac{q_{max}}{\sqrt{q_{min}}}\sqrt{\phi m} \sum_{e\in S} q(e, S) y_e \geq c_S. \tag{11}$$

We will prove (11) by using both y^1 and y^2. The main idea is to use the second solution as a *back-up*: whenever on some set $S \in \mathcal{S}$, y^1 is not high enough we will prove that y^2 is sufficiently high. We define a new solution y as a convex combination of y^1, y^2: $y_e = \frac{1}{2}\left(y_e^1 + y_e^2\right)$, for each $e \in U$; also, define z as $z_S = z_S^2$ for each $S \in \mathcal{S}$.

Proving (11). Suppose first that $S' \in \mathcal{S} \setminus \mathcal{P}$. The reason that S' has not been included in solution $\mathcal{P}$ must be an $e \in SAT_\mathcal{P}$ such that $e \in S'$. This means that adding set S' to solution $\mathcal{P}$ would violate constraint (2). Let $\mathcal{E} \subseteq \mathcal{P}$ be the family of all sets in the solution that contain element e. Observe, that $|\mathcal{E}| \geq 1$.

Lower-Bounding y^1. Using our greedy selection rule, and lower-bounding appropriately σ, we are able to show the following bound:

$$\sum_{e' \in S'} q(e', S') y_{e'}^1 \geq \frac{c_{S'}}{m\gamma} \cdot \left(\sum_{S \in \mathcal{E}} q_S\right) \cdot \sqrt{\sum_{e' \in S'} \frac{\gamma q(e', S')}{b_{e'}}}. \qquad (12)$$

Lower-Bounding y^2. Again using our greedy selection rule, a well known inequality between the arithmetic and harmonic means [16], we can lower-bound appropriately parameters Δ to obtain ($|\mathcal{E}| = p$):

$$\sum_{e' \in S'} q(e', S') y_{e'}^2 \geq \frac{c_{S'}}{b_e} \frac{q_{min} p^2}{\left(\sum_{S \in \mathcal{E}} q_S\right) \sqrt{\sum_{e' \in S'} \frac{\gamma q(e', S')}{b_{e'}}}}. \qquad (13)$$

Lower-Bounding y. Using (12) and (13) we can write

$$\sum_{e' \in S'} q(e', S') y_{e'} = \frac{1}{2} \sum_{e' \in S'} \left(q(e', S') y_{e'}^1 + q(e', S') y_{e'}^2\right) \geq$$

$$\frac{1}{2} \left(\frac{c_{S'}}{m\gamma} \cdot \left(\sum_{S \in \mathcal{E}} q_S\right) \cdot \sqrt{\sum_{e' \in S'} \frac{\gamma q(e', S')}{b_{e'}}} + \frac{c_{S'} q_{min} p^2}{b_e \left(\sum_{S \in \mathcal{E}} q_S\right) \sqrt{\sum_{e' \in S'} \frac{\gamma q(e', S')}{b_{e'}}}} \right)$$

$$\frac{c_{S'}}{2} \left(\frac{x}{m\gamma} + \frac{q_{min} p^2}{b_e x} \right), \quad \text{where} \quad x = \left(\sum_{S \in \mathcal{E}} q_S\right) \cdot \sqrt{\sum_{e' \in S'} \frac{\gamma q(e', S')}{b_{e'}}}.$$

Consider now function $f(x) = \frac{x}{m\gamma} + \frac{q_{min} p^2}{b_e x}$ for $x > 0$. We can show that $f(x) \geq 2\sqrt{\frac{q_{min} p^2}{\gamma m b_e}}$ for all $x \in (0, \infty]$. Observing that $|\mathcal{E}| = p \geq \frac{b_e}{q_{max}}$, and $\frac{\gamma}{b_e} \leq \phi$, we obtain $f(x) \geq 2\sqrt{\frac{q_{min}}{q_{max}^2 \phi m}}$ for all $x \in (0, \infty]$. This proves claim (11) when set $S' \in \mathcal{S} \setminus \mathcal{P}$. If $S' \in \mathcal{P}$, then claim (11) follows from the definition of $z_{S'}$.

Finishing the Proof. We have shown that the dual solution $(\frac{q_{max}}{\sqrt{q_{min}}} \sqrt{\phi m} \cdot y, z)$ is feasible for the dual linear program and so by weak duality $\sum_{S \in \mathcal{S}} z_S + \frac{q_{max}}{\sqrt{q_{min}}} \sqrt{\phi m} \cdot$

$\sum_{e \in U} b_e y_e$ is an upper bound on the value of the optimal integral solution to our problem. By (10), we have that $\sum_{e \in U} b_e y_e \leq \frac{1}{\sqrt{q_{min}}} \sum_{S \in \mathcal{P}} c_S$. Thus, we obtain that

$$opt \leq \sum_{S \in \mathcal{S}} z_S + \frac{q_{max}}{\sqrt{q_{min}}} \sqrt{\phi m} \cdot \sum_{e \in U} b_e y_e \sum_{S \in \mathcal{P}} z_S + \frac{q_{max}}{\sqrt{q_{min}}} \sqrt{\phi m} \cdot \sum_{e \in U} b_e y_e$$

$$\leq \sum_{S \in \mathcal{P}} c_S + \frac{q_{max}}{q_{min}} \sqrt{\phi m} \cdot \left(\sum_{S \in \mathcal{P}} c_S \right) \left(\frac{q_{max}}{q_{min}} \sqrt{\phi m} + 1 \right) \cdot \left(\sum_{S \in \mathcal{P}} c_S \right). \quad \square$$

Other Greedy Selection Rules. The next two theorems – Theorem 2 and 3 can easily be shown by using our generic proof (of Theorem 1).

$$\text{Given a set } S \in \mathcal{S}, \text{ let: } \rho_2(S) = \frac{c_S}{\sqrt{\sum_{e \in S} q(e, S)}}. \tag{14}$$

Greedy-2 is Greedy with the greedy selection rule ρ_2.

Theorem 2. *Algorithm* Greedy-2 *has an approximation ratio of* $\frac{q_{max}}{q_{min}} \sqrt{\frac{\sum_{e \in U} b_e}{b_{min}}} + 1$ *for the generalized set packing problem, and for (PIP), assuming that* $b_{min} = \min\{b_e : e \in U\}$, *and for each* $S \in \mathcal{S}$, $e \in S$, *we have* $q_{min} \leq q(e, S) \leq q_{max}$ *or* $q(e, S) = 0$.

Suppose that $q_{max}/q_{min} = 1$. We improve the best known approximation ratio for a combinatorial (greedy) algorithm for (PIP) from $\sqrt{\sum_{e \in U} b_e/q_{min}}$ (Gonen & Lehmann [10]) to $\min\left\{\sqrt{\sum_{e \in U} b_e/b_{min}} + 1, \sqrt{\phi m} + 1\right\}$ (Theorems 1, 2). This ratio is always better than [10], since w.l.o.g. $q_{min} \leq 1$ and $b_{min} \geq 1$ (see Srinivasan [31]). Also if $\phi = 1$, then our ratio is $O(\sqrt{m})$, and that of [10] is still $\sqrt{\sum_{e \in U} b_e}$ (for $q_{min} = 1$).

$$\text{Given a set } S \in \mathcal{S}, \text{ let: } \rho_3(S) = c_S/|S|.$$

We call algorithm Greedy with the greedy selection rule ρ_3, Greedy-3.

Theorem 3. *Algorithm* Greedy-3 *has an approximation ratio of* $\frac{q_{max}}{q_{min}} d + 1$ *for the generalized set packing problem, and for (PIP), assuming that* $|S| \leq d$ *for each* $S \in \mathcal{S}$. *Moreover, for each* $S \in \mathcal{S}$, $e \in S$, *we have* $q_{min} \leq q(e, S) \leq q_{max}$ *or* $q(e, S) = 0$.

Observe that the approximation ratio is close to best possible. Let $\frac{q_{max}}{q_{min}} = 1$ and all $q(e, S) \in \{0, 1\}$; then our ratio is $d+1$. The considered PIP can express the unweighted set packing problem for which obtaining a ratio of $O(\frac{d}{\log d})$ is NP-hard [18].

2.2 The Lower Bound

In this section we consider (0,1)-PIP. By Theorem 1 we obtain the following.

Corollary 1. *Algorithm* Greedy-1 *is an oblivious greedy* ($\sqrt{\phi m} + 1$)-*approximation algorithm for the (0,1)-PIP problem, where* $\phi = \max\{b_e/b_f : e, f \in S, S \in \mathcal{S}\}$.

We show below, by modifying an argument in [10], that this upper bound can be matched by a lower bound in the class of oblivious greedy algorithms.

Proposition 1. *Let us consider $(0,1)$-PIP problem, assuming that $b_e = b \in \mathsf{N}_{\geq 1}$ for each $e \in U$. Then any oblivious greedy polynomial time algorithm for this problem has an approximation ratio of at least $\sqrt{m} - \varepsilon$, for any $\varepsilon > 0$.*

Note, that this lower bound above is meaningful, since there is a polynomial time $O(m^{1/(b+1)})$-approximation to the described problem via LP randomized rounding.

3 An Application to Truthful Combinatorial Auctions

We will use now our results to give truthful approximation mechanisms for combinatorial auctions (CAs) with single-minded bidders. A seller (auctioneer) wants to sell m kinds of goods U, to n potential customers (bidders). A good $e \in U$ is available in $b_e \in \mathsf{N}_{\geq 1}$ units (supply). Suppose each bidder j can valuate subsets of goods: a valuation $v_j(S) \in \mathsf{R}_{\geq 0}$ for a subset $S \subseteq U$ means the maximum amount of money j is willing to pay for getting S. For simplicity, assume that bidders can bid for 0 or single unit of a good, i.e., $q(e,S) \in \{0,1\}$ for all $e \in U$, and $S \subseteq U$. An allocation of goods to bidders is a packing $S^1, \ldots, S^n \subseteq U$ w.r.t. the defined (PIP) with $\mathcal{S} = 2^U$, i.e., bidder j gets set S^j, and e appears at most b_e times in $S^1, \ldots, S^n$. The objective is to find an allocation with maximum *social welfare*, $\sum_j v_j(S^j)$. Each v_j is only known to bidder j. Our bidders are *single-minded* [24], i.e., for each bidder j there exists a set $S_j \subseteq U$ she prefers and a $v_j^* \geq 0$, such that $v_j(S) = v_j^*$ if $S_j \subseteq S$ and $v_j(S) = 0$ otherwise.

An *auction mechanism* (seller) is an algorithm which first collects the bids from the bidders, i.e., their declarations (S_j', v_j'), where S_j' is supposed to mean the preferred set S_j, and v_j' the valuation v_j for j. The mechanism then determines the allocation and a payment p_j for each bidder j. The utility of bidder j is $u_j = v_j(S) - p_j$ for winning set S. Note that v_j is the true valuation function. We assume the mechanism is *normalized*, i.e., $p_j = 0$, when bidder j is not allocated any set. Our allocation problem is to maximize the social welfare; this corresponds to approximating our (PIP) problem.

Each bidder aims at maximizing her own utility. It may be profitable for bidder j to lie and report $v_j' \neq v_j$ and $S_j' \neq S_j$ to increase her utility. A mechanism is *truthful (incentive compatible)* if declaring truth, i.e., $v_j' = v_j$ and $S_j' = S_j$, is a *dominant* strategy for each bidder j. That is for any fixed set of bids of all bidders except j, if j does not declare the truth, then this may only decrease j's utility. Our goal is a truthful approximate mechanism.

An allocation algorithm is *monotone* if whenever bidder j declares (S_j', v_j') (given other bidders' declarations) and wins, i.e., gets set S_j' allocated, then declaring (S_j'', v_j'') s.t. $S_j'' \subseteq S_j'$, $v_j' \leq v_j''$ results also in winning set S_j''. It is well known that if an allocation (approximate) algorithm is monotone and exact (i.e., a bidder gets exactly her declared set or nothing), then there is a payment scheme which together with the allocation algorithm is a truthful (approximate) mechanism (see, e.g., [26,24]). It is easy to see that all our greedy algorithms are monotone and exact. We can also modify the payment scheme in [24] to serve our purposes. Thus, using Theorems 1, 2 and 3, this gives the following result.

Theorem 4. *Suppose we have m kinds of goods U, each good $e \in U$ available in $b_e \in \mathsf{N}_{\geq 1}$ units, and $b_{min} = \min\{b_e : e \in U\}$. Suppose each bidder bids only on at most*

$d \in \mathsf{N}_{\geq 1}$ goods, and for each bid (S'_j, v'_j) we have $\max\{b_e/b_f : e, f \in S'_j\} \leq \phi$. There is a truthful mechanism for CAs with single-minded bidders with an approximation ratio

$$\min\left\{1 + \sqrt{\phi m},\ 1 + \sqrt{\sum_{e \in U} b_e/b_{min}},\ 1 + d\right\}.$$

Note, that we assume d is known to the mechanism. Theorem 4 improves on the mechanism of Lehmann et al. [24], where they assume $b_e = 1$ for each $e \in U$, and their ratio is $\sqrt{m}$. We achieve the same ratio (+1) for a more general setting, where the supplies of the goods are given arbitrary numbers. The best known truthful mechanism for the problem with $b_e = b\ \forall e \in U$, is $5.44 \cdot (m)^{1/b}$-approximate, see Briest, Krysta and Vöcking [6]. Thus, our ratio for the same problem is slightly better for $b = 2$. Our ratio is also very good if d is small – a very natural assumption for bidders.

4 An Application to the Unsplittable Flow Problem

We show that our dual fitting analysis can be extended to deal with a difficult routing problem – multicommodity unsplittable flow problem (UFP).

Let $G = (V, E)$ be a given graph ($|E| = m$), and $C = \{(s_i, t_i) : s_i, t_i \in V, i = 1, \ldots, k\}$ be k source-sink pairs, or commodities. For each commodity $i \in \{1, \ldots, k\}$, we are given a demand $d_i \in \mathsf{N}_{\geq 1}$, and a profit $p_i \in \mathsf{R}_{\geq 0}$. For each edge $e \in E$, $b_e \in \mathsf{N}_{\geq 0}$ denotes its capacity. Given $i \in \{1, \ldots, k\}$, let C_i be the set of all simple s_i-t_i-paths in G, such that all edge capacities on these paths are at least d_i. The multicommodity unsplittable flow problem (UFP) is to route for each commodity i demand d_i along a single path in C_i, respecting edge capacities. The objective is to maximize the sum of the profits of all commodities that can be simultaneously routed.

We give now an LP relaxation of this problem (see Guruswami et al. [12]). The ground set is $U = E$, and the set family is $\mathcal{S} = \cup_{i=1}^{k} C_i$. Each set $S \in \mathcal{S} \cap C_i$, that is an s_i-t_i-path, has cost $c_S = c_i = p_i$. The LP relaxation of UFP is:

$$\max \sum_{i=1}^{k} c_i \cdot \left(\sum_{S \in C_i} x_S\right) \tag{15}$$
$$\text{s.t.} \sum_{S: S \in \mathcal{S}, e \in S} d_S x_S \leq b_e \quad \forall e \in U \tag{16}$$
$$\sum_{S \in C_i} x_S \leq 1 \quad \forall i \in \{1, \ldots, k\} \tag{17}$$
$$x_S \geq 0 \quad \forall S \in \mathcal{S}, \tag{18}$$

where $d_S = d_i$ iff $S \in C_i$. The corresponding dual linear program reads then:

$$\min \sum_{e \in U} b_e y_e + \sum_{i=1}^{k} z_i \tag{19}$$
$$\text{s.t.} \ z_i + \sum_{e \in S} d_i y_e \geq c_i \quad \forall i \in \{1, \ldots, k\}\ \forall S \in C_i \tag{20}$$
$$z_i, y_e \geq 0 \quad \forall i \in \{1, \ldots, k\}\ \forall e \in U. \tag{21}$$

In this dual linear program, dual variable z_i corresponds to the constraint (17).

We will use algorithm **Greedy**-1 to approximate UFP. We will say how to realize it in the case of UFP. Given a commodity i, the greedy selection rule is:

$$\max_{S \in C_i} \rho_1(S) = \max_{S \in C_i} \frac{c_i}{\sqrt{\sum_{e \in S} \frac{d_i}{b_e}}}. \tag{22}$$

We replaced $q(e, S)$ with d_i, since $S \in C_i$. (22) is same as $\min_{S \in C_i} (\sum_{e \in S} 1/b_e) d_i/c_i^2$, and such an $S \in C_i$ can be found by a shortest path computation. The implementation of **Greedy**-1 is as follows. We maintain the current edge capacities b'_e. We declare all commodities *unsatisfied*, and put $b'_e := b_e$ for each $e \in E$. Perform the following until all commodities are *satisfied*. Find a commodity i and $S \in C_i$ that minimize $(\sum_{e \in S} 1/b_e) d_i/c_i^2$ among all unsatisfied commodities: when computing the shortest path for i use only edges e with $d_i \le b'_e$. If s_i and t_i are disconnected by such edges, then declare i satisfied. Otherwise, let i_0 and $S_0 \in C_{i_0}$ be the shortest path and commodity. Route demand d_{i_0} along S_0, put $b'_e := b_e - d_{i_0}$ for all $e \in S_0$, and declare i_0 satisfied. (We define dual variable $z_{i_0} := c_{i_0}$ as in our generic proof.)

The definitions of other dual variables remain the same as in the generic proof (proof of Theorem 1). This proof basically goes through. Observe, that now $q_{min} = \min\{d_i : d_i > 0, i = 1, \ldots, k\} = d_{min}$ and $q_{max} = \max\{d_i : i = 1, \ldots, k\} = d_{max}$. Thus, by employing the generic proof we obtain the following theorem for the UFP problem.

Theorem 5. *There is a greedy $(\frac{d_{max}}{d_{min}}\sqrt{\phi m} + 1)$-approximation algorithm for the unsplittable flow problem (UFP), where $\phi = \max\{b_e/b_f : \exists i\ \exists S \in C_i\ s.t.\ e, f \in S\}$.*

Guruswami et al. [12] prove that a repeated use of a similar greedy algorithm gives a $(\frac{2d_{max}}{d_{min}}\sqrt{m})$-approximation for unweighted UFP (i.e., $c_i = 1, \forall i$). We now present some applications of Theorem 5 but the proofs are omitted, due to the lack of space.

Theorem 5 gives a $(\sqrt{m}+1)$-approximation when specialized to the weighted edge-disjoint paths (EDP) problem: this is just UFP problem with $b_e = 1$ for each $e \in E$, and $d_i = 1$ for each $i \in \{1, \ldots, k\}$. Our greedy when specialized to this case, reduces to the greedy algorithm of Kolliopoulos & Stein [21]. They prove a similar approximation ratio for this algorithm for the unweighted EDP problem. There is also a greedy $O(\min\{\sqrt{m}, |V|^{2/3}\})$-approximation for the same problem by Chekuri and Khanna [7]. Such a result is not possible for (0,1)-PIP, since obtaining ratio $O(d/\log d)$ is NP-hard [18]; note that d in (0,1)-PIP corresponds to $|V|$. Notice, that our $(\sqrt{m}+1)$-approximation holds also for the weighted and more general (than EDP) problem where we allow each edge to be used up to b times, where $b_e = b$.

The previous best known algorithm for general UFP, assuming $d_{max} \le b_{min}$, is a combinatorial $32\sqrt{m}$-approximation algorithm of Azar and Regev [3]. If $\phi = 1$, then Theorem 5 implies a combinatorial $(2 + \epsilon)\sqrt{m}$-approximation to the UFP, assuming $d_{max} \le b_{min}$, for any $\epsilon > 0$. On uniform capacity networks (a stronger assumption than $\phi = 1$), there is a combinatorial $O(\min\{\sqrt{m}, |V|^{2/3}\})$-approximation [7].

Another application is a $(\sqrt{|V|} + 1)$-approximation algorithm for the weighted vertex-disjoint paths problem, improving on the previous best known ratio of roughly $18\sqrt{|V|}$ due to Kolliopoulos & Stein [21]. We get this result by using ideas from [21] and our algorithm from Theorem 5 as a subroutine.

Note, that results in [22,23] hold, unlike ours, for the unweighted UFP, i.e., unit profits problem. We would like to point out that we are able to extend our analysis to UFP thanks to the use of LP duality theory. For instance, it seems difficult to extend the combinatorial analysis in [10] to such problems.

References

1. R. Aharoni, P. Erdös and N. Linial. Optima of dual linear programs. *Combinatorica*, **8**, pp. 13–20, 1988.
2. A. Archer, C.H. Papadimitriou, K. Talwar and É. Tardos. An approximate truthful mechanism for combinatorial auctions with single parameter agents. In the *Proc. 14th ACM-SIAM Symposium on Discrete Algorithms (SODA)*, 2003.
3. Y. Azar and O. Regev. Strongly polynomial algorithms for the unsplittable flow problem. In the *Proc. 8th Conference on Integer Programming and Combinatorial Optimization (IPCO)*, Springer LNCS, 2001.
4. Y. Bartal, R. Gonen, and N. Nisan. Incentive Compatible Multi-Unit Combinatorial Auctions. In the *Proc. 9th conference on Theoretical Aspects of Rationality and Knowledge (TARK)*, Bloomington, IN, USA, June, 2003.
5. P. Berman and P. Krysta. Optimizing misdirection. In the *Proc. 14th ACM-SIAM Symposium on Discrete Algorithms (SODA)*, pp. 192–201, 2003.
6. P. Briest, P. Krysta and B. Vöcking. Approximation Techniques for Utilitarian Mechanism Design. In the *Proc. 37th ACM Symposium on Theory of Computing (STOC)*, 2005.
7. C. Chekuri and S. Khanna. Edge-disjoint paths revisited. In the *Proc. 14th ACM-SIAM Symposium on Discrete Algorithms (SODA)*, 2003.
8. V. Chvátal. A greedy heuristic for the set-covering problem. *Mathematics of Operations Research*, **4**: 233–235, 1979.
9. G. Dobson. Worst-case analysis of greedy heuristics for integer programming with nonnegative data. *Mathematics of Operations Research*, **7**, pp. 515–531, 1982.
10. R. Gonen, D.J. Lehmann. Optimal solutions for multi-unit combinatorial auctions: branch and bound heuristics. In *Proc. 2nd ACM Conference on Electronic Commerce (EC)*, 2000.
11. R. Gonen, D.J. Lehmann. Linear Programming Helps Solve Large Multi-Unit Combinatorial Auctions. In the *Proceedings of INFORMS 2001*, November 2001.
12. V. Guruswami, S. Khanna, R. Rajagopalan, B. Shepherd, and M. Yannakakis. Near-optimal hardness results and approximation algorithms for edge-disjoint paths and related problems. In the *Proc. 31st ACM Symposium on Theory of Computing (STOC)*, 1999.
13. M.M. Halldórsson, J. Kratochvíl, and J.A. Telle. Independent sets with domination constraints. In the *Proc. ICALP*. Springer LNCS, 1998.
14. M.M. Halldórsson. A survey on independent set approximations. In the *Proc. APPROX*, Springer LNCS, **1444**, pp. 1–14, 1998.
15. M.M. Halldórsson. Approximations of Weighted Independent Set and Hereditary Subset Problems. *Journal of Graph Algorithms and Applications*, 4(1), pp. 1–16, 2000.
16. G. Hardy, J.E. Littlewood, G. Polya. *Inequalities.* 2nd Edition, Cambridge Univ. Press, 1997.
17. J. Hastad. Clique is hard to approximate within $n^{1-\epsilon}$. In the *Proc. IEEE FOCS*, 1996.
18. E. Hazan, S. Safra and O. Schwartz. On the hardness of approximating k-dimensional matching. In the *Proc. APPROX*, Springer LNCS, 2003.
19. D.S. Hochbaum. Efficient bounds for the stable set, vertex cover, and set packing problems. *Discrete Applied Mathematics*, **6**, pp. 243–254, 1983.
20. J. Kleinberg. *Approximation algorithms for disjoint paths problems.* PhD thesis, MIT, 1996.
21. S.G. Kolliopoulos and C. Stein. Approximating disjoint-path problems using greedy algorithms and packing integer programs. In *Proc. 6th IPCO*, LNCS, **1412**, pp. 153–162, 1998.
22. P. Kolman. A note on the greedy algorithm for the unsplittable flow problem. *Information Processing Letters*, **88**(3), pp. 101–105, 2003.
23. P. Kolman and Ch. Scheideler. Improved bounds for the unsplittable flow problem. In the *Proc. 13th ACM-SIAM Symposium on Discrete Algorithms (SODA)*, 2002.

24. D. Lehmann, L. Ita O'Callaghan and Y. Shoham. Truth revelation in rapid, approximately efficient combinatorial auctions. In the *Proc. 1st ACM Conference on Electronic Commerce (EC)*, 1999. Journal version in: *Journal of the ACM*, **49**(5): 577–602, 2002.
25. L. Lovász. On the ratio of optimal integral and fractional covers. *Discrete Mathematics*, **13**, pp. 383–390, 1975.
26. A. Mu'alem and N. Nisan. Truthful Approximation Mechanisms for Restricted Combinatorial Auctions. In the *Proc. 18th National AAAI Conference on Artificial Intelligence*, 2002.
27. P. Raghavan. Probabilistic construction of deterministic algorithms: Approximating packing integer programs. *Journal of Computer and System Sciences*, **37**, pp. 130–143, 1988.
28. P. Raghavan and C.D. Thompson. Randomized rounding: a technique for provably good algorithms and algorithmic proofs. *Combinatorica*, **7**: 365–374, 1987.
29. S. Rajagopalan and V.V. Vazirani. Primal-dual RNC approximation algorithms for set cover and covering integer programs. *SIAM Journal on Computing*, **28**(2), 1998.
30. A. Srinivasan. A extension of the Lovász Local Lemma and its applications to integer programming. In the *Proc. 7th ACM-SIAM Symposium on Discrete Algorithms (SODA)*, 1996.
31. A. Srinivasan. Improved Approximation Guarantees for Packing and Covering Integer Programs, *SIAM Journal on Computing*, **29**, pp. 648–670, 1999.

Tight Approximability Results for the Maximum Solution Equation Problem over Z_p

Fredrik Kuivinen

Department of Computer and Information Science,
Linköping University, S-581 83 Linköping, Sweden
freku045@student.liu.se

Abstract. In the maximum solution equation problem a collection of equations are given over some algebraic structure. The objective is to find an assignment to the variables in the equations such that all equations are satisfied and the sum of the variables is maximised. We give tight approximability results for the maximum solution equation problem when the equations are given over groups of the form Z_p, where p is prime. We also prove that the weighted and unweighted versions of this problem have equal approximability thresholds. Furthermore, we show that the problem is equally hard to solve even if each equation is restricted to contain at most three variables and solvable in polynomial time if the equations are restricted to contain at most two variables. All of our results also hold for a generalised version of maximum solution equation where the elements of the group are mapped arbitrarily to non-negative integers in the objective function.

1 Introduction

Problems related to solving equations over various algebraic structures have been studied extensively during a large time frame. The most fundamental problem is, perhaps, EQN which is the problem of: given an equation, does it have a solution? That is, is it possible to assign values to the variables in the equation such that the equation is satisfied? Goldmann and Russell [8] studied this problem over finite groups. They showed that EQN is **NP**-complete for all non-solvable groups and solvable in polynomial time for nilpotent groups.

A problem related to EQN is EQN*. In EQN* a collection of equations are given and the question is whether or not there exists an assignment to the variables such that all equations are satisfied. For finite groups Goldmann and Russell [8] have shown that this problem is solvable in polynomial time if the group is abelian and **NP**-complete otherwise. Moore et al. [13] have studied this problem when the equations are given over finite monoids. The same problem have been studied for semigroups [11,16] and even universal algebras [12].

Another problem is the following: given a over-determined system of equations, satisfy as many equations as possible simultaneously. This problem have been studied with respect to approximability by Håstad [9]. He proved optimal inapproximability bounds for the case when the equations are given over a finite abelian group. Håstad's result has later on been generalised by Engebretsen et al. [6] to cover non-abelian groups as well. Those results uses the PCP theorem [1] which has been used to prove a number of inapproximability results. Other problems that have been studied which are related to this area is #EQN* (counting the number of solutions to a system of equations) [15] and EQUIV-EQN* and ISO-EQN* (deciding whether two systems of equations are equivalent or isomorphic, respectively) [14].

In this paper we study the following problem: given a system of equations over a group of the form $\mathbf{Z}_p$ where p is prime, find the best solution. With "best solution" we mean a solution (an assignment to the variables that satisfies all equations) that maximises the sum of the variables. We call this problem MAXIMUM SOLUTION EQUATION (here after called MAX SOL EQN).

A problem that is similar to MAX SOL EQN is NEAREST CODEWORD.[1] In this problem we are given a matrix A and a vector $\bar{b}$. The objective is to find a vector $\bar{x}$ such that the hamming weight (the number of positions in the vector that differs from 0) of $A\bar{x} - \bar{b}$ is minimised. The decision version of a restricted variant[2] of this problem was proved to be **NP**-complete by Bruck and Noar [4]. Later on Feige and Micciancio [7] proved inapproximability results for the same restricted problem. Arora et al. [2] proved that NEAREST CODEWORD over $GF(2)$ is not approximable within $2^{\log^{1-\epsilon} n}$ for any $\epsilon > 0$ unless **NP** $\subseteq DTIME(n^{poly(\log n)})$. NEAREST CODEWORD is interesting because it has practical applications in the field of error correcting codes.

MAX SOL EQN is parametrised on the group we are working with and a map from the elements of the group to non-negative integers. The map is used in the objective function to compute the measure of a solution. Our main result give tight approximability results for MAX SOL EQN for every group of the form $\mathbf{Z}_p$ where p is prime and every map from the elements of the group to non-negative integers. That is, we prove that for every group of the form $\mathbf{Z}_p$ and every map from group elements to non-negative integers there is a constant, α, such that MAX SOL EQN is approximable within α but not approximable within $\alpha - \epsilon$ in polynomial time for any $\epsilon > 0$ unless **P = NP**. As a special case of the main result we show that for the most natural map, where every group element x in $\mathbf{Z}_p$ is mapped to the integer x, it is not possible to approximate MAX SOL EQN within $2 - \epsilon$ in polynomial time for any $\epsilon > 0$ unless **P = NP**. Furthermore, we show that it is possible to approximate this variant of MAX SOL EQN within 2. We also show that the weighted and the unweighted versions of this problem are asymptotically equally hard to approximate. All our hardness results holds even if the instances are restricted to have at most three variables per equation. We also prove that this is tight since with two variables per equation the problems are solvable to optimum in polynomial time.

Our work may be seen as a generalisation of Khanna et al.'s [10] work on the problem MAX ONES($\mathcal{F}$) in the sense that we study larger domains. However, their work is not restricted to equations over finite groups which the results in this paper are. Nevertheless, they give a 2-approximation algorithm for MAX ONES($\mathcal{F}$) when $\mathcal{F}$ is affine. We prove that, unless **P = NP**, this is tight. (MAX ONES($\mathcal{F}$) when $\mathcal{F}$ is an affine constraint family is equivalent to a specific version of MAX SOL EQN.)

---

[1] This problem is sometimes called MLD for MAXIMUM LIKELIHOOD DECODING.
[2] The problem we are referring to is NEAREST CODEWORD with preprocessing. See [4] for a definition.

The structure of this paper is as follows: we will begin with constructing an approximation algorithm for MAX SOL EQN. Our approximation algorithm is a modification of Khanna et al.'s [10] approximation algorithm for MAX ONES($\mathcal{F}$) where $\mathcal{F}$ is an affine constraint family. After that we will prove an inapproximability result for the weighted version of MAX SOL EQN. Our proof uses Håstad's [9] amazing inapproximability results for MAX-Ek-LIN-p. The basic idea is to transform an instance of MAX-Ek-LIN-p to an instance of MAX SOL EQN where it is easy to simultaneously satisfy every equation, but it is hard to do so and at the same time assign the identity element to certain variables. In the last section we will prove that the approximability thresholds for the weighted and unweighted versions of MAX SOL EQN are asymptotically equal. The proof of this result is based upon works by Khanna et al. [10] and Crescenzi et al. [5].

2 Preliminaries

We assume that the reader has some basic knowledge of complexity theory. We will briefly state some fundamental definitions of optimisation problems and approximation, see e.g. [3] for a more detailed presentation.

An *optimisation problem* has a set of admissible input data, called the *instances* of the problem. Each instance has a set of *feasible solutions*. The optimisation problem also has a function of two variables, an instance and a feasible solution, that associates an integer with each such pair. This function denotes the *measure* of the solution. The *goal* of an optimisation problem is to find a feasible solution that either maximises or minimises the measure for a given instance.

An **NPO** problem is an optimisation problem where instances and feasible solutions can be recognised in polynomial time, feasible solutions are polynomially bounded in the input size and the measure can be computed in polynomial time. We will only study **NPO** maximisation problems in this paper.

We will denote the measure of our problems with $m(I, s)$, where I is an instance and s is a feasible solution. The optimum for an instance I of some problem (which problem we are talking about will be clear from the context) is designated by $\text{OPT}(I)$. We say that a maximisation problem Π is r-approximable if there exists a (possibly probabilistic) polynomial time algorithm A such that for every instance I of Π, $m(I, A(I)) \geq \text{OPT}(I)/r$, in the case of a probabilistic algorithm we require that $\text{E}\left[m(I, A(I))\right] \geq \text{OPT}(I)/r$.

In reductions we will work with two different problems simultaneously. The objects associated with the problem that the reduction is from will be denoted by symbols without $'$ and objects associated with the other problem will be denoted by symbols with $'$. Thus, for example, the measuring function of the problem that the reduction starts with will be denoted by $m(I, s)$ and the measuring function of the other problem will be denoted by $m'(I', s')$.

Let f be a function such that $f : X \to \mathbf{N}$. We define the following quantities

$$f_{\max} = \max_{x \in X} f(x); \quad \text{and} \quad f_{\text{sum}} = \sum_{x \in X} f(x).$$

Those notations will only be used when they are well defined. Variables named p will denote prime numbers in this paper. For a group G we denote its identity element with 0_G. We will use addition as the group operator.

2.1 Definitions and Results

In this section we define the problems that will be studied in this paper and we state our results.

Definition 1. WEIGHTED MAXIMUM SOLUTION EQUATION(G, g) where $G = (D, +)$ is a group and $g : D \to \mathbf{N}$ is a function, is denoted by W-MAX SOL EQN(G, g). An instance of W-MAX SOL EQN(G, g) is defined to be a triple (V, E, w) where,

- V is a set of variables.
- E is a set of equations of the form $w_1 + \ldots + w_k = 0_G$, where each w_i is either a variable, an inverted variable or a group constant.
- w is a weight function $w : V \to \mathbf{N}$.

The objective is to find an assignment $f : V \to D$ to the variables such that all equations are satisfied and the sum

$$\sum_{v \in V} w(v) g(f(v))$$

is maximised.

Note that the function g and the group G are not parts of the input. Thus, W-MAX SOL EQN(G, g) is a problem parametrised by G and g. The function g is introduced to make the set of problems more general, we call the function $g(x) = x$ the *natural* choice of g. If MAX SOL EQN is to be studied over e.g. general abelian groups then something like g will be required, because there is no obvious map from elements of a general abelian group to non-negative integers. It is therefore, in our opinion, better to study the more generalised set of problems where an arbitrary map from group elements to non-negative integers is allowed than to chose one of those maps and argue that the chosen one is the most interesting one.

We will also study the unweighted problem, MAX SOL EQN(G, g), which is equivalent to W-MAX SOL EQN(G, g) with the additional restriction that the weight function is equal to 1 for every variable in every instance.

Due to Goldmann and Russell's result [8] that solving systems of equations over non-abelian groups is **NP**-hard, it is **NP**-hard to find feasible solutions to MAX SOL EQN(G, g) if G is non-abelian. It is therefore sufficient to only study MAX SOL EQN(G, g) where G is abelian.

The main result of this paper is the following theorem about the approximability of MAX SOL EQN($\mathbf{Z}_p$, g).

Theorem 1. *For every prime p and every function $g : \mathbf{Z}_p \to \mathbf{N}$, MAX SOL EQN($\mathbf{Z}_p$, g) is approximable within α where*

$$\alpha = p \frac{g_{\max}}{g_{\text{sum}}}.$$

Furthermore, for every prime p and every non-constant function $g : \mathbf{Z}_p \to \mathbf{N}$ MAX SOL EQN($\mathbf{Z}_p$, g) is not approximable within $\alpha - \epsilon$ for any $\epsilon > 0$ unless $\mathbf{P} = \mathbf{NP}$.

Note that if g is a constant function then every feasible solution has the same measure and finding an optimum is solvable in polynomial time. As a consequence of Theorem 1 we get the following result for the natural choice of g.

Corollary 1. *If $g(x) = x$ then* MAX SOL EQN($\mathbf{Z}_p$, g) *is approximable within 2. Furthermore, it is not possible to approximate* MAX SOL EQN($\mathbf{Z}_p$, g) *within $2 - \epsilon$ for any $\epsilon > 0$ unless $\mathbf{P} = \mathbf{NP}$.*

Proof. If $g(x) = x$ then $g_{\max} = p - 1$ and $g_{\text{sum}} = p(p-1)/2$. This gives us

$$p\frac{g_{\max}}{g_{\text{sum}}} = p\frac{p-1}{p(p-1)/2} = 2.$$

The desired result follows from Theorem 1. □

We will also prove that the approximability threshold for W-MAX SOL EQN($\mathbf{Z}_p$, g) is asymptotically equal to the approximability threshold for MAX SOL EQN($\mathbf{Z}_p$, g). I.e., we will prove that W-MAX SOL EQN($\mathbf{Z}_p$, g) is approximable within $\alpha + o(1)$ where the $o(\cdot)$-notation is with respect to the size of the instance. Furthermore, we will prove that W-MAX SOL EQN($\mathbf{Z}_p$, g) is not approximable within $\alpha - \epsilon$ for any $\epsilon > 0$, unless $\mathbf{P} = \mathbf{NP}$.

3 Approximability

In this section we will prove our approximability results. We begin with giving a basic definition and proving a fundamental lemma about the modular sum of independent and uniformly distributed discrete random variables.

Definition 2. *Let $U(a, b)$ denote the uniform discrete distribution over*

$$\{a, a+1, \ldots, b-1, b\}.$$

We will write $X \sim U(a, b)$ to denote that the random variable X is distributed according to $U(a, b)$.

Lemma 1. *If X and Y are independent, $X \sim U(0, n-1)$ and $Y \sim U(0, n-1)$ then $X + Y \pmod{n} \sim U(0, n-1)$.*

Proof. Let k be constant such that $0 \leq k \leq n - 1$, then

$$\Pr[X + Y \pmod{n} = k] = \sum_{i,j : i+j \pmod{n} = k} \Pr[X = i]\Pr[Y = j] = n\frac{1}{n^2} = 1/n.$$

The second equality holds because given constants c_1, c_2 such that $0 \leq c_1, c_2 \leq n - 1$ then the equation $x + c_1 \pmod{n} = c_2$ has exactly one solution such that $0 \leq x \leq n - 1$. □

The following theorem is the main approximability result of this paper.

Theorem 2. *For every prime p and every $g : \mathbf{Z}_p \to \mathbf{N}$, MAX SOL EQN($\mathbf{Z}_p, g$) has an α-approximate algorithm, where*

$$\alpha = p\frac{g_{\max}}{g_{\text{sum}}}.$$

Proof. To find a feasible solution is equivalent to solving a system of equations in $\mathbf{Z}_p$. Let the system of equations be given by $A\bar{x} = \bar{b}$. Where A is a $m \times n$ matrix, $\bar{b}$ is a $m \times 1$ vector and $\bar{x}$ is a $n \times 1$ vector.

Assume without loss of generality that the rows in A are linearly independent. By elementary row operations and reordering of the variables we can, if there are any feasible solutions, transform the system to the form $[I|A']\bar{x} = \bar{b}'$. If no feasible solutions exists this will be detected during the transformation procedure. Let $\bar{x}' = (x_1, \ldots, x_m)$ and $\bar{x}'' = (x_{m+1}, \ldots, x_n)$. Then, the solutions to the system of equations are given by the set

$$\left\{ (\bar{x}', \bar{x}'') \mid \bar{x}'' \in \{0, \ldots, p-1\}^{n-m} \wedge \bar{x}' = -A'\bar{x}'' + \bar{b}' \right\}. \tag{1}$$

Choose a random element from this set by assigning values to the variables in $\bar{x}''$, uniformly at random from $\mathbf{Z}_p$. The variables in $\bar{x}'$ are then assigned values according to $\bar{x}' = -A'\bar{x}'' + \bar{b}'$.

We claim that the procedure described above is a α-approximate algorithm. Note that for every $m + 1 \leq i \leq n$ and every $0 \leq k \leq p - 1$ the following holds: $\Pr[x_i = k] = 1/p$. We will now prove that for $x_i, 1 \leq i \leq m$, one of the following two cases holds:

1. $x_i = b'_i$ in every solution, or
2. $x_i = k$ with probability $1/p$, for every $0 \leq k \leq p-1$.

For every $x_i, 1 \leq i \leq m$, we see that

$$x_i = b'_i - \sum_{j=1}^{n-m} a'_{ij} x_{j+m} = b'_i - \sum_{\{j \mid a'_{ij} \neq 0\}} a'_{ij} x_{j+m}. \tag{2}$$

If, for some i, we have $a'_{ij} = 0_G$ for all j, then $x_i = b'_i$ and we have case 1.

Let us introduce fresh variables t_j for each j such that $a'_{ij} \neq 0_G$. Let $t_j = a'_{ij} x_{j+m}$. Equation (2) is now equivalent to

$$x_i = b'_i - \sum_{\{j \mid a'_{ij} \neq 0\}} t_j. \tag{3}$$

Since $x_{j+m} \sim U(0, p-1)$ and $a'_{ij} \neq 0_G$, we have $t_j \sim U(0, p-1)$ and according to Lemma 1 $x_i \sim U(0, p-1)$. Let B denote the set of indices i such that $x_i \sim U(0, p-1)$ and let $C = \{1, \ldots, n\} \setminus B$ denote the set with indices of the variables that are fixed in all feasible solutions. Furthermore, let

$$S^* = |B|g_{\max} + \sum_{i \in C} g(b'_i).$$

It is easy to see that $S^* \geq \text{OPT}$. Let S denote the solution produced by the algorithm. The expected value of S is

$$\mathrm{E}\left[S\right] = \mathrm{E}\left[\sum_{i \in B} g(x_i) + \sum_{i \in C} g(b'_i)\right] = \sum_{i \in B} \mathrm{E}\left[g(x_i)\right] + \sum_{i \in C} g(b'_i)$$

$$= |B|\frac{g_{\text{sum}}}{p} + \sum_{i \in C} g(b'_i) \geq \frac{g_{\text{sum}}}{p} \frac{S^*}{g_{\max}} \geq \frac{g_{\text{sum}}}{p g_{\max}} \text{OPT}. \qquad \square$$

4 Inapproximability

In this section we will prove our inapproximability results. We begin with the definition of MAX-Ek-LIN-p which will be used in the main reduction later on.

Definition 3 (MAX-Ek-LIN-p [9]). *An instance of* MAX-Ek-LIN-p *is defined to be* (V, E) *where*

- *V is a set of variables, and*
- *E is a set of linear equations over the group $\mathbf{Z}_p$ with exactly k variables in each equation.*

The objective is to find an assignment $f : V \to \mathbf{Z}_p$ such that the maximum number of equations in E are satisfied.

The following theorem can be deduced from the proof of Theorem 1 in [6], which is a generalisation of a similar theorem about abelian groups by Håstad [9].

Theorem 3. *For every problem Π in* **NP** *there is a polynomial time reduction from instances I of Π to instances $I' = (V, E)$ of* MAX-E3-LIN-p *such that*

- *if I is a* YES *instance then at least $(1 - \delta)|E|$ equations can be satisfied, and*
- *if I is a* NO *instance then no assignment satisfies more than $|E|(1+\delta)/p$ equations*

where δ is an arbitrary constant such that $0 < \delta < \frac{p-1}{p+1}$.

We will prove our inapproximability results with a special kind of reduction, namely a gap-preserving reduction introduced by Arora in [1]. The definition is as follows.

Definition 4 (Gap-preserving reduction [1]). *Let Π and Π' be two maximisation problems and $\rho, \rho' > 1$. A gap-preserving reduction with parameters c, ρ, c', ρ' from Π to Π' is a polynomial time algorithm f. For each instance I of Π, f produces an instance $I' = f(I)$ of Π'. The optima of I and I', satisfy the following properties:*

- *if $\text{OPT}(I) \geq c$ then $\text{OPT}(I') \geq c'$, and*
- *if $\text{OPT}(I) \leq c/\rho$ then $\text{OPT}(I') \leq c'/\rho'$.*

Gap-preserving reductions are useful because if for every language in **NP** there is a polynomial time reduction to the maximisation problem Π such that YES instances are mapped to instances of Π of measure at least c and NO instances to instances of measure at most c/ρ, then a gap-preserving reduction from Π to Π' implies that finding ρ'-approximations to Π' is **NP**-hard. [1]

Note that Theorem 3 implies that MAX-E3-LIN-p is a suitable problem to do gap-reductions from.

Theorem 4. *For every prime p and every non-constant $g : \mathbf{Z}_p \to \mathbf{N}$, it is not possible to approximate W-MAX SOL EQN$(\mathbf{Z}_p, g)$ within $\alpha - \epsilon$ where*

$$\alpha = p\frac{g_{\max}}{g_{\text{sum}}},$$

for any $\epsilon > 0$ unless $\mathbf{P} = \mathbf{NP}$.

Proof. We will prove the theorem with a gap-preserving reduction from MAX-E3-LIN-p. Given an instance $I = (V, E)$ of MAX-E3-LIN-p we will construct an instance $I' = (V', E', w')$ of W-MAX SOL EQN$(\mathbf{Z}_p, g)$. Let a be the element in $\mathbf{Z}_p$ that maximises g, i.e., $g(a) = g_{\max}$, and let $s = g_{\text{sum}} - g_{\max}$.

Let

$$V' = V \cup \{z_j \mid 1 \le j \le |E|\} \cup \left\{z_j^{(i)} \mid 1 \le j \le |E|, 1 \le i \le p-1\right\}.$$

Every equation e_j in E is of the form $w_{j_1} + w_{j_2} + w_{j_3} = c_j$ where w_{j_k} is either a variable or an inverted variable and c_j is a group constant. For each such equation add the equation $w_{j_1} + w_{j_2} + w_{j_3} = c_j + z_j$ to E_1', where z_j are fresh variables.

Construct the following set of equations

$$E_2' = \left\{z_j^{(i)} = iz_j + a \mid 1 \le j \le |E|, 1 \le i \le p-1\right\}.$$

Note that i is an integer and the expression iz_j should be interpreted as the sum $z_j + \ldots + z_j$ with i terms.

Let $E' = E_1' \cup E_2'$, and let

$$w\left(z_j^{(i)}\right) = 1, \quad 1 \le i \le p-1 \text{ and } 1 \le j \le |E|$$

and $w(\cdot) = 0$ otherwise.

We claim that the procedure presented above is a gap-preserving reduction from MAX-E3-LIN-p to W-MAX SOL EQN$(\mathbf{Z}_p, g)$ with parameters

$$c = (1-\delta)|E|,$$
$$c' = (1-\delta)|E|(p-1)g_{\max} + \delta|E|s, \text{ and}$$
$$\rho = p\frac{1-\delta}{1+\delta}.$$

Where δ is the constant from Theorem 3. The last parameter, ρ', is specified below. According to Theorem 3 we know that either OPT$(I) \ge |E|(1-\delta)$ or OPT$(I) \le |E|(1+\delta)/p$.

Case 1: (OPT$(I) \ge (1-\delta)|E|$) Let f be an assignment such that $m(I, f) \ge (1-\delta)|E|$. We can then construct an assignment, f', to I' as follows: if $x \in V$ then let $f'(x) = f(x)$, for every $1 \le j \le |E|$, let $f'(z_j)$ be the value such that equation e_j is satisfied and finally let

$$f'\left(z_j^{(i)}\right) = if'(z_j) + a, \quad \text{for } 1 \le i \le p-1 \text{ and } 1 \le j \le |E|.$$

Note that $f'(z_j) = 0_G$ iff equation e_j is satisfied by f. It is easy to verify that every equation in E' is satisfied by f'.

Before we prove a bound on $m'(I', f')$, note that the sum

$$\sum_{i=1}^{p-1} g\left(f'\left(z_j^{(i)}\right)\right)$$

is either equal to $(p-1)g_{\max}$ (if $f'(z_j) = 0_G$) or equal to s (if $f'(z_j) \neq 0_G$). The latter case follows from the fact that $\mathbf{Z}_p$ is cyclic group and every element except 0_G is a generator. Hence, every equation in E will either contribute $(p-1)g_{\max}$ to $m'(I', f')$ or it will contribute s. We also know that at least $(1-\delta)|E|$ equations are satisfied and each one of them will therefore contribute $(p-1)g_{\max}$ to $m'(I', f')$. This argument gives us the following bound on $m'(I', f')$,

$$m'(I', f') = \sum_{j=1}^{|E|}\sum_{i=1}^{p-1} g\left(f'\left(z_j^{(i)}\right)\right) \geq (1-\delta)|E|(p-1)g_{\max} + \delta|E|s = c'.$$

Hence, if $\text{OPT}(I) \geq c$ then $\text{OPT}(I') \geq c'$.

Case 2: ($\text{OPT}(I) \leq |E|(1+\delta)/p$) If $\text{OPT}(I) \leq |E|(1+\delta)/p = c/\rho$, then any assignment to I' cannot have assigned more than $|E|(1+\delta)/p$ of the z_i variables the value 0_G. If there is such an assignment, then it can be used to construct an assignment f to I such that $m(I, f) > |E|(1+\delta)/p$ which contradicts our assumption. Hence, at most $|E|(1+\delta)/p$ variables of the form z_i has been assigned the value 0_G in any assignment to I'.

Let f' be an assignment to I' such that $\text{OPT}(I') = m'(I', f')$. >From the argument above we know that

$$m'(I', f') = \sum_{i=1}^{|E|}\sum_{j=1}^{p} g\left(f'\left(z_j^{(i)}\right)\right) \leq |E|\frac{1+\delta}{p}g_{\max}(p-1) + |E|\left(1 - \frac{1+\delta}{p}\right)s.$$

Let h denote the quantity on the right hand side of the inequality above. We want to find the largest ρ' that satisfies $\text{OPT}(I') \leq c'/\rho'$. If we choose ρ' such that $c'/\rho' = h$ then $\text{OPT}(I') \leq c'/\rho'$ because of $\text{OPT}(I') = m'(I', f') \leq h$.

$$\rho' = \frac{c'}{h} = \frac{(1-\delta)|E|(p-1)g_{\max} + \delta|E|s}{|E|\frac{1+\delta}{p}g_{\max}(p-1) + |E|\left(1 - \frac{1+\delta}{p}\right)s}$$

$$= \frac{pg_{\max}}{g_{\max} + s\frac{p-(1+\delta)}{p-1} + \delta g_{\max}} + p\delta\frac{s/(p-1) - g_{\max}}{g_{\max} + s\frac{p-(1+\delta)}{p-1} + \delta g_{\max}}$$

Now, given a fixed but arbitrary $\epsilon > 0$ we can choose $0 < \delta < \frac{p-1}{p+1}$ such that

$$\rho' > p\frac{g_{\max}}{g_{\max} + s} - \epsilon = p\frac{g_{\max}}{g_{\text{sum}}} - \epsilon = \alpha - \epsilon.$$

The gap-preserving reduction implies that it is **NP**-hard to find ρ'-approximations to W-MAX SOL EQN($\mathbf{Z}_p, g$), and since $\rho' > \alpha - \epsilon$ we have the desired result. □

5 Proof of Main Theorem

Before we can prove our main result we need the lemma below.

Lemma 2. *If* MAX SOL EQN($\mathbf{Z}_p$, g) *is approximable within in r, then* W-MAX SOL EQN($\mathbf{Z}_p$, g) *is approximable within $r + o(1)$, where the $o(\cdot)$-notation is with respect to the size of the instance.*

This lemma can be proved using techniques similar to those used in the proof of Lemma 3.11 in [10] and the proof of Theorem 4 in [5]. Due to this fact and lack of space we omit the proof. We are now ready to prove Theorem 1.

Proof (of Theorem 1). The approximation algorithm in Theorem 2 is the first part of Theorem 1.

Lemma 2 says that if we can find r-approximate solutions for MAX SOL EQN($\mathbf{Z}_p$, g), then we can find $(r + o(1))$-approximate solutions for W-MAX SOL EQN($\mathbf{Z}_p$, g). Hence, if we can find $\alpha - \delta$ approximations for some $\delta > 0$ for MAX SOL EQN($\mathbf{Z}_p$, g) then we can find $(\alpha - \delta + o(1))$-approximate solutions for W-MAX SOL EQN($\mathbf{Z}_p$, g). However, as the sizes of the instances grow we will, at some point, have $-\delta + o(1) < 0$ which means that we would be able to find $(\alpha - \epsilon)$-approximate solutions, where $\epsilon > 0$, for W-MAX SOL EQN($\mathbf{Z}_p$, g). But Theorem 4 says that this is not possible. Therefore, MAX SOL EQN($\mathbf{Z}_p$, g) is not approximable within $\alpha - \delta$, for any $\delta > 0$, unless $\mathbf{P} = \mathbf{NP}$. □

The situation is almost the same for W-MAX SOL EQN($\mathbf{Z}_p$, g). We have an α-approximate algorithm for MAX SOL EQN($\mathbf{Z}_p$, g) (Theorem 2) therefore, due to Lemma 2 we have a $(\alpha + o(1))$-approximate algorithm for W-MAX SOL EQN($\mathbf{Z}_p$, g). Furthermore, it is not possible to approximate W-MAX SOL EQN($\mathbf{Z}_p$, g) within $\alpha - \epsilon$ for any $\epsilon > 0$ (Theorem 4).

All our hardness results holds for equations with at most three variables per equation. If we are given an equation with n variables where $n > 3$, we can reduce this equation to one equation with $n - 1$ variables and one equation with 3 variables in the following way: Given the equation $x_1 + \ldots + x_n = c$ where each x_i is either a variable or an inverted variable and c is a group constant, introduce the equation $z = x_1 + x_2$ where z is a fresh variable. Furthermore replace the original equation with the equation $z + x_3 + \ldots + x_n = c$. Let the weight of z be zero. Those two equations are clearly equivalent to the original equation in the problem W-MAX SOL EQN($\mathbf{Z}_p$, g). The proof of Lemma 2 do not introduce any equations with more than two variables, so we get the same result for MAX SOL EQN($\mathbf{Z}_p$, g).

If the instances of W-MAX SOL EQN($\mathbf{Z}_p$, g) are restricted to have at most two variables per equation then the problem is tractable. The following algorithm solves this restricted problem in polynomial time.

A system of equations where there are at most two variables per equation can be represented by a graph in the following way: let each variable be a vertex in the graph and introduce an edge between two vertices if the corresponding variables appear in the same equation. It is clear that the connected components of the graph are independent subsystems of the system of equations. Hence, finding the optimum of the system of equations is equivalent to finding the optimum of each of the subsystems that corresponds to the connected components. To find the optimum of one such subsystem, choose a variable, x, and assign a value to it. This assignment will force assignments of values to every other variable in the subsystem. The optimum can be found by testing every possible assignment of values to x. If this is done for every independent subsystem the optimum for the entire system of equations will be found in polynomial time.

6 Concluding Remarks

We have given tight approximability results for some maximum solution equation problems. One natural generalisation of our work might be to investigate the problem where the set of equations are replaced by a set of constraints over some constraint family. The problem will then be a proper generalisation of the problem MAX ONES studied by Khanna et al [10]. A start for such results might be to try to classify which families of constraints give rise to tractable optimisation problems.

References

1. Sanjeev Arora. *Probabilistic checking of proofs and hardness of approximation problems.* PhD thesis, 1995.
2. Sanjeev Arora, Laszlo Babai, Jacques Stern, and Z. Sweedyk. The hardness of approximate optima in lattices, codes, and systems of linear equations. *J. Comput. Syst. Sci.*, 54(2):317–331, 1997.
3. Daniel Pierre Bovet and Pierluigi Crescenzi. *Introduction to the theory of complexity.* Prentice Hall International (UK) Ltd., 1994.
4. Jehoshua Bruck and Moni Naor. The hardness of decoding linear codes with preprocessing. *IEEE Transactions on Information Theory*, 36(2):381–385, 1990.
5. Pierluigi Crescenzi, Riccardo Silvestri, and Luca Trevisan. To weight or not to weight: Where is the question? In *ISTCS 96': Proceedings of the 4th Israeli Symposium on Theory of Computing and Systems*, pages 68–77. IEEE, 1996.
6. Lars Engebretsen, Jonas Holmerin, and Alexander Russell. Inapproximability results for equations over finite groups. *Theor. Comput. Sci.*, 312(1):17–45, 2004.
7. Uriel Feige and Daniele Micciancio. The inapproximability of lattice and coding problems with preprocessing. *J. Comput. Syst. Sci.*, 69(1):45–67, 2004.
8. Mikael Goldmann and Alexander Russell. The complexity of solving equations over finite groups. *Inf. Comput.*, 178(1):253–262, 2002.
9. Johan Håstad. Some optimal inapproximability results. *J. ACM*, 48(4):798–859, 2001.
10. Sanjeev Khanna, Madhu Sudan, Luca Trevisan, and David P. Williamson. The approximability of constraint satisfaction problems. *SIAM J. Comput.*, 30(6):1863–1920, 2000.
11. Ondrej Klíma, Pascal Tesson, and Denis Thérien. Dichotomies in the complexity of solving systems of equations over finite semigroups. Technical Report TR04-091, Electronic Colloq. on Computational Complexity, 2004.
12. Benoit Larose and Laszlo Zadori. Taylor terms, constraint satisfaction and the complexity of polynomial equations over finite algebras. Submitted for publication.
13. Cristopher Moore, Pascal Tesson, and Denis Thérien. Satisfiability of systems of equations over finite monoids. In *Mathematical Foundations of Computer Science 2001, 26th International Symposium, MFCS 2001*, volume 2136 of *Lecture Notes in Computer Science*, pages 537–547. Springer, 2001.

14. Gustav Nordh. The complexity of equivalence and isomorphism of systems of equations over finite groups. In *Mathematical Foundations of Computer Science 2004, 29th International Symposium, MFCS 2004*, volume 3153 of *Lecture Notes in Computer Science*, pages 380–391. Springer, 2004.
15. Gustav Nordh and Peter Jonsson. The complexity of counting solutions to systems of equations over finite semigroups. In *Computing and Combinatorics, 10th Annual International Conference, COCOON 2004*, volume 3106 of *Lecture Notes in Computer Science*, pages 370–379. Springer, 2004.
16. Pascal Tesson. *Computational Complexity Questions Related to Finite Monoids and Semigroups*. PhD thesis, 2003.

The Complexity of Model Checking Higher Order Fixpoint Logic

Martin Lange[1] and Rafał Somla[2]

[1] Institut für Informatik, University of Munich, Germany
mlange@informatik.uni-muenchen.de
[2] IT department, Uppsala University, Sweden
rsomla@it.uu.se

Abstract. This paper analyzes the computational complexity of the model checking problem for Higher Order Fixpoint Logic – the modal μ-calculus enriched with a typed λ-calculus. It is hard for every level of the elementary time/space hierarchy and in elementary time/space when restricted to formulas of bounded type order.

1 Introduction

Temporal logics are well-established tools for the specification of correctness properties and their verification in hard- and software design processes. One of the most famous temporal logics is Kozen's modal μ-calculus $\mathcal{L}_\mu$ [16] which extends multi-modal logic with extremal fixpoint quantifiers. $\mathcal{L}_\mu$ subsumes many other temporal logics like CTL* [10], and with it CTL [9] and LTL [19], as well as PDL [12].

$\mathcal{L}_\mu$ is equi-expressive to the bisimulation-invariant fragment of Monadic Second Order Logic over trees or graphs [11,15]. Hence, properties expressed by formulas of the modal μ-calculus are only regular. There are, however, many interesting correctness properties of programs that are not regular. Examples include *uniform inevitability* [8] which states that a certain event occurs globally at the same time in all possible runs of the system; counting properties like "at any point in a run of a protocol there have never been more *send-* than *receive-* actions"; formulas saying that an unbounded number of data does not lose its order during a transmission process; etc.

When program verification was introduced to computer science, programs as well as their correctness properties were mainly specified in temporal logics. Hence, verification meant to check formulas of the form $\varphi \to \psi$ for validity, or equally formulas of the form $\varphi \wedge \psi$ for satisfiability. An intrinsic problem for this approach and non-regular properties is undecidability. Note that the intersection problem for context-free languages is already undecidable [1].

One of the earliest attempts at verifying non-regular properties of programs was *Non-Regular PDL* [13] which enriches ordinary PDL by context-free programs. Non-Regular PDL is highly undecidable, hence, the logic did not receive much attention for program verification purposes. Its model checking problem, however, remains decidable on finite transition systems.

Although the theoretical complexity of the model checking problem is normally below that of its satisfiability problem, it often requires a lot more time or space to do model checking. This is simply because the input to a model checker is usually a lot bigger compared to that of a satisfiability checker. Hence, the feasibility of model checking is very much limited by the state space explosion problem: real-world examples result in huge transition systems that are very hard to model check simply because of their sheer size. However, in recent years various clever techniques have been invented that can cope with huge state spaces, starting with *local model checking*, and resulting in symbolic methods like *BDD-based* [4] or *bounded model checking* [6]. They are also a reason for the shift in importance from the satisfiability checking to the model checking problem for program verification.

More expressive power naturally comes with higher complexities. But with good model checking techniques at hand, verifying non-regular properties has become worthwhile again. This is for example reflected in the introduction of *Fixpoint Logic with Chop*, FLC, [18] which extends $\mathcal{L}_\mu$ with a sequential composition operator. It is capable of expressing many non-regular – and even non-context-free – properties, and its model checking problem on finite transition systems is decidable in deterministic exponential time [17].

Another logic capable of expressing non-regular properties is the *Modal Iteration Calculus*, MIC, [7] which extends $\mathcal{L}_\mu$ with inflationary fixpoint quantifiers. Similar to FLC, the satisfiability checking problem for MIC is undecidable but its model checking problem is decidable in deterministic polynomial space [7].

In order to achieve non-regular effects in FLC, the original $\mathcal{L}_\mu$ semantics is lifted to a function from sets of states to sets of states. This idea has been followed consequently in the introduction of *Higher Order Fixpoint Logic*, HFL, [23] which incorporates a typed λ-calculus into the modal μ-calculus. This gives it even more expressive power than FLC. HFL is, for example, capable of expressing *assume-guarantee-properties*. Still, HFL's model checking problem on finite transition systems remains decidable. This has been stated in its introductory work [23] but no analysis of its computational complexity has been done so far.

Here we set out to answer the open question concerning the complexity of HFL's model checking problem. We start by recalling the logic and giving a few examples in Section 2. Section 3 presents a reduction from the satisfiability problem for First Order Logic over finite words to HFL's model checking problem. Consequently, the latter is hard for every level of the elementary time/space hierarchy. I.e. there is no model checking algorithm for HFL that runs in time given by a tower of exponentials whose height does not depend on the input formula. This is not too surprising because HFL incorporates the typed λ-calculus for which the problem of deciding whether a given term can be transformed into another given one, is also non-elementary [21]. This can be reduced to the model checking problem for HFL.

Here we provide a more fine-grained analysis of HFL's model checking problem. When restricted to terms of type order k, it is hard for $(k-3)$ExpSpace and included in $(k+1)$ExpTime. It remains to be seen whether this gap can be closed.

2 Preliminaries

Let $\mathcal{P} = \{p, q, \ldots\}$ be a set of atomic propositions, $\mathcal{A} = \{a, b, \ldots\}$ be a finite set of action names, and $\mathcal{V} = \{X, Y, X_1, \ldots\}$ a set of variable names. For simplicity, we fix $\mathcal{P}$, $\mathcal{A}$, and $\mathcal{V}$ for the rest of the paper.

A $v \in \{-, +, 0\}$ is called a variance. The set of HFL-types is the smallest set containing the atomic type Prop and being closed under function typing with variances, i.e. if σ and τ are HFL-types and v is a variance, then $\sigma^v \to \tau$ is an HFL-type. Formulas of HFL are given by the following grammar:

$$\varphi ::= q \mid X \mid \neg\varphi \mid \varphi \vee \varphi \mid \langle a \rangle \varphi \mid \varphi\,\varphi \mid \lambda(X^v : \tau).\varphi \mid \mu(X : \tau).\varphi \;.$$

We use the standard abbreviations: $\mathsf{tt} := q \vee \neg q$ for some $q \in \mathcal{P}$, $\mathsf{ff} := \neg\mathsf{tt}$, $\varphi \wedge \psi := \neg(\neg\varphi \wedge \neg\psi)$, $[a]\psi := \neg\langle a \rangle \neg\psi$, and $\nu X.\varphi := \neg\mu X.\neg\varphi[\neg X/X]$. We will assume that any variable without an explicit type annotation is of the ground type Prop. Also, if a variance is omitted it is implicitly assumed to be 0.

A sequence Γ of the form $X_1^{v_1} : \tau_1, \ldots, X_n^{v_n} : \tau_n$ where X_i are variables, τ_i are types and v_i are variances is called a *context* (we assume all X_i are distinct). An HFL-formula φ has type τ in context Γ if the statement $\Gamma \vdash \varphi : \tau$ can be inferred using the rules of Figure 1. We say that φ is *well-formed* if $\Gamma \vdash \varphi : \tau$ for some Γ and τ.

For a variance v, we define its complement v^- as $+$ if $v = -$, as $-$, if $v = +$, and 0 otherwise. For a context $\Gamma = X_1^{v_1} : \tau_1, \ldots, X_n^{v_n} : \tau_n$, the complement Γ^- is defined as $X_1^{v_1^-} : \tau_1, \ldots, X_n^{v_n^-} : \tau_n$.

A (labeled) transition system is a structure $\mathcal{T} = (\mathcal{S}, \{\xrightarrow{a}\}, L)$ where $\mathcal{S}$ is a finite set of states, $\xrightarrow{a}$ is a binary relation on states for each $a \in \mathcal{A}$, and

$$(var)\;\; \Gamma, X^v : \tau \vdash X : \tau \quad \text{if } v \in \{0, +\} \qquad (neg)\;\; \frac{\Gamma^- \vdash \varphi : \mathsf{Prop}}{\Gamma \vdash \neg\varphi : \mathsf{Prop}}$$

$$(or)\;\; \frac{\Gamma \vdash \varphi : \mathsf{Prop} \quad \Gamma \vdash \psi : \mathsf{Prop}}{\Gamma \vdash \varphi \vee \psi : \mathsf{Prop}} \qquad (mod)\;\; \frac{\Gamma \vdash \varphi : \mathsf{Prop}}{\Gamma \vdash \langle a \rangle \varphi : \mathsf{Prop}}$$

$$(abs)\;\; \frac{\Gamma, X^v : \sigma \vdash \varphi : \tau}{\Gamma \vdash \lambda(X^v : \sigma).\varphi : (\sigma^v \to \tau)} \qquad (fix)\;\; \frac{\Gamma, X^+ : \tau \vdash \varphi : \tau}{\Gamma \vdash \mu(X : \tau).\varphi : \tau}$$

$$(app^+)\;\; \frac{\Gamma \vdash \varphi : (\sigma^+ \to \tau) \quad \Gamma \vdash \psi : \sigma}{\Gamma \vdash (\varphi\,\psi) : \tau} \qquad (app^-)\;\; \frac{\Gamma \vdash \varphi : (\sigma^- \to \tau) \quad \Gamma^- \vdash \psi : \sigma}{\Gamma \vdash (\varphi\,\psi) : \tau}$$

$$(app^0)\;\; \frac{\Gamma \vdash \varphi : (\sigma^0 \to \tau) \quad \Gamma \vdash \psi : \sigma \quad \Gamma^- \vdash \psi : \sigma}{\Gamma \vdash (\varphi\,\psi) : \tau} \qquad (prop)\;\; \frac{}{\Gamma \vdash p : \mathsf{Prop}}$$

Fig. 1. Type inference rules for HFL

$$\mathcal{T}[\![\Gamma \vdash q : \mathsf{Prop}]\!]\eta = \{s \in \mathcal{S} \mid q \in L(s)\}$$
$$\mathcal{T}[\![\Gamma \vdash X : \tau]\!]\eta = \eta(X)$$
$$\mathcal{T}[\![\Gamma \vdash \neg\varphi : \mathsf{Prop}]\!]\eta = \mathcal{S} - \mathcal{T}[\![\Gamma^- \vdash \varphi : \mathsf{Prop}]\!]\eta$$
$$\mathcal{T}[\![\Gamma \vdash \varphi \vee \psi : \mathsf{Prop}]\!]\eta = \mathcal{T}[\![\Gamma \vdash \varphi : \mathsf{Prop}]\!]\eta \cup \mathcal{T}[\![\Gamma \vdash \psi : \mathsf{Prop}]\!]\eta$$
$$\mathcal{T}[\![\Gamma \vdash \langle a\rangle\varphi : \mathsf{Prop}]\!]\eta = \{s \in \mathcal{S} \mid s \xrightarrow{a} t \text{ for some } t \in \mathcal{T}[\![\Gamma \vdash \varphi : \mathsf{Prop}]\!]\eta\}$$
$$\mathcal{T}[\![\Gamma \vdash \lambda(X^v : \tau).\varphi : \tau^v \to \tau']\!]\eta = F \in \mathcal{T}[\![\tau^v \to \tau']\!] \text{ s.t. } \forall d \in \mathcal{T}[\![\tau]\!]$$
$$F(d) = \mathcal{T}[\![\Gamma, X^v : \tau \vdash \varphi : \tau']\!]\eta[X \mapsto d]$$
$$\mathcal{T}[\![\Gamma \vdash \varphi\,\psi : \tau']\!]\eta = (\mathcal{T}[\![\Gamma \vdash \varphi : \tau^v \to \tau']\!]\eta)(\mathcal{T}[\![\Gamma' \vdash \psi : \tau]\!]\eta)$$
$$\mathcal{T}[\![\Gamma \vdash \mu(X:\tau)\varphi : \tau]\!]\eta = \bigsqcap\nolimits_{\mathcal{T}[\![\tau]\!]} \{d \in \tau \mid$$
$$\mathcal{T}[\![\Gamma, X^+ : \tau \vdash \varphi : \tau]\!]\eta[X \mapsto d] \leq_{\mathcal{T}[\![\tau]\!]} d\}$$

Fig. 2. Semantics of HFL

$L : \mathcal{S} \to 2^{\mathcal{P}}$ is a labeling function denoting the set of propositional constants that are true in a state.

The semantics of a type w.r.t. a transition system $\mathcal{T}$ is a complete lattice, inductively defined on the type as

$$\mathcal{T}[\![\mathsf{Prop}]\!] = (2^{\mathcal{S}}, \subseteq)\,, \qquad \mathcal{T}[\![\sigma^v \to \tau]\!] = (\mathcal{T}[\![\sigma]\!])^v \to \mathcal{T}[\![\tau]\!]\,.$$

Here, for two partial orders $\bar{\tau} = (\tau, \leq_\tau)$ and $\bar{\sigma} = (\sigma, \leq_\sigma)$, $\bar{\sigma} \to \bar{\tau}$ denotes the partial order of all monotone functions ordered pointwise, and, $\bar{\tau}^v$ denotes $(\tau, \leq_\tau^v)$. $\leq_\tau^+$ is $\leq_\tau$, $a \leq_\tau^- b$ iff $b \leq_\tau a$, and $\leq_\tau^0 = \leq_\tau \cap \leq_\tau^-$.

An environment η is a possibly partial map on the variable set $\mathcal{V}$. For a context $\Gamma = X_1^{v_1} : \tau_1, \ldots, X_n^{v_n} : \tau_n$, we say that η respects Γ, denoted by $\eta \models \Gamma$, if $\eta(X_i) \in \mathcal{T}[\![\tau_i]\!]$ for $i \in \{1,\ldots,n\}$. We write $\eta[X \mapsto a]$ for the environment that maps X to a and otherwise agrees with η. If $\eta \models \Gamma$ and $a \in \mathcal{T}[\![\tau]\!]$ then $\eta[X \mapsto a] \models \Gamma, X : \tau$, where X is a variable that does not appear in Γ.

For any well-typed term $\Gamma \vdash \varphi : \tau$ and environment $\eta \models \Gamma$, Figure 2 defines the semantics of φ inductively to be an element of $\mathcal{T}[\![\tau]\!]$. In the clause for function application $(\varphi\,\psi)$ the context Γ' is Γ if $v \in \{+, 0\}$, and is Γ^- if $v = -$.

The model checking problem for HFL is the following: Given an HFL sentence $\varphi : \mathsf{Prop}$, a transition system $\mathcal{T}$ and a set of states A decide whether or not $\mathcal{T}[\![\varphi]\!] = A$.

We consider fragments of formulas that can be built using restricted type orders only. Let

$$\mathrm{ord}(\mathsf{Prop}) := 0\,, \quad \mathrm{ord}(\sigma \to \tau) := \max\{1 + \mathrm{ord}(\sigma), \mathrm{ord}(\tau)\}$$

and $\mathrm{HFL}^k := \{\varphi \in \mathrm{HFL} \mid \vdash \varphi : \mathsf{Prop} \text{ using types } \tau \text{ with } \mathrm{ord}(\tau) \leq k \text{ only}\}$.

Example 1. The following HFL formula expresses the non-regular property "on any path the number of a's seen at any time never exceeds the number of b's seen so far."

$$\nu X.[a]\mathsf{ff} \wedge [b]\big((\nu(Z : \mathsf{Prop} \to \mathsf{Prop}).\lambda Y.([a]Y \wedge [b]\,(Z\,(Z\,Y)))\,X\big)\,.$$

Note how function composition is used to "remember" in the argument to Z how many times a b-action has been seen along any path. Every b-action gives the potential to do another a-action later on which is remembered in the additional application of Z. a-action "uses up" one Z. If there have been as many a's as b's then the current state must be in the semantics of X again, hence, cannot do another a-action, etc.

Example 2. Let $\mathbf{2}_0^n := n$ and $\mathbf{2}_{m+1}^n := 2^{2_m^n}$. For any $m \in \mathbb{N}$, there is an HFL formula φ_m expressing the fact that there is a maximal path of length $\mathbf{2}_m^1$ (number of states on this path) through a transition system. It can be constructed using a typed version of the Church numeral 2. Let $P_0 = \mathsf{Prop}$ and $P_{i+1} = P_i \to P_i$. For $i \geq 1$ define ψ_i of type P_{i+1} as $\lambda(F : P_i).\lambda(X : P_{i-1}).F\,(F\,X)$. Then

$$\varphi_m := \psi_m\,\psi_{m-1}\,\ldots\,\psi_1\,\bigl(\lambda X.\langle -\rangle X\bigr)\,\bigl([-]\,\mathsf{ff}\bigr)\,.$$

Note that for any $m \in \mathbb{N}$, φ_m is of size polynomial in m. This indicates that HFL is able to express computations of Turing Machines of arbitrary elementary complexity. The next section shows that this is indeed the case.

3 The Lower Complexity Bound

Let Σ be a finite alphabet. Formulas of FO in negation normal form over words in Σ^* are given by the following grammar.

$$\varphi \;::=\; x \leq y \mid x < y \mid P_a(x) \mid \varphi \vee \varphi \mid \varphi \wedge \varphi \mid \exists x.\varphi \mid \forall x.\varphi$$

where x, y are variables and $a \in \Sigma$.

A word $w \in \Sigma^*$ of length n is a function of type $\{0, \ldots, n-1\} \to \Sigma$. Thus, $w(i)$ denotes the i-th letter of w. FO formulas are interpreted over words in the usual way, written $w \models_\eta \varphi$ for a word w, a formula φ and an environment η evaluating the free variables in φ by positions in w.

Let Σ_0 and Π_0 be the set of all quantifier-free formulas of FO. Σ_{k+1} is the closure of $\Sigma_k \cup \Pi_k$ under the boolean operators and existential quantification. Similarly, Π_{k+1} is constructed from $\Sigma_k \cup \Pi_k$ using universal quantification instead.

Let $\mathrm{DTime}(f(n))$ and $\mathrm{DSpace}(f(n))$ be the classes of languages that can be decided by a deterministic Turing Machine in time, resp. space, $f(n)$ where n measures the length of the input word to the machine. The k-th levels of the elementary time/space hierarchy are

$$k\textsc{ExpTime} \;=\; \bigcup_{c\in\mathbb{N}} \mathrm{DTime}(\mathbf{2}_k^{c\cdot n}), \qquad k\textsc{ExpSpace} \;=\; \bigcup_{c\in\mathbb{N}} \mathrm{DSpace}(\mathbf{2}_k^{c\cdot n})\,.$$

Furthermore, the elementary time/space hierarchy is

$$\textsc{ElTime} \;:=\; \bigcup_{k\in\mathbb{N}} k\textsc{ExpTime} \;=\; \bigcup_{k\in\mathbb{N}} k\textsc{ExpSpace}\,.$$

The standard translation of an FO formula into a finite automaton [3] and the encoding of space-bounded Turing Machine computations in FO [22,20] yield the following results.

Theorem 1. *An FO sentence $\varphi \in \Sigma_{k+1}$ of length n has a model iff it has a model of length 2_k^n.*

Theorem 2. *For all $k \geq 1$: The satisfiability problem for FO formulas in Σ_{k+1} is hard for kExpSpace.*

We will use these results to prove kExpSpace hardness of HFL. Our first step is to translate the problem of deciding whether a given binary word w of length 2_k^n is a model of an FO sentence φ into the HFL model checking problem. Let us fix n and define $\mathcal{T}_n$ to be a transition system with states $S_n = \{0, 1, \ldots, n-1\}$, an empty labeling (we will not use propositional constants) and a cyclic next state relation $\rightarrow \subseteq S_n \times S_n$ given by $0 \rightarrow n-1$ and $i \rightarrow i-1$ for $i = 1, \ldots, n-1$.

We represent the Boolean values *false* and *true* by the two elements of $\mathcal{B} = \{\varnothing, S_n\} \subset \mathcal{T}_n[\text{Prop}]$. In the sequel we will implicitly assume that all semantic interpretations are w.r.t $\mathcal{T}_n$ and omit it in front of semantic brackets. Hence we can write $[\![\text{ff}]\!]$ and $[\![\text{tt}]\!]$ for the representations of *false* and *true*, respectively.

Let $\mathcal{N}_0^n = \{\{0\}, \{1\}, \ldots, \{n-1\}\}$ and let $\mathcal{N}_{k+1}^n = \mathcal{N}_k^n \rightarrow \mathcal{B}$. Clearly, each $\mathcal{N}_k^n$ has exactly 2_k^n elements which we will use to represent numbers in the range $0, \ldots, 2_k^n - 1$. We have also $\mathcal{N}_k^n \subseteq [\![N_k]\!]$ where $N_0 = \text{Prop}$ and $N_{k+1} = N_k \rightarrow \text{Prop}$. Note that the elements of $\mathcal{N}_{k+1}^n = \mathcal{N}_k^n \rightarrow \mathcal{B}$ can be equivalently viewed as predicates over $\mathcal{N}_k^n$, subsets of $\mathcal{N}_k^n$ or binary words of length 2_k^n.

For a number $j \in \{0, 1, \ldots, 2_k^n - 1\}$ let $[\![j]\!]^k$ be the element of $\mathcal{N}_k^n$ representing j, defined inductively by $[\![j]\!]^0 = \{j\}$ and

$$[\![j]\!]^k(x) := \begin{cases} [\![\text{tt}]\!] & \text{if } x = [\![i]\!]^{k-1} \text{ and } j_i = 1, \\ [\![\text{ff}]\!] & \text{o.w.} \end{cases}$$

where j_i is the i-th bit in the binary representation of j. For a binary word w of length 2_k^n let $[\![w]\!]^k := [\![\sum_i w_i \cdot 2^i]\!]^{k+1}$.

The possibility of a compact encoding of FO logic in HFL relies on the existence of polynomial size HFL formulas describing basic operations on numbers represented as elements of $\mathcal{N}_k^n$. We define

$$\text{inc}_k : N_k \rightarrow N_k \ , \quad \text{eq}_k : N_k \rightarrow N_k \rightarrow \text{Prop} \ , \quad \text{search}_k : (N_k \rightarrow \text{Prop}) \rightarrow \text{Prop}$$

adhering to the following specifications

$$[\![\text{inc}_k]\!] \, [\![j]\!]^k = [\![j+1]\!]^k \ , \quad [\![\text{eq}_k]\!] \, [\![j]\!]^k \, [\![i]\!]^k = \begin{cases} [\![\text{tt}]\!] \ , & \text{iff } j = i \\ [\![\text{ff}]\!] \ , & \text{o.w.} \end{cases}$$

For a predicate $p \in \mathcal{N}_{k+1}^n$,

$$[\![\text{search}_k]\!] \, p = \begin{cases} [\![\text{tt}]\!] & \text{iff exists } x \in \mathcal{N}_k^n \text{ s.t. } p(x) = [\![\text{tt}]\!] \\ [\![\text{ff}]\!] & \text{o.w.} \end{cases}$$

The search function search_k can be implemented using inc_k and recursion. A helper function $\text{search}'_k \, P \, x$ applies P to the successive numbers, starting from x, taking the union of the results.

$$\text{search}_k := \lambda(P : N_k \to \text{Prop}).\text{search}'_k \ P \perp^k \ ,$$
$$\text{search}'_k := \lambda(P : N_k \to \text{Prop}).\ \mu(Z : N_k \to \text{Prop}).$$
$$\lambda(X : N_k).\ (P \ X) \lor (Z \ (\text{inc}_k \ X)) \ .$$

Formula $\perp^k : N_k$ represents 0 and is defined as

$$\perp^0 := [-] \ \text{ff} \ , \qquad \perp^k := \lambda(X : N_{k-1}).\ \text{ff} \quad \text{for } k > 0 \ .$$

Functions eq_k and inc_k are defined by induction on k. For $k = 0$ we set

$$\text{eq}_0 := \lambda X.\lambda Y.(X \leftrightarrow Y) \ , \qquad \text{inc}_0 := \lambda X.\langle - \rangle X \ .$$

For $k > 0$, function eq_k is implemented by searching for an argument at which two number representations differ:

$$\text{eq}_k := \lambda(X : N_k).\lambda(Y : N_k).\ \neg\bigl(\text{search}_{k-1} \ \lambda(I : N_{k-1}).\neg(X \ I \leftrightarrow Y \ I)\bigr) \ .$$

Function inc_k is the usual incrementation of a number in binary representation. The helper function $\text{inc}'_k \ x \ i$ adds one to the i-th bit of n and possibly the following bits if the carry-over occurs.

$$\text{inc}_k := \lambda(X : N_k).\ \text{inc}'_k \ X \perp^{k-1} \ .$$

The value of $[\![\text{inc}'_k]\!] \ [\![x]\!]^k \ [\![i]\!]^{k-1}$ is a function which for each j returns the j-th bit of $x + 2^i$ (encoded as $[\![\text{tt}]\!]$ or $[\![\text{ff}]\!]$). For $j = i$ the corresponding bit is $\neg x_i$. If there is no carry-over ($x_i = 0$) then the remaining bits are unchanged. Otherwise the remaining bits are the same as in $x + 2^{i+1}$.

$$\text{inc}'_k := \lambda(X : N_k).\ \mu(Z : N_{k-1} \to N_k).\ \lambda(I : N_{k-1}).$$
$$\lambda(J : N_{k-1}).\ \text{if } (\text{eq}_{k-1} \ J \ I)$$
$$(\neg(X \ I))$$
$$\bigl(\text{if } \neg(X \ I) \ (X \ J) \ (Z \ (\text{inc}_{k-1} \ I) \ J)\bigr)$$

where $\text{if} := \lambda P.\lambda Q.\lambda R.\ (P \land Q) \lor (\neg P \land R)$.

Note that the lengths of inc_k, eq_k and search_k as strings can be exponential in k. However, the number of their subformulas is only polynomial in k.

Lemma 1. *For any $k \geq 0$, any $i \in \{0, \ldots, 2^n_k - 1\}$, and any $p \in N^n_k \to \mathcal{B}$ we have:* $[\![\text{search}'_k]\!] \ p \ [\![i]\!]^k = [\![\text{tt}]\!]$ *iff* $p([\![j]\!]^k) = [\![\text{tt}]\!]$ *for some $i \leq j < 2^n_k$.*

Proof. Simply because $[\![\text{search}'_k]\!] \ p \ [\![i]\!]^k \equiv \bigcup_{j=i}^{2^n_k - 1} p([\![j]\!]^k)$. □

Proposition 1. *For any k, $\text{eq}_k, \text{inc}_k \in \text{HFL}^{k+1}$ and $\text{search}_k, \text{search}'_k \in \text{HFL}^{k+2}$.*

Let φ be an FO sentence. For given k we translate φ into an HFL^{k+2} formula $tr_k(\varphi) : N_{k+1} \to \text{Prop}$ s.t. for any word w of length 2^n_k, w is a model of φ iff $(\mathcal{T}_n[\![tr_k(\varphi)]\!]) \ (\mathcal{T}_n[\![w]\!]^k) = \mathcal{T}_n[\![\text{tt}]\!]$.

$$tr_k(x \leq y) := \lambda(w : N_{k+1}).\, \text{search}'_k \, x \, (\lambda(u : N_k).\, \text{eq}_k \, u \, y)\,,$$
$$tr_k(P_0(x)) := \lambda(w : N_{k+1}).\, \neg(w \, x)\,,$$
$$tr_k(P_1(x)) := \lambda(w : N_{k+1}).\, (w \, x)\,,$$
$$tr_k(\exists x.\varphi) := \lambda(w : N_{k+1}).\, \text{search}_k \, (\lambda(x : N_k).\, tr_k(\varphi) \, w)\,,$$
$$tr_k(\neg\varphi) := \lambda(w : N_{k+1}).\, \neg(tr_k(\varphi) \, w)\,,$$
$$tr_k(\varphi \vee \psi) := \lambda(w : N_{k+1}).\, (tr_k(\varphi) \, w) \vee (tr_k(\psi) \, w)\,.$$

Conjunctions and universal quantifiers can be translated using negation and the above formulas. Note that free variables of φ become free variables of type N_k and variance 0 in $tr_k(\varphi)$.

Lemma 2. *For any FO sentence φ the translation $tr_k(\varphi)$ is a predicate. That is, $\mathcal{T}_n[\![tr_k(\varphi)]\!]$ is an element of $\mathcal{N}^n_{k+2}$ – a function which returns either $\mathcal{T}_n[\![\text{tt}]\!]$ or $\mathcal{T}_n[\![\text{ff}]\!]$ when applied to an argument from $\mathcal{N}^n_{k+1}$.*

Proof. This follows from the fact that, by their specifications, search_k and search'_k are predicates. Hence $tr_k(\varphi)$ is a predicate as a Boolean combination of predicates.

Proposition 2. *For any k and φ, $tr_k(\varphi) \in \text{HFL}^{k+2}$.*

Lemma 3. *Let φ be an FO formula with variables $x_1, \ldots, x_l$. For any word w of length 2^n_k and FO-environment η we have*

$$w \models_\eta \varphi \quad \text{iff} \quad (\mathcal{T}_n[\![tr_k(\varphi)]\!]\rho)(\mathcal{T}_n[\![w]\!]^k) = \mathcal{T}_n[\![\text{tt}]\!]$$

where ρ is an HFL environment given by $\rho(x_i) = \mathcal{T}_n[\![\eta(x_i)]\!]^k$.

Proof. By induction on the structure of φ. We fix n and k and as before omit $\mathcal{T}_n$ in front of semantic brackets.

Case $\varphi = x_i \leq x_j$: Then $[\![tr_k(\varphi)]\!]\rho = [\![\text{search}'_k]\!] \, [\![a]\!]^k \, p$ where $a = \eta(x_i)$, $b = \eta(x_j)$ and predicate p is given by $p(x) = [\![\text{eq}_k]\!] \, x \, [\![b]\!]^k$. We have $w \models_\eta \varphi$ iff $a \leq b$ iff exists $a \leq c < 2^n_k$ s.t. $p([\![c]\!]^k) = [\![\text{tt}]\!]$ iff $[\![\text{search}'_k]\!] \, [\![a]\!]^k \, p = [\![\text{tt}]\!]$, by Lemma 1.

Case $\varphi = \exists x.\psi$: Then $[\![tr_k(\varphi)]\!]\rho = [\![\text{search}_k]\!] \, p$ where $p([\![i]\!]^k) = [\![tr_k(\psi)]\!]\rho[x \mapsto [\![i]\!]^k] \, [\![w]\!]^k$. By the specification of search_k and induction hypothesis we have $([\![tr_k(\varphi)]\!]\rho) \, [\![w]\!]^k = [\![\text{tt}]\!]$ iff $p([\![i]\!]^k) = [\![\text{tt}]\!]$ for some i iff $w \models_{\eta[x \mapsto i]} \psi$ iff $w \models_\eta \varphi$.

Case $\varphi = P_0(x_i)$: Then $w \models_\eta \varphi$ iff $w_{\eta(x_i)} = 0$ iff $[\![w]\!]^k \, [\![\eta(x_i)]\!]^k = [\![\text{ff}]\!]$ iff $([\![tr_k(\varphi)]\!]\rho) \, [\![w]\!]^k = [\![\text{tt}]\!]$.

All other cases are either analogous or follow immediately from the induction hypothesis when negation is used in formulas. □

Lemma 4. *The satisfiability problem for FO sentences in Σ_k is polynomially reducible to the model checking problem for HFL^{k+2}.*

Proof. First note that we can restrict ourselves to the satisfiability problem for FO^k over the binary alphabet $\{0, 1\}$ because any other alphabet can be encoded in it at a logarithmic expense only.

Given a Σ_k formula φ of length n we can construct in polynomial time and space an instance of the HFL^{k+2} model checking problem consisting of a formula $\varphi' = \mathsf{search}_k\ tr_{k-1}(\varphi)$, transition system $\mathcal{T}_n$ and the set $\mathcal{T}_n[\![\mathsf{tt}]\!] = \{0, 1, \ldots, n-1\}$. Note that $\varphi' \in \mathrm{HFL}^{k+2}$ by Propositions 1 and 2. From Lemmas 1, 2 and 3 it follows that $\mathcal{T}_n[\![\varphi']\!] = \mathcal{T}_n[\![\mathsf{tt}]\!]$ iff φ has a model of size $\mathbf{2}_{k-1}^n$ which, by Theorem 1, is equivalent to φ having a model. □

Theorem 2 together with the reduction of Lemma 4 yields the following result.

Theorem 3. *The model checking problem for* HFL^{k+3} *is hard for* kExpSpace *under polynomial time reductions.*

Corollary 1. *The model checking problem for* HFL *is not in* ElTime.

Note that the reduction only uses modal formulas $\langle - \rangle \varphi$ and $[-]\varphi$ because $\mathbf{2}_k^n$ is an upper bound on a minimal model for an FO sentence in Σ_{k+1} of length n. However, $\mathbf{2}_k^n = \mathbf{2}_{k+\log^* n}^1$. This enables us to use modality-free formulas and transition systems of fixed size 1 in the reduction. The price to pay is that, in order to achieve kExpSpace-hardness, one needs formulas with unrestricted types. But it shows that HFL model checking is not in ElTime for fixed and arbitrarily small transition systems already.

4 The Upper Complexity Bound

In the following we will identify a type τ and its underlying complete lattice induced by a transition system $\mathcal{T}$ with state set $\mathcal{S}$. In order to simplify notation we fix $\mathcal{T}$ for the remainder of this section.

Suppose $|\mathcal{S}| = n$ for some $n \in \mathbb{N}$. We define the size $\#(\tau)$ of an HFL type τ, as well as $rp(\tau)$ – a space measure for a representation of one of its elements.

$$\#(\mathsf{Prop}) := 2^n, \qquad \#(\sigma \to \tau) := \#(\tau)^{\#(\sigma)},$$

$$rp(\mathsf{Prop}) := n, \qquad rp(\sigma \to \tau) := \#(\sigma) \cdot rp(\tau).$$

Lemma 5. *For all* HFL *types τ we have:*
(a) *There are only $\#(\tau)$ many different elements of τ.*
(b) *An element x of τ can be represented using space $O(rp(\tau))$.*
(c) $\#(\tau) \leq 2_{ord(\tau)+1}^{O(n)}.$
(d) $rp(\tau) \leq 2_{ord(\tau)}^{O(n)}.$

Proof. Part (a) is standard. Part (b) is proved by induction on the structure of τ. The claim is easily seen to be true for $\tau = \mathsf{Prop}$. Let $\tau = \tau_1 \to \tau_2$, i.e. any element of τ is a function. Such a function can be represented by cascading tables where a table for type Prop consists of one entry only. For $\tau_1 \to \tau_2$ the table must contain for every element of τ_1 one entry of τ_2. Since $\#(\tau_1)$ is finite, we can assume to have a standard enumeration for all elements of τ_1. This enumeration can be used to determine the order in which the function values are stored in the table. With such an enumeration at hand, one does not need to write down the function arguments, and by (a) and the induction hypothesis the overall space needed is $O(rp(\tau_1 \to \tau_2))$.

We also prove (c) by induction on the structure of τ. For $\tau =$ Prop this is true. Let $\tau = \tau_1 \to \tau_2$. Then we have

$$\#(\tau) = \#(\tau_2)^{\#(\tau_1)} \leq (2^{O(n)}_{ord(\tau_2)+1})^{2^{O(n)}_{ord(\tau_1)+1}} = 2^{2^{O(n)}_{ord(\tau_2)} \cdot 2^{O(n)}_{ord(\tau_1)+1}}$$

$$\leq 2^{2^{2 \cdot O(n)}_{\max\{ord(\tau_2), ord(\tau_1)+1\}}} = 2^{O(n)}_{ord(\tau)+1}.$$

Again, the claim in (d) is easily seen to be true for $\tau =$ Prop. Let $\tau = \tau_1 \to \tau_2$, i.e. $ord(\tau) \geq 1$. Note that $ord(\tau_1) \leq ord(\tau) - 1$, and $ord(\tau_2) \leq ord(\tau)$. Then

$$rp(\tau) = 2^{O(n)}_{ord(\tau_1)+1} \cdot 2^{O(n)}_{ord(\tau_2)} \leq 2^{O(n)}_{ord(\tau)} \cdot 2^{O(n)}_{ord(\tau)} \leq 2^{O(n)}_{ord(\tau)}$$

by (c) and the hypothesis. □

Theorem 4. *For all $k \in \mathbb{N}$, the model checking problem for HFL^k and finite transition systems is in $(k+1)$EXPTIME.*

Proof. For a finite transition system $\mathcal{T} = (\mathcal{S}, \{\xrightarrow{a} \mid a \in \mathcal{A}\}, L)$ with $|\mathcal{S}| = n$ and an HFL formula φ of type τ, we describe an alternating procedure for finding the denotation of φ. The existential player $\exists$ proposes an element of τ as $[\![\varphi]\!]$ and the universal player $\forall$ challenges her choice. The game proceeds along the structure of φ in the following way.

If $\varphi = \lambda(X : \sigma).\psi$ then $\forall$ chooses an entry in the table written by $\exists$ as a value of φ. The row in which the entry is found determines the argument to φ, i.e. a value for X. $\forall$ can now invoke the verification protocol to check that this is the correct value of $\psi(X)$ when X has the value given by the entry.

For $\varphi = \psi_1 \psi_2$, first player $\exists$ has to provide a table for ψ_1 and a denotation for ψ_2. The player $\forall$ can either check that the value in ψ_1's table in the row corresponding to ψ_2 is the previously guessed value for φ. Or he can proceed to challenge the denotation of ψ_2.

To verify a guess x of the value of $\varphi = \mu(X : \tau).\psi$ first $\exists$ writes a table of a function $\lambda(X : \tau).\psi$. Furthermore, she names a row in this table that determines a value for X. If the value in this row is not the same as the value of X she looses because she has not found a fixpoint. Then player $\forall$ can either challenge the whole table as above or name another smaller table row that defines a fixpoint. Note that it is always possible to require the entries in a table of type τ to respect the order $\leq_\tau$.

In all other cases φ is of type Prop and its value is a bit vector of length n. Correctness of Boolean operations can be easily verified by $\forall$ using additional $O(n)$ space for storing the values of the operands.

Clearly the space needed to perform the above procedure is bounded by the maximal $rp(\tau)$ where τ is a type of a subformula of φ. This includes using the enumeration function to find corresponding table rows. It is fair to assume that in order to enumerate $2_k^{O(n)}$ elements one does not need more space than $2_k^{O(n)}$. Hence, by Lemma 5 the model checking problem is in alternating $2_k^{O(n)}$-space which equals $(k+1)$ExpTime [5]. □

5 Conclusion

We have shown that the model checking problem for HFL is hard for every kExpTime and consequently of non-elementary complexity. It is tempting to dismiss HFL as a specification formalism of any practical use. But the same argument would also rule out any practical implementation of a satisfiability checker for Monadic Second Order Logic over words or trees (MSO) since this problem has non-elementary complexity, too. However, the verification tool Mona [14] shows that in many cases satisfiability of MSO formulas can be checked efficiently. This is mainly because of the use of efficiently manipulable data structures like BDDs [2], and the fact that formulas used in practical cases do not coincide with those that witness the high complexity.

Thus, the theoretical complexity bounds proved in this paper need to be seen as a high alert warning sign for someone building a model checking tool based on HFL. This will certainly require the use of efficient data structures as well as other clever optimizations. However, only such an implementation will be able to judge the use of HFL as a specification formalism properly. Furthermore, Theorems 3 and 4 show that in order to reach high levels in ElTime, one needs formulas of high type order. But such types might not be needed in order to formulate natural correctness properties, cf. Example 1.

It remains to be seen whether the gap between $(k-3)$ExpSpace-hardness and inclusion in $(k+1)$ExpTime can be reduced and finally closed.

Acknowledgments. We would like to thank Martin Leucker for discussing HFL's model checking problem and an anonymous referee for pointing out Statman's result on the typed λ-calculus.

References

1. Y. Bar-Hillel, M. Perles, and E. Shamir. On formal properties of simple phrase structure grammars. *Zeitschrift für Phonologie, Sprachwissenschaft und Kommunikationsforschung*, 14:113–124, 1961.
2. R. E. Bryant. Graph-based algorithms for boolean function manipulation. *IEEE Transactions on Computers*, 35(8):677–691, August 1986.
3. J. R. Büchi. On a decision method in restricted second order arithmetic. In *Proc. Congress on Logic, Method, and Philosophy of Science*, pages 1–12, Stanford, CA, USA, 1962. Stanford University Press.
4. J. R. Burch, E. M. Clarke, K. L. McMillan, D. L. Dill, and L. J. Hwang. Symbolic model checking: 10^{20} states and beyond. *Information and Computation*, 98(2):142–170, June 1992.
5. A. K. Chandra, D. C. Kozen, and L. J. Stockmeyer. Alternation. *Journal of the ACM*, 28(1):114–133, January 1981.

6. E. M. Clarke, A. Biere, R. Raimi, and Y. Zhu. Bounded model checking using satisfiability solving. *Formal Methods in System Design*, 19(1):7–34, 2001.
7. A. Dawar, E. Grädel, and S. Kreutzer. Inflationary fixed points in modal logic. In L. Fribourg, editor, *Proc. 15th Workshop on Computer Science Logic, CSL'01*, LNCS, pages 277–291, Paris, France, September 2001. Springer.
8. E. A. Emerson. Uniform inevitability is tree automaton ineffable. *Information Processing Letters*, 24(2):77–79, January 1987.
9. E. A. Emerson and J. Y. Halpern. Decision procedures and expressiveness in the temporal logic of branching time. *Journal of Computer and System Sciences*, 30:1–24, 1985.
10. E. A. Emerson and J. Y. Halpern. "Sometimes" and "not never" revisited: On branching versus linear time temporal logic. *Journal of the ACM*, 33(1):151–178, January 1986.
11. E. A. Emerson and C. S. Jutla. Tree automata, μ-calculus and determinacy. In *Proc. 32nd Symp. on Foundations of Computer Science*, pages 368–377, San Juan, Puerto Rico, October 1991. IEEE.
12. M. J. Fischer and R. E. Ladner. Propositional dynamic logic of regular programs. *Journal of Computer and System Sciences*, 18(2):194–211, April 1979.
13. D. Harel, A. Pnueli, and J. Stavi. Propositional dynamic logic of nonregular programs. *Journal of Computer and System Sciences*, 26(2):222–243, April 1983.
14. J. G. Henriksen, J. Jensen, M. Joergensen, and N. Klarlund. MONA: Monadic second-order logic in practice. *LNCS*, 1019:89–110, 1995.
15. D. Janin and I. Walukiewicz. On the expressive completeness of the propositional μ-calculus with respect to monadic second order logic. In U. Montanari and V. Sassone, editors, *Proc. 7th Conf. on Concurrency Theory, CONCUR'96*, volume 1119 of *LNCS*, pages 263–277, Pisa, Italy, August 1996. Springer.
16. D. Kozen. Results on the propositional μ-calculus. *TCS*, 27:333–354, December 1983.
17. M. Lange and C. Stirling. Model checking fixed point logic with chop. In M. Nielsen and U. H. Engberg, editors, *Proc. 5th Conf. on Foundations of Software Science and Computation Structures, FOSSACS'02*, volume 2303 of *LNCS*, pages 250–263, Grenoble, France, April 2002. Springer.
18. M. Müller-Olm. A modal fixpoint logic with chop. In C. Meinel and S. Tison, editors, *Proc. 16th Symp. on Theoretical Aspects of Computer Science, STACS'99*, volume 1563 of *LNCS*, pages 510–520, Trier, Germany, 1999. Springer.
19. A. Pnueli. The temporal logic of programs. In *Proc. 18th Symp. on Foundations of Computer Science, FOCS'77*, pages 46–57, Providence, RI, USA, October 1977. IEEE.
20. K. Reinhardt. The complexity of translating logic to finite automata. In E. Grädel, W. Thomas, and Th. Wilke, editors, *Automata, Languages, and Infinite Games*, volume 2500 of *LNCS*, pages 231–238. Springer, 2002.
21. R. Statman. The typed λ-calculus is not elementary recursive. *Theoretical Computer Science*, 9:73–81, 1979.
22. L. Stockmeyer. *The Computational Complexity of Word Problems*. PhD thesis, MIT, 1974.
23. M. Viswanathan and R. Viswanathan. A higher order modal fixed point logic. In Ph. Gardner and N. Yoshida, editors, *Proc. 15th Int. Conf. on Concurrency Theory, CONCUR'04*, volume 3170 of *LNCS*, pages 512–528, London, UK, 2004. Springer.

An Efficient Algorithm for Computing Optimal Discrete Voltage Schedules

Minming Li[1,*] and Frances F. Yao[2]

[1] State Key Laboratory of Intelligent Technology and Systems,
Dept. of Computer Science and Technology, Tsinghua Univ., Beijing, China
liminming98@mails.tsinghua.edu.cn
[2] Department of Computer Science, City University of Hong Kong
csfyao@cityu.edu.hk

Abstract. We consider the problem of job scheduling on a variable voltage processor with d discrete voltage/speed levels. We give an algorithm which constructs a minimum energy schedule for n jobs in $O(dn \log n)$ time. Previous approaches solve this problem by first computing the optimal continuous solution in $O(n^3)$ time and then adjusting the speed to discrete levels. In our approach, the optimal discrete solution is characterized and computed directly from the inputs. We also show that $O(n \log n)$ time is required, hence the algorithm is optimal for fixed d.

1 Introduction

Advances in processor, memory, and communication technologies have enabled the development and widespread use of portable electronic devices. As such devices are typically powered by batteries, energy efficiency has become an important issue. With dynamic voltage scaling techniques (DVS), processors are able to operate at a range of voltages and frequencies. Since energy consumption is at least a quadratic function of the supply voltage (hence CPU speed), it saves energy to execute jobs as slowly as possible while still satisfying all timing constraints.

We refer to the associated scheduling problem as min-energy DVS scheduling problem (or DVS problem for short); the precise formulation will be given in Section 2. The problem is different from classical scheduling on fixed-speed processors, and it has received much attention from both theoretical and engineering communities in recent years. One of the earliest theoretical models for DVS was introduced in [1]. They gave a characterization of the min-energy DVS schedule and an $O(n^3)$ algorithm [1] for computing it . No special assumption was made on the power consumption function except convexity. This optimal schedule has

---

* This work is supported by National Natural Science Foundation of China (60135010), National Natural Science Foundation of China (60321002) and the Chinese National Key Foundation Research & Development Plan (2004CB318108).
[1] The complexity of the algorithm was said to be further reducible in [1], but that claim has since been withdrawn.

An Efficient Algorithm for Computing Optimal Discrete Voltage Schedules   653

been referenced widely, since it provides a main benchmark for evaluating other scheduling algorithms in both theoretical and simulation work.

In the min-energy DVS schedule mentioned above, the processor must be able to run at *any* real-valued speed s in order to achieve optimality. In practice, variable voltage processors only run at a finite number of speed levels chosen from specific points on the power function curve. For example, the Intel *SpeedStep* technology [2] currently used in Intel's notebooks supports only 3 speed levels, although the new *Foxon* technology will soon enable Intel server chips to run at as many as 64 speed grades. Thus, an accurate model for min-energy scheduling should capture the discrete, rather than continuous, nature of the available speed scale. This consideration has motivated our present work.

In this paper we consider the discrete version of the DVS scheduling problem. Denote by $s_1 > s_2 > \ldots > s_d$ the clock speeds corresponding to d given discrete voltage levels. The goal is to find, under the restriction that only these speeds are available for job execution, a schedule that consumes as little energy as possible. (It is assumed that the highest speed s_1 is fast enough to guarantee a feasible schedule for the given jobs.) This problem was considered in [3] for a single job (i.e. $n = 1$), where they observed that minimum energy is achieved by using the immediate neighbors s_i, s_{i+1} of the ideal speed s in appropriate proportions. It was later extended in [4] to give an optimal discrete schedule for n jobs, obtained by first computing the optimal continuous DVS schedule, and then individually adjusting the speed of each job appropriately to adjacent levels as done in [3].

The question naturally arises: Is it possible to find a direct approach for solving the optimal discrete DVS scheduling problem, without first computing the optimal continuous schedule? We answer the question in the affirmative. For n jobs with arbitrary arrival-time/deadline constraints and d given discrete supply voltages (speeds), we give an algorithm that finds an optimal discrete DVS schedule in $O(dn \log n)$ time. We also show that this complexity is optimal for any fixed d. We remark that the $O(n^3)$ algorithm for finding continuous DVS schedule (cf Section 2) computes the highest speed, 2nd highest speed, etc for execution in a strictly sequential manner, and may use up to n different speeds in the final schedule. Therefore it is unclear a priori how to find shortcuts to solve the discrete problem. Our approach is different from that of [4] which is based on the continuous version and therefore requires $O(n^3)$ time.

Our algorithm for optimal discrete DVS proceeds in two stages. In stage 1, the jobs in J are partitioned into d disjoint groups $J_1, J_2, \ldots, J_d$ where J_i consists of all jobs whose execution speeds in the continuous optimal schedule S_{opt} lie between s_i and s_{i+1}. We show that this multi-level partition can be obtained without determining the exact optimal execution speed of each job. In stage two, we proceed to construct an optimal schedule for each group J_i using two speeds s_i and s_{i+1}. Both the separation of each group J_i in stage 1, and the subsequent scheduling of J_i using two speed levels in stage 2 can be accomplished in time $O(n \log n)$ per group. Hence this two-stage algorithm yields an optimal discrete voltage schedule for J in total time $O(dn \log n)$. The algorithm admits a simple implementation although its proof of correctness and complexity analysis are non-trivial. Aside from its theoretical value, we also expect our algorithm to be useful in generating optimal discrete DVS schedules for simulation purposes as in the continuous case.

We briefly mention some additional theoretical results on DVS, although they are not directly related to the problem considered in this paper. In [1], two on-line heuristics AVR (Average Rate) and OPA (Optimal Available) were introduced for the case that jobs arrive one at a time. AVR was shown to have a competitive ratio of at most 8 in [1]; recently a tight competitive ratio of 4 was proven for OPA in [5]. For jobs with fixed priority, the scheduling problem is shown to be NP-hard and an FPTAS is given in [6]. In addition, [7] gave efficient algorithms for computing the optimal schedule for job sets structured as trees. (Interested reader can find further references from these papers.)

The remainder of the paper is organized as follows. We give the problem formulation and review the optimal continuous schedule in Sections 2. Section 3 discusses some mathematical properties associated with EDF (earliest deadline first) scheduling under different speeds. Section 4 and Section 5 give details of the two stages of the algorithm as outlined above. The combined algorithm and a lower bound are presented in Section 6. Finally some concluding remarks are given in Section 7. Due to the page limit, many of the proofs are omitted in this version.

2 Problem Formulation

Each job j_k in a job set J over $[0,1]$ is characterized by three parameters: arrival time a_k, deadline b_k and required number of CPU cycles R_k. A schedule S for J is a pair of functions $(s(t), job(t))$ defining the processor speed and the job being executed at time t. Both functions are piecewise constant with finitely many discontinuities. A *feasible* schedule must give each job its required number of cycles between arrival time and deadline (with perhaps intermittent execution). We assume that the power P, or energy consumed per unit time, is a convex function of the processor speed. The total energy consumed by a schedule S is $E(S) = \int_0^1 P(s(t))dt$. The goal of the min-energy scheduling problem is to find, for any given job set J, a feasible schedule that minimizes $E(S)$. We refer to this problem as *DVS* scheduling (or sometimes *Continuous DVS* scheduling to distinguish it from the discrete version below).

In the discrete version of the problem, we assume d discrete voltage levels are given, enabling the processor to run at d clock speeds $s_1 > s_2 > \ldots > s_d$. The goal is to find a minimum-energy schedule for a job set using only these speeds. We may assume that, in each problem instance, the highest speed s_1 is always fast enough to guarantee a feasible schedule for the given jobs. We refer to this problem as *Discrete DVS* scheduling.

For the continuous DVS scheduling problem, the optimal schedule S_{opt} can be characterized based on the notion of a *critical interval* for J, which is an interval I in which a group of jobs must be scheduled at maximum constant speed $g(I)$ in any optimal schedule for J. The algorithm proceeds by identifying such a critical interval I, scheduling those 'critical' jobs at speed $g(I)$ over I, then constructing a subproblem for the remaining jobs and solving it recursively. The optimal $s(t)$ is in fact unique, whereas $job(t)$ is not always so. The details are given below.

Definition 1. *Define the intensity of an interval $I = [z, z']$ to be $g(I) = \frac{\sum R_j}{z'-z}$ where the sum is taken over all jobs j_ℓ with $[a_\ell, b_\ell] \subseteq [z, z']$.*

The interval $[c, d]$ achieving the maximum $g(I)$ will be the critical interval chosen for the current job set. All jobs $j_\ell \in J$ satisfying $[a_\ell, b_\ell] \subseteq [c, d]$ can be feasibly scheduled at speed $g([c, d])$ by EDF principle. The interval $[c, d]$ is then removed from $[0, 1]$; all remaining intervals $[a_j, b_j]$ are updated (compressed) accordingly and the algorithm recurses. The complete algorithm is give in Algorithm 1.

```
Input: a job set J
Output: Optimal Voltage Schedule S
repeat
    Select I* = [z, z'] with s = max g(I)
    Schedule j_i ∈ J_{I*} at s over I* by Earliest Deadline First policy
    J ← J - J_{I*}
    for all j_k ∈ J do
        if b_k ∈ [z, z'] then
            b_k ← z
        else if b_k ≥ z' then
            b_k ← b_k - (z' - z)
        end if
        Reset arrival times similarly
    end for
until J is empty
```

Algorithm 1: *OS (Optimal Schedule)*

Let $CI_i \subseteq [0, 1]$ be the ith critical interval of J. Denote by Cs_i the execution speed during CI_i, and by CJ_i those jobs executed in CI_i. We take note of a basic property of critical intervals which will be useful in later discussions.

Lemma 1. *A job $j_\ell \in J$ belongs to $\bigcup_{k=1}^{i} CJ_k$ if and only if the interval $[a_\ell, b_\ell]$ of j_ℓ satisfies $[a_\ell, b_\ell] \subseteq \bigcup_{k=1}^{i} CI_k$.*

3 EDF with Variable Speeds

The EDF (earliest deadline first) principle defines an ordering on the jobs according to their deadlines. At any time t, among jobs j_k that are available for execution, that is, j_k satisfying $t \in [a_k, b_k)$ and j_k is not yet finished by t, it is the job with minimum b_k that will be executed during $[t, t + \epsilon]$. The EDF is a natural scheduling principle and many optimal schedules (such as the continuous min-energy schedule described above) in fact conform to it. All schedules considered in the remainder of this paper are EDF schedules. Hence we assume the jobs in $J = \{j_1, \ldots, j_n\}$ are indexed by their deadlines.

We introduce an important tool for solving Discrete DVS scheduling problem: an EDF schedule that runs at some constant speed s (except for periods of idleness).

Definition 2. *An s-schedule for J is a schedule which conforms to EDF principle and uses constant speed s in executing any job of J.*

As long as there are unfinished jobs available at time t, an s-schedule will select a job by EDF principle and execute it at speed s. An s-schedule may contain periods of idleness when there are no jobs available for execution. An s-schedule may also yield an unfeasible schedule for J since the speed constraint may leave some jobs unfinished by deadline.

Definition 3. *In any schedule S, a maximal subinterval of $[0, 1]$ devoted to executing the same job j_k is called an execution interval (for j_k with respect to S). Denote by $I_k(S)$ the collection of all execution intervals for j_k with respect to S. With respect to the s-schedule for J, any execution interval will be called an s-execution interval, and the collection of all s-execution intervals for job j_k will be denoted by I_k^s.*

Notice that for any EDF schedule S, it is always true that $I_i(S) \subseteq [a_i, b_i] - \cup_{k=1}^{i-1} I_k(S)$. For a given J, we observe some interesting monotone relations that exist among the EDF schedules of J with respect to different speed functions. These relations will be exploited by our algorithms later. They may also be of independent interest in studying other types of scheduling problems.

Lemma 2. *Let S_1 and S_2 be two EDF schedules whose speed functions satisfy $s_1(t) > s_2(t)$ for all t whenever S_1 is not idle.*
1) For any t and any job j_k, the workload of j_k executed by time t under S_1 is always no less than that under S_2.
2) $\cup_{k=1}^{i} I_k(S_1) \subseteq \cup_{k=1}^{i} I_k(S_2)$ for any i, $1 \leq i \leq n$.
3) Any job of J that can be finished under S_2 is always finished strictly earlier under S_1.
4) If S_2 is a feasible schedule for J, then so is S_1.

Note that as a special case, Lemma 2 holds when we substitute s_1-schedule and s_2-schedule, with $s_1 > s_2$, for S_1 and S_2 respectively.

Lemma 3. *The s-schedule for J contains at most $2n$ s-execution intervals and can be computed in $O(n)$ time if the arrival times and deadlines are sorted.*

4 Partition of Jobs by Speed Level

We will consider the first stage of the algorithm in this section. Clearly, to obtain an $O(dn \log n)$-time partition of J into d groups corresponding to d speed levels, it suffices to give an $O(n \log n)$ algorithm which can properly separate J into two groups according to any given speed s.

Definition 4. *Given a job set J and any speed s, let $J^{\geq s}$ and $J^{<s}$ denote the subset of J consisting of jobs whose executing speed are $\geq s$ and $< s$ respectively in the (continuous) optimal schedule of J. We refer to the partition $\langle J^{\geq s}, J^{<s} \rangle$ as the s-partition of J.*

Let $T^{\geq s} \subseteq [0,1]$ be the union of all critical intervals CI_i with $Cs_i \geq s$. By Lemma 1, a job i is in $J^{\geq s}$ if and only if its interval $[a_i, b_i] \subseteq T^{\geq s}$. Thus $J^{\geq s}$ is uniquely determined by $T^{\geq s}$ and we can focus on computing $T^{\geq s}$ instead. Let $T^{<s} = [0,1] - T^{\geq s}$ and we refer to $\langle T^{\geq s}, T^{<s} \rangle$ as the s-partition of time for J.

An example of J with 11 jobs is given in Figure 1, together with the optimal speed function $S_{opt}(t)$. The portion of $S_{opt}(t)$ lying above the horizontal line $Y = s$ projects to $T^{\geq s}$ on the time-axis. In general, $T^{\geq s}$ may consist of a number of connected components.

In the remainder of this section, we will show that certain features existing in the s-schedule of J can be used for identifying connected components of $T^{<s}$. This then leads to an efficient algorithm for computing the s-partition of time $\langle T^{\geq s}, T^{<s} \rangle$.

Definition 5. *In the s-schedule for J, we say a deadline b_i is tight if job j_i is either unfinished at time b_i, or it is finished just on time at b_i. An idle interval $g = [t, t']$ in the s-schedule is called a gap.*

Figure 2 depicts the s-schedule for the sample job set J considered in Figure 1. All tight deadlines and gaps have been marked along the time axis. By overlaying the s-partition of time $\langle T^{\geq s}, T^{<s} \rangle$ for J, we notice that 1) tight deadlines only exist in $T^{\geq s}$, and 2) each connected component of $T^{\geq s}$ ends with a tight deadline. We prove below that these properties always hold for any job set.

Lemma 4.

1) Tight deadlines in an s-schedule can only exist in $T^{\geq s}$.
2) The rightmost point of each connected component of $T^{\geq s}$ must be a tight deadline.

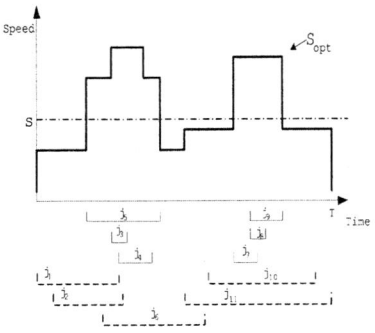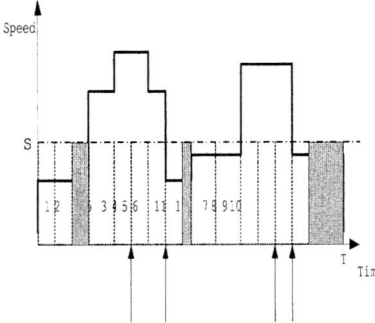

Fig. 1. The s-partition for a sample J. The jobs are represented by their intervals only, and indexed according to deadline. Solid intervals represent jobs belonging to $J^{\geq s}$, while dashed intervals represent jobs belonging to $J^{<s}$.

Fig. 2. The s-execution intervals for the same J in Fig. 1 are illustrated, where the number indicates which job is being executed. Shaded blocks correspond to gaps (idle time), while arrows point to tight deadlines.

Property 2) of Lemma 4 gives a necessary condition for identifying the right boundary of each connected component of $T^{\geq s}$. The corresponding left boundary of such a component can also be identified through left-right symmetry of the scheduling problem with respect to time.

Definition 6. *Given a job set J, the reverse job set J^{rev} consists of jobs with the same workload but time intervals $[1 - b_i, 1 - a_i]$. The s-schedule for J^{rev} is called the reverse s-schedule for J. We call an arrival time a_i (for the original job set J) tight if $1 - a_i$ corresponds to a tight deadline in the reverse s-schedule for J.*

One may also view the reverse s-schedule as a schedule which runs backwards: starting from time 1 and executing jobs of J by the LAF principle (Latest Arrival time First) at constant speed s whenever possible.) Lemma 5 is the symmetric analogue of Lemma 4.

Lemma 5.

1) Tight arrival times in an s-schedule can only exist in $T^{\geq s}$.
2) The leftmost point of each connected component of $T^{\geq s}$ must be a tight arrival time.

Lemmas 4 and 5 are not sufficient by themselves to enable an efficient separation of $T^{\geq s}$ from $T^{<s}$. Fortunately, we have an additional useful property related to $T^{<s}$. Observe that in Figure 2 all gaps of the s-schedule fall within $T^{<s}$. This is in fact true in general and furthermore, a gap must exist in $T^{<s}$.

Lemma 6. *Gaps in an s-schedule can only exist in $T^{<s}$; furthermore a gap must exist in $T^{<s}$.*

Finally we collect the properties that will be used by the partition algorithm in the following theorem. We first give a definition.

Definition 7. *Given a gap $[x, y]$ in an s-schedule, we define the expansion of $[x, y]$ to be the smallest interval $[b, a]$ satisfying 1) $[b, a] \supseteq [x, y]$, and 2) b and a are tight deadline and tight arrival time respectively of the s-schedule. (Note: we adopt the convention that 0 is considered a tight deadline while 1 is considered a tight arrival time.)*

Theorem 1.

1) A gap always exists in an s-schedule if $T^{<s} \neq \emptyset$.
2) The expansion $[b, a]$ of a gap $[x, y]$ defines the connected component in $T^{<s}$ containing $[x, y]$.

Proof. Properties 1) comes from Lemma 6, while Property 2) follows from Lemma 4 and Lemma 5. □

Notice that, although Theorem 1 guarantees that one can always find a gap and then use it to identify a connected component C of $T^{<s}$ (provided $T^{<s} \neq \emptyset$), it is not true that *all* connected component of $T^{<s}$ must contain gaps and can

be identified simultaneously. However, once a component C is found, by deleting the s-execution intervals of all jobs whose interval $[a_i, b_i]$ intersects with C, gaps can surely be found (provided $T^{<s} - C \neq \emptyset$) and the process can continue. This is true because, by reasoning similar to that of Lemma 6, the total workload of the remaining jobs in $J^{<s}$ over $T^{<s} - C$ is less than $s \cdot |T^{<s} - C|$, hence a gap must exist.

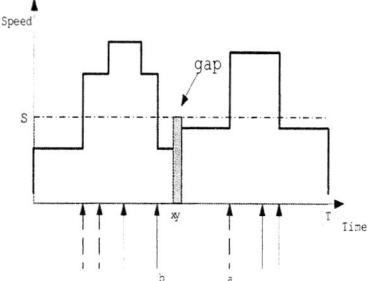

Fig. 3. Gap expansion: the indicated gap will be expanded into $[b, a]$, a connected component of $T^{<s}$

The detailed algorithm for generating the s-partition is given in Algorithm 2 below.

Input: job set J and speed s
Output: s-partition of J
Sort arrival times and deadlines
Generate the s-schedule and reverse s-schedule for J
$J^{\geq s} \leftarrow J$
$J^{<s} \leftarrow \emptyset$
$T^{\geq s} \leftarrow [0, 1]$
$T^{<s} \leftarrow \emptyset$
$Gaps$ = sorted list of gaps in s-schedule
while $Gaps \neq \emptyset$ do
   1. Choose any gap $[x, y]$ from $Gaps$. Find the expansion $[b, a]$ of $[x, y]$.
   2.$J^{<s}_{new}$ = {all jobs in $J^{\geq s}$ whose interval $[a_j, b_j]$ intersects with $[b, a]$ }
   3.$J^{\geq s} \leftarrow J^{\geq s} - J^{<s}_{new}$
   4.$J^{<s} \leftarrow J^{<s} \cup J^{<s}_{new}$
   5.$T^{\geq s} \leftarrow T^{\geq s} - [b, a]$
   6.$T^{<s} \leftarrow T^{<s} \cup [b, a]$
   7.$Gaps$ = $Gaps \cup$ { s-execution intervals of jobs in $J^{<s}_{new}$}
   8.Delete all gaps that are contained in $[b, a]$
end while
Return $J^{<s}$ and $J^{\geq s}$

Algorithm 2: Bi-Partition

Theorem 2. *Algorithm 2 finds the s-partition $\langle J^{\geq s}, J^{<s} \rangle$ for a job set J in $O(n \log n)$ time.*

Proof. The correctness of the algorithm is based on Theorem 1 and the discussions following the theorem. For the complexity part, sorting and generating s-schedules take $O(n \log n)$ time. We now analyze individual steps inside the for loop. For Step 1, finding the expansion of a gap only takes $O(\log n)$ time by binary search; with at most n expansions (to find at most n connected components) the total cost is $O(n \log n)$. Step 2 can be done, with standard data structures such as interval trees, in time $O(\log n) + |J_{new}^{<s}|$ which amounts to total time $O(n \log n)$ since $\sum |J_{new}^{<s}| = O(n)$. It remains to consider the cost of Steps 7 and 8. Since each individual gap is added to and deleted from the sorted list $Gaps$ only once, and there are at most $2n$ s-execution intervals (hence gaps), the cost is at most $O(n \log n)$. This shows that the total running time of Algorithm 2 is $O(n \log n)$. □

We next use Algorithm 2 as a subroutine to obtain Algorithm 3.

Input:
job set J and speed $s_1 > \ldots > s_d > s_{d+1} = 0$
Output:
Partition of J into $J_1, \ldots, J_d$ corresponding to speed levels
  for $i = 1$ to d do
    Obtain $J^{\geq s_i+1}$ from J using Algorithm 2
    $J_i \leftarrow J^{\geq s_i+1}$
    $J \leftarrow J - J_i$
    Update J as in Algorithm 1
  end for

Algorithm 3: Multi-level Partition

Theorem 3. *Algorithm 3 partitions job set J into d subsets corresponding to d speed levels in time $O(dn \log n)$.*

5 Two-Level Schedule

After Algorithm 3 completes the multi-level partition of J into subsets $J_1, \ldots, J_d$, we can proceed to schedule the jobs in each subset J_i with two appropriate speed levels s_i and s_{i+1}. We will present a two-level scheduling algorithm whose complexity is $(n \log n)$ for a set of n jobs. For this purpose, it suffices to describe how to schedule the subset J_1 with two available speeds s_1 and s_2 where $s_1 > s_2 > 0$. We will schedule each connected component of J_1 separately. Thus, the two-level scheduler only deals with 'eligible' input job sets, i.e., those with a continuous optimal schedule speed $s_{opt}(t)$ satisfying $s_1 \geq s_{opt}(t) \geq s_2$ for all t. (Clearly, this condition is satisfied by each connected component of $J_1 = J^{\geq s_2}$ output from Algorithm 3.) We give an alternative and equivalent definition of 'eligibility' in the following. This definition does not make reference to $s_{opt}(t)$ and hence is more useful for the purpose of deriving a two-level schedule directly.

Definition 8. *For a job set J over $[0,1]$, a two-level schedule with speeds s_1 and s_2 (or (s_1, s_2)-schedule for short) for J is a feasible schedule $s(t)$ for J, which is piecewise constant over $[0,1]$ with either $s(t) = s_1$ or $s(t) = s_2$ for any t.*

In other words, an (s_1, s_2)-schedule for J is a schedule using only speeds s_1 and s_2 which finishes every job and leaves no idle time.

Lemma 7. *For a job set J over $[0,1]$, an (s_1, s_2)-schedule exists if and only if*
1) the s_1-schedule for J is a feasible schedule, and
2) the s_2-schedule for J contains no idle time in $[0,1]$.

An Efficient Algorithm for Computing Optimal Discrete Voltage Schedules   661

In view of the preceding lemma, we give the following definition of eligibility for input job sets to two-level scheduling.

Definition 9. *A job set J over $[0,1]$ is said to be eligible for (s_1, s_2)-scheduling if 1) the s_1-schedule for J is a feasible schedule, and 2) the s_2-schedule for J contains no idle time in $[0,1]$.*

We will consider only eligible job sets in discussing two-level scheduling in the remainder of this section. An (s_1, s_2)-schedule for J is said to be *optimal* if it consumes minimum energy among all (s_1, s_2)-schedules for J.

Lemma 8. *All (s_1, s_2)-schedules for an eligible job set J consume the same amount of energy and hence are optimal.*

The two-level schedule as described in the proof of Lemma 7, which first computes the continuous optimal schedule and then rounds the execution speed of each job up and down appropriately [4], requires $O(n^3)$ computation time. We now describe a more efficient algorithm which directly outputs a two-level schedule without first computing the continuous optimal schedule. The algorithm runs in $O(n)$ time if the input jobs are already sorted by deadline (as obtained via Multi-level Partition), and $O(n \log n)$ time in general.

The two-level scheduling algorithm (Algorithm 4) proceeds as follows. It first computes the s_2-schedule for J which in general does not provide a feasible schedule. We then transform it into a feasible schedule by suitably adjusting the execution speed of each job from s_2 to s_1, and possibly extending its execution interval if necessary. These adjustments are done in an orderly and systematic manner to ensure overall feasibility. The algorithm needs to consult the corresponding s_1-schedule of J in making the transformation. An (s_1, s_2)-schedule for J is produced at the end which by lemma 8 is an optimal two-level schedule.

Input:
speeds s_1, s_2 where $s_1 > s_2$
An eligible job set J for (s_1, s_2)-scheduling
Variables:
Committed: the list of allocated time intervals.
Committed(i): the time intervals allocated to job j_i.

Output:
Optimal (s_1, s_2)-schedule for J
  Compute s_1-schedule for J to obtain $I_k^{s_1}$ for $k = 1, \ldots, n$.
  Compute s_2-schedule for J to obtain $I_k^{s_2}$ for $k = 1, \ldots, n$.
  $Committed \leftarrow \emptyset$
  for $i = n$ downto 1 do
    1. $I = I_i^{s_2} - Committed$
    2. Take $I' \subseteq I_i^{s_1}$ of appropriate length (possibly 0) from the right end of $I_i^{s_1}$
      to obtain an (s_1, s_2)-schedule for j_i over $I \cup I'$
    3. $Committed(i) = I \cup I'$
    4. $Committed \leftarrow Committed \cup Committed(i)$
  end for

Algorithm 4: Two-Level Schedule

In the remainder of this section, we consider the correctness and complexity of Algorithm 4.

Let J be an eligible job set for (s_1, s_2)-scheduling. Assume the jobs in J are sorted in increasing order by their deadlines as $j_1, j_2, \ldots, j_n$. After computing the s_1-schedule and s_2-schedule for J, the algorithm then allocates appropriate execution time and speed for each job j_i, in the order $i = n, \ldots, 1$. Step 2 of the for loop carries out the allocation for job j_i. We examine this step in more detail in the following lemma.

Lemma 9. *In step 2 of the for loop, by choosing an appropriate interval $I' \subseteq I_i^{s_1}$ (assuming $I_i^{s_1} \cap Committed = \emptyset$), an (s_1, s_2)-schedule for job j_i over $I \cup I'$ can be found where $I = I_i^{s_2} - Committed$.*

We next show that the assumption $I_i^{s_1} \cap Committed = \emptyset$ in Lemma 9 is indeed satisfied when step 2 is encountered in the i-th iteration (see property 3) below). In fact, we show by induction on i that the following induction hypotheses are maintained by the algorithm at the start of the i-th iteration for $i = n, \ldots, 1$.

Lemma 10. *At the beginning of the i-th iteration of the for loop, the following are true:*
1) $Committed(i+1) \subseteq I_{i+1}^{s_1} \cup I_{i+1}^{s_2}$
2) $\cup_{k=i+1}^{n} I_k^{s_2} \subseteq Committed \subseteq (\cup_{k=i+1}^{n} I_k^{s_1}) \cup (\cup_{k=i+1}^{n} I_k^{s_2})$
3) $Committed \cap (\cup_{k=1}^{i} I_k^{s_1}) = \emptyset$.

Theorem 4. *Given an eligible job set J for (s_1, s_2)-scheduling, Algorithm 4 generates an (s_1, s_2)-schedule for J.*

Proof. Each job j_i is feasibly executed, with no idle time, over $Committed(i)$ at speeds $\{s_1, s_2\}$ as specified in Lemma 9. By the time the algorithm terminates, $Committed = \cup_{k=1}^{n} Committed(k) \supseteq \cup_{k=1}^{n} I_k^{s_2} = [0, 1]$ by Property 2) of Lemma 10. Hence there is no idle time in $[0, 1]$. The resulting schedule thus satisfies the requirements of an (s_1, s_2)-schedule for J. □

Theorem 5. *Algorithm 4 computes an optimal two-level schedule for J in $O(n)$ time if the jobs in J are sorted by deadline (as output by Algorithm 3), and in $O(n \log n)$ time otherwise.*

6 Optimal Discrete Voltage Schedule

Theorem 6. *Algorithm 5 generates a min-energy Discrete DVS schedule with d voltage levels in time $O(dn \log n)$ for n jobs.*

Proof. This is a direct consequence of Theorem 3, Theorem 4 and Theorem 5. □

We next show that the running time of Algorithm 5 is optimal by proving an $\Omega(n \log n)$ lower bound in the algebraic decision tree model.

Theorem 7. *Any deterministic algorithm for computing a min-energy Discrete DVS schedule (MDDVS) with $d \geq 2$ voltage levels will require $\Omega(n \log n)$ time for n jobs.*

```
Input:
  job set J
  speed levels: s_1 > s_2 > ... > s_d > s_{d+1} = 0
Output:
  Optimal Discrete DVS Schedule for J
    Generate J_1, J_2, ..., J_d by Algorithm 3
    for i = 1 to d do
      Schedule jobs in J_i using Algorithm 4 with speeds s_i and s_{i+1}
    end for
    The union of the schedules give the optimal Discrete DVS schedule for J
```

Algorithm 5: Optimal Discrete DVS Schedule

7 Conclusion

In this paper we considered the problem of job scheduling on a variable voltage processor with d discrete voltage/speed levels. We give an algorithm which constructs a minimum energy schedule for n jobs in $O(dn \log n)$ time, which is optimal for fixed d. The min-energy discrete schedule is obtained without first computing the continuous optimal solution. Our algorithm consists of two stages: a multi-level partition of J into d disjoint groups J_i, followed by finding a two-level schedule for each J_i using speeds s_i and s_{i+1}. The individual modules in our algorithm, such as the multi-level partition and two-level scheduling, may be of interest in themselves aside from the main result. Our algorithm admits a simple implementation although its proof of correctness and complexity analysis are non-trivial. We have also discovered some nice fundamental properties associated with EDF scheduling under variable speeds. Some of these properties are stated as lemmas in Section 3 for easy reference. Our results may provide some new insights and tools for the problem of min-energy job scheduling on variable voltage processors. Aside from the theoretical value, we also expect the algorithm to be useful in generating optimal discrete schedules for simulation purposes as in the continuous case.

References

1. F. Yao, A. Demers and S. Shenker, *A Scheduling Model for Reduced CPU Energy*, IEEE Proc. FOCS 1995, 374-382.
2. Intel Corporation, *Wireless Intel SpeedStep Power Manager - Optimizing Power Consumption for the Intel PXA27x Processor Family*, Wireless Intel SpeedStep(R) Power Manager White Paper, 2004.
3. T. Ishihara and H. Yasuura, *Voltage Scheduling Problem for Dynamically Variable Voltage Processors*, ISLPED, 1998.
4. W. Kwon and T. Kim, *Optimal Voltage Allocation Techniques for Dynamically Variable Voltage Processors*, 40th Design Automation Conference, 2003.
5. N. Bansal, T. Kimbrel and K. Pruhs, *Dynamic Speed Scaling to Manage Energy and Temperature*, IEEE Proc. FOCS 2004, 520-529.
6. H. S. Yun and J. Kim, *On Energy-Optimal Voltage Scheduling for Fixed-Priority Hard Real-Time Systems*, ACM Trans. Embedded Comput. Syst. 2(3):393-430, 2003.
7. M. Li, J. B. Liu and F. F. Yao, *Min-Energy Voltage Allocation for Tree-Structured Tasks*, to appear in COCOON 2005.
8. A. C. Yao, *Lower Bounds for Algebraic Computation Trees with Integer Inputs*, SIAM J. Comput. 20(1991):308-313.

Inverse Monoids: Decidability and Complexity of Algebraic Questions

Markus Lohrey and Nicole Ondrusch

Universität Stuttgart, FMI, Germany
{lohrey, ondrusch}@informatik.uni-stuttgart.de

Abstract. The word problem for inverse monoids generated by a set Γ subject to relations of the form $e = f$, where e and f are both idempotents in the free inverse monoid generated by Γ, is investigated. It is shown that for every fixed monoid of this form the word problem can be solved in polynomial time which solves an open problem of Margolis and Meakin. For the uniform word problem, where the presentation is part of the input, EXPTIME-completeness is shown. For the Cayley-graphs of these monoids, it is shown that the first-order theory with regular path predicates is decidable. Regular path predicates allow to state that there is a path from a node x to a node y that is labeled with a word from some regular language. As a corollary, the decidability of the generalized word problem is deduced. Finally, it is shown that the Cayley-graph of the free inverse monoid has an undecidable monadic second-order theory.

1 Introduction

The decidability and complexity of algebraic questions in various kinds of structures is a classical topic at the borderline of computer science and mathematics. The most basic algorithmic question concerning algebraic structures is the word problem, which asks whether two given expressions denote the same element of the underlying structure. Markov and Post proved independently that the word problem for finitely presented monoids is undecidable in general. This result can be seen as one of the first undecidability results that touched real mathematics. Later, Novikov and Boone extended the result of Markov and Post to finitely presented groups, see [9] for references.

In this paper, we are interested in a class of monoids that lies somewhere between groups and general monoids: inverse monoids [15]. In the same way as groups can be represented by sets of permutations, inverse monoids can be represented by sets of partial injections [15]. Algorithmic questions for inverse monoids received increasing attention in the past, and inverse monoid theory found several applications in combinatorial group theory, see e.g. the survey [11]. In [10], Margolis and Meakin presented a large class of finitely presented inverse monoids with decidable word problems. An inverse monoid from that class is of the form $\mathrm{FIM}(\Gamma)/P$, where $\mathrm{FIM}(\Gamma)$ is the free inverse monoid generated by the set Γ and P is a presentation consisting of a finite number of identities between idempotents of $\mathrm{FIM}(\Gamma)$; we call such a presentation idempotent. In fact, in [10] it is shown that even the uniform word problem for idempotent presentations is decidable. In this problem, also the presentation is part of the input.

The decidability proof of Margolis and Meakin uses Rabin's seminal tree theorem [16], concerning the decidability of the monadic second-order theory of the complete binary tree. From the view point of complexity, the use of Rabin's tree theorem is somewhat unsatisfactory, because it leads to a nonelementary algorithm for the word problem, i.e., the running time is not bounded by an exponent tower of fixed height. Therefore, in [1,10] the question for a more efficient approach was asked. In Section 6 we show by using tree automata techniques that for every fixed idempotent presentation the word problem for $\text{FIM}(\Gamma)/P$ can be solved in polynomial time. For the uniform word problem for idempotent presentations we prove completeness for EXPTIME (deterministic exponential time). Similarly to the method of Margolis and Meakin, we use results from logic for the upper bound. But instead of translating the uniform word problem into monadic second-order logic over the complete binary tree, we exploit a translation into the modal μ-calculus, which is a popular logic for the verification of reactive systems. Then, we can use a result from [5,19] stating that the model-checking problem of the modal μ-calculus over context-free graphs [13] is EXPTIME-complete.

In Section 7 we study Cayley-graphs of inverse monoids of the form $\text{FIM}(\Gamma)/P$. The Cayley-graph of a finitely generated monoid $\mathcal{M}$ w.r.t. a finite generating set Γ is a Γ-labeled directed graph with node set $\mathcal{M}$ and an a-labeled edge from a node x to a node y if $y = xa$ in $\mathcal{M}$. Cayley-graphs of groups are a fundamental tool in combinatorial group theory [9] and serve as a link to other fields like topology, graph theory, and automata theory, see, e.g., [12,13]. Here we consider Cayley-graphs from a logical point of view, see [6,7] for previous results in this direction. More precisely, we consider an expansion of the Cayley-graph G that contains for every regular language L over the generators of $\mathcal{M}$ a binary predicate reach_L. Two nodes u and v of G are related by reach_L if there exists a path from u to v in the Cayley-graph, which is labeled with a word from the language L. Our main result of Section 7 states that this structure has a decidable first-order theory, whenever the underlying monoid is of the form $\text{FIM}(\Gamma)/P$ for an idempotent presentation P (Theorem 6). An immediate corollary of this result is that the generalized word problem of $\text{FIM}(\Gamma)/P$ is decidable. The generalized word problem asks whether for given elements $w, w_1, \ldots, w_n \in \text{FIM}(\Gamma)/P$, w belongs to the submonoid of $\text{FIM}(\Gamma)/P$ generated by $w_1, \ldots, w_n$. Our decidability result for Cayley-graphs should be also compared with the undecidability result for the existential theory of the free inverse monoid $\text{FIM}(\{a,b\})$ [17], which consists of all true statements over $\text{FIM}(\{a,b\})$ of the form $\exists x_1 \cdots \exists x_m : \varphi$, where φ is a boolean combination of word equations (with constant). Finally, we complement our decidability result for Cayley-graphs by showing that already the Cayley-graph of the free inverse monoid $\text{FIM}(\{a,b\})$ has an undecidable monadic second-order theory (Theorem 7).

Proofs that are omitted in this extended abstract can be found in the full version [8].

2 Preliminaries

For a finite alphabet Γ, we denote with $\Gamma^{-1} = \{a^{-1} \mid a \in \Gamma\}$ a disjoint copy of Γ. For $a^{-1} \in \Gamma^{-1}$ we define $(a^{-1})^{-1} = a$; thus, $^{-1}$ becomes an involution on the alphabet $\Gamma \cup \Gamma^{-1}$. We extend this involution to words from $(\Gamma \cup \Gamma^{-1})^*$ by setting $(b_1 b_2 \cdots b_n)^{-1} = b_n^{-1} \cdots b_2^{-1} b_1^{-1}$, where $b_i \in \Gamma \cup \Gamma^{-1}$. The set of all regular languages over an alphabet Γ is denoted by REG(Γ). We assume that the reader has some basic background in complexity theory. An *alternating Turing-machine* [2] $T = (Q, \Sigma, \delta, q_0, q_f)$ is a nondeterministic Turing-machine (where Q is the state set, Σ is the tape alphabet, δ is the transition relation, q_0 is the initial state, and q_f is the unique accepting state), where the set of nonfinal states $Q \setminus \{q_f\}$ is partitioned into two sets: $Q_\exists$ (existential states) and $Q_\forall$ (universal states). We assume that T cannot make transitions out of the final state q_f. A configuration C with current state q is accepting, if (i) $q = q_f$, or (ii) $q \in Q_\exists$ and there exists a successor configuration of C that is accepting, or (iii) $q \in Q_\forall$ and every successor configuration of C is accepting. An input word w is accepted by T if the corresponding initial configuration is accepting. It is known that EXPTIME (deterministic exponential time) equals APSPACE (the class of all problems that can be accepted by an alternating Turing-machine in polynomial space) [2].

3 Relational Structures and Logic

See any text book on logic for more details on the subject of this section. A signature is a countable set $\mathcal{S}$ of relational symbols, where each relational symbol $R \in \mathcal{S}$ has an associated arity n_R. A (relational) structure over the signature $\mathcal{S}$ is a tuple $\mathcal{A} = (A, (R^\mathcal{A})_{R \in \mathcal{S}})$, where A is a set (the universe of $\mathcal{A}$) and $R^\mathcal{A}$ is a relation of arity n_R over the set A, which interprets the relational symbol R. We will assume that every signature contains the equality symbol $=$ and that $=^\mathcal{A}$ is the identity relation on the set A. As usual, a constant $c \in A$ can be encoded by the unary relation $\{c\}$. Usually, we denote the relation $R^\mathcal{A}$ also with R. For $B \subseteq A$ we define the restriction $\mathcal{A}{\upharpoonright}B = (B, (R^\mathcal{A} \cap B^{n_R})_{R \in \mathcal{S}})$; it is again a structure over the signature $\mathcal{S}$.

Next, let us introduce *monadic second-order logic (MSO-logic)*. Let $\mathbb{V}_1$ (resp. $\mathbb{V}_2$) be a countably infinite set of *first-order variables* (resp. *second-order variables*) which range over elements (resp. subsets) of the universe A. First-order variables (resp. second-order variables) are denoted x, y, z, x', etc. (resp. X, Y, Z, X', etc.). *MSO-formulas* over the signature $\mathcal{S}$ are constructed from the atomic formulas $R(x_1, \ldots, x_{n_R})$ and $x \in X$ (where $R \in \mathcal{S}$, $x_1, \ldots, x_{n_R}, x \in \mathbb{V}_1$, and $X \in \mathbb{V}_2$) using the boolean connectives $\neg, \wedge,$ and $\vee$, and quantifications over variables from $\mathbb{V}_1$ and $\mathbb{V}_2$. The notion of a free occurrence of a variable is defined as usual. A formula without free occurrences of variables is called an *MSO-sentence*. If $\varphi(x_1, \ldots, x_n, X_1, \ldots, X_m)$ is an MSO-formula such that at most the first-order variables among $x_1, \ldots, x_n$ and the second-order variables among $X_1, \ldots, X_m$ occur freely in φ, and $a_1, \ldots, a_n \in A$, $A_1, \ldots, A_m \subseteq A$, then $\mathcal{A} \models \varphi(a_1, \ldots, a_n, A_1, \ldots, A_m)$ means that φ evaluates to true in $\mathcal{A}$ if the free variable x_i (resp. X_j) evaluates to a_i (resp. A_j). The *MSO-theory* of $\mathcal{A}$, denoted by MSOTh($\mathcal{A}$), is the set of all MSO-sentences φ such that $\mathcal{A} \models \varphi$. A *first-order formula* over the signature $\mathcal{S}$ is an MSO-formula that does not contain any occurrences of second-order variables. In particular, first-order formulas do not contain atomic subformulas of the form $x \in X$. The *first-order theory* FOTh($\mathcal{A}$) of $\mathcal{A}$ is the set of all first-order sentences φ such that $\mathcal{A} \models \varphi$.

Several times, we will use implicitly the well-known fact that reachability in graphs can be expressed in MSO. More precisely, there exists an MSO-formula reach(x, y) (over the signature containing a binary relation symbol E) such that for every directed graph $G = (V, E)$ and all nodes $s, t \in V$ we have $G \models \text{reach}(s, t)$ iff $(s, t) \in E^*$. Another important fact is that finiteness of a subset of a finitely-branching tree can be expressed in MSO: There is an MSO-formula $\text{fin}(X)$ (over the signature containing a binary relation symbol E) such that for every finitely-branching (and downward-directed) tree $T = (V, E)$ and all subsets $U \subseteq V$ we have $T \models \text{fin}(U)$ iff U is finite (by König's lemma, U is infinite iff the upward-closure of U contains an infinite path), see also [16, Lemma 1.8].

In Section 6 we will make use of the *modal μ-calculus*, which is a popular logic for the verification of reactive systems. Formulas of this logic are interpreted over edge-labeled directed graphs. Let Σ be a finite set of edge labels. The syntax of the modal μ-calculus is given by the following grammar (we only introduce those operators that are needed later; other operators like $\neg\varphi$ or $[a]\varphi$ are defined as usual): $\varphi ::= \text{true} \mid X \mid \varphi \vee \varphi \mid \varphi \wedge \varphi \mid \langle a \rangle \varphi \mid \mu X.\varphi$. Here $X \in \mathbb{V}_2$ is a second-order variable ranging over sets of nodes and $a \in \Sigma$. Variables from $\mathbb{V}_2$ are bounded by the μ-operator. We define the semantics of the modal μ-calculus w.r.t. an edge-labeled graph $G = (V, (E_a)_{a \in \Sigma})$ ($E_a \subseteq V \times V$ is the set of all a-labeled edges) and a valuation $\sigma : \mathbb{V}_2 \to 2^V$. To each formula φ we assign the set $\varphi^G(\sigma) \subseteq V$ of nodes where φ evaluates to true under the valuation σ. For a valuation σ, a variable $X \in \mathbb{V}_2$, and a set $U \subseteq V$ define $\sigma[U/X]$ as the valuation with $\sigma[U/X](X) = U$ and $\sigma[U/X](Y) = \sigma(Y)$ for $X \neq Y$. Now we can define $\varphi^G(\sigma)$ inductively as follows:

- $\text{true}^G(\sigma) = V$, $X^G(\sigma) = \sigma(X)$ for every $X \in \mathbb{V}_2$,
- $(\varphi \vee \psi)^G(\sigma) = \varphi^G(\sigma) \cup \psi^G(\sigma)$, $(\varphi \wedge \psi)^G(\sigma) = \varphi^G(\sigma) \cap \psi^G(\sigma)$,
- $(\langle a \rangle \varphi)^G(\sigma) = \{u \in V \mid \exists v \in V : (u, v) \in E_a \wedge v \in \varphi^G(\sigma)\}$,
- $(\mu X.\varphi)^G(\sigma)$ is the smallest fixpoint of the monotonic function $U \mapsto \varphi^G(\sigma[U/X])$

Note that only the values of the valuation σ for free variables is important. In particular, if φ is a sentence (i.e., a formula where all variables are bounded by μ-operators), then the valuation σ is not relevant and we can write φ^G instead of $\varphi^G(\sigma)$, where σ is an arbitrary valuation. For a sentence φ and a node $v \in V$ we write $(G, v) \models \varphi$ if $v \in \varphi^G$.

A *context-free graph* [13] is the transition graph of a pushdown automaton, i.e., nodes are the configurations of a given pushdown automaton, and edges are given by the transitions of the automaton. A more formal definition is not necessary for the purpose of this paper. We will only need the following result:

Theorem 1 ([5,19]). *The following problem is in EXPTIME:*
INPUT: A pushdown automaton A defining a context-free graph $G(A)$, a node v of $G(A)$, and a formula φ of the modal μ-calculus
QUESTION: $(G(A), v) \models \varphi$?
Moreover, there exists already a fixed formula φ for which this question is EXPTIME-complete.

4 Word Problems and Cayley-Graphs

Let $\mathcal{M} = (M, \circ, 1)$ be a finitely generated monoid with identity 1 and let Σ be a finite generating set for $\mathcal{M}$, i.e., there exists a surjective monoid homomorphism $h : \Sigma^* \to$

$\mathcal{M}$. The *word problem* for $\mathcal{M}$ w.r.t. Σ is the computational problem that asks for two given words $u, v \in \Sigma^*$, whether $h(u) = h(v)$. It is well-known that if Σ_1 and Σ_2 are two finite generating sets for $\mathcal{M}$, then the word problem for $\mathcal{M}$ w.r.t. Σ_1 is logspace reducible to the word problem for $\mathcal{M}$ w.r.t. Σ_2. Thus, the computational complexity of the word problem does not depend on the underlying set of generators.

The *Cayley-graph* of the monoid $\mathcal{M}$ w.r.t. the generating set Σ is the relational structure $\mathcal{C}(\mathcal{M}, \Sigma) = (M, (\{(u, v) \in M \times M \mid u \circ h(a) = v\})_{a \in \Sigma}, 1)$. It is a rooted directed graph, where every edge has a label from Σ and $\{(u, v) \mid u \circ h(a) = v\}$ is the set of a-labeled edges. Since Σ generates $\mathcal{M}$, every $u \in M$ is reachable from the root 1. Cayley-graphs of groups play an important role in combinatorial group theory [9]. On the other hand, only a few papers deal with Cayley-graphs of monoids. Combinatorial aspects of Cayley-graphs of monoids are studied in [4,20]. In [18], Cayley-graphs of automatic monoids are investigated.

The *free group* $\mathrm{FG}(\Gamma)$ generated by the set Γ is the quotient $(\Gamma \cup \Gamma^{-1})^*/\delta$, where δ is the smallest congruence on $(\Gamma \cup \Gamma^{-1})^*$ that contains all pairs (bb^{-1}, ε) for $b \in \Gamma \cup \Gamma^{-1}$. Let $\gamma : (\Gamma \cup \Gamma^{-1})^* \to \mathrm{FG}(\Gamma)$ denote the canonical morphism mapping a word $u \in (\Gamma \cup \Gamma^{-1})^*$ to the group element represented by u. It is well known that for every $u \in (\Gamma \cup \Gamma^{-1})^*$ there exists a unique word $r(u) \in (\Gamma \cup \Gamma^{-1})^*$ (the reduced normalform of u) such that $\gamma(u) = \gamma(r(u))$ and $r(u)$ does not contain a factor of the form bb^{-1} for $b \in \Gamma \cup \Gamma^{-1}$. The word $r(u)$ can be calculated from u in linear time. It holds $\gamma(u) = \gamma(v)$ iff $r(u) = r(v)$. The Cayley-graph of $\mathrm{FG}(\Gamma)$ w.r.t. the standard generating set $\Gamma \cup \Gamma^{-1}$ will be denoted by $\mathcal{C}(\Gamma)$; it is a finitely-branching tree and a context-free graph [13].

Similarly to the word problem, if Σ_1 and Σ_2 are finite generating sets for the same monoid $\mathcal{M}$, then $\mathrm{FOTh}(\mathcal{C}(\mathcal{M}, \Sigma_1))$ is logspace reducible to $\mathrm{FOTh}(\mathcal{C}(\mathcal{M}, \Sigma_2))$ and the same holds for the MSO-theories, see [7]. It is easy to see that the decidability of the first-order theory of the Cayley-graph implies the decidability of the word problem. On the other hand, there exists a finitely presented monoid for which the word problem is decidable, but the first-order theory of the Cayley-graph is undecidable [7]. When restricting to groups, the situation is different: The Cayley-graph of a finitely generated group has a decidable first-order theory iff the group has a decidable word problem [6]. Moreover, the Cayley-graph of a finitely generated group has a decidable MSO-theory iff the group is virtually free (i.e., has a free subgroup of finite index) [6,13]. We will only need this result for the Cayley-graph $\mathcal{C}(\Gamma)$ of the free group $\mathrm{FG}(\Gamma)$:

Theorem 2 ([13]). *For every finite Γ, $\mathrm{MSOTh}(\mathcal{C}(\Gamma))$ is decidable but nonelementary.*

5 Inverse Monoids

A monoid $\mathcal{M}$ is called an *inverse monoid* if for each $m \in \mathcal{M}$ there is a unique $m^{-1} \in \mathcal{M}$ such that $m = mm^{-1}m$ and $m^{-1} = m^{-1}mm^{-1}$. For detailed reference on inverse monoids see [15]; here we only recall the basic notions. Since the class of inverse monoids forms a variety it follows from universal algebra that *free inverse monoids* exist. The free inverse monoid generated by a set Γ is denoted by $\mathrm{FIM}(\Gamma)$; it is isomorphic to $(\Gamma \cup \Gamma^{-1})^*/\rho$, where ρ is the smallest congruence on the free monoid $(\Gamma \cup \Gamma^{-1})^*$ which contains for all words $v, w \in (\Gamma \cup \Gamma^{-1})^*$ the pairs

$(w, ww^{-1}w)$ and $(ww^{-1}vv^{-1}, vv^{-1}ww^{-1})$ (which are also called the Vagner equations). Let $\alpha : (\Gamma \cup \Gamma^{-1})^* \to \text{FIM}(\Gamma)$ denote the canonical morphism mapping a word $u \in (\Gamma \cup \Gamma^{-1})^*$ to the element of $\text{FIM}(\Gamma)$ represented by u. Obviously, there exists a morphism $\beta : \text{FIM}(\Gamma) \to \text{FG}(\Gamma)$ such that $\gamma = \beta \circ \alpha$. The free inverse monoid $\text{FIM}(\Gamma)$ can be also represented via *Munn trees*: The Munn tree $\text{MT}(u)$ of $u \in (\Gamma \cup \Gamma^{-1})^*$ is $\text{MT}(u) = \{\gamma(v) \in \text{FG}(\Gamma) \mid \exists w \in (\Gamma \cup \Gamma^{-1})^* : u = vw\}$; it is a finite and connected subset of the Cayley-graph $\mathcal{C}(\Gamma)$ of the free group $\text{FG}(\Gamma)$. In other words, $\text{MT}(u)$ is the set of all nodes along the unique path in $\mathcal{C}(\Gamma)$ that starts in 1 and that is labeled with the word u. We identify $\text{MT}(u)$ with the subtree $\mathcal{C}(\Gamma)\restriction_{\text{MT}(u)}$ of $\mathcal{C}(\Gamma)$. Munn's theorem [14] states that $\alpha(u) = \alpha(v)$ for $u, v \in (\Gamma \cup \Gamma^{-1})^*$ iff $r(u) = r(v)$ (i.e., $\gamma(u) = \gamma(v)$) and $\text{MT}(u) = \text{MT}(v)$. It is well known that for a word $u \in (\Gamma \cup \Gamma^{-1})^*$, the element $\alpha(u) \in \text{FIM}(\Gamma)$ is an idempotent element, i.e., $\alpha(uu) = \alpha(u)$, iff $r(u) = \varepsilon$, i.e., $\gamma(u) = 1$.

For a finite set $P \subseteq (\Gamma \cup \Gamma^{-1})^* \times (\Gamma \cup \Gamma^{-1})^*$ define $\text{FIM}(\Gamma)/P = (\Gamma \cup \Gamma^{-1})^*/\tau$ to be the inverse monoid with the set Γ of generators and the set P of relations, where τ is the smallest congruence on $(\Gamma \cup \Gamma^{-1})^*$ generated by $\rho \cup P$. Then the canonical morphism $\mu_P : (\Gamma \cup \Gamma^{-1})^* \to \text{FIM}(\Gamma)/P$ factors as $\mu_P = \nu_P \circ \alpha$ with $\nu_P : \text{FIM}(\Gamma) \to \text{FIM}(\Gamma)/P$. For the rest of the paper, the meaning of the morphisms α, γ, μ_P, and ν_P will be fixed. We say that $P \subseteq (\Gamma \cup \Gamma^{-1})^* \times (\Gamma \cup \Gamma^{-1})^*$ is an *idempotent presentation* if for all $(e, f) \in P$, $\alpha(e)$ and $\alpha(f)$ are both idempotents of $\text{FIM}(\Gamma)$, i.e., $r(e) = r(f) = \varepsilon$. In this paper, we are concerned with inverse monoids of the form $\text{FIM}(\Gamma)/P$ for a finite idempotent presentation P. To solve the word problem for such a monoid, Margolis and Meakin [10] constructed a closure operation for Munn trees. We shortly review the ideas here. As remarked in [10], every idempotent presentation P can be replaced by the presentation $P' = \{(e, ef), (f, ef) \mid (e, f) \in P\}$, i.e., $\text{FIM}(\Gamma)/P \cong \text{FIM}(\Gamma)/P'$. Since $\text{MT}(e) \subseteq \text{MT}(ef) \supseteq \text{MT}(f)$ if $r(e) = r(f) = \varepsilon$, we can restrict in the following to idempotent presentations P such that $\text{MT}(e) \subseteq \text{MT}(f)$ for all $(e, f) \in P$. Let $V \subseteq \text{FG}(\Gamma)$. Define sets $V_i \subseteq \text{FG}(\Gamma)$ ($i \geq 1$) inductively as follows: (i) $V_1 = V$ and (ii) for $n \geq 1$ let

$$V_{n+1} = V_n \cup \bigcup_{(e,f) \in P} \{u \circ v \mid u \in V_n, \forall w \in \text{MT}(e) : u \circ w \in V_n, v \in \text{MT}(f)\},$$

where $\circ$ refers to the multiplication in the free group $\text{FG}(\Gamma)$. Finally, define the closure of V w.r.t. the presentation P as $\text{cl}_P(V) = \bigcup_{n \geq 1} V_n$.

Theorem 3 ([10]). *Let P be an idempotent presentation and let $u, v \in (\Gamma \cup \Gamma^{-1})^*$. Then $\mu_P(u) = \mu_P(v)$ iff $r(u) = r(v)$ (i.e., $\gamma(u) = \gamma(v)$) and $\text{cl}_P(\text{MT}(u)) = \text{cl}_P(\text{MT}(v))$.*

The result of Munn for $\text{FIM}(\Gamma)$ mentioned above is a special case of this result.

Example 1. Let $\Gamma = \{a, b\}$, $u = aa^{-1}bb^{-1}$, and $P = \{(aa^{-1}, a^2a^{-2}), (bb^{-1}, b^2b^{-2})\}$. The Munn trees for the words in the presentation P and u are shown on the right; the bigger circle represents the 1 of $\text{FG}(\Gamma)$. Then $\text{cl}_P(\text{MT}(u)) = \{a^n \mid n \geq 0\} \cup \{b^n \mid n \geq 0\} \subseteq \text{FG}(\Gamma)$.

In the next section, instead of specifying a word $w \in (\Gamma \cup \Gamma^{-1})^*$ (that represents an idempotent in $\mathrm{FIM}(\Gamma)$, i.e., $r(w) = \varepsilon$) explicitly, we will only show its Munn tree, where as above the 1 of $\mathrm{FG}(\Gamma)$ is drawn as a bigger circle. In fact, one can replace w by any word that labels a path from the circle back to the circle and that visits all nodes; the resulting word represents the same element of $\mathrm{FIM}(\Gamma)$ as the original one.

Margolis and Meakin used Theorem 3 in order to decide the word problem for $\mathrm{FIM}(\Gamma)/P$. More precisely, they have shown that from a finite and idempotent presentation P one can effectively construct an MSO-formula $\mathrm{CL}_P(X, Y)$ over the signature of the Cayley-graph $\mathcal{C}(\Gamma)$ such that for all words $u \in (\Gamma \cup \Gamma^{-1})^*$ and all subsets $A \subseteq \mathrm{FG}(\Gamma)$: $\mathcal{C}(\Gamma) \models \mathrm{CL}_P(\mathrm{MT}(u), A)$ iff $A = \mathrm{cl}_P(\mathrm{MT}(u))$. The decidability of the word problem for $\mathrm{FIM}(\Gamma)/P$ is an immediate consequence of Theorem 2 and Theorem 3. But the application of Theorem 2 results in a nonelementary algorithm.

6 Complexity of the Word Problem

Using an efficient translation into tree automata over finite trees, we can prove:

Theorem 4. *For every finite idempotent presentation $P \subseteq (\Gamma \cup \Gamma^{-1})^* \times (\Gamma \cup \Gamma^{-1})^*$ the word problem for $\mathrm{FIM}(\Gamma)/P$ can be solved in deterministic polynomial time.*

In the uniform case, where the presentation P is part of the input, the complexity increases considerably:

Theorem 5. *The following problem is EXPTIME-complete:*
  INPUT: A finite alphabet Γ, words $u, v \in (\Gamma \cup \Gamma^{-1})^*$, and a finite idempotent presentation $P \subseteq (\Gamma \cup \Gamma^{-1})^* \times (\Gamma \cup \Gamma^{-1})^*$
  QUESTION: $\mu_P(u) = \mu_P(v)$?

Proof. For the lower bound we use the fact that EXPTIME equals APSPACE. Thus, let $T = (Q, \Sigma, \delta, q_0, q_f)$ be a fixed alternating Turing machine that accepts an EXPTIME-complete language. Assume that T works in space $p(n)$ for a polynomial p on an input of length n. W.l.o.g. we assume that:

- T alternates in each state, i.e., it either moves from a state of $Q_\exists$ to a state from $Q_\forall \cup \{q_f\}$ or from a state of $Q_\forall$ to a state from $Q_\exists \cup \{q_f\}$.
- The initial state q_0 belongs to $Q_\exists$.
- For each pair $(q, a) \in (Q \setminus \{q_f\}) \times \Sigma$, the machine T has precisely two choices according to δ, which we call choice 1 and choice 2.
- If T terminates in the final state q_f, then the symbol that is currently read by the head is some distinguished symbol $\$ \in \Sigma$.

Define $\Gamma = \Sigma \cup (Q \times \Sigma) \cup \{a_1, a_2, b_1, b_2, \#\}$, where all unions are assumed to be disjoint. A configuration of T is encoded as a word from $\#\Sigma^*(Q \times \Sigma)\Sigma^*\# \subseteq \Gamma^*$. Now let $w \in \Sigma^*$ be an input of length n and let $m = p(n)$. Then a configuration of T is a word from $\bigcup_{i=0}^{m-1} \#\Sigma^i(Q \times \Sigma)\Sigma^{m-i-1}\# \subseteq \Gamma^{m+2}$. Clearly, the symbol at position $1 < i < m+2$ at time $t+1$ in a configuration only depends on the symbols at the positions $i-1$, i, and $i+1$ at time t. Assume that $c, c_1, c_2, c_3 \in \Sigma \cup (Q \times \Sigma) \cup \{\#\}$

such that $c_1c_2c_3 \in \{\varepsilon, \#\}\Sigma^*(Q \times \Sigma)\Sigma^*\{\varepsilon, \#\}$. We write $c_1c_2c_3 \xrightarrow{j} c$ for $j \in \{1,2\}$ if the following holds: If three consecutive positions $i-1, i$, and $i+1$ of a configuration contain the symbol sequence $c_1c_2c_3$, then choice j of T results in the symbol c at position i. We write $c_1c_2c_3 \xrightarrow{\exists} (d_1, d_2)$ for $c_1, c_2, c_3, d_1, d_2 \in \Sigma \cup (Q \times \Sigma) \cup \{\#\}$ if one of the following two cases holds: (i) $c_1c_2c_3 \in \{\varepsilon, \#\}\Sigma^*(Q_\exists \times \Sigma)\Sigma^*\{\varepsilon, \#\}$ and $c_1c_2c_3 \xrightarrow{j} d_j$ for $j \in \{1,2\}$ or (ii) $c_1c_2c_3 \in \{\varepsilon, \#\}\Sigma^*\{\varepsilon, \#\}$ and $d_1 = d_2 = c_2$. The notation $c_1c_2c_3 \xrightarrow{\forall} (d_1, d_2)$ is defined analogously, except that in the first case we require $c_1c_2c_3 \in \{\varepsilon, \#\}\Sigma^*(Q_\forall \times \Sigma)\Sigma^*\{\varepsilon, \#\}$.

We encode a configuration $\#c_1c_2 \cdots c_m\#$, where the current state is from $Q_\exists$ by the subtree of $\mathcal{C}(\Gamma)$ on the right, where $i = 1$ or $i = 2$. If the current state is from $Q_\forall$, then we take the same subgraph, except that b_i replaces a_i.

The idempotent presentation $P \subseteq (\Gamma \cup \Gamma^{-1})^* \times (\Gamma \cup \Gamma^{-1})^*$ is constructed in such a way from the machine T that building the closure from a Munn tree that represents the initial configuration (in the above sense) corresponds to generating the whole computation tree of the Turing machine T starting from the initial configuration. We will describe each pair $(e, f) \in P$ by the Munn trees of e and f.

For all $x \in \{a_1, a_2, b_1, b_2\}$ put the identity on the right into P, which propagates the end-marker $\#$ along intervals of length $m+2$ (here, the x^m-labeled edge abbreviates a path consisting of m many x-labeled edges). Successor configurations of the current configuration are generated by the equations below, where $i \in \{1,2\}, 0 \le k \le m-1$, and $c_1c_2c_3 \xrightarrow{\exists} (d_1, d_2)$ (resp. $c_1c_2c_3 \xrightarrow{\forall} (d_1, d_2)$) for the left (resp. right) equation:

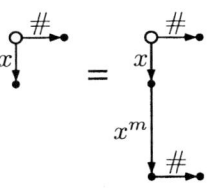

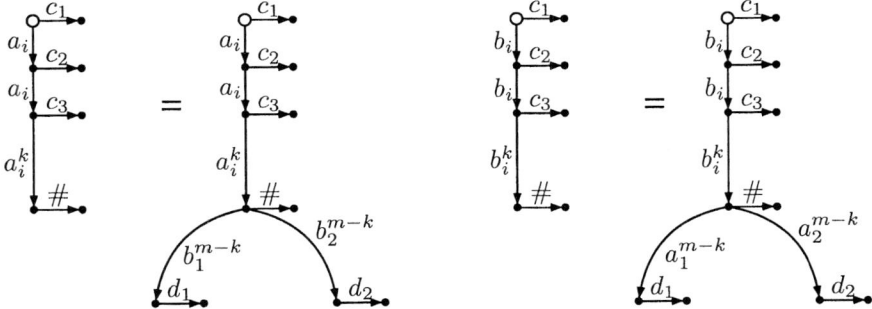

The remaining equations propagate acceptance information back to the initial Munn tree. Here the separation of the state set into existential and universal states becomes important. Let $f = (q_f, \$)$; recall that $\$$ is the symbol under the head of T when T terminates in state q_f. For all $i, j \in \{1, 2\}$ and all $x \in \{a_1, a_2, b_1, b_2\}$ we put the following equations into P:

This concludes the description of the presentation P. Now define the words $u, v \in (\Gamma \cup \Gamma^{-1})^*$ as follows: Assume that the input word for our alternating Turing machine w is of the form $w = w_1 w_2 \cdots w_n$ with $w_i \in \Sigma$. For $n+1 \leq i \leq m$ define $w_i = \Box$, where $\Box$ is the blank symbol of T. Then the Munn trees of u and v are shown on the right (we assume $r(u) = r(v) = \varepsilon$).

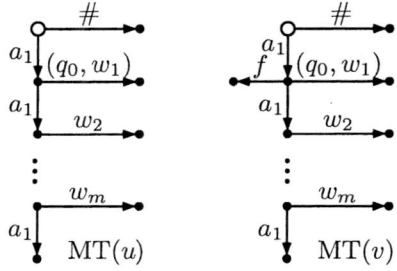

We claim that $\mu_P(u) = \mu_P(v)$ iff the machine T accepts the word w. From the construction of u, v, and P it follows that T accepts the word w iff $\mathrm{MT}(v) \subseteq \mathrm{cl}_P(\mathrm{MT}(u))$. Since $\mathrm{MT}(u) \subseteq \mathrm{MT}(v)$ this is equivalent to $\mathrm{cl}_P(\mathrm{MT}(v)) = \mathrm{cl}_P(\mathrm{MT}(u))$, i.e., $\mu_P(u) = \mu_P(v)$ due to Theorem 3 (note that $r(u) = r(v) = \varepsilon$). This proves the EXPTIME lower bound.

For the upper bound let $P \subseteq (\Gamma \cup \Gamma^{-1})^* \times (\Gamma \cup \Gamma^{-1})^*$ be an idempotent presentation and $u, v \in (\Gamma \cup \Gamma^{-1})^*$. Since $r(u) = r(v)$ can be checked in linear time, it suffices by Theorem 3 to show that we can verify in EXPTIME whether $\mathrm{MT}(v) \subseteq \mathrm{cl}_P(\mathrm{MT}(u))$ (note that $\mathrm{cl}_P(\mathrm{MT}(v)) = \mathrm{cl}_P(\mathrm{MT}(u))$ iff $\mathrm{MT}(u) \subseteq \mathrm{cl}_P(\mathrm{MT}(v))$ and $\mathrm{MT}(v) \subseteq \mathrm{cl}_P(\mathrm{MT}(u))$). Let G be the graph that results from the Cayley-graph $\mathcal{C}(\Gamma)$ by taking a new edge label $\#$, adding a new node v_0, and adding a $\#$-labeled edge from node 1 (i.e., the origin) of $\mathcal{C}(\Gamma)$ to the new node v_0. Since $\mathcal{C}(\Gamma)$ is a context-free graph, also G is context-free. By Theorem 1 it suffices to construct in polynomial time a modal μ-calculus formula $\varphi_{u,v,P}$ such that $\mathrm{MT}(v) \subseteq \mathrm{cl}_P(\mathrm{MT}(u))$ iff $(G, 1) \models \varphi_{u,v,P}$.

For $w = a_1 a_2 \cdots a_m$ ($a_i \in \Gamma \cup \Gamma^{-1}$) and two positions $i, j \in \{1, \ldots, m\}$, $i \leq j$, let $w[i, j] = a_i \cdots a_j$. If $i > j$, then set $w[i, j] = \varepsilon$. Moreover, we use $\langle w \rangle \phi$ as an abbreviation for $\langle a_1 \rangle \langle a_2 \rangle \cdots \langle a_m \rangle \phi$. Assume that $P = \{(e_i, f_i) \mid 1 \leq i \leq n\}$, where $\mathrm{MT}(e_i) \subseteq \mathrm{MT}(f_i)$. First, let

$$\varphi_{u,P} = \mu X. \left(\bigvee_{i=0}^{|u|} \langle u[1,i]^{-1} \rangle \langle \# \rangle \mathrm{true} \vee \bigvee_{i=1}^{n} \bigvee_{j=0}^{|f_i|} \langle f_i[1,j]^{-1} \rangle (\bigwedge_{k=0}^{|e_i|} \langle e_i[1,k] \rangle X) \right).$$

Then $(G, x) \models \varphi_{u,P}$ iff $x \in \mathrm{cl}_P(\mathrm{MT}(u))$. The disjunction $\bigvee_{i=0}^{|u|} \langle u[1,i]^{-1} \rangle \langle \# \rangle \mathrm{true}$ expresses $\mathrm{MT}(u) \subseteq \mathrm{cl}_P(\mathrm{MT}(u))$, whereas $\bigvee_{i=1}^{n} \bigvee_{j=0}^{|f_i|} \langle f_i[1,j]^{-1} \rangle (\bigwedge_{k=0}^{|e_i|} \langle e_i[1,k] \rangle X)$ defines all nodes such that via the inverse of some prefix of some word f_i a node x can be reached such that the whole path starting in x and labeled with e_i already belongs to X. Finally, set $\varphi_{u,v,P} = \bigwedge_{i=0}^{|v|} \langle v[1,i] \rangle \varphi_{u,P}$. □

The following result was conjectured in [19].

Corollary 1. *There is a fixed context-free graph, for which the model-checking problem of the modal μ-calculus (restricted to formulas of μ-nesting depth 1) is EXPTIME-complete.*

Proof. We can reuse the constructions from the previous proof. Note that the generating set Γ from the lower bound proof is a fixed set; thus, the Cayley-graph $\mathcal{C}(\Gamma)$ is a fixed context-free graph. Hence, also the graph G constructed in the upper bound proof by adding a $\#$-labeled edge that leaves the origin 1 is a fixed context-free graph. For the input word w for the Turing machine T let u, v, and P be the data constructed in the lower bound proof. Then w is accepted by T iff $\mathrm{MT}(v) \subseteq \mathrm{cl}_P(\mathrm{MT}(u))$ iff $(G, 1) \models \varphi_{u,v,P}$. This proves the corollary. □

7 Cayley-Graphs of Inverse Monoids

Let $\mathcal{M} = (M, \circ, 1)$ be a monoid with a finite generating set Σ and let $h : \Sigma^* \to \mathcal{M}$ be the canonical morphism. We define the following expansion $\mathcal{C}(\mathcal{M}, \Sigma)_{\mathrm{reg}}$ of the Cayley-graph $\mathcal{C}(\mathcal{M}, \Sigma)$: $\mathcal{C}(\mathcal{M}, \Sigma)_{\mathrm{reg}} = (M, (\mathrm{reach}_L)_{L \in \mathrm{REG}(\Sigma)}, 1)$ with $\mathrm{reach}_L = \{(u, v) \in M \times M \mid \exists w \in L : u \circ h(w) = v\}$. Thus, $\mathcal{C}(\mathcal{M}, \Sigma) = (M, (\mathrm{reach}_{\{a\}})_{a \in \Sigma}, 1)$. The main result of this section is:

Theorem 6. *Let $P \subseteq (\Gamma \cup \Gamma^{-1})^* \times (\Gamma \cup \Gamma^{-1})^*$ be a finite and idempotent presentation. Then the first-order theory of the structure $\mathcal{C}(\mathrm{FIM}(\Gamma)/P, \Gamma \cup \Gamma^{-1})_{\mathrm{reg}}$ is decidable.*

The following undecidability result contrasts Theorem 6. It is easy to see that the decidability of $\mathrm{MSOTh}(\mathcal{C}(\mathcal{M}, \Gamma))$ implies the decidability of $\mathrm{FOTh}(\mathcal{C}(\mathcal{M}, \Gamma)_{\mathrm{reg}})$.

Theorem 7. $\mathrm{MSOTh}(\mathcal{C}(\mathrm{FIM}(\{a,b\}), \{a, b, a^{-1}, b^{-1}\}))$ *is undecidable.*

Theorem 7 can be shown by identifying an infinite grid as a minor of $\mathcal{C}(\mathrm{FIM}(\{a,b\}))$.

Before we prove Theorem 6, let us first state a corollary. The *generalized word problem* for $\mathcal{M}$ asks whether for given words $u, u_1, \ldots, u_n \in \Sigma^*$ the monoid element $h(u)$ belongs to the submonoid of $\mathcal{M}$ that is generated by the elements $h(u_1), \ldots, h(u_n)$. Theorem 6 easily implies:

Corollary 2. *Let $P \subseteq (\Gamma \cup \Gamma^{-1})^* \times (\Gamma \cup \Gamma^{-1})^*$ be a finite and idempotent presentation. Then the generalized word problem for $\mathrm{FIM}(\Gamma)/P$ is decidable.*

To prove Theorem 6 we first need some lemmas.

Lemma 1. *There exists a fixed MSO-formula $\varphi(x, y)$ (over the signature consisting of a binary relation symbol E) such that for every finite directed graph $G = (V, E)$ and all nodes $s, t \in V$ we have: $G \models \varphi(s, t)$ iff there is a path in G with initial vertex s and terminal vertex t visiting all vertices from V.*

For the proof of Lemma 1 one defines a partial order $\prec$ on the set of strongly connected components of G: $U \prec V$ for two different strongly connected components U and V if and only if there is a (directed) path from a node of U to a node of V. Then there is a path in G with initial vertex s and terminal vertex t visiting all vertices from V iff $\prec$ is a total order and s (resp. t) belongs to the minimal (resp. maximal) strongly connected component of G. These conditions can be easily formalized in MSO-logic.

Lemma 2. *Let Σ be a finite alphabet and let $L \in \mathrm{REG}(\Sigma)$. Then one can construct an MSO-sentence ψ_L (over a signature consisting of binary relation symbols E_a ($a \in \Sigma$) and two constants s and t) such that for every finite structure $G = (V, (E_a)_{a \in \Sigma}, s, t)$ we have $G \models \psi_L$ iff there exists a path $p = (v_1, a_1, v_2, a_2, \ldots, v_n)$ ($v_i \in V$, $a_i \in \Sigma$) such that: $v_1 = s$, $v_n = t$, $(v_i, v_{i+1}) \in E_{a_i}$ for all $1 \leq i < n$, $a_1 a_2 \cdots a_{n-1} \in L$, and $V = \{v_1, v_2, \ldots, v_n\}$.*

Let us just give a brief sketch of the proof of Lemma 2. Let $A = (Q, \Sigma, \delta, q_0, F)$ be a deterministic finite automaton with $L(A) = L$. W.l.o.g. $Q = \{1, \ldots, m\}$. Define the structure $f_A(G)$ by $f_A(G) = (V \times Q, E, \Delta, I_s, F_t)$, where

$$E = \{((u,i),(v,j)) \mid \exists a \in \Sigma : (u,v) \in E_a \land \delta(i,a) = j\},$$
$$\Delta = \{((v,1),\ldots,(v,m)) \mid v \in V\}, I_s = \{(s, q_0)\}, \text{ and } F_t = \{t\} \times F.$$

Then one can show that f_A is an MSO-transduction in the sense of [3]. Thus, there exists a backwards translation $f_A^\sharp$ such that for every MSO-sentence ϕ over the signature of $f_A(G)$ we have: $f_A(G) \models \phi$ iff $G \models f_L^\sharp(\phi)$ [3]. Now, using Lemma 1 and the relation Δ it is easy to write down an MSO-sentence ϕ over the signature of $f_A(G)$ expressing that there exists a path from (s, q_0) to a node in F_t such that the set of first components of nodes along that path is precisely V. Then the sentence $f_A^\sharp(\phi)$ is the desired sentence.

Lemma 2 easily implies the next lemma.

Lemma 3. *Let Σ be a finite alphabet and let $L \in \text{REG}(\Sigma)$. Then one can construct an MSO-formula $\theta_L(X)$ (over a signature consisting of binary relation symbols E_a ($a \in \Sigma$) and two constants s and t) such that for every finite structure $G = (V, (E_a)_{a \in \Sigma}, s, t)$ and every finite set $U \subseteq V$ we have $G \models \theta_L(U)$ iff there exists a path $p = (v_1, a_1, v_2, a_2, \ldots, v_n)$ ($v_i \in V$, $a_i \in \Sigma$) such that: $v_1 = s$, $v_n = t$, $(v_i, v_{i+1}) \in E_{a_i}$ for all $1 \leq i < n$, $a_1 a_2 \cdots a_{n-1} \in L$, and $U \subseteq \{v_1, v_2, \ldots, v_n\}$.*

We are now able to finish the proof of Theorem 6. Let $P \subseteq (\Gamma \cup \Gamma^{-1})^* \times (\Gamma \cup \Gamma^{-1})^*$ be a finite and idempotent presentation. We want to show that the first-order theory of the structure $\mathcal{A} = \mathcal{C}(\text{FIM}(\Gamma)/P, \Gamma \cup \Gamma^{-1})_{\text{reg}}$ is decidable. For this, we use Theorem 3 and translate each first-order sentence φ over $\mathcal{A}$ into an MSO-sentence $\|\varphi\|$ over the Cayley graph $\mathcal{C}(\Gamma)$ of the free group $\text{FG}(\Gamma)$ such that $\mathcal{A} \models \varphi$ iff $\mathcal{C}(\Gamma) \models \|\varphi\|$. Together with Theorem 2 this will complete the proof of Theorem 6.

To every variable x (ranging over $\text{FIM}(\Gamma)/P$) in φ we associate two variables in $\|\varphi\|$: (i) an MSO-variable X' representing $\text{cl}_P(\text{MT}(u))$, where u is any word representing x, and (ii) an FO-variable x' representing $\gamma(u) \in \text{FG}(\Gamma)$. The relationship between x' and X' is expressed by the MSO-formula (over the signature of $\mathcal{C}(\Gamma)$) $\text{MT}(x', X') = \exists X : \Theta(x', X, X')$, where $\Theta(x', X, X') = (1, x' \in X \land X$ is connected and finite $\land \text{CL}_P(X, X'))$. Recall that finiteness and connectedness of a subset of the finitely-branching tree $\mathcal{C}(\Gamma)$ can be expressed in MSO, see the remarks in Section 3. Here $\text{CL}_P(X, X')$ is the MSO-formula constructed by Margolis and Meakin in [10], see the remark at the end of Section 5. Next, note that by Lemma 3 for every language $L \in \text{REG}(\Gamma \cup \Gamma^{-1})$ there exists an MSO-formula $\xi_L(x', X, y', Y)$ over the signature of $\mathcal{C}(\Gamma)$ such that for all finite sets $U, V \subseteq \text{FG}(\Gamma)$ and all nodes $u', v' \in \text{FG}(\Gamma)$ we have: $\mathcal{C}(\Gamma) \models \xi_L(u', U, v', V)$ iff $U \subseteq V$ and there is a path from u' to v' in $\mathcal{C}(\Gamma)\restriction_V$ that visits all vertices of $V \backslash U$ and which is labeled with a word from the language L. Now let φ be an FO-formula over the signature of $\mathcal{A}$. We define $\|\varphi\|$ inductively: $\|\text{reach}_L(x, y)\| = \exists X, Y : \Theta(x', X, X') \land \Theta(y', Y, Y') \land \xi_L(x', X, y', Y)$, $\|\neg \psi\| = \neg \|\psi\|$, $\|\psi_1 \land \psi_2\| = \|\psi_1\| \land \|\psi_2\|$, and $\|\forall x : \psi\| = \forall x' \forall X' : \text{MT}(x', X') \Rightarrow \|\psi\|$. It is straight-forward to verify that $\mathcal{A} \models \varphi$ iff $\mathcal{C}(\Gamma) \models \|\varphi\|$. This concludes the proof of Theorem 6.

8 Open Problems

We plan to investigate for which monoids $\mathcal{M}$ the structure $\mathcal{C}(\mathcal{M}, \Gamma)_{\text{reg}}$ has a decidable first-order theory. In particular, the group case is interesting. It is easy to see that the decidability of the MSO-theory of $\mathcal{C}(\mathcal{M}, \Gamma)$ implies the decidability of the first-order theory of $\mathcal{C}(\mathcal{M}, \Gamma)_{\text{reg}}$. Thus, the class of groups $\mathcal{G}$ for which $\mathcal{C}(\mathcal{G}, \Gamma)_{\text{reg}}$ is decidable lies somewhere between the virtually-free groups (i.e., those groups for which the MSO-theory of the Cayley-graph is decidable) and the groups with a decidable word problem (i.e., those groups for which the first-order theory of the Cayley-graph is decidable).

References

1. J.-C. Birget, S. W. Margolis, and J. Meakin. The word problem for inverse monoids presented by one idempotent relator. *Theoretical Computer Science*, 123(2):273–289, 1994.
2. A. K. Chandra, D. C. Kozen, and L. J. Stockmeyer. Alternation. *Journal of the Association for Computing Machinery*, 28(1):114–133, 1981.
3. B. Courcelle. The expression of graph properties and graph transformations in monadic second-order logic. In *Handbook of graph grammars and computing by graph transformation, Volume 1 Foundations*, pages 313–400. World Scientific, 1997.
4. A. V. Kelarev and S. J. Quinn. A combinatorial property and Cayley graphs of semigroups. *Semigroup Forum*, 66(1):89–96, 2003.
5. O. Kupferman and M. Y. Vardi. An automata-theoretic approach to reasoning about infinite-state systems. In *Proc. CAV 2000*, LNCS 1855, pages 36–52. Springer, 2000.
6. D. Kuske and M. Lohrey. Logical aspects of Cayley-graphs: the group case. *Annals of Pure and Applied Logic*, 131(1–3):263–286, 2005.
7. D. Kuske and M. Lohrey. Logical aspects of Cayley-graphs: the monoid case. *International Journal of Algebra and Computation*, 2005. to appear.
8. M. Lohrey and N. Ondrusch. Inverse monoids: decidability and complexity of algebraic questions. Technical Report 2005/2, University of Stuttgart, Germany, 2005. Available via ftp://ftp.informatik.uni-stuttgart.de/pub/library/ ncstrl.ustuttgart_fi/TR-2005-02/
9. R. C. Lyndon and P. E. Schupp. *Combinatorial Group Theory*. Springer, 1977.
10. S. Margolis and J. Meakin. Inverse monoids, trees, and context-free languages. *Trans. Amer. Math. Soc.*, 335(1):259–276, 1993.
11. S. Margolis, J. Meakin, and M. Sapir. Algorithmic problems in groups, semigroups and inverse semigroups. In *Semigroups, Formal Languages and Groups*, 147–214. Kluwer, 1995.
12. D. E. Muller and P. E. Schupp. Groups, the theory of ends, and context-free languages. *Journal of Computer and System Sciences*, 26:295–310, 1983.
13. D. E. Muller and P. E. Schupp. The theory of ends, pushdown automata, and second-order logic. *Theoretical Computer Science*, 37(1):51–75, 1985.
14. W. Munn. Free inverse semigroups. *Proc. London Math. Soc.*, 30:385–404, 1974.
15. M. Petrich. *Inverse semigroups*. Wiley, 1984.
16. M. O. Rabin. Decidability of second-order theories and automata on infinite trees. *Transactions of the American Mathematical Society*, 141:1–35, 1969.
17. B. V. Rozenblat. Diophantine theories of free inverse semigroups. *Siberian Mathematical Journal*, 26:860–865, 1985. English translation.
18. P. V. Silva and B. Steinberg. A geometric characterization of automatic monoids. *Quarterly J. Mathematics*, 55:333–356, 2004.
19. I. Walukiewicz. Pushdown processes: Games and model-checking. *Information and Computation*, 164(2):234–263, 2001.
20. B. Zelinka. Graphs of semigroups. *Casopis. Pest. Mat.*, 27:407–408, 1981.

Dimension Is Compression

María López-Valdés and Elvira Mayordomo*

Dept. de Informática e Ing. de Sistemas, Universidad de Zaragoza,
María de Luna 1, 50018 Zaragoza, Spain
{marlopez, elvira}@unizar.es

Abstract. Effective fractal dimension was defined by Lutz (2003) in order to quantitatively analyze the structure of complexity classes. Interesting connections of effective dimension with information theory were also found, in fact the cases of polynomial-space and constructive dimension can be precisely characterized in terms of Kolmogorov complexity, while analogous results for polynomial-time dimension haven't been found.

In this paper we remedy the situation by using the natural concept of reversible time-bounded compression for finite strings. We completely characterize polynomial-time dimension in terms of polynomial-time compressors.

1 Introduction

Effective fractal dimension was defined in [13] in order to quantitatively analyze the structure of complexity classes. See [12,16] for a summary of the main results.

In parallel, the connections of this effective dimension with algorithmic information started being patent. The cases of constructive, recursive and polynomial-space dimension were characterized precisely as the best case asymptotic compression rate when using plain, recursive, and polynomial-space-bounded Kolmogorov complexity, respectively [15,14,6].

But the case of polynomial-time bounds remained elusive [8]. This is not strange since computing even approximately the value of time-bounded Kolmogorov complexity seems to require an exponential search. The main difference with space-bounded Kolmogorov complexity is reversibility, in this later case the encoding phase can be performed within similar space-bounds.

In this paper we look at the usual notion of compression algorithm for finite strings. A polynomial-time compression scheme is just a pair of encoder and decoder algorithms, both working in polynomial-time. We consider encoders that do not completely start from scratch when working on an extension of a previous input. This last condition is formalized in Sect. 3 with a conditional-entropy like inequality.

We exactly characterize polynomial-time (or p) -dimension as the best case asymptotic (that is, i.o.) compression ratio attained by these polynomial-time

---

* This research was supported in part by Spanish Government MEC project TIC 2002-04019-C03-03.

compression schemes. Dually, strong polynomial-time-dimension [2] corresponds to the worst case asymptotic compression ratio.

Several results on the polynomial-time dimension of complexity classes can be now interpreted as compressibility results. For example, the (characteristic sequences of) languages in a class of p-dimension 1 cannot be i.o. compressed by more that a sublinear amount. Here we obtain results on the compressibility of complete and autoreducible languages.

Buhrman and Longprè have given a characterization of p-measure in terms of compressibility in [4], but in that case the compressors are restricted to extenders and the encoder is required to give several alternatives, one of them being the correct output. In the light of our present results we can view effective dimension as an information content measure for infinite strings, whereas resource-bounded measure can only distinguish the extreme case of non measure 0 classes that are the most incompressible ones.

2 Preliminaries

The Cantor space **C** is the set of all infinite binary sequences. If $w \in \{0,1\}^*$ and $x \in \{0,1\}^* \cup \mathbf{C}$, $w \sqsubseteq x$ means that w is a prefix of x. For $0 \leq i \leq j$, we write $x[i \ldots j]$ for the string consisting of the i-th through the j-th bits of x. We use λ for the empty string.

Let p be the set of polynomial-time computable functions. Let E= DTIME($2^{O(n)}$).

Definition 1. *Let $s \in [0, \infty)$.*

1. *An s-gale is a function $d : \{0,1\}^* \to [0, \infty)$ satisfying*

$$d(w) = 2^{-s}[d(w0) + d(w1)]$$

   for all $w \in \{0,1\}^*$.
2. *A martingale is a 1-gale, that is, a function $d : \{0,1\}^* \to [0, \infty)$ satisfying*

$$d(w) = \frac{d(w0) + d(w1)}{2}$$

   for all $w \in \{0,1\}^*$.

Definition 2. *Let $s \in [0, \infty)$ and d be an s-gale.*

1. *We say that d succeeds on a sequence $S \in \mathbf{C}$ if*

$$\limsup_{n \to \infty} d(S[0 \ldots n]) = \infty$$

   The success set of d is $S^\infty[d] = \{S \in \mathbf{C} \mid d \text{ succeeds on } S\}$
2. *We say that d succeeds strongly on a sequence $S \in \mathbf{C}$ if*

$$\liminf_{n \to \infty} d(S[0 \ldots n]) = \infty$$

   The strong success set of d is $S^\infty_{\text{str}}[d] = \{S \in \mathbf{C} \mid d \text{ succeeds strongly on } S\}$.

Definition 3. *Let* $X \subseteq \mathbf{C}$,

1. *The p-dimension of X is*

$$\dim_{\mathrm{p}}(X) = \inf \left\{ s \in [0, \infty) \;\middle|\; \begin{array}{l} \text{there is a p-computable s-gale d s.t.} \\ X \subseteq S^{\infty}[d] \end{array} \right\}$$

2. *The strong p-dimension of X is*

$$\mathrm{Dim}_{\mathrm{p}}(X) = \inf \left\{ s \in [0, \infty) \;\middle|\; \begin{array}{l} \text{there is a p-computable s-gale d s.t.} \\ X \subseteq S^{\infty}_{\mathrm{str}}[d] \end{array} \right\}$$

For a complete introduction and motivation of effective dimension and effective strong dimension see [12].

3 Compressors That Do Not Start from Scratch

In this section we develop the notion of compressors that "do not start from scratch" in the sense that when encoding successively longer extensions of an input, the outputs are restricted in the way we make precise below. The extreme case of this behavior is when the compressor is a mere extender, that is, $C(w)$ is always a prefix of $C(wu)$. We consider here a much weaker restriction than extension.

Definition 4. *A pair of functions (C, D) (C the encoder, D the decoder) $C, D : \{0,1\}^* \to \{0,1\}^*$ is a polynomial-time compressor if:*

(i) C and D can be computed in polynomial-time on their corresponding input length.
(ii) For all $w \in \{0,1\}^*$, $D(C(w), |w|) = w$.

In this paper, we could make all codes prefix-free, that is, $C(\{0,1\}^n)$ is a prefix set for each n. For the asymptotic compression rates the difference is not significant.

Notice that in the previous definition there is no restriction whatsoever on the behavior of C, the encoder, when working on two inputs that are one an extension of the other. For instance, we can have $|C(wu)| \ll |C(w)|$ and $C(wu)$ can have no common prefix with $C(w)$. In definition 5 we introduce a restriction on the compressor that has an effect on the variety of $C(wu)$ for different u, that will be controlled by $|C(w)|$.

Definition 5. *A polynomial-time compressor (C, D) does not start from scratch if $\forall \epsilon > 0$ and for almost every $w \in \{0,1\}^*$ there exists $k = O(\log(|w|))$, $k > 0$, such that*

$$\sum_{|u| \leq k} 2^{-|C(wu)|} \leq 2^{\epsilon k} 2^{-|C(w)|}. \tag{1}$$

We will consider only compressors that do not start from scratch.

Notice that when there is a constant k such that $\sum_{|u| \leq k} 2^{-|C(wu)|} \leq 2^{-|C(w)|}$, condition (1) is trivial, while in general $\sum_{|u| \leq k} 2^{-|C(wu)|}$ can be as large as 1, so condition (1) is a proper restriction on compressors.

We first remark that if $C(w)$ and $C(wu)$ have a long common prefix then C fulfills condition (1).

Lemma 1. *Polynomial-time compressors for which $C(w)$ and $C(wu)$ have a common prefix of length at least $|C(w)| - O(\log(|w|))$ ($\forall w, u \in \{0,1\}^*$) don't start from scratch.*

We next present two easy examples of compressors not starting from scratch, including Lempel-Ziv algorithms.

Example 1. For the following polynomial-time compressor condition (1) holds

1. An extender, that is, $\forall w, w' \in \{0,1\}^*$

$$w \sqsubseteq w' \Rightarrow C(w) \sqsubseteq C(w').$$

2. Lempel-Ziv data compression algorithm for its three most common variants (notice that it is not an extender. See [10,11] for details).
    In fact, Lempel-Ziv compression algorithm verifies the common-prefix condition lemma 1. Let $w \in \{0,1\}^*$. If $w = w_1 w_2 \ldots w_n v$ where $w_1, w_2, \ldots, w_n$ are the phrases obtained by the Lempel-Ziv parsing, then $LZ(w)$ and $LZ(wu)$ have a common prefix of length at least $|LZ(w)| - \log n \geq |LZ(w)| - \log(|w|)$.

We leave for the complete version of this paper an analysis of the case of Grammar-based compressors, that generalize Lempel-Ziv methods [9].

Polynomial-time compressors (C, D) that are length increasing in the encoder C and for which we can control, for all w and all $i \geq 0$, the number of strings u satisfying $|C(wu)| = |C(w)| + i$, don't start from scratch. More formally,

Lemma 2. *Polynomial-time compressors (C, D) that satisfy both of the following conditions don't start from scratch.*

i) For all $w, u \in \{0,1\}^*$, $|C(wu)| \geq |C(w)|$

ii) For all $\epsilon > 0$ and for almost every $w \in \{0,1\}^*$ there exists $k = O(\log(|w|))$ such that $\forall i \geq 0$

$$N_i = N_i(w, k) = \#\left\{u \in \{0,1\}^{\leq k} \,\Big|\, |C(wu)| = |C(w)| + i\right\} \leq 2^{i + \epsilon k - \log k}.$$

4 Main Theorem

In the main theorem we obtain an exact characterization of polynomial-time dimension in terms of polynomial-time compressors that don't start from scratch.

Our characterization holds both for the best and worst asymptotic compression ratio, corresponding to p-dimension and strong p-dimension.

We formalize the notion of a.e. (almost everywhere) and i.o. (infinitely often) compressibility for sets of infinite sequences as the asymptotic best (respectively worse) compression ratio.

Definition 6. *For $\alpha \in [0,1]$ and $X \subseteq \mathbf{C}$,*

1. *X is α-i.o. polynomial-time compressible if there is a polynomial-time compressor (C, D) that does not start from scratch and such that for every $A \in X$*

$$\liminf_n \frac{|C(A[0\ldots n-1])|}{n} \leq \alpha.$$

2. *X is α-a.e. polynomial-time compressible if there is a polynomial-time compressor (C, D) that does not start from scratch and such that for every $A \in X$*

$$\limsup_n \frac{|C(A[0\ldots n-1])|}{n} \leq \alpha.$$

Definition 7. *Let $X \subseteq \mathbf{C}$,*

1. *X is i.o. polynomial-time incompressible if for every (C, D) polynomial-time compressor that does not start from scratch, there exist $A \in X$ such that*

$$\liminf_n \frac{|C(A[0\ldots n-1])|}{n} = 1.$$

2. *X is a.e. polynomial-time incompressible if for every (C, D) polynomial-time compressor that does not start from scratch, there exist $A \in X$ such that*

$$\limsup_n \frac{|C(A[0\ldots n-1])|}{n} = 1.$$

We next state our main theorem.

Theorem 1. *Let $X \subseteq \mathbf{C}$,*

$$\dim_\mathrm{p}(X) = \inf\{\,\alpha\,|\,X \text{ is } \alpha\text{-i.o. polynomial-time compressible}\}$$
$$\mathrm{Dim}_\mathrm{p}(X) = \inf\{\,\alpha\,|\,X \text{ is } \alpha\text{-a.e. polynomial-time compressible}\}$$

The proof of theorem 1 will be split between sections 5 and 6. In section 5 we transform each gale into a compressor that requires only a time increase of a linear factor. In section 6 we show that compression is an upper bound on dimension.

Hitchcock showed in [7] that p-dimension can be characterized in terms of on-line prediction algorithms, using the well-studied log-loss prediction ratio. Our result can thus be interpreted as a joining bridge between the performance of polynomial-time prediction and compression algorithms, both in the best and the worse case.

5 Compression Is at Most Dimension

Proposition 1. *Let $X \in \mathbf{C}$,*

$$\dim_\mathrm{p}(X) \geq \inf\{\,\alpha\,|\,X \text{ is } \alpha\text{-i.o. polynomial-time compressible}\}.$$

We first transform each gale into a simple version that requires little accuracy. Then we apply a generalization of arithmetic coding to this new gale.

For the first part we will need the following lemma stating that very simple gales characterize p-dimension.

Lemma 3. *Let $X \subseteq \mathbf{C}$. If $\dim_p(X) = \alpha$ then $\forall s > \alpha$ there exists an s-gale d with $X \subseteq S^\infty[d]$ such that for all $w \in \{0,1\}^*$, there exists $m_w, n_w \in \mathbb{N}$ with $n_w \leq |w| + 1$ and*
$$d(w)2^{-|w|s} = m_w 2^{-(n_w + |w|)}$$

That is, if $\dim_p(X) < s$, then there exists a p-computable s-gale d as in the previous lemma. We define a polynomial-time compressor that doesn't start from scratch using this s-gale. Roughly speaking, the idea for the encoder C is associate to each $w \in \{0,1\}^*$ an interval of size proportionally related to $d(w)$. By the properties of d, the extreme points of such interval are dyadic rational numbers. By using the following lemma, we codify each interval with a string z. We will take $C(w) = z$.

Lemma 4. *Let a, b be dyadic numbers. and let $I = [a, b)$ be an interval of length $r \in [0,1)$, then there exists a string z of length $-\lfloor \log(r) \rfloor + 1$ such that $a < 0.z < b$ and z can be computed in time polynomial in $|z|$.*

Proof sketch of Proposition 1. Let s be such that $\dim_p(X) < s$, then there exists a p-computable s-gale d as in Lemma 3 with $d(\lambda) = 1$, $X \subseteq S^\infty[d]$.

Let $h : \{0,1\}^* \to \mathbb{R}$ be defined as follows.
$$h(w) := \sum_{|y|=|w|, y<w} d(y) 2^{(1-s)|y|-|w|}$$

where $y < w$ means that y precedes x in lexicographic order. Denote by $succ(w)$ the successor of w in lexicographic order. Notice that $h(w)$ is a dyadic number $m2^{-n}$ with $n \leq 2|w| + 1$, therefore there is a $z \in \{0,1\}^*$ such that $|z| \leq 2|w| + 2$ and $h(w) < 0.z < h(succ(w))$. Let z_w be the first shortest string such that $h(w) < 0.z < h(succ(w))$. We define the encoder as $C(w) = z_w$.

It can be shown that the encoder C and the corresponding decoder form a polynomial-time compressor that does not start from scratch.

Finally, let us see that (C, D) compresses X. Notice that for each w the interval $[h(w), h(succ(w)))$ has length exactly $d(w)2^{-s|w|}$. Then by lemma 4, there is a string z of length $-\lfloor \log(2^{-s|w|}d(w)) \rfloor + 1 \leq |w| - \lfloor \log(2^{(1-s)|w|}d(w)) \rfloor + 1$ such that $h(w) < 0.z < h(succ(w))$. So,
$$|z_w| \leq |w| - \lfloor \log(2^{(1-s)|w|}d(w)) \rfloor + 1.$$

For all $A \in X$, as $X \subseteq S^\infty[d]$ then $d(A[0 \ldots n-1]) > 1$ i.o. n and
$$|C(A[0 \ldots n-1])| = |z_{A[0 \ldots n-1]}|$$
$$\leq n - \lfloor \log(d(A[0 \ldots n-1])) \rfloor + 1$$
$$\leq n - \log(2^{(1-s)n}) + 1$$
$$= sn + 1$$

□

6 Dimension Is at Most Compression

Next we prove that compressibility is an upper bound on dimension.

Proposition 2. *Let $X \in \mathbf{C}$,*

$$\dim_{\mathrm{p}}(X) \leq \inf\{\,\alpha \mid X \text{ is } \alpha\text{-i.o. polynomial-time compressible}\}$$

Proof. Let $s' > s$ and $\epsilon > 0$ such that $s' - s > \epsilon$. Let N be such that condition (1) is true for each $w \in \{0,1\}^{\geq N}$. For each of these w, let $k = k(w,\epsilon) = O(\log(|w|))$ be the smallest one such that

$$\sum_{|u| \leq k} 2^{-|C(wu)|} \leq 2^{\epsilon k} 2^{-|C(w)|}$$

Let $w = w_1 \ldots w_n$ with $|w_1| = N$ and $|w_i| = k(w_1 \ldots w_i - 1, \epsilon)$ for $i > 0$.
We define an s' gale d as follows

$$d(wu) := d(w) \frac{2^{-|C(wu)|}}{\sum_{|v| \leq k} 2^{-|C(wv)|}} 2^{s'|u|} \qquad \text{if } |u| = k(w,\epsilon)$$

$$d(wr) := \sum_{r \sqsubseteq u, |u|=k} d(wu) 2^{s'(|r|-|u|)} \qquad \text{if } |r| < k(w,\epsilon)$$

d is computable in polynomial-time. Notice that, for induction, if $w = w_1 w_2 \ldots w_n$ with $|w_1| = N$ and $|w_i| = k(w_1 \ldots w_{i-1}, \epsilon)$, then

$$d(w) = d(w_1) 2^{s'(|w|-N)} \prod_{h=1}^{n-1} \frac{2^{-|C(w_1 \ldots w_{h+1})|}}{\sum_{|v| \leq k(w_1 \ldots w_h, \epsilon)} 2^{-|C(w_1 \ldots w_h v)|}}$$

By condition (1),

$$d(w) \geq d(w_1) 2^{(\epsilon-s')N} 2^{|C(w_1)|} 2^{(s'-\epsilon)|w|} 2^{-|C(w)|}$$
$$\geq a 2^{(s'-\epsilon)|w|} 2^{-|C(w)|}$$

where a is the minimum of

$$d(w_1) 2^{|C(w_1)|} 2^{(\epsilon-s')N}$$

for $w_1 \in \{0,1\}^N$.

For all $A \in X$,

$$\liminf_n \frac{|C(A[0 \ldots n-1])|}{n} \leq s$$

so there exists $(b_n)_{n \in \mathbb{N}}$ a sequence of natural numbers such that

$$\lim_n \frac{|C(A[0 \ldots b_n - 1])|}{b_n} \leq s$$

Let $(a_n)_{n \in \mathbb{N}}$ be defined as $a_1 = N$, $a_i = k(A[0 \ldots a_{i-1} - 1])$ for $i > 1$.
Then
$$d(A[0 \ldots a_{i-1} - 1]) \geq a 2^{(s'-\epsilon)a_i} 2^{-|C(A[0 \ldots a_i - 1])|}$$

For each n, let $a_m < b_n \leq a_{m+1}$, by condition (1)

$$|C(A[0 \ldots a_m - 1])| < |C(A[0 \ldots b_n - 1])| + O(\log b_n) \leq b_n(s + \epsilon/2) \leq a_{m+1}(s + \epsilon/2)$$

for all but finitely n.
Then,
$$d(A[0 \ldots a_m - 1]) \geq a 2^{(s'-\epsilon)a_m} 2^{-a_{m+1}(s+\epsilon/2)}$$
$$\geq a 2^{a_m \epsilon/2} 2^{-O(\log(a_m))}$$

And d succeeds on X.

Notice that in the last proof we didn't need the decoder so we have that for each polynomial-time encoder satisfying the condition of not starting from scratch we automatically get a polynomial-time decoder.

Corollary 1. *Let C be a polynomial-time encoder that satisfies inequality (1). Then there exist (C', D') a polynomial-time compressor that does not start from scratch and such that for every $A \in \mathbf{C}$*

$$\liminf_n \frac{C'(A[0 \ldots n-1])}{n} \leq \liminf_n \frac{C(A[0 \ldots n-1])}{n}$$

$$\limsup_n \frac{C'(A[0 \ldots n-1])}{n} \leq \limsup_n \frac{C(A[0 \ldots n-1])}{n}$$

7 Applications of the Main Result

In this section we obtain interesting consequences of our characterization for the polynomial-time compressibility of complete and autoreducible sets from previously known p-dimension results.

Notice that in this section we identify each language A with its characteristic sequence χ_A, therefore compressibility of a class always means compressibility of the corresponding characteristic sequences.

We start by showing that no polynomial-time compressor works on all many-one complete sets.

Theorem 2. *The class of polynomial-time many-one complete sets for E is i.o. polynomial-time incompressible.*

Proof. Ambos-Spies et al. prove in [1] that the class has p-dimension 1.

Next we consider $\deg_m^P(A)$, the class of sets that are equivalent to A by $\leq_m^P$-reductions. The compression ratio of $\deg_m^P(A)$ and $\deg_m^P(B)$, for $A \leq_m^P B$, is related by the following theorem.

Theorem 3. *Let A, B be sets in E such that $A \leq_m^P B$, then*

1. *The i.o. p-compression ratio of $\deg_m^P(A)$ is at most the i.o. p-compression ratio of $\deg_m^P(B)$.*
2. *The a.e. p-compression ratio of $\deg_m^P(A)$ is at most the a.e. p-compression ratio of $\deg_m^P(B)$.*

Proof. Ambos-Spies et al. prove 1. in [1] for p-dimension. Athreya et al. prove in [2] the strong dimension result for 2.

We next consider the property of autoreducibility. A set A is autoreducible if A can be decided by using A as an oracle but without asking query x on input x. We obtain incompressibility results both in the case of polynomial-time many-one autoreducibility and for the *complement* of i.o. p-Turing autoreducible sets. Therefore for each polynomial-time bound there are i.o. incompressible sets that are $\leq_m^P$-autoreducible and other that are not even i.o. $\leq_T^P$-autoreducible.

Theorem 4. *The class of polynomial-time many-one autoreducible sets are i.o. polynomial-time incompressible.*

Proof. Ambos-Spies et al. prove in [1] that the class has p-dimension 1.

Theorem 5. *The class of sets that are NOT i.o. polynomial-time Turing autoreducible are i.o. polynomial-time incompressible.*

Proof. Beigel et al. prove in [3] that the class has p-dimension 1.

We next show that there exist polynomial-time many-one degrees with every possible value for both a.e. and i.o. compressibility.

Theorem 6. *Let x, y be computable reals such that $0 \leq x \leq y \leq 1$. Then there is a set A in E such that the i.o. p-compression ratio of $\deg_m^P(A)$ is x and the a.e. p-compression ratio of $\deg_m^P(A)$ is y.*

Proof. Athreya et al. prove in [2] the result for p-dimension and strong p-dimension.

This last theorem includes the extreme case for which the i.o. compression ratio is 0 whereas the a.e. ratio is 1.

Finally, the hypothesis "NP has positive p-dimension" can be interpreted in terms of incompressibility. This hypothesis has interesting consequences on the approximation algorithms for MAX3SAT.

Theorem 7. *If for some $\alpha > 0$ NP is not α-i.o.-compressible in polynomial-time then any approximation algorithm $\mathcal{A}$ for MAX3SAT must satisfy at least one of the following*

1. *For some $\delta > 0$, $\mathcal{A}$ uses time at least 2^{n^δ}*
2. *For all $\epsilon > 0$, $\mathcal{A}$ has performance ratio less than $7/8 + \epsilon$ on an exponentially dense set of satisfiable instances.*

Proof. Hitchcock proves in [5] that the consequence follows from NP having positive p-dimension.

References

1. K. Ambos-Spies, W. Merkle, J. Reimann, and F. Stephan. Hausdorff dimension in exponential time. In *Proceedings of the 16th IEEE Conference on Computational Complexity*, pages 210–217, 2001.
2. K. B. Athreya, J. M. Hitchcock, J. H. Lutz, and E. Mayordomo. Effective strong dimension in algorithmic information and computational complexity. In *Proceedings of the Twenty-First Symposium on Theoretical Aspects of Computer Science*, volume 2996 of *Lecture Notes in Computer Science*, pages 632–643. Springer-Verlag, 2004.
3. R. Beigel, L. Fortnow, and F. Stephan. Infinitely-often autoreducible sets. In *Proceedings of the 14th Annual International Symposium on Algorithms and Computation*, volume 2906 of *Lecture Notes in Computer Science*, pages 98–107. Springer-Verlag, 2003.
4. H. Buhrman and L. Longpré. Compressibility and resource bounded measure. *SIAM Journal on Computing*, 31(3):876–886, 2002.
5. J. M. Hitchcock. MAX3SAT is exponentially hard to approximate if NP has positive dimension. *Theoretical Computer Science*, 289(1):861–869, 2002.
6. J. M. Hitchcock. *Effective Fractal Dimension: Foundations and Applications*. PhD thesis, Iowa State University, 2003.
7. J. M. Hitchcock. Fractal dimension and logarithmic loss unpredictability. *Theoretical Computer Science*, 304(1–3):431–441, 2003.
8. J. M. Hitchcock and N. V. Vinodchandran. Dimension, entropy rates, and compression. In *Proceedings of the 19th IEEE Conference on Computational Complexity*, pages 174–183, 2004.
9. J C. Kieffer and En hui Yang. Grammar based codes: A new class of universal lossless source codes. *IEEE Transactions on Information Theory*, 46:737–754, 2000.
10. A. Lempel and J. Ziv. A universal algortihm for sequential data compression. *IEEE Transaction on Information Theory*, 23:337–343, 1977.
11. A. Lempel and J. Ziv. Compression of individual sequences via variable rate coding. *IEEE Transaction on Information Theory*, 24:530–536, 1978.
12. J. H. Lutz. Effective fractal dimensions. *Mathematical Logic Quarterly*. To appear. Preliminary version appeared in *Computability and Complexity in Analysis*, volume 302 of Informatik Berichte, pages 81-97. FernUniversität in Hagen, August 2003.
13. J. H. Lutz. Dimension in complexity classes. *SIAM Journal on Computing*, 32:1236–1259, 2003.
14. J. H. Lutz. The dimensions of individual strings and sequences. *Information and Computation*, 187:49–79, 2003.
15. E. Mayordomo. A Kolmogorov complexity characterization of constructive Hausdorff dimension. *Information Processing Letters*, 84(1):1–3, 2002.
16. E. Mayordomo. Effective Hausdorff dimension. In *Classical and New Paradigms of Computation and their Complexity Hierarchies, Papers of the conference "Foundations of the Formal Sciences III"*, volume 23 of *Trends in Logic*, pages 171–186. Kluwer Academic Press, 2004.

Concurrent Automata vs. Asynchronous Systems

Rémi Morin

Laboratoire d'Informatique Fondamentale de Marseille,
39 rue F. Joliot-Curie, F-13453 Marseille cedex 13, France
remi.morin@lif.univ-mrs.fr

Abstract. We compare the expressive power of two automata-based finite-state models of concurrency. We show that Droste's and Kuske's coherent stably concurrent automata and Bednarczyk's forward-stable asynchronous systems describe the same class of regular event structures. This connection subsumes a previous study by Schmitt which relates Stark's trace automata to asynchronous systems. This work relies on Zielonka's theorem and some unrecognized result due to Arnold.

1 Introduction

In a seminal paper [12] Nielsen, Plotkin and Winskel introduced prime event structures as natural unfoldings of 1-safe Petri nets. This semantics can be decomposed into several steps by means of intermediate models which are prefix-closed Mazurkiewicz trace languages [9] and asynchronous systems [2]. More general automata-based models of concurrency were later related to more general event structures [19], namely trace automata [15] and concurrent automata [5]. Interestingly three classes of automata-based models are known to describe exactly prime event structures: Forward-stable asynchronous systems, stable trace automata, and the more general model of coherent stably concurrent automata.

More recently the problem of characterizing the unfoldings of *finite* concurrent automata has been investigated [14,18,11]. In [14], Schmitt established that all unfoldings of finite stable trace automata are also unfoldings of finite forward-stable asynchronous systems. In [18], Thiagarajan proved with the help of Zielonka's theorem [20] that all unfoldings of finite forward-stable asynchronous systems are also unfoldings of finite 1-safe Petri nets.

In this paper we improve both approaches and show that all unfoldings of finite coherent stably concurrent automata are unfoldings of finite 1-safe Petri nets. We proceed in two steps. With the help of some unrecognized difficult work by Arnold [1] we prove that if a prime event structure is the unfolding of a finite coherent stably concurrent automaton then it is also the unfolding of a finite coherent asynchronous system. Next we use Zielonka's theorem to establish that if a prime event structure is the unfolding of a finite coherent asynchronous system then it is also the unfolding of a finite forward-stable asynchronous system. This step is more technical so we sketch the construction in more details.

To simplify the presentation of this paper we consider particular domains as semantical objects instead of prime event structures. These domains are known to be equivalent to prime event structures so that we shall sketch in the conclusion how our results can be rephrased in that setting.

2 Background and Results

In this section we present the main framework of this study and compare the contribution of this paper to some known results from the literature.

First we introduce the very general automata-based model of concurrency known as automata with concurrency relations [5]. The latter appear as a generalization of several other models such as asynchronous systems, trace automata, and Mazurkiewicz traces.

DEFINITION 2.1. *An* automaton with concurrency relations *over the alphabet Σ is a structure $\mathcal{A} = (Q, \imath, \Sigma, \longrightarrow, (\|_q)_{q \in Q})$ such that*
1. *Q is a non-empty (possibly infinite) set of states, with an initial state $\imath \in Q$;*
2. *$\longrightarrow \subseteq Q \times \Sigma \times Q$ is a set of transitions;*
3. *if $q \xrightarrow{a} q_1$ and $q \xrightarrow{a} q_2$ then $q_1 = q_2$;*
4. *$(\|_q)_{q \in Q}$ is a family of binary, irreflexive, and symmetric relations on Σ;*
5. *if $a\|_q b$ then there exist $q \xrightarrow{a} q_1$, $q \xrightarrow{b} q_2$, $q_1 \xrightarrow{b} q_3$ and $q_2 \xrightarrow{a} q_3$.*

We say that $\mathcal{A}$ is finite *if Q and Σ are finite.*

The *language $L(\mathcal{A})$ of sequential computations* of $\mathcal{A}$ consists of all words $u = a_1...a_n \in \Sigma^*$ for which there are some states $q_0, ..., q_n \in Q$ such that $\imath = q_0$ and for each $i \in [1, n]$, $q_{i-1} \xrightarrow{a_i} q_i$. For short, these conditions will be denoted by $q_0 \xrightarrow{u} q_n$. Now the independence relations $\|_q$ yield a natural equivalence relation over the set of sequential computations $L(\mathcal{A})$ as follows. The *trace equivalence* $\sim_\mathcal{A}$ associated with $\mathcal{A}$ is the least equivalence over $L(\mathcal{A})$ such that for all words $u, v \in \Sigma^*$ and all actions $a, b \in \Sigma$ if $\imath \xrightarrow{u} p \xrightarrow{ab} q \xrightarrow{v} r$ and $a\|_p b$ then $u.ab.v \sim_\mathcal{A} u.ba.v$. Conditions 3 and 5 ensure that if w and w' are two trace equivalent words then they lead from the initial state to the same state.

For any word $u \in L(\mathcal{A})$, the trace $[u]$ consists of all words $v \in L(\mathcal{A})$ that are trace equivalent to u: Formally we put $[u] = \{v \in \Sigma^* \mid v \sim_\mathcal{A} u\}$. The *trace language* $\mathcal{L}(\mathcal{A}) = L(\mathcal{A})/\sim_\mathcal{A}$ consists of all traces. The latter are partially ordered in the following way: We put $[u] \sqsubseteq [v]$ if there exists some word $z \in \Sigma^*$ such that $u.z \sim_\mathcal{A} v$. The *trace domain* of $\mathcal{A}$ is the partial order $(\mathcal{L}(\mathcal{A}), \sqsubseteq)$.

EXAMPLE 2.2. Let $D = \{(n, m) \in \mathbb{N}^2 \mid n \leqslant m\}$ be equipped with the partial order $\sqsubseteq$ for which $(n, m) \sqsubseteq (n', m')$ if $n \leqslant n'$ and $m \leqslant m'$. Then $(D, \sqsubseteq)$ is (isomorphic to) the trace domain of the automaton with concurrency relations $\mathcal{A} = (D, (0,0), \{a, b\}, \longrightarrow, (\|_q)_{q \in Q})$ where $(n, m) \xrightarrow{a} (n', m')$ if $n' = n + 1$ and $m' = m$; $(n, m) \xrightarrow{b} (n', m')$ if $n' = n$ and $m' = m + 1$; and $a\|_{(n,m)} b$ if $n < m$. Noteworthy it is easy to see that no *finite* automaton with concurrency relations admits the partial order $(D, \sqsubseteq)$ as trace domain.

2.1 Coherent Stably Concurrent Automata

In the literature, a particular attention was devoted to automata whose concurrency relations $\|_q$ depend locally on each other. In the two following definitions, for all actions $a, b, c \in \Sigma$ and all states q, we write $a\|_{q.c} b$ if there exists a state $q' \in Q$ such that $q \xrightarrow{c} q'$ and $a\|_{q'} b$.

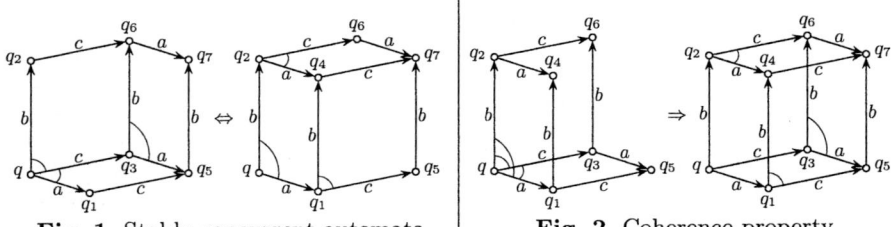

Fig. 1. Stably concurrent automata **Fig. 2.** Coherence property

DEFINITION 2.3. *An automaton with concurrency relations $\mathcal{A}$ is called a* stably concurrent automaton *if for all states $q \in Q$ and all actions $a, b, c \in \Sigma$:*
$a\|_q c \wedge b\|_q c \wedge a\|_{q.c} b$ *if and only if* $a\|_q b \wedge b\|_{q.a} c \wedge a\|_{q.b} c$

This requirement is depicted in Fig. 1. In this paper we are interested in stably concurrent automata that satisfy an additionnal coherence condition that ressembles the requirement (C) of the generalized trace languages from [13].

DEFINITION 2.4. *A stably concurrent automaton is* coherent *if for all states $q \in Q$ and all actions $a, b, c \in \Sigma$: $a\|_q b \wedge a\|_q c \wedge b\|_q c$ implies $a\|_{q.c} b$.*

This requirement is depicted in Fig. 2. The trace language of such a coherent stably concurrent automaton can be viewed as a generalized trace language [13]. Consequently, it corresponds also to a labeled event structure with a binary conflict. As we will explain in the conclusion, most results used or established in this paper can be rephrased and applied in the framework of event structures.

2.2 Forward-Stable Asynchronous Systems

Automata with concurrency relations are a generalization of several other models of concurrency, in particular Bednarczyk's asynchronous automata [2] and Stark's trace automata [15]. These models are known to have close relationships to event structures, too. As opposed to automata with concurrency relations, both models involve a single independence relation.

DEFINITION 2.5. *Let Σ be some alphabet and $\|$ be a binary, symmetric, and irreflexive relation over Σ. Let $\mathcal{A} = (Q, \iota, \Sigma, \longrightarrow, \|)$ be a structure satisfying Conditions 1, 2, and 3 of Definition 2.1. Then $\mathcal{A}$ is called an* asynchronous system *[2] if we have*

ID: $q_1 \xrightarrow{a} q_2 \wedge q_2 \xrightarrow{b} q_3 \wedge a\|b$ *implies* $q_1 \xrightarrow{b} q_4 \wedge q_4 \xrightarrow{a} q_3$ *for some* $q_4 \in Q$.

On the other hand, $\mathcal{A}$ is called a trace automaton *[15] if we have*

FD: $q_1 \xrightarrow{a} q_2 \wedge q_1 \xrightarrow{b} q_3 \wedge a\|b$ *implies* $q_2 \xrightarrow{b} q_4 \wedge q_3 \xrightarrow{a} q_4$ *for some* $q_4 \in Q$.

Finally, if $\mathcal{A}$ satisfies both conditions ID *and* FD *then it is called a* forward-stable asynchronous system *[2]*.

Asynchronous systems and trace automata can be seen as automata with concurrency relations by putting $a\|_q b$ if $a\|b$, $q \xrightarrow{a} q_1 \xrightarrow{b} q_2$, and $q \xrightarrow{b} q_3 \xrightarrow{a} q_2$. Consequently they are associated with a set of sequential computations $L(\mathcal{A})$, a trace language $\mathcal{L}(\mathcal{A})$, and a trace domain $(\mathcal{L}(\mathcal{A}), \sqsubseteq)$.

Observe that any trace automaton satisfies the coherence property (Fig. 2). Moreover it fulfills also half of the requirement to be a stably concurrent automaton (Fig. 1) namely the implication from left to right. We say that a trace automaton is *stable* if it is a stably concurrent automaton.

2.3 Comparisons of Expressive Power

Observe now that any asynchronous automaton is a stably concurrent automaton. Furthermore, it is coherent as soon as it is forward-stable. Thus the trace domain of any forward-stable asynchronous automaton is obviously the trace domain of a coherent stably concurrent automaton. The next theorem expresses the converse property.

THEOREM 2.6. *For any coherent stably concurrent automaton $\mathcal{A}$, there exists a forward-stable asynchronous system $\mathcal{A}'$ such that the trace domains $(\mathcal{L}(\mathcal{A}), \sqsubseteq)$ and $(\mathcal{L}(\mathcal{A}'), \sqsubseteq)$ are isomorphic.*

This theorem summarizes some results from [2,8]. In [2] it is shown that the trace domains of forward-stable asynchronous systems can be identified with event structures by means of a coreflection between the two models. This connection can be extended to coherent stably concurrent automata [8].

Consider now again the trace domain $(D, \sqsubseteq)$ of Example 2.2. As explained above, this partial order is isomorphic to the trace domain of some coherent stably concurrent automaton over the finite alphabet $\{a, b\}$. However, as noticed by Husson [7], any asynchronous automaton whose trace domain is isomorphic to $(D, \sqsubseteq)$ admits some *infinite* alphabet. This shows that Theorem 2.6 fails if one considers finite alphabets only.

In this paper, we focus on finite automata. We show in Corollary 4.6 that Theorem 2.6 remains valid if we restrict to *finite* stably concurrent automata. Since stable trace automata are coherent stably concurrent automata, Corollary 4.6 subsumes the following theorem due to Schmitt.

THEOREM 2.7. *[14, Th. 3.12] For any finite stable trace automaton $\mathcal{A}$, there exists a finite forward-stable asynchronous system $\mathcal{A}'$ such that the trace domains $(\mathcal{L}(\mathcal{A}), \sqsubseteq)$ and $(\mathcal{L}(\mathcal{A}'), \sqsubseteq)$ are isomorphic.*

Organization of the Paper. In the rest of this paper we consider only *finite* alphabets. In the following section we relate stably concurrent automata to the theory of regular consistent sets of pomsets [1] and Mazurkiewicz traces [4]. This first step allows us to transform a finite stably concurrent automaton into a finite asynchronous system with the same trace domain. Moreover this process preserves coherence. Next we state our main result (Cor. 4.6) and sketch its proof by means of Zielonka's theorem [20]. This second step shows how to build a forward-stable finite asynchronous system from a coherent one while preserving the trace domain. In order to improve [14] and [18], the main difficulty here is to ensure Property **FD** from the assumption of coherence. In the conclusion we explain how this work applies to the setting of regular event structures [18].

3 Consistent Sets of Pomsets

A *pomset* over an alphabet Σ is a triple $t = (E, \preccurlyeq, \xi)$ where $(E, \preccurlyeq)$ is a finite partial order and ξ is a mapping from E to Σ *without autoconcurrency*: $\xi(x) = \xi(y)$ implies $x \preccurlyeq y$ or $y \preccurlyeq x$ for all $x, y \in E$. A pomset can be seen as an abstraction of an execution of a concurrent system. In this view, the elements e of E are *events* and their label $\xi(e)$ describes the action that is performed in the system by the event $e \in E$. Furthermore, the order $\preccurlyeq$ describes the causal dependence between events. We denote by $\mathbb{P}(\Sigma)$ the class of all pomsets over Σ.

An *order extension* of a pomset $t = (E, \preccurlyeq, \xi)$ is a pomset $t' = (E, \preccurlyeq', \xi)$ such that $\preccurlyeq \subseteq \preccurlyeq'$. A *linear extension* of t is an order extension that is linearly ordered. It corresponds to a sequential view of the concurrent execution t. Linear extensions of a pomset t over Σ can naturally be regarded as words over Σ. By $\mathrm{LE}(t) \subseteq \Sigma^*$, we denote the set of linear extensions of a pomset t over Σ. For any subset of pomsets $\mathcal{L} \subseteq \mathbb{P}(\Sigma)$, we put $\mathrm{LE}(\mathcal{L}) = \bigcup_{t \in \mathcal{L}} \mathrm{LE}(t)$.

Two isomorphic pomsets admit the same set of linear extensions. Noteworthy the converse property holds [17]: If $\mathrm{LE}(t) = \mathrm{LE}(t')$ then t and t' are two isomorphic pomsets. In the sequel of this paper we do not distinguish between isomorphic pomsets any longer because they are used as representative of sets of words. In particular, $\mathrm{LE}(t) = \mathrm{LE}(t')$ implies $t = t'$.

An *ideal* of a pomset $t = (E, \preccurlyeq, \xi)$ is a subset $H \subseteq E$ such that $x \in H \wedge y \preccurlyeq x \Rightarrow y \in H$. The restriction $t' = (H, \preccurlyeq \cap (H \times H), \xi \cap (H \times \Sigma))$ is then called a *prefix* of t and we write $t' \leqslant t$. For any set of pomsets $\mathcal{L}$, $\mathrm{Pref}(\mathcal{L})$ denotes the set of prefixes of pomsets from $\mathcal{L}$. We say that $\mathcal{L}$ is *prefix-closed* if $\mathrm{Pref}(\mathcal{L}) = \mathcal{L}$.

3.1 Regular Consistent Sets of Pomsets

We borrow now the notion of consistent sets of pomsets from [1]. Although we deal mainly with prefix-closed sets of pomsets, we present here the general definition of a regular consistent set of pomsets. Intuitively, consistency means that concurrency is determined by any sequential ordering of events.

DEFINITION 3.1. *A set of pomsets $\mathcal{L}$ is called* consistent *if*
$$\forall t_1, t_2 \in \mathrm{Pref}(\mathcal{L}) : \mathrm{LE}(t_1) \cap \mathrm{LE}(t_2) \neq \emptyset \Rightarrow t_1 = t_2.$$

Let $\mathcal{L}$ be a consistent set of pomsets. The *pomset equivalence* $\sim_\mathcal{L}$ over $\mathrm{LE}(\mathcal{L})$ is such that $w \sim_\mathcal{L} w'$ if and only if $\{w, w'\} \subseteq \mathrm{LE}(t)$ for some $t \in \mathcal{L}$. Note that $\sim_\mathcal{L}$ is an equivalence relation over $\mathrm{LE}(\mathcal{L})$ because $\mathcal{L}$ is consistent.

For any two words $w, w' \in \Sigma^*$, we put $w \equiv_\mathcal{L} w'$ if for all words $u, v \in \Sigma^*$ it holds: $w.u \sim_\mathcal{L} w.v \Leftrightarrow w'.u \sim_\mathcal{L} w'.v$. It is easy to see that $\equiv_\mathcal{L}$ is a right-congruence over Σ^*. Relation $\equiv_\mathcal{L}$ appeared in [1] in the definition of regular, complete, consistent, and prefix-closed sets of pomsets.

DEFINITION 3.2. *A consistent set of pomsets $\mathcal{L}$ is regular if $\equiv_{\mathcal{L}}$ is of finite index.*

Regularity satisfies several natural properties. In particular if $\mathcal{L}$ is a regular consistent set of pomsets then $\mathrm{Pref}(\mathcal{L})$ is consistent and regular, too.

3.2 Relationships with Stably Concurrent Automata

The connection between prefix-closed consistent sets of pomsets and stably concurrent automata originates from a pomset description of traces.

THEOREM 3.3. *[3, Th. 4.6] Let $\mathcal{A}$ be a stably concurrent automaton over Σ. Each trace $[u] \in \mathcal{L}(\mathcal{A})$ is the set of linear extensions of a (unique) pomset.*

This result shows that the trace language $\mathcal{L}(\mathcal{A})$ of a stably concurrent automaton can be represented by a set of pomsets. *We adopt this dual view in the rest of this paper.* Clearly $L(\mathcal{A}) = \mathrm{LE}(\mathcal{L}(\mathcal{A}))$ and $\sim_{\mathcal{A}} = \sim_{\mathcal{L}(\mathcal{A})}$. Noteworthy $[u] \sqsubseteq [v]$ means in this setting that the pomset $[u]$ is a prefix of the pomset $[v]$. Moreover the trace language of any stably concurrent automaton is prefix-closed [3, Cor. 4.11]. It follows that $\mathcal{L}(\mathcal{A})$ is a consistent set of pomsets.

The mapping from stably concurrent automata to prefix-closed and consistent sets of pomsets is actually onto: *Any prefix-closed and consistent set of pomsets is the trace language of a stably concurrent automaton.* This connection specializes into a correspondance between *finite* stably concurrent automata and *regular*, prefix-closed, and consistent sets of pomsets.

3.3 From Consistent Sets of Pomsets to Mazurkiewicz Traces

Basically Arnold's result relates regular consistent sets of pomsets to regular Mazurkiewicz trace languages [4]. The latter can be viewed as the trace language of a particular asynchronous system. Consider some independence relation $(\Sigma, \|)$ and some asynchronous system $\mathcal{A}$ that has a single state q such that $q \xrightarrow{a} q$ for all $a \in \Sigma$. Then the set of Mazurkiewicz traces $\mathbb{M}(\Sigma, \|)$ may be defined as the trace language $\mathcal{L}(\mathcal{A})$. Since $\mathcal{A}$ is a stably concurrent automaton, for each $u \in \Sigma^{\star}$ there exists a (unique) pomset t over Σ such that $[u] = \mathrm{LE}(t)$ (Th. 3.3). That is why subsets of Mazurkiewicz traces are particular cases of consistent sets of pomsets. Noteworthy a subset of Mazurkiewicz traces $\mathcal{L} \subseteq \mathbb{M}(\Sigma, \|)$ is regular (Def. 3.2) if and only if $\mathrm{LE}(\mathcal{L})$ is a regular set of words.

Let Σ and Γ be two alphabets and $\pi : \Gamma \to \Sigma$ a mapping from Γ to Σ. This mapping extends in a natural way into a function that maps each pomset $t = (E, \preccurlyeq, \xi)$ over Γ to the structure $\pi(t) = (E, \preccurlyeq, \pi \circ \xi)$. The latter might not be a pomset over Σ in case some autoconcurrency appears in it. This situation can occur if $\pi(a) = \pi(b)$ for two distinct actions $a, b \in \Sigma$ while there are two events e and f that are labelled by a and b and that are not causally related. The next notion of refinement allows to relate two sets of pomsets that are identical up to some relabeling.

DEFINITION 3.4. *Let $\mathcal{L}$ and $\mathcal{L}'$ be two consistent sets of pomsets over Σ and Γ respectively. A mapping $\pi : \Gamma \to \Sigma$ from Γ to Σ is a refinement from $\mathcal{L}$ to $\mathcal{L}'$ if π is a bijection from $\mathcal{L}'$ onto $\mathcal{L}$ and from $\mathrm{Pref}(\mathcal{L}')$ onto $\mathrm{Pref}(\mathcal{L})$.*

Now one main contribution of [1] can be slightly extended as follows.

THEOREM 3.5. *[1, Th. 6.16] For any regular consistent set of pomsets $\mathcal{L}$ over Σ, there is a refinement from $\mathcal{L}$ to a regular set of Mazurkiewicz traces $\mathcal{L}' \subseteq \mathrm{M}(\Gamma, \|)$.*

4 From Coherence to Forward-Stability

In this section we want to apply Theorem 3.5 in order to build a finite *forward-stable* asynchronous system from a finite coherent stably concurrent automaton (Cor. 4.6). The requirement that the asynchronous system should be forward-stable is the main difficulty tackled in this section: Without this requirement, the result would follow directly from Th. 3.5 because any regular prefix-closed set of Mazurkiewicz traces is the trace language of a finite asynchronous system.

4.1 Coherent and Forward-Stable Mazurkiewicz Trace Languages

It is easy to characterize the trace languages associated with the stably concurrent automata we are intererested in.

DEFINITION 4.1. *A prefix-closed and consistent set of pomsets $\mathcal{L}$ over Σ is coherent if for all words $u \in \Sigma^\star$, all distinct actions $a, b, c \in \Sigma$:*
$u.ab \sim_\mathcal{L} u.ba \wedge u.bc \sim_\mathcal{L} u.cb \wedge u.ca \sim_\mathcal{L} u.ac$ *implies* $u.abc \sim_\mathcal{L} u.acb \sim_\mathcal{L} u.cab$.

Clearly, the trace language of a coherent stably concurrent automaton is coherent. Conversely, we can show that any coherent prefix-closed consistent set of pomsets is the trace language of some coherent stably concurrent automaton.

Since we deal also with forward-stable asynchronous systems (Def. 2.5 and Cor. 4.6), we focus also on forward-stable Mazurkiewicz trace languages.

DEFINITION 4.2. *A prefix-closed set of Mazurkiewicz traces $\mathcal{L} \subseteq \mathrm{M}(\Sigma, \|)$ is forward-stable w.r.t. $(\Sigma, \|)$ if for all words $u, v \in \Sigma^\star$ and all actions $a, b \in \Sigma$:*
$[u.a] \in \mathcal{L} \wedge [u.b] \in \mathcal{L} \wedge a\|b$ *implies* $[u.ab] \in \mathcal{L}$.

This condition is well-known. A forward-stable Mazurkiewicz trace language is called *safe-branching* in [16], *forward independence closed* in [10], *ideal* in [2], and *proper* in [9]. Clearly the trace language of a forward-stable asynchronous system is forward-stable. Actually, the converse property holds. As expressed by the next basic lemma, this connection specializes into a correspondance between finite asynchronous systems and regular sets of Mazurkiewicz traces.

LEMMA 4.3. *Any regular, forward-stable, and prefix-closed set of Mazurkiewicz traces is the trace language of some finite forward-stable asynchronous system.*

Observe now that any forward-stable prefix-closed set of Mazurkiewicz traces is coherent. It is easy to see that the converse property does not hold. However we can represent coherent set of Mazurkiewicz traces by forward-stable sets of Mazurkiewicz traces by means of a refinement (Def. 3.4). This is expressed for regular languages in the next useful result whose proof will be sketched in Subsection 4.3 and relies on Zielonka's theorem.

THEOREM 4.4. *Let $\mathcal{L}$ be a regular, coherent, and prefix-closed set of Mazurkiewicz traces. There exists a refinement from $\mathcal{L}$ to a regular, forward-stable, and prefix-closed set of Mazurkiewicz traces.*

In order to apply Theorem 4.4 together with Theorem 3.5, we observe that coherence of consistent sets of pomsets is preserved by refinements.

LEMMA 4.5. *Let $\mathcal{L}_1$ and $\mathcal{L}_2$ be two consistent sets of pomsets over Σ_1 and Σ_2 respectively. Let $\pi : \Sigma_1 \to \Sigma_2$ be a refinement from $\mathcal{L}_2$ to $\mathcal{L}_1$. If $\mathcal{L}_2$ is prefix-closed and coherent then $\mathcal{L}_1$ is prefix-closed and coherent, too.*

We come now to the statement of our main result.

COROLLARY 4.6. *For any finite coherent stably concurrent automaton $\mathcal{A}$, there exists some finite forward-stable asynchronous system $\mathcal{A}'$ such that the trace domains $(\mathcal{L}(\mathcal{A}), \sqsubseteq)$ and $(\mathcal{L}(\mathcal{A}'), \sqsubseteq)$ are isomorphic.*

Proof. The trace language $\mathcal{L}(\mathcal{A})$ is a regular, coherent, prefix-closed, and consistent set of pomsets. By Th. 3.5 there exists a refinement from $\mathcal{L}(\mathcal{A})$ to a regular and prefix-closed set of Mazurkiewicz traces $\mathcal{L}'$. By Lemma 4.5, $\mathcal{L}'$ is coherent, too. By Theorem 4.4, we get a refinement from $\mathcal{L}'$ to a regular, forward-stable, and prefix-closed set of Mazurkiewicz traces $\mathcal{L}''$. By Lemma 4.3, $\mathcal{L}''$ is the trace language of a finite forward-stable asynchronous system $\mathcal{A}''$. Since we can compose refinements, we get a refinement from $\mathcal{L}(\mathcal{A})$ to $\mathcal{L}(\mathcal{A}'')$. It follows that the trace domains $(\mathcal{L}(\mathcal{A}), \sqsubseteq)$ and $(\mathcal{L}(\mathcal{A}''), \sqsubseteq)$ are isomorphic. ∎

4.2 Zielonka's Theorem

Let $\mathcal{S} = (\mathcal{P}_i)_{i \in I}$ be a family of finite automata $\mathcal{P}_i = (Q_i, \iota_i, \Sigma_i, \longrightarrow_i)$ where Q_i is a non-empty finite set of states, $\iota_i \in Q_i$ is the initial state, Σ_i is an alphabet of actions and $\longrightarrow_i \subseteq Q_i \times \Sigma_i \times Q_i$ is a set of *deterministic* transitions: If $q \stackrel{a}{\longrightarrow} q'$ and $q \stackrel{a}{\longrightarrow} q''$ then $q' = q''$. The global behaviour of such a system can be modelled by a single automaton which is the *mixed product* of its components [6]: $\prod \mathcal{S} = (\prod_{i \in I} Q_i, (\iota_i)_{i \in I}, \bigcup_{i \in I} \Sigma_i, \longrightarrow)$ where $(q_i)_{i \in I} \stackrel{a}{\longrightarrow} (q'_i)_{i \in I}$ if and only if for all $i \in I$ it holds $a \in \Sigma_i \Rightarrow q_i \stackrel{a}{\longrightarrow}_i q'_i$ and $a \notin \Sigma_i \Rightarrow q_i = q'_i$. We can enrich the mixed product of $\mathcal{S}$ by explicitly modelling concurrency: We put $a \| b$ if $\{a,b\} \not\subseteq \Sigma_i$ for all $i \in I$. In that way we provide the mixed product $\prod \mathcal{S}$ with an independence relation $\|$ and turn it into a *forward-stable* asynchronous system. The latter is associated with a trace language $\mathcal{L}(\prod \mathcal{S})$.

Let us now formulate a particular version of Zielonka's theorem [20,10,16] in terms of mixed products and refinements. Let $\mathcal{A}$ be an asynchronous system over the independence alphabet $(\Sigma, \|)$ with set of states Q and initial state $\iota \in Q$. A finite family $\delta = (\Sigma_i)_{i \in I}$ of subsets of Σ is called a *distribution of* $(\Sigma, \|)$ if for all actions $a, b \in \Sigma$ we have $a \not\| b \Leftrightarrow \exists i \in I, \{a, b\} \subseteq \Sigma_i$. Given a subset of states $F \subseteq Q$, we let $\mathcal{L}_F(\mathcal{A})$ denote the subset of traces $[u]$ such that $\iota \stackrel{u}{\longrightarrow} q \in F$.

THEOREM 4.7. *Let $\delta = (\Delta_i)_{i \in I}$ be a distribution of some independence alphabet $(\Sigma, \|)$. Let $\mathcal{L} \subseteq \mathbb{M}(\Sigma, \|)$ be a regular, forward-stable, and prefix-closed set of*

Mazurkiewicz traces. There exists a family of finite automata $\mathcal{S} = (\mathcal{P}_i)_{i \in I}$ with local alphabets $(\Sigma_i)_{i \in I}$ and a refinement $\pi : \bigcup_{i \in I} \Sigma_i \to \Sigma$ from $\mathcal{L}$ to $\mathcal{L}(\prod \mathcal{S})$ such that for all $i \in I$ and all $a \in \bigcup_{i \in I} \Sigma_i$ it holds $a \in \Sigma_i \Leftrightarrow \pi(a) \in \Delta_i$.

4.3 Proof of Theorem 4.4

In this section we fix a regular, coherent, and prefix-closed set of Mazurkiewicz traces $\mathcal{L} \subseteq \mathbb{M}(\Gamma_1, \|_1)$. There exists a finite forward-stable asynchronous system $\mathcal{A}_1 = (Q_1, \imath_1, \Gamma_1, \longrightarrow_1, \|_1)$ together with a subset of states $F \subseteq Q_1$ such that $\mathcal{L} = \mathcal{L}_F(\mathcal{A}_1)$. Clearly we can assume that all states of $q \in Q_1$ are reachable from the initial state $\imath_1$. Consequently if $q \xrightarrow{a}_1 q' \in F$ then $q \in F$ because $\mathcal{L}$ is prefix-closed.

We build from $(\Gamma_1, \|_1)$ an extended independence alphabet $(\Gamma_2, \|_2)$ such that $\Gamma_2 = \Gamma_1 \uplus \{\{a, b\} \subseteq \Gamma_1 \mid a\|_1 b\}$ and the independence relation $\|_2$ is defined as follows:

– for all $a, b \in \Gamma_1$, $a\|_2 b$ if $a\|_1 b$;
– for all $\{a, b\} \in \Gamma_2 \setminus \Gamma_1$, for all $c \in \Gamma_1$, $c\|_2 \{a, b\}$ if $c\|_1 a$ and $c\|_1 b$;
– for all $\{a, b\}, \{c, d\} \in \Gamma_2 \setminus \Gamma_1$, $\{a, b\}\|_2 \{c, d\}$ if $a\|_1 c$, $a\|_1 d$, $b\|_1 c$, and $b\|_1 d$.

We fix some arbitrary distribution $\delta = (\Delta_i)_{i \in I}$ of $(\Gamma_1, \|_1)$. For each $i \in I$, we define an extended subset of actions $\Delta_i' = \Delta_i \uplus \{x \in \Gamma_2 \setminus \Gamma_1 \mid x \cap \Delta_i \neq \emptyset\}$. We can check easily that $\delta' = (\Delta_i')_{i \in I}$ is a distribution of $(\Gamma_2, \|_2)$.

We build also a new structure $\mathcal{A}_2 = (Q_2, \imath_2, \Gamma_2, \longrightarrow_2, \|_2)$ where $Q_2 \subseteq (Q_1)^{2^{\Gamma_1}}$, that is, a state $\sigma \in Q_2$ is a map that associates each subset of actions $A \subseteq \Gamma_1$ with some state $\sigma(A) \in Q_1$. The initial state $\imath_2 \in Q_2$ maps each subset $A \subseteq \Gamma_1$ to the initial state $\imath_1$. The transition relation $\longrightarrow_2$ is defined as follows: Consider two states $\sigma : 2^{\Gamma_1} \to Q_1$ and $\sigma' : 2^{\Gamma_1} \to Q_1$

– we put $\sigma \xrightarrow{a}_2 \sigma'$ for some action $a \in \Gamma_1$ if for all $A \subseteq \Gamma_1$,
  • if $a \in A$ then $\sigma'(A) = \sigma(A)$;
  • if $a \notin A$ then $\sigma(B) \xrightarrow{a}_1 \sigma'(A)$ where $B = \{c \in A \mid c\|_1 a\}$.
– we put $\sigma \xrightarrow{x}_2 \sigma'$ with $x = \{a, b\} \in \Gamma_2 \setminus \Gamma_1$ if $\sigma = \sigma'$, $\sigma(A) \xrightarrow{a}_1 q_a \in F$, $\sigma(A) \xrightarrow{b}_1 q_b \in F$, and $\sigma(A) \xrightarrow{ab}_1 q \notin F$ where $A = \{c \in \Gamma_1 \mid c\|_1 a \land c\|_1 b\}$.

Finally let Q_2 be the subset of states that are reachable from $\imath_2$. By an immediate induction, it is clear that for all words $u \in \Gamma_1^*$ and all states $\sigma \in Q_2$, if $\imath_2 \xrightarrow{u}_2 \sigma$ then $\imath_1 \xrightarrow{u}_1 \sigma(\emptyset)$. We shall prove a useful converse property in Lemma 4.8.

The product of two Mazurkiewicz traces $[w], [w'] \in \mathbb{M}(\Gamma_1, \|_1)$ is defined as usual by $[w] \cdot [w'] = [w.w']$. For all $A \subseteq \Gamma_1$ and all traces $[u] \in \mathbb{M}(\Gamma_1, \|_1)$ we denote by $[u]/A$ the least trace $[v]$ such that $[u] = [v] \cdot [z]$ for some $z \in A^*$. If $u \in L(\mathcal{A}_1)$ then we define the map $\sigma_u : 2^{\Gamma_1} \to Q_1$ as follows: For all $A \subseteq \Gamma_1$, we let $\sigma_u(A)$ be the state from Q_1 such that $\imath_1 \xrightarrow{v}_1 \sigma_u(A)$ for all $v \in [u]/A$. In particular $\imath_1 \xrightarrow{u}_1 \sigma_u(\emptyset)$. Note also that $\imath_2 = \sigma_\varepsilon$ and $u \sim v$ implies $\sigma_u = \sigma_v$.

LEMMA 4.8. *For all $u \in \Gamma_1^*$, $\imath_1 \xrightarrow{u}_1 q_1$ in $\mathcal{A}_1$ if and only if $\imath_2 \xrightarrow{u}_2 \sigma_u$ in $\mathcal{A}_2$.*

Proof. The proof follows by induction by means of the next two key properties. For all words $u \in \Gamma_1^\star$, for all actions $a \in \Gamma_1$, and for all subsets $A \subseteq \Gamma_1$
1. if $u.a \in L(\mathcal{A}_1)$ and $a \notin A$ then $[u.a]/A = [v.a]$ where $[v] = [u]/\{c \in A \mid c\|_1 a\}$;
2. if $u.a \in L(\mathcal{A}_1)$ and $a \in A$ then $\sigma_{u.a}(A) = \sigma(A)$. ∎

This lemma enables us to check easily that the structure $\mathcal{A}_2$ is a forward-stable asynchronous system. Now $\mathcal{L}(\mathcal{A}_2)$ is a regular, prefix-closed, and forward-stable set of Mazurkiewicz traces $\mathcal{L}(\mathcal{A}_2) \subseteq \mathbb{M}(\Gamma_2, \|_2)$. We can apply Zielonka's theorem (Th. 4.7) to $\mathcal{L}(\mathcal{A}_2)$ with the distribution $\delta' = (\Delta'_i)_{i \in I}$ from above. We get a new independence alphabet $(\Gamma_3, \|_3)$, a mapping $\pi : \Gamma_3 \to \Gamma_2$, and a finite system of finite automata $\mathcal{S}_3 = (\mathcal{P}_i)_{i \in I}$ with alphabets Σ_i such that $(\Sigma_i)_{i \in I}$ is a distribution of $(\Gamma_3, \|_3)$, π is a refinement from $\mathcal{L}(\mathcal{A}_2)$ to $\mathcal{L}(\prod \mathcal{S}_3)$, and for all $i \in I$ and all $a \in \bigcup_{i \in I} \Sigma_i$ we have $a \in \Sigma_i \Leftrightarrow \pi(a) \in \Delta_i$. Then for all $a, b \in \Gamma_3$, we have $a \|_3 b \Leftrightarrow \pi(a) \|_1 \pi(b)$. We put $\mathcal{A}_3 = (Q_3, \imath_3, \Gamma_3, \longrightarrow_3, \|_3) = \prod \mathcal{S}_3$. We can assume that in each component automaton $\mathcal{P}_i$, each action occurs in at most one transition.

We consider now a new forward-stable asynchronous system $\mathcal{A}_4$ by restricting the alphabet and the independence relation over $\mathcal{A}_3$. We let $(\Gamma_4, \|_4)$ be the independence alphabet such that $\Gamma_4 = \Gamma_3 \cap \pi^{-1}(\Gamma_1)$ and for all $a, b \in \Gamma_4$, $a \|_4 b$ if $a \|_3 b$ or there exists some action $x \in \Gamma_3 \setminus \Gamma_4$ such that $\pi(x) = \{\pi(a), \pi(b)\}$ and the next condition is satisfied for all $i \in I$ and all $q_i \in Q_i$:
$$\forall c \in \{a, b\} : \left(c \in \Sigma_i \wedge \exists q'_i \in Q_i, q_i \xrightarrow{c}_i q'_i\right) \Rightarrow \left(x \in \Sigma_i \wedge \exists \widetilde{q}_i \in Q_i, q_i \xrightarrow{x}_i \widetilde{q}_i\right)$$
As transitions, we simply restrict to transitions carrying actions from Γ_4: We put $q \xrightarrow{a}_4 q'$ if $q \xrightarrow{a}_3 q'$ and $\pi(a) \in \Gamma_1$. Obviously $\mathcal{A}_4 = (Q_3, \imath_3, \Gamma_4, \longrightarrow_4, \|_4)$ is also a forward-stable asynchronous system.

We build a last structure $\mathcal{A}_5$ by synchronizing $\mathcal{A}_1$ and $\mathcal{A}_4$ with a restriction to the global states $F \subseteq Q_1$. We put $\mathcal{A}_5 = (Q_5, (\imath_1, \imath_3), \Gamma_4, \longrightarrow_5, \|_4)$ where $Q_5 \subseteq F \times Q_3$ and the transition relation is defined as follows: We put $(q_1, q_3) \xrightarrow{a}_5 (q'_1, q'_3)$ if $q_1 \xrightarrow{\pi(a)}_1 q'_1$ and $q_3 \xrightarrow{a}_3 q'_3$. Now a pair $(q_1, q_3) \in F \times Q_3$ belongs to Q_5 if it is reachable, that is: There exists some $u \in \Gamma_4^\star$ such that $(\imath_1, \imath_3) \xrightarrow{u}_5 (q_1, q_3)$. It is easy to check that $\mathcal{A}_5$ is a finite asynchronous system.

We can use now the hypotheses that $\mathcal{L}$ is coherent and each action appears locally in at most one transition to show the crucial following fact.

LEMMA 4.9. *The finite asynchronous system $\mathcal{A}_5$ is forward-stable.*

Proof. Assume $(\imath_1, \imath_3) \xrightarrow{u}_5 (q_1, q_3) \xrightarrow{a}_5 (q'_1, q'_3)$ and $(q_1, q_3) \xrightarrow{b}_5 (q''_1, q''_3)$ with $a\|_4 b$. Since $a\|_3 b$, $q'_3 \xrightarrow{b}_3 q'''_3$ and $q''_3 \xrightarrow{a}_3 q'''_3$ for some state $q'''_3 \in Q_3$ because $\mathcal{A}_3$ is forward-stable. Similarly there exists some state $q'''_1 \in Q_1$ such that $q'_1 \xrightarrow{\pi(b)}_1 q'''_1$ and $q''_1 \xrightarrow{\pi(a)}_1 q'''_1$. It is sufficient to prove that $q'''_1 \in F$. We proceed by contradiction and assume $q'''_1 \notin F$. Let $A = \{c \in \Gamma_1 \mid c\|_1 \pi(a) \wedge c\|_1 \pi(b)\}$. We consider $[v] = [\pi(u)]/A$. We have $\imath_1 \xrightarrow{v}_1 \sigma_{\pi(u)}(A) \xrightarrow{z}_1 q_1 = \sigma_{\pi(u)}(\emptyset)$ with $z \in A^\star$. Since $q'_1 \in F$, $q''_1 \in F$, $q'''_1 \notin F$, and $\mathcal{L}$ is coherent we have $\sigma_{\pi(u)} \xrightarrow{x}_2 \sigma_{\pi(u)}$ with $x = \{\pi(a), \pi(b)\}$. There exists $x' \in \Gamma_3 \setminus \Gamma_4$ such that $u.x' \in L(\mathcal{A}_3)$ and $\pi(x') = x$. We show now that $a \|_4 b$ (which is the expected contradiction). Let $i \in I$ and $q_i \in Q_i$ be such that $a \in \Sigma_i$ and $q_i \xrightarrow{a} q_i'$. Then $\pi(a) \in \Delta_i'$ and $x \in \Delta_i'$. It follows that $x' \in \Sigma_i$ hence $q_i \xrightarrow{x'}_i \widetilde{q}_i$ for some local state $\widetilde{q}_i \in Q_i$ because $u.a \in L(\mathcal{A}_3)$, $u.x' \in L(\mathcal{A}_3)$, and each action appears locally in at most one transition. Similarly, for all $j \in I$ and all $q_j \in Q_j$ if $b \in \Sigma_j$ and $q_j \xrightarrow{b} q_j''$ then $x' \in \Sigma_j$ and $q_j \xrightarrow{x'}_j \widetilde{q}_j$ for some state $\widetilde{q}_j \in Q_j$. Thus $a \not\|_4 b$. ∎

We can easily check that the mapping $\pi : \Gamma_4 \to \Gamma_1$ induces a bijection from $L(\mathcal{A}_5)$ to $\text{LE}(\mathcal{L})$. Moreover $u \sim_{\mathcal{A}_5} v$ implies $\pi(u) \sim_{\mathcal{L}} \pi(v)$. To complete the proof and show that $\pi : \Gamma_4 \to \Gamma_1$ is a refinement from $\mathcal{L}$ to $\mathcal{L}(\mathcal{A}_5)$ it is sufficient to establish the converse property. Assume $u.ab.v \sim_{\mathcal{L}} u.ba.v$ where $u, v \in \Gamma_1^\star$ and $a\|_1 b$. There are $u', v' \in \Gamma_3^\star$ and $a', b' \in \Gamma_3$ such that $\pi(u') = u$, $\pi(v') = v$, $\pi(a') = a$, $\pi(b') = b$, and $u'.a'b'.v' \in L(\mathcal{A}_3)$. Moreover $u'.a'b'.v' \sim_{\mathcal{A}_3} u'.b'a'.v'$ because $a'\|_3 b'$. We need just to show that $a'\|_4 b'$. We proceed by contradiction and assume $a' \not\|_4 b'$. There exists some $x \in \Gamma_3 \setminus \Gamma_4$ such that $\pi(x) = \{a, b\}$ and the next condition is satisfied for all $i \in I$ and all $q_i \in Q_i$:
$$\forall c \in \{a', b'\} : \left(c \in \Sigma_i \wedge \exists q_i' \in Q_i, q_i \xrightarrow{c}_i q_i' \right) \Rightarrow \left(x \in \Sigma_i \wedge \exists \widetilde{q}_i \in Q_i, q_i \xrightarrow{x}_i \widetilde{q}_i \right)$$
We have $\iota_3 \xrightarrow{u'}_3 q_3 \xrightarrow{a'}_3 q_3' \xrightarrow{b'}_3 q_3''$ and $q_3 \xrightarrow{b'}_3 q_3''' \xrightarrow{a'}_3 q_3''$. If $x \in \Sigma_i$ then $\pi(x) \in \Delta_i'$ hence it holds $a \in \Delta_i$ or $b \in \Delta_i$, which implies that $a' \in \Sigma_i$ or $b' \in \Sigma_i$. Therefore $q_3 \xrightarrow{x}_3 \widetilde{q}_3$ in $\mathcal{A}_3$. It follows that $u'.x \in L(\mathcal{A}_3)$ and $u.\pi(x) \in L(\mathcal{A}_2)$. Hence $u.ab \notin \text{LE}(\mathcal{L}_F(\mathcal{A}_1)) = \text{LE}(\mathcal{L})$, a contradiction.

5 Related Works

To conclude we wish to sketch some connections between this paper and the theory of regular event structures [18]. Due to the page limit, most definitions are omitted here. However we believe that the following arguments can clarify how our results apply to that setting.

Prefix-closed, coherent, and consistent sets of pomsets are the trace languages of coherent stably concurrent automata. They can be regarded as generalized trace languages [13]. The latter are closely related to event structures by means of a coreflection whose units are labelings [13]. Event structures are a classical semantical model in concurrency theory since they appeared as the unfoldings of (possibly infinite) 1-safe Petri nets [12]. This strong relationship can be extended to forward-stable asynchronous systems [2] and coherent stably concurrent automata [8].

The results presented in this paper allow us to claim that for an event structure $\mathcal{E}$ the following conditions are equivalent:
 (i) $\mathcal{E}$ is the unfolding of a finite 1-safe Petri net;
 (ii) $\mathcal{E}$ admits a regular forward-stable Mazurkiewicz labeling;
 (iii) $\mathcal{E}$ is the unfolding of a finite forward-stable asynchronous system;
 (iv) $\mathcal{E}$ is the unfolding of a finite coherent asynchronous system;
 (v) $\mathcal{E}$ is the unfolding of a finite coherent stably concurrent automaton;
 (vi) $\mathcal{E}$ admits a regular labeling.

In [18], the equivalence between (i) and (ii) is established by means of Zielonka's theorem. The equivalences (ii) ⇔ (iii) and (v) ⇔ (vi) are easy consequences of the corresponding definitions. The implications (iii) ⇒ (iv) and (iv) ⇒ (v) are trivial. As explained in the proof of Corollary 4.6, Arnold's result (Th. 3.5) was used here to get (v) ⇒ (iv). Then we used Zielonka's theorem to prove (iv) ⇒ (iii) by means of Theorem 4.4. This work subsumes a difficult work by Schmitt who established that (iii) holds if and only if $\mathcal{E}$ is the unfolding of a finite stable trace automaton [14]. Note here that a direct proof of (vi) ⇒ (ii) would provide us easily with an alternative proof of Theorem 3.5 which is a difficult result.

A very interesting conjecture by Thiagarajan [18] characterizes the unfoldings of finite 1-safe Petri nets. It asserts that *any regular event structure is the unfolding of a finite 1-safe Petri net*. Since (vi) ⇒ (i), our contribution reduces this conjecture to proving that *any regular event structure admits a regular labeling*. We are investigating at present this issue by adapting to the general setting the techniques developped in [11]. We stress finally that any direct proof of Thiagarajan's conjecture would show that (vi) ⇒ (ii) because (vi) implies that $\mathcal{E}$ is a regular event structure. It would lead to a new proof of Theorem 3.5. The latter is thus a natural ingredient to answer this question.

References

1. Arnold A.: *An extension of the notion of traces and asynchronous automata*. RAIRO, Theoretical Informatics and Applications **25** (Gauthiers-Villars, 1991) 355–393
2. Bednarczyk M.A.: *Categories of Asynchronous Systems*. PhD thesis in Computer Science (University of Sussex, 1988)
3. Bracho F., Droste M., Kuske D.: *Representations of computations in concurrent automata by dependence orders*. Theoretical Computer Science **174** (1997) 67–96
4. Diekert V. and Rozenberg G.: *The Book of Traces*. (World Scientific, 1995)
5. Droste M.: *Concurrency, automata and domains*. LNCS **443** (1990) 195–208
6. Duboc C.: *Mixed product and asynchronous automata*. Theoretical Computer Science **48** (1986) 183–199
7. Husson J.-Fr.: *Modélisation de la causalité par des relations d'indépendances*. PhD thesis (Université Paul Sabatier de Toulouse, 1996)
8. Kuske D.: *Nondeterministic automata with concurrency relations and domains*. CAAP, LNCS **787** (1994) 202–217
9. Mazurkiewicz A.: *Trace theory*. LNCS **255** (1987) 279–324
10. Mukund M.: *From global specifications to distributed implementations*. In *Synthesis and Control of Discrete Event Systems*, Kluwer (2002) 19–34
11. Nielsen M., Thiagarajan P.S.: *Regular Event Structures and Finite Petri Nets: The Conflict-Free Case*. ICATPN, LNCS **2260** (2002) 335–351
12. Nielsen M., Plotkin G and Winskel G.: *Petri nets, event structures and domains I*. Theoretical Computer Science **13** (1980) 86–108
13. Sassone V., Nielsen M., Winskel G.: *Deterministic Behavioural Models for Concurrency (Extended Abstract)*. MFCS, LNCS **711** (1993) 682–692
14. Schmitt V.: *Stable trace automata vs. full trace automata*. Theoretical Computer Science **200** (1998) 45–100

15. Stark E.W.: *Connections between a Concrete and an Abstract Model of Concurrent Systems.* LNCS **442** (1990) 53–79
16. Ştefănescu A., Esparza J., and Muscholl A.: *Synthesis of distributed algorithms using asynchronous automata.* CONCUR, LNCS **2761** (2003) 20–34
17. Szpilrajn E.: *Sur l'extension de l'ordre partiel.* Fund. Math. **16** (1930) 386–389
18. Thiagarajan P.S.: *Regular Event Structures and Finite Petri Nets: A Conjecture.* Formal and Natural Computing, LNCS **2300** (2002) 244–256
19. Winskel G.: *Event structures.* Advances in Petri Nets, LNCS **255** (1987) 325–392
20. Zielonka W.: *Notes on finite asynchronous automata.* RAIRO, Theoretical Informatics and Applications **21** (Gauthiers-Villars, 1987) 99–135

Completeness and Degeneracy in Information Dynamics of Cellular Automata

Hidenosuke Nishio

Iwakura Miyake-cho 204, Sakyo-ku,
Kyoto, 606-0022 Japan
YRA05762@nifty.ne.jp

Abstract. This paper addresses an algebraic problem which arises from our study on the information dynamics of cellular automata (CA). The state set of a cell is assumed to be a polynomial ring $Q[X]$ modulo $X^q - X$ over a finite field $\mathrm{GF}(q)$, where X is the indeterminate called the information variable. When a CA starts with an initial configuration containing a cell with state X, the information of X is transmitted to neighboring cells by cellular computation. In such a computation, every cell of a global configuration takes a polynomial in $Q[X]$. Generally denote such a configuration by c_X and let G_{c_X} be the set of polynomials appearing in c_X. Our problem is to ask *how much information* of X is contained by G_{c_X}. For G_{c_X} we define the *degree of completeness* $\lambda(G_{c_X}) = \log_q |\langle G_{c_X} \rangle|$, where $\langle G_{c_X} \rangle$ is the subring of $Q[X]$ generated by G_{c_X} and investigate its relation to the *degree of degeneracy* $m(c_X)$ introduced before. We note here that $m(c_X) = q - |V(G_{c_X})|$, where $|V(G_{c_X})|$ is the cardinality of the value set of G_{c_X}. Then, we prove that $\lambda(G_{c_X}) = |V(G_{c_X})|$ and in turn that $\lambda(G_{c_X}) + m(c_X) = q$. This result suggests that the computation of the size of subrings is reduced to that of the value size, which is executed much easier than subring generation.

1 Introduction

The information dynamics of cellular automata (CA for short) is formulated exploiting the theory of polynomials over finite fields [1]. First, the state set of a cell is assumed to be a polynomial ring $Q[X]$ modulo $X^q - X$, where X is the indeterminate called the information variable. When a CA starts with an initial configuration containing a cell having X, the information of X is transmitted to neighboring cells by cellular computation. In such a computation cells take polynomial states in X. Suppose that at a certain time t, the global configuration c_X contains cells having polynomials in X. Denote such a set of polynomials by G_{c_X}. Then we ask a question *how much information of X is contained by c_X* or G_{c_X}. If we can restore (compute) X from G_{c_X} by ring operations, then c_X is said to preserve the information of X completely or c_X is called *informationally complete*. If not, we ask next *how much information* is preserved in c_X. For answering the question, in the previous paper [1], we defined the *degree of degeneracy* of the configuration and proved for instance that it generally increases in time. In this paper, we newly define the *degree of completeness* based on the subring generation and discuss its relationship to the degree of degeneracy. Particularly we prove a theorem that the degree of completeness and the degree of degeneracy makes the number of states (q). Its corollaries and examples are also given. Finally we generalize the theorem to several inderminates.

In the next two sections we extract the definitions and results from [1] as far as they are relevant to the present topics. For finite fields and polynomials, we refer to [2].

2 Preliminaries

2.1 Cellular Automaton over a Finite Field

One-dimensional CA over a finite field is defined by a 4-tuple $(Q, \mathbb{Z}, N, f)$, where Q is the set of the states of a cell, $\mathbb{Z}$ is the set of the integers, N is the neighborhood index, and f is the local state transition function. In this paper, we assume the basic neighborhood $N = \{-1, 0, +1\}$, though the theory works for general settings. Since the space $\mathbb{Z}$ and the neighborhood N are understood, CA is denoted by (Q, f).

The set of states Q is assumed to be a finite field GF(q), where $q = p^n$ with prime p and positive integer n.

The local function $f : Q \times Q \times Q \to Q$ is uniquely expressed by the polynomial form:

$$f(x, y, z) = u_0 + u_1 x + u_2 y + \cdots + u_i x^h y^j z^k + \cdots$$
$$+ u_{q^3-2} x^{q-1} y^{q-1} z^{q-2} + u_{q^3-1} x^{q-1} y^{q-1} z^{q-1},$$
$$\text{where } u_i \in Q \ (0 \le i \le q^3 - 1). \quad (1)$$

x, y and z assume the state values of the neighboring cells -1(left), 0(center) and $+1$(right), respectively.

The global map $F : C \to C$ is defined on the set of configurations $C = Q^{\mathbb{Z}}$; For any cell $i \in \mathbb{Z}$, $F(c)(i) = f(c(i-1), c(i), c(i+1))$, where $c(i)$ is the state of cell $i \in \mathbb{Z}$ of $c \in C$. For a configuration $c \in C$, the dynamics of CA is defined by $F^{t+1}(c) = F(F^t(c))$, $t \ge 0$, where $F^0(c) = c$.

2.2 Information Dynamics of Cellular Automata

Let X be a symbol different from those used in the polynomial form (1). It stands for an *unknown* state or the *information* about a cell of CA which is supposed to be in an (unknown) state from the state set. It is often called the *information variable*. In order to investigate the dynamics of information X in CA space, we consider another polynomial form, which generally defines the cell state of the extended CA.

$$g(X) = a_0 + a_1 X + \cdots + a_i X^i + \cdots + a_{q-1} X^{q-1},$$
$$\text{where } a_i \in Q, 0 \le i \le q - 1. \quad (2)$$

The polynomial form g uniquely defines a function $Q \to Q$ and the set of such functions is denoted by $Q[X]$. $Q[X]$ is a polynomial ring modulo $X^q - X$. Note that the cardinality of $Q[X]$ equals q^q and $pX = 0$, $X^q - X = 0$ in $Q[X]$. $Q[X]$ contains constant polynomials such as $g \equiv a, a \in Q$. They are often called *constants* or trivial polynomials.

Based upon CA=(Q, f) we define its extension CA$[X]$=$(Q[X], f_X)$, where the set of cell states is $Q[X]$ and the local function f_X is defined on the same basic neighborhood $\{-1, 0, 1\}$ and expressed by the same polynomial form (1) as f. The variables x, y and z, however, move in $Q[X]$ instead of Q. That is, $f_X : Q[X] \times Q[X] \times Q[X] \to Q[X]$. The global map is given by $F_X : C_X \to C_X$, where $C_X = Q[X]^{\mathbb{Z}}$. For any $i \in \mathbb{Z}$, $F_X(c_X)(i) = f(c_X(i-1), c_X(i), c_X(i+1))$, where $c_X(i)$ is the state of cell $i \in \mathbb{Z}$ of $c_X \in C_X$. For a configuration $c_X \in C_X$, the dynamics of CA$[X]$ is defined by $F_X^{t+1}(c_X) = F_X(F_X^t(c_X))$, $t \geq 0$, where $F_X^0(c_X) = c_X$. The global behavior(dynamics) of CA$[X]$ is called the *information dynamics* of CA$[X]$.

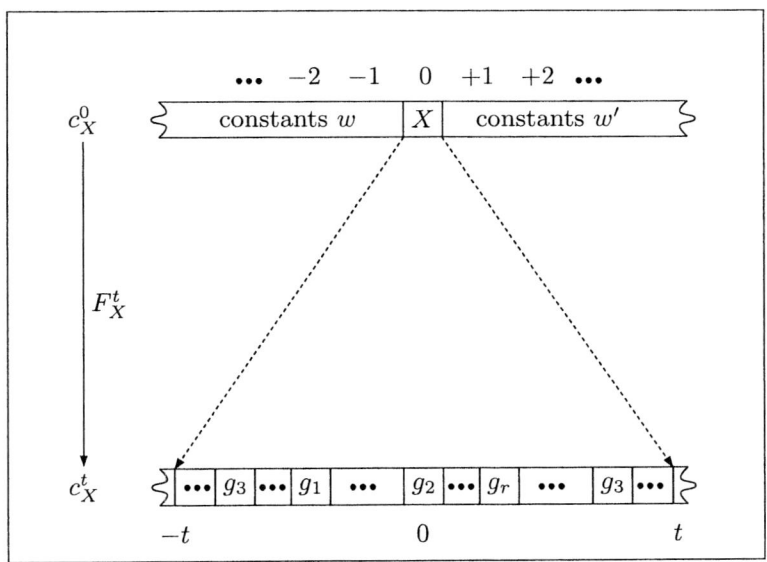

Fig. 1. Information dynamics of CA$[X]$

Fig.1. illustrates an elementary information dynamics, which begins with the initial configuration $c_X^0 = wXw'$, where $c_X^0(0) = X$ and w and w' are semi infinite strings of constant polynomial functions. Obviously w and w' do not contain any information about X. Then, by repeated application of F_X, the information X spreads in time t among cells $-t, -t+1, ..., 0, ..., t-1, t$ and at time t we observe a configuration c_X^t, which might contain some polynomials in X. Note that a same polynomial may appear at several cells. Such a set of polynomials is generally expressed by $G_{c_X^t} = \{g_1, g_2, ..., g_r\}$. $G_{c_X^t}$ is considered to preserve some amount of information of X contained by the initial configuration c_X^0.

3 Completeness and Degeneracy

As is shown in the preceding section, the information X may spread in CA space in the form of polynomials in X. >From here on we generally suppose a configuration c_X which contains polynomials $g_1, g_2, ..., g_r$ and denote the set of them by G_{c_X}. If we can retrieve or *compute* X from G_{c_X}, then we would be able to say that c_X completely preserves the information of X. Such a case trivially occurs, when there is the identity polynomial X or a permutation polynomial in X. Generally, however, some or all of information are lost during CA dynamics and we can not retrieve X from G_{c_X}. The case trivially occurs, if c_X contains only constant polynomials. In order to deepen such an observation, we have formulated the notion of *completeness* based on the ring operation as is shown in the following section.

Definition 1 (Complete configuration). *A subset $G \subseteq Q[X]$ is called complete, if G generates $Q[X]$. Any constant is allowed to be used at the computation. For any configuration $c_X \in C_X$, define the set of polynomials $G_{c_X} = \{c_X(i) | i \in \mathbb{Z}\} \subseteq Q[X]$. A configuration c_X is called complete, if G_{c_X} is complete.*

The notion of completeness has turned out to be closely related to the following notion of *degeneracy*, which is more tractable in studying the informational behavior of CA. For defining the degeneracy we need to define *substitution*.

Definition 2 (Substitution). *Evaluation of a function $g(X) \in Q[X]$ for $a \in Q$ is called substitution of $g(X)$. Symbolically we denote it as $\psi_a(g) = g(a)$. If g is a constant polynomial such that $g \equiv a, a \in Q$, then by definition $\psi_b(g) = a$ for any $b \in Q$. For any configuration $c_X \in Q[X]^{\mathbb{Z}}$ and $a \in Q$, we define the substitution of configuration $\psi_a(c_X)$ by the mapping $\psi_a : Q[X]^{\mathbb{Z}} \to Q^{\mathbb{Z}}$ such that $\psi_a(c_X)(i) = \psi_a(c_X(i)), i \in \mathbb{Z}$. Note that ψ_a is generally a many to one mapping. Throughout this paper $\psi_a(c_X)$ is often written as c_a.*

Proposition 1. *Substitution and ring operations of polynomials commute each other. That is, for any polynomials g and h, we have the following formulae.*

$$(A) \quad \psi_a(g+h) = \psi_a(g) + \psi_a(h) \quad (3)$$
$$(B) \quad \psi_a(g \cdot h) = \psi_a(g) \cdot \psi_a(h) \quad (4)$$

Proposition 2. *Substitution and global maps commute each other. That is, for any $c_X \in C_X$ and $a \in Q$,*

$$\begin{array}{ccc} c_X & \xrightarrow{F_X} & F_X(c_X) \\ \psi_a \downarrow & & \downarrow \psi_a \\ c_a & \xrightarrow{F} & F(c_a) \end{array} \quad (5)$$

Definition 3 (Degeneracy). *For a configuration $c_X \in Q[X]^{\mathbb{Z}}$, let $\psi_a(c_X)$ be a substitution of a in c_X. Then, c_X is called m-degenerate, if*

$$|\{\psi_a(c_X) \mid a \in Q\}| = |Q| - m. \tag{6}$$

It is easily seen that $0 \le m \le |Q| - 1$. Such m will be called the degree of degeneracy of c_X and denoted as m_{c_X}. A configuration c_X is simply called degenerate if $m_{c_X} \ne 0$.

Theorem 1. *A configuration is complete if and only if it is not degenerate.*

Proof. The proof has been given in [1].

4 Degree of Completeness and Value Size of G

It is generally interesting to investigate the lattice structure (set inclusion) of subrings of $Q[X]$, see [3]. In this paper, however, we focus on the cardinality (size) of subrings which is an important quantity in the information dynamics of CA.

For any subset $G \subseteq Q[X]$, we define two kinds of entity concerning G, the *degree of completeness* and the *value size*.

4.1 Degree of Completeness

Taking into account the fact that the cardinality of any (nontrivial) subring of $Q[X]$ is a power of q, we define the *degree of completeness* of G. It is also called the *log-ring size* of G.

Definition 4. *For any subset $G \subseteq Q[X]$, let $\langle G \rangle$ denote the subring generated by G and $|\langle G \rangle|$ its cardinality. Then, the degree of completeness $\lambda(G)$ is defined by*

$$\lambda(G) = \log_q |\langle G \rangle|. \tag{7}$$

Note that $1 \le \lambda(G) \le q$. Obviously, if $\lambda(G) = q$ then G is complete. That is, G preserves the information without any loss. See Section 3. If not, we claim that the greater $\lambda(G)$ is, the greater is the information contained by G. If $\lambda(G) = 1$ or G contains only constant polynomial functions (elements of Q), G does not contain any information about X.

4.2 Value Size of G

The following definition is a reformulation of Definition 3.

Definition 5. *Suppose that a subset $G \subseteq Q[X]$ consists of r polynomials: $G = \{g_1, g_2, ..., g_r \mid g_i \in Q[X], 1 \le i \le r\}$. Then an r-tuple of values $(g_1(a), g_2(a), ..., g_r(a))$ for $a \in Q$ is called the value vector of G for a and denoted by $G(a)$. Note that $G(a) \in Q^r$. The value set $V(G)$ of G is defined by*

$$V(G) = \{G(a) \mid a \in Q\}. \tag{8}$$

Finally we define the *value size* of G by $|V(G)|$. Note that $1 \le |V(G)| \le q$. It is seen that

$$|V(G_{c_X})| = q - m(c_X). \tag{9}$$

5 Results

We state and prove here the main theorem of this paper and its corollaries. The claim of the theorem has appeared, in a different form, without proof in the concluding remarks of [1]. It also provides another proof of Theorem 1 as Corollary 3 below. The theorem is generalized to polynomials in several indeterminates in the next section.

Theorem 2. *For any subset $G \subseteq Q[X]$, the degree of completeness is equal to the value size. That is,*

$$\lambda(G) = \log_q |\langle G \rangle| = |V(G)|. \tag{10}$$

Proof. For given G, we first assume $m = q - |V(G)| > 0$. Then there exist $q - m$ different value vectors $\gamma_1, \gamma_2, .., \gamma_{q-m} \in V(G)$ such that

$$G^{-1}(\gamma_i) = \{a \in Q | G(a) = \gamma_i\}, 1 \leq i \leq q - m. \tag{11}$$

$$\bigcup_{i=1}^{q-m} G^{-1}(\gamma_i) = Q. \tag{12}$$

It is seen that $\{G^{-1}(\gamma_i) | 1 \leq i \leq q - m\}$ constitutes a partition of Q, which is denoted by $\pi(G)$ and called a partition *induced by* G.

Let $\pi(g)$ be a partition of Q induced by g or a partition by the evaluation of g: $a \equiv b$ if and only if $g(a) = g(b)$. For two partitions π and π', we define a relation $\pi \succeq \pi'$ such that an equivalence class of π is a union of equivalence classes of π'. That is π is *coarser* than or equal to π' or π' is *finer* than or equal to π.

We prove the theorem using the following three lemmas.

Lemma 1. *If $g \in G$, then $\pi(g) \succeq \pi(G)$.*

Proof. From the definitions of $\pi(G)$ and the value vector, if $G(a) = G(b)$ then $g(a) = g(b)$ for any $a, b \in Q$. That is $\pi(g)$ is coarser than $\pi(G)$. □

We note here the commutativity of substitution and ring operations given in Proposition 1: $(g + h)(a) = g(a) + h(a)$ and $(g \cdot h)(a) = g(a)h(a)$.

Then we have the following lemma.

Lemma 2. *If $\pi(g) \succeq \pi(G)$ and $\pi(h) \succeq \pi(G)$, then $\pi(g+h) \succeq \pi(G)$ and $\pi(g \cdot h) \succeq \pi(G)$.*

Proof. Assume that $\pi(g) \succeq \pi(G)$ and $\pi(h) \succeq \pi(G)$, then, for any $a, b \in Q$, $G(a) = G(b)$ implies $g(a) = g(b)$ and $h(a) = h(b)$. Then, we see that $(g+h)(a) = g(a) + h(a) = g(b) + h(b) = (g+h)(b)$. Therefore we have $\pi(g + h) \succeq \pi(G)$. As for $g \cdot h$ the similar calculation applies and the lemma holds. □

Now we have the final lemma.

Lemma 3.
$$\langle G \rangle = \{h \in Q[X] \mid \pi(h) \succeq \pi(G)\}. \tag{13}$$

Proof. Since the subring $\langle G \rangle$ is a set of all polynomials that are obtained by ring operations on the polynomials of G and any constant, from the above lemmas $h \in \langle G \rangle$, $\pi(h) \succeq \pi(G)$ holds for any polynomial. Conversely, any polynomial function h such that $\pi(h) \succeq \pi(G)$ is obtained by ring operations on the polynomials of G with constants. In fact, assume that $\pi(h) = \{H_1, H_2, ..., H_s\}, 1 \leq s \leq q - m$, where every equivalence class H_k is a union of those of $\pi(G)$ and h takes a value $h_k \in Q$ on H_k such that $h_k \neq h_{k'}$ for $k \neq k'$. Then the following system of equations is solved for coefficients αs and powers βs.

$$h(a_i) = \sum_{\alpha \in Q} \alpha \prod_{j=1}^{r} g_j^{\beta_j}(a_i), \quad 0 \leq i \leq q-1, \tag{14}$$

where $\alpha \in Q$ and $0 \leq \beta_j \leq q-1, 1 \leq j \leq r$. That is h is expressed by a linear combination of products of powers of polynomials in G.

$$h(X) = \sum_{\alpha \in Q} \alpha \prod_{j=1}^{r} g_j^{\beta_j}(X). \tag{15}$$

Thus we have proved the lemma. □

By Lemma 3, $|\langle G \rangle|$ equals the number of polynomials h such that $\pi(h) \succeq \pi(G)$. It is easily seen to be equal to the number of all assignments of q elements from Q to each of $q - m$ equivalence classes of $\pi(G)$. Therefore, we see that $|\langle G \rangle| = q^{q-m} = q^{|V(g)|}$. Taking $\log_q$ of both sides, we have the theorem. When $m = 0$, every value vector of G is different, G generates $Q[X]$ and therefore $|\langle G \rangle| = q^q$. So, taking $\log_q$ we have the theorem. □

Example 1. Let $Q = GF(3) = \{0, 1, 2\}$ and $G = \{g\} = \{X^2\}$. Since $g(0) = 0, g(1) = g(2) = 1$, we see that $m = q - |V(g)| = 1$ and have a partition $\pi_I = \{\{0\}, \{1, 2\}\}$. There are 3^2 assignments of values from $\{0, 1, 2\}$ to each of two equivalence classes. Therefore, there are 9 polynomials each of which induces a partition coarser than or equal to π_I:

$$0, 1, 2, X^2, 2X^2, 1 + X^2, 2 + X^2, 1 + 2X^2, 2 + 2X^2.$$

For another generator set $G = \{1 + X + X^2\}$ we have another partition $\pi_{II} = \{\{0, 2\}, \{1\}\}$ and also $|\langle 1 + X + X^2 \rangle| = 3^2$.

Example 2. Let $Q = GF(5) = \{0, 1, 2, 3, 4\}$. Consider two subsets of $Q[X]$ $G_1 = \{X^2\}$ and $G_2 = \{X + X^2\}$. It is seen that $\pi(X^2) = \{\{0\}, \{1, 4\}, \{2, 3\}\}$ with $V(G_1) = \{0, 1, 4\}$ and $\pi(X + X^2) = \{\{0, 4\}, \{1, 3\}, \{2\}\}$ with $V(G_2) = \{0, 1, 2\}$. $|V(G_1)| = |V(G_2)| = 3$. According to the theorem, we have $\lambda(G_1) = \lambda(G_2) = 2$, while $\lambda(G_1 \cup G_2) = 5$ since $\langle G_1 \cup G_2 \rangle = Q[X]$ or $G_1 \cup G_2$ is complete.

Example 3. Concerning the proof of Lemma 3, we consider again $GF(5)$ and $G_1 = \{X^2\}$. Assume a function h which induces a coarser partition such that $\pi(h) = \{\{0\}, \{1, 2, 3, 4\}\} \succeq \pi(X^2)$. If $h(0) = 0$ and $h(1) = h(2) = h(3) = h(4) = 1$, then by solving the system of equations (14) we have $h(X) = X^4 = (X^2)^2$ a multiplication of X^2 by itself. If $h(0) = 1$ and $h(1) = h(2) = h(3) = h(4) = 0$, then we see that $h(X) = 1 + 4X^4$, which is also obtained by ring operations on X^2 with constants 1 and 4.

Using Theorem 2 and its proof we have the following corollaries.

Corollary 1. *For $0 \leq m \leq q-1$ and a partition π of Q, we can compute all (q^{q-m}) polynomial functions (subring) that are m-degenerate with respect to π.*

Proof. By assigning values from Q to equivalence classes of π, we obtain a function $h : Q \to Q$ such that $h(a) = h(b)$ if and only if a and b are equivalent. Then, by using Lagrange interpolation formula[1], we have a polynomial g such that $g(a) = h(a), \forall a \in Q$. □

The following corollary is a special case of Corollary 1.

Corollary 2. *For any $1 \leq i \leq q-1$, there exists a subring R such that $|R| = q^i$.*

Proof. Consider a polynomial function which has the value size i. For example, we can take a function h such that

$$h(a_0) = a_0, h(a_1) = a_1, h(a_2) = a_2, \cdots,$$
$$h(a_{i-1}) = a_{i-1} = h(a_i) = h(a_{i+1}) = \cdots = h(a_{q-1}). \quad (16)$$

Then, using the interpolation formula, we obtain a polynomial g such that $g(c) = h(c)$, for any $c \in Q$. Therefore we see $|V(g)| = |V(h)|$. Then by Theorem 2 we have $|\langle g \rangle| = |V(g)| = |V(h)| = q^i$. □

Theorem 1 is proved as a corollary to Theorem 2.

Corollary 3. *A configuration is complete if and only if it is not degenerate.*

Proof. c_X is complete, if and only if $\langle G_{c_X} \rangle = Q[X]$. Therefore, $\lambda(G_{c_X}) = q$. Then by Theorem 2 we see $|V(G_{c_X}| = q$, which implies $m_{c_X} = 0$. □

6 Polynomials in Several Indeterminates

In the information dynamics of CA, we can consider a CA with n *mutually independent* information variables $X_1, X_2, ..., X_n$ [4] [5]. For instance, we can investigate informational interactions between two cells, say $i, j \in \mathbb{Z}$, by considering a CA with X_1, X_2, which starts with an initial configuration $w'X_1wX_2w"$, where the state of cell i is X_1 and that of cell j is X_2 while the other cells have constant states.

---
[1] Equation (7.20), page 369, [2].

To formalize it, let $Q[X_1, X_2, ..., X_n]$ be the polynomial ring in $X_1, X_2, ..., X_n$ over Q modulo $(X_1^q - X_1)(X_2^q - X_2)\cdots(X_n^q - X_n)$. In the sequel the n-tuple of indeterminates $X_1, X_2, ..., X_n$ is denoted by $\mathbf{X^n}$. The basic one-dimensional CA with $\mathbf{X^n}$ is defined by $(Q[\mathbf{X^n}], f_{\mathbf{X^n}})$ in the same way as one-variable case. The local function is defined by $f_{\mathbf{X^n}} : Q[\mathbf{X^n}] \times Q[\mathbf{X^n}] \times Q[\mathbf{X^n}] \to Q[\mathbf{X^n}]$ and the global map induced by $f_{\mathbf{X^n}}$ is $F_{\mathbf{X^n}} : Q[\mathbf{X^n}]^{\mathbb{Z}} \to Q[\mathbf{X^n}]^{\mathbb{Z}}$.

Among others, we note here that Theorem 2 and its corollaries are generalized to polynomial rings in $\mathbf{X^n}$. The degree of completeness and the value size of $G \subseteq Q[\mathbf{X^n}]$ are defined in the same way as one indeterminate case. Note, however, that $1 \leq \lambda(G) \leq q^n$ and $1 \leq |V(G)| \leq q^n$. Then, we have the following theorem which is proved in the same manner as one variable case.

Theorem 3. *For any subset* $G \subseteq Q[\mathbf{X^n}]$,

$$\lambda(G) = \log_q |\langle G \rangle| = |V(G)|. \tag{17}$$

Proof. Omitted.

7 Concluding Remarks

Lemma 3 shown in the proof of Theorem 2 means that the subring generated by any subset G is equal to the set of polynomial functions which induces coarser partitions than that of G. Owing to Theorem 2 the computation of the size of subring $|\langle G \rangle|$ is reduced to that of the value size $|V(G)|$, which is much easier to execute. The complexity of subalgebra generation is generally discussed (proved to be **P**-complete) in [6], but the present result will serve the computational algebra in another direction.

Many thanks are due to Friedrich von Haeseler for his discussions on this topics by and his submitted paper [3], which provides a comprehensive solution for our problem about the structure of subrings generated by subsets of polynomials appearing in information dynamics of CA.

References

1. Nishio, H., Saito, T.: Information Dynamics of Cellular Automata I : An Algebraic Study. Fundamenta Informaticae **58** (2003) 399–420
2. Lidl, R., Niederreiter, H.: Finite Fields. Second edn. Cambridge University Press (1997)
3. von Haeseler, F.: On a Problem in Information Dynamics of Cellular Automata. (Int. Journal of Unconventional Computing, submitted 2004)
4. Nishio, H., Saito, T.: Information Dynamics of Cellular Automata II: Completeness, Degeneracy and Entropy (2002) 8th Workshop IFIP WG1.5, Prague, Sept. 2002.
5. Nishio, H., Saito, T.: Information Dynamics of Cellular Automata : CA Computation and Information Theory. Technical Report (kokyuroku) vol. 1325, RIMS, Kyoto University (2003) Proceedings of LA Symposium, Feb. 2003.
6. Bergman, C., Slutzki, G.: Computational Complexity of Generators and Nongenerators in Algebra. International Journal of Algebra and Computation **12** (2002) 719–735

Strict Language Inequalities and Their Decision Problems

Alexander Okhotin*

Department of Mathematics, University of Turku, Turku FIN–20014, Finland
http://www.cs.queensu.ca/home/okhotin/

Abstract. Systems of language equations of the form $\{\varphi(X_1,\ldots,X_n) = \varnothing, \psi(X_1,\ldots,X_n) \neq \varnothing\}$ are studied, where φ, ψ may contain set-theoretic operations and concatenation; they can be equivalently represented as strict inequalities $\xi(X_1,\ldots,X_n) \subset L_0$. It is proved that the problem whether such an inequality has a solution is Σ_2-complete, the problem whether it has a unique solution is in $(\Sigma_3 \cap \Pi_3) \setminus (\Sigma_2 \cup \Pi_2)$, the existence of a regular solution is a Σ_1-complete problem, while testing whether there are finitely many solutions is Σ_3-complete. The class of languages representable by their unique solutions is exactly the class of recursive sets, though a decision procedure cannot be algorithmically constructed out of an inequality, even if a proof of solution uniqueness is attached.

1 Introduction

Language equations are equalities of the form $\varphi(X_1,\ldots,X_n) = \psi(X_1,\ldots,X_n)$, where the variables $X_1,\ldots,X_n$ assume values of languages, while the expressions φ and ψ use language-theoretic operations from some predefined set. In a more general sense, a language equation is any formally specified relationship between sets of strings that contains unknowns. This definition includes some particular variants, such as language inequalitites $\varphi \subseteq \psi$ [8,10], inequations $\varphi \neq \psi$, proper inequalities $\varphi \subset \psi$, as well as mixed systems of equations of these four types.

The origins of language equations can be traced to a paper by Ginsburg and Rice [5] on the specification of context-free languages, and to the monographs on automata theory by Salomaa [18] and by Conway [4]. The automata-theoretic direction in the study of language equations led to an important concept of an alternating finite automaton [3]. The results of Ginsburg and Rice have been extended to conjunctive grammars [11], and finally led to a definition of Boolean grammars [13], which depart from the generative paradigm.

What is perhaps the most important about language equations, is that language-theoretic problems arising in different areas can be reduced to their decision problems. The important turn towards understanding this role of language equations was made by Baader and Narendran [1] and by Baader and Küsters [2], who reduced term unification in several description logics to certain

---

* Supported by the Academy of Finland under grant 206039.

decision problems for language equations with one-sided concatenation, and determined the complexity of those problems.

The next step towards a computational theory of language equations was undertaken by the author [12], who began a systematic study of language equations with Boolean operations and unrestricted concatenation. An important result was their computational universality [12], which has subsequently been extended to more restricted cases of language equations [9,15]. One of these cases led to an unexpected negative solution, due to Kunc [9], to the long-standing Conway's problem on the commutation of languages [4,7].

This paper continues the study of the computational properties of language equations [12], investigating new problems and more general types of equations. Important decision problems, such as whether a system has a finite or a regular solution [1,2], are for the first time considered for a general type of language equations with unrestricted concatenation and all Boolean operations. Another novelty is a further extension of the model: strict inequalities and inequations make their first appearance in the systems. Let us start with defining the general form of language equations to be studied in the following.

2 Forms of Equations

We shall study all four types of language equations mentioned in the introduction, which are:

- *equations* (in the narrow sense) $\varphi(X_1, \ldots, X_n) = \psi(X_1, \ldots, X_n)$,
- *inequalities* $\varphi(X_1, \ldots, X_n) \subseteq \psi(X_1, \ldots, X_n)$,
- *inequations* $\varphi(X_1, \ldots, X_n) \neq \psi(X_1, \ldots, X_n)$ and
- *strict inequalities* $\varphi(X_1, \ldots, X_n) \subset \psi(X_1, \ldots, X_n)$.

Both sides of each of these types of equations may contain concatenation, union, intersection and complement, as well as any regular constant languages over a fixed alphabet Σ, while the variables assume values of languages over Σ. Actually, the constants can be restricted to $\{\varepsilon\}$ and $\{a\}$ ($a \in \Sigma$) without decreasing the expressive power [12,15], or relaxed to arbitrary recursive languages without changing any of the constructions of this paper.

We shall consider mixed systems of language equations of all four types. A vector of languages $L = (L_1, \ldots, L_n)$ is a solution of such a system, if a substitution of L_i for X_i in all equations yields a tautology in each. Such systems can be represented in two normal forms; the first of these forms is a system of an equation and an inequation, each with the empty set in the right-hand side:

$$\varphi(X_1, \ldots, X_n) = \varnothing \tag{1a}$$
$$\psi(X_1, \ldots, X_n) \neq \varnothing \tag{1b}$$

Any inequality $\xi \subseteq \eta$ can be equivalently rewritten as $\xi \setminus \eta = \varnothing$. Any equality $\xi = \eta$ holds if and only if $\xi \triangle \eta = \varnothing$ (where $\triangle$ denotes the symmetric difference of sets), and, similarly, an inequation $\xi \neq \eta$ can rewritten as $\xi \triangle \eta \neq \varnothing$. A proper inequality $\xi \subset \eta$ is equivalent to a system $\{\xi \setminus \eta = \varnothing, \eta \setminus \xi \neq \varnothing\}$. Finally, a system with multiple equations of each type (1a,1b), $\{\varphi_1 = \varnothing, \ldots, \varphi_m = \varnothing, \psi_1 \neq \varnothing, \ldots, \psi_n \neq \varnothing\}$ can be simplified to $\{\varphi_1 \cup \ldots \cup \varphi_m = \varnothing, \psi_1 \cdot \ldots \cdot \psi_n \neq \varnothing\}$. These transformations can be applied to convert any system to the form (1).

The other form serving as a universal representation is a single strict inequality $\xi(X_1, \ldots, X_n) \subset L_0$, where L_0 is a regular constant language. It is easy to see that (1) holds if and only if $a\varphi \cup b\overline{\psi} \subset b\Sigma^*$. The form (1) will be used in this paper for all results on arbitrary systems of language equations, while the examples of systems will utilize all four types of equations.

Example 1. The following system over $\Sigma = \{a\}$

$$Y \subseteq aY \cup \varepsilon \tag{2a}$$
$$Y \neq a^* \tag{2b}$$
$$X \subseteq Y \tag{2c}$$
$$aX \setminus Y \neq \varnothing \tag{2d}$$

has the set of solutions $\{(L, L') \mid L \text{ is finite}, L' \text{ is the substring closure of } L\}$.

The first equation (2a) states that Y is suffix-closed, which means that it is either $a^{<n}$ for some $n \geqslant 0$, or a^*; the latter possibility is ruled out by (2b). So Y must be a finite language of the form $a^{<n}$, while X is its subset by (2c). However, Y cannot be *any* superset of X: according to (2d), if a is appended to the longest string in X, the resulting string should not be in Y. This limits Y to the substring closure of X.

Example 2. The system from Example 1 can be equivalently rewritten as a system of an equation and an inequality

$$(Y \setminus (aY \cup \varepsilon)) \cup (X \setminus Y) = \varnothing \tag{3a}$$
$$(Y \bigtriangleup a^*)(aX \setminus Y) \neq \varnothing \tag{3b}$$

or as a proper inequality $a\big[(Y \setminus (aY \cup \varepsilon)) \cup (X \setminus Y)\big] \cup b\overline{(Y \bigtriangleup a^*)(aX \setminus Y)} \subset ba^*$.

The language of valid accepting computations of a Turing machine, VALC(T) [6], has proved to be a very important tool in the study of language equations [12,15,16]. In short, for every TM T over an input alphabet Σ one can construct an alphabet Γ and an encoding of computations $C_T : \Sigma^* \to \Gamma^*$, such that

$$\text{VALC}(T) = \{w \# C_T(w) \mid C_T(w) \text{ is an accepting computation}\}, \tag{4}$$

is an intersection of two linear context-free languages, and hence can be specified by a system of language equations.

Example 3 ([15]). For every Turing machine T, there exists and can be effectively constructed a two-variable language equation $\upsilon_T(Y, Z) = \varnothing$ which uses all Boolean operations and linear concatenation, and has a unique solution of the form (VALC(T), L'), where L' is a certain auxiliary language.

Language equations with Boolean operations and concatenation are expressive enough to extract the language *recognized* by a Turing machine out of the language of its computations [12]. This makes such equations computationally universal and binds their study to the arithmetical hierarchy.

Let us recall the definition of this key notion of the classical recursion theory [17]. The arithmetical hierarchy consists of the classes Σ_k and Π_k (for all $k \geqslant 1$). A language L is said to be in Σ_k if it can be represented as $\{w \mid \exists x_1 \forall x_2 \ldots Q_k x_k \ R(w, x_1, \ldots, x_k)\}$ for some recursive predicate R, where $Q_k = \exists$ if k is odd, $Q_k = \forall$ if k is even. Similarly, L is in Π_k if its complement is in Σ_k, or, in other words, if it is of the form $\{w \mid \forall x_1 \exists x_2 \ldots Q_k x_k \ R(w, x_1, \ldots, x_k)\}$. There are complete sets in each Σ_k and Π_k ($k \geqslant 1$). It is easy to see that $\Sigma_1 = \mathrm{RE}$ and $\Pi_1 = \mathrm{co\text{-}RE}$, and their complete sets are the TM halting problem and its complement. For all k, the inclusions $\Sigma_k, \Pi_k \subset \Sigma_{k+1}$ and $\Sigma_k, \Pi_k \subset \Pi_{k+1}$ are known to be proper, while Σ_k and Π_k are incomparable. The intersection of Σ_k and Π_k is the class of languages decidable using an oracle for Σ_{k-1}.

3 Existence of Finite and Regular Solutions

A vector $(L_1, \ldots, L_n)$ is said to be a finite (a regular) solution of a system of language equations if it is a solution and all languages $L_1, \ldots, L_n$ are finite (regular, respectively). These special types of solutions have an advantage of being effectively representable, and algorithms to compute finite [1] and regular solutions [2] for some restricted types of language equations have been constructed.

However, that turns out to be impossible in our more general case:

Theorem 1. *The set of systems of language equations* $\{\varphi(X_1, \ldots, X_n) = \varnothing, \psi(X_1, \ldots, X_n) \neq \varnothing\}$ *that have a finite solution (a regular solution) is RE-complete. Both problems remain RE-complete for individual equations* $\varphi(X_1, \ldots, X_n) = \varnothing$, *in which the concatenation is restricted to linear.*

Proof. **Membership in RE.** It suffices to consider all vectors of n finite languages $(L_1, \ldots, L_n)$ (all vectors of n finite automata $A_1, \ldots, A_n$ in the case of regular solutions) substituting each into both equations and determining whether the system is satisfied. If the system has a finite (regular, resp.) solution, such a vector will eventually be found. If there are no finite (regular, resp.) solutions, the computation will never terminate.

RE-hardness. Reduction from the Post Correspondence Problem for nonempty strings, stated as *"Given an alphabet Σ and a finite set of pairs $(u_1, v_1), \ldots, (u_m, v_m)$, where $u_i, v_i \in \Sigma^+$, determine whether there exists a finite sequence of numbers $i_1, \ldots i_n$ ($n \geqslant 1$, $1 \leqslant i_j \leqslant m$), such that $u_{i_1} \ldots u_{i_n} = v_{i_1} \ldots v_{i_n}$"*.

Let $\{x_1, \ldots, x_m\}$ be a block code over Σ (i.e., $|x_1| = \ldots = |x_m|$ and $x_i \neq x_j$ for all $i \neq j$). Define a set of variables $\{X, Y, Z, Y_1, \ldots, Y_m, Z_1, \ldots, Z_m\}$ and construct the following system of language equations over $\Sigma \cup \{\#\}$ ($\# \notin \Sigma$):

$$X = Y \cap Z \cap \Sigma^* \# \Sigma^+ \tag{5a}$$
$$Y = x_1 Y_1 u_1 \cup \ldots \cup x_m Y_m u_m \cup \# \tag{5b}$$
$$Y = Y_1 \cup \ldots \cup Y_m \cup X \tag{5c}$$
$$Z = x_1 Z_1 v_1 \cup \ldots \cup x_m Z_m v_m \cup \# \tag{5d}$$
$$Z = Z_1 \cup \ldots \cup Z_m \cup X \tag{5e}$$

Note that if the nonterminals Y_i are replaced with Y in (5b), while (5c) is altogether removed, and the same is done with respect to Z, we obtain the usual encoding of PCP as an intersection of two linear context-free languages, in which Y and Z assume nonregular values regardless of the solvability of the PCP instance. In our case, each variable Y_i means the set of those strings from Y that are prolonged with x_i and u_i. The equation (5c) means: "every string w in Y, unless it is also in X, must be prolonged to $x_i w y_i$ at least for one i". If any strings get into X, this process can be stopped, which allows us to obtain a finite solution when PCP is solvable.

It is easy to see that if a vector L satisfies (5), then every $w \in Y(L)$ must be of the form $x_{i_k} \ldots x_{i_1} \# u_{i_1} \ldots u_{i_k}$, for some $k \geqslant 0$ and $1 \leqslant i_j \leqslant m$; similarly, every $w \in Z(L)$ is of the form $x_{i_k} \ldots x_{i_1} \# v_{i_1} \ldots v_{i_k}$. Accordingly, if $x_{i_1} \ldots x_{i_n} \# w \in X(L)$, then PCP has a solution $u_{i_1} \ldots u_{i_n} = v_{i_1} \ldots v_{i_n} = w$.

Suppose the given instance of PCP is not solvable, and let us prove that all solutions of the system (5) are infinite. Suppose L is a finite solution and let w be the longest string in $Y(L)$. Since $X(L) = \varnothing$, by (5c), there exists i, such that $w \in Y_i(L)$. Then $u_i w x_i \in Y(L)$ by (5b), and hence w is not the longest string. The contradiction obtained proves that $Y(L)$ is not finite. Furthermore, using the pumping lemma it can be proved that $Y(L)$ is not regular.

Now let the instance of PCP be solvable, and let $u_{i_1} \ldots u_{i_n} = v_{i_1} \ldots v_{i_n} = w$ be the shortest string that meets its specification. Then (5) has the following finite solution: ($X = \{x_{i_n} \ldots x_{i_1} \# w\}$, $Y = \{x_{i_k} \ldots x_{i_1} \# u_{i_1} \ldots u_{i_k} \mid 0 \leqslant k \leqslant n\}$, $Y_i = \{x_{i_k} \ldots x_{i_1} \# u_{i_1} \ldots u_{i_k} \mid 0 \leqslant k < n, i_{k+1} = i\}$, $Z = \{x_{i_k} \ldots x_{i_1} \# v_{i_1} \ldots v_{i_k} \mid 0 \leqslant k \leqslant n\}$, $Z_i = \{x_{i_k} \ldots x_{i_1} \# v_{i_1} \ldots v_{i_k} \mid 0 \leqslant k < n, i_{k+1} = i\}$). □

4 Existence of a Solution

Let us now study the question of the existence of solutions of an arbitrary form. Some further terminology is required.

Two languages, L' and L'', are said to be *equal modulo* a third language M, which is substring-closed (i.e., contains all substrings of each of its strings), if $L' \cap M = L'' \cap M$ [12]. This definition is extended to vectors of languages, which are said to be equal modulo M, if their corresponding components are equal modulo M. These definitions are also naturally extended to inequalities, strict inequalities and inequations. Note that if two vectors of languages L', L'' are equal modulo a substring-closed M, then $\varphi(L') = \varphi(L'') \pmod{M}$.

A vector of languages is a *solution modulo M* of a system of language equations, if each equation holds modulo M under the substitution of these languages for variables. A given vector $(L_1, \ldots, L_n)$ can be tested for being a solution modulo M by substituting $X_i = L_i \cap M$ into the equations and computing, modulo M, the value of each subexpression.

Let $\varphi(X_1, \ldots, X_n) = \varnothing$ be an equation. Its solution L_M modulo M is said to be *extendable to M'*, for a given $M' \supseteq M$, if there exists a solution modulo M' that coincides with L_M modulo M. The vector L_M is said to be *extendable to a solution*, if the equation has a solution that equals L_M modulo M.

Lemma 1 ([12]). *Let $\varphi(X_1, \ldots, X_n) = \varnothing$ be an equation and let M be a finite substring-closed language. Then there exists a finite substring-closed language $M' \supseteq M$, such that all solutions modulo M extendable to M' are extendable to solutions.*

Theorem 2 ([12]). *A language equation $\varphi(X_1, \ldots, X_n) = \varnothing$ has a solution if and only if for every finite substring-closed language M there exists a solution of $\varphi(X) = \varnothing$ modulo M. The decision problem is co-RE-complete.*

Let us now establish an analogous necessary and sufficient condition of solution existence for more general systems involving inequations.

Theorem 3. *A system $\{\varphi(X_1, \ldots, X_n) = \varnothing, \psi(X_1, \ldots, X_n) \neq \varnothing\}$ has a solution if and only if there exists a finite substring-closed language M_0, such that for every finite substring-closed language $M \supseteq M_0$ there exists a solution of $\varphi = \varnothing$ modulo M that is a solution of $\psi \neq \varnothing$ modulo M_0.*

Proof. $\Leftarrow$ Let $L = (L_1, \ldots, L_n)$ be a solution of the system. Since $\psi(L_1, \ldots, L_n) \neq \varnothing$, there exists a finite substring-closed language, such that this inequality is satisfied modulo that language. Denote it by M_0 and consider an arbitrary finite substring-closed superset $M \supseteq M_0$. The vector $(L_1 \cap M, \ldots, L_n \cap M)$ satisfies $\varphi = \varnothing$ modulo M and $\psi \neq \varnothing$ modulo M_0.

$\Rightarrow$ Given a finite substring-closed M_0, apply Lemma 1 to the equation $\varphi(X_1, \ldots, X_n) = \varnothing$ and the modulus M_0, to obtain a greater finite substring-closed modulus $M \supseteq M_0$.

By assumption, for this M there exists a vector L_M, such that $\varphi(L_M) = \varnothing$ (mod M) and $\psi(L_M) \neq \varnothing$ (mod M_0). Let L_{M_0} be L_M taken modulo M_0; then we know that $\psi(L_{M_0}) \neq \varnothing$ (mod M_0) and that L_{M_0} is extendable to M. The latter, by the choice of M according to Lemma 1, implies that the equation $\varphi = \varnothing$ has a solution L that coincides with L_{M_0} modulo M_0. Hence L also satisfies $\psi \neq \varnothing$, and therefore is a solution of the system. □

Theorem 4. *The set of systems of language equations $\{\varphi(X_1, \ldots, X_n) = \varnothing, \psi(X_1, \ldots, X_n) \neq \varnothing\}$ that have solutions is Σ_2-complete.*

Proof. A necessary and sufficient condition of having solutions given by Theorem 3 is of the form $\exists M_0 \forall M \ R(\varphi, \psi, M_0, M)$, where the quantifiers range over countable sets and R is a recursive predicate. A set thus defined is in Σ_2 by definition.

In order to prove Σ_2-hardness, let us use a reduction from the complement of the Π_2-complete Turing machine universality problem. This Σ_2-complete problem can be stated as "Given a TM T over Σ, determine whether $L(T) \neq \Sigma^*$".

$$\upsilon_T(Y, Z) = \varnothing \tag{6a}$$
$$Y \subseteq X \# \Gamma^* \tag{6b}$$
$$X \subseteq \Sigma^* \tag{6c}$$
$$X \neq \Sigma^* \tag{6d}$$

The equation (6a) expresses $Y = \mathrm{VALC}(T)$ and $Z = L'$, as in Example 3. The next equation (6b) specifies that every string that begins an accepting computation (that is, every string accepted by M) should be in X. Together with (6c), this means that

$$L(T) \subseteq X \subseteq \Sigma^* \tag{7}$$

If $L(T) = \Sigma^*$, then the bounds (7) are tight, and the only candidate for being a solution of (6) is $X = \Sigma^*$. However, it is ruled out by the inequation (6d), and hence there are no solutions.

If $L(T) \neq \Sigma^*$, then $X = L(T)$ fits into the bounds (7) and at the same time satisfies the inequation (6d), and, therefore, $(L(T), \mathrm{VALC}(T), L')$ is a solution of the system (6). This proves the correctness of the reduction. □

5 Uniqueness of a Solution

In order to study the systems that have exactly one solution, let us first characterize the following property:

Theorem 5. *Let $k \geqslant 1$. A system $\{\varphi(X_1, \ldots, X_n) = \varnothing, \psi(X_1, \ldots, X_n) \neq \varnothing\}$ has at most k solutions if and only if for every finite substring-closed language M, there exists a finite substring-closed language $M' \supseteq M$, such that all solutions of $\varphi(X_1, \ldots, X_n) = \varnothing$ modulo M' that are solutions of $\psi(X_1, \ldots, X_n) \neq \varnothing$ modulo M have at most k distinct images modulo M.*

In the case $k = 1$, the statement reads: "... all solutions of $\varphi(X) = \varnothing$ modulo M' that are solutions of $\psi(X) \neq \varnothing$ modulo M coincide modulo M".

Proof. ⇐ Fix an M, and let M' be the language defined for the equation $\varphi = \varnothing$ and the modulus M by Lemma 1. Suppose that there exist $k+1$ vectors modulo M', which are distinct modulo M, which satisfy $\varphi = \varnothing$ modulo M', and which remain solutions of $\psi \neq \varnothing$ modulo M. Let $L^{(1)}, \ldots, L^{(k+1)}$ be these vectors taken modulo M. According to Lemma 1, each of them can be extended to a solution of $\varphi = \varnothing$, which will at the same time remain a solution modulo $\psi \neq \varnothing$. Therefore, the system has at least $k+1$ solutions, which yields a contradiction.

⇒ Suppose the vectors $L^{(1)}, \ldots, L^{(k+1)}$ are pairwise distinct solutions of the system. Then there exists a finite M closed under substring, such that these vectors are pairwise distinct modulo M, and each of them satisfies $\psi \neq \varnothing$ modulo M. By assumption, for this particular M there exists a finite substring-closed M', such that all solutions of $\varphi = \varnothing$ modulo M' that are solutions of $\psi \neq \varnothing$ have at most k distinct images modulo M. Since each of $L^{(1)}, \ldots, L^{(k+1)}$ fits this description, they are not pairwise distinct modulo M, which contradicts the choice of M. □

Now the property of having a unique solution can be expressed as a conjunction of two conditions: the Σ_2-condition of having at least one solution (Theorem 3), and the Π_2-condition of having at most one solution (Theorem 5 with $k = 1$). This can be represented as a Σ_3- or as a Π_3-formula, which puts the decision problem to $\Sigma_3 \cap \Pi_3$. Actually, this is the optimal representation.

Theorem 6. *For any $k \geqslant 1$, the set of systems $\{\varphi(X_1, \ldots, X_n) = \varnothing, \psi(X_1, \ldots, X_n) \neq \varnothing\}$ that have exactly k solutions is Σ_2-hard, Π_2-hard and recursive in Σ_2 (i.e., it belongs to $\Sigma_3 \cap \Pi_3$).*

Proof. Σ_2-**hardness.** Though the solution existence problem is Σ_2-complete, Theorem 4 does not imply the Σ_2-hardness of our case. Taking a look at its proof, it is easy to see that, unless $L(T) = \Sigma^* \setminus \{w\}$ for some $w \in \Sigma^*$, the solution of the system (6) is not unique. Hence, a different proof is needed.

Let us use a reduction from another Σ_2-complete problem, the Turing machine finiteness, stated as "Given a TM T over a unary alphabet $\{a\}$, determine whether $L(T)$ is finite". It is claimed that the following system has a unique solution if and only if $L(T)$ is finite.

$$\upsilon_T(Y, Z) = \varnothing \quad (8a)$$
$$Y \subseteq X \# \Gamma^* \quad (8b)$$
$$X \subseteq aX \cup \varepsilon \quad (8c)$$
$$(aY \cup \#) \setminus X \# \Gamma^* \neq \varnothing \quad (8d)$$

As in the proof of Theorem 4, the equations (8a, 8b) specify that $Y = \text{VALC}(T)$, $Z = L'$ and $L(T) \subseteq X$. The inequality (8c), cf. (2a) in Example 1, requires that $X \subseteq a^*$ and that X is closed under suffix. Hence, the system (8a, 8b, 8c) has the set of solutions $\{(L, \text{VALC}(T), L') | L(T) \subseteq L \subseteq a^*$ and L is suffix-closed$\}$.

If $L(T) = \varnothing$, then $\text{VALC}(T) = \varnothing$, and X must be $\varnothing$ as well. Supposing the contrary, that X is a nonempty suffix-closed language $\{\varepsilon, a, aa, \ldots, a^\ell\}$ ($\ell \geqslant 0$), the left-hand side of the inequation (8d) takes form $(\varnothing \cup \#) \setminus \{\varepsilon, \ldots\} \# \Gamma^* = \varnothing$, and hence the inequation is not satisfied. It can similarly be proved that if $L(T)$ is a finite nonempty set and a^m is the longest string in it, then X must be equal to $a^{\leqslant m}$.

If $L(T)$ is infinite, then a^* is the only potential value of $X \supseteq L(T)$, since this is the only infinite suffix-closed language over $\{a\}$. But then $a\text{VALC}(T) \cup \{\#\} \subseteq X \# \Gamma^* = a^* \# \Gamma^*$, and therefore the inequation (8d) does not hold.

Π_2-**hardness.** Follows from the Π_2-completeness of the same problem for equations $\varphi(X_1,\ldots,X_n) = \varnothing$ [12]. Its proof uses a reduction from the Turing machine universality problem, and uses exactly the system (6a, 6b 6c).

Recursiveness in Σ_2. As noted above, the Σ_2 condition of solution existence (Theorem 3) and the Π_2 condition of having at most one solution (Theorem 5) have to be checked. Hence, the problem is Turing-reducible to Σ_2. □

Theorem 7. *Let $\{\varphi(X_1,\ldots,X_n) = \varnothing, \psi(X_1,\ldots,X_n) \neq \varnothing\}$ be a system of an equation and an inequality that has a unique solution $(L_1,\ldots,L_n)$. Then each component L_i is recursive.*

Proof. Since $\psi(L_1,\ldots,L_n) \neq \varnothing$, there exists a string $w_0 \in \psi(L_1,\ldots,L_n)$. Using this string, construct the following algorithm:

Input: $w \in \Sigma$
Let $M = substrings(w_0) \cup substrings(w)$
For all finite substring-closed $M' \supseteq M$
    Let $L^{(1)},\ldots,L^{(k)}$ be all vectors (mod M')
    that satisfy $\varphi = \varnothing$ modulo M' and $\psi \neq \varnothing$ modulo M
  If $L^{(1)} = \ldots = L^{(k)}$ (mod M)
    Accept if $w \in L_i^{(1)}$, reject if $w \notin L_i^{(1)}$

By Theorem 5, for the language M used by the algorithm there exists a finite substring-closed language $M' \supseteq M$, such that all solutions of $\varphi = \varnothing$ modulo M' that satisfy $\psi \neq \varnothing$ modulo M coincide modulo M. This M' will be eventually reached by the algorithm's loop, the condition in the *if* statement will become true, and the algorithm will terminate.

It remains to argue that the algorithm accepts w if and only if it is in L_i. The actual solution $(L_1,\ldots,L_n)$ satisfies $\varphi = \varnothing$ modulo M' and $\psi \neq \varnothing$ modulo M, so this vector, taken modulo M', must be among $L^{(1)},\ldots,L^{(k)}$. Therefore, $L^{(1)} = (L_1,\ldots,L_n)$ (mod M) and $w \in L_i^{(1)}$ is equivalent to $w \in L_i$. □

Since it is already known that every recursive language can be specified by a unique solution of a language equation [12,15], the following can be concluded:

Corollary 1. *The class of languages representable as components of unique solutions of systems of the form $\{\varphi(X_1,\ldots,X_n) = \varnothing, \psi(X_1,\ldots,X_n) \neq \varnothing\}$ is exactly the class of recursive languages.*

Given an individual language equation $\varphi(X_1,\ldots,X_n) = \varnothing$ with a unique solution, the algorithm for determining the membership of strings in this solution can be effectively constructed [12]. This turns out to be different in our more general case involving inequations:

Theorem 8. *There is no algorithm that, given a system of language equations $\{\varphi(X_1,\ldots,X_n) = \varnothing, \psi(X_1,\ldots,X_n) \neq \varnothing\}$ with an attached proof that it has a unique solution, determines this unique solution modulo $\{\varepsilon\}$.*

Proof. Suppose such an algorithm exists. Consider an arbitrary Turing machine T, and use the language equation $v_T(Y,Z) = \varnothing$ from Example 3 to construct the following system of language equations:

$$v_T(Y, Z) = \varnothing \tag{9a}$$
$$X \subseteq \varepsilon \tag{9b}$$
$$XY = \varnothing \tag{9c}$$
$$X \cup Y \neq \varnothing \tag{9d}$$

The equation (9a) requires that $Y = \mathrm{VALC}(T)$ and $Z = L'$, where L' is a certain irrelevant auxilliary language (see Example 3). By (9b), X can assume one of the two possible values: $\varnothing$ or $\{\varepsilon\}$. The next two equations (9c, 9d) specify that exactly one of the languages X and $\mathrm{VALC}(T)$ is nonempty. Since $\mathrm{VALC}(T)$ is nonempty if and only if $L(T)$ is nonempty, it can be concluded that if $L(T) = \varnothing$, then $X = \{\varepsilon\}$, and if $L(T) \neq \varnothing$, then $X = \varnothing$. The system has a unique solution in either case, it is either $(\{\varepsilon\}, \varnothing, L')$ or $(\varnothing, \mathrm{VALC}(T), L')$.

This argument can be properly formalized and attached to the system (9) to produce an input for the supposed algorithm. The solution modulo $\{\varepsilon\}$ it computes will be $(\{\varepsilon\}, \varnothing, L' \cap \{\varepsilon\})$ if $L(T) = \varnothing$, or $(\varnothing, \varnothing, L' \cap \{\varepsilon\})$ if $L(T) \neq \varnothing$. Therefore, the supposed algorithm can be used to solve the Turing machine emptiness problem, which is known to be undecidable. □

In particular, Theorem 8 implies that, though for every system with a unique solution there exists an algorithm for testing the membership of strings in the components of that solution, this algorithm cannot be algorithmically constructed. This reveals a principal difference between these systems and the earlier studied computationally universal types of language equations [9,12,15].

6 Equations with Finitely Many Solutions

Let us consider the problem of testing whether a given system of language equations has finitely many solutions. This property can naturally be represented as "there exists a number $k \geqslant 0$, such that there are exactly k solutions", and we shall now see that this representation is optimal with respect to the number of quantifiers.

Theorem 9. *The set of systems of language equations $\{\varphi(X_1, \ldots, X_n) = \varnothing, \psi(X_1, \ldots, X_n) \neq \varnothing\}$ that have finitely many solutions is Σ_3-complete. It remains Σ_3-complete for individual equations $\varphi(X_1, \ldots, X_n) = \varnothing$.*

Proof. **Membership in Σ_3.** A system has finitely many solutions if and only if there exists a number $k \geqslant 1$, such that the system has at most k solutions. According to Theorem 5, the condition of having at most k solutions is representable as a Π_2 formula $\forall M \exists M'\, R(k, M, M')$, where R is a certain recursive predicate stated by the theorem. This predicate R can be used to characterize our problem as follows: $\exists k \forall M \exists M'\, R(k, M, M')$. This is a Σ_3-formula.

Σ_3**-hardness.** Reduction from the co-finiteness problem for Turing machines, stated as "Given a TM T over Σ, determine whether $\Sigma^* \setminus L(T)$ is finite", which is known to be Σ_3-complete [17, Corollary 14-XVI].

Construct the system (6a, 6b, 6c), as in Theorem 4, which has the set of solutions
$$\{(L, \mathrm{VALC}(T), L') \mid L(T) \subseteq L \subseteq \Sigma^*\} \tag{10}$$
If $L(T)$ is co-finite, then there are finitely many values L that fit within the bounds in (10), and hence the set of solutions has finitely many elements. If $\Sigma^* \setminus L(T)$ is infinite, then (10) is infinite as well. This proves the reduction. □

We already know that if a system has a unique solution, then the components of this solution are recursive languages (see Theorem 7). Let us extend this recursiveness result to the case of systems with any finite number of solutions.

Theorem 10. *Let $\{\varphi(X_1, \ldots, X_n) = \varnothing, \psi(X_1, \ldots, X_n) \neq \varnothing\}$ be a system of an equation and an inequality that has finitely many solutions. Then all components of all these solutions are recursive.*

Proof. Let $L^{(1)}, \ldots, L^{(k)}$ ($k \geqslant 0$) be all solutions of the system. If $k = 0$, the result trivially holds, and if $k = 1$, it holds by Theorem 7. Consider the case $k \geqslant 2$ and let us construct a new system that would have $L^{(1)}$ as its unique solution. Since $L^{(1)}$ is different from each $L^{(i)}$ ($2 \leqslant i \leqslant k$), for every such i there exists a variable X_{j_i}, such that $L_{j_i}^{(1)} \neq L_{j_i}^{(i)}$; let w_i be any string in their symmetric difference. Now construct the following system of language equations:

$$\varphi(X_1, \ldots, X_n) = \varnothing \tag{11a}$$

$$\psi(X_1, \ldots, X_n) \neq \varnothing \tag{11b}$$

$$\begin{cases} X_{j_i} \cap w_i \neq \varnothing, & \text{if } w_i \in L_{j_i}^{(1)} \\ X_{j_i} \cap w_i = \varnothing, & \text{if } w_i \notin L_{j_i}^{(1)} \end{cases} \quad \text{(for all } 2 \leqslant i \leqslant k\text{)} \tag{11c}$$

Every solution of the constructed system satisfies (11a, 11b), i.e., is a solution of the original system. Therefore, $L^{(1)}, \ldots, L^{(k)}$ are the only candidates for being solutions of (11). While each $L^{(i)}$ ($i \geqslant 2$) does not satisfy the i-th equation (11c), $L^{(1)}$ satisfies all of them, which makes it the unique solution of the system (11). Therefore, by Theorem 7, all components of $L^{(1)}$ are recursive. □

7 Conclusion

The complexity of main decision problems for language equations with equality only and for systems involving proper inequalities and inequations is shown in Table 1. The results are fairly disparate, but they have one thing in common: undecidability.

The close relation of language equations of the general form to the recursion theory and to logic has been noted before [12,16], and the extensions introduced in this paper (namely, inequations and strict inequalities) yield new interesting undecidabilities that manifest themselves in Theorem 8. This particular turn of the theory of language equations suggests further mathematical problems to study. For instance, how hard is the following decision problem: "Given a systemof language equations of one of the two forms, determine whether its set

Table 1. Complexity of decision problems. Expressive power of unique solutions.

	$\varphi(X_1,\ldots,X_n) = \varnothing$	$\begin{cases}\varphi(X_1,\ldots,X_n) = \varnothing \\ \psi(X_1,\ldots,X_n) \neq \varnothing\end{cases}$
Does there exist a finite solution?	Σ_1	Σ_1
Does there exist a regular solution?	Σ_1	Σ_1
Do there exist any solutions?	Π_1 [12]	Σ_2
Does there exist a unique solution?	Π_2 [12]	$\Sigma_3 \cap \Pi_3$
Are there finitely many solutions?	Σ_3	Σ_3
Class of languages	recursive [12]	recursive

of solutions is countable"? For systems with countably many solutions, what is the class of languages that can occur in those solutions?

References

1. F. Baader, P. Narendran, "Unification of concept terms in description logic", *Journal of Symbolic Computation*, 31 (2001), 277–305.
2. F. Baader, R. Küsters, "Unification in a description logic with transitive closure of roles", *LPAR 2001* (Havana, Cuba), LNCS 2250, 217–232.
3. J. A. Brzozowski, E. L. Leiss, "On equations for regular languages, finite automata, and sequential networks", *Theoretical Computer Science*, 10 (1980), 19–35.
4. J. H. Conway, *Regular Algebra and Finite Machines*, Chapman and Hall, 1971.
5. S. Ginsburg, H. G. Rice, "Two families of languages related to ALGOL", *Journal of the ACM*, 9 (1962), 350–371.
6. J. Hartmanis, "Context-free languages and Turing machine computations", *Proceedings of Symposia in Applied Mathematics*, Vol. 19, AMS, 1967, 42–51.
7. J. Karhumäki, I. Petre, "Two problems on commutation of languages", in: Gh. Păun, G. Rozenberg, A. Salomaa (Eds.), *Current Trends in Theoretical Computer Science: The Challenge of the New Century, vol. 2*, World Scientific, 2004, 477–494.
8. M. Kunc, "Regular solutions of language inequalities and well quasi-orders", *ICALP 2004*, LNCS 3142, 870–881.
9. M. Kunc, "The power of commuting with finite sets of words", *STACS 2005*.
10. M. Kunc, "Largest solutions of left-linear language inequalities", *AFL 2005*.
11. A. Okhotin, "Conjunctive grammars and systems of language equations", *Programming and Computer Software*, 28 (2002), 243–249.
12. A. Okhotin, "Decision problems for language equations with Boolean operations", *ICALP 2003*, LNCS 2719, 239–251.
13. A. Okhotin, "Boolean grammars", *Information and Computation*, 194:1 (2004), 19–48.
14. A. Okhotin, "The dual of concatenation", *MFCS 2004*, LNCS 3153, 698–710.
15. A. Okhotin, "On computational universality in language equations", *MCU 2004* (St. Petersburg, Russia), LNCS 3354, 292–303.
16. A. Okhotin, "A characterization of the arithmetical hierarchy by language equations", *DCFS 2004* (London, Ontario, Canada), 225–237.
17. H. Rogers, Jr., *Theory of Recursive Functions and Effective Computability*, McGraw-Hill, 1967.
18. A. Salomaa, *Theory of Automata*, Pergamon Press, Oxford, 1969.

Event Structures for the Collective Tokens Philosophy of Inhibitor Nets*

G. Michele Pinna

Dipartimento di Matematica e Informatica - Università di Cagliari
`pinna@sc.unica.it`

Abstract. In recent years the collective tokens philosophy for Petri Nets has gained again the stage, but it is commonly set in opposition to the individual tokens philosophy. In this paper we investigate what can be an adequate event structures to capture the collective tokens philosophy of nets when inhibitor arcs are taken into account.

1 Introduction

When describing the computations of a concurrent system each step in the computation is determined by some previous happenings, determining a causal dependency between the step and the previous ones. As noticed in many papers (e.g. [12,1,2,3,4,5]) there are several kinds of causality arising from the various different situations. In [6,12] and [7] a new relation between events is proposed that is able to model several different situations. The *disabling/enabling* relation has been introduced to take into account the causality arising in Petri nets with read and inhibitor arcs under the so called *individual tokens philosophy*. This relation can be used to model dependencies between activities in a much broader sense, as we will try to point out in the following. Consider the following two nets:

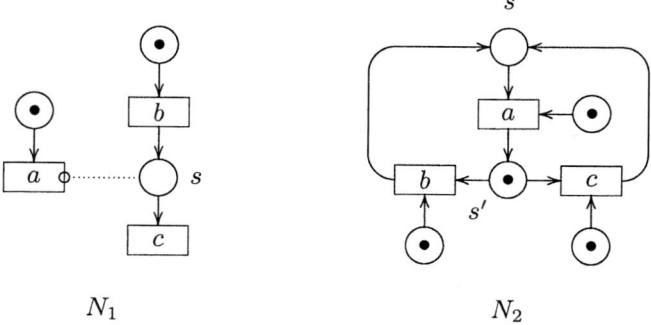

In the net N_1 the transition a fires either at the initial state or, in case the transition b fires, after the firing of c because there is an *inhibiting* arc between s and a. This is the typical case where a is inhibited by something caused by b,

---

* Work partially supported by the SYBILLA progetto "Systems Biology: modellazione, linguaggi e analisi".

but the inhibition can be removed. In the other net, a is enabled by the firing of b or c, and the firing of a re-enables the other (putting some condition consumed by b or c). This is the typical case where we do not care too much about who enabled a, rather we focus on the fact that a is enabled. It is worth to notice that the net N_1 is an inhibitor occurrence net as defined in [6], whereas the second is an occurrence net as defined in [8].

The dependencies between these events can be easily modeled using (a generalization of) the disabling/enabling relation as introduced in [6] and [12]: for N_1 we write $\mathfrak{I}_1(\{b\}, a, \{\{c\}\})$, $\mathfrak{I}_1(\emptyset, c, \{\{b\}\})$ and for N_2 we have $\mathfrak{I}_2(\{c\}, b, \{\{a\}\})$, $\mathfrak{I}_2(\{b\}, c, \{\{a\}\})$, $\mathfrak{I}_2(\emptyset, a, \{\{b\}, \{c\}\})$ (we give here only the relevant ones), where the intuition behind this relation is the following one: given $\mathfrak{I}(a, e, A)$, the event e can be added to a state, seen as the set of events happened so far, if either a is not contained in the state or also some set of event of A is present at the state. Thus the relation is general enough to model several different kinds of dependencies. In particular, the individual tokens philosophy which advocates the interpretation that the happening of an event (step in the computation) has a unique history (hence the set of its causes is unique) and the collective tokens philosophy which consent a broader interpretation, namely that events can happen with different histories, can be both modeled using the same relation. The claim that events may have different history has been made in several other situations: in [9] it has been used to introduce the notion of *event automata* to solve a problem of compositionality and specification involving prime event structures, in [8] it has been used to introduce *configuration structures*, and in [10] it has been used to give an event structure semantics to Place/Transition nets.

We investigate on an adequate event structure semantics for nets with inhibitor arcs under the *collective tokens philosophy*, namely when it is irrelevant the history of an event. As the *disabling/enabling* relation seems to be well suited to represent the dependencies among events in the case of the collective tokens philosophy, we place this notion as the starting point of this investigation. The final target would be a unique framework for considering the individual and the collective tokens philosophy. A similar perspective in considering the collective tokens philosophy is adopted in [11], where the semantics based on configuration structures ([8]) is compared with an algebraic model based on monoidal categories.

Thus, we first generalize the disabling/enabling relation introduced in [6,12] to ease the modeling of the nets tokens game, showing that this is indeed a conservative extension of the previous notion. We then show that even recent proposals of event structures (e.g. *event structure with resolvable conflicts*, defined in [13]) cannot capture the dependencies arising from the inhibition, even under the individual tokens philosophy. Following the Gunawardena's approach to the study of the semantics of a net ([14]), where he characterize in a logical way the firings of a transition, we use the disabling/enabling relation to characterize the conditions under which the n-th occurrence of a transition can happen. This approach is easily extended to the case of net with inhibitor arc, as it is obvious that the characterization of the *positive causes* follows the Gunawardena's intuition, whereas the characterization of *negative causes*, i.e. these arising from inhibitor arcs, follows the intuition behind the disabling/enabling relation.

The paper is organized as follows: in the next section we introduce and discuss the disabling/enabling relation and introduce the notion of *Disabling/Enabling* event structure. We then compare it with other brands of event structure (also in section 3 where we show that the DE-relation is substantially different from the relation introduced by van Glabbeek and Plotkin in their event structures with resolvable conflicts). In section 4 we show how our event structure capture the collective tokens philosophy and finally, in section 5 we extend our approach to the case of inhibitor arcs.

2 DE-Event Structures

In [6,12] and [7] the *disabling/enabling* relation has been introduced to take into account the causalities arising in Petri nets with read and inhibitor arcs. This relation can be used to model dependencies between activities in a much broader sense, as shown in [15]. Here we conservatively extend the definition of *disabling/enabling*. Let us fix some notational conventions. The powerset of a set X is denoted by $\mathbf{2}^X$, while $\mathbf{2}^X_{fin}$ denotes the set of finite subsets of X and $\mathbf{2}^X_1$ the set of subsets of X of cardinality at most one (singletons or the empty set). With $\mathbb{N}$ we denote the set of natural numbers, and with $\mathbb{Z}$ the set of relative numbers.

Definition 1. *A* disabling/enabling *relation over a set E (DE-relation for short) is a ternary relation* $\mathfrak{I} \subseteq \mathbf{2}^E_{fin} \times E \times \mathbf{2}^{\mathbf{2}^E_{fin}}$.

Informally, if $\mathfrak{I}(a, e, A)$ then the events in a inhibits the event e, which can be enabled again by one of the sets of events in A. The first argument of the relation can be also the empty set $\emptyset$, $\mathfrak{I}(\emptyset, e, A)$ meaning that the event e is inhibited in the initial state of the system, thus some other event should happen before e. Moreover the third argument (the set of set of events A) can be empty, $\mathfrak{I}(a, e, \emptyset)$ meaning that there are no events that can re-enable e after it has been disabled by a.

In the papers [6,12] and [15] the triples considered are further constrained. The first component can be at most a singleton and also the set of events in the third components are singletons. Thus the disabling/enabling relation is in this case as follows: $\vdash\!\circ \,\subseteq\, \mathbf{2}^E_1 \times E \times \mathbf{2}^E$.[1] As shown in [6], this relation is sufficient to represent both causality and asymmetric conflict and thus it is the only relation one need in the individual tokens philosophy approach. However, this relation cannot take into account more complex notion of *consistency* as it is done in general event structures. Here the consistency relation is induced by the asymmetric conflict relation (a set of events is consistent if there is no chain of asymmetric conflicts, see [12]), thus there is no event structure with

---

[1] To be more precise, we should write $\vdash\!\circ \,\subseteq\, \mathbf{2}^E_1 \times E \times \mathbf{2}^{\mathbf{2}^E_1}$, but here we retain the notation of [6,12].

this relation that is able to model the situation where the set of events $\{a, b, c\}$ is not consistent, but $\{a, b\}, \{a, c\}$ and $\{b, c\}$ are consistents and there is no dependency among these three events. The generalization of the relation can model this situation. In fact, adding the triples $\mathfrak{I}(\{a, b\}, c, \emptyset)$, $\mathfrak{I}(\{a, c\}, b, \emptyset)$ and $\mathfrak{I}(\{c, b\}, a, \emptyset)$, we can avoid that the configuration $\{a, b, c\}$ is reached. General event structures by Winskel ([16]) model this situation by forbidding that $\{a, b, c\}$ is consistent. It is easy to see that $\vdash\!\!\circ$ relation is an instance of the more general definition of disabling/enabling relation we have introduced here. We can now introduce the notion of DE-event structure.

Definition 2. *A DE-event structure (*DE-ES*) is a pair* $\mathbf{E} = \langle E, \mathfrak{I} \rangle$, *where* E *is a set of events and* $\mathfrak{I} \subseteq 2^E_{fin} \times E \times 2^{2^E_{fin}}$ *is a ternary relation, called disabling-enabling relation, such that if* $\mathfrak{I}(a, e, A)$ *and* $A \neq \emptyset$ *then* $\emptyset \notin A$.

The DE-relation is used to represent the conditions under which an event can happen. Differently from other notions of event structures, it could be that no computation is actually represented by such relation. As usual a computation is presented as a *configuration*, i.e. a set of events, and the DE-relation is used to *extend* the set of events.

Definition 3. *Let* $\mathbf{E} = \langle E, \mathfrak{I} \rangle$ *be a* DE-ES. *Given* $C, C' \subseteq E$, *we say that* C' *extends* C, *notation* $C \sqsubset C'$, *iff (a)* $C' \setminus C = \{e\}$, *and (b) for all* $\mathfrak{I}(a, e, A)$, *if* $a \subseteq C$ *then exists* $a' \in A$ *and* $a' \subseteq C$.

Using this definition we can define what a configuration of a DE-ES event structure is.

Definition 4. *Let* $\mathbf{E} = \langle E, \mathfrak{I} \rangle$ *be a* DE-ES. *A subset* C *of* E *is a configuration iff there exists a sequence* $C_0, \ldots, C_n, \ldots$ *of subsets of* E *with* $C_0 = \emptyset$ *and* $\bigcup_{i \in \mathbb{N}} C_i = C$ *such that for all* $i \in \mathbb{N}$, $i > 0$, *it holds that* $C_{i-1} \sqsubset C_i$. *The set of configurations of a DE event structure will be denoted by* $Conf(\mathbf{E})$.

Before comparing this notion of event structure with the generalization of general event structures developed by van Glabbeek and Plotkin in [13], we briefly discuss the capabilities of the DE-relation. In [6] and [12] the event structures considered had the disabling/enabling relation of the form of $\vdash\!\!\circ$ and were further constrained by auxiliary generalized causality and conflict (asymmetric and symmetric) relations (denoted with $< \subseteq 2^E \times E$, $\nearrow \subseteq E \times E$ and $\# \subseteq 2^E_{fin}$ respectively) defined by the following set of rules:

$$\frac{\vdash\!\!\circ (\emptyset, e, A) \quad \#_p A}{A < e} \ (<1) \qquad \frac{A < e \quad \forall e' \in A. \ A_{e'} < e' \quad \#_p(\cup\{A_{e'} \mid e' \in A\})}{(\cup\{A_{e'} \mid e' \in A\}) < e} \ (<2)$$

$$\frac{\vdash\!\!\circ (\{e'\}, e, \emptyset)}{e \nearrow e'} \ (\nearrow 1) \qquad \frac{e \in A < e'}{e \nearrow e'} \ (\nearrow 2) \qquad \frac{\#\{e, e'\}}{e \nearrow e'} \ (\nearrow 3)$$

$$\frac{e_0 \nearrow \ldots \nearrow e_n \nearrow e_0}{\#\{e_0, \ldots, e_n\}} \ (\#1) \qquad \frac{A' < e \quad \forall e' \in A'. \ \#(A \cup \{e'\})}{\#(A \cup \{e\})} \ (\#2)$$

where $\#_p A$ means that the events in A are pairwise conflicting, namely $\#\{e, e'\}$ for all $e, e' \in A$ with $e \neq e'$. We will use the infix notation for the binary conflicts, writing $e \# e'$ instead of $\#\{e, e'\}$. Moreover we will write $e < e'$ to indicate $\{e\} < e'$.

For a detailed discussion on the interpretation of these relations, the interested reader can consult [6] and [12]. Here it is enough to summarize as follows: $A < e$ means that in every computation where e is executed, there is exactly one event $e' \in A$ which is executed and it precedes e; $e' \nearrow e$ means that in every computation where both e and e' are executed, e' precedes e; and $\# A$ means that there are no computations where all events in A are executed. These relations are used to define the notion of inhibitor event structure, by suitably constraining the triples that are allowed.

Definition 5. *An* inhibitor event structure (IES) *is a pair* $I = \langle E, \vdash\!\circ \rangle$ *satisfying, for all* $e \in E$, $a \in 2_1^E$ *and* $A \subseteq E$, *(1) if* $\vdash\!\circ (a, e, A)$ *then* $\#_p A$ *and* $\forall e' \in a.\ \forall e'' \in A.\ e' < e''.$; *(2) if* $A < e$ *then* $\vdash\!\circ (\emptyset, e, A)$; *(3) if* $e \nearrow e'$ *then* $\vdash\!\circ (\{e'\}, e, \emptyset)$.

This definition captures the behavior of a (weakly) safe Petri nets with inhibitor and read arcs under the individual tokens philosophy, as it is shown in [6] and [12]. Furthermore there it is shown that Winskel's *prime event structures* ([17]), *asymmetric event structures* by Baldan, Corradini and Montanari ([1]) and *event structures with possible events* by Pinna and Poigné ([9]) can be viewed as IES. In fact the causality and the symmetric conflict can be derived by these triples, as well as asymmetric conflict (the $\nearrow$ relation). Possible events can be viewed as asymmetric conflicts. For what regards the treatment of disjunctive or-causality (relation $<$) the presented rules resembles also the equivalence rules for *(extended) bundle event structures* by Langerak ([18]), thus also this brand of event structures can be seen as a special case of IES. Using the fact that *flow event structures* by Boudol ([19]) can be represented by event structures with possible events, it is quite clear that IES, and henceforth DE-ES, can capture all these brands of event structures. We show that general event structures can be seen as DE-ES. An even structure is the triple $E = \langle E, Con, \vdash \rangle$ such that (a) E is a set of events, (b) $Con \subseteq 2_{fin}^E$ such that $\emptyset \in Con$ and $Y \subseteq X$, $X \in Con \Rightarrow Y \in Con$, and (c) $\vdash \subseteq Con \times E$ such that $X \vdash e$ and $X \subseteq Y \Rightarrow Y \vdash e$. A configuration of such event structure is a subset C of E which is (1) *consistent*, i.e. every finite subset of C belongs to Con, and (2) *secured*, i.e. $\forall e \in C, \exists e_0, \ldots, e_n \in C.\ e_n = e$ and $\forall i \leq n.\ \{e_0, \ldots, e_{i-1}\} \vdash e_i$. To an event structure $\langle E, Con, \vdash \rangle$ we associate a DE-ES $\mathbf{E} = \langle E, \mathfrak{I} \rangle$ where $\mathfrak{I}$ is as follows:(a) $\mathfrak{I}(a, e, \emptyset)$ if $a \cup \{e\} \notin Con$ and $a \in Con$, and (b) $\mathfrak{I}(a, e, \{a\})$ if $a \vdash e$. Now assume that C is a configuration of E, then it is consistent and secured. Assume that it is not a configuration of the associated DE-ES, then for all possible orderings of subsets, there is an index i where the $\sqsubset$ is violated. But this means that either $\mathfrak{I}(a, e, \emptyset)$ cannot be used and $a \cup \{e\}$ is in C, contradicting the consistency, or that $\mathfrak{I}(a, e, \{a\})$ cannot be used, contradicting the securedness. Thus we can say that DE-ES are able to represent also Winskel's event structures. Using the fact that a $\vdash\!\circ$ is just a special case of the $\mathfrak{I}$ relation, we have the following theorem.

Theorem 1. *General, prime, flow, bundle event structures and event structures with possible events can be modeled using the $\mathfrak{I}$ relation.*

3 Event Structure with Resolvable Conflict

In [13] van Glabbeek and Plotkin propose a generalization of Winskel's event structures [16,17]. We briefly review their definition and compare it with our.

An *event structure* is a pair $E = \langle E, \vdash \rangle$ with: (a) E a set of events, and (b) $\vdash \subseteq 2^E \times 2^E$, the *enabling* relation.

No particular constraint is placed on the enabling relation $\vdash$, and it intuitively says that for *all* the events in Y to occur, for some set X with $X \vdash Y$, the events in X have to happen first. Configurations are defined according to the *step transition relation* defined as follows: $X \longrightarrow_E Y \Leftrightarrow (\forall X \subseteq Y \wedge \forall Z \subseteq Y. \exists W \subseteq X. W \vdash Z)$. For the single action transition relation it is further required that $Y \setminus X$ contains at most one element. Let $E = \langle E, \vdash \rangle$ be an event structure. Its set of left closed configurations, denoted with $L(E)$, is $L(E) = \{X \subseteq E \mid X \longrightarrow_E X\}$. The configurations defined in this way are not necessarily reachable from a given initial state (as in the case of DE-ES). Thus a notion of secured configuration is added. Let $E = \langle E, \vdash \rangle$ be an event structure. A configuration $X \in L(E)$ is *secured* if there is an infinite sequence $\emptyset = X_0 \longrightarrow_E X_1 \longrightarrow_E \ldots \longrightarrow_E X_n \longrightarrow_E \ldots$ and $X = \bigcup_{i \in \mathbb{N}} X_i$. The set of secured configuration is denoted with $S(E)$.

We limit our attention to event structures satisfying the following conditions: (1) $\vdash \subseteq 2^E_{fin} \times 2^E_{fin}$, (2) *rooted*: if $\emptyset \vdash \emptyset$, (3) *singular*: if $X \vdash Y$ implies $X = \emptyset$ or Y as singleton, (4) *binary conflict*: if $|X| > 2$ implies $\emptyset \vdash X$, and (5) the set of secured configurations coincides with the set of left closed configurations.

The requirement of *finite cause* formulated as if $X \vdash Y$ implies X finite, is subsumed by the restriction we place on the $\vdash$ relation. This class of event structure is clearly contained in the more general one of the definition of van Glabbeek and Plotkin, but it is enough for our purposes. The enabling of the form $\emptyset \vdash X$ with $|X| > 1$ are used to model consistency of events, whereas the other to express how event are added. These kinds of event structures can be seen as DE-ES, using the same schema of translation we used for Winskel's general event structures: $X \vdash \{e\}$ is translated into $\mathfrak{I}(\emptyset, e, \{X\})$ whereas the consistency is $\emptyset \vdash X$ is translated with $\mathfrak{I}(X, e, \emptyset)$ provided that $\neg(\emptyset \vdash X \cup \{e\})$, but not the vice versa. Consider the following situation (a net with inhibitor arcs):

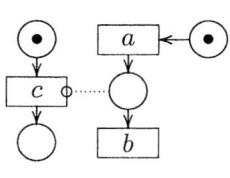

the net

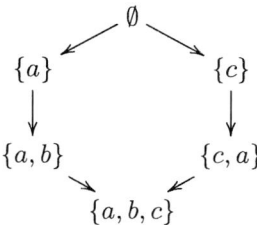

its configurations

This situation is modeled by the triples $\Im(\emptyset, b, \{\{a\}\})$ and $\Im(\{a\}, c, \{\{b\}\})$ but there is no possibility of modeling it using the $\vdash$ relation, even dropping the assumptions we made before. The reason is the following: in van Glabbeek and Plotkin event structures (as in Winskel's general event structures), enabling is monotonic, once that an event is *enabled*, there is no way to state that it is *temporarily disabled* as it is in this case (hence an enabled event cannot be added if adding the event would lead to a set which is not consistent).

Thus, there are specific situations where event structures with resolvable conflicts cannot capture the behaviour of (extensions of) Petri nets, whereas on the contrary, it seems not so easy to overcome the general requirement that $\vdash \subseteq 2^E \times 2^E$.

4 Nets and Event Structures

We investigate now an event structure semantics for nets with inhibitor arcs, defining it firstly on nets without inhibitor arcs and showing then how to extend to this case. We first recall some notions about nets. A net is a tuple $N = (S, T, F, m_0)$ where S is a set of *places*, T is a set of transitions, $F : (S \times T) \cup (T \times S) \to \mathbb{N}$ is a flow relation and $m : S \to \mathbb{N}$ is the initial marking. The evolution of a net is described as usually with the tokens game. Let $m : S \to \mathbb{N}$ be a marking of a net, a finite multiset $U : T \to \mathbb{N}$ of transitions is *enabled* under m if $\sum_{t \in T} U(t) \cdot F(s, t) \leq m(s)$ and the reached marking is $m'(s) = m(s) + \sum_{t \in T} U(t) \cdot (F(t, s) - F(s, t))$. We write $m \, [U\rangle \, m'$ to express the firing of a step enabled at the marking m and yielding the marking m'. A *firing sequence* is defined as follows:

- m_0 is a firing sequence,
- if $m_0 \, [U_1\rangle \, m_1 \, [U_2\rangle \, m_2 \ldots m_{n-1} \, [U_n\rangle \, m_n$ is a firing sequence and $m_n \, [U_{n+1}\rangle \, m_{n+1}$, then $m_0 \, [U_1\rangle \, m_1 \, [U_2\rangle \, m_2 \ldots m_{n-1} \, [U_n\rangle \, m_n \, [U_{n+1}\rangle \, m_{n+1}$ is a firing sequence.

A marking m is reachable if there is a firing sequence $m_0 \, [U_1\rangle \, m_1 \, [U_2\rangle \, m_2 \ldots m_{n-1} \, [U_n\rangle \, m_n$ and $m = m_n$.

Given a net $N = (S, T, F, m)$, with $^{\bullet N} F$ ($F^{\bullet N}$) we denote the multiset $S \to \mathbb{N}$ defined as $^{\bullet N} F(t) = F(\_, t)$ ($F^{\bullet N}(t) = F(t, \_)$), and with $^{\bullet N} s$ ($s^{\bullet N}$) we denote the sets $F(\_, s)$ ($F(s, \_)$). We will omit the index N in $^{\bullet N}$, $^{\bullet N}$ when it is clear from the context what is the referring net. Over multisets a sum is defined as usual: $(U \oplus U')(x) = U(x) + U'(x)$. When a multiset is a set we often identify it with its elements. We end this brief review recalling that a *safe* net is a net where $F : (S \times T) \cup (T \times S) \to \{0, 1\}$ and each reachable marking is a set, i.e. $M(s) \leq 1$ for all $s \in S$. In this paper, for simplicity, we consider safe nets. The behaviour of a *safe* net (with inhibitor arcs) under the individual tokens philosophy can be captured using the notion of i-occurrence net ([12]) and to such i-occurrence net a DE-ES can be easily associated (indeed in [12] it is shown how to associate an IES to an i-occurrence net, and we have shown that an IES is a subclass of DE-ES). To capture the behaviour of a net under the collective tokens philosophy, van Glabbeek and Plotkin proposed the notion of 1-occurrence net [8].

Definition 6. Let $N = (S, T, F, m)$ be a Petri net, a configuration of a net is any finite multiset X of transitions with the property that the function $m_X : S \to \mathbb{Z}$ given by $m_X(s) = m(s) + \sum_{t \in T} X(t) \cdot (F(t,s) - F(s,t))$ is a reachable marking of the net.

We can now define the notion of 1-occurrence net.

Definition 7. Let $N = (S, T, F, m)$ be a Petri net, N is a 1-occurrence net (1-ON for short) if every configuration is a set.

Given a net, its behaviour according to the collective tokens philosophy can be captured by the following definition ([8]).

Definition 8. Let $N = (S, T, F, m)$ be a Petri net. Its 1-unfolding $N' = (S', T', F', m')$ into a 1-occurrence net is given by (a) $T' = T \times (\mathbb{N} \setminus \{0\})$, (b) $S' = S \cup (T' \times \{*\})$, (c) $F'(s, (t, n)) = F(s, t)$ and $F'((t, n), s) = F(t, s)$, (d) $F'((t, n), ((t, n+1), *)) = 1$ and $F'((u, *), u) = 1$, and (e) $m'(s) = m(s)$, $m'((t, 1), *) = 1$ and $m'((t, n), *) = 0$ if $n > 1$.

With respect to the definition in [8] we add the requirement that the various occurrence of the same transition are ordered. In [8] it is shown how to associate a configuration structures. To relate in a more precise way the notion of 1-unfolding of a net and the net itself, we use the mapping $\imath : [T' \to \mathbb{N}] \to [T \to \mathbb{N}]$ as follows: $\imath(U)(t)$ is $X(t) = |\{(t, i) | U((t, i)) > 0\}|$. It is easy to see that the following proposition holds.

Proposition 1. Let $N = (S, T, F, m)$ be a Petri net and $N' = (S', T', F', m')$ its 1-unfolding. Let C be a configuration of N' and $\hat{m}$ the reached marking, then there exists a firing sequence $m_0 \, [U_1 \rangle \, m_1 \, [U_2 \rangle \, m_2 \ldots m_{n-1} \, [U_n \rangle \, m_n$ such that $m_n(s) = \hat{m}(s)$ for all $s \in S$ and $\imath(\sum_{i \leq n} U_i) = C$.

An event structure with resolvable conflicts can be associated to the 1-unfolding of a net without inhibitor arcs defining the $\vdash$ relation as $C \vdash C'$ where C is a configuration of the 1-unfolding and C' is a configuration that can be reached by C, and $\emptyset \vdash C$ for all possible configurations of the net.

A different approach to the characterization of the firings of a net under the collective tokens philosophy is taken by Gunawardena in [14], where he characterizes the firings of the transitions of a safe net by counting the occurrences of the firings and relating them in an appropriate way. We illustrate the approach using the following (part) of a (safe) net:

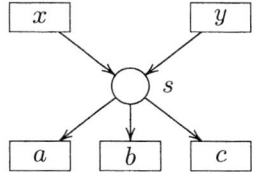

The transition x and y put tokens in s which are consumed by one in a, b or c, hence the i-th firing of a depends on the presence of a token in s and this is present only when the sum of the firings of x and y is k and the sum of the firings of a, b and c is j with $i = k - j + 1$. When s is initially marked we require that $i = k - j$.

Gunawardena gives, for each happening of an event (namely the firing of a transition, where the subscript identifies how many times the transition has fired before), the following formula, which must hold at a certain configuration in order to add the event:

$$\rho(a_n) = \bigwedge_{s \in {}^\bullet a} \ [\bigvee_{i+j=n+k+l+\phi(s)} \ ((x_i \wedge y_j) \wedge \neg(b_k \vee c_l))] \tag{1}$$

with ${}^\bullet s = \{x, y\}$ and $s^\bullet = \{a, b, c\}$, and $\phi(s) = 1$ if the place contains a token and 0 otherwise.

This formula can be rephrased as follows:

$$\rho'(a_n) = a_{n-1} \wedge \bigwedge_{s \in {}^\bullet a} \ [\bigwedge_{i+j+\psi(s)=n+k+l} \ (b_k \wedge c_l) \Rightarrow (x_i \wedge y_j)] \tag{2}$$

where $\psi(s) = 0$ if the place s is initially marked, and $\psi(s) = 1$ otherwise.

The formula is interpreted on configurations, with the intuitive meaning that a propositional letter is true if the corresponding occurrence of the transition is in the configuration. We assume that $\emptyset \models \top$, and in the case of the formula (2) we assume that the subscript 0 means that the propositional letter is $\top$. As usual we write $C \models \rho'(a_n)$ to say that the formula $\rho'(a_n)$ is true at the given configuration. The following proposition relates the configurations of a net with the formulas exemplified above, provided that a_n is interpreted on a net as (a, n).

Proposition 2. *Let $N' = (S', T', F', m')$ be the 1-unfolding of $N = (S, T, F, m)$ and C a configuration. If $C \models \rho'(t_n)$ then (1) $(m' + \sum_{t \in T'} C(t) \cdot (F'^\bullet(t) - {}^\bullet F'(t))) \, [(t, n)\rangle \, m''$, and (2) $C \oplus (t, n)$ is a configuration of N'.*

Thus these formulas describe exactly the enabling of a transition, giving a logical account of the firing of a net.

The formulas above define a DE-ES as well: to each $\rho'(a_n)$ the triples

$$\mathfrak{I}(\{(b, k), (c, l)\}, (a, n), \{\{(x, i), (y, j)\} | i + j = n + k + l\})$$

are associated. Henceforth we can associate a DE-ES to each 1-unfolding of a net. To formalize this notion, we need some more notation. Let $N = (S, T, F, m)$ be a net, with $Pre_N(\mathscr{Y}, s)$ the set $\{b_k | b \in s^{\bullet_N} \text{ and } k = \mathscr{Y}(b)\}$, where $\mathscr{Y} : s^{\bullet_N} \to \mathbb{N}$ is a finite multiset, and with $Post_N(\mathscr{Z}, s)$ the set $\{b_k | b \in {}^{\bullet_N} s \text{ and } k = \mathscr{Z}(b)\}$, where $\mathscr{Z} : {}^{\bullet_N} s \to \mathbb{N}$ is a again a finite multiset. $\mathscr{Y}$ and $\mathscr{Z}$ are multisets restricted to the preset and postset of s representing possible (parts of) configurations of the net N.

Definition 9. *Let $N' = (S', T', F', m')$ be the 1-unfolding of $N = (S, T, F, m)$. With $\mathscr{E}(N')$ we denote the DE-ES $\langle T', \mathfrak{I} \rangle$ defined as follows: for all $t \in T'$ and for all $s \in {}^{\bullet_N} t$ we introduce the triples (where $\mathscr{Y}(t) = 0$):*

$$\mathfrak{I}(Pre_N(\mathscr{Y}, s), (t, n), \{Post_N(\mathscr{Z}, s) | \sum_{s' \in s^{\bullet_N}} \mathscr{Y}(s') + n + \psi(s) = \sum_{s' \in {}^{\bullet_N} s} \mathscr{Z}(s')\})$$

Theorem 2. *Let $N' = (S', T', F', m')$ be the 1-unfolding of $N = (S, T, F, m)$ and $\mathscr{E}(N')$ be the associated* DE-ES. *Then C is a configuration of N' iff $C \in Conf(\mathscr{E}(N'))$.*

Thus, DE-ES events structures are able to capture the behaviour of a net under the collective tokens philosophy.

5 Nets with Inhibitor Arcs and Event Structures

We turn now our attention to nets with inhibitor arcs. A net with inhibitor arcs is the tuple $N = (S, T, F, I, m_0)$ where (S, T, F, m_0) is a net and $I \subseteq T \times S$. With $^\odot{}^Nt$ we denote the places inhibiting t, i.e. those such that $I(t,s)$. The notion of enabling changes accordingly adding that $(t,s) \in I$ implies $M(s) = 0$ and $\forall t, t' \in U$ such that $t = t' \Rightarrow U(t) \geq 2$ we have that $F(s,t) > 0$ implies $(t',s) \notin I$. The notion of firing sequence does not change. We rephrase quite easily the notion of 1-occurrence net and of 1-unfolding adapting to the case of inhibitor arcs.

Definition 10. *Let $N = (S, T, F, I, m)$ be a Petri net with inhibitor arcs, N is a 1-i-occurrence net (1-I-ON for short) if every configuration is a set.*

Definition 11. *Let $N = (S, T, F, I, m)$ be a Petri net. Its 1-unfolding $N' = (S', T', F', I', m')$ into a 1-i-occurrence net is defined as $N' = (S', T', F', m')$ and 1-occurrence net and $I'(s, (t,n))$ if $I(s,t)$.*

Let us see how to associate a formula to the n-firing of a transition of a net N with inhibitor arcs. Consider the following (part) of a (safe) net with inhibitor arcs:

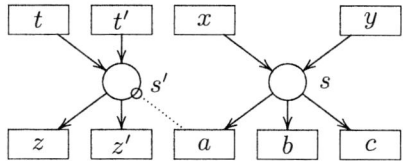

In order to fire the transition a should have the place s' empty. Thus to ρ' we have to add a piece corresponding to the left part of the net, stating that if t, t' have occurred each a number of times, then z, z' have to be occurred one time more:

$$\bigwedge_{s \in {}^\odot a} [\bigwedge_{i+j=k+l+\phi(s)} (t_k \wedge t'_l) \Rightarrow (z_i \wedge z'_j)] \tag{3}$$

where $\phi(s) = 1$ if the place s is initially marked. The whole formula is

$$\rho'(a_n) = a_{n-1} \wedge \bigwedge_{s \in {}^\bullet a} [\bigwedge_{i+j=n+k+l+\psi(s)} (b_k \wedge c_l) \Rightarrow (x_i \wedge y_j)] \wedge$$
$$\wedge \bigwedge_{s \in {}^\odot a} [\bigwedge_{i+j=k+l+\phi(s)} (t_k \wedge t'_l) \Rightarrow (z_i \wedge z'_j)]$$

Proposition 3. Let $N = (S, T, F, I, m)$ be a Petri net with inhibitor arcs and $N' = (S', T', F', I', m')$ its 1-unfolding. Let C be a configuration of N' and $\hat{m}$ the reached marking, then there exists a firing sequence $m_0 [U_1\rangle m_1 [U_2\rangle m_2 \ldots m_{n-1} [U_n\rangle m_n$ such that $m_n(s) = \hat{m}(s)$ for all $s \in S$ and $\iota(\sum_{i \leq n} U_i) = C$.

The following proposition relates the holding of a formula to the firing of a transition.

Proposition 4. Let $N' = (S', T', F', I', m')$ be the 1-unfolding of $N = (S, T, F, I, m)$ and C a configuration. If $C \models \rho'(t_n)$ then (a) $(m' + \sum_{t \in T'} C(t) \cdot (F'^\bullet(t) - {}^\bullet F'(t))) [(t, n)\rangle m''$, and (b) $C \oplus (t, n)$ is a configuration of N'.

The DE-ES is easily associated to the 1-unfolding of a net, as the part of the formula concerning the inhibition *generates* a number of different triples but does not influence the others (the positive enabling part).

Definition 12. Let $N' = (S', T', F', I', m')$ be the 1-unfolding of $N = (S, T, F, I, m)$. With $\mathscr{E}(N')$ we denote the DE-ES $\langle T', \mathfrak{I}\rangle$ defined as follows: for all $s \in {}^\bullet_N t$ we introduce the triples

$$\mathfrak{I}(Pre_N(\mathscr{Y}, s), (t, n), \{Post_N(\mathscr{Z}, s) | \sum_{s' \in s^\bullet_N} \mathscr{Y}(s') + n + \psi(s) = \sum_{s' \in {}^\bullet_N s} \mathscr{Z}(s')\})$$

with $\mathscr{Y}(a) = 0$, and for all $s \in {}^{\circledcirc}_N t$ we introduce the triples

$$\mathfrak{I}(Pre_N(\mathscr{Y}, s), (t, n), \{Post_N(\mathscr{Z}, s) | \sum_{s' \in s^\bullet_N} \mathscr{Y}(s') + n = \phi(s) + \sum_{s' \in {}^\bullet_N s} \mathscr{Z}(s')\})$$

The following theorem relates the two notion of configuration in the net and in DE-ES, showing the adequateness of DE-ES to give a faithful account of the behaviour of a net with inhibitor arcs under the collective tokens philosophy.

Theorem 3. Let $N' = (S', T', F', I', m')$ be the 1-unfolding of $N = (S, T, F, I, m)$ and $\mathscr{E}(N')$ be the associated DE-ES. Then C is a configuration of N' iff $C \in Conf(\mathscr{E}(N'))$.

6 Conclusions

In this paper we have shown how to generalize a disabling/enabling relation in order to give an event based faithful account of the collective tokens philosophy of nets with inhibitor arcs. We have limited our attention to safe nets, but the generalization to step semantics is rather obvious.

The proposed generalization seems to cover all other event based models in literature, but this should be further investigated. The event structures we have introduced are quite general, and indeed suitable subclasses have already been identified, e.g. IES, and these subclasses are obtained putting suitable constraints on the triple of the DE-relation. We will investigate on a more precise characterization of the triples in the case of the collective tokens philosophy.

Another line that should be pursued is on which kind of morphisms can be defined on this new relation. Usually morphisms identify conflicting computations, here the situation is more subtle as it is not so easy to identify conflicting computations on the basis of the triples alone. Thus it should be investigated whether a different notion of causality and conflict could arise in the spirit of the calculus presented in [6] and [12].

We do not consider here a relevant issue, namely a characterization of the theory that give rise to the model we consider, along the lines of what have been started in [13] and [11], leaving this topics to further investigations.

Acknowledgment. I would like to thank the referees for their useful suggestions that helped in improving the paper.

References

1. Baldan, P., Corradini, A., Montanari, U.: Contextual Petri nets, asymmetric event structures and processes. Information and Computation **171** (2001) 1–49
2. Busi, N., Pinna, G.M.: Process semantics for Place/Transition nets with inhibitor and read arcs. Fundamenta Informaticae **40** (1999) 165–197
3. Degano, P., Priami, C.: Causality for mobile processes. In Fülöp, Z., Gécseg, F., eds.: Proceedings of ICALP'95. (1995) 660–671
4. Janicki, R., Koutny, M.: Semantics of inhibitor nets. Information and Computation **123** (1995) 1–16
5. Pinna, G.M., Poigné, A.: On the nature of events. In Havel, I.M., Koubek, V., eds.: MFCS'92 Conference Proceedings. LNCS 629, Springer Verlag (1992) 430–441
6. Baldan, P.: Modelling concurrent computations: from contextual Petri nets to graph grammars. PhD thesis, University of Pisa (2000).
7. Busi, N., Pinna, G.: Contextual event structure and Petri nets with read and inhibitor arcs. manuscript (1999)
8. Glabbeek, R.J.v., Plotkin, G.D.: Configuration structures. In Kozen, D., ed.: Proceedings of 10th Annual IEEE Symposium on Logic in Computer Science, IEEE Computer Society Press (1995) 199–209
9. Pinna, G.M., Poigné, A.: On the nature of events: another perspective in concurrency. Theoretical Computer Science **138** (1995) 425–454
10. Hoogers, P.W., Kleijn, H.C.M., Thiagarajan, P.S.: An event structure semantics for general Petri nets. Theoretical Computer Science **153** (1996) 129–170
11. Bruni, R., Meseguer, J., Montanari, U., Sassone, V.: A comparison of petri net semantics under the collective tokens philosophy. In Hsiang, J., Ohori, A., eds.: ASIAN98. LNCS 1538, Springer Verlag (1998) 225–244
12. Baldan, P., Busi, N., Corradini, A., Pinna, G.: Domain and event structure semantics for Petri nets with read and inhibitor arcs. Theoretical Computer Science **323** (2004) 129–189
13. Glabbeek, R.J.v., Plotkin, G.D.: Event structures for resolvable conflict. In: Proceedings of MFCS '04. LNCS 3153, Springer (2004) 550–561
14. Gunawardena, J.: A generalized event structure for the muller unfolding of a safe net. In Best, E., ed.: CONCUR'93 Conference Proceedings. LNCS 715, Springer Verlag (1993)
15. Pinna, G.: Event structures with disabling/enabling relation and event automata. submitted (2005)
16. Winskel, G.: Event Structures. In: Petri Nets: Applications and Relationships to Other Models of Concurrency. LNCS 255, Springer Verlag (1987) 325–392

17. Nielsen, M., Plotkin, G., Winskel, G.: Petri Nets, Event Structures and Domains, Part 1. Theoretical Computer Science **13** (1981) 85–108
18. Langerak, R.: Transformation and Semantics for LOTOS. PhD thesis, University of Twente (1992)
19. Boudol, G.: Flow Event Structures and Flow Nets. In: Semantics of System of Concurrent Processes. LNCS 469, Springer Verlag (1990) 62–95

An Exact 2.9416^n Algorithm for the Three Domatic Number Problem*

Tobias Riege and Jörg Rothe

Institut für Informatik, Heinrich-Heine-Universität Düsseldorf,
40225 Düsseldorf, Germany
{riege, rothe}@cs.uni-duesseldorf.de

Abstract. The three domatic number problem asks whether a given undirected graph can be partitioned into at least three dominating sets, i.e., sets whose closed neighborhood equals the vertex set of the graph. Since this problem is NP-complete, no polynomial-time algorithm is known for it. The naive deterministic algorithm for this problem runs in time 3^n, up to polynomial factors. In this paper, we design an exact deterministic algorithm for this problem running in time 2.9416^n. Thus, our algorithm can handle problem instances of larger size than the naive algorithm in the same amount of time. We also present another deterministic and a randomized algorithm for this problem that both have an even better performance for graphs with small maximum degree.

Keywords: Exact algorithms, domatic number problem

1 Introduction

In this paper, we design a deterministic algorithm for the three domatic number problem, which is one of the standard NP-complete problems. This problem asks, given an undirected graph G, whether or not the vertex set of G can be partitioned into three dominating sets. A dominating set is a subset of the vertex set that "dominates" the graph in that its closed neighborhood covers the entire graph. Motivated by the tasks of distributing resources in a computer network and of locating facilities in a communication network, this problem and the related problem of finding a minimum dominating set in a given graph have been thoroughly studied.

The exact (i.e., deterministic) algorithm designed in this paper runs in exponential time. However, its running time is better than that of the naive exact algorithm for this problem. That is, we improve the trivial $\tilde{\mathcal{O}}(3^n)$ time bound to a time bound of $\tilde{\mathcal{O}}(2.9416^n)$, where the $\tilde{\mathcal{O}}$ notation neglects polynomial factors as is common for exponential-time algorithms. The point of such an improvement is that a $\tilde{\mathcal{O}}(c^n)$ algorithm, where $c < 3$ is a constant, can deal with larger instances than the trivial $\tilde{\mathcal{O}}(3^n)$ algorithm in the same amount of time before the exponential growth rate eventually hits and the running time becomes infeasible. For example, if $c = \sqrt{3} \approx 1.732$ then we have $\tilde{\mathcal{O}}\left(\sqrt{3}^{2n}\right) = \tilde{\mathcal{O}}(3^n)$, so one can deal with inputs twice as large as before. Doubling the size of inputs that can be handled by some algorithm can

---

* Work supported in part by the DFG under Grant RO 1202/9-1.

make quite a difference in practice. Exact exponential-time algorithms with improved running times have been designed for various other important NP-complete problems. Comprehensive surveys on this subject have been written by Woeginger [Woe03] and Schöning [Sch05].

In designing domatic number algorithms, it might be tempting to exploit known results for the graph three colorability problem, which resembles the three domatic number problem in that both are partitioning problems. However, as Cockayne and Hedetniemi [CH77] point out, the theory of domination is dual to the theory of coloring in the following sense. Coloring is based on the hereditary property of independence. A graph property is *hereditary* if whenever some set of vertices has the property then so does every subset of it. In contrast, domination is an *expanding* property in that every superset of a dominating set also is a dominating set of the graph. Further, graph colorability is a minimum problem, whereas the domatic number problem is a maximum problem. Independence (and thus colorability) can be seen as a *local* property, since it suffices to check the immediate neighborhood of a set of vertices to determine whether or not it is independent. In contrast, dominance is a *global* property, since in order to check it one has to consider the relation between the given set of vertices and the entire graph. In this sense, determining the domatic number of a graph intuitively appears to be harder than computing its chromatic number, notwithstanding that both problems are NP-complete. More to the point, the algorithms developed for graph coloring seem to be of no help in designing algorithms for dominating set or domatic number problems.

After introducing some definitions and notation in Section 2, we describe and analyze our algorithm in Section 3. In Section 4, we give another deterministic and a randomized algorithm, which have an even better running time for graphs with small maximum degree. Finally, we summarize and discuss our results in Section 5.

2 Preliminaries and Simple Observations

We start by introducing some graph-theoretical notation. We only consider simple, undirected graphs without loops in this paper. Let $G = (V, E)$ be a graph. Unless stated otherwise, n denotes the number of vertices in G. The *neighborhood of a vertex* v *in* V is defined by $N(v) = \{u \in V \mid \{u,v\} \in E\}$, and the *closed neighborhood of* v is defined by $N[v] = N(v) \cup \{v\}$. For any subset $S \subseteq V$ of the vertices of G, define $N[S] = \bigcup_{v \in S} N[v]$ and $N(S) = N[S] - S$. The *degree of a vertex* v *in* G is the number of vertices adjacent to v, i.e., $deg_G(v) = ||N(v)||$. If the graph G is clear from the context, we omit the subscript G. Define the *minimum degree in* G by $min\text{-}deg(G) = \min_{v \in V} deg(v)$, and the *maximum degree in* G by $max\text{-}deg(G) = \max_{v \in V} deg(v)$. A path $P_k = u_1 u_2 \cdots u_k$ of length k is a sequence of k vertices, where each vertex is adjacent to its successor, i.e., $\{u_i, u_{i+1}\} \in E$ for $1 \leq i \leq k-1$. If, in addition, $\{u_k, u_1\} \in E$, then path P_k is said to be a *cycle*, and we write C_k instead of P_k.

Definition 1. *Let* $G = (V, E)$ *be a graph. A subset* $D \subseteq V$ *is a* dominating set *of* G *if and only if* $N[D] = V$, *i.e., if and only if every vertex in* G *either belongs to* D *or has some neighbor in* D. *The* domination number *of* G, *denoted* $\gamma(G)$, *is the* minimum size of a dominating set of G. The domatic number of G, denoted $\delta(G)$, is the maximum number of disjoint dominating sets of G, i.e., $\delta(G)$ is the maximum k such that $V = V_1 \cup V_2 \cup \ldots \cup V_k$, where $V_i \cap V_j = \emptyset$ for $1 \leq i < j \leq k$, and each V_i is a dominating set of G. The dominating set problem *asks, given a graph G and a positive integer k, whether or not* $\gamma(G) \leq k$. The domatic number problem *asks, given a graph G and a positive integer k, whether or not* $\delta(G) \geq k$.

For fixed $k \geq 3$, both the dominating set problem and the domatic number problem are known to be NP-complete, see Garey and Johnson [GJ79]. Thus, they are not solvable in deterministic polynomial time unless P = NP, and all we can hope for is to design an exponential-time algorithm having a better running time than the trivial exponential time bound. For exponential-time algorithms, it is common to drop polynomial factors, as indicated by the $\tilde{\mathcal{O}}$ notation: For functions f and g, we write $f \in \tilde{\mathcal{O}}(g)$ if and only if $f \in \mathcal{O}(p \cdot g)$ for some polynomial p. The naive deterministic algorithm for the dominating set problem runs in time $\tilde{\mathcal{O}}(2^n)$. Fomin, Kratsch, and Woeginger [FKW04] improved this trivial upper bound to $\tilde{\mathcal{O}}(1.93782^n)$. For various restricted graph classes, they achieve even better bounds.

The naive deterministic algorithm for the domatic number problem runs in time $\tilde{\mathcal{O}}(k^n)$. A better result can be achieved via the dynamic programming across the subsets technique, which was introduced by Lawler [Law76] to compute the chromatic number of a graph by exploiting the fact that every minimum chromatic partition contains at least one maximum independent set. By suitably modifying this technique, one can compute the domatic number of a graph in time $\tilde{\mathcal{O}}(3^n)$.

One tempting way of designing an improved algorithm for the domatic number problem might be to exploit the result for the dominating set problem mentioned above. However, we observe that no such useful connection between the two problems exists in general.[1] Thus, for solving the domatic number problem, one cannot use in any obvious way the exact $\tilde{\mathcal{O}}(1.93782^n)$ algorithm for the dominating set problem by Fomin et al. [FKW04].

For the three domatic number problem, no algorithm with a running time better than $\tilde{\mathcal{O}}(3^n)$ is known. We improve this trivial upper bound to $\tilde{\mathcal{O}}(2.9416^n)$.

We now define some technical notions suitable to measure how "useful" a vertex is to achieve domination of the graph $G = (V, E)$. Intuitively, the vertex degree is a good (local) measure, since the larger the neighborhood of a vertex is, the more vertices are potentially dominated by the set to which it belongs. The technical notions introduced in Definition 2 will be used later on to describe our algorithm.

Definition 2. *Let* $G = (V, E)$ *be a graph with n vertices, and let* $\mathcal{P}(D_1, D_2, D_3, R)$ *be a partition of V into four sets, D_1, D_2, D_3, and R. The subsets D_i of V will eventually yield a partition of V into the three dominating sets (if they exist) to be constructed, and the subset* $R \subseteq V$ *collects the remaining vertices not yet assigned at the current point in the computation of the algorithm. Let* $r = ||R||$ *be the number of these remaining vertices, and let* $d = n - r$ *be the number of vertices already assigned to some set D_i. The area of G covered by $\mathcal{P}$ is defined as* $\mathrm{area}_\mathcal{P}(G) = \sum_{i=1}^{3} ||N[D_i]||$.

---

[1] Examples of graphs with maximum domatic partitions not including a minimum dominating set are presented in the upcoming full version of this paper.

Note that $\text{area}_{\mathcal{P}}(G) = 3n$ if and only if D_1, D_2, and D_3 are dominating sets of G. For a partition $\mathcal{P}$, we also define the *surplus* of graph G as $\text{surplus}_{\mathcal{P}}(G) = \text{area}_{\mathcal{P}}(G) - 3d$.

Some of the vertices in R may be assigned to three, not necessarily disjoint, auxiliary sets A_1, A_2, and A_3 arbitrarily. Let $\mathcal{A} = (A_1, A_2, A_3)$. For each vertex $v \in R$ and for each i with $1 \leq i \leq 3$, define the *gap* of vertex v with respect to set D_i by

$$\text{gap}_{\mathcal{P},\mathcal{A}}(v, i) = \begin{cases} \|N[v]\| - \|\{u \in N[v] \mid (\exists w \in N[u])\, [w \in D_i]\}\| & \text{if } v \notin A_i \\ \bot & \text{otherwise,} \end{cases}$$

where $\bot$ is a special symbol that indicates that $\text{gap}_{\mathcal{P},\mathcal{A}}(v, i)$ is undefined for this v and i. (Our algorithm will make sure to properly handle the cases of undefined gaps.) Additionally, given $\mathcal{P}$ and $\mathcal{A}$, define for all vertices $v \in R$:

$$\text{maxgap}_{\mathcal{P},\mathcal{A}}(v) = \max\{\text{gap}_{\mathcal{P},\mathcal{A}}(v, i) \mid 1 \leq i \leq 3\},$$
$$\text{mingap}_{\mathcal{P},\mathcal{A}}(v) = \min\{\text{gap}_{\mathcal{P},\mathcal{A}}(v, i) \mid 1 \leq i \leq 3\},$$
$$\text{sumgap}_{\mathcal{P},\mathcal{A}}(v) = \sum_{i=1}^{3} \text{gap}_{\mathcal{P},\mathcal{A}}(v, i).$$

Given G, $\mathcal{P}$, and $\mathcal{A}$, define the *maximum gap* of G and the *minimum gap* of G by taking the maximum and minimum gaps over all vertices in G not yet assigned:

$$\text{maxgap}_{\mathcal{P},\mathcal{A}}(G) = \max\{\text{maxgap}_{\mathcal{P},\mathcal{A}}(v) \mid v \in R\},$$
$$\text{mingap}_{\mathcal{P},\mathcal{A}}(G) = \min\{\text{mingap}_{\mathcal{P},\mathcal{A}}(v) \mid v \in R\}.$$

Let $\mathcal{P}$ be given. A vertex $u \in V$ is called an *open neighbor* of $v \in V$ if $u \in N[v]$ and u has not been assigned to any set D_1, D_2, or D_3 yet. A potential dominating set D_i, $1 \leq i \leq 3$, is called an *open set* of $v \in V$ if its closed neighborhood does not include v, i.e., v is not dominated by D_i. The *balance* of $v \in V$ is defined as the difference between the number of open vertices and the number of open sets. Formally, define

$$\text{openNeighbors}_{\mathcal{P}}(v) = \{u \in N[v] \mid u \in R\},$$
$$\text{openSets}_{\mathcal{P}}(v) = \{i \in \{1, 2, 3\} \mid v \notin N[D_i]\},$$
$$\text{balance}_{\mathcal{P}}(v) = \|\text{openNeighbors}_{\mathcal{P}}(v)\| - \|\text{openSets}_{\mathcal{P}}(v)\|.$$

We call a vertex $v \in V$ *critical* if and only if $\text{balance}_{\mathcal{P}}(v) \leq 0$ and $\|\text{openSets}_{\mathcal{P}}(v)\| > 0$.

The proof of the next proposition is straightforward. Once $\text{balance}_{\mathcal{P}}(v) = 0$, no two vertices remaining in $N[v] \cap R$ can be assigned to the same dominating set D_i, $1 \leq i \leq 3$, since $\text{balance}_{\mathcal{P}}(v)$ would then be negative.

Proposition 1. *Let $\mathcal{P} = (D_1, D_2, D_3, R)$ be given as in Definition 2, and $v \in V$ be a critical vertex for this partition. The only way to modify $\mathcal{P}$ so as to contain three dominating sets is to assign all vertices $u \in N[v] \cap R$ to distinct dominating sets D_i.*

3 The Algorithm

Our strategy is to recursively assign the vertices $v \in V$ to obtain a correct potential solution consisting of a partition into three dominating sets, D_1, D_2, and D_3. Once a previous assignment of v to some set D_i turns out to be wrong, we remember this by adding this vertex to A_i. More precisely, the basic idea is to first pick those vertices with the highest maximum gap. While the algorithm is progressing, it dynamically updates the gaps for every vertex in each step. We now state our main result.

Theorem 1. *The three domatic number problem can be solved by a deterministic algorithm running in time $\tilde{\mathcal{O}}(2.9416^n)$.*

Proof. Let $G = (V, E)$ be the given graph. The algorithm seeks to find a partition of V into three disjoint dominating sets. Note that every vertex $v \in V$ is contained in one of these sets and is dominated by the remaining two sets, i.e., it is adjacent to at least one of their elements. The algorithm is described in pseudo-code in Figures 1, 2, and 3. (Function RECALCULATE-GAPS, which updates all gaps with respect to the current $\mathcal{P}$ and $\mathcal{A}$ in polynomial time, is omitted here due to space constraints.) Since $\delta(G) \leq \textit{min-deg}(G) + 1$, we may assume that $\textit{min-deg}(G) \geq 2$.

The algorithm starts by initializing the potential dominating sets D_1, D_2, and D_3 and the auxiliary sets A_1, A_2, and A_3, setting each to the empty set. The initial partition thus is $\mathcal{P}(\emptyset, \emptyset, \emptyset, V)$ and the initial triple of auxiliary sets is $\mathcal{A} = (\emptyset, \emptyset, \emptyset)$.

Algorithm for the Three Domatic Number Problem
 Input: Graph $G = (V, E)$ with vertex set $V = \{v_1, v_2, \ldots, v_n\}$ and edge set E
 Output: Partition of V into three dominating sets $D_1, D_2, D_3 \subseteq V$ or "failure"
 Set each of D_1, D_2, D_3, A_1, A_2, and A_3 to the empty set;
 Set $R = V$;
 Set $\mathcal{P} = (D_1, D_2, D_3, R)$;
 Set $\mathcal{A} = (A_1, A_2, A_3)$;
 DOMINATE$(G, \mathcal{P}, \mathcal{A})$; // Start recursion
 output "failure" and **halt**;

Fig. 1. Algorithm for the Three Domatic Number Problem

Then, the recursive function DOMINATE is called for the first time. It is always invoked with graph G, a partition $\mathcal{P} = (D_1, D_2, D_3, R)$, and a triple $\mathcal{A} = (A_1, A_2, A_3)$ of not necessarily disjoint auxiliary sets. $\mathcal{P}$ and $\mathcal{A}$ represent a situation in which the vertices in $V - R$ have been assigned to D_1, D_2, and D_3, and $v \in A_i$ means that in some previous recursive call to function DOMINATE the vertex v has been assigned to D_i without successfully changing $\mathcal{P}$ to contain three dominating sets.

Function DOMINATE starts by calling RECALCULATE-GAPS, which calculates all gaps with respect to $\mathcal{P}$ and $\mathcal{A}$. Additionally, openNeighbors$_\mathcal{P}(v)$, openSets$_\mathcal{P}(v)$, and balance$_\mathcal{P}(v)$ are determined for every vertex $v \in V$. Four trivial cases can occur.

Case 1: The sets D_1, D_2, and D_3 are dominating sets of graph G. In this case, we are done and may add the remaining vertices $v \in R$ to any set D_i, say to D_1.

Function DOMINATE($G, \mathcal{P}, \mathcal{A}$) {   // $\mathcal{P}$ is a partition of graph G,
                                                                                          // $\mathcal{A}$ is a triple of auxiliary sets
    RECALCULATE-GAPS($G, \mathcal{P}, \mathcal{A}$);   // recalculate all gaps, open neighbors, etc.
    if (each D_i is a dominating set) {
        $D_1 = D_1 \cup R$;
        output D_1, D_2, D_3;
    }
    if (not HANDLE-CRITICAL-VERTEX($G, \mathcal{P}, \mathcal{A}$)) {
        select vertex $v \in R$ with
            maxgap$_{\mathcal{P},\mathcal{A}}(v) = $ maxgap$_{\mathcal{P},\mathcal{A}}(G)$ and
            sumgap$_{\mathcal{P},\mathcal{A}}(v) = \max\{$sumgap$_{\mathcal{P},\mathcal{A}}(u) \mid u \in R \wedge $ maxgap$_{\mathcal{P},\mathcal{A}}(u) = $ maxgap$_{\mathcal{P},\mathcal{A}}(G)\}$;
        find i with gap$_{\mathcal{P},\mathcal{A}}(v, i) = $ maxgap$_{\mathcal{P},\mathcal{A}}(v)$;
        ASSIGN($G, \mathcal{P}, \mathcal{A}, v, i$);   // First recursive call
        $A_i = A_i \cup \{v\}$;   // If recursion fails, put v in A_i and try again
        DOMINATE($G, \mathcal{P}, \mathcal{A}$);   // Second recursive call
    }
    return;
}

Fig. 2. Recursive function to dominate graph G

Case 2: For some vertex $v \in V$, we have balance$_\mathcal{P}(v) < 0$. That is, there are less vertices in $R \cap N[v]$ than dominating sets with $v \notin N[D_i]$. Thus, no matter how the vertices in $R \cap N[v]$ are assigned, $\mathcal{P}$ won't contain three dominating sets. We have run into a dead-end and return to the previous level of the recursion.

Case 3: There exists a vertex $v \in R$ that is also a member of two of the auxiliary sets A_1, A_2, and A_3. Hence, vertex v was previously assigned to two distinct sets D_i and D_j, $1 \leq i < j \leq 3$, but the recursion returned without success. We assign v to the only possible set D_k left, with $i \neq k \neq j$.

Case 4: For some vertex $v \in V$, we have balance$_\mathcal{P}(v) = 0$ and $||$openSets$_\mathcal{P}(v)|| > 0$. That is, v is a critical vertex, since it is not dominated by all three sets D_1, D_2, and D_3 contained in the current $\mathcal{P}$, and there are as many open neighbors as open sets left for it. Note that this is the case for each vertex v with $deg(v) = 2$ and $N[v] \cap R \neq \emptyset$, as v and its two neighbors have to be assigned to three different dominating sets. We select one of the at most three vertices left in $N[v] \cap R$, say u, and call function ASSIGN($G, \mathcal{P}, \mathcal{A}, u, i$) for all i with $u \notin A_i$.

Function HANDLE-CRITICAL-VERTEX deals with the latter three of these trivial cases. After they have been ruled out, one of the remaining vertices $v \in R$ is selected and assigned to one of the three sets D_i, under the constraint that a vertex $v \in R$ cannot be added to D_i if it is already a member of A_i. This case occurs whenever the recursion returns because no three dominating sets could be found with this combination. The recursion continues by calling ASSIGN($G, \mathcal{P}, \mathcal{A}, v, i$), which adds v to D_i, and then calls DOMINATE($G, \mathcal{P}, \mathcal{A}$). If no three dominating sets are found by this choice, we remember this by adding v to the set A_i. A final call to DOMINATE is made without assigning a vertex to one potential dominating set D_i. If this call fails, the recursion

Function boolean HANDLE-CRITICAL-VERTEX$(G, \mathcal{P}, \mathcal{A})$ {
  for all (vertices $v \in V$) {
    if (balance$_\mathcal{P}(v) < 0$) {   // impossible to three dominate v with $\mathcal{P}$
      return true;
    } **else if** ($||\{i \in \{1,2,3\} \mid v \in A_i\}|| == 2$) {   // one choice for v remaining
      select i with $v \notin A_i$;
      ASSIGN$(G, \mathcal{P}, \mathcal{A}, v, i)$;
      return true;
    } **else if** (balance$_\mathcal{P}(v) == 0$ and $||\text{openSets}_\mathcal{P}(v)|| > 0$) {   // v is critical
      select $u \in N[v] \cap R$;
      for all (i with $u \notin A_i$ and v not dominated by D_i)
        ASSIGN$(G, \mathcal{P}, \mathcal{A}, u, i)$;
      return true;
    }
  }
  return false;   // no critical vertices were found
}

Function ASSIGN$(G, \mathcal{P}, \mathcal{A}, v, i)$ {
  $D_i = D_i \cup \{v\}$;
  $R = R - \{v\}$;
  DOMINATE$(G, \mathcal{P}, \mathcal{A})$;
}

Fig. 3. Functions to handle the critical vertices and to assign vertex v to set D_i

returns to the previous level. This completes the description of the algorithm. We now argue that it is correct and estimate its running time.

To see that the algorithm works correctly, note that it outputs three sets D_1, D_2, and D_3 only if they each are dominating sets of G. It remains to prove that these sets are definitely found in the recursion tree. All drop-backs within the recursion occur when, for the current $\mathcal{P} = (D_1, D_2, D_3, R)$, we have balance$_\mathcal{P}(v) < 0$ for some vertex $v \in V$. Thus, $\mathcal{P}$ cannot be modified so as to contain a correct partition into three dominating sets on this branch of the recursion tree. Since the algorithm checks every possible partition of G into three sets, unless it is stopped by such a drop-back, a partition into three dominating sets will be found, if it exists. If the algorithm does not find three dominating sets, it eventually terminates when returning from the first recursive call of function DOMINATE. It reports the failure, and thus always yields the correct output.

To estimate the running time of the algorithm, an important observation is that the recalculation of the gaps takes no more than quadratic time in n, the number of vertices of the graph G. Thus, in terms of the $\tilde{\mathcal{O}}$-notation, the running time of the algorithm depends solely on the number of recursive calls. Let $T(m)$ be the number of steps of the algorithm, where m is the number of potential dominating sets left for all vertices that have not been selected as yet. Initially, every vertex may be a member of any of the three dominating sets to be constructed (if they exist), hence $m = 3n$.

There are two scenarios where the algorithm calls function DOMINATE recursively. If HANDLE-CRITICAL-VERTEX detects a vertex $v \in V$ as being critical, it selects a vertex $u \in N[v] \cap R$ and calls function ASSIGN (and thus DOMINATE) for each i with $u \notin A_i$. Function HANDLE-CRITICAL-VERTEX will be called until all vertices in $N[v] \cap R$ have been assigned to any of D_1, D_2, and D_3. Since $\|\text{openSets}_\mathcal{P}(v)\| \leq 3$, at most three vertices in the closed neighborhood of v have not been assigned when v turns critical. By Proposition 1, all vertices in $N[v] \cap R$ have to be assigned to different dominating sets. In the worst case we have $T(m) \leq 6T(m-6)$, as we will handle three vertices for which at least two choices for dominating sets are left, and we have at most six different choices that lead to a partition into three disjoint dominating sets. With $m = 3n$, it follows that $T(m) \leq 6^{m/6} = 6^{n/2}$, i.e., $T(m) = \tilde{\mathcal{O}}(2.4495^n)$.

The only other branching into two different recursive calls happens in the main body of function DOMINATE, when selecting a vertex v with the currently highest maximum gap. Two cases might occur. On the one hand, we might have considered a correct dominating set D_i for v. If v had not been looked at so far, i.e., if v is not contained in any set A_j, $1 \leq j \leq 3$, $j \neq i$, we have eliminated all three possible sets for v to belong to. Thus, in this case, $T(m) = T(m-3)$. On the other hand, if the algorithm returns from the recursion and thus did not make the right choice for v, we have $T(m) = T(m-1)$, since v is added to A_i, and function DOMINATE is called without assigning any vertex. In the second case, we have already visited vertex v in a previous stage of the algorithm and unsuccessfully tried to assign it to some set D_j, with $1 \leq j \leq 3$. There are only two dominating sets for v left. Either way, if we put v into the correct dominating set right away or fail the first time, we have $T(m) = T(m-2)$. Suppose that the first and the second case occur equally often, i.e., the algorithm considers every vertex twice. It then follows that $T(m) \leq \frac{1}{2}(T(m-1) + T(m-3)) + \frac{1}{2}(2T(m-2))$ with $m = 3n$. Thus, we have $T(m) = \tilde{\mathcal{O}}(3^n)$, and the trivial time bound cannot be beaten. To improve this running time, we have to make sure that the recursion tree will not reach its full depth, i.e., not all vertices are considered by the algorithm or function HANDLE-CRITICAL-VERTEX will be called for a sufficiently large portion of the vertices. It is clear that the algorithm has found three dominating sets once $\text{area}_\mathcal{P}(G) = 3n$ (recall the notions from Definition 2). By selecting the maximum gap possible for a partition $\mathcal{P}$, we try to reach this goal as fast as possible. For every vertex $v \in R$ that we assign to one of the potential dominating sets D_i, $1 \leq i \leq 3$, we increase $\text{area}_\mathcal{P}(G)$ by $\text{gap}_{\mathcal{P},\mathcal{A}}(v,i)$, and additionally we add $(\text{gap}_{\mathcal{P},\mathcal{A}}(v,i) - 3)$ to $\text{surplus}_\mathcal{P}(G)$.

Since the vertices of degree two are critical, they and their neighbors can be handled in time $\tilde{\mathcal{O}}(2.4495^n)$, as argued above. So assume that $min\text{-}deg(G) \geq 3$. Then, we have $\text{maxgap}_{\mathcal{P},\mathcal{A}}(G) > 3$ at the start of the algorithm. If this condition remains to hold for at least $3n/4$ steps, we have reached $\text{area}_\mathcal{P}(G) = 3n$, and the algorithm terminates successfully. To make use of more than $3n/4$ vertices, $\text{maxgap}_{\mathcal{P},\mathcal{A}}(G)$ has to drop below four at one point of the computation. We exploit the fact that up to this point, the surplus has grown sufficiently large with respect to n. Decreasing it will force $\text{maxgap}_{\mathcal{P},\mathcal{A}}(G)$ to drop below three, and this condition can hold only for a certain portion of the remaining vertices until the algorithm terminates. To see this, we now analyze the remaining steps of the algorithm after the given graph G has reached a certain maximum gap with respect to the current $\mathcal{P}$ and $\mathcal{A}$.

If $\text{maxgap}_{\mathcal{P},\mathcal{A}}(G) = 0$, the recursion stops immediately. Either we have already found three disjoint dominating sets (in which case we put the remaining vertices $v \in R$ into set D_1 and halt), or we have $\text{balance}_\mathcal{P}(v) < 0$ for some vertex $v \in V$. The question is how many vertices are left in R when we reach $\text{maxgap}_{\mathcal{P},\mathcal{A}}(G) = 0$.

Lemma 1. *Let $G = (V,E)$ be a graph and $\mathcal{P} = (D_1, D_2, D_3, R)$ be a partition of V as in Definition 2. Let $r = ||R||$ and $\text{maxgap}_{\mathcal{P},\mathcal{A}}(G) = 3$. Then, for at least $r/64$ vertices in R, the algorithm will not recursively call function DOMINATE.*

Proof of Lemma 1. Let $\text{maxgap}_{\mathcal{P},\mathcal{A}}(G) = k$ with $k > 0$. Since $\text{gap}_{\mathcal{P},\mathcal{A}}(v,i) \leq k$ for each $v \in R$ and for each i, $1 \leq i \leq 3$, we have $\sum_{v \in R} \text{sumgap}_{\mathcal{P},\mathcal{A}}(v) \leq 3kr$. Every vertex v that is selected for a set D_i with $\text{gap}_{\mathcal{P},\mathcal{A}}(v,i) = k$ decreases at least k gaps of the vertices in $R - \{v\}$ by one. Otherwise, HANDLE-CRITICAL-VERTEX would have found a critical vertex $u \in N[v]$ with $N[u] \cap R = \{v\}$. Then, either $||\text{openSets}_\mathcal{P}(u)|| > 1$ (which implies $\text{balance}_\mathcal{P}(u) < 0$ and we abort), or $||\text{openSets}_\mathcal{P}(u)|| = 1$, in which case v is added to the appropriate set D_i without further branching of function DOMINATE. Thus, if no critical vertex is detected, selecting a vertex $v \in R$ for some set D_i decreases at least k gaps, and since v does not belong to R anymore, additionally all gaps previously defined for v are now undefined.

Now suppose that $\text{maxgap}_{\mathcal{P},\mathcal{A}}(G) = 3$ and $\text{sumgap}_{\mathcal{P},\mathcal{A}}(v) = 9$ for all vertices $v \in R$. As long as there exists a vertex $v \in R$ with $\text{gap}_{\mathcal{P},\mathcal{A}}(v,i) = 3$ for all i, it will be selected by the algorithm. After calling function RECALCULATE-GAPS, the number of gaps equal to three will be decreased at least by six. If exactly three other gaps of vertices in $R - \{v\}$ decrease by one in every step, it takes at least $r/4$ vertices until $\text{sumgap}_{\mathcal{P},\mathcal{A}}(v) < 9$ for all $v \in R$. Another $1/4$ of the $3r/4$ vertices remaining have to be selected until $\text{sumgap}_{\mathcal{P},\mathcal{A}}(v) < 8$. Adding $1/4$ of the $9r/16$ vertices left in R, we have reached $\text{maxgap}_{\mathcal{P},\mathcal{A}}(G) = 2$ with $\text{sumgap}_{\mathcal{P},\mathcal{A}}(v) = 6$ for all vertices $v \in R$. This implies that every defined gap is equal to two. Summing up, we have selected $\frac{1}{4} \cdot r + \frac{1}{4} \cdot \frac{3}{4}r + \frac{1}{4} \cdot \frac{9}{16}r = \frac{37}{64}r$ vertices until $\text{maxgap}_{\mathcal{P},\mathcal{A}}(G) = 2$, under the constraint that a minimum number of gaps is reduced in each step, while simultaneously trying to reduce the maximum summation gap in the fastest possible way. This way we reach level $\text{maxgap}_{\mathcal{P},\mathcal{A}}(G) = 0$ with as few vertices left in R as possible, which describes the worst case that might happen.

Analogously, we can show that $\text{maxgap}_{\mathcal{P},\mathcal{A}}(G)$ drops from 2 to 1 after selecting another $19r/64$ vertices. And once we have $\text{maxgap}_{\mathcal{P},\mathcal{A}}(G) = 1$, it takes $7r/64$ vertices to get to $\text{maxgap}_{\mathcal{P},\mathcal{A}}(G) = 0$. Now, there are $r/64$ vertices remaining in R, which do not have to be processed recursively. ∎ Lemma 1

Continuing the proof of Theorem 1, note that we assumed $min\text{-}deg(G) \geq 3$, so when the gaps are initialized for graph G, we have $\text{mingap}_{\mathcal{P},\mathcal{A}}(v) \geq 4$ for each vertex $v \in V$. Thus, more than three vertices are dominated by the selected set D_i for vertex v. As long as $\text{maxgap}_{\mathcal{P},\mathcal{A}}(G) > 3$ is true, $\text{surplus}_\mathcal{P}(G)$ is increasing. The only way to lower the surplus is by adding vertices v to a set D_i with $\text{gap}_{\mathcal{P},\mathcal{A}}(v,i) < 3$. The surplus decreases by one when $\text{gap}_{\mathcal{P},\mathcal{A}}(v,i) = 2$, and it decreases by two when $\text{gap}_{\mathcal{P},\mathcal{A}}(v,i) = 1$.

Let $S = \text{surplus}_\mathcal{P}(G)$ be the surplus collected for a partition $\mathcal{P}$ until we reach a point where $\text{maxgap}_{\mathcal{P},\mathcal{A}}(G) = 3$. To make use of the most recursive calls and to even out the surplus completely, there have to be at least $r = ||R||$ vertices remaining with

$0 \cdot \frac{37r}{64} + 1 \cdot \frac{19r}{64} + 2 \cdot \frac{7r}{64} = S$, so $r \geq 64S/33$. A fraction of $1/64$ of these vertices will be handled by the algorithm without branching into more than one recursive call, which is at least $S/33$. The question is how big the surplus S might grow and how many vertices are left in R before $\text{maxgap}_{\mathcal{P},\mathcal{A}}(G) = 3$ is reached. The lowest surplus with as few vertices in R as possible occurs if $\text{min-deg}(G) = \text{max-deg}(G) = 3$. Surplus S is increased by one in each step until we arrive at $\text{maxgap}_{\mathcal{P},\mathcal{A}}(G) = 3$. When selecting a vertex v of degree 3 for a set D_i, the gap of its neighbors $u \in N(v)$ and the gaps of the neighbors of every u might be decreased. Summing up, at most 10 vertices can have decreased their gaps for some i. After selecting at least $n/10$ vertices for each i, we have $\text{mingap}_{\mathcal{P},\mathcal{A}}(G) = 3$ (in the worst case). From this point on, we cannot be sure if the next vertex selected for some D_i satisfies $\text{gap}_{\mathcal{P},\mathcal{A}}(v,i) > 3$. But so far we have already collected a surplus of $S = 3n/10$, and applying this we obtain $64n/110 \leq r \leq 7n/10$. Thus, for at least $n/110$ vertices we never branch into two different recursive calls. Setting $m = 3(109n/110)$, we obtain a running time of $\tilde{O}(2.9416^n)$. ∎

4 Graphs with Bounded Maximum Degree

As seen in the last section, the running time of the algorithm crucially depends on the degrees of the vertices of G. If we restrict ourselves to graphs G with bounded maximum degree (say $\Delta = \text{max-deg}(G)$), we can optimize our strategy in finding three disjoint dominating sets. In this section, we present a simple deterministic algorithm, which has a better running time than the algorithm from Theorem 1, provided that Δ is low. By using randomization, we can further improve the running time for graphs G with low maximum degree.

Before stating the two results, note that graphs with maximum degree two can trivially be partitioned into three dominating sets, if such a partition exists. Every component of such a graph is either an isolated vertex, a path, or a cycle, and each such property can be recognized in polynomial time.

Proposition 2. *Let $G = (V, E)$ be a given graph with $\text{max-deg}(G) = 2$. There exists a partition of the vertices of G into three dominating sets if and only if every component of G is a cycle of length k such that 3 divides k.*

We use the terms from Definition 2 in Section 3 to describe a snapshot within the algorithm. The auxiliary sets $\mathcal{A} = (A_1, A_2, A_3)$ will not be needed in this section. Only connected graphs are considered, as it is possible to treat every connected component separately, producing the desired output within the same time bounds.

Table 1 lists the running times of both the deterministic and the random algorithm, where the maximum degree of the input graph is bounded by Δ, $3 \leq \Delta \leq 9$. Note that the exact deterministic algorithm from Theorem 1 in Section 3 beats the deterministic algorithm from Theorem 2 whenever $\Delta \geq 7$.

Theorem 2. *Let $G = (V, E)$ be a graph with $\text{max-deg}(G) = \Delta$, where $\Delta \geq 3$. There exists a deterministic algorithm solving the three domatic number problem in time $\tilde{O}(d^{\frac{n}{\Delta}})$, where*

Table 1. Results for $max\text{-}deg(G) = k$, where $3 \leq k \leq 9$

Δ	3	4	5	6	7	8	9
deterministic	2.2894^n	2.6591^n	2.8252^n	2.9058^n	2.9473^n	2.9697^n	2.9823^n
randomized	2^n	2.3570^n	2.5820^n	2.7262^n	2.8197^n	2.8808^n	2.9210^n

$$d = \sum_{a=0}^{\Delta-2} \left[\binom{\Delta}{a} \sum_{b=1}^{\Delta-a-1} \binom{\Delta-a}{b} \right]. \tag{4.1}$$

Proof. The algorithm works as follows. We start with an arbitrary vertex $v \in V$ and assign it to the first set D_1. In each step, we first check whether we found a partition $\mathcal{P} = (D_1, D_2, D_3, R)$ into dominating sets D_1, D_2, and D_3. If not, one vertex $v \in V$ is selected that is not dominated by all three sets D_1, D_2, and D_3, and additionally has a vertex $u \in N[v]$ in its closed neighborhood that has already been added to some set D_i, $1 \leq i \leq 3$. It follows that $1 \leq ||\text{openSets}_{\mathcal{P}}(v)|| \leq 2$.

If $\text{balance}_{\mathcal{P}}(v) < 0$, we return within the recursion. Otherwise, we try all combinations to partition the vertices in $N[v] \cap R$, so that after this step vertex v is dominated by all three potential dominating sets. If no such combination leads to a valid partition, we again return within the recursion.

Suppose now that $\text{balance}_{\mathcal{P}}(v) \geq 0$, $||\text{openSets}_{\mathcal{P}}(v)|| = 2$, and $N[v] \cap D_1 \neq \emptyset$. To obtain three disjoint dominating sets, at least one vertex in $N[v]$ has to be assigned to D_2, and at least one vertex in $N[v]$ has to be added to D_3. This limits our choices, especially if the degree of v is bounded by some constant Δ.

To measure the running time of the algorithm, we consider the worst case with the most possible combinations that might yield a partition into three dominating sets. This occurs when only one vertex $u \in N[v]$ has already been added to one set, i.e., $||N[v] \cap (D_1 \cup D_2 \cup D_3)|| = 1$. If $N[v] \cap D_1 \neq \emptyset$, then any number between 0 and $\Delta - 2$ of vertices in $N[v] \cap R$ may be assigned to the same set D_1. Let this number be a. It follows that from one to $\Delta - a - 1$ vertices remaining in $N[v] \cap R$ are allowed to be in the next potential dominating set D_2. This is how Equation 4.1 for d is derived. After assigning the last vertices in $N[v] \cap R$ to the dominating set D_3, exactly Δ vertices have been removed from R. Thus, we have a worst case running time of $\tilde{\mathcal{O}}(d^{\frac{n}{\Delta}})$. Table 1 lists the running time for graphs with maximum degree from three to nine. ∎

In the next theorem, randomization is used to speed up this procedure. Instead of assigning all vertices in the closed neighborhood of some vertex $v \in V$ in one step, only one or two vertices in $N[v] \cap R$ are added to the potential dominating sets D_1, D_2, and D_3. The goal is to dominate one vertex by all three sets in one step. We will select the one or two vertices that are missing for this goal at random.

Due to space limitations, the proof of Theorem 3 is omitted. Here is a rough sketch of the idea: In the case $||\text{openSets}_{\mathcal{P}}(v)|| = 2$, we randomly assign two vertices in the neighborhood of $v \in V$ to the sets D_1, D_2, and D_3. Since $deg(v) \leq \Delta$, there are at most $3^{\Delta-2}$ valid choices for the vertices left in $N[v] \cap R$. Thus, the success probability is greater than $3^{\Delta-2}/d$, where d is the number from Equation (4.1) in Theorem 2. The case $||\text{openSets}_{\mathcal{P}}(v)|| = 1$ is treated analogously.

Theorem 3. *Let $G = (V, E)$ be a graph with max-deg$(G) = \Delta$, where $\Delta \geq 3$, and let d be defined as in Equation (4.1) in Theorem 2. For each constant $c > 0$, there exists a randomized algorithm solving the three domatic number problem with error probability at most e^{-c} in time $\tilde{O}(r^{\frac{n}{2}})$, where $r = d/3^{\Delta-2}$.*

5 Conclusion

We have shown that the three domatic number problem can be solved by a deterministic algorithm in time $\tilde{O}(2.9416^n)$. Furthermore, we presented two algorithms solving the three domatic number problem for graphs with bounded maximum degree, improving the above time bound for graphs with small maximum degree. Although our running times seem to be not too big of an improvement of the trivial $\tilde{O}(3^n)$ bound, they are to our knowledge the first such algorithms breaking this barrier. For $k > 3$, the decision problem of whether $\delta(G) \geq k$ can be solved in time $\tilde{O}(3^n)$ by Lawler's dynamic programming algorithm for the chromatic number, appropriately modified for the domatic number problem. Therefore, it would not be reasonable to use our gap approach of Section 3 to decide if $\delta(G) \geq k$ for a graph G and $k > 3$.

Acknowledgement. We thank Dieter Kratsch for pointing us to Lawler's algorithm.

References

[CH77] E. Cockayne and S. Hedetniemi. Towards a theory of domination in graphs. *Networks*, 7:247–261, 1977.

[FKW04] F. Fomin, D. Kratsch, and G. Woeginger. Exact (exponential) algorithms for the dominating set problem. In *Proceedings of the 30th International Workshop on Graph-Theoretic Concepts in Computer Science (WG 2004)*, pages 245–256. Springer-Verlag *Lecture Notes in Computer Science #3353*, 2004.

[GJ79] M. Garey and D. Johnson. *Computers and Intractability: A Guide to the Theory of NP-Completeness*. W. H. Freeman and Company, New York, 1979.

[Law76] E. Lawler. A note on the complexity of the chromatic number problem. *Information Processing Letters*, 5(3):66–67, 1976.

[Sch05] U. Schöning. Algorithmics in exponential time. In *Proceedings of the 22nd Annual Symposium on Theoretical Aspects of Computer Science*, pages 36–43. Springer-Verlag *Lecture Notes in Computer Science #3404*, 2005.

[Woe03] G. Woeginger. Exact algorithms for NP-hard problems. In M. Jünger, G. Reinelt, and G. Rinaldi, editors, *Combinatorical Optimization: "Eureka, you shrink!"*, pages 185–207. Springer-Verlag *Lecture Notes in Computer Science #2570*, 2003.

D-Width: A More Natural Measure for Directed Tree Width

Mohammad Ali Safari

Department of Computer Science,
University of British Columbia, Vancouver, BC, Canada
safari@cs.ubc.ca

Abstract. Due to extensive research on tree-width for undirected graphs and due to its many applications in various fields it has been a natural desire for many years to generalize the idea of tree decomposition to directed graphs, but since many parameters in tree-width behave very differently in the world of digraphs, the generalization is still in its preliminary steps.

In this paper, after surveying the main work that has been done on this topic, we propose a new simple definition for directed tree-width and prove a special case of the min-max theorem (duality theorem) relating haven order, bramble number, and tree-width on digraphs. We also compare our definition with previous definitions and study the behavior of some tree-width like parameters such as brambles and havens on digraphs.

1 Introduction

The notion of tree-width is considered as a generalization of trees (trees have tree-width 1) and many intractable problems are efficiently solvable on bounded tree-width graphs. Examples include *Hamiltonian cycle*, *graph isomorphism*, *vertex coloring*, *edge coloring*, and so on. Such problems arise in various fields including (but not restricted to) expert systems, telecommunication network design, VLSI design, Choleski factorization, natural language processing, etc. See [Bod93] for an overview of some such applications.

In 1996 Reed et al. [RRST96] proved Youngers's conjecture [You73] roughly saying that every directed graph has either a large set of disjoint directed circuits or a small set of vertices that cover all directed circuits. They considered an analogous definition of *well-linked* sets for directed graphs and since the size of the largest well-linked set in undirected graphs has close relationship with tree-width[Ree00] they suggested the idea that the analogous definition of tree-width for directed graphs might be very useful, as pointed out in [Ree97]. A proper definition should ideally measure the global connectivity of a digraph, for example the tree-width of a directed acyclic graph, DAG, is expectedly small because it is lowly connected.

Unfortunately finding an analogous definition for directed tree-width is not easy at all, and almost all concepts related to undirected tree-width behave differently in directed graphs. For example, the bramble number is equal to the haven order in undirected graphs, while they may differ by a factor of 2 in directed graphs. There is not even a single agreed-upon definition of tree-width for directed graphs.

For the first time Johnson, Robertson, Seymour, and Thomas[JRST01] gave a formal definition of directed tree-decomposition (called *arboreal-decomposition* in their paper) and directed tree-width, and proved some theorems relating directed tree-width and haven order.

Theorem 1. *[JRST01] For any digraph D, $H(D)-1 \leq$ tree-width$(D) \leq 3H(D) - 1$, where tree-width(D) and $H(D)$ are the tree-width and the haven order of D, respectively.*

They also show how their definition agrees width tree-width on undirected graphs in the sense that if we obtain a digraph D from an undirected graph G by replacing every edge (u,v) of G by two edges (u,v) and (v,u) in D, then the tree-width of G equals the directed tree-width of D. Their other results in that paper include relating the tree-widths of an Eulerian digraph and its underlying undirected graph and proposing a general algorithm for solving many hard problems like Hamiltonian cycle on digraphs of bounded tree-width. Finally they conjectured that digraphs with large tree-width have large grids as minors by defining directed variants of *grid* and *minor*. It is worth mentioning that the latter conjecture is a well-known fact for undirected graphs[RS86].

The other main work on the topic was by Reed[Ree00] who was also among those who proved Younger's conjecture in 1996. In his paper, Reed presents various global connectivity measures, such as bramble number($BN(D)$), $link(D)$, and $wlink(D)$, and proves that they are within a constant factor of each other in order to justify that all these terms are essentially measuring the same thing. Then, he presents another definition for directed tree-width which is very close to the definition of Johnson et al.[JRST01] (the two values differ by at most one for any digraph). The other parts of Reed's paper mainly discuss the hardness of obtaining results for directed graphs similar to those for undirected graphs.

In this paper we propose a new definition for directed tree-width which resembles the undirected version of tree-width in a natural way, and it seems to have the potential to form the basis for more efficient algorithms on some hard problems on digraphs. We have also studied, as an example to show how the close relationship of our definition and the undirected definition is useful, the situation in which our definition and the existing definition of Johnson et al.[JRST01] are equivalent by proving a min-max theorem relating haven order and d-width under certain conditions.

In the next section we state some preliminary definitions that are related to this paper. In section 3 we present our definition of directed tree-width and discuss its properties as well as its relationship with the definition of Johnson et al. [JRST01]. Section 4 is devoted to the min-max theorem on directed graphs. In that section we prove a weak version of the min-max theorem which is correct if the graph has a special separator property called *the augmenting condition*.

Although we believe that the augmenting condition holds for all digraphs, it still remains as an open part of our work.

2 Preliminaries

For general concepts of tree-decomposition and tree-width on undirected graphs including various related terms and algorithms the reader is referred to [Bod93]. Here we define some related concepts that we use in this paper.

A **tree-decomposition** of an undirected graph $G = (V, E)$ is a a pair (X, T), where $T = (I, F)$ is a tree and $X = \{X_i | i \in I\}$ is a family of subsets of V such that

- $\cup_{i \in I} X_i = V$
- For every edge $(u, v) \in E$ there exists some node i such that $u, v \in X_i$.
- For any vertex $u \in V$ the set of nodes $r \in I$ such that $u \in X_r$ induce a connected subtree in T.

The width of a T is $max_{i \in I} |X_i| - 1$, and the **tree-width** of G is the minimum width over all tree-decompositions of G.

A **haven** of order w in D (for integer w) is a function β which assigns to every subset X of less than w vertices of D a strongly connected component of $D \setminus X$ with the extra condition that if X and Y are two subsets of size less than w and X is a subset of Y, then $\beta(Y)$ is a subset of $\beta(X)$. The haven order of a digraph D, represented by $H(D)$, is the maximum w such that D has a haven of order w.

A **bramble** in a digraph D is a family of strongly connected subsets of D any two of which touch, that is, either have a vertex in common or there are edges from one to the other in both directions. The *order* of a bramble is the size of the minimum set cover of those strongly connected subsets. The *bramble number* of a graph D, represented by $BN(D)$, is the maximum bramble order over all brambles of D.

For example the digraph D which is depicted in Fig. 1 has 9 vertices. The three parts A, B, and C are undirected cliques (i.e. there is an edge between any two of their vertices in both directions) and all other edges are of the form (a_i, b_j), (b_i, c_j), or (c_i, a_j), for $i, j = 1, 2, 3$. The bramble $\varphi = \{\{a_1\}, \{a_2\}, \{a_2, b_2, c_2\}, \{a_3, b_3, c_3\}\}$ has order 3 because $\{a_1, a_2, a_3\}$ is its minimum cover. We can also compute a haven of order 3 by defining $\beta(Z)$ to be the strongly connected component in $D \setminus Z$ that has nonempty intersection with A. In Section 4 we show that $H(D) = 6$ whereas $BN(D) = 3$.

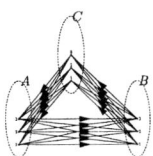

Fig. 1. A digraph with bramble number 3 and haven order 6

Notice that all definitions above have a corresponding version for undirected graphs.

2.1 Arboreal Decomposition

Since we use the definition of tree-width by Johnson et al.[JRST01] many times and compare our definition with theirs, it is worth mentioning their definition here.

An **arborescence** is formed from a rooted tree by directing all edges in the root-leaf direction, that is, for every vertex r in the arborescence there exists a unique path from the root to r.

If r and r' are two vertices of an arborescence R, we say $r > r'$ if $r \neq r'$ and there is a path in R from r' to r. For a vertex r and an edge $e = (t, r')$ we say $r > e$ if $r = r'$ or $r > r'$.

For two disjoint subsets Z and S of $V(D)$, S is called Z-**normal** if every walk in $D\backslash Z$ with first and last vertices in S contains no vertex of $D\backslash\{Z \cup S\}$.

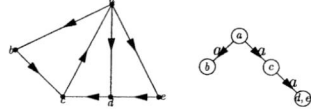

Fig. 2. A digraph with its arboreal decomposition

An **arboreal decomposition** of a digraph D is a triple (R, X, W), where R is an arborescence, and $X = \{X_e | e \in E(R)\}$ and $W = \{W_r | r \in V(R)\}$ are two families of subsets of $V(D)$, i.e. we assign to every edge and vertex of R a subset of vertices of D, with the following two conditions:

A1. W is a partition of $V(D)$ into nonempty sets.
A2. For any $e \in E(R)$, $\bigcup\{W_r | r > e\}$ is X_e-normal.

The width of (R, X, W) is the minimum w such that

$$\left| W_r \cup \bigcup_{e \sim r} X_e \right| \leq w + 1$$

for all vertices r, where $e \sim r$ means r is either the head or the tail of e. The *tree-width* of D is defined as the minimum width over all arboreal decompositions of D. Fig. 2 shows a digraph with an arboreal decomposition of it with width 2. To verify the condition A2, we need to verify that the sets $\{d, e\}$, $\{c, d, e\}$, and $\{b\}$ are $\{a\}$-normal.

3 Introducing D-Width

Let the triple $T = (R, X, W)$ be an arboreal decomposition of a digraph D. For any node $r \in R$ let $W'_r = W_r \cup \bigcup_{e \sim r} X_e$. Let S be a strongly connected set of D,

R_s be the set of vertices r of R such that $W_r \cap S \neq \emptyset$, and E_s be the minimum edges in R that are required to connect all vertices of R_s in R (ignore the edge directions for the moment). It is easy to verify that condition A2 implies all edges $e \in E_s$ have $X_e \cap S \neq \emptyset$. Let E'_s be the set of all edges $e = (r, r')$ in R for which $W'_r \cap W'_{r'} \cap S \neq \emptyset$. Obviously $E_s \subseteq E'_s$, so E'_s connects all the vertices of R_s. In general, E'_s induces one or more connected subtrees in R (ignore the direction of edges), but all vertices of R_s lie in exactly one of these components.

After considering many directed graphs we noticed that we might be able to restrict E'_s to form exactly one connected subtree without affecting the tree-width of the digraph. More formally, we define the d-decomposition and d-width of a graph as follows.

A **d-decomposition** of a digraph D is a pair (T, X), where T is a tree and X is a function that assigns to every node of T a subset of vertices of D such that:

B1 For any vertex v of D there exists some node i of T such that $v \in X_i$, i.e. $\bigcup_{i \in V(T)} X_i = V(D)$.

B2 For any strongly connected subset S of D, the nodes of T containing vertices of S form a connected subtree, i.e. if we select every node of T that contains a vertex of S and every edge $e = (i, j)$ of T such that $S \cap X_i \cap X_j \neq \emptyset$, then the result is a connected subtree of T.

Notice that the second condition yields the fact that the nodes of T containing a single vertex u of D form a subtree because every vertex of a digraph is a strongly connected set by itself.

The width of a d-decomposition is the minimum w such that $|X_i| \leq w+1$ for all $i \in T$, and the **d-width** of D is the minimum width over all d-decompositions of D.

A d-decomposition of width 1 of the directed graph in Fig. 2 is depicted in Fig. 3. The strongly connected sets of the digraph in Fig. 2 are $\{a\}$, $\{b\}$, $\{c\}$, $\{d\}$, $\{e\}$, $\{a, b, c\}$, $\{a, c, d\}$, $\{a, c, d, e\}$, $\{a, b, c, d\}$, and $\{a, b, c, d, e\}$, and it is easy to verify that condition B2 holds for all of them.

It's worth mentioning that if we replace 'strongly connected' in condition B2 above with 'connected', then the definition reduces to the tree-width definition for undirected graphs.

From now on we use the phrases tree-decomposition and tree-width to refer to the arboreal decomposition and tree-width defined in [JRST01] and the terms d-decomposition and d-width to refer to the concepts that we defined above.

For any digraph D, d-width(D) is at least as large as tree-width(D) because the digraph D must satisfy a more restrictive condition in order to have a given d-width than to have a given tree-width.

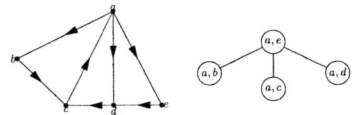

Fig. 3. A d-decomposition of width 1

Corollary 1. *For any digraph D, d-width(D) $\geq$ tree-width(D).*

d-width is a generalization of undirected tree-width in the sense that if we make a digraph D out of an undirected graph G in the most obvious way, i.e. replace every edge of G by two edges in D in both directions, then the d-width of D equals the tree-width of G. The reason is that every strongly connected set in D corresponds to a connected set of G, and vice versa. Johnson et al. [JRST01] also proved that tree-width(D) is equal to tree-width(G); thus, for the special class of undirected graphs, both definitions are equivalent. For this special case, the two decompositions can be efficiently transformed to each other in time $O(mn^2)$[Saf03], where m and n are the number of edges and vertices of G, respectively.

Corollary 2. *Let G be an undirected graph with (undirected) tree-width w and D be the digraph obtained by replacing every edge of G by two edges in both directions. Then, tree-width(D) = d-width(D) = w.*

If X, Y, and $X \cup Y$ are all strongly connected sets and $X \cap Y \neq \emptyset$, then the correctness of condition B2 for X and Y implies its correctness for $X \cup Y$. This means we only need to verify condition B2 for minimal strongly connected sets, that is, those sets S for which there is no strongly connected sets X and Y such that $S = X \cup Y$ and $X \cap Y \neq \emptyset$. That's why in the case of undirected tree-width it suffices to verify condition B2 for just edges and vertices because they are the only minimal connected sets in an undirected graph.

It seems to us that d-width and tree-width are equal on every digraph, though we are not able to prove this at the moment. In the following sections we relate tree-width and d-width and show some evidence that d-width seems to be a proper global measure of connectivity.

3.1 Properties

D-width is a nice measure because of its resemblance to the undirected tree-width, and is algorithmically useful at least because of its being a restricted version of directed tree-width defined by [JRST01].

Given a digraph D with a d-decomposition of width w, one can compute an undirected graph with tree-width at most w by connecting two vertices iff there exists some node in the d-decomposition that contains both the vertices.

Theorem 2. *Let $D = (V, E)$ be a digraph of d-width w. There exists an undirected graph $G = (V, E')$ whose tree-width is at most w such that every strongly connected set in D is a connected set in G.*

In regard to the algorithmic aspects of bounded d-width graphs, d-decomposition is at least as restrictive as tree-decomposition, so at least as many problems are efficiently solvable on digraphs of bounded d-width as on digraphs with bounded tree-width. This includes the linkage problem for fixed number of terminals, the Hamiltonian cycle and the Hamiltonian path problems, the Hamiltonian path problem with prespecified ends, the even cycle problem through a specified vertex, etc.

Since the suggested algorithm in [JRST01] for the above problems is not practically efficient[1], d-decomposition seems to be a good alternative in finding more efficient algorithms. However, the algorithmic properties of d-decomposition are not deeply studied yet.

One basic question related to our work is whether d-width equals tree-width or not. Since tree-width is believed to be equal to the haven order (minus one)[JRST01], we focus on the equality of haven order (minus one) and d-width in the next section. The weak min-max theorem which has been proved in the next section is a very nice example of how the close relationship between d-width and undirected tree-width is useful.

4 The Min-Max Theorem

In the case of undirected graphs there exists a very important theorem that relates haven order, bramble number, and tree-width together. This theorem is called the min-max theorem in [ST93] and the duality theorem in [Ree00].

Theorem 3. *[ST93, Ree00, BD02] For every undirected graph G, the tree-width of G equals the haven order of G minus one, and equals its bramble number minus one.*

Notice that tree-width is a minimized parameter, whereas haven order (or bramble number) is a maximized parameter. According to the above theorem the haven order (or the bramble number) of a digraph is at most w if and only if its tree-width is at least $w - 1$, which has a min-max flavor similar to the max-flow-min-cut theorem.

The original proof appears in [ST93] but other authors like Ballenbaum and Diestel in [BD02] give shorter and cleaner proofs. The above equality relation between haven order, bramble number, and tree-width not only convinces us of the properness of tree-width as a measure of global connectivity, but also has some algorithmic and theoretical consequences. For the case of directed graphs, the story is very different. We know that the bramble number may not be equal to the haven order and may differ by a factor of 2 (see the next section) but we suspect that directed tree-width is equal to the haven order, though there is no proof or disproof at the moment. Johnson et al. [JRST01] were able to prove Theorem 1 which only bounds the two terms, directed tree-width and haven order, within a constant factor of each other.

Here we talk about the relations between bramble number, haven order, and d-width. In section 4.1, we present a tight inequality relating haven order and bramble number. Then, in section 4.2, we present a theorem, similar to the min-max theorem, for directed graphs.

---

[1] It is $\Omega(n^w)$ where w is the tree-width.

4.1 Brambles and Havens in Digraphs

In this section we prove that the haven order and the bramble number of a digraph are not necessarily equal, but are within a constant factor of each other.

Lemma 1. *For every digraph D, $BN(D) \leq H(D)$ and there exists some D for which the equality holds.*

Proof. Let $\varphi = \{\varphi_1, \varphi_2, \cdots, \varphi_m\}$ be a bramble of order w in D. We show that D has a haven of order at least w. For every subset Z of vertices of D such that $|Z| < w$, there exists some $\varphi_k \in \varphi$ such that Z fails to cover φ_k. Let $\beta(Z)$ be the strongly connected component of $D \setminus Z$ that contains φ_k. β has order at least w. By the way, if D is undirected, then the equality holds.

Lemma 2. *For any digraph D, $\left\lceil \frac{H(D)}{2} \right\rceil \leq BN(D)$.*

Proof. Let $\varphi = \{\beta(Z) | Z \subset V(D) \text{ and } |Z| \leq \left\lceil \frac{w}{2} \right\rceil - 1\}$, where β is a haven of order w in D. φ is a bramble of order at least $\left\lceil \frac{w}{2} \right\rceil$.

Lemma 3. *There exist digraphs D for which $\left\lceil \frac{H(D)}{2} \right\rceil = BN(D)$.*

Proof. Let D be a digraph similar to the one depicted in Fig. 1 in which A, B, and C are undirected cliques, i.e. there is an edge between any two of their vertices in both directions, $|A| = |B| = |C| = k$, and all other edges are of the form (a_x, b_y), (b_x, c_y), or (c_x, a_y), for $x, y = 1, 2, \cdots, k$. It can be shown that its haven order is at least $2k$, whereas its bramble number is at most k. For the haven order, if we define $\beta(Z)$ to be the strongly connected component of $D \setminus Z$ containing $A - Z$ if $A - Z \neq \emptyset$, and the strongly connected component of $D \setminus Z$ that contains $B - Z$, otherwise, then β has order at least $2k$.

For the bramble number, it is easy to show that A or B or C are the cover for any bramble in D.

Our goal is now achieved as a direct consequence of the above lemmas.

Theorem 4. *For any digraph D, $\left\lceil \frac{H(D)}{2} \right\rceil \leq BN(D) \leq H(D)$ and both inequalities are tight.*

4.2 The Min-Max Theorem on Directed Graphs

For the proof of Theorem 3, Seymour and Thomas [ST93] use powerful properties of brambles and separators in undirected graphs. They prove the following lemma which reduces to Theorem 3 in the special case $\varphi = \emptyset$.

Lemma 4. *[ST93] Let G be an undirected graph with bramble number at most k and φ be a bramble in G. There exists a tree-decomposition T of G such that every node of size more than k fails to cover φ, that is, for every node X such that $|X| > k$ there exists some non-empty $\varphi_i \in \varphi$ such that $X \cap \varphi_i = \emptyset$.*

To prove lemma 4, Seymour and Thomas find a minimum covering of φ, say X, and examine the components of $G\backslash X$. For any component C_i of $G\backslash X$ they find a tree-decomposition T_i of $G[X \cup C_i]$ that satisfies the condition in lemma 4 with the additional property that there is some node in T_i that contains X. They finally join all of these tree-decompositions to form the final tree-decomposition for G. The join process is simply adding a new node containing X and connecting it to a node of T_i that contains X, for all i's.

The above proof does not work for the case of directed graphs. First, unlike for undirected graphs the bramble number of a digraph is not equal to its d-width. Second, the behavior of separators on directed graphs is very different. There may be minimal strongly connected subsets of D that do not appear as strongly connected subsets of $D[X \cup C_i]$ for any strongly connected component C_i of $D\backslash X$. For example in Fig. 4, where $X = \{a\}$, the strongly connected component $\{a, b, c, d\}$ does not show up in $D[\{a, b\}]$ or $D[\{a, c\}]$ or $D[\{a, d\}]$, so the resulting d-decomposition may not be valid because the strongly connected set $\{a, b, c, d\}$ violates property B2.

In order to be able to prove a theorem similar to Theorem 3, we need to resolve the bad behavior of separators on digraphs and to use brambles properly. For separators, before we obtain the components of $D\backslash X$ we add (or remove) some edges to the graph, and obtain a new digraph D' so that all minimal (recall the concept of minimal from Section 3) strongly connected sets of D show up in some components of $D'\backslash X$. For example in Fig. 5 the original digraph of Fig. 4 has been changed to a newer digraph in which the set of minimal strongly connected components is $\{\{a, b\}, \{a, c\}, \{a, d\}\}$. Notice that every d-decomposition of this new digraph is a d-decomposition of the original digraph as well.

Our goal is to transform the original digraph D to a a digraph D' with the same haven order such that D' preserves all strongly connected sets of D', i.e., every strongly connected set in D is strongly connected in D', and D' has no bad separator.

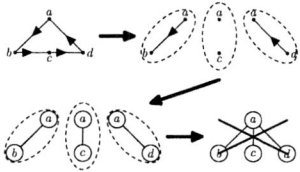

Fig. 4. The bad behavior of separators in digraphs

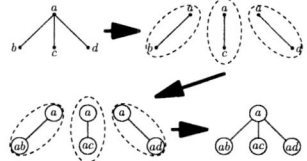

Fig. 5. Making a good separator

Notice that if d-width$(D) = H(D) - 1$, then such a digraph D' exists according to Theorem 2. In what follows we prove that the reverse is also true: The existence of such a digraph D' implies the min-max theorem.

A set X in a digraph D is called a **good-separator** if and only if for every minimal strongly connected set H of D, $V(H) \subseteq V(D[X \cup C])$, for some strongly connected component C of $D\backslash X$.

Augmenting Condition: A digraph D satisfies this condition if for any bramble ϕ and any minimum cover X of ϕ, there exists a digraph D' such that

- D and D' have the same haven order.
- X is a good separator in D'.
- Every strongly connected set of D is a strongly connected set of D'.

The latter condition guarantees that every d-decomposition of D' is a d-decomposition of D as well.

Notice that the augmenting condition requires only minimum bramble covers to be good separators rather than all subsets of vertices. This weaker condition may make it easier to establish the augmenting condition for some directed graphs.

We are now ready to prove the following version of min-max theorem.

Theorem 5. *Let D be a digraph with haven order w. If D satisfies the augmenting condition, then d-width$(D) \leq w - 1$.*

Proof. Our proof has a flavor very similar to the proof of Seymour and Thomas for min-max theorem on undirected graphs [ST93]. First we prove the following lemma.

Lemma 5. *Let D be a digraph of haven order at most w and φ be a bramble of D. Assume that D satisfies the augmenting condition. Then, there exists a d-decomposition T of D such that every node of size more than w fails to cover φ.*

Let X be a cover of minimum size for φ. Since D satisfies the augmentation condition, there exists some digraph D' with haven order at most w in which X is a good separator. Let $C_1, C_2, \cdots, C_k$ be the components of $D'\backslash X$. Assume there is no edge between a vertex of C_i and a vertex of C_j for any i, j such that $i \neq j$; otherwise, we can remove them without affecting any required condition. Now we make a d-decomposition for $D'[X \cup C_i]$ by using the following lemma.

Lemma 6. *For any i, there exists a d-decomposition T_i for $D'[X \cup C_i]$ such that any node of size more than w fails to cover φ and there exists a node r in T_i that contains X, i.e., $X \subseteq X_r$.*

Proof. There are two cases:

Case 1: C_i does not touch some $\varphi_j \in \varphi$. In this case, a d-decomposition T_i with two nodes u and v such that $X_u = X$ and $X_v = (C_i \cup X) - \varphi_j$ has the necessary conditions. First, $(C_i \cup X) - \varphi_j$ fails to cover φ_j and so fails to cover φ. Second, let S be a strongly connected subset in $(C_i \cup X)$. We have two cases:

1. $(X - \varphi_j) \cap S \neq \emptyset$. In this case, the vertices of S clearly form a connected subtree in T_i.
2. $(X - \varphi_j) \cap S = \emptyset$. In this case either $S \subset C_i$ or $S \subset \varphi_j$.

Case 2: C_i touches every member of φ. So $\varphi \cup C_i$ is a bramble and its order is at most w.

By the induction hypothesis (The induction is based on the number of connected sets in φ), there exists a d-decomposition T' for D' such that every node of size more than w fails to cover $\varphi \cup C_i$. If every node of size more than w fails to cover φ, then it directly implies Lemma 5. Otherwise, let y be a node of T' such that $Y = X_y$ has size more than w and fails to cover C_i, but covers every member of φ.

We now build an undirected graph G from T' by adding edge (u, v) if and only if $u, v \in X_r$ for some node r in T' and both are vertices of the same subgraph $D'[X \cup C_j]$, for some j. Since X is a good separator, it can be easily seen that every strongly connected subset of T, and in particular every member of φ, forms a connected subset in G. We now use the following useful lemma whose proof is similar to the undirected version proof in [ST93] and is stated in the appendix.

Lemma 7. *There are at least $|X|$ vertex-disjoint paths from X to Y in G.*

Let $X = \{x_1, x_2, \cdots, x_m\}$, and $\{P_1, P_2, \cdots, P_m\}$ be a set of m vertex-disjoint paths from X to Y such that P_j begins with x_j and ends with a vertex of Y. We assume, by truncating the path at the first vertex in Y, that P_j has exactly one vertex from Y.

Let $C = C_i \cup X$. Let T'' be obtained from T' by replacing every set X_t by $X_t' = (X_t \cap C) \cup \{x_i | P_i \cap X_t \neq \emptyset\}$, for every node t of T'. T'' has the following properties:

1. $X_y' = X$.
2. For any node t, $|X_t'| \leq |X_t|$.
3. Every node r in T'' of size more than w fails to cover φ.

The remaining steps of the proof are straightforward. We join all the T_j's and attach them to a new node that contains only X. The resulting d-decomposition satisfies all the requirements of Lemma 5.

□

Corollary 3. *For any digraph D that satisfies the augmentation condition tree-width$(D) = $ d-width$(D) = H(D) - 1$.*

Proof. This is an easy consequence of Theorem 1, Corollary 1, and Theorem 5.

5 Conclusions and Future Work

Since the topic is very new there exists a huge number of open questions. Any research topic related to undirected tree-width may have a corresponding problem on directed graphs. Here we list just some of these:

- Corresponding definitions for separators, path-width, and branch-width.
- Identifying the class of digraphs whose tree-width is k for a constant k, $k = 2, 3, \cdots$. For $k = 0$, the answer is the class of directed acyclic graphs.
- Identifying the class of digraphs for which the augmenting condition holds. We believe all digraphs satisfy the augmenting condition, but proving it for some special classes of digraphs would also be a big step forward.
- Finding d-decomposition of small d-width for bounded d-width graphs.

Acknowledgement

The main part of this work was done when I was a graduate student at the school of Computer Science at the University of Waterloo. I should thank my supervisor, Prabhakar Ragde, for his many helps and comments during this research.

My sincere thanks to Will Evans for his contribution in preparing this paper. I would like to thank him for his invaluable hints on this paper.

References

[BD02]   P. Bellenbaum and R. Diestel. Two short proofs concerning tree-decompositions. *Combinatoric, Probability and Computing*, 11:1–7, 2002.

[Bod93] H. L. Bodlaender. A tourist guide through treewidth. *Acta Cybernetica*, pages 1–21, 1993.

[JRST01] T. Johnson, N. Robertson, P. D. Seymour, and R. Thomas. Directed Tree-Width. *Journal of Combinatorial Theory(Series B)*, 82:128–154, 2001.

[Ree97] B. Reed. Tree Width and Tangles: A New Connectivity Measure And Some Applications. *Suervey in Combinatorics*, 241:87–158, 1997.

[Ree00] B. Reed. Introducing directed tree width. In H.J. Broersma, U. Faigle, C. Hoede, and J.L. Hurink, editors, *Electronic Notes in Discrete Mathematics*, volume 3. Elsevier, 2000.

[RRST96] B. Reed, N. Robertson, P. Seymour, and R. Thomas. On packing directed circuits. *Combinatorica*, 16:535–554, 1996.

[RS86]   N. Robertson and P. D. Seymour. Graph minors II: algorithmic aspects of tree-width. *Journal Algorithms*, 7:309–322, 1986.

[Saf03] M.A. Safari. Directed tree-width. Master's thesis, School of Computer Science, University of Waterloo, 2003.

[ST93]   P. Seymour and R. Thomas. Graph searching, and a min-max theorem for treewidth. *Journal of Combinatorial Theory (Series B)*, 58:239–257, 1993.

[You73] D. H. Younger. Graphs with interlinked directed circuits. In *Proceedings of the Mid-west Symposium on Circuit Theory*, volume 2, pages XVI 2.1 – XVI 2.7, 1973.

On Beta-Shifts Having Arithmetical Languages

Jakob Grue Simonsen

Department of Computer Science, University of Copenhagen (DIKU),
Universitetsparken 1, DK-2100 Copenhagen Ø, Denmark
simonsen@diku.dk

Abstract. Let β be a real number with $1 < \beta < 2$. We prove that the language of the β-shift is Δ_n^0 iff β is a Δ_n-real. The special case where n is 1 is the independently interesting result that the language of the β-shift is decidable iff β is a computable real. The "if" part of the proof is non-constructive; we show that for Walters' version of the β-shift, no constructive proof exists.

1 Introduction

Symbolic dynamics is a vast and varied field of research originating with Morse's work in the 1920ies [12], and has a wide variety of applications [6,1,11]. A well-known class of symbolic dynamical systems is that of the *β-shifts* introduced by Renyi [16], developed by Parry in the seminal paper [15], and studied intensely [7,19,22,10,2,5,20,21]. From the vantage point of the computer scientist, the class of β-shifts is also interesting because of the following fact concerning its *topological entropy*, a quantity of major importance in dynamical systems theory also having connections to data compression [6]:

Theorem 1 ([15,16]). *If β is a non-integral real number > 1, then the topological entropy of the β-shift is $\log(\beta)$.*

The computability of the topological entropy of various dynamical systems has been studied closely [8,9,4]. For none of the studied classes of systems, is it known whether, for each *computable* real number α, there exists a system having topological entropy equal to α. As log is a computable function, Theorem 1 thus offers a tantalizing opportunity to have a class of dynamical systems with this property. Ideally, such a correspondence should be effective, ie. we would like to have an *algorithm* that transformed any computable real β, in some suitable representation, to some suitable representation of the β-shift.

As we shall show, β is a computable real iff the socalled "language" of the β-shift is decidable. Therefore, the "suitable representation" of the β-shift is an algorithm for deciding its language. However, we also show that an algorithm as is asked for above does not exist. Our methods are not particular to the setting of decidable sets, but can be recast to fit effective procedures with access to oracles. Consequently, we prove our results for all Δ_n^0 in the Arithmetical Hierarchy. This proof establishes a surprising correspondence with the elegant notion of Δ_n-reals introduced by Weihrauch and Zheng [27].

For ease of notation, we prove our results for reals in the open interval $(1;2)$. The extension of our results to non-integral βs greater than 2 is certainly possible, but requires some awkward encoding.

2 Preliminaries

For ease of notation, we use the computability notions of recursion theory. The reader in need of intuitive understanding may substitute "program" for "partial recursive function" and "program that always halts" for "total recursive function". Good introductions to recursion theory are [18,14]. Familiarity with computable analysis or any of the varieties of constructive mathematics will be an advantage, but not a prerequisite; Weihrauch's monograph [25] is recommended.

Throughout the paper, $\mathbb{R}$ denotes the usual set of real numbers from classical mathematics, as does any use of the term "real number". As usual, the greatest integer less than or equal to a real number β is denoted by $\lfloor \beta \rfloor$. We denote the set of positive reals by $\mathbb{R}^{>0}$.

We set $\mathbf{2} \triangleq \{0,1\}$. The set of right-infinite binary sequences is denoted by $\mathbf{2}^{\mathbb{N}}$, the set of bi-infinite such by $\mathbf{2}^{\mathbb{Z}}$; if b is a finite binary string, we denote by b^ω the right-infinite string consisting of an infinite number of concatenations of b. If M is a language of finite binary strings and $k \in \mathbb{N}$, M^k denotes the set of all finite strings obtained by $k-1$ successive concatenations of k elements of M (with $M^1 = M$ as a special case). As usual, we set $M^* \triangleq \{\lambda\} \cup \bigcup_{k=1}^{\infty} M^k$ where λ is the empty string.

The (strict) *lexicographic order* on $\mathbf{2}^{\mathbb{N}}$ (or $\mathbf{2}^k$ for any $k \in \mathbb{N}$) is defined by $\alpha <_{\text{lex}} \gamma$ iff there is an $n \in \mathbb{N}$ such that $\alpha(n) = 0$, $\gamma(n) = 1$, and $\alpha(k) = \gamma(k)$ for all $k < n$. The non-strict lexicographic order is then defined in the obvious way.

We set $\mathbb{N} \triangleq \{1,2,\ldots\}$, $\mathbb{N}_0 \triangleq \{0\} \cup \mathbb{N}$, and define $\mathbb{Z}$ and $\mathbb{Q}$ as usual. For computability purposes, we assume elements of $\mathbb{N}_0$, $\mathbb{Z}$ and $\mathbb{Q}$ to have suitable representations as elements of $\mathbb{N}$, whence comparison under $<$, $>$ and $=$ are decidable in these sets. Indices k, m, n, s ranges over $\mathbb{N}$.

2.1 The Beta-Shift

For any finite alphabet Σ, the *one-sided shift* map on $\Sigma^{\mathbb{N}}$, denoted σ, is defined by $\sigma(b_1 b_2 \cdots) \triangleq b_2 b_3 \cdots$. The *two-sided shift* on $\Sigma^{\mathbb{Z}}$, also denoted σ, is defined by $\sigma(b)_i = b_{i+1}$ for all $i \in \mathbb{Z}$.

Definition 1. *Let β be a non-integral real number > 1. The (greedy) expansion of 1 in powers of β^{-1} is the sequence $a = (a_k)_{k=1}^{\infty}$ where $a_1 = \lfloor \beta \rfloor$, and $a_k = \lfloor \beta^k - \sum_{i=1}^{k-1} a_i \beta^{k-i} \rfloor$ for $k > 1$.*

If there is an $m \in \mathbb{N}$ such that $k \geq m$ implies $a_k = 0$, then the expansion is said to be finite.

It is easy to see that $1 = \sum_{n=1}^{\infty} a_n \beta^{-n}$, and that β is the unique positive solution to $1 = \sum_{k=1}^{\infty} a_k x^{-k}$. Observe that if $k = \lfloor \beta \rfloor + 1$, then $0 \leq a_n \leq k-1$ for all $n \in \mathbb{N}$, and thus $a = (a_n)_{n=1}^{\infty}$ is an element of the full shift on k letters (i.e.

the set of all right-infinite sequences of words from a k-letter alphabet—this set is unique up to injective renaming of the letters). As we restrict our attention to the open interval $(1; 2)$, we may take $\Sigma = 2$ in the remainder of the paper.

Note that $\sigma^n(a) \leq_{\text{lex}} a$ for all $n \in \mathbb{N}$; this gives rise to the standard definition of the β-shift:

Definition 2. *Let β be a real number with $1 < \beta < 2$, and let $a = (a_n)_{n=1}^{\infty}$ be the expansion of 1 in powers of β^{-1}. The* one-sided W-β-shift *is the subset $\tilde{X}_\beta$ of $\mathbf{2}^{\mathbb{N}}$ containing exactly those b such that, for all $n \in \mathbb{N}_0$, we have $\sigma^n(b) \leq_{\text{lex}} a$.*

The one-sided β-shift, *denoted X_β, is defined to be $\tilde{X}_\beta$ if a is not finite. If a is finite, i.e. $a = a_1 a_2 \ldots a_k 0^\omega$ such that $a_k = 1$, define $a' \triangleq (a_1 a_2 \cdots a_{k-1} 0)^\omega$. Then, X_β is defined to be the subset of $\mathbf{2}^{\mathbb{N}}$ such that, for all $n \in \mathbb{N}_0$, we have $\sigma^n(b) \leq_{\text{lex}} a'$*

The two-sided W-β-shift *is the subset of $\mathbf{2}^{\mathbb{Z}}$ containing exactly those b such that, for all $i \in \mathbb{Z}$, we have we have $b_i b_{i+1} b_{i+2} \cdots \in \tilde{X}_\beta$. The two-sided β-shift is defined analogously, using X_β.*

It is easy to see that both the one- and two-sided (W-)β-shifts are shift-invariant subsets of $\{0, \ldots, \lfloor \beta \rfloor\}^{\mathbb{N}}$ and $\{0, \ldots, \lfloor \beta \rfloor\}^{\mathbb{Z}}$, ie., $\sigma(\tilde{X}_\beta) = \tilde{X}_\beta$ and $\sigma(X_\beta) = X_\beta$.

The term "W-β-shift" is short for "Walters-β-shift", since $\tilde{X}_\beta$ is studied in Walters' book [24] (a point of confusion is that the W-β-shift is occasionally called the β-shift in the literature). The special case where the definition of the W-β-shift differs from the β-shift (i.e. with finite a) stems from the original research of the β-shift [15] where it was necessary to consider the special case to study aspects of number theory. Both the W-β-shift and the β-shift satisfy Theorem 1.

A fundamental concept in the study of shift spaces is that of *language*:

Definition 3. *Let β be a real number with $1 < \beta < 2$. The* language *of the W-β-shift, denoted $\mathcal{L}(\tilde{X}_\beta)$, is the set of all finite binary strings occurring in elements of $\tilde{X}_\beta$, ie. $\mathcal{L}(\tilde{X}_\beta) \triangleq \{b_i b_{i+1} \cdots b_j \mid b \in \tilde{X}_\beta \wedge i, j \in \mathbb{N} \wedge i \leq j\}$. $\mathcal{L}(X_\beta)$ is defined analogously.*

Define the shift map σ_{fin} on *finite* strings by $\sigma_{\text{fin}}(b_1 b_2 \cdots b_k) \triangleq b_2 \cdots b_k$ and note that $|\sigma_{\text{fin}}(a)| + 1 = |a|$. Extend the map to sets of finite strings by letting σ_{fin} act on each string in the set. We have:

Proposition 1. *For all $j, k \in \mathbb{N}$, we have $\mathcal{L}(\tilde{X}_\beta) \cap \mathbf{2}^{jk} \subseteq (\mathcal{L}(\tilde{X}_\beta) \cap \mathbf{2}^k)^j$, and $\mathcal{L}(X_\beta) \cap \mathbf{2}^{jk} \subseteq (\mathcal{L}(X_\beta) \cap \mathbf{2}^k)^j$*

Proof. As $\sigma(\tilde{X}_\beta) = \tilde{X}_\beta$, we see that $\sigma_{\text{fin}}(\mathcal{L}(\tilde{X}_\beta)) = \mathcal{L}(\tilde{X}_\beta)$. From the above, we see that $\sigma^k(\mathcal{L}(\tilde{X}_\beta) \cap \mathbf{2}^{jk}) = \mathcal{L}(\tilde{X}_\beta) \cap \mathbf{2}^{(j-1)k}$ (where $\mathbf{2}^0 = \{\lambda\}$). Hence, $\mathcal{L}(\tilde{X}_\beta) \cap \mathbf{2}^{jk} \subseteq (\mathcal{L}(\tilde{X}_\beta) \cap \mathbf{2}^k) \cdot (\mathcal{L}(\tilde{X}_\beta) \cap \mathbf{2}^{(j-1)k})$, and the result follows by a simple induction on j. The proof for X_β is completely analogous. □

2.2 Computable Reals

There are several definitions of "computable reals" in the literature, but these are all equivalent [23,13,17,25,26]. The definition that will be easiest to work with in this paper is, essentially, that of [25]:

Definition 4. *A sequence* $(I_s)_{s \in \mathbb{N}} = ([p_s; q_s])_{s \in \mathbb{N}}$ *of closed intervals with endpoints in* $\mathbb{Q}$ *is said to be computable if there is a total recursive function* $\phi : \mathbb{N} \longrightarrow \mathbb{Q}$ *where, for all* $s \in \mathbb{N}$, *we have* $\phi(2s) = p_s$ *and* $\phi(2s+1) = q_s$. *A computable name is a computable sequence* $(I_s)_{s \in \mathbb{N}}$ *of closed intervals with endpoints in* $\mathbb{Q}$ *such that, for all* $s \in \mathbb{N}$, *we have* $I_{s+1} \subseteq I_s$ *such that* $\bigcap_{s \in \mathbb{N}}$ *is a singleton.*

A real number α *is said to be* computable *if there is a computable name* $(I_s)_{s \in \mathbb{N}}$ *with* $\{\alpha\} = \bigcap_{s \in \mathbb{N}} I_s$.

From any computable name $(I_s)_{s \in \mathbb{N}}$ of some real α, we may effectively obtain a computable name $(I'_s)_{s \in \mathbb{N}}$ of α such that $|I'_s| \leq 2^{-s}$ for all $s \in \mathbb{N}$: Since we know that $|I_s| \to 0$ for $s \to \infty$ and we can, in finite time, check the length of an interval I_s, we may simply wait for $(I_s)_{s \in \mathbb{N}}$ to produce sufficiently small intervals.

Definition 5. *Let* α *be a real number. Then,* α *is said to be* left-computable *(resp.* right-computable*) if there is a total recursive function* $\phi : \mathbb{N} \longrightarrow \mathbb{Q}$ *such that* $\sup_s \phi(s) = \alpha$ *(resp.* $\inf_s \phi(s) = \alpha$*).*

It is well-known that a real number is computable iff it is both left- and right-computable. Also:

Proposition 2. *For each fixed computable name of some real* α, *the following problem is undecidable:*
  Given: *A computable name* $(I_n)_{n \in \mathbb{N}}$ *of some computable real* β.
  To decide: *Is* $\beta < \alpha$?

Proof. Standard. See e.g. [3,25]. □

We use the above proposition in Section 6, specialized to the case where α is the Golden Mean $(1 + \sqrt{5})/2$.

We shall need an effective way of finding the unique positive root of equations of the form $1 = \sum_{j=1}^{k} c_j x^{-j}$ where all $c_j \in \mathbf{2}$ and at least one of the c_j equals 1.

Lemma 1. *There is a total recursive function* $\psi : \mathbb{N} \longrightarrow \mathbb{N}$ *such that, for each* $k \in \mathbb{N}$, $\phi_{\psi(k)} : \mathbf{2}^k \longrightarrow \mathbb{N}$ *is a partial recursive function such that, if* $c_1, \ldots, c_k \in \mathbf{2}$ *with at least one* $c_j = 1$, *then* $\phi_{\psi(k)}(c_1, \ldots, c_k)$ *is defined and* $\phi_{\phi_{\psi(k)}(c_1,\ldots,c_k)} :$ $\mathbb{N} \longrightarrow \mathbb{Q}$ *is a computable name of the unique positive solution to* $1 = \sum_{j=1}^{k} c_j x^{-j}$.

Proof. The positive solution of $1 = \sum_{j=1}^{k} c_j x^{-j}$ is an isolated zero of $f(x) \triangleq \sum_{j=1} c_j x^{-j} - 1$, which is a computable function in the sense of Weihrauch [25]. The result now follows from standard root-finding algorithms, indeed from the fact that every isolated zero of a computable function is a computable real, and that there is an effective way of finding a computable name for it [25, Ch. 6]. □

In the above lemma, ψ is merely a way of getting the right arity, and $\phi_{\psi(k)}$ an "algorithm" for converting the relevant "coefficients" to a computable name of the solution.

2.3 The Arithmetical Hierarchy of Reals

We briefly summarize a few notions from recursion theory:

Definition 6. *Let $A \subseteq \mathbb{N}$. We let $(\phi_i^A)_{i \in \mathbb{N}}$ be an effective enumeration of all partial functions from $\mathbb{N} \longrightarrow \mathbb{N}$ that are recursive-in-A (ie., computable by Turing Machines with access to an oracle for A). Observe that $A = \emptyset$ gives the usual partial recursive functions, and we write ϕ_i in place of $\phi_i^\emptyset$. We will usually suppress the index i if it is not necessary for the exposition.*

We overload the ϕ_i^A to denote partial recursive-in-A functions with domain or codomain any of the sets $\mathbf{2}, \mathbb{N}, \mathbb{Z}, \mathbb{Q}$ (using suitable representations). If B is any of these sets, observe that $C \subseteq B$ is decidable iff there exists a total recursive function $\phi_i : B \longrightarrow \mathbf{2}$ such that $\phi_i(x) = 1$ iff $x \in C$.

Let, for each $n \in \mathbb{N}$, $\langle \cdot, \ldots, \cdot \rangle : \mathbb{N}^n \longrightarrow \mathbb{N}$ be a total recursive pairing function, e.g. the one obtained by repeated use of the Cantor pairing function $\langle i, j \rangle \triangleq (i+j)(i+j+1)/2 + j$ and its accompanying projections.

Using the pairing function, we may extend the concepts introduced above to finite Cartesian products of any of these sets. If $\phi : \mathbb{N} \longrightarrow \mathbb{N}$ is a *total* function, we say that $\psi : \mathbb{N} \longrightarrow \mathbb{N}$ is recursive-in-ϕ if it is recursive-in-$\{\langle n, \phi(n) \rangle \mid n \in \mathbb{N}\}$.

Definition 7. *For any $A \subseteq \mathbb{N}$, the jump, A' is defined by $A' \triangleq \{i \in \mathbb{N} \mid \phi_i^A(i) \text{ is defined}\}$. For $n \in \mathbb{N}$, the nth jump $A^{(n)}$ is defined by $A^{(1)} \triangleq A'$, and $A^{(n+1)} \triangleq \left(A^{(n-1)}\right)'$. For convenience, we set $A^{(0)} \triangleq A$.*

Define $\Sigma_0^0 = \Pi_0^0 = \Delta_0^0$ to be the set of decidable subsets of $\mathbb{N}$. For any $n \in \mathbb{N}$, define the sets of subsets of $\mathbb{N}$ called Σ_n^0, Π_n^0, and Δ_n^0 as follows: a set $A \subseteq \mathbb{N}$ satisfies $A \in \Sigma_n^0$ iff there is a decidable set $R \subseteq \mathbb{N}$ such that, for any $i \in \mathbb{N}$: $i \in A$ iff $(\exists m_1)(\forall m_2)(\exists m_3) \cdots (Q m_n).(\langle i, m_1, \ldots, m_n \rangle \in R)$ where Q is '$\exists$' if n is odd and '$\forall$' otherwise. $A \subseteq \Pi_n^0$, iff the complement $\overline{A} \in \Sigma_n^0$, and we define $\Delta_n^0 \triangleq \Sigma_n^0 \cap \Pi_n^0$.

It is easy to see that Σ_1^0 contains precisely the recursively enumerable (henceforth "r.e.") subsets of $\mathbb{N}$, and Π_1^0 precisely the co-r.e. sets. It is a standard result that, for $n \in \mathbb{N}$, $A \in \Delta_n^0$ iff there is a total recursive-in-$\emptyset^{(n-1)}$ function $\phi : \mathbb{N} \longrightarrow \mathbf{2}$ such that $\phi(j) = 1$ iff $j \in A$.

Recognizing the similarity between alternating quantifiers in the usual notion of arithmetical hierarchy for $\mathbb{N}$ and the alternating uses of inf and sup in certain generalizations of the computable reals, Weihrauch and Zheng introduced the *arithmetical hierarchy of reals* [27]. Each class in the hierarchy constitutes a closed subfield of $\mathbb{R}$ corresponding to a degree of unsolvability.

A full introduction to the arithmetical hierarchy of reals is beyond the scope of this paper; we shall only need to recapitulate a few facts. The lemma below may be taken as a definition of the classes.

Lemma 2 (Lemma 7.2 of [27]). *With the convention $\emptyset^{(0)} = \emptyset$, the following hold for all $n \in \mathbb{N}_0$ and all $x \in \mathbb{R}$:*

1. *$x \in \Sigma_{n+1}$ iff there is a recursive-in-$\emptyset^{(n)}$ total function $\phi_i : \mathbb{N} \longrightarrow \mathbb{Q}$ with $x = \sup_s \phi_i(s)$.*
2. *$x \in \Pi_{n+1}$ if there is a recursive-in-$\emptyset^{(n)}$ total function $\phi_i : \mathbb{N} \longrightarrow \mathbb{Q}$ with $x = \inf_s \phi_i(s)$.*
3. *$x \in \Delta_{n+1}$ if there is a total function as above such that $x = \lim_{s \to \infty} \phi_i(s)$ that converges effectively, ie. there is a recursive-in-$\emptyset^{(n)}$ total function $\xi : \mathbb{N} \longrightarrow \mathbb{N}$ such that for all $s, j \in \mathbb{N}$, we have $s \geq \xi(j) \Rightarrow |x - \phi_i(s)| \leq 2^{-j}$.*
4. *$x \in \Delta_{n+2}$ if there is a total function as above such that $x = \lim_{s \to \infty} \phi_i(s)$.*

In [27], the lemma is stated only for $n \geq 1$, but the case $n = 0$ is proved elsewhere *loc. cit.*

From the above lemma, it is not hard to see that $\Delta_n = \Sigma_n \cap \Pi_n$ for all $n \in \mathbb{N}$, that Δ_1 coincides with the set of computable reals, and Σ_1 (resp. Π_1) coincides with the set of left-computable (resp. right-computable) reals.

Proposition 3 (First part of Prop. 7.6 of [27]). *For any $n \in \mathbb{N}$, Δ_n is an algebraic field, ie. is closed under the arithmetical operations of addition, subtraction, multiplication and division.*

Examination of the proof in [27] and the standard proof of algebraic closure of the computable reals [25] yields that the closure under algebraic operations is effective. For example, if $\phi_i, \phi_j : \mathbb{N} \longrightarrow \mathbb{Q}$ are total recursive-in-$\emptyset^{(n-1)}$ functions with $\lim_{s \to \infty} \phi_i(s) = \alpha$ and $\lim_{s \to \infty} \phi_j(s) = \beta$ (where the convergence is effective in both cases), then there is a total recursive-in-$\emptyset^{(n-1)}$ function $\psi : \mathbb{N} \longrightarrow \mathbb{Q}$ such that $\lim_{s \to \infty} \psi(s) = \alpha + \beta$, effectively.

We now prove a series of ancillary propositions and lemmas.

Proposition 4. *For any $n \in \mathbb{N}$, if α is a Π_n-real, then so is 2^α.*

Proof. As α is Π_n, there is, by Lemma 2, a total recursive-in-$\emptyset^{(n-1)}$ function $\phi : \mathbb{N} \longrightarrow \mathbb{Q}$ such that $\alpha = \inf_k f(k)$. Using standard methods from computable analysis, it is easy to show that there is a total recursive function $\xi : \mathbb{N} \times \mathbb{Q} \longrightarrow \mathbb{Q}$ such that, for each $k \in \mathbb{N}$ and $p/q \in \mathbb{Q}$, we have $0 \leq \xi(k, p/q) - 2^{p/q} < 2^{-k}$. Hence, $0 \leq \xi(k, f(k)) - 2^{f(k)} < 2^{-k}$ for all $k \in \mathbb{N}$. The function $\zeta : \mathbb{N} \longrightarrow \mathbb{Q}$ defined by $\zeta(k) \triangleq \xi(k, f(k))$ is thus recursive-in-$\emptyset^{(n-1)}$ and, since $x \mapsto 2^x$ is an increasing map, satisfies $\inf_k \zeta(k) = 2^\alpha$. Thus, $2^\alpha \in \Pi_n$. □

We need the concept of Δ_n^0-good sequences to make some of the subsequent proofs more readable:

Definition 8. *Let $n \in \mathbb{N}$. A sequence $(x_s)_{s \in \mathbb{N}}$ of computable reals is called Δ_n^0-good if there is a $\emptyset^{(n-1)}$-computable total function $\psi : \mathbb{N} \longrightarrow \mathbb{N}$ such that, for each $s \in \mathbb{N}$, $\phi_{\psi(s)} : \mathbb{N} \longrightarrow \mathbb{Q}$ is a computable name of x_s.*

Taking the sup or inf of such sequences does not force us into a higher level of the arithmetical hierarchy:

Proposition 5. *Let $n \in \mathbb{N}$, and let $(x_s)_{s \in \mathbb{N}}$ be a Δ_n^0-good, convergent sequence of computable reals. Then:*

1. *If $\forall s \in \mathbb{N}.x_s \leq \lim_s x_s$, then $\lim_{s \to \infty} x_s = \sup_s x_s \in \Sigma_n$.*
2. *If $\forall s \in \mathbb{N}.x_s \geq \lim_s x_s$, then $\lim_{s \to \infty} x_s = \inf x_s \in \Pi_n$.*

Proof. We prove (1); the proof of (2) is similar.

As we have $\forall s \in \mathbb{N}.x_s \leq \lim_{s \to \infty} x_s$, we immediately get $\lim_{s \to \infty} x_s = \sup_s x_s$. As $(x_s)_{s \in \mathbb{N}}$ is Δ_n^0-good, there is a total recursive-in-$\emptyset^{(n-1)}$ function ψ with the properties of Definition 8. For each s, $\phi_{\psi(s)}(2s)$ is a left endpoint of an interval a name of x_s; there is clearly a total recursive-in-$\emptyset^{(n-1)}$ function $\xi : \mathbb{N} \longrightarrow \mathbb{Q}$ such that $\xi(s) = \phi_{\psi(s)}(2s)$, for all $s \in \mathbb{N}$.

By the comments after Definition 4, we may assume wlog. that for each $s \in \mathbb{N}$, we have $|x_s - \phi_{\psi(s)}(2s)| \leq 2^{-s}$, Furthermore, for each $s \in \mathbb{N}$, $\phi_{\psi(s)}(2s)$ is a *left* endpoint of a name of x_s, and we thus have $x_s \geq \phi_{\psi(s)}(2s)$ for all $s \in \mathbb{N}$, and thus $\lim_s \phi_{\psi(s)}(2s) = \sup_s \phi_{\psi(s)}(2s) = \sup_s \xi(s) \in \Sigma_n$, as desired. □

3 Beta-Shifts Having Arithmetical Languages

In this and the remaining sections, we assume a $\beta \in \mathbb{R}$ with $1 < \beta < 2$. Furthermore, we freely refer to $(a_k)_{k \in \mathbb{N}}$ as the expansion of 1 in powers of β^{-1}.

Let log be the logarithm to base 2; we now establish a sufficient condition for $\log(\beta)$ to be in Π_n:

Proposition 6. *Let $\mathcal{L}(\tilde{X}_\beta)$ be Δ_n^0. Then, the quantity*

$$\log(\beta) = h_{top}\left(\tilde{X}_\beta\right) = \lim_{k \to \infty}\left(\frac{\log(|\mathcal{L}(\tilde{X}_\beta) \cap 2^k|)}{k}\right)$$

is a Π_n-real. The result holds with $\mathcal{L}(\tilde{X}_\beta)$ replaced by $\mathcal{L}(X_\beta)$.

Proof. The limit always exists and equals $\log(\beta)$ by the standard theory of the β-shift [24]. We want to use Proposition 5 and proceed as follows:

- If $\mathcal{L}(\tilde{X}_\beta)$ is Δ_n^0, then there is a total recursive-in-$\emptyset^{(n-1)}$ function $\zeta : 2^* \longrightarrow 2$ such that $\zeta(a) = 1$ iff $a \in \mathcal{L}(\tilde{X}_\beta)$; hence, there is a total recursive-in-$\emptyset^{(n-1)}$ function $\xi : \mathbb{N} \longrightarrow \mathbb{N}$ such that $\xi(k) = |\mathcal{L}(\tilde{X}_\beta) \cap 2^k|$ for all $k \in \mathbb{N}$. For each $k \in \mathbb{N}$, $\log(|\mathcal{L}(\tilde{X}_\beta) \cap 2^k|)/k$ is a computable real, and we can effectively find a computable name for it given the natural number $|\mathcal{L}(\tilde{X}_\beta) \cap 2^k|$ as input. Thus, there is a total recursive-in-$\emptyset^{(n-1)}$ function $\psi : \mathbb{N} \longrightarrow \mathbb{N}$ such that $\phi_{\psi(k)} : \mathbb{N} \longrightarrow \mathbb{Q}$ is a computable name of $\log(\mathcal{L}(\tilde{X}_\beta) \cap 2^k)/k$ for all $k \in \mathbb{N}$, proving that $(\log(\mathcal{L}(\tilde{X}_\beta) \cap 2^k)/k)_{k \in \mathbb{N}}$ is a Δ_n^0-good sequence.
- For all $j, k \in \mathbb{N}$, Proposition 1 entails that $\mathcal{L}(\tilde{X}_\beta) \cap 2^{kj} \subseteq (\mathcal{L}(\tilde{X}_\beta) \cap 2^k)^j$, hence that $|\mathcal{L}(\tilde{X}_\beta) \cap 2^{kj}| \leq |(\mathcal{L}(\tilde{X}_\beta) \cap 2^k)^j| = |\mathcal{L}(\tilde{X}_\beta) \cap 2^k|^j$.

Thus:
$$\frac{\log(|\mathcal{L}(\tilde{X}_\beta) \cap \mathbf{2}^{kj}|)}{kj} \leq \frac{\log(|\mathcal{L}(\tilde{X}_\beta) \cap \mathbf{2}^{k}|^j)}{kj} = \frac{\log(|\mathcal{L}(\tilde{X}_\beta) \cap \mathbf{2}^{k}|)}{k}.$$

The rightmost expression above does not depend on j, whence we have, for each $k \in \mathbb{N}$:
$$\lim_{j \to \infty} \left(\frac{\log(|\mathcal{L}(\tilde{X}_\beta) \cap \mathbf{2}^{j}|)}{j} \right) = \lim_{j \to \infty} \left(\frac{\log(|\mathcal{L}(\tilde{X}_\beta) \cap \mathbf{2}^{kj}|)}{kj} \right) \leq \frac{\log(|\mathcal{L}(\tilde{X}_\beta) \cap \mathbf{2}^{k}|)}{k}.$$

Thus, for each $k \in \mathbb{N}$, $\log(|\mathcal{L}(\tilde{X}_\beta) \cap \mathbf{2}^k|)/k$ is an upper bound on $h_{\text{top}}\left(\tilde{X}_\beta\right)$.

Finally, Proposition 5 yields $h_{\text{top}}\left(\tilde{X}_\beta\right) \in \Pi_n$. The proof for $\mathcal{L}(X_\beta)$ can be carried out by copying the arguments for $\mathcal{L}(\tilde{X}_\beta)$ verbatim. □

The following lemma establishes a useful correspondence between $\mathcal{L}(\tilde{X}_\beta)$ and $\{k \mid a_k = 1\}$.

Lemma 3. $\mathcal{L}(\tilde{X}_\beta)$ *is* Δ_n^0 *iff* $\{k \in \mathbb{N} \mid a_k = 1\}$ *is* Δ_n^0.

Proof. Let, for each $k \in \mathbb{N}$, $D_k \triangleq \{d \in \mathbf{2}^k \mid \forall j \in \{0, \ldots, k-1\}. \sigma^j(d) \leq_{\text{lex}} a_1 \cdots a_{k-j}\}$. Observe that if $d \in D_k$, then $d \cdot 0^\omega \in \tilde{X}_\beta$, and thus $D_k \subseteq \mathcal{L}(\tilde{X}_\beta) \cap \mathbf{2}^k$. Conversely, if $d \in \mathcal{L}(\tilde{X}_\beta) \cap \mathbf{2}^k$, then $\sigma_{\text{fin}}^j(d) \leq_{\text{lex}} a_1 \cdots a_{k-j}$ for $j \in \{0, \ldots, k-1\}$, ie. $d \in D_k$. Hence, $D_k = \mathcal{L}(\tilde{X}_\beta) \cap \mathbf{2}^k$.

If $\mathcal{L}(\tilde{X}_\beta)$ is Δ_n^0, then we can obviously establish a total recursive-in-$\emptyset^{(n-1)}$ function $\phi : \mathbb{N} \longrightarrow \mathbf{2}$ such that $\phi(k) = b_k$ where $b_1 \cdots b_k$ is the lexicographically greatest element of D_k. By definition of the β-shift, the lexicographically greatest element of D_k is the prefix of length k of $a_1 a_2 \cdots$. But then $\phi(k) = 1$ iff $a_k = 1$, ie. $\{k \in \mathbb{N} \mid a_k = 1\}$.

Conversely, if $\{k \in \mathbb{N} \mid a_k = 1\}$ is Δ_n^0, we can recursively-in-$\emptyset^{(n-1)}$ establish $a_1 \cdots a_k$ for each $k \in \mathbb{N}$. With $a_1 \cdots a_k$ in hand, we can effectively establish D_k. For a given $d \in \mathbf{2}^*$, to decide whether $d \in \mathcal{L}(\tilde{X}_\beta)$, we need only examine whether $d \in D_{|d|}$, which is thus recursive-in-$\emptyset^{(n-1)}$, ie. there is a total recursive-in-$\emptyset^{(n-1)}$ function $\psi : \mathbf{2}^* \longrightarrow \mathbf{2}$ such that $\psi(d) = 1$ iff $d \in \mathcal{L}(\tilde{X}_\beta)$. □

Observe that the proof is constructive, ie. we have an effective way of producing decision procedures for $\{k \mid a_k = 1\}$ given decision procedures for $\mathcal{L}(\tilde{X}_\beta)$ as input, and vice versa.

Proposition 7. $\mathcal{L}(X_\beta)$ *is* Δ_n^0 *iff* $\mathcal{L}(\tilde{X}_\beta)$ *is* Δ_n^0.

Proof. If a is not finite, we have $\mathcal{L}(X_\beta) = \mathcal{L}(\tilde{X}_\beta)$, and the result follows. If a is finite, then $\{k \mid a_k = 1\}$ is Δ_1^0 (there are only a finite number of 1s), whence Lemma 3 furnishes that $\mathcal{L}(\tilde{X}_\beta)$ is Δ_1^0. Also, we have that $\mathcal{L}(X_\beta)$ is Δ_1^0, since we can use the same construction as in the second part of the proof of Lemma 3 applied to the sequence $a' = (a_1 a_2 \cdots a_{k-1} 0)^\omega$ where k is the largest integer with $a_k = 1$. □

Let $s \in \mathbb{N}$, $a_1 = 1$ and $a_j \in \mathbf{2}$ for $j \in \{2, \ldots, s\}$. Consider the map $f_s : \mathbb{R}^{>0} \longrightarrow \mathbb{R}^{>0}$ defined by $f_s(x) = \sum_{j=1}^{s} a_j x^{-j}$. Now, $f_s(x)$ is strictly decreasing, continuous and onto, whence $1 = f_s(x)$ has a unique positive real solution for all s. We now show that this solution is a computable real, and that there is an effective way to find it given $a_1, \ldots, a_s$ as input:

Proposition 8. *If $\{k \in \mathbb{N} \mid a_k = 1\}$ is Δ_n^0, then the sequence $(\alpha_s)_{s \in \mathbb{N}}$ of positive solutions to $1 = \sum_{j=1}^{s} a_j x^{-j}$ is a Δ_n^0-good sequence of computable reals, convergent with limit β, and satisfying $\forall s \in \mathbb{N}. \alpha_s \leq \beta$.*

Proof. Observe that we always have $a_1 = 1$. By Lemma 1, there is an effective procedure yielding a computable name of the unique positive real solution to $1 = \sum_{j=1}^{s} a_j x^{-j}$, when given $(a_1, \ldots, a_s)$ as input. Let the notation and names of recursive functions be as in Lemma 1; Then $\phi_{\phi_{\psi(s)}(a_1, \ldots, a_s)} : \mathbb{N} \longrightarrow \mathbb{Q}$ is a computable name of the unique positive solution, and the function $\psi : \mathbb{N} \longrightarrow \mathbb{N}$ is total recursive. As $\{k \in \mathbb{N} \mid a_k = 1\}$ is Δ_n^0, there is a total recursive-in-$\emptyset^{(n-1)}$ function $\xi : \mathbb{N} \longrightarrow \mathbf{2}$ with $\xi(k) = 1$ iff $a_k = 1$, and hence a total recursive-in-$\emptyset^{(n-1)}$ function $\zeta : \mathbb{N} \longrightarrow \mathbf{2}$ such that $\zeta(k) = a_k$ for all $k \in \mathbb{N}$. Hence, there is a total recursive-in-$\emptyset^{(n-1)}$ function mapping $s \in \mathbb{N}$ to an index of $\phi_{\psi(s)}(\zeta(s), \ldots, \zeta(1))$, whence $(\alpha_s)_{s \in \mathbb{N}}$ is a Δ_n^0-good sequence of computable reals. The sequence is non-decreasing, since $\alpha_{s+1} = \alpha_s$ if $a_{s+1} = 0$ and $\alpha_{s+1} > \alpha_s$ if $a_{s+1} = 1$. Now, β is the unique positive solution to $1 = \sum_{j=1}^{\infty} a_j x^{-j}$, and clearly all of the α_s are less than or equal to this solution. Hence, $\forall s \in \mathbb{N}. \alpha_s \leq \beta$. Proving that $\lim_{s \to \infty} \alpha_s = \beta$ is a standard exercise in undergraduate (classical) mathematics. □

We now have the following key lemma:

Lemma 4. *If $\mathcal{L}(\tilde{X}_\beta)$ is Δ_n^0, then β is a Δ_n-real.*

Proof. Propositions 6 and 4 furnish that $\beta \in \Pi_n$. Furthermore, Lemma 3, and Propositions 8 and 5 furnish that $\beta \in \Sigma_n$, whence the result. □

4 Arithmetical Betas

In the first lemma of this section, we give a sufficient condition for $\{k \mid a_k = 1\}$ to be Δ_n^0.

Lemma 5. *Let $n \in \mathbb{N}$, and assume that, for all $k \in \mathbb{N}$, we have $1 \neq \beta^k - \sum_{j=1}^{k-1} a_j \beta^{k-j}$. Then there is a total recursive-in-$\emptyset^{(n-1)}$ function $\xi : \mathbb{N} \longrightarrow \mathbf{2}$ such that $\xi(n) = 1$ iff $\beta^k - \sum_{j=1}^{k-1} a_j \beta^{k-j} \geq 1$, ie. $\{k \mid a_k = 1\}$ is a Δ_n^0 subset of $\mathbb{N}$.*

Proof. By Lemma 2, there is a recursive-in-$\emptyset^{(n-1)}$ total function $f : \mathbb{N} \longrightarrow \mathbb{Q}$ such that $\beta = \lim_{i \to \infty} f(i)$ effectively (that is, there is an $\emptyset^{(n-1)}$-computable total function $\psi : \mathbb{N} \longrightarrow \mathbb{N}$ such that, for all $m \in \mathbb{N}$, $|\beta - f(i)| < 2^{-m}$ for all $i \geq \psi(m)$). By Proposition 3, Δ_n is an algebraic field, and we thus have $\beta^k - \sum_{j=1}^{k-1} a_j \beta^{k-j} \in \Delta_n$ for all $k \in \mathbb{N}$. By the comments after the proposition, the algebraic operations are recursive, and there is thus a total recursive-in-$\emptyset^{(n-1)}$ function $\xi : \mathbb{N} \times \mathbb{N} \longrightarrow \mathbb{Q}$ such that, for all $k, m \in \mathbb{N}$, $|\beta^k - \sum_{j=1}^{k-1} a_j \beta^{k-j} - \xi(k,m)| < 2^{-m}$.

Consider the recursive-in-ξ procedure that does the following: For each $k \in \mathbb{N}$, run $\xi(k,i)$ on successively greater i until an i is found for which $|1 - \xi(k,i)| > 2^{-(i-1)}$ (the assumption $1 \neq \beta^k - \sum_{j=1}^{k-1} a_j \beta^{k-j}$ implies existence of such an i). As $|\beta^k - \sum_{j=1}^{k-1} a_j \beta^{k-j} - \xi(k,i)| < 2^{-i}$, we have $\xi(k,i) > 1$ iff $\beta^k - \sum_{j=1}^{n-1} a_j \beta^{k-j} > 1$.

This procedure can clearly be made into a total recursive-in-$\emptyset^{(n-1)}$-function $h : \mathbb{N} \longrightarrow \mathbf{2}$ such that $h(n) = 1$ iff $\beta^k - \sum_{j=1}^{n-1} a_j \beta^{k-j} > 1$. □

The next lemma is a counterpart to Lemma 4.

Lemma 6. *Let $\beta \in \Delta_n$. Then, $\{k \mid a_k = 1\}$ is Δ_n^0.*

Proof. Consider $(a_k)_{k \in \mathbb{N}}$. Either there is a $k \in \mathbb{N}$ such that $1 = \beta^k - \sum_{i=1}^{k-1} a_i \beta^{k-i}$, or there is not[1]. If there is no such k, then Lemma 5 furnishes the result. If there is no such k, the a_i, for $1 \leq i \leq k-1$, are the initial coefficients of the expansion of 1 in negative powers of β. Hence, $a_k = \lfloor \beta^k - \sum_{i=1}^{k-1} a_i \beta^{k-i} \rfloor = \beta^k - \sum i = 1^{k-1} a_i \beta^{k-i} = 1$, showing that $1 = \sum_{i=1}^{k} a_i \beta^{-i}$ is the β-expansion of 1, all further coefficients therefore being 0. Thus, there is a total *recursive* function $\phi : \mathbb{N} \longrightarrow \mathbf{2}$ such that $\phi(k) = 1$ iff $a_k = 1$. □

5 The Correspondence Theorem

We now prove our main result:

Theorem 2. *Let β be a real number with $1 < \beta < 2$, and let $n \in \mathbb{N}$. The following are equivalent:*

1. *β is a Δ_n-real.*
2. *$\{k \mid a_k = 1\}$ is a Δ_n^0 subset of $\mathbb{N}$.*
3. *$\mathcal{L}(\tilde{X}_\beta)$ is a Δ_n^0 subset of $\mathbf{2}^*$.*
4. *$\mathcal{L}(X_\beta)$ is a Δ_n^0 subset of $\mathbf{2}^*$.*

Proof. (1) ⇒ (2) is Lemma 6, (2) ⇒ (3) is one-half of Lemma 3, and (3) ⇒ (1) is Lemma 4. Finally, Proposition 7 furnishes equivalence of (3) and (4). □

The case where n is 1 is of particular interest:

Corollary 1. *Let β be a real number with $1 < \beta < 2$. The following are equivalent:*

1. *β is a computable real.*
2. *The set $\{k \mid a_k = 1\}$ is a decidable subset of $\mathbb{N}$.*
3. *$\mathcal{L}(\tilde{X}_\beta)$ is a decidable subset of $\mathbf{2}^*$.*
4. *$\mathcal{L}(X_\beta)$ is a decidable subset of $\mathbf{2}^*$.*

---

[1] This use of the Law of the Excluded Middle is the essential non-constructive part of the proof: We are asking for an answer to the undecidable problem of whether such a k exists.

6 Absence of a Constructive Proof

Inspection of the proof of Lemma 4 reveals that it is constructive and thus yields an effective procedure for converting a decision procedure for $\mathcal{L}(\tilde{X}_\beta)$ to a computable name of β. Hence, (3) $\Rightarrow$ (1) of Theorem 2 is effective in the case where n equals 1.

Unfortunately, that fact is not very interesting; what we *really* want is for (1) $\Rightarrow$ (3) to be constructive, ie. we desire a *program* to generate a decision procedure for $\mathcal{L}(\tilde{X}_\beta)$ when given a computable name of a computable real β as input. Alas, this is impossible:

Theorem 3. *There is no partial recursive function $\psi : \mathbb{N} \longrightarrow \mathbb{N}$ such that if $\phi_i : \mathbb{N} \longrightarrow \mathbb{Q}$ is a computable name of a computable real $\beta \in (1;2)$, then $i \in dom(\psi)$ and $\phi_{\psi(i)} : \mathbf{2}^* \longrightarrow \mathbf{2}$ is a total recursive function such that $\phi_{\psi(i)}(c) = 1$ iff $c \in \mathcal{L}(\tilde{X}_\beta)$ for all $c \in \mathbf{2}^*$.*

Proof. Observe that for any $\beta \in (1,2)$, we have $a_1 = 1$. Also, $a_2 = 0$ iff $\lfloor \beta^2 - \beta \rfloor = 0$ iff $\beta^2 - \beta < 1$ iff $\beta < (1 + \sqrt{5})/2$. If ψ existed, we could, by Lemma 3 and the comments thereafter, effectively establish the sequence $(a_n)_{n \in \mathbb{N}}$. Thus, we could decide whether $a_2 = 0$ or $a_2 = 1$, and hence decide whether $\beta < (1 + \sqrt{5})/2$, which is impossible by Proposition 2. $\square$

In other words, the proof of the theorem shows that there is no program converting computable names to decision procedures for the associated shifts. Note also that the proof can immediately be adapted to show that (1) $\Rightarrow$ (2) in Theorem 2 cannot be made effective. As $x \mapsto 2^x$ is a computable function on the computable reals, another adaptation of the proof yields:

Corollary 2. *There is no partial recursive function $\psi : \mathbb{N} \longrightarrow \mathbb{N}$ such that if $\phi_i : \mathbb{N} \longrightarrow \mathbb{Q}$ is a computable name of a computable real $\beta \in (0;1)$, then $i \in dom(\psi)$ and $\phi_{\psi(i)} : \mathbb{N} \longrightarrow \mathbf{2}$ is a total recursive function with $\phi_{\psi(i)}(c) = 1$ iff $c \in \mathcal{L}(\tilde{X}_\beta)$.*

Thus, there is no effective way to find decision procedures for the W-β-shift given its topological entropy $\log(\beta)$.

Whether the corresponding result holds for X_β is still open; we strongly conjecture that it does.

References

1. M.F. Barnsley. *Fractals Everywhere*. Morgan Kaufmann, 1993.
2. F. Blanchard. β-Expansions and Symbolic Dynamics. *Theoretical Computer Science*, 65:131–141, 1989.
3. D.S. Bridges. *Computability: A Mathematical Sketchbook*, volume 146 of *Graduate Texts in Mathematics*. Springer-Verlag, 1994.
4. J.-C. Delvenne and V.D. Blondel. Quasi-Periodic Configurations and Undecidable Dynamics for Tilings, Infinite Words and Turing Machines. *Theoretical Computer Science*, 319:127–143, 2004.

5. C. Frougny and B. Solomyak. Finite Beta-Expansions. *Ergodic Theory and Dynamical Systems*, 12:713–723, 1992.
6. G. Hansel, D. Perrin, and I. Simon. Compression and Entropy. In *Proceedings of the 9th Annual Symposium on Theoretical Aspects of Computer Science (STACS '92)*, volume 577 of *Lecture Notes in Computer Science*, pages 515–528. Springer-Verlag, 1992.
7. F. Hofbauer. β-Shifts have Unique Maximal Measure. *Monatshefte für Mathematik*, 85:189–198, 1978.
8. L. Hurd, J. Kari, and K. Culik. The Topological Entropy of Cellular Automata is Uncomputable. *Ergodic Theory and Dynamical Systems*, 12:255–265, 1992.
9. P. Koiran. The Topological Entropy of Iterated Piecewise Affine Maps is Uncomputable. *Discrete Mathematics and Theoretical Computer Science*, 4(2):351–356, 2001.
10. D. Lind. The Entropies of Topological Markov Shifts and a Related Class of Algebraic Integers. *Ergodic Theory and Dynamical Systems*, 4:283–300, 1984.
11. D. Lind and B. Marcus. *An Introduction to Symbolic Dynamics and Coding*. cambridge University Press, 1995.
12. M. Morse. Recurrent Geodesics on a Surface of Negative Curvature. *Transactions of the American Mathematical Society*, 22:84–110, 1921.
13. J. Myhill. Criteria of Constructivity for Real Numbers. *Journal of Symbolic Logic*, 18:7–10, 1953.
14. P. Odifreddi. *Classical Recursion Theory*, volume 129 of *Studies of Logic and the Foundations of Mathematics*. North-Holland, 1989.
15. W. Parry. On the β-Expansion of Real Numbers. *Acta Math. Acad. Sci. Hung.*, pages 401–416, 1960.
16. A. Renyi. Representations for Real Numbers and their Ergodic Properties. *Acta Math. Acad. Sci. Hung.*, 8:477–493, 1957.
17. H.G. Rice. Recursive Real Numbers. *Proceedings of the American Mathematical Society*, 5:784–791, 1954.
18. H. Rogers Jr. *Theory of Recursive Functions and Effective Computability*. The MIT Press, paperback edition, 1987.
19. K. Schmidt. On Periodic Expansions of Pisot Numbers and Salem Numbers. *Bulletin of the London Mathematical Society*, 12:269–278, 1980.
20. N. Sidorov. Almost Every Number has a Continuum of Beta-Expansions. *American Mathematic Monthly*, 110:838–842, 2003.
21. N. Sidorov. Arithmetic Dynamics. In *Topics in Dynamics and Ergodic Theory*, volume 310 of *London Mathematical Society Lecture Notes Series*, pages 145–189. London Mathematical Society, 2003.
22. Y. Takahashi. Shift with Free Orbit Basis and Realization of One-Dimensional Maps. *Osaka Journal of Mathematics*, 20:599–629, 1983.
23. A.M Turing. On Computable Numbers with an Application to the "Entscheidungsproblem". *Proceedings of the London Mathematical Society*, 42(2):230–265, 1936.
24. P. Walters. *An Introduction to Ergodic Theory*, volume 79 of *Graduate Texts in Mathematics*. Springer-Verlag, 1981.
25. K. Weihrauch. *Computable Analysis: An Introduction*. Springer, 1998.
26. X. Zheng. Recursive Approximability of Real Numbers. *Mathematical Logic Quarterly*, 48:131–156, 2002.
27. X. Zheng and K. Weihrauch. The Arithmetical Hierarchy of Real Numbers. *Mathematical Logic Quarterly*, 47(1):51–65, 2001.

A BDD-Representation for the Logic of Equality and Uninterpreted Functions

Jaco van de Pol[1,2] and Olga Tveretina[2]

[1] Centrum voor Wiskunde en Informatica, Dept. of Software Engineering,
P.O.-Box 94.079, 1090 GB Amsterdam, The Netherlands
Jaco.van.de.Pol@cwi.nl
[2] Department of Computer Science, TU Eindhoven,
P.O. Box 513, 5600 MB Eindhoven, The Netherlands
o.tveretina@tue.nl

Abstract. The logic of equality and uninterpreted functions (EUF) has been proposed for processor verification. This paper presents a new data structure called Binary Decision Diagrams for representing EUF formulas (EUF-BDDs). We define EUF-BDDs similar to BDDs, but we allow equalities between terms as labels instead of Boolean variables. We provide an approach to build a reduced ordered EUF-BDD (EUF-ROBDD) and prove that every path to a leaf is satisfiable by construction. Moreover, EUF-ROBDDs are logically equivalent representations of EUF-formulae, so they can also be used to represent state spaces in symbolic model checking with data.

1 Introduction

Binary Decision Diagrams (BDDs) are one of the biggest breakthroughs in computer-aided design. Reduced ordered BDDs [1] form a canonical representation of Boolean formulas, making testing of equivalence straightforward. Unfortunately, their power is mostly restricted to propositional logic, which is often not sufficiently expressive for verification. The equality logic with uninterpreted functions (EUF) has been proposed for verifying hardware [2]. EUF formulae have been successfully applied for the verification of pipelined processors [2], and translation validation [3].

Using uninterpreted functions simplifies proofs as the only retained information about a function is the property of *functional consistency*, i.e. if $x = y$ then $f(x) = f(y)$. The abstraction process does not preserve validity and may transform a valid formula into an invalid one, e.g. $x + y = y + x$ is valid but $f(x,y) = f(y,x)$ is not. However, in some application domains the process of abstraction is justified.

The original approach to decide this logic was to solve equalities while maintaining *congruence closure* with respect to the uninterpreted functions [4]. This is mainly applied to the conjunction of equalities. Disjunctions can be treated by case splitting [5]. Another approach is based on work of Ackermann [6], who has shown that deciding the validity of EUF formulae can be reduced to checking the satisfiability of pure *equality logic* formulae. Such reduction can be performed by replacing each application of an uninterpreted function symbol with a new variable and for each pair of function applications to add a constraint which enforces the property of functional consistency, i.e. while replacing any two subterms of the form $F(x)$ and $F(y)$ by new variables f_1 and f_2, we have to add a constraint of the form $x = y \rightarrow f_1 = f_2$.

Due to the *finite domain property*, which states that an equality logic formula is satisfiable if and only if it is satisfiable over a finite domain, Pnueli et al. [3] find a small domain for each variable, which is large enough to maintain satisfiability. Goel et al. [7] proposed to decide equality logic formulae by replacing all equalities with new Boolean variables. Similarly, in [8] a BDD-based decision procedure for combinations of theories is presented. As a result of both approaches, BDDs are not a canonical representation for formulas anymore. Also, there can be paths to a leaf which are not satisfiable. Hence, all paths must be checked, for instance if they satisfy transitivity of equality. Therefore, the constraint solver can be invoked exponentially many times because of the Boolean structure of the formula.

Bryant et al. [9] reduce an equality formula to a propositional one by adding *transitivity constraints*. In that approach it is analyzed which transitivity properties may be relevant. Tveretina et al [10] proposed a resolution-based approach to check satisfiability of equality logic formulae.

In [11], equational BDDs (EQ-BDDs) are defined, in which all paths are satisfiable by construction. That approach extends the notion of orderedness to capture the properties of reflexivity, symmetry, transitivity, and substitutivity. The advantage of the method is that satisfiability checking for a given ordered EQ-BDD can be done immediately. However, it is restricted to the case when equalities do not contain function symbols. EQ-BDDs have been extended in [12]. Here some *interpreted* functions, viz. natural numbers with zero and successor were added. In [13] an alternative solution was provided, with a different orientation of the equations.

Contribution. We introduce EUF-BDDs, which are BDDs with internal nodes labelled by equalities between ground terms. We introduce reduced ordered EUF-BDDs, and prove that these have no contradictory paths. This makes them suitable for theorem proving and satisfiability checking. Moreover, contrary to the approaches to EUF mentioned above, we obtain a representation which is logically equivalent to the original formula. This method extends the approach introduced in [11]. However, the changed orientation of [13] is essential for the completeness of our method. So technically, the EQ-BDDs of [11] are not a special case of our EUF-BDDs, because the orientation of the guards is reversed.

Application. We have made a prototype implementation of our EUF-BDDs in the special purpose theorem prover for the μCRL toolset [14]. The prover is used to discharge proof obligations generated in protocol verifications, and it is also used in the symbolic model checker with data, proposed in [15]. In the latter application it is essential to have a concise representation of formulas, which is provided by EUF-BDDs.

2 Basic Definitions

2.1 Syntax

In this section we define a syntax for formulae. A *signature* is a tuple $\Sigma =$ (Fun, ar), where Fun $= \{f, g, h, \ldots\}$ is an enumerable set of *function symbols* and ar : Fun $\to \mathbf{N}$ is a function describing the arity of the function symbols. Function symbols with the arity 0 are called constants (typically $a, b, c, \ldots$). The set of constant symbols is denoted by Const. The set Term of terms is defined inductively: for $n \geq 0$, $f(t_1, \ldots, t_n)$ is a term if $t_1, \ldots, t_n$ are terms, $f \in$ Fun, and $\text{ar}(f) = n$. For $n = 0$, we write a instead of $a()$. In the following, we use the lower case letters s, t, and u to denote terms. The set SubTerm(t) of *subterms* of a term t is defined inductively: for $n \geq 0$, $\text{SubTerm}(f(t_1, \ldots, t_n)) = \{f(t_1, \ldots, t_n)\} \cup \bigcup_{i=1}^{n} \text{SubTerm}(t_i)$. A subterm of a term t is called *proper* if it is distinct from t. The set of proper subterms of a term t is denoted by $\text{SubTerm}_p(t)$.

Definition 1. *(Equalities) An equality is a pair of terms* $(s, t) \in (\text{Term} \times \text{Term})$. *We write an equality as* $s \approx t$. *The set of equalities over Σ is defined by* Eq(Σ) *or if it is not relevant by* Eq.

Here we write '$\approx$' for equality, and we use $\equiv$ to denote syntactical identity between two elements. We define the set of subterms occurring in an equality $s \approx t$ as $\text{SubTerm}(s \approx t) = \text{SubTerm}(s) \cup \text{SubTerm}(t)$, and the set of proper subterms occurring in $s \approx t$ as $\text{SubTerm}_p(s \approx t) = \text{SubTerm}_p(s) \cup \text{SubTerm}_p(t)$.

Definition 2. *Formulae (denoted by* For(Σ)*) are expressions satisfying the following syntax.*
$$\varphi := \text{true} \mid \text{false} \mid \text{Eq} \mid \text{ITE}(\varphi, \varphi, \varphi)$$
In the following, the abbreviation $\neg \varphi$ *stands for* ITE(φ, false, true), $\varphi \wedge \psi$ *stands for* ITE(φ, ψ, false), $\varphi \vee \psi$ *stands for* $\neg(\neg \varphi \wedge \neg \psi)$, $\varphi \to \psi$ *stands for* $\neg \varphi \vee \psi$, *and* $\varphi \leftrightarrow \psi$ *stands for* $(\varphi \to \psi) \wedge (\psi \to \varphi)$.

We write $s \not\approx t$ as an abbreviation of $\neg(s \approx t)$. For a given formula φ, the set of all equalities occurring in φ is denoted by Eq(φ).

We define the set of subterms occurring in a formula φ as $\text{SubTerm}(\varphi) = \bigcup_{e \in \text{Eq}(\varphi)} \text{SubTerm}(e)$, and the set of proper subterms occurring in a formula φ as $\text{SubTerm}_p(\varphi) = \bigcup_{e \in \text{Eq}(\varphi)} \text{SubTerm}_p(e)$. We define the set Lit of *literals* as Lit $= \{l \mid l \in \text{Eq}\} \cup \{\neg l \mid l \in \text{Eq}\}$. Given a conjunction of literals φ, by Lit(φ) we denote the set of all literals occurring in it.

2.2 Semantics

A *structure* $\mathcal{D}$ over a signature $\Sigma =$ (Fun, ar) is defined to consist of a non-empty set D called the *domain*, and for every $f \in$ Fun, with $\text{ar}(f) = n$, a map

$f_D : D^n \to D$. The *interpretation* $[\![t]\!]_D : \mathsf{Term}(\Sigma) \to D$ of a term t is inductively defined as follows. For $n \geq 0$, $[\![f(t_1, \ldots, t_n)]\!]_D = f_D([\![t_1]\!]_D, \ldots, [\![t_n]\!]_D)$, where $t_1, \ldots, t_n \in \mathsf{Term}$. The interpretation $[\![\varphi]\!]_D : \mathsf{For}(\Sigma) \to \{\mathsf{true}, \mathsf{false}\}$ of a formula φ is defined as usual, i.e. $[\![\mathsf{true}]\!]_D = \mathsf{true}$, $[\![\mathsf{false}]\!]_D = \mathsf{false}$, $[\![s \approx t]\!]_D = \mathsf{true}$, if $[\![s]\!]_D = [\![t]\!]_D$, and false otherwise.

$$[\![\mathsf{ITE}(\varphi, \psi, \chi)]\!]_D = \begin{cases} [\![\psi]\!]_D & \text{if } [\![\varphi]\!]_D = \mathsf{true} \\ [\![\chi]\!]_D & \text{otherwise} \end{cases}$$

Definition 3. *A structure D satisfies a formula φ if $[\![\varphi]\!]_D = \mathsf{true}$. A formula φ is called* satisfiable *if there exists a satisfying structure. Otherwise φ is called* a contradiction. *If each structure D satisfies φ then φ is a* tautology. *We say that a formula φ is* logically equivalent *to a formula ψ if for every structure D, $[\![\varphi]\!]_D = [\![\psi]\!]_D$.*

3 Binary Decision Diagrams for EUF-Logic

This paper presents a new data structure called an EUF-BDD for representing and manipulating formulas containing *equalities* and *uninterpreted functions*. We consider EUF-BDDs as a restricted subset of formulas.

Definition 4. *We define the set B of EUF-BDDs as follows.*

$$\mathsf{B} := \mathsf{true} \mid \mathsf{false} \mid \mathsf{ITE}(\mathsf{Eq}, \mathsf{B}, \mathsf{B})$$

It is straightforward to show that every formula defined above is equivalent to at least one EUF-BDD.

EUF-BDDs are nested ITE formulas which are represented in implementations as directed acyclic graphs. The difference between BDDs representing Boolean formulae and EUF-BDDs is, that in the latter case internal nodes are labelled with equalities. An EUF-BDD can be represented as a rooted, directed acyclic graph with nodes of out-degree zero labelled by true and false, and a set of nodes of out-degree two labelled by equalities between ground terms. For a node l the two outgoing edges are given by two functions $low(l)$ and $high(l)$.

Throughout the paper we use T and S to denote EUF-BDDs.

Example 5. The EUF-BDD representing the property of functional consistency $a \approx b \to f(a) \approx f(b)$ can be depicted as in Figure 1. The EUF-BDD can be written as $\mathsf{ITE}(a \approx b, \mathsf{ITE}(f(a) \approx f(b), \mathsf{true}, \mathsf{false}), \mathsf{true})$.

In order to define *ordered* EUF-BDDs, we need a total well-founded order on equalities. This is built from a total well-founded order on terms. To ensure structural properties of ordered EUF-BDDs, this total order should satisfy certain properties.

Definition 6. (**Order on terms**) *We define a simplification order on the set Term as satisfying the following conditions:*

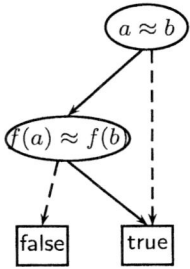

Fig. 1. Dashed lines represent low/false edges, and solid ones represent high/true edges

1. For all $s, t \in$ Term, $s \prec t$ if $s \in \mathsf{SubTerm}_p(t)$.
2. For each $f \in$ Fun and for all $1 \leq i, j \leq n$, and $s_i, t \in$ Term, if $s_j \prec t$ then $f(s_1, \ldots, s_j, \ldots, s_n) \prec f(s_1, \ldots, t, \ldots, s_n)$.
3. The order is total and well-founded.
4. $\succ$ is the reverse of $\prec$.

In the sequel, we work with an arbitrary but fixed simplification order $\prec$. An example of such an order is the recursive path order [16], which we also used in our implementation.

Definition 7 (Order on equalities). *Given a simplification order $\prec$ on terms, the total well-founded order on the set* Eq *is defined as follows.*

$$(s \approx t) \prec (u \approx v) \text{ if either } s \prec u \text{ or } s \equiv u \text{ and } t \prec v.$$

We use terminology from *term rewrite systems* (TRS). In particular, by a *normal form* with respect to some TRS we mean a term to which no rules of the TRS are applicable. A system is *terminating* if no infinite rewrite sequence exists.

A first operation on EUF-BDDs is simplification of equalities as defined below.

Definition 8 (Simplified equalities and EUF-BDDs). *An equality $s \approx t$ is called* simplified, *if $s \succ t$. In order to simplify all equalities in a BDD, we introduce the following rewrite rules:*

- $s \approx t \to t \approx s$, for all $s, t \in$ Term such that $s \prec t$.
- $\mathsf{ITE}(t \approx t, T_1, T_2) \to T_1$.

Suppose T is an EUF-BDD. By $T \downarrow$ we mean the normal form of T obtained after applying these rules. An EUF-BDD T is called simplified if $T \equiv T \downarrow$.

In the following, by $t[s]$ we mean a term t such that $s \in \mathsf{SubTerm}(t)$, by $e[s]$ we mean an equality such that $s \in \mathsf{SubTerm}(e)$, and by $T[s]$ we mean an EUF-BDD T such that there is a node l in T associated with an equality $e(l)$ and $s \in \mathsf{SubTerm}(e(l))$.

Given the order on equalities, we can define a system of reduction rules as in [11], but now the equations are oriented differently, as in [13]. Now starting with an arbitrary simplified EUF-BDD, we can transform it by repeatedly applying the following reduction rules.

Definition 9 (Reduction rules on simplified EUF-BDDs). *We define a TRS* Reduce-Order *as follows.*

1. $\mathsf{ITE}(e, T, T) \to T$
2. $\mathsf{ITE}(e, T_1, \mathsf{ITE}(e, T_2, T_3)) \to \mathsf{ITE}(e, T_1, T_3)$
3. $\mathsf{ITE}(e, \mathsf{ITE}(e, T_1, T_2), T_3) \to \mathsf{ITE}(e, T_1, T_3)$
4. $\mathsf{ITE}(e_1, \mathsf{ITE}(e_2, T_1, T_2), T_3) \to \mathsf{ITE}(e_2, \mathsf{ITE}(e_1, T_1, T_3), \mathsf{ITE}(e_1, T_2, T_3))$, *if* $e_1 \succ e_2$.
5. $\mathsf{ITE}(e_1, T_1, \mathsf{ITE}(e_2, T_2, T_3)) \to \mathsf{ITE}(e_2, \mathsf{ITE}(e_1, T_1, T_2), \mathsf{ITE}(e_1, T_1, T_3))$, *if* $e_1 \succ e_2$.
6. $\mathsf{ITE}(s \approx t, T_1[s], T_2) \to \mathsf{ITE}(s \approx t, T_1[t] \downarrow, T_2)$, *if* $s \succ t$.

Rules 1–5 are the rules for simplifying BDDs for propositional logic, eliminating redundant tests and ensuring the right ordering. Rule 6 allows to substitute equals for equals. Note that we immediately apply simplification after a substitution. The transformation by the reduction rules yields a logically equivalent EUF-BDD.

Definition 10 (EUF-ROBDDs). *We define an EUF-ROBDD to be a simplified EUF-BDD which is a normal form with respect to the TRS* Reduce-Order.

It follows from Definition 10 that in a *reduced ordered* EUF-BDD (EUF-ROBDD) all equalities labelling the nodes are *oriented*, i.e. for a given order $\prec$ on terms, if a node l is associated with an equality $s \approx t$ then $s \succ t$; the equalities along a path appear only in a fixed order; and for each EUF-ROBDD of the form $\mathsf{ITE}(s \approx t, T_1, T_2)$, s doesn't occur in T_1.

Example 11. Consider $\varphi \equiv (x \approx y \wedge y \approx z) \to f(x) \approx f(z)$. For a given order $x \prec y \prec z$, the derivation of an EUF-ROBDD is depicted in Figure 2. The EUF-ROBDD consists of one node true. In the picture, we combined several steps in one arrow. Note that intermediate EUF-BDDs should always be kept simplified. In the middle arrow of the picture, we explicitly show a simplification step.

An EUF-ROBDD is a normal form with respect to the TRS Reduce-Order. The following theorem states that the system of reduction rules is terminating. As a consequence, for each EUF-BDD there exists a logically equivalent EUF-ROBDD.

Theorem 12. *The rewrite system* Reduce-Order *is terminating.*

The proof is based on the *recursive path order* (RPO) [16] to prove termination. The details can be found in a full technical report [17].

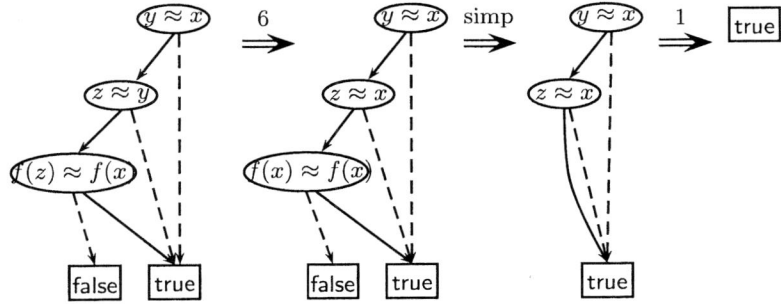

Fig. 2. The derivation of the EUF-ROBDD for $(x \approx y \land y \approx z) \to f(x) \approx f(z)$

4 Satisfiability of Paths in EUF-ROBDDs

Checking equivalence of two Boolean functions can be done by comparing their ROBDD representation: equivalent formulas have identical ROBDDs. Unfortunately, the canonicity property of EUF-ROBDDs is violated as is shown for the plain equality case in [11]. In this section we prove that if the EUF-ROBDD corresponding to a formula φ consists of one node true then φ is a tautology, if the EUF-ROBDD consists of one node false then φ is a contradiction, and in all other cases φ is satisfiable. As a consequence, our approach allows to check whether φ and ψ are equivalent. It can be done by verifying whether $\varphi \leftrightarrow \psi$ is a tautology.

When BDDs are used to represent formulas including equalities and uninterpreted functions, a path to the true leaf in the BDD might not be consistent, i.e. the set of literals occurring along the path does not have a model. We show that each path in an EUF-ROBDD is satisfiable by construction.

For proving satisfiability of a path, we see it as a conjunction of literals occurring along the path, where $\land$ is considered modulo associativity and commutativity. We use letters α and β to denote finite sequences of literals, ϵ for the empty sequence and $\alpha.\beta$ for the concatenation of sequences α and β.

Definition 13 (EUF-BDD paths).

- We define the set Path(T) of all paths contained in an EUF-BDD T inductively as follows.
  - Path(true) = Path(false) = ϵ,
  - Path(ITE(e, T_1, T_2)) = $\{e.\alpha \mid \alpha \in$ Path$(T_1)\} \cup \{\neg e.\alpha \mid \alpha \in$ Path$(T_2)\}$.
- For a given path $\alpha \equiv l_1.\ldots.l_n$, we use an abbreviation φ_α to denote a formula $l_1 \land \cdots \land l_n$.
- The formula φ_α corresponds to a path α.
- We say that α is a satisfiable path if φ_α is satisfiable.

Example 14. Consider an EUF-BDD
ITE$(a \approx b, \text{true}, \text{ITE}(f(a) \approx g(c), \text{ITE}(b \approx c, \text{false}, \text{true}), \text{false})$. The EUF-BDD and the path $a \not\approx b.f(a) \approx g(c).b \not\approx c$ are depicted in Figure 3.

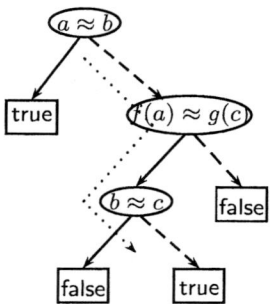

Fig. 3. The path $a \not\approx b.f(a) \approx g(c).b \not\approx c$

4.1 Satisfiability of Reduced Formulas

To prove satisfiability of paths in EUF-ROBDDs we use a satisfiability criterium from [18]. Before turning to a proof that every path in an EUF-ROBDD is satisfiable, we need to give a definition of a non-propagated equality and a definition of a reduced formula. In [18] the definition of non-propagated equalities is given for CNFs. Here, for sake of simplicity, we rather speak of formulas, but actually we are interested in the case when a formula is a conjunction of literals. Since we see a path as a conjunction of literals, where $\wedge$ is considered modulo associativity and commutativity, this corresponds to a set of unit clauses, as in [18].

Definition 15 (Non-propagated equality). *An equality $s \approx t$ is called* non-propagated *in a formula φ if the following holds.*

- $\varphi \equiv (s \approx t) \wedge \psi$ *for some formula ψ, and*
- $s, t \in \mathsf{SubTerm}(\psi)$.

The set of all non-propagated equalities in φ is denoted by $\mathsf{NPEq}(\varphi)$.

Definition 16 (Reduced formula). *We say that $\varphi \equiv l_1 \wedge \cdots \wedge l_n$, where $l_i \in \mathsf{Lit}$, for all $1 \leq i \leq n$, is* reduced *if the following holds.*

- $\mathsf{NPEq}(\varphi) = \emptyset$, *and*
- *for each $t \in \mathsf{Term}$, $(t \not\approx t) \notin \mathsf{Lit}(\varphi)$.*

In the following Red *is used to denote the set of reduced formulas.*

Theorem 17. *Every $\varphi \in \mathsf{Red}$ is satisfiable.*

Proof. See [18]. □

4.2 Satisfiability of EUF-ROBDD Paths

In this section we prove that every path in an EUF-ROBDD is satisfiable. For a given path α, we transform φ_α into the logically equivalent reduced formula

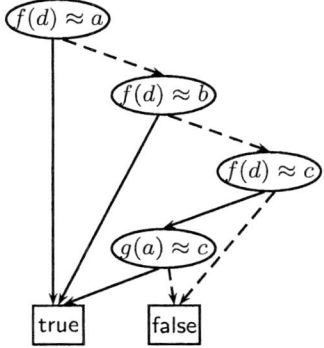

Fig. 4. The EUF-ROBDD representation of a formula $\mathsf{ITE}(f(d) \approx a, \mathsf{true}, \mathsf{ITE}(f(d) \approx b, \mathsf{true}, \mathsf{ITE}(f(d) \approx c, \mathsf{ITE}(g(a) \approx c, \mathsf{true}, \mathsf{false}), \mathsf{false})))$.

φ_α^{red}. The idea of the proof is that a path in a (simplified) EUF-ROBDD contains segments of the form $s \not\approx t_0. \cdots .s \not\approx t_n.s \approx t$. The term s doesn't occur as a subterm in any t_i, nor in any of the other segments. We obtain a path corresponding to $\varphi_\alpha^{red} \in \mathsf{Red}$ by propagating $s \approx t$, i.e. replacing the segment by $t \not\approx t_0. \cdots .t \not\approx t_n.s \approx t$. Note that this operation doesn't introduce new subterms, so propagated equalities in other segments remain propagated. The result is an equivalent formula in Red, hence it is satisfiable.

Example 18. For a given order $a \prec b \prec c \prec f(d) \prec g(a)$, the EUF-ROBDD representation of $\mathsf{ITE}(f(d) \approx a, \mathsf{true}, \mathsf{ITE}(f(d) \approx b, \mathsf{true}, \mathsf{ITE}(f(d) \approx c, \mathsf{ITE}(g(a) \approx c, \mathsf{true}, \mathsf{false}), \mathsf{false})))$ is depicted in Figure 4.

Consider the path $f(d) \not\approx a.f(d) \not\approx b.f(d) \approx c.g(a) \approx c$. The formula corresponding to the path contains one non-propagated equality $f(d) \approx c$. By propagating this equality, i.e. replacing $f(d) \not\approx a \wedge f(d) \not\approx b$ with $c \not\approx a \wedge c \not\approx b$, we obtain a reduced formula $c \not\approx a \wedge c \not\approx b \wedge f(d) \approx c \wedge g(a) \approx c$.

The formula corresponding to the path and the reduced formula are logically equivalent. By Theorem 17, the reduced formula is satisfiable. Therefore, the path is also satisfiable.

Theorem 19. *Every path in an EUF-BDD is satisfiable.*

The proof can be found in a full technical report [17].

Corollary 20. *From Theorem 19*

- *The only tautological EUF-ROBDD is* true.
- *The only contradictory EUF-ROBDD is* false.
- *All other EUF-ROBDDs are satisfiable.*

5 Implementation and Applications

We implemented our proposal for EUF-BDDs within the special purpose theorem prover for the μCRL toolset [14]. The language μCRL combines abstract data types with process algebra. The prover is used to discharge proof obligations generated in protocol verifications, in particular to prove process invariants and confluence of internal computation steps [19]. It is also used in the symbolic model checker with data, proposed in [15]. In the latter application it is essential to have a concise representation of formulas, which is provided by EUF-BDDs.

Usually, symbolic model checking uses ordered binary decision diagrams to provide a compact representation of the transition system. BDD-based model checking performs an exhaustive traversal of the model by considering all possible behaviors in a compact way. Such exhaustive exploration allows BDD based model checking algorithms to conclude whether a given property is satisfied. In a similar way, the symbolic model checker with data represents a possibly infinite state space by BDDs extended with equalities and function symbols. In this case, the main operation for the model checker is to compute the sequence of EUF-ROBDDs $\Phi^n(\bot)$ [for some operator Φ]. A fixed point has been reached as soon as the formula $\Phi^n(\bot) \iff \Phi^{n+1}(\bot)$ is a tautology, which can be checked by our method.

As input, the prover takes a data specification consisting of a signature of constructor and defined symbols, and a set of equations. It also takes a quantifier-free formula as input, and it returns a logically equivalent EUF-ROBDD. If the result is either true or false, we know for sure that the formula is a tautology or a contradiction, respectively. For the other cases, we would like to conclude that both the formula and its negation are satisfiable. However, this is only possible for certain fragments. We call the prover complete for such fragments.

The previous implementation of this theorem prover [20] was based on EQ-BDDs [11], and consequently it was only complete for the case of equality logic with equations between variables. The implementation also used plain term rewriting with equations from the abstract datatype. It was sound for any data specification, but not complete.

The current implementation is based on the observations in this paper. Consequently, it is now complete for the theory with equality and uninterpreted function symbols. It is sound – but incomplete – for the case that functions denote constructors, or when they are specified by means of equations.

In the current implementation, we use the reversed equation order as introduced in [13]. Moreover, we use the lexicographic path order to compare terms; this order satisfies the conditions of Definition 6. Given a formula φ, we find the smallest equation $t \approx s$ in it, then recursively compute the EUF-BDDs A of $\varphi[t := s]$ and B of $\varphi[t \approx s := \mathsf{false}]$, and return $\mathsf{ITE}(t \approx s, A, B)$. This procedure must be repeated in order to obtain an EUF-ROBDD.

The new prover was applied to many existing case studies (see for instance [19,14] for a description). It is confirmed that the new prover can handle more formulas. Moreover, it was never slower than the version of [20].

The resulting EUF-ROBDD can be visualized (using graphviz/dot) for small formulas. Figure 5 shows the EUF-ROBDDs for the formulas $f(f(f(f(f(a))))) \approx b \land f(f(a)) \approx f(a)$ and $(f(b) \approx f(c) \Rightarrow a \approx b) \Leftrightarrow (f(b) \approx f(c) \Rightarrow a \approx c)$, respectively.

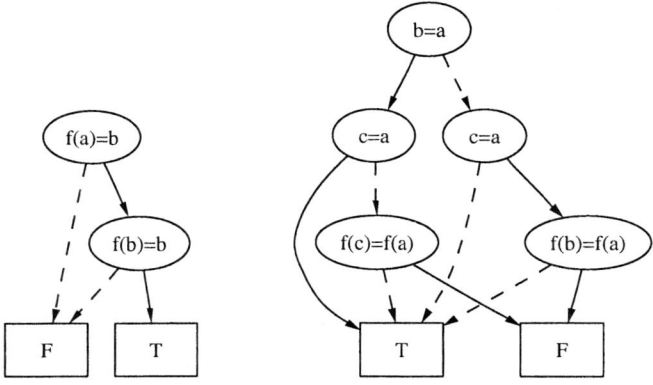

Fig. 5. EUF-ROBDDs obtained by the implementation

6 Conclusions

We have extended the approach from [11] in the presence of uninterpreted function symbols, and the changed orientation of [13] is essential for the completeness of our method. Starting from the EUF-BDD representing an arbitrary EUF formula and applying rewrite rules of a rewrite system Reduce-Order, a normal form, called an EUF-ROBDD, can be calculated. We proved that all paths in a EUF-ROBDD are satisfiable by construction. Our approach is suitable for checking tautology, satisfiability, and equivalence of formulas. A prototype implementation of this method works within the special purpose theorem prover for the μCRL toolset [14].

Future Work. We have not yet studied strategies for choosing an ordering on equalities. A good ordering is crucial since it yields a compact representation: for some Boolean functions, the ROBDD sizes are linear in the number of variables for one ordering, and exponential for another. It is interesting to extend our methods beyond the EUF fragment. The current implementation handles any data specified by a TRS, but is in general incomplete. The study in [12] shows that finding complete extensions may be hard.

References

1. Bryant, R.: Symbolic boolean manipulation with ordered binary decision diagrams. ACM Computing Surveys **24** (1992) 293–318
2. Burch, J., Dill, D.: Automated verification of pipelined microprocesoor control. In Dill, D., ed.: Computer-Aided Verification (CAV'94). Volume 818 of LNCS., Springer-Verlag (1994) 68–80
3. Pnueli, A., Rodeh, Y., Shtrichman, O., Siegel, M.: The small model property: how small can it be? Information and Computation **178** (2002) 279 – 293

4. Nelson, G., Oppen, D.: Fast decision procedures based on congruence closure. Journal of the ACM **27 (2)** (1980) 356 – 364
5. Shostak, R.: An algorithm for reasoning about equality. Communications of the ACM **21** (1978) 583–585
6. Ackermann, W.: Solvable cases of the decision problem. Studies in Logic and the Foundations of Mathematics. North-Holland, Amsterdam (1954)
7. Goel, A., Sajid, K., Zhou, H., Aziz, A., Singhal, V.: BDD based procedures for a theory of equality with uninterpreted functions. In Hu, A.J., Vardi, M.Y., eds.: Computer-Aided Verification (CAV'98). Volume 1427 of LNCS., Springer-Verlag (1998) 244–255
8. Fontaine, P., Gribomont, E.P.: Using BDDs with combinations of theories. In Baaz, M., Voronkov, A., eds.: Logic for Programming and Reasoning (LPAR'2002). Volume 2514 of LNCS., Springer-Verlag (2002) 190–201
9. Bryant, R., Velev, M.: Boolean satisfiability with transitivity constraints. ACM Transactions on Computational Logic **3** (2002) 604–627
10. Tveretina, O., Zantema, H.: A proof system and a decision procedure for equality logic. In Farach-Colton, M., ed.: LATIN 2004: Theoretical Informatics. Volume 2976 of LNCS. (2004) 530–539
11. Groote, J., van de Pol, J.: Equational binary decision diagrams. In Parigot, M., Voronkov, A., eds.: Logic for Programming and Reasoning (LPAR'2000). Volume 1955 of LNAI. (2000) 161–178
12. Badban, B., van de Pol, J.: Zero, sucessor and equality in BDDs. Annals of Pure and Applied Logic **133/1-3** (2005) 101–123
13. Badban, B., van de Pol, J.: An algorithm to verify formulas by means of (0,s,=)-BDDs. In: Proceedings of the 9th Annual Computer Society of Iran Computer Conference (CSICC 2004), Tehran, Iran (2004)
14. Blom, S., Groote, J., van Langevelde, I., Lisser, B., van de Pol, J.: New developments around the μCRL tool set. In: Proceedings of FMICS 2003. ENTCS volume 80 (2003)
15. Groote, J., Willemse, T.: Parameterised boolean equation systems (extended abstract). In Gardner, P., Yoshida, N., eds.: Proceedings of CONCUR 2004. LNCS 3170 (2004) 308–324
16. Baader, F., Nipkow, T.: Term Rewriting and All That. Cambridge University Press (1998)
17. van de Pol, J., Tveretina, O.: A BDD-representation for the logic of equality and uninterpreted functions (a full version with proofs). Technical Report SEN-R0509, Centrum voor Wiskunde en Informatica, Amsterdam (2005)
18. Tveretina, O.: A decision procedure for equality logic with uninterpreted functions. In Buchberger, B., Campbell, J., eds.: Artificial Intelligence and Symbolic Mathematical Computation. Volume 3249 of LNAI., Springer-Verlag (2004) 63–76
19. Blom, S., van de Pol, J.: State space reduction by proving confluence. In Brinksma, E., Larsen, K., eds.: Proceedings of CAV'02. LNCS 2404, Springer (2002) 596–609
20. van de Pol, J.: A prover for the μCRL toolset with applications – Version 0.1. Technical Report SEN-R0106, CWI, Amsterdam (2001)

On Small Hard Leaf Languages

Falk Unger

Centrum voor Wiskunde en Informatica (CWI), Kruislaan 413,
P.O. Box 94079, 1090 GB Amsterdam, The Netherlands
falk.unger@cwi.nl

Abstract. This paper deals with balanced leaf language complexity classes, introduced independently in [1] and [14]. We propose the seed concept for leaf languages, which allows us to give "short" representations for leaf words. We then use seeds to show that leaf languages A with $NP \subseteq BLeaf^P(A)$ cannot be polylog-sparse (i.e. $census_A \in O(\log^{O(1)})$), unless PH collapses.

We also generalize balanced $\leq_m^{P,bit}$-reductions, which were introduced in [6], to other bit-reductions, for example (balanced) truth-table- and Turing-bit-reductions. Then, similarly to above, we prove that NP and Σ_2^P cannot have polylog-sparse hard sets under those balanced truth-table- and Turing-bit-reductions, if the polynomial-time hierarchy is infinite.

Keywords: Computational Complexity, Leaf Languages, Seeds, Sparseness.

1 Introduction

The leaf language formalism, introduced in [1] and independently in [14], can be used to describe polynomial-time complexity classes in the following way: Let M be some non-determistic polynomial-time Turing machine (short: NPTM). For input x let each path in the computation tree of $M(x)$ be marked with a word from $\{0,1\}^*$, characterizing the non-deterministic choices on that path. $M(x)$ produces output bit 0 or 1 on each path. If we concatenate all output bits of $M(x)$ in the lexicographical order of their path marks, we get the leaf word $leaf^M(x)$. For a given leaf language A we define the complexity class $Leaf^P(A)$ as the set of all languages L for which there is an NPTM M such that for all inputs x it holds: $x \in L \leftrightarrow leaf^M(x) \in A$.

In this paper we additionally require M to be balanced. An NPTM is *balanced* if for any input its computation tree is balanced. A computation tree is balanced if there is a maximal path mark y with the property that all path marks x with $|x| = |y|$ and $x \leq_{lex} y$ exist and further no other path marks exist in that computation tree. $BLeaf^P$ is defined analogously to $Leaf^P$ with the only difference that now M has to be balanced. See Section 2.1 for further details.

This formalism allows to define many known complexity classes in a uniform way. If we set $A = \Sigma^*\{1\}\Sigma^*$ and $B = \{0\}^*$ we get for example $NP =$

$BLeaf^P(A) = Leaf^P(A)$, $coNP = BLeaf^P(B) = Leaf^P(B)$. For these particular leaf languages the balanced leaf language classes are the same as the non-balanced ones. In general that does not need to be the case (see [7] and [8]).

It is well-known that SAT contains exponentially many instances of each size, that is $census_{SAT}(n) \in \Omega(2^{cn})$ for some $c > 0$. In this paper we tackle the question whether NPTMs can reduce the exponentially many instances of SAT to less than exponentially many words, or more precisely: How many words are necessary in a leaf language A with $NP \subseteq BLeaf^P(A)$? We first show that sparse sets are sufficient, i.e. $NP \subseteq BLeaf^P(SPARSE)$ (see Proposition 1). This yields an upper bound. In our main result we then give a lower bound by showing that no polylog-sparse leaf language A can satisfy $NP \in BLeaf^P(A)$, unless PH collapses. We call a leaf language A *polylog-sparse* if there is a polynomial p such that $census_A(n) \leq p(\log n)$. Thus, leaf languages for NP can probably not be arbitrarily "empty".

1.1 Generalized Bit-Reductions and Seeds

In [6] the notation $L \leq_m^{P,bit} A$ was introduced for $L \in BLeaf^P(A)$. This is very intuitive, since $L \in BLeaf^P(A)$ means that there is some balanced NPTM M such that $leaf^M$ is a reduction from L to A, i.e. $x \in L \leftrightarrow leaf^M(x) \in A$. However, we prefer the notation $L \leq_m^{P,Bbit} A$ to emphasize the fact that the reduction is carried out by a balanced NPTM. Later in 2.3 we will generalize this to $\leq_{tt}^{P,Bbit}$, $\leq_T^{P,Bbit}$ and other balanced bit-reductions.

Similar to our main result we then show in Sections 4 and 5 the following: There are no polylog-sparse leaf languages which are hard for NP or Σ_2^P under $\leq_m^{P,Bbit}$-, $\leq_T^{P,Bbit}$- or $\leq_{btt}^{P,Bbit}$-reductions, unless PH collapses. An overview of our results is given in Section 3.

In the proofs and for the definition of general $\leq^{P,Bbit}$-reductions we need seeds. A set S is called a *seed set* of a leaf language A if there is a balanced NPTM M such that $leaf^M(S) = A$. The crucial feature of seeds is that they can be much smaller than the corresponding leaf words they induce.

1.2 Relation to Other Work

Our work is related to the following results about sparse hard sets for NP from [10], [9], [11] and [2]. For $S \in SPARSE$ it holds

[10]	$NP \leq_m^P S$	$\Longrightarrow$	$P{=}NP$
[9]	$NP \leq_T^P S \wedge S \in NP$	$\Longrightarrow$	$PH{=}\Theta_2^P$
[11]	$NP \leq_{btt}^P S$	$\Longrightarrow$	$P{=}NP$
[2]	$NP \leq_T^P S$	$\Longrightarrow$	$PH{=}S_2^P$.

Let us look at the theorem from [10]. It implies that there is probably no deterministic polynomial-time Turing-machine that can manyone-reduce the $\Omega(2^{cn})$ satisfiable boolean formulae of size n to only a polynomial number of words. Our main result implies the same conclusion for *non-deterministic* polynomial-time machines: If PH does not collapse, then there is no NPTM that can manyone-reduce (i.e. $\leq_m^{P,Bbit}$) all satisfiable boolean formulae of size n to only a polynomial number of words. Thus, in this respect non-determinism does not seem to be more powerful than determinism.

We also use similar ideas as in [10]. The main new ingredient in our proof is the use of seeds. Our other results relate similarly to the other theorems above.

2 Notation and Definitions

Before we start, let us agree on the following conventions: We fix the alphabet $\Sigma = \{0, 1\}$ and add symbols (e.g. delimiters) when necessary. Furthermore, we extend a function $f : A \to B$ to $f : 2^A \to 2^B$ by setting $f(X) = \bigcup_{x \in X}\{f(x)\}$. Let $\leq_{lex}$ be the lexicographical ordering of words and $\sqsubseteq$ the prefix relation, that is: For $x, y \in \Sigma^*$ we write $x \sqsubseteq y$ if and only if $\bigvee_{z \in \Sigma^*} xz = y$. For a set A its census function is defined as $census_A(n) := |\{w : w \in A \wedge |w| \leq n\}|$. Call a set A sparse if $census_A \in O(n^{O(1)})$ and let $SPARSE$ be the set of all sparse sets. Analogously, a set A is called polylog-sparse if $census_A \in O(\log^{O(1)})$. Let Pol be the set of all polynomials. FP is the class of all functions computable by polynomial-time Turing-machines.

The class $\Theta_2^P = P^{NP[\log]}$ consists of all languages that can be accepted by a polynomial-time Turing-machine with at most $O(\log n)$ adaptive queries to an NP-oracle.

The complexity class S_2^P was introduced in [3] and [13]. A language L is in S_2^P iff there is a polynomial-time predicate P such that: If $x \in L$ then there exists a y, such that for all z the predicate $P(x, y, z)$ is true. Conversely, if $x \notin L$ then there exists a z, such that for all y the predicate $P(x, y, z)$ is false. The lengths of y and z are in both cases polynomially bounded in the length of x. From [13] it is known that $P^{NP} \subseteq S_2^P$.

2.1 Leaf Language Complexity Classes

We first formalize balanced NPTMs.

Definition 1. BFP *is the set of all functions f such that there exist $f_1, f_2 \in FP$ with $f_1 : \Sigma^* \times N \to \Sigma$, $f_2 : \Sigma^* \to N$ and for all $x \in \Sigma^*$*

$$f(x) = f_1(x, 1) \ldots f_1(x, f_2(x)).$$

Such functions f are called polynomial-time bit-computable.

If in the following we are given a function $f \in BFP$, we always assume that f_1 and f_2 denote the respective FP-functions as in Definition 1. For $f \in BFP$ we say: f produces leaf word $f(x)$ on input x.

It is easy to see that $BFP = \{leaf^M : M$ is a balanced NPTM$\}$. The inclusion $\subseteq$ is trivial and the inclusion $\supseteq$ follows by noting that if M is balanced, the i-th bit of $leaf^M(x)$ can be computed in polynomial time: Let $b \in \{0,1\}^*$ be the binary representation of i. First compute the maximal path mark y in the computation of $leaf^M(x)$. The i-th bit exists iff $|b| \leq |y|$ and $0^{|y|-|b|}b \leq_{lex} y$. It can then be computed by following the path with mark $0^{|y|-|b|}b$.

Note that the same procedure cannot be applied for unbalanced M, since the path computing the i-th bit does not necessarily have mark $0^{|y|-|b|}b$. In fact, computing the i-th bit if M is unbalanced seems to be much harder. See [7] for more on balanced vs. unbalanced computation trees. However, in this paper we are only concerned with balanced computations.

We now formally define balanced leaf language complexity classes.

Definition 2. *For* $A \subseteq \Sigma^*$ *we define:*

$$L \in BLeaf^P(A) \leftrightarrow_{def} \bigvee_{f \in BFP} [x \in L \leftrightarrow f(x) \in A]$$

For $A, R \subseteq \Sigma^*$, $A \cap R = \emptyset$ *we also define promise classes:*

$$L \in BLeaf^P(A, R) \leftrightarrow_{def}$$
$$\bigvee_{f \in BFP} \left[(x \in L \to f(x) \in A) \land (x \notin L \to f(x) \in R) \right].$$

2.2 Seeds

We now introduce the notion of seeds.

Definition 3. *For* $S, A \subseteq \Sigma^*$ *we say that* S *is a* seed set *of the leaf language* A, *if there is a function* $f \in BFP$ —*the* generating function *or simply* generator— *with* $f(S) = A$.

We say further: x *is the* seed *of* $f(x)$ *and* $f(x)$ *is the* f-expansion *of seed* x.

The crucial feature of seeds is that they are (often) much smaller than the leaf word they represent. If we know the seed s of a leaf word b we can compute each bit of b in polynomial time in $|s|$ whenever needed. If S is a seed set for A with generator $f \in BFP$ then B is, in a way, effectively constructible from S via f.

2.3 Balanced Bit-Reductions

In [6] it was pointed out, that we can interpret $L \in BLeaf^P(A)$ as a bit reduction from L to A. Recall from Definition 2 that $L \in BLeaf^P(A)$ means, that there is some $r \in BFP$ such that $x \in L \leftrightarrow r(x) \in A$. This resembles very much the definition of $\leq_m^P$-reductions with the only difference that now r is in BFP and not in FP. That is why the notation $L \leq_m^{P,bit} A$ was introduced. Instead of $\leq_m^{P,bit}$ as in [6] we rather use $\leq_m^{P,Bbit}$ to emphasize that the reduction must be realized by a balanced NPTM.

Using seeds, we now define tt-, T- and other bit-reductions.

Definition 4. If for $X = L, NL, P, NP, RP, \ldots$ and $Y = m, tt, T, \ldots$ the reduction $\leq_Y^X$ is already defined, we define for $L, A \subseteq \Sigma^*$

$$L \leq_Y^{X,Bbit} A \leftrightarrow_{def} \bigvee_{S \subseteq \Sigma^*} \bigvee_{f \in BFP} [f(S) \subseteq A \land f(\bar{S}) \subseteq \bar{A} \land L \leq_Y^X S.] \quad (1)$$

For such $\leq_Y^{X,Bbit}$-reductions we call the $\leq_Y^X$-reduction from A to S the corresponding seed reduction, S the corresponding seed oracle, f the access function and A the leaf oracle.

For $L, A, R \subseteq \Sigma^*$ with $A \cap R = \emptyset$ we also define promise reductions

$$L \leq_Y^{X,Bbit} (A, R) \leftrightarrow_{def}$$
$$\bigvee_{S_1, S_2 \subseteq \Sigma^*} \bigvee_{f \in BFP} [f(S_1) \subseteq A \land f(S_2) \subseteq R \land L \leq_Y^X (S_1, S_2)]$$

where $L \leq_Y^X (S_1, S_2)$ means, that there is a $\leq_Y^X$-reduction r from L to S_1 with the promise that r only asks queries from $S_1 \cup S_2$. Analogously we call (S_1, S_2) the seed oracle with accepting/rejecting parts S_1/S_2 and (B_1, B_2) the leaf oracle with accepting/rejecting parts B_1/B_2.

Our $\leq_m^{P,Bbit}$ coincides with $\leq_m^{P,bit}$ in [6].

Note that the answer of a $\leq_Y^{X,Bbit}$-oracle A on a query x is not $\chi_A(x)$ (as for a $\leq_Y^X$-oracle) but rather $\chi_A(f(x))$, where f is the access function to A and χ_A is the characteristic function of A. It is obvious how to extend Definition 4 to unbalanced bit-reductions, but in this paper we do not need that.

Access functions in the context of bit-reductions are quite similar to generating functions in the context of seed sets. We have deliberately chosen a different name to reflect the different approach: Whereas a generating function "generates" a leaf language from a *given seed set*, an access function is used to check whether the expansion of some seed yields a leaf that is contained in some *given leaf language*. Thus, an access function "gives access" to a leaf oracle.

Now, the following fact is easy to see: $L \leq_Y^X A \to L \leq_Y^{X,Bbit} A$. This is because if r is a $\leq_Y^X$-reduction from L to A, then the same r with access function $f = id$ is a $\leq_Y^{X,Bbit}$-reduction from L to A. That means that $\leq_Y^{X,Bbit}$-reductions can be seen as extensions of $\leq_Y^X$-reductions.

3 Overview of Results

In the next sections we show the following results for polylog-sparse sets K:

$$\begin{array}{lll} \text{Theorem 1} & NP \leq_m^{P,Bbit} K & \Longrightarrow PH = \Theta_2^P \\ \text{Theorem 3} & NP \leq_T^{P,Bbit} (K, 0^*) & \Longrightarrow PH = S_2^P \\ \text{Theorem 4} & \Sigma_2^P \leq_{btt}^{P,Bbit} K & \Longrightarrow PH = \Delta_2^P \\ \text{Theorem 5} & \Sigma_2^P \leq_T^{P,Bbit} K & \Longrightarrow PH = \Sigma_4^P \end{array}$$

In the following interpretation of the theorems we assume that PH does not collapse to Σ_4^P. Theorem 1 then suggests that there are no polylog-sparse $\leq_m^{P,Bbit}$-hard leaf languages for NP.

In the other theorems we try to relax the $\leq_m^{P,Bbit}$-reduction in the supposition to $\leq_{tt}^{P,Bbit}$- and $\leq_T^{P,Bbit}$-reductions. But then we can prove a collapse of PH only under some other additional assumptions.

In Theorem 3 for example, we additionally require that the $\leq_T^{P,Bbit}$-reduction is a particular promise reduction. Theorem 3 can be interpreted as saying that NP does probably not have polylog-sparse leaf languages which are NP-typical, i.e. accepting leaf words contain at least one 1 and rejecting leaf words contain only 0's.

Theorems 4 and 5, on the other hand, require additionally that not only NP but also Σ_2^P can be reduced to some polylog-sparse leaf language. These theorems suggest that there are no polylog-sparse $\leq_{btt}^{P,Bbit}$- or $\leq_T^{P,Bbit}$-hard leaf languages for Σ_2^P.

Looking at the related results about sparse hard sets (see Section 1.2), some readers might wonder why we sometimes have to require that Σ_2^P and not merely NP reduces to K and also why the resulting collapse of PH is on a higher level. Informally, the reason is that in order to check that two seeds y_1 and y_2 yield the same leaf word can only be checked by expanding them and checking if $f(y_1) = f(y_2)$, which requires a $coNP$-computation.

4 Lower Bounds for NP-Hard Leaf Languages

In this section we demonstrate, that no polylog-sparse leaf language A can be $\leq_m^{P,Bbit}$-hard for NP, unless PH collapses. This clearly suggests a lower bound for the number of words needed in leaf languages which characterize NP. Nevertheless, we first give an upper bound, by showing that every set has a $\leq_m^{P,Bbit}$-hard, sparse leaf language.

Proposition 1. *For all $L \subseteq \Sigma^*$ it holds*

$$L \leq_m^{P,Bbit} SPARSE.$$

Proof. Assume $L \subseteq \Sigma^*$. Define padding function $f \in BFP$ as $f(x) = 0^{2^{|x|}}x$. Define the set A by $a \in A \leftrightarrow \bigvee_{x \in L} f(x) = a$. Clearly, A is sparse. Then obviously $x \in L \leftrightarrow f(x) \in A$ and thus $L \in BLeaf^P(A)$. □

Now, what are the lower bounds for $\leq_m^{P,Bbit}$-hard leaf languages? In this section we tackle that question for the class NP. Firstly, it is obvious that finite sets cannot be $\leq_m^{P,Bbit}$-hard for NP, unless $P = NP$. Assuming that PH does not collapse the following theorem gives a better bound:

Theorem 1. *For a polylog-sparse set K it holds*

$$NP \leq_m^{P,Bbit} K \implies PH = \Theta_2^P.$$

Proof. Choose arbitrary $L \in \Pi_2^P$. We will show $L \in \Theta_2^P$.

$L \in \Pi_2^P$ means that there must be polynomials p, q and $r \in FP$ with

$$x \in L \leftrightarrow \bigwedge_y^p r(x,y) \in SAT \tag{2}$$

and $\bigwedge_y^p |r(x,y)| \leq q(|x|)$. By assumption, there is an $f \in BFP$ and a polylog-sparse K such that for all boolean formulae ϕ

$$\phi \in SAT \leftrightarrow f(\phi) \in K. \tag{3}$$

Choose a polynomial t such that $census_K(n) \leq t(\log n)$ for all n.

We now construct a Θ_2^P-machine M_L that decides L. On input x with $|x| = n$ it first uses oracle A

$$A := \left\{ (0^n, 1^m) : \bigvee_{y_1 \in SAT}^{|y_1| \leq q(n)} \cdots \bigvee_{y_m \in SAT}^{|y_m| \leq q(n)} \bigwedge_{1 \leq i < j \leq m} f(y_i) \neq f(y_j) \right\}$$

to compute the number of leaves in K that have seeds of length $\leq q(n)$. We call that number $\hat{m}(n)$. Clearly, to compute $\hat{m}(|x|)$ at most $2\lceil \log \hat{m}(|x|) \rceil \leq 2\lceil \log t(\log f_2(q(|x|))) \rceil \in O(\log |x|)$ queries to A are needed. Note that A is in NP since an A-accepting machine M_A only needs to guess $y_1, \ldots, y_m$ and paths $p_{ij, 1 \leq i,j \leq m} \in \Sigma^{O(Pol(n))}$ such that for all $i \neq j$ it holds $f_1(y_i, p_{ij}) \neq f_1(y_j, p_{ij})$ or $f_2(y_i) \neq f_2(y_j)$.

After having computed $\hat{m}(n)$, M_L makes one last query $(x, \hat{m}(n))$ to another $coNP$-machine M_B, which will accept iff $x \in L$. We finish the proof by showing how such a $coNP$-machine M_B works.

M_B on input $(x, \hat{m}(n))$ starts like M_A on input $(0^n, 1^{\hat{m}(n)})$. On non-accepting M_A-paths M_B stops and rejects, too. Note that on accepting paths of the M_A-computation M_B knows seeds $y_1, \ldots, y_{\hat{m}(n)}$ for all leaves in K which can be constructed from seeds of length $\leq q(n)$. With (3) this means that for all boolean formulae ϕ with $|\phi| \leq q(n)$ it therefore holds

$$\phi \in SAT \leftrightarrow \bigvee_i^{i \leq \hat{m}(n)} f(y_i) = f(\phi). \tag{4}$$

On accepting M_A-paths M_B then checks whether

$$\bigwedge_y^p \bigvee_i^{i \leq \hat{m}(n)} f(y_i) = f(r(x,y)). \tag{5}$$

By (2) and (4) this is equivalent to $x \in L$. Thus, the rest of the construction of M_B is straightforward if we show that (5) is a $coNP$-predicate. But this is true because (5) is equivalent to

$$\bigwedge_y^p \bigvee_i^{i \leq \hat{m}(n)} \left[f_2(r(x,y)) = f_2(y_i) \wedge \bigwedge_v^{v \leq f_2(r(x,y))} f_1(y_i, v) = f_1(r(x,y), v) \right], \tag{6}$$

which is a $coNP$-predicate since $\bigvee_i$ quantifies only over polynomially many elements.

Thus, M_L is a Θ_2^P-machine deciding L. □

If we require $NP \leq_{dtt}^{P,Bbit} K$ instead of $NP \leq_m^{P,Bbit} K$ in Theorem 1, almost the same proof establishes $PH = \Theta_2^P$.

In the next Theorem 3 we will achieve a collapse of PH by assuming that NP is only $\leq_T^{P,Bbit}$-reducible to some polylog-sparse leaf oracle $(K, \{0\}^*)$. In order to do so we additionally require the leaf reduction to be "NP-typical", that is positive queries produce leaf words containing at least one 1 and negative queries produce leaf words from $\{0\}^*$. In the proof we need a result from Cai.

Theorem 2. *[2] For $S \in SPARSE$*

$$NP \leq_T^P S \implies PH = S_2^P.$$

Theorem 3. *For polylog-sparse K it holds*

$$NP \leq_T^{P,Bbit} (K, \{0\}^*) \implies PH = S_2^P.$$

Proof. Let K be given as in the assumption with $census_K(n) \leq t(\log n)$ for some polynomial t. Let M be a deterministic polynomial-time oracle Turing machine (short: DPOM) that can decide SAT with the help of the $\leq_T^{P,Bbit}$-oracle $(K, \{0\}^*)$. Let $f \in BFP$ be the access function of that reduction and (S_1, S_2) the corresponding seed oracle, that is $f(S_1) \subseteq K$ and $f(S_2) \subseteq \{0\}^*$. Let the computation time of M be bounded by the polynomial p.

Because of Theorem 2 it suffices to construct a sparse oracle $Path$ with $NP \subseteq P^{Path}$. We first define an auxiliary oracle $Path'$ that contains all $(0^n, \hat{w})$ for which

$$\bigvee_{s \in S_1}^{|s|=n} \left[\hat{w} \leq f_2(s) \wedge f_1(s, \hat{w}) = 1 \wedge \bigwedge_{w'}^{w' \leq \hat{w}} f_1(s, w') = 0 \right]. \quad (7)$$

Note that for each possible query s with $f(s) \in K$, the smallest $\hat{w}$ with $f_1(s, \hat{w}) = 1$ is in $Path'(0^{|s|})$. Hence, if $Path'(0^{|s|})$ is known, it can be decided in polynomial time whether a given formula s is in SAT. We extend $Path'$ to

$$Path := \{(0^n, w) : \bigvee_{\hat{w} \in Path'(0^n)} w \sqsubseteq \hat{w}\}. \quad (8)$$

$Path(0^n)$ contains all prefixes of $Path'(0^n)$. Using the oracle $Path$ it is possible to compute $Path'(0^n)$ in polynomial time in n. Since K is polylog-sparse it is clear that $|Path'(0^n)| \in O(n^{O(1)})$. Hence $Path' \in SPARSE$ and also $Path \in SPARSE$.

A SAT-deciding DPOM M' with oracle $Path$ could basically work like M with the only difference: Whenever M asks the leaf oracle K if $f(s) \in K$, M' computes $Path'(0^{|s|})$ and then checks whether $f_1(s, \hat{w}) = 1$ for some $\hat{w} \in Path'(0^{|s|})$. If that is the case, M' proceeds as M on answer "Yes". Otherwise M' proceeds as M on answer "No". Thus, $SAT \leq_T^P Path$, which together with Theorem 2 establishes the result. □

Note that in [13] it was proven that $P^{NP} \subseteq S_2^P$. Looking at Theorems 1 and 3 a natural question is whether it is possible to conclude something stronger than $PH = S_2^P$ by combining both assumptions (i.e. $NP \leq_m^{P,Bbit} (K, \{0\}^*)$). That is still open.

We now want to see if similar results are possible for $coNP$.

Proposition 2. *There is a polylog-sparse set K with*

$$coNP \leq_m^{P,Bbit} K.$$

For example $K = \{0^{2^n} : n \in N\}$ proves this proposition. Thus, a theorem that is analogous to Theorem 1 for $coNP$ would immediately imply a collapse of PH, which is assumed to be highly unlikely. Similarly, we also see why Theorem 1 can (probably) not be extended to $\leq_{btt/tt/T}^{P,Bbit}$-reductions, since that would also imply a collapse of PH. The relaxation to $\leq_T^{P,Bbit}$-reductions in Theorem 3 was only possible because we additionally required the leaf oracle to be "NP-typical".

5 Lower Bounds for Σ_2^P-Hard Leaf Languages

Following on from what was said at the end of the previous section there is another way of extending Theorem 1 to $\leq_{btt}^{P,Bbit}$- and $\leq_T^{P,Bbit}$-reductions. But again, we also have to accept an additional requirement, which now is that Σ_2^P (and not only NP) must be balanced bit-reducible to some polylog-sparse leaf language K. In this section we show two theorems of that kind.

The first of them is an adaption of a result by Ogihara-Watanabe (see [11]), which originally stated, that there cannot be sparse $\leq_{btt}^P$-hard sets for NP, unless $P = NP$.

Theorem 4. *For polylog-sparse K it holds*

$$\Sigma_2^P \leq_{btt}^{P,Bbit} K \implies PH = P^{NP} = \Delta_2^P.$$

The proof is quite long and technical. Due to space restrictions it is deferred to the final journal version. However, for readers knowing the original proof from [11] the following lines might give some idea of how it works. We also use the left-set technique. The main difference is the following: In the original proof it is easy to check whether two queries y_1 and y_2 to the oracle are the same (which is the case iff $y_1 = y_2$). Now, in our case that is more difficult. Even for queries y_1 and y_2 with $y_1 \neq y_2$ it is still possible that they are in fact the same question, namely if $f(y_1) = f(y_2)$. So, checking whether two queries ask for the same leaf word now requires a $coNP$-oracle. Since that test already needs a $coNP$-oracle, it is not sufficient to require $NP \leq_{btt}^{P,Bbit} K$ in the assumption but rather $\Sigma_2^P \leq_{btt}^{P,Bbit} K$.

Our last theorem assumes only the weaker supposition $\Sigma_2^P \leq_T^{P,Bbit} K$ but then PH only collapses to Σ_4^P.

Theorem 5. *For polylog-sparse K it holds*

$$\Sigma_2^P \leq_T^{P,Bbit} K \to PH = \Sigma_4^P.$$

Proof. Assume $L \in \Pi_4^P$. We show $L \in \Sigma_4^P$.

Suppose K as in the assumption and let t be a polynomial such that $census_K(n) \leq t(\log n)$. Choose $Q \in P$ and polynomials $p_1, \ldots, p_4$ such that

$$x \in L \leftrightarrow \bigwedge_{x_1}^{p_1} \cdots \bigvee_{x_4}^{p_4} Q(x, x_1, \ldots, x_4).$$

We define L' as the set containing all (x, x_1, x_2) with

$$|x_1| \leq p_1(|x|) \wedge |x_2| \leq p_2(|x|) \wedge \bigwedge_{x_3}^{p_3} \bigvee_{x_4}^{p_4} Q(x, x_1, \ldots, x_4). \tag{9}$$

Obviously, $L' \in \Pi_2^P$ and thus by the assumption $\overline{L'} \leq_T^{P,Bbit} K$. But then also $L' \leq_T^{P,Bbit} K$, since $\leq_T^{P,Bbit}$ is a deterministic Turing-reduction. Hence, there must be a DPOM M, with associated access function $f \in BFP$, which realizes a $\leq_T^{P,Bbit}$-reduction from L' to K. Choose polynomial q with $|f(x)| \leq 2^{q(|x|)}$. Because of the constraint $\bigwedge_{i=1,2} |x_i| \leq p_i(|x|)$ in (9), we can also bound the computation time of $M(x, x_1, x_2)$ by a polynomial in $|x|$, instead of $|(x, x_1, x_2)|$. Let r be such a polynomial.

We now define a predicate O, that is true for given $n \in \mathbb{N}$ and $S = \{s_1, \ldots, s_m\}$ iff: M decides with $\leq_T^{P,Bbit}$-leaf oracle $f(S)$ for all inputs (x, x_1, x_2), $|x| \leq n$, correctly whether $(x, x_1, x_2) \in L'$.

$$O(S, 0^n) \leftrightarrow \bigwedge_x^{|x| \leq n} \bigwedge_{x_1}^{p_1} \bigwedge_{x_2}^{p_2} \tag{10}$$

$$\bigvee_{c \in \mathbb{N}}^{0 \leq c \leq r(|x|)} \bigvee_{l_i \in \Sigma^{\leq r(|x|)}}^{i=1,\ldots,c} \bigvee_{o_i \in \Sigma}^{i=1,\ldots,c} \tag{11}$$

$$\begin{bmatrix} M(x, x_1, x_2) \text{ makes queries } l_1, \ldots, l_c \\ \text{and gets answers } o_1, \ldots, o_c \end{bmatrix} \tag{12}$$

$$\wedge \left[\bigwedge_i^{1 \leq i \leq c} o_i = 1 \leftrightarrow \bigvee_{u \in S} f(l_i) = f(u) \right] \tag{13}$$

$$\wedge \left[M(x, x_1, x_2) \text{ rejects} \rightarrow \bigvee_{x'_3}^{p_3} \bigwedge_{x_4}^{p_4} \neg Q(x, x_1, x_2, x'_3, x_4) \right] \tag{14}$$

$$\wedge \left[M(x, x_1, x_2) \text{ accepts} \rightarrow \bigwedge_{x_3}^{p_3} \bigvee_{x_4}^{p_4} Q(x, x_1, x_2, x_3, x_4) \right]. \tag{15}$$

(For the time being ignore the arrow.) Let us have a closer look at a each line. Line (10) is clear: We want to know if M gives correct answers for all (x, x_1, x_2) with $|x| \leq n$, $|x_1| \leq p_1(|x|)$ and $|x_2| \leq p_2(|x|)$. In line (11) it is guessed, how many queries ($= c$) and which queries ($= \{l_1, \ldots, l_c\}$) are made by $M(x, x_1, x_2)$ and what the answers to these queries are ($= \{o_1, \ldots, o_c\}$). In (12) it is checked whether the queries were guessed correctly: Is l_1 the first query of $M(x, x_1, x_2)$? If M gets answer o_1 on that query, is the next query l_2? And so on. In line (13) it is tested, if the guessed answers o_i are correct with respect to $f(S)$. In lines (14)-(15) it is checked, if $M(x, x_1, x_2)$ actually computes the correct answer.

Now let us see what the complexity of O is. Line (12) can be decided in P. Line (13) can be rewritten as

$$\left[\bigwedge_i^{i \leq c} o_i = 1 \to f_2(l_i) = f_2(u) \wedge \bigwedge_v^{v \leq f_2(l_i)} f_1(l_i, v) = f_1(u, v) \right] \quad (16)$$

$$\wedge \left[\bigwedge_i^{i \leq c} o_i = 0 \to \bigwedge_{u \in S} \left[f_2(l_i) \neq f_2(u) \vee \bigvee_v^{v \leq f_2(l_i)} f_1(l_i, v) \neq f_1(u, v) \right] \right]. \quad (17)$$

Line (16) can be checked in $coNP$. Line (17) can be decided in NP since $\bigwedge_i^{1 \leq i \leq c}$ and $\bigwedge_{u \in S}$ quantify only over polynomially many choices. Lines (14) and (15) are in Σ_2^P and Π_2^P respectively. The longest chain of alternating quantifiers is thus $\bigwedge_{x, x_1, x_2} \bigvee_{c, l_i, o_i} \bigwedge_{x_3} \bigvee_{x_4}$ from lines (10), (11) and (15). Therefore, O seems to be a Π_4^P-predicate.

But we can improve that to Π_3^P. Just move the $\bigwedge_{x_3}^{P3}$-quantifier from line (15) to the beginning of line (11), as is indicated by the arrow. Certainly, the resulting predicate is in Π_3^P. Since the computation $M(x, x_1, x_2)$ implicitly depends on the values o_i, this shift seems to change the semantics of O. The following argument shows that it does not:

If we move the $\bigwedge_{x_3}$-quantifier from line (15) to the end of line (11) the semantics are not changed since x_3 does not occur in between. Now, the crucial observation is that for fixed S, n, x, x_1, x_2 there is *exactly one* choice for c, l_i, o_i that makes lines (12) and (13) true. We call them $c(S, n, x, x_1, x_2)$, $l_i(S, n, x, x_1, x_2)$, and $o_i(S, n, x, x_1, x_2)$. Thus, we can replace the existential quantifiers $\bigvee_c, \bigvee_{l_i}, \bigvee_{o_i}$ in line (11) with

$$c = c(S, n, x, x_1, x_2), o_i = o_i(S, n, x, x_1, x_2), l_i = l_i(S, n, x, x_1, x_2). \quad (18)$$

Since that term does not depend on x_3 we can further move the $\bigwedge_{x_3}$-quantifier past it, i.e. from its current position at the end of line (11) to the beginning of that line. Substituting $\bigvee_c, \bigvee_{l_i}, \bigvee_{o_i}$ back for (18) establishes that O is indeed a Π_3^P-predicate.

Since K is polylog-sparse and the computation time of M is bounded by r, we conclude: For each input length n there must be a set S, $\bigwedge_{s \in S} |s| \leq r(n)$ and $|S| \leq t(q((r(n)))$, such that $O(S, 0^n)$ is satisfied. Thus, sets S with $O(S, 0^n)$ can be non-deterministically guessed in polynomial time. We get

$$x \in L \leftrightarrow \bigvee_{\substack{S=\{s_1,\ldots,s_{t(q(r(|x|)))}\} \\ |s_i| \leq r(|x|)}} O(S, 0^{|x|})$$

$$\wedge \bigwedge_{x_1}^{p_1} \bigvee_{x_2}^{p_2} [\text{ lines } (11) - (13)$$

$$\wedge\, M(x, x_1, x_2) \text{ accepts}],$$

which proves $L \in \Sigma_4^P$. □

It is conceivable that a different approach in the proof collapses the polynomial hierarchy to a lower level than Σ_4^P.

Acknowledgements

I would like to thank Gerd Wechsung for supervising my Diploma Thesis on leaf languages and Harry Buhrman for hints on this paper.

References

[1] D. P. Bovet, P. Crescenzi, R. Silvestri, *A Uniform Approach to Define Complexity Classes*, Theoretical Computer Science 104, 1992, pp. 263-283.
[2] J.-Y. Cai, $S_2^P \subseteq ZPP^{NP}$, FOCS 2001, pp. 620-629
[3] R. Canetti, *More on BPP and the polynomial-time hierarchy*, Information Processing Letters 57, 1996, pp. 237-241
[4] J.-Y. Cai, M. Ogihara, *Sparse Sets versus Complexity Classes*, Complexity Theory Retrospective II by L.A. Hemaspaandra, A.L. Selman (Eds.), 1997, pp. 53-80
[5] S. Homer, L. Longpré *On reductions of NP sets to sparse sets*, Journal of Computer and System Sciences 48(2), 1994, pp. 324-336
[6] U. Hertrampf, C. Lautemann, T. Schwentick, H. Vollmer, K. Wagner, *On the Power of Polynomial Time Bit-Reductions*, Proc. 8th Structure in Complexity Theory Conference, 1993, pp. 200-207.
[7] U. Hertrampf, H. Vollmer, K. Wagner, *On Balanced vs. Unbalanced Computation Trees*, Mathematical Systems Theory 29, 1996, pp. 411-421.
[8] B. Jenner, P. McKenzie, D. Therien, *Logspace and Logtime Leaf Languages*, Information and Computation 141, 1996, pp. 21-33
[9] J. Kadin $P^{NP[\log n]}$ *and sparse Turing-complete sets for NP*, Journal of Computer and System Sciences 39, 1989, pp. 282-298
[10] S. Mahaney, *Sparse complete sets for NP: Solution of a conjecture of Berman and Hartmanis*, Journal of Computer and System Sciences 25(2), 1982, pp. 130-143
[11] M. Ogiwara, O. Watanabe, *On polynomial time bounded truth-table reducibility of NP sets to sparse sets*, SIAM JC 20 1991, 471-483
[12] C. H. Papadimitriou, *Computational Complexity*, Addison Wesley 1994
[13] A. Russell, R. Sundaram, *Symmetric alternation captures BPP*, Computational Complexity 7(2), 1998, pp. 152-162
[14] N. Vereshchagin, *Relativizable and Nonrelativizable Theorems in the Polynomial Theory of Algorithms*, Russian Acad. Sci. Izv. Math. 42, 1994, pp. 261-298

Explicit Inapproximability Bounds for the Shortest Superstring Problem

Virginia Vassilevska[*]

Carnegie Mellon University, Pittsburgh PA 15213, USA
virgi@cs.cmu.edu

Abstract. Given a set of strings $S = \{s_1, \ldots, s_n\}$, the *Shortest Superstring* problem asks for the shortest string s which contains each s_i as a substring. We consider two measures of success in this problem: the *length* measure, which is the length of s, and the *compression* measure, which is the difference between the sum of lengths of the s_i and the length of s. Both the length and the compression versions of the problem are known to be MAX-SNP-hard. The only explicit approximation ratio lower bounds are by Ott: 1.000057 for the length measure and 1.000089 for the compression measure. Using a natural construction we improve these lower bounds to 1.00082 for the length measure and 1.00093 for the compression measure. Our lower bounds hold even for instances in which the strings are over a binary alphabet and have equal lengths. In fact, we show a somewhat surprising result, that the Shortest Superstring problem (with respect to both measures) is as hard to approximate on instances over a binary alphabet, as it is over any alphabet.

1 Introduction

Given a set of strings over some alphabet, the Shortest Superstring problem asks for the shortest string over the same alphabet which contains each of the given strings as a substring. The problem was first shown to be NP-hard by Maier and Storer [7]. As an optimization problem, it has two optimization measures: the length of the resulting superstring, and the compression, which is the difference between the sum of lengths of the given strings and the length of the superstring. The Shortest Superstring problem was shown by Blum et al. [4] to be MAX-SNP-hard with respect to both measures (over an unbounded alphabet), which implies that unless $P = NP$, there exists some $\epsilon > 0$ for which it is hard to approximate the optimal superstring to within a factor better than $(1 + \epsilon)$. Ott gave explicit approximation ratio lower bounds (assuming $P \neq NP$) for Shortest Superstring instances over a binary alphabet [12]. These ratios (1.000057 for the length measure and 1.000089 for the compression measure) are far from the best known upper bounds for the problem: 2.5 for the length measure by Sweedyk [13] and 1.625 for the compression measure by Bläser [3].

Using a natural reduction, we show a relationship between the approximability of the Vertex Cover and Shortest Superstring problems. Given a constant

---
[*] Supported by the NSF ALADDIN Center (NSF Grant No. CCR-0122581).

approximation ratio lower bound for a class of graphs for which the optimal vertex cover is linear in the number of edges, we can obtain an inapproximability constant for the Shortest Superstring problem on equal length strings. Berman and Karpinski [2] and Karpinski [8] gave a series of inapproximability results for Vertex Cover on bounded degree graphs. Here we use the result for graphs of degree at most 5 to get that, unless $P = NP$, the Shortest Superstring problem is not 1.00082-approximable with respect to the length measure, and not 1.00093-approximable with respect to the compression measure. Notice that these constants, although small, improve on Ott's result by an order of magnitude. Moreover, these results have potential for much improvement if different inapproximability results for Vertex Cover are used.

Most hardness results for the Shortest Superstring problem, except Ott's result, are for instances over an unbounded alphabet. In fact, Ott [12] stresses that their result is the first APX-hardness result for instances over a binary alphabet. Small size alphabet instances are of interest because of their immediate relation to DNA sequencing, where an alphabet of size 4 (A,T,G,C) is used. The hardness of Shortest Superstring over a binary alphabet does not imply, however, that Shortest Superstring is not easier to approximate on smaller alphabet instances than in general. In this paper we show that the problem on a binary alphabet is just as hard to approximate as the general case, $i.e.$ if one can approximate Shortest Superstring over a binary alphabet by a factor α in polynomial time, then the problem over any (finite) alphabet can be approximated by a factor α in polynomial time.

2 Preliminaries

In this section we define some terminology we will need.

Definition 1. *Given an alphabet Σ, a string over Σ is an element of Σ^*. Given two strings $s = s_1 \ldots s_m$ and $t = t_1 \ldots t_k$ over Σ, s is said to be a substring of t, if $|s| \leq |t|$ and there exists a j: $0 \leq j \leq k - m$ so that for every i: $1 \leq i \leq m$, $s_i = t_{j+i}$. t is said to be a superstring of s iff s is a substring of t. s is said to overlap to the right with t if there exists a j: $0 \leq j \leq m - 1$ so that for every i: $1 \leq i \leq m - j$, $s_i = t_{j+i}$. Then t is said to overlap to the left with s.*

Now consider the following procedure, $Induced(\pi, S)$, which given a set of n strings, $S = \{s_1, \ldots, s_n\}$, and a permutation $\pi \in S_n$, greedily builds a superstring of S:

$Induced(\pi, S)$:
    $s \leftarrow s_{\pi(1)}$
    for i from 2 to n
        $s \leftarrow$ string obtained by maximally overlapping s to the right with $s_{\pi(i)}$

Definition 2. *Let Σ be an alphabet, $S = \{s_1, \ldots, s_n\} \subset \Sigma^*$ be a set of n strings over Σ, and $\pi \in S_n$. We say π induces a superstring s_π on S if $s_\pi =$*

$Induced(\pi, S)$. Define $ov(\pi, S)$ to be the amount of overlap induced by π on S, i.e. $ov(\pi, S)(\sum_{i=1}^{n}|s_i|) - |Induced(\pi, S)|$.

Intuitively, the superstring induced on S by π is obtained by sequentially overlapping the strings in the order given by π.

Definition 3. *Given an alphabet Σ and a set of strings $S = \{s_1, \ldots, s_n\} \subset \Sigma^*$ such that no string in S is a substring of another string in S, the Shortest Superstring problem asks for the shortest string s which is a superstring of every $s_i \in S$. In terms of optimization, the* length *measure minimizes $|s|$, and the* compression *measure maximizes $(\sum_{i=1}^{n}|s_i|) - |s|$. When the compression measure is used, the problem is often referred to as the* maximum compression *problem.*

Note that the shortest superstring is the shortest length superstring over all superstrings induced by a permutation from S_n on S. Since the Shortest Superstring problem is defined on finite strings, here we consider the alphabet for the superstring instance to consist solely of characters occurring in the strings. With this definition, the alphabet size is always bounded by the sum of the string lengths.

3 Binary Alphabet Shortest Superstring

Until now the size of the underlying alphabet has been assumed to make a difference in the approximability of the Shortest Superstring problem. This may be related to the fact that a related problem, Shortest Common Supersequence, seems to be easier on instances over a small alphabet. For example, Jiang and Li [6] give an algorithm for Shortest Common Supersequence with an approximation ratio directly related to the size of the alphabet. In the next theorem we show that in the case of the Shortest Superstring problem, the alphabet size does not affect the approximability of the problem.

Theorem 1. *Suppose the Shortest Superstring problem can be approximated by a factor α on instances over a binary alphabet (with respect to either measure). Then the Shortest Superstring problem can be approximated by a factor α on instances over any alphabet.*

Proof. Given an alphabet $A = \{a_1, \ldots, a_k\}$ of size k, associate with a_i the binary string $s_i = 0^i(01)^{(k+1-i)}1^i$. Notice that if $i \neq j$, s_i does not overlap with s_j, and that the only way s_i overlaps with itself is by its whole length. Furthermore, all s_i have the same length, $2(k+1)$.

Consider an instance $T = \{t_1, \ldots, t_n\}$ over A and the instance $T' = \{t'_1, \ldots, t'_n\}$ obtained by substituting s_i for a_i. As noted earlier, we take A to contain only the characters present in the strings of T, so $|A| = k \leq \sum_i |t_i|$.

For all permutations $\pi \in S_n$ let s_π and s'_π be the superstrings induced by π on T and T' respectively. Then $|s'_\pi| = 2(k+1)|s_\pi|$. In particular, $|s'_{opt}| =$

$2(k+1)|s_{opt}|$ for the optimal permutation opt (in terms of the length measure for T). And if $|s'_\pi| \leq \alpha |s'_{opt}|$, we have

$$|s_\pi| = \frac{|s'_\pi|}{2(k+1)} \leq \frac{\alpha |s'_{opt}|}{2(k+1)} = \alpha |s_{opt}|.$$

For all permutations $\pi \in S_n$, let $ov_\pi = ov(\pi, T)$ and $ov'_\pi = ov(\pi, T')$. Then $ov'_\pi = 2(k+1) \cdot ov_\pi$. As above we have $ov'_{opt} = 2(k+1) \cdot ov_{opt}$ for the optimal (now in terms of overlaps) permutation opt. For $ov'_\pi \leq \alpha \cdot ov'_{opt}$,

$$ov_\pi = \frac{ov'_\pi}{2(k+1)} \leq \frac{\alpha \cdot ov'_{opt}}{2(k+1)} = \alpha \cdot ov_{opt}.$$

Hence if we have an α-approximation algorithm for the Shortest Superstring problem on binary strings, then we can use it to get an α-approximation for Shortest Superstring instances over any alphabet. The running time of the algorithm is polynomial since k is at most linear in the length of the input and the transformation can be carried out in polynomial time. □

4 Approximation Ratio Lower Bounds

In this section we derive explicit approximation ratio lower bounds (assuming $P \neq NP$) for the Shortest Superstring problem restricted to instances with equal length strings. We do this by a reduction from Vertex Cover, and by using the following theorem of Berman and Karpinski [2]:

Theorem 2 ([2]). *For any $0 < \epsilon < \frac{1}{2}$ it is NP-hard to decide whether an instance of Vertex Cover with $140n$ nodes and maximum degree at most 5 has its optimum above $(73 - \epsilon)n$ or below $(72 + \epsilon)n$.*

Moreover we will need the following fact concerning the reduction:

Claim. The Vertex Cover instances in Theorem 2 have at most $286n$ edges.

Proof of Claim: The instances in the reduction used in the proof of the theorem above have at most $12n$ nodes of degree 5, and the rest of the nodes have degree at most 4. Hence the instances considered have at most $30n + 256n = 286n$ edges. □

We are now prepared to derive the inapproximability bounds.

Theorem 3. *For any $\epsilon > 0$, unless $P = NP$, Shortest Superstring on instances with equal length strings is not approximable in polynomial time within a factor of*
- *$1.00082 - \epsilon$ with respect to the length measure, and*
- *$1.00093 - \epsilon$ with respect to the compression measure.*

Proof. Suppose we are given an instance of Vertex Cover $G = (V, E)$ with $|E| = m$. Let our alphabet contain a letter a for each vertex $a \in V(G)$, and our strings be $abab$ and $baba$ for each edge $e = (a, b)$.

Suppose G has a vertex cover S of size k. Then, assign each edge to one of its end points which is in S. If $e = (a, b)$ was assigned to a, then overlap $abab$ (to the left) with $baba$ to obtain $ababa$. Otherwise overlap them in the opposite order to obtain $babab$.

For every $b \in S$, consider all edges (a_i, b) assigned to b. For each one of these edges we have a string of the form ba_iba_ib. By consecutively overlapping these strings by 1 letter, we obtain a string s_b. By concatenating the s_b strings for all $b \in S$ we obtain a string s.

Claim. The string s has length $4m + k$.

Proof of Claim: The sum of the lengths of the original strings is $8m$ and each two strings corresponding to the same edge are overlapped by 3 symbols. This gives a total of $5m$ for the strings of the form $babab$. If all of these were overlapped by one letter we would get a compression of $(m - 1)$ since there are m of these strings. However, since the vertex cover is of size k, there are k groups with no overlap between them. So the length of the superstring is actually longer by $(k - 1)$ symbols. We have $|s| = 5m - (m - 1) + (k - 1) = 4m + k$ □

Now suppose for some $k \geq 1$ we have a superstring of length $4m + k$. First we show that wlog we may assume for every $(a, b) \in E$ that either $abab$ is overlapped to the right with $baba$ or vice-versa.

Suppose some $abab$ and $baba$ are not overlapped with each other. We will construct a new superstring of length $\leq 4m + k$ such that they do overlap.

Consider the permutation π of the strings which induces the superstring. Wlog, $abab$ occurs before $baba$, in π. In the worst case, there is a string $ba'ba'$ right after $abab$ overlapping with $abab$ to the right, and there is a string $a''ba''b$ right before $baba$ overlapping with $baba$ to the left (where clearly $a' \neq a \neq a''$). We can break these two overlaps moving all strings between $abab$ and $baba$ to the end of the permutation (without breaking any other overlaps), and then overlapping $abab$ with $baba$, for an overall gain of 1 letter overlap. After doing this for all edges we get a superstring of no greater length in which for each $(a, b) \in E$ either $abab$ is overlapped to the left with $baba$ or vice-versa.

After this transformation, the superstring s' is a concatenation of strings of the form

$$a_s \prod_i (b_i a_s b_i a_s),$$

where $\prod$ stands for iterated concatenation.

For an edge $(a, b) \in E$, if $abab$ overlaps to the left with $baba$, then put a in the vertex cover S. S is clearly a vertex cover since we used a vertex for each edge. If s' consists of t strings of the above form, $S = \bigcup_s a_s$ and $|S|$ is at most t. The length of the superstring is

$$5m - (m - 1) + (t - 1) = 4m + t$$

by an argument similar to the one given earlier. Since the length of the superstring did not increase due to our manipulations,

$$4m + t \leq 4m + k,$$

which yields $t \leq k$.

Therefore G has a vertex cover of size k iff the string set has a superstring of size $4m + k$.

Now we prove the inapproximability bounds. By Theorem 2, we have that for any $0 < \epsilon < \frac{1}{2}$ it is NP-hard to decide whether an instance of Vertex Cover with $140n$ nodes and at most $286n$ edges has its optimum above $(73 - \epsilon)n$ or below $(72 + \epsilon)n$.

Hence for Shortest Superstring on $2m \leq 572n$ strings of length 4 it is NP-hard to distinguish whether there is a superstring of length below $4m + (72+\epsilon)n$ or above $4m + (73 - \epsilon)n$. So if Shortest Superstring can be approximated within an α factor, then

$$\alpha \geq \frac{4m + (73 - \epsilon)n}{4m + (72 + \epsilon)n}.$$

Taking limits on both sides we get

$$\alpha \geq \lim_{\epsilon \to 0} \frac{4m + (73 - \epsilon)n}{4m + (72 + \epsilon)n} = \frac{4m + 73n}{4m + 72n} = 1 + \frac{1}{4\frac{m}{n} + 72}$$

But $4\frac{m}{n} \leq 286 \times 4 = 1144$ and so

$$\alpha \geq 1 + \frac{1}{1216} \geq 1.00082$$

Therefore, for any $\epsilon > 0$, Shortest Superstring on instances with equal length strings cannot be approximated within a factor of $1.00082 - \epsilon$, with respect to the length measure, unless $P = NP$.

When the length of the superstring is $4m+k$, the compression is $8m - (4m+k) = 4m - k$. So for the maximum compression on the strings from our reduction it is NP-hard to decide whether the optimum compression is above $4m - (72+\epsilon)n$ or below $4m - (73 - \epsilon)n$. If the compression can be approximated by a factor β, then

$$\beta \geq \frac{4m - (72 + \epsilon)n}{4m - (73 - \epsilon)n}$$

Taking limits on both sides,

$$\beta \geq \lim_{\epsilon \to 0} \frac{4m - (72 + \epsilon)n}{4m - (73 - \epsilon)n} = \frac{4m - 72n}{4m - 73n} =$$

$$= 1 + \frac{1}{\frac{4m}{n} - 73} \geq 1 + \frac{1}{1071} \geq 1.00093$$

Hence for any $\epsilon > 0$, Shortest Superstring on instances with equal length strings cannot be approximated within a factor of $1.00093 - \epsilon$, with respect to the compression measure, unless $P = NP$. □

Using Theorems 1 and 3 we also get

Corollary 1. *For any $\epsilon > 0$, unless $P = NP$, Shortest Superstring on instances with equal length binary strings is not approximable in polynomial time within a factor of*
- *$1.00082 - \epsilon$ with respect to the length measure, and*
- *$1.00093 - \epsilon$ with respect to the compression measure.*

5 Conclusion

We have derived explicit approximation ratio lower bounds for the Shortest Superstring problem, when restricted to instances with equal length strings. These bounds are far from the best known upper bounds for the Shortest Superstring problem. The reduction given in this paper presents a promising avenue for improving the lower bounds further, since any better bounds for a class of Vertex Cover instances with an optimum linear in the number of edges immediately improves our result. This of course would only give lower bounds for the restricted version of Shortest Superstring, which may be weaker than the best lower bounds for the general problem. It is an interesting question whether Shortest Superstring on equal length strings is easier in terms of approximation than the general Shortest Superstring problem.

We have also shown that the alphabet size does not affect the approximability of Shortest Superstring. It is an open problem whether a similar result can be obtained for the related Shortest Common Supersequence problem, which is also known to be MAX-SNP-hard over a binary alphabet [11]. Our reduction exploited a property of the Shortest Superstring problem which is not present in the Shortest Common Supersequence problem. Hence, if the alphabet size does not affect the approximability of Shortest Common Supersequence, then proving this may require very different ideas from ours.

Our result on the alphabet importance for Shortest Superstring is significant since the main application of the Shortest Superstring problem is in DNA sequencing where the alphabet has only 4 symbols. Until now it was assumed that because of this restriction the real-world applications of the problem may be much better approximable. In this paper we have refuted this hope. But we also shed light on a very natural relation between Shortest Superstring and Vertex Cover, a problem which has been well-studied. Moreover, we conjecture that the relation between the two problems is much tighter than our reduction indicates, and that if Vertex Cover is not 2-approximable, then neither is Shortest Superstring.

Acknowledgments

I would like to thank Ryan Williams, Maverick Woo and my advisor Guy Blelloch for numerous helpful discussions, Avrim Blum for suggesting the Shortest Superstring problem, Uriel Feige for several insightful observations, and the three anonymous reviewers for their comments.

References

1. C. Armen, C. Stein, Short Superstrings and the Structure of Overlapping Strings. J. Comput. Biol. **2** (2) (1995) 307–332
2. P. Berman, M. Karpinski, On Some Tighter Inapproximability Results (Extended Abstract). ICALP (1999) 200–209
3. M. Bläser, An 8/13–Approximation Algorithm for the Asymmetric Maximum TSP. SODA (2002) 64–73
4. A. Blum, T. Jiang, M. Li, J. Tromp, and M. Yannakakis, Linear Approximation of Shortest Superstrings. STOC (1991) 328–336
5. D. Breslauer, T. Jiang and Z. Jiang, Rotations of Periodic Strings and Short Superstrings. J. Algorithms. **24** (2) (1997) 340–353
6. T. Jiang, M. Li, On the Approximation of Shortest Common Supersequences and Longest Common Subsequences. SIAM J. Comput. **24** (5) (1995) 1122–1139
7. D. Maier and J.A. Storer. A Note on the Complexity of the Superstring Problem. Princeton University Technical Report 233, Department of Electrical Engineering and Computer Science (1977)
8. M. Karpinski, Approximating Bounded Degree Instances of NP-hard Problems. Proc. 13th Symp. on Fundamentals of Computation Theory, LNCS 2138, Springer **10** (2001) 24–34
9. H. Kaplan, N. Shafrir, The Greedy Algorithm for Shortest Superstrings. Information Processing Letters **93** (2005) 13–17
10. M. Middendorf, More on the Complexity of Common Superstring and Supersequence Problems. Theoretical Computer Science **125** (2) (1994) 205–228
11. M. Middendorf, On Finding Various Minimal, Maximal, and Consistent Sequences over a Binary Alphabet. Theoretical Computer Science **145** (1995) 317–327.
12. S. Ott, Lower Bounds for Approximating Shortest Superstrings over an Alphabet of Size 2. WG (1999) 55–64
13. Z. Sweedyk, A $2\frac{1}{2}$–Approximation Algorithm for Shortest Superstring. SIAM J. Comput. **29** (3) (1999) 954–986
14. J. Tarhio and E. Ukkonen, A Greedy Approximation Algorithm for Constructing Shortest Common Superstrings. Theoretical Computer Science **57** (1988) 131–145
15. J.S. Turner, Approximation Algorithms for the Shortest Common Superstring Problem. Information and Computation **83** (1989) 1–20

Stratified Boolean Grammars

Michał Wrona

Institute of Computer Science, University of Wrocław

Abstract. We study Boolean grammars. We introduce stratified semantics for Boolean grammars. We show, how to check, if a Boolean grammar generates a language according to this semantics. We show, that stratified semantics covers a class of important and natural languages. We introduce a recognition algorithm for Boolean grammars compliant to this semantics.

1 Introduction

As it is widely known, context free grammars are too weak to define many important formal languages and in consequence a real programming language. Therefore many various generalizations of them were introduced, inter alia, Boolean grammars of Alexander Okhotin [5]. BG are usual context-free grammars, where in productions, operations of intersection and completion are allowed. Rules of Boolean grammars look as follows:

$$X \to \alpha_1 \& \ldots \& \alpha_m \& \neg \beta_1 \& \ldots \& \neg \beta_n \tag{1}$$

In the case of CFG, a language, generated by a grammar G is an element of a vector of the least solution of a corresponding system of language equations. But after introducing negation such a system need not to have the least solution. Therefore other semantics are necessary. In [5], *semantics of a naturally reachable solution* and *semantics of a unique solution in the strong sense* were introduced. However, for both of them the problem, whether a grammar generates a language according to this semantics is undecidable (co-**RE**-complete).

In Sect. 2, we introduce new semantics, for which such a problem can be decided in linear time. A grammar is compliant to the *stratified semantics*, if there exists a function, that orders linearly a set of nonterminals in the following way. If a nonterminal X occurs negatively in the production, that defines a nonterminal Y, then a nonterminal Y cannot occur in a production, that defines X. A Boolean grammar that is compliant to the stratified semantics is called a *stratified Boolean grammar*. Let us note, that stratification is a natural way of dealing with negation. See for example [2]. In this section, we also argue, that a class of stratified Boolean grammars generates a set of natural and important languages. As an example we show a stratified Boolean grammar, that generates $\{ww|w \in \Sigma^*\}$. To our knowledge, the only language, for which a stratified Boolean grammar is not known, is a quite artificial one namely $\{a^{2^n}|n >= 0\}$ [5]. Moreover, in [4], there is shown a Boolean grammar, that generates proper programs of a simple programming language. As it is not difficult to check, this grammar is compliant to the stratified semantics.

In Sect. 3, we show, that there is a close relation between the new semantics and the old ones. In particular we show, that every stratified Boolean grammar is also compliant to the semantics of a naturally reachable solution and for such a grammar, languages defined by both of these semantics are equal. We also see, that if a stratified Boolean grammar is compliant to the semantics of a unique solution in the strong sense, then the languages defined by both of these semantics are equal. Of course not every grammar, that is compliant to one of the old semantics is also compliant to the stratified one.

In [5] there is also presented a recognition algorithm for Boolean grammars. If the entry Boolean grammar is in a binary normal form – a generalization of Chomsky normal form for Boolean grammars – then this algorithm works in usual time $\Theta(|G| |w|^3)$. In the other case the transformation to this normal form, also presented there, can blow up the size of the entry grammar exponentially. So, for an arbitrary Boolean grammar this algorithm together with the transformation works in time $\Omega(2^{|G|} |w|^3)$.

Fortunately the stratified semantics gives us an opportunity to avoid the exponentially blowup. To obtain this, in Sect. 4 we introduce a new normal form and develop a recognition algorithm for stratified Boolean Grammars. It appears, that transformation to our normal form can increase size of an entry grammar only $|G| \log(|G|)$ times and our algorithm works in usual time $O(|G| |w|^3)$. So, our algorithm together with the transformation works in time $O(|G| \log(|G|) |w|^3)$, thus avoiding the exponentially blowup.

2 Stratified Semantics

In this section, we define a notion of a Boolean grammar, stratified semantics for Boolean grammars and show an algorithm, that for a Boolean grammar, checks if it is compliant to this semantics.

2.1 Stratified Boolean Grammars

First, we define notions of a language formula and of a value of a language formula. Against as it was defined in [5], we extend these definitions by introducing a set of language symbols and their interpretation. Language symbol is a generalization of a terminal i.e., it is a symbol, that can stand not just for one letter, but for an arbitrary language. We do not see any practical purposes for this extension, but as we see below it simplifies some reasonings connected with stratified semantics.

Definition 1. *(A Language formula) Let Σ be a finite nonempty alphabet, $N = (X_1, ..., X_n)$ a finite nonempty set of variables (nonterminals) and $\Phi = (\mathcal{L}_1, ..., \mathcal{L}_m)$ a finite set of language symbols. Language formula over Σ and Φ in variables N is defined inductively as follows:*

- the empty string ϵ is a formula
- any symbol from $\Sigma \cup \Phi$ is a formula
- any variable from X is a formula
- if ϕ and ψ are formulas, then $(\phi \cdot \psi)$, $(\phi \& \psi)$, $(\phi \vee \psi)$ and $(\neg \phi)$ are formulae.

Definition 2. (*Value of a formula*) Let ψ be a formula over $\Sigma \cup \Phi$ in variables $N = (X_1, ..., X_n)$. Let $L = (L_1, ..., L_n)$ be a set of languages over Σ and $\mathcal{I} : \Phi \to 2^{\Sigma^*}$ an interpretation of language symbols from Φ. The value of the formula ψ on the set of languages L, denoted as $\eta(L)$ is defined inductively on the structure of η:

- $\epsilon(L) = \epsilon$
- $a(L) = a$ for every $a \in \Sigma$
- $L(\mathcal{L}) = I(\mathcal{L})$ for every $\mathcal{L} \in \Phi$
- $X_i(L) = L_i$ for every $(1 \leq i \leq n)$
- $(\psi \phi)(L) = \psi(L) \cdot \phi(L)$
- $(\psi \vee \phi)(L) = \psi(L) \cup \phi(L)$
- $(\psi \& \phi)(L) = \psi(L) \cap \phi(L)$
- $(\neg \psi)(L) = \Sigma^* \setminus \psi(L)$.

The value of a vector of formulae $\psi = (\psi_1, ..., \psi_l)$ on a vector of languages $L = (L_1, ..., L_n)$ is a vector of languages $\psi(L) = (\psi_1(L), ..., \psi_l(L))$.

Now we define a special normal form of a language formula i.e., a concatenation normal form.

Definition 3. (*A concatenation normal form*) We say, that a language formula ψ is in a concatenation normal form, if it is one of the following:

1. ϵ, a terminal, a nonterminal, a language symbol,
2. concatenation of symbols from point 1,
3. a Boolean formula, where every Boolean variable is replaced by an expression from point 2.

If it is not stated differently, from now on, we assume, that every considered formula is in a concatenation normal form. Now, we introduce a notion of a (new) Boolean grammar. We name it new, because productions of our Boolean grammar are a little bit different (more general) from those of the Boolean grammar, introduced in [5].

Definition 4. (*A (new) Boolean grammar*) A (new) Boolean grammar is $G = (\Sigma, N, P, S, \Phi, I)$, where Σ is an alphabet, N a set of nonterminals, P is a set of productions of the form $X \to \psi_X(N)$, where $X \in N$ and $\psi_X(N)$ is a language formula over Σ and Φ in variables N. S is a start symbol.

A (new) Boolean grammar without language symbols is sometimes denoted by $G = (\Sigma, N, P, S)$.

Because of the fact, that we allow the operator $\vee$, we can assume, that for every nonterminal $X \in N$, there is only one production $\psi_X(N)$. Now, we define a system of language equations corresponding to a (new) Boolean grammar.

Definition 5. *(A system of equations corresponding to a grammar)* Let $G = (\Sigma, N, P, S, \Phi, I)$ be a (new) Boolean grammar, then a system of language equations corresponding to G is defined as follows: for every production of the form $X \to \psi_X(N)$ we have an equation of the form $X = \psi_X(N)$.

Let $N' \subseteq N$, then by $N' = \psi(N)$, we denote a system $N = \psi(N)$, restricted to equations, such that their left side is a variable, that belongs to N'.

Roughly speaking, a vector of languages is a solution of a system of language equations if and only if an application of this vector to left-hand sides is equal to an application of this vector to right-hand sides of this system (see Definition 2).

Definition 6. *(Solution of a system of equations)* We say that a vector of languages $L = (L_1, \ldots, L_n)$ is a solution of a system of language equations $N = \psi(N)$, where $N = (X_1, \ldots, X_n)$ and $\psi(N) = (\psi_{X_1}(N), \ldots, \psi_{X_n}(N))$ if and only if $L(N) = L(\psi(N))$.

Definition 7. *(Solution modulo)* We say that a vector of languages $L = (L_1, \ldots, L_N)$ is a solution of a system of language equations $N = \psi(N)$ modulo some set M if and only if $L_i(X_i) \cap M = L_i(\psi(X_i)) \cap M$ for every $1 \leq i \leq n$.

Now, after one more auxiliary definition, we introduce notions of stratified semantics and of a stratified Boolean grammar.

Definition 8. Let $G = (\Sigma, N, P, S, \Phi, I)$ be a (new) Boolean grammar. Let $\psi_X(N)$ be a language formula and let $X \in N$. Then $Y \in N$ appears positively in $\psi_X(N)$ if and only if every occurrence of Y is in the scope of even number of negations. If Y occurs in $\psi_X(N)$ and does not occur positively, then it occurs negatively.

Definition 9. *(Stratified Semantics)* We say that a (new) Boolean grammar $G = (\Sigma, N, P, S, \Phi, I)$ is compliant to the stratified semantics if there exists function $\mathcal{F}: N \to \mathbf{Nat}$, such that for each production $X \to \psi_X(N)$ holds the following:

- For every nonterminal Y, that occurs positively in $\psi_X(N)$, it holds: $\mathcal{F}(X) \geq \mathcal{F}(Y)$.
- For every nonterminal Y, that occurs negatively in $\psi_X(N)$, it holds: $\mathcal{F}(X) > \mathcal{F}(Y)$.

Then, we name $\mathcal{F}$ a stratifying function for G and $G = (\Sigma, N, P, S, \Phi, I, \mathcal{F})$ a stratified Boolean grammar.

For every $1 \leq i \leq max(\{k | \mathcal{F}^{-1}(k) \neq \emptyset\})$ a set of nonterminals $\mathcal{F}^{-1}(i)$ we name a stratum.

Now, we give a definition of a stratified solution of a system of language equations. We define it by dual induction with a notion of a system of equations, that defines k-th stratum.

Definition 10. *(Stratified solution of a system of equations)* Let $N = \psi(N)$ be a system of language equations corresponding to a stratified Boolean grammar $G = (\Sigma, N, P, S, \Phi, I, \mathcal{F})$. Then $N = \psi(N)$ has a unique stratified solution L_{st}, defined by induction as follows:
- $L_{st}(\mathcal{F}^{-1}(1))$ is the least solution of the system $\mathcal{F}^{-1}(1) = \psi(N)$.
- The system that defines k-th stratum $\mathcal{F}^{-1}(k) = \psi_D(N)$ can be obtained from $\mathcal{F}^{-1}(k) = \psi(N)$ by replacing:
  - each occurrence of $Y \in \mathcal{F}(\{0, ..., k-1\})$ by $\mathcal{L}$, where $\mathcal{L}$ is a new language symbol and $I(\mathcal{L}) = L_{st}(Y)$,
- $L_{st}(\mathcal{F}^{-1}(k))$ is the least solution of the system, that defines k-th stratum.

Definition 11. *(A language generated by a stratified Boolean grammar)* Let $G = (\Sigma, N, P, S, \Phi, I, F)$ be a stratified Boolean grammar and L_{st} be a stratified solution of a system of language equations corresponding to G, then a language generated by G according to stratified semantics, we denote by $L_{st}(G)$ and there holds, that $L_{st}(G) = L_{st}(S)$.

If for some language L, there exists a stratified Boolean grammar G, such that $L(G) = L$, then we call L a stratified language.

An Example of a Stratified Language. Let $G = (\{a,b\}, N, P, S, \mathcal{F})$ be a stratified Boolean grammar, where $N = \{S, A, B, C, D\}$, P consist of the following productions:

$$S \to \neg BA \& \neg AB \& C$$
$$A \to DAD \lor a$$
$$B \to DBD \lor b$$
$$D \to a \lor b$$
$$C \to DDC \lor \epsilon \qquad (2)$$

and $\mathcal{F}$ is defined as follows: $\mathcal{F}(A) = \mathcal{F}(B) = \mathcal{F}(C) = \mathcal{F}(D) = 0$ and $\mathcal{F}(S) = 1$.

Now, we show, how to find a language generated by G. We do it in a way suggested by definitions 10 and 11. Let $N = \psi(N)$ be a system of equations corresponding to G. First, we have to obtain the least solution of the system: $\mathcal{F}^{-1}(1) = \psi(N)$ i.e., $\{A, B, C, D\} = \psi(N)$. It is as follows:

$$L_{st}(D) = \{a, b\}$$
$$L_{st}(C) = \{vw|v, w \in \Sigma^*\}$$
$$L_{st}(A) = \{vaw|v, w \in \Sigma^*\}$$
$$L_{st}(B) = \{vbw|v, w \in \Sigma^*\} \qquad (3)$$

Thereafter, we have to obtain the least solution of a system, that defines a second stratum, in that case, it is a system $S = \psi_D(N)$, i.e,

$$S = \neg \mathcal{L}_A \mathcal{L}_B \& \neg \mathcal{L}_B \mathcal{L}_A \& \mathcal{L}_C, \text{ where } H \in \{A, B, C, D\} \text{ and } I(\mathcal{L}_H) = L_{st}(H) \quad (4)$$

Finally we get:
$$L_{st}(S) = L(G) = \{ww|w \in \{a,b\}^*\} \qquad (5)$$

2.2 Checking If a Boolean Grammar is Compliant to Stratified Semantics

Here, we show, that a set of (new) Boolean grammars (see Definition 4), compliant to the stratified semantics, is recognizable in linear time. The algorithm below is based on the one for partioning set of states of Weakly Alternating Automata [3]. To state this algorithm transparently, we introduce a special structure, induced by a (new) Booleann grammar. But let us first notice, that not every Boolean grammar generates a language. As an example consider a grammar, that contains a production: $A \to B\&\neg B$.

Definition 12. *Let $G = (\Sigma, N, P, S)$ be a (new) Boolean grammar. Then by D, we denote a digraph with set of vertices N and two different set of arcs E^+ and E^-, defined as follows:*

- $E^+ = \{(X,Y) \in N \times N | Y \text{ occurs positively in } X \to \psi_X(N)\}$
- $E^- = \{(X,Y) \in N \times N | Y \text{ occurs negatively in } X \to \psi_X(N)\}$

Algorithm 1. *Let $G = (\Sigma, N, P, S)$ be a (new) Boolean grammar.*

If G is compliant to the stratified semantics, then algorithm returns a stratifying function $\mathcal{F}$ for G, if it is not, then it answers: NO.

1. *Create a structure D corresponding to G, as it is stated in Definition 12.*
2. *Find a digraph $D_{MSCC} = (V_{MSCC}, E_{MSCC})$ of maximal strongly connected components of D.*
3. *If D has a cycle, that contains an arc from E^-, then answer NO.*
4. *If there are no such cycles in D, then compute $\mathcal{F}$ as follows:*
    (a) *Let $(V_{MSCC}, \prec)$ be a partial order induced by D_{MSCC} i.e., $\forall V_1, V_2 \in V_{MSCC}$ we have, that $V_1 \prec V_2$ if and only if $(V_1, V_2) \in E_{MSCC}$.*
    (b) *Return $\mathcal{F}$ compliant to $\prec$, defined as follows:*
        – *for every $X, Y \in V$, where $V \in V_{MSCC}$, it holds, that: $\mathcal{F}(X) = \mathcal{F}(Y)$.*
        – *for every $X \in V_1$ and $Y \in V_2$, where $V_1 \prec V_2$, it holds, that: $\mathcal{F}(X) < \mathcal{F}(Y)$*

Lemma 1. *Algorithm 1 works in time $O(|G|)$.*

Proof. The point 1 can be of course done in $O(|G|)$. Points 2 and 3 can be done by a Tarjan algorithm [6] also in $O(|G|)$. Point 4 in the same time by breadth-first search.

Lemma 2. *A (new) Boolean grammar G is compliant to the stratified semantics if and only if Algorithm 1 returns $\mathcal{F}$.*

Theorem 1. *Set of Boolean grammars compliant to the stratified semantics is recognizable in linear time.*

Proof. It follows simply from lemmas 1 and 2.

3 Relations Between Semantics

In the previous section we introduced new semantics. Here we state some facts about relations between semantics from [5] and the stratified semantics.

3.1 Semantics of a Naturally Reachable Solution

Definition 13. *(A naturally reachable solution)* Let $N = \psi(N)$ be a system of language equations corresponding to some Boolean grammar $G = (\Sigma, N, P, S)$. A solution $L_{nr}(N)$ of a system is called a naturally reachable solution if for every finite set M closed under substring and for every string $u \notin M$ (such that all proper substring of u are in M) every sequence of vectors of the form

$$L^{(0)}, L^{(1)}, ..., L^{(i)}, ... \tag{6}$$

(where $L^{(0)} = (L_1 \cap M, ..., L_n \cap M)$) and every next vector $L^{(i+1)} \neq L^{(i)}$ in the sequence is obtained from the previous vector $L^{(i)}$ by substituting some j-th component with $\psi_j(L^i) \cap (M \cup \{u\})$ converges to $(L_{nr}(N_1) \cap (M \cup \{u\}), ..., L_{nr}(N_n) \cap (M \cup \{u\}))$ in finitely many steps regardless of the choice of components.

Definition 14. *(A naturally reachable language)* We say that a Boolean grammar $G = (N, \Sigma, P, S, \Phi, I)$ generates a language according to the semantics of a naturally reachable solution $L_{nr}(G)$, if a system of language equations $N = \psi(N)$ corresponding to G has a naturally reachable solution $L_{nr}(N)$. Then, it holds, that $L_{nr}(G) = L_{nr}(S)$.

If for a language L, there exists such a Boolean grammar G, compliant to the semantics of a naturally reachable solution, such that $L_{nr}(G) = L$, then we call L a naturally reachable language.

Theorem 2. Let $G = (\Sigma, N, P, S, \mathcal{F})$ be a stratified Boolean grammar, and let $L_{st}(G)$ be a stratified language generated by G. Then G generates also a naturally reachable language $L_{nr}(G)$ and $L_{st}(G) = L_{nr}(G)$.

Because of the fact, that a set of Boolean grammars, compliant to the stratified semantics is recognizable in linear time (see Algorithm 1) and set of Boolean grammars, compliant to the semantics of a naturally reachable solution is co-**RE**-complete, we have the following.

Proposition 1. *Not every Boolean grammar, that is compliant to the semantics of a naturally reachable solution is also compliant to the stratified semantics.*

Let us also note, that from Theorem 2 comes the following proposition.

Proposition 2. *Every stratified language is also a naturally reachable one.*

3.2 Semantics of a Unique Solution in the Strong Sense

Definition 15. *(A unique language in the strong sense)* A Boolean grammar $G = (\Sigma, N, P, S)$ is said to be compliant to the semantics of a unique solution in the strong sense, if for every finite M closed under substrings, the system of language equations $N = \psi(N)$ corresponding to G has a unique solution modulo M. Such a solution, we denote by $L_u(N)$. Then, it holds, that $L_u(G) = L_u(S)$

If for a language L, there exists a Boolean grammar G, compliant to the semantics of a unique solution in the strong sense and $L(G) = L$, we name L a unique language in the strong sense.

From [5] comes also the following propositions.

Proposition 3. *A system of language equations corresponding to a Boolean grammar has a unique solution modulo every language closed under substrings if and only if it has unique solution.*

Proposition 4. *A language L is a naturally reachable language if and only if it is a unique language in the strong sense.*

Lemma 3. *Every stratified language is also a unique language in the strong sense.*

Proof. Lemma comes simply from propositions 2 and 4.

Unfortunately, not every Boolean grammar compliant to the stratified semantics is also compliant to the semantics of a unique solution in the strong sense. As an example consider a grammar with productions: $A \to \neg B$ and $B \to B$. A stratifying function $\mathcal{F}$ for G can be defined as follows: $\mathcal{F}(B) = 0$ and $\mathcal{F}(A) = 1$. But the system of language equations corresponding to this grammar has, of course, many different solutions. Thus, G is not compliant to the semantics of a unique solution in the strong sense. However, we can prove the following theorem.

Theorem 3. *If a stratified Boolean grammar G is compliant to the semantics of a unique solution in the strong sense, then $L_{st}(G) = L_u(G)$*

4 Recognition Algorithm

In this section, we first remind an approach to recognition from [5] and then we show a recognition algorithm for stratified Boolean grammars.

4.1 Previous Approach

In [5] there was shown a recognition algorithm for Boolean grammars, based on the Cocke-Kasami-Younger algorithm and the transformation of an arbitrary Boolean grammar to a binary normal form, that is a generalization of Chomsky normal form. A scheme of this transformation is also based on a scheme of the analogous transformation for CFG. If a Boolean grammar G is in a binary normal form, then the recognition algorithm works in usual time $O(|G|\,|w|^3)$, where w is the entry word. But if G is arbitrary, then during transformation to a binary normal form, its size can blow up exponentially. Therefore, in our opinion, treating a size of a grammar as a constant, like it was done in [5], is not properly. First, we remind a part of transformation, shown in [5] and argue, that only this part can increase a size of a grammar expotentially.

Let us remind, that productions of Boolean grammars in apprehension of [5] are of the form (1).

Definition 16. *(A binary normal form)* We say that a Boolean grammar $G = (\Sigma, N, P, S)$ is in a binary normal form if every production from P is of the form:

1. $X \to a$, where $a \in \Sigma$
2. $X \to Y_1 Z_1 \& ... \& Y_m Z_m \& \neg V_1 Q_1 \& ... \& \neg V_n Q_n$, where for every $1 \le i \le m$ and for every $1 \le j \le n$, we have $Y_i, Z_i, V_j, Q_j \in N$

and G generates a vector of languages $L = (L_1 \ldots L_k)$, such that for every $1 \le l \le k$, we have $\epsilon \notin L_l$.

Here, we show only a part of transformation to a binary normal form i.e., removing unit conjuncts. We say, that a Boolean grammar contains unit conjunct if there exists a production, in which at least one of its conjuncts is a single nonterminal. So, for a grammar G it has to be generated a grammar G', that generates the same language and has no unit conjuncts.

First, it is defined $R = (\Sigma \cup N)^* \setminus N$. Then, for every assignment $W : R \to \{0, 1\}$, it is sought a solution of the system of Boolean equations, where every $\alpha \in R$, is replaced with 0 or 1, in a way compliant to W and every nonterminal is treated like a Boolean variable.

If such a system has more than one solution, this situation is treated like an artificial one and this case is omitted. If there is only one solution L_B of this system and $N' = \{X \in N | L_B(X) = 1\}$. Then for every $X \in N'$, we add to G a production of the form: $X \to \mu_1 \& ... \& \mu_k \& \neg \nu_1 \& ... \& \neg \nu_l$, where $W(\mu_1) = ... = W(\mu_k) = 1$, and $W(\nu_1) = ... = W(\nu_l) = 0$.

Because of the fact, that we have 2^R assignments, and there exists such grammars, that for every such assignment the systems obtained (in the above way) from the systems corresponding to them, have one solution, we have the following.

Proposition 5. *Let G be a Boolean grammar in apprehension of [5] and G' be a grammar, returned by the transformation ,presented above, that removes unit conjuncts, then $|G'| = \Omega(2^{|G|})$.*

4.2 New Approach

Stratified semantics gives an opportunity to develop an algorithm of recognition, that is polynomial against to a size of an entry word as well as to a size of a grammar and works for an arbitrary stratified Boolean grammar.

First, let us introduce a notion of a conjunctive/disjunctive normal form.

Definition 17. *We say that a stratified Boolean grammar $G = (\Sigma, N, P, S, \mathcal{F})$ is in a conjunctive/disjunctive normal form if for every production $X \to \psi_X(N)$, the formula $\psi_X(N)$ is:*

1. *a terminal, a nonterminal, a concatenation of two nonerminals,*
2. *a negation of an expression from point 1,*
3. *a conjunction or a disjunction of expressions from points 1 and 2.*

Moreover, for every nonterminal X it holds, that $\epsilon \notin L_{st}(X)$ i.e., we do not allow generalized ϵ-productions.

To transform a stratified Boolean grammar into a disjunctive/conjunctive normal form, we do the following.

- For every $a \in \Sigma$ we introduce a new production of the form $N_a \to a$, fix $\mathcal{F}(N_a) = 0$ and in every other production, we replace each occurrence of terminal a by a nonterminal N_a.
- We get rid of ϵ-productions. To obtain this, we compute $Epsilon = \{X \in N | \epsilon \in L_{st}(X)\}$. (As we argue later, $Epsilon$ can be computed in linear time). Then for every occurence of XY, where $X \in Epsilon$, we replace XY by $(XY \vee Y)$ and if $Y \in Epsilon$ by $(XY \vee X)$.
- We reduce every production to a form, such that every negation is attached to nonterminals or to concatenations of two nonterminals.
- We flatten right-hand sides of productions. For example, for a production: $A \to B \vee (\neg C \& D)$, we introduce a new nonterminal E and two new productions, namely: $A \to B \vee E$ and $E \to \neg C \& D$. We also fix $\mathcal{F}(E) = \mathcal{F}(D)$.
- As longs as there exists any production containing subformula of the form: XYZ, we replace YZ by a new nonterminal V, introduce a new production: $V \to YZ$ and fix $\mathcal{F}(V) = \max(\mathcal{F}(Y), \mathcal{F}(Z))$.

Lemma 4. *For a stratified Boolean grammar G, the transformation to a conjunctive/disjunctive normal form can increase its size at most $O(|G| \log(|G|))$ times.*

Let us note, that our conjunctive/disjunctive normal form allows unit conjuncts. Now, we take a first look on a recognition algorithm, that works for stratified Boolean grammars in a conjunctive/disjunctive normal form. Algorithm 2 uses a definition of the procedure *ComputeModulo*, that is described below.

Algorithm 2. Let $G = (\Sigma, N, P, S, \mathcal{F})$ be a stratified Boolean grammar in a conjunctive/disjunctive normal form. Let $w = a_1...a_n \in \Sigma^+$ ($n \geq 1$) be the input string.
For every $1 \leq i \leq j \leq n$, compute $T_{i,j} = \{X \in N | w_{i,j} \in L_{st}(X)\}$ and check if $S \in T_{1,n}$.

1. for $d = 0$ to $n - 1$
2.   for $i = 1$ to $n - d$
3.   let $j = i + d$
4.   if $d = 0$ then
5. $R_{i,i} = \{X | X \to a_i\}$
6. $T_{i,i} = \textbf{\textit{ComputeModulo}}(R_{i,i})$
7.   else
8.   for $l = i$ to $j - 1$
9. $R_{i,j} = R_{i,j} \cup T_{i,l} \times T_{l+1,j}$
10. $T_{i,j} = \textbf{\textit{ComputeModulo}}(R_{i,j})$

As it is not difficult to see Algorithm 2 is an usual recognition algorithm for CFG with some extra lines. We consider them now. Because of the fact, that we allow unit conjuncts, productions of the form $X \to a_i$ are not the only ones, that can produce single terminal a_i, where $1 \leq i \leq n$. As an example consider a grammar, that contains the following productions: $A \to B \vee \neg C$, $B \to a$ and $C \to b$. We have, that $a \in L_{st}(B)$ as well as $a \in L_{st}(A)$.

So, first in line 5 we gather in $R_{i,i}$ all nonterminals X, such that $X \to a_i$ and then we pass the set of nonterminals $R_{i,i}$ to the procedure *ComputeModulo*, that computes the set of all nonterminals X, such that $a_i \in L_{st}(X)$. Similarly, in the induction step, we first in line 9 gather concatenations of pairs of nonterminals XY, such that $w_{i,j} \in L_{st}(XY)$ and then pass the set $R_{i,j}$ to the procedure *ComputeModulo*, that computes the set of nonterminals X, such that $w_{i,j} \in L_{st}(X)$. So, we want the procedure *ComputeModulo*, to compute stratified solution modulo $w_{i,j}$ of the system $N = \psi(N)$, that is corresponding to G.

ComputeModulo. We consider here only the induction step. The base case is, in fact, very similar. Stratified solution modulo some $w_{i,j}$, we denote by $L_{stm}^{i,j}$. Further, let us assume, that we have $L_{stm}^{i,j}(\mathcal{F}^{-1}(1,\ldots,k-1))$. Now, we show, how to compute $L_{stm}^{i,j}$ for k-th stratum. A conjunctive/disjunctive normal form does not allow ϵ-productions, so when we pass $R_{i,j}$ ($1 \leq i < j \leq n$) to *ComputeModulo*, it contains all pairs XY, such that $w_{i,j} \in L_{st}(XY)$. Therefore, to achieve $L_{stm}^{i,j}(k)$, it is enough to compute the least solution of the system of Boolean equation, that can be obtained from system of language equations, that defines k-th stratum ($\mathcal{F}^{-1}(k+1) = \psi_D(N)$) as follows.

1. Replace each occurence of every terminal with a Boolean value *false* and every occurence of the concatenation of two nonterminals XY with *true* if $XY \in R_{i,j}$ and with *false* in the opposite case.
2. Replace each occurence of a nonterminal $Z \in \mathcal{F}^{-1}(1,\ldots,k-1)$ with *true* if $w_{i,j} \in L_{stm}^{i,j}(Z)$ and with *false* in opposite case.

3. Replace each occurence of a nonterminal $Z \in \mathcal{F}^{-1}(k)$ with a Boolean variable.

The least solution of every such system of Boolean equations in conjunctive/disjunctive normal form can be find in linear time by reducing it to the Boolean graph [1]. Thus, we have the following lemma.

Lemma 5. *The procedure* ComputeModulo *works in time* $O(|G|)$.

Proof. ComputeModulo computes the least solutions of systems of Boolean equations of total size $O(|G|)$. □

Now let us go back to the set *Epsilon*, that was being computed during transformation to a conjunctive/disjunctive normal form. As we noted earlier it can be computed in linear time and a procedure like *ComputedModulo* can be used to it. It is enough to compute stratified solution of the system of language equations modulo ϵ.

Theorem 4. *Let* $G = (\Sigma, N, P, S, \mathcal{F})$ *be a stratified Boolean grammar in a conjunctive/disjunctive normal form, let* $w = a_1 \ldots a_n \in \Sigma^+$. *Let* L_{st} *be a stratified solution of a system of language equations corresponding to* G. *Then for every* $X \in N$ *and* $1 \leq i, j \leq n$ *we have* $w_{i,j} \in L_{st}(X)$ *if and only if* $X \in T_{i,j}$, *where* $T_{i,j}$ *is computed by Algorithm 2.*

Theorem 5. *Let* G *be a stratified Boolean grammar. Let* $w \in \Sigma^+$. *Then we have the following:*

1. *If* G *is in a conjunctive/disjunctive normal form, then Algorithm 2 works in time* $O(|w|^3 |G| + |w|^2 |G|) = O(|w|^3 |G|)$.
2. *If* G *is arbitrary, then Algorithm 2 together with the transformation to a conjunctive/disjunctive normal form works in time* $O(|w|^3 |G| \log(|G|))$.

Proof. Point 1 comes from the fact, that CKY-algorithm for CFG works in time $O(|G| |w|^3)$ and from Lemma 5. Point 2 comes from point 1 and Lemma 4. □

References

1. Andersen, H. R.: Model Checking and Boolean Graphs. Theor. Comput. Sci. **1(126)** (1994) 3–30
2. Chandra, A. K., Harel, D.: Horn Clauses and the Fixpoint Query Hierarchy. Proceedings of the ACM Symposium on Principles of Database Systems, March 29-31, 1982, Los Angeles, California (1982) 158–163
3. Kupferman, O., Vardi, M. Y., Wolper, P.: An automata-theoretic approach to branching-time model checking. Journal of ACM (2000) **47(2)** 312–360
4. Okhotin, A.: A Boolean grammar for a simple programming language. Technical Report 2004-478, School of Computing, Queen's University, Kingston, Ontario, Canada
5. Okhotin, A.: Boolean grammars. Developments in Language Theory (2003) 398 – 410
6. Tarjan, R.E.: Depth first search and linear graph algorithms. SIAM Journal of Computing **1(2)** (1972) 146–160

Author Index

Allender, Eric 71
Alonso, César L. 83
Àlvarez, Carme 95
Amano, Kazuyuki 107

Bauland, Michael 71, 119
Berthé, Valérie 131
Bienkowski, Marcin 1
Blesa, María J. 144
Boros, E. 556
Böttcher, Julia 156

Calzada, Daniel 144
Carayol, Arnaud 168, 180
Cervelle, J. 192
Chalopin, Jérémie 212
Chan, Ho-Leung 224
Chan, Wun-Tat 236
Cheng, Ho-lun 248
Chubarov, D. 260

Dawar, Anuj 495
De Marco, Gianluca 271
Doty, David 283

Elbassioni, K. 556
Epstein, Leah 295

Faliszewski, Piotr 308
Fatès, Nazim 316
Fernández, Antonio 144
Fertin, Guillaume 328
Fiala, Jiří 340
Fishkin, Aleksei V. 352
Fomin, Fedor V. 364
Formenti, E. 192
Fraigniaud, Pierre 364
Freivalds, Rūsiņš 15

Gabarró, Joaquim 95
Gaintzarain, J. 376
Gargano, Luisa 271
Gerber, Olga 352
Glaßer, Christian 387, 399
Goldsmith, Judy 410

Grohe, Martin 422
Grunsky, Igor 435
Gu, Xiaoyang 283
Gurevich, Yuri 26
Gurvich, V. 556
Gurvits, Leonid 447

Hagen, Matthias 410
Hansen, Uffe Flarup 459
Heinemann, Bernhard 471
Hemaspaandra, Edith 119
Hermo, M. 376
Hernich, André 483
Hunter, Paul 495

Ibarra, Oscar H. 39
Immerman, Neil 71

Janicki, Ryszard 507
Jansen, Klaus 352
Jansson, Jesper 224, 520
Jurdziński, Tomasz 532

Kapoutsis, Christos 544
Khachiyan, L. 556
Kneis, Joachim 568
Kolman, Petr 580
Korovin, Konstantin 591
Král', Daniel 603
Kranakis, Evangelos 271
Kreutzer, Stephan 422
Krizanc, Danny 271
Krysta, Piotr 615
Kuivinen, Fredrik 628
Kurganskyy, Oleksiy 435

Lam, Tak-Wah 224, 236
Lange, Martin 640
Levy, Meital 295
Li, Minming 652
Liu, Kin-Shing 236
Lohrey, Markus 664
López, Luis 144
López-Valdés, María 676
Lutz, Jack H. 283

Martínez, Andrés L. 144
Maruoka, Akira 107
Masson, B. 192
Mayordomo, Elvira 283, 676
Meer, Klaus 459
Métivier, Yves 212
Meyer auf der Heide, Friedhelm 1
Meyer, Antoine 180
Mölle, Daniel 568
Montaña, José L. 83
Morin, Rémi 686
Morvan, Michel 316
Moser, Philippe 283
Mundhenk, Martin 410

Navarro, M. 376
Nickelsen, Arfst 483
Nishio, Hidenosuke 699
Nisse, Nicolas 364

Ogihara, Mitsunori 308, 387
Okhotin, Alexander 708
Ondrusch, Nicole 664
Otto, Friedrich 532

Pangrác, Ondřej 603
Pardo, Luis M. 83
Paulusma, Daniël 340
Pavan, A. 387
Pelc, Andrzej 271
Peng, Zeshan 520
Pinna, G. Michele 720
Pol, Jaco van de 769
Potapov, Igor 435

Richter, Stefan 568
Riege, Tobias 733
Rigo, Michel 131
Rizzi, Romeo 328

Rödl, Vojtěch 52
Rossmanith, Peter 568
Rothe, Jörg 733
Ruciński, Andrzej 52

Safari, Mohammad Ali 745
Santos, Agustín 144
Schabanel, Nicolas 316
Schnoor, Henning 71
Schweikardt, Nicole 422
Selman, Alan L. 387, 399
Serna, Maria 95, 144
Simonsen, Jakob Grue 757
Solis-Oba, Roberto 352
Somla, Rafał 640
Szemerédi, Endre 52

Tan, Tony 248
Telle, Jan Arne 340
Thierry, Éric 316
Tveretina, Olga 769

Unger, Falk 781

Vaccaro, Ugo 271
Vassilevska, Virginia 793
Vialette, Stéphane 328
Vollmer, Heribert 71
Voronkov, Andrei 260, 591

Wong, Prudence W.H. 236
Wrona, Michał 801

Yao, Andrew C. 57
Yao, Frances F. 652
Yiu, Siu-Ming 224

Zhang, Liyu 387, 399
Zielonka, Wiesław 58

Lecture Notes in Computer Science

For information about Vols. 1–3561

please contact your bookseller or Springer

Vol. 3687: S. Singh, M. Singh, C. Apte, P. Perner (Eds.), Pattern Recognition and Image Analysis, Part II. XXV, 809 pages. 2005.

Vol. 3686: S. Singh, M. Singh, C. Apte, P. Perner (Eds.), Pattern Recognition and Data Mining, Part I. XXVI, 689 pages. 2005.

Vol. 3672: C. Hankin, I. Siveroni (Eds.), Static Analysis. X, 369 pages. 2005.

Vol. 3671: S. Bressan, S. Ceri, E. Hunt, Z.G. Ives, Z. Bellahsène, M. Rys, R. Unland (Eds.), Database and XML Technologies. X, 239 pages. 2005.

Vol. 3664: C. Türker, M. Agosti, H.-J. Schek (Eds.), Peer-to-Peer, Grid, and Service-Orientation in Digital Library Architectures. X, 261 pages. 2005.

Vol. 3663: W. Kropatsch, R. Sablatnig, A. Hanbury (Eds.), Pattern Recognition. XIV, 512 pages. 2005.

Vol. 3662: C. Baral, G. Greco, N. Leone, G. Terracina (Eds.), Logic Programming and Nonmonotonic Reasoning. XIII, 454 pages. 2005. (Subseries LNAI).

Vol. 3660: M. Beigl, S. Intille, J. Rekimoto, H. Tokuda (Eds.), UbiComp 2005: Ubiquitous Computing. XVII, 394 pages. 2005.

Vol. 3659: J.R. Rao, B. Sunar (Eds.), Cryptographic Hardware and Embedded Systems – CHES 2005. XIV, 458 pages. 2005.

Vol. 3654: S. Jajodia, D. Wijesekera (Eds.), Data and Applications Security XIX. X, 353 pages. 2005.

Vol. 3653: M. Abadi, L.d. Alfaro (Eds.), CONCUR 2005 – Concurrency Theory. XIV, 578 pages. 2005.

Vol. 3649: W.M.P. van der Aalst, B. Benatallah, F. Casati, F. Curbera (Eds.), Business Process Management. XII, 472 pages. 2005.

Vol. 3648: J.C. Cunha, P.D. Medeiros (Eds.), Euro-Par 2005 Parallel Processing. XXXVI, 1299 pages. 2005.

Vol. 3645: D.-S. Huang, X.-P. Zhang, G.-B. Huang (Eds.), Advances in Intelligent Computing, Part II. XIII, 1010 pages. 2005.

Vol. 3644: D.-S. Huang, X.-P. Zhang, G.-B. Huang (Eds.), Advances in Intelligent Computing, Part I. XXVII, 1101 pages. 2005.

Vol. 3642: D. Ślezak, J. Yao, J.F. Peters, W. Ziarko, X. Hu (Eds.), Rough Sets, Fuzzy Sets, Data Mining, and Granular Computing, Part II. XXIV, 738 pages. 2005. (Subseries LNAI).

Vol. 3641: D. Ślezak, G. Wang, M.S. Szczuka, I. Düntsch, Y. Yao (Eds.), Rough Sets, Fuzzy Sets, Data Mining, and Granular Computing, Part I. XXIV, 742 pages. 2005. (Subseries LNAI).

Vol. 3639: P. Godefroid (Ed.), Model Checking Software. XI, 289 pages. 2005.

Vol. 3638: A. Butz, B. Fisher, A. Krüger, P. Olivier (Eds.), Smart Graphics. XI, 269 pages. 2005.

Vol. 3637: J. M. Moreno, J. Madrenas, J. Cosp (Eds.), Evolvable Systems: From Biology to Hardware. XI, 227 pages. 2005.

Vol. 3636: M.J. Blesa, C. Blum, A. Roli, M. Sampels (Eds.), Hybrid Metaheuristics. XII, 155 pages. 2005.

Vol. 3634: L. Ong (Ed.), Computer Science Logic. XI, 567 pages. 2005.

Vol. 3633: C. Bauzer Medeiros, M. Egenhofer, E. Bertino (Eds.), Advances in Spatial and Temporal Databases. XIII, 433 pages. 2005.

Vol. 3632: R. Nieuwenhuis (Ed.), Automated Deduction – CADE-20. XIII, 459 pages. 2005. (Subseries LNAI).

Vol. 3629: J.L. Fiadeiro, N. Harman, M. Roggenbach, J. Rutten (Eds.), Algebra and Coalgebra in Computer Science. XI, 457 pages. 2005.

Vol. 3628: T. Gschwind, U. Aßmann, O. Nierstrasz (Eds.), Software Composition. X, 199 pages. 2005.

Vol. 3627: C. Jacob, M.L. Pilat, P.J. Bentley, J. Timmis (Eds.), Artificial Immune Systems. XII, 500 pages. 2005.

Vol. 3626: B. Ganter, G. Stumme, R. Wille (Eds.), Formal Concept Analysis. X, 349 pages. 2005. (Subseries LNAI).

Vol. 3625: S. Kramer, B. Pfahringer (Eds.), Inductive Logic Programming. XIII, 427 pages. 2005. (Subseries LNAI).

Vol. 3624: C. Chekuri, K. Jansen, J.D.P. Rolim, L. Trevisan (Eds.), Approximation, Randomization and Combinatorial Optimization. XI, 495 pages. 2005.

Vol. 3623: M. Liśkiewicz, R. Reischuk (Eds.), Fundamentals of Computation Theory. XV, 576 pages. 2005.

Vol. 3621: V. Shoup (Ed.), Advances in Cryptology – CRYPTO 2005. XI, 568 pages. 2005.

Vol. 3620: H. Muñoz-Avila, F. Ricci (Eds.), Case-Based Reasoning Research and Development. XV, 654 pages. 2005. (Subseries LNAI).

Vol. 3619: X. Lu, W. Zhao (Eds.), Networking and Mobile Computing. XXIV, 1299 pages. 2005.

Vol. 3618: J. Jędrzejowicz, A. Szepietowski (Eds.), Mathematical Foundations of Computer Science 2005. XVI, 814 pages. 2005.

Vol. 3615: B. Ludäscher, L. Raschid (Eds.), Data Integration in the Life Sciences. XII, 344 pages. 2005. (Subseries LNBI).

Vol. 3614: L. Wang, Y. Jin (Eds.), Fuzzy Systems and Knowledge Discovery, Part II. XLI, 1314 pages. 2005. (Subseries LNAI).

Vol. 3613: L. Wang, Y. Jin (Eds.), Fuzzy Systems and Knowledge Discovery, Part I. XLI, 1334 pages. 2005. (Subseries LNAI).

Vol. 3612: L. Wang, K. Chen, Y. S. Ong (Eds.), Advances in Natural Computation, Part III. LXI, 1326 pages. 2005.

Vol. 3611: L. Wang, K. Chen, Y. S. Ong (Eds.), Advances in Natural Computation, Part II. LXI, 1292 pages. 2005.

Vol. 3610: L. Wang, K. Chen, Y. S. Ong (Eds.), Advances in Natural Computation, Part I. LXI, 1302 pages. 2005.

Vol. 3608: F. Dehne, A. López-Ortiz, J.-R. Sack (Eds.), Algorithms and Data Structures. XIV, 446 pages. 2005.

Vol. 3607: J.-D. Zucker, L. Saitta (Eds.), Abstraction, Reformulation and Approximation. XII, 376 pages. 2005. (Subseries LNAI).

Vol. 3606: V. Malyshkin (Ed.), Parallel Computing Technologies. XII, 470 pages. 2005.

Vol. 3604: R. Martin, H. Bez, M. Sabin (Eds.), Mathematics of Surfaces XI. IX, 473 pages. 2005.

Vol. 3603: J. Hurd, T. Melham (Eds.), Theorem Proving in Higher Order Logics. IX, 409 pages. 2005.

Vol. 3602: R. Eigenmann, Z. Li, S.P. Midkiff (Eds.), Languages and Compilers for High Performance Computing. IX, 486 pages. 2005.

Vol. 3599: U. Aßmann, M. Aksit, A. Rensink (Eds.), Model Driven Architecture. X, 235 pages. 2005.

Vol. 3598: H. Murakami, H. Nakashima, H. Tokuda, M. Yasumura, Ubiquitous Computing Systems. XIII, 275 pages. 2005.

Vol. 3597: S. Shimojo, S. Ichii, T.W. Ling, K.-H. Song (Eds.), Web and Communication Technologies and Internet-Related Social Issues - HSI 2005. XIX, 368 pages. 2005.

Vol. 3596: F. Dau, M.-L. Mugnier, G. Stumme (Eds.), Conceptual Structures: Common Semantics for Sharing Knowledge. XI, 467 pages. 2005. (Subseries LNAI).

Vol. 3595: L. Wang (Ed.), Computing and Combinatorics. XVI, 995 pages. 2005.

Vol. 3594: J.C. Setubal, S. Verjovski-Almeida (Eds.), Advances in Bioinformatics and Computational Biology. XIV, 258 pages. 2005. (Subseries LNBI).

Vol. 3593: V. Mařík, R. W. Brennan, M. Pěchouček (Eds.), Holonic and Multi-Agent Systems for Manufacturing. XI, 269 pages. 2005. (Subseries LNAI).

Vol. 3592: S. Katsikas, J. Lopez, G. Pernul (Eds.), Trust, Privacy and Security in Digital Business. XII, 332 pages. 2005.

Vol. 3591: M.A. Wimmer, R. Traunmüller, Å. Grönlund, K.V. Andersen (Eds.), Electronic Government. XIII, 317 pages. 2005.

Vol. 3590: K. Bauknecht, B. Pröll, H. Werthner (Eds.), E-Commerce and Web Technologies. XIV, 380 pages. 2005.

Vol. 3589: A M. Tjoa, J. Trujillo (Eds.), Data Warehousing and Knowledge Discovery. XVI, 538 pages. 2005.

Vol. 3588: K.V. Andersen, J. Debenham, R. Wagner (Eds.), Database and Expert Systems Applications. XX, 955 pages. 2005.

Vol. 3587: P. Perner, A. Imiya (Eds.), Machine Learning and Data Mining in Pattern Recognition. XVII, 695 pages. 2005. (Subseries LNAI).

Vol. 3586: A.P. Black (Ed.), ECOOP 2005 - Object-Oriented Programming. XVII, 631 pages. 2005.

Vol. 3584: X. Li, S. Wang, Z.Y. Dong (Eds.), Advanced Data Mining and Applications. XIX, 835 pages. 2005. (Subseries LNAI).

Vol. 3583: R.W. H. Lau, Q. Li, R. Cheung, W. Liu (Eds.), Advances in Web-Based Learning – ICWL 2005. XIV, 420 pages. 2005.

Vol. 3582: J. Fitzgerald, I.J. Hayes, A. Tarlecki (Eds.), FM 2005: Formal Methods. XIV, 558 pages. 2005.

Vol. 3581: S. Miksch, J. Hunter, E. Keravnou (Eds.), Artificial Intelligence in Medicine. XVII, 547 pages. 2005. (Subseries LNAI).

Vol. 3580: L. Caires, G.F. Italiano, L. Monteiro, C. Palamidessi, M. Yung (Eds.), Automata, Languages and Programming. XXV, 1477 pages. 2005.

Vol. 3579: D. Lowe, M. Gaedke (Eds.), Web Engineering. XXII, 633 pages. 2005.

Vol. 3578: M. Gallagher, J. Hogan, F. Maire (Eds.), Intelligent Data Engineering and Automated Learning - IDEAL 2005. XVI, 599 pages. 2005.

Vol. 3577: R. Falcone, S. Barber, J. Sabater-Mir, M.P. Singh (Eds.), Trusting Agents for Trusting Electronic Societies. VIII, 235 pages. 2005. (Subseries LNAI).

Vol. 3576: K. Etessami, S.K. Rajamani (Eds.), Computer Aided Verification. XV, 564 pages. 2005.

Vol. 3575: S. Wermter, G. Palm, M. Elshaw (Eds.), Biomimetic Neural Learning for Intelligent Robots. IX, 383 pages. 2005. (Subseries LNAI).

Vol. 3574: C. Boyd, J.M. González Nieto (Eds.), Information Security and Privacy. XIII, 586 pages. 2005.

Vol. 3573: S. Etalle (Ed.), Logic Based Program Synthesis and Transformation. VIII, 279 pages. 2005.

Vol. 3572: C. De Felice, A. Restivo (Eds.), Developments in Language Theory. XI, 409 pages. 2005.

Vol. 3571: L. Godo (Ed.), Symbolic and Quantitative Approaches to Reasoning with Uncertainty. XVI, 1028 pages. 2005. (Subseries LNAI).

Vol. 3570: A. S. Patrick, M. Yung (Eds.), Financial Cryptography and Data Security. XII, 376 pages. 2005.

Vol. 3569: F. Bacchus, T. Walsh (Eds.), Theory and Applications of Satisfiability Testing. XII, 492 pages. 2005.

Vol. 3568: W.-K. Leow, M.S. Lew, T.-S. Chua, W.-Y. Ma, L. Chaisorn, E.M. Bakker (Eds.), Image and Video Retrieval. XVII, 672 pages. 2005.

Vol. 3567: M. Jackson, D. Nelson, S. Stirk (Eds.), Database: Enterprise, Skills and Innovation. XII, 185 pages. 2005.

Vol. 3566: J.-P. Banâtre, P. Fradet, J.-L. Giavitto, O. Michel (Eds.), Unconventional Programming Paradigms. XI, 367 pages. 2005.

Vol. 3565: G.E. Christensen, M. Sonka (Eds.), Information Processing in Medical Imaging. XXI, 777 pages. 2005.

Vol. 3564: N. Eisinger, J. Małuszyński (Eds.), Reasoning Web. IX, 319 pages. 2005.

Vol. 3562: J. Mira, J.R. Álvarez (Eds.), Artificial Intelligence and Knowledge Engineering Applications: A Bioinspired Approach, Part II. XXIV, 636 pages. 2005.